For Reference

Not to be taken from this room

W0408
W0512
W0515
W0218

GEO-DATA

PROPERTY OF CLPL

GEO-DATA

The World Geographical Encyclopedia

THIRD EDITION

John F. McCoy, Project Editor

First and second editions edited by George Kurian

Detroit • New York • San Diego • San Francisco • Cleveland • New Haven, Conn. • Waterville, Maine • London • Munich

GEO-DATA: THE WORLD GEOGRAPHICAL ENCYCLOPEDIA, THIRD EDITION

Project Editor
John F. McCoy

Editorial
Mary Rose Bonk, Pamela A. Dear, Rachel J. Kain, Lynn U. Koch, Michael D. Lesniak, Nancy Matuszak, Michael T. Reade

Imaging and Multimedia
Randy Bassett, Christine O'Bryan, Barbara J. Yarrow

Product Design
Cindy Baldwin, Tracey Rowens

Manufacturing
Nekita McKee

© 2003 by Gale. Gale is an imprint of The Gale Group, Inc., a division of Thomson Learning, Inc.

Gale and Design™ and Thomson Learning™ are trademarks used herein under license.

For more information contact
The Gale Group, Inc.
27500 Drake Rd.
Farmington Hills, MI 48331–3535
Or you can visit our Internet site at
http://www.gale.com

ALL RIGHTS RESERVED
No part of this work covered by the copyright hereon may be reproduced or used in any form or by any means--graphic, electronic, or mechanical, including photocopying, recording, taping, Web distribution, or information storage retrieval systems--without the written permission of the publisher.

For permission to use material from this product, submit your request via Web at http://www.gale-edit.com/permissions, or you may download our Permissions Request form and submit your request by fax or mail to:

Permissions Department
The Gale Group, Inc.
27500 Drake Rd.
Farmington Hills, MI 48331–3535
Permissions Hotline:
248–699–8006 or 800–877–4253; ext. 8006
Fax: 248–699–8074 or 800–762–4058

Cover photographs reproduced by permission of Digital Stock (Taj Mahal, center left), Digital Vision (Fall trees, top left; Waterfall, top right; Water carriers, center; City in smog, center right), and PhotoDisc (Pink buttes, top center; Mountains, bottom).

While every effort has been made to ensure the reliability of the information presented in this publication, The Gale Group, Inc. does not guarantee the accuracy of the data contained herein. The Gale Group, Inc. accepts no payment for listing; and inclusion in the publication of any organization, agency, institution, publication, service, or individual does not imply endorsement of the editors or publisher. Errors brought to the attention of the publisher and verified to the satisfaction of the publisher will be corrected in future editions.

Library of Congress Cataloging-in-Publication Data

Geo-data : the world geographical encyclopedia / John F. McCoy, project editor.-- 3rd ed.
p. cm.
Includes bibliographical references (p.) and index.
ISBN 0-7876-5581-3 (hardcover : alk. paper)
1. Gazetteers. I. McCoy, John, 1976-
G103.5 .G36 2002
910'.3--dc21
2002010926

Printed in the United States of America
10 9 8 7 6 5 4 3 2

CONTENTS

PREFACE

Gale is pleased to present the third edition of *Geo-Data: The World Geographical Encyclopedia.* This is the first new edition of *Geo-Data* since 1989, and it represents a complete revision and updating of that work. The purpose of the book remains unchanged: to provide the reader with the most detailed and comprehensive descriptions available for the physical geography of countries.

Geo-Data's focus and design is unique. Many encyclopedias, atlases, and other books provide some information on geography, but it is usually secondary to history, current events, or other topics. In these sorts of references physical geography receives a few paragraphs of coverage per country at best, compared to pages in *Geo-Data.* Among the few books that can rival *Geo-Data* coverage, none can match its ease of use. There is no need to hunt from one end of a dictionary or gazetteer to another, searching for details on a country's many different features. In *Geo-Data* everything is presented in one place. Each of *Geo-Data's* entries gives the reader a complete portrait of a country's mountains, fields, forests, deserts, rivers, seas, wetlands, and other features, with additional information on population distribution and natural resources. Key facts and statistics are highlighted at the beginning of each entry. In addition to the country entries, *Geo-Data* features a World Rankings Appendix, listing the world's tallest mountains, longest rivers, deepest oceans, largest lakes, most populous countries, and other outstanding features.

Selection Criteria

Geo-Data features 207 entries, each describing a single country or dependency. Every entry includes a relief map depicting the country or dependency. There is a separate entry for each of the world's 192 countries, including the new nation of East Timor. In addition, the autonomous island of Taiwan and the continent of Antarctica have their own entries, as do 13 dependencies: Anguilla, Aruba, Bermuda, the British Virgin Islands, the Cayman Islands, French Guiana, Greenland, Guadeloupe, Martinique, Netherlands Antilles, Puerto Rico, the Turks and Caicos Islands, and the U.S. Virgin Islands. Smaller, less populated, and more remote dependencies, such as the Cook Islands, Palmyra Atoll, and Svalbard, are discussed in the entries of their parent states.

Although all countries and dependencies are described in detail, entries can vary greatly in length. This is often due to the obvious reason that some countries are much larger than others. However, countries with particularly complicated geography, such as Nepal, may have longer entries than much larger countries with fewer outstanding features, such as Libya. Countries with a large number of dependencies, such as France or the United Kingdom, also have longer entries than comparably sized countries.

Names and Measurements

Geography is an imprecise science. Knowledgeable experts, most of whom have lived, worked, or studied extensively in the countries about which they wrote, prepared the entries for this new edition of *Geo-Data.* Gale editors have relied on authoritative sources such as *Merriam Webster's Geographical Dictionary, 3rd ed., The World Factbook 2001* of the CIA, *The Columbia Gazetteer of the World,* and the statistics and geographic divisions of the United Nations and various countries to check our facts. However, the reader should be aware that it is difficult if not impossible to measure exactly such large and irregular features as mountains, lakes, and rivers. Even authoritative sources often vary slightly in their information. Population information can be even more difficult to determine accurately, as many countries do not measure this scientifically and on a regular basis. While basic population statistics are present in every entry, detailed tables and other breakdowns are present only in those entries for which accurate and reasonably current information was available.

It is common for a single geographical feature or country to be called by many different names, depending on the source and context. In *Geo-Data* we have attempted to use the names that will be most familiar to the majority of our readers. These are most commonly English-language names, but widely used names in other languages and names which have no English equivalent also appear frequently. In addition, the reader will often find alternative names for the same feature given in parentheses after it is first mentioned in the text.

In the case of country and dependency names, the editors have again opted for using the most easily recognized versions of their names. Thus the reader will find Brazil under B, not under F for Federative Republic of Brazil or R for República Federative do Brasil. Full, formal, names are used with the two Congos and the two Koreas, but these countries still appear under C and K, respectively.

When discussing geography it is impossible to avoid controversy. There are many regions around the world whose ownership is disputed, sometimes violently. Although Gale does not support or seek to legitimize the views of any of the parties of these disputes, for the sake of completeness it was necessary to include these disputed regions in one entry or another. The editors have chosen to discuss disputed regions within the entries of the countries that physically control them, while at the same time carefully noting in those entries that the region is claimed by other countries and its status is not determined. Thus the West Bank can be found under Israel, Jammu and Kashmir within the entries on India and Pakistan, and Taiwan in its own separate entry.

How Each Profile is Organized

Geo-Data's 207 entries are organized alphabetically. All entries follow the same basic structure, beginning with a collection of Key Facts. The purpose of the Key Facts is to gather together in one easy to find place the basic information about each country that readers most often need. These facts always include a country's: area (total and ranked relative to other entries), location relative to the rest of the world, coordinates, border length (overall and with individual countries), coastline length, territorial seas, highest point, lowest point, population (total and ranked relative to other entries), largest city, and capital city. When available and applicable, the longest distances across the country, its longest river, and its largest lake are also provided.

Following the Key Facts are eight rubrics. First, the overall geography and geology of the country is described in the Overview. Next, Mountains and Hills describes the major elevations of the country. Inland Waterways then describes the country's lakes, rivers, and wetlands. The Coast, Islands, and the Ocean gives information on the bodies of water that border on the country, the country's coasts, and its major islands. Climate and Vegetation describes the country's temperature and rainfall patterns, as well as its forests, grasslands, deserts, and other ground cover. Human Population outlines the major concentrations of people in the country. Then Natural Resources identifies the most important resources found within the country. Finally, the Further Readings rubric gives an interested reader suggestions on where to find more information.

All eight rubrics appear in every entry, in the order given above, making it easy to navigate within entries and to compare information across countries. Entries are written as a narrative, so the reader looking for the complete picture on a country should find the entire entry an enjoyable read. Rubrics are clearly denoted in the text, however, so a reader looking for specific information can find it easily. Throughout the text, major features are identified by name, and their height, length, or other dimensions are provided whenever possible, both in imperial and metric units. Sidebars highlight particularly unusual and interesting facts.

Geo-Data's entries are greatly enhanced by their maps. Every entry in the book includes a relief map. These maps were created specifically for their entries and give the user an extremely detailed view of the geography they are reading about. Since they were created especially for *Geo-Data,* the reader can be assured that all but the smallest of features mentioned in the text also appear on the map. In addition to the maps, most entries feature one or two tables. These tables list the country's largest cities and their populations and the country's provinces, their capitals, area, and populations.

Additional Features

Readers looking for comparative information on the world's most remarkable features will find the World Rankings Appendix useful. The 19 tables in this section include lists of the world's tallest mountains by continent and country; the world's longest borders; the world's largest bays and gulfs; the world's tallest volcanoes; the world's deepest oceans; and much more. *Geo-Data's* exhaustive subject index can help the reader find more information in the text on these and other features. A glossary is also provided, for those readers who are unfamiliar with some of the terms used throughout the book.

Acknowledgements

Earlier edition of *Geo-Data* were produced by George Kurian and associates, and distributed by Gale. Gale thanks Mr. Kurian entrusting us with *Geo-Data* and allowing us to continue his excellent work. Many assisted Gale in producing *Geo-Data,* for which Gale thank's them. Eastword Publications, Inc., of Cleveland, OH, contributed most of the research and writing required for the book's text and tables, as well as typesetting the book. Their many contributors are acknowledged below, but Gale would like to give special thanks to Eastword's Susan Gall. All maps appearing in Geo-Data were created by XNR Productions of Madison, WI (Tanya Buckingham, Jon Daugherity, Laura Exner, Paul Exner, Andy Grosvold, Cory Johnson, Rob McCaleb, and Paula Robbins). Copy-editing support was provided by Cathy Dybiec Holm, John Krol, Christopher "Rollo" Romig, and Carrie Snyder. The index was prepared by Synapse Corporation of Franktown, CO.

Contributors

T. Anne Dabb
Karen Ellicott
Susan Gall
Robert J. Groelsema, Ph.D.
Jim Henry
Jeneen Hobby, Ph.D.
Tara Hohne
Hendrick Isengingo
Sarah Kunz
David H. Long
Daniel Lucas
John McCoy
Edith Mirante
Christine Sciulli
Jeanne Marie Stumpf, Ph.D.
Michelle Tackla
Max White
Rosalie Wieder

We Welcome Your Suggestions

Gale appreciates your comments and suggestions. Please direct all correspondence to:

Editor
Geo-Data: The World Geographical Encyclopedia
Gale Group
27500 Drake Rd.
Farmington Hills, MI 48331-3535

Afghanistan

- **Area:** 250,001 sq mi (647,500 sq km) / World Rank: 42
- **Location:** Eastern and Northern Hemispheres, Southern Asia; bordering Turkmenistan, Uzbekistan, and Tajikistan to the north; China to the east; Pakistan to the east and south; and Iran to the west
- **Coordinates:** 33°N, 65°E
- **Borders:** 3,428 mi (5,529 km) total boundary length / China, 47 mi (76 km); Iran, 582 mi (936 km); Pakistan, 1,511 mi (2,430 km); Tajikistan, 750 mi (1,206 km); Turkmenistan, 463 mi (744 km); Uzbekistan, 85 mi (137 km)
- **Coastline:** None
- **Territorial Seas:** None
- **Highest Point:** Mt. Nowshāk, 24,558 ft (7,485 m)
- **Lowest Point:** Amu Dar'ya River, 846 ft (258 m)
- **Longest Distances:** 770 mi (1,240 km) NE-SW / 350 mi (560 km) SE-NW
- **Longest River:** Amu Dar'ya, 1,654 mi (2,661 km)
- **Natural Hazards:** Flooding, droughts, earthquakes
- **Population:** 26,813,057 (July 2001 est.) / World Rank: 39
- **Capital City:** Kabul, east-central Afghanistan
- **Largest City:** Kabul, 1,780,000 (2000 est.)

OVERVIEW

Afghanistan is a landlocked nation in south-central Asia. Strategically located at the crossroads of major north-south and east-west trade routes, it has attracted a succession of invaders ranging from Alexander the Great, in the fourth century B.C., to the Soviet Union in the twentieth century A.D.

Almost as large as the state of Texas, Afghanistan is bounded by six different countries. Afghanistan's longest border—accounting for its entire southern boundary and most of its eastern one—is with Pakistan. The shortest one, bordering China's Xinjiang province, is a mere 47 mi (76 km) at the end of the Vākhān corridor, a narrow sliver of land 150 mi (241 km) long that extends eastward between Tajikistan and Pakistan. At its narrowest point it is only 7 mi (11 km) wide.

The Hindu Kush mountains, running northeast to southwest across the country, divide it into three major regions: 1) the Central Highlands, which form part of the Himalayan Mountains and account for roughly two-thirds of the country's area; 2) the Southwestern Plateau, which accounts for one-fourth of the land; and 3) the smaller Northern Plains area, which contains the country's most fertile soil.

Land elevations generally slope from northeast to southwest, following the general shape of the Hindu Kush massif, from its highest point in the Pamir Mountains near the Chinese border to the lower elevations near the border with Iran. To the north, west, and southwest there are no mountain barriers to neighboring countries. The northern plains pass almost imperceptibly into the plains of Turkistan. In the west and southwest, the plateaus and deserts merge into those of Iran.

The greater part of the northern border and a small section of the border with Pakistan are marked by rivers; the remaining boundary lines are political rather than natural. The northern frontier extends approximately 1,050 mi (1,689 km) southwestward, from the Pamir Mountains in the northeast to a region of hills and deserts in the west, at the border with Iran. The border with Iran runs generally southward from the Harirud River across swamp and desert regions before reaching the northwestern tip of Pakistan. Its southern section crosses the Helmand River.

The border with Pakistan runs eastward from Iran through the Chagai Hills and the southern end of the Rīgestān Desert, then northward through mountainous country. It then follows an irregular northeasterly course some 281 km (175 mi) before reaching the Durand Line, established in 1893 by agreement with British authorities. This line, which defines the border from this point on, continues on through mountainous regions to the Khyber Pass area. Beyond this point it rises to the crest of the Hindu Kush, which it follows eastward to the Pamir Mountains. The Durand Line divides the Pashtun tribes of the region between Afghanistan and Pakistan. Its creation has caused much dissatisfaction among Afghans and has given rise to political tensions between the two countries.

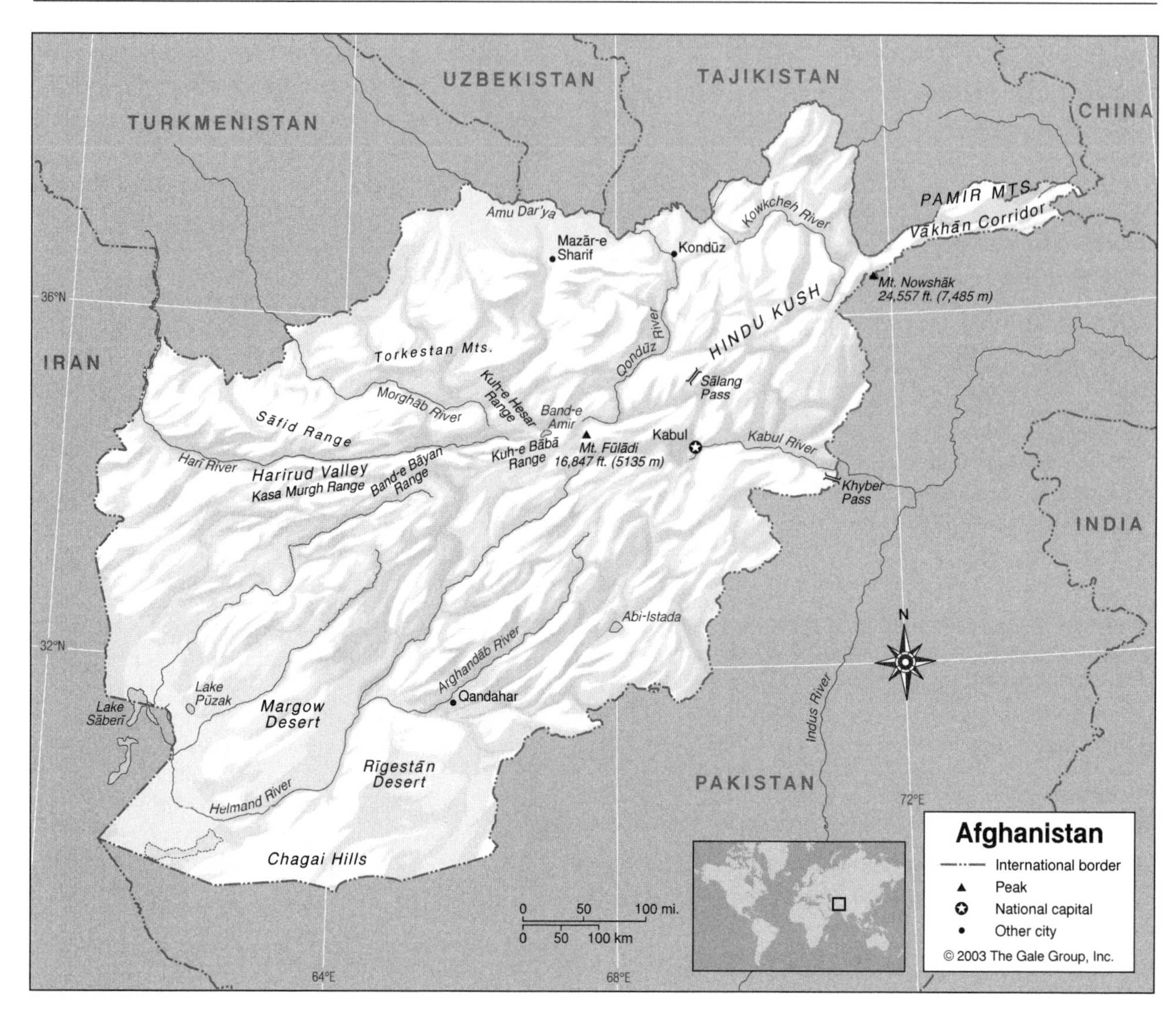

Afghanistan is located on the Eurasian Tectonic Plate. The Vākhān corridor and the rest of northeastern Afghanistan, including Kabul, are situated in a geologically active area. Over a dozen earthquakes occurred there during the twentieth century.

MOUNTAINS AND HILLS

Mountains

The mountainous Central Highlands formed by the Hindu Kush and its subsidiary ranges—extensions of Himalayan mountain chain—are Afghanistan's dominant physical feature. Traversing the country for 600 mi (965 km) from east to west and covering an area of approximately 160,000 sq mi (414,400 sq km), the towering peaks alternate with steep gorges and barren slopes. This mountain system is composed of three high ridges with altitudes descending toward Iran.

The main ridge begins in China and runs southwestward some 300 mi (482 km) as the Pamir Mountains and the Eastern Hindu Kush, with peaks over 21,000 ft (6,400 m) high and mountain passes at altitudes between 12,000 and 15,000 ft (3,657 m and 4,572 m). The very highest peaks are in the Vākhān corridor, including the country's highest peak, Mt. Nowshāk (24,558 ft / 7,485 m).

West of the approximately 12,000 ft (3,681 m) high Sālang Pass, connecting the Kabul area with the northern plains, the Eastern Hindu Kush becomes the Central Hindu Kush. Although not as high and desolate as the Eastern Hindu Kush, terrain is still very rugged and there are many high mountains, including Mt. Fūlādī (16,843 ft / 5,134 m). The Central Hindu Kush continues southeast roughly to the center of the country.

The Bābā Range, with peaks at 15,000 ft (4,572 m), runs parallel to and south of the western end of the Central Hindu Kush, to which it is connected by two ridges. Other important ranges include the Hesār—extending northward from the upper reaches of the Morghāb River in the northwest—and the Sāfid mountains situated north of the broad Harirud Valley and south of the Morghāb. The Torkestan Mountains run parallel to the Sāfid, north of the Morghāb River. They are broken by

GEO-FACT

Afghanistan is a country of extreme weather conditions. Besides the great heat experienced in the summers and the icy cold of its high mountains, there is the effect known as the "Winds of 120 Days," so-called because of strong winds that blow between June and September with a velocity of up to 108 mph (180 kmph). Another example of extreme weather is "Allah's minesweepers," a term Afghan resistance fighters gave to hailstorms so heavy that they would set off some of the thousands of landmines planted in the country during its long wars.

deep river valleys that start in the Bābā Range or in the western part of the Central Hindu Kush.

Several mountain chains fan out southwestward from the Bābā Range, Bāyan Mountains, and Kasa Murgh Ranges. These decrease in altitude as they approach the Southwestern Plateau region, where they yield to gently undulating deserts. Southeast of the Central Hindu Kush, a series of lower ridges enclose long valleys that run parallel to the boundary with Pakistan. The valley region that is home to the capital city of Kabul is bounded by this range system. The Khyber Pass (approximately 3518 ft / 1070 m high) gives access across the border into Pakistan through these ridges.

Plateaus

The Southwestern Plateau, situated southwest of the Central Highlands, is an arid region of deserts and semidesert extending into Pakistan in the south and into Iran in the west. With an average altitude of about 3,000 ft (914 m), it slopes gently to the southwest. It is crossed by a few large rivers, among which the Helmand and its major tributary the Arghandāb are the most important. The region covers approximately 129,500 sq km (50,000 sq mi) and includes the Rīgestān Desert.

INLAND WATERWAYS

Lakes

There are few lakes in Afghanistan. The major ones are Lake Puzak and Lake Sāberī (Lake Helmand), situated on the southwestern border. Lake Sāberī has most of its surface area in Iran. Lake Zorkul is located in the Vākhān corridor near the border with Tajikistan. Abi-Istada, situated on a plateau about 120 mi (193 km) northeast of Qandahār, is a salt lake. The Band-e Amir is a group of five small lakes in the Central Highlands that owe their unique coloration—ranging from a filmy white to a deep green—to the bedrock beneath them.

Rivers

Afghanistan's drainage system is essentially landlocked. Most of the rivers and streams end in shallow desert lakes or oases inside or outside the country's boundaries. Nearly half of the country's total area is drained by watercourses south of the Hindu Kush–Sāfīd ridge line, and half of this area is drained by the Helmand and its tributaries alone. The Amu Dar'ya on the northern border, the country's other major river, has the next-largest drainage area.

The 1,654 mi (2,661 km) long Amu Dar'ya originates in the glaciers of the Pamir Mountains in the northeast. Some 600 mi (965 km) of its upper course constitutes Afghanistan's border with the former Soviet states of Turkmenistan, Uzbekistan, and Tajikistan. Flowing in rapid torrents in its upper course, the Amu Dar'ya becomes calmer below the mouth of the Kowkcheh, 60 mi (96 km) west of Fey ābād. The Qondūz River is another major tributary. During its flood period the upper course of the Amu Dar'ya, swollen by snow and melting ice, carries along much gravel and large boulders.

The Helmand is the principal river in the southwest, bisecting the entire region. Starting some 50 mi (80 km) west of Kabul in the Bābā Mountains, the Helmand is approximately 870 mi (1,400 km) long, making it the longest river situated entirely within Afghanistan. With its many tributaries, the most important of which is the Arghandāb, it drains more than 100,000 sq mi (258,998 sq km).

The Kabul River, 320 mi (515 km) long, is a vital source of water in the Bābā Mountains and for Kabul itself, which it flows through. The Kabul and its tributaries are among the few in Afghanistan that eventually reach the sea, as it flows east into the Indus River in Pakistan.

In the west the sandy deserts along most of the Iranian frontier have no watercourses. However, in the northwest, the Harī and Morghāb Rivers flow into Iran and Turkmenistan.

THE COAST, ISLANDS, AND THE OCEAN

Afghanistan is landlocked, with the closest seacoast roughly 300 mi (483 km) away in Pakistan, on the shores of the Arabian Sea.

CLIMATE AND VEGETATION

Temperature

Afghanistan has a semiarid to arid climate with wide variations in temperature, both between seasons and

Population Centers – Afghanistan

(2002 POPULATION ESTIMATES)

Name	Population	Name	Population
Kabul (capital)	2,142,300	Herat	166,600
Qandahar	339,200	Jalalabad	158,800
Mazār-e Sharif	239,800	Baghlan	117,700
Charikar	196,700	Kondūz	114,600

SOURCE: Data based on United Nations and other international organization estimates; as of 2002, relief organizations estimate that 25–65% of urban dwellers have left the cities for rural areas, or are refugees in neighboring nations. There were no accurate population figures available for Afghanistan in 2002; the last census was undertaken in 1988.

between different times of day. Its summers are hot and dry, but its winters are bitterly cold. Recorded temperatures have ranged as high as 128°F (53°C) in the deserts, and as low as -15°F (-26°C) in the central highlands, which have a subarctic climate. Summertime temperatures in the capital city of Kabul can vary from 61°F (16°C) at sunrise to 100°F (38°C) by noon. The mean January temperature in Kabul is 32°F (0°C).

Rainfall

In much of the country, rainfall is sparse and irregular, averaging 10 to 12 in (25 to 30 cm) and mostly falling between October and April. However, a record 53 in (135 cm) annually has been recorded in the Hindu Kush mountains, and rainfall is generally heavier in the eastern part of the country than in the west. Indian summer monsoons can bring heavy rains in the southeastern mountains in July and August. Otherwise, Afghan summers are generally dry, cloudless, and hot. Humid air from the Persian Gulf sometimes produces summer showers and thunderstorms in the southwest.

Northern Plains

North of the mountainous Central Highlands are the Northern Plains. Afghanistan's smallest natural region, they stretch from the Iranian border in the west to the foothills of the Pamir Mountains in the east. The eastern half of this area, which forms a part of the Central Asia steppe, is bounded by the Amu Dar'ya River in the north; in the west it extends into Turkmenistan. Covering an area of approximately 40,000 sq mi (103,600 sq km), the Northern Plains are situated at an average elevation of 2,000 ft (609 m), except for the Amu Dar'ya valley floor, where it drops to as low as 600 ft (183 m). A considerable portion of the area consists of fertile, loess-covered plains with rich natural gas resources.

Deserts

The Rīgestān Desert at the country's southern border occupies roughly one-quarter of the Southwestern Plateau. It is bounded by the Helmand and Arghandāb Rivers to the north and continues into Pakistan to the south and east. Sand ridges and dunes alternate with wide desert plains, devoid of vegetation and covered with windblown sand changing here and there into barren gravel and clay areas. West of the Rīgestān Desert lies the Margow Desert, a desolate steppe with salt flats.

Administrative Divisions – Afghanistan

2002 POPULATION ESTIMATES

Name	Population	Area (sq mi)	Area (sq km)	Capital
Badakhshan	965,000	17,589	45,556	Fayzabad
Badghis	645,000	8,547	22,136	Qala-i-Naw
Baghlan	828,000	7,005	18,144	Baghlan
Balkh	1,473,000	4,756	12,319	Mazār-e Sharif
Bamian	458,000	6,963	18,033	Bamian
Farah	434,000	22,741	58,900	Farah
Faryab	893,000	7,850	20,331	Maymana
Ghazni	1,053,000	8,517	22,059	Ghazni
Ghowr	609,300	15,539	40,247	Chaghcharan
Helmand	876,600	23,591	61,420	Lashkargah
Herat	1,309,500	16,645	43,334	Herat
Jowzjan	695,300	4,016	10,456	Shibirghan
Kabul (Kabol)	2,846,000	1,800	4,685	Kabul
Kandahar	1,406,700	18,679	48,630	Qandahar
Kapisa	442,000	2,077	5,407	Mahmud Raqi
Kunar (Konar)	523,000	3,900	10,153	Asadabad
Kunduz (Kondoz)	736,100	3,294	8,576	Kondūz
Laghman	610,000	2,745	7,147	Mihtarlam
Lowgar	315,000	1,864	4,853	Puli Alam
Nangarhar	1,410,000	2,938	7,650	Jalalabad
Nimruz	210,000	15,710	40,902	Zaranj
Nurestan (Nuristan)	135,000	3,256	8,477	- -
Paktia (Paktiya)	500,000	2,461	6,408	Gardez
Paktika	443,000	7,337	19,101	Sharan
Parvan	770,000	2,445	6,365	Charikhar
Samangan	535,300	4,956	12,902	Aybak
Sar-e Pul (Sar-e Pol or Saripul)	644,000	9,404	24,484	Sar-e Pul
Takhar	757,100	4,970	12,939	Taluqan
Urruzgan	289,000	10,955	28,522	Tirin Kot
Vardak	357,000	3,750	9,762	Maidanshar
Zabul (Zabol)	389,500	6,723	17,503	Qalat

SOURCE: Compiled and projected from United Nations and United States Agency for International Development statistics; these figures should be considered very rough estimates only. International organizations projected that up to 65% of the population was displaced as of 2002.

A flat strip of desert and steppe extends along the banks of the Amu Dar'ya. Desert and desert-like steppe areas are also found west of Badakhshan along the foothills of the Central Hindu Kush, and also west of Mazār-e Sharif. The southernmost fringe of the area passes gradually into elevated plains.

Forests and Jungles

A light forest cover of oak, pine, cedar, walnut, and other species grows in the Central Hindu Kush and the Safid range. Willows and poplars are found in the mountain valleys.

HUMAN POPULATION

Since much of the country is covered by deserts and receives little precipitation, water has been a dominant factor in determining the location and distribution of human settlement. Many of the historically important towns are located near rivers and streams. Kabul, the country's capital and a crossroads of trade routes from east, west, north, and south, lies on the well-watered plains of the river with the same name. At least 80 percent of Afghanistan's population is rural, and about 20 percent are nomadic.

Afghanistan has seen almost constant war since the 1980s, including a long civil war and an invasion by the Soviet Union. This has resulted in enormous population shifts, with as many as six million Afghans thought to have sought refuge in Pakistan and Iran.

NATURAL RESOURCES

Afghanistan's major mineral resource that has been exploited so far is its natural gas reserves in the Northern Plains. Large deposits of high-grade iron ore remain unmined due to difficulty of access. Other mineral resources include coal, copper, petroleum, sulfur, lead, zinc, chromite, talc, barites, salt, and precious and semi-precious stones.

FURTHER READINGS

Edwards, David B. *Heroes of the Age: Moral Fault Lines on the Afghan Frontier.* Berkeley: University of California Press, 1996.

Elliot, Jason. *An Unexpected Light: Travels in Afghanistan.* London: Picador, 1999.

Ellis, Deborah. *Women of the Afghan War.* Westport, Conn.: Praeger, 2000.

Ewans, Martin. *Afghanistan: A New History.* Richmond: Curzon, 2001.

Giustozzi, Antonio. *War, Politics, and Society in Afghanistan, 1978-1992.* Washington, D.C.: Georgetown University Press, 2000.

"The Most Dangerous Place on Earth: A Look Inside Afghanistan." Special report. *Current Events.* Nov. 30, 2001, pp.S1-5.

QUAZICO. *Afghanistan Online.* http://www.afghan-web.com (accessed February 14, 2002).

Albania

- **Area:** 17,864 sq mi (28,748 sq km) / World Rank: 142
- **Location:** Northern and Eastern Hemispheres. Southeastern Europe bordering the Adriatic and Ionian Seas to the west and southwest, Yugoslavia to the north, the Former Yugoslav Republic of Macedonia to the east, Greece to the southeast
- **Coordinates:** 41°00′ N, 20°00′ E
- **Borders:** 447 mi (720 km) total boundary length; Former Yugoslav Republic of Macedonia 94 mi (151 km), Yugoslavia 179 mi (287 km) [Serbia 71 mi (114 km), Montenegro 108 mi (173 km)], Greece 175 mi (282 km)
- **Coastline:** 225 mi (362 km)
- **Territorial Seas:** 12 NM (22 km)
- **Highest Point:** Mt. Korabit, 9,033 ft (2,753 m)
- **Lowest Point:** Sea level
- **Longest Distances:** 211 mi (340 km) N-S / 92 mi (148 km) E-W
- **Longest River:** Drin River, 177 mi (285 km)
- **Largest Lake:** Lake Scutari, 149 sq mi (385 sq km)
- **Natural Hazards:** Earthquakes; tsunamis occur along southwestern coast; periodic drought
- **Population:** 3,510,484 (July 2001 est.) / World Rank: 127
- **Capital City:** Tiranë, located in the center of the country
- **Largest City:** Tiranë (population 270,000 in 2000)

OVERVIEW

Albania is one of the smallest countries in Europe. It is located on the west coast of the Balkan peninsula in southeastern Europe along the Strait of Otranto, which connects the Adriatic and Ionian seas. More than 70 percent of Albania's terrain is rugged and mountainous, with mountains running the length of the country from north to south. The remainder consists mostly of coastal lowlands, of which a large portion is former marshland that was reclaimed during the Communist era and is now agriculturally productive land. The largest lake in the Balkans—Lake Shkodër or Scutari—is found in Albania, as well as the deepest (Lake Ohrid).

Albania is located on the Eurasian Tectonic Plate. Shifting of the earth along the fault line that roughly defines the western edge of the central uplands causes frequent and occasionally severe earthquakes. Major damage occurred over wide areas in 1967 and 1969.

MOUNTAINS AND HILLS

Mountains

Albania's mountains, located to the north, east, and south of the coastal lowlands, can be divided into three groups. The northernmost group, the North Albanian

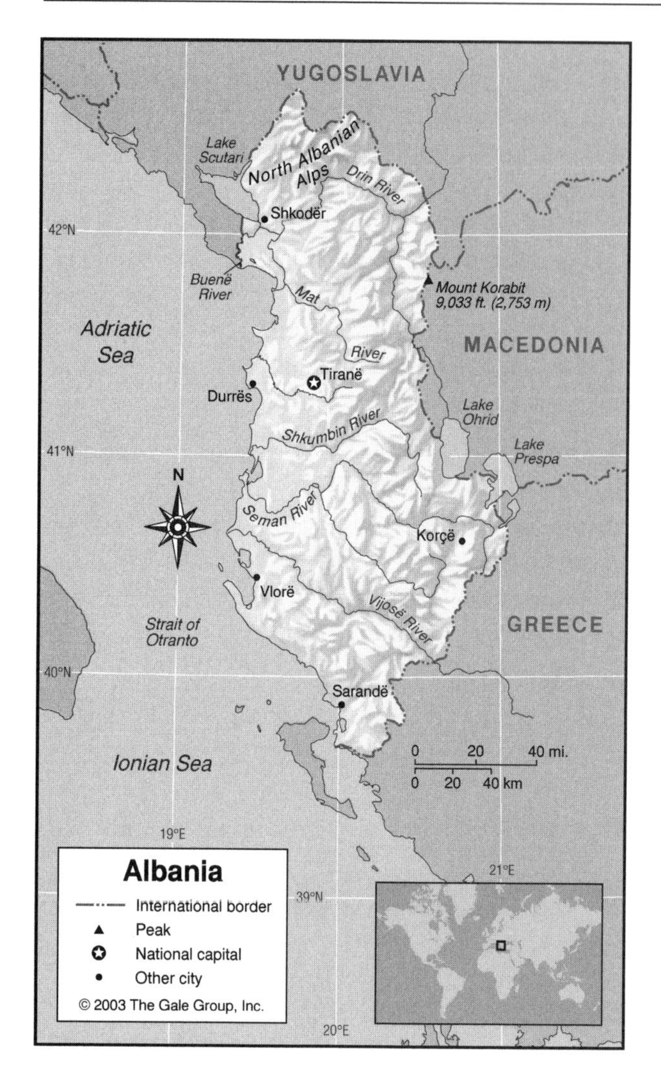

Alps, are an extension of the Dinaric Alpine chain and the Montenegrin limestone plateau. Some of the mountains in this region reach heights greater than 8,800 ft (2,700 m). These limestone peaks, popularly known as "the accursed mountains," are the country's most rugged.

The central uplands region extends south along the Macedonian border from the Drin River valley, which marks the southern boundary of the North Albanian Alpine area, to the southern mountains. Although the central uplands are generally lower than the North Albanian Alps, Albania's highest peak, Mt. Korabit, is found in this region.

The southern highlands are lower and more rounded than the mountains to the north. At the southernmost end of Albania, south of Vlorë, the mountains reach all the way across the country, meeting the Ionian Sea.

INLAND WATERWAYS

Lakes

Albania has three large tectonic lakes, which it shares with neighboring countries: Skadarsko Jezero (Lake Scutari) with Yugoslavia; Ohridsko Jezero (Lake Ohrid) with Macedonia; and Prespansko Jezero (Lake Prespa) with Greece. Ohridsko Jezero is the deepest lake, not only in Albania but in the Balkans, with a depth of 965 ft (294 m).

Rivers

Albania's major rivers are the Drin, the Buenë, the Mat, the Shkumbin, the Seman, and the Vijosë. They all empty into the Adriatic Sea. The Drin is the longest river in the country, while the Buenë is Albania's only navigable river.

THE COAST, ISLANDS, AND THE OCEAN

Oceans and Seas

Albania lies on the southeastern shore of the Adriatic Sea and is also bordered by the Ionian Sea to the south. The Strait of Otranto lies between southern Albania and Italy, serving to connect the Adriatic, Ionian, and Mediterranean seas.

The Coast and Beaches

Albania has no good natural harbors. Albania's Ionian coastal area is known for its great natural beauty; the area between Vlorë and Sarandë is called the "Riviera of Flowers."

CLIMATE AND VEGETATION

Temperature

Albania is located in a transition zone between a coastal Mediterranean climate to the west and a continental climate to the east, so its climate varies. The coastal plain has mild, rainy winters and hot, dry summers, moderated by sea breezes. In the mountains, continental air masses produce warm to hot summers and very cold winters with heavy snowfall; summer rainfall is also heavier in this region than on the coast. Albania's average annual temperature is 59°F (15°C).

Rainfall

Average annual rainfall ranges from about 40 in (100 cm) on the coastal plain to more than 100 in (250 cm) in the mountains.

Grasslands

Albania's coastal plain stretches south from the northern border to Vlorë, stretching 124 mi (200 km) from north to south and extending as much as 31 mi (50 km) inland. Wetlands covered much of this area until the Communist era, when they were drained to create productive agricultural land. However, flooding still occurs there. Citrus fruits, maize (corn), and wheat are grown in these lowlands.

Population Centers – Albania

(2000 POPULATION ESTIMATES)

Name	Population	Name	Population
Tiranë (Tirana, the capital)	244,000	Elbasan	69,900
Durrës (Durazzo)	72,400	Vlorë (Vlone, Valona)	61,100
Shkodër (Scutari)	71,200	Korçë (Koritsa)	57,100

SOURCE: Projected from United Nations Statistics Division and Albania Institute of Statistics data.

Provinces – Albania

2001 POPULATION ESTIMATES

Name	Population	Area (sq mi)	Area (sq km)	Capital
Berat	128,000	353	915	Berat
Bulqize	43,000	277	718	Bulqize
Delvine	11,000	142	367	Delvine
Devoll	35,000	166	429	Bilisht
Diber	86,000	294	761	Peshkopi
Durrës	183,000	176	455	Durrës
Elbasan	225,000	498	1,290	Elbasan
Fier	200,000	328	850	Fier
Gjirokaster	56,000	439	1,137	Gjirokaster
Gramsh	36,000	268	695	Gramsh
Has	20,000	144	374	Krum
Kavaje	78,000	152	393	Kavaje
Kolonje	17,000	311	805	Erseke
Korçë	143,000	676	1,752	Korçë
Kruje	64,000	144	372	Kruje
Kucove	36,000	43	112	Kucove
Kukes	112,000	916	2,373	Kukes
Kurbin	55,000	91	235	Lac
Lezhe	68,000	185	479	Lezhe
Librazhd	73,000	425	1,102	Librazhd
Lushnje	144,000	275	712	Lushnje
Malesi e Madhe	37,000	346	897	Koplik
Mallakaster	40,000	125	325	Ballsh
Mat	62,000	397	1,028	Burrel
Mirdite	37,000	335	867	Rreshen
Peqin	33,000	74	191	Peqin
Permet	26,000	359	930	Permet
Pogradec	71,000	280	725	Pogradec
Puke	34,000	399	1,033	Puke
Sarande	35,000	281	730	Sarande
Shkodër	186,000	630	1,631	Shkodër
Skrapar	30,000	299	775	Corovoda
Tepelene	32,000	315	817	Tepelene
Tiranë	523,000	461	1,193	Tiranë
Tropoje	28,000	403	1,043	Bajram Curri
Vlorë	147,000	621	1,609	Vlorë

SOURCE: Institute of Statistics, Albania.

Forests

Maquis (chaparral) is predominant in the coastal lowlands, giving way to oak and other deciduous trees on the country's lower slopes and beeches, chestnuts, and conifers on the higher ones. About 40 percent of Albania is forested, much of it with pine, oak, and beech trees. There are six national forests and 24 nature reserves.

HUMAN POPULATION

Albania is one of the most densely populated countries in Europe. The ban on birth control during the Communist era, coupled with increased life expectancy, has led to a high population growth rate. Between 30 percent and 40 percent of Albanians live in urban areas, compared to 15 percent before World War II. The capital city of Tiranë is by far the country's most populous city.

NATURAL RESOURCES

In addition to its forests, Albania is rich in mineral resources including copper, iron, phosphates, chromium, coal, petroleum, and natural gas. The Drin River has been dammed to produce hydroelectric energy.

FURTHER READINGS

CARE International in Albania Web site. *Better Economic and Social Development.* http://www.care.org.al/mission.htm. accessed Jan. 27, 2002.)

Carver, Robert. *The Accursed Mountains: Journeys in Albania.* London: John Murray, 1998.

Costa, Nicolas J. *Albania: A European Enigma.* New York: East European Monographs, 1995; distributed by Columbia University Press.

Senechal, Marjorie, and Stan Sherer. *Long Life to Your Children!: A Portrait of High Albania.* Amherst: University of Massachusetts Press, 1997.

Zickel, Raymond E., and Walter R. Iwaskiw, eds. *Albania, A Country Study.* 2nd ed. Federal Research Division, Library of Congress. Washington, D.C.: U.S. G.P.O., 1994.

Algeria

- **Area:** 919,590 sq mi (2,381,740 sq km) / World Rank: 12
- **Location:** Northern Hemisphere, straddles Eastern and Western Hemispheres, Northern Africa bordering the Mediterranean Sea to the north, Tunisia and Libya to the east, Niger to the southeast, Mali to the southwest, Mauritania and Western Sahara to the west, Morocco to the northwest
- **Coordinates:** 28°00′ N, 3°00′ E
- **Borders:** 3,933 mi (6,343 km) total boundary length; Tunisia 598 mi (965 km), Libya 610 mi (982 km), Niger 594 mi (956 km), Mali 855 mi (1,376 km), Mauritania 288 mi (463 km), Morocco 993 mi (1,601 km)
- **Coastline:** 620 mi (998 km)

- **Territorial Seas:** 12 NM (22 km)
- **Highest Point:** Mt. Tahat, 9,853 ft (3,003 m)
- **Lowest Point:** Chott Melrhir, 131 ft (40 m) below sea level
- **Longest Distances:** 1,500 mi (2,400 km) E-W/ 1,300 mi (2,100 km) N-S
- **Longest River:** Chelif, 143 mi (230 km)
- **Natural Hazards:** Desert regions subject to hot winds, called sirocco, which cause sandstorms, most common in summer; sporadic torrential rains cause flooding and mudslides (most recent November 2001); mountainous regions subject to earthquakes
- **Population:** 31,736,053 (July 2001 est.) / World Rank: 34
- **Capital City:** Algiers, centrally located on the northern coast of the Mediterranean Sea
- **Largest City:** Algiers (population estimated at 3 million)

OVERVIEW

The largest of the three countries (Algeria, Morocco, and Tunisia) that form the Maghreb region of northwest Africa, Algeria is the second-largest country on the continent, surpassed in size only by Sudan. It is a little less than 3.5 times the size of Texas and as large as the whole of Western Europe.

More than 80 percent of Algeria's land is part of the Sahara Desert, and almost completely uninhabited. To the north of the Sahara, roughly paralleling the country's Mediterranean border, lies the Tell region. It comprises a narrow strip of coastal plains and the two Algerian sections of the Atlas Mountains, as well as a plateau region that separates them. The Atlas Mountains cover much of Morocco and extend eastward into Tunisia. Within Algeria they are known as the Tell Atlas and Saharan Atlas systems. The northeastern corner of Algeria, where a compact massif area is broken up into mountains, plains, and basins, differs from the Tell region in that its prominent topographic features do not parallel the coast.

Algeria lies on the African Tectonic Plate. Major seismic activity periodically occurs in the northern Tell ranges, threatening the lives of inhabitants with earthquakes and mudslides. About 20,000 Algerians were killed by major earthquakes in 1790. Thousands of people were killed by earthquakes at Chlef in 1954 and 1980.

A land boundary dispute between Algeria and Tunisia was settled in 1993 but Libya still claims land in a portion of southeast Algeria.

GEO-FACT

Some of Algeria's desert oases rely on water supplied by man-made foggaras. A foggara is constructed by finding a spring or other underground source of water located in high ground, and digging a tunnel slanting downwards from the spring until it reaches level ground. Once finished, gravity forces water down the foggara and out into the desert where it forms an oasis. Foggaras are an ancient technology, with some dating back thousands of years, and are used throughout the Sahara and Asia.

MOUNTAINS AND HILLS

Mountains

The Tell Atlas is the more northerly of Algeria's Atlas mountain systems. Its peaks, which rise to heights of over 6,000 ft (1,830 m), include the ranges of the Greater and Lesser Kabylie, as well as the Tlemcen and Madjera ranges.

The Saharan Atlas Mountains are higher and more continuous than the Tell Atlas Mountains. They consist of three mountain chains: the Ksour near the Moroccan border, the Amour, and the Ouled Nail south of Algiers. The mountains are better watered than the plateaus to their north and the highland topography includes some good grazing land.

Dominating the southeastern section of Algeria's Saharan desert region are the

Ahaggar Mountains, with irregular heights reaching above 6,561 ft (2,000 m). The country's highest peak, Mt. Tahat (9,853 ft / 3,003 m), is found in this range.

Plateaus

Stretching more than 372 mi (600 km) eastward from the Moroccan border, the High Plateaus consist of a steppe-like tableland lying between the Tell and Saharan Atlas ranges. Averaging between 3,609 and 4,265 ft (1,100 and 1,300 m) in elevation in the west, the plateaus drop to 1,312 ft (400 m) in the east. So dry that they are sometimes thought of as part of the Sahara, they are covered by alluvial debris formed by erosion of the mountains, with an occasional ridge projecting through the alluvial cover to interrupt the monotony of the landscape.

The plains are broken up by shallow basins in which water collects during the rainy season but which become

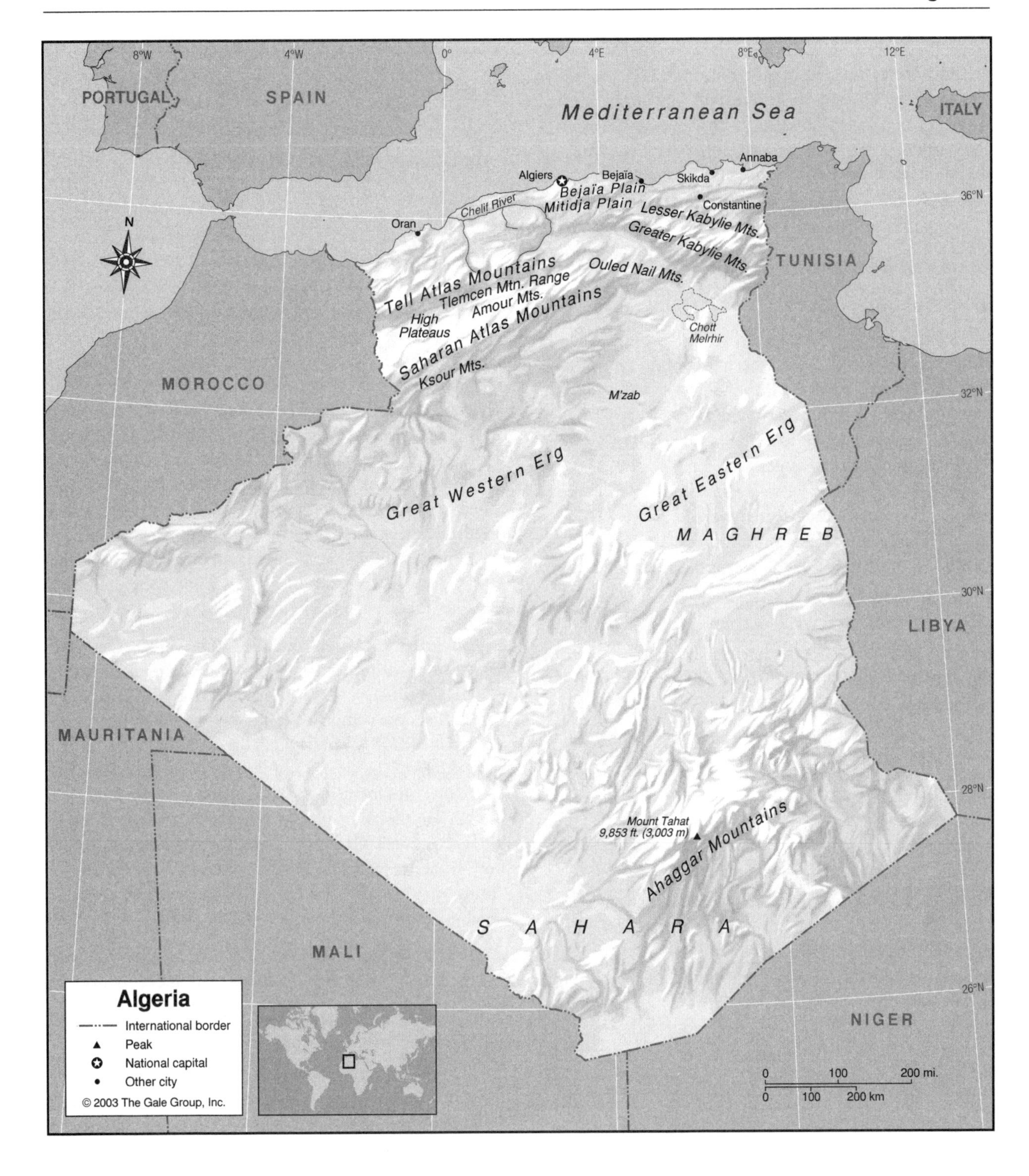

dry lake beds and salt flats (chotts) during the hottest months.

INLAND WATERWAYS

Because precipitation is intermittent and often scanty, Algeria has few permanent inland bodies of water and no navigable rivers—almost all of them flow only seasonally or irregularly. The longest and best known is the Chelif, which wanders for 143 mi (230 km) from its source in the Tell Atlas to the Mediterranean. Most of the Tell streams diminish to trickles or go dry in summer, but in the west reservoirs have been developed in the Chelif and Hamiz river basins for irrigation purposes. Rivers and streams on the high plateaus flow irregularly, and the region contains salt marshes and shallow salt lakes. The land in the southernmost Saharan region is largely arid but has some date-palm oases.

THE COAST, ISLANDS, AND THE OCEAN

Algeria borders the Mediterranean Sea on the north. The major cities of Algiers, Oran, and Annaba are located on Algeria's narrow, intermittent coastal plains, as well as the port cities of Bejaia and Skikda. These plains alternate with massifs along much of the coast, except for the easternmost section, where the coast is mostly mountainous.

CLIMATE AND VEGETATION

Temperature

Algeria's geographical diversity makes for a range of climatic conditions. The northern part of the country has a Mediterranean climate with mild, wet winters and hot, dry summers. Temperatures range from 50° to 54°F (10° to 12°C) in winter to 75° to 79°F (24° to 26°C) in summer. The plateau region has a semiarid climate, with a greater range of temperatures between the summer and winter months. Summer temperatures average 79° to 82°F (26° to 28°C) and winter temperatures average 39° to 43°F (4° to 6°C), with frost and occasional snow on the massifs.

Temperatures vary the most in the Sahara Desert region, which has an arid climate. Days with temperatures of 95°F (35°C) or higher can be accompanied by nights with temperatures below freezing. Temperatures range from 14° to 93°F (-10° to 34°C), and reaching extreme highs of 120°F (49°C). The hot, dusty wind called the sirocco is common in the summer.

Rainfall

Like its temperatures, Algeria's rainfall varies in its different geographical regions. The northern Tell region receives 15 to 27 in (38 to 69 cm) of rain annually, with up to 40 in (100 cm) in its easternmost section. Heavy rains can cause flooding, but evaporate too soon to irrigate the land. In the plateau region rainfall is seasonal due to changes to wind direction (northerly and westerly in winter, easterly and northeasterly in summer). Precipitation is heaviest between September and December, and lighter from January to August with very little rainfall in the summer months. Rainfall in the country's Saharan region is sparse and irregular, averaging less than 4 in (10 cm) annually, and drought conditions are common.

Grasslands

Most of Algeria's flatlands are desert or semi-arid. The best agricultural areas in the country lie in the coastal plains. The only areas of the country that are relatively flat and well-watered are the gentle hills extending 62 mi (100 km) westward from Algiers, the Mitidja Plain, which was a malarial swamp before its clearing by the French, and the Bejaïa Plain. They support much of Algeria's agriculture. The area around Algiers and Oran is also heavily cultivated.

There are areas of scrub and grassland in the Tell Atlas region, where cedar and pine forests have been destroyed by fire and logging. Esparto grass, a needlegrass common to African deserts, is abundant in the western part of the High Plateaus, where brushwood is also found.

Deserts

South of the Saharan Atlas, the Algerian portion of the Sahara Desert extends southward 931 mi (1,500 km) to the Niger and Mali frontiers, with an average elevation of about 1,500 ft (460 m). Immense areas of sand dunes, called ergs, occupy about one-fourth of the region. Among these areas, the two major ones are the Grand Erg Occidental (Great Western Erg) and the larger Grand Erg Oriental (Great Eastern Erg), where enormous dunes 6.5 to 16.5 ft (two to five meters) high are spaced about 130 ft (40 m) apart. Much of the remainder of the desert is covered by bare, rocky platforms called hamada that are elevated above the sand dunes. Almost the entire southeastern quarter is taken up by the Ahaggar Mountains. Surrounding the Ahaggar are sandstone plateaus cut by deep gorges, and to the west a flat, pebble-covered expanse stretches to the Mali frontier.

The desert can be divided into two sectors. The northern one extends a little less than half the distance to Algeria's southern borders. Less arid than the area to the south, it supports most of the few people who live in the region, and most of the region's oases are found here. The sand dunes of the Grand Erg Oriental and the Grand Erg Occidental are its most prominent topographical feature, but between them lie plateaus, including a complex limestone structure called the M'zab, where the M'zabite Berbers have settled. The southern zone of the Sahara is almost totally arid. Barren rock predominates, and its most prominent feature is the Ahaggar range.

Vegetation in the desert, which includes acacia, jujube, and desert grasses, is sparse and unevenly distributed.

Forests

Due to extensive deforestation, Algeria's forests have receded to the upper reaches of the Tell Atlas and Saharan Atlas mountains. The evergreen forests of the Tell Atlas mountains contain Aleppo pine, juniper, cedar, and evergreen oak trees, as well as some deciduous varieties including cork oak and several other species of oak.

HUMAN POPULATION

Most of Algeria's cities and over 90 percent of its population are concentrated in the northern Tell region nearthe Mediterranean coast, while the plateau and desert regions to the south are sparsely populated. Urban dwellers account for about 60 percent of the population.

Population Centers – Algeria

(2000 POPULATION ESTIMATES)

Name	Population
Algiers (El-Djezair, capital)	1,700,000
Oran (Ouahran)	700,000
Constantine (Qacentina)	500,000
Batna	300,000

SOURCE: Statistics Algeria, National Office of Statistics.

NATURAL RESOURCES

Natural gas and petroleum dominate Algeria's economy. Its natural gas reserves, concentrated in the northeastern and north-central parts of the country, are among the largest in the world, as are its phosphate deposits. Algeria's petroleum reserves are among the largest in Africa. Algeria is also the only African country that produces mercury. Other natural resources include iron ore, bentonite, uranium, lead, zinc, kaolin, barites, sulfur, strontium, fuller's earth, and salt. Forestry products include cork, firewood, charcoal, and wood for industrial use.

FURTHER READINGS

ArabNet. *Algeria.* http://www.arab.net/algeria/algeria_contents.html. (Accessed Feb.3, 2002.)

Fromentin, Eughne. *Between Sea and Sahara: An Algerian Journal.* Translated by Blake Robinson. Athens: Ohio University Press, 1999.

McLaughlan, Anne, and Keith McLaughlin. *Morocco & Tunisia Handbook, 1996.* Lincolnwood, IL: Passport Books, 1995.

Metz, Helen Chapin, ed. *Algeria, A Country Study.* 5th ed. Federal Research Division, Library of Congress. Washington, D.C.: U.S. Government Printing Office, 1994.

Ruedy, John. *Modern Algeria: The Origins and Development of a Nation.* Bloomington: Indiana University Press, 1992.

Stone, Martin. *The Agony of Algeria.* New York: Columbia University Press, 1997.

Andorra

- **Area:** 180 sq mi (468 sq km) / World Rank: 184
- **Location:** Land-locked nation in the Eastern and Northern Hemispheres, lies on the southern slopes of Pyrenees Mountains of Europe south of France and north of Spain
- **Coordinates:** 42°25′ to 42°40′ N, 1°30′E

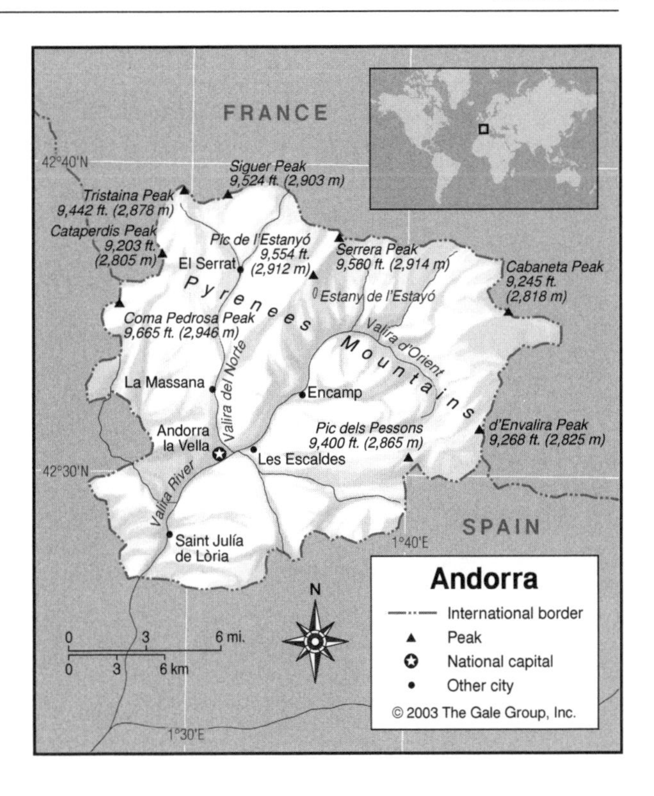

- **Borders:** 74.6 mi (120.3 km) total boundary length / France, 35.1 mi (56.6 km); Spain, 39.5 mi (63.7 km)
- **Coastline:** None. Andorra is landlocked.
- **Territorial Seas:** None
- **Highest Point:** Coma Pedrosa Peak, 9,665 ft (2,946 m)
- **Lowest Point:** Runer River, 2,755 ft (840 m), where the Runer River and Valera River meet.
- **Longest Distances:** 18.7 mi (30.1 km) east to west / 15.8 mi (25.4 km) north to south.
- **Natural Hazards:** Severe winters
- **Population:** 67,627 (July 2001 est.) / World Rank: 191
- **Capital City:** Andorra la Vella, located in the southwestern part of Andorra
- **Largest City:** Andorra la Vella, 25,000 (2000)

OVERVIEW

Andorra is one of the smallest independent countries on earth. Located in the Pyrenees Mountains between Spain and France, Andorra's terrain consists of gorges, narrow valleys, and defiles surrounded by mountain peaks rising higher than 9,500 ft (2,900 m). There is little level ground. All the valleys are at least 3,000 ft (900 m) above sea level and the mean altitude is over 6,000 ft (1,800 m).

MOUNTAINS AND HILLS

All of Andorra is mountainous. The highest mountain peak is Pic de Coma Pedrosa (Coma Pedrosa Peak), which rises to 9,665 ft (2,946 m) near where the western border of Andorra and the borders of France and Spain meet. Along the northwestern border with France, Pic de Cataperdis (9,203 ft/2,805 m) and Pic de Tristaina (9,442 ft/2,878 m) can be found. Pic de Siguer (9,524 ft/2,903 m) and Pic de la Serrera (9,560 ft/2,914 m) lie along the northern border, and Pic de la Cabaneta (9,245 ft/2,818 m) is in the east. Near the southeastern point where the borders of the three countries meet lies Pic d'Envalira (9,268 ft/2,825 m) and Pic dels Pessons (9,400 ft/2,865 m). A lake, Estany de l'Estanyó, and a mountain peak, Pic de l'Estanyó (9,564 ft/2,915 m) lie just east of El Serrat and are accessible only by hiking trail.

INLAND WATERWAYS

Andorra is drained by a single basin whose main river, Valira River (Riu Valira), has two branches and six smaller open basins. These basins gave the name by which the region was traditionally known, The Valleys (Les Valls). The Valira del Norte is the northwest branch of the main river, flowing through the cities of La Massana, Ordino, and El Serrat. The Valira d'Orient is the northeast branch, flowing through Les Escaldes, Encamp, Canillo, Soldeu, and Pas de la Casa.

CLIMATE AND VEGETATION

Temperature

Andorra has a temperate climate, but winters are severe because of the high elevation. Snow completely fills the northern valleys for several months. During the April to October rainy season, rainfall can be heavy and it is reported to exceed 44 in (122 cm) per year in the most mountainous regions.

HUMAN POPULATION

The population is concentrated in the valleys where cities have developed; about 95% of the population lives in these cities, including the capital Andorra la Vella, Les Escaldes, Sant Julía de Lòria, Encamp, and La Massana.

NATURAL RESOURCES

Deposits of iron ore, lead, alum, and building stones are among the resources exploited in Andorra, although the economy depends to a much greater extent on tourism. Andorra's mountainous terrain attracts about 12 million tourists annually, primarily for skiing and hiking.

Parishes – Andorra

1999 POPULATION ESTIMATES

Name	Population	Area (sq mi)	Area (sq km)	Capital
Andorra la Vella	21,200	22	59	Andorra la Vella
Canillo	2,700	47	121	Canillo
Encamp	10,600	29	74	Encamp
La Massana	6,300	25	65	La Massana
Les Escaldes-Engordany	15,300	*	*	-
Ordino	2,300	33	85	Ordino
Sant Julia de Uria	7,600	23	60	Sant Julia de Uria

* included in Andorra la Vella

SOURCE: Government of Andorra.

FURTHER READINGS

De Cugnac, Pascal. *Pyrenees & Gascony: Including Andorra.* London: Hachette UK, 2000.

Duursma, Jorri. *Self-Determination, Statehood, and International Relations of Micro-states.* New York: Cambridge University Press, 1996.

Angola

- **Area:** 481,226 sq mi (1,246,700 sq km) including the exclave of Cabinda. / World Rank: 24
- **Location:** Located in the Eastern and Southern Hemispheres on the west coast of the African continent,south and southeast of the Democratic Republic of Congo (DROC), west of Zambia, north of Namibia, east of the Atlantic Ocean. Cabinda region is separated from the rest of Angola by the DROC, and is completely surrounded by that country and the Republic of the Congo.
- **Coordinates:** 12° 30′S and 18° 30′E
- **Borders:** 3,233 mi (5,198 km) total boundary length; the border of DROC, 1,557 mi (2,511 km, 136 mi (220 km) of which is the boundary of the discontiguous Cabinda province; Republic of the Congo, 125 mi (201 km); Namibia, 853 mi (1,376 km); Zambia, 688 mi (1,110 km)
- **Coastline:** 992 mi (1,600 km)
- **Territorial Seas:** 12 NM (22 km) from the coast; exclusive economic zone is 200 NM (360 km)
- **Highest Point:** Mount Moco, 8,596 ft (2,620 m)
- **Lowest Point:** Sea level
- **Longest Distances:** 1,092 mi (1,758 km) SE-NW / 926 mi (1,491 km) NE-SW; Cabinda: 103 mi (166 km) NNE-SSW / 39 mi (62 km) ESE-WNW
- **Longest River:** Congo, 2,693 mi (4344 km)

- **Natural Hazards:** Occasional heavy rainfall with accompanying floods
- **Population:** 10,366,031 (July 2001) / World Rank: 72
- **Capital City:** Luanda, located on the Atlantic Coast.
- **Largest City:** Luanda, 2,665,000 (2000 estimate).

OVERVIEW

A land of many broad tablelands above 3,300 ft (1,000 m), Angola also has both central and southern high plateaus that range up to 7,900 ft (2,400 m). The country as a whole is relatively dry, especially in the south and along the coast. The interior is the source of many rivers but is predominately savanna. Land abuse, such as desertification, forest loss, and water impurity are significant environmental problems throughout the country.

The Angolan province of Cabina lies somewhat to the north of Angola, separated from the rest of the country by the Democratic Republic of the Congo (DROC). It receives more rainfall than most of Angola and parts of it are covered by rain forest.

MOUNTAINS AND HILLS

Mountains

Angola's tallest mountains are found along the edges of the coastal plain where it meets the plateaus that make up most of the interior. Its highest peak, Mount Moco (Morro de Moco), can be found here near the city of Huambo. Other major peaks include Mount Mejo (Morro de Mejo) at 8,474 ft (2,583 m) in the Benguela region and Morro de Vavéle (Mount Vavéle) at 8,133 ft (2,479 m) in Kuanza Sul. Running through the center of the country and into Zambia is the Lunda Divide, a set of low ridges marking the division between north flowing and southeast flowing rivers.

Plateaus

Most of Angola's interior is part of the great central plateaus that make up much of southern Africa. In the west central part of the country is the Bié Plateau, in the south central region the Huila Plateau, and in the north the Malanje Highlands. The Bié Plateau and the Huila Plateau are the most elevated regions, with Angola's tallest mountains. They drop off sharply to the coastal plains. In the north, the Malanje Highlands are of lower elevation and reach the coastal fringe in a gradual drop. Most of the eastern half of Angola is relatively flat and open plateau characterized by sandy soils.

Hills and Badlands

The northeastern part of the Cabinda exclave is a hilly region known as the Mayombé Hills. They were once under rainforest cover but have been heavily cut.

GEO-FACT

The Quicama National Park covers an area between the Atlantic Ocean to the west, the Cuanza River to the north, and the Longa River to the south in northwestern Angola. Quicama was established as a game reserve in 1938 and became a national park in 1957. The park incorporates several habitats, including dense thicket, savanna, and grasslands, and once provided sanctuary for a wide variety of African animals. Due to Angola's civil war and illegal poaching, since the 1980s many of the animal herds have been seriously threatened. International observers report sightings of red buffalo, antelope, eland, bushbuck, and waterbuck, but no reliable information is available; a vast variety of bird species are believed to be thriving in the park. The numbers of elephant, rhinoceros, and giant black sable surviving are not known.

INLAND WATERWAYS

Rivers

Angola has a diverse system of rivers, including some large constantly fed rivers, such as the Congo; seasonally fed rivers; and temporary rivers and streams. Most of the country's many rivers originate in central Angola, but the pattern of flow and ultimate outlets is varied.

The enormous Congo River (or Cuango), which drains much of central Africa, flows along a small part of the border with the DROC before emptying into the Atlantic Ocean. This is one of the few navigable rivers in Angola. Most of the other rivers in northeastern Angola, those north of the Lunda Divide, flow into the Congo basin.

Another great river, the Zambezi, flows through a portion of eastern Angola. Many of the rivers to the south of the Lunda Divide and east of the coastal plains are tributaries of this river. The Okavango (Cubango) River has its source in central Angola and flows southeast into Namibia, running for 998 mi (1610 km) before emptying into the Okavango Swamp in Botswana.

The Cuanza River (or Kuanza or Kwanza), at 600 mi (966 km), is the longest river that flows entirely within

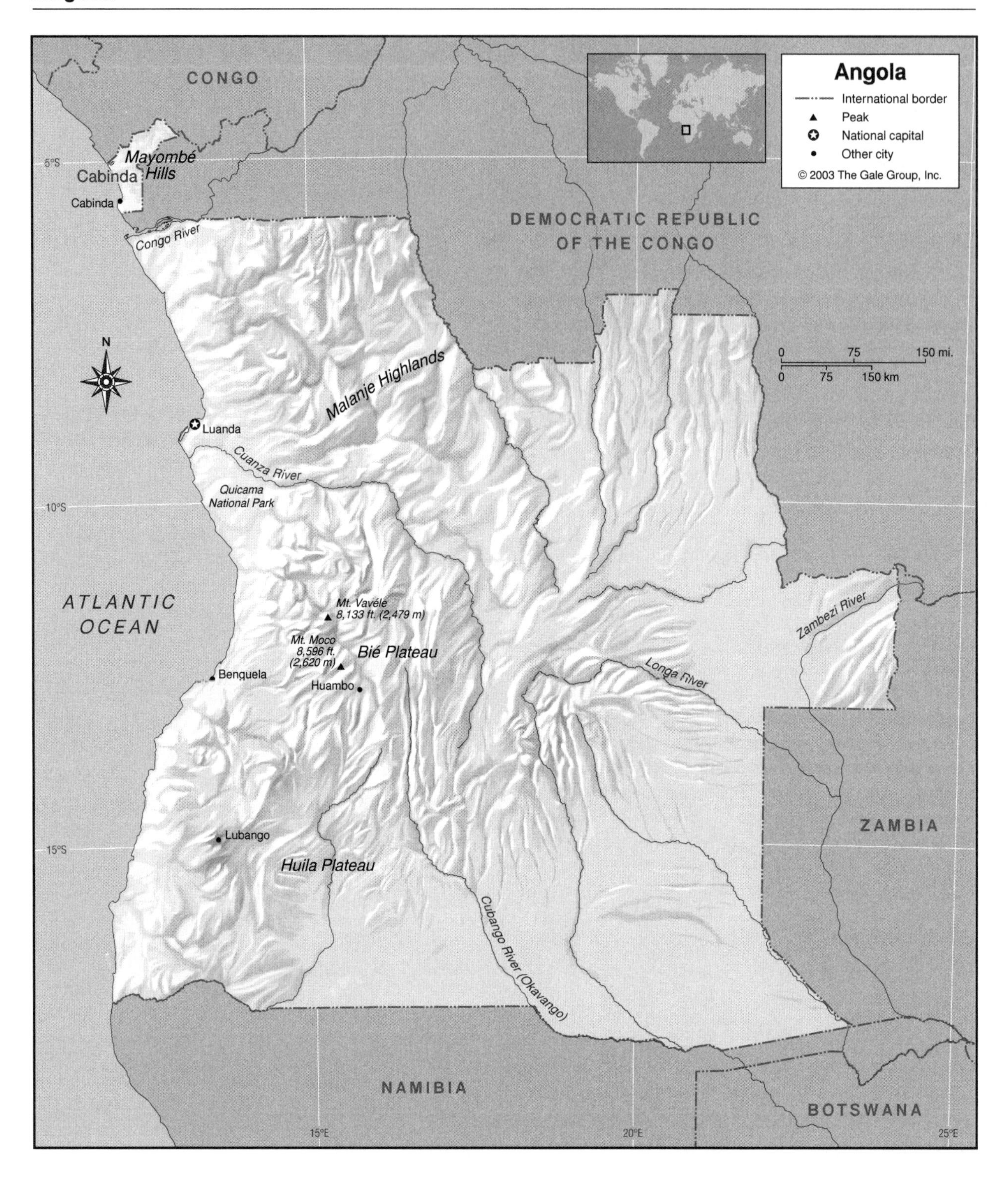

Angola's borders, and is the only one besides the Congo that is navigable, although only for 126 mi (200 km). The Cambembe Dam on the Cuanza River provides power to Luanda. Several other smaller rivers flow from the plateaus westward into the Atlantic and provide both irrigation for the otherwise dry coastal strip and hydroelectric power. The country has six dams, but as of 2002 only three were functioning. The southernmost rivers flowing into the Atlantic are seasonal—many are completely dry much of the year.

THE COAST, ISLANDS, AND THE OCEAN

The Coast and Beaches

Angola's Atlantic coast is an arid strip that is irrigated by the westward flowing rivers. The Benguela Current that runs north along much of the coast is cold, reducing precipitation, although it does support a food fishing industry that contributes to export income.

CLIMATE AND VEGETATION

Temperature

Angola's temperatures average 68°F (20°C), but there is considerable variation from the warmer coastal region to the cooler central plateau. The north—from Ambriz to Cabinda—has a wet, tropical climate; the eastern strip, starting slightly north of Luanda and stretching to Namibe, has a moderate tropical climate; the southern central strip between the plateau and the border with Namibia, as well as along the coast as far north as Namibe, has hot, dry desert conditions.

There are two seasons in Angola. The dry and cool winter lasts from June to late September. The hot, rainy summer extends from October into May.

Rainfall

Coastal regions are arid or semiarid because of the effects of the north-flowing cold Benguela Current in the Atlantic Ocean. The coastal region south of Benguela is made up of the most northern reaches of the Namib Desert of Namibia. The most southerly areas, near the border with Namibia, are characterized by sand dunes. In the middle coastal region around Benguela, low-growing scrubby plants are found; thicker brush develops further north along the coast. The interior, while not as arid as the southern coast, is still fairly dry. Only in the north is rainfall at all plentiful.

The annual average rainfall is 2 in (5 cm) on the coast at Namibe; 13 in (34 cm) along the northern coast at Luanda; and as high as 59 in (150 cm) in the northeast. The rainy season is October through May, but drought is not uncommon.

Grasslands

The coastal regions, while they receive relatively little rainfall, are well irrigated because of the drainage of the rivers from the higher central plateaus and serve as good cropland. The coastal plains vary in width from about 15 mi (25 km) to more than 93 mi (150 km) south of Luanda. The flat interior plateaus are predominantly savanna. Meadows and pastures constitute about 23 percent of the total land area. Elephant grass and scrubby forest cover the surface of the sandy floodplains.

Deserts

The southern part of the country is sandy and dry and has sparse vegetation, except along the courses of major rivers. This is less true in the southeastern parts of the country, but even there much of the vegetation disappears during the dry season. Due to vagaries in precipitation the far south is marked by sand dunes, which give way to dry scrub in the central portions.

Provinces – Angola

ESTIMATED 1995 POPULATION

Name	Population	Area (sq mi)	Area (sq km)	Capital
Bengo	170,000	14,173	36,708	Caxito
Benguela	700,000	15,115	39,151	Benguela
Bié	1,130,000	27,149	70,317	Kuito
Cabinda	170,000	2,744	7,107	Cabinda
Huambo	1,600,000	12,796	33,141	Huambo
Huíla	875,000	30,499	78,992	Lubango
Cuando Cubango	132,000	76,671	198,577	Menongue
Cuanza Norte	390,000	7,717	19,988	N'Dalatando
Cuanza Sul	680,000	21,281	55,117	Sumbe
Cunene	240,000	29,327	75,956	N'Giva
Luanda	1,650,000	570	1,477	Luanda
Lunda Norte	300,000	39,685	102,784	Lucapa
Lunda Sul	160,000	29,860	77,336	Saurimo
Malanje	920,000	33,686	87,247	Malanje
Moxico	330,000	77,870	201,683	Lwena
Namibe	125,000	22,043	57,090	Namibe
Uíge	925,000	23,728	61,455	Uige
Zaire	202,000	14,281	36,989	M'Banza Kongo

SOURCE: Projected from data compiled by Instituto Nacional de Estatística, Angola.

Forests and Jungles

Precipitation at the highest points in the central plateaus permit the growth of deciduous forest, although it has been much depleted from timber and fuel demands. The Mayombé Hills in Cabinda have also been heavily cut of their former cover of rain forest. The northeastern part of Angola also has some rain forest.

HUMAN POPULATION

A July 2001 figure of 10,366,061 includes a 2.15 percent growth rate estimate. It is also estimated that 34 percent of the population live in urban areas. Ethnic groups include Ovimbundu (37 percent), Kimbundu (25 percent), Bakongo (13 percent), mestiço (mixed European and Native African, 2 percent), European (1 percent), and other (22 percent).

NATURAL RESOURCES

Natural resources in Angola are abundant and include: petroleum, diamonds, iron ore, phosphates, copper, feldspar, gold, bauxite, and uranium. Angola's war for independence and ongoing civil strife has hampered the economic potential that resides in the land. Both petroleum production and diamond mining lead the industrial sector.

FURTHER READINGS

Burness, Don. *On the Shoulder of Martí: Cuban Literature of the Angolan War.* Colorado Springs, CO: Three Continents Press, 1996.

Cushman, Mary Floyd. *Missionary Doctor, The Story of Twenty Years in Africa.* New York: Harper & Brothers, 1944.

Ekwe-Ekwe, Herbert. *Conflict and Intervention in Africa: Nigeria, Angola, Zaire.* New York: St. Martin's Press, 1990.

Embassy of The Republic of Angola. *Welcome to the Public of Angola.* http://www.angola.org (accessed February 22, 2002.)

Maier, Karl. *Angola: Promise and Lies.* London: Serif, 1996.

University of Pennsylvania, African Studies Program. *Angola Page.* http://www.sas.upenn.edu/African_Studies/Country_Specific/Angola.html (accessed February 22, 2002).

U.S. Department of State. *Angola, 1996 Post Report.* Washington, DC: The Department of State, 1996.

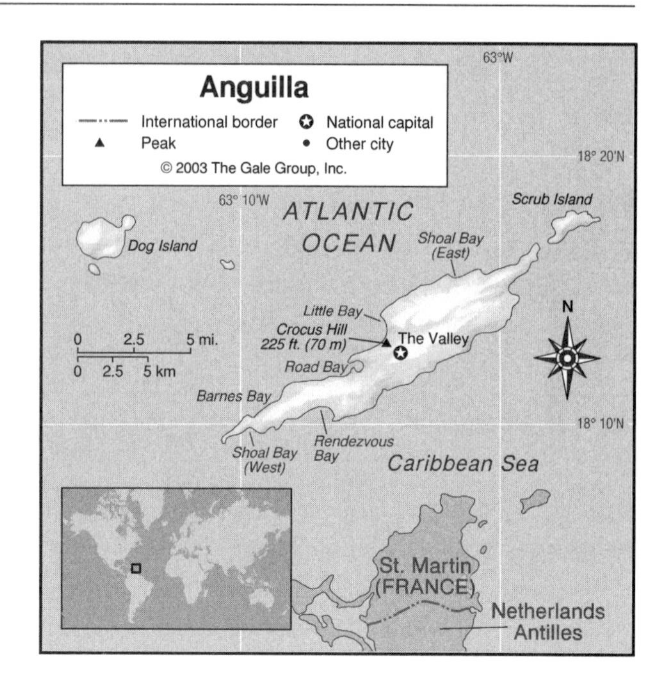

Anguilla

Territory of the United Kingdom

- **Area:** 35 sq mi (90 sq km), in addition to Sombero Island (2 sq mi /5 sq km) / World Rank: 201
- **Location:** Northern and Western Hemispheres, northernmost of the Leeward Islands
- **Coordinates:** 18° 15′ N and 63° 10′ W
- **Borders:** Entirely bounded by ocean, no international boundaries
- **Coastline:** 38 mi (61 km)
- **Territorial Seas:** 3 NM (5 km)
- **Highest Point:** Crocus Hill, 213 ft (65 m)
- **Lowest Point:** Sea level
- **Longest Distances:** Main island: 13 mi (21 km) long / 3 mi (4.8 km) wide
- **Longest River:** None
- **Natural Hazards:** Subject to hurricanes and severe tropical storms
- **Population:** 13,132 (mid-2001 est.) / World Rank:203
- **Capital City:** The Valley, located at mid-point of the north coast (administrative center; Anguilla has no capital)
- **Largest City:** The Valley

OVERVIEW

Anguilla is one of the Leeward Islands, which lie between the Caribbean Sea in the west and the open Atlantic Ocean in the east. It is a long, flat, dry, scrub-covered coral island, south and east of Puerto Rico and north of the Windward chain. It is an island of no significant elevations with its terrain consisting entirely of beaches, dunes, and low limestone bluffs.

MOUNTAINS AND HILLS

Mountains

Anguilla's highest elevation, Crocus Hill, is 225 ft (70 m). Crocus Hill is among the cliffs that line the northern shore.

INLAND WATERWAYS

Anguilla has no inland waterways of any significant size.

THE COAST, ISLANDS, AND THE OCEAN

Major Islands

Sombrero, a 1-mi (1.6-km) long rock island, lies about 35 mi (56 km) to the northwest of the main island.

Other, smaller, islands lie close by Anguilla, including Scrub Island and Dog Island.

The Coast and Beaches

The numerous bays—Barnes, Little, Rendezvous, Shoal, and Road—lure many vacationers to this tropical island. The coast and the beautiful, pristine beaches are integral to the tourism-based economy of Anguilla. Because of Anguilla's warm climate, the beaches can be used year-round.

GEO-FACT

Although sighted by Columbus in 1496, Europeans did not colonize Anguilla until 1650 when the British arrived from neighboring St. Kitts. Anguilla became a separate dependency from the Leeward Islands, which include St. Kitts and Nevis, in 1980.

CLIMATE AND VEGETATION

Temperature

Northeastern trade winds keep this tropical island cool and dry. Average annual temperature is 81°F (27°C). July–October is its hottest period, December–February, its coolest.

Rainfall

Rainfall averages 35 in (89 cm) annually, although the figures vary from season to season and year to year. The island is subject to both sudden tropical storms and hurricanes, which occur in the period from July to October. The island suffered damage in 1995 from Hurricane Luis.

Vegetation

Anguilla's coral and limestone terrain provide no subsistence possibilities for forests, woodland, pastures, crops, or arable lands. Its dry climate and thin soil hamper commercial agricultural development.

HUMAN POPULATION

Estimated population for Anguilla in mid-2001 was 12,132 with a growth rate of 2.68 percent. Anguillans are primarily of African descent, with an European (especially Irish) ancestral presence. The population is overwhelmingly Christian. Most residents are involved in fishing and subsistence farming, raising such crops as pigeon peas, sweet potatoes, Indian corn, and beans.

NATURAL RESOURCES

Anguilla's natural resources are the waters and coral reefs that surround this island nation. Deep sea and lobster fishing provides not only food for its natives, but is an important part of the tourism industry on the island. The island has no forests, pastures, woodland, or arable land, but does harvest salt from commercial salt ponds.

FURTHER READINGS

Blanchard, Melinda, and Robert Blanchard. *A Trip to the Beach.* New York: Clarkson Potter Publishers, 2000.

Brisk, William J. *The Dilemma of a Ministate: Anguilla.* Columbia, SC: Institute of International Studies, University of South Carolina, 1969.

Browne, Whitman T. *From Commoner to King; Robert L. Bradshaw, Crusader for Dignity and Justice in the Caribbean.* Lanham, MD: University Press of American, 1992.

Westlake, Donald E. *Under an English Heaven.* New York: Simon and Schuster, 1972.

Antarctica

- **Area:** 5,405,430 sq mi (14,000,000 sq km, approximation) / World Rank: 2
- **Location:** Covers the South Pole in the Southern Hemisphere. Its territory forms the southernmost portions of the Eastern and Western Hemispheres.
- **Coordinates:** 90°00′ S / 0°00′ E
- **Borders:** None
- **Coastline:** 11,164 mi (17,968 km)
- **Territorial Seas:** None
- **Highest Point:** Vinson Massif, 16,864 ft (5,140 m)
- **Lowest Point:** Bentley Subglacial Trench, 8,333 ft (2,540 m) below sea level
- **Longest Distances:** Longest distance traversing the South Pole 3,337 mi (5,339 km); shortest distance traversing the South Pole 771 mi (1,234 km)
- **Longest River:** Onyx River, 20 mi (25 km)
- **Largest Lake:** Lake Vostok, estimated size 10,000 sq mi (26,000 sq km), but buried under 2.8 mi (3.5 km) of ice.
- **Natural Hazards:** Extreme cold; blizzards; cyclones; volcanoes; icebergs; high radiation due to ozone depletion
- **Population:** No indigenous population; numbers of scientists and other research personnel from nearly 30 nations range seasonally from 1,000 to 4,000 or more. / World Rank: 207
- **Capital City:** None
- **Largest City:** McMurdo Station, in East Antarctica, summer population estimated at 1,200; winter, 250.

OVERVIEW

The continent of Antarctica is almost entirely south of the Antarctic Circle (66.5°S), with the South Pole located on it. Antarctica ranks fifth in size among the world's continents, being larger than Australia or Europe. It is surrounded by the Southern Ocean, and islands in

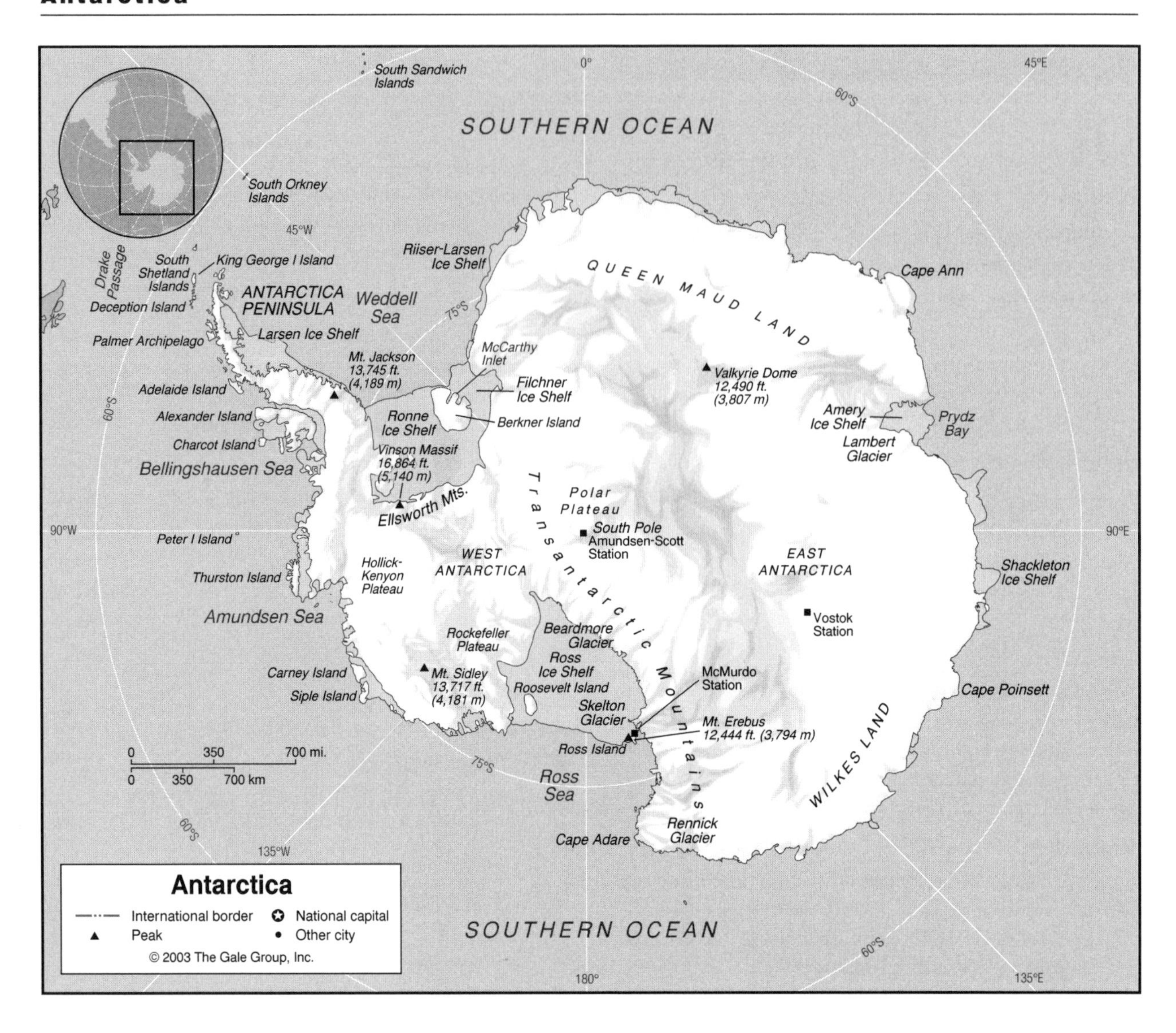

this ocean (south of 60° latitude) are considered part of Antarctica.

Antarctica is generally described as having two parts, West Antarctica and East Antarctica. West Antarctica lies directly south of the South American continent, and includes the Antarctic Peninsula, which extends farther north than any other part of the continent. East Antarctica is the larger part of the continent, and lies south of the southern tips of Africa and Australia. East and West Antarctica are separated by the Transantarctic Mountains. Explorers and research scientists have given names to almost every stretch of the Antarctic coast, and to the notable geographic features they have discovered.

Located as it is in the far south, Antarctica is extremely cold, even during the summer. About 98 percent of the land area is permanently covered with ice sheet. The remainder is exposed barren rock. Antarctica is generally mountainous, with elevations typically ranging from 6,600 to13,200 ft (2,000 to 4,000 m). Mountain peaks rise to heights in excess of 16,500 ft (5,000 m).

Antarctica is unique in that it does not belong to any nation. Parts of Antarctica are claimed by seven nations: Argentina, Australia, Chile, France, New Zealand, Norway, and the United Kingdom. However, their claims are not recognized by the international community, and are in abeyance under the terms of the Antarctic Treaty. This treaty went into effect in 1961, when it was first signed by 12 nations. The Treaty attempts to clarify issues related to territorial claims and to provide a framework for this unique and complex international cooperation. It specifies that "Antarctica shall be used for peaceful purposes only." Other nations that sponsor research in Antarctica have since become consulting members of the Treaty. As of 2002, there are 27 nations with consulting member status in the treaty agreements for Antarctica.

MOUNTAINS AND HILLS

Mountains

Dividing Antarctica into two regions, East Antarctica (Greater Antarctica) and West Antarctica (Lesser Antarctica), is the continent's major mountain range, the Transantarctic Mountains. Lying between the mountain peaks are Victoria Valley, Wright Valley, and Taylor Valley. These large, relatively ice-free areas, are known collectively as the McMurdo Dry Valleys. They account for about 1,733 sq mi (4,800 sq km) of territory free of ice in an area measuring approximately 48 by 60 mi (60 by 75 km). The areas are ice-free because the mountains impede the flow of the sheet of ice that covers most of the continent. The valleys are filled with sandy, spongy gravel. Glaciers flow into the deepest parts of the valleys from the surrounding mountains.

The Antarctic Peninsula, a finger of land jutting into the ocean from the mainland of West Antarctica, is also mountainous, with underlying volcanic activity. The Ellsworth Mountains of West Antarctica include the territory's highest point, the Vinson Massif (16,864 ft / 5,140 m). Other notable peaks in West Antarctica are Mount Sidley (13,717 ft / 4,181 m), Mount Jackson (13,745 ft / 4,189 m), and Mount Berlin 11,543 ft (3,518 m).

East Antarctica features at least two active volcanoes, and scientists believe they will likely discover more with peaks buried beneath the ice. Mount Erebus (12,444 ft/ 3,794 m), one of the active volcanoes, is on Ross Island. Other notable peaks in East Antarctica are Mount Melbourne at 9,016 ft (2,732 m), peaks lying beneath the Beardmore Glacier 14,942 ft (4,528 m), and the Gamgurtsev subglacial mountains 13,300 ft (4,030 m).

Plateaus

Even where it is not mountainous, Antarctica's elevations are high. It's average elevation of roughly 8,000 ft (2,440 m) is greater than that of any other continent. As a consequence, most of the land outside of the mountain ranges can be considered plateau. Covered by thick ice, most of these plateaus have no names. The Hollick-Kenyon and Rockefeller Plateaus can be found in West Antarctica. The Polar Plateau lies over the South Pole in East Antarctica. The elevation of the South Pole is 9,355 ft (2,835 m).

Glaciers

In Antarctica, glaciers (ice sheets) completely cover the land forms beneath them, allowing only the most dramatic mountain peaks to poke through. Antarctic ice represents 90 percent of the world's total. Over land it averages 1.5 mi (2 km) thick, and is about 3 mi (3.5 km) thick at its thickest point. The East Antarctic glacier (ice sheet) is slightly larger that the West Antarctic ice sheet. Some coastal areas support a few lichens during the summer months, but the ice sheets are otherwise barren.

Glaciers move over the land at a slow and steady pace. Dramatic formations and striations (stripes, believed to be remnants of volcanic ash) may be observed in the glaciers. The advancing edge of the glacier becomes a high sheer cliff as the top levels of ice push forward. The polar ice cap moves an average of 33 ft (10 m) each year.

In East Antarctica, the continent's largest valley glacier, the Lambert Glacier, lies over several mountain peaks that rise to 3,355 ft (1,017 m). Massive sections of ice discharge from the Lambert Glacier to become part of the floating Amery Ice Shelf each year. Other noteworthy glaciers include the Skelton Glacier, Rennick Glacier, Recovery Glacier, and Beardmore Glacier. The Bentley Subglacial Trench, a canyon extending 8,333 ft (2,540 m) below sea level, is covered by solid ice, making it the lowest point on earth not underwater.

INLAND WATERWAYS

Lakes

While a large portion of the world's fresh water is located on Antarctica, it is mostly in the form of ice. Antarctica's largest known lake, Lake Vostok, is approximately the size of North America's Lake Erie. Other lakes found in the McMurdo Dry Valleys, the valleys that lie between mountain peaks around the McMurdo Research Station in East Antarctica, include Lake Vanda, Lake Brownworth, Lake Fryxell, Lake Bonney, and Lake Hoare. These lakes are fed by glacial melt from the glaciers that lie in the deepest mountain valleys. During the summer, the air temperatures warm to about freezing (32°F / 0°C), causing the glaciers to melt slightly and to send water flowing into small streams for a few weeks, before the temperature again drops below freezing. The stream flow feeds the lakes, which lie beneath 10 ft (3 m) of permanent ice cover. Non-frozen water exists in the lakes beneath the ice. These lakes are believed to be 100 ft (30 m) deep or deeper. Scientists are studying these lakes to determine whether they support any marine life. To conduct their experiments, they must use exceptionally sterile methods to collect specimens, to avoid contaminating the glacial environment.

Rivers

The only river of any significance on Antarctica is the Onyx River. It is the largest of the streams that flow during the summer months. The Onyx River flows into Lake Vanda.

THE COAST, ISLANDS, AND THE OCEAN

The Southern Ocean

In 2000, the International Hydrographic Association delimited a new ocean, called the Southern Ocean, that encompasses all of the water south of 60° latitude. Since this decision Antarctica has been completely surrounded by the Southern Ocean. Before it was considered to border on the South Atlantic, South Pacific, and Indian Oceans. Due to the great temperature differences between the ice and the open ocean, as well as the lack of

GEO-FACT

Ice shelves are generally thought of as permanent, unlike the sea ice that forms around Antarctica in the winter only to melt again in the summer. However, it is common for parts of an ice shelf to break away, forming icebergs in a process called "calving." Most icebergs are small, but occasionally huge sections of the Antarctic ice shelves have been known to break away. An iceberg estimated to be the size of Belgium (208 mi / 260 km long and 60 mi / 75 km wide)—the largest recorded as of 2002—was sighted off Antarctica in November, 1956. In October, 1987, an iceberg measuring 1,750 sq mi (more than 4,500 sq km—nearly the size of Delaware) broke away from the Ross Ice Shelf.

In 1995 and then again in 2002, huge chunks of the Larsen Ice Shelf collapsed into the Weddell Sea, forming many icebergs. The area that collapsed in 2002 was roughly the size of Rhode Island, and contained approximately 500 million billion tons of sheet ice. The remaining ice shelf is only 40 percent of its pre-1995 size. These and other dramatic collapses have stimulated concerns about global warming.

any land to impede them, powerful winds blow across the Southern Ocean and the southernmost parts of the surrounding oceans.

The Southern Ocean is home to the Antarctic Circumpolar Current. This ocean current flows east completely around the earth in a great circle to the north of Antarctica. The current is the most powerful on earth, and unique in that it is unimpeded by landforms in its passage around the globe. The current tends to keep cold water to the south, near Antarctica, and holds warmer water back to the north, with a relatively sharp boundary flowing down the middle of the current known as the Antarctic Convergence.

Sea Ice

Even during the summer only a few coastal areas are ever free of ice, including parts of Wilkes Land in East Antarctica and parts of the Antarctic Peninsula. During the winter the ocean around Antarctica freezes, surrounding the continent with ice that expands far out to sea. In winter the ice surrounding the Antarctic land mass grows at the rate of about 40,000 sq mi (103,600 sq km) per day. By the heart of winter it is roughly six times as extensive as normal, expanding the effective size of the continent to 13,000,000 sq mi (33,000,000 sq km).

Coastal Waters

The Bellingshausen Sea lies off the west coast of the Antarctic Peninsula; it is named for Russian explorer Fabian von Bellingshausen, the first to sail completely around Antarctica in 1819–21; his expedition also gave names to Queen Maud Land and Peter I island. Off of West Antarctica is the Amundsen Sea, named for the first man to reach the South Pole (on December 14, 1911), the Norwegian explorer Roald Amundsen.

The Ross Sea lies off the coast of the Ross Ice Shelf directly south of New Zealand; both are named for Sir James C. Ross, an explorer in the region in 1839–43 from the United Kingdom. Also named for a British explorer—James Weddell—is the Weddell Sea, the body of water east of the Antarctic Peninsula. Weddell conducted his expedition to Antactica in 1823.

The whales that inhabited the waters around Antarctica were hunted without controls in the late 1800s and the early decades of the 1900s, until the International Whaling Commission imposed protection for most species to prevent their extinction. A similar pattern developed for fur seals: they were hunted to the point of near-extinction, but since the 1978 promulgation of a treaty to protect them, populations have been thriving in Antarctica. Many species of penguin are also native to Antarctic waters.

Major Islands

Alexander Island (16,700 sq mi/43.200 sq km), Antarctica's largest island, is separated from the Antarctic Peninsula by the George VI sound, although thick ice sheets connect the two land masses. There are dozens of smaller islands in the Bellingshausen Sea and Amundsen Seas, including Thurston, Siple, Carney, and Charcot Islands. Further north along the Antarctic Peninsula is Adelaide Island and the Palmer Archipelago. Most of these islands are connected to the main land mass by ice.

Berkner Island (1,500 sq mi/3,880 sq km), covered by the Ronne and Filchner Ice Shelves, lies in the McCarthy Inlet of the Weddell Sea. Roosevelt Island is the largest found within the Ross Sea, but it is covered by the Ross Ice Shelf. Ross Island is smaller, but has access to the ocean in the summer months.

The South Shetland Islands, lying between Antarctica and the southern tip of South America, include Deception Island and King George I island, among others. Deception Island, which lies in an active volcanic field known as the Branfield Rift, is horseshoe shaped, with a caldera (surface area of about 10 sq mi/26 sq km) that is breached at one end and accessible from the open sea. The water of the caldera is heated by underground volcanic activity, and has at times reached the boiling point. Destructive volcanic eruptions occurred in 1967, 1969, 1970, and 1991; more eruptions are predicted for the future. Also lying in the ocean between Antarctica and South America are the South Orkney Islands, South Georgia, and the South Sandwich Islands. Zavadovski Island in the South Sandwich Islands is home to one of the largest penguin colonies in the world—estimated at two million penguins.

The Coast and Beaches

Ice shelves—thick fields of ice formed by glaciers that last year round—cover almost half the coastal regions. These glaciers move slowly toward the sea at speeds of less than one mile per hour (2,950 to 4,250 ft / 900 to1300 m per year), and in some places extend out into the water for hundreds of miles.

The Ross Sea indents Antarctica on one side, the Weddell Sea on the other. Both have enormous ice shelves covering the parts of them closest to the shore and the South Pole. The Ross Ice Shelf, in the sea of the same name, is the largest with an area of roughly 130,100 sq mi (336,770 sq km). The Ronne, Filchner, Larsen, and Riiser-Larsen Ice Shelves are all found in the Weddell Sea.

West Antarctica has a highly irregular coastline, with many small peninsulas and inlets, most of them ice-covered. The S-shaped Antarctic Peninsula extends far to the northeast. It comes closer to another continent than any other part of Antarctica; South America is hundreds of miles to the north across the Drake Passage. Away from the Weddell and Ross Seas East Antarctica has a much more regular coastline than the western part of the continent. It arcs in a rough half circle from one sea to the other. For most of its length the coast is much closer to the Antarctic Circle than in West Antarctica, and as a consequence the ice shelves are smaller. Prydz Bay, the only significant indentation on the middle of the East Antarctic coast, is mostly covered by the Amery Ice Shelf. East Antarctica extends north slightly beyond the Antarctic Circle at Cape Ann and Cape Poinsett. The Shackleton Ice Shelf lies not far from the second of these capes. Cape Adare marks the point where the East Antarctic coast curves sharply inwards to form one side of the Ross Sea.

CLIMATE AND VEGETATION

Temperature

About 97 percent of the surface of Antarctica is permanently covered by ice, more than 15,000 ft (4500 m) thick at its thickest point. The mean annual temperature in the interior is a frigid -71°F (-57°C), with the mean summer temperature -40°F (-40°C) and mean winter temperature -90°F (-68°C). In the coastal areas the mean summer temperature is 32°F (0°C). McMurdo Station near the Ross Ice Shelf in East Antarctica has the most moderate climate, with a mean winter temperature of 16°F (-9C°). The lowest temperature ever recorded on Earth was at Vostok, East Antarctica, where a reading of -129°F (-89°C) was taken in 1983.

Since the 1950s, scientists have recorded an overall increase in temperature on Antarctica of about 4°F (2°C), which is much more than the increase in overall temperature elsewhere in the world. Five of the largest ice shelves have shrunk in size during this time period. Some scientists speculate that this is an early sign of global warming caused by human activity, but this has not been proven.

Sunlight

Antarctica has continuous daylight from mid-September to mid-March, when the continent receives more solar radiation than equatorial regions; and six months of continuous darkness from mid-March to mid-September.

Observation has shown that the layer of high atmosphere ozone that helps reflect harmful solar radiation away from the Earth's surface is thin to nonexistent over Antarctica. The ozone hole varies in size from season to season but appears to be expanding. Many blame human activity for this hole in the ozone, but the exact causes are unknown.

Rainfall

Most of the continent receives less than 2 in (5 cm) of precipitation (in the form of snow) annually. Due to this lack of precipitation the entire continent is technically considered a desert, despite the fact that it holds more than two-thirds of the world's fresh water.

HUMAN POPULATION

There is no native human population. During the summer months, as many as several thousand scientists may be in residence at the scientific research stations in Antarctica; in addition, several thousand tourists travel to Antarctica by ship or airplane. Few scientists remain year round. McMurdo Station, a U.S. research station, is the largest settlement. Amundsen-Scott—another U.S. research station—is located at the South Pole.

NATURAL RESOURCES

Antarctica is believed to be rich in mineral resources, but the severe climate combined with the complexities of the international relationships has made exploration and exploitation impractical. Deposits of copper, lead, zinc, gold, and silver and extensive deposits of coal have been identified on the Antarctic Peninsula, and iron ore has been found in East Antarctica. A 1991 treaty restricts mineral exploration to scientific purposes only.

FURTHER READINGS

Campbell, David G. *The Crystal Desert: Summers in Antarctica.* Boston: Houghton Mifflin, 1992.

Mastro, Jim. Antarctica: *A Year at the Bottom of the World.* Boston: Little, Brown, 2002.

Mount Erebus Volcano Observatory. http://www.ees.nmt.edu/Geop/mevo/mevo.html (accessed June 12, 2002).

Public Broadcasting Service. *Antarctic Almanac.* http://www.pbs.org/wgbh/nova/warnings/almanac.html (accessed June 12, 2002).

Stewart, John. *Antarctica: An Encyclopedia.* Jefferson, N.C.: McFarland, 1990.

Wheeler, Sara. *Terra Incognita: Travels in Antarctica.* London: Vintage, 1997.

Antigua and Barbuda

Dependency of the United Kingdom

- **Area:** 170 sq mi (440 sq km): Antigua, 108 sq mi (280 sq km); Barbuda, 62 sq mi (161 sq km); Redonda, 5 sq mi (1.3 sq km). / World Rank: 187
- **Location:** Part of the Leeward Islands, Caribbean Sea, in the Northern and Western Hemispheres,261 mi (420 km) southeast of Puerto Rico, 110 mi (180 km) north of Montserrat and Guadeloupe
- **Coordinates:** 17° 03′ N, 61° 48′ W
- **Borders:** No international boundaries
- **Coastline:** 95 mi (153 km)
- **Territorial Seas:** 12 NM (exclusive economic zone: 200 NM)
- **Highest Point:** Boggy Peak, 1,319 ft (402 m)
- **Lowest Point:** Sea level
- **Longest Distances:** 14 mi (22.4 km) N-S / 9 mi (14.4 km) E-W
- **Longest River:** None
- **Natural Hazards:** Subject to hurricanes and periodic drought
- **Population:** 66,970 (2001 est.) / World Rank: 192
- **Capital City:** St John's in the northwest on the island of Antigua
- **Largest City:** St. John's, 24,000 (2000)

OVERVIEW

Antigua is both partly volcanic and partly coral in make-up, giving it deeply indented shores lined by shoals, reefs, and natural beaches. It is the largest of the British Leeward Islands. Its northeastern coast is lined by many islets and its central area is a fertile plain. Barbuda is a coral island that has a large natural lagoon in the northwest. Redonda is a rocky, low-lying islet.

MOUNTAINS AND HILLS

Despite Antigua being a partly volcanic island, there have been no eruptions in recent history. The highest elevations are in the southwestern part of the island. This is where Boggy Peak (1,319 ft / 402 m), the tallest mountain on the island, is located.

Neither Barbuda nor Redonda have any significant elevations.

INLAND WATERWAYS

Antigua and Barbuda lack any large rivers or lakes of significant size.

THE COAST, ISLANDS, AND THE OCEAN

Oceans and Seas

Antigua and Barbuda are located in the eastern Caribbean Sea. The open Atlantic Ocean lies to the north and east. There are many coral reefs in the vicinity of Antigua and Barbuda. The island of Guadeloupe lies to the south, on the far side of the Guadeloupe Passage from Antigua.

The Coast and Beaches

Antigua and Barbuda is famous for its beaches, estimated at 365, particularly those on Antigua itself. The most noteworthy feature of Barbuda's coastline is the natural lagoon on the western side of the island.

CLIMATE AND VEGETATION

Temperature

Temperatures average 84°F (29°C) in July and 75°F (24°C) in January, a result of the cooling trade winds from the east and northeast.

Rainfall

Rainfall averages 46 in (117 cm) per year with September through November being the wettest months. The islands are subject to both the occasional summer drought and autumn hurricanes, although the low humidity makes it one of the most temperate climates in the world.

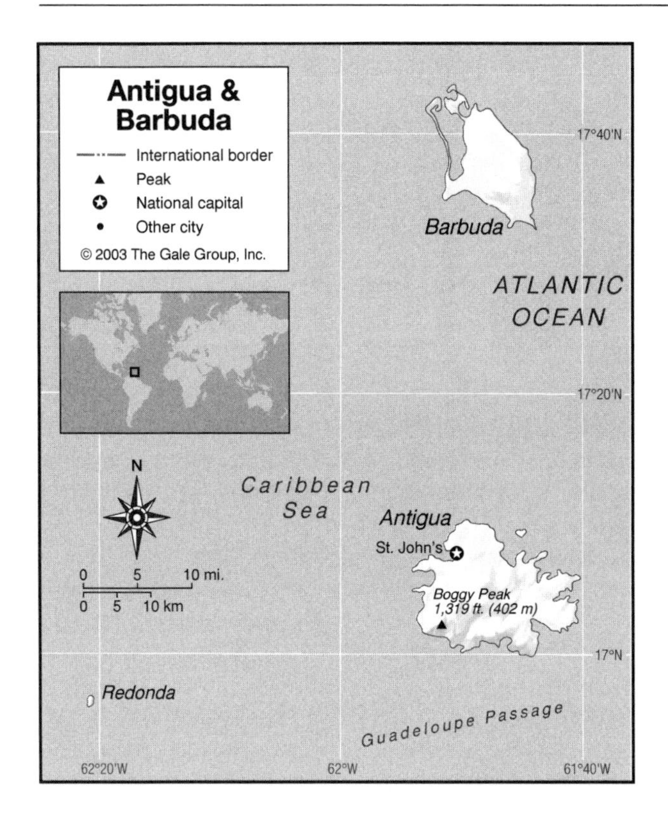

Vegetation

The sandy soil on much of the islands has only scrub vegetation. Some parts of Antigua are more fertile–most notably the central plain–due to the volcanic ash in the soil. These areas support some tropical vegetation, and agricultural uses. The planting of acacia, mahogany, and red and white cedar on Antigua has led to as much as 11 percent of the land becoming forested, helping to conserve soil and water.

HUMAN POPULATION

The mid-July 2001 estimate is 66,970 for both of the inhabited islands combined, although the greatest number live on Antigua. An estimated 37 percent of the total are urban dwellers, and claim African descent, although persons of European and Middle Eastern origins also live here. The Anglican Church claims about 45 percent of the population; other Protestant denominations, 42 percent; and the Roman Catholics, 8.7 percent. Minority religions include Islam and Baha'i.

NATURAL RESOURCES

Most fishing is for local consumption, but the US and neighboring islands receive lobster from Antigua and Barbuda. Recent exploitation of mineral resources—limestone on Antigua, salt on Barbuda, and phosphate on Redonda—has opened up new manufacturing possibilities for local usage.

Parishes – Antigua and Barbuda

POPULATIONS FROM 1991 CENSUS

Name	Population	Area (sq mi)	Area (sq km)
Barbuda	1,200	62.0	160.6
Redonda	N/A	0.5	1.3
Saint George	4,500	10.2	26.4
Saint John	35,600	26.2	67.9
Saint Mary	5,300	25.1	65.0
Saint Paul	6,100	17.7	45.8
Saint Peter	3,600	12.8	33.2
Saint Phillip	3,000	16.0	41.4

SOURCE: *Geo-Data* 1989 ed. and; Johan van der Heyden. Geohive, http://www.geohive.com (accessed July 2002).

FURTHER READINGS

Corum, Robert. *Caribbean Time Bomb; the United States' Complicity in the Corruption of Antigua.* New York: Morrow, 1993.

Dyde, Brian. *Antigua and Barbuda; the Heart of the Caribbean.* London: M Caribbean, 1993.

U.S. Department of State, Bureau of Western Hemisphere Affairs. *Background Notes. Antigua and Barbuda.* Washington, D.C., 2001.

Vaitilingham, Adam. *Antigua; the Mini Rough Guide* / compiled by Adam Vaitilingam. New York: Penguin Books, 1998.

Argentina

- **Area:** 1,068,302 sq mi (2,766,890 sq km) / World Rank: 9
- **Location:** Southern and Western Hemispheres, in the southern region of the South American continent; bordered by Bolivia and Paraguay on the north; Brazil, Uruguay, and the South Atlantic Ocean on the east; and Chile on the south and west.
- **Coordinates:** 34°00′S, 64°00′W
- **Borders:** 6,006 mi (9,665 km) / Bolivia, 517 mi (832 km); Brazil, 761 mi (1,224 km); Chile, 3,200 mi (5,150 km); Paraguay, 1,168 mi (1,880 km); Uruguay, 360 mi (579 km)
- **Coastline:** 3,100 mi (4,989 km)
- **Territorial Seas:** 12 NM
- **Highest Point:** Cerro Aconcagua, 22,835 ft (6,960 m)
- **Lowest Point:** Salinas Chicas, 131 ft (40 m) below sea level

- **Longest Distances:** 2,268 mi (3,650 km) N-S; 889 mi (1,430 km) E-W
- **Longest River:** Paraná, 3,060 mi (4,900 km)
- **Largest Lake:** Lago Buenos Aires, 860 sq mi (2,240 sq km)
- **Natural Hazards:** Earthquakes, violent windstorms known as *pamperos*, periodic heavy flooding, volcanic activity
- **Population:** 37,384,816 (July 2001 est.) / World Rank: 31
- **Capital City:** Buenos Aires, on the northernmost point of the Atlantic coast
- **Largest City:** Buenos Aires, population 12,955,300 (2002 est.)

OVERVIEW

Argentina is the eighth largest country in the world. The terrain varies dramatically across the country's different regions, as both altitude and latitude play a major role in Argentina's geography. The country's four major physiographic provinces are the Andean region, the lowland North, the central Pampas, and Patagonia. Patagonia includes Tierra del Fuego, the southernmost point of the South American continent, shared by Argentina and Chile.

The Andean region makes up 30 percent of Argentina. Home to the Western Hemisphere's highest point, Cerro Aconcagua, the Andes Mountains are broad and lofty in the north, and narrower and progressively lower in the south. This area also contains half of the Lake District (the other half is in Chile).

The lowland north, the Gran Chaco and Mesopotamia regions, consists of tropical and subtropical lowlands. The landscape ranges from dry savannas to swamps. The province of Misiones, the northeasternmost extension of Argentina, is rich with both mountain and forest areas and is home to the majestic Iguazú Falls.

The central Pampas region forms the heartland of Argentina. This grassland area is oval-shaped, and extends more than 500 mi (800 km) both north and south and east to west. The Pampas also has a natural division into humid and dry subregions.

Patagonia, in the southern region of Argentina, has a geography that ranges from a vast, windy, treeless plateau to glacial regions in the southern area of Tierra del Fuego. Patagonia extends more than 1,200 mi (2,000 km) from the Colorado River in the north to Cape Horn, the southernmost tip of the continent.

Eons ago, Tierra del Fuego existed under the sea. The land slowly raised and mountains formed as the South American and Scotia Tectonic Plates pushed together. The Ice Ages occurred, and most of what is now the Patagonian continental shelf was land. The waters of the Strait of Magellan eventually broke through the tip of the continent approximately 9,000 years ago, separating Tierra del Fuego from the Patagonia mainland. This region is also within the Subantarctic Zone.

GEO-FACT

The region of Patagonia takes it name from the word *pantagon*, meaning "big feet." This name referred to the Tehuelche Indians who first entered the fertile plains wearing oversized boots.

MOUNTAINS AND HILLS

Mountains

Stretching more than 4,500 mi (7,000 km), the Andes Mountains form the western border of Argentina, which is nearly parallel with the coast of the Pacific Ocean. First formed by tectonic movement approximately 70 million years ago, the mountain range is the highest in the Western Hemisphere. Its peaks reach nearly 23,000 ft (7,015 m) and stretch to form a natural border with Chile for more than 2,000 miles (3,219 km). The San Miguel de Tucumán and Mendoza areas are still subject to earthquakes due to tectonic plate movement.

The Argentinean Andes contain some of the tallest mountains in South America, including Cerro Mercedario (22,205 ft / 6,768 m), and Cerro Aconcagua, which at 22,834 ft (6,960 m) is the tallest peak on the continent and in the entire Western Hemisphere. Both of these peaks are located near the Chile border southwest of San Juan. The Andes region is also home to arid basins, lush foothills covered with grape vineyards, glacial mountains, and half of the Lake District (the other half is in Chile). The Lake District, named for the many glacial lakes carved out of the mountains and subsequently filled by melt-water and rain, is located in the southern Andes and boasts a diverse natural landscape of glaciers, native old growth forests, lakes, rivers, fjords, volcanoes, and sentinel mountains. Throughout the Andes that separate Chile and Argentina there are more than 1,800 volcanoes, 28 of which are considered to be active. These account for approximately one-fifth of the earth's active volcanoes.

Patagonia, the southern region of Argentina, is a combination of pastoral steppes and glacial regions. Located in this region near the Chilean border is Parc Nacional Los Glaciares (Glacier National Park), where some 300 glaciers make up part of the Patagonian Ice Cap (8,400 sq mi / 21,760 sq km). The ice cap, flowing into the

Pacific oceans from the Andes, is the largest in the Southern Hemisphere outside of Antarctica. Thirteen of the glaciers feed lakes in the region. The Upsala glacier, at 37 mi (60 km) long and 6 mi (10 km) wide, is the largest in South America and can only be reached by boat, since it floats in Lago Argentino. The next largest is the 3-mi (4.8-km) wide Perito Moreno, which stretches about 22 mi (35 km) to Lago Argentino, where it forms a natural dam in the lake. Jagged mountain peaks formed from granite include Cerro Fitz Roy (11,236 ft / 3,405 m), Cerro Torre (10,346 ft / 3,102 m), and Cerro Pináculo (7,128 ft / 2,160 m).

Smaller mountain ranges also exist in central South America. These ranges cut across the center of the country and serve as the divider between the southern Patagonia region and the northeastern Pampas. From west to east these ranges are the Sierra Lihuel-Calel, the Sierra de la Ventana, and the Sierra del Tandil.

Plateaus

The Somuncurá Plateau is a basalt plateau with alternating hills and depressions. It stretches across the Río Negro and Chubut provinces, or the area from the Chubut River north to the Negro River. The region undergoes severe climate changes between the winter and summer months. The area has lava formations and contains many fruit and alfalfa plantations. Cattle ranchers find this area to be ideal for raising their livestock. A smaller plateau, the Atacama Plateau, occupies the region just east of the Andes Mountains in northern Argentina and extends east to the city of San Miguel de Tucumán.

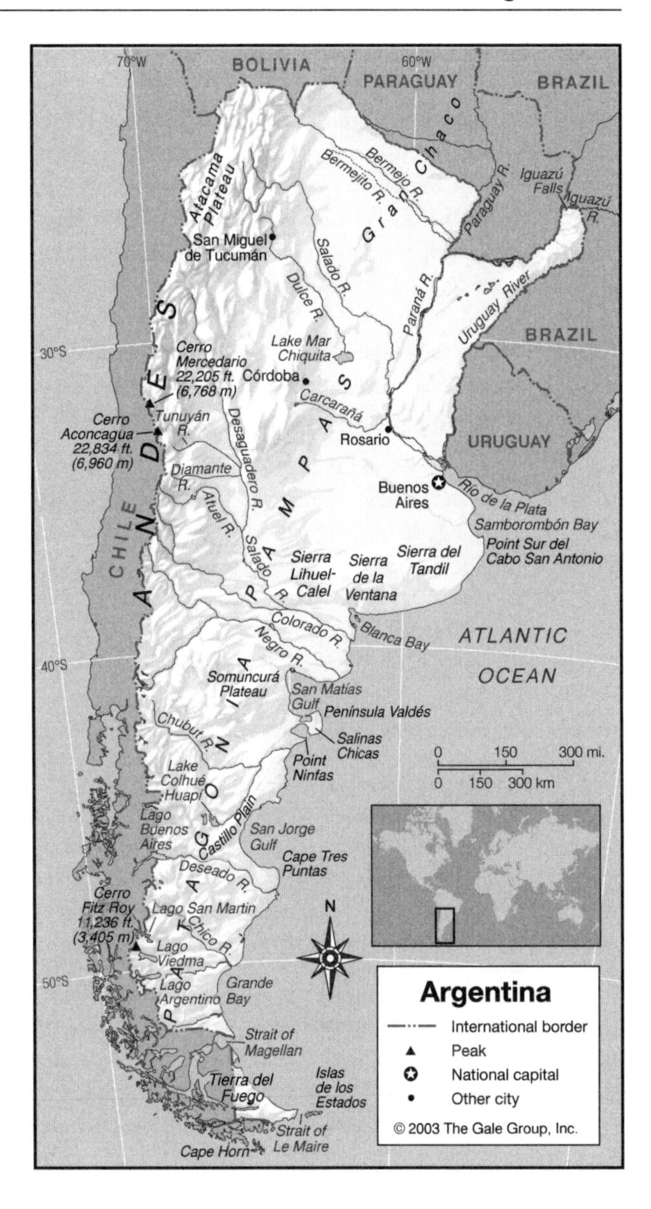

INLAND WATERWAYS

Lakes

The Lake District, on the border of Chile and Argentina in the Andes mountain region, contains many glacial lakes that are carved out of the mountains then filled by melt-water and rain. The most significant of these is Lago Buenos Aires, also known as General Carrera, located in southern Argentina and shared with Chile. It is the largest lake in the country and the fifth largest in all of South America with an average surface area of 860 sq mi (2,240 sq km). Moving south along the border one would encounter Lago San Martín, Lago Viedma, and finally Lago Argentino, the second largest lake in this region with a surface area of 566 sq mi (1466 sq km). Not far from Lake Buenos Aires on the Castillo Plain near Comodoro Rivadavia is Lake Colhué Huapí.

One of the world's largest salt lakes, and the second largest lake in Argentina, is Lake Mar Chiquita (Little Sea), located in central Argentina. Its surface area varies from year to year and season to season, but has in it wettest periods spanned 2,228 sq mi (5,770 sq km). The reservoir created by the Chocón dam, located on the Río Negro, is one of the country's largest manmade lakes.

Rivers

Except in the Northeast there are few large rivers, and many have only seasonal flows. Nearly all watercourses drain eastward toward the Atlantic, but a large number terminate in lakes and swamps or become lost in the thirsty soils of the Pampas and Patagonia. The four major rivers systems are those that feed into the Río de la Plata estuary, those made up of the Andean streams, those of the central river system, and those of the southern system.

The Paraná, the second-longest river in South America after the Amazon, flows approximately 3,060 mi (4,900 km) and forms part of the borders between Brazil and Paraguay, and Paraguay and Argentina. The Paraná is navigable only to Rosário. Its upper reaches feature many waterfalls. It is joined by the Iguazú River (Río Iguaçu) where it enters Argentina in the northeast. This area is well known throughout the world for the spectacular Iguazú Falls (Cataratas Iguaçu, meaning "great water"). One of the world's great natural wonders, they are located on the border between Argentina and Brazil with two-thirds of the falls in Argentina. They include approximately 275 falls, ranging between 197 and 262 ft (60 and 80 m) high. These falls are higher and wider than Niagara Falls in the United States. Other tributaries of the Paraná, which feed in from the west, are the Bermejo, Bermejito, Salado, and Carcarañá.

The Uruguay River (1,000 mi/1,600 km) forms part of the borders between Argentina and Brazil and Argentina and Uruguay. It is navigable for about 190 mi (300 km) from its mouth to Concordia. The 1,594-mile long (2,550 km) Paraguay forms part of the border between Paraguay and Argentina, and flows into the Paraná north of Corrientes and Alto Paraná. These all join to flow into the Río de la Plata, and eventually into the Atlantic Ocean in northern Argentina. Where these rivers meet, a wide estuary is formed, which can reach a maximum width of 138 mi (222 km).

In north central Argentina, Lake Mar Chiquita is supplied with its water by several rivers. The Dulce River originates near San Miguel de Tucumán and flows southwest into the lake. From the southwest it is also fed by the Primero and Segundo Rivers.

In the northern Patagonia region, the major rivers are the Colorado and Negro Rivers, both of which rise in the Andes and flow to the Atlantic Ocean. The Colorado is fed by the Salado River, which flows from Pico Ojos del Salado in a southeasterly direction to the Colorado. Tributaries of the Salado include the Atuel, Diamante, Tunuyán, Desaguadero, and the San Juan, all of which originate in the northwest Andes. The Negro also has two main tributaries of its own, the Neuquén and the Limay. In the central Patagonia region the Chubut rises in the Andes and flows east to form a sizable lake before making its way to the ocean. The Lake District is also coursed by its share of rivers, all originating in the mountains and flowing to the Atlantic. These include the Deseado, Chico, Santa Cruz, and Gallegos Rivers.

Wetlands

Iberá, in the northeast of Argentina, is a biologically rich region, with more than sixty ponds joined to marshes and swampland. The area is extremely humid, and is home to hundreds of bird species and thousands of insects, including a wide variety of butterflies. The area hosts a diverse array of flora and fauna, notably the royal water lily, silk-cotton trees, alligators, and capybara, the largest rodent species in the world.

THE COAST, ISLANDS, AND THE OCEAN

Oceans and Seas

The Atlantic Ocean comprises Argentina's eastern border. Argentina has one of the largest ports on the Atlantic Ocean in Buenos Aires. The area of the South Atlantic around the Valdés Peninsula is home to one of the world's largest concentrations of the Atlantic Right Whale (*Eubalaena glacialis*). The Argentine coast is also known for being home to the Magellanic Penguins.

Major Islands

Argentina shares the offshore island territory of Tierra del Fuego with Chile, and also owns Isla de los Estados, separated from Tierra del Fuego by the Strait of Le Maire. Argentina claims the Falkland Islands (Islas Malvinas), a large archipelago to the southeast of the country that is under the control of the United Kingdom.

The Coast and Beaches

The Atlantic coast, curving generally from northeast to southwest, features a number of gulfs, bays, and inlets. Starting in the north, the bay on which Buenos Aires sits is Samborombón Bay. The coast then juts out at Point Sur del Cabo San Antonio before curving due west. At Bahía Blanca the coast abruptly turns southward, forming Blanca Bay. This pattern is then repeated not too far to the south forming the San Matías Gulf. The Península Valdés (Valdes Peninsula), with miles of beaches and tall cliffs, forms the southern rim of the San Matías Gulf at about the midpoint of the Atlantic coast. This area is home to large colonies of marine mammals, including penguins and the southern elephant seal, which mate in the protected lagoons of the peninsula. It is also where Salinas Chicas, Argentina's lowest elevation—131 feet (40 m) below sea level—is found. Just south of Valdés Peninsula there is a tiny bay which is bordered to the south by Point Ninfas. Following a similar pattern twice more in the south, the coastline sweeps in at San Jorge Gulf, culminating in Cape Tres Puntas, and then finally at the mouth of the Chico River at Grande Bay. The Strait of Magellan serves as the divider between the mainland and Tierra del Fuego, the southernmost tip of the country.

A popular destination for tourists and Argentineans alike is the Mar del Plata, a city on the Atlantic coast known for its sprawling beaches, which cover about 5 mi (8 km). This area boasts more than 140 bird species, including flamingos.

CLIMATE AND VEGETATION

Temperature

Argentina's climate ranges from subtropical in the north to humid in the central regions, to subantarctic in the south. The average temperatures for the summer months—January, February, and March—are highs of 75° to 95°F (24° to 35°C) and lows of 60° to 75°F (16° to

Population Centers – Argentina

(2000 POPULATION ESTIMATES)

Name	Population
Buenos Aires (capital)	12,024,000
Córdoba	1,368,000
Rosario	1,279,000
Mendoza	934,000
San Miguel de Tucumán	792,000

SOURCE: "Table A.12. Population of Urban Agglomerations with 750,000 Inhabitants or More in 2000, by Country 1950–2015. Estimates and Projections: 1950–2015." United Nations Population Division, World Urbanization Prospects: The 2001 Revision.

24°C), with January being the warmest month. The winter months, May through August, are the driest period of the year, and the coldest months are June and July. The average winter temperatures are highs of 58° to 65°F (15° to 8°C) and lows of 47° to 53°F (8° to 12°C). Climate variations are due the country's range in altitude as well as latitude.

Rainfall

Average rainfall declines from east to west. Buenos Aires receives an average of 37 in (94 cm) of rain annually, experiencing light snow during the winter months. Areas north of Negro River experience little precipitation during winter. The Pampas receives enough rainfall to support its crops, but is also subject to flooding. The northeastern region bordering Brazil and Uruguay known as Mesopotamia also receives sufficient rainfall. The Chaco region north of the Pampas receives an average of 30 in (76 cm) of rainfall per year. The Andes region is subject to intense changes in weather, including flash floods during the summer months. Puna de Atacama expects an annual rainfall of only 2 in (5 cm).

Grasslands

The Pampas comprises fertile grasslands that cover much of the central Argentinean region. Stretching south, west, and north in a radius of 600 mi (970 km) from Buenos Aires, the capital city, the eastern half of the Pampas is humid, with fertile agricultural lands well suited for the cultivation of wheat. The western region approaching the Andes mountain range is dry open land, providing grazing land for Argentina's famous horse and sheep ranches. This region—along with the northeastern Chaco region—is subject to violent windstorms known as pamperos.

Deserts

Thin areas of desert extend eastward from the glacial mountains down into the Patagonian plains of Argentina. The land is dry, wind-eroded, and marked by sparse scrub vegetation and remnants of a petrified forest.

Forests and Jungles

Lumbering in the north-central part of Argentina, called the Chaco, has become a major industry; its forests contain valuable hardwood trees. Harvesting of the quebracho tree is one of the region's most important economic activities. The tree produces a resin used in the tanning of leather, which is an important by-product of the cattle industry.

Iguazú National Park is an unspoiled jungle on the border of Brazil and Argentina, which protects more than 400 bird and wild animal species, and 2,000 flora species. It is also home to most of the well-known Iguazú Falls.

HUMAN POPULATION

More than one-third of Argentina's population lives within Greater Buenos Aires, an area that includes the capital's 22 suburbs. Only 4.5 percent of the population lives in the southern region of Patagonia. The central Pampas region, which includes Buenos Aires, forms the heartland of Argentina, which houses 80 percent of the Argentine people.

NATURAL RESOURCES

Argentina's natural resources, in addition to the fertile plains of the Pampas, are lead, zinc, tin, copper, iron ore, manganese, petroleum, and uranium. Iron ore is mainly found in the province of Río Negro. Argentina is one of the world's leading agricultural exporters; major crops include wheat, rice, cotton, and fruit.

FURTHER READINGS

Argentina. London: APA Publications, 1997.

Argentina in Pictures. Minneapolis: Lerner, 1988.

Argentina Travel Net. http://www.argentinatravelnet.com/indexE.htm (Accessed June 13, 2002).

Bernhardson, Wayne. *Argentina, Uruguay, and Paraguay.* Oakland: Lonely Planet, 1999.

Crane, Jonathan. "Eubalaena glacialis: Atlantic Right Whale," *The Animal Diversity Web: The Regents of the University of Michigan.* http://animaldiversity.ummz.umich.edu/accounts/eubalaena/e._glacialis$narrative.html (Accessed June 24, 2002).

Frank, Nicole. *Argentina.* Milwaukee, Wis.: Gareth Stevens, 2000.

Mayell, Hillary. "Patagonia Penguins Make a Comback," *National Geographic.* http://news.nationalgeographic.com/news/2001/12/1221_patapenguins.html (Accessed June 24, 2002).

Nickles, Greg. *Argentina: The Land.* New York: Crabtree, 2001.

Armenia

- **Area:** 11,500 sq mi (29,800 sq km) / World Rank: 141
- **Location:** Northern and Eastern Hemispheres, in southwestern Asia; bordered on the north by Georgia, on the east by Azerbaijan, on the southwest by Azerbaijan's Naxçivan exclave, on the south by Iran, and on the west by Turkey
- **Coordinates:** 40°N, 45°E
- **Borders:** 778 mi (1,254 km) / Azerbaijan 488 mi (789 km; 137 mi / 221 km of this is with the Naxçivan enclave); Georgia 102 mi (164 km); Iran 22 mi (35 km); Turkey 166 mi (268 km)
- **Coastline:** None
- **Territorial Seas:** None
- **Highest Point:** Mt. Aragats, 13,425 ft (4095 m)
- **Lowest Point:** Debed River valley, 1,320 ft (400 m)
- **Longest River:** Aras, 568 mi (914 km)
- **Largest Lake:** Lake Sevan, 480 sq mi (1,244 sq km)
- **Natural Hazards:** Earthquakes and droughts
- **Population:** 3,336,100 (July 2001 est.) / World Rank: 129
- **Capital City:** Yerevan, located on the Hrazdan River in west-central Armenia
- **Largest City:** Yerevan, 1,322,000 (2000 est.)

OVERVIEW

Armenia is located in the mountainous Transcaucasia region, southwest of Russia between the Black Sea and the Caspian Sea. A small, landlocked country, Armenia's terrain is largely that of plateaus and rugged mountain ranges, with the exception of a few fertile river valleys and the area around Lake Sevan. Half of the country is above 6,090 ft (2,000 m) in elevation. The Aras River and the Debed River valleys in the far north are the lowest points, with elevations of 1,158 ft (380 m) and 1,310 ft (430 m) respectively.

The Armenian Plateau was formed in a geological upheaval of earth's crust twenty-five million years ago. The northern mountain chain began forming in the Jurassic period; the western chain dates from the Tertiary period. Geological instability still causes major earthquakes in Armenia. Gyumri, the second largest city in the republic, was devastated by a massive quake that killed more than 25,000 people in December 1988.

The Armenian climate ranges between subtropical and sub-temperate. The rich soils of the arable river valleys contain vineyards and orchards; the flat tablelands are primarily pastoral.

MOUNTAINS AND HILLS

Mountains

The Lesser Caucasus Mountains enter into Armenia in the north and extend across the entire country along the border with Azerbaijan and into Iran. The Lesser Caucasus includes the P'ambaki, Geghama, Vardenis, and Zangezur ranges of the Lesser Caucasus system. Composed largely of granite and crystalline rock, the mountains are high, rugged, and include some extinct volcanoes and many glaciers. The terrain is particularly rugged in the extreme southeast.

Plateaus

The Armenian Plateau occupies the western part of the country. It slopes down from the Lesser Caucasus Mountains toward the Aras River valley. The plateau features some smaller mountain ranges and extinct volcanoes including Mount Aragats (Aragats Lerr), which at 13,425 ft (4095 m) is the highest point in Armenia.

INLAND WATERWAYS

Lakes

Lake Sevan lies 6,200 ft (2,070 m) above sea level on the Armenian Plateau. With an area of 480 sq mi (1,244 sq km), it is the country's largest lake—and one of the largest high-elevation lakes in the world. At its widest point, Lake Sevan measures 58 mi (72.5 km) across; it is 301 mi (376 km) long. The lake's greatest depth is about

Population Centers – Armenia

(1995 POPULATION ESTIMATES)

Name	Population
Yerevan	1,254,400
Gyumri	206,600
Vanadzor	170,200

SOURCE: Armenian Statistical Yearbook.

272 ft (83 m). Many tributaries flow into the lake from the south and southeast, but the Hrazdan River is its only outlet.

Rivers

The Aras River, which is 568 mi (914 km) long, is Armenia's largest and longest river. It originates in southwestern Asia near Erzurum, Turkey, then flows east and southeast, forming part of the border between Turkey and Armenia. It then continues into Azerbaijan before joining the Kura River and flowing into the Caspian Sea. Its chief tributary in Armenia is the Hrazdan. The Debed River in the north of the country flows northeast into Georgia. The Bargushat River drains the southeastern part of Armenia.

THE COAST, ISLANDS, AND THE OCEAN

Armenia is landlocked and has no coast.

CLIMATE AND VEGETATION

Temperature

Although Armenia lies not far from several seas, its high mountains block their effects and give it a continental highland climate. It has cold, dry winters and hot, dusty summers. Temperature and precipitation depend greatly on elevation, with colder and wetter seasons in the high north and northeast. The widest variation in temperature between winter and summer occurs in the central Armenian Plateau, where in midwinter the mean temperature is 32° F (0°C); in midsummer the mean temperature is over 77° F (25°C). Overall, Armenia is a sunny country.

Rainfall

Precipitation rates depend on altitude and location, but are heaviest during autumn. In the lower Aras River valley, the average annual precipitation is 10 in (250 mm). It can reach 32 in (800 mm) in the mountains.

Forests and Jungles

Armenia's natural forest land was largely cleared in the early 1990s for use as firewood.

HUMAN POPULATION

With a population of 3,336,100 (July 2001 estimate), Armenia has a population density of 298 persons per sq mi (115 per sq km). A highly urbanized country, 69 percent of the country's residents live in cities or towns. Most of these are concentrated along the river valleys in the west and northwest, primarily that of the Hrazdan River, where Yerevan, the capital and largest city, is located.

NATURAL RESOURCES

Agriculture employs more than a third of Armenia's population. Major food crops include wine grapes, citrus fruits, wheat, barley, potatoes, and sugar beets. Tobacco and cotton comprise the primary industrial crops. Lake Sevan supports a fishing industry. Armenia also produces modest amounts of copper, molybdenum, zinc, gold, perlite (a lightweight aggregate used in concrete and plaster), granite, bauxite, lead, iron, pyrites, manganese, chromite, and mercury. Health resorts have prospered from the abundance of salts and other minerals.

FURTHER READINGS

Lang, David Marshall. *Armenia: Cradle of Civilization.* London: Allen and Unwin, 1980.

Lynch, H. F. B. *Armenia: Travels and Studies* (2 vols.). Beirut: Khayats, 1965.

Suny, Ronald G. *Armenia in the Twentieth Century.* Chico, CA: Scholars Press, 1983.

———. *Looking Toward Ararat: Armenia in Modern History.* Bloomington: Indiana University Press, 1993.

Walker, Christopher J. "Armenia: A Nation in Asia," *Asian Affairs*, 19, February 1988, pp. 20-35.

Aruba

Dependency of the Netherlands

- **Area:** 75 sq mi (193 sq km) / World Rank: 197
- **Location:** Northern and Western Hemispheres, in the southern Caribbean Sea off the north coast of the South American continent
- **Coordinates:** 12° 30′ N, 69°58 W
- **Borders:** No international boundaries
- **Coastline:** 42.6 mi (68.5 km)
- **Territorial Seas:** 12 NM (22 km)

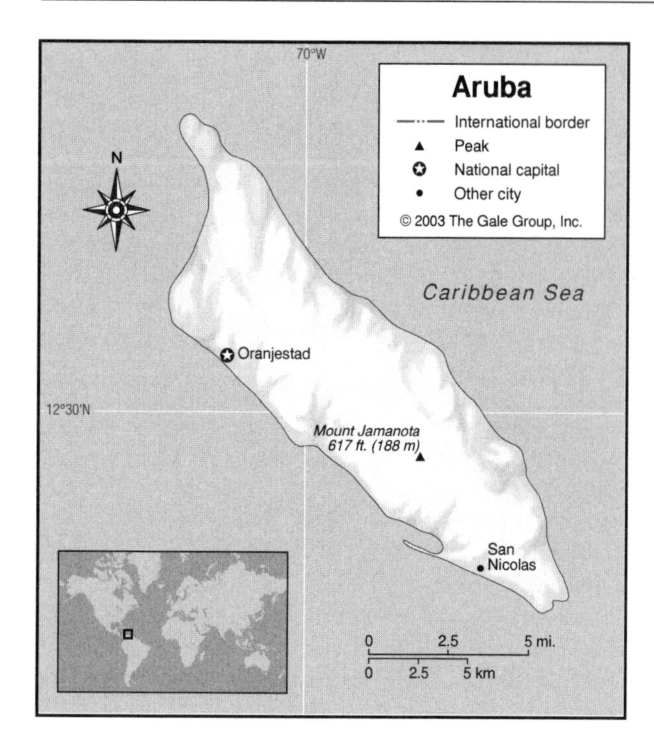

GEO-FACT

Aruba was part of the Netherlands Antilles until 1986, when it seceded and became a separate dependency. The island was on its way to full independence by 1996—until deciding to turn it down. In 1990 it took back its request for independence, electing to remain a dependency.

- **Highest Point:** Mount Jamanota, 617 ft (188 m)
- **Lowest Point:** Sea level
- **Longest Distances:** 18.6 mi (30 km) NW-SW / 5 mi (8 km) SW-NE
- **Longest River:** None
- **Natural Hazards:** None
- **Population:** 70,007 (July 2001 est.) / World Rank: 190
- **Capital City:** Oranjestad, on northwest coast
- **Largest City:** Oranjestad (20,046, 1991 est.)

OVERVIEW

Aruba is a Caribbean island about the size of Washington, D.C., located 15 mi (25 km) north of the coast of Venezuela and 42 mi (68 km) northwest of Curaçao, the largest island of the Netherlands Antilles. Aruba's terrain is mostly flat with a few hills. There is little in the way of vegetation or outstanding physical features and no inland water. Aruba's best-known geographical feature is its white-sand beaches, which are the basis of an active tourism industry that is the mainstay of the island's economy.

Aruba is situated on the Caribbean Tectonic Plate. The island is made up of limestone-capped hills and ridges, with cliffs on the northern and northeastern coasts and coral reefs on the southern coast.

MOUNTAINS AND HILLS

Aruba's terrain is almost entirely flat. The highest elevation, so-called Mount Jamanota, is only 617 ft (188 m) above sea level. Rock formations characterize the interior of the island.

INLAND WATERWAYS

Aruba has no inland waterways.

THE COAST, ISLANDS, AND THE OCEAN

The Coast and Beaches

Aruba has three deepwater harbors located at Oranjestad, Barcadera, and San Nicolas (Sint Nicolaas). The coastal area is known for its white-sand beaches and the calm waters surrounding Aruba are clear, making it a popular tourist destination.

CLIMATE AND VEGETATION

Temperature

Aruba's tropical marine climate varies little seasonally, with an average annual temperature of 81°F (27°C), varying from about 78°F (26°C) in January to 84°F (29°C) in July.

Rainfall

Most rain brought by the prevailing easterly winds of the region falls on the Windward Islands of the Lesser Antilles, leaving Aruba with a very dry climate. Rainfall averages 20 in (51 cm) or less annually, and the island's residents rely on one of the world's largest desalination plants for most of their drinking water. The rainy season occurs between October and December.

Forests and Other Vegetation

Aruba has little vegetation. Due to the island's scant rainfall only hardy, drought-resistant tree, shrub, and cactus species can survive, and there is no arable land.

HUMAN POPULATION

Most Arubans live in the island's major cities and work in the tourism industry or at the island's oil refinery.

The average estimated life expectancy for 2001 was 78.5 years, with an estimated population growth rate of 0.64 percent and an estimated birthrate of 12.64 per 1,000 population. Over two-thirds of the population is aged 15–64.

NATURAL RESOURCES

Aruba has few natural resources. Its major economic activities are tourism and refining crude oil imported from Venezuela.

FURTHER READINGS

Brushaber, Susan, and Arnold Greenberg. *Aruba, Bonaire and Curacao Alive!* 2nd ed. Edison, N.J.: Hunter, 2002.

Fine, Brenda. "Aruba: Let's Go Dutch." *Travel-Holiday* (February 2000): 74.

Fodor's Pocket Aruba. New York: Fodor's Travel Publications, 1998.

Garrett, Echo, and Kevin Garrett. "Dutch Treats." *Chicago* (November 2000): 75-8.

Official Web site of the Aruba Tourism Authority. *Bon Bini: Welcome to Aruba.* http://www.interknowledge.com/aruba/index.htm (accessed January 26, 2002).

Australia

- **Area:** 2,966,200 sq mi (7,682,300 sq km), including island state of Tasmania / World Rank: 7
- **Location:** Southern and Eastern Hemispheres, southeast of Asia in the region known as Oceania, south of the Timor and Arafura Seas, between the Pacific Ocean and the Indian Ocean.
- **Coordinates:** 27° 00′ S, 133° 00′ E
- **Borders:** No international borders.
- **Coastline:** 22,831 mi (36,735 km)
- **Territorial Seas:** 12 NM, exclusive economic zones extends 200 NM.
- **Highest Point:** Mt. Kosciusko, 7,310 ft (2,228 m), highest point on the mainland / Mawson Peak, 9,006 ft (2,745 m), on Heard Island, tallest point in all Australian territory
- **Lowest Point:** Lake Eyre, 52 ft (16 m) below sea level
- **Longest River:** Darling River, 1,702 mi (2,739 km)
- **Largest Lake:** Lake Eyre, 3,668 sq mi (9,500 sq km)
- **Natural Hazards:** Coastal areas subject to cyclones; periodic severe drought
- **Population:** 19,357,594 (July 2001) / World Rank: 53
- **Capital City:** Canberra, located in off the coast in the southeast
- **Largest City:** Sydney, 3,665,000 (2000)

OVERVIEW

Australia is the smallest continent in the world and the only one occupied by a single country, the Commonwealth of Australia. It is the lowest, the flattest, and the driest continent, and is one of the oldest landmasses, with exposed bedrock more than 3 billion years old. It lies in the Southern and Eastern Hemispheres, southeast of Asia, between the Pacific and Indian Oceans. The sixth largest country on earth, Australia is slightly smaller than the 48 contiguous United States.

The Commonwealth of Australia has six states, including the island of Tasmania, and two territories: the Australian Capital Territory and the Northern Territory. It also has External Territories, including the Coral Sea Islands, the Heard and McDonald Islands, Christmas Island, Norfolk Island, the Cocos (Keeling) Islands, Ashmore and Cartier Island, and the Australian Antarctic Territory.

The continent of Australia is made up of four general topographic regions: the Eastern Coastal Plain, the Eastern Highlands, the Central Plains, and the Western Plateau. The Eastern Coastal Plain is a low, sandy region following along that coast. The Eastern Highlands generally parallel the east coast, extending from Cape York in the north to the southern edge of the continent, and reemerging from the sea to form the island of Tasmania. Most of Australia's mountains are in this region, although it is far from mountainous by world standards with altitudes ranging only from 1,000 ft (300 m) to more than 7,000 ft (2,100 m). The Central Plains, an area just west of the Eastern Highlands, is made up of horizontal sedimentary rock and contains the Great Artesian Basin, one of the largest areas of internal drainage in the world. It underlies nearly one-fifth of the continent. The Western Plateau, a desert region underlain by ancient rock shield, is the largest region, encompassing the approximately two-thirds of the continent west of the Central Plains. Escarpments rim the Plateau.

All of Australia is on the Australian Tectonic Plate. There are no major fault lines and no active volcanoes on the mainland.

MOUNTAINS AND HILLS

Mountains

Australia is the flattest continent in the world; only six percent of its total area lies over 2,000 ft (610 m). Most of the country's mountains lie in the Eastern Highlands in the Great Dividing Range. Other mountain ranges

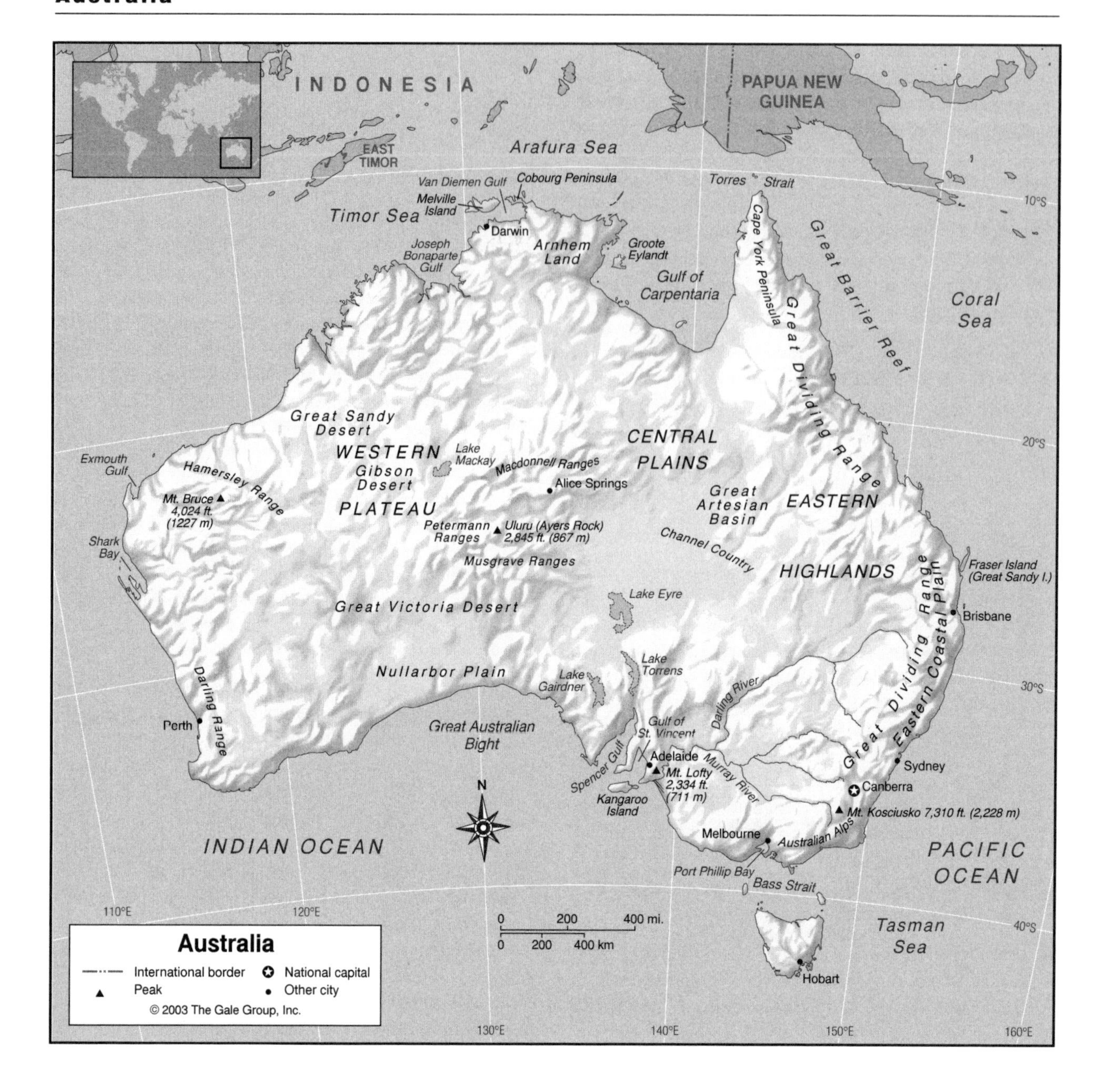

exist as part of the Western Plateau. Running north-south through the Eastern Highlands, the Great Dividing Range is significant more for the drainage patterns it creates than its height, although the terrain there can be quite rugged. The range consists of eroded plateaus and cones and plugs of long-extinct volcanoes. The ten highest mountains in Australia are part of this range, which includes the so-called Australian Alps. Located in New South Wales, they are less than half the height of their European counterparts. Mt. Kosciusko, at 7,310 ft (2,228 m), is the highest peak on mainland Australia, yet it is possible to drive a car to the top of it. The ironically named Mt. Lofty, near Adelaide, reaches only 2,334 ft (711 m). There are no active volcanoes on the continent, although some exist in Australia's territories. The highest point on an Australian territory is Mawson Peak, 9,006 ft (2,745 m), on Big Ben, a volcano on Heard Island near the Antarctic.

Plateaus

The Western Plateau covers approximately two-thirds of the continent. It averages about 1,000 feet above sea level (305 m) and is broken by widely separated small mountains. The Hamersley Range, in the northwest, features Mt. Bruce, with a peak of just over 4,000 ft (1,219 m). On the eastern side of the plateau lie three mountain ranges—the Macdonnell, the Musgrave, and the Petermann. They run east-west, are cut through by gorges, and have peaks of more than 4,900 ft (1,493 m). The Plateau continues unbroken across much of central Australia, however, with only occasional rock outcroppings. The best known of these is the impressive Uluru, formerly known as Ayers Rock. It is a sandstone monolith rising abruptly to 1,100 ft (335 m), more than 5.6 miles (9 km) around its base. The rock is renowned for changing color as the angle of the sun changes throughout the day. Escarpments rim the Western Plateau. In the south, it drops to the Nullarbor Plain, a flat, barren limestone lowland, riddled by numerous underground caves. This plain extends to the coast all along the Great Australian Bight. The Darling Scarp in the southwest separates the Plateau from the western coastal plain.

Hills and Badlands

The Western Plateau gives way to rolling hills on the west coast near Perth. There are also hills in the Eastern Highlands, adjacent to the Great Dividing Range, and some hilly areas in South Australia near the City of Adelaide. Australia has no badlands.

INLAND WATERWAYS

Lakes

Since much of Australia is arid, and rainfall is highly variable, the country does not have many large, permanent lakes. The largest lake, Lake Eyre, (3,668 sq mi / 9,500 sq km) in South Australia, is a salt lake that is often dry, appearing as a vast sheet of salt. It was full only three times in the twentieth century. Lake Torrens (2,218 sq mi / 5,745 sq km) and Lake Gairdner (1,680 sq mi / 4,351 sq km), also in South Australia, are intermittent salt lakes as well. These lakes receive drainage from several intermittent streams, principally Cooper's Creek and the Warburton-Diamantina and Georgina rivers. Together, these inland waterways make up what is known as the Channel Country, which is dry for most of the year except for scattered waterholes. Lake Mackay (1,349 sq mi / 3,494 sq km), in Western Australia, is the largest natural fresh water lake.

Each major population center in the county has a reservoir for a reliable water supply and for hydroelectric power generation. The largest reservoir, Lake Gordon, in Tasmania, holds 16,284 million cu yds (12,450 million cu m) of water, behind the Gordon Dam. Second in size, Lake Argyle in Western Australia holds 7,582 million cu yds (5,797 million cu m) of water behind the Ord River Dam. New South Wales has Lake Eucumbene, with 6,276 million cu yds (4,798 million cu m) of water behind the Eucumbene Dam.

Rivers

Only a few of Australia's rivers are permanent. The most important river system in the country is the Murray-Darling, largely in New South Wales, emptying into Lake Alexandrina, near Adelaide, in South Australia. The catchment area of this drainage system is 1.063 million sq km (410,318 sq mi), and it is the main source of water for 80 percent of Australia's irrigated land. This diversion of water for irrigation has caused an increased salinity in the water downstream. This system contains the country's two longest rivers: the Darling River (1,702 mi / 2,739 km) and the Murray (1,609 mi / 2,589 km) The Darling is a tributary of the Murray. Other Murray tributaries include the Murrumbidgee, the Lachlan, and the Goulburn, which drain the western slopes of the southeastern highlands. The rivers that flow northward into the Gulf of Carpentaria from the northern part of the Central Eastern Lowlands form numerous small drainage systems, of which only the Gregory River is permanent.

Australia's somewhat rugged Eastern Highlands have three significant waterfalls. The Wallaman Waterfall, in Queensland, has a drop of 1,000 (305 m). The Wollomombi, in New South Wales, has a gentler gradient fall of 722 ft (220 m), with a single drop of 328 ft (100 m). The Ellenborough, also in New South Wales, drops 656 ft (200 m).

GEO-FACT

Western Australia is home to the enormous State Barrier Fence. When first constructed in 1907 it ran 1,139 miles (1,834 km) from the north coast of Australia to the south. It proved unable to stop the spread of European rabbits into Western Australia, its original purpose, but was later rebuilt along a 731-mile (1170 km) stretch to keep kangaroos and emus out of Western Australia's farmland.

THE COAST, ISLANDS, AND THE OCEAN

Oceans and Seas

Australia lies between the Pacific and Indian Oceans. To the northwest is the Timor Sea, and to the north is the Arafura Sea. The Coral Sea is off the northeast coast, the Tasman Sea off the southeast. The Bass Strait separates the island of Tasmania from the rest of Australia. The Torres Strait separates Australia from Papua New Guinea.

The coastal shelf of Australia extends all around the continent, encompassing Tasmania, widening at the Great Australian Bight, and extending under the Arafura Sea and much of the Timor Sea. The Great Barrier Reef—the largest living reef system on earth at 1,243 miles long (2,000 km)—lies on the coastal shelf off of Queensland. This coral reef encloses an area of 79,902 sq mi (207,000 sq km) that is an important marine ecosystem.

Major Islands

Tasmania is an island state approximately 150 miles (241 km) off of mainland Australia to the southeast, across the Bass Strait. The island is 26,200 sq mi (67,800 sq km) in area. Australia has tens of thousands of other, smaller, islands surrounding it. The largest are: Melville Island, off the Northern Territory, at 2,234 sq mi (5,786 sq km); Kangaroo Island, off South Australia, at 1,705 sq mi (4,416 sq km); Groote Eylandt, also off the Northern Territory, at 882 sq mi (2,285 sq km); Bathurst Island, off the Northern Territory, at 654 sq mi (1,693 sq km); and Fraser Island, the world's largest sand island, off Queensland, at 638 sq mi (1,693 sq km).

External Territories:

Australia possess or administers a number of islands and islands groups located away from the continent itself. The Coral Sea Islands, covering an area totaling just 1 sq mi (3 sq km), are scattered over 2.6 million sq mi (1 million sq km) in the Coral Sea east of Australia. A meteorological station on Willis Islets supports the only human inhabitants. Norfolk Island, at 13.3 sq mi (34.6 sq km), lies in the Tasman Sea about 869 mi (1,390 km) east of Brisbane on mainland Australia. Ashmore and Cartier Islands, at 2 sq mi (5 sq km), lie to the northwest in the Indian Ocean near the Timor Sea, between mainland Australia and Indonesia.

Christmas Island (52 sq mi / 135 sq km) also lies directly south of the western tip of the Indonesian island, Java, in the Indian Ocean; its location is 1,450 mi (2,330 km) northwest of Perth. Christmas Island is almost completely surrounded by reef. Its 33-mi (54-km) coast features steep cliffs that rise abruptly to a central plateau. The capital, The Settlement, is found on the island's northeast coast.

The Cocos (Keeling) Islands, a group of 27 coral atolls covering 5.4 sq mi (14 sq km), lie about halfway between Australia and Sri Lanka in the Indian Ocean, about 1,720 mi (2,752 km) northwest of Perth. The two inhabited islands are West Island, home to the capital, and Home Island; both support coconut palms and dense tropical vegetation.

The uninhabited Heard and the McDonald Islands, with a combined area of 158 sq mi (412 sq km), lie in the Southern Ocean approaching Antarctica. Heard Island, home to the islands' highest point, the inactive volcano Big Ben (9150 ft / 2,745 m) features mountainous terrain. The much smaller McDonald Islands are rocky and provide a haven for a large seal population.

The Coast and Beaches

The mainland of Australia has 22,831 mi (36,735 km) of coastline, and along that coastline are many, many miles of beautiful beaches. Bondi Beach in New South Wales, Surfers Paradise in the Gold Coast of Queensland, Bell's Beach in Victoria, and the Margaret River region of Western Australia all attract surfers and swimmers from around the world.

The tip of Cape York Peninsula is the northernmost point on the continent. The Cobourg Peninsula, extending north from Arnhem Land, reaches nearly the same latitude as Cape York. In the south, the Yorke Peninsula and Eyre Peninsula protect the coastal city of Adelaide. The west coast has several long, narrow peninsulas, forming Shark Bay and Exmouth Gulf.

The southern coastline of the continent has many bays and inlets, including Port Philips Bay, near Melbourne; Spencer Gulf and the Gulf of St. Vincent, near Adelaide; the broad Great Australian Bight; and numerous smaller inlets. In the northwest are the Van Diemen Gulf, near Darwin, and the Joseph Bonaparte Gulf, opening into the Timor Sea, and the large Gulf of Carpentaria. Many of Australia's cities, including Brisbane, Sydney, Melbourne, Adelaide, and Darwin, were built on natural bays and harbors.

CLIMATE AND VEGETATION

Temperature

The country is generally warm, hotter toward the interior. Winters are mild. Mean temperatures in July (Australian winter) on the southeast coast average 48°F (9°C). In January (summer) the coastal temperatures average from 68°F (20°C) along the southeast to 86°F (30°C) at Darwin on the tropical north coast. In the deserts of the Western Plateau, temperatures are often 100°F (38°C) or higher, and may exceed 115°F (46°C) in the central core of the Plateau.

Rainfall

Australia is the driest continent in the world, with a large portion of its area covered by desert. Except for some areas along the coast, the Western Plateau is all desert or semi-desert. The center of the Plateau is especially arid and it, as well as the semi-arid plains to the north, are referred to generally as "the Outback." Rainfall averages only 4.9 in (125 mm) per year in the interior. Even outside of the Western Plateau, conditions are dry. Only about 20 percent of the country gets more than 30 in (762 mm) of rain annually. Generally, rainfall increases concentrically from the arid core, getting the heaviest along the north and east coasts. Only the coastal areas of Victoria, Queensland, parts of New South Wales, and the island of Tasmania regularly receive sufficient rainfall year round. Tasmania has the most frequent rains, averaging 250 days of rain per year.

Australia has often suffered serious floods and droughts. In 1974 a cyclone and subsequent flooding in Darwin resulted in 49 deaths and 20,000 people were left homeless. In the early 1980s severe drought led to dust

Population Centers – Australia

(2000 POPULATION ESTIMATES)

Name	Population
Sydney	4,041,400
Melbourne	3,417,200
Brisbane	1,601,400
Perth	1,364,200
Adelaide	1,092,900

SOURCE: Australian Bureau of Statistics, 30 June 2000.

States and Territories – Australia

2001 POPULATION ESTIMATES

Name	Population	Area (sq mi)	Area (sq km)	Capital
Australian Capital Territory	322,600	900	2,400	Canberra
New South Wales	6,642,900	309,500	801,600	Sydney
Northern Territory	199,900	519,800	1,346,200	Darwin
Queensland	3,670,500	666,900	1,727,200	Brisbane
South Australia	1,518,900	379,900	984,000	Adelaide
Tasmania	473,300	26,200	67,800	Hobart
Victoria	4,854,100	87,900	227,600	Melbourne
Western Australia	1,918,800	975,100	2,525,500	Perth

SOURCE: Australian Bureau of Statistics. "3101.0 Australian Demographic Statistics."

storms, fires, and multibillion-dollar crop losses. Serious droughts occurred again in 1994–1995.

Plains

The Central Plains have large areas of grasslands, with some steppe and some wooded savannah. Much of this area is used for grazing livestock, most famously sheep but also much cattle. The northern parts of the Western Plateau, while semi-arid, also have grasslands and some grazing. This is also the case on the coastal plain southwest of the plateau around Perth.

Deserts

Deserts cover approximately 18 percent of the land area of Australia. The Great Sandy Desert and the Gibson Desert cover a large part of Western Australia. The Great Victoria Desert, extending over parts of Western Australia and South Australia, and the Simpson Desert, covering parts of the Northern Territory, Queensland, and South Australia, are also of significant size. Smaller deserts exist scattered throughout the rest of the country, excepting Tasmania. Even outside of the deserts, much of the rest of the country is semi-arid. All together, the deserts and semi-arid areas combined cover nearly two-thirds of the continent.

Forests and Jungles

In the Eastern Highlands there are areas of forests, ranging from tropical rainforest in the north, particularly on Cape York Peninsula near the coast, to lighter tropical forest, to broadleaf and mixed broadleaf coniferous in the mid-latitudes of New South Wales and Victoria. Tasmania continues with the broadleaf and mixed broadleaf coniferous forest. There are also scattered coastal forests in the northwest and the southwest around Perth.

HUMAN POPULATION

Australia has a very small population relative to its size, with only 19,357,594 (2001 est.) people living on the entire continent and Tasmania. The population is mostly concentrated in and around the major coastal cities. The largest of these, Sydney, Melbourne, and Brisbane, are all located in the east and southeast of the country as is another large city, Adelaide. The national capital of Canberra is also located in the southeast, but inland; it is the only large inland city in Australia. There is another, smaller concentration of people along the southwestern edge of the continent, where the large city of Perth is located. Away from these two areas, the population is very sparse, especially in the desert outback, although there is one sizable town, Alice Springs, located near the center of the continent.

NATURAL RESOURCES

Australia is rich in mineral resources, with bauxite, coal, iron ore, copper, tin, silver, uranium, tungsten, mineral sands, lead, zinc, diamonds, and gold. Mining is an important industry for the county and the coal and iron have made possible a steel industry.

FURTHER READINGS:

Bechervaise, John. *Australia: World of Difference.* Adelaide: Rigby Limited, 1967.

Cousteau, Jean-Michel. *Cousteau's Australia Journey.* New York: Abrams, 1993.

Flannery, Tim. *The Explorers: Stories of Discovery and Adventure from the Australian Frontier.* New York: Grove Press, 2000.

Krannich, Ronald L. *The Treasures and Pleasures of Australia: Best of the Best.* Manassas Park, VA: Impact, 2000.

Robinson, K. W. *Australia, New Zealand and the Southwest Pacific,* 2nd edition. London: University of London Press Ltd., 1968.

Smith, Roff Martin. *Australia: Journey Through a Timeless Land.* Washington, D.C.: National Geographic, 1999.

Austria

- **Area:** 32,378 sq mi (83,858 sq km) / World Rank: 115
- **Location:** Northern and Eastern Hemispheres, Central Europe, bordering the Slovak Republic and Hungary to the east; Slovenia and Italy to the south; Switzerland to the west and southwest; Liechtenstein to the west; Germany to the northwest; the Czech Republic to the north
- **Coordinates:** 47°20′N, 13°20′E
- **Borders:** 1,588 mi (2,562 km) / Czech Republic, 224 mi (362 km); Germany, 486 mi (784 km); Hungary, 227 mi (366 km); Italy, 267 mi (430 km); Liechtenstein, 22 mi (35 km); Slovakia, 56 mi (91 km); Slovenia, 205 mi (330 km); Switzerland, 102 mi (164 km)
- **Coastline:** None; landlocked
- **Territorial Seas:** None
- **Highest Point:** Grossglockner, 12,461 ft (3,798 m)
- **Lowest Point:** Neusiedler See, 377 ft (115 m)
- **Longest Distances:** 356 mi (573 km) E-W / 183 mi (294 km) N-S
- **Longest River:** Danube, 1,775 mi (2,857 km)
- **Largest Lake:** Neusiedler See, 124 sq mi (320 sq km)
- **Natural Hazards:** None
- **Population:** 8,150,835 (July 2001 est.) / World Rank: 85
- **Capital City:** Vienna, northeastern Austria on the Danube
- **Largest City:** Vienna, 2,072,000 (2000)

OVERVIEW

Situated at the heart of Central Europe and bordering eight different countries, Austria has historically been a political, economic, and cultural crossroads. The Brenner Pass and the Danube River have provided crucial links between the Mediterranean and Balkan lands to the south and east and the Germanic countries to the north. For hundreds of years, the now small, landlocked, country of Austria was at the center of the great Hapsburg Empire that ruled much of Europe until World War I.

Austria's topography is dominated by the Alpine mountains that extend eastward from Switzerland, covering the western two-thirds of the country. Austria's two other major geographical regions are the Bohemian highlands bordering the Czech Republic to the north and the eastern lowlands, which include the Vienna Basin, home to the nation's capital city of the same name, located on the shores of the Danube, Europe's second-longest river. Austria is located on the Eurasian Tectonic Plate.

GEO-FACT

Although the composer Johann Strauss Jr. immortalized the Danube River in his famous waltz entitled "On the Beautiful Blue Danube," the Danube River is not blue—its waters appear either greenish or brown.

MOUNTAINS AND HILLS

Mountains

More than three-fourths of Austria's terrain is mountainous. The Alpine mountains spread across the western and southern parts of the country, with numerous ranges dividing into three major groups as they fan out across the land.

The limestone peaks of the northern Alps begin in Switzerland and Germany and extend west into Austria, continuing all the way to the Vienna Woods in the east. They lie to the north of a longitudinal depression that follows the valleys of the Inn, Salzach, and Enns rivers in an eastward direction from the Arlberg Pass (5881 ft / 1792 m above sea level) and crosses the Schober Saddle to the Mur and Mürz river valleys, ending at the Semmering Pass (3232 ft / 985 m above sea level) and the Vienna Basin. Many of its peaks rise above 8,000 ft (2,400 m); the highest point in the range, the Zugspitz (9,721 ft; 2,963 m), is located in Germany.

The central group of mountains is the largest and has the highest elevations, including the highest point in Austria, the Grossglockner (12,461 ft / 3,798 m). Many of its crystalline peaks top 10,000 ft (3,000 m). The major ranges of the central Alps include the Hohe Tauern and Niedere Tauern, and the Noric, Ötztal, Zillertal, Lechtal, and Kitzbühel Alps. The Pasterze, the largest of numerous glaciers found among the Austrian Alps, is also located in the central Alpine system, not far from the Grossglockner. It provides a venue for skiers as late as mid-June. The Brenner Pass (4497 ft / 1371 m above sea level) lies in the Ötztal Alps on the border with Italy. One of the largest and the lowest passes running through the Alps, it has been an important route for north-south travel through the mountains since ancient times.

The central group's southern boundaries are demarcated by the Drava (Drau) River valley, south of which lie Austria's southern Alps. These limestone peaks lie mostly in northern Italy. Within Austria, they occupy a relatively narrow strip in the southeast, along Austria's borders with Italy and Slovenia, lying within the province of Carinthia. They include the Carnic and Karawanken ranges.

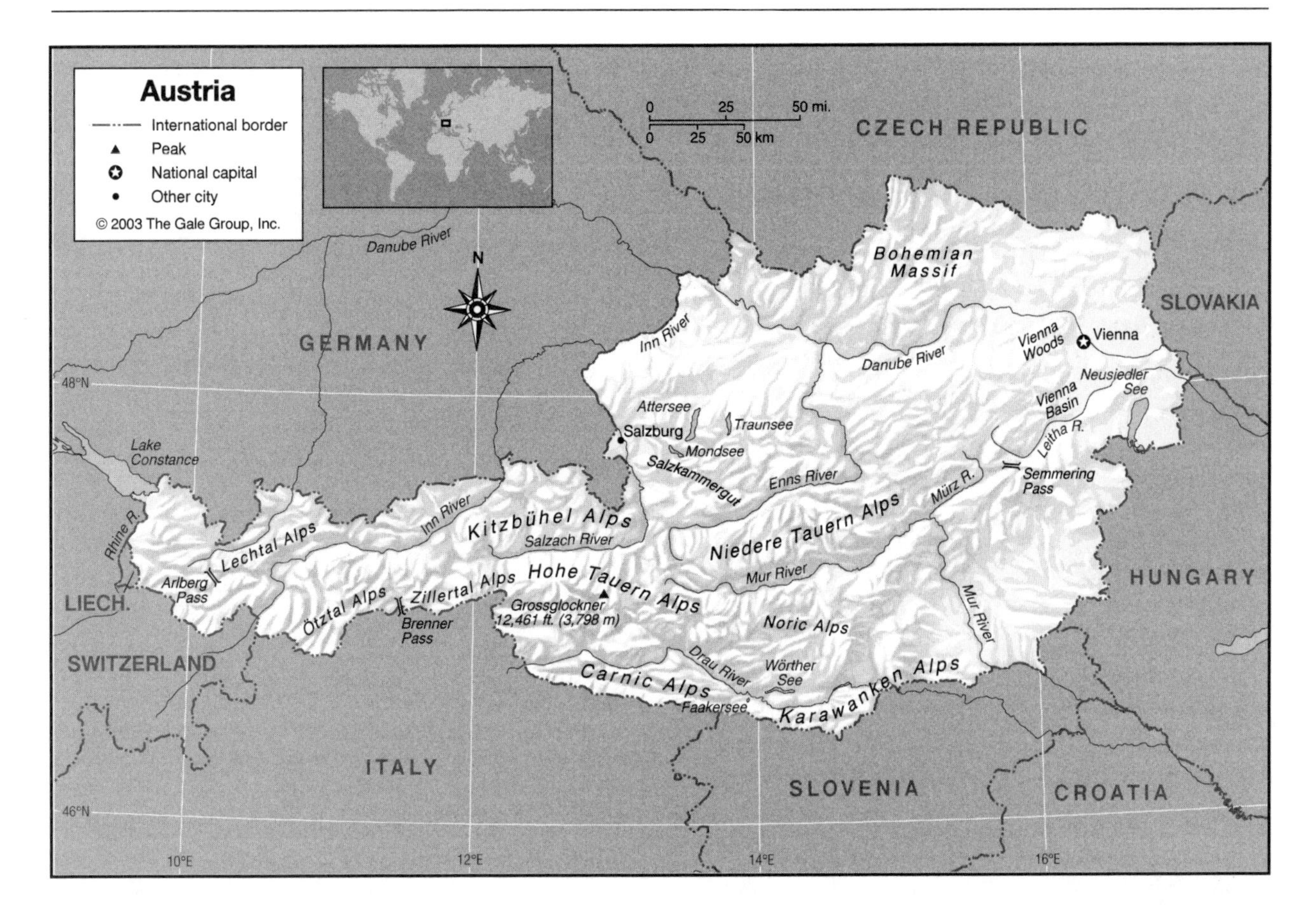

Plateaus

North of the Danube River and northwest of Vienna lie the granite and gneiss highlands of the Bohemian Massif, a plateau region that extends northward into the Czech Republic at elevations of up to 4,000 ft (1,200 m). These highlands account for roughly one-tenth of Austria's total area.

Hills and Badlands

The Northern Alpine Forelands is a region of foothills and valleys that lies between Austria's northern Alpine ranges and the Danube River valley. There are also foothills at the southeastern edge of the Alpine system, leading to the plains region bordering Hungary. Other hilly regions include the Waldviertel (wooded quarter) and Mühlviertel (mill district), rugged forested areas near the borders with Germany and the Czech Republic.

INLAND WATERWAYS

Lakes

The many lakes in Austria's mountain valleys contribute to the country's scenic beauty. Neusiedler See (Neusiedler Lake) lies mostly in Austria but also straddles the border between Austria and Hungary. It is over 20 mi (32 km) long and about 5 mi (8 km) wide. At the opposite end of Austria, at its furthest northwestern tip, is a small part of Lake Constance (also known as the Bodensee), which lies along the course of the Rhine River, where Austria, Switzerland, and Germany meet. It is one of the largest lakes in Western Europe.

The Salzkammergut region near Salzburg is known for a district that contains about seventy lakes, of which the largest include the Attersee, the Mondsee, and the Traunsee. The southern province of Carinthia, which alone boasts a total of over 1,200 lakes, is home to five famous lakes known as the Five Sister Lakes *(Funf Schwesterseen)*, of which the largest is the Wörther See. The Drau River Valley, in which Carinthia is located, is known for other picturesque lakes, including the Faakersee.

Rivers

Austria's principal river is the Danube, the second-longest river in Europe, which originates in Germany and flows southeastward to the Black Sea. The Danube flows eastward for 217 mi (350 km) within Austria's borders, through the northern part of the country; Vienna, the Austrian capital, is situated on its banks. Three of Austria's other major rivers—the Inn, Salzach, Enns—are tributaries of the Danube, flowing eastward through the central part of the country.

The major rivers in the southeast are the Mur and Mürz, in the industrial province of Styria. The Leitha flows northeast, draining the area from the Semmering

Pass to the Hungarian border. The Rhine River flows along part of Austria's western borders, and forms Lake Constance in Austria's extreme northwest.

THE COAST, ISLANDS, AND THE OCEAN

Austria is a landlocked nation.

CLIMATE AND VEGETATION

Temperature

Austria is in the Central European transitional climatic zone, and its climate varies by region. Atlantic maritime influences are felt in the northern and western provinces, where northwest winds from the North Atlantic moderate temperatures and bring moisture. Annual temperatures range between 20°F to 30°F (–7°C to –1°C) in winter and 65°F to 75°F (18°C to 24°C) in summer.

Continental influences are stronger in the eastern provinces, which have less precipitation and a greater range of temperatures, with colder winters, hot, humid summers, and mild, cloudy weather in the spring and fall. Average temperatures in Vienna are 25°F to 34°F (–4°C to 1°C) in January and 59°F to 77°F (15°C to 25°C) in July.

The mountain regions in the south and west have an Alpine climate, with warm but short summers and frequent storms. Winters are generally long with clear, sunny days. On the highest mountains, summertime temperatures often remain below the freezing point.

In the fall and spring, a warm, dry southern wind called the *föhn* moderates temperatures in the Alpine regions; it can also bring fog, and contributes to avalanches by causing snow to melt suddenly and fall from high elevations.

Rainfall

Precipitation is heaviest in the mountainous western regions (as high as 40 in / 102 cm annually) and lighter in the eastern plains regions (under 30 in / 76 cm), especially the area east of Neusiedler See, the driest region in the country. Average annual rainfall is 34 in (86 cm) at Innsbruck in the mountainous Tyrol region, and 26 in (66 cm) in Vienna, to the east.

Grasslands

East of the Alpine mountains is a region of low hills and level plains that forms part of the Hungarian Plain and constitutes Austria's lowland region. Even here, however, the land is often hilly, with elevations averaging 500 to 1,300 ft (150 to 400 m). The Vienna Basin in the north contains the most productive agricultural land in the country. The basin itself is not completely flat, but the terrain is gentle. The Vienna Basin extends into the Leitha River valley in a southeasterly direction toward the Semmering Pass and is separated from the Neusiedler Lake (Neusiedler See) by the Leitha Mountains. In the hilly Alpine foreland region, forests give way to arable land. Characteristic vegetation of the eastern lowlands includes scrub, heathland, and deciduous forests. Plants typical of salt steppe regions are found east of Neusiedler See.

Population Centers – Austria

(2001 CENSUS)

Name	Population	Name	Population
Vienna (capital)	1,562,676	Klagenfurt	90,255
Graz	226,424	Villach	57,740
Linz	186,298	Wels	56,516
Salzburg	144,816	Sankt Pölten	49,272
Innsbruck	113,826	Dornbirn	42,337

SOURCE: Österreichischen Statistischen Zentralamt, as cited on Geohive. Available http://www.geohive.com (accessed May 2002).

States – Austria

2001 POPULATION ESTIMATES

States	Population	Area (sq mi)	Area (sq km)	Capital
Burgenland	278,600	1,531	3,965	Eisenstadt
Kärnten	561,114	3,681	9,533	Klagenfurt
Niederösterreich	1,549,640	7,403	19,173	Sankt Pölten
Oberösterreich	1,382,017	4,625	11,980	Linz
Salzburg	518,580	2,762	7,154	Salzburg
Steiermark	1,185,911	6,327	16,388	Graz
Tirol	675,063	4,883	12,648	Innsbruck
Vorarlberg	351,565	1,004	2,601	Bregenz
Wien	1,562,676	160	415	Vienna

SOURCE: Population data from Österreichischen Statistischen Zentralamt. As cited on Geohive. http://www.geohive.com (accessed May 2002).

Forests and Jungles

Austria has the highest percentage of forestland of any country in Central Europe; forests cover nearly half its total area. Oak, beech, birch, and other deciduous varieties are found in the country's river valleys and on its lower slopes, while conifers, including spruce, larch, pine, and stone pine, grow at higher elevations, which are also home to common varieties of Alpine flowers.

HUMAN POPULATION

Most Austrians live in the country's eastern lowland regions and river valleys, particularly the Danube Valley. Urban dwellers account for roughly two-thirds of the population, and their numbers are growing. In the final decades of the twentieth century, thousands of foreign workers from southern European countries, including

Turkey and the former Yugoslavia, came to Austria to find work in its cities.

NATURAL RESOURCES

Austria's most important natural resources are its forests and the potential for hydroelectric power from its numerous streams and steep mountain slopes. Strict conservation laws protect the nation's forests, mandating replanting when trees are cut down to provide timber for paper and wood pulp production. Important mineral resources include lignite, iron ore, crude oil, magnesite, lead, and copper.

FURTHER READINGS

Austria. Oakland, Calif.: Lonely Planet Publications, 1996.

Austrian Press and Information Service. *Austria.* http://www.austria.org (accessed February 17, 2002)

Brook-Shepherd, Gordon. *The Austrians: A Thousand-year Odyssey.* New York: Carroll & Graf Publishers, 1997.

Frommer's Austria. New York: Macmillan, 1997.

Lichtenberger, Elisabeth. *Austria: Society and Regions.* Translated by Lutz Holzner. Vienna: Austrian Academy of Sciences, 2000.

Rice, Christopher, and Melanie Rice. *Essential Austria.* Lincolnwood (Chicago), Ill.: Passport Books, 2000.

Azerbaijan

- **Area:** 33,400 sq mi (86,600 sq km) / World Rank: 114
- **Location:** Northern and Eastern Hemispheres; northern Azerbaijan in Europe, the rest in Asia, bordering the Caspian Sea, between Iran and Russia; Armenia lies to the west and separates Azerbaijan from its Naxçivan exclave.
- **Coordinates:** 40°30′N, 47°30′E
- **Borders:** 1,251 mi (2,013 km) / Armenia, 490 mi (787 km); Georgia 200 mi (322 km); Iran (with Azerbaijan-proper) 379 mi (611 km); Russia 176 mi (284 km); Turkey 6 mi (9 km)
- **Coastline:** 500 mi (800 km, estimated) on Caspian Sea
- **Highest Point:** Mount Bazardyuze, 14,714 ft (4,485 m)
- **Lowest Point:** Caspian Sea, 92 ft (28 m) below sea level
- **Longest Distances:** Approximately 320 mi (510 km) E-W / 240 mi (380 km) N-S
- **Longest River:** Kura, 941 mi (1514 km)
- **Largest Lake:** Mingäçevir Reservoir, 234 sq mi (605 sq km)

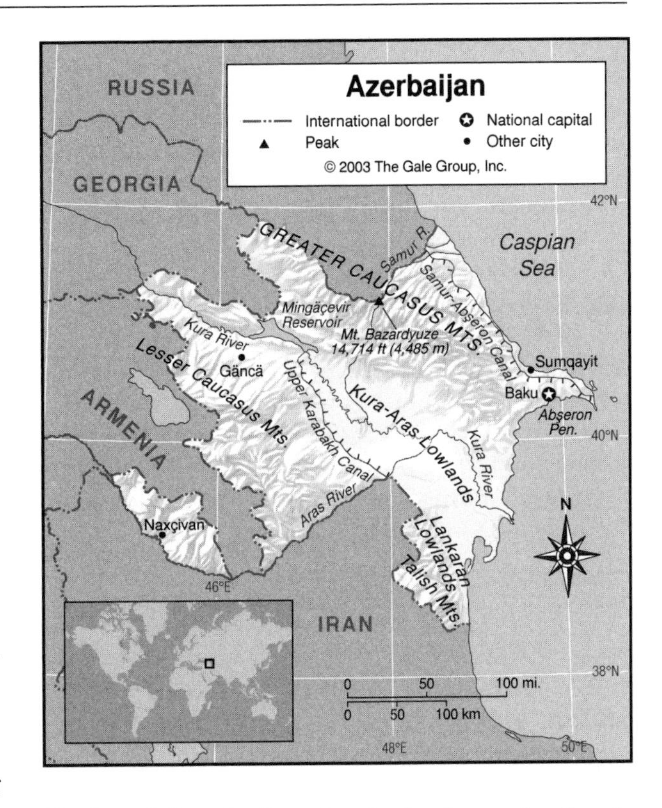

- **Natural Hazards:** Droughts; lowlands may be threatened by rising levels of the Caspian Sea
- **Population:** 7,771,092 (July 2001 est.) / World Rank: 86
- **Capital City:** Baku, on the coast of the Caspian Sea
- **Largest City:** Baku, 1.7 million (1993 est.)

OVERVIEW

Azerbaijan is the easternmost country of Transcaucasia (the southern portion of the region of Caucasia). It lies within the southern part of the isthmus between the Black and Caspian seas. The country is bordered on the north by Russia, on the east by the Caspian Sea, on the south by Iran, on the west by Armenia, and on the northwest by Georgia. An exclave called Naxçivan lies to the west, separated from the rest of Azerbaijan by Armenia. Naxçivan borders on Iran and Turkey in the west. Control of the Nagorno-Karabakh region, located within Azerbaijan's borders, east of Armenia, is disputed with that country.

About half of Azerbaijan is covered by mountain ranges; these mountains run along most of the borders, surrounding the Kura-Aras lowlands on three sides. These lowlands are an essentially flat region, much of which is below sea level. The coastline along the Caspian Sea is also essentially flat. The rise in elevation, from lowlands to highlands, occurs over a relatively small area.

The climate is varied for an area of only 33,400 sq mi (86,600 sq km): in the center and east are dry, semiarid steppe; in the southeast, it is subtropical; high mountain elevations to north are cold; and along the Caspian Sea coast, weather is temperate.

Population Centers – Azerbaijan

(1997 POPULATION ESTIMATES)

Name	Population
Baku (capital)	1,725,500
Gäncä (Gyandzha)	294,100
Sumqayit (Sumgait)	273,200
Mingäçevir (Mingechavir or Mingachevir)	97,400

SOURCE: State Statistical Committee of Azerbaijan Republic.

MOUNTAINS AND HILLS

Mountains

Azerbaijan is nearly surrounded by mountains. The Greater Caucasus range, with the country's highest elevations, lies in the north along the border with Russia and run southeast to the Abşeron Peninsula on the Caspian Sea. The country's highest peak, Bazardyuze, rises to 14,800 ft (4,485 m) in this range near the Azerbaijan-Russia border. The Lesser Caucasus range, with elevations up to 11,500 ft (3500 m), lies to the west along the border with Armenia. The Talish Mountains form part of the border with Iran at the southeast tip of the country.

Kobustan Mountain, located near Baku, is carved by deep ravines, from which bubble mud volcanoes and mineral springs.

INLAND WATERWAYS

Lakes

The Mingäçevir Reservoir, a 234 sq mi (605 sq km) reservoir formed by a dam built in 1953 on the Kura River, is the largest body of water in Azerbaijan.

Rivers and Canals

Most of the country's rivers flow down from the Caucasus ranges into the central Kura-Aras lowlands. The Kura River flows through Turkey, Georgia, and Azerbaijan; and at 940 mi (1500 km) it is the longest river of the Transcaucasia Region. The lower 300 mi (480 km) in Azerbaijan is navigable. It enters the Caspian Sea in a delta south of Baku. The Araks (Aras) River, which is 568 mi (914 km) long, flows along much of the border with Iran before joining the Kura River in south-central Azerbaijan.

The Kura-Aras lowlands west of the Caspian shoreline form an area of alluvial flatlands and low seacoast deltas. However, much of the time water here is supplied through irrigation; the area is naturally arid. The Upper Karabakh Canal channels water from the Mingäçevir Reservoir in northwestern Azerbaijan to the Kura and Araks further south, bringing water to farms in the central lowland during the dry summer months. In the east, the Samur-Abşeron Canal redirects water from the Samur River on Azerbaijan's northeastern border to the Abşeron Peninsula, an arid area near the capital, Baku.

THE COAST, ISLANDS, AND THE OCEAN

The Caspian Sea

Azerbaijan has a 500 mi (800 km) coastline on the Caspian Sea, which is essentially a giant saltwater lake in southeastern Europe and southwestern Asia. With an area of 143,000 sq mi (371,000 sq km) it is the largest inland body of water in the world. It has no outlet to the ocean. Baku is Azerbaijan's chief port on the Caspian.

The Coast and Beaches

The irregular Caspian coastline features the Abşeron Peninsula in the center of Azerbaijan's coast. Pollution from agricultural chemicals (especially pesticides), industry, and oil drilling has had a serious impact on the Caspian Sea coastline environment.

CLIMATE AND VEGETATION

Temperature

Climate varies from subtropical and humid in the southeast to subtropical and dry in central and eastern Azerbaijan. Along the shores of the Caspian Sea it is temperate, while the higher mountain elevations are generally cold. Baku, on the Caspian, enjoys mild weather that averages 39°F (4°C) in January and 77°F (25°C) in July.

Rainfall

Most of Azerbaijan receives little rainfall, only 6 to 10 inches (152 to 254 millimeters) annually on average. As a result, agricultural areas require irrigation. Approximately 3,861 sq mi (10,000 sq km) of the land is irrigated (1993 est.). The greatest precipitation falls in the highest elevations of the Caucasus but also in the Länkäran Lowlands of the extreme southeast. The yearly average in these areas can exceed 39 inches (1,000 millimeters).

Kura-Aras Lowlands

The Kura-Aras lowlands occupy the center of Azerbaijan, between the mountain ranges and the Caspian Sea. Much of this area is below true sea level, as are the shores of the Caspian. Only 18 percent of the land is considered arable, with permanent crops comprising 5 percent of the total.

HUMAN POPULATION

With a population of 7,771,092 (July 2001 estimate), Azerbaijan has a population density of 214 people per sq mi (83 people per sq km). Azerbaijan is one of the most densely populated of the Transcaucasian states. The population is concentrated in a few urban centers and in the Kura-Aras agricultural regions. Baku, the capital, is the most populated city. As of 1998 Baku's population exceeded 1.7 million people, including a mid-1990s surge of an estimated 300,000 to 500,000 war refugees.

NATURAL RESOURCES

Crude oil and natural gas are the most important of Azerbaijan's natural resources. The oil fields are located offshore near the Abşeron Peninsula, beneath the Caspian Sea. Iron ore, aluminum, copper, lead, zinc, limestone, and salt are the most abundant mineral resources.

Despite decades of severe air, water, and soil pollution, highly productive Caspian fisheries provide valuable catches of sturgeon (for caviar), salmon, perch, herring, and carp.

FURTHER READINGS

Economic Review: Azerbaijan. Washington, D.C.: International Monetary Fund., 1992.

Edwards-Jones, Imogen. *The Taming of Eagles: Exploring the New Russia*. London: Weidenfeld & Nicolson, 1993.

International Petroleum Encyclopedia, 1992. Edited by Jim West. Tulsa, OK: Penn Well, 1992.

Maggs, William Ward. "Armenia and Azerbaijan: Looking Toward the Middle East," *Current History*, 92, January 1993, pp. 6-11.

Richards, Susan. *Epics of Everyday Life: Encounters in a Changing Russia*. New York: Viking, 1991.

Streissguth, Thomas. *The Transcaucasus*. San Diego, CA: Lucent Books, 2001.

Weekes, Richard W., ed. *The Muslim Peoples: A World Ethnographic Survey*. 2nd edition. Westport, CT: Greenwood Press, 1984.

The Bahamas

- **Area:** 5,382 sq mi (13,940 sq km) / World Rank: 158
- **Location:** Northern and Western Hemispheres; North Atlantic Ocean southeast of Florida
- **Coordinates:** 24°15′N, 76°00′W
- **Borders:** No international boundaries
- **Coastline:** 2,201 mi (3,542 km)
- **Territorial Seas:** 12 NM (22 km)
- **Highest Point:** Mt. Alvernia, 206 ft (63 m)
- **Lowest Point:** Sea level
- **Longest Distances:** 590 mi (950 km) SE-NW / 185 mi (298 km) NE-SW
- **Longest River:** None
- **Natural Hazards:** Hurricanes and severe tropical storms cause flooding and wind damage
- **Population:** 297,852 (July 2001 est.) / World Rank: 171
- **Capital City:** Nassau, on New Providence Island in the center of the island group
- **Largest City:** Nassau, 195,000 (2000 est.)

OVERVIEW

The Commonwealth of the Bahamas occupies an archipelago that lies across the Tropic of Cancer at the northwestern end of the West Indies, about 50 mi (80 km) off the southeast coast of Florida. Roughly 700 islands are included in the chain, as well as some 2,000 rock formations, islets, and cays (keys). Nassau, the capital, is located on New Providence, which occupies a central position in the archipelago and is the most densely populated island. Collectively, the rest of the inhabited Bahamas islands are known as the Family Islands.

The islands, most of which are long, narrow, and fringed by coral reefs, originated as surface outcroppings of two submerged limestone banks, today known as the Great Bahama Bank and the Little Bahama Bank. The Bahamas lie on the North American Tectonic Plate, and the nation exercises maritime claims over the continental shelf to a depth of 656 ft (200 m).

MOUNTAINS AND HILLS

The terrain of the Bahamas is mostly flat and low, only a few feet above sea level in most places. There are no true mountains, and only a few hills. The tallest point is Mt. Alvernia on Cat Island (206 ft / 63 m).

INLAND WATERWAYS

None of the Bahamas are large enough to support significant rivers or lakes, although there are many small streams and ponds. Coastal wetlands and mangrove swamps are common throughout the archipelago.

THE COAST, ISLANDS, AND THE OCEAN

Oceans and Seas

The Bahamas are spread over approximately 90,000 sq mi (233,000 sq km) of water in the southwestern portion of the North Atlantic Ocean. They are separated from Florida in the United States by the Straits of Florida. The Mayaguana Passage and the Caicos Passage, respectively north and south of the island of Mayaguana, allow

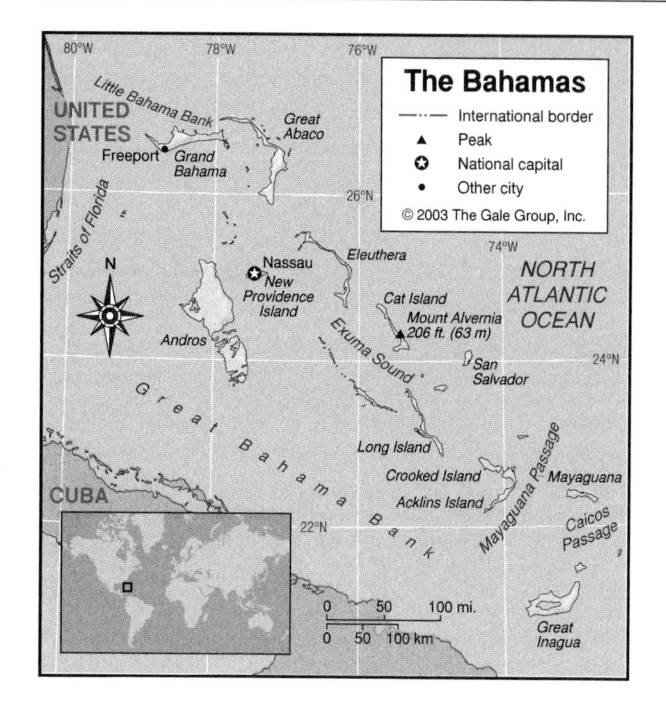

GEO-FACT

Christopher Columbus made his first landfall in the New World in the Bahamas on October 12, 1492, on the island he named San Salvador.

clear sailing from east to west through the chain. A relatively large area of open water in the middle of the archipelago is known as Exuma Sound.

Major Islands

The most important island is New Providence, home to the capital city of Nassau; it has an area of 80 sq mi (207 sq km). Next in importance are Andros (at 1,600 sq mi / 4,144 sq km, the largest island), Great Abaco, and Eleuthera. Other major islands include Grand Bahama, Cat Island, San Salvador, Long Island, Great Exuma, Crooked Island, Acklins Island, Mayaguana, and Great Inagua.

The Coast and Beaches

The numerous coral reefs on the shorelines of the Bahamas combine with iron compounds found in the limestone terrain to produce rare and beautiful colors in the shallow seas surrounding the islands. This, as well as the many coral reefs to be found throughout the archipelago, make the Bahamas a popular destination for beach lovers and divers.

CLIMATE AND VEGETATION

Temperature

The Bahamas have a subtropical marine climate moderated by warm breezes from the Gulf Stream and the Atlantic Ocean. Average temperatures are 73°F (23°C) in winter (December–April) and 81°F (27°C) in summer (May–November).

Rainfall

Rainfall averages 50 in (127 cm) annually, with some variation among the different islands. Occasional hurricanes occur between mid-July and mid-November. Hurricanes can cause major damage from winds and flooding, but their effects are limited since the islands are so widely scattered, with a reduced incidence of risk for any one island.

Forests

Trees found in the Bahamas include the cork and black olive tree, as well as casuarina, cascarilla, mahogany, cedar, and several species of palm. Some islands, including Andros, Grand Bahama, and Great Abaco, have pine forests.

Tropical vegetation found on the islands includes jasmine, bougainvillea, and oleander.

HUMAN POPULATION

Only 30 to 40 of the islands are inhabited. The greatest population concentration is on the island of New Providence, which has both the largest population and the highest population density. More than half the population of the Bahamas lives on this island.

The islands with the next-largest population concentrations are Grand Bahama (home to Freeport, the second-largest city) and Great Abaco.

NATURAL RESOURCES

Commercially significant minerals include salt, aragonite, and limestone. Other natural resources include fish from the surrounding water, and modest amounts of timber and arable land. However, the Bahamas' economy relies more heavily on tourism than it does these resources.

FURTHER READINGS

Dulles, Wink, and Marael Johnson. *Fielding's Bahamas.* Redondo Beach, Calif.: Fielding Worldwide, 1997.

Geographia Tourist Guide to the Bahamas. *The Islands of The Bahamas.* http://www.geographia.com/bahamas/ (accessed Feb. 7, 2002).

Jenkins, Olga Culmer. *Bahamian Memories: Island Voices of the Twentieth Century.* Gainesville Fl.: University Press of Florida, 2000.

Johnson, Howard. *The Bahamas from Slavery to Servitude, 1783-1933.* Gainesville, Fl.: University of Florida Press, 1996.

Lloyd, Harvey. *Isles of Eden: Life in the Southern Family Islands of the Bahamas.* Akron, Oh.: Benjamin Publishing, 1991.

Permenter, Paris, and John Bigley. *The Bahamas: A Taste of the Islands.* Edison, N.J.: Hunter, 2000.

Bahrain

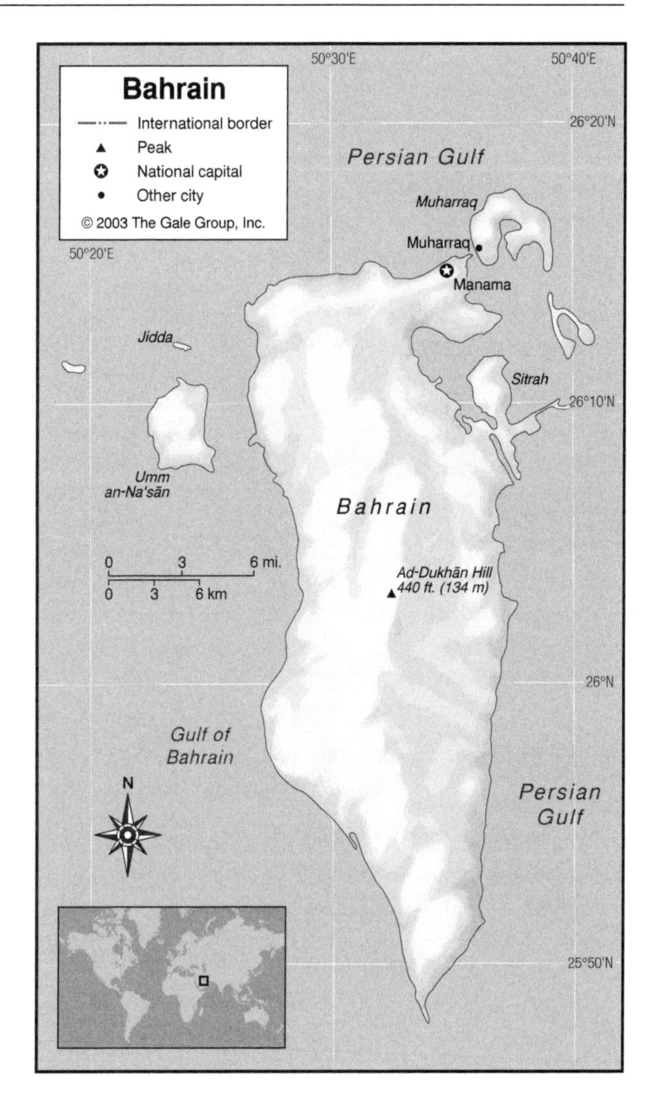

- **Area:** 239 sq mi (620 sq km) total area of 33 islands. The main island, Bahrain, accounts for 85 percent of total land mass. / World Rank: 182
- **Location:** Northern and Eastern Hemispheres, lies at the entrance to the Gulf of Bahrain in the western Persian Gulf, 18 mi (29 km) northwest of Qatar and east of Saudi Arabia
- **Coordinates:** 26°00′N, 50°33′E
- **Borders:** No international boundaries
- **Coastline:** 78 mi (126 km)
- **Territorial Seas:** 26 NM in a contiguous zone, territorial seas of 12 NM. continental shelf boundaries have yet to be determined.
- **Highest Point:** Ad-Dukhān Hill, 440 ft (134 m)
- **Lowest Point:** Sea level
- **Longest Distances:** Archipelago extends 30 mi (48 km) N-S /12 mi (19m km) E-W
- **Longest River:** None
- **Natural Hazards:** Subject to period droughts and dust storms
- **Population:** 645,361 (2001 est.) / World Rank: 158
- **Capital City:** Manama, located on the northeastern coast of the main island Bahrain
- **Largest City:** Manama (133,784, 1990 est.)

OVERVIEW

Bahrain's six inhabited islands—Bahrain, Al Muharraq, Sitrah, Umm an-Na'sān, Nabih Salih, and Jidda—and their position in an inlet of the Persian Gulf have given this multi-island nation a regional importance as a trade and transportation center. The low rolling hills, rocky cliffs, and wadis comprise the majority of this barren land, although along the north coast of the island of Bahrain is a narrow strip of land that is irrigated by natural springs and artesian wells.

Most of the lesser islands are flat and sandy, although date groves cover the island of Nabih Salih. Bahrain also possesses the Hawār Islands, off the coast of Qatar.

MOUNTAINS AND HILLS

Hills and Badlands

On the main island of Bahrain, the land gradually rises from the shoreline to the center, where rocky cliffs surround a basin. Near the center of this basin is the country's highest elevation, Ad-Dukhān Hill, which rises only 440 ft (134 m) above sea level.

INLAND WATERWAYS

Comprised of mostly barren, un-arable land, Bahrain has little fresh water, and no rivers or lakes. On the main island in this group (Bahrain) are about 6.2 sq mi (10 sq km) of irrigated land.

THE COAST, ISLANDS, AND THE OCEAN

Oceans and Seas

Located in an inlet of the Persian Gulf, the Gulf of Bahrain, the island of Bahrain is connected to other major islands by bridges, and to Saudi Arabia by the King Fahd Causeway.

Major Islands

The country is comprised of 33 islands, the 6 major islands being: Bahrain, the largest; Al Muharraq; Sitrah; Umm an-Na'sān; Nabih Salih; and Jidda. In 2001 the International Court of Justice awarded the Hawār Islands, long disputed with Qatar, to Bahrain. The remaining islands are little more than exposed rock and sandbar.

The Coast and Beaches

Unfortunately, coastal degradation (damage to coral reefs and sea vegetation) resulting from oil spills and other discharges from large tankers, oil refineries, and distribution stations has adversely affected Bahrain's coastline and beaches.

CLIMATE AND VEGETATION

Temperature

Summers are very hot and humid with southwest winds raising dust storms and drought conditions. Winters are mild, cool, and pleasant. Average temperatures in July range from 84°F (29°C) to 79°F (37°C) , and in January from 68°F (20°C) to 57°F (14°C).

Rainfall

Prevailing southwest winds that sweep this low-lying desert plain contribute to dust storms and occasional drought. Rainfall averages less than 4 in (10 cm) annually and occurs primarily from December to March.

Deserts

Due to its location and weather patterns, the State of Bahrain is primarily desert, with only a narrow strip of land on the main island that is irrigated by natural springs and artesian wells.

HUMAN POPULATION

Bahrain's estimated population in 2001 was 645,361, but 228,424 of these were non-nationals in the country as a temporary work force. The growth rate for 2001 was 1.73%. The vast majority of the population lives in Manama and the other major cities in the north of the country.

NATURAL RESOURCES

Bahrain depends on the exploitation of oil, natural gas, fish, and pearls to support its economy. There is little arable land and no forests.

Population Centers – Bahrain

(1992 POPULATION ESTIMATES)

Name	Population
Manama (capital)	140,401
Ar Rifá	45,956
Muharraq	45,337

SOURCE: "Population of Capital Cities and Cities of 100,000 and More Inhabitants." United Nations Statistics Division.

Regions – Bahrain

1991 POPULATION ESTIMATES

Name	Population	Area (sq mi)	Area (sq km)
Al-Hadd	8,610	2.0	5.2
Al Mintaqah al Gharbiyah	22,034	60.2	156.0
Al Mintaqah al Wustal	34,304	13.6	35.2
Al Mintaqah ash Shamaliyah	33,763	14.2	36.8
Al-Muharraq	74,245	5.9	15.2
Jidd Hafs	44,769	8.4	21.6
Ar Rifa	49,752	112.6	291.6
Madinat Hamad	29,055	5.1	13.1
Madinat 'Isá	34,509	4.8	12.4
Manama	136,999	9.8	25.5
Sitrah	36,755	11.0	28.6

SOURCE: *Geo-Data*, 1989 ed., and Johan van der Heyden, Geohive, http://www.geohive.com (accessed June 2002).

FURTHER READINGS

Crawford, Harriet E. W. *Dilmun and Its Gulf Neighbors.* New York: Cambridge University Press, 1998.

Jenner, Michael. *Bahrain, Gulf Heritage in Transition.* New York: Longman, 1984.

Khuri, Fuad I. *Tribe and State in Bahrain: the Transformation of Social and Political Authority in an Arab State.* Chicago: University of Chicago, 1980.

U.S. Dept. of State. Bureau of Public Affairs. Office of Public Communication. *Background Notes: Bahrain.* Washington, D.C., 1991.

Vine, Peter. *Pearls in Arabian Waters: The Heritage of Bahrain.* London: Immel Publications, 1986.

Bangladesh

- **Area:** 55,598 sq mi (143,998 sq km) / World Rank: 93
- **Location:** South Asia, between the Himalayan foothills and the Bay of Bengal, Northern and Eastern Hemispheres
- **Coordinates:** 24° 00′N, 90° 00′E
- **Borders:** 2,638 mi (4,246 km) total border length; India, 2,518 mi (4,053 km); Myanmar, 120 mi (193 km)
- **Coastline:** 357 mi (574 km) long, on the Bay of Bengal of the Indian Ocean
- **Territorial Seas:** 12 NM (22 km)
- **Highest Point:** Keokradong, 4,034 ft (1,230 m)
- **Lowest Point:** Sea level
- **Longest Distances:** 477 mi (767 km) SSE-NNW/ 267 mi (429 km) ENE-WSW
- **Longest River:** Brahmaputra River, total length 1,700 mi (2,900 km)
- **Largest Lake:** Karnaphuli Reservoir, 253 sq mi (655 sq km)
- **Natural Hazards:** Severe seasonal flooding, cyclonic storms, tidal bores, tornadoes, ground water arsenic contamination, hailstorms, moderate earthquake risk
- **Population:** 131,269,860 (2001 est.) / World Rank: 8
- **Capital City:** Dhaka, located in the center of Bangladesh.
- **Largest City:** Dhaka (2002 population estimated at over 10 million)

OVERVIEW

Bangladesh is dominated by a vast multi-river basin and delta. The huge Brahamputra (or Jamuna) and Ganges (or Padma) Rivers meet in the center of the country and countless connecting streams and rivers, some of them very large, run between the two rivers or flow south into the Indian Ocean. A small coastal region (Chittagong) in the southeast is the only part of the country outside this network of rivers. Small hill regions in the northeast and southeast are the only topological variation of the land's flat alluvial plains. This geography is both a gift, creating bountiful rice-growing land by the seasonal flux of great river systems, and a curse, making the population vulnerable to flooding. With some 90 percent of Bangladesh's land within 33 ft (10 m) of sea level, the country is extraordinarily vulnerable to flooding due to its many great rivers and heavy rainfall. Bangladesh also has one of the world's highest populations and is believed to be the most densely populated country on Earth.

MOUNTAINS AND HILLS

Hills and Badlands

Bangladesh's significant hill regions are the Chittagong and Bandarban Hill Tracts, which are a series of ridges along the Myanmar (Burma) frontier, and Sylhet District. Rising near the intersection of the borders of Myanmar, India, and Bangladesh, Keokradong at 4,034 ft (1230 m) is the country's highest peak. The countryside in the northeast corner of Bangladesh, near the town of Sylhet, features sedimentary hills, some of which exceed 300 ft (90 m) in elevation. Also near Sylhet, but south of the Kusiyara River, six hill ranges connect to the Tripura Hills of India. In these ranges the maximum elevation, near the Bangladesh-Indian border, is about 1,100 ft (335 m).

INLAND WATERWAYS

Lakes

Bangladesh's one major lake is largely an artificial one: Karnaphuli Reservoir (also known as Kaptai Lake), in the Chittagong Hill Tracts. It was for the most part formed by damming the Karnaphuli River in 1963 for hydroelectric power. Much smaller lakes called "mils" or "haors" are formed within the network of rivers winding across Bangladesh's plains. The region around Sylhet is known as the Sylhet Plain, or the Surma River Plain. There and throughout the upper Meghna-Surma drainage area, the most distinguishing feature is the profusion of large and small lakes. This whole lake-dotted northeast quadrant of the Bangladesh Plain is even more vulnerable to flooding than the other parts of the country.

Rivers

Bangladesh is formed by the great rivers that move the snow-melt of the Himalayas through India to their outlets in Bangladesh's delta region. Rich soil is brought along by the rivers as old soil is carried away, and the rivers provide fish and transportation for the people of Bangladesh. The rivers also cause hardship due to seasonal flooding and shrinking land for settlement. The balance of water input and output is intrinsic to Bangladesh's survival, and environmental policies in India, upstream, can have huge effects on Bangladesh.

The river network of Bangladesh's plain moves generally from north to south but includes untold numbers of feeder and effluent streams, like capillaries, flowing east, west, southeast, and southwest. Within Bangladesh's riverine labyrinth, change is constant as rivers modify or shift their channels.

The Ganges River of India, flowing southeastward, comes to the boundary of Bangladesh in the northwest of the country. For about 90 mi (145 km) the Ganges is the

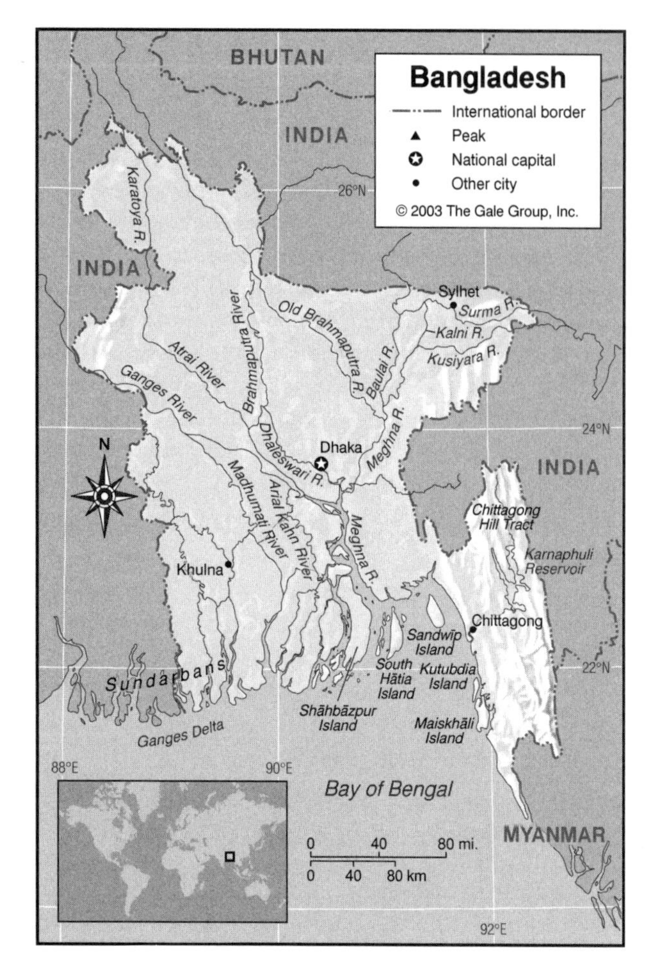

boundary between India and Bangladesh; it then continues to the southeast across the alluvial plain. In Bangladesh the Ganges is commonly called the Padma. The area south of the Padma is of very low elevation, and the hundreds of rivers and streams in this true delta segment of the plain are virtually all distributary from the Padma. Principal among them are the Madhumati River and the Arial Khan (also known as the Bhubanswar). Most flow south, and many of them reach the Indian Ocean. The result is a network of channels entering the Bay of Bengal through a crumbled seacoast that stretches for hundreds of miles, including part of India. It is often referred to as "the many mouths of the Ganges."

Like the Ganges, the Brahmaputra River rises in the high Himalayas. It flows east across Tibet and the Assam Valley of India. Upon reaching the northern border of Bangladesh, the Brahmaputra turns south and enters the country in multiple, narrowly separated channels and then becomes known as the Jamuna River. The wide Jamuna flows south, joining with the Karatoya River and the Atrai River. Near the center of the country, it merges with the Padma. After junction, the combined waters of the Brahmaputra-Jamuna and Ganges-Padma continue southeast for about 60 mi (96 km) to the even wider junction with a third great river, the Meghna.

GEO-FACT

The Grameen Bank, founded in 1976 in Bangladesh, has pioneered the concept of "micro-credit" as a way of helping the poor, especially village women, by granting them small collateral-free loans with which to start up their own businesses. This unusual form of banking is based on mutual trust and community participation.

The Dhaleswari River, a distributary of the Brahmaputra-Jamuna, leaves the parent river above its junction with the Padma. It flows southeast, below Dhaka and roughly parallel to the lower Padma, receives the Lakhya River, and then joins the Meghna a few miles above its junction with the Padma. This is only one of the largest of the countless distributary streams and rivers that branch off from or into the Padma, Jamuna, and Meghna.

In the northeast corner of Bangladesh, two branches of the Barak River, the Surma and the Kusiyara, enter the country from India. These rivers, with smaller tributaries, form the Kalni River, which soon becomes the upper Meghna and is then reinforced by a major tributary from the north, the Baulai. From this point the Meghna continues southwest in a twisting, multi-channeled course to the junction with the lower Padma about 65 mi (105 km) away, south of Dhaka.

From the wide Padma-Meghna junction, the three combined rivers, here called the Meghna, move south in an S-shaped stem channel for about 40 mi (64 km) and then spread out into the Bay of Bengal through one of the largest estuaries in the world, roughly 100 mi (160 km) across, from Khulna to Chittagong.

Wetlands

About 3,861 sq mi (10,000 sq km) of Bangladesh are covered with water under normal conditions. The rivers flowing though Bangladesh often silt up to form marshlands. Western Bangladesh's Kulna Division is as much as two-thirds marsh and mangrove forest (a tidal wetland environment of low trees and salt bog.) In Kulna the delta land and islands along the coast from the Indian border east to the Padma-Meghna estuary and extending inland are called the Sundarbans. This is a forested, tidal-flushed, saltmarsh region, much of it so shifting, low, and swampy as to prevent permanent human habitation. Another important wetlands area is Rajshahi Division, the northwest segment of the Bangladesh Plain between the Padma and Jamuna rivers. This has been called the

Population Centers – Bangladesh

(2000 POPULATION ESTIMATES)

Name	Population
Dhaka (capital)	12,519,000
Chittagong	3,651,000
Khulna	1,442,000
Rajshahi	1,035,000

SOURCE: "Table A.12. Population of Urban Agglomerations with 750,000 Inhabitants or More in 2000, by Country 1950–2015. Estimates and Projections: 1950–2015." United Nations Population Division, World Urbanization Prospects: The 2001 Revision.

Divisions – Bangladesh

POPULATIONS ESTIMATED IN 1998

Name	Population	Area (sq mi)	Area (sq km)	Administrative center
Barisal	8,678,000	5,134	13,297	Barisal
Chittagong	24,609,000	17,535	45,415	Chittagong
Dhaka	39,020,000	11,881	30,772	Dhaka
Khulna	14,944,000	12,963	33,574	Khulna
Rajshahi	30,937,000	13,219	34,237	Rajshahi
Sylhet	8,021,000	4,863	12,596	Sylhet

SOURCE: *Bangladesh Data Sheet, 1999.* Bangladesh Bureau of Statistics, National Data Bank.

"paradelta" by geographers. It is an extensive plain falling from about 300 to 350 ft (91 to 106 m) general elevation in the north to about 100 ft (30 m) centrally and down to about 30 ft (9 m) in the south. It is cut by many old river courses as well as by newer, active rivers and, like the rest of the country, is subject to disastrous flooding.

THE COAST, ISLANDS AND THE OCEAN

Oceans and Seas

Bangladesh sits at the apex of the Bay of Bengal, a part of the Indian Ocean. Despite this, Bangladesh is a river nation rather than an ocean-oriented nation. Its ports, including the largest, Chittagong, are all river ports. The ocean often threatens catastrophe for Bangladesh, in the form of cyclonic storms and tidal bores. The tidal waves called bores recur from time to time in Bangladesh's estuary areas and its stem rivers. Tidal bores tend to form high waves with abrupt fronts when the incoming surge of water at flood tide encounters a resistance, such as a sandbar. The funnel shape of the Padma-Meghna River estuary and the many channels between the islands are highly favorable to bore formation. A typical tidal bore rushes in with thunderous noise and a wall of water. Velocity and height are magnified if the tidal bore is backed by strong winds from the south, as occurs during cyclonic storms.

Major Islands

Bangladesh is a country of countless and changing islands In the Padma-Meghna estuary triangle there are a number of permanent islands, many islands that surface only at low tide, and also many temporary "chars," land forms built up by silting that may become permanent or erode away again. "Chars" also occur in many places along the larger rivers and have frequently been the subject of dispute as to ownership. A series of flat islands lie not far offshore in the Bay of Bengal, with many inhabited by fishing communities despite extreme cyclone danger. The largest of the permanent islands are Shāhbāzpur, North Hātia, South Hātia, and Sandwīp. Along the Chittagong coast in the south lie Kutubdia and Maiskhāl islands.

The Coast and Beaches

Bangladesh's coastline in the delta region is characterized by its fragmentation by rivers and streams. In contrast, in the southeast (Chittagong Division) the coastline includes a vast uninterrupted 75 mi (120 km) stretch of sand near Cox's Bāzār that is promoted as "the world's longest beach."

CLIMATE AND VEGETATION

Temperature

Bangladesh experiences three seasons. The winter, from October to early March, has temperatures from 41°F to 72°F (5°C to 22°C). Then the hot season arrives in March with temperatures averaging 90°F (32°C). During the monsoon season, from May to September, temperatures average 88°F (31°C) with 90 to 95 percent humidity.

Rainfall

With a tropical monsoon climate, Bangladesh has a heavy average annual rainfall of approximately 47 in (119 cm) to 57 in (145 cm). About 80 percent of Bangladesh's rainfall occurs in the monsoon season from May to September. Hailstorms of Himalayan origin frequently affect Bangladesh in that season. Cyclones also often sweep in from the Bay of Bengal during April-June and October-November. Particularly devastating cyclones hit Bangladesh in 1970 (as many as half a million people killed) and 1991 (at least 140,000 people killed).

Forests and Jungles

Large scale deforestation has been caused by land-clearing for agriculture, plus logging and firewood use. Less than 7.8 percent of Bangladesh remained forested as of 1999. Pockets of rainforest, mostly secondary, still exist in the eastern regions. The mangrove forests of the delta regions have been somewhat protected from encroachment, although they face threats from economic interests. The officially protected 1,390 sq mi (3,600 sq km) Sundarbans forest region of Bangladesh is a wildlife refuge for endangered Bengal tigers.

HUMAN POPULATION

Usually estimated to be the most densely populated country on Earth, Bangladesh has over 2,460 people per sq mi (950 people per sq km). Most Bangladeshis live in rural rice-growing communities. However, Dhaka grew enormously in the late 20th Century to become a "mega-city" of over 10 million; some 20 percent of all Bangladeshis now live in Dhaka. Despite significant gains in slowing the birth rate through family planning programs–Bangladesh had achieved a 1.59 percent population growth rate in 2001 estimate–the nation's population expansion still outpaces its economic growth and its sustainable living space.

Population pressure leads to deforestation, pollution, epidemics, vulnerability to cyclones and flooding, and persistant poverty in Bangladesh.

NATURAL RESOURCES

Bangladesh's economic resources are primarily agricultural. Bangladesh is a major rice producer, ranked 5th in the world in 1999-2000. The rice crop is largely used for domestic consumption. Jute (a fiber crop) and fishing are other significant economic resources.

Important reserves of natural gas and some oil exist in northeast Bangladesh and offshore in the Bay of Bengal. Coal is found in western Bangladesh.

FURTHER READINGS

Bornstein, David. *The Price of a Dream: The Story of the Grameen Bank.* Chicago: University of Chicago Press, 1997.

Heitzman, James, ed. *Bangladesh: A Country Study.* Washington, D.C.: Library of Congress, 1988.

Jahan, Rounaq, ed. *Bangladesh: Promise and Performance.* London: Zed Books, 2001.

Novak, James. *Bangladesh: Reflections on the Water.* Bloomington: Indiana University Press, 1993.

USAID Bangladesh. *USAID Bangladesh, Making a Difference.* http://www.usaid.gov/bd (accessed February 22, 2002).

Virtual Bangladesh. *Welcome to Bangladesh.* http://www.virtualbangladesh.com (accessed February 22, 2002).

Barbados

- **Area:** 166 sq mi (430 sq km) / World Rank:188
- **Location:** Northern and Western Hemispheres, Caribbean Sea, northeast of Trinidad and Tobago
- **Coordinates:** 13°10′N, 59°32W
- **Borders:** None
- **Coastline:** 60 mi (97 km)
- **Territorial Seas:** 12 NM (22 km)
- **Highest Point:** Mt. Hillaby, 1,102 ft (336 m)
- **Lowest Point:** Sea level
- **Longest Distances:** 21 mi (34 km) N-S / 14 mi (23 km) E-W
- **Longest River:** None of significant size
- **Natural Hazards:** Landslides and occasional hurricanes
- **Population:** 275,330 (July 2001 est.) / World Rank: 173
- **Capital City:** Bridgetown, on the southwestern coast
- **Largest City:** Bridgetown, 126,000 (2000 est.)

OVERVIEW

The second-smallest independent country in the Western Hemisphere and the easternmost Caribbean Island, Barbados lies between the Caribbean Sea and the North Atlantic Ocean. It is located east of the Windward Islands and roughly 200 mi (320 km) north-northeast of Trinidad and Tobago. The low-lying island is composed of limestone and coral and almost totally ringed with undersea coral reefs. Barbados is situated on the Caribbean Tectonic Plate. Numerous inland cliffs were created by past seismic activity.

MOUNTAINS AND HILLS

Barbados is mostly flat, but a series of terraces rises from the western coast to a central ridge, culminating in the highest point, Mt. Hillaby (1,102 ft / 336 m), in the north-central part of the island. Hackleton's Cliff at the eastern edge of the island's central plateau rises to 1,000 ft (305 m) above sea level and extends over several miles. South and east of this elevated area is the smaller Christ Church ridge. They are separated by the St. George Valley. At one time this valley was covered by a shallow sea, with each ridge forming a separate island.

INLAND WATERWAYS

Barbados has no rivers and little surface water of any other kind, but a few springs are fed by underground water stored in limestone beds and some water courses are temporarily filled by heavy rains.

THE COAST, ISLANDS, AND THE OCEAN

The western coast of Barbados borders the Caribbean Sea, and its eastern coast borders the North Atlantic Ocean. The port city of Bridgetown is located on Barbados's only natural harbor, Carlisle Bay, at the southwestern end of the island. The coast is ringed by flat land and

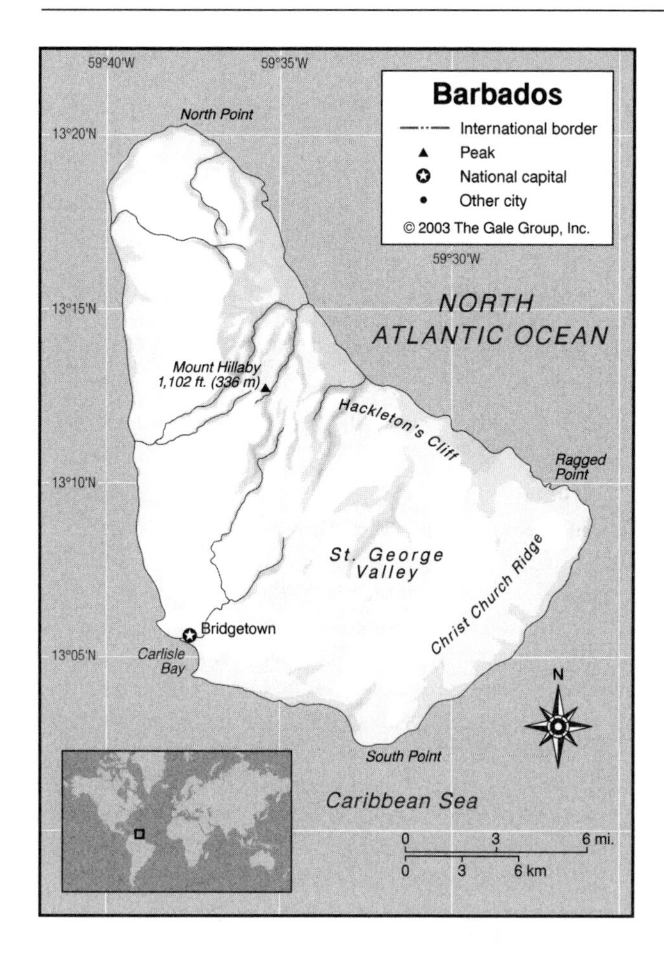

Parishes – Barbados

Name	Area (sq mi)	Area (sq km)
Christ Church	22	57
St. Andrew	14	36
St. George	17	44
St. James	12	31
St. John	13	34
St. Joseph	10	26
St. Lucy	14	36
St. Michael	15	39
St. Peter	13	34
St. Phillip	23	60
St. Thomas	13	34

SOURCE: *Geo-Data: The World Geographical Encyclopedia,* 2nd ed. Detroit: Gale Research, 1989.

wide strips of sandy beach. At Ragged Point at the eastern end of the island, flat rocks form a low, jagged rim to the ocean. The southern and northern ends of the island are known as South Point and North Point, respectively.

CLIMATE AND VEGETATION

Temperature

The northeasterly trade winds that blow across Barbados's eastern coast, which faces the Atlantic Ocean, moderate the island's tropical maritime Caribbean climate. The weather is cool and dry between December and May, and hotter and humid during the rainy season between June and December. Temperatures in the capital city of Bridgetown range from 70–82°F (21–28°C) in February to 73–86°F (23–30°C) in June and September.

Rainfall

Rainfall is heaviest between June and December but falls throughout the year. Average annual precipitation varies from about 40 in (100 cm) in coastal areas to 90 in (230 cm) at higher elevations.

Vegetation

Although the clearing of land for sugarcane plantations has left Barbados without substantial forested areas; palm, mahogany, frangipani, Poinciana, and other tropical trees still grow on the island. There is an abundance of flowering shrubs, including lilies, roses, carnations, and various types of cactus.

HUMAN POPULATION

Barbados is one of the world's most densely populated countries. According to 2000 estimates, roughly half the population of the island is urban, and the great majority of these urban dwellers live in the capital city of Bridgetown.

NATURAL RESOURCES

Barbados's most important natural resource is sugarcane; the sugarcane industry was the island's most important economic sector until the 1960s, when it was surpassed by tourism. Barbados also has modest natural gas and petroleum resources and supports a fishing industry.

FURTHER READINGS

Barbados Daily Nation. http://www.nationnews.com (accessed February 18, 2002).

Beckles, Hilary. *A History of Barbados.* Cambridge: Cambridge University Press, 1990.

Forde, G. Addington, Sean Carrington, Henry Fraser, and John Gilmore. *The A-Z of Barbadian Heritage.* Bridgetown: Heinemann Caribbean, 1990.

Handler, Jerome S. *Plantation Slavery in Barbados: An Archaeological and Historical Investigation.* Cambridge: Harvard University Press, 1978.

Spark, Debra. *The Ghost of Bridgetown.* Saint Paul, Minn.: Graywolf Press, 2001.

Stow, Lee Karen. *Essential Barbados.* Lincolnwood, Ill.: Passport Books, 2001.

Belarus

- **Area:** 80,154 sq mi (207,600 sq km) / World Rank: 85
- **Location:** Northern and Eastern Hemispheres, in Eastern Europe, east of Poland, west of Russia, south of Latvia and Lithuania, north of Ukraine
- **Coordinates:** 53°00′N, 28°00′E
- **Borders:** 1,925 mi (3,098 km) total / Latvia, 88 mi (141 km); Lithuania, 312 mi (502 km); Poland, 376 mi (605 km); Russia, 596 mi (959 km); Ukraine, 554 mi (891 km)
- **Coastline:** Landlocked country, Belarus has no coastline
- **Highest Point:** Dzerzhinskaya Mountain, 1,135 ft (346 m)
- **Lowest Point:** Neman River, 295 ft (90 m)
- **Longest Distances:** 400 mi (640 km) SW-NE / 310 mi (490 km) N-S
- **Longest River:** Dnieper, 1,420 mi (2290 km)
- **Largest Lake:** Lake Naroch, 30 sq mi (80 sq km)
- **Natural Hazards:** None
- **Population:** 10,350,194 (July 2001 est.) / World Rank: 74
- **Capital City:** Minsk, located in the center of the country
- **Largest City:** Minsk, 1.7 million (2002 est.)

OVERVIEW

The Republic of Belarus is a landlocked country in east central Europe, about 161 mi (260 km) southeast of the Baltic Sea coastline. The topography is relatively flat (average elevation 100 ft / 162 m), and Belarus has no natural borders. The country features thousands of lakes, areas of marshland, and forests.

Although its topography is chiefly flat-to-hilly, the country does have five distinct geographic regions. In the north is the Polotsk Lowland, an area of lakes, hills, and forests. The Neman Lowland in the northwest is similar. The lowlands are separated from each other and the rest of the country by the Belorussian Ridge and smaller uplands. Plains and grasslands lie in the east and central part of the country. The south is dominated by the Polesye Marshes, a vast swampy area that extends into Ukraine. The swampy plains of the south, the northern lakes, and the gently sloping ridges were all the work of glaciers.

MOUNTAINS AND HILLS

Although its terrain is generally level, the Belorussian Ridge, a region of highlands, runs across the center of the country from the southwest to the northeast. The highest elevation is Dzerzhinskaya Mountain (Dzyarzhynskaya Hara; 1135 ft / 346 m).

INLAND WATERWAYS

Lakes

Belarus has over 4,000 lakes. Lakes Drisvyaty and Osveyskoye are near the northern border. The largest is Lake Naroch (Narach) covering 50 sq mi (80 sq km) in the northwest.

Rivers

The Dnieper is the longest river in Belarus at 1420 mi (2290 km). It is the third longest river in Europe; only the Volga and Danube are longer. After crossing the Russian border southwest of the Belorussian Ridge, the river bends south and flows across most of eastern Belarus, passing through the city of Mahilyow before entering Kiev. Main tributaries are the Berezina in the central region and the Pripyat in the south.

The Berezina River in east central Belarus originates in the marshes near the town of Barysaw. It flows southeast for 365 mi (587 km) then joins the Dnieper River, which continues to the Black Sea. The Berezina is important for transporting timber.

The Pripyat and its tributaries flow eastward across the southern part of Belarus before curving southwest and meeting the Dnieper just inside of Ukraine. They are surrounded by the Polesye (or Pripyat) Marshes. The Bug River, which flows north along part of the border with Poland, is connected to the Pripyat-Dnieper system by the Dnieper-Bug Canal.

The major rivers in the north of the country are the Western Dvina and the Neman. The Western Dvina enters the country from Russia and flows across the northern tip of Belarus into Latvia. The Neman has its source in the center of the country and flows west before turning north and entering Lithuania. Canals link both these rivers with the Dnieper helping make it one of the main waterways linking the Black and the Baltic seas.

Wetlands

About 25 percent of Belarus is covered in peat bogs and marshes. The Polesye Marshes are a poorly drained lowland around the Pripyat River, featuring some low hills, that dominate the southern part of Belarus and northern Ukraine. Roughly 300 mi (485 km) across from east to west and 140 mi (225 km) from north to south, they are the largest wetland in Europe. Forests cover about a third of the marshes. The marsh soils are predominantly sandy, and about 70 percent of the soil in Belarus is acidic with fairly large amounts of iron oxides, a type of soil called podzolic. The Polesye Marsh area was once covered by a glacial lake.

THE COAST, ISLANDS, AND THE OCEAN

Belarus is landlocked and has no coast.

CLIMATE AND VEGETATION

Temperature

The Belarusian climate is considered transitional between continental and maritime. Cool temperatures and high humidity predominate, with helpful influence from the nearby Baltic Sea. In January the average temperature is 23°F (-5°C); in July, 67°F (19°C). Sometimes in the north, frosts of below -40°F (-40°C) have been recorded. Summer lasts up to 150 days, while winter ranges between 105 and 145 days.

Rainfall

Precipitation ranges between 22.5 and 26.5 in (570 and 610 mm) in an average year; the central region generally receives the highest amount. It is said, with some truth, that in Belarus it either rains or snows every two days.

Population Centers – Belarus

(1999 POPULATION ESTIMATES)

Name	Population	Name	Population
Minsk (Meinsk)	1,725,100	Baranovichi (Baranovichy)	173,800
Hoyel' (Gomel or Homiel)	503,700	Borisov (Barysau)	153,500
Mahilyow (Mogilev or Mahilou)	371,300	Pinsk	133,500
Vitsyebsk (Vitebsk or Viciebsk)	358,700	Orsha (Vorsha)	124,300
Grodno (Horadnia)	308,900	Mozyr (Mazyr)	110,000
Brest (Bierascie)	300,400	Soligorsk	101,700
Bobruysk	228,000	Lida	99,600

SOURCE: Ministry of Statistics and Analysis, Belarus

Grasslands

Outside of the Belorussian Ridge highlands most of the country is flat and well-watered, and although substantial portions are forested or marshland, Belarus still has vast areas of grassland. Roughly a third of the country is suitable for farming.

Forests

Forests and woodlands cover 34 percent of Belarus's land. The forests are scattered and variable in size. In the north, pine is the principal tree, but spruce, oak, birch, alder, and ash trees also are common. A significant part of the Polesye Marshes is wooded. In the southwest, the Belovezhskaya Pushcha Reserve is part of the oldest existing European forest, safe haven of the nearly extinct European bison, or wisent, as well as other birds and animals that have become extinct elsewhere. The reservation extends into Poland, and both countries administer it. Belarus's forests shelter more than 70 mammal and 280 bird species.

HUMAN POPULATION

Some 10,350,194 people (130 persons per sq mi / 50 persons per square kilometer) live in Belarus (July 2001 estimate). Of that number, 68 percent live in urban areas. Roughly four-fifths of the population is ethnically Belarusian, with Russians making up the largest minority. Belarus was a part of the Soviet Union for many years, and Russia before that, and much of the population still uses that language.

NATURAL RESOURCES

Belarus's major natural resources are its forests and peat deposits. The country also has small quantities of oil and natural gas. Manufacturing and commerce are the most important parts of the economy.

FURTHER READINGS

Glover, Jeffrey. "Outlook for Belarus." *Review and Outlook for the Former Soviet Union.* Washington: PlanEcon, August 1995, pp. 89-104.

"In the Slav Shadowlands," *Economist*, 335, No. 7915, May 20, 1995, pp. 47-49.

World Bank. *Belarus: Energy Sector Review.* Washington: April 21, 1995.

Zaprudnik, Jan. *Belarus: At a Crossroads in History.* Boulder, Colorado: Westview Press, 1993.

Belgium

- **Area:** 11,780 sq mi (30,510 sq km) / World Rank: 139
- **Location:** Northern and Eastern Hemispheres; Western Europe bordering the North Sea between France and the Netherlands
- **Coordinates:** 50°50′N, 4°00′E
- **Borders:** 859 mi (1,385 km) total boundary length; France, 385 mi (620 km), Germany, 104 mi (167 km), Luxembourg, 92 mi (148 km), Netherlands, 280 mi (450 km)
- **Coastline:** 41 mi (66 km)
- **Territorial Seas:** 12 NM (22 km)
- **Highest Point:** Mt. Botrange, 2,277 ft (694 m)
- **Lowest Point:** Sea level
- **Longest Distances:** 174 mi (280 km) SE-NW/ 137 mi (222 km) NE-SW
- **Longest River:** Meuse, 580 mi (933 km)
- **Natural Hazards:** Flooding threatens reclaimed coastal areas that are protected by dikes.
- **Population:** 10,258,762 (July 2001 est.) / World Rank: 77
- **Capital City:** Brussels, located in the center of the country
- **Largest City:** Brussels, 1,122,000 (2000 est.)

GEO-FACT

Man has heavily influenced the geography of Belgium. Almost one-fifth of present-day Belgium was reclaimed from the North Sea between the 8th and the 13th centuries. The reclamation added a coastal strip 30 mi (48 km) wide to the country. During the 1930s the 80 mi (130 km) Albert Canal was constructed to connect the Meuse River in eastern Belgium with the Scheldt River. The canal links the eastern industrial city of Liège with Antwerp, where there is access to the North Sea and international shipping.

OVERVIEW

A small but densely populated country, Belgium is located in a part of northwestern Europe once called the Low Countries and today known as the Benelux region (primarily due to Belgium's economic partnership with its neighbors Luxembourg and the Netherlands). Situated at the southern tip of the North Sea, it is bordered by the Netherlands to the northeast, Germany to the east, Luxembourg to the southeast, and France to the west and southwest.

With its central location and with few natural frontiers, Belgium has been called the crossroads of Europe. For much of its history it was a battleground for the major European powers of France, Britain, and Germany. Its capital, Brussels, is the seat of both the North Atlantic Treaty Organization (NATO) and the European Union (EU), and is within 621 mi (1,000 km) of most other West European capitals. Belgium's central location has also made it an important Western European financial and commercial center.

Belgium lies on the Eurasian Tectonic Plate. It can be divided into three major geographic regions: coastal plains to the northwest, a low central plateau region, and the Ardennes highlands to the southeast. An extensive system of dikes has allowed to Belgians to reclaim substantial amounts of land from the sea. Rivers flow throughout the country, and there are many canals in the plains and central plateau.

MOUNTAINS AND HILLS

Mountains

The heavily forested Ardennes highlands extend south of the Meuse River Valley, continuing into France. They range in elevation from 1,300 ft (400 m) in the broken hill country of the Condroz Plateau to between 1,900 and 2,300 ft (580 to 700 m) in the Hautes Fagnes near the German border. This last region includes Belgium's highest peak, Mt. Botrange (Signal de Botrange), 2,277 ft (694 m) above sea level. The hilly southernmost area of the Ardennes, called the Belgian Lorraine, is sometimes viewed as a separate region.

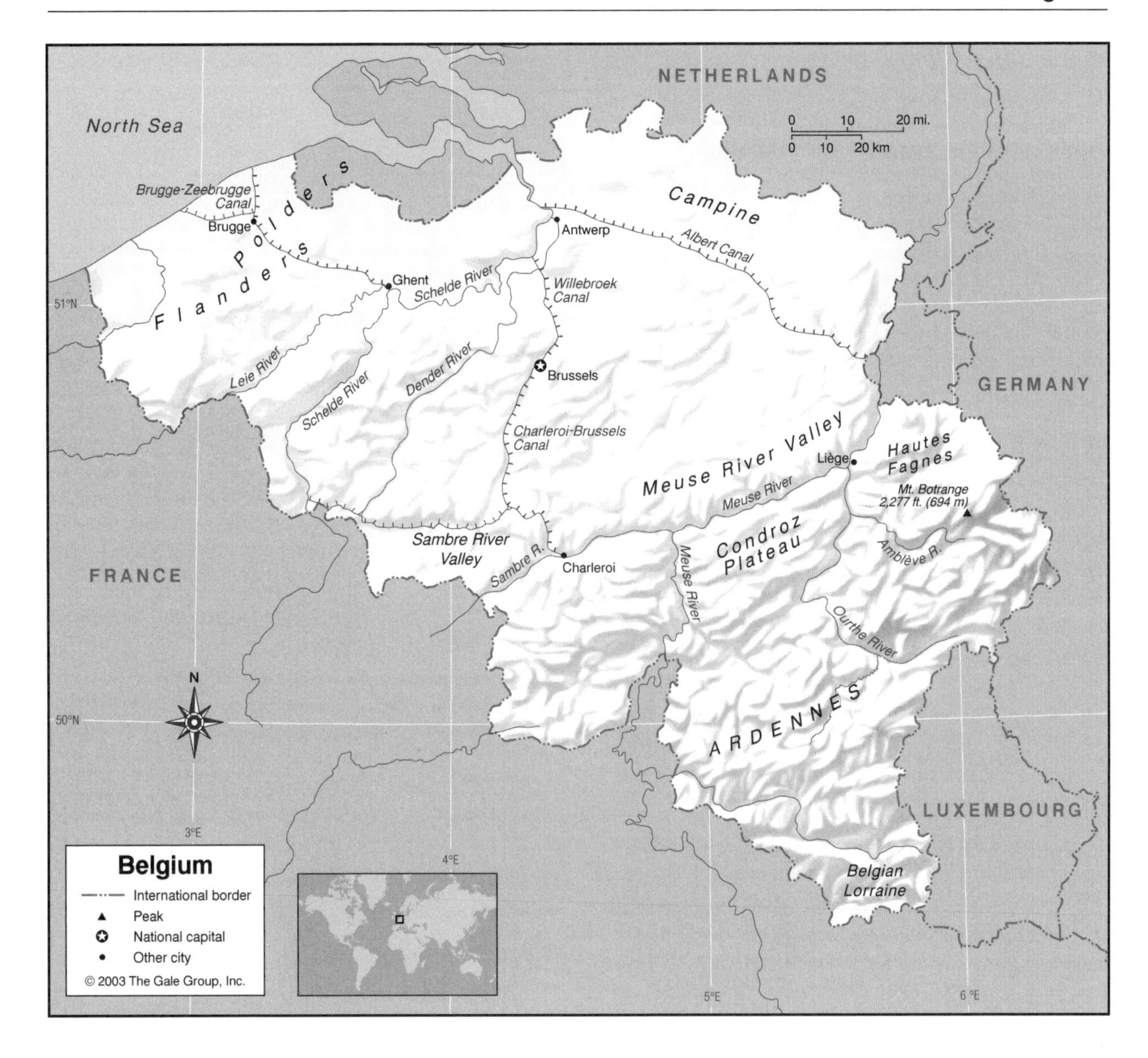

Plateaus

Located between the northern lowlands and the Ardennes highlands to the south, Belgium's central plateau region extends across the middle of the country, from the Borinage area in the west to the Brabant region near the southeastern Dutch border. Its elevations range from 65 ft (20 m) near the coastal lowlands to 650 ft (200 m) near the southern highlands. The capital city of Brussels is located in this central region. Tributaries of the Schelde (Escaut) River flow in the plateau region, and its southern boundary is formed by the Meuse and Sambre river valleys.

INLAND WATERWAYS

Lakes

Belgium has relatively few natural lakes, and none of any great size. The largest complex of lakes is located in the Ardennes Region.

Rivers

Belgium has two major rivers, the Schelde and the Meuse, both of which have their source in France, flow east across Belgium to the Netherlands, and then drain into the North Sea. The valleys of the central plateau region are irrigated by many canals and by the tributaries of the Schelde River. Some of the largest of these rivers are the Leie and Dender. In the south, the Sambre, Semois, Ourthe, and Amblève flow into the Meuse.

There is also an extensive network of canals running throughout the coastal plains and central plateau region, connecting Belgium's major cities and rivers to the sea. Chief among them are the Brugge-Zeebrugge, Charleroi-Brussels, Willebroek, and Albert Canals.

Wetlands

Maritime Flanders includes polders (reclaimed land) that were formerly marshland. The salt marshes of the region were transformed into rich farmland behind a barrier of dikes.

THE COAST, ISLANDS, AND THE OCEAN

Belgium is situated at the southern tip of the North Sea. The coastline is nearly straight, and the beach is white. It is practically free of pebbles and stabilized by fences called groins, which reach from the higher beach into the water. Behind the beach lie dunes and behind them the polders (wetlands reclaimed for agricultural use in the Middle Ages).

CLIMATE AND VEGETATION

Temperature

Belgium has a temperate maritime climate with moderate temperatures in both summer and winter, with wet westerly and southwesterly winds. Except for the mountainous Ardennes region, its climate resembles those of northern France, the Netherlands, northwestern Germany, and the British Isles. The mean temperature in Brussels ranges from 37°F (3°C) in January to 64°F (18°C) in July. The higher Ardennes and Campine regions have more of a continental climate, with winter temperatures below 32°F (0°C).

Rainfall

Rainfall averages between 28 and 40 in (70 and 100 cm) per year and is evenly spread out over the year. The elevated Ardennes region can receive as much as 55 in (140 cm) of rain annually.

Grasslands

Outside of the Ardennes region, most of Belgium is flat, well watered, and agriculturally suitable land, although much of it is now urbanized. These lowlands are part of the Great European Plain stretching across much of the continent. The western part of the Belgian lowlands is occupied by Flanders, which can be divided into two regions. Maritime Flanders, reaching 5 to 10 mi (8 to 16 km) inland from the sea, is a totally flat fringe of land protected from floods and tides by sand dunes and dikes. Interior Flanders is composed of terrain that gently rises to about 150 ft (46 m) above sea level before giving way to the central plateau near Brussels.

The eastern region of Belgium's lowlands, northeast of Antwerp, is known as the Kempenland (or Campine). Belonging to the delta of the Meuse and Rhine rivers, it consists of sparsely populated and barren heathlands.

Forests and Jungles

About half the Ardennes region is forested, with spruce and pine predominating. Other trees common in Belgium include beech and oak.

Population Centers – Belgium

(1990 CENSUS OF POPULATION)

Name	Population	Name	Population
Brussels (metropolitan area, capital)	960,324	Liège	484,518
Antwerp (metropolitan area)	668,125	Charleroi	294,962
		Ghent	250,666
		Mons	175,290
		Brugge (Bruges)	117,100

SOURCE: "Population of Capital Cities and Cities of 100,000 and More Inhabitants." United Nations Statistics Division.

Provinces – Belgium

2000 POPULATION ESTIMATES

Name	Population	Area (sq mi)	Area (sq km)	Capital
Antwerp	1,643,972	1,107	2,867	Antwerp (Antwerpen)
Brussels Region	959,318	63	162	Brussels (Bruxelles)
East Flanders	1,361,623	1,151	2,982	Ghent
Flemish Brabant	1,014,704	813	2,106	Leuven
Hainaut	1,279,467	1,462	3,787	Mons
Liège	1,019,442	1,491	3,862	Liège
Limburg	791,178	935	2,422	Hasselt
Luxembourg	246,820	1,715	4,441	Arlon
Namur	443,903	1,415	3,665	Namur
Walloon Brabant	349,884	421	1,091	Wavre
West Flanders	1,128,774	1,210	3,134	Brugge

SOURCE: Statistics Belgium, Ministry of Economic Affairs.

HUMAN POPULATION

Belgium is one of the most densely populated countries in Europe, and has distinctive ethnic and linguistic divisions. The Flemings, who speak Flemish (a form of Dutch) and live in the northern part of the country, account for over half the population. The French-speaking Walloons, who live in the south, account for only about one-third. German speakers in the east of the country and other minorities make up the remainder of the population. The Ardennes region in the south is the most sparsely populated part of the country. With birth and death rates that are roughly equal, Belgium has a low population growth rate, and this rate is especially low for the Walloons. Since 1945, Belgium's foreign-born population has had a higher growth rate than its native population.

NATURAL RESOURCES

Although once a significant coal producer, Belgium now has to import most of its basic raw materials and is heavily dependent on crude-oil imports. There are still some coal deposits in the Sambre-Meuse Valley and in the Campine region. Belgium has a thriving diamond cutting industry and is known as the diamond capital of the world. There are small deposits of lead, copper, zinc, and iron ore in the Ardennes, but Belgium has no metal ore deposits that are commercially exploitable. Chalk, limestone, and stones used in construction, such as granite, are plentiful.

FURTHER READINGS

Belgian Federal Govt Online. http://belgium.fgov.be/en_index.htm (accessed Feb. 8, 2002).

Blom, J.C.H., and E. Lamberts, eds. *History of the Low Countries.* Translated by James C. Kennedy. Providence, RI: Berghahn Books, 1999.

Blyth, Derek. *Belgium.* 9th ed. New York: W.W. Norton, 2000.

Fielding's Belgium: The Most In-Depth and Entertaining Guide to the Charms and Pleasures of Belgium. Redondo Beach, CA: Fielding Worldwide, 1994.

Fox, Renie C. *In the Belgian Château: The Spirit and Culture of a European Society in an Age of Change.* Chicago: I.R. Dee, 1994.

Frommer's Belgium, Holland & Luxembourg. New York: Macmillan USA, 1997.

Belize

- **Area:** 8,867 sq mi (22,966 sq km) / World Rank: 149
- **Location:** Located in the Northern and Western Hemispheres, Belize is bounded on the north by Mexico, on the east by the Caribbean Sea, and on the south and west by Guatemala.
- **Coordinates:** 17° 15′N and 88° 45′W.
- **Borders:** 320 mi (516 km) / Guatemala, 165 mi (266 km); Mexico, 155 mi (250 km)
- **Coastline:** 239 mi (386 km)
- **Territorial Seas:**. Territorial seas vary from 12 NM in the north to the south 3 NM.
- **Highest Point:** Victoria Peak, 3,680 ft (1,122 m)
- **Lowest Point:** Sea level
- **Longest Distances:** 174 mi (280 km) N-S; 68 mi (109 km) E-W
- **Longest River:** Belize River, 180 mi (288 km)
- **Natural Hazards:** Hurricanes; coastal flooding
- **Population:** 256,062 (2001 estimate) / World Rank: 174
- **Capital City:** Belmopan, located in the center of the country

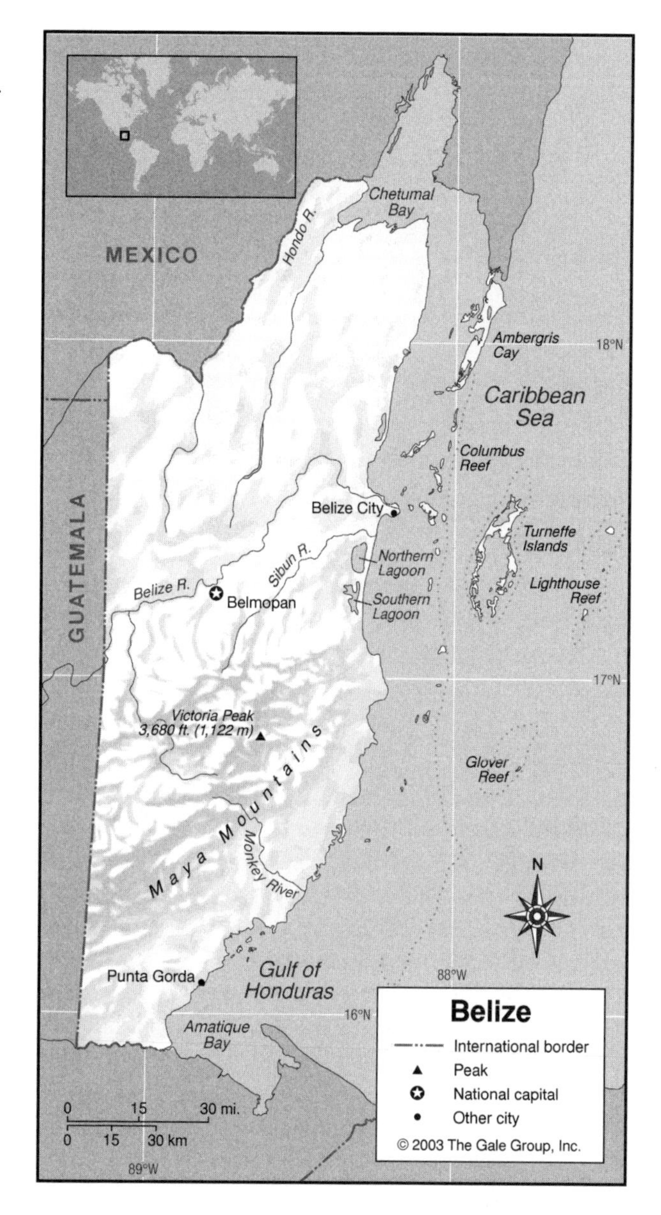

- **Largest City:** Belize City, on the Caribbean coast, population 46,342 (2000 est.)

OVERVIEW

The Belize River effectively cuts the country into northern and southern halves. The north features predominantly level landscape, interrupted only by the Manatee Hills, while the south, containing the Maya and Cockscomb mountain ranges, is generally elevated and contains plateaus and basins. The coastlines are flat and swampy and are marked by numerous lagoons. Just beyond the shoreline is the Belize Barrier Reef, the second largest in the world, in which numerous islands known as cays are located.

Belizean geology consists largely of varieties of limestone, with the notable exception of the Maya Mountains,

GEO-FACT

As British Honduras, Belize was the last British colony on the American mainland. Belize became independent in 1981.

a large intrusive block of granite and other Paleozoic sediments. Several major faults rive the Mayan highlands, but much of Belize lies outside the tectonically active zone that underlies most of Central America.

MOUNTAINS AND HILLS

Mountains

The Maya and Cockscomb mountain ranges form the backbone of the country and dominate the southern landscape. Mountains in the Maya range rise to a height of 3,400 ft (1,100 m), and run northeast to southwest across the central and southern parts of the country. The highest elevation, Victoria Peak, is located in the Cockscomb range, located just below the Maya range.

Plateaus

The area north of Belize City is mostly flat, broken occasionally by the Manatee Hills. During the Cretaceous period, what is now the western part of the Maya Mountains stood above sea level, creating the oldest land surface in Central America, the Mountain Pine Ridge plateau. This plateau is covered in pine trees and houses interesting bird life, including the stigeon owl, pine siskin, eastern bluebird, and orange breasted falcon. Some areas, such as Badly Beacon, are infertile due to excessive erosion.

INLAND WATERWAYS

Lakes

Located on the central coast of Belize, the Northern and Southern lagoons are rich in marine wildlife. The population of manatees, a large marine mammal, is particularly large here, and the lagoons are also known as the Manatee Lagoons. Limestone hills, marshes, and mangroves surround the area.

Rivers

Seventeen rivers, among them the Belize, drain the countryside. The Belize runs across the center of the country, draining into the Caribbean Sea near Belize City. The city itself is bisected by Haulover Creek, one of the river's tributaries. Monkey River is located in the south of the country, emptying into the Caribbean near the Gulf of Honduras. The Sibun River, which drains the Maya Mountain range, carries large amounts of forest debris as it flows down to the ocean. In the north, the Hondo River marks the border with Mexico.

Hidden Valley Falls, aptly known as the Thousand Foot Falls for their 1,000-ft (323-m) drop, are located near the Mountain Pine Ridge. These scenic falls are the highest in Central America.

Wetlands

The coastal regions are particularly swampy due to their proximity to the Caribbean Sea and their susceptibility to hurricanes and flooding. Crocodiles are common in the heavily vegetated swamplands.

THE COAST, ISLANDS, AND THE OCEAN

Oceans and Seas

Belize's eastern border lies on the Caribbean Sea. The central coast is on the open sea, but in the north it is one side of Chetumal Bay, and in the south it fronts on the Gulf of Honduras and Amatique Bay.

The Coast and Beaches

The coastline of Belize is full of indented areas providing for a dynamic coastline with many beaches, as well as swamplands and lagoons. Belize's barrier reef, the second largest in the world, stretches over the entire coastline. Within this expansive reef, smaller coral reefs and cays can be found.

Major Islands

More than 1,000 small islands dot the coastline of Belize. Laughing Bird Cay, 11 miles (17.6 km) off the coast of Placentia Village, is found within a faro, a steep sided coral island on the continental shelf. Laughing Bird Cay contains several smaller reefs and a lagoon, which house an abundant variety of coral, the main attraction of the cay. Beyond the barrier reef, numerous cays—Ambergris Cay, Columbus Reef, and Glover's Reef—line the coast of Belize. The Turneffe Islands, just east of Belize City, comprise about 200 small islands covered in mangroves. These trees nourish the surrounding waters, providing for unique marine life.

CLIMATE AND VEGETATION

Temperature

Belize has a humid tropical climate that is tempered by its northeast trade winds. Temperatures generally remain between 16° and 32° C (61° and 90° F) along the coastal regions but are higher inland. The southern mountain regions have an average annual temperature of 72°F (22°C), with temperatures generally cooler between November and January. It is the level of humidity, rather than actual changes in temperature, which differentiates the seasons. The average humidity level is 83 percent.

Population Centers – Belize

(POPULATION AT 2000 CENSUS)

Name	Population	Name	Population
Belize City	49,050	Dangriga	8,814
Orange Walk	13,483	Belmopan (capital)	8,130
San Ignacio	13,260	Corozal Town	7,888

SOURCE: Government of Belize

Districts – Belize

POPULATIONS FROM 2000 CENSUS

Name	Population	Area (sq mi)	Area (sq km)	Capital
Belize	68,197	1,624	4,206	Belize City
Cayo	52,564	2,061	5,338	San Ignacio
Corozal	32,708	718	1,860	Corozal
Orange Walk	38,890	1,829	4,737	Orange Walk
Stann Creek	24,548	840	2,176	Dangriga
Toledo	23,297	1,795	4,649	Punta Gorda

SOURCE: Government of Belize.

Rainfall

Rainfall increases dramatically from north to south, ranging from 50 in (127cm) in the northern regions to more than 150 in (380 cm) in the south. Hurricanes are prevalent between July and October. There is a generally dry season lasting from February to May, when rainfall is predictable throughout the country.

Forests and Jungles

About half of the country is covered by forest areas, which are quickly disappearing. The tropical forests harbor a wide variety of wildlife, including jaguars, pumas, macaws, crocodiles, and the endangered black howler monkey that is found only in Belize. Many other endangered species, such as the peregrine falcon, the iguana, and several types of hawks and parrots, are found in the lush forest areas.

HUMAN POPULATION

In July 2001 the population was estimated at 256,062, with approximately half the population living in urban areas. Overall population density is 28.9 people per sq mi (11.6 people per sq km), with the density highest on agricultural land. Ethnic groups are the Mestizo (43.7 percent), Creole (29.8 percent), Mayan (10 percent), Garifuna (6.2 percent), and others (10.3 percent).

NATURAL RESOURCES

The valleys of the Belize, Sibun, and Monkey Rivers are mining sites for clay, gravel, and limestone. Although only 10 percent of total land area is currently arable, there is great potential for development. Additional natural resources are timber, fish, and hydropower.

FURTHER READINGS

Bradbury, Alex; updated by Peter Hutchinson. *The Bradt Travel Guide. Belize.* Guilford, Conn.: Bradt Publications, 2000.

Fisk, Erma J. *Parrot's Wood.* New York: Norton, 1985.

Hoffman, Eric. *Adventuring in Belize: the Sierra Club Travel Guide to the Islands, Waters, and Inland Parks of Central America's Tropical Paradise.* San Francisco: Sierra Club Books, 1994.

Norton, Natasha. *Belize.* Old Saybrook, Conn.: Globe Pequot Press, 1997.

Virtual Belize Tour. http://www.belizeexplorer.com/ (Accessed June 27, 2002)

Wright, Peggy, and Brian E. Coutts, comp. *Belize.* Oxford: Clio Press, 1993.

Benin

- **Area:** 43,483 sq mi (112,620 sq km) / World Rank: 101
- **Location:** Northern and Eastern Hemispheres, in West Africa, bordering Nigeria to the east, the South Atlantic Ocean to the east, Togo to the west, Burkina Faso to the northwest, Niger to the north
- **Coordinates:** 9°30′N, 2°15′E
- **Borders:** 1,233 mi (1,989 km) total / Burkina Faso, 190 mi (306 km); Niger 165 mi (266 km); Nigeria, 479 mi (773 km); Togo, 399 mi (644 km)
- **Coastline:** 75 mi (121 km)
- **Territorial Seas:** 12 NM (22 km)
- **Highest Point:** Mt. Sokbaro, 2,159 ft (658 m)
- **Lowest Point:** Sea level
- **Longest Distances:** 413 mi (665 km) N-S / 207 mi (333 km) E-W
- **Longest River:** Niger River, 2,600 mi (4,184 km)
- **Largest Lake:** Lake Ahémé, 39 sq mi (100 sq km)
- **Natural Hazards:** Harmattan winds in the north
- **Population:** 6,590,782 (July 2001 est.) / World Rank: 94
- **Capital City:** Porto-Novo, southeastern Benin
- **Largest City:** Cotonou, 400,000 (2000 est.) on the southern coast

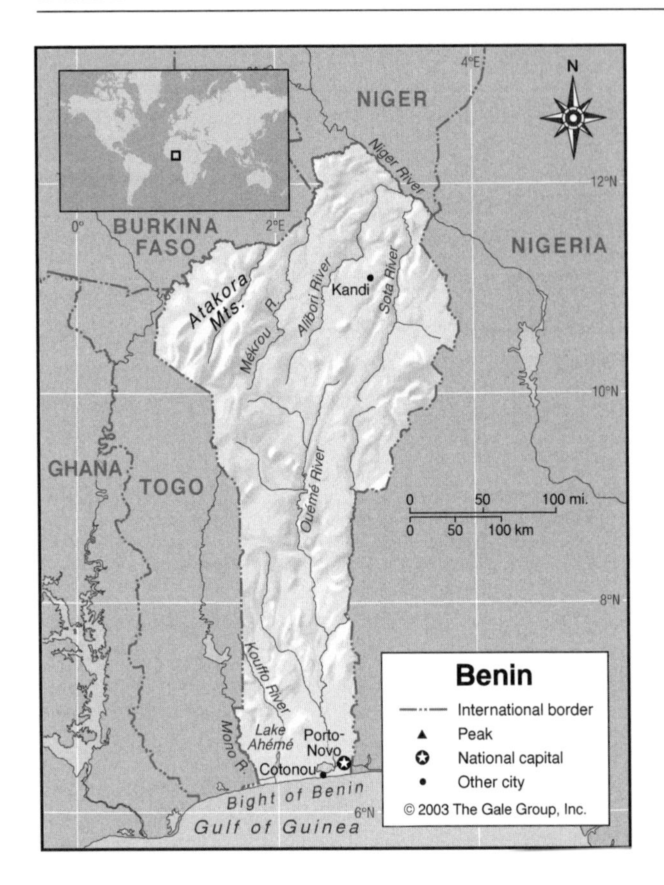

OVERVIEW

Formerly a French colony known as Dahomey, Benin is a small country on the coast of West Africa, between Togo and Nigeria. It is bounded on the north by the Niger River, and on the south by the Bight of Benin, which forms part of the Gulf of Guinea. From south to north, Benin's major geographical divisions consist of a coastal belt that includes sandbanks and lagoons; a savanna-covered clay plateau; and, in the northern two-thirds of the country, a higher plateau region that includes the Atakora Mountains and the Niger Plains. Benin is situated on the African Tectonic Plate.

MOUNTAINS AND HILLS

Mountains

The Atakora Mountains extend northeast to southwest across the plateau of Upper Benin, in the northwestern part of the country, at elevations of 1,000 to around 2,000 ft (300 to 600 m). Heavily forested, they belong to the same system as the Togo Mountains to the south.

Plateaus

North of the coastal region, 300 to 750 ft (90 to 230 m) above sea level, lies a fertile, savanna-covered clay plateau called the *terre de barre*, composed of lateritic clay and bisected by the swampy Lama depression.

INLAND WATERWAYS

Lakes

Benin's principal lake is Lake Ahémé, in the southern part of the country.

Rivers

Benin's longest river is the Niger River, which forms part of its border with Niger in the northeast and is navigable for 55 mi (89 km) in Benin. The longest river located entirely within Benin's borders is the Ouémé, which is 285 mi (459 km) long.

Most of Benin's rivers flow in a generally north-south direction. Those in the north, including the Alibori, the Mékrou, and the Sota, drain into the Niger and are prone to flooding. The Ouémé flows southward through about two-thirds of Benin, starting at about the center and winding its way southeast to the Porto-Novo lagoon. The river is navigable for about 125 mi (200 km) of its length. To the southwest, the Mono River forms part of the border with Togo and is navigable for 62 mi (100 km). Other than the Ouémé, the major river in the southern part of the country is the Kouffo.

Wetlands

A large, swampy, depression called the Lama Marsh extends across the terre de barre plateau in south-central Benin.

THE COAST, ISLANDS, AND THE OCEAN

Benin is bounded on the south by a wide, natural indentation on the Gulf of Guinea called the Bight of Benin. Benin has no natural harbors, and access to its coast is further impeded by the sandbanks that form part of its coastal belt. Benin's coastal belt includes four lagoons: Grand Popo, Ouidah, Cotonou, and Porto Novo.

CLIMATE AND VEGETATION

Temperature

Southern Benin, which lies near the equator, has a hot, humid tropical climate, with average temperatures around 80°F (27°C). The north has a semiarid climate with greater variability, ranging from 56°F (13°C) in June to 104°F (40°C) in January.

Rainfall

Southern Benin's primary rainy season occurs from March to July, with a secondary rainy period between September and November. The hot, dry harmattan wind blows during the dry season between December and March.

Average annual rainfall is highest (53 in / 135cm) in the central part of the country and lower in the north

GEO-FACT

The area of low precipitation in southwest Benin—a dramatic exception to the high rainfall elsewhere in this tropical region—is called the "Benin window." It is thought to have resulted from the destruction of the native rainforest, which decreased the evaporation of moisture into the air that results in local "convection rains."

(38 in / 97 cm). The driest part of Benin is the southwest, which averages 32 in (82 cm) of rain per year.

Grasslands

Most of Upper Benin (above about 9°N) has a sparse covering of savanna and is mainly infertile, except for the northeastern plains around Kandi that descend to the Niger River valley.

Forests and Jungles

Mahogany, ebony, and various species of palm have been cultivated in the southernmost part of Benin, but tracts of original rainforest are still found north of Abomey, where they alternate with savanna.

HUMAN POPULATION

Close to half of Benin's inhabitants are urban dwellers, and about 75 percent of the population lives in the southern half of the country.

NATURAL RESOURCES

Mineral resources include marble, limestone, gold, and modest offshore oil deposits. Other natural resources are timber and hydropower from the Nangbeto Dam on the Mono River, which has supplied most of Benin's electricity since 1988.

Population Centers – Benin

(2002 POPULATION ESTIMATES)

Name	Population
Cotonou	536,827
Porto-Novo (capital)	179,138
Djougou	134,099
Parakou	103,577

SOURCE: Bureau Central du Recensement, Institut National de la Statistique et de l'Analyse Économique, Ministère du Plan et de la Restructuration Économique, Benin.

Provinces – Benin

POPULATIONS FROM 2002 CENSUS

Name	Population	Area (sq mi)	Area (sq km)	Capital
Atacora	649,308	12,050	31,200	Natitingou
Atlantique	1,066,373	1,250	3,200	Cotonou
Borgou	827,925	19,700	51,000	Parakou
Mono	676,377	1,450	3,800	Lokossa
Quémé	876,574	1,800	4,700	Porto-Novo
Zou	818,998	7,200	18,700	Abomey

SOURCE: Bureau Central du Recensement, Institut National de la Statistique et de l'Analyse Économique, Ministère du Plan et de la Restructuration Économique, Benin.

FURTHER READINGS

Ben-Amos, Paula Girshick. *Art, Innovation, and Politics in Eighteenth-Century Benin.* Bloomington: Indiana University Press, 1999.

Eades, J. S., and Chris Allen. *Benin.* Santa Barbara, Calif.: CLIO Press, 1996.

Edgerton, Robert B. *Women Warriors: The Amazons of Dahomey and the Nature of War.* Boulder, Colo.: Westview Press, 2000.

Manning, Patrick. *Slavery, Colonialism, and Economic Growth in Dahomey, 1640–1960.* Cambridge. Eng.: Cambridge University Press, 1982.

Scholefield, Alan. *The Dark Kingdoms: The Impact of White Civilization on Three Great African Monarchies.* New York: Morrow, 1975.

World Desk Reference. *Benin.* http://www.travel.dk.com/wdr/BJ/mBJ_Intr.htm# (Accessed February 21, 2002)

Bermuda

Overseas Territory of the United Kingdom

- **Area:** 22.7 sq mi (58.8 sq km) / World Rank: 203
- **Location:** Located in the North Atlantic Ocean, in the Northern and Western Hemispheres, 580 mi (933 km) east of North Carolina, U.S.A.
- **Coordinates:** 32°20′N, 64°45′W
- **Borders:** No international boundaries
- **Coastline:** 64 mi (103 km)
- **Territorial Seas:** 12 NM, Bermuda exercises exclusive fishing rights extending 200 NM (370 km)

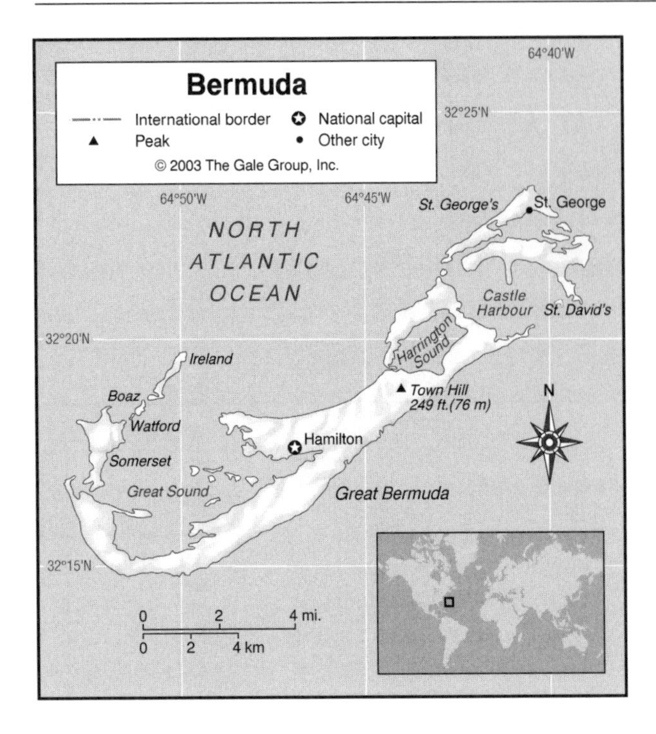

■ **Highest Point:** Town Hill, 249 ft (76 m)
■ **Lowest Point:** Sea level
■ **Longest Distances:** Main island: 24 mi (39 km) long / 1 mi (1.6 km) average width
■ **Longest River:** None
■ **Largest Lake:** None
■ **Natural Hazards:** Subject to hurricanes and severe tropical storms June through November
■ **Population:** 63,503 (July 2001 est.) / World Rank: 193
■ **Capital City:** Hamilton, located at the midpoint of the northwest shore on the main island
■ **Largest City:** Hamilton (3,800, 2000 est.)

OVERVIEW

Bermuda, Britain's oldest colony, is an archipelago consisting of roughly 130 to 150 small coral islands in the western North Atlantic Ocean east of Cape Hatteras, off the coast of North Carolina. Central to the archipelago, and by far the largest island in the chain, is the island of Bermuda itself, also called Great Bermuda or Main Island. It is about 14 mi (23 km) long and has an average width of 1 mi (1.6 km). Together with the six next largest islands—Ireland, Watford, Boaz, Somerset, St. David's, and St. George's—it forms a fishhook-shaped curve about 22 mi (35 km) long and less than 1 mi (1.6 km) wide. These seven islands are connected by a network of bridges and causeways.

Bermuda's islands constitute the exposed portion of a submerged, extinct volcanic mountain with a limestone cap 200 ft (60 m) thick, fringed with coral reefs. These are the northernmost coral reefs in the world. Bermuda is located on the North American Tectonic Plate.

GEO-FACT

The Bermuda Triangle, categorized by the U.S. Board of Geographic Names as an imaginary area, is located in the western Atlantic Ocean that lies between Bermuda and Florida. Since the 1800s it has been notorious as the site of many unexplained disappearances of ships and aircraft. The points or apexes of the triangle are considered to be Bermuda, the southern tip of Florida, and San Juan, Puerto Rico. The Bermuda Triangle is also known as the "Devil's Triangle." There is some reason to believe that the area is unusually dangerous. Special conditions in the region cause compasses to work somewhat differently there than normal. The Gulf Stream current is strong and turbulent, and there are many reefs and shoals as well as many deep trenches within the Triangle. All of this means that not only is it easy to get lost or wrecked within the Triangle, but that evidence of a wreck can easily sink into deep water or be borne away by the currents, leaving no traces.

MOUNTAINS AND HILLS

The Bermuda islands are rocky but mostly flat. There are a few hills, the tallest of which is Town Hill (249 ft / 76 m).

INLAND WATERWAYS

Bermuda has no rivers or freshwater lakes.

THE COAST, ISLANDS, AND THE OCEAN

Bermuda's northwest coast is heavily indented. The capital city of Hamilton has a deepwater harbor; other major harbors are St. George's and Castle Harbors. Most of Bermuda's smaller islands are in these harbors, the

Great Sound, Harrington Sound, or off the North and South Shores.

The sand on Bermuda's beaches is not volcanic but formed from pulverized shell remains and the skeletons of invertebrates including clams, corals, and forams. Shells of the pink foram make Bermuda one of the only places on earth with coral reefs that have pink sand.

Population Centers – Bermuda

(2002 POPULATION ESTIMATES)

Name	Population
St. George	2,200
Hamilton (capital)	1,400

SOURCE: Geo-Data 1989 ed., and Bermuda Online, http://www.bermuda-online.com (accessed June 2002).

CLIMATE AND VEGETATION

Bermuda has abundant semitropical vegetation. Its limited arable land produces flowers, fruits, herbs, and vegetables year round. There are no forests, and no tall trees like oaks, maples, and sycamore. Some of the more common trees include the Bermuda cedar, Bermuda olivewood, and Bermuda palmetto. Widely seen types of ground cover include ajuga, lily turf, tea plant, and wandering jew.

Temperature

Bermuda has a subtropical climate. Its location in the Gulf Stream makes the weather mostly mild and humid; however brisk winds are common in the winter, from December to April. The mean annual temperature is about 70°F (21°C), ranging from below 60°F (16°C) in winter to above 80°F (27°C) in summer.

Rainfall

Rainfall averages 48 in (147 cm) and is evenly distributed throughout the year. The hurricane season runs from June to November.

Parishes – Bermuda

2002 POPULATION ESTIMATES

Name	Population	Area (sq mi)	Area (sq km)
Devonshire	8,100	1.9	4.9
Hamilton	4,500	2.0	5.1
Paget	5,300	2.0	5.3
Pembroke	12,400	2.1	5.4
St. George's	3,400	3.8	9.3
Sandy's	7,500	2.5	6.7
Smith's	5,300	1.9	4.9
Southampton	5,500	2.2	5.8
Warwick	8,300	2.2	5.7

SOURCE: *Geo-Data* 1989 ed. and Bermuda Online, http://www.bermuda-online.org/ (accessed June 2002).

HUMAN POPULATION

Only about 20 of Bermuda's islands are inhabited. The two major population centers are Hamilton (the capital) and St. George. The island of Great Bermuda is the most densely inhabited, with roughly even population distribution over its entire area.

NATURAL RESOURCES

Bermuda's natural resources include limestone and enough marine life to support a modest fishing industry. Its pleasant climate is also a resource, as it is the basis for the tourism industry. International finance is another major sector of Bermuda's economy.

FURTHER READINGS

Bermuda. Oakland, Calif.: Lonely Planet Publications, 1997.

Bermuda Online. *Welcome to Bermuda.* http://Bermuda-online.org (accessed Jan. 26, 2002).

Gaffron, Norma. *The Bermuda Triangle: Opposing Viewpoints.* San Diego: Greenhaven Press, 1995.

Philpott, Don. *Bermuda.* Landmark Visitors Guides. Edison, N.J.: Hunter Publications, 2000.

Wilkinson, Henry Campbell. *Bermuda from Sail to Steam: The History of the Island from 1784 to 1901.* London: Oxford University Press, 1973.

Bhutan

- **Area:** 18,147 sq mi (47,000 sq km) / World Rank: 131
- **Location:** Northern and Eastern Hemispheres, Southern Asia, bordering China on the north and northwest and India on the east, south, and west
- **Coordinates:** 27°30′N, 90°30′E
- **Borders:** 668 mi (1,075 km) / China, 292 mi (470 km); India 376 mi (605 km)
- **Coastline:** Bhutan is landlocked
- **Territorial Seas:** none
- **Highest Point:** Kula Kangri, 24,781 ft (7,553 m)
- **Lowest Point:** Drangme River, 318 ft (97 m)
- **Longest Distances:** 190 mi (306 km) E-W / 90 mi (145 km) N-S

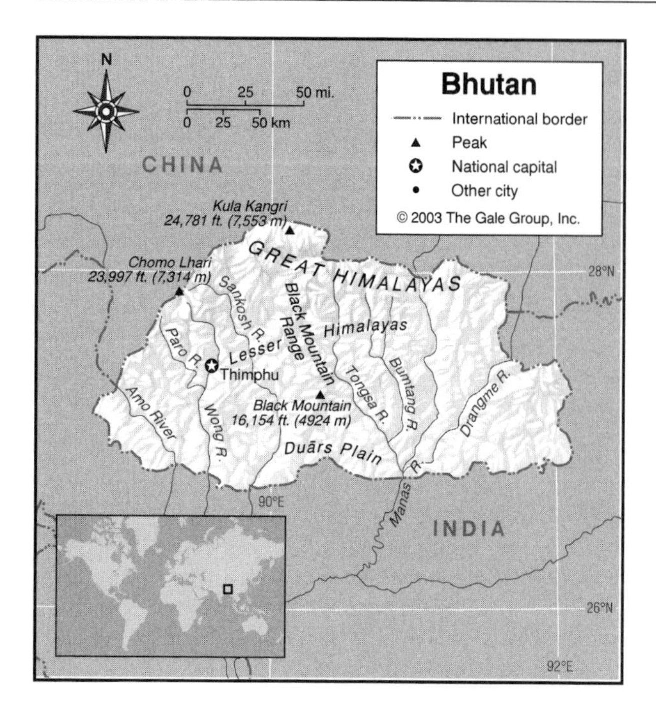

- **Longest River:** Tongsa River, 220 mi (350 km est.)
- **Natural Hazards:** Landslides and severe storms
- **Population:** 2,049,412 (July 2001 est.) / World Rank: 140
- **Capital City:** Thimphu, west-central Bhutan
- **Largest City:** Thimphu, 30,000 (mid-1990s est.)

OVERVIEW

Bhutan is a small, landlocked country in the Himalayan Mountains, between China and India in Southern Asia. It is situated on the Indo-Australian Tectonic Plate. To the north and northwest it borders the Chinese autonomous region of Tibet (Xizang Zizhiqu); to the south and southwest, the Indian states of West Bengal and Assam; and to the east, the Indian state of Arunachal Pradesh (formerly the North-East Frontier Agency).

All of Bhutan is mountainous except for narrow fringes of land at the southern border where the Duārs Plain, the lowlands of the Brahmaputra River, protrude northward over the border with India. The rest of Bhutan can be divided into two mountain regions: the Lesser, or Inner, Himalayas, which rise from the Duārs Plains through the central part of the country, and the snow-capped peaks of the Great Himalayas at the far north.

MOUNTAINS AND HILLS

Mountains

Bhutan is known for the sometimes dramatic irregularity of its mountainous terrain. Elevations vary from approximately 1,000 ft (305 m) in the south to almost 25,000 ft (7,620 m) in the north—in some places within distances of 60 mi (under 100 km) from each other.

The snowcapped Great Himalayas rise along the Tibetan border to the north, spreading across Bhutan in a belt about 10 mi (16 km) wide. Four peaks in this range have elevations above 20,000 ft (6,096 m). The highest is Kula Kangri, north of Gasa Dzong, at 24,781 ft (7,553 m). Next in height is the country's most famous peak, picturesque Chomo Lhari northwest of Punakha, towering over the Chumbi Valley at an elevation of 23,997 ft (7,314 m). The Great Himalayas have an arctic climate in their highest areas and are permanently snow-covered in many places, with valleys at elevations of 12,000 to 18,000 ft (3,700 to 5,500 m) sloping down from vast glaciers. At lower elevations, yaks graze in pastureland during the summer months.

Spurs extending southward from the Great Himalayas make up the north-south ranges of Bhutan's Inner, or Lesser, Himalayas. Fertile valleys lie between their peaks, which form the watersheds of Bhutan's major rivers. The dominant range in this system is the Black Mountain Range, which runs north to south and divides the country almost equally down the middle. It forms the watershed between the Sankosh and Drangme Chhu rivers. Its highest peak is Black Mountain at 16,514 ft (5,033 m) above sea level. Picturesque gorges are found at its lower elevations. Another major spur in the eastern half of the country is the Donga range.

Several strategically important passes, accessible through the Duārs Plain to the south, follow the major river courses through the valleys of Bhutan's Himalayan mountains. They were formerly of great significance for trade. Since Bhutan stopped trading with Tibet in 1953 to impede the spread of Communist influence, the passes have lost their earlier importance. They now serve as escape routes for Tibetan refugees, and Bhutanese authorities regard them with concern as potential invasion routes for Chinese Communist forces. With elevations ranging from approximately 15,000 ft (4,572 m) to more than 20,000 ft (6,096 m), the passes are negotiable only by pack animals or porters. The three most important are those on routes leading from Paro in Bhutan across the northwestern frontier into the Chumbi Valley. Other important passes include those that lead across the mountain spurs of the Inner Himalayan Range. Tashigang in eastern Bhutan and Paro in the west are connected by the country's only lateral communication route, which must cross a series of valleys and ridges.

Hills and Badlands

At the southern edge of the Inner Himalayas, sloping down to the Duārs Plain, are low, densely forested foothills called the Siwalik, or Southern, Hills.

GEO-FACT

Bhutan is home to the extremely rare blue poppy, its national flower, which grows only at elevations of about 13,000 ft (4,000 m). The poppy, which can grow as high as 3 ft (1 m) tall, has lavender-colored petals and a bright orange center. A single plant lasts from three to five years, flowering only once each summer.

INLAND WATERWAYS

Rivers

All of Bhutan's numerous rivers flow southward through gorges and narrow valleys and eventually drain into the Brahmaputra River, some 50 mi (80 km) south of the boundary with India. Except in the east and in the west, the headwaters of the streams are in the regions of permanent snow along the Tibetan border. None of the rivers are navigable, but many are potential sources of hydroelectric power.

Bhutan is drained by four main river systems. The area east of the Black Mountain watershed is drained by the Tongsa River and its tributaries, the Bumtang and Drangme Rivers (river names in Bhutan are often followed by *Chu* or *Chhu*, which means river). The Tongsa River (Tongsa Chhu) is known as the Manas River further south, where it enters the Duārs Plain and continues on into India. The eastern area of Bhutan drained by this system is known as the Drangme River Basin (Drangme Chhu Basin).

West of the Black Mountain Range the drainage pattern changes to a series of parallel streams, beginning with the Sankosh (or Puna Tsang) River and its tributaries, the Mo River (Mo Chhu) and Pho River (Pho Chhu). These tributaries, originating in northwestern Bhutan and fed by melting snow from the Great Himalayas, flow southward to Punakha, where they join the main river, continuing their southward course into the Indian state of West Bengal.

Farther west is the third major system, the Wong River (Wong Chhu) and its tributaries, including the Paro River (Paro Chhu). They flow through west-central Bhutan, joining to form the Raigye River (Raigye Chhu) before flowing into West Bengal. Still farther west is the smallest system, the Torsa River (Torsa Chhu) (called the Amo Chhu farther north), which flows through the Chumbi Valley and the major urban center of Phuntsholing before entering India.

THE COAST, ISLANDS, AND THE OCEAN

Bhutan is completely landlocked.

CLIMATE AND VEGETATION

Temperature

Bhutan has three distinct climates, corresponding to its three topographical divisions. The Duārs Plain areas in the south have a hot, humid, subtropical climate, with heavy rainfall. Temperatures generally average between 59°F (15°C) and 86°F (30°C) year round, although temperatures in the valleys of the southern foothills of the Himalayas may rise as high as 101°F (40°C) in the summer.

The central Inner Himalayan region has a temperate climate, with hot summers, cool winters, and moderate rainfall. Temperatures in the capital city of Thimphu, located in the western part of this region, are generally between about 59°F (15°C) and 79°F (26°C) between June and September (the monsoon season), falling to between 25°F (–4°C) and 61°F (16°C) in January. The high mountains of the Greater Himalayas in the north have more severe weather than the regions to the south, with cool summers and cold winters. At their highest elevations, they are snow-covered year round, with an arctic climate.

Rainfall

Like other aspects of Bhutan's climate, rainfall varies by region. The northern Himalayas are relatively dry, with most precipitation falling in the form of snow. The Inner Himalayan slopes and valleys in the central part of the country have moderate rainfall, averaging between 39 and 59 in (100 and 150 cm) annually. Rainfall in the subtropical southern regions averages between about 197 in and 295 in (500 cm and 750 cm) per year.

Bhutan has distinct dry and rainy seasons. The greatest amount of rain falls during the summer monsoon season from late June to the end of September, accompanied by high humidity, flash flooding, and landslides. The weather during this period is generally overcast. Days become bright and sunny during the dry autumn season, which lasts from late September to around the end of November. During winter, which lasts from late November to March, frost occurs in many areas, and snow falls at elevations above 9,843 ft (3,000 m).

In addition to its summer monsoons, Bhutan gets a winter monsoon from the northeast. The name by which Bhutan is known to its own people—Drak Yul, or the Land of the Thunder Dragon—comes from the high-velocity winds of this storm, which thunder down from the mountains. Bhutan's weather becomes drier again in the spring, from early March until mid-April, when summer weather begins. Summer showers are only occasional until the onset of the monsoon season in June.

Forest and Jungles

The southernmost part of the Duārs Plain region is composed of savanna, bamboo, and dense jungle vegetation.

Bhutan's Inner Himalayan slopes are densely covered with deciduous forests. Species found at elevations between 5,000 ft to 8,000 ft (1,500 m to 2,400 m) include ash, birch, beech, cypress, maple, and yew. Between 8,000 and 9,000 ft (2,400 to 2,700 m), these give way to oak and rhododendron, with spruce, fir, and juniper trees growing beyond that point, up to the tree line.

Bhutan has a total of more than 5,000 plant species, including many varieties of orchid, the giant rhubarb, magnolias, over 300 species of medicinal plants, and carnivorous plants.

Grasslands

In the far north, livestock graze on pastureland in the alpine valleys of the Greater Himalayas. In the south is the Duārs Plain. It lies mostly in India but extends northward across Bhutan's border in strips 6 to 9 mi (10 to 15 km) wide. The northern edges of these plains, which border the Himalayan foothills, have rugged terrain and porous soil. Fertile flatlands are found farther south.

HUMAN POPULATION

Most Bhutanese live in small rural villages in the Inner Himalayan region. Settlements are spread out among the valleys in this region, with farmers living in houses on the lower mountain slopes above their farmland. At higher elevations, population distribution is more concentrated because the lack of level land forces inhabitants to cluster together in smaller areas. The upper reaches of the Himalayas are largely uninhabited except for scattered Buddhist monasteries in valleys.

The capital city of Thimphu is located at the northern edge of the Inner Himalayas, in the western part of the country. The major commercial centers of Phuntsholing, Geylegphug, and Samdrup Jongkhar are located near the southern border.

NATURAL RESOURCES

Bhutan's most productive forestlands are found in the central Inner Himalayan mountain region. Oak, pine, and the tropical hardwoods found on the Duārs Plain are the main types of timber harvested. With its abundant rivers and steep mountain slopes, Bhutan has great hydroelectric power potential, although only a small fraction of it is currently being exploited. Further hydroelectric development is being planned. Other natural resources include gypsum and calcium chloride.

FURTHER READINGS

Apte, Robert Z. *Three Kingdoms on the Roof of the World: Bhutan, Nepal, and Ladakh.* Berkeley, Calif.: Parallax Press, 1990.

Bhutan Tourism Corporation Web site. *Kingdom of Bhutan.* http://www.kingdomofbhutan.com/ (accessed February 20, 2002)

Dompnier, Robert. *Bhutan, Kingdom of the Dragon.* Boston: Shambhala, 1999.

Hellum, A. K. *A Painter's Year in the Forests of Bhutan.* Edmonton: University of Alberta Press, 2001.

Savada, Andrea Matles, ed. *Nepal and Bhutan: Country Studies.* Federal Research Division, Library of Congress. 3rd edition. Washington, D.C.: U.S. G.P.O., 1993.

Swift, Hugh. *Trekking in Nepal, West Tibet, and Bhutan: The Sierra Club Travel Guide to the Eastern Himalayas.* San Francisco: Sierra Club Books, 1989.

Zeppa, Jamie. *Beyond the Sky and the Earth: A Journey into Bhutan.* New York: Riverhead Books, 1999.

Bolivia

- **Area:** 424,164 sq mi (1,098,580 sq km) / World Rank: 29
- **Location:** Southern and Western Hemispheres, bordering Brazil to the north and northeast, Paraguay to the southeast, Argentina to the south, Chile to the southwest, and Peru to the northwest.
- **Coordinates:** 17°00′S, 65°00′W
- **Borders:** 4,190 mi (6,743 km) / Argentina, 517 mi (832 km); Brazil, 2,113 mi (3,400 km); Chile, 535 mi (861 km); Paraguay, 466 mi (750 km); Peru, 559 mi (900 km)
- **Coastline:** Landlocked
- **Territorial Seas:** Landlocked
- **Highest Point:** Nevado Sajama, 21,463 ft (6,542 m)
- **Lowest Point:** Rio Paraguay, 295 ft (90 m)
- **Longest Distances:** 950 mi (1,530 km) N-S; 900 mi (1,450 km) E-W
- **Longest River:** Mamoré, 1,200 mi (1,931 km)
- **Largest Lake:** Titicaca, 9,660 sq mi (25,086 sq km)
- **Natural Hazards:** Subject to localized flooding in spring
- **Population:** 8,300,463 (July 2001 est.) / World Rank: 84

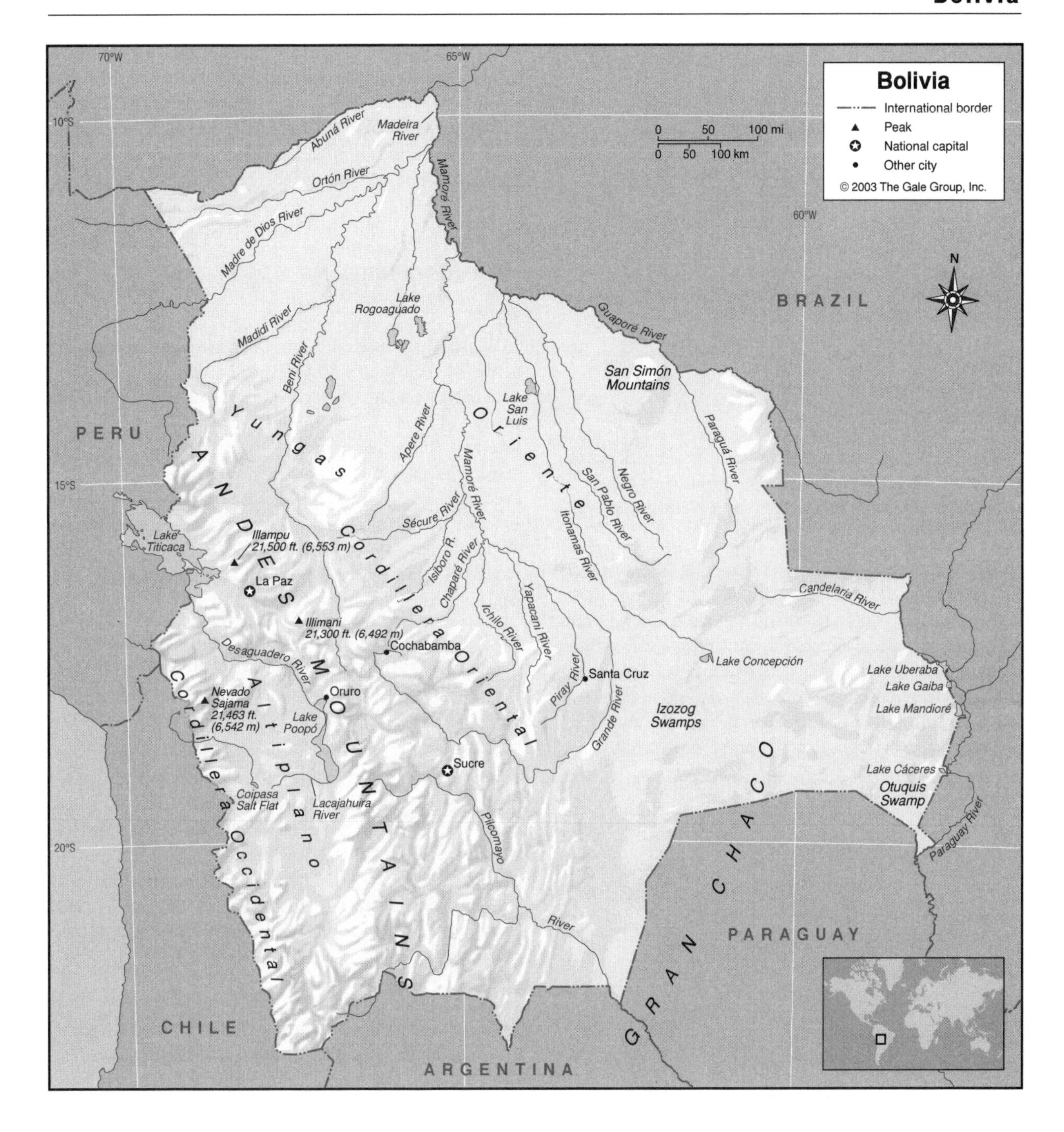

■ **Capital City:** Sucre, southwestern Bolivia

■ **Largest City:** La Paz, west-central Bolivia, population 1,458,000 (2000 metropolitan est.)

OVERVIEW

Home to the world's highest capital city—La Paz, which is the governmental capital—and highest commercially navigable lake, Bolivia has been called the "rooftop of the world." This landlocked country in central South America is the continent's fifth-largest nation.

The Andean highlands of southwest Bolivia, made up of the Cordillera Occidental and Cordillera Oriental Mountain Ranges separated by a high plateau called the Altiplano, constitute roughly one-third of the country. The remaining two-thirds belong to the Oriente, the country's northern and eastern tropical lowlands, which encompass forestland, savannas, and marshes. At the far southeastern corner of the country lies the Bolivian portion of the Gran Chaco, a thinly populated plain that continues southward into Paraguay and northern Argentina.

Bolivia is situated on the South American Tectonic Plate. The area north of Sucre, near the center of the

GEO-FACT

The Andean Condor, found in the mountains of Bolivia, is the largest flying bird in the Americas. It nests at elevations above 10,000 ft (3,048 m).

country, is seismically active. A 1958 earthquake inflicted major damage in this area.

MOUNTAINS AND HILLS

Mountains

Rising to both their greatest average elevations and greatest width in Bolivia, the Andes Mountains consist of two chains separated by the lofty high plateau that is the country's heartland.

On the west, the Cordillera Occidental (Western Cordillera), which forms the border with Chile, has crests higher than 19,000 ft (5,800 m) above sea level. This range includes Mt. Sajama, Bolivia's loftiest peak with an incredible height of 21,463 ft (6,542 m). The few passes through these perpetually snowcapped peaks are at elevations of 13,000 ft (4,000 m) or more, and the chain includes a number of both active and inactive volcanoes.

The eastern arm of the Bolivian Andes is called either the Cordillera Oriental or Cordillera Real, although the latter name is often reserved for only that section of the range that extends northwestward from the environs of Cochabamba and Oruro. This region, in which the governmental capital city of La Paz is located (Sucre is the legal capital), includes the country's most dramatic Andean peaks, with average heights of over 18,000 ft (5,486 ft) for more than 200 mi (322 km). The best known of the crests are Illampu, at 21,500 ft (6,553 m) and the triple crown of Illimani, which rises to 21,300 ft (6,492 m) behind the city of La Paz.

The eastern slopes of the northern Cordillera Oriental, called the Yungas, are rugged, steep, and densely forested, descending swiftly to the eastern plains. South of the Yungas is an area of valleys and mountain basins called the Valles. Bolivia also has the San Simón Mountains on its northeastern border with Brazil but, compared to the majestic Andes, these mountains are relatively only hills.

Plateaus

The forbidding lunar-like landscape of the Altiplano extends southward for a distance of 500 mi (804 km), with an average width of 80 mi (50 km) and altitudes varying between 12,000 and 14,000 ft (3,657 and 4,267 m). It tilts upward from the center toward both the eastern and western Cordillera systems and descends gradually from north to south. The plateau floor is made up of sedimentary debris washed down from the adjacent mountains. The material frequently appears to consist of rock, but it is in fact made up of compressed sandy materials, clays, and gravels, and it is highly susceptible to erosion.

INLAND WATERWAYS

Lakes

Lake Titicaca straddles the Peruvian border near the center of the country, just west of La Paz. With a surface area of 9,660 sq mi (25,086 sq km) at 12,484 ft (3,805 m) above sea level, it is both South America's largest inland lake and the world's highest body of navigable water. The remainder of a much larger ancient body of water, Titicaca has a length of 138 mi (222 km) and a width of 70 mi (113 km) and contains twenty-five islands, which played an important role in Inca mythology. Lake Titicaca has depths of up to 700 ft (213 m), and its icy waters are only slightly saline.

Southeast of Lake Titicaca and connected to it by the Desaguadero River, Lake Poopó is a shallow, salty body of brackish water with depths of 10 ft (3 m) or less and an area of around 1,000 sq mi (386 sq km) when its waters are low.

Several large lakes are found in the plains drained by the Beni and Mamoré Rivers, including Lake Rogoaguado and Lake San Luis. Shallow lakes on the eastern border near the Paraguá River and Candelaria River include Lakes Cáceres, Mandioré, Gaiba, and Uberaba. The only significant lake in the eastern interior is Lake Concepción, which forms part of the origin of the Itonamas River.

Rivers

Bolivia is drained by three different river systems. Flowing down from the Yungas area of the Cordillera Real, the Beni and Mamoré River and the Mamoré's tributaries—including the Apere, Sécure, Isiboro, Chaparé, Ichilo, Yapacani, Piray, and Grande—form part of the Amazon River system. Eventually joined by rivers draining the Orient lowlands to the east (including the Paraguá, Guaparé, Itonamas, and Negro), and the Abuná, Ortón, and Madidi in the northwest, these Amazon headwaters flow north to join at the Madeira River on the northernmost part of the border with Brazil.

At Bolivia's western border Lake Titicaca, fed by mountain streams rushing down the Cordillera Occidental from the Altiplano, is drained to the south by the Desaguadero River. Flowing southward for 322 km (200 mi) to Lake Poopó, Bolivia's other major lake, the river is the only major stream on the surface of the Altiplano.

Lake Poopó, in turn, drains into the Lacajahuira River, which flows south to the Coipasa Salt Flat.

Farther south, the Pilcomayo River rises in the heart of the Yungas and flows southward to the border with Argentina and Paraguay, to join the Paraguay River (and the Rio de la Plata system) near Asunción in Paraguay.

Wetlands

North of the lakes in the Paraguá River region are the Xarayes swamps. South of these lakes in the southeast corner along the borders with Paraguay and Brazil is the Otuquis Swamp, an area of palm trees, aquatic vegetation, and shallow soil. Near Lake Concepción, to the southwest surrounding the head of the Parapetí River, the Izozog Swamps is a large area of swampland. The plains drained by the Beni and Mamoré regions also include swampland, as well as lakes and lagoons.

THE COAST, ISLANDS, AND THE OCEAN

Bolivia is landlocked.

CLIMATE AND VEGETATION

Temperature

Although Bolivia is a tropical country, its climate varies widely with differences in elevation and terrain. The high peaks of the Cordillera Occidental to the west have a cool, arctic climate, with days that are often clear and sunny and night temperatures that frequently fall below the freezing point. Cold winds bring cool weather to the Altiplano, where afternoon thunderstorms are common in summer (from December to January), and snow sometimes falls in the winter. Nights in this region are cold year round. In the northern Altiplano, however, the climate is moderated by Lake Titicaca, and temperatures on sunny winter days may reach 70°F (21°C).

The valleys of the lower Cordillera Oriental have a semiarid Mediterranean climate, while the climate becomes semitropical in the Yungas region on the eastern slopes of these mountains, and tropical in the eastern lowlands. The mean annual temperature in La Paz, at the edge of the Altiplano, is about 46°F (8°C), compared with mean temperatures of 60° to 68°F (16° to 19°C) in the Yungas region, and 79°F (26°C) in the city of Trinidad, in the eastern plains.

A strong wind from the Argentine pampas, called the surazo, can bring fierce storms and plunging temperatures in the winter months (June through August).

Rainfall

Like climate conditions in general, rainfall in Bolivia varies greatly by region, ranging from 5 in (13 cm) or less in the southwest to more than 60 in (152 cm) in the Amazon basin to the northeast. Rainfall in the Yungas region on the eastern slopes of the Cordillera Oriental averages 30 in (76 cm) to 50 in (127 cm) annually—it is heaviest between December and February but falls year round. The southern part of the country has a long summer dry season that can last from four to six months (up to nine in the Gran Chaco Region), while the dry season in the northern areas is shorter. Flooding often occurs in the northeast in March and April.

Grasslands

Savanna grasslands cover much of the lowland Oriente region, which encompasses the eastern and northern two-thirds of Bolivia, or all the land east of the Cordilleras Ranges of the Andes Mountains. The region slopes gradually from south to north, and from elevations of 2,000 to 2,500 ft (610 to 762 m) at the foot of the Andes in the west to as little as 300 ft (91 m) along parts of the Brazilian border in the east. The southeastern portion, which is a continuation of the Gran Chaco of Paraguay, is virtually rainless for nine months of the year, receiving torrents of rain in the winter. This extreme variation in rainfall supports only sparse plant life including cacti, prickly scrub, scorched grass, and the quebracho tree.

Forests and Jungles

The fertile alluvial soils along the courses of the many rivers that flow northward through the Oriente lowlands support high forest growth. Tropical Amazon rainforest predominates in the northernmost *llanos,* or plains regions of the Oriente lowlands, mostly in the extreme northern regions. Rubber, mahogany, and Brazil nut are a few of the many different tree species found here. Stretches of tropical forest also line riverbanks and foothills farther south and at the eastern and southeastern edge of the country. The rainforests are also home to vanilla, saffron, sarsaparilla, palm trees, and many types of fruit.

The mountain jungles of the Yungas region contain tree species including green pine, laurel, cedar, and the quina, or cinchona, tree, which is used for producing quinine, as well as a variety of medicinal plants. Farther south, the deciduous forests of the Valles region include walnut and quebracho trees.

Other vegetation

Ichu, a coarse species of bunch grass grazed on by llamas, is the most widespread form of vegetation on the Altiplano. Cacti also grow in this region, as do a hardy form of scrub called the *tola* and the mossy *yareta*. Reeds grow along the shores of Lake Titicaca.

HUMAN POPULATION

The most densely populated region in Bolivia is the northern part of Altiplano, home to the governmental capital city of La Paz. The pleasant climate, moderated by the proximity of Lake Titicaca, and the fertility of the land

Population Centers – Bolivia

(1997 POPULATION ESTIMATES)

Name	Population	Name	Population
Santa Cruz	914,795	El Alto	523,280
La Paz (administrative capital)	758,141	Oruro	202,548
		Sucre (legal capital)	163,563
Cochabamba	560,284	Potosí	122,962

SOURCE: INE (National Institute of Statistics), Bolivia.

Departments – Bolivia

2000 POPULATION ESTIMATES

Name	Population	Area (sq mi)	Area (sq km)	Capital
Chuquisaca	589,948	19,893	51,524	Sucre
Cochabamba	1,524,724	21,479	55,631	Cochabamba
Beni	366,047	82,458	213,564	Trinidad
La Paz	2,406,377	51,732	133,985	La Paz
Oruro	393,991	20,690	53,588	Oruro
Pando	57,316	24,644	63,827	Cobija
Potosí	774,696	45,644	118,218	Potosí
Santa Cruz	1,812,522	143,098	370,621	Santa Cruz
Tarija	403,079	14,526	37,623	Tarija

SOURCE: INE (National Institute of Statistics), Bolivia.

have long made this the most heavily settled part of the country. However, the building of new roads—especially the Cochabamba-Santa Cruz highway in the 1950s—and the discovery of oil and gas have led to population growth in the lowland region around Santa Cruz. Otherwise most of the lowland plains are uninhabited.

NATURAL RESOURCES

Although Bolivia has rich mineral resources—including abundant quantities of tin, antimony, and tungsten, as well as gold, silver, copper, zinc, and lead—its minerals industry remains underdeveloped, and its economy is primarily based on agriculture. The nation also has a wealth of forest resources, including deciduous hardwoods, evergreens, and more than 2,000 species of tropical hardwoods, but this sector, too, has yet to tap its full potential. The recent discovery of natural gas and petroleum in the foothills of the Andes near Santa Cruz holds potential for the future.

FURTHER READINGS

Bolivia Web. http://www.boliviaweb.com (Accessed Feb. 25, 2002).

Bradt, Hilary. *Peru and Bolivia: Backpacking and Trekking.* Old Saybrook, Conn.: Globe Pequot Press, 1999.

Klein, Herbert S. *Bolivia: The Evolution of a Multi-ethnic Society.* 2nd ed. New York: Oxford University Press, 1992.

Morales, Waltraud Q. *Bolivia: Land of Struggle.* Boulder: Westview Press, 1992.

Murphy, Alan. *Bolivia Handbook.* Lincolnwood, Ill.: Passport Books, 1997.

Swaney, Deanna. *Bolivia: A Lonely Planet Travel Survival Kit.* 3rd ed. Oakland, Calif.: Lonely Planet Publications, 1996.

Bosnia and Herzegovina

- **Area:** 19,741 sq mi (51,129 sq km) / World Rank: 127
- **Location:** Northern and Western Hemispheres, in southeastern Europe, bordering the Adriatic Sea, east of Croatia, west of Yugoslavia.
- **Coordinates:** 44°00′N, 18°00′E
- **Borders:** 907 mi (1,459 km) / Croatia, 579 mi (932 km); Yugoslavia, 327 mi (527 km)
- **Coastline:** 12 mi (20 km)
- **Territorial Seas:** none
- **Highest Point:** Mt. Maglic, 7,828 ft (2,386 m)
- **Lowest Point:** Sea level
- **Longest Distances:** 202 mi (325 km) N-S; 202 mi (325 km) E-W
- **Longest River:** Sava, 589 mi (947 km)
- **Largest Lake:** Buško Blato, 21.5 sq mi (55.8 sq km).
- **Natural Hazards:** Earthquakes
- **Population:** 3,922,205 (July 2001 est.) / World Rank: 120
- **Capital City:** Sarajevo, in east-central Bosnia and Herzegovina.
- **Largest City:** Sarajevo, 434,000 (2002 estimate).

OVERVIEW

The Republic of Bosnia and Herzegovina is located in the Balkan Peninsula of Southern Europe. Most of the county lies inland, but a short, narrow, corridor extends to the Adriatic Sea. The country as a whole is roughly shaped like an isosceles triangle, with each side of the right angle measuring about 185 mi / 300 km (this shape is symbolized in the gold triangle of the national flag). The Republic consists of joint administrative divisions roughly equal in size: the Muslim-Croat area known as the Federation of Bosnia and Herzegovina (FBH), and the Republic of Srpska (RS). The FBH mostly occupies the central portion of the country, while the RS arcs from the northern border with Croatia eastward and southward along the border with Yugoslavia.

The Republic has three main geographic types: a thin band of high plains and plateau running along the northern border with Croatia, roughly from Bosanska Gradiška to Bijeljina; low mountains in the center between Banja Luka, Zenica, and Sarajevo; and the higher Dinaric Alps, which cover the rest of the country.

Tectonic fault lines run through the central part of the country, from Bosanska Gradiška to Sarajevo, and also in the northwest corner between the Sana and Unac rivers. A thrust fault also runs through southern Bosnia and Herzegovina in the vicinity of Mostar. These structural seams in the earth's crust periodically shift, causing earth tremors and occasional destructive earthquakes. Bosnia and Herzegovina is located on the Eurasian Tectonic Plate.

MOUNTAINS AND HILLS

Mountains

About two-thirds of Bosnia and Herzegovina is mountainous, running from the northwest to the southeast. There are 64 mountains with peaks of more than 4,922 ft (1,500 m) above sea level. Mt. Maglic, in the southeast on the Yugoslav border, is the highest peak in the country at 7,828 ft (2,386 m). Nearby are the country's second and third highest mountains, Volujak (7,664 ft / 2,336 m) and Velika Ljubuša (7,343 ft / 2,238 m), respectively.

The Dinaric Alps chain consists of ridges that run parallel to the coast. The limestone ranges of the Dinaric Alps are frequently referred to as karst or karstland, and are distinctive because of the underground drainage channels that have been formed by the long-term seepage of water down through the soluble limestone. This action leaves the surface dry and over the years has formed many large depressions. The central mountains are lower and less rugged than the Dinaric Alps. They include the Kozara Mountains, Vlašić Mountains, and Majevica Mountains.

Canyons

At 4,265 ft (1,300 m), Tara Canyon is Europe's deepest canyon. The canyon follows the Tara River along the southeastern border with Yugoslavia.

INLAND WATERWAYS

Lakes

Buško Blato, in the southwest near the Croatian border, is the country's largest lake, with a surface area of 21.5 sq mi (55.8 sq km). The lake lies within the Dinaric Alps; its surface is 2351.2 ft (716.6 m) above sea level, the maximum depth is 56.8 ft (17.3 m) with a total volume of 27.6 billion cu ft (782 cu m) of water. There are thermal springs along the Sava River valleys and mineral springs near Sarajevo.

Rivers

The rivers of Bosnia and Herzegovina typically flow from south to north, because the slope of the mountains gradually rises towards the south. From west to east, the main rivers are: the Una and its tributary, the Sana; the Vrbas; the Bosna; and the Drina. The Drina River forms part of the border with Yugoslavia. All these rivers eventually flow into the Sava River, the largest river in the country, which forms the northern border with Croatia. The Sava is itself a tributary of the Danube River. In the southeastern part of the country there are some rivers that flow south into the Adriatic. The largest of these is the Neretva.

THE COAST, ISLANDS, AND THE OCEAN

Bosnia and Herzegovnia only has about 12 mi (20 km) of coastline on the Adriatic Sea. Neum is the country's only coastal town, but it is not conducive to shipping.

Population Centers – Bosnia and Herzegovina

(1992 POPULATION)

Name	Population	Name	Population
Sarajevo	434,000*	Mostar	127,034
Banja Luka	179,200*	Prijedor	112,635
Zenica	145,837	Doboj	102,624
Tuzla	131,866		

*Sarajevo and Banja Luka populations estimated for 2002.

SOURCE: Federal Office of Statistics, Bosnia and Herzegovina.

CLIMATE AND VEGETATION

Temperature

In Sarajevo, the average daily high temperature in summer is 64.6°F (18.1°C), while in winter it is 32.5°F (0.3°C). The overall average annual temperature in Sarajevo is 49.1°F (9.5°C). Mostar averages 73.8°F (23.2°C) in the summer and 42.4°F (5.8°C) in the winter. Mostar also holds the highest recorded temperature, at 115.2°F (46.2°C) on July 31, 1901.

Pannonian Plain

Occupying the north near the Croatian border is a narrow segment of the Pannonian Plain. The plain is an ancient seabed, now filled with sediment that makes it the country's most fertile area and the site of many farms. Low and flat, it features rolling hills and wide valley basins.

Forests

Beech forests cover much of the mountainous areas, with mixed forests of beech, fir, and spruce on higher mountains. The Sutjeska National Park is the country's oldest national park; its 43,250 acres (17,500 ha) includes Mt. Maglic. The park contains the old-growth Perucica Forest.

HUMAN POPULATION

The estimated 2002 population was 4,339,600, of which some 2,689,700 lived in the FBH and approximately 1,649,800 lived in the RS. Sarajevo, located in the FBH, is the country's largest city. Bosnia and Herzegovina was the scene of extensive warfare and ethnic conflict during the 1990s. As a result, as much as half of the country's pre-war population was still living as refugees in 2000, either in Bosnia and Herzegovina's cities or in neighboring countries.

NATURAL RESOURCES

Bosnia and Herzegovina has an abundance of metallurgical resources. Iron ore is mined in the east, near Vareš, Lubija, and Radovan. Bauxite is mined in the southwest, near Posusje and Lištica. Lead, manganese, copper, chromium, and zinc are other mineral resources. Lignite and brown coal are mined near Tuzla in the northeast.

FURTHER READINGS

Brân, Zoë. *After Yugoslavia.* Oakland, Calif.: Lonely Planet, 2001.

Bosnia and Herzegovina. http://www.fbihvlada.gov.ba/engleski/bosna/index.html (accessed July 1, 2002).

Lovrenovic, Ivan. *Bosnia: A Cultural History.* New York: New York University Press, 2001.

Malcolm, Noel. *Bosnia: A Short History.* New York: New York University Press, 1996.

Botswana

- **Area:** 231,802 sq mi (600,370 sq km) / World Rank: 46
- **Location:** Southern and Eastern hemispheres, located in southern Africa, bordered on the northeast by Zambia, on the south and southeast by South Africa, on the west and north by Namibia
- **Coordinates:** 22°00′S, 24°00′E
- **Borders:** 2,488 mi (4,013 km) total length; Zimbabwe, 504 mi (813 km); South Africa, 1,141 mi (1,840 km); Namibia, 843 mi (1,360 km)
- **Coastline:** Botswana is landlocked.
- **Territorial Seas:** None
- **Highest Point:** Tsodilo Hills, 4,884ft (1,489 m)
- **Lowest Point:** Junction of the Limpopo and Shashe Rivers at1,683 ft (513 m)
- **Longest Distances:** 690 mi (1,110 km) NNE-SSW / 597 mi (960 km) EWE-WNW
- **Longest River:** Limpopo River, 1,000 mi (1,600 km)
- **Largest Lake:** Lake Ngami, 401 sq mi (1,040 sq km).
- **Natural Hazards:** Much of the western part of the country is desert. Seasonal winds blow from the west, carrying sand and dust across the country, which can obscure visibility and contribute to periodic droughts.
- **Population:** 1,586,119 (2001 est.) / World Rank: 145
- **Capital City:** Gaborone, located in the southeast corner near the South African border.
- **Largest City:** Gaborone, 182,000 (2000 est.)

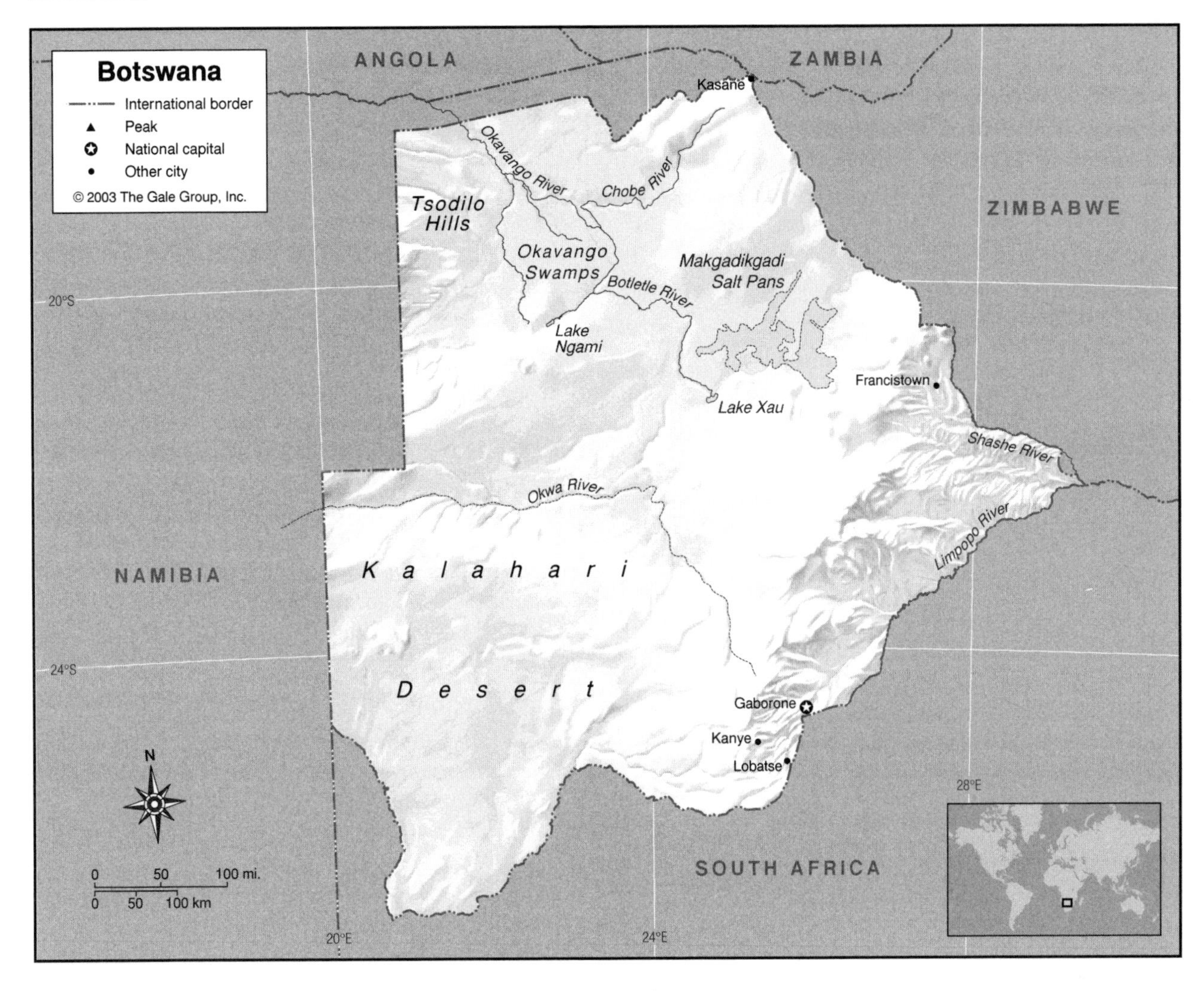

OVERVIEW

Botswana is a landlocked country located in southern Africa. It is a vast tableland with a mean altitude of 1,000 m (3,300 ft). A gently undulating plateau running from the South African border near Lobatse to a point west of Kanye and from there northward to Bulawayo in Zimbabwe border forms the watershed between the two main natural divisions of Botswana. The fertile land to the south and east of this plateau is hilly bush country and grassland, or veld. To the west of the plateau, stretching over the border into Namibia, is the Kalahari (or Kgalagadi) Desert. In the north lies the area known as Ngamiland. It is dominated by the Okavango Swamps, a great inland delta of some 6,500 sq mi (16,835 sq km), and the Makgadikgadi Salt Pans. Around the swamps and along the northeastern border from Kasane to Francistown there is forest and dense bush.

MOUNTAINS AND HILLS

Mountains

There are no mountains in this elevated but relatively flat country.

Plateaus

All of Botswana is located on a broad tableland with an average altitude of 3,300 ft (1,000 m). Dividing the country into two distinct topographical regions is a vast plateau, about 4,000 ft in height, that extends from near Kanye in the south corner northeast to the border with Zimbabwe.

Hills and Badlands

The Tsodilo Hills are granite cliffs on the northwest fringe of the Kalahari Desert and are the highest elevations in the country. The hills form a fortress-like ridge 12 mi (20 km) in length and have long been considered sacred by the native population. At their highest point they reach 4,884 ft (1,489 m) above sea level.

INLAND WATERWAYS

Lakes

Temporary lakes form in the Okavango Swamps and the Makgadikgadi Salt Pans during seasons of heavy rainfall. Lakes Ngami and Xau are more permanent, but also rely on the floodwaters that rush down the high plateaus.

Rivers

There are few permanent rivers in Botswana, and its temporary rivers never reach the sea. The Chobe River in the north is permanent, and a major tributary of the Zambezi River. The Zambezi itself forms a brief portion of Botswana's border. The Limpopo River in the east is a large river, and marks the border with South Africa. The Okavango River enters the country in the northwest and ends in the Okavango Swamps. The Boteti River flows south from these swamps into Lake Xau.

Wetlands

In the heart of the Kalahari Desert in the western portion of Botswana, the Okavango River spreads out into a seasonally flooded wetland the size of Massachusetts, comprising swamps, channels, lagoons, and flood plains. The Okavango Delta is one of the world's largest wetlands, and provides a unique ecosystem and habitat for an astounding abundance of African mammals, birds, fish, amphibians, and reptiles. The Okavango Delta depends on the annual floods from central Africa, doubling in extent after the floods, then receding during the dry months, leaving behind vast salt pans. The swamps are world-renowned and often a destination for safari tourists.

THE COAST, ISLANDS, AND THE OCEAN

Botswana is completely landlocked.

CLIMATE AND VEGETATION

Temperature

A subtropical climate is experienced by most of the country, while the higher altitudes have cooler temperatures. Winter days are warm with cool nights, although the desert is commonly covered in heavy frost. Temperatures range from 91°F (33°C) in January to 72°F (22°C) in July. The August seasonal winds that blow from the west carry sand and dust across the landscape, often contributing to droughts.

Rainfall

Normal averages of 18 in (45 cm) occur throughout most of the country, except for the Kalahari, in the south, which has less than 10 in (25 cm), and 27 in (69 cm) in the northern plateau regions.

Grasslands

Although Botswana is 90 percent covered by some kind of savanna, most areas are too arid to sustain agriculture. The most fertile region is the veld and hilly bush country in the eastern portion of the country, which serves as pastureland for Botswana's cattle.

Population Centers – Botswana

(2001 POPULATION ESTIMATES)

Name	Population
Gaborone (capital)	224,286
Francistown	106,553
Selebi-Phikwe	50,312

SOURCE: Central Statistics Office, Botswana.

Deserts

The Kalahari Desert lies in the western portion of the country. It is a large, dry sandy basin that covers about 190,000 square miles (500,000 square km). The Kalahari reaches from the Orange River in South Africa north to Angola, in the west to Namibia, and in the east to Zimbabwe. The erosion of soft stone formations created the sand masses that characterize the terrain. The dominant vegetation is grasses, thorny shrubs, and Acacia trees that can survive the long drought periods of more than ten months every year.

Forests and Jungles

Most of Botswana is too dry to sustain true forests, but there are extensive tracts of scrub and brush, as well as some forested lands in the northeast.

HUMAN POPULATION

The July 2001 estimated population of 1,586,119 is concentrated in the eastern portion of the country. Very few people live in the desert or swamps. Ethnic groups are comprised of the Tswana (or Setswana), 79 percent; Kalanga, 11 percent; Basarwa, 3 percent; and others, including Kgalagadi and white, 7 percent. Religious practitioners are equally divided between Christians and those who practice indigenous beliefs.

NATURAL RESOURCES

Botswana is rich in mineral resources—particularly diamonds, copper, nickel, salt, soda ash, potash, coal, iron ore, and silver. Mining makes up about 33 percent of the Gross Domestic Product (GDP) of the country and 50 percent of government's revenue. Cattle ranching is another major economic activity.

FURTHER READINGS

Alverson, Marianne. *Under African Sun.* Chicago: University of Chicago Press, 1987.

Augustinus, Paul. *Botswana: A Brush with the Wild.* Randburg, South Africa: Acorn Books, 1987.

Chirenje, J. Mutero. *Chief Kgama and His Times: The Story of a Southern African Ruler.* London: R. Collings, 1978.

Maylam, Paul. *Rhodes, the Tswana, and the British: Colonialism, Collaboration, and Conflict in the Bechuanaland Protectorate, 1885–1899*. Westport, CT: Greenwood Press, 1980.

Picard, Louis A., ed. *Politics and Rural Development in South Africa: The Evolution of Modern Botswana*. Lincoln, NE: University of Nebraska Press, 1986.

Brazil

- **Area:** 3,286,469 sq mi (8,511,965 sq km) / World Rank: 6
- **Location:** Southern, Northern, and Western Hemispheres, in eastern South America; bordering the Atlantic Ocean to the east; French Guiana, Suriname, Guyana, and Venezuela in the north; Colombia in the northwest; Peru in the west; Bolivia, Argentina, and Paraguay in the southwest; Uruguay in the south
- **Coordinates:** 10°00′S, 55°00′W
- **Borders:** 9,108 mi (14,691 km) total / Argentina, 759 mi (1,224 km); Bolivia, 2,108 mi (3,400 km); Colombia, 1,019 mi (1,643 km); French Guiana, 417 mi (673 km); Guyana, 694 mi (1,119 km); Paraguay, 800 mi (1,290 km); Peru, 967 mi (1,560 km); Suriname, 307 mi (597 km); Uruguay, 611 mi (985 km); Venezuela, 1,364 mi (2,200 km)
- **Coastline:** 4,644 mi (7,491 km)
- **Territorial Seas:** 12 NM
- **Highest Point:** Neblina Peak, 9,888 ft (3,014 m)
- **Lowest Point:** Sea level
- **Longest Distances:** 2,689 mi (4,328 km) N-S / 2,684 mi (4,320 km) E-W
- **Longest River:** Amazon, 3,900 mi (6,280 km; approximation)
- **Largest Lake:** Lagoa dos Patos, 3,920 sq mi (10,153 sq km)
- **Natural Hazards:** frequent river flooding; recurring droughts in northeast; floods and occasional frost in south; vulnerability to severe erosion
- **Population:** 176,274,000 (2002 est.) / World Rank: 5
- **Capital City:** Brasília, south-central Brazil
- **Largest City:** São Paulo, located on southeastern coast along the Atlantic Ocean, 10,057,700 (est.)

OVERVIEW

Brazil is an immense country that makes up about half of the landmass of South America, is the home of half of its population, and borders all but two of its countries. Its territory is larger than that of the forty-eight contiguous states of the United States.

The Amazon River, the second longest in the world and the world's largest flow of water, cuts laterally across the country's northern flank, and countless tributary streams drain a vast flat to rolling lowland basin that encompasses three-fifths of the national territory. The entire basin, including fringes in neighboring countries, supports a tropical forest that provides natural replacement for 15 percent or more of the world's oxygen.

Although there are no high mountains—the highest elevations are at less than 10,000 ft (3,048 m)—most of the territory outside the Amazon Basin consists of a great highland block, part of the South American Tectonic Plate. The highlands drop precipitously to a narrow Atlantic coastal plain. Brazil's entire coastline measures 4,495 mi (7,491 km), and its continental shelf extends some 200 NM into the Atlantic. With fewer high elevations than any other country of South America except Uruguay and Paraguay, not more than one-fifth of Brazil's terrain is beyond the limits of agricultural usefulness. There are, however, many low mountain systems, rounded hills, and deep valleys. The drainage is generally good, but much of the landscape is highly vulnerable to erosion.

MOUNTAINS AND HILLS

Highlands and Plateaus

Brazil defines its highlands as areas with elevations in excess of 656 ft (200 m). Some 59 percent of the national territory is in the highlands according to this definition, but only 0.5 percent is at more than 3,937 ft (1,200 m).

The principal highland zone, the Brazilian Highlands, is an enormous block covering almost all of Brazil south of the Amazon Basin. It is tilted almost imperceptibly westward and northward so that rivers rising near its eastern rim, almost within sight of the Atlantic, flow inland for hundreds of miles before veering northward or southward to join larger streams.

The northern, western, and central parts of the Highlands are made up of broad, rolling terrain punctuated irregularly by low, rounded hills. Frequently these hills are formed into systems that are given range names but are not high enough to be considered mountains. The Planalto Central is a large plateau in this region; there are many smaller ones.

Further south the terrain becomes much more rugged. Basic elevations are no loftier than those to the north and west, but the crystalline rock and its cover of softer materials have been folded and eroded into a complex mass with ridges and ranges extending in all directions. Gradients are precipitous, and passage through them is

GEO-FACT

The Amazon's awesome tidal bore is called *pororoca*. It can reach heights of up to 12 ft (3.7 m), and sometimes travels as much as 500 mi (800 km) upstream. The river's immense silt-laden discharge is noticeable some 200 miles (320 km) out to sea.

tortuous and difficult. Roads frequently traverse many miles to reach destinations only short linear distances apart. In the far south and east of the Highlands the terrain rises to form true mountain ranges.

Mountains

The Serra do Mar parallels the coast in the southeast for 1,000 mi (1,609 km) from Santa Catarina to Rio de Janeiro and continues northward as the Serra dos Orgaos. This extended range has a mean crest of about 5,000 ft (1,524 m) topped by peaks above 7,000 ft (2,133 m) including Pedra Acu, which rises to 7,605 ft (2,318 m) just west of Rio de Janeiro. The Serra do Mar is so near the tidewater in many places that it rises almost directly from the shore. In others it recedes to leave a narrow littoral varying from twenty to forty miles in width. There are passes below 3,000 ft (914 m) only in two places, where rivers have cut their way through the coastal escarpment north of the city of Rio de Janeiro. The valleys of these streams, however, are blocked off from the interior plateau by a second range of mountains, the Serra da Mantiqueira. This range is the highest and most rugged of the Central Highlands; it includes the Pico da Bandeira, which at 9,495 ft (2,890 m) is the highest elevation in the Central Highlands and is frequently but incorrectly believed to be the highest in Brazil.

A third significant range of mountains runs from north to south behind the Serra da Mantiqueira. Appropriately named the Serra do Espinhaço, which means Spine Mountains, its range determines the drainage divide between the Sao Francisco River to the west and short streams that tumble eastward to the Atlantic. It is important because of the great wealth of minerals that it contains. Sometimes the Serra do Espinhaço and the Serra da Mantiqueira together with that range's southward extending spurs are referred to collectively as the Serra Geral.

The Guiana Highlands lie along Brazil's northern borders. They form part of an immense plateau extending into Venezuela, Guyana, Suriname, and French Guiana, and are much higher on average than the Central Highlands. The crests of its ranges constitute the divide between drainage northward to the Orinoco River in Venezuela and southward to the Amazon and define the national borders. With an elevation of 9,888 ft (3,013 m), the Pico da Neblina in the Imeri range is Brazil's highest mountain. Its location, immediately to the north of the equator in a zone of heavy rainfall, makes it the source of countless rivers and streams that descend in rushing falls and rapids to the Amazon. None of these watercourses is navigable very far upstream, and they contribute little to the development of the considerable mineral and forest wealth believed to exist near their headwaters.

INLAND WATERWAYS

The Amazon River

Brazil's river systems are among the world's most extensive, and the Amazon in particular is the world's mightiest river. Only the Nile River is longer, and no other three rivers of the world combined equal the flow of 80 million gallons of water per second that the Amazon discharges into the Atlantic.

The Amazon rises high in the Peruvian Andes and flows for a considerable distance before entering Brazil at the northwest corner of their border. From here until it receives its tributary the Negro, the Brazilians refer to the river as Solimões. During its 2,000 mi (3,218 km) course eastward across northern Brazil, it drops only about 215 ft (65 m) in elevation. It is navigable by oceangoing vessels as far as Iquitos in Peru. Manaus, in the middle of Brazil, is a major seaport. Smaller craft can reach Pôrto Velho, near the Bolivian frontier, on the tributary Madeira River. During most of its course the river is slow flowing, but at Monte Alegre about 400 mi (643 km) from its mouth it is constricted by hills to a width of about one mile, and the flow reaches six miles per hour.

Altogether there are more than 200 rivers and 500 smaller tributaries in the Amazon system, which drains about 59 percent of Brazil and 35 percent of all of South America. Approximately 20 percent of all the world's fresh water flows through the Amazon basin. Some of the largest tributaries are the Juruá, Purus, Madeira, Tapajós, Xingu, and Tocantins, which flow into the Amazon from the south. Major tributaries entering from the north are the Jari, Japurá, Negro, and Branco Rivers. All of these tributaries are major rivers in their own right; each of them carries more water than the Mississippi, for example. Slow-flowing like the Amazon in their lower courses, the tributary rivers meander intricately, and oxbow lakes and islands are numerous.

Other Rivers

Outside of the Amazon basin, Brazil's other major rivers arise in the Central Highlands and flow east or south. The São Francisco River is the longest contained entirely in Brazil at 1,988 mi (3,199 km) in length. It rises

near Belo Horizonte in eastern Brazil and flows northeastward along a line parallel to the coast for a great distance before turning eastward and flowing into the Atlantic. It drops 265 ft (80 m) at the spectacular Paulo Afonso Falls about 150 mi (241 km) from the coast, but is navigable for about 1,000 mi (1,609 km) in its middle reaches. The major Sobradinho Reservoir makes up part of its course.

Only two other major rivers of the Central Highlands cut through the escarpments of the Atlantic coastal ranges. The Doce River also begins near Belo Horizonte but flows more directly east to the ocean. The Parnaíba River has its source near the Atlantic but first flows northwest, away from the ocean, in a rift between two coastal mountain ranges, before turning sharply to the east and eventually entering the sea at Cape São Tomé. Its valley forms the best line of communication between Rio de Janeiro and São Paulo.

Although many other rivers of the Central Highlands originate close to the sea, the coastal mountains and the westward inclination of the plateau cause them to flow westward to join major streams of the Río de la Plata drainage basin in the southwest of the country. Of the three major rivers forming part of the Río de la Plata basin, the Paraná is the largest (3,030 mi / 4,870 km). It receives most of the tributary streams, and is formed by the confluence of the Parnaíba and the Rio Grande. It then flows south and forms parts of the borders with Paraguay. The Itaipu Dam and reservoir are located at this point, after which the Paraná flows southwest into Argentina on its way to the Río de la Plata estuary.

The magnificent Iguazú Falls, which eclipse Niagara Falls in magnitude, are located on the Iguazú River close to the point at which it joins the Paraná. The complex is three miles wide and 270 ft (82 m) high and consists of some 300 cataracts.

The second of the Río de la Plata basin rivers, the Uruguay, is fed by streams originating in the Serra do Mar in the far south of Brazil. It flows northwest away from the Atlantic in a broad arc, eventually turning southwest and forming much of the border with Argentina. The third major Brazilian river in this system is the Paraguay (Paraguai), which has its source in the Pantanal wetlands in west central Brazil. It flows southward along the border with Paraguay, eventually cutting across that country and into and Argentina, where it joins the Paraná.

Wetlands

The Pantanal is a lowland area, with an average elevation of 500 ft (152 m) above sea level, in west-central Brazil near the Bolivian border. The floor of this lowland is largely swamp and marshland; it is the largest system of wetlands in the world. The area is too wet to support forest, except for lightwoods on patches of higher ground. Away from its many streams, which make up the headwaters of the Paraguay River, sedimentary deposits have left a soil suitable for varied agriculture. There are smaller swampy areas, called *varzea*, scattered throughout the Amazon basin along the courses of the major rivers. These are often flooded during the rainy season.

THE COAST, ISLANDS, AND THE OCEAN

Major Islands

Countless islands are found throughout Brazil's river systems, the most noteworthy being Tupinambarama and Bananal Islands. Countless islands can be found in the huge delta of the Amazon, formed by alluvial deposits. By far the largest of the delta islands is Marajó, around which the Amazon splits into two principal channels. The southern outlet, also called the Pará River, is the smaller of the two, but it receives the Tocantins River and has the important port of Belém.

Outside of the Amazon delta, there are few coastal islands. The largest are Maracá, which lies north of the mouths of the Amazon, São Luis Island in São Marcos Bay, and São Sebastião and Santa Catarina Islands off the southeast coast. In addition, Fernando de Noronha, comprising Ilha Fernando de Noronha and the nearby Rocas Atoll, lies 250 mi (402 km) off the eastern bulge of the continent.

The Coast and Beaches

Brazil's entire coastline is on the Atlantic Ocean. Off the northeast seaboard, the waters of the continental shelf are extremely shallow, and the shoreline is rimmed by reefs and sandbars. The port of Belém is hemmed in by sandbars that prevent the entry of the largest vessels, and the ports of Salvador and Recife are flanked by reefs. The name of the latter city means reef.

Brazil's beaches are among the most famous in the world, including Copacabana and Ipanema, found near Rio de Janeiro on the Baia de Guanabara, immortalized in song and film. There are a number of excellent harbors. Besides those already mentioned, there is Vitória somewhat to the north of Rio de Janiero; Santos, the port of São Paulo, one of the greatest coffee ports in the world; and Pôrto Alegre in the south. In the northeast there is the Amazon delta, with the associated Marajó Bay. Further east is São Marcos Bay. The coastline then continues southeastward until it reaches Cape São Roque, at which point it turns south. It continues in this direction, with only gentle curves, until Cape São Tomé, with the exception of All Saints Bay. After Cape São Tomé it angels more to the southwest, with its most significant features being the Lagoa Dos Patos and Lagoa Mirim, two vast coastal lagoons at the extreme southern end of the country.

CLIMATE AND VEGETATION

Temperature

Brazil's geographical diversity makes for a range of climatic conditions, but the country is predominantly tropical, with the equator crossing through the northern part of the country. Brazil is so vast that the southernmost regions lie outside the tropics and have a temperate climate. May to September are the coolest months, and the higher elevations in the south may receive snow at this time. Further north in Rio de Janeiro, the average high temperature in February is 84°F (29°C) and the average low in July is 63°F (17°C). In contrast to the tropical humidity of the coast, the upland interior is relatively dry and moderate. It has been suggested that the moderate climate resulting from São Paulo's 2,500 ft (762 m) elevation has contributed substantially to that city's spectacular prosperity and growth.

Rainfall

Rainfall varies widely across the country. In the southern and central part of the country it generally ranges between 58 in to 78 in a year (150 cm to 200 cm), but it can be much higher in certain areas. Rainfall is heavier in the Amazon basin, reaching as much as 117 in (300 cm) annually. It is also more seasonal in this region, with different parts of the basin experiencing dry spells of three months or more each year. The northeast is the driest and hottest parts of the country, with lengthy droughts a regular occurrence.

Grasslands

Grasslands characterize much of southern and west-central Brazil. Between the Paraná River and the Serra do Mar Mountains is Brazil's famously fertile *terra roxa* (purple earth) area, where its most productive farmland is found. Further south the highlands dip downward, creating a somewhat less fertile pampa related topographically more to Uruguay than to Brazil.

The narrow strip of coastal plain that extends along the Atlantic seaboard from French Guiana to Uruguay constitutes another, highly developed, lowland. At one time much of this narrow plain was covered by tropical rainforest, but almost all of this has been cleared to make room for farms and cities. North of São Marcos Bay it is fairly wide and is more or less an extension of the Amazon basin. To the south it initially remains tropical, merging into subtropical. Beyond Cape São Roque it gradually narrows to a mere ribbon at the foot of the highland escarpment. In some places, particularly between the cities of Rio de Janeiro and Santos, it disappears entirely, and at no place in this region does it offer any large level areas except at the deltas of the Doce and Parnaíba Rivers. This northeastern coastal plain receives less rainfall than anywhere else in the country and experiences frequent droughts. As a result the plains are semi-arid by nature, although they remain fertile and have been extensively irrigated. In the far south the coastal plain broadens again, eventually merging into the rolling grasslands of the pampas that continue into Uruguay.

In the enormous Amazon basin the predominant form of ground cover is *selva* (rainforest) but there are some natural grasslands and savanna areas, particularly in the northwest. Large areas of the rainforest have been cleared for agricultural purposes. However the soil of the rainforest is generally not very fertile and once cleared will not support farming for long. As a result farmers are constantly pushing deeper into the forest, leaving their abandoned farmland behind them. An exception is the stretch of land south of the Amazon between the Xingu and Tapajós rivers, where the soil is similar to the *terra roxa* of the south and better able to support farming.

Forests and Jungles

The Amazon basin contains the world's largest tropical rain forest. It is a region of incredible diversity. Here are found Brazil nut trees, brazilwood, myriad palms, kapok-bearing ceiba trees, rosewood, orchids, the wild rubber tree, and numerous humid tropical forest species. The exact number of species in the forest is unknown, but about one-fourth of the world's known plant species are found in Brazil.

Population Centers – Brazil

(POPULATIONS FROM 2000 CENSUS)

Name	Population	Name	Population
São Paulo	10,057,700	Curitiba	1,642,300
Rio de Janeiro	6,029,300	Manaus	1,524,600
Salvador	2,539,500	Recife	1,464,100
Belo Horizonte	2,307,800	Pôrto Alegre	1,355,100
Fortaleza	2,230,800	Belém	1,344,900
Brasilía (capital)	2,089,500	Goiânia	1,132,600

SOURCE: Instituto Brasileiro de Geografia e Estatística (IBGE—Brazilian Institute of Geography and Statistics).

States – Brazil

POPULATIONS ESTIMATED IN 2002

Name	Population	Area (sq mi)	Area (sq km)	Capital
Acre	597,700	58,915	152,589	Rio Branco
Alagoas	2,914,400	10,707	27,731	Maceió
Amapá	530,100	54,161	140,276	Macapá
Amazonas	3,088,400	604,035	1,564,445	Manaus
Bahia	13,341,800	216,613	561,026	Salvador
Ceará	7,743,100	58,159	150,630	Fortaleza
Distrito Federal	2,043,169	2,245	5,814	Brasilía
Espírito Santo	3,254,200	17,605	45,597	Vitória
Goiás	5,254,200	247,913	642,092	Gioânia
Maranhão	5,859,600	126,897	328,663	São Luís
Mato Grosso	2,644,300	340,156	881,001	Cuiabá
Mato Grosso do Sul	2,155,800	135,347	350,548	Campo Grande
Minas Gerais	18,449,600	226,708	587,172	Belo Horizonte
Pará	6,548,800	482,906	1,250,722	Belém
Paraíba	3,504,100	21,765	56,372	João Pessoa
Paraná	9,855,700	77,048	199,554	Curitiba
Pernambuco	8,175,300	37,946	98,281	Recife
Piauí	2,926,600	96,886	250,934	Teresina
Rio Grande do Norte	2,883,600	20,469	53,015	Natal
Rio Grande do Sul	10,468,100	108,952	282,184	Pôrto Alegre
Rio de Janeiro	14,920,000	17,092	44,268	Rio de Janeiro
Rondônia	1,462,700	93,840	243,044	Pôrto Velho
Roraima	363,900	88,844	230,104	Boa Vista
Santa Catarina	5,582,700	37,060	95,985	Florianópolis
São Paulo	38,505,500	95,714	247,898	São Paulo
Sergipe	1,866,200	8,492	21,994	Aracaju
Tocantins	1,212,300	10,749	278,420	Palmas

SOURCE: Instituto Brasileiro de Geografia e Estatística (IBGE—Brazilian Institute of Geography and Statistics).

South of the Amazonian forest is a mixture of semi-deciduous and scrub trees and shrubs (mata), which opens to grasses in the south and to plantations along the coast. Southern Brazil is known for exotic flowering trees such as the ipê tree with yellow petals lining the streets of Sao Paulo. A mixture of pine and broadleaf species is found in the temperate south.

Commercial farming, ranching, logging, mining, roads, railways, and other development account for the clearing and loss of nearly 50 million acres of forest a year. Scientists predict a negative impact on climate and global warming. For Brazil, deforestation and erosion pose significant threats to forest-based livelihoods, agriculture, and ultimately to reforestation once valuable topsoil is lost.

HUMAN POPULATION

Brazil is the most populous country in South America, and has the fifth highest population in the world. The people are diverse in origin, and the amalgam of African, European, and indigenous strains has led some to speak of a "Brazilian race." Brazil's origins as a country can be traced to Portuguese colonies along its coast, and Portuguese is the official language.

Brazil's population is concentrated along the coast and in the areas just to the west of the coastal mountain ranges. In 2000 approximately 81 percent of the population was urban. The southeast is the most densely populated region. Brazil's two huge metropolitan areas of São Paolo and Rio de Janiero, both with populations well over 10 million, can be found here. The population is less dense further to the south, but there are still many large cities. At least 50 percent of the population is in the south and southeast, collectively.

The northeast is another area of dense settlement, dating back to the earliest colonies. About a quarter of the people can be found here, and there are many cities and farms, but its population is in decline relative to the rest of the country. The vast interior of Brazil is home to as much as a third of the population, but is sparsely populated relative to its size. Some large cities can be found, particularly on the Amazon. The area north of the Amazon is the least settled part of the country. Most of the estimated 150,000 indigenous peoples (chiefly of Tupí or Guaraní linguistic stock) are found in the rain forests of the Amazon River basin.

NATURAL RESOURCES

Brazil has vast stores of natural resources including: bauxite, iron ore, manganese, nickel, phosphates, platinum, tin, uranium, and petroleum. Hydropower, timber, and fresh water are also extremely plentiful. There are extensive coal fields in the south, and gold can be found in the north and northeast. The Amazon region produces timber, rubber, and other forest products such as Brazil nuts and pharmaceutical plants. Brazil's many rivers give it abundant reserves of fresh water and hydropower; its farms produce a wide variety of tropical and temperate crops. In addition to these abundant resources, Brazil's cities are major centers for manufacturing and commerce.

FURTHER READINGS

"Amazonia: A World Resource at Risk." *National Geographic Magazine*, August 1992.

Anderson, Anthony B., ed. *Alternatives to Deforestation: Steps Toward Sustainable Use of the Amazon Rain Forest.* New York: Columbia University Press, 1990.

Burns, E. Bradford. *A History of Brazil.* 3rd edition. New York: Columbia University Press, 1993.

Malingreau, J. P., and C. J. Tucker. "Large scale deforestation in the southeastern Amazon basin of Brazil." *Ambio* 17, pp. 49-55, 1988.

Moran, Emílio. *Through Amazonian Eyes: The Human Ecology of Amazonian Populations.* Iowa City: University of Iowa Press, 1993.

Poor, D. *Ecological Guidelines for Development in Tropical Rainforests.* Geneva: IUCN, 1976.

Skole, D.L., et al. "Physical and human dimensions of deforestation in Amazonia." *Bioscience*, 44, pp. 314-22, 1994.

World Resources Institute and World Bank. *Tropical Forests: A Call for Action.* Washington D.C.: World Resources Institute and World Bank, 1985.

British Virgin Islands

Dependency of the United Kingdom

- **Area:** 58 sq mi (150 sq km) / World Rank: 200
- **Location:** Northern and Western Hemispheres, between the North Atlantic Ocean and the Caribbean Sea, east of Puerto Rico, at the north end of the Leeward Islands
- **Borders:** No international boundaries
- **Coordinates:** 18°25′N 64°30′W
- **Coastline:** 50 mi (80 km)
- **Territorial Seas:** 3 NM (5.5 km), with 200 NM (364 km) exclusive fishing zone
- **Highest Point:** Mount Sage, 1,709 ft (521 m)
- **Lowest Point:** Sea level
- **Longest River:** None
- **Largest Lake:** None
- **Natural Hazards:** Subject to seasonal hurricanes and tropical storms.
- **Population:** 20,812 (2001 est.) / World Rank: 200
- **Capital City:** Road Town, on the central southern coast of the island of Tortola.
- **Largest City:** Road Town, 2,500 (1992 est.)

GEO-FACT

The Rhone was a 310-foot Royal Mail Ship that was dashed against the rocks off the British Virgin Islands' Salt Island's southwest coast during a hurricane in 1867. Its remains are extensive and have become a fascinating underwater habitat for marine life. It is part of the national park system and is a popular dive site.

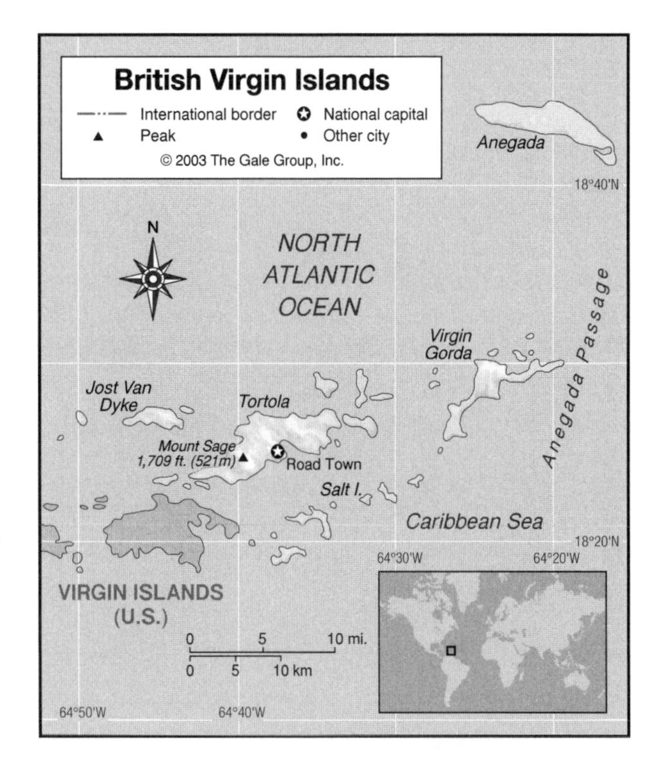

OVERVIEW

Consisting of some 40 islands and 20 other islets, of which only 15 are inhabited, the British Virgin Islands are the northern half of the Virgin Island chain, the southern portion of which is the U.S. Virgin Islands. The islands are comprised of flat coral reefs; steeper, hillier volcanic landscapes; and beaches.

MOUNTAINS AND HILLS

Mountains

Although the coral islands are relatively flat, there are also volcanic islands in the group that are steep and hilly. None of the volcanoes have been active recently. At only 1,709 ft (521 m), Mount Sage, on Tortola, is the highest point on the islands. The island of Anegada is especially flat, the entire island being no more than a few feet above sea level.

INLAND WATERWAYS

There are no large bodies of fresh water on the islands.

THE COAST, ISLANDS, AND THE OCEAN

The principal islands of this British dependency are Tortola, Virgin Gorda, Anegada, and Jost Van Dyke. The beaches of the British Virgin Islands are among their most valuable assets, bringing many tourists to the islands. Many of the British Virgin Islands rest on and are surrounded by coral reefs. The Anegada Passage lies between the British Virgin Islands and Anguilla and St. Martin to the east.

CLIMATE AND VEGETATION

Temperature

The British Virgin Islands have a sub-tropical, humid climate moderated by trade winds with summer temperatures varying from 79°F (26°C) to 89°F (31°C), and winter temperatures ranging from 72°F (22°C) to 82°F (28°C).

Rainfall

Located in an area of seasonable hurricanes and tropical storms—the last major storm being Lenny in November of 1999—rainfall varies for this island group. Over the entire chain rainfall averages 39–40 in (94–96 mm) per year. In 1998, the islands experienced a record rainfall of about 60 in (1,524 mm).

HUMAN POPULATION

The July 2001 estimate was 20,812, indicating a population growth rate of 2.22 percent. The primary ethnic group is black (87 percent) with whites and Asians making up the difference. Most Islanders are Christians (86 percent Protestant).

NATURAL RESOURCES

Natural resources are negligible; the beaches attract vacationers, but yield little else economically.

FURTHER READINGS

Connett, Eugene V., ed. *Virgin Islands.* Princeton, NJ: VanNostrand, 1959.

Maurer, Bill. *Recharting the Caribbean: Land, Law, and Citizenship in the British Virgin Islands.* Ann Arbor, MI: University of Michigan Press, 1997.

White, Robb. *In Privateer's Bay.* New York: Harper & Brothers, 1939.

Brunei

- **Area:** 2,228 sq mi (5,770 sq km) / World Rank: 166
- **Location:** Northern and Eastern Hemispheres, in the northwest of the island of Borneo, in Southeast Asia
- **Coordinates:** 4°30′N, 114°40′E
- **Borders:** 237 mi (381 km)
- **Coastline:** 100 mi (160 km)
- **Territorial Seas:** 12 NM
- **Highest Point:** Mt. Pagon, 6,070 ft (1,850 m)
- **Lowest Point:** Sea level
- **Longest River:** Belait River, 130 mi (209 km)
- **Largest Lake:** Tasek Merimbun, 0.4 sq mi (1.2 sq km)
- **Natural Hazards:** Earthquakes and typhoons (rare)
- **Population:** 343,653 (2001 est.) / World Rank: 169
- **Capital City:** Bandar Seri Begawan, located in the north along an inlet of the Brunei River
- **Largest City:** Bandar Seri Begawan (75,000, 2001 est.)

OVERVIEW

The small country of Brunei is an enclave on the northern coast of the island of Borneo, which it shares with the Malaysian state of Sarawak, and with Indonesia. The country consists of distinct eastern and western segments, separated by Malaysia's Limbang River valley, but linked by the waters of Brunei Bay. The terrain in both the eastern segment (the Temburong District) and the more populated western segment is composed of coastal plain rising gradually to hills and cut through by rivers running north to the sea.

MOUNTAIN AND HILLS

Hills and Badlands

In the west of Brunei, hills lower than 295 ft (90 m) rise towards an escarpment and higher hills approaching the Sarawak border. Brunei's highest peak, Mt. Pagon (6,070 ft / 1,850 m) is located in this region. Brunei's eastern sector is also covered with low hills, which gain height close to the border with Sarawak.

INLAND WATERWAYS

Lakes

There are a few lakes in Brunei. In Tutong District, the unusual S-shaped Tasek Merimbun is surrounded by a 30 sq mi (77 sq km) nature park. Wong Kadir and Teraja Lakes are in Belait District.

Rivers

Four indigenous river systems and one originating in the Malaysian state of Sarawak flow north through and between the segments of Brunei to the South China Sea. The Belait River, Brunei's longest, flows through western Brunei, as does the Tutong River. The Brunei River runs southwest from an inlet of Brunei Bay (where Bandar Seri Begawan is located). In the eastern segment of Brunei, the Temburong River provides drainage for the entire Temburong District. The Limbang River valley, which belongs to Malaysian Sarawak, splits Brunei in two.

Wetlands

The mangrove forests of Brunei's estuaries are an ecological treasure, considered among the most intact in Southeast Asia. Mangrove forests take up an estimated 3.2 percent of Brunei's land. Brunei's ecologically intact peat swamps (rare in the north of Borneo) are found in western Brunei, along sections of the Belait River and the Tutong River.

THE COAST, ISLANDS, AND THE OCEAN

Oceans and Seas

The Sultanate of Brunei originally arose as a trading state strategically located on shipping lanes linking the trade routes of the Indian Ocean and Pacific Ocean through the South China Sea. The immensely valuable hydrocarbon deposits that have produced Brunei's petroleum export boom lie mainly under the South China Sea off Brunei's coast, and along the coastline itself in Belait District.

Major Islands

Brunei has 33 islands, comprising 1.4 percent of its land area. Two islands are in the South China Sea. The others are river islands, or like Pulau Muara Besar, are in Brunei Bay. The islands are important wildlife habitats for species including proboscis monkeys, and are mostly uninhabited by humans, although there is some recreational usage. Brunei has demarcated a fishing zone in an area of the Spratly Islands (which are contested by the Philippines, Malaysia, China, Taiwan and Vietnam) but has not made a formal claim on the Spratly Islands.

The Coast and Beaches

The western section of Brunei has a coastline on the South China Sea, where sandbars lie between estuaries

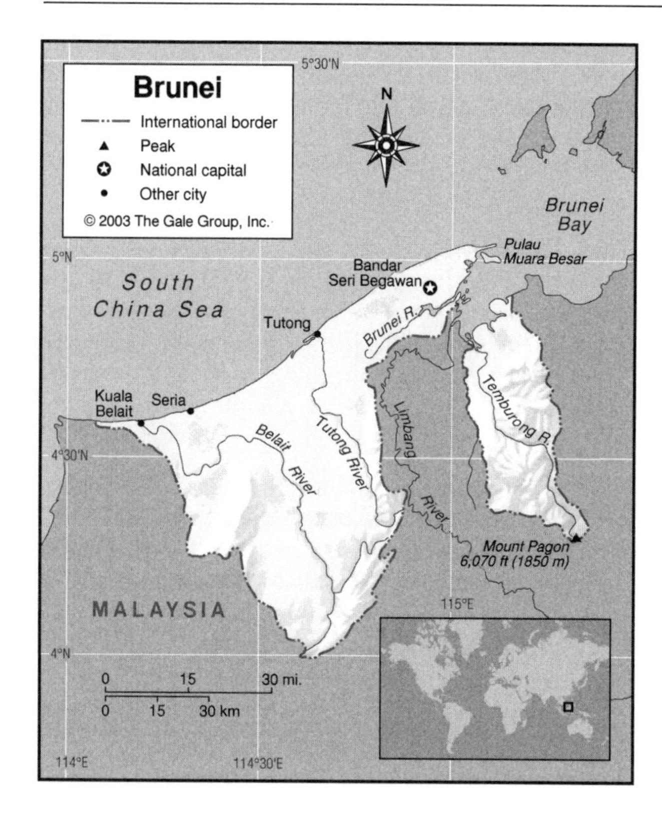

and the open ocean. Belait, Tutong, and Brunei Districts have three river estuaries and significant mangrove forests. In Temburong District (the east) the steep muddy banks of Brunei Bay and its inlets form a major wildlife habitat.

CLIMATE AND VEGETATION

Temperature

The temperature of Brunei, a tropical country, averages from 73° to 89°F (23°C to 32°C) year-round. Humidity stays around 80 percent.

Rainfall

The northeast monsoon affects Brunei with heavy rains in November and December. On Brunei's coast the annual rainfall averages around 110 in (275 cm) while inland rainfall amounts to 200 in (500 cm) or more. Brunei is out of the path of most ocean storms such as typhoons, although it can be affected by tidal surges.

Forests and Jungles

Much of Brunei is still covered by exceptionally biodiverse dipterocarpaceous rainforest, in a mixture of primary and secondary growth. Almost all of the interior of Brunei is forested, roughly 75 percent of the entire country. Logging is strictly limited, in contrast to the rampant deforestation inflicted on the rest of Borneo in the last two decades. Only 130,795 cu yd (100,000 cu m) of timber is allowed to be cut each year, for Brunei's domestic use only.

Districts – Brunei

Name	Area (sq mi)	Area (sq km)	Capital
Belait	1,053	2,727	Kuala Belait
Brunei and Muara	220	570	Bandar Seri Begawan
Temburong	503	1,303	Bangar
Tutong	450	1,165	Tutong

SOURCE: *Geo-Data: The World Geographical Encyclopedia,* 2nd ed. Detroit: Gale Research, 1989.

HUMAN POPULATION

Brunei is sparsely populated, with an estimated density of just 53 people per sq mi (20 per sq km) in 1997. Some 85 percent of the country's population lives in the coastal areas, particularly the capital city Bandar Seri Begawan and the coastal towns of Seria, Tutong, and Kuala Belait. Seria and Kuala Belait are centers for the oil and natural gas industry. There is some migration from neighboring countries, and Brunei hosts tens of thousands of foreign workers, most employed in the petroleum sector.

NATURAL RESOURCES

Brunei has been economically dependent on vast hydrocarbon reserves along its coastline, producing oil, natural gas, and liquified natural gas. These petroleum deposits are expected to be depleted by the second or third decade of the twenty-first century. Agriculture is a minor part of Brunei's economic picture (most food is imported) and there are no other significant economic resources. Brunei's forests have been protected from economic exploitation but are a resource for scientific study and eco-tourism.

FURTHER READINGS

Cleary, Mark. *Oil, Economic Development and Diversification in Brunei Darussalam.* New York: Palgrave, 1994.

Edwards, David S. *A Tropical Rainforest: The Nature of Biodiversity in Borneo at Belalong, Brunei.* Torrance, CA: Heian International, 1995.

Pelton, Robert Young. *Fielding's Borneo.* Redondo Beach, CA: Fielding Worldwide, 1995.

Thia-Eng, Chua. *Brunei Darussalam: Coastal Environmental Profile of Brunei Darussalam.* Washington, DC: U.S. Agency for International Development, 1987.

Bulgaria

- **Area:** 42,811 sq mi (110,910 sq km) / World Rank: 104
- **Location:** Northern and Eastern Hemispheres, part of the Balkan Peninsula in southern Europe, south of Romania, west of the Black Sea, northwest of Turkey, north of Greece, and east of The Former Yugoslav Republic of Macedonia and Yugoslavia.
- **Coordinates:** 43° 00′N and 25° 00′E.
- **Borders:** 1,343 mi (1,808 km) / The Former Yugoslav Republic of Macedonia, 92mi (148 km); Greece, 307 mi (494 km); Romania, 378 mi (608 km); Turkey, 149 mi (240 km); Yugoslavia, 197 mi (318 km).
- **Coastline:** 214 mi (354 km)
- **Territorial Seas:** 12 NM
- **Highest Point:** Musala, 9,596 ft (2,925 m)
- **Lowest Point:** Sea level
- **Longest Distances:** 205 mi (330 km) N-S; 323 mi (520 km) E-W
- **Longest River:** Danube, 1,770 mi (2,850 km).
- **Largest Lake:** Popovo Lake, 30.7 acres (12.4 ha)
- **Natural Hazards:** Earthquakes and landslides
- **Population:** 7,707,495 (July 2001 estimate) / World Rank: 87
- **Capital City:** Sofia, located in the west central part of the country.
- **Largest City:** Sofia, 1,188,000 (2000 est.)

OVERVIEW

Bulgaria occupies a relatively small area, but is nevertheless a land of unusual scenic beauty, having picturesque mountains, wooded hills, sheltered valleys, grain-producing plains, and a seacoast along the Black Sea that has both rocky cliffs and long sandy beaches.

In the north of the country is the Danubian Plain, the central portion houses the Balkan Mountains, and south of them is the Maritsa River. The Rhodope Mountains are found in the south and southwest of the country. Located on the Eurasian Tectonic Plate, Bulgaria is crossed by fault lines and earthquakes are not infrequent.

MOUNTAINS AND HILLS

Mountains

The Balkan Mountains (Stara Planina) comprise the biggest and longest mountain chain. An extension of the Carpathian Mountains, the Balkans stretch 435 mi (700 km) across the central portion of the entire country, declining in altitude towards the east. The range's highest peak is Botev at 7,793 ft (2,376 m). Just to the south of the central part of this range are the Sredna Mountains (Sredna Gora), a 100 mi (160 km) long ridge that runs almost directly from east to west, at an average height of 5,249 ft (1,600 m).

The other major mountain range is the Rhodope. These mountains mark the southern and southwestern borders of Bulgaria, and include the Vitosha, Rila, and Pirin Mountains. These last two ranges are largely volcanic in origin, and are the highest mountains on the Balkan Peninsula. Musala in the Rila Mountains is the tallest peak in the country (9,596 ft / 2,925 m).

Plateaus

The Danubian Plain extends from the Yugoslav border to the Black Sea. The plateau rises from cliffs along the Danube and extends south to the Balkan Mountains at elevations as high as 1,500 ft (457 m). On the southern side of the Balkan Mountains is another plateau, the Thracian Plain, which is drained by the Maritsa River. Both plateaus are fertile regions of hills and plains, gradually declining in elevation as they approach the Black Sea.

Canyons and Caves

A long geological trench that contains the Valley of Roses lies between the Balkan and Sredna Mountains. The north-flowing rivers have cut deep valleys through the Balkan Mountains and the Danubian Plain. More than 2,000 caves are scattered amidst the limestone layers of the Pirin and the Balkan Mountains.

INLAND WATERWAYS

Lakes

Glacial lakes are numerous in Bulgaria. There are about 280 of them, located in the higher zones of the Rila and Pirin Mountains. Most are located at altitudes of 7,216 to 7,872 ft (2,200 to 2,400 m); the lake lying at the highest elevation—Ledenika Lake in the Rila Mountains—lies at an altitude of 8,905 ft (2,715 m). Located in the Pirin Mountains, Popovo Lake, also known as the "Pirin Sea," is the largest lake in the country. It covers an area of 30.7 acres (12.4 hectares) and is 1,575 ft (480 m) long and 1,102 ft (336 m) wide.

Rivers

The Danube (Dunav), which forms the majority of Bulgaria's border with Romania, is by far the largest river in the country and is the second longest in Europe. It is navigable by ocean vessels throughout Bulgaria. Most of the northern part of the country drains into the Black Sea via the Danube and its tributaries. All but one of these tributaries rise in the Balkan Mountains, including the Yantra and the Osŭm. The one exception is the Iskŭr, which rises in the Rila Mountains and flows northward, passing through Sofia's eastern suburbs, and then cuts a valley through the Balkan Mountains.

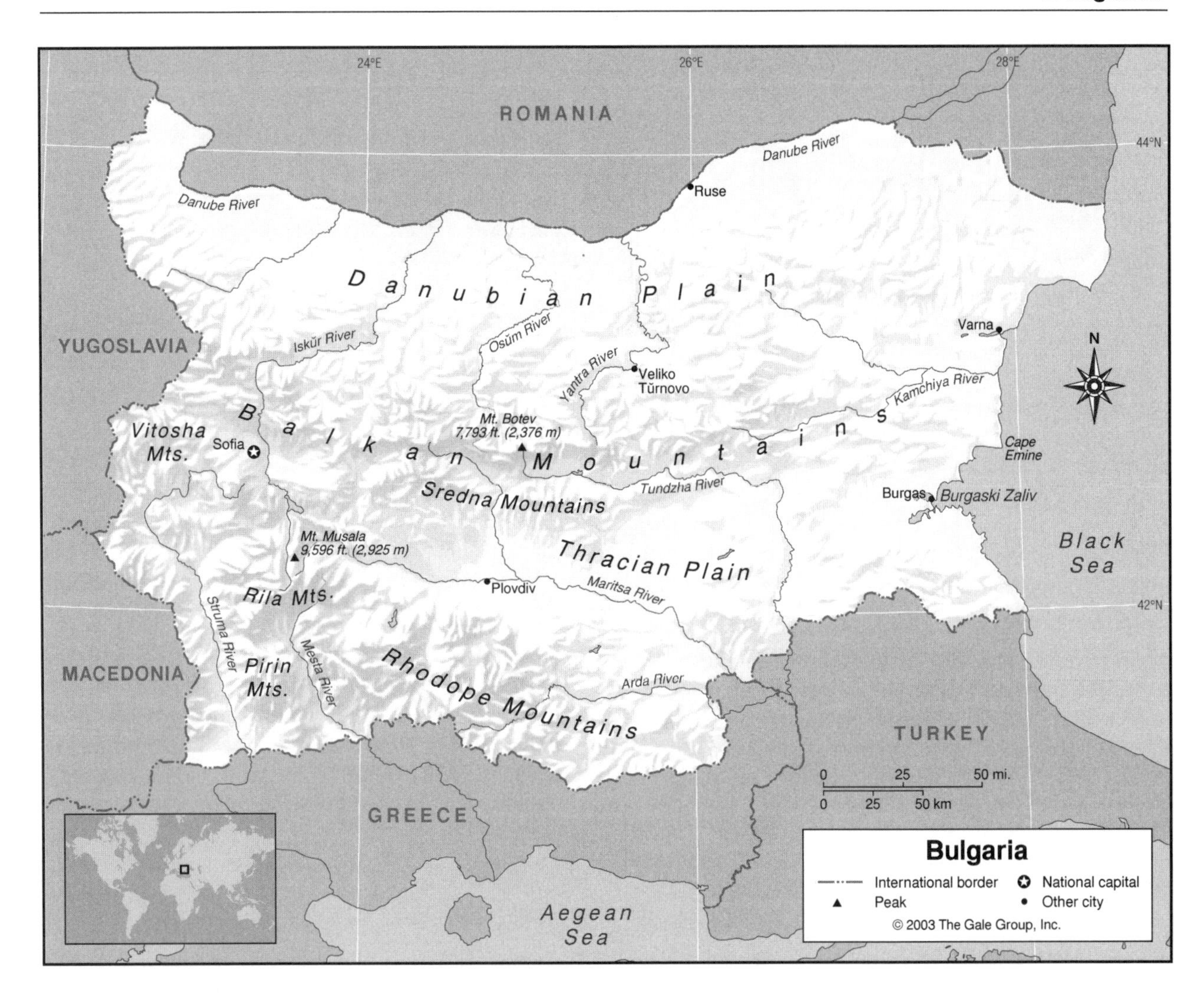

South of the Balkan Mountains most rivers flow south into the Aegean Sea. Most notable among these are the Mesta, the Struma, the Maritsa, and the Maritsa's tributaries the Tundzha and Arda. Together, they provide drainage for most of the Thracian Plain. The Kamchiya in the northeast is the only large river to flow directly into the Black Sea.

THE COAST, ISLANDS, AND THE OCEAN

Oceans and Seas

Located east of Bulgaria, the Black Sea contains calm waters that are free of tides and dangerous marine life. Called the "Hospitable Sea" by the ancient Greeks, the Black Sea is half as saline as the Mediterranean Sea and has gentle sandy slopes, making it ideal for swimming.

The Coast and Beaches

Bulgaria's eastern coast on the Black Sea is curved, providing for many beaches along the 214 mi (354 km) of shoreline. Many of the country's beaches have been awarded for their environmental excellence by the European Union. The coastline is a varied, with coves, rugged shores, wooded hills, orchards, and fishing villages dotting the expansive area. Burgaski Zaliv indents the coast deeply in the south, with Cape Emine to the north.

CLIMATE AND VEGETATION

Temperature

Lying along the southern margins of the continental climate of Central and Eastern Europe, Bulgaria experiences cold winter winds from the north in the Danubian Plain and a modified Mediterranean climate in the Thacian Plain because of the protection offered by the Balkan Mountains. January temperatures average 32–36°F (0–2°C) in the north, and colder in the mountainous regions; July temperatures range from 72 to 75°F (22 to 24°C). Overall, Bulgaria's climate is temperate with cold, damp winters and hot, dry summers.

Rainfall

Rainfall is generally light in the plateaus, averaging about 25 in (65 cm) per year, and higher in the mountain

Population Centers – Bulgaria

(MARCH 2001 POPULATION ESTIMATES)

Name	Population	Name	Population
Sofia (capital)	1,096,389	Ruse (Rousse)	162,128
Plovdiv	340,638	Stara Zaeora	143,989
Varna	314,539	Pleven	122,149
Burgas (Bourgas)	193,316	Sliven	100,695

SOURCE: National Statistical Institute, Bulgaria.

ranges, where it can reach up to 60 in (152 cm). Most rainfall occurs during the winter months.

Grasslands

The lower peaks of the Pirin and Rila Mountain ranges are covered in Alpine meadows. The Thracian Plain and Danubian Plain have great varieties of vegetation. They are both densely populated and cultivated.

Forests

More than one third of the country's territory (38 percent) is forested. The densest forests are in the mountainous regions. The Balkans and Rhodopes are covered by broadleaf forests in the low areas and conifers in the higher elevations. Broadleaf forests are the predominant forest type throughout the country.

HUMAN POPULATION

Bulgaria's population was 7,707,495 in 2001 (estimated). Almost 75 percent lived in urban areas. The Danubian Plain, Thracian Plain, Black Sea coastline, and the area around Sofia, the capital, were the most densely populated.

NATURAL RESOURCES

Arable land comprises 83 percent of the total land area, making it one of Bulgaria's most valuable natural resources. Mining of bauxite, copper, lead, zinc, and coal meets domestic needs, but the future of the minerals industry is uncertain because of a decline in production. Covering more than 9 million acres (3.7 million hectares), timber harvesting and reforestation is difficult because more than half of the forests are on mountain slopes. Exploitation and neglect have also contributed to the deterioration of this natural resource.

FURTHER READINGS

Bell, John D., ed. *Bulgaria in Transition: Politics, Economic, Society, and Culture after Communism.* Boulder, Colo.: Westview Press, 1998.

Cary, William. *Bulgaria Today: The Land and the People, a Voyage of Discovery.* New York: Exposition Press, 1965.

Detrez, Raymond. *Historical Dictionary of Bulgaria.* Lanham, Md.: Scarecrow Press, 1997.

Hoddinott, Ralph F. *Bulgaria in Antiquity: An Archaeological Introduction.* New York: St. Martin's Press, 1975.

McIntyre, Robert J. *Bulgaria: Politics, Economics, and Society.* New York: Pinter Publishers, 1988.

Pettifer, James. *Bulgaria.* New York: W.W. Norton, 1998.

Burkina Faso

- **Area:** 105,869 sq mi (274,200 sq km) / World Rank: 74
- **Location:** Western Africa, in Northern Hemisphere, and split between the Eastern and Western Hemispheres, west of Niger, northwest of Benin, north of Mali, Togo, Ghana, and Côte d'Ivoire, and on the east and south of Mali
- **Coordinates:** 13°00′N, 2°00′W
- **Borders:** 1,983 mi (3,192 km) / Niger, 389 mi (628 km); Benin, 190 mi (306 km); Mali, 621 mi (1,000 km); Togo, 78 mi (126 km); Ghana, 340 mi (548 km); Ivory Coast 636 mi (584 km)
- **Coastline:** Burkina Faso is landlocked.
- **Territorial Seas:** None
- **Highest Point:** Tena Kourou, 2,451 ft (747 m)
- **Lowest Point:** Black Volta River Valley, 650 ft (198 m)
- **Longest Distances:** 542 mi (873 km) from ENE-WSW / 295 mi (474 km) SSE-NNW
- **Longest River:** Black Volta, 1,000 mi (1610 km)
- **Largest Lake:** None
- **Natural Hazards:** Subject to recurring droughts
- **Population:** 12,272,289 (July 2001 est.) / World Rank: 65
- **Capital City:** Ouagadougou, located in the center of the country
- **Largest City:** Ouagadougou, 1,131,000 (2000 est.)

OVERVIEW

Located in West Africa, Burkina Faso (formerly Upper Volta) is a single, vast plateau that is carved into three valleys by the Black, White, and Red Volta rivers, and their main tributary, the Sourou. The rivers are either in flood or dry, making the terrain of this savanna arid and poor. This wild bush country does have a mixture of grasslands and small trees, which vary in degree from the dry (harmattan) to rainy seasons.

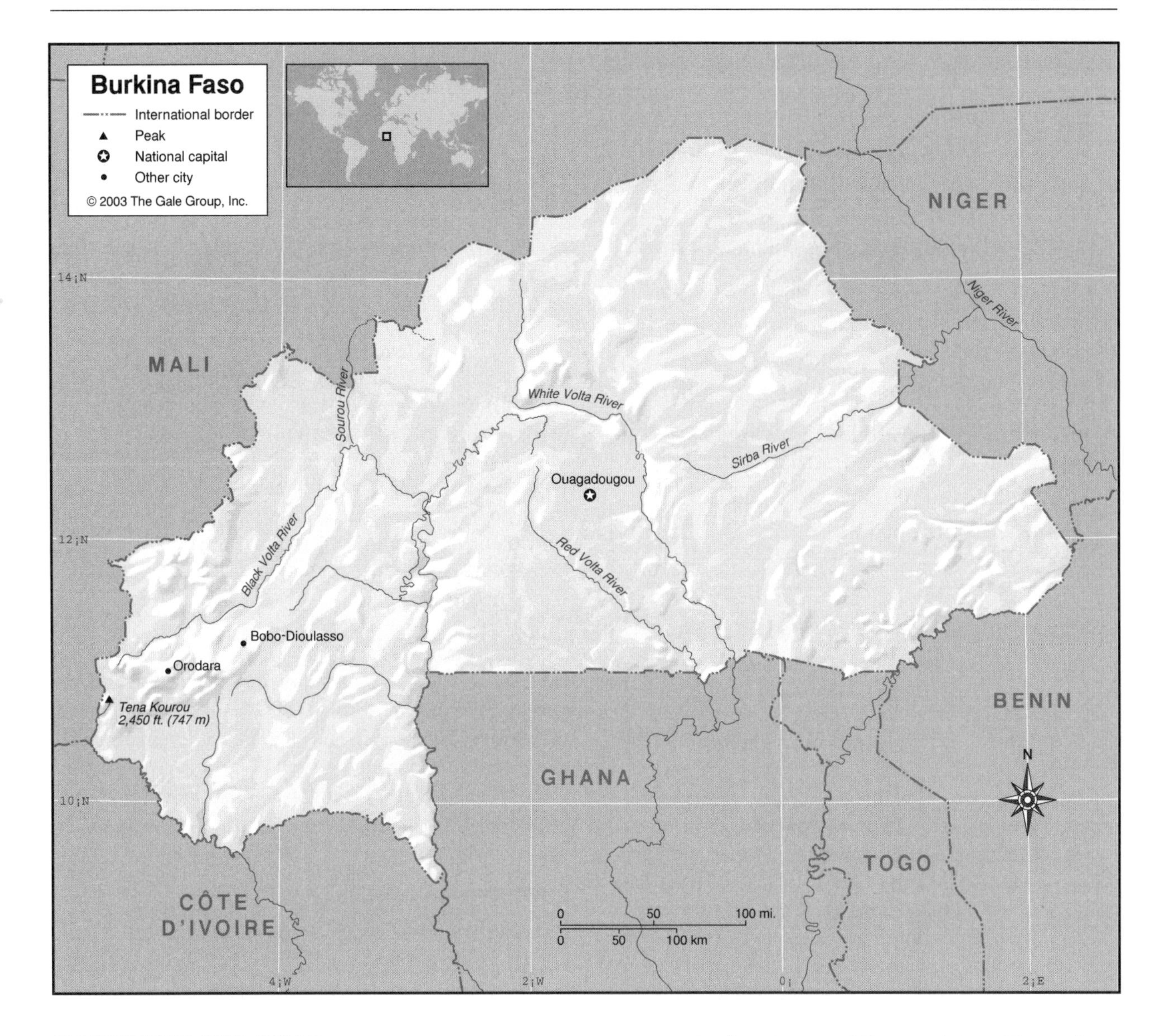

MOUNTAINS AND HILLS

Plateaus

The country consists, for the most part, of a vast plateau in the West African savanna, approximately 650-1000 ft (198-305 m) above sea level. This plateau is slightly inclined toward the south. There are some hills in the west and southeast, but no mountains of significance. The highest elevation is Tena Kourou at 2,451 ft (747 m), near the Mali border, east of Orodara.

Canyons

The plateau of Burkina Faso is notched by shallow valleys formed by the three Volta Rivers, and their tributaries.

INLAND WATERWAYS

Rivers

Burkina Faso has three principal rivers: the Black, White, and Red Voltas. All three rivers have their headwaters in Burkina Faso, and flow south into Ghana where they eventually meet to form the Volta River. Their only major tributary within the country is the Sourou. The Sirba River flows east into Niger. These rivers are alternately nearly dry or in flood and none are navigable. Other temporary rivers may form during the rainy season.

THE COAST, ISLANDS, AND THE OCEAN

A completely landlocked nation, Burkina Faso has no coastland and no island dependencies.

CLIMATE AND VEGETATION

Temperature

Located near the equator, high temperatures are typical in Burkina Faso, especially during the dry season. From March to May, the harmattan, a dry east wind, contributes to considerably hot temperatures that range from

GEO-FACT

Ruled by the Mossi from the eleventh to the nineteenth centuries, Burkina Faso accepted French domination late in the nineteenth century to protect itself from hostile incursions from its neighbors. In 1960 the territory received its independence as the Volta Democratic Union. It was in 1984 that the present name of Burkina Faso was given to the nation by its then leader Captain Thomas Sankara. The name may be loosely translated as "Land of Upright Men" or "the land (or house) of the incorruptible people."

104°–119°F (40°–48°C). From May to October, the weather is hot, but wet, and from November to March, more comfortable. January temperatures vary from 44°–55°F (7° 13°C).

Rainfall

Average annual rainfall varies from 45 inches (115 cm) in the southwest to a low of 10 inches (25 cm) in the extreme north and northeast portion of the country. The rainy season, only four months in the north, lasts upwards of six months (from May through October) in the southwest. Humidity increases as one moves from the north to the south, ranging from a winter low of 12 percent to 45 percent, to a rainy season high of 68 percent to 96 percent. The country suffers from recurring droughts.

Vegetation

Burkina Faso is essentially a large savanna. This savanna is primarily grassland during the rainy season, dotted with small forests. This terrain supports what is probably the widest variety of animal life on the continent. During the hot and dry harmattan season, the savannah becomes a semi-desert. Burkina Faso's primitive forestland has been all but decimated in order to provide farmland and fuel resources. No real program of reforestation was begun until 1973. About 50 percent of the total land area has been considered forest or woodland. In the years 1990 through 1995, deforestation proceeded at a rate of 0.7 percent per year.

HUMAN POPULATION

A total population of 12,272,289 (July 2001 estimate) is directly affected by the excess mortality rates because of the AIDS epidemic, which lowers life expectancy, causes high infant mortality and death rates, causes lower growth rates, and changes the distribution of population by age and sex. Ethnic groups in Burkina Faso are represented by Mossi (over 40 percent), Gurunsi, Senufo, Lobi, Bobo, Mande, and Fulani. Religions represented are indigenous beliefs, 40 percent; Muslim, 50 percent; and Christian (mainly Roman Catholic), 10 percent. Population density is .88 per sq mi (2.3 per sq km).

Population Centers – Burkina Faso

(1991 POPULATION ESTIMATES)

Name	Population
Ouagadougou (capital)	634,479
Bobo-Dioulasso	268,926

SOURCE: Population of Capital Cities and cities of 100,000 and More Inhabitants." United Nations Statistics Division.

NATURAL RESOURCES

Burkina Faso is a poor country, lacking much mineral resources, water, or fertile soil. There are deposits of manganese, limestone, and marble, as well as small deposits of precious metals such as gold. A large portion of the male population migrates to surrounding countries for seasonal employment.

FURTHER READINGS

Anderson, Samantha, translator. *Thomas Sankara Speaks: The Burkina Faso Revolution, 1983-87*. New York: Pathfinder Press, 1988.

Baxter, Joan, and Keith Sommerville. *Burkina Faso*. New York: Pinter Publishers, 1989.

Englebert, Pierre. *Burkina Faso: Unsteady Statehood in West Africa*. Boulder CO: Westview Press, 1996.

McFarland, Daniel Miles. *Historical Dictionary of Upper Volta*. Metuchen, NJ: Scarecrow Press, 1978.

Shepard, Steve. *Elvis Hornbill, International Business Bird*. New York: Holt, 1991.

Burundi

- **Area:** 10,745 sq mi (27,830 sq km) / World Rank: 145
- **Location:** Southern and Eastern Hemispheres in east-central Africa, between the Democratic Republic of the Congo, Rwanda, and Tanzania
- **Coordinates:** 3°30′S, 30°00′E

- **Borders:** 605 mi (974 km) / Rwanda, 180 mi (290 km); Tanzania, 280 mi (451 km); Democratic Republic of the Congo, 145 mi (233 km)
- **Coastline:** None
- **Territorial Seas:** None
- **Highest Point:** Mt. Heha, 8,760 ft (2,670 m)
- **Lowest Point:** Lake Tanganyika, 2,533 ft (772 m)
- **Longest Distances:** 163 mi (263 km) NNE-SSW / 121 mi (194 km) ESE-WNW
- **Longest River:** Muragarazi River, 348 mi (560 km)
- **Largest Lake:** Lake Tanganyika, 12,700 sq mi (33,020 sq km)
- **Natural Hazards:** Subject to periodic flooding accompanied by landslides, and drought
- **Population:** 6,223,897 (July 2001 est.) / World Rank: 98
- **Capital City:** Bujumbura, located on the northeast coast of Lake Tanganyika
- **Largest City:** Bujumbura, 278,000 (2000 est.)

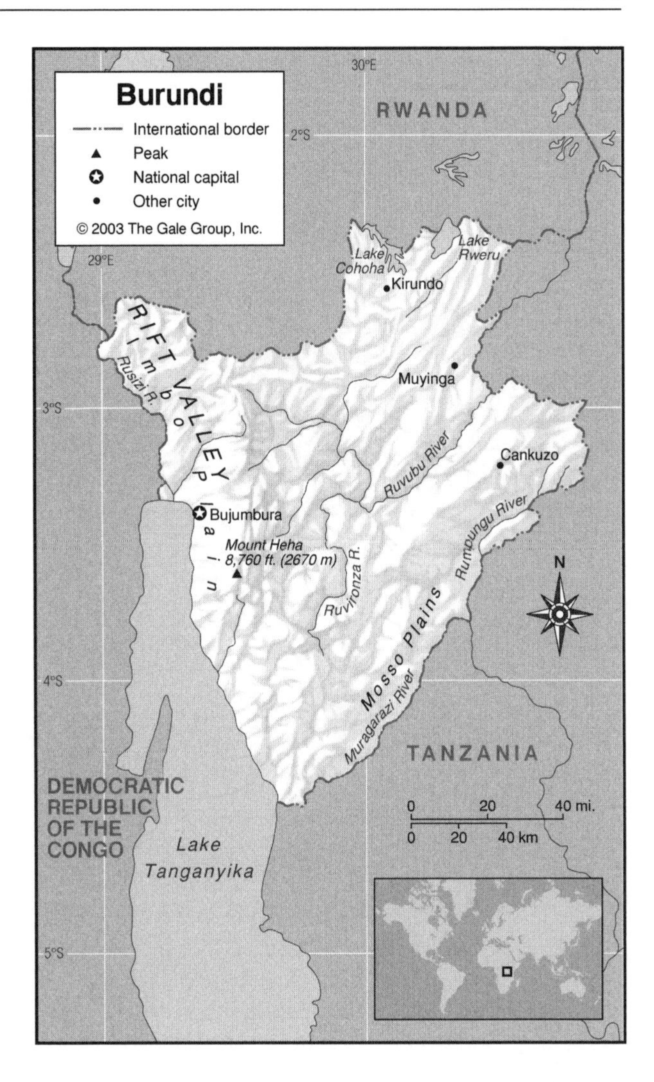

OVERVIEW

Burundi is a small, densely populated, landlocked country in east-central Africa, slightly larger than the state of Maryland. It has three major natural regions: the Rift Valley area in the west, consisting of the narrow plains along the Rusizi (Ruzizi) River and the shores of Lake Tanganyika, together with the belt of foothills on the western face of the divide between the Congo and Nile rivers; the range of peaks that form this divide; and the extensive central and eastern plateaus, separated by wide valleys and sloping into the warmer, drier plains of the eastern and southeastern borders.

Burundi is located on the African Tectonic Plate. The Great Rift Valley system on which it lies has moderate geological activity, including periodic tremors and earthquakes.

MOUNTAINS AND HILLS

Mountains

Burundi's mountains form part of the divide between the basins of the Nile and Congo Rivers. Located in the western half of Burundi, they extend the entire length of the country from north to south, forming an elongated series of ridges that are generally less than 10 mi (16 km) wide, with an average elevation of about 8,000 ft (2,438 km). The tallest peak in the country, Mt. Heha (8,760 ft / 2,670 m), is located in this range.

Plateaus

East of the rugged Congo-Nile divide lies a large central plateau with an average elevation of 5,000 to 6,500 feet (1,525 to 2,000 m). This pleasant highland, inhabited by farmers and cattle herders, is heavily farmed and grazed. Coffee and cotton are the most important commercial crops.

Hills and Badlands

Above the flat western plains that border the Rusizi River and Lake Tanganyika, a belt of foothills and steeper slopes forms the western face of the Congo-Nile Divide. This region includes valleys and farmland.

INLAND WATERWAYS

Lakes

The shores of Lake Tanganyika form Burundi's southeastern border, extending for over 100 mi (161 km). Lake Tanganyika is the second-deepest freshwater lake in the world and is home to over 100 species of fish, many of which are found nowhere else in the world. Burundi has a number of other lakes, of which Lake Rweru and Lake Cohoha in the north along the border with Rwanda are among the largest.

Provinces – Burundi

1990 CENSUS OF POPULATION

Name	Population	Area (sq mi)	Area (sq km)	Capital
Bubanza	223,000	422	1,093	Bubanza
Bujumbura	608,900	515	1,334	Bujumbura
Bururi	385,500	971	2,515	Bururi
Cankuzo	142,700	749	1,940	Cankuzo
Cibitoke	279,800	633	1,639	Cibitoke
Gitega	565,200	768	1,989	Gitega
Karuzi	287,900	563	1,459	Karuzi
Kayanza	443,100	475	1,229	Kayanza
Kirundo	401,100	661	1,711	Kirundo
Makamba	223,800	761	1,972	Makamba
Muramvya	441,700	591	1,530	Muramvya
Muyinga	373,400	705	1,825	Muyinga
Ngozi	482,200	567	1,468	Ngozi
Rutana	195,800	733	1,898	Rutana
Ruyigi	238,600	913	2,365	Ruyigi

SOURCE: ISTEEBU, Bujumbura, Burundi. Cited by Johan van der Heyden, GeoHive, http://www.geohive.com (accessed June 2002).

Rivers

West of the Congo-Nile Divide, runoff waters drain down Burundi's narrow western watershed into the Rusizi River and Lake Tanganyika. The major rivers of the central plateaus include the Ruvironza (or Luvironza) and the Ruvubu, whose river basin is the southernmost extension of the White Nile River. In the east, the two principal rivers on the border with Tanzania are the Rumpungu and the Muragarazi (Malagarasi), which forms most of Burundi's southern border.

THE COAST, ISLANDS, AND THE OCEAN

Oceans and Seas

Burundi is a landlocked nation.

CLIMATE AND VEGETATION

Temperature

Burundi lies fairly close to the equator, but its tropical climate is moderated by elevation, keeping its temperatures at a comfortable level. However, humidity is high. The average annual temperature in the western plains (including the capital city of Bujumbura) is 73°F (23°C). Temperatures average 68°F (20°C) in the plateau region and 60°F (16°C) in the mountains.

Rainfall

Average annual rainfall in most of Burundi is 51–63 in (130–160 cm); on the plains bordering the Rusizi River and Lake Tanganyika it is 30–40 in (75–100 cm). Dry seasons occur from June to August and December to January, and rainy seasons from February to May and September to November. Flooding and landslides occur during heavy rains; during years with below-average rainfall, Burundi may experience periods of drought.

Grasslands

The eastern and central plateau regions of Burundi, flat and well-watered, supports much of the country's agriculture and population. Savannas are found on the eastern border, at elevations of under 5,000 ft (1,524 m). On the southeastern border, the Mosso plains lie along the Muragarazi, Rumpungu, and Rugusi rivers.

At the westernmost edge of the country, the narrow Imbo plain extends southward along the Rusizi River from the Rwanda border through Bujumbura at the north corner of Lake Tanganyika, then extends southward for another 30 mi (48 km) along the eastern shore of the lake. All of this plain, which belongs to the western branch of the Great Rift Valley, is below 3,500 ft (1,066 m) in elevation.

Forests and Jungles

Deforestation in Burundi has been among the most severe in Africa. The remaining tree species include eucalyptus, acacia, and fig, as well as palms along the shores of Lake Tanganyika.

HUMAN POPULATION

Burundi is one of Africa's most densely populated countries. The central and eastern plateaus, from the Congo-Nile divide to the towns of Kirundo, Muyinga, and Cankuzo, are the most heavily populated part of the country. At least half the population lives in this region. Until the mid-twentieth century, Burundi's western plain, on the shores of the Rusizi River and Lake Tanganyika, was mostly uninhabited. However, after 1950, resettlement programs brought farmers to the region.

NATURAL RESOURCES

Burundi's natural resources include cobalt, copper, uranium, nickel, peat, and vanadium. In 2001 most of these had not yet been exploited. Burundi also has a good supply of arable land and water.

FURTHER READINGS

Forster, Peter G., Michael Hitchcock, and Francis F. Lyimo. *Race and Ethnicity in East Africa.* New York: St. Martin's Press, 2000.

Nyankanzi, Edward L. *Genocide : Rwanda and Burundi.* Rochester, Vt.: Schenkman Books, 1998.

Ould Abdallah, Ahmedou. *Burundi on the Brink, 1993-95: A UN Special Envoy Reflects on Preventive Diplomacy.* Washington, D.C. : United States Institute of Peace Press, 2000.

Weinstein, Warren. *Historical Dictionary of Burundi.* Metuchen, N.J. : Scarecrow Press, 1976.

University of Pennsylvania African Studies Program. *Burundi—Geography.* http://www.sas.upenn.edu/African_Studies/NEH/br-geog.html (February 10, 2002).

Cambodia

- **Area:** 69,900 sq mi (181,040 sq km)/ World Rank: 89
- **Location:** Northern and Eastern Hemispheres; in Southeast Asia; bordering Laos to the northeast; Vietnam to the east and southeast; the Gulf of Thailand to the southwest; and Thailand to the west and northwest
- **Coordinates:** 13°N, 105°E
- **Borders:** 1,598 mi (2,572 km) total / Laos, 336 mi (541 km); Thailand, 499 mi (803 km); Vietnam, 763 mi (1,228 km)
- **Coastline:** 275 mi (443 km)
- **Territorial Seas:** 12 NM (22 km)
- **Highest Point:** Phnom Aôral, 5,939 ft (1,810 m)
- **Lowest Point:** Gulf of Thailand, 0 ft (0 m)
- **Longest Distances:** 454 mi (730 km) NE-SW / 318 mi (512 km) SE-NW
- **Longest River:** Mekong River, 2,600mi (4,184 km)
- **Largest Lake:** Tonle Sap, 800 sq mi (24,605 sq km) at the height of the flood season
- **Natural Hazards:** monsoons, floods, and drought
- **Population:** 12,491,501 (2001 est.) / World Rank: 64
- **Capital City:** Phnom Penh, south-central Cambodia
- **Largest City:** Phnom Penh, 862,000 (1998 est.)

OVERVIEW

Cambodia is located in the southwestern part of the Indochina Peninsula. It lies completely within the tropics—its southernmost points are only a little more than 10° above the equator.

It is bounded by highlands to the east and northeast and by the Krâvanh (Cardamom) and Dâmrei (Elephant) mountain ranges to the southwest, with a flat central basin between them. The regular flooding of this plain irrigates the land for the cultivation of rice and other crops. Cambodia's other dominant physical feature is the Mekong River, which traverses the country from north to south.

The mountain ranges that mark the southwestern edge of the central plains are bordered on the Gulf of Thailand side by a narrow coastal plain.

Sections of Cambodia's border with Vietnam, which is an artificially created political boundary rather than a natural one, are disputed. Cambodia is situated on the Eurasian Tectonic Plate and exercises maritime rights over the continental shelf to a depth of 200 nm (370 km).

MOUNTAINS AND HILLS

Mountains

The Krâvanh Range, extending in a northwest-southeast direction, has elevations rising to over 5,000 ft (1,524 m); Phnom Aural, an eastern spur of this range, is the highest point in the country. The Dâmrei Range, running south and southeastward from the Krâvanh, has elevations above 3,000 ft (914 m).

The Dangrek range at the northern rim of the basin consists of a steep escarpment with an average elevation of about 1,600 ft (487 m). The escarpment faces southward and constitutes the southern edge of the Khorat Plateau, which extends northward into Thailand. The watershed along the escarpment, which marks the boundary between Thailand and Cambodia, impedes easy communication between the two countries.

Plateaus

East of the Mekong River, mountains and plateaus extend eastward into the central highlands of Vietnam, at an average elevation of 1,200 ft (360 m).

INLAND WATERWAYS

Lakes

Cambodia's largest lake is the Tonle Sap, or Great Lake. Connected to the Mekong River by the Tonle Sap River, it acts as a natural reservoir during the Mekong's flood period. During this time the Mekong's waters back up, reversing the flow of the Tonle Sap River and enlarging the area of the lake from a low of about 1,000 sq mi (2,590 sq km) in the dry season to nearly 9,500 sq mi (24,605 sq km) at the height of the flooding. When the Mekong's level lowers, the Tonle Sap River reverses direction again and discharges the waters of the Tonle Sap. The annual drainage of these excess river waters into the lake has made it one of the world's most plentiful sources of freshwater fish.

Rivers

Cambodia's central basin is drained primarily by the Mekong River and the Tonle Sap. In the southwest, the Krâvanh and Dâmrei ranges form a separate drainage divide. To the east of this divide, the rivers flow into the Tonle Sap; those to the west drain into the Gulf of Thailand. The Mekong in Cambodia flows southward for

about 315 mi (505 km), from the Cambodia-Laos border to below the provincial capital of Krâchéh, where it turns westward and then southwestward to Phnom Penh. There are extensive rapids above Krâchéh. Elevations below Kâmpóng Cham are extremely low, and areas along the river are inundated during the flood season between June and November. From Phnom Penh the river flows generally southeastward. It divides at this point into two principal channels, the new one being known as the Tonle Basak River, which flows independently from there on through the delta area into the South China Sea.

THE COAST, ISLANDS, AND THE OCEAN

Cambodia's southwestern corner borders the Gulf of Thailand. Numerous islands dot the waters off the Cambodian coast. The largest include Krŏng Kaôh Kŏng and Kaôh Rŭng.

Cambodia has a short coastline whose most important feature is the natural bay at the port of Kompong Som (Kâmpóng Saôm; formerly Sihanoukville). Before the construction of a connecting road and rail line from Phnom Penh, and the opening of the port of Kompong Som, this coastal strip was for the most part an isolated region.

CLIMATE AND VEGETATION

Temperature

Cambodia has a humid, tropical climate. There is little seasonal variation in temperatures, which generally range from 68°F to 97°F (10°C to 38°C). The two main seasons are determined by the monsoons in the region.

Southwestern winds bring the rainy season, which lasts from April or May to November; northeast monsoon winds bring trigger a drier season for the remainder of the year, characterized by lower rainfall, less humidity, and variable skies.

Rainfall

Rainfall varies from 50 to 55 in (127 to 140 cm) in the great central basin to 200 in (508 cm) or more in the southwestern mountains, especially on their westward-facing slopes.

Grasslands

Savanna grasslands are found in the transitional areas around the central lowlands, with grasses growing as high as five feet in some places.

The heart of Cambodia, occupying three-quarters of the country, is the large drainage basin and flood plains of the Tonle Sap Lake and the Mekong River. Located in the center of the country, it consists chiefly of alluvial plains with elevations generally under 300 ft (91 m) above sea level.

Forests and Jungles

Forests cover about half of Cambodia's land area. The evergreen forests of the Dangrek Mountains in the north have a thick, varied undergrowth that includes vines, rattan, and palms, and its taller trees reach heights of 100 ft (30 m).

There are rainforests on the seaward-facing western slopes of the southwestern Krâvanh and Dâmrei Mountains, which include both native tree species and forests of cultivated teak. Pine forests grow at higher elevations. At the lower elevations along the coast, vegetation varies from mangrove forests to evergreens. Deciduous forest is found in the eastern highlands.

HUMAN POPULATION

Roughly three-quarters of Cambodians are rural dwellers, and of these, some 90 percent live in the central plains region. The vast majority of urban dwellers live in the capital city of Phnom Penh. The coastal strip is thinly populated outside the port city of Kompong Som.

NATURAL RESOURCES

Cambodia's forests have been its most heavily exploited natural resource, resulting in substantial deforestation in the latter part of the twentieth century. Other resources include phosphates, which are used in processing fertilizer; rubies, sapphires, and other gemstones; modest deposits of iron ore and manganese; and substantial reserves of both onshore and offshore oil and gas.

Population Centers – Cambodia

(1998 CENSUS OF POPULATION)

Name	Population
Phnom Penh (capital)	862,000
Battambang	150,000

SOURCE: 1998 Population Census of Cambodia, National Institute of Statistics, Ministry of Planning, Cambodia

Provinces – Cambodia

POPULATIONS FROM 1998 CENSUS

Name	Population	Area (sq mi)	Area (sq km)	Capitals
Banteay Méan Cheay	577,772	2,578	6,679	...
Batdâmbâng	793,129	7,407	19,184	Battambang
Kâmpóng Cham	1,608,914	3,783	9,799	Kâmpóng Cham
Kâmpóng Chhnang	417,693	2,132	5,521	Kâmpóng Chhnang
Kâmpóng Spoe	598,882	2,709	7,017	Kâmpóng Spoe
Kâmpóng Thum	569,060	10,657	27,602	Kâmpóng Thum
Kâmpôt	528,405	2,320	6,008	Kâmpôt
Kândal	1,075,125	1,472	3,812	...
Kaôh Kong	132,106	4,309	11,161	Krong Kaôh Kong
Krâchéh	263,175	4,283	11,094	Krâchéh
Kêb, Krong	28,660	130	336	Kêb
Pailin, Krong	22,906	310	803	Pailin
Preah Sihanouk, Krong	155,690	335	868	Preah Sihanouk
Môndól Kiri	32,407	5,517	14,288	Senmonorom
Phnom Penh	999,804	18	46	Phnom Penh
Pouthisat	360,445	4,900	12,692	Pouthisat
				Phnum Tbéng
Preah Vihéar	119,261	5,324	13,788	Meanchey
Prey Vêng	946,042	1,885	4,883	Prey Vêng
Rôtânak Kiri	94,243	4,163	10,782	Lumphat
Siem Réab	696,164	6,354	16,457	Siem Réab
Stoeng Tëng	81,074	4,283	11,092	Stoeng Tëng
Svay Rieng	478,252	1,145	2,966	Svay Rift
Takêv	790,168	1,376	3,563	Takêv

SOURCE: 1998 Population Census of Cambodia, National Institute of Statistics, Ministry of Planning, Cambodia.

FURTHER READINGS:

Asian Studies Virtual Library Web site. *Cambodia.* http://www.iias.nl/wwwvl/southeas/cambodia.html (accessed Feb. 26, 2002).

Downie, Susan. *Down Highway One: Journeys Through Vietnam and Cambodia.* North Sydney: Allen & Unwin, 1993.

Fifield, Adam. *A Blessing Over Ashes: The Remarkable Odyssey of My Unlikely Brother.* New York: W. Morrow, 2000.

Gray, Spalding. *Swimming to Cambodia.* New York: Theatre Communications Group, 1985.

Lafreniere, Bree. *Music Through the Dark : A Tale of Survival in Cambodia.* Honolulu: University of Hawaii, 2000.

Livingston, Carol. *Gecko Tails: A Journey Through Cambodia.* London: Weidenfeld and Nicolson, 1996.

Wurlitzer, Rudolph. *Hard Travel to Sacred Places.* Boston: Shambhala, 1994.

Cameroon

- **Area:** 183,567 sq mi (475,440 sq km) / World Rank: 54
- **Location:** Western Africa, bordering the Bight of Biafra, between Equatorial Guinea and Nigeria
- **Coordinates:** 6°N, 12°E
- **Borders:** 2,846 mi (4,591 km) / Chad, 678 mi (1,094 km); Central African Republic, 494 mi (797 km); Republic of the Congo, 324 mi (523 km); Gabon, 185 mi (298 km); Equatorial Guinea, 117 mi (189 km); Nigeria, 1,048 mi (1,690 km)
- **Coastline:** 249 mi (402 km)
- **Territorial Seas:** 50 NM
- **Highest Point:** Mount Cameroon, 13,353 ft (4,070 m)
- **Lowest Point:** Sea level
- **Longest Distances:** 748 mi (1,206 km) N-S / 446 mi (717 km) E-W
- **Longest River:** Sanaga, 570 mi (920 km)
- **Largest Lake:** Lake Chad (shared, varies regionally from c.4,000 to c.10,000 sq mi / 10,360-25,900 sq km)
- **Natural Hazards:** Volcanic activity with release of poisonous gases
- **Population:** 15,906,500 (2002 est.) / World Rank: 60
- **Capital City:** Yaoundé, located in south-central Cameroon
- **Largest City:** Douala, south-western Cameroon along the Bight of Biafra coast, 1,239,100 (2002 est.)

OVERVIEW

Cameroon (Cameroun), the hinge between west and central Africa, forms an irregular wedge extending northeastward from a coastline on the Gulf of Guinea, an arm of the Atlantic Ocean, to Lake Chad, 700 mi (1,126 km) inland. Behind the swamps and the lowlands generally referred to as the southwestern coastal zone, the land rises to mountains and plateaus extending more than 500 mi (804 km) inland before descending to a flat plain of moderate elevation in the far north. It is one of the most physically and socially diverse countries on the African continent.

Situated in the African Plate, Cameroon is made up of four loosely defined regions: the northern plains, the central and southern plateaus, the western highlands and mountains, and the lowlands along the coast. Near the coast are volcanic peaks, dominated by Mt. Cameroon (Fako), the highest point in the country. Beyond the coastal marshes and plains, the land rises to a densely forested plateau. The interior of the country is a dissected plateau, which forms a barrier between the agricultural south and the pastoral north. The Northern Plains extend to Lake Chad, where the borders of Nigeria, Chad, and Cameroon intersect.

The Bakassi Peninsula and its extended maritime boundaries have significant oil reserves and have been the source of boundary disputes between Cameroon and Nigeria.

MOUNTAINS AND HILLS

Mountains

The Cameroon Mountains, the highest range in the country, extend southeastward from the Cameroon/Nigeria border area at about 7°N latitude to Mt. Cameroon on the coast. The major mountain range and the upland areas on its eastern and western slopes were built up by volcanic activity associated with a series of faults in the granite substructures underlying the African continent. All of the ancient volcanoes in this complex had subsided before the dawn of recorded history except for Mt Cameroon, which has been active on seven occasions since 1900. In 1922 and 1959 molten lava flowed several miles, destroying plantations n the lower slopes. During the 1999 eruption, several hundred people were evacuated from their villages.

Mt. Cameroon is a complex of several connected fissures and cones, one of which reaches 13,353 ft (4,070 m) above sea level, more than half again as high as any other peak in the country. Elsewhere in the Cameroon Mountains, elevations range between 5,500 ft (1,676 m) and 8,000 ft (2,438 m). Other ranges of lower elevation stand in the north near the western border of Northern Province. The most important of these are the Alantika Mountains, which mark the border for a short distance at about 8°30′N latitude, and the Mandara Hills, which extend northward from the town of Garoua and the Bénoué River to about 11°N latitude.

Plateaus

The Adamawa (Adamaoua) Plateau, lying between 7°N and 9°N latitude, extends from the eastern to the western border of Cameroon at elevations that are more than 3,000 ft (914 m) above sea level and average about 4,500 ft (1,371 m). Surface features in the central parts of

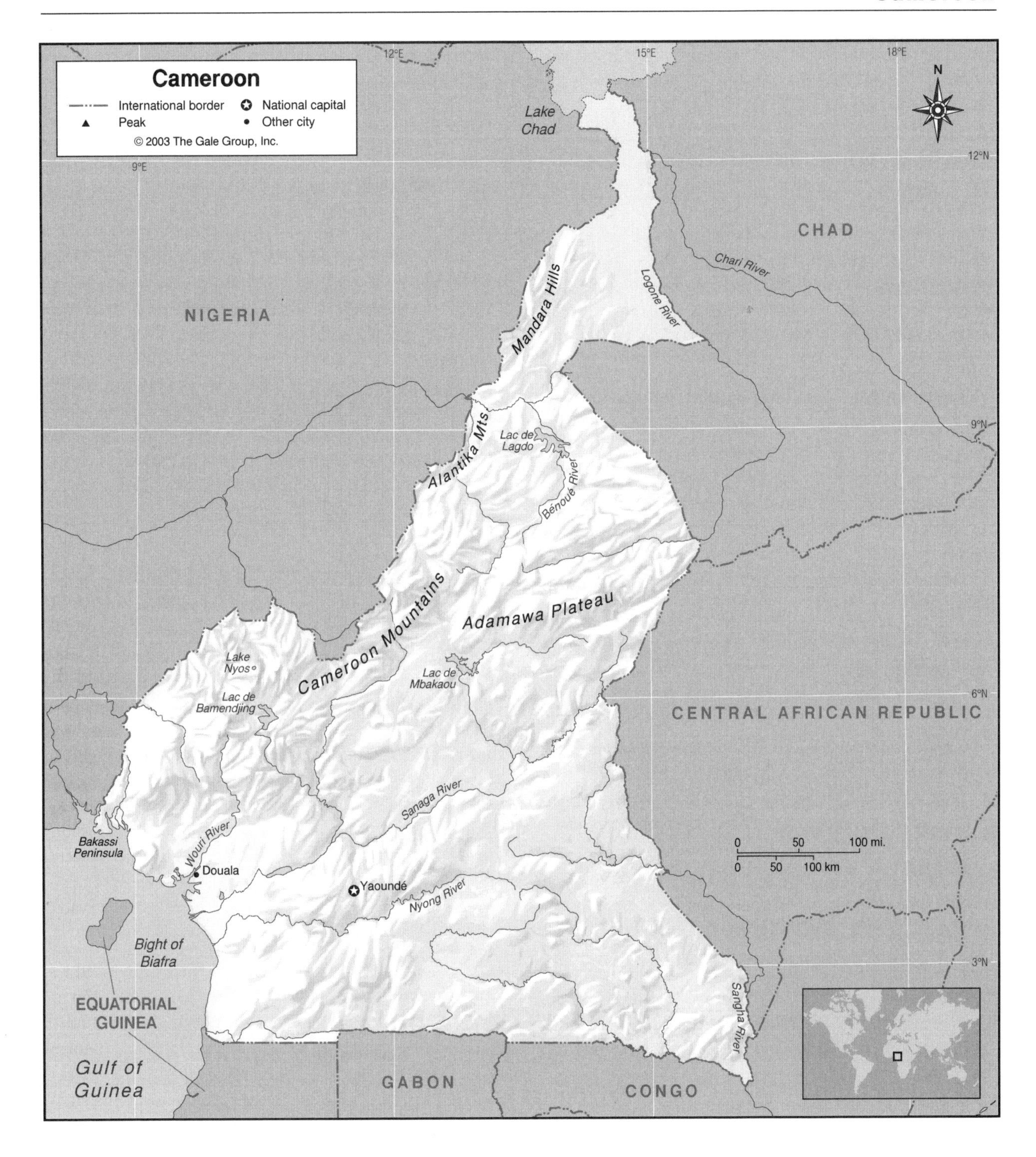

this high plateau include small hills or mounds capped by erosion-resistant granite or gneiss. Along the western and, to a lesser degree, the eastern borders, old eruptions from fissures and volcanoes have covered thousands of square miles of the underlying granite with lava.

Hills and Badlands

The Mandara Hills present a moonscape marked by scattered outcrops or hills of resistant rock rising above the general erosion level. West of Maroua the scattered rocky mounds and minor ridges are more numerous, rising westward to hills and elongated ridges. In this area—one of Cameroon's most scenic—volcanic plugs dot rugged hills like sentinels watching over deep undulating valleys. South of the Adamawa Plateau begins a series of lower plateaus that extend throughout most of South Central and Eastern provinces at elevations averaging about 3,000 ft (914 m) but descending gradually southward to the border and westward toward a series of terraces leading downward to the coastal plain.

INLAND WATERWAYS

Lakes

Lake Chad, shared with Niger, Nigeria, and Chad, is the largest body of water in the Sahel. The size of the lake varies seasonally from c.4,000 to c.10,000 sq mi (10,360-25,900 sq km). It is divided into north and south basins, neither of which is generally more than 25 ft (7.6 m) deep, although the lake was formerly much larger and attained a depth of c.930 ft (285 m) in the 19th century. Its chief tributary, the Chari River, has no outlets. The low salinity of the lake has led scientists to explore if lake water is seeping into the deep water table rather than evaporating.

In 1986 the build-up and sudden release of toxic gas in Lake Nyos near the Bamenda highlands killed 1,600 people on mountain slopes. In January 2001 a team of environmental experts began work on a filtering device that would release carbon gases from the volcanic lakes of Nyos and Monoun slowly into the atmosphere.

Rivers

Three primary watersheds drain the country into the tributaries of the Niger and Congo Rivers, and into the Atlantic Ocean. Rivers in Northern Province exhibit major seasonal fluctuations in volume while practically all rivers in the other provinces carry a heavy flow for most of the year.

Runoff originating on the Adamawa Plateau flows northward into the upper tributaries of the Bénoué River, or feeds into the Sanaga, the largest river in the south-western part of the country. Three other major rivers—the Wouri, Dibamba, and Nyong—also feed into the tangled complex of deltas on the central Atlantic coast. Farther south, near the borders with Equatorial Guinea and Gabon, the Campo River watershed extends inland for about 200 mi (321 km). Both the Sanaga and the Nyong rivers collect runoff from parts of Eastern province, but most of this wet forest area is drained by various tributaries of the Sangha River, which for a short distance marks the border with the Congo and then flows southward into the Congo River.

Wetlands

The low coastal plain in the south is covered by equatorial rain forests and swamp lands along its edges. The Logone and Chari river systems along the northeastern border annually inundate a broad area before emptying into Lake Chad.

THE COAST, ISLANDS, AND THE OCEAN

Along its west coast, Cameroon borders the Gulf of Guinea and the Bight of Biafra in the Atlantic Ocean. Most of the coastal zone is a flat area of sedimentary soils that front on the Gulf of Guinea for about 160 mi (257 km). Along its seaward edges the central segment of the coastal zone is a series of many adjoining deltas. The beaches of Limbe at the base of Mt. Cameroon are renowned for their black volcanic sand.

CLIMATE AND VEGETATION

Temperature

Cameroon has a tropical climate which varies from equatorial in the south to sahelian (semiarid and hot) in the north. Average temperature ranges in Yaoundé are from 64°-84°F (18°-29°C).

Rainfall

The Sahelian north has a wet season between April and September while the rest of the year is dry. Average annual precipitation for this region is between 39 in (100 cm) and 69 in (175 cm). The equatorial south has two wet seasons and two dry seasons. One wet season is between March and June, and the great wet season is between August and November. One dry season is between June and August, and the great dry season is between November to March. In the south, the average annual precipitation reaches 159 in (403 cm).

Grasslands

Vast stretches of grassland are typical in the Bamenda area. The central plateau is a transition zone where slash and burn agriculture has left fire resistant trees and thorny scrub interspersed throughout expanses of prairie grass. There are no deserts, but the northern plains between Maroua and Lake Chad are sub-arid.

Forests and Jungles

Equatorial rain forests, mangroves, and swamps cover much of the low coastal plain in the south. The south to central areas have patches of rain forest interspersed with cultivation and coffee plantations. The western mountains are mostly forest-covered except lower altitudes. Upland wooded savanna marks the east-central part of the country becoming steadily drier mixed forests approaching the Adamawa Plateau. Thorn trees and scrub cover the semi-arid northern plains.

HUMAN POPULATION

Population densities are highest in the west, south-central, and Sudan savannah zone of the north. Cameroon Highlanders constitute 31 percent of the population; Equatorial Bantu, 19 percent; Kirdi, 11 percent; Fulani, 10 percent; Northwestern Bantu, 8 percent; Eastern Nigritic, 7 percent; other African, 13 percent; and non-African, less than 1 percent. Peoples of the southwest are largely agricultural and influenced by Christianity, whereas those of the north are more pastoral and Muslim. Furthermore, the densely populated Anglophone northwest and southwest provinces provide a

Population Centers – Cameroon

(1991 POPULATION ESTIMATES)

Name	Population
Douala	884,000
Yaoundé (captial)	750,000
Garoua	177,000
Maroua	143,000

SOURCE: "Länderbericht Kamerun," Statistisches Bundesamt, Germany.

Provinces – Cameroon

Name	Area (sq mi)	Area (sq km)	Capital
Adamoua	23,979	62,105	Ngaoundéré
Centre	26,655	69,035	Yaoundé
Est	42,086	109,002	Bertoua
Extreme-Nord	12,477	32,316	Maroua
Littoral	7,810	20,229	Douala
Nord	26,134	67,686	Garoua
Nord-Ouest	6,722	17,409	Bamenda
Ouest	5,360	13,883	Bafoussam
Sud	18,200	47,137	Ebolowa
Sud-Ouest	9,540	24,709	Buea

SOURCE: *Geo-Data: The World Geographical Encyclopedia,* 2nd ed. Detroit: Gale Research, 1989.

striking set of cultural, trade, and legal differences with the rest of the country (former East Cameroon), which is Francophone.

NATURAL RESOURCES

Cameroon has one of the best-endowed primary commodity economies in sub-Saharan Africa thanks to agriculture, petroleum, iron ore (Southern Province), timber, and hydropower. Very large beds of bauxite are located in the Northern Province. Deposits of tin ore, manganese, gold, asbestos, mica, and diamonds are either small or of poor quality and have little economic value. Deforestation, over-grazing, and over-fishing are among the major environmental threats.

FURTHER READINGS

Africa South of the Sahara 2002: Cameroon. London: Europa Publications Ltd., 2002.

Debel, Anne. *Le Cameroun Aujourd'hui.* Paris: Les Editions Jeune Afrique, 1985.

DeLancy, Mark W., and Mark Dike DeLancey. *Historical Dictionary of Cameroon.* African Historical Dictionaries, No. 81. Lanham, MD and London: The Scarecrow Press, Inc., 2000.

Europa World Yearbook 2000: Cameroon. London: Europa Publications, Ltd., 2000.

Gaillard, P. *Le Cameroun.* Paris: Editions L'Harmattan, 1989.

Wo Yaa! *Cameroon.* http://www.woyaa.com/ (accessed March 23, 2002).

Canada

- **Area:** 3,851,809 sq mi (9,976,185 sq km) / World Rank: 3
- **Location:** Northern and Western Hemispheres, northern North America, bordering the North Atlantic Ocean and North Pacific Ocean, north of the conterminous US.
- **Coordinates:** 60°00′N, 95°00′W
- **Borders:** 5,526 mi (8,893 km) / United States: conterminous, 3,987 mi (6,416 km); Alaska, 1,539 mi (2,477 km)
- **Coastline:** 151,485 mi (243,791 km)
- **Territorial Seas:** 12 NM
- **Highest Point:** Mount Logan, 19,551 ft (5,959 m)
- **Lowest Point:** Sea level
- **Longest Distances:** 3,223 mi (5,187 km) E-W; 2,875 mi (4,627 km) N-S
- **Longest River:** Mackenzie River, 2,635 mi (4,290 km)
- **Largest Lake:** Lake Superior, 31,802 sq mi (82,367 sq km)
- **Natural Hazards:** Continuous permafrost in north; cyclonic storms east of the Rocky Mountains
- **Population:** 31,592,805 (July 2001 est.) / World Rank: 35
- **Capital City:** Ottawa, located in the southeast on the Ottawa River
- **Largest City:** Toronto, located on the northern shore of Lake Ontario, population 4,657,000

OVERVIEW

Canada occupies all of the North American continent north of the United States except for Alaska and the small French islands of St. Pierre and Miquelon. The most striking geographical characteristic of Canada is its immense size. It is the largest country in the Western Hemisphere and the second-largest in the world, next to Russia. Canada's size is about the same as that of the continent of Europe. Canada also encompasses the Canadian continental margin, including Hudson Bay, the Gulf of St. Lawrence, the Pacific Coast Straits, and the channels of

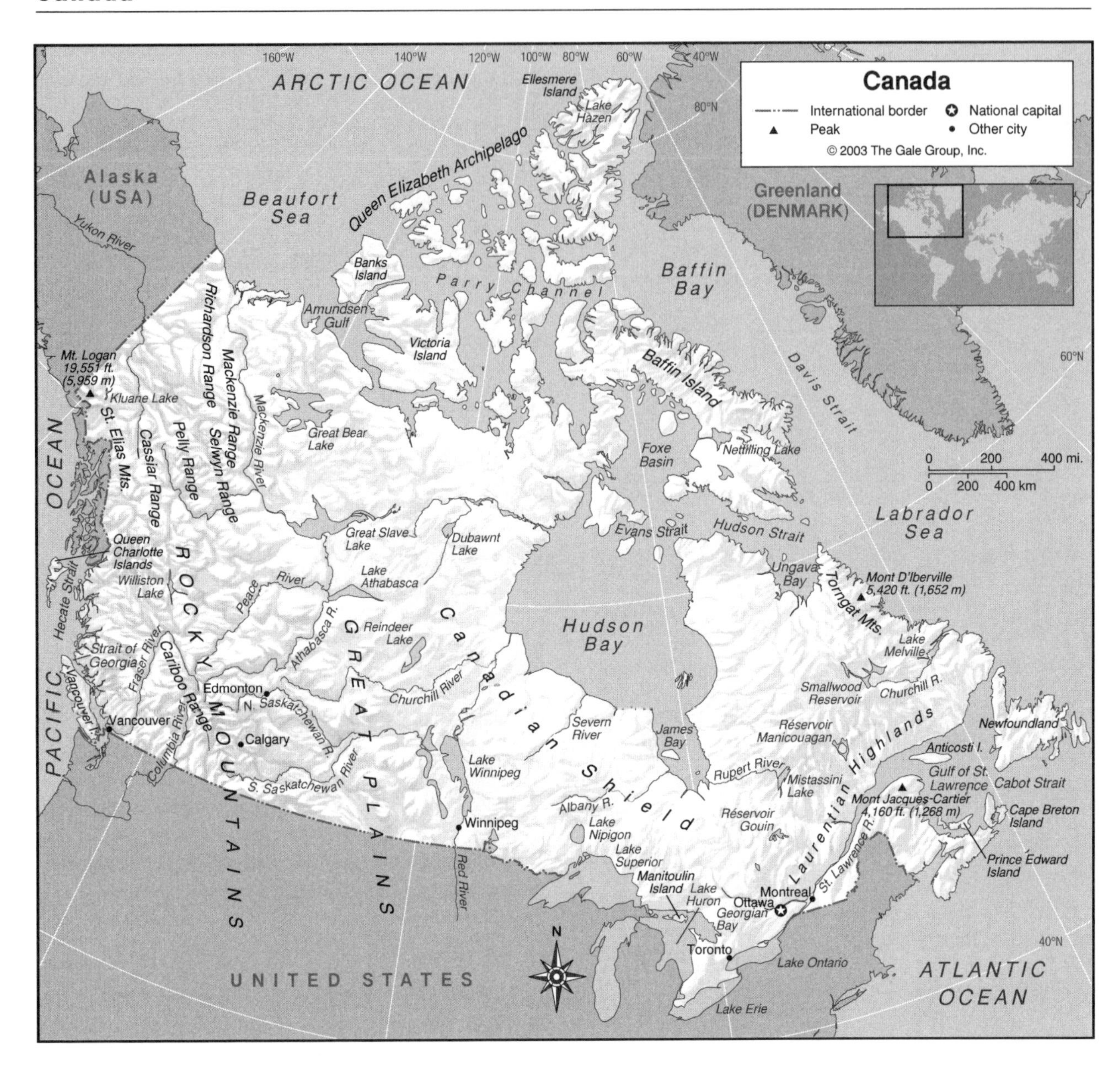

the Arctic Archipelago. Canada's longitudinal extent is such that it requires six time zones.

Topographically, Canada is divided into the Atlantic provinces, the Great Lakes-St. Lawrence Lowlands, the Canadian Shield, the Interior Plains, the Western Cordillera, and the Northwest Territories, which is a political rather than a geographical term. The Territories cover the region east of the Western Cordillera and north of the Interior Plains and the Canadian Shield. Within this large area there are two distinct sub-regions: the sub-arctic Mackenzie River Valley to the west, and the arctic area of the islands and north-central mainland.

The Canadian Shield is surrounded by a series of lowlands: the Atlantic region and the Great Lakes-St. Lawrence Lowlands to the east, the Interior Plains to the west, and the Arctic Lowlands to the north. The Atlantic provinces have rugged, indented coasts. The Great Lakes-St. Lawrence Lowlands constitute the heartland of the country's population. This region has the largest area of level land easily accessible by water from the east.

Canada is located on the North American Tectonic Plate.

MOUNTAINS AND HILLS

Mountains

The principal mountainous region is the Western Cordillera, or Cordilleran, Mountain system located in the westernmost portion of Canada. The Cordilleran is colloquially known as the Canadian Rockies, and is composed of relatively young, folded and faulted mountains and plateaus. The Cordilleran Chain is made up of several ranges, including the Richardson, Mackenzie, Selwyn, Pelly, Cassiar, and Cariboo Ranges. The chain is much narrower than in the United States, with less extensive interior plateaus. However, the mountains are generally much higher in Canada and contain some of the most beautiful scenery in the world. Most peaks in the Canadian Rockies are over 14,765 ft (4,500 m) high with 24 peaks over 13,123 ft (4,000 m). Canada's highest point is Mount Logan (19,551 ft; 5,959 m) in the St. Elias Mountains of Yukon Territory near the Alaskan border. The only other parts of Canada with comparable spectacular mountains are Baffin and Ellesmere Islands in the northeastern Arctic.

A second major mountain system is located along the north-eastern seaboard from Ellesmere Island down through the Torngat Mountains of Quebec, Newfoundland and Labrador. A third and less significant system is the Appalachian Chain, which crosses parts of Eastern Canada. The highest point in Quebec is Mont D'Iberville at 5,420 ft. (1,652 m) in the Torngat range. In the Appalachians, the highest peak is Mont Jacques-Cartier at 4,160 ft. (1,268 m).

Plateaus

The foundation of Canadian geology is the Canadian Shield (sometimes called the Precambrian Shield or the Laurentian Plateau), which takes up almost half of Canada's total area. It extends beyond the Canadian boundary into the United States in two limited areas: at the head of Lake Superior and in the Adirondack Mountains. Structurally, the shield may be thought of as a huge saucer, the center of which is occupied by Hudson Bay and James Bay, which have breached the northeastern rim to drain into the Atlantic Ocean through the Hudson Strait. Most of the shield is relatively level and less than 2,000 ft (612 m) above sea level.

Only along the dissected rim of the saucer are there major hills and mountains: the Torngat Mountains in northeastern Labrador, the Laurentian Highlands, and along the northern shores of Lake Superior. Except for the plains, the rest of the shield is composed of undulating terrain with rocky, knoblike hills, the hollows between which are occupied by lakes interconnected by rapid streams. A second and far less extensive plateau supports the Western Cordillera.

INLAND WATERWAYS

Lakes

Canada has 31,752 lakes, more than a third of which are in the northern half of the country in the Northwest Territories and Nunavut. In the east, Quebec has more than 8,000 lakes and Ontario almost 4,000, while in the west British Columbia has only about 800. Fully 7.6 percent of Canada's total area is covered by lakes and rivers, making surface water the source of 90 percent of freshwater. Indeed, Canada's lakes play a critical role in their ecosystems as natural regulators of river flow, smoothing out peak flows during flooding and sustaining the flow during dry seasons.

The largest freshwater bodies in the world are the Great Lakes, of which 36 percent lie in Canada. Of the five Great Lakes, only Lake Michigan is completely outside Canadian borders. Lake Superior, Canada's largest lake in terms of volume (shared with the United States), has a surface area of 31,802 sq mi (82,367 sq km) and is the world's largest freshwater lake. Also in the east near Lake Superior is Lake Nipigon (1,700 sq mi / 4,500 sq km) which is known for being surrounded by towering cliffs and its green-black beaches. Further east in Quebec are Mistassini Lake (835 sq mi / 2,164 sq km) and Réservoir Gouin (480 sq mi/1,240 sq km). Mistassini Lake is the source of the Rupert River, which flows into James Bay. It is rather deep, and is home to abundant wildlife and fish. Since its waters open to many rivers and waterways that reach Montreal, in the 1800s its shores were chosen as the site for a Hudson Bay Company fur trading post.

In Canada's northern provinces—the Northwest Territories, Yukon Territory, and Nunavut—there are two of the greatest lakes in the country. The Great Slave Lake in the Northwest Territories is the deepest and fourth-largest of Canada's lakes with a depth of 2,014 ft (614 m) and a surface area of 11,030 sq mi (28,568 sq km). Also in this region is the Great Bear Lake, located in a largely uninhabited part of northwestern Canada. It is third in size but the largest lake wholly within Canada; its surface area is 12,095 sq mi (31,328 sq km). Further west and near the Alaskan border is Kluane Lake (156 sq mi / 537 sq km). Kluane Lake is located in Kluane National Park at the foot of the St. Elias Mountains, which include Mt. Logan. The extreme elevation difference between the lake and the surrounding mountain crests presents a variety of geographical research opportunities in a small area. Besides offering hiking, fishing, and tours of icefield ranges, there is also an Arctic Institute of North America station in the park. In the northeast, the province of Nunavut offers two notable lakes. Nettilling Lake, the country's tenth-largest lake (1,956 sq mi / 5,066 sq km), is fed by the slightly smaller Amadjuak Lake that helps to drain Baffin Island into Foxe Basin. The lake is frozen most of the year. Far to the north on Ellesmere Island is Lake Hazen (210 sq mi / 540 sq km), the largest lake in the world to lie completely north of the Arctic Circle. This lake helps function as a "thermal oasis," catching the sun's energy and heating the surrounding land to temperatures which are anomalous for such an altitude. However, the lake itself still remains frozen all year except in especially warm years.

In the eastern coastal regions the important lakes are Smallwood Reservoir (2,500 sq mi / 6,460 sq km) and Lake Melville (1,160 sq mi / 3,000 sq km). Smallwood Reservoir was formed by connecting many smaller lakes and wetlands to create the opportunity not only for hydroelectric energy but also for fishing. It is the largest reservoir in Canada. Lake Melville is connected to Smallwood Reservoir by Churchill River. It is a large coastal lake that is linked to the Atlantic Ocean by Hamilton Inlet. Most of the coastal province of Newfoundland drains via Lake Melville. A similar coastal lake is located on Cape Breton Island in the southeast. Bras d'Or Lake (425 sq mi / 1,100 sq km) is a deeply indented arm of the Atlantic Ocean occupying much of the island. Its name means "Arms of Gold," named because its sheltered waters are clear and clean and it is surrounded by a beautiful natural landscape and lush vegetation.

Canada's lower central and western region includes the remainder of the country's largest lakes. Lake Winnipeg, in the province of Manitoba, is the country's sixth-largest lake with a surface area of 9,174 sq mi (23,760 sq km). Lake Winnipeg drains much of the Great Plains region, being fed by many rivers including the Saskatchewan, the Red, and Winnipeg Rivers. The lake then empties to the northeast into Hudson Bay via the Nelson River. Lake Winnipeg offers many ocean-like sandy beaches and attracts large numbers of tourists. Not far to the northwest is Reindeer Lake, the ninth-largest lake (2,185 sq mi / 5,660 sq km) in Canada. This lake also helps to drain the Great Plains region, and is itself home to many fishing tournaments and lodges. Heading further northwest toward the Great Slave Lake one comes across Lake Athabasca, Canada's eighth-largest lake (3,030 sq mi / 7,850 sq km). This lake is surrounded by many other lakes—including Lake Claire (546 sq mi / 1,415 sq km) in Wood Buffalo National Park—and rivers draining the Canadian Shield region. Lake Athabasca is also known for its plentiful trout, producing one of the world's largest lake trout ever—a 102-pound (46-kg) fish caught in a gillnet in 1961. In western Canada British Columbia has one important lake, Williston Lake (680 sq mi / 1,761 sq km). Williston Lake is the largest artificial lake in Canada.

Rivers

Canada's rivers drain into five major ocean outlets: The Pacific, Arctic, and Atlantic Oceans; Hudson Bay; and the Gulf of Mexico. On an average annual basis, Canadian rivers discharge roughly 1.152 billion sq ft (107 million sq m) per second. This is nearly 9 percent of the world's renewable water supply, equivalent to 60 percent of Canada's mean annual precipitation. The Yukon and Mackenzie in the west, the North Saskatchewan, South Saskatchewan, Saskatchewan, Peace, and the Athabasca Rivers in central Canada, and the Ottawa and St. Lawrence Rivers in the east comprise Canada's main rivers.

GEO-FACT

Visitors to Gros Morne National Park in Newfoundland can see plate tectonics in action. Geologists believe that North America and Europe were once a single landmass 600 million years ago. As the two continents pulled apart, magma from deep inside the Earth oozed up between them. This solidifed magma—as well 500-million-year old fossils preserved in the sedimentary rock—are visible to visitors of Gros Morne, who can also just enjoy some of eastern Canada's most dramatic scenery.

The Central Canadian Shield is drained by the Nelson-Saskatchewan, Churchill, Severn and Albany Rivers, flowing into Hudson Bay. The 2,635 mi long (4,290 km long) Mackenzie River, with its tributaries and three large lakes—Great Bear Lake, Great Slave Lake, and Lake Athabasca—drains an area of more than 1 million sq mi (2.6 million sq km) into the Arctic Ocean. The Columbia, Fraser and Yukon rivers are the principal drainage systems of western Canada. The Great Lakes drain into the broad St. Lawrence River, which flows into the Gulf of St. Lawrence. The river in Canada with the greatest annual discharge is the St. Lawrence River at 347,606 cubic feet per second (9,850 cubic meters per second).

In the prairies, groundwater is the principal source of water for streams during the frequent dry weather periods. In hot summer months, melting glaciers may contribute up to 25 percent of the flow of the Saskatchewan and Athabasca Rivers.

Wetlands

Canada possesses 24 percent of the world's wetlands, more than 314 million acres (127 million hectares). Most of the wetlands are located in the boreal peat bogs in arctic and subarctic regions, and the Prairie pothole region across south central Canada and the United States, which contains more than four million wetlands and ponds.

More than one-seventh of wetland areas that existed before European settlement have been converted through agriculture and commercial use. In the Great Lakes region, 83 percent of wetlands have been destroyed, and along the shore of Lake Ontario, 90 percent of the wetlands have been sacrificed. Despite the loss of wetlands, 14 percent of Canada is still covered by these lands in their natural states in the form of ponds, marshes, flood plains, and water-logged ground.

THE COAST, ISLANDS, AND THE OCEAN

Oceans and Seas

Canada borders three oceans: The Pacific on the west, the Arctic to the north, and the Atlantic on the east. The Yukon Territory and Banks Island face the oil-rich Beaufort Sea in the Arctic Ocean. To the east of the Beaufort Sea and south of Banks Island is the Amundsen Gulf. A series of gulfs, straits, and channels—comprising Viscount Melville Sound, M'Clintock Channel, Queen Maud Gulf, the Gulf of Boothia, and Lancaster Sound to the south of Parry Channel, and M'Clure Strait, Peary Channel, Norwegian Bay, Jones Sound, and Smith Sound north of Parry Channel—winds through the Artic Archipelago, but is locked in ice most of the year.

East of the Queen Elizabeth Islands, Baffin Bay separates Baffin Island from Greenland opening to the Davis Strait and then to the Labrador Sea, which lies off the southeastern tip of Greenland. Turning south around Newfoundland—the easternmost point of the North American continent—and skirting the greater North Atlantic Ocean, the Cabot Strait separates Newfoundland and Nova Scotia, and provides a channel to the Atlantic for the Gulf of St. Lawrence. Inland, the Hudson Bay and its southern arm James Bay, as well as the Foxe Basin, connect to the Labrador Sea through Evans Strait and the Hudson Strait. On the west coast, fronting the North Pacific is a labyrinth of straits and sounds extending from Vancouver Island in the south and winding through the Alexander Archipelago in the north.

The Arctic Archipelago lies on a submerged plateau whose floor varies from flat to gently undulating. From the Alaskan border eastward to the mouth of the Mackenzie River the shelf is shallow and continuous, with its outer edge at a depth of 210 ft (64 m) about 40 naut mi (69 naut km) from the shore. Near the western edge of the Mackenzie River delta it is indented by the Mackenzie Trough (formerly known as the Herschel Sea Canyon), whose head comes within 15 naut mi (24 naut km) of the coast. The submerged portion of the Mackenzie Delta forms a pock-marked undersea plain, most of it less than 180 ft (55 m) deep and up to 75 naut mi (121 naut km) wide and 250 naut mi (402 naut km) long. Most of the continental shoulder is over 1,801 ft (549 m) deep, sloping to the abyssal Canada Basin at 12,002 ft (3,658 m). The deeply submerged continental shelf runs along the entire western coast of the Arctic Archipelago from Banks Island to Greenland.

Major Islands

Canada has more than 52,000 islands with all but a few hundred of them considered 'minor' in size—less than 49.81 sq mi (129 sq km) in area. By far, the largest islands are those in the Arctic Archipelago, extending from James Bay to Ellesmere Island. Baffin is larger than 193,050 sq mi (500,000 sq km), Victoria contains 83,783 sq mi (217,000 sq km), and Banks Island covers some 27,027 sq mi (70,000 sq km).

The Queen Elizabeth Archipelago surrounding the north magnetic pole has 35 islands larger than 49.81 sq mi (129 sq km) in size. Ellesmere, the northernmost of Canada's islands, is the largest of the Elizabeth group with more than 75,675 sq mi (196,000 sq km).

The largest islands on the western coast are Vancouver Island (12,079 sq mi / 31,285 sq km) and the Queen Charlotte Islands. On the eastern coast are Newfoundland (42,030 sq mi /108,860 sq km); Prince Edward Island (2,170 sq mi /5,620 sq km); Cape Breton Island (3,981 sq mi /10,311 sq km); Grand Manan and Campobello Islands of New Brunswick; and Anticosti Island (3,066 sq mi /7,941 sq km) and the Iles de la Madeleine of Quebec. Manitoulin Island in Lake Huron is the world's largest island in a freshwater lake covering some 1,068 sq mi (2,765 sq km).

The Coast and Beaches

Canada's coastlines of nearly 151,647 mi (244,000 km) on the mainland and offshore islands are among the largest of any country in the world. On the Atlantic coast the submerged continental shelf has great width and diversity of relief. From the coast of Nova Scotia its width varies from 60 to 100 naut mi (97 to 161 naut km), from Newfoundland 100 to 280 naut mi (161 to 451 naut km) at the entrance of Hudson Strait, and northward it merges with the submerged shelf of the Arctic Ocean. The outer edge varies in depth from 620 to 10,201 ft (189 to 3,110 m). The overall gradient is slight, but the shelf is studded with shoals, ridges and banks. Hudson Bay is a shallow inland sea, 317,417 sq mi (822,325 sq km) in area, having an average depth of 422 ft (128 m). Hudson Strait separates Baffin Island from the continental coast and connects Hudson Bay with the Atlantic Ocean. It is 495 mi (796 km) long and from 43 to 138 mi (69 to 222 km) wide.

The Pacific coast is strikingly different and is characterized by bold, abrupt relief—a repetition of the mountainous landscape. Numerous inlets penetrate the coast for up to 89 mi (140 km), usually 1 naut mi (1.6 naut km) wide with deep side canyons. From the islet-strewn coast the continental shelf extends from 50 to 100 naut mi (80 to 161 naut km) except on the western slopes of Vancouver Island and the Queen Charlotte Islands, where the seafloor drops rapidly. These two island groups contain the relatively shallow Queen Charlotte Sound as well as two Straits, Hecate Strait and the Strait of Georgia.

CLIMATE AND VEGETATION

Temperature

Canada's continental climate is sub-artic to arctic in the north, while near the US border a narrow strip has a temperate climate with cold winters; the east and west coasts are maritime and more temperate. The north Canadian coast is permanently icebound except for Hudson Bay which is only frozen for nine months of the year. Canada's greatest temperature variation is found in the Northwest Territories where at Fort Good Hope temperatures range from -24°F (-31°C) in January to 61°F (16°C) in July.

Temperatures on the west coast range from about 39°F (4°C) in January to 61°F (16°C) in July. On the Atlantic coast the winter temperatures are warmer than those of the interior, but summer temperatures are lower. Much of the southern interior of Canada has high summer temperatures and long cold winters. Average temperature ranges in Ottawa are from 5° to 21° F (-15° to -6°C) in January to 59° to 79°F (15° to 26°C) in July.

Rainfall

The west coast and some inland valleys have mild winters and mild summers with rainfall occurring throughout the year. The west coast receives between 60 and 120 in (150 cm to 300 cm) of rain annually while the maritime provinces receive 45 to 60 in (115 to 150 cm) annually. The driest area is the central prairie where less than 20 in (50 cm) of rain falls each year. The region to the east of Winnipeg is considerably wetter than the western prairie, receiving 20–40 in (50–100 cm) yearly.

Grasslands

Between the Western Cordillera and the Canadian Shield is the region broadly known as the West, including the Manitoba and Mackenzie lowlands. The Manitoba Lowland (leading to the Saskatchewan and Alberta plains) is one of only a few parts of Canada that is as flat as a tabletop. The boundary between the Manitoba Lowland and the Saskatchewan Plain is marked by the Manitoba Escarpment. The Saskatchewan and Alberta plains are divided in the south by the Missouri Couteau. The landscape of the two plains is similar to that of the U.S: great plains, with prairie and rolling plains; deeply incised rivers; water-filled depressions called sloughs; dry streambeds called coulees; and, in the drier areas, mesas, buttes, and badlands. To the south and southeast of the Shield lies a triangular, flat and fertile plain bounded by Georgian Bay in Lake Huron, the St. Lawrence River, and Lake Ontario. Grasslands made of up many different types of stunted bushes and grasses extend over much of the southern Canadian Great Plains.

Tundra

The Tundra is situated on the northern Canadian Shield, which is an area of Precambrian rock with moss covered, frozen subsoil. Low-growing grasses and small bushes thrive in this arctic region. Altogether, the tundra makes up a significant portion of the 27.4 percent of Canadian territory that lies north of the tree line.

Forests

Canada's great boreal forest is the largest of its woodlands, occupying 35 percent of the total Canadian land area and 77 percent of Canada's total forest land. Named for the Greek god of the north wind, Boreas, this forest stretches between northern tundra and southern grassland and mixed hardwood trees, and constitutes a band 600 mi (1,000 km) wide. The boreal forest is characterized by the predominance of coniferous trees, which first occurred during the Miocene Epoch, from 12 to 15 million years ago, and now is an important source of paper products, jack pine railway ties, and logs.

Population Centers – Canada

(2001 POPULATION ESTIMATES)

Name	Population	Name	Population
Toronto, Ontario	4,682,897	Calgary, Alberta	951,395
Montréal, Quebec	3,426,350	Edmonton, Alberta	937,845
Vancouver, British Columbia	1,986,965	Québec, Quebec	682,757
Ottawa-Hull, Ontario–Quebec	1,063,664	Winnipeg, Manitoba	671,274
		Hamilton, Ontario	662,401
		London, Ontario	432,451

SOURCE: "Census Metropolitan Area Populations and Growth Rates," *2001 Census Analysis Series: A Profile of the Canadian Population: Where We Live,* Statistics Canada.

Provinces and Territories – Canada

2001 POPULATION ESTIMATES

Name	Population	Area (sq mi)	Area (sq km)	Capital
Alberta	3,064,249	248,800	644,390	Edmonton
British Columbia	4,095,934	358,971	929,730	Victoria
Manitoba	1,150,034	211,723	548,360	Winnipeg
New Brunswick	757,077	2,834	72,090	Fredericton
Newfoundland and Labrador	533,761	145,510	371,690	Saint John's
Nova Scotia	942,691	20,402	52,840	Halifax
Ontario	11,874,436	344,090	891,190	Toronto
Prince Edward Island	138,514	2,185	5,660	Charlottetown
Quebec	7,410,504	523,859	1,356,790	Québec
Saskatchewan	1,015,783	220,348	570,700	Regina
Northwest Territories	40,860	452,478	1,171,198	Yellowknife
Nunavut Territory	28,159	870,424	2,254,402	Iqaluit
Yukon Territory	29,885	186,660	483,450	Whitehorse

SOURCE: Statistics Canada.

The Rocky Mountain area boasts alpine fir, Engelmann spruce, lodgepole pine, aspen, and mountain hemlock. Additionally, the west is known for western hemlock, red cedar, Douglas fir, Sirka spruce, and western white pine. Further east, aspen, bur oak, balm of Gilead, cottonwood, balsam poplar, white birch, and other deciduous trees dot the great prairies. In the Great Lakes region, the flora is characterized by forests of white pine, hemlock, sugar and red maples, yellow birch, and beech trees. Balsam fir, white cedar, tamarack, white birch, and aspen dominate eastern Canada. Red spruce has colonized the Maritime region, black spruce the eastern Laurentian zone, and white spruce the western areas.

HUMAN POPULATION

Due to extremely harsh winters and adverse climatic, geographic, and soil conditions, approximately 85 percent of the population is concentrated within 180 mi (300 km) of the United States. Most of the northern half of Canada is completely uninhabited except for scattered settlements of mainly indigenous Amerindian population. People of British Isles origin account for 28 percent of the Canadian population, French origin 23 percent, other European 15 percent, Amerindian 2 percent, and other—mostly Asian, African, and Arab—6 percent. People of mixed background comprise 26 percent, making Canada an increasingly interracial society. Forty-two per cent of Canadians profess to be Roman Catholic, 40 percent Protestant, and 18 percent other faiths. English and French are the official languages. Fifty-nine percent of the population speaks English, and 23.2 percent speaks French. However, Canada is a country of immigrants where as much as 17.5 percent of the population speaks other languages.

NATURAL RESOURCES

Large deposits of oil and potash are found on the interior plains. The country also has significant deposits of iron ore, nickel, zinc, copper, gold, lead, molybdenum, potash, silver, coal. There is a large supply of petroleum and natural gas buried under and around the Beaufort Sea. In addition, Canada has great potential in hydropower and water resources, and is rich in fish, timber, and wildlife. Five percent of the land is arable, 3 percent is utilized for meadows and pastures, and some 35 percent is forest and woodland.

FURTHER READINGS

Atlas of Canada. *Facts about Canada.* http://atlas.gc.ca/site/english/facts/index.html (Accessed June 2002).

CyberNatural Software. *Canada's Aquatic Environments.* http://www.aquatic.uoguelph.ca (Accessed June 2002).

Lightbody, Mark, Thomas Huhti, and Ryan Ver Berkmoes. *Canada.* 7th ed. Oakland, Calif.: Lonely Planet, 1999.

Lonely Planet World Guide. *Canada.* http://www.lonelyplanet.com/destinations/north_america/canada/ (Accessed June 2002).

MacLean, Doug, comp. *Canadian Geographic Quizbook: Over 1000 Questions on All Aspects of Canadian Geography.* Markham, Ontario: Fitzhenry and Whiteside, Ltd., 2000.

Rayburn, Alan. *Naming Canada: Stories about Canadian Place Names.* Toronto: University of Toronto Press, 2001.

Cape Verde

- **Area:** 1,557 sq mi (4,033 sq km) / World Rank: 168
- **Location:** Northern and Western Hemispheres in Atlantic Ocean, about 370 mi (595 km) west of Dakar, Senegal
- **Coordinates:** 16°00′N, 24°00′W
- **Borders:** None
- **Coastline:** 598 mi (965 km)
- **Territorial Seas:** Exclusive economic zone of 200 NM and territorial seas of 12 NM
- **Highest Point:** Mt. Fogo, 9,281 ft (2,829 m)
- **Lowest Point:** Sea level
- **Longest Distances:** 206 mi (332 km) SE-NW / 186 mi (299 km) NE-SW
- **Longest River:** No rivers of importance, only four islands have year-round running streams
- **Largest Lake:** None
- **Natural Hazards:** Subject to extended droughts, harmattan winds, volcanic and seismic activity
- **Population:** 405,163 (2001 est.) / World Rank: 167
- **Capital City:** Praia, located on São Tiago Island, on the southeastern coast
- **Largest City:** Praia 68,000 (2000 est.)

OVERVIEW

Cape Verde is an island nation located some distance off the coast of West Africa. The islands are of volcanic origin and most are mountainous. These mountains support trees that are typical of both temperate and tropical climates. This flora is possible because of the varying elevations found on the islands. The high ground and slopes of the southwesterly mountain faces allow growth of lush vegetation because of moisture condensation.

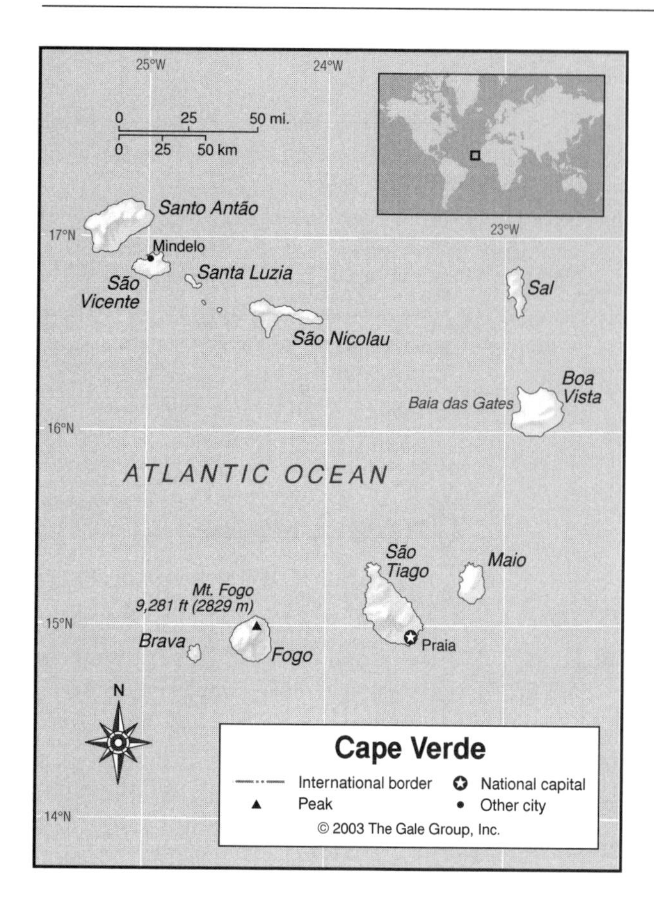

MOUNTAINS AND HILLS

Mountains

Except for the low-lying islands of Sal, Boa Vista, and Maio, the Cape Verde islands are quite mountainous with both rugged cliffs and deep ravines. Mount Fogo (Pico da Cano) (9,281 ft / 2,829 m), located on Fogo, is the highest peak and the only active volcano.

INLAND WATERWAYS

There are no large bodies of fresh water in the Cape Verde islands, and only four islands possess year-round running streams.

THE COAST, ISLANDS, AND THE OCEAN

Oceans and Seas

The Cold Canary Current, which runs adjacent to the islands, is thought to provide an ideal environment for a fishing industry that has yet to be fully exploited for its economic potential. An estimated 50,000 tons of fish, lobster, and additional marine products are available for harvest, yet only some 1,500 tons reach markets annually.

Major Islands

Cape Verde consists of 10 islands and five islets that are divided into a northern windward group (Barlavento)—Santo Antão, São Vicente, Santa Luzia (uninhabited), São Nicolau, Sal, Boa Vista, and two islets—and a southern leeward group (Sotavento)—Brava, Fogo, São Tiago, Maio, and three islets. The two districts of Barlavento and Sotavento owe their names to the direction of the prevailing northeasterly winds.

The Coast and Beaches

The beaches at Baia das Gates on Boa Vista support considerable tourist traffic. Cape Verde has several fine harbors, with Mindelo on São Vicente being the principal one.

CLIMATE AND VEGETATION

Temperature

There are only two seasons in Cape Verde due to the cold Atlantic current that produces an arid atmosphere in the islands. December through June is cool and dry, with temperatures at sea level averaging 70°F (21°C). July through November is warmer, with temperatures averaging 81°F (27°C).

Rainfall

Rainfall is scarce overall in the islands, although some precipitation occurs in the latter half of the year. Generally accumulations are 5 in (13 cm) in the northern islands, and 12 in (30 cm) in the southern ones. The country is subject to cyclical droughts, often lasting years and devastating the environment.

Grasslands

There is little in the way of lowlands and grasslands in Cape Verde. Only three of the islands of Cape Verde are low-lying—Sal, Boa Vista, and Maio—and only nine percent of the land of Cape Verde is arable. The well-watered regions of Cape Verde are the high grounds and southwestern slopes of the mountains. This is not due to rainfall, but to condensation of moisture, which accumulates off the mountains from the Atlantic currents.

Deserts

Much of Cape Verde's scant flatlands are effectively semi-arid. This is due not only to scant rainfall, but also to storms, insect infestation, and overuse of the land, all which make erosion a constant problem.

HUMAN POPULATION

The Cape Verde islands were uninhabited when they were discovered and colonized by Portugal in the fifteenth century. The islands were subsequently used as a shipping center for the slave trade, and most of the current population of 405,163 (July 2001 est.), is descended from Portuguese settlers or Africans brought there as slaves. More than half of the country's population lives on the island of São Tiago. The primary language spoken is Portuguese.

Population Centers – Cape Verde

(1990 CENSUS OF POPULATION)

Name	Population
Praia (capital)	61,644
Mindelo	47,100
São Felipe	5,600

SOURCE: United Nations Statistics Division.

Counties – Cape Verde

2000 POPULATION ESTIMATES

Name	Population	Area (sq mi)	Area (sq km)	Capital
Boa Vista	4,193	239	620	Sal Rei
Brava	6,820	26	67	Nova Sintra
Fogo	37,409	184	476	São Filipe
Maio	6,742	104	269	Porto Inglês
Paúl	8,325	21	54	Pombas
Porto Novo	17,239	214	558	Porto Novo
Praia	106,052	153	396	Praia
Ribeira Grande	21,560	64	167	Ponta do Sol
Sal	14,792	83	216	Santa Maria
Santa Catarina	49,970	94	243	Assomeda
Santa Cruz	32,822	58	149	Pedra Badejo
Santiago	236,352	383	991	
Santo Antão	47,124	301	779	
Sao Nicolau	13,536	150	388	Riberia Brava
Sao Vincente	67,844	88	227	Mindelo
Tarrafal	18,059	78	203	Tarrafal

SOURCE: INE (Instituto Nacional de Estatística de Cabo Verde—National Institute of Cape Verde Statistics).

NATURAL RESOURCES

Cape Verde has a very weak economy. There is very little farmland, and most food is imported. Nor do the islands have many mineral resources, although deposits of salt and pozzuolana (a siliceous volcanic ash used to produce hydraulic cement) were mined in the 1990s. Fishing holds the greatest economic potential for Cape Verde, but has yet to realize its full potential as an industry.

FURTHER READINGS

Duncan, T. Bentley. *Atlantic Islands: Madeira, the Azores, and the Cape Verdes in Seventeenth Century Commerce and Navigation.* Chicago: University of Chicago Press, 1972.

Halter, Marilyn. *Between Race and Ethnicity: Cape Verdean American Immigrants, 1880–1965.* Urbana, IL: University of Illinois Press, 1993.

Irwin, Aisling, and Colum Wilson. *Cape Verde Islands: The Bradt Travel Guide.* Old Saybrook, CT: Globe Pequot Press, 1998.

U.S. Dept. of State, Bureau of Public Affairs, Office of Public Communication. *Background Notes, Cape Verde.* Washington, D.C.: Superintendent of Documents, 1998.

Cayman Islands

Overseas Territory of the United Kingdom

- **Area:** 100 sq mi (259 sq km); Grand Cayman 76 sq mi (97 sq km) / Little Cayman, 10 sq mi (26 sq km); Cayman Brac, 14 sq mi (36 sq km) / World Rank: 196
- **Location:** Caribbean Sea in the Northern and Western Hemispheres; Grand Cayman lies about 180 mi (290 km) northwest of Jamaica and 150 mi (240 km) south of Cuba; Little Cayman and Cayman Brac lie about 90 mi (145 km) further northeast
- **Coordinates:** 19°30′N, 80°30′W
- **Borders:** No international borders
- **Coastline:** 100 mi (160 km)
- **Territorial Seas:** 12 NM (22 km)
- **Highest Point:** The Bluff, Cayman Brac, 141 ft (43 m)
- **Lowest Point:** Sea level
- **Longest River:** None of significance
- **Natural Hazards:** Hurricanes
- **Population:** 35,527 (July 2001 est.) / World Rank: 196
- **Capital City:** George Town, located on the western shore of Grand Cayman Island
- **Largest City:** George Town, 13,000 (2001 est.)

OVERVIEW

The three islands that make up the Cayman Islands are Grand Cayman, Cayman Brac, and Little Cayman. The low-lying islands are outcroppings of the underwater mountain range known as the Cayman Ridge that extends from the southeast area of Cuba west-southwest toward Belize in Central America. Coral reefs surround the Cayman Islands.

Two types of limestone make up the islands. The older type, known as bluff limestone, formed the central core of each island about 30 million years ago. The limestone surrounding this core, known as "ironshore" or coastal limestone, was formed from limestone compacted with coral and mollusk shells around 120,000 years ago. The Cayman Islands are situated on the Caribbean Plate.

MOUNTAINS AND HILLS

On Grand Cayman Island, the highest elevation is about 60 ft (18 m). On Cayman Brac, The Bluff, the highest point in the Cayman Islands, rises along the 12 mi (19 km) length of the island, reaching a height of 141 ft (43

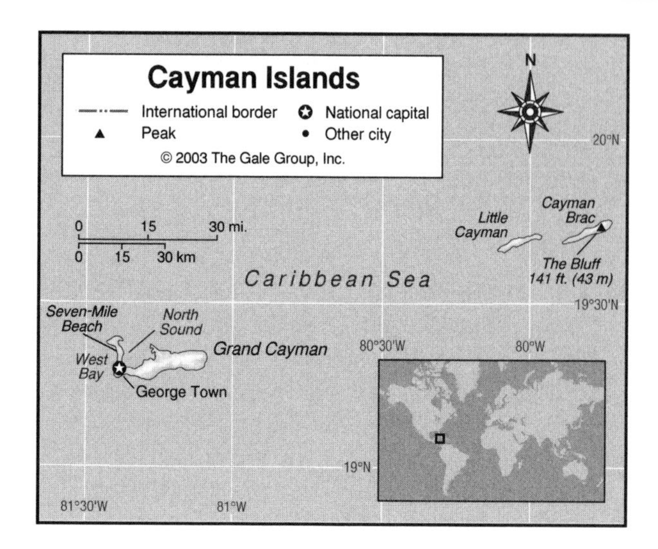

m) at the far eastern end of the island, where it forms a dramatic cliff at the edge of the sea. Little Cayman has little variance in elevation.

INLAND WATERWAYS

The central part of Grand Cayman Island features some minor wetlands where the mangrove, a dense tropical plant, thrives. There are no inland waterways on the Cayman Islands.

THE COAST, ISLANDS, AND THE OCEAN

Oceans and Seas

The deepest part of the Caribbean Sea lies between the Cayman Islands and Jamaica. Known as the Cayman Trough, the sea is over 4 mi (6 km) deep in this area. Another deep area, known as the Barlett Deep, lies between the Cayman Islands and Honduras to the southwest. A channel 7 mi (11 km) wide separates Little Cayman from Cayman Brac.

Major Islands

Grand Cayman spans about 25 miles (40 km) from east to west, and lies 150 miles (240 km) south of Cuba and about 180 miles (290 km) west of Jamaica. Little Cayman and Cayman Brac lie 80 miles (130 km) and 90 miles (144 km) to the east of Grand Cayman, respectively; both extend about 10 mi (16 km) from east to west, and about a mile (less than 2 km) from north to south.

The Coast and Beaches

At the western end of Grand Cayman Island, the north coast features the North Sound, a large lagoon measuring about 35 sq mi (120 sq km). Grand Cayman is also the site of the Seven-Mile Beach, a long, uninterrupted stretch of sandy beach, actually measuring 5 mi (8 km), along the island's westernmost coast along West Bay.

Islands – Cayman Islands

Name	Area (sq mi)	Area (sq km)
Grand Cayman	76	197
Cayman Brac	14	36
Little Cayman	10	26

SOURCE: *Geo-Data: The World Geographical Encyclopedia,* 2nd ed. Detroit: Gale Research, 1989.

The Cayman Bank is a shallow area 10 mi (16 km) west of Grand Cayman. It measures about 5 mi (8 km) long by one-half mile (800 m) wide.

CLIMATE AND VEGETATION

Temperature

The average high temperature in summer (May to October is 85°F (29°C); the average high temperature in winter (November to April) is 75°F (24°C). The record low temperature of 58°F (24°C) was set January 19, 2000.

Rainfall

The rainy season extends from May to October with May-June and September-October usually being the wettest months. The western side of the islands receives the most rainfall since the prevailing winds blow toward the west; the record rainfall was November 30, 1993, with 7.8 in (198 mm). Tropical storms and even hurricanes occasionally hit the Cayman Islands and their neighbors in the Caribbean and Central America.

HUMAN POPULATION

The majority of Caymanians reside on Grand Cayman. Cayman Brac, the most easterly island, has the fewest permanent residents. Over 600,000 tourists visit the Cayman Islands annually.

NATURAL RESOURCES

Tourism is the primary economic activity, attracting sport fishers and divers. The government has established several marine parks, bird sanctuaries, and other nature reserves.

FURTHER READINGS

Frink, Stephen. *The Cayman Islands Dive Guide.* New York: Abbeville Press Publishers, 1999.

Permenter, Paris. *Adventure Guide to the Cayman Islands.* Edison, NJ: Hunter, 2000.

Philpott, Don. *Cayman Islands.* Edison, NJ: Hunter, 2000.

Smith, Martha K. *The Cayman Islands: The Beach and Beyond.* Edison, NJ: Hunter, 1995.

Smith, Roger C. *The Maritime Heritage of the Cayman Islands.* Gainesville: University Press of Florida, 2000.

World Travel Guide.net. *Cayman Islands.* http://www.travel-guides.com/data/cym/cym.asp (accessed March 5, 2002).

Central African Republic

- **Area:** 240,534 sq mi (622,984 sq km) / World Rank: 44
- **Location:** Northern and Eastern Hemispheres, Central Africa, bordering Sudan to the northeast and east, Republic of the Congo to the southwest, Democratic Republic of the Congo to the southeast, Cameroon to the west, Chad to the northwest
- **Coordinates:** 7°N, 21°E
- **Borders:** 3,233 mi (5,203 km) total / Cameroon, 495 mi (797 km); Chad, 744 mi (1,197 km); Democratic Republic of the Congo, 980 mi (1,577 km); Republic of the Congo, 290 mi (467 km); Sudan, 724 mi (1,165 km)
- **Coastline:** None
- **Territorial Seas:** None
- **Highest Point:** Mont Ngaoui, 4,659 ft (1,420 m)
- **Lowest Point:** Ubangi River, 1,099 ft (335 m)
- **Longest Distances:** 893 mi (1,437 km) E-W / 480 mi (772 km) N-S
- **Longest River:** Ubangi 1,400 mi (2,253 km; including the Uele River)
- **Natural Hazards:** Subject to flooding, harmattan winds in the north
- **Population:** 3,576,884 (July 2001 est.) / World Rank: 126
- **Capital City:** Bangui, southwestern Central African Republic
- **Largest City:** Bangui, 553,000 (2000 est.)

OVERVIEW

In accordance with its name, the landlocked Central African Republic, in equatorial Africa, lies roughly at the center of the African continent, more than 375 mi (603 km) from the Atlantic Ocean. Most of the country consists of a large plateau that separates the basin of Lake Chad, to the north, from that of the Congo River to the south. The dominant features of the landscape are the Bongo Mountains in the eastern part of the country and the Karre Mountains (Yadé Massif) to the west. The Central African Republic is located on the African Tectonic Plate.

MOUNTAINS AND HILLS

Mountains

The country's central plateau rises to the Bongo Mountains near the border with Sudan in the northeast, and to the Karre Mountains near the borders with Cameroon and Chad in the northwest. The Bongo Mountains rise to elevations as high as 4,488 ft (1,368 m) and extend into the Sudan. The granite escarpment of the Karre Mountains in the northwest, a continuation of Cameroon's Adamawa Plateau, includes Mont Ngaoui (4,659 ft / 1,420 m), the Central African Republic's highest peak.

Plateaus

An undulating plateau with elevations roughly between 2,000 ft and 2,500 ft (610 m and 762 m) extends across the center of the country, covered with grass and scattered groups of trees, and broken in places by river valley, ridges, and isolated granite peaks called *kaga.* Its eastern portion slopes southward toward the Mbomou and Ubangi rivers. A large expanse of sandstone plateau is located in the southwestern part of the country, in the vicinity of Berbérati and Bouar.

INLAND WATERWAYS

Lakes

Many of the country's lakes are seasonal, filling during the rainy season and drying up when the rains stop.

Rivers

The Central African Republic is drained by two river systems: one flowing south, the other flowing north. Of the southward-flowing rivers, the Chinko, Mbari, Kotto, Ouaka, and Lobaye are tributaries of the Ubangi River, which forms most of the country's southern border with the Democratic Republic of the Congo; the Mambéré and Kadei, which join in the southwest to form the Sangha, are tributaries of the Congo River. Two northern rivers, the Ouham and Bamingui, are tributaries of the Chari River, which flows northward to the Chad Basin.

From the conjunction of the Uele and Mbomou rivers, the Ubangi flows westward along the Congo border from Bangassou, turning south after Bangui to form the border between the Republic of the Congo and the Democratic Republic of the Congo, and draining into the Congo River Basin. Draining frequently flood-swollen water, the Ubangi transports large volumes of water, discharging at least 30,000 cu ft of water (849 cu m) per second at Bangui during the rainy seasons.

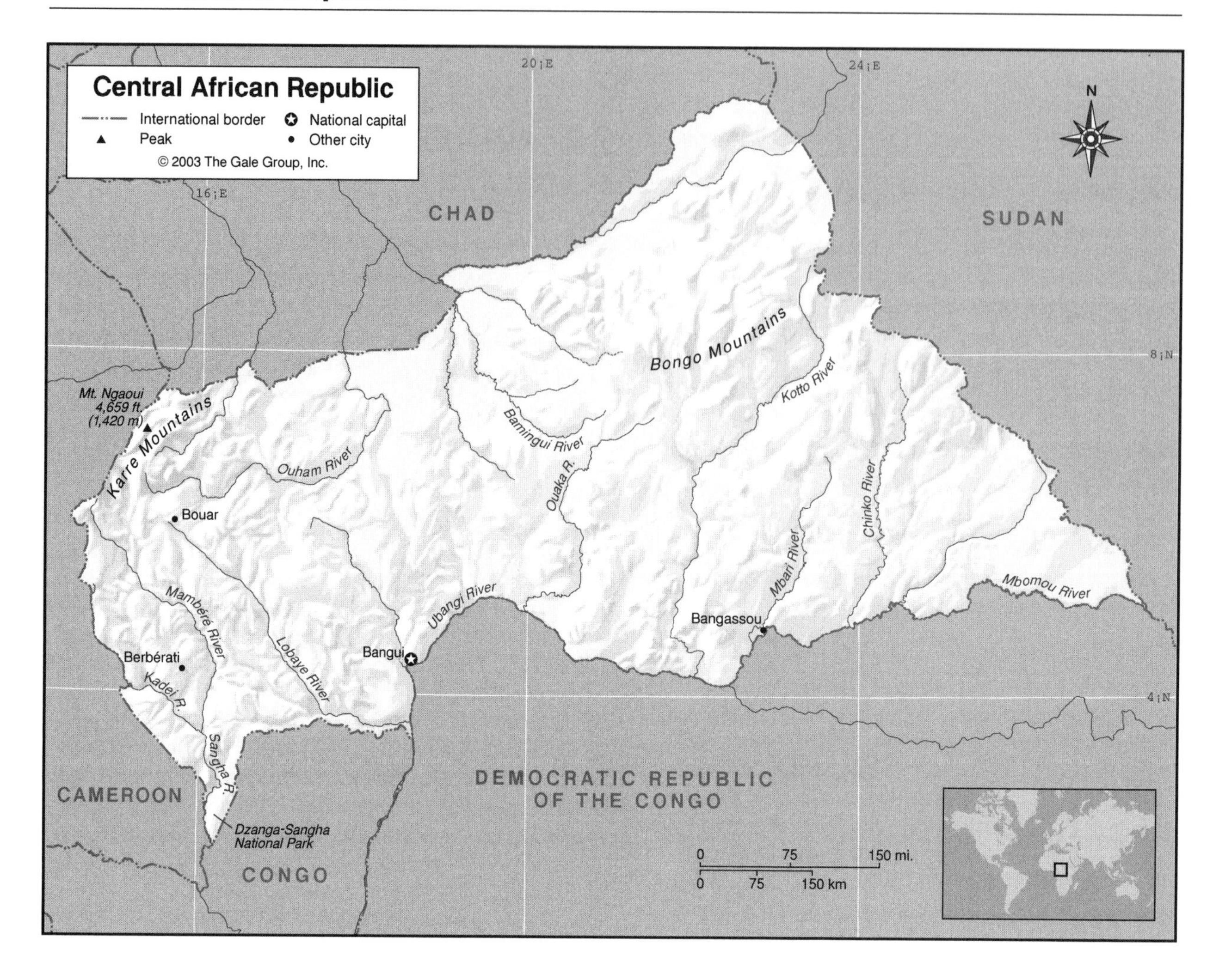

THE COAST, ISLANDS, AND THE OCEAN

The Central African Republic is landlocked.

CLIMATE AND VEGETATION

Temperature

The climate is tropical, but temperatures are moderated by rainfall and altitude, and temperatures average around 80°F (27°C) all year. Temperatures in Bangui average 70°-84°F (21-29°C) in July and August, and 70°-93°F (21-34°C) in February. The harmattan, a hot, dry Saharan wind, affects the climate during the summer months.

Rainfall

Rainfall varies, increasing from north to south. The northern part of the country, influenced by proximity to the sub-Saharan Sahel region, is relatively dry, with an annual average rainfall of about 30 in (76 cm) and a six-month dry season from November to April. The northeast, with a semiarid climate, is the driest part of the country.

The dry season is shorter in most of the central plateau region, which receives up to 60 in (152 cm) of rain per year, sometimes falling in very heavy showers. The southern part of the country has only a very brief dry season of about two months, and in some parts of the region

GEO-FACT

In the southwest region of the country around the Sangha River, the Central African Republic established the Dzanga-Sangha nature reserve. It protects the last of the country's rainforests, a habitat for such wildlife as lowland gorillas, forest elephants, bongos, crowned eagles, waterbuck, warthogs, chimpanzees, and many monkey species (including the white-bearded DeBrazza monkey).

Population Centers – Central African Republic

(2000 POPULATION ESTIMATES)

Name	Population
Bangui (capital)	553,000
Berwrati	125,000

SOURCE: Projected from United Nations Statistics Division data.

Prefectures – Central African Republic

Name	Area (sq mi)	Area (sq km)	Capital
Bamingui-Bangoran	22,471	58,200	Ndélé
Bangui	26	67	Bangui
Basse-Kotto	6,797	17,604	Mobaye
Gribingui-Économique	7,720	19,996	Kaga-Bandoro
Haut-Mbomou	21,440	55,530	Obo
Haute-Kotto	33,456	86,650	Bria
Haute-Sangha	11,661	30,203	Berbérati
Kemo-Gribingui	6,642	17,204	Sibut
Lobaye	7,427	19,235	Mbaïki
Mbomou	23,610	61,150	Bangassou
Nana-Mambere	10,270	26,600	Bouar
Ombella-Mpoko	12,292	31,835	Bimbo
Ouaka	19,266	49,900	Bambari
Ouham	19,402	50,250	Bossangoa
Ouham-Pendé	12,394	32,100	Bozoum
Sangha-Économique	7,495	19,412	Nola
Vakaga	17,954	46,500	Birao

SOURCE: *Geo-Data: The World Geographical Encyclopedia,* 2nd ed. Detroit: Gale Research, 1989.

rain falls year round. Annual rainfall here averages 70 in (178 cm) per year or more, and river levels can rise several feet in a few hours during the rainy season.

Grasslands

Much of the plateau region is savanna grassland, with forest growth along the rivers. Ground cover varies with the seasons. Foliage is lush during the rainy season, but the leaves turn brown and fall during dry periods.

Deserts

The country's northeastern tip, which borders the Sahel, has species of vegetation that are supported by a semiarid climate, including various hardy grasses and shrubs, as well as shea and acacia trees.

Forests and Jungles

More than half the country is wooded, with thick tropical rainforest growing along the Ubangi River and in the country's southwestern tip. This gives way to savanna woodlands and grassland further north. The trees of the rainforest, which can grow as high as 150 ft (46 m), include sapele mahogany and obeche.

HUMAN POPULATION

Most of the country is sparsely inhabited, with populated areas concentrated along the rivers. Urban dwellers account for about half the population, and this percentage is on the increase.

NATURAL RESOURCES

The tropical hardwoods that grow in the country's rainforests are a substantial source of potential wealth. Rubber is another natural resource of the rainforest. Diamond mining is carried out in the sandstone plateau in the southwest, and other mineral resources include uranium and gold. The numerous rivers and waterfalls are a rich potential source of hydropower.

FURTHER READINGS

Fung, Karen. *Africa South of the Sahara.* http://www-sul.stanford.edu/depts/ssrg/africa/centralafr.html (accessed March 4, 2002).

Hagmann, Michael. "On the Track of Ebola's Hideout?" *Science,* October 22, 1999, p. 654.

Kalck, Pierre. *Central African Republic: A Failure in Decolonisation.* Translated by Barbara Thomson. New York: Praeger, 1971.

O'Toole, Thomas. *Central African Republic in Pictures.* Minneapolis: Lerner Publications, 1989.

Sillery, Bob. "Urban Rainforest: An African Jungle Comes to Life on New York's West Side." *Popular Science,* March 1998, pp. 70-1.

Titley, Brian. *Dark Age: The Political Odyssey of Emperor Bokassa.* Montreal: McGill-Queen's University Press, 1997.

Chad

- **Area:** 495,752 sq mi (1,284,000 sq km); World Rank: 22
- **Location:** Central Africa, south of Libya
- **Coordinates:** 15° N, 19° E
- **Borders:** 3,700 mi (5,968 km) / Libya 655 mi (1,055 km); Sudan 845 mi (1,360 km); Central African Republic 745 mi (1,197 km); Cameroon 651 mi (1,094 km); Nigeria 55 mi (87 km); Niger 730 mi (1,175 km)
- **Coastline:** None
- **Territorial Seas:** None
- **Highest Point:** Emi Koussi 11,204 ft (3,415 m)
- **Lowest Point:** Bodele Depression 525 ft (160 m)

- **Longest Distances:** 1,094 mi (1,765 km) N-S; 639 mi (1,030 km) E-W
- **Longest River:** Chari River 744 mi (1,200 km)
- **Largest Lake:** Lake Chad (shared, varies regionally from c.4,000 - c.10,000 sq mi (10,360-25,900 sq km)
- **Natural Hazards:** hot, dry, dusty harmattan winds occur in north; periodic droughts; locust plagues
- **Population:** 7,114,400 (2002 est.) World Rank: 92
- **Capital City:** N'Djamena, located at the confluence of the Logone and Chari Rivers on Chad's western border with Cameroon
- **Largest City:** N'Djamena, 601,500 (2002 est.)

OVERVIEW

A relatively large country in north-central Africa, Chad extends north-south for more than 1,000 mi (1,609 km) from the Tropic of Cancer, which crosses the heart of the Sahara Desert at 23.5° N, through broad transitional zones of sub-arid and humid savanna to the edge of the tropical rain forest at about 7.5° N. The nation is landlocked, having no easy or direct access to the sea. N'Djamena, the nation's capital and the only major city, lies 700 mi (1,126 km) from the nearest seaport—Douala, a Cameroon port on the Atlantic Ocean.

The most important structural features are a broad, shallow central bowl and Lake Chad, together with the lake's major water source—the Chari-Logone river system. This drainage network collects considerable flow from the uplands along the southern border and adjacent areas in the Central African Republic and Cameroon. Part of this great volume of water is retained in swampy flood plains along the way; some is used for irrigation; much of the annual flow reaches the lake. The water is not highly mineralized, and the lake has continued to be an economically important reservoir of fresh water. The lake is an inland basin with no outlet to the sea and, in the sub-arid climate, there is a high rate of surface evaporation.

Northward from Lake Chad and the other depressions and swamps of the western border area, the basin extends for more than 500 mi (804 km) to the plateaus, mountain ranges, and extinct volcanoes associated with the Tibesti Massif (Mountains) in northern Chad, a major landmark of the Sahara Desert. Eastward and southward from the lake, the relatively flat sedimentary basin extends for several hundred miles before rising gently to the rolling plateaus and scattered low mountains of the eastern and southern border areas. Central Chad is an area of mixed farming and grazing, a transition zone between the well-watered south and the barren north. Chad is situated in the African Plate.

MOUNTAINS AND HILLS

Mountains

The highest mountains in Chad are found in the Tibesti Massif, a northern range with—a defunct volcano at Emi Koussi rising to an altitude of more than 11,204 ft (3,415 m), nearly 10,000 ft (3,048 m) above the surrounding plateau. Elsewhere the uplands and lesser mountains in northern, eastern, and southern areas of the country average from 1,000 to 3,000 ft (305 to 914 m) in elevation, sloping toward the central basin and the western border area. The Ennedi and Biltine highlands of the eastern border form the divide between this great inland basin and the Nile River drainage system in Sudan.

Plateaus

From the central bowl to southern Chad, the land slopes upward almost imperceptibly to rolling plateaus, which for the most part are less than 2,000 ft (610 m) above sea level. The plateaus are marked here and there by mountains, such as the Guera Massif near Mongo, which has at least one peak above 4,900 ft (1,493 m).

Hills and Badlands

Isolated hills (inselbergs) generally found in the Chari-Baguirmi (Bagirmi) and Mayo (Mao)-Kebbi regions do not exceed for the most part 1,500 ft (457 m). These rocky outcroppings, which resemble piles of boulders, rise unexpectedly over the flat and gentle rolling landscape, but support only sparse vegetation.

INLAND WATERWAYS

Lake Chad

Lake Chad is an internal basin having no outlet to the sea, making it the seventh-largest permanent lake in the world, and Africa's fourth-largest freshwater lake. It is, however, less than six feet deep during some dry seasons and less than twelve feet deep in most areas during the annual flood stage. The surface area varies greatly by season ranging from c.4,000 to 10,000 sq mi (12,950 to 25,900 sq km), and it is estimated that the lake has shrunk to one-twentieth of its size 40 years ago.

The area covered by the shallow waters depends primarily upon the balance between the rate of evaporation—about eight inches annually from the surface of the lake—and the flow of the Chari-Logone river system. Much of the 40 billion cubic meters (volume) carried by this river system eventually reaches the lake. However, overgrazing and irrigation are mainly to blame for the shrinking of the lake. Losses into the shore areas, especially into the dunes on the northeastern shores, may account for no more than one foot of depth a year. Rivers other than the Chari and the Logone originate in semi-arid areas and carry relatively small inputs; rainfall over Lake Chad adds about fifteen inches annually.

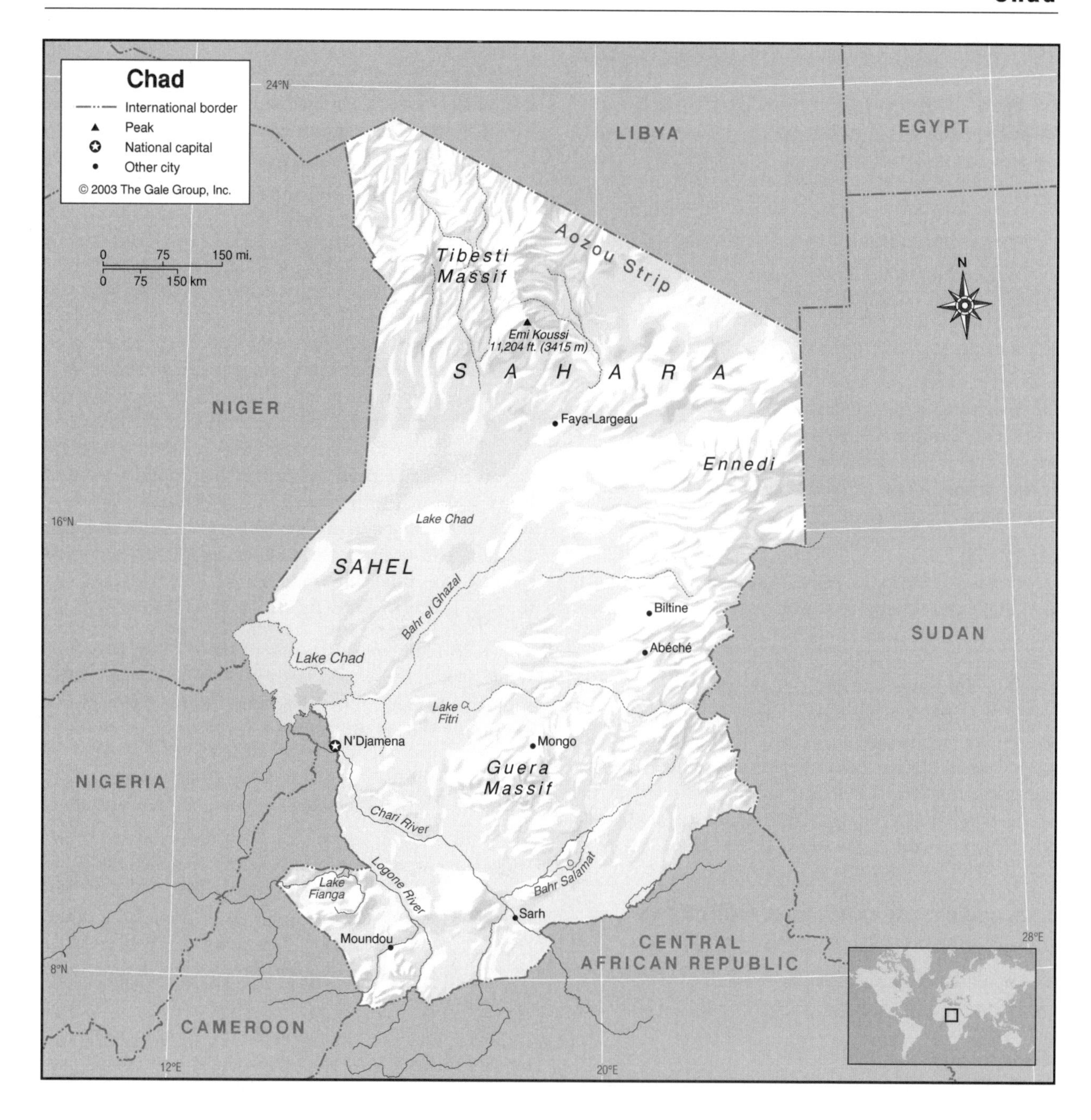

Other Lakes

Other very shallow bowls, similar to Lake Chad in geologic origin, are scattered across the flat plains northeast, east, and southeast of the lake. Almost all of these sandy depressions are dry before the end of the annual dry (winter) season. One of the largest, the Bahr el Ghazal, receives some overflow from Lake Chad during its flood stage. Lake Fitri to the southeast holds water year round and is a major supplier of fish in the area.

The only other lakes of consequence are in the southwest in Mayo-Kebbi. Lake Fianga is a shallow body of water that expands and contracts with the fluctuations of the wet and dry seasons.

Rivers

The Chari and Logone Rivers in southern Chad flow throughout the year, although they become shallow and sluggish toward the end of the dry season. Their headwaters are in the equatorial rain belt in Cameroon and the Central African Republic, where rainfall averages more than fifty inches per year. A huge volume of water is carried from these upland areas into southern Chad, where the annual rainfall ranges between thirty and fifty inches. Tributaries, such as the Mayo-Kebbi in southwestern Chad, the Bahr Salamat in the southeast, and innumerable smaller streams, add to the flow as the six-month rainy season progresses.

GEO-FACT

As of 2002, Lake Chad had shrunk to about 1/20 of its 1960s size. About half of the water loss can be attributed to the implementation of extensive irrigation systems. The countries that border the lake—Chad, Nigeria, Niger, and Cameroon—drain water from the lake for irrigation.

There are no permanent streams in northern or central Chad. Summer rainfall collected by the various shallow wadis (seasonally dry streambeds) flows toward inland basins; most of these streams disappear in the sands soon after the end of the brief rainy season.

Wetlands

Chad's wetlands principally occur from seasonal flooding during the rainy season, especially in the lower reaches of the Chari and Logone Rivers. These two major rivers join at N'Djamena and inundate the flat delta area for more than 50 mi (80 km) between the capital city and Lake Chad. Much of the land around N'Djamena and for a considerable distance upstream (southeastward) is also under water during the average autumn flood season. Swampy areas are also found farther upstream in the areas of Sarh and Moundou, on tributary streams, and in low-lying areas within the Lake Chad basin.

THE COAST, ISLANDS, AND THE OCEAN

Chad is completely landlocked.

CLIMATE AND VEGETATION

Temperature

From north to south, Chad has three climate zones. In the north the Saharan climate offers extreme temperatures between day and night. In the Sahel where N'Djamena is located the average daily maximums and minimums are 108° F (42° C) and 73° F (28° C) in April, and 91° F (33° C) and 57° F (14° C) in December. In the south, the Sudanic region, temperatures are more moderate. The most extreme temperatures in the country range from 10° F (-12° C) to 122° F (50° C).

Rainfall

Like temperatures, rainfall varies considerably from north to south. In the Sahara at Faya-Largeau, rain averages only 1 in (2.5 cm) annually. The rains last from April (in the south) or July (farther north) through October. At N'Djamena, average annual rainfall is about 30 in (76 cm). In the far south it is as much as 48 in (122 cm).

Grasslands

Bodele (Bodéle, Djourab) Depression is a low area, northeast of Lake Chad and south of the Tibesti Mountains; it is historically important for its forage grasses.

The southern Sudanic climatic zone consists of broad grasslands or prairies and the Sahel covers itself with a carpet of brilliant green grass following the first rains of the season. Grasses exist for several months of the year in the Sahel with thorn trees interspersed throughout.

Deserts

Desert covers roughly one-half of the country beginning with the Sahelian zone and extending to Libya and the Aozou strip in the north. The southern reaches of the desert extend north and northeast of Lake Chad within the Chad basin for more than 500 mi (800 km). The region is characterized by great rolling dunes separated by very deep depressions. In some of these are found oases with date-palm groves. Although vegetation holds the dunes in place in the Kanem region, farther north they are bare, fluid, and rippling.

Forests

The middle zone, Sahel, consists of thorn trees and scrub. Palms and acacia trees grow in this region. The southern, Sudanic zone, is savanna country with mixed dry forests and grasslands.

HUMAN POPULATION

Chad is one of Africa's least densely populated countries (6 per sq mi/16 per sq km), but roughly half of the population lives in the southwestern 10%. In 2001, the growth rate was estimated at 3.29%. In 2000, 24% of the population was urban, and the capital, N'Djamena, had nearly six times the population of next largest city, Moundou.

The population is divided into the Islamic north and peoples of the south, the five southernmost prefectures. The Muslim grouping includes Arabs, Toubou, Hadjerai, Fulbe, Kotoko, Kanembou, Baguirmi, Boulala, and others. Among the non-Muslim indigenous peoples are the Sara (30%) living in the valleys of the Chari and Logone

Population Centers – Chad

(2000 POPULATION ESTIMATES)

Name	Population	Name	Population
N'Djamena, (capital, metropolitan area)	1,044,000	Koumra	115,000
		Kéla	98,000
		Bongor	136,000
Moundou	117,500	Sarh	129,600

SOURCE: United Nations Statistics Division estimates.

Rivers and the Ngambaye, Mbaye, Goulaye, Moundang, Moussei, and Massa.

Chad is located at a rich linguistic crossroads where more than 100 different languages and dialects are spoken. Arabic predominates throughout the northern two-thirds of the country and Sara and Sango are spoken in the south. French, along with Arabic, is an official language.

NATURAL RESOURCES

Chad's land use comprises 2% arable; 36% meadows and pastures; 11% forest and woodland; 51% other. In addition to the potentially mineral-rich Aozou Strip, north of Tibesti and along the Libyan border, Chad has significant petroleum reserves in the south. The Doba oil fields between Sahr and Moundou are expected to come on-stream in 2004 after completion of a 630 mi (1,050 km) buried pipeline through Cameroon to the port of Kribi on the Atlantic coast. Chad also has deposits of uranium, natron, kaolin, and fish (Lake Chad). Environmental threats include desertification, the shrinking of Lake Chad, inadequate supplies of potable water, and improper waste disposal in rural areas, which contributes to soil and water pollution.

FURTHER READINGS

Cabot, Jean. *Atlas practique du Tchad.* Paris: Institut de géographie national, 1972.

Cabot, Jean, and Christian Bouquet. *Le Tchad.* Paris: Presses universitaires de France, 1973.

Collelo, Thomas. *Chad: a country study.* Washington, D.C.: U.S. Government Printing Office, 1990.

Cordell, Dennis D. "The Savannas of North-Central Africa." Pages 30-74 in David Birmingham and Phyllis Martin (eds.), *History of Central Africa.* London: Longman, 1983.

Decalo, Samuel. *Historical Dictionary of Chad.* (2d ed.) Metuchen, N.J.: Scarecrow Press, 1987.

National Geographic. http://www.nationalgeographic.com/ (accessed March 25, 2002).

Chile

- **Area:** 292,260 sq mi (756,950 sq km) / World Rank: 39
- **Location:** Southern and Western Hemispheres, on the southwestern coast of South America, bordering Bolivia to the northeast, Argentina to the east, the Pacific Ocean to the south and west, Peru to the northwest
- **Coordinates:** 30°00′S, 71°00′W
- **Borders:** 3,835 mi (6,171 km) total / Argentina, 3,200 mi (5,150 km); Bolivia, 535 mi (861 km); Peru, 99 mi (160 km)
- **Coastline:** 3,999 mi (6,435 km)
- **Territorial Seas:** 12 NM
- **Highest Point:** Ojos del Salado, 22,573 ft (6,880 m)
- **Lowest Point:** Sea level
- **Longest Distances:** 2,653 mi (4,270 km) N-S / 221 mi (356 km) E-W
- **Longest River:** Loa, 275 mi (442 km)
- **Largest Lake:** General Carrera, 865 sq mi (2,240 sq km)
- **Natural Hazards:** earthquakes, tsunamis, volcanoes, floods, avalanches, landslides, severe storms
- **Population:** 15,328,467 (July 2001 est.) / World Rank: 61
- **Capital City:** Santiago, central Chile
- **Largest City:** Santiago, 5,261,000 (2000 est.)

OVERVIEW

Chile is a long, narrow country fringing the southwestern edge of South America, between the Pacific Ocean and the Andes Mountains. It reaches to Cape Horn, the southernmost edge of the continent, and touches the Atlantic Ocean at the Strait of Magellan, which separates it from Tierra del Fuego, an archipelago that it shares with Argentina.

Covering 2,653 mi (4,270 km) between its northern and southern extremities, Chile averages not much more than 100 mi (161 km) in width, making it the world's longest, narrowest country. Its 38-degree longitudinal span gives it an extremely varied climate and range of vegetation. Although Chile lies along the western edge of the South American continent, its capital city, Santiago, is located almost due south of New York.

Chile has three dominant topographical features, which are parallel to each other and span nearly the entire country from north to south. These include a low coastal range to the west; the Andes mountains to the east along the border with Argentina; and, lying between them, a structural depression whose composition varies at changing latitudes, ranging from a desert plateau in the north to the fertile Central Valley in the country's midsection, to land submerged under the fjords and channels of the south.

The country is commonly divided into regions by latitude. These include the northernmost desert region, the Norte Grande; a semiarid region, Norte Chico, immedi-

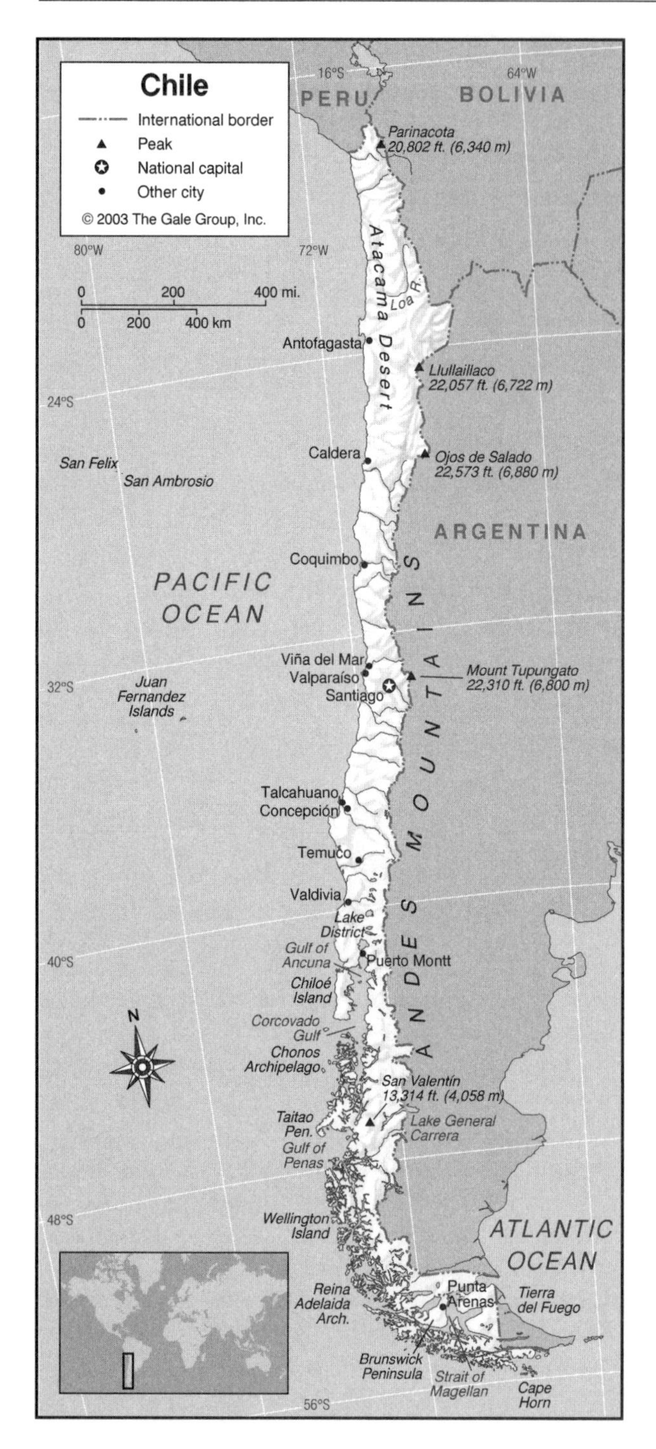

ately to the south; the temperate Central Valley that is the country's heartland; the south-central region that contains dense rain forest and the picturesque Lake District; and the cold, windswept southern region, with a coastline including thousands of islands extending down to Cape Horn. In addition, Chile also has several island dependencies in the Pacific Ocean including Easter Island, which is over 2,000 mi (3,218 km) west of the mainland. Chile is also one of several nations that claim land in Antarctica.

Chile is situated on the South American Tectonic Plate. The geologically young Andes form a seismically active environment. The country's Andean peaks include hundreds of volcanoes, of which several dozen are active. Well over 100 major earthquakes have been recorded since compilation of records began in 1575; many have been accompanied by fires and tidal waves.

MOUNTAINS AND HILLS

Mountains

The Andes Mountains reach their greatest elevations in Chile, where they span nearly the entire length of the country from north to south, starting with the peaks of the Atacama Desert in the north. The Andes chain forms most of Chile's border with Argentina to the east. Its valleys are carved deep into the often snow-capped peaks.

The crests of the Andean range are highest in the northern half of the country, where even the passes are at elevations of more than 10,000 ft (3,048 m). Those at heights above 14,700 ft (4,500 m) are permanently snowcapped. In this northern sector is Ojos del Salado, Chile's loftiest peak, and—at 22,573 ft (6,880 m)—the second-highest point in the Western Hemisphere. Many other peaks are over 20,000 ft (6,096 m) in height, including the volcanoes Parinacota (20,802 ft / 6,340 m) and Llullaillaco (22,057 ft / 6,722 m). Mount Tupungato rises to 22,310 ft (6,800 m) to the east of Santiago. It is the world's tallest active volcano.

South of Santiago the peaks of the Andes become progressively lower, with passes as low as 5,000 ft (1,524 m). In the far south, the Andes continue to decline in elevation, merging into the lowlands of Chilean Patagonia on both sides of the Strait of Magellan. Even here, however, high peaks are found, including San Valentín, at 13,314 ft (4,058 m). The system makes a final appearance at Cape Horn, which is the crest of a submerged mountain.

The peaks and plateaus of the coastal range are lower than those of the Andes, ranging from 1,000 to 7,000 ft (300 to 2,100 m) in the northern half of the country. They form a series of rounded forms with flat summits, broken occasionally by gorges and rivers. The system declines in elevation south of Valparaíso and plunges into the sea in the far south, although its peaks reappear as the islands of the southern archipelagoes.

Plateaus

In northern Chile, the central structural depression between the eastern and western mountain ranges consists of dry, barren, nitrate-rich plateau basins at elevations of between 2,000 ft and 4,000 ft (610 m and 1,219 m). In the north-central part of the country, much of this plateau land gives way to the ridges of Andean spurs, with fertile valleys in between.

INLAND WATERWAYS

Lakes

In south-central Chile, at the eastern edge of the Central Valley between Concepción and Puerto Montt, lies a district of lakes, hills, and waterfalls whose picturesque scenery has made it a popular tourist attraction. In the southern part of the district lies Lake Llanquihue, the largest lake found entirely within Chile. It has a maximum length of 22 mi (35 km), a maximum width of 25 mi (40 km), and depths of 5,000 ft (1,500 m). Further south, high in the Andes, is Lake General Carrera (Lake Buenos Aires). Shared with Argentina, it has an area of 865 sq mi (2,240 sq km).

Rivers

Because most of Chile's rivers flow across the narrow country in a westward direction—down the Andes and into the Pacific—they are short. Nevertheless, their steep path down the mountainsides makes them a good source for hydroelectric power. There are around thirty rivers, including the Loa, Aconcagua, Huasco, Coquimbo, Limarí, Mapocho, Maule, Maipo, Bío-Bío, Copiapó, and Toltén. The longest is the Loa River in the north. Other rivers in this parched northern region have smaller volumes and often dry up before reaching the sea. Elsewhere the rivers are regular and have a greater flow, fed by the permanent snowcaps atop the Andean peaks.

THE COAST, ISLANDS, AND THE OCEAN

Oceans and Seas

Chile borders the South Pacific Ocean, and the curved southernmost portion of its coast reaches to the Atlantic Ocean at the Straight of Magellan. South of Cape Horn lies the Drake Passage, separating from Antarctica far to the south. The Humboldt Current, flowing northward from Antarctica, keeps the waters of the Pacific off the Chilean coast very cold.

Major Islands

Chile exercises sovereignty over several islands in the South Pacific. The one farthest from the South American mainland is Easter Island, also known by its Polynesian name of Rapa Nui. Lying 2,330 mi (3,749 km) due west from the port of Caldera in northern Chile, it is the most remote possession of any Latin American country. It is a volcanic island with an area of 45 mi (117 km) and a subtropical climate.

Like Easter Island, several of Chile's other island possessions are located between the Tropic of Capricorn (23°27'S) and 30°S. From the farthest to the closest, they are Sala y Gómez, San Felix, and San Ambrosio. Farther south, about 360 mi (579 km) west of Valparaíso, are the Juan Fernández Islands. Like Easter Island, these islands are a national park.

Stretching for some 700 mi (1,130km), the southern part of the coast, made up of a series of submerged mountaintops separated by thin channels, consists of an extensive series of islands and archipelagoes stretching in a long chain from Chiloé Island slightly south of Puerto Montt all the way to the southernmost tip of the country, Tierra del Fuego, which Chile shares with Argentina, and the Diego Ramírez Islands, 60 mi (100 km) southwest of Cape Horn. The formations in between include the Chonos Archipelago, Wellington Island, and the Reina Adelaida Archipelago.

The Coast and Beaches

There are few beaches and natural harbors along Chile's long, narrow coast. In the north the coastal mountains rise close to the shoreline in steep cliffs. However, rocky outcroppings provide good protection from the sea at the harbors of Valparaíso and Talcahuano.

South of Puerto Montt is the Gulf of Ancua. South of here, stretching along over one-third of the coast to the foot of the continent, are thousands of uninhabited islands separated by channels and fjords reaching all the way to the Tierra del Fuego archipelago. Corcovado Gulf, the Taitao Peninsula, and the Gulf of Penas mark the coast in the northern part of this region. Further south is the Brunswick Peninsula, the southernmost point on mainland South America. The Strait of Magellan separates it from the Tierra del Fuego Archipelago. It is here that Cape Horn, the southernmost point in South America, with the Drake Passage to the south.

CLIMATE AND VEGETATION

Temperature

Due to its great length and latitudinal extension, Chile's climate varies widely, with temperatures steadily cooling as the country extends southward away from the equator and toward Antarctica. The mean average temperature at Arica, in the far north, is 64°F (18°C), while that of Santiago, in the central section of the country is 57°F (14°C), and Punta Arenas in the extreme south averages 43°F (6°C). In spite of the polar front from the south, winter temperatures are moderated by winds off the Pacific Ocean, and sea winds also temper the heat in summer.

GEO-FACT

The early nineteenth-century shipwreck of Alexander Selkirk on Chile's Juan Fernandez Islands inspired the British author Daniel Defoe to write *Robinson Crusoe.*

Central Chile, where most of the country's population and agricultural production are concentrated, has a pleasant Mediterranean climate with well-differentiated seasons; its winters are mild, and its summers are warm and dry.

The southern part of the country is subject to frequent storms.

Rainfall

While temperatures in Chile steadily drop with increasing latitude, the amount of rainfall gradually rises, varying from virtually no precipitation north of 27°S to around 160 in (406 cm) annually at 48°S (the heaviest precipitation for any region outside the tropics). Santiago, in the center of the country, averages 13 in (33 cm) of precipitation, and Puerto Montt, in the Lakes region, averages 73 in (185 cm). In the far south, precipitation once again decreases to 18 in (46 cm) at Punta Arenas. Snow and sleet are common in the southern third of the country, and the coastal archipelagos are among the world's rainiest regions.

Grasslands

The characteristic native vegetation of Chile's central temperate region is a mix called *matorral* that includes grasses, shrubs, cacti, and hardwoods. However, overcutting has destroyed much of the native ground cover, and farms have replaced much of it as well.

In southernmost Chile, the country extends to the east into the lowlands of Patagonia. Grasses and various types of herbs grow there.

Deserts

The Atacama Desert, which extends for some 800 mi (1,300 km) from the northern border as far southward as the Aconcagua River, consists largely of dry river basins and salt flats, with a few rivers and oases. It is both the warmest and driest part of the country, and is said to be the world's driest desert. Large stretches of the desert have no vegetation at all. Plant life found in other areas includes shrubs, brambles, flowering herbs, cactus species, a type of acacia called the tamarugo, and, on the slopes of the northern Andes, ichu and tola grasses.

North-central Chile, the region immediately south of the Atacama Desert, is a semiarid region with cacti, shrubs, and some hardwoods, but fruits and vegetables are cultivated in its valleys and there is an active fishing industry along its coast.

Forests and Jungles

South of the Bío-Bío River, the Central Valley is thickly forested with both evergreen and deciduous tree species including laurels, magnolias, conifers, and beeches. The forests of this region contain some one-of-a-kind species including the southern cedar and the evergreen laurel. The Chile pine, also called the monkey puzzle tree, is found on the seaward-facing Andean slopes of this region. Dense rainforests are found in the south-central lake district, and dwarf beeches grow on the islands to the far south, where strong winds limit the growth of other forms of plant life. Other vegetation in this cold, rainy region includes lichens and sphagnum moss.

Population Centers – Chile

(1997 POPULATION ESTIMATES)

Name	Population
Santiago (capital)	4,641,000
Concepción	363,000
Viña del Mar	331,000
Valparaiso	283,000
Talcahuano	269,000

SOURCE: Instituto Nacional de Estadísticas (INE), Chile National Institute of Statistics.

HUMAN POPULATION

The Mediterranean climate, even terrain, and rich soils of central Chile, contrasting with the inhospitable climate and geography in thc north and thc south, have attracted a heavy concentration of population to that region, which is home to over three-fourths of the country's population although it accounts for only about one-fourth of the country's area. After the capital city of Santiago, the next-largest population center is the area encompassing the port of Valparaíso and nearby Viña del Mar, a popular resort city.

The major population centers of north-central Chile include the neighboring port cities of Serena and Coquimbo, and Copiapó, which is located in the central valley. The largest city in the lake district of south-central Chile is Valdivia.

NATURAL RESOURCES

Chile's mountains and deserts are rich in minerals. The country is one of the world's largest copper producers, and its other mineral wealth—much of it in the Atacama Desert—includes iron ore, gold, silver, salt, nitrates, and lithium. Coal, copper, and manganese are found in the semiarid north-central region. Other important natural resources include the country's forests, which are exploited for timber, wood products, and paper, and the hydropower generated by the rivers rushing down the steep slopes of the Andes.

FURTHER READINGS

Allende, Isabel. *Portrait in Sepia.* Translated by Margaret Sayers Peden. New York: Harper, 2001.

Bernhardson, Wayne. *Chile & Easter Island: A Lonely Planet Travel Atlas.* Hawthorn, Victoria, Australia: Lonely Planet Publications, 1997.

Caistor, Nick. *Chile: A Guide to the People, Politics, and Culture.* New York: Interlink Books, 1998.

Keenan, Brian, and John McCarthy. *Between Extremes.* London: Black Swan, 2000.

Hickman, John. *News from the End of the Earth: A Portrait of Chile.* New York: St. Martin's Press, 1998.

Wheeler, Sara. *Travels in a Thin Country: A Journey Through Chile.* New York: Modern Library, 1999.

China

- **Area:** 3,705,407 sq mi (9,596,960 sq km) / World Rank: 5
- **Location:** Northern and Eastern hemispheres; Eastern Asia, bordering East China Sea, Korea Bay, Yellow Sea, and South China Sea, west of the Democratic People's Republic of Korea (North Korea), north of Vietnam, Laos, Myanmar, and Bhutan, northeast of Nepal, east of India, Pakistan, and Tajikistan, southeast of Kyrgyzstan, and south of Russia and Mongolia
- **Coordinates:** 35°00′N, 105°00′E
- **Borders:** 13,743 mi (22,166 km) / Afghanistan, 47 mi (76 km); Bhutan, 292 mi (470 km); Myanmar, 1,358 mi (2,185 km); Hong Kong, 19 mi (30 km); India, 2,100 mi (3,380 km); Kazakhstan, 953 mi (1,533 km); Kyrgyzstan, 533 mi (858 km); Laos, 263 mi (423 km); Mongolia, 2,906 mi (4,677 km); Nepal, 768 mi (1,236 km); North Korea, 880 mi (1,416 km); Pakistan, 325 mi (523 km); Russia, 2,265 mi (3,645 km); Tajikistan, 257 mi (414 km); Vietnam, 796 mi (1,281 km)
- **Coastline:** 9,010 mi (14,500 km)
- **Territorial Seas:** 12 NM
- **Highest Point:** Mount Everest, 29,035 ft (8,850 m)
- **Lowest Point:** Turpan Pendi, 505 ft (154 m) below sea level
- **Longest Distances:** 525 mi (845 km) ENE-WSW; 2,082 mi (3,350 km) SSE-NNW
- **Longest River:** Chang Jiang, 3,434 mi (5,525 km)
- **Largest Lake:** Qinghai, 1,625 sq mi (4,209 sq km)
- **Natural Hazards:** Damaging floods, tsunamis, earthquakes, droughts, typhoons
- **Population:** 1,273,111,290 (July 2001 est.) / World Rank: 1
- **Capital City:** Beijing, in the northeast at the foot of the Mongolian Uplands
- **Largest City:** Shanghai, on the East China Sea coast, population 12,900,000 (2000 est.)

OVERVIEW

The vast territory of China exhibits great variation in terrain and vegetation. The elevation drops from west to east. The highest elevations are found in the far southwest, in the Plateau of Tibet (Xizang Gaoyuan) and in the Himalayas, the highest mountain range on Earth. The high elevations of the western portion of the country, making up more than half of the overall territory, combined with cold temperatures and generally arid conditions have prevented the development of agriculture. Thus, the western region is more isolated and much more sparsely populated than the east.

The eastern quarter of the country is mostly lowlands, and may be divided roughly into northern China and the slightly larger southern China, separated from each other by the Huang He (Yellow River) and the Qinling Shandi (Ch'in Ling Shan) Mountain Range. In the northeastern region is the large Manchurian Plain. Separated from the Manchurian Plain by the Da Hinggan Ling (Great Khingan Mountains) is the Gobi Desert, which occupies north-central China straddling the China-Mongolia border. To the southeast, stretching from Beijing to Nanjing across the valley of China's second-largest river, the Huang He, lies the heavily populated Loess Plateau.

China lies entirely on the Eurasian Tectonic Plate. However, the Tibetan region in the southwest sits at the boundary of the Indian and Eurasian Plates, and seismic fault lines also run north to south through the eastern region of China and the Manchurian Plain—the region along the northeastern coast that includes Beijing and is densely populated. Consequently, both the northeast and southwest regions are centers of seismic activity and experience periodic earthquakes, some of which have been devastating. In July, 1976, Tangshan, about 102 mi (165 km) east of Beijing, was leveled by an earthquake resulting in more than 500,000 dead, according to estimates by international sources.

When Communists took control of mainland China in 1949, the former government of China fled to the island of Taiwan, off the southeast coast. It has continued to govern Taiwan as an effectively independent entity since that time, but has long since ceased to be recognized as the government of China itself. China maintains that Taiwan is an integral part of its territory that has been occupied by rebels, and strongly refutes the idea that Taiwan is an independent country. China has disputed claims on many other smaller islands off its coasts, as well as territory along its border with India.

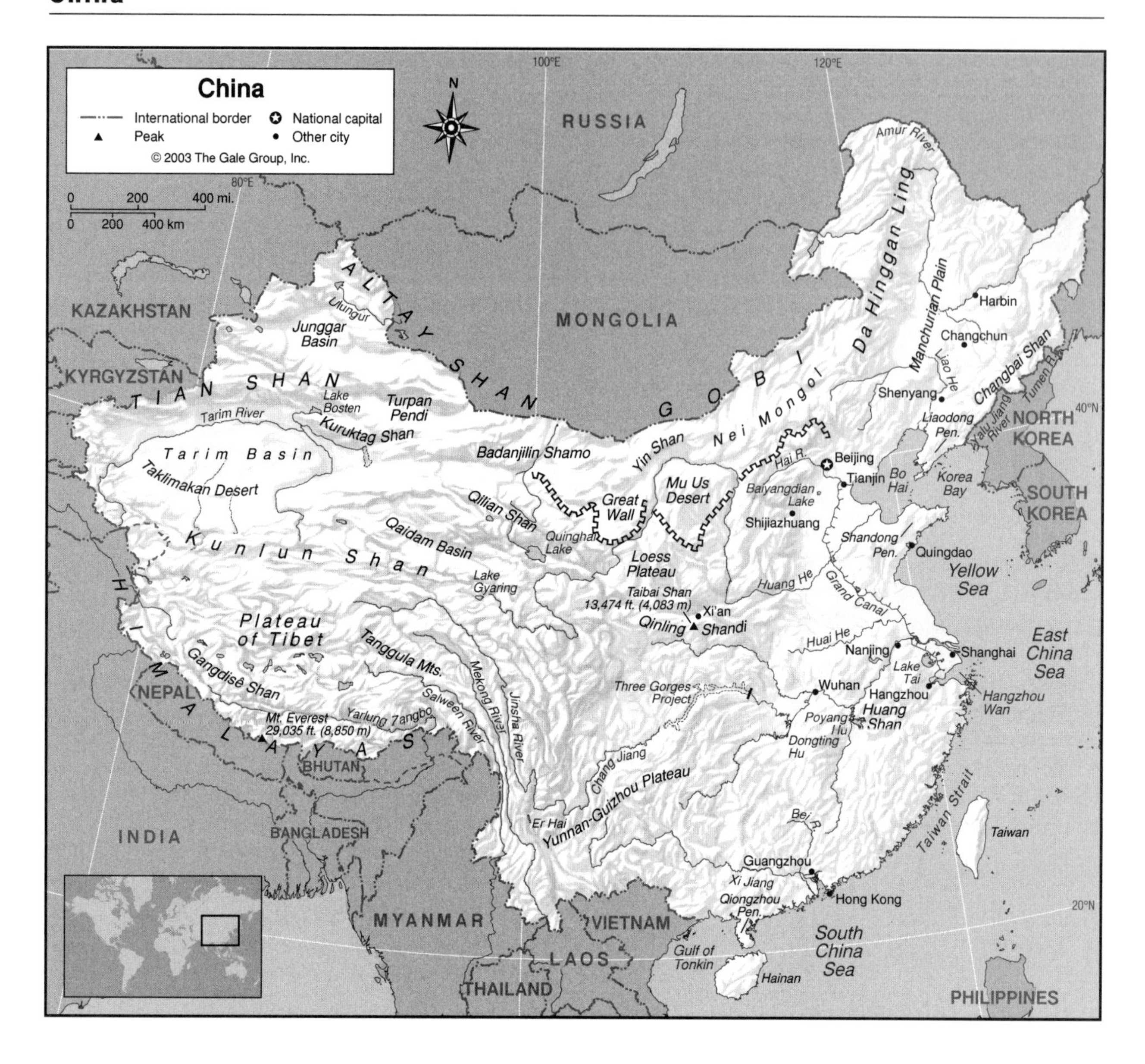

MOUNTAINS AND HILLS

Mountains

Mountains cover more than two-thirds of the nation's territory, impeding communication and leaving only limited areas of level land for agriculture. Most ranges, including all the major ones and some of the most significant mountains in Asia, trend east-west. In the west, the Himalayas on the border of Tibet form a natural boundary with countries on the southwest. Similarly, the Altay Shan (Altai Mountains) form the extreme northwest border with Mongolia.

The Himalayas are the highest mountains on Earth. They extend along a 1,500-mile (2,414-km) arc from Jammu and Kashmir (disputed by India, Pakistan, and China) in the northwest to where the Brahmaputra River cuts south through the mountains near the Myanmar border. This range forms much of China's western and all of its southwestern international borders. Mount Everest (Zhumulangma Feng), the world's highest mountain at 29,035 ft (8,850 m), is found in this region on the border between Nepal and China, as are seven of the world's 19 peaks of over 23,000 ft (7,000 m).

Moving north from the Himalayas, several ranges also run west-east, including the Gangdisê Shan (Kailas Mountains), Tanggula Mountains, the Kunlun Shan, the Kuruktag Shan, the Qilian Shan, and the Tian Shan. The Tian Shan stretch across China between Kyrgyzstan and Mongolia and stand between two great basins.

The Qinling Shandi (Ch'in Ling Shan), a continuation of the Kunlun Shan, divide the Loess Plateau from the Chang Jiang (Yangtze River) Delta. The Qinling Shandi form both a geographic boundary between the two great parts of China, and have served to create a cultural boundary as well. The highest peak is Taibai Shan, which rises to a height of 13,474 ft (4,083 m). To the south lie the densely populated and highly developed areas of the lower and middle plains of the Chang Jiang. To the north are more remote, more sparsely populated areas.

The Yin Shan, a system of mountains with average elevations of 1,364 m (4,475 ft), extends east-west through the center of the Chinese section of the Gobi Desert and steppe peneplain.

In the far northeast, north of the Great Wall, the Da Hinggan Ling (Greater Khingan Range) forms a barrier along the border with Mongolia, extending from the Amur River (Heilong Jiang) to the Liao He (Liao River) in a north-south orientation, with elevations reaching 5,660 ft (1,715 m). To the east, along the border with Korea lie the Changbai Shan (Forever White Mountains), where snow covers the peaks year round.

The Huang Shan (Yellow Mountains), southwest of Shanghai, comprise 72 peaks, the tallest of which is Lianhua Feng (Lotus Flower Peak) at 6,151 ft (1,864 m). The Huang Shan region also includes hot mineral springs, where the water temperature is constant at 108°F (42°C).

Notable Peaks of China

Nine mountain peaks are designated as sacred by observers of either Taoism or Buddhism. The sacred Taoist peaks are Bei Heng Shan (10,095 ft / 3,060 m); Nan Heng Shan (4,232 ft / 1,282 m), in the southeast; Hua Shan (6,552 ft / 1,985 m) and Song Shan (4,900 ft / 1,485), along the Huang He; and Tai Shan (5,069 ft / 1,530m) on the Shandong Peninsula. The sacred Buddhist peaks are Emei Shan (10,095 ft / 3,060 m), in south-central China southwest of Chengdu; Jiuhua Shan (4,340 ft / 130 m); Putuo Shan (932 ft / 282 m); and Wutai Shan (10,003 ft / 3,031 m), west of Beijing.

Plateaus

About 25 percent of China's total area may be characterized as plateau; the principal plateaus are the Plateau of Tibet (Xizang Gaoyuan) in southwestern China; the Nei Mongol (Inner Mongolia) plateau in northern China; the Loess Plateau in the east-central part of the country south of the Great Wall; and the Yunnan-Guizhou plateau in the southwest.

The Plateau of Tibet is in China's southwest, enclosed by the Himalayas and the Kunlun Shan. It is the highest and most extensive plateau in the world, incorporating some 888,000 sq mi (2.3 million sq km) with elevations that average more than 13,123 ft (4,000 m) above sea level. The loftiest summits rise to over 23,622 ft (7,200 m). It is referred to as the "roof of the world," and the land there continues to rise, gaining an average of 0.04 in (10 mm) per year in elevation. North of Tibet in the northwest are two more plateau basins, the Tarim Basin and the Junggar Basin. In these regions the elevation averages 15,000 ft (4,600 m). The two plateaus are separated by the Tian Shan Range.

The Nei Mongol (Inner Mongolia) Plateau, China's second-largest plateau, lies in the northeast near the border with Mongolia. It covers an area of about 386,100 sq mi (1,000,000 sq km), stretching 1,250 mi (2,000 km) from east to west, and 300 mi (500 km) from north to south. The elevation averages 3,300–6,600 ft (1,000–2,000 m).

To the south is Loess Plateau, the largest loess plateau in the world and the third-largest plateau in China, covering 308,881 sq mi (600,000 sq km). The plateau is veneered by a layer of loess—a yellowish soil blown in from the deserts of Nei Mongol (Inner Mongolia); the loess layer ranges from 330–660 ft (100–200 m) in depth. The Loess Plateau covers about 154,400 sq mi (400,000 sq km), and rises to elevations that range from 2,640–6,600 ft (800–2,000 m). The Loess Plateau experiences some of the most severe soil erosion conditions of anywhere in the world.

The last notable plateau in China is the Yunnan-Guizhou plateau in the southwest. This plateau region has areas of the highest altitude in the northwest. The smallest plateau in China, it features unusual geology with dramatic stone outcroppings and overhangs.

Canyons

The Grand Yaluzangbu Canyon is largest canyon in the world: 316 mi (505 km) long and 10,830 ft (6,009 m) deep. It is located in southwestern China, and was carved by the Yarlung Zangbo, the river that eventually becomes the Bramaputra.

The Three Gorges, a famous 200-mi (322-km) canyon on the Chang Jiang (Yangtze River), will be submerged when the Three Gorges Dam becomes operational in 2009. One of the world's deepest canyons, the 9,900-ft (3,000-m) deep Hutiaojian ("Tiger Leaping") Canyon, lies along the Jinsha River, an upper tributary of the Chang Jiang.

Hills and Badlands

Being so mountainous, China has many hill regions between and at the feet of the various ranges. However, there are some notable hilly regions in the south, along the coastline of the South China Sea (Nan Hai), where farmers must terrace the land to grow rice.

INLAND WATERWAYS

Lakes

The Qaidam Basin, a sandy and swampy basin, contains many salt lakes. Qinghai Lake (formerly Koko Nor)—China's largest lake and the third-largest salt lake in the world, with an area of 1,625 sq mi (4,209 sq km)—is located here, but is drying up and shrinking in size

GEO-FACT

The Great Wall of China is the largest structure ever built by humans. It stretches across 3,729 mi (6,000 km) of mountainous and desert terrain in northeastern China. Most of the Great Wall along the country's northern flank, the east-west extent of which is more than 2,050 mi (3,300 km), was built about 220 B.C.E. as a barrier against invasion.

each year. Lakes Ngoring and Gyaring are also located in this basin.

Poyang Hu is the largest freshwater lake in China with a surface area of 1,073 sq mi (2,779 sq km); it is found on the south Chang Jiang (Yangtze River) in southeast China. Dongting Hu is a large, shallow lake also south of the Chang Jiang upstream from Poyang Hu in China's southeastern region. About 40 percent of the Chang Jiang's water passes through several channels into the lake. Lake Tai, located at the base of Mt. Yu Shan (12,956 ft / 3,949 m) on the other side of the Great Canal, lies just inland from Shanghai. Baiyangdian Lake (140 sq mi / 360 sq km), a water source for the region just to the southwest of Beijing which is home to hundreds of thousands of people, is drying up due to overuse for industrial and agricultural production and drinking water, compounded by drought.

There are several other notable lakes in China, many of which are located in the various mountain ranges catching water from the many mountain streams. Erhai Lake is a freshwater lake on the plateau of Yunnan. Tianchi Lake (Heavenly Lake) lies in the Tian Shan Mountains in the northwest, about 70 mi (115 km) northeast of Ürümqi. Formerly known as Yaochi Lake (Jake Lake), it is a major attraction for tourists. Also in the northwest between the Tian Shan and Kuruktag Shan Mountains is Lake Bosten, which receives the Kaidu River and other streams.

Three Gorges Dam

In 1994, work began on the 17-year project to construct the world's largest dam on the Chang Jiang (Yangtze), the world's third-largest river. The Three Gorges Dam will be the largest hydroelectric dam in the world, measuring just over a mile (about 2 km) across and 610 ft (185 m) high when it is completed (projected for 2009). Its reservoir was expected to extend more than 350 mi (560 km) upstream, flooding the towns and villages that are home to an estimated 2 million people, all of whom will be forced to relocate when the dam is completed.

Rivers

China's most important rivers lie in the eastern and northeastern part of the country. The three major river systems here are the Chang Jiang (Yangtze River)—which is the longest in China at 3,434 mi (5,525 km)—the Huang He (Yellow River), and the Hai, all flowing generally east. The Kunlun and Qinling Mountains form the chief watersheds between these rivers. The country's longest waterway, the Chang Jiang, is found south of these mountains. It is navigable over much of its length and offers significant hydroelectric power generation potential. The Chang Jiang begins on the Plateau of Tibet and flows east through the heart of the country, draining an area of 694,000 million sq mi (1.8 million sq km) before emptying into the East China Sea. A principal feature of the south-central part of China is the fertile plain that is home to the Chang Jiang. The large Jinsha River is a major tributary of the upper Chang Jiang.

Flowing initially northeast from its source in the Kunlun Shan, the Huang He (Yellow River) follows a winding path, measuring 2,903 mi (4,671 km), as it courses toward the sea through the Loess Plateau, the historic center of Chinese expansion and influence. It is China's second-longest river. Over the centuries the Huang He has become choked with silt as it brings down a heavy load of sand and mud from the upper reaches, much of which is deposited on the flat plain. The flow is channeled by artificial embankments that require constant repair. After years of these repairs, the river now actually flows on a raised ridge, the riverbed having risen 164 ft (50 m) or more above the plain.

The Hai River flows west to east north of the Huang He. Its upper course consists of five rivers that converge near Tianjin, then flow 43 mi (70 km) before emptying into Bo Hai (Gulf of Chihli). Another major river, the Huai He, rises southwest of Beijing and flows through several lakes before joining the Chang Jiang (Yangtze).

Other significant rivers in northeastern China include the Amur River (Heilong Jiang), which flows a total 2,719 mi (4,350 km) through Russia and China; the Liao He; and the Yalu Jiang, which, along with the Tumen River, forms the border with North Korea. The largest river flowing in the southeast is the Zhu Jiang (Pearl River). It is formed at Guangzhou, where the Xi Jiang (West River) and Bei River join. The Zhu Jiang then flows to form the large Boca Tigris estuary between Hong Kong and Macao, linking Guangzhou to the South China Sea. The estuary must be dredged frequently to keep it open for shipping. The Xi Jiang is an important commercial waterway and one of the most important rivers in China. The total length of this entire system from its origin is about 1,200 mi (1,930 m). All of these rivers eventually empty into the seas that form China's eastern border, merging into the Pacific Ocean.

Central and Western China also has many important rivers as well. Between the high mountains of the north and northwest the rivers have no outlet to the sea. This area of inland drainage by smaller rivers and streams involves a number of upland basins, and accounts for less than 40 percent of the country's total drainage area. Many such waterways terminate in lakes, or else diminish in the desert. A few are useful for irrigation. The largest of these rivers are the Konqi, the Kaidu, the Ulungur, and the Tarim. Its length of 1,354 mi (2,179 km) makes the Tarim River China's longest river without an outlet to the sea.

South of the Kunlun Shan the major rivers flow south, unlike most of the rivers in China, and all eventually reach the ocean. The easternmost of these rivers—the Tongtian, the Jinsha, the Yalong, and the Dadu—are the major tributaries of the Chang Jiang. Further west are the Lancang and the Nu. These rivers flow south out of China and become the major rivers of Southeast Asia, known as the Mekong and Salween, respectively.

The Yarlung Zangbo has its headwaters in the southwesternmost part of China, and runs east for 1,300 mi (2,075 km) across the Tibetan plateau. It then curves sharply to the south and breaches the Himalayas, flowing into India. In India it takes its better known name, the Brahmaputra. It then flows south and west into the Ganges, eventually reaching the Indian Ocean, with a total length of 1,800 mi (2,900 km).

The Grand Canal (Dayun He), running from Beijing in the north to Hangzhou in the south, is the longest (1,126 ft / 1,801 km) and oldest artificial canal in the world. It links five rivers: the Hai He, Huang He, Huai He, Chang Jiang, and Qiantang Jiang. It was dug by hand over a period that stretched from 486 B.C.E. to C.E. 1293.

Wetlands

There are wetlands areas along most of China's major rivers. Management of water resources and flood control has posed a challenge to the government for decades, and mismanagement has exacerbated the water supply problems. In early 2002 the government announced plans to allocate the equivalent of one billion U.S. dollars for a ten-year program of wetlands conservation, and designated 200 new wetlands areas for protection. Also, the coastal area of Bo Hai (Gulf of Chihli) has extensive wetlands, including riverine wetland, marshes, and salt marshes.

THE COAST, ISLANDS, AND THE OCEAN

Oceans and Seas

China's extensive territorial waters are principally marginal seas of the Western Pacific, washing a long and much-indented coastline and having many islands. The Yellow Sea (Huang Hai), East China Sea (Dong Hai), and South China Sea (Nan Hai) are marginal seas of the Pacific Ocean. The South China Sea features a deep ocean floor. Elsewhere, the continental shelf supports coastal fish farms and also contains substantial oil deposits.

The most northerly coastal waters of the Yellow Sea—Korea Bay and Bo Hai (Gulf of Chihli)—are the most southerly ocean regions where substantial amounts of sea ice are found. The waters of Bo Hai, relatively shallow at 70 ft (20 m), are very turbulent. Korea Bay separates the Liaodong Peninsula from North Korea. The Taiwan Strait lies between the mainland and the island of Taiwan, which maintains a separate government. The Gulf of Tonkin lies off the coast of Guangxi, the extreme southeastern province of China, between Hainan Island and Vietnam.

Major Islands

There are more than 5,000 islands lying off the eastern coast of China. Taiwan (about 22,500 sq mi / 36,000 square km) is the largest, although it considers itself the independent Republic of China and has a separate democratic government. Hainan Island (about 21,250 sq mi / 34,000 sq km) is the second-largest, but it is the largest island fully under the jurisdiction of China. Disputed islands include the Spratly Islands, the Diaoyutai Islands, the Paracel Islands, and the Pescadores.

Coast and Beaches

China's coastline extends more than 11,000 mi (16,000 km). More than half the coastline (predominantly in the south) is rocky; most of the remainder is sandy. The Hangzhou Wan (Bay of Hangzhou), just south of Shanghai, roughly divides the two types of shoreline.

The Shandong Peninsula juts out at the northernmost reach of the Yellow Sea. It features the dramatic and sacred peak, Tai Shan (5,069 ft / 1,530 m). North of the Shandong Peninsula the coastline curves around another peninsula, the Liaodong Peninsula. This peninsula separates Korea Bay from Bo Han. In the south, separating the Gulf of Tonkin from the South China Sea, the narrow Qiongzhou Peninsula extends out from the mainland at China's southernmost point and almost touches Hainan Island.

CLIMATE AND VEGETATION

Temperature

China's climate varies from region to region. Since the country is so large with such variations in altitude, many extremes in climate exist although the climate in most of the country is temperate. At the highest elevations in southwestern China, there are only fifty frost-free days per year. The hottest spot in China is in northwestern China in the Turpan Pendi, where summer highs can reach as much as 116°F (47°C). Winter temperatures in northern China drop to as low as -17°F (-27°C), and in summer reach just 54°F (12°C). In the Chang Jiang River valley, the mean temperature in summer is 85°F (29°C).

Rainfall

Most of the country's rainfall occurs during the summer months. Rainfall is heaviest in the southeast, averaging 80 in (200 cm) per year. In the northeastern region near Beijing, annual rainfall averages about 25 in (60 cm). In the far northwest, the annual rainfall averages 4 in (10 cm), although some desert regions may go a year or longer with no precipitation. Along the southern coast, severe storms are common with destructive typhoons occasionally occurring.

Grasslands

Only about 12 percent of China's land area may be classified as grasslands. However, because of the country's size, there are still some significant plains regions. A principal feature of the south-central part of China is the fertile plain that is home to the Chang Jiang (Yangtze River). To the south of the river a large plate-shaped section of the plain surrounds Lake Tai.

The Loess Plateau is mainly a large plain, known as the North China Plain. It is actually a continuation of the central Manchurian Plain to the northeast, but separated from it by Bo Hai, an extension of the Yellow Sea. Han people, China's largest ethnic group, have farmed the rich alluvial soils of the plain since ancient times, constructing the Grand Canal (Dayun He) for north-south transport.

There are also grasslands in the massive Tarim Basin and the Junggar Basin in China's northwest corridor. Rich deposits of coal, oil, and metallic ores lie in the area. The Tarim is China's largest inland basin, measuring 932 mi (1,500 km) from east to west and 373 mi (600 km) from north to south at its widest parts.

Desert

One of the significant problems facing China is desertification; the total desert area comprises more than 1 million sq mi (2.6 million sq km), or about 30 percent of the country's total land area. In the extreme west of the country, between two east-west mountain ranges, lies the Tarim Basin, where Asia's driest desert, the Taklimakan Desert, is found. Brutal sandstorms, arid conditions, extreme temperatures, and the remoteness of the area have prevented any significant exploitation of the vast petroleum reserves of this desert region. The Gobi Desert lies along the northern border with Mongolia; in China, the Badanjilin Shamo forms the southern limit of the Gobi. Much of the Gobi is mountainous, stark terrain. The Mu Us (or Ordos) Desert is the extension of the Gobi that lies along the southern edge of Nei Mongol (Inner Mongolia)

Population Centers – China

(2000 POPULATION ESTIMATES)

Name	Population	Name	Population
Shanghai	14,173,000	Nanjing	5,323,000
Beijing (capital)	12,033,000	Chengdu	5,293,000
Tianjin	10,239,000	Hefei	4,260,000
Shijiazhuang	8,672,700	Kunming	3,896,000
Wuhan	7,317,900	Nanchang	3,896,000
Qingdao	6,995,700	Tai'an	3,825,000
Changchun	6,868,700	Suzhou	3,273,000
Shenyang	6,748,600	Yantai	3,205,000
Guangzhou	6,741,000	Zaozhuang	3,192,000
Xi'an	6,682,000	Guiyang	3,157,000
Hong Kong	6,200,000	Taiyuan	2,957,000
Hangzhou	6,116,400	Nanning	2,846,300
Zhengzhou	6,060,000	Lanzhou	2,839,000
Fuzhou	5,780,000	Zibo	2,484,000
Chongqing	5,771,000	Anshan	2,479,000
Changsha	5,770,000	Liupanshui	1,844,000
Jinan	5,535,000	Linyi	1,590,000
Harbin	5,475,000	Ürümqi	1,549,000
Dalian	5,432,000	Tangshan	1,485,000
Ningbo	5,353,000	Qiqihar	1,401,000

SOURCE: National Bureau of Statistics of China.

Forests and Jungles

Forests covered much of the territory, especially in western China, in the early twentieth century. Along the Yarlung Zangbo (Brahmaputra) River in Tibet lie dense virgin forests. In the 1960s and 1970s, forest cover had fallen to less than 10 percent of the total land area. Trees were being cut down in state forests in western China at double the rate of natural growth. The government responded in 1989 by launching a reforestation program, known as the Great Green Wall of China. The forests of the northeast near the Tumen River include larch, fir, pine, cypress, juniper, birth, and walnut.

HUMAN POPULATION

China is the most populous nation on earth. Its population comprises dozens of ethnic groups, but more than 90 percent of the people are Han Chinese, the country's largest ethnic group. The people of the minority ethnic groups are concentrated in southern China. The coastal areas of China are the most densely populated regions, with a density of more than 1,036 people per sq mi (400 people per sq km). Bustling port cities lie along the coast—from Shanghai near the Chang Jiang delta to Guangzhou (Canton) where the Xi Jiang (West River) and Bei River join to become the Zhu Jiang (Pearl River). Major commercial development and high population density also characterize the regions around the mouths of the major rivers, including the Zhu Jiang, Chang Jiang (Yangtze River), and the Huang He (Yellow River). The central regions have an average population density of 518 people per sq mi (200 people per sq km). The western

Provinces – China

POPULATIONS ESTIMATED IN 2000

Name	Population	Area (sq mi)	Area (sq km)	Capital
Anhui (Anhwei)	59,860,000	54,000	139,900	Hefei (Ho-fei)
Fujian (Fukien)	34,710,000	47,500	123,100	Fuzhou
Hainan	7,870,000	13,243	34,300	Haikou
Hebei (Hopch)	67,440,000	78,200	202,700	Shijiazhuang (Shih-chia-chuang)
Heilongjiang (Heilungkiang)	36,890,000	179,000	463,600	Harbin
Henan (Honan)	92,560,000	64,500	167,000	Zhengzhou (Cheng-chou)
Hubei (Hupeh)	60,280,000	72,400	187,500	Wuhan
Hunan	64,400,000	81,300	210,500	Changsha (Ch'ang-sha)
Gansu (Kansu)	25,620,000	141,500	366,500	Lanzhou (Lanchou)
Guangdong (Kwangtung)	86,420,000	89,300	231,400	Guangzhou (Canton)
Guizhou (Kweichow)	35,250,000	67,200	174,000	Guiyang (Kuei-yang)
Jiangsu (Kiangsu)	74,380,000	39,600	102,600	Nanjing (Nanking)
Jiangxi (Kiangsi)	41,400,000	63,600	164,800	Nanchang (Nan ch'ang)
Jilin (Kirin)	27,280,000	72,200	187,000	Changchun (Ch'ang-ch'un)
Liaoning	42,380,000	58,300	151,000	Shenyang (Shen-yang)
Shaanxi (Shensi)	36,050,000	75,600	195,800	Xi'an (Sian)
Shandong (Shantung)	90,790,000	59,200	153,300	Jinan (Tsinan)
Shanxi (Shansi)	31,720,000	60,700	157,100	Taiyuan
Sichuan (Szechwan)	83,290,000	219,700	569,000	Chengdu (Ch'eng-tu)
Qinghai (Tsinghai)	5,180,000	278,400	721,000	Xining (Hsi-ning)
Yunnan	42,880,000	168,400	436,200	Kunming (K'un ming)
Zhejiang (Chekiang)	46,770,000	39,300	101,800	Hangzhou (Hangchow)
Autonomous regions				
Guangxi Zhuangzu (Kwangsi Chuang)	44,890,000	85,100	220,400	Nanning
Inner Mongolia (Nei Monggol)	23,760,000	454,600	1,177,500	Hohhot (Hu-ho-hao-t'e)
Ningxia Huizu (Ningsia Hui)	5,620,000	25,600	66,400	Yinchuan (Yin-ch'uan)
Tibet (Xizang)	2,620,000	471,700	1,221,600	Lhasa
Xinjiang Uygur (Sinkiang Uighur)	19,250,000	635,900	1,646,900	Ürümqi
Municipalities				
Beijing (Peking)	13,820,000	6,500	16,800	-
Chongqing	30,900,000	31,660	82,000	-
Shanghai	16,740,000	2,400	6,200	-
Tianjin (Tientsin)	10,010,000	4,400	11,300	-

SOURCE: National Bureau of Statistics of China.

plateaus are sparsely populated, with densities averaging less than 26 people per sq mi (10 people per sq km). Approximately 30 percent of the inhabitants live in urban areas.

NATURAL RESOURCES

Extensive petroleum deposits have been discovered in the remote, arid Taklimakan Desert in western China; they have yet to be fully exploited. There are coal deposits underlying all regions of the country. Especially rich deposits of coal, oil, and metallic ores lie in the Tarim Basin. China is the world's largest producer of tin, and is a major producer of iron ore (although generally low-grade), along with antimony, tungsten, barite, and magnetite. Because of the mountainous terrain, China's rivers represent the world's largest potential for producing hydroelectric power.

FURTHER READINGS

China in Brief. http://www.chinaguide.org/e-china/index.htm (Accessed June 4, June).

China National Tourism Administration. *Welcome to China.* http://www.cnta.com/lyen/index.asp (Accessed June 13, 2002).

Gamer, Robert E., ed. *Understanding Contemporary China.* Boulder, Colo: Lynne Rienner Publishers, 1999.

Gargan, Edward A. *A River's Tale: A Year on the Mekong.* New York: Alfred A. Knopf, 2002.

Harper, Damian. *The National Geographic Traveler: China.* Washington, D.C.: National Geographic Society, 2001.

Leeming, Frank. *The Changing Geography of China.* Cambridge, Mass.: Blackwell, 1993.

Ma, Jian. *Red Dust: A Path through China, tr. Flora Drew.* New York: Pantheon Books, 2001.

Paine. S.C.M. *Imperial Rivals: China, Russia, and Their Disputed Frontier.* Armonk, N.Y.: M.E. Sharpe, 1996.

Riboud, Marc. "China's Magic Mountain." *Life,* 7 (March 1984): 48+.

Shaughnessy, Edward L., ed. *China: Empire and Civilization.* New York: Oxford University Press, 2000.

Smith, Christopher J. *China: People and Places in the Land of One Billion.* Boulder, CO: Westview Press, 1991.

Spence, Jonathan D. *The Chan's Great Continent: China in Western Minds.* New York: Norton, 1998.

Starr, John Bryan. *Understanding China: A Guide to China's Economy, History, and Political Structure.* New York: Hill & Wang, 1997.

Colombia

- **Area:** 439,736 sq mi (1,138,910 sq km) / World Rank: 27
- **Location:** Northern and Western Hemispheres, in northwest South America, southeast of Panama, south of the Caribbean Sea, west of Venezuela, northwest of Brazil, north of Peru, northeast of Ecuador, and east of the North Pacific Ocean.
- **Coordinates:** 4°00′N, 72°00′W
- **Borders:** 3,731 mi (6,004 km) / Brazil, 1,021 mi (1,643 km); Ecuador, 367 mi (590 km); Panama, 140 mi (225 km); Peru, 930 mi (1,496 km); Venezuela, 1,274 mi (2,050 km)
- **Coastline:** 1,993 mi (3,208 km) / Caribbean Sea, 1,100 mi (1,760 km); North Pacific Ocean, 905 mi (1,448 km)
- **Territorial Seas:** 12 NM
- **Highest Point:** Pico Cristóbal Colón, 18,947 ft (5,775 m)
- **Lowest Point:** Sea level
- **Longest Distances:** 1,056 mi (1,700 km) NNW-SSE; 752 mi (1,210 km) NNE-SSW
- **Longest River:** Amazon, 4,080 mi (6,570 km)
- **Natural Hazards:** Occasional earthquakes; periodic drought; highlands subject to volcanic eruptions
- **Population:** 40,349,388 (July 2001 est.) / World Rank: 28
- **Capital City:** Bogotá, at the center of the country
- **Largest City:** Bogotá, population 6,834,000 (2000 metropolitan est.)

OVERVIEW

Located in the northwest corner of the South American continent, Colombia is the only country of South America with both Atlantic (Caribbean) and Pacific Ocean coastlines. It is the fifth largest in size of the Latin American countries. The country consists of four main geographic regions: the central highlands, surrounded by the three Andean ranges and the lowlands between them; the Atlantic lowlands; the Pacific lowlands and coastal regions; and the eastern plain. Colombia's northwest border follows the Isthmus of Panama.

Colombia sits on the extreme edge of the South American Tectonic Plate. Just to the east is the Nazca Plate, and immediately to the north is the Caribbean Plate. Subduction at these plate boundaries has pushed up the rock, resulting in the mountains that exist on Colombia's coasts. Volcanoes were also formed, many of which remain active. Folding and faulting of the earth's crust resulted in seismic fault lines between the mountain ranges, and continued movement of the plates subjects Colombia to frequent earthquakes, some of which are destructive.

MOUNTAINS AND HILLS

Mountains

Beginning near the border with Ecuador, the Andes Mountains divide into three distinct cordilleras (mountain chains) that extend northward almost to the Caribbean Sea. The Cordillera Occidental in the west roughly follows the Pacific coast; slightly inland, the Cordillera Central extends parallel to the Cordillera Occidental; and the Cordillera Oriental lies furthest east. Altitudes reach almost 19,000 ft (5,791 m) and mountain peaks are permanently covered with snow. The elevated basins and plateaus of these ranges have a moderate climate that provides pleasant living conditions and enables farmers in many places to harvest twice a year.

The Cordillera Occidental is separated from the Cordillera Central by the deep rift of the Cauca River valley; this range is the lowest and the least populated of the three and supports little economic activity. A pass about 5,000 ft (1,524 m) above sea level provides the major city of Cali with an outlet to the Pacific Ocean. The relatively low elevation of the cordillera permits dense vegetation, which on the western slopes is truly tropical.

To the west of the Atrato River along the Pacific Coast and the Panama border rises the Serranía de Baudó, an isolated chain that occupies a large part of the coastal plain. Its highest elevation is less than 6,000 ft (1,829 m).

The Cordillera Central, also called the Cordillera del Quindío, is the loftiest of the mountain systems. Its crystalline rocks form a 500-mi (805-km) long towering wall dotted with snow-covered volcanoes, several of which reach elevations above 18,000 ft (5,500 m). There are no plateaus in this range and no passes under 11,000 ft (3,352 m). The highest peak, the Nevado del Huila, rises 18,865 ft (5,750 m) above sea level. Toward its northern end this cordillera separates into several branches that descend toward the Atlantic coast, including the San Jerónimo Mountains, the Ayapel Mountains, and the San Lucas Mountains.

The Cordilla Oriental is the longest of the three systems, covering more than 745 mi (1200 km). In the far north, where the Cordillera Oriental makes an abrupt turn to the northwest near the Venezuela border, lies the Sierra Nevada de Cocuy; the highest point of this range rises to 18,310 ft (5,581 m) above sea level. The northernmost region of the range, around Cúcuta and Ocaña, is so rugged that historically it has been easier to

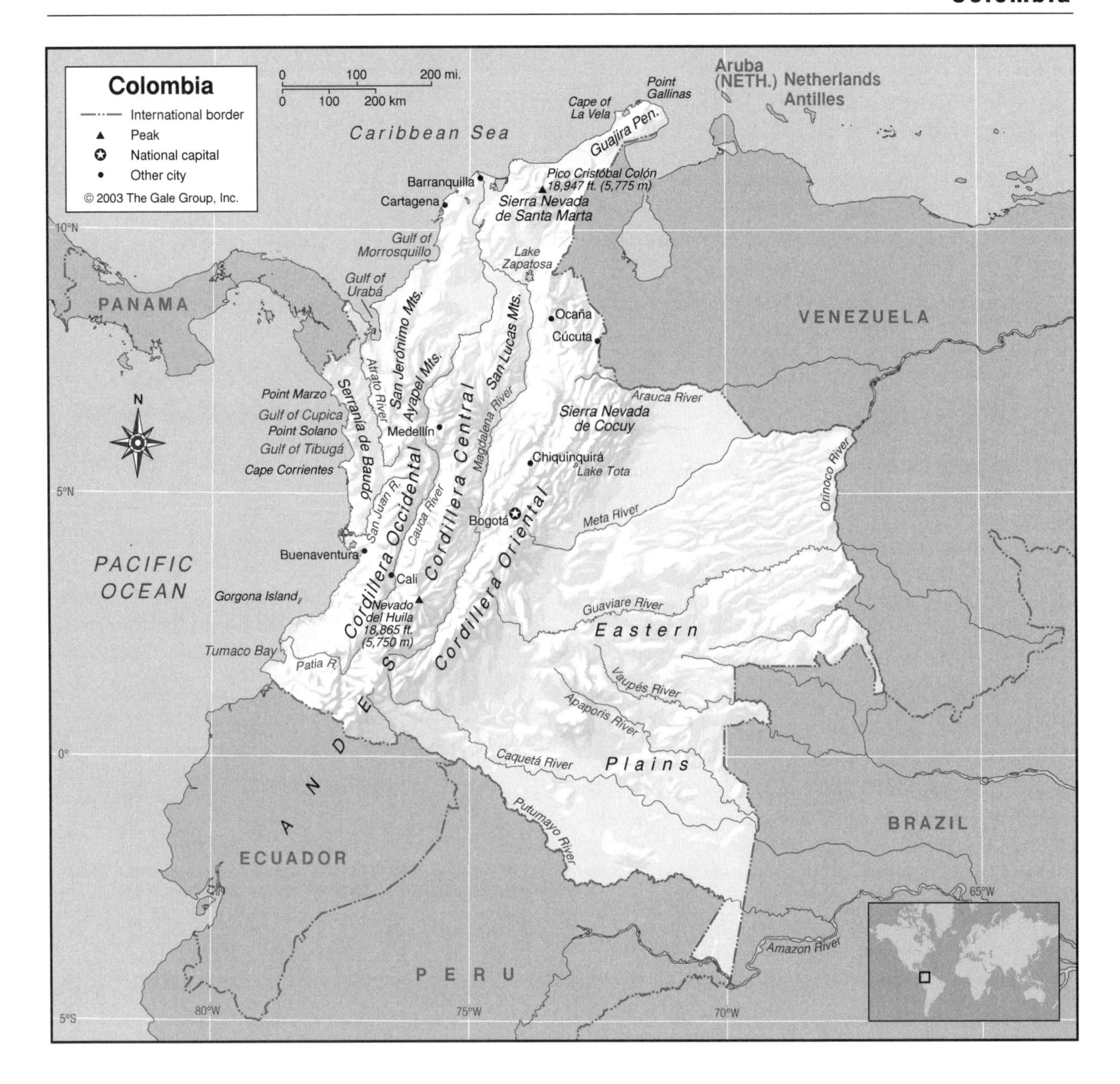

maintain communication and transportation toward Venezuela from this area than toward the adjacent parts of Colombia.

The semiarid Guajira Peninsula in the extreme north bears little resemblance to the rest of the region. In the southern part of the peninsula, the Sierra Nevada de Santa Marta rise to a height of 18,947 ft (5,775 m) at Pico Cristóbal Colón, the highest peak in Colombia. The Sierra Nevada de Santa Marta is an isolated mountain system with slopes generally too steep for cultivation.

Plateaus

North of Bogotá, the densely populated plateaus of Chiquinquirá and Boyacá feature fertile fields, rich mines, and large industrial establishments. The average elevation in this area is about 8,000 feet.

INLAND WATERWAYS

Lakes

Lake Tota near Bogotá supports tourism with abundant resources for fishing and boating. The largest lake in the north, in the Sierra Nevada de Cocuy Mountain Range near the border with Venezuela is Laguna de la Plaza, with a shore lined with rock formations. Another lake in the area is Laguna Grande de los Verdes. Lake Fúquene (11 sq mi/30 sq km) is a shallow lake that lies in the Cordillera Oriental. Lake Zapatosa is the largest of the many lakes of northern Colombia.

Rivers

Torrential rivers on the slopes of the Andes mountain ranges that make up the central highlands produce a large hydroelectric power potential and add their volume to the navigable rivers in the valleys. The two rivers that separate the lines of the Andean trident have formed fertile floodplains in valleys that reach deep into the highlands. The Magdalena River rises near a point some 110 mi (177 km) north of Ecuador where the Cordillera Oriental and the Cordillera Central diverge. Its spacious drainage area is fed by numerous mountain torrents originating high in the snowfields, where for millennia glaciers have planed the surface of folded and stratified rocks. The Magdalena is navigable from the Caribbean Sea as far as the town of Neiva, deep in the interior, but is interrupted at the midpoint of the country by rapids at the town of Honda. The valley floor is very deep; at nearly 500 mi (805 km) from the river's mouth the elevation is no more than about 1,000 ft (305 m).

Running parallel to the Magdalena and separated from it by the Cordillera Central, the Cauca River has headwaters not far from those of the Magdalena, which it eventually joins in swamplands of the Atlantic (Caribbean) Coast region. The Cordillera Occidental is separated from the Cordillera Central by the deep rift of the Cauca River valley. The Atlantic region, the second-most important region after the Central Highlands in terms of population and economic activity—merges into and is connected with the cordilleras of the Central Highlands through the two river valleys. Further west, the navigable Atrato River flows northward to the Gulf of Urabá, a circumstance that makes the river settlements accessible to the major Atlantic ports and commercially related primarily to the Atlantic Lowlands hinterland.

There are no great rivers in western Colombia, as the mountains lie close to the coastline. The longest rivers in this region are the San Juan and the Patia. Conversely, east of the Andes there are many large rivers, including many that are navigable. The Orinoco River flows north along part of the border with Venezuela. It is the major river of that country and many of Colombia's eastern rivers flow into it. The Guaviare River and the rivers to its north, the Arauca and the Meta, are the Orinoco's major Colombian tributaries. The Guaviare serves as a border for five political subdivisions, and it divides eastern Colombia into the Eastern Plains subregion in the north and the Amazonas subregion in the south.

The rivers south of the Guaviare—the Vaupés, Apaporis, Caquetá, and the Putumayo—flow southeast into the basin of the Amazon, which is the longest river in South America and the second-longest river in the world. The Amazon touches the southernmost part of Colombia in the course of its 4,080-mi (6,570-km) journey east to the Atlantic Ocean. In this southern region of Colombia the plains give way to largely unexplored tropical jungle.

Wetlands

The narrow region along the Pacific coast, known as the Pacific Lowlands, is swampy, heavily forested, and sparsely populated. Along the Atlantic coast, the Atlantic Lowlands also consist largely of open, swampy land, but there are cattle ranches and plantations there, and settlements centered on the port cities.

The Atrato swamp is a 40-mi (64-km) wide area adjoining the Panama frontier. For years, engineers seeking to complete the Pan-American Highway were challenged by this inhospitable stretch of terrain, known as the Tapón del Chocó (Chocó Plug).

GEO-FACT

The Sierra Nevada de Santa Marta is the tallest coastal mountain range in the world. The range includes many tall peaks as well as active volcanoes. The town of Arboletes is especially known for its pungent mud volcanoes, which, instead of molten rock, bubble and spatter a mixture of hot water and clay or mud from deep within the earth's surface. One of its volcanoes has a large crater that fills with a lake of mud. Locals and tourists alike swim and bathe in the lake.

THE COAST, ISLANDS, AND THE OCEAN

Oceans and Seas

The waters along the Caribbean Coast are attractive to snorkelers and scuba divers from around the world since the water is clear and the coastal areas are lined with extensive coral reefs.

Rich marine life fills the Pacific Ocean waters along Colombia's western coast, influenced by the Humbolt Current. It is common to see dolphins, and deep-sea fishing is a popular tourist activity. During the period from July through September, humpback whales populate the waters for their mating season.

Major Islands

Colombia possesses a few islands in the Caribbean and some in the Pacific Ocean, the combined areas of which do not exceed 25 sq mi (65 sq km). Off Nicaragua about 400 mi (644 km) northwest of the Colombian coast lies the San Andrés y Providencia Intendency, an archipelago of thirteen small cays grouped around the two larger islands of San Andrés and Providencia. Other islands in the same area—the sovereignty of which has been in dispute—are the small islands, cays, or banks of Santa Catalina, Roncador, Quita Sueno, Serrana, and Serranilla. Off the coast south of Cartagena are several small islands, among them the islands of Rosario, San Bernardo, and Fuerte.

The island of Malpelo lies in the Pacific Ocean about 270 mi (434 km) west of Buenaventura. Nearer the coast a prison colony is located on Gorgona Island. Gorgonilla Cay is off its southern shore.

The Coast and Beaches

The Atlantic Lowlands consist of all Colombia north of an imaginary line extending northeastward from the Gulf of Urabá to the Venezuelan frontier at the northern extremity of the Cordillera Oriental. The region corresponds generally to one often referred to by foreign writers as the Caribbean Lowland or Coastal Plain; in Colombia, however, it is consistently identified as Atlantic rather than Caribbean. The Atlantic Lowlands region is roughly the shape of a triangle, the longest side of which is the coastline. Inland from coastal cities are swamps, hidden streams, and shallow lakes that support banana and cotton plantations, countless small farms and, in higher places, cattle ranches. The northernmost extension of the Atlantic Coast is Point Gallinas. The coastline curves around the Cape of La Vela and continues in a southwestern direction until it gets to Cartagena. Here it turns to the south, cutting back to the west at the Gulf of Morrosquillo. It then continues uneventful until it cuts sharply into the mainland just before the Isthmus of Panama, forming the Gulf of Urabá.

The Pacific Lowlands are a thinly populated region of jungle and swamp with considerable but little-exploited potential in minerals and other resources. Buenaventura, at about the midpoint of the 800-mi (1,287-km) coast, is the only port of any size. On the east the Pacific Lowlands are bounded by the Cordillera Occidental, from which run numerous streams. The peaks of the Cordillera Occidental provide a barrier to rainclouds, and the rainfall along the coast is heavy as a result. The rainforest that lines the coast is dense, with a rich diversity of plant, animal, and bird life. The coast of the Pacific is very irregular, featuring many alternating bays and capes. From north to south they are Point Marzo, the Gulf of Cupica, Point Solano, the Gulf of Tibugá, and Cape Corrientes, and, at the southernmost point, Tumaco Bay.

Population Centers – Colombia

(2001 POPULATION ESTIMATES)

Name	Population	Name	Population
Bogotá (capital)	6,712,247	Barranquilla	1,305,334
Cali	2,264,256	Cartagena	952,523
Medellín	2,026,789	Cúcuta	682,325

SOURCE: Departamento Administrativo Nacional de Estadística (National Statistics Administration—DANE), Columbia.

Departments – Colombia

POPULATIONS ESTIMATED IN 2000

Name	Population	Area (sq mi)	Area (sq km)	Capital
Amazonas	70,489	42,342	109,665	Leticia
Antioquia	5,377,854	24,561	63,612	Medellín
Arauca	240,190	9,196	23,818	Arauca
Atlántico	2,127,567	1,308	3,388	Barranquilla
Bolivar	1,996,906	10,030	25,978	Cartagena
Boyacá	1,365,110	8,953	23,189	Tunja
Caldas	1,107,627	3,046	7,888	Manizales
Caquetá	418,998	34,349	88,965	Florencia
Casanare	285,416	17,236	44,640	Yopal
Cauca	1,255,333	11,316	29,308	Popaygn
César	961,535	8,844	22,905	Valledupar
Chocó	407,255	17,965	46,530	Quibdó
Córdoba	1,322,852	9,660	25,020	Monterfa
Cundinamarca	2,142,260	8,735	22,623	Bogotá
Guainía	37,162	27,891	72,238	Puerto Inírida
La Guajira	483,106	8,049	20,848	Riohacha
Guaviare	117,189	16,342	42,327	Guaviare
Huila	924,968	7,680	19,890	Neiva
Magdalena	1,284,135	8,953	23,188	Santa Marta
Meta	700,506	33,064	85,635	Villavicencio
Nariño	1,632,093	12,845	33,268	Pasto
Norte de Santander	1,345,697	8,362	21,658	Cúcuta
Putumayo	332,434	9,608	24,885	Mocoa
Quindío	562,156	712	1,845	Armenia
Risaralda	944,298	1,598	4,140	Pereira
San Andrés y Providencia	73,465	17	44	San Andrés
Santander	1,964,361	11,790	30,537	Bucaramanga
Sucre	794,631	4,215	10,917	Sincelejo
Tolima	1,296,942	9,097	23,562	Ibagué
Valle de Cauca	4,175,515	8,548	22,140	Cali
Vaupés	29,942	25,200	65,268	Mitú
Vichada	83,467	38,703	100,242	Puerto Carreño
Special District				
Santa Fe de Bogotá	6,437,842	613	1,587	Bogotá

SOURCE: Departamento Administrativo Nacional de Estadística (National Statistics Administration—DANE), Columbia.

CLIMATE AND VEGETATION

Temperature

Temperatures throughout the country are dependent more on altitude than by a change in seasons. The hottest area, also known as *tierra caliente*, is a tropical zone that extends vertically from sea level to about 3,500 ft (1,100 m). In this area, the temperature is usually between 75°F and 81°F (24 and 27°C), with a maximum near 100°F (38°C) and a minimum of 64°F (18°C). A temperate zone, or *tierra templada*, exists in elevations between 3,500 and 6,500 ft (1,100 and 2,000 m), with an average temperature of 64°F (18°C). Rising to elevations between 6,500 and 10,000 ft (2,000 and 3,000 m), one encounters the *tierra fría*, or cold country, which has yearly temperatures averaging 55°F (13°C). Above 10,000 ft (3,000 m), one encounters more frigid temperatures, often between 1°F and 55°F and (-17°C and 13°C).

Rainfall

Seasons are determined more by changes in rainfall than by changes in temperature. Areas in the north generally experience only one rainy season, lasting from May through October. Other areas of the country, particularly on the western coast and near the Andes, experience alternating three-month cycles of wet and dry seasons. Annual rainfall averages 42 in (107 cm).

Grasslands

The tropical Cauca Valley, a fertile sugar zone that includes the best farmland in the country, follows the course of the Cauca River for about 150 mi (241 km) southward from a narrow gorge at about its midpoint near the town of Cartago. The cities of Cali and Palmira are situated on low terraces above the floodplain of the Cauca Valley.

The area east of the Andes includes about 270,000 sq mi (699,297 sq km)—three-fifths of the country's total area, but contains only a small percentage of the population. The entire area is known as the Eastern Plains, and is crossed from east to west by many large rivers. The Spanish term for plains (*llanos*) can be applied only to the open plains in the northern part where cattle raising is practiced, particularly in piedmont areas near the Cordillera Oriental.

Forests and Jungles

Some 45 percent of the country is forested, about half of it in exploitable timber. In the south, east of the Andes, lies a plains region crossed from east to west by many large rivers and covered with dense tropical rain forest. There are few settlements east of the Andes, and the region, from north to south, remains largely undeveloped. There is also dense tropical rainforest along the Pacific Coast.

HUMAN POPULATION

About three-quarters of the population lives on the plateaus and in the basins scattered among the mountainous regions of the Andes, close to two-thirds in urban areas. The south and *Llanos* regions are very sparsely populated. Before the arrival of Spanish explorers, the Chibchas Indians living in the area near modern-day Bogotá (elevation of 8,660 ft / 2,639 m) had developed a culture almost as complex and elaborate as that of the Incas to the south and the Aztecs to the north. There are also significant population centers in the valleys of the two great rivers that lie between the Cordillera Mountain Ranges. Almost two-thirds of the population are mestizo (mixed European and indigenous descent).

NATURAL RESOURCES

Colombia is a major producer of coal, nickel, and quarried materials (limestone, sand, marble, gravel, gypsum, etc.). The country is South America's only producer of platinum, and is a leading producer of precious metals and gemstones, such as gold, silver, and emeralds. Even though much of its terrain is mountainous, Colombia also has considerable agricultural resources due to its favorable climate. It is one of the leading producers of coffee and bananas, as well as the world's largest producer of cocaine. The nation also has large supplies of petroleum and natural gas buried beneath the eastern plains region and the northern mountainous region.

FURTHER READINGS

Dydynski, Krzysztof. *Colombia: A Travel Survival Kit.* Hawthorn, Vic., Australia: Lonely Planet Publications, 1995.

Lessard, Marc. *Colombia.* Montréal, Que.: Ulysse, 1999.

Pollard, Peter. *Colombia Handbook.* Lincolnwood, Ill.: Passport Books, 1998.

Ruiz, Bert. *The Colombian Civil War.* Jefferson, N.C.: McFarland, 2001.

Williams, Raymond L., et al. *Culture and Customs of Colombia.* Westport, Conn.: Greenwood Press, 1999.

Comoros

- **Area:** 838 sq mi (2,170 sq km) / World Rank: 171
- **Location:** Southern and Eastern Hemispheres; group of islands at the northern point of the Mozambique Channel, east of Mozambique and west of the island of Madagascar
- **Coordinates:** 12°10′S, 44°15′E
- **Borders:** No international borders
- **Coastline:** 211 mi (340 km)
- **Territorial Seas:** 12 NM
- **Highest Point:** Mt. Karthala, 7,743 ft (2,360 m)
- **Lowest Point:** Sea level
- **Longest Distances:** 110 mi (180 km) ESE-WNW / 60 mi (110 km) NNE-SSW
- **Longest River:** None of significance size
- **Natural Hazards:** Cyclones, volcanoes
- **Population:** 596,202 (July 2001 est.) / World Rank: 159
- **Capital City:** Moroni, located at the western edge of Grande Comore
- **Largest City:** Moroni, 36,000 (2000 est.)

OVERVIEW

Comoros includes four main islands: Grande Comore (Ngazidja), Mohéli (Mwali), Anjouan (Nzwani), Mayotte (Maore) (144 sq mi / 374 sq km), and several smaller islands. Mayotte is claimed Comoros, but remains under French administrative control. Each island is of varying age with distinct topographical characteristics.

Located in the African Plate, the archipelago is the result of volcanic action along a fissure in the seabed running west-northwest to east-southeast in the western Indian Ocean. The center of Grande Comore is a desert lava field. The black basalt relief rises 3,950 to 5,250 ft (1,200 to 1,600 m) on Anjouan and 1,650 to 2,600 ft (500 to 800 m) on Mohéli.

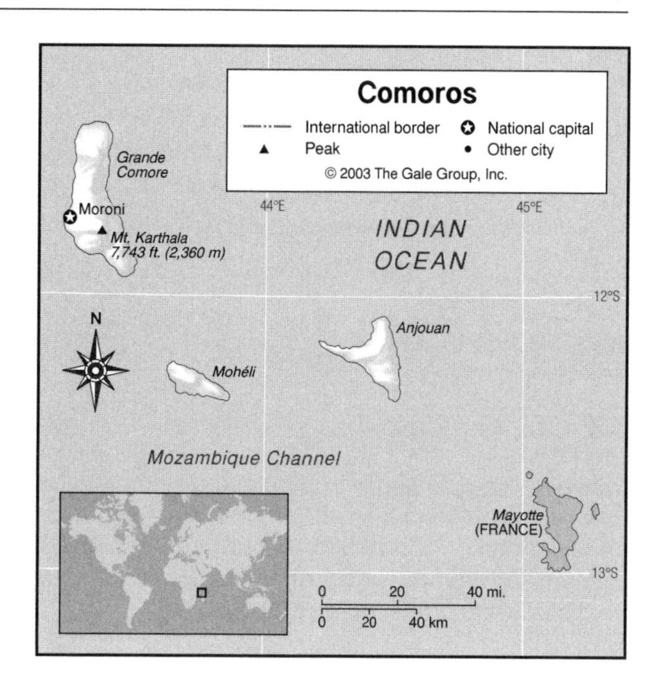

MOUNTAINS AND HILLS

Mt. Karthala (Mt. Kartala, Le Kartala) is an active volcano located on the southern tip of the island of Grande Comore. A number of other peaks rise from a plateau, nearly 2,000 ft (600 m), on Grande Comore.

The island of Anjouan has steep hills that rise nearly 5,000 ft (1,500 m) from a volcanic massif in the center of the island. A ridge lies on Mohéli in the center of a plain that reaches 1,900 ft (580 m) above sea level.

INLAND WATERWAYS

None of the islands have any rivers of note. Mangrove swamps can be found along the coastal zones of the islands.

THE COAST, ISLANDS, AND THE OCEAN

The largest islands are Grande Comore, 443 sq mi (1,148 sq km); Mohéli ,112 sq mi (290 sq km); and Anjouan, 164 sq mi (424 sq km). There are also several smaller islands of lesser size. All of the islands lie in the Indian Ocean, north of Mozambique Channel. The sandy beaches of the islands have the potential to become an important resource for the tourism industry in Comoros.

CLIMATE AND VEGETATION

Temperature

Located a little more than 10 degrees below the equator in the western Indian Ocean, the islands have a maritime tropical climate. In the wet season from October to April, the predominant northerly winds of the Indian Ocean bring moist, warm air to the region. The temperature averages 82°F (28°C) in March, the hottest month. From May to September southerly winds dominate the region. These winds are cooler and drier, and temperatures in the islands average around 66°F (19°C).

Rainfall

Heaviest rainfall occurs during the period from December to April; January rainfall averages 16.5 in (420 mm). Rainfall and temperature vary from island to island during any month and even vary on an island due to the topography. The central, higher areas of an island are often cooler and moister than the coastal regions. This variation results in microecologies on the islands with distinct flora and fauna.

Grasslands

There are large tracts of fertile soil on these volcanic islands, yet because of the dense population, farming has been forced upwards on the hills, leading to deforestation and erosion.

Deserts

A desert lava field lies in the central interior of the island of Grande Comore.

Forests and Jungles

The rich volcanic soils enable the growth of plentiful vegetation. Mangroves predominate in the coastal areas,

Population Centers – Comoros

(1991 CENSUS)

Name	Population
Moroni (capital)	30,365
Mutsamudu	16,785
Domoni	10,400

SOURCE: Direction de la Statistique, Comores.

with palms, bananas, and mangoes further inland. Above this area is a forest zone, home to various tropical hardwoods.

Demand for firewood has put the country's forests at risk for deforestation. The forest zones are in the interior of the islands, lying above the terraced farms and coastal areas.

HUMAN POPULATION

The population, numbering 596,202 (July 2001 estimate), is made up of people of mixed descent. Maritime commerce before the entry of Europeans into the Indian Ocean brought Comorians into contact with peoples from southern Africa to southeast Asia. The majority are Sunni Muslim (98 percent), with a small Roman Catholic minority (2 percent). The activist population of Anjouan (Nzwani) has moved to secede from Comoros.

NATURAL RESOURCES

Having few natural resources, Comoros is one of the world's poorest countries. Agriculture, including fishing, hunting, and forestry, is the leading sector of the economy. Vanilla, cloves, perfume essences, copra, coconuts, bananas, and cassava (tapioca) account for the majority of Comoros' agricultural products.

FURTHER READINGS

ArabNet. *Comoros Islands.* http://www.arab.net/comoros/comoros_contents.html (accessed March 5, 2002).

Madagascar and Comoros: A Travel Survival Kit. Berkeley, CA: Lonely Planet Publications, 1989.

Ottenheimer, Martin, and Harriet Ottenheimer. *Historical Dictionary of the Comoros Islands.* Metuchen, NJ: Scarecrow Press, 1994.

U.S. Department of State, Bureau of Public Affairs. *Background Notes, Comoros.* Washington, DC: Office of Public Communications, 1992.

Weinberg, Samantha. *Last of the Pirates: The Reach for Bob Denard.* New York: Pantheon Books, 1994.

Democratic Republic of the Congo

- **Area:** 905,562 sq mi (2,345,410 sq km) / World Rank: 13
- **Location:** Central Africa; crossed by the equator, in the Northern, Southern, and Eastern Hemispheres; borders the Republic of the Congo in the west; Central African Republic and Sudan in the north; Uganda, Rwanda, Burundi, and Tanzania in the east; Zambia in the southeast; and Angola in the southwest
- **Coordinates:** 0°00′N, 25°00′E
- **Borders:** 6,661mi (10,744 km) total / Angola, 1,559 mi (2,511 km); Burundi, 145 m (233 km); Central African Republic, 979 mi (1,577 km); Republic of the Congo, 1,494 mi (2,410 km); Rwanda, 135 mi (217 km); Sudan, 390 mi (624 km); Tanzania, 295 mi (473 km); Uganda, 474 mi (765 km); Zambia, 1,197 mi (1,930 km)
- **Coastline:** 23 mi (37 km)
- **Territorial Seas:** 12 NM
- **Highest Point:** Margherita Peak, 16,756 ft (5,110 m)
- **Lowest Point:** Sea level
- **Longest Distances:** 1,414 mi (2,276 km) SSE-NNW / 1,389 mi (2,236 km) ENE-WSW
- **Longest River:** Congo River, 2,700 mi (4,344 km)
- **Largest Lake:** Lake Tanganyika, 12,480 sq mi (32,000 sq km)
- **Natural Hazards:** Drought, volcanic activity
- **Population:** 53,624,718 (2002 est.) / World Rank: 23
- **Capital City:** Kinshasa, in the southwest on the south bank of the Congo River
- **Largest City:** Kinshasa, 6,301,100 (2002 est.)

OVERVIEW

Slightly less than one-fourth the size of the United States of America, the Democratic Republic of the Congo (DROC, also formerly known as Zaire) dominates central southern Africa. This part of Africa is made up of many plateaus, and almost all of the DROC lies more than 1,300 (400 m) above sea level. This includes its many river valleys, for the DROC is a wet country lying almost entirely within the Congo River Basin and constitutes about two-thirds of it. The Congo River itself and the many streams, large and small, draining into it (and ultimately into the Atlantic Ocean) provide the most significant system of inland waterways in Africa. In places where navigation is not possible, the rapids and falls that prevent it furnish hydroelectric potential surpassing that of any country in the world.

There are four major geographic regions in the DROC, defined in terms of terrain and patterns of natural vegetation. The core region is the Congo Basin, a large depression (average elevation of 1,312 ft / 400 m) with the Congo River at its heart, often referred to simply as the Basin. The region's roughly 312,000 sq mi (800,000 sq km) constitute about a third of Congo's territory, and it

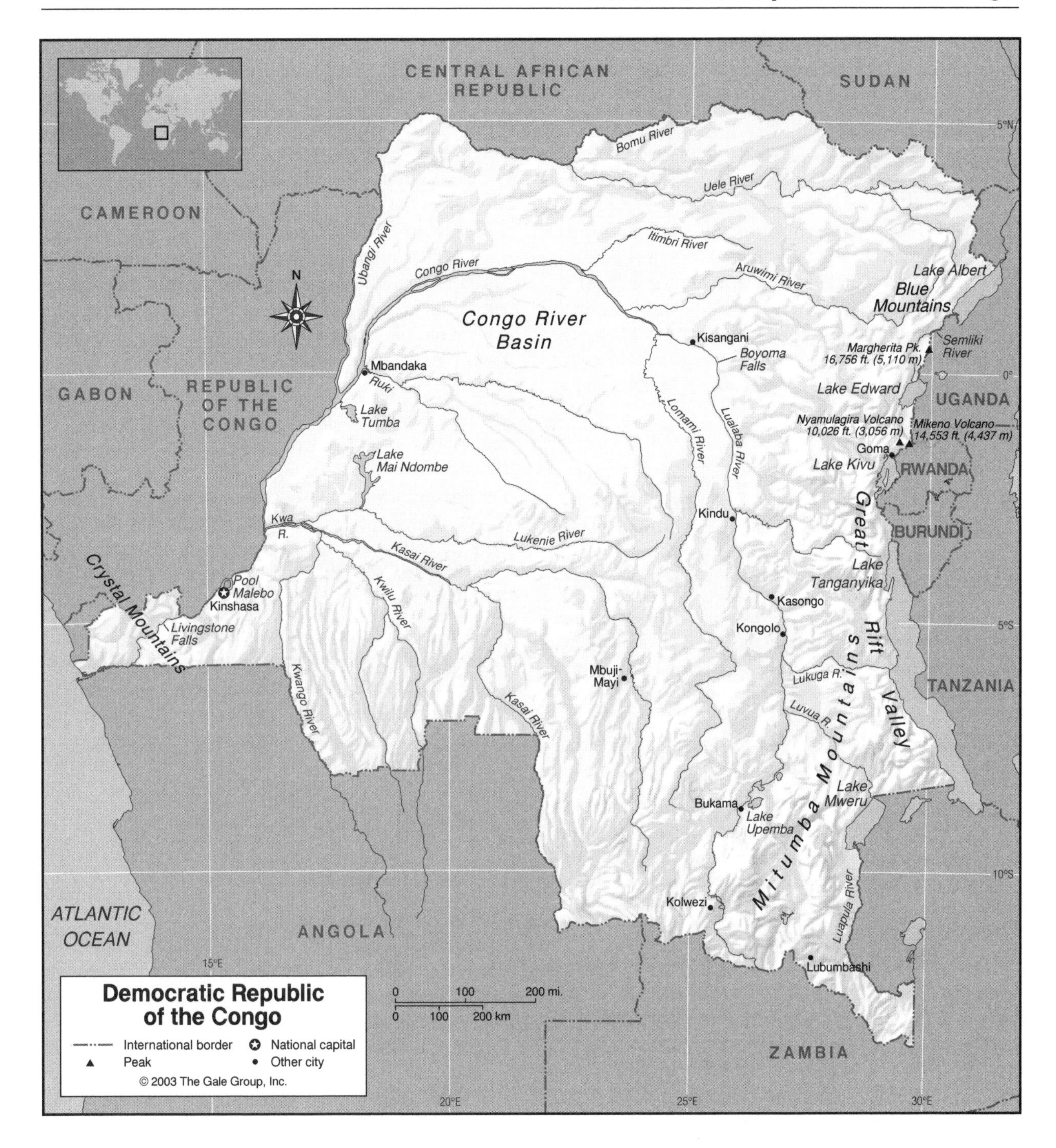

encompasses most of the country's more than 390,000 sq mi (1 million sq km) of tropical rainforest. A substantial proportion of the forest within the Basin is swamp, and still more of it consists of a mixture of marshy and firm land.

North of the Basin are the northern uplands, a region of higher plains and occasional hills covered with varying mixtures of savanna grasses and woodlands. This region slopes from south to north, starting at about 3,280 ft (1,000 m) and falling to about 1,640 ft (500 m) as it approaches the Basin. To the south of the Basin are the larger southern uplands, which constitute about a third of the DROC's territory. Its vegetation cover is somewhat more varied than that of the northern uplands, with savanna and scattered woodlands still dominant, but also extensive gallery forests. In the far southeast there are somewhat higher plateaus and low mountains. The westernmost part of the DROC, a partly forested panhandle reaching the Atlantic Ocean, is an extension of the southern uplands that drops sharply to a very narrow shore.

Along the eastern border there are high mountains associated with the western part of the Great Rift Valley (or East African Rift), as well as many very large lakes. It is an area of significant volcanic activity. The DROC is located on the African Tectonic Plate.

MOUNTAINS AND HILLS

The Great Rift Valley

The relief of the eastern part of the country is primarily characterized by the Great Rift Valley. The Great Rift Valley is a lengthy depression, the result of volcanic and tectonic activity, which stretches from north to south across most of eastern Africa and into Asia. In the DROC, Lakes Albert, Edward, Kivu, and Tanganyika occupy most of the bottom of this valley. On either side of the valley are mountain ranges.

Mountains

There are many mountain ranges making up the chain that borders the Great Rift Valley in the DROC. In the north are the Blue Mountains around Lake Albert. They reach heights of up to around 6,600 ft (2,000 m), separating the Congo and Nile River Basins.

The Ruwenzori Mountains between Lakes Albert and Edward are the highest mountain range in the country, and include Albert Peak (16,830 ft / 5,100 m) and Margherita Peak (16,896 ft / 5,120 m), also the highest elevation in the country. Margherita, the highest point in the DROC, is perpetually covered by snow despite being located practically on the equator.

To the south are the Ngoma Mountains, which extend to the Lukuga River. Their average altitude is 6,600 ft (2,000 m) with their highest point at Sambrini Peak (7,425 ft / 2,250 m). Continuing south, one finds the Kundelungu Mountains. Between 5,280 ft and 5,610 ft (1,600 m and 1,700 m) in height on average, they lie between the Lufira River in the west and the Luapula River and Mweru Lake in the east. These are older and more heavily eroded mountains, with poor soil. The Mitumba Mountains, with heights up to 7,260 ft (2,200 m) border Lake Tanganyika in the extreme southeast.

On the far side of the country, near the Atlantic shore, are the Mayumbe Mountains, part of the Crystal Mountain range. These are old mountains, strongly attacked by erosion, resembling a hilly plateau. Its lower elevations of the range are at 1,980 ft (600 m) and it reaches maximum elevation at Uia Mountain (3,465 ft / 1,050 m). The Congo River drains into the Mayumbe Mountains through a valley broken with rapids and falls. A significant portion of the Bas-Congo region presents a relief of hills, and the old mountain represents a very small part of the region.

Volcanoes

The Virunga Mountains, north of Kivu Lake between the Ruwenzori and Ngoma Mountains, consist of a series of volcanoes. Many of them are active, such as Karisimbi (14,873 ft / 4,507 m), Nyamulagira (10,026 ft / 3,068 m), and Nyiragongo (11,365 ft / 3,465 m). Others, such as Mikeno (14,642 ft / 4,437 m), Visoke (12,246ft / 3,711 m), and Sabinio (12,035 ft / 3,647 m), are now dormant.

Nyiragongo is one of Africa's most active volcanoes. On January 10, 1977 the lava lake at the summit poured over the countryside at speeds of up to 40 miles per hour (60 km / hr), killing about 70 people. It was also active in 1982 and again in June of 1994. Its most recent eruption began in January 17, 2002 when its lava flows forced the evacuation of refugee-filled Goma, before filling its downtown streets with pumice several feet thick.

In March 2002 scientists were monitoring significant seismic activity from Mt. Nyamulagira, located 9 mi (14 km) northwest of Nyiragongo. Since 1882 Nyamulagira had erupted 34 times, although only the 1912-13 eruption was serious enough to cause fatalities. An eruption of Nyamulagira began on July 4, 1994 from a fissure on the west flank, spewing ash, lava fountains, and flows. Ash and Pele's hair fell 12 mi (20 km) from the volcano. The eruption ended July 27, 1995.

Plateaus

Most of the DROC is a low plateau, dropping in elevation only as it nears the Atlantic Ocean, and rising to mountains in the east. The southeastern part of the country was once all mountainous, but the effect of erosion has leveled much of these mountains. The result is Upemba, a hilly plateau with an altitude greater than 4,950 ft (1,500 m).

INLAND WATERWAYS

The Congo River System

The Congo River is second only to the Amazon River in volume of flow, and is the sixth longest river in the world (2,700 mi / 4,344 km). The basin of the Congo River covers almost all of the DROC, and extends into many of the surrounding countries. There are hundreds of tributaries, many of them significant rivers in their own right. Including its tributaries, the Congo River offers a 9,106 mi (14,500 km) navigable path.

The Congo River can be divided into three portions. It begins with its main tributary, the Lualaba River, which has its source at 5,115 ft (1,550 m) elevation west of Lumbubashi, close to the Zambian border. It then flows north, and is navigable between Bukama to Kongolo. During this stretch it receives many tributaries. The most important are the Luvua and Luapula Rivers, which drain waters from Lakes Bangwelo (in Zambia) and Mweru, and the Lukuga River, which drains the waters from the lakes Tanganyika and Kivu.

Past Kongolo there are falls which block river traffic. North of this the river is again navigable for the 69 mi

GEO-FACT

The town of Goma, 11 mi (18 km) south of the summit of Nyiragongo and on the shore of Lake Kivu, provided refuge to nearly a million refugees from the civil war in Rwanda. On January 17, 2002, lava from Nyiragongo flowed on the eastern and southern flanks of the volcano at a rate of 1.2 to 1.8 km / hour toward Goma. As lava several feet thick flowed down city streets, 400,000 people were evacuated for three days and 14 villages were damaged by the lava flows. The eruption killed more than 45 people and left 12,000 families homeless.

(110 km) between Kasongo and Kibomho, has another waterfall, and is once again navigable for the 195 mi (310 km) between Kindu and Ubundu. Beyond that point the navigation is stopped by the Boyoma (Stanley) Falls, located directly upstream of Kisangani. After Kisangani, the river is considered to be the Congo River proper, and is known as the Upper Congo (Haut-Congo). It also changes direction, gradually curing west and then southwest.

The Congo receives countless tributaries in this stretch, and soon widens considerably. The Lomami (800 mi / 1,285 km, estimate) enters the Congo from the south, after having paralleled the Lualaba for much of its course. The Aruwimi and Itimbiri enter from the northeast, but the greatest of the Congo's northern tributaries is the Ubangi. Including its headstreams the Uele and Bomu, it is 1,400 mi (2,255 km) long and forms most of the DROC's border with the Central African Republic, as well as part of the border with the Republic of the Congo. Still yet more tributaries enter the Congo from the east, after it has completed its curve and is flowing southwest. These include the Ruki and its many tributaries and the Kwa, with its large tributaries of Kasai, Lulua, Lukenie, Kwilu, and Kwango.

The Upper Congo, navigable for its entire length, terminates in Pool Malebo. Downstream from the pool the river is known as the Lower Congo. Here the river digs a deep, narrow, and winding pass through the Crystal Mountains, and is broken by the 32 rapids and falls known as the Livingstone Falls. These terminate at Matadi, and past this city the river is once again navigable as it flows west into the nearby Atlantic Ocean. The Congo is 7 mi (11 km) wide at its mouth.

Great Lakes

DROC is home to several of the Great Lakes of Africa, which fill basins in the Great Rift Valley along the DROC's eastern border. The northernmost of these lakes is Lake Albert. Lying at 2,030 ft (619 m) above sea level, Lake Albert has an area of 2,075 sq mi (5,374 sq km). To the south lies Lake Edward, at an elevation of 3,023 ft (916 m) and an area of 830 sq mi (2,150 sq km). Lake Edward drains into Lake Albert through the Semliki River. The outflow of Lake Albert is the Albert Nile in Uganda, making these two lakes part of the Nile River Basin. They are the only sizable bodies of water in the country not connected to the Congo River.

Lake Kivu, the highest of the Great Lakes at an altitude of 4,851 ft (1,470 m), has an area of 1,042 sq mi (2,699 sq km). Located amidst the volcanic Virunga Mountains, the lake has a high methane content. It is connected to Lake Tanganyika to the south by the Ruzizi River. Lake Tanganyika is one of the largest lakes in the world, with a surface area of 12,480 mi (32,000 sq km). It is the world's longest freshwater lake; 408 mi (650 km) long and 50 mi (80 km) wide at its greatest extent. It is also the second deepest lake in the world, with a maximum depth of 4,851 ft (1,470 m). This vast lake drains its waters into the Congo River through the Lukuga River.

The southern-most lake of the Great Lakes chain (excepting Lake Malawi, which is outside DROC) is Lake Mweru (Moero). It has an area of 1,770 sq mi (4,584 sq km). Lake Mweru straddles the border between Congo and Zambia, and is drained by the Luvua River, a tributary of the Lualaba and Congo Rivers.

Other Lakes and Wetlands

Lakes Tumba and Mai-Ndombe can be found in the western part of the country, part of the Congo Basin. Vestiges of an ancient sea, their shores are generally swampy. Another swampy depression surrounds Lake Upemba on the southeastern plateau of the same name. Pool Malebo (Stanley Pool) is a river lake on the lower Congo. The capitals of both the DROC and the Republic of the Congo are located on its shores. There are many lesser lakes as well.

THE COAST, ISLANDS, AND THE OCEAN

Oceans and Seas

The DROC claims a narrow corridor of land north of the Congo River, extending west from the heartland of the country to reach the Atlantic Ocean. The narrow coastline is only 23 mi (37 km) long.

Major Islands

There are no coastal islands, though countless alluvial islands are found throughout the river systems and interspersed along the Congo River between Kisangani and Mbandaka. Idjwi Island is located on Lake Kivu.

CLIMATE AND VEGETATION

Temperature

Temperatures can vary widely across this vast country. Near the equator in the northern DROC, it is hot and very humid. Only at the highest elevations does the temperature ever drop below 68°F (20°C). It is cooler and less humid in the southern highlands, and cooler and wetter in the eastern highlands and mountains. The average temperature is 77°F (25°C) around the central Basin; 79°F (26°C) on the coastline. Altitude also has a major impact on climate. Temperatures average around 65°F to 68°F (18°C to 20°C) at an altitude of 1,500m; from 61°F to 63°F (16°C to 17°C) at 6,600 ft (2,000 m); 59°F (11°C) at 9,840 ft (3,000 m); and 43°F (6°C) at 13,200 ft (4,000 m).

Rainfall

To the north of the equator there are two rainy seasons, one from April to June and another in September to October. The long dry season falls between November and March, although it does not necessarily last for the entire period and can be as short as one month. There is a short dry season during July and August. To the south of the equator, this cycle is reversed. In the mountainous regions, two short dry seasons only last a month each (January and July). In the south and southeast, the rainy season starts in mid-October and continues until mid-May. In the southern central part of the country the rainy season starts in early October and ends in late April, but a short dry season occurs generally in January.

Grasslands

The DROC's natural ground cover is predominately forest and woodland. Natural grasslands can be found in marshy areas in the central Congo Basin, in the savanna of the northern region, and also cover sizeable portions of Upemba Plateau and the far western region. Many areas throughout the county have been deforested for farming and settlement purposes.

Forests and Jungles

Despite logging, mining, and agricultural practices, tropical and sub-tropical forests cover over 390,000 sq mi (1 million sq km) of the DROC. The immense equatorial rainforest within Congo's borders is by far the largest forest on the continent. It harbors countless species of plants and animals. In the extreme southwest, scattered remnants of the largely cleared Mayumbe Forest can be found. Gallery forests are situated throughout the country.

HUMAN POPULATION

There are over 200 ethno-linguistic groups native to the DROC. The majority are of the Bantu type. The four largest groups—Mongo, Luba, Kongo (Bantu), and the Mangbetu-Azande (Hamitic)—make up about 45 percent of the population.

Population Centers – Democratic Republic of the Congo

(2002 POPULATION ESTIMATES)

Name	Population	Name	Population
Kinshasa (capital)	6,301,100	Kolwezi	803,900
Lubumbashi	1,074,600	Kananga	539,600
Mbuji-Mayi	905,800	Kisangani	510,300

SOURCE: Government of the Democratic Republic of the Congo.

Provinces – Democratic Republic of the Congo

POPULATIONS ESTIMATED IN 2002

Name	Population	Area (sq mi)	Area (sq km)	Capital
Bandundu	6,680,000	114,154	295,658	Bandundu
Bas-Congo	3,694,000	20,819	53,920	Matadi
Équateur	6,202,000	155,712	403,293	Mbandaka
Kasaï Occidental	4,399,000	60,605	156,967	Kananga
Kasaï Oriental	5,412,000	64,949	168,216	Mbuji-Mayi
Katanga	4,306,000	191,831	496,877	Lubumbashi
Kinshasa	6,301,000	3,848	9,965	Kinshasa
Maniema	1,609,000	51,057	132,250	Kindu
Nord-Kivu	4,738,000	22,966	59,483	Goma
Orientale	6,789,000	194,301	503,239	Kisangani
Sud-Kivu	3,773,000	25,124	65,070	Bukavu

SOURCE: Projections of the Government of Democratic Republic of the Congo.

Nearly a third of the population is concentrated in and around the capital of Kinshasa, either in the city itself or in the surrounding Lower Congo area. The remainder is distributed throughout the rest of the country. There are a few densely populated regions—the southeast around Lubumbashi, the area between the Kwilu and Kasai Rivers, and the mountainous east near Lakes Kivu and Edward. The rest of the country is sparsely populated, with a density below the overall national average of 13 per sq km. Even though the population is largely rural, labor there is relatively scarce due to the flow of youth to the urban areas.

The DROC's population was greatly affected by warfare in and around the country throughout the 1990s and early 2000s. More than 1 million refugees fled from Rwanda into the eastern DROC following the 1994 genocide in that country. After war broke out within the DROC in August 1998, some 1.8 million Congolese were estimated to have become displaced by 2002, and at least 300,000 had fled to neighboring countries.

NATURAL RESOURCES

Congo holds many important mineral reserves. Some of the world's largest deposits of cobalt are in southeastern DROC. Along with neighboring northern Zambia, this region also contains the "Copper Belt," a wide strip of land hundreds of miles long, rich in copper, zinc and manganese. Besides these minerals, southeastern DROC has abundant resources of gold, tin, silver, coal, uranium, cadmium, and tungsten. Further north, in the Lake Kivu region, there are additional major deposits of titanium, gold, diamonds, tin, and, in Lake Kivu itself, methane. While some oil production occurs off the country's narrow Atlantic shoreline, a greater hydrocarbon potential exists in the oil and coal deposits in the east. The Congo River and its tributaries present great opportunities for hydropower, especially in the Livingstone Falls. Agricultural products include coffee, tea, quinine (in the east), sugar, palm oil (in the west), cassava (tapioca), rubber, bananas, root crops, corn, and fruits. The DROC's vast forests represent major reserves of timber. Instability and warfare in the DROC and the surrounding area limited the development of these resources during the 1990s and early 2000s.

FURTHER READINGS:

Bechky, Allen. *Adventuring in East Africa: The Sierra Club Travel Guide to Kenya, Tanzania, Rwanda, Zaire, and Uganda.* San Francisco: Sierra Club Books, 1990.

Bobb, F. Scott. *Historical Dictionary of Democratic Republic of the Congo (Zaire).* Lanham, Md.: Scarecrow Press, 1999.

Diallo, Siradiou. *Le Zaire Aujourdhui.* 2nd Edition. Paris: Les Editions Jeune Afrique, 1984.

Henry-Biabaud, Chantal. *Living in the Heart of Africa.* Translated by Vicki Bogard. Ossining, NY: Young Discovery Library, 1991.

Meditz, Sandra W., and Tim Merrill, eds. *Zaire: A Country Study.* Federal Research Division, Library of Congress. Foreign Area Studies. Washington, DC.: American University, 1993.

Myers, N. "Tropical forests: present status and future outlook." *Climatic Change,* 19, 1991, pp. 3-32.

Simkin, T., and Siebert, L. *Volcanoes of the World.* Tucson, Arizona: Geoscience Press, 1994.

Smithsonian Institution. *Bulletin of the Global Volcanism Network.* Vol. 20, No. 1, Washington, D.C., 1995, pp. 11-12.

University of North Dakota. *Volcano World.* http://volcano.und.nodak.edu/ (accessed March 16, 2002).

University of Pennsylvania. *Democratic Republic of Congo (Zaire) Page.* http://www.sas.upenn.edu/African_Studies/Country_Specific/Zaire.html (accessed March 16, 2002).

White, F. *The Vegetation Map of Africa.* Paris: UNESCO, 1983.

World Resources Institute and World Bank. *Tropical Forests: A call for Action.* Washington, D.C.: World Resources Institute and World Bank, 1985.

Republic of the Congo

- **Area:** 132,047 sq mi (342,000 sq km) / World Rank: 64
- **Location:** Western Africa, crossed by the equator, in the Eastern, Southern, and Northern Hemispheres, bordering the South Atlantic Ocean, between Angola and Gabon
- **Coordinates:** 1°S, 15°E
- **Borders:** 3,413 mi (5,504 km) / Cameroon, 324 mi (523 km); Central African Republic 290, mi (467 km); Democratic Republic of the Congo, 1,494 mi (2,410 km); Angola, 125 mi (201 km); Gabon, 1,180 mi (1,903 km)
- **Coastline:** 105 mi (169 km)
- **Territorial Seas:** 200 NM
- **Highest Point:** Mount Berongou 2,963 ft (903 m)
- **Lowest Point:** Sea level
- **Longest Distances:** 798 mi (1,287 km) NNE-SSW / 249 mi (402 km) ESE-WNW
- **Longest River:** Congo River, 2,678 mi (4,320 km)
- **Natural Hazards:** Seasonal flooding
- **Population:** 3,258,400 (2002 est.) / World Rank: 130
- **Capital City:** Brazzaville, southeastern Republic of the Congo on north bank of the Congo River
- **Largest City:** Brazzaville, 1,133,800 (2002 est.)

OVERVIEW

The Republic of the Congo is an irregularly shaped equatorial country located on the west coast of Africa and may be divided into four topographical regions. Along the seaboard lies a relatively treeless coastal plain, flanked by the forested Mayombé escarpment (2,000-3,000 ft / 600-900 m). The escarpment is bordered by a vast plateau region to the north and east covering some 50,000 sq mi (129,000 sq km). Still further to the northeast lies an expansive lowland covering some 60,000 sq mi (155,000 sq km) flooded seasonally by multiple crisscrossing tributaries of the Congo River. The Republic of the Congo is situated in the African Plate.

MOUNTAINS AND HILLS

Mountains

Inland from the seacoast the land rises somewhat abruptly to a series of eroded hills and plateaus, which run parallel to the coastline. From the lower reaches of

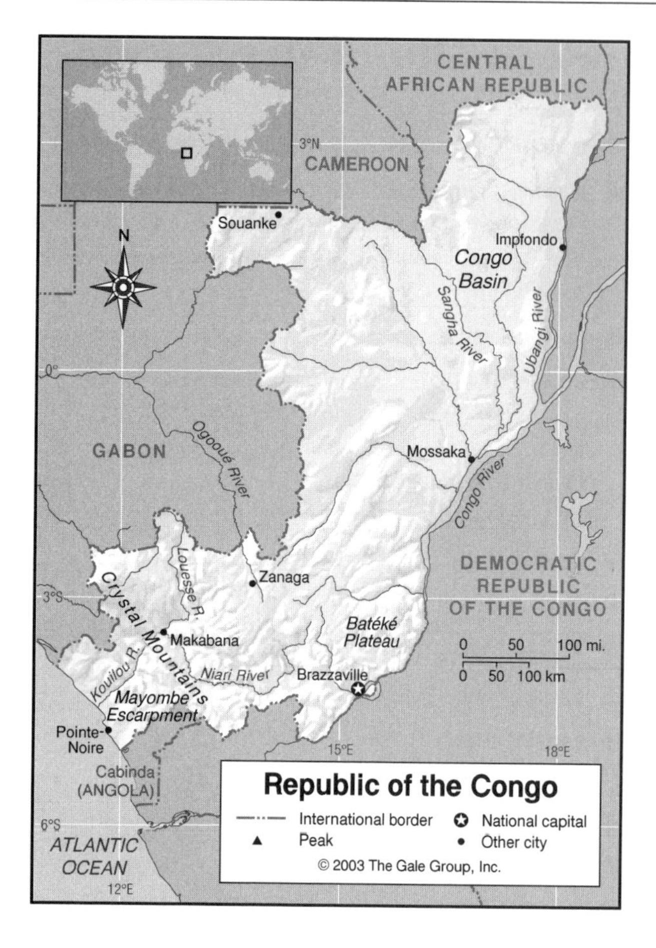

the Crystal Mountains on the Gabon border, this area rises southeasterly in a succession of sharp ridges of the Mayombé range that reach elevations of 1,600 to 2,000 ft (487 to 610 m). Deep gorges have been cut in these ridges by the swift Kouilou River or its tributaries. Mount Berongou, 2,963 ft (903 m), Republic of the Congo's highest point, is located in the upper reaches of the Crystal Mountains on the border with Gabon.

Plateaus and Hills

The Central Highlands encompass the area generally known as the Batéké Plateau and extend for approximately 50,000 sq mi (129,500 sq km) over the south-central portion of the country. This region is characterized predominantly by rounded, low hills of less than 1,000 ft (305 m) elevation and scattered rolling plains. In the northern part of this sector, however, toward the lower Gabon border, the hills are more peaked, and crests rise as high as 2,700 ft (823 m) above sea level. For the most part the lower hills are denuded and the plateaus grass covered.

To the northwest along the Gabon border and running almost to the equator, a region of hills and plateaus forms the western rim of the Congo Basin. The plateaus are separated from each other by deep valleys that carry the eastward-flowing tributaries of the Congo River.

INLAND WATERWAYS

Lakes and Rivers

Several lakes, lagoons, and swamps mark the coastal plain. Republic of the Congo has two river systems: the coastal rivers, which flow into the Kouilou River, and—by far the larger and more extensive—the Congo River and its tributaries. The plateau region divides the watershed between the Niari and Ogooué river systems. The Congo and Ubangi (Oubangui) Rivers provide 670 mi (1,120 km) of commercially navigable water transport, and form the main trade artery for the country's 4,000 miles (6,400 km) of navigable streams. The Congo River constitutes a 496 mi (800 km) border between the two Congo republics, is the sixth longest river in the world (2,678 mi / 4,320 km), and second in volume world-wide.

Wetlands

The northeast section of the country, covering an area of approximately 60,000 sq mi (155,400 sq km) lies within the Congo Basin (cuvette). It consists of flat, swampy valleys and low divides descending east and southeast from the western hills to the Congo River. Large portions lying northeast and southwest of the Sangha River are permanently inundated. Flooding occurs seasonally almost everywhere, and in areas south of Mossaka, along the Congo River, extensive marshland covered with swamp vegetation exists.

The confluence of the Niari River with the Kouilou and the Louessé Rivers permanently inundates much of the low-lying land south of Makabana. In addition, extensive swampy areas exist both to the northwest and southeast of the mouth of the Kouilou River.

THE COAST, ISLANDS, AND THE OCEAN

Republic of the Congo has a seaboard less than 100 mi long (160 km) on the Atlantic Ocean. There are no coastal islands, though countless alluvial islands are found throughout the river systems, especially the Congo and Ubangi Rivers and their tributaries. The coastal area, lying to the southwest between Gabon and the enclave of Cabinda, is an undulating plain fringed with sandy shores. It stretches for about 100 mi (160 km) along the south Atlantic coast and reaches inland approximately 40 miles to the Mayombé escarpment to the northwest and to the foothills of the Crystal Mountains. The area is bisected by the Kouilou River, which drains the area between these two mountain ranges from Makabana to the sea.

The effect of the Benguela (Antarctic) Current flowing from the south enhances the formation of sand spits along the coastal plain, which is virtually treeless except in scattered areas. In addition to the mangrove-fringed lagoons, the area is marked by lakes and rivers with accompanying marshland and heavy vegetation in low-lying areas.

CLIMATE AND VEGETATION

Temperature

The Republic of the Congo has constantly high temperatures and humidity and the climate is particularly enervating astride the Equator. At Brazzaville in the south, the average daily maximum temperature is 86°F (30°C) and the average minimum temperature 68°F (20°C). At Souanké, in the far north, the extremes are 84°F (29°C) and 64°F (18°C).

Rainfall

Republic of the Congo enjoys two wet and two dry seasons. In the south there is a rainy season from October to December with a short dry season in January followed by another rainy season from March to June and a long dry season June to October. The equator region receives rain throughout the year, but further north wet and dry seasons become more pronounced and are reversed from those in the south. Annual rainfall varies from 41 in (105 cm) at Pointe-Noire in the southwest, to 73 in (185 cm) at Impfondo in the northeast.

Grasslands

Savanna and grasslands cover much of the central plateau, plains, and deforested hills and valleys. The Niari Valley for almost 200 mi (322 km) was originally covered with tall grasses and savanna, but has been extensively cleared to permit a great variety of agricultural pursuits and diversified industrial activity.

Forests and Jungles

Approximately 60 percent of Republic of the Congo is lowlands covered by forest savanna. Tropical rain forest is interspersed throughout the entire northwest area along the Gabon border and running almost to the Equator forming part of the continent's great equatorial rain forest. A considerable portion of the area north of the Niari River and extending as far west as the vicinity of Zanaga is covered with dense tropical forest

GEO-FACT

During 1981-1985 deforestation in the Congo proceeded at a rate of 54,400 acres (22,000 ha) per year. As of the mid-1990s, 12 of 198 species of mammals were endangered. 3.4 percent of the nation's natural areas were protected.

Population Centers – Republic of the Congo

(2002 POPULATION ESTIMATES)

Name	Population
Brazzaville (capital)	1,133,800
Pointe-Noire	650,000

SOURCE: Projected from United States Agency for International Development and "Population of Capital Cities and Cities of 100,000 and More Inhabitants." United Nations Statistics Division.

About half the land area is covered by okoumé, limba, and other trees of the heavy rainforest. On the plateaus, the forest gives way to savanna broken by patches of bushy undergrowth. The Republic of the Congo's forests, which cover about 62 percent of the land area, are endangered by fires set to clear land for cultivation. They are also used for fuel. The most accessible forest, that of the Kouilou-Mayombé Mountains, has been over-exploited.

HUMAN POPULATION

At least four-fifths of the population live in the southern third of the country. About 70 percent of the people live in Brazzaville, Pointe-Noire, or along the railroad between them. Considerable portions of the Congo Basin are virtually uninhabited where hunting, fishing, trading, and subsistence agriculture are practiced. The average population density in 1998 was 21 per sq mi (8 per sq km) with a population growth rate of 2.8 percent between 1995 and 2000.

NATURAL RESOURCES

Republic of the Congo utilizes only 2 percent of its land for agriculture; 29 percent of the land is meadows and pastures; 62 percent forest and woodland; and 7 percent other. Natural resources include petroleum, timber, potash, lead, zinc, uranium, copper, phosphates, natural gas, and hydropower. Deposits of iron ore, bauxite, diamonds, and titanium have been discovered. Threats to the environment come from air pollution from vehicle emissions, and water pollution from the dumping of raw sewage and deforestation.

FURTHER READINGS

Africa South of the Sahara 2000: Congo. London: Europa Publications Ltd, 1999.

Bernier, Donald W. *Area Handbook for the People's Republic of the Congo.* Area Handbook Series. Washington, D.C.: U.S. Government Printing Office, 1971.

Chevron. *Republic of Congo; A Nation in Transition.* http://www.chevron.com/community/whats_new_stories/congo.shtml (accessed March 18, 2002).

Decalo, Samuel, Virginia Thompson, and Richard Adloff. "Historical Dictionary of the Congo." *African Historical Dictionaries #69*. Landham, MD: Scarecrow Press, Inc., 1996.

Ecological Science Based Forest Conservation Advocacy. *Forest Conservation Portal*. http://forests.org/ (accessed March 18, 2002).

Europa World Yearbook 2000: Congo. London: Europa Publications, Ltd., 2000.

Rainforest Action Network. http://www.ran.org/ran/ (accessed March 18, 2002).

Costa Rica

- **Area:** 19,560 sq mi (50,660 sq km) / World Rank: 128
- **Location:** Central America, between the North and South American continents, in the Northern and Western Hemispheres; bordered by Nicaragua on the north, the Caribbean Sea on the east, Panama on the southeast, the Pacific Ocean on the southwest and west / Its major dependency, Cocos Island, is approximately 480 km (300 mi) off the Pacific Coast
- **Coordinates:** 10°00′N, 84°00′W
- **Borders:** 399 mi (639 km) total / Nicaragua, 193 mi (309 km); Panama, 206 mi (330 km)
- **Coastline:** 805 mi (1,290 km) / Caribbean Sea, 132 mi (212 km); Pacific Ocean, 633 mi (1,016 km); Isla de Coco (Cocos Island) 40 mi (62 km)
- **Territorial Seas:** 12 NM
- **Highest Point:** Chirripó Grande 12,530 ft (3,819 m)
- **Lowest Point:** Sea level
- **Longest Distances:** 288 mi (464 km) N-S / 170 mi (274 km) E-W
- **Longest River:** San Juan, 140m (220 km)
- **Natural Hazards:** Earthquakes, hurricanes, flooding, active volcanoes
- **Population:** 3,773,057 (July 2001 est.) / World population ranking/ World Rank: 123
- **Capital City:** San José, located in the center of Costa Rica.
- **Largest City:** San José, 1,063,000 (2000 est.)

OVERVIEW

Costa Rica is the second-smallest Central American nation after El Salvador. Nicaragua and Panama comprise Costa Rica's land borders, and the Pacific Ocean and Caribbean Sea form its western and eastern coastline borders. Although the country lies completely within the tropics, elevation plays a role in the variations of its climate. The landscape varies from seasonally snow-capped mountains to seasonal marshlands to lush rain forests. Costa Rica has many rivers, but only few lakes, and Laguna de Arenal (Lake Arenal), the largest lake in the country, is man-made.

The country sits at the boundary where the Pacific's Cocos Plate—a piece of the earth's crust about 316 mi (510 km) wide—meets the crustal plate underlying the Caribbean Sea. The two plates continue to converge as the Cocos Plate moves east at a rate of about 4 in (10 cm) per year, causing occasional earthquakes in the country. Costa Rica also lies at the heart of one of the most active volcanic regions on Earth.

MOUNTAINS AND HILLS

Mountains

Extending the length of Costa Rica, there are several distinct cordilleras: the Cordillera de Guanacaste, Cordillera Central, Cordillera de Tilarán, and Cordillera de Talamanca. They are the principal mountain ranges, which are part of the Andean-Sierra Madre chain that runs along the western shore of the Americas.

The Cordillera de Guanacaste, volcanic in origin, stretches for 70 mi (112 km) from the western border with Nicaragua to the Cordillera Central, from which it is separated by low mountains. The highest peak in the Guanacaste chain is the Miravalles volcano at 6,640 ft (2,024 m). To the southeast, the Cordillera de Tilarán is home to the Arenal volcano, one of the world's most active volcanoes. To the east lies Cordillera Central, which contains four volcanoes and the Meseta Central (which is also home to the capital city). Cordillera de Talamanca rises in the south, housing the country's highest point, Chirripó Grande (Mount Chirripó).

Lying at the heart of one of the most active volcanic regions on Earth, Costa Rica is home to seven active volcanoes, and 60 dormant or extinct ones. Four volcanoes, two of which are active, Irazú and Poás, rise near the capital city of San José, along with the Barba and Turrialba volcanoes. The remaining active to semi-active volcanoes are: Arenal, Miravalles, and Rincon de la Vieja.

Plateaus

The most important area of Costa Rica is the Meseta Central, which is two upland basins separated by low volcanic hills, and is home to half of the Costa Rican population. About 3,500 sq mi (9,065 sq km) in area and located in the temperate country, it lies between the Cordillera Central to the north and low mountains and hills to the south. The land surface of the Meseta is generally level or rolling, which is acceptable for agriculture, except near the headwaters of rivers, where it is hilly and occasionally too steep for agriculture.

The slightly higher and smaller eastern basin, called the Cartago basin, is drained by the headwaters of the Reventazón River, which flows through its deeply gorged valley to the Caribbean Sea.

The General Valley, drained by the General River, which becomes the Grande de Térraba River and empties into the Bay of Coronado, lies between the granitic Cordillera de Talamanca to the north and the coastal mountains of the southwest. Almost as large as the Meseta Central, the General Valley is a relatively isolated structural depression that ranges from 600-3,500 ft (183-1,066 m). River flood plains, terraces, rolling hills, and savanna dominate the landscape.

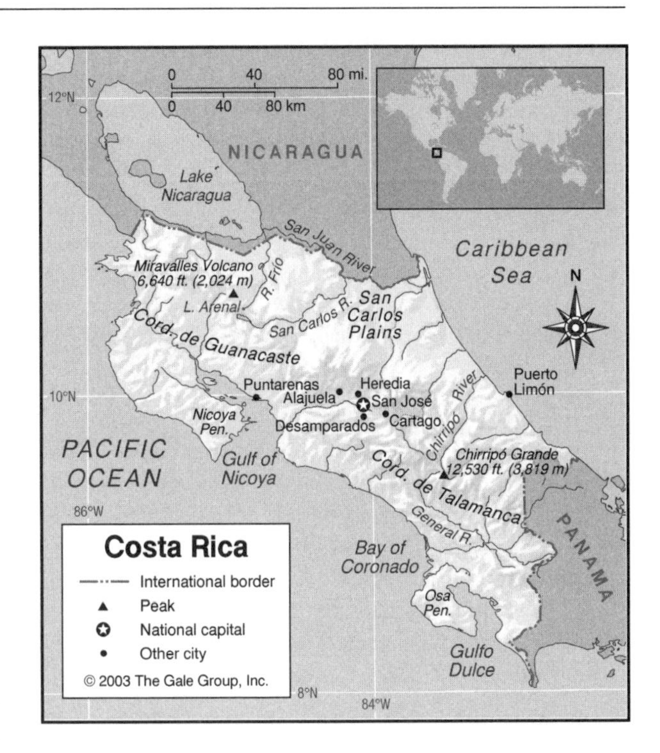

INLAND WATERWAYS

Lakes

Lake Arenal, located near Arenal volcano in the Arenal National Park, is the largest lake in Costa Rica. A man-made lake, it was formed in 1973 when a dam was built to provide hydroelectric power. Year-round water temperatures remain at 65-70°F (18-21°C), and winds of 60 knots or greater. Lake Coter is a small lake located near Lake Arenal.

Lake Cachí was also created by a dam across the Río Reventazón at its eastern end. Its dam supplies hydroelectric power to San José, the capital city. Lake Hule, 4 mi (6 km) south of San Miguel, is set in a dormant volcanic crater. Fed by the fresh waters of the Río Frío, Lake Caño Negro is a seasonal lake (appearing during the wet season) near Costa Rica's northern border.

Rivers

The northern part of Costa Rica's lowlands is drained by the San Juan River and its three main tributaries, which rise in the volcanic highlands. An extensive delta has built up around the mouth of the San Juan, which is in flood stage from September through November. Although the San Juan River lies within Nicaraguan territory, Costa Rica has, by treaty, full rights of navigation. The lower reaches of the river are shallow, although it is navigable all the way from the Caribbean to Lake Nicaragua. The remaining rivers that drain the highlands and the Caribbean lowland south of the San Juan are relatively short. They drop precipitously from the highlands, but are not long enough to have built up extensive flood plains in their lower courses.

Wetlands

Near the Nicaraguan border, the San Carlos and Chirripó rivers commonly flood during the wet season, turning the surrounding landscape into swampy marshlands. Along the coasts, mainly where river mouths meet the ocean, there are extensive mangrove forests and swamps.

THE COAST, ISLANDS, AND THE OCEAN

Oceans and Seas

Costa Rica is bordered on the west by the Pacific Ocean, and on the east by the Caribbean Sea.

Major Islands

Coco Island (Isla del Coco), a dependency of Costa Rica, is heavily forested, with a maximum elevation of 2,788 ft (850 m). The island, 300 mi (480 km) southwest of Costa Rica in the Pacific Ocean, is uninhabited, with no permanent population.

The Coast and Beaches

Bordered by water on both sides, Costa Rica is home to numerous beaches of varying lengths and textures. The Caribbean coast of Costa Rica is flat and open, with gray or black sand beaches, while the Pacific coast is irregular with hilly or mountainous peninsulas, coastal lowlands, bays, and deep gulfs—the Golf of Nicoya (Goflo de Nicoya) in the north and Golfo Dulce in the south.

CLIMATE AND VEGETATION

Temperature

Most of Costa Rica has two seasons: the wet season from May to November (winter months), and the dry season from December to April (summer months). Climatic conditions are dependent mostly on altitude, as well as proximity to one or the other of the coasts. The area known as the tierra caliente (hot country) in the coastal and northern plains, extends from sea level to about 1,500 ft (457 m) and experiences daytime temperatures of 85-90°F (29-32°C). The tierra templada (temperate country), including the central valleys and plains, extends from 1,500-5,000 ft (457-1,524 m), with average daytime temperatures of 75-80°F (24-27°C). The tierra fría (cold country) composes the land above 5,000 ft (1,524 m) and experiences average daytime temperatures of 75-80°F (24-27°C), and nighttime temperatures of 50-55°F (10-13°C).

Rainfall

A series of three mountain ranges, flanked by lower hilly sections, bisects the country from northwest to southeast and is partly responsible for the different climatic conditions of the two coasts. The mountains block the rain-bearing northeast trade winds, which cause the heavy and continual rainfall of the Caribbean coastal area. The Pacific coast receives its rain from May through October when the southwest winds blow on shore.

Because of the greater rainfall on the Caribbean side and the warmer waters of the Caribbean, which affect the coastal air temperatures, the hot country and the temperate country climatic zones extend to higher altitudes on the Caribbean side than on the Pacific side.

The tierra caliente (hot country) is characterized by heavy rains, the tierra templada (temperate country) receives regular rains from April through November, and the tierra fría (cold country) is less rainy but more windy than the temperate regions. The average rainfall for Costa Rica is more than 100 in (250 cm).

Grasslands

The northern lowlands are broad, flat, wedge-shaped, and, in some areas, are cut off from the more densely populated highlands by a virtually impassible hardwood forest. The region is made up of two separate llanuras (low-lying plains); the Llanura de los Guatusos in the west and the San Carlos Plains (Llanura de San Carlos) farther east. The llanuras make up one-fifth of Costa Rica's land area, and extend along the entire length of the Río San Juan, whose course runs along the Costa Rica / Nicaragua border.

Forests and Jungles

Once nearly completely covered by forests, Costa Rica's deforestation for agricultural purposes and cattle ranching has greatly reduced its pristine forests to only 25 percent of the total area. The country lost an additional 5.2 percent of its forests and woodlands between 1983 and 1993.

The lush tropical evergreen rainforest of the Caribbean lowlands gives way on the Pacific side to a seasonally dry evergreen forest in the well-watered south, and tropical dry forest in the northwest.

Population Centers – Costa Rica

(2000 POPULATION ESTIMATES)

Name	Population	Name	Population
San José	1,063,000	Puntarenas	92,360
Alajuela	158,276	Puerto Limón	67,784
Cartago	108,958	Heredia	67,387

SOURCE: Instituto Nacional de Estadistica y Censos (National Institute for Statistics and Censuses), Costa Rica.

HUMAN POPULATION

More than half of the population is concentrated in urban areas in the temperate highland valley called the Meseta Central (Central Basin), which contains San José (the capital city), most of the country's large cities, as well as densely populated areas. The remaining population is distributed throughout other larger cities near the capital, such as Alajuela, Cartago, and Heredia. Puerto Limón, another larger city, is located on the Caribbean coast and Puntarenas is located on the Gulf of Nicoya.

NATURAL RESOURCES

Costa Rica's natural resources include hydroelectric power, bananas, coffee, the country's forests, and cattle ranching. Intel opened two side-by-side assembly plants in Costa Rica in 1997, which currently accounts for over 35 percent of the country's exports.

FURTHER READINGS

Baker, Christopher. *Costa Rica Handbook.* 3rd edition. Chico, CA: Moon Publications, Inc., 1999.

Dunlop, Fiona. *Fodor's Exploring Costa Rica.* 3rd edition. New York: Fodor's Travel Publications, 2001.

Greenspan, Eliot. *Frommer's Costa Rica.* New York: Hungry Minds, 2001.

In Costa Rica.Net. http://www.incostarica.net/centers/visitor/ (accessed February 27, 2002).

Lonely Planet.com. *Costa Rica.* http://www.lonelyplanet.com/destinations/central_america/costa_rica/ (accessed Feb. 26, 2002).

Tourism-Costa Rica.com. *Costa Rica.* http://www.tourism-costarica.com/tourism-costaricacom/html/index.html (accessed February 27, 2002).

Côte d'Ivoire

- **Area:** 124,502 sq mi (322,460 sq km) / World Rank: 69
- **Location:** Southern and Eastern Hemispheres in western Africa, bordering Liberia and Guinea in the west, Mali and Burkina Faso in the north, Ghana in the east, the Atlantic Ocean to the south
- **Coordinates:** 8°00′N, 5°00′W
- **Borders:** 1,932 mi (3,110 km) / Burkina Faso, 363 mi (584 km); Ghana, 415 mi (668 km); Guinea, 379 mi (610 km); Liberia, 445 mi (716 km); Mali, 330 mi (532 km)
- **Coastline:** 322 mi (515 km)
- **Territorial Seas:** 12 NM
- **Highest Point:** Mt. Nimba, 5,748 ft (1,752 m)
- **Lowest Point:** Sea level
- **Longest Distances:** 502 mi (808 km) SE-NW / 485 mi (780 km) NE-SW
- **Longest River:** Bandama, 500 mi (800 km)
- **Natural Hazards:** Subject to flooding in the rainy season
- **Population:** 16,393,221 (July 2001 est.) / World Rank: 57
- **Capital City:** Yamoussoukro, in central Côte d'Ivoire, just south of Lake Kossou
- **Largest City:** Abidjan, located on the southeastern coast, 2,793,000 (2000 est.)

OVERVIEW

From its southern coast on the Atlantic's Gulf of Guinea, Côte d'Ivoire's terrain slopes gently to elevations of about 1,400 ft (426 m) along the northern border. The four parallel drainage basins formed by the four main rivers of the country run generally north to south, but except for the westernmost, the divides between them are not sharply defined.

There are no great rivers, mountain barriers, or marked climatic differences dividing the land into distinctive geographic regions. More than by any other physical feature, the land is differentiated by zones of natural vegetation, extending roughly east and west across the entire country, parallel to the coastline. Three main regions corresponding to these zones are commonly recognized: the Lagoon Region, Dense Forest Region, and Savanna Woodland Region.

MOUNTAINS AND HILLS

Viewed as a whole, almost all of the country is little more than a wide plateau, sloping gradually southward to the sea. The only mountain masses of any consequence are along the western border and in the northwest where some of the higher peaks exceed 914 m (3,000 ft) in elevation. This includes Mount Nimba near the borders with Liberia and Guinea, which at 5,748 ft (1,752 m) is the highest peak in the country.

INLAND WATERWAYS

Lakes

The largest lakes in Côte d'Ivoire are all reservoirs. Lake Kossou in the central part of the nation is the largest, with the smaller Lake Taabo directly south. Lake Buyo is in the southwest and Lake Ayamé in the southeast are the other large reservoir lakes.

Rivers

There are four main rivers draining the vast plateau that comprises Côte d'Ivoire. All of them run roughly parallel from the north to the south. These rivers are the Cavalla (on the border with Liberia for over half its length), Sassandra, Bandama, and Komoé (the easternmost river). The Bandama is the longest and largest, draining about half the country. Although these are permanent streams with a good volume of water, they are commercially navigable only for short distances inland from the coast because of rocky ledges and rapids. Uncontrolled and given to seasonal flooding, they are not only of little use as communications lines but are also obstacles to east-west travel.

Wetlands

The Lagoon Region is a narrow coastal belt extending east along the Gulf of Guinea from the Ghana border almost to the mouth of the Sassandra River. In this region, sand bars and islands line the entire coast, forming many shifting lagoons. The region varies in width from a few hundred feet to 3 or 4 mi (5 or 6 km) and seldom rises more than 100 ft (30.4 m) above sea level. On the seaward side, the smooth, steep beaches are pounded by surf, heavy at all seasons, but particularly so in July and November. Behind the beaches, the sandy soil supports a luxuriant growth of coconut palm and salt-resistant coastal shrubs. Mudbanks, sandbars, and wooded islands dot the sheltered surface of the lagoons; their landward shores are indented with little forested bays and steep rocky headlands.

Most of the lagoons are narrow, salty, and shallow and are parallel to the coastline. They are linked to one another by small watercourses or canals, built by the French during the 1800s. Where the larger rivers empty, broad estuaries that may extend 10 or 15 mi (16 or 24 km) inland are formed. These are the only permanent natural exits from the region into the open sea.

THE COAST, ISLANDS, AND THE OCEAN

Oceans and Seas

Côte d'Ivoire borders the broad Gulf of Guinea—part of the Atlantic Ocean—in the south. Its coastline and the surrounding waters were named the Ivory Coast by European explorers centuries ago, and it is from this that the country takes its name (Côte d'Ivoire is French for Ivory Coast).

The Coast and Beaches

A narrow coastal belt of lagoons extends along nearly two-thirds of the coastal border, from the Sassandra River to Ghana. For its entire length, the coast of this region is fringed by a strip of low, sandy islands or sandbars, known as the *cordon littoral.* Built by the combined action of the heavy surf and the ocean current which sets eastward, the sand barrier has closed all but a few of the river mouths and formed a series of lagoons between itself and the true continental shore.

CLIMATE AND VEGETATION

Temperature

Côte d'Ivoire has a warm, humid climate that transitions from equatorial to tropical. Temperatures average between 75°F (25°C) and 90°F (32°C) and range from

50°F (10°C) to 104°F (40°C), dependent upon the time of year and the area of the country. In January, temperatures along the coast range from 75° to 90°F (24° to 32°C); it is slightly cooler in July, with temperatures ranging from 72° to 83°F (22° to 28°C). In the middle of the county, November temperatures (usually the highest for the year) range from 70° to 95°F (21° to 35°C) and July temperatures (usually the lowest for the year) range from 68° to 84°F (20° to 29°C). In the highlands in the north, temperatures are coolest, ranging from 63° to 86°F (17° to 30°C) in November.

Rainfall

In the north, heavy rains occur between June and October averaging 43 in (110 cm) annually. Along the equatorial coast and the southwest, some rain falls in most months, but is heaviest between May and July and August and September; the annual rainfall averages 43 to 87 in (110 to 200 cm) annually. The major dry season lasts from December to April.

Grasslands

In the southeast, the area behind the Lagoon region has been largely cleared of forest and is now farm and pastureland. Much land has also been cleared over the years in the central part of the country. Natural savanna exists in the most northern region of the country, the Savanna Woodland. Scattered trees and shrubs dot the grasslands, although their size and frequency diminish progressively from south to north. This is the driest region of the country.

Forests and Jungles

There are three types of forests in Côte d'Ivoire: rainforest, deciduous, and the secondary forest of the savanna in the far north. Dense forest characterizes the southern third of the country, but farther inland the woodlands become more and more sparse and grassy, with the heaviest growth bordering the rivers or dispersed in isolated pockets.

The dense forest region forms a broad belt that covers roughly a third of the country north of the lagoon region and extends from Ghana on the east to Liberia on the west. West of the Sassandra River, it reaches all the way to the sea. The Taï National Park is located here, protecting one of the largest remaining areas of West African rain forest. It also provides a habitat to many species of endangered wildlife, including the pygmy hippopotamus. The dense forest region's northern boundary, although well-defined, is very irregular, descending in the form of a wide V from points on the Ghana and Guinea borders some 200 mi (322 km) inland to within about 75 mi (121 km) of the sea north of Grand Lahou.

The heavy tropical forest flourishes throughout the region (and once stretched south to the Gulf of Guinea), except where it has been disturbed by man. Its northern limit is marked by a transition to open, grassy woodlands. This division is, only in small part, caused by climatic differences; the forest gives way to grassland primarily as the result of persistent cutting and burning by man encroaching on the forest from the north. No such limits exist on the east and west where the forest continues into the adjacent countries.

Population Centers – Côte d'Ivoire

(2000 POPULATION ESTIMATES)

Name	Population
Abidjan	2,793,000
Bouaké	462,000
Yamoussoukro (capital)	177,000
Daloa	171,000

SOURCE: Projected from United Nations Statistics Division data.

Departments – Côte d'Ivoire

Name	Area (sq mi)	Area (sq km)	Capital
Abengourou	2,664	6,900	Abengourou
Abidjan	5,483	14,200	Abidjan
Aboisso	2,413	6,250	Aboisso
Adzopé	2,019	5,230	Adzopé
Agboville	1,486	3,850	Agboville
Biankouma	1,911	4,950	Biankouma
Bondoukou	6,382	16,530	Boundoukou
Bongouanou	2,151	5,570	Bongouanou
Bouaflé	2,189	5,670	Bouaflé
Bouaké	9,189	23,800	Bouaké
Bouna	8,290	21,470	Bouna
Boundiali	3,048	7,895	Boundiali
Dabakala	3,734	9,670	Dabakala
Daloa	4,483	11,610	Daloa
Danané	1,776	4,600	Danané
Dimbokro	3,293	8,530	Dimbokro
Divo	3,058	7,920	Divo
Ferkessedougou	6,845	17,728	Ferkessdedougou
Gagnoa	1,737	4,500	Gagnoa
Guiglo	5,463	14,150	Guiglo
Issia	1,386	3,590	Issia
Katiola	3,637	9,420	Katiola
Korhogo	4,826	12,500	Korhogo
Lakota	1,054	2,730	Lakota
Man	2,722	7,050	Man
Mankono	4,116	10,660	Mankono
Odienné	7,954	20,600	Odienné
Oumé	927	2,400	Oumé
Sassandra	6,768	17,530	Sassandra
Séguéla	4,340	11,240	Séguéla
Soubré	3,193	8,270	Soubré
Tingréla	849	2,200	Tingréla
Touba	3,367	8,720	Touba
Zuénoula	1,093	2,830	Zuénoula

SOURCE: *Geo-Data: The World Geographical Encyclopedia,* 2nd ed. Detroit: Gale Research, 1989.

HUMAN POPULATION

The July 2001 estimate of 16,393,221 indicates a growth rate of 2.51 percent. The population is about evenly split between urban and rural, but has been becoming more urban for years. From 1950 to 2000 the percentage of the population living in cities grew from under 15 percent to nearly 50 percent.

NATURAL RESOURCES

Petroleum, diamonds, manganese, iron ore, cobalt, bauxite, and copper constitute the major mineral resources of Côte d'Ivoire. All of the major rivers have been harnessed for hydropower. The economy is primarily agricultural, Côte d'Ivoire is one of the world's largest producers of palm oil, cocoa, and coffee.

FURTHER READINGS

Fuchs, Regina. *Ivory Coast.* New Jersey: Hunter Publications, 1991.

Gottlieb, Alma, and Philip Graham. *Parallel Worlds: An Anthropologist and a Writer Encounter Africa.* New York: Crown Publishers, 1993.

Harshé, Rajen. *Pervasive Entente: France and Ivory Coast in African Affairs.* Atlantic Highlands, NJ: Humanities Press, 1984.

Mundt, Robert J. *Historical Dictionary of Côte d'Ivoire (the Ivory Coast).* Metuchen, NJ: Scarecrow Press, 1995.

Weiskel, Timothy C. *French Colonial Rule and the Baule Peoples: Resistance and Collaboration, 1889–1911.* New York: Oxford University Press, 1980.

Croatia

- **Area:** 21,831 sq mi (56,542 sq km) / World Rank: 126
- **Location:** Northern and Western Hemispheres, in southeastern Europe, bordering the Adriatic Sea, south of Hungary, west of Yugoslavia, northwest of Bosnia and Herzegovina, east of Slovenia.
- **Coordinates:** 45°10′N, 15°30′E
- **Borders:** 1,260 mi (2,028 km) / Bosnia and Herzegovina, 579 mi (932 km); Hungary, 204 mi (329 km); Slovenia, 311 mi (501 km); Yugoslavia, 165 mi (266 km)
- **Coastline:** 3,626 mi (5,835 km)
- **Territorial Seas:** 12 NM
- **Highest Point:** Mt. Dinara, 6,004 ft (1,830 m)
- **Lowest Point:** Sea level
- **Longest Distances:** N-S 310 mi (499 km) / E-W 288 mi (463 km)
- **Longest River:** Danube, 1771 mi (2,850 km)
- **Largest Lake:** Lake Vrana, 11.6 sq mi (30 sq km).
- **Natural Hazards:** Earthquakes are frequent in the east
- **Population:** 4,334,142 (July 2001 est.) / World Rank: 116
- **Capital City:** Zagreb, in north central Croatia
- **Largest City:** Zagreb, 765,200 (2002 estimate)

OVERVIEW

Croatia sprawls along eastern side of the Adriatic Sea, on the western side of the Balkan Peninsula. Its long coastal region stretches from the Istria Peninsula in the north to the Gulf of Kotor (Boka Kotorska) in the south, becoming increasingly narrow as it goes. For a short distance it is interrupted by a branch of neighboring Bosnia and Herzegovina. In the north, between Bosnia and Slovenia, Croatia extends inland as far as the Danube River.

Croatia has three main geographic types: the Pannonian and Peri-Pannonian Plains of eastern and northwestern Croatia, the hilly and mountainous central area, and the Adriatic coastal area that extends down to Dalmatia in the south.

Tectonic fault lines are widespread in north central Croatia, and also run through the Dinaric Alps down to Dalmatia. These structural seams in the earth's crust periodically shift, causing earth tremors and occasional destructive earthquakes.

MOUNTAINS AND HILLS

Much of Croatia lies at an altitude of over 1,640 ft (500 m). The Dinaric Alps are the most significant mountain range in the country, and Mt. Dinara (6,004 ft / 1,830 m), Croatia's highest peak, lies in them near the border with Bosnia and Herzegovina. They run across the central region of the country, forming the boundary between the coastal area and the eastern plains, then extend southeastward along the border with Bosnia.

Subsidiary ranges of the Dinaric Alps in Croatia include the Velika Kapela, Plješevica, and Velebit Mountains, with the high peaks of Kame Plješevica (5,437 ft / 1657 m), Velika Kapela (5,030 ft / 1,533 m), and Risnjak (5,013 ft / 1,528 m). In eastern Croatia the Psunj Mountains, Papuk Mountains, and Zagorje Hills can be found.

The limestone ranges of the Dinaric Alps are frequently referred to as karst or karstland, and are distinctive because water seeping through the soluble limestone has formed underground drainage channels. This leaves the mountains dry and rocky, with their surface pockmarked by depressions and caves.

INLAND WATERWAYS

Lakes and Waterfalls

Croatia's largest lake is Vrana, near Biograd, which only has a surface area of 11.6 sq mi (30 sq km). The

Plitvička Lakes are a string of 16 lakes within a national park of the same name. Croatia's most notable waterfall is the series of cascades between these lakes, the tallest of which has a 275 ft (72 m) drop. The cascade beds are fairly recent features—the oldest parts of the beds are only about 4,000 years old. Croatia's interior area has 14 thermal springs, including seven mineral springs.

Rivers

Croatia has two types of rivers. In the coastal region there are many short rivers and streams that run quickly down the steep mountains into the Adriatic Sea to the south and west. Among the largest of these are the Krka, Rasa, and the Neretva, which flows through the country for a short distance after entering from Bosnia.

In the interior east, rivers are wider and calmer. Blocked from the Adriatic by the Dinaric Alps, they instead flow east towards the Danube River and, ultimately, the Black Sea. The largest of these rivers form much of Croatia's borders in this region. The Drava and Mura make up almost all of the northwest border with Hungary. The Sava, after flowing across the country from Slovenia, forms the southern border with Bosnia and Herzegovina. The Kupa and Una are tributaries of the Sava. In the east the mighty Danube River, second longest in Europe, flows between Croatia and Yugoslavia. Both the Sava and the Drava are tributaries of this river.

THE COAST, ISLANDS, AND THE OCEAN

Oceans and Seas

Off Croatia's coast in the north by Slovenia, the Adriatic Sea is shallow, only reaching 75 ft (23 m) in the Gulf of Venice. The waters off southern Croatia, however, reach to depths of more than 3,900 ft (1,200 m).

The Coast and Beaches

Most of Croatia's coast is covered by rocks rather than sandy beaches. In the north is the Istria Peninsula and Kvarner Bay, while the Gulf of Kotor marks the far south. The southern half of Croatia's coastline is called Dalmatia, after the ancient Roman name for this region.

Major Islands

Croatia has a total of 1,185 islands, of which only 66 are inhabited. Croatia's coastal islands are mountainous, as they are extensions of the Dinaric Alps. The largest islands are Krk (157 sq mi/406 sq km), Cres (157 sq mi/406 sq km), Brač (153 sq mi/395 sq km), Pag (116 sq mi/300 sq km) and Korčula (110 sq mi/285 sq km).

CLIMATE AND VEGETATION

Temperature

In Zagreb, the average daily high temperature in July is 80°F (27°C), while in January it is 35°F (2°C). The overall average annual temperature in Zagreb is 52.9°F (11.6°C). The Adriatic coast has a more moderate, Mediterranean, climate. The average annual temperatures for Split and Dubrovnik are 61.9°F (16.6°C) and 62.8°F (17.1°C), respectively. The prevailing northeast winds include the maestral (mistral), which mitigates the heat in the summer, and the cold, dry bora.

Rainfall

Zagreb's annual precipitation is 36.4 in (924 mm); in the winter there are an average of 49 days with snow cover of greater than 0.4 inch (1 cm). The narrow Adriatic coastal belt has very dry summers. Neither Split nor Dubrovnik typically experience snow accumulation in the winter and each averages more than 100 clear days per year. Split averages 37.1 in (943 mm) of precipitation annually; Dubrovnik, 40.1 in (1,020 mm).

Plains

Occupying the east and northeast is the Pannonian Plain, a lowland that is the best farmland in the country. The plain was once occupied by an ancient sea, which was gradually filled by silt until it formed a fertile basin, marked by low hills and broad flood plains. The plains of Slavonia extend through the eastern arm of Croatia near Yugoslavia.

Forests and Jungles

Croatia's eight National Parks cover 196,000 acres (79,320 ha). The government also has 22 forest parks cov-

Population Centers – Croatia

(2001 CENSUS OF POPULATION)

Name	Population	Name	Population
Zagreb (capital)	691,724	Osijek	90,411
Split	175,140	Zadar	69,556
Rijeka	143,800	Slavonski Brod	58,642

SOURCE: "Population of the Twenty Largest Cities, 2001 Census." Central Bureau of Statistics, Republic of Croatia.

ering 12,159 acres (4,921 ha). The government has extended special protection to Velebit National Park in the north, one of the few remaining old growth forests in the Mediterranean region. The islands of Lokrum, Mljet, and Korčula contain densely wooded regions.

HUMAN POPULATION

Croatia's 21 counties had 422 municipalities and 123 towns in 1999. Most of the people live in the interior, although the coastal region is also well populated. The capital city of Zagreb is by far the largest in the country.

NATURAL RESOURCES

Oil fields in Slavonia and coalmines on the Istria peninsula are Croatia's main energy resources. Bauxite is mined at Obravac, Drni, and Rovinj. Croatia is mostly self-sufficient in mining and producing industrial minerals for construction such as cement, clays, and lime.

FURTHER READINGS

Brân, Zoë. *After Yugoslavia.* Oakland, Calif.: Lonely Planet, 2001.

Carmichael, Cathie. *Croatia.* Santa Barbara, Calif.: Clio Press, 1999.

Department of Telecommunications, Faculty of Electrical Engineering and Computing, University of Zagreb. *Republic of Croatia Homepage.* http://www.hr/hrvatska/geography/shtml (accessed 29 April 2002).

Foster, Jane. *Croatia.* London: APA, 2001.

Sabo, Alexander. *Croatia, Adriatic Coast.* Munich: Nelles, 1999.

Cuba

- **Area:** 42,803 sq mi (110,860 sq km) / World Rank: 105
- **Location:** Northern and Western Hemispheres, bordered by the Atlantic Ocean, Caribbean Sea, and the Gulf of Mexico, south of the Florida Keys and east of the Yucatan Peninsula.
- **Coordinates:** 21°30′ N, 80°00′ W
- **Borders:** None
- **Coastline:** 2,316 mi (3,735 km)
- **Territorial Seas:** 12 NM
- **Highest Point:** Pico Turquino, 6,578 ft (2,005 m)
- **Lowest Point:** Sea level
- **Longest Distances:** 55 mi (89 km) N-S; 760 mi (1,223 km) E-W
- **Longest River:** Cauto River, 213 mi (343 km)
- **Natural Hazards:** Hurricanes; earthquakes; floods; drought
- **Population:** 11,184,023 (July 2001 est.) / World Rank: 67
- **Capital City:** Havana, on the northwestern shore
- **Largest City:** Havana, population 2,300,000 (2000 est.)

OVERVIEW

The long, narrow island of Cuba has a shape that has been compared to a cigar caught between the fingers of Florida and the Yucatan. It is flanked by Jamaica on the south, Hispaniola on the southeast, and the Bahamas on the northeast. The principal trade routes to the Gulf of Mexico skirt its northern and southern coasts, and in the sixteenth century the island received from the Spanish monarchy the designation of "key to the Gulf of Mexico." This strategic location is memorialized at the top of the national coat of arms by a key that hangs suspended between the two headlands: Florida and the Yucatan Peninsula.

Cuba is slightly smaller than Pennsylvania and extends some 746 mi (1,200 km) from Cape Maisí on the east to Cape San Antonio on the west, about the distance from New York to Chicago. The largest of the West Indian islands, its territory almost matches that of all the other islands combined. In addition to the main island, the Cuban archipelago includes the Isla de la Juventud (Isle of Pines) near the south coast in the Gulf of Batabanó and some 1,600 coastal cays and islets. The main island occupies 94.7 percent of the national territory, and the Isla de la Juventud and the other cays and islets occupy respectively 2.0 percent and 3.3 percent of the total. Well more than half of the terrain consists of flat or rolling plains with a great deal of rich soil well suited to the cultivation of sugarcane, the dominant crop. There are rugged hills and mountains in the southeast and the most extensive mountainous zone of Cuba lies near its eastern extremity. Smaller mountain zones with lower elevations occur near the midsection and in the far west.

Cuba was raised from the seafloor by geological action occurring about 20 million years ago and at one time was connected with other Antillean islands. The

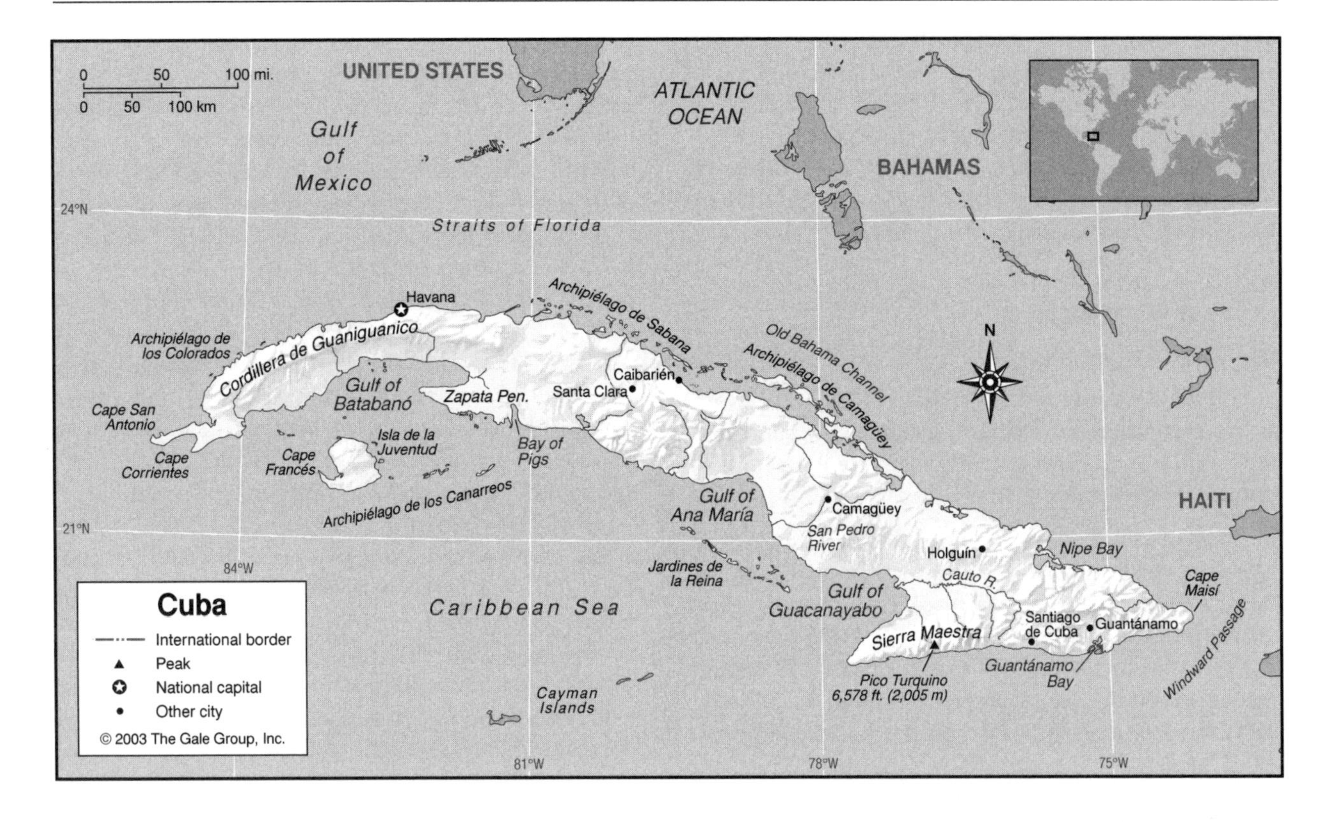

mountains of southeastern Cuba are related to those of southern Mexico, Jamaica, and Hispaniola; the limestone formations that make up much of the island resemble those of Florida, Jamaica, and the Yucatan Peninsula. Cuba's topography has resulted from the interaction of constructive forces that determined the basic structure and alignment of landforms and the destructive forces of wind and water that sculpted the structure into its present configurations. Soil erosion, however, has been less severe than on most other Antillean islands. The North American and Caribbean tectonic plates meet in the 23,622-ft (7,200-m) deep Cayman Trench between Jamaica and Cuba, and the region is thus prone to earthquakes. Consequently, the island is still subject to some crustal instability and its history has been marked by earthquakes of varying intensity. The zone of maximum instability occurs in the southeastern mountains. Light quakes are recorded frequently in the southeast, and a severe tremor suffered by Santiago de Cuba in 1578 was followed by another exactly one century later. Havana recorded several strong disturbances during the nineteenth century, and a severe tremor struck Pinar del Río Province in 1880. None, however, occurred in the twentieth century.

MOUNTAINS AND HILLS

Mountains

The Oriental, Central, and Occidental mountain ranges cover 25 percent of the country, while the loftiest mountain system is the Sierra Maestra. Sierra Maestra is the most heavily dissected and steepest of the Cuban ranges, and its peaks include Pico Turquino, which reaches 6,578 ft (2,005 m), the country's highest elevation. It skirts the southeastern coastline west of Guantánamo Bay except where it is broken by the small lowland depression on which Santiago de Cuba is located. On the east, the Sierra Maestra terminates in a low area around the United States Naval Base at Guantánamo Bay. The lowlands around Guantánamo mark the termination of the Central Valley, which is some 60 mi (96 km) in length and merges with plains to the west.

Most of the island east of a line from north to south between Nipe Bay and Santiago de Cuba is mountainous and includes such ranges as the Sierra de Nipe, the Sierra de Nicaro, the Sierra del Cristal, and the Cuchillas de Toa. The Escambray Mountains are the principal mountains of central Cuba. They are located in the southern part of that region, and are separated by the Agabama River into two ranges: the Sierra de Trinidad in the west and the Sierra de Sancti Spíritus in the east. The principal ranges of the western highlands are the Sierra del Rosario and the Sierra de los Organos. The Sierra del Rosario Range commences near the town of Guanajay, west of Havana, and extends southwestward along the spine of the island for about 60 mi (96 km). The Sierra de los Organos, continues in the same direction almost to the tip of the island. These western highlands, known collectively as the Cordillera de Guaniguanico, are limestone formations weathered into strange shapes. Ranks of tall erosion-resistant limestone columns resembling organ pipes gave the Sierra de los Organos its name. The numerous shapes, sinkholes, and underground caverns and streams are limestone developments known as karst. Karst landscape is most characteristic of the western highlands but is found all over the island.

INLAND WATERWAYS

Rivers

About 200 rivers run northward or southward from an interior watershed and are predominantly short and rapid. They provide good drainage but are not generally suitable for navigation.

Most of Cuba's rivers originate in the interior near the island's watershed and flow northward or southward to the sea. Smaller streams and arroyos that remain dry during most of the year are also numerous. River levels rise significantly during the rainy season when 80 percent of their flow occurs, and seasonal flooding is common. Most watercourses are not navigable while the potential to use them for hydroelectric power has yet to be tapped. One of the major rivers of Cuba, the San Pedro, runs from the city of Camagüey southwest to the Gulf of Ana María. The Cauto River of the Oriente Province, which flows for 213 mi (343 km), is the longest and heaviest flowing river in Cuba, rising in the Sierra Maestra near Santiago de Cuba and flowing westward to the Gulf of Guacanayabo. However, only small boats can navigate it. Smaller rivers can be found throughout the eastern part of the country, but are less common in the west.

Seven subterranean river basins constitute the sources of many surface rivers, and there are extensive reservoirs of fresh and brackish groundwater. Sulfide mineral springs are located near Pinar del Río and Matanzas, and there are radioactive thermal springs on the Isla de la Juventud.

Wetlands

There are no large lakes in Cuba, but many coastal swamplands extend throughout the country. Zapata Swamp, the largest on the island, covers more than 1,700 sq mi (4,403 sq km) on the Zapata Peninsula. Much of the southern coast has mangrove swamps that support small fish and birdlife. While the majority of the northern coast is bordered by rugged beaches, swamps still occur there and on the Isla de la Juventud.

THE COAST, ISLANDS, AND THE OCEAN

Oceans and Seas

Cuba is cradled between the Caribbean Sea to its south, the North Atlantic Ocean to its northeast, and the Gulf of Mexico to its northwest. It is separated from Florida to the north by the Straits of Florida, and from Hispaniola to the southeast by the narrow Windward Passage. The main island of Cuba rests on a subsurface shelf from which the numerous cays, coral islets, and reefs rise. Submerged about 300 to 600 ft (91 to 182 m) below sea level, the shelf varies in breadth off the north coast, is almost nonexistent off the southeast coast, and attains its maximum breadth off the remainder of the southern coastline where it extends to the limits of the Gulfs of Batabanó, Ana Maria, and Guacanayabo. Its outer rim is flanked on the southeast by the deep Bartlett Trough, which separates Cuba and Jamaica, and on the southwest by the Cayman, or Yucatan, depression. The two troughs are separated by the shallows of the submerged Cayman Ridge, a continuation of the Sierra Maestra range that reemerges as the Cayman Islands. Off the central northern coast the sea-lane of the Old Bahama Channel at some points is only ten miles wide as it passes between the Cuban shelf and the shallows of the Great Bahama Bank. The warm waters off the coast of Cuba are populated more than 900 species of fish and crustaceans. Coral reefs, the most complex and variable community of organisms in the world, surround Cuba.

Major Islands

The 570 sq-mi (220 sq-km) Isla de la Juventud is the westernmost island in a chain of smaller islands, the Archipiélago de los Canarreos, which extend 68.32 mi (110 km) across the Gulf of Batabanó wrapping around the Zapata Peninsula to the Bay of Pigs. This island is very circular, yet features the large indentation of Cape Francés, which points toward Cape Corrientes on the mainland. Farther east, beneath east-central Cuba, tiny coral cays, each with a beach, sit a mere 13–16 ft (4–5 m) above the sea. The extreme northwestern coast is flanked by the Archipiélago de los Colorados extending from Cape San Antonio to Havana. Offshore from the center of the island to the north of Sagua la Grande lie the islands of the Archipiélago de Sabana. East of those islands, stretching around the coast from Morón to Neuyitas is the Archipiélago de Camagüey, the largest of the archipelagos surrounding Cuba. Overall, about 4,200 coral cays and islets surround Cuba, most of them low-lying and uninhabited.

The Coast and Beaches

Satellite cays and islets are strung along both the northern and the southern coasts of Cuba and, except near its western tip, a wealth of excellent harbors indent the shoreline. The submerged shelf on which the island rests, however, is capped by extensive coral reef development that poses a serious hazard to navigation. The coastline includes more than 289 natural beaches. In the north, beaches tend to be longer and whiter with rolling surf and undertow while the southern beaches are darker, feature sea urchins, and are rockier or more swampy.

GEO-FACT

Desembarco del Granma National Park, a park in southwest Cuba near Cabo Cruz, features dramatic cliffs lining the shore of the Atlantic Ocean, as well as limestone terraces uplifted by geological forces. It was designated a World Heritage Site in 1999 by UNESCO.

Since the southern coast is closer to the equator, it is warmer in winter and less likely to show the effects of cold fronts that move down from the north.

Except where the precipitous cliffs of the Sierra Maestra plunge into the sea, most of the Cuban shoreline is fringed with coral reefs and archipelagos of cays. In the north an almost unbroken chain of cays extends from Cárdenas to Nueyitas. In the south, the Isla de la Juventud is the largest member of the Canarreos Archipelago, and the Jardines de la Reina chain flanks the Gulf of Ana Maria. Coral reefs are interlaced with many of the cays, clog the gulfs of Ana Maria and Guacanayabo, and form a chain of unborn islets off the western extremity of the island. Cuba's approximately 2,200-mile (3,540-km) coastline is indented by some of the world's finest natural harbors. There are about 200 in all, and many are of the pouch or bottleneck kind with narrow entrances that broaden into spacious deepwater anchorages. Ports on the north coast with these kinds of harbors include Mariel, Havana, Nueyitas, Manati, Puerto Padre, Gibara, and Antilla. South coast bottleneck ports include Guantánamo, Santiago de Cuba, and Crenfuegos. The principal open bay ports, Cárdenas and Matanzas, are located close to one another on the north coast of Matanzas Province. These were developed primarily to export sugar.

There are no important harbors west of Crenfuegos on the south coast or Mariel on the north. Shallow waters, coral formations, and a lack of good natural harbors are characteristic of the coasts of western Cuba, but since sugarcane is not grown in the west the need for ports is not pressing. Good ports along the rest of the coastline are lacking only along the 150 mi (241 km) between Caibarién and Neuyitas, where cays are numerous and there is extensive coral development.

CLIMATE AND VEGETATION

Temperature

Cuba has a pleasant subtropical climate strongly influenced by gentle northeast trade winds, which shift slightly to the east in the summer. The island's long, tapered shape allows the moderating sea breezes to have their effect and there are no pronounced seasonal variations in temperature. The average temperature in July and August, the warmest months, is 81°F (27°C) and in February, the coolest month, it is 72°F (22°C).

Rainfall

The wet summer season is between May and October, and the drier winter season runs from November through April. On average, rain falls on Cuba 85 to 100 days per year with three-quarters of it falling during the wet season. The humidity varies between 75 percent and 95 percent year round. The eastern coast is subject to hurricanes from August to October and the country averages about one hurricane every year. Droughts are also common.

Grasslands

Almost two-thirds of the Cuban landscape consists of flatlands and rolling plains. Cattle graze on these fertile flatlands, and sugarcane, coffee, and tobacco are grown. Grasslands, with hills and the lower and gentler slopes of the mountains, make up as much as three-fourths of the national territory. The generally easy gradients minimize the hazards of land erosion and facilitate both development of the transportation network and land tillage, including the use of mechanized equipment.

Forests and Jungles

Although much of the native forest was harvested in the first half of the twentieth century, about 1.8 million seedlings (eucalyptus, mahogany, and cedar, among others) were planted from 1960 to 1985 during a reforestation program. About one-fifth of the land area is made up of state forests.

There are more than 6,000 plant species in Cuba, about half of which are endemic. The royal palm *(Reistonea regia)* found on the country's coat of arms is said to number 20 million in Cuba. Also, Cuba offers the rare and prehistoric cork palm *(Microcycas calocoma),* descending from the Cretaceous Period; the jaguey, a fig with aerial roots; the big belly palm *(Palma barrigona)*; the ceiba; the sacred silk-cotton tree; and the mariposa (butterfly jasmine), the white national flower.

HUMAN POPULATION

An island of more than 11 million inhabitants, Cuba has an overall population density of 38.6 people per sq mi (100 people per sq km). However, 20 percent of its population (2.3 million), resides in its capital city, Havana. In an effort to curb the influx, the government has required that Cubans have special permission to migrate to Havana since May 1998. The most heavily populated areas are around Havana, between Cienfuegos and Santa Clara, and in the east.

Population Centers – Cuba

(1995 POPULATION ESTIMATES)

Name	Population
Havana	2,184,990
Santiago de Cuba	432,396
(Victoria de) las Tunas	324,011
Camagüey	296,601

SOURCE: Cámara de Comercio de la República de Cuba, as cited on GeoHive, 2002. http://www.geohive.com (accessed July 2002).

NATURAL RESOURCES

Cuba is rich with natural resources. The climate and rich soil make it a wonderful place to grow crops such as sugarcane, citrus fruit, coffee, rice, and tobacco. Its chief export is sugar, grown nearly everywhere on the island except for the western third. Cuba also has a large tobacco industry, producing many products including cigars that are considered among the best in the world. The fertile flatlands are well suited for livestock raising. Before being severely depleted, Cuba's forests provided an excellent source of timber, and the reforestation programs are increasing this resource every year. The island nation is also endowed with mineral resources including cobalt, nickel, iron ore, copper, manganese, salt, silica, and petroleum.

FURTHER READINGS

Baker, Christopher. *Moon Handbooks: Cuba.* Emeryville, Calif.: Avalon Travel Publishing, 2000.

Coe, Andrew. *Cuba.* Lincolnwood, Ill.: Passport Books, 1997.

Latimer Clarke Corporation Pty Ltd. *Atlapedia Online: Countries A to Z.* "Cuba." http://www.atlapedia.com/online/countries/cuba.htm (Accessed June 10, 2002).

Stanley, David. *Cuba.* Oakland, Calif.: Lonely Planet, 2000.

Cyprus

- **Area:** 3,571 sq mi (9,250 sq km) / World Rank: 164
- **Location:** Northern and Eastern Hemispheres, eastern Mediterranean Sea, south of Turkey
- **Coordinates:** 35°N, 33°E
- **Borders:** No international borders
- **Coastline:** 403 mi (648 km)
- **Territorial Seas:** 12 NM
- **Highest Point:** Olympus, 6,401 ft (1,951 m)
- **Lowest Point:** Sea level
- **Longest Distances:** 141 mi (227 km) ENE-WSW / 60 mi (97 km) SSE-NNW
- **Longest River:** Pedieos, 62 mi (100 km)
- **Natural Hazards:** Earthquakes and drought
- **Population:** 762,887 (July 2001 est.) / World Rank: 156
- **Capital City:** Nicosia, north-central Cyprus
- **Largest City:** Nicosia, 195,300 (1999 est.)

OVERVIEW

The largest Mediterranean island after Sicily and Sardinia, Cyprus is located in the extreme northeastern corner of the Mediterranean Sea, 44 mi (71 km) south of Turkey, 65 mi (105 km) west of Syria, and 230 mi (370 km) north of Egypt. Its average width is between 35 mi and 45 mi (56 km and 72 km) and it includes, at its northeasternmost tip, the small island outposts of Cape Andreas known as the Klidhes. The long, narrow Karpas peninsula extends 46 mi (74 km) northeastward from the broad heartland of the island to form Cape Andreas, leading to the frequent description of the island as being shaped like a skillet or frying pan.

Cyprus's topography is dominated by two mountain ranges and the central plain they encompass, called the Mesaoria. The Troodos Mountains cover the southern and western half of the country except for the coastal plains. The narrower and less elevated Kyrenia Range extends along the northern coastline. Coastal lowlands, varying in width, surround the island. The two mountain systems generally run parallel to the Taurus Mountains, and Cyprus, located on the Eurasian Tectonic Plate, is geologically a part of Asia Minor.

Since 1974 Cyprus has been divided into autonomous northern and southern sectors, separated by what is known as the Green Line. The Turkish sector north of the line, whose self-proclaimed government is recognized only by Turkey, comprises 37 percent of the island. The Greek sector, whose government is recognized internationally, takes up 59 percent, and the remainder belongs to a UN-controlled buffer zone.

MOUNTAINS AND HILLS

Mountains

The jagged slopes of the narrow Kyrenia Range stretch along the country's northern coast for some 100 mi (161 km), giving way to foothills as they extend into the Karpas Peninsula in the east. This mountain range, also known as the Pentadaktylos range because its most famous peak has a five-fingered shape, is a limestone ridge that belongs geologically to the Alpine-Himalayan system. Its highest peak is Kyparissouvouno (3,359 ft / 1,024 m); other notable peaks include St. Hilarion and Buffavento.

The rugged Troodos mountain range, an extensive massif formed of molten igneous rock, is the single most conspicuous feature of the landscape. It dominates the southwestern part of the island, with spurs going off in all directions. The island's highest peak, Mt. Olympus (6,401 ft / 1,951 m), is centrally located in the heart of these mountains, which extend across the southwestern portion of Cyprus from the Akamas Peninsula at the island's northwestern tip. The landscape of the Troodos range is more elevated but less dramatic and steep than that of the Kyrenia range to the north, with rolling valleys and foothills. To the southwest, the mountains descend in a series of stepped foothills to the coastal plain.

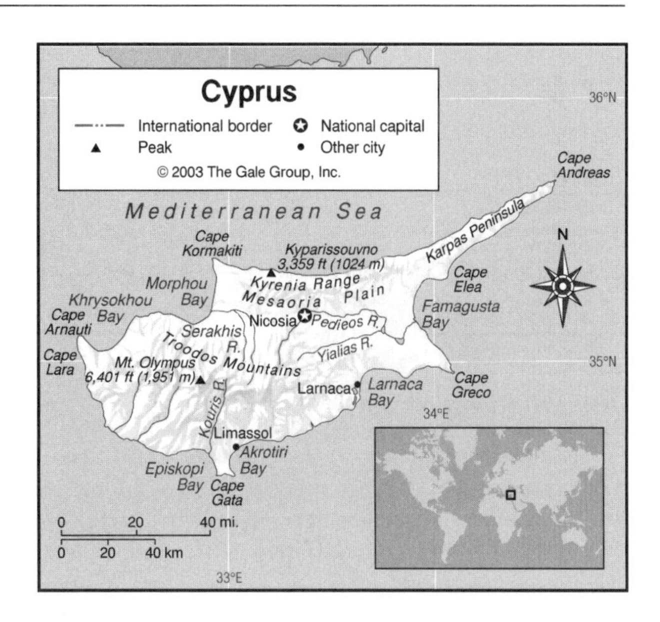

INLAND WATERWAYS

Lakes

Cyprus has few permanent lakes. Two large saltwater lagoons near Larnaca and Limassol on the southern coast dry up every summer and are filled by the winter rains.

Rivers

A network of rivers flows in all directions down the Troodos Mountains. Even the largest of these, the Pedieos, which drains eastward across the Mesaoria to empty into Famagusta Bay, is a winter river that becomes a dry course in the summer. So do Cyprus's other major rivers, including the Kouris, which flows south into Episkopi Bay; the Serakhis, which flows northwest to Morphou Bay; and the Yialias, which, like the Pedieos, flows eastward to Famagusta Bay.

THE COAST, ISLANDS, AND THE OCEAN

Cyprus is located at the far northeastern corner of the Mediterranean Sea. The island's coastline is rocky and heavily indented, with many bays and capes. The former include Famagusta Bay and Larnaca Bay in the east, the Akrotiri and Episkopi bays to the south, and the Khrysokhou and Morphou bays to the northwest. Capes include Andreas to the northeast (at the end of the Karpas Peninsula), Elea and Greco to the east (enclosing Famagusta Bay), Gata to the south, Lara to the west, and Arnauti and Kormakiti to the northwest.

The coast is fringed by sandy beaches.

CLIMATE AND VEGETATION

Temperature

The climate is Mediterranean, with sharply defined seasons. There are hot, dry summers between June and September; rainy winters from November to March; with short spring and autumn seasons in between. Average temperatures in Nicosia range from 70°F (21°C) to 98°F (37°C) in summer and 41°F (5°C) to 59°F (15°C) in winter.

Rainfall

Annual rainfall averages around 20 in (50 cm). Precipitation is highest in the Troodos Mountains, and lowest in the Mesaoria Plain.

Grasslands

The Mesaoria Plain, whose name ("Between the Mountains") describes its location between the island's northern and southern mountain ranges, stretches from Morphou Bay in the west to Famagusta Bay in the east. This flat, low expanse is the country's agricultural heartland and home to the capital city of Nicosia. Flowering bushes and shrubs grow between the autumn and spring seasons, and acacia, cypress, eucalyptus, and pine trees are found as well in wooded patches.

Forests and Jungles

Evergreens grow along the narrow coastal plain, as well as typical Mediterranean trees such as citrus and olive. Cypress, pine, cedar, and dwarf oak grow in the Troodos Mountains.

GEO-FACT

In ancient times Cyprus was famous for its copper mines, so much so that the modern word copper is thought to have evolved from Cyprus's ancient Greek name: Kypros.

Population Centers – Cyprus

(1994 POPULATION ESTIMATES)

Name	Population
Nicosia (capital)	188,800
Limassol	146,200

SOURCE: "Population of Capital Cities and Cities of 100,000 and More Inhabitants." United Nations Statistics Division.

HUMAN POPULATION

Steady urbanization took place throughout the twentieth century. By 2000 urban dwellers accounted for over half the population. Settlement patterns were affected by the partition of the island into Greek and Turkish sectors in 1974, when some 180,000 Greek Cypriots were obliged to relocate on the other side of the Green Line, settling mostly in the environs of Nicosia. In spite of a corresponding transfer to Turkish Cypriots to the north and an influx of Turkish immigrants, the Turkish sector is still more sparsely populated than the Greek one.

NATURAL RESOURCES

Cyprus's mineral resources include copper, asbestos, gypsum, pyrites, salt, and marble. The economies of the two parts of the island are separate, but both rely heavily on agriculture and tourism.

FURTHER READINGS

Borowiec, Andrew. *Cyprus: A Troubled Island.* Westport, Conn.: Praeger, 2000.

Bulmer, Robert. *Essential Cyprus.* Lincolnwood, Ill.: Passport Books, 1998.

Durrell, Lawrence. *Bitter Lemons.* New York: Marlowe, 1996.

Hellander, Paul D. *Cyprus.* London: Lonely Planet, 2000.

Thubron, Colin. *Journey into Cyprus.* New York: Atlantic Monthly Press, 1990.

Zaphiris, Panayiotis. *Kypros-Net Home Page.* http://www.kypros.org/Cyprus/root.html (accessed March 10, 2002).

Czech Republic

- **Area:** 30,450 sq mi (78,866 sq km) / World Rank: 117
- **Location:** Northern and Eastern Hemispheres, Central Europe, bordering Poland to the northeast, Slovakia to the southeast, Austria to the south, Germany to the southwest, west, and northwest
- **Coordinates:** 49°45′N, 15°30′E
- **Borders:** 1,169 mi (1,881 km) / Austria, 225 mi (362 km); Germany, 401 mi (646 km); Poland, 409 mi (658 km); Slovakia, 134 mi (215 km)
- **Coastline:** Landlocked.
- **Territorial Seas:** None
- **Highest Point:** Mt. Sněžka, 5,256 ft (1,602 m)
- **Lowest Point:** Elbe River, 377 ft (115 m)
- **Longest Distances:** 307 mi (494 km) E-W / 167 mi (269 km) N-S
- **Longest River:** Elbe River, 724 mi (1165 km)
- **Largest Lake:** Lake Rozmberk, 1,235 acres (500 hectares)
- **Natural Hazards:** None
- **Population:** 10,264,212 (July 2001 est.) / World Rank: 76
- **Capital City:** Prague, west-central Czech Republic
- **Largest City:** Prague, 1,233,000 (2000 est.)

OVERVIEW

Located in the heart of Central Europe, the landlocked Czech Republic is one of two nations formed by the breakup of Czechoslovakia in 1993. It consists of two major regions—Bohemia to the west and Moravia to the east. In addition, its northwestern corner is part of Silesia, a region that lies mostly in southwestern Poland. Bohemia, the larger of the two main regions, consists of highlands bordered by low mountains, while Moravia, although also surrounded by mountains, is composed of hilly lowlands.

The Czech Republic is located on the Eurasian Tectonic Plate. The mountains and hills that enclose the region of Bohemia are part of the north-central European uplands that run northeast through Germany and into Belgium. These uplands, which are distinct from the Alps to the south and the Carpathians to the east, are known geologically as the Hercynian Massif.

MOUNTAINS AND HILLS

Mountains

Mountain ranges ring much of the country and also separate its two major regions. Part of the border with Poland, to the north, is formed by the Krkonoše (or Great) Mountains, which are part of the Sudety Mountains and also form the northern border of Bohemia. The country's highest peak, Mt. Sněžka, is found in these mountains. Farther east, the Hrubý Jeseník Mountains

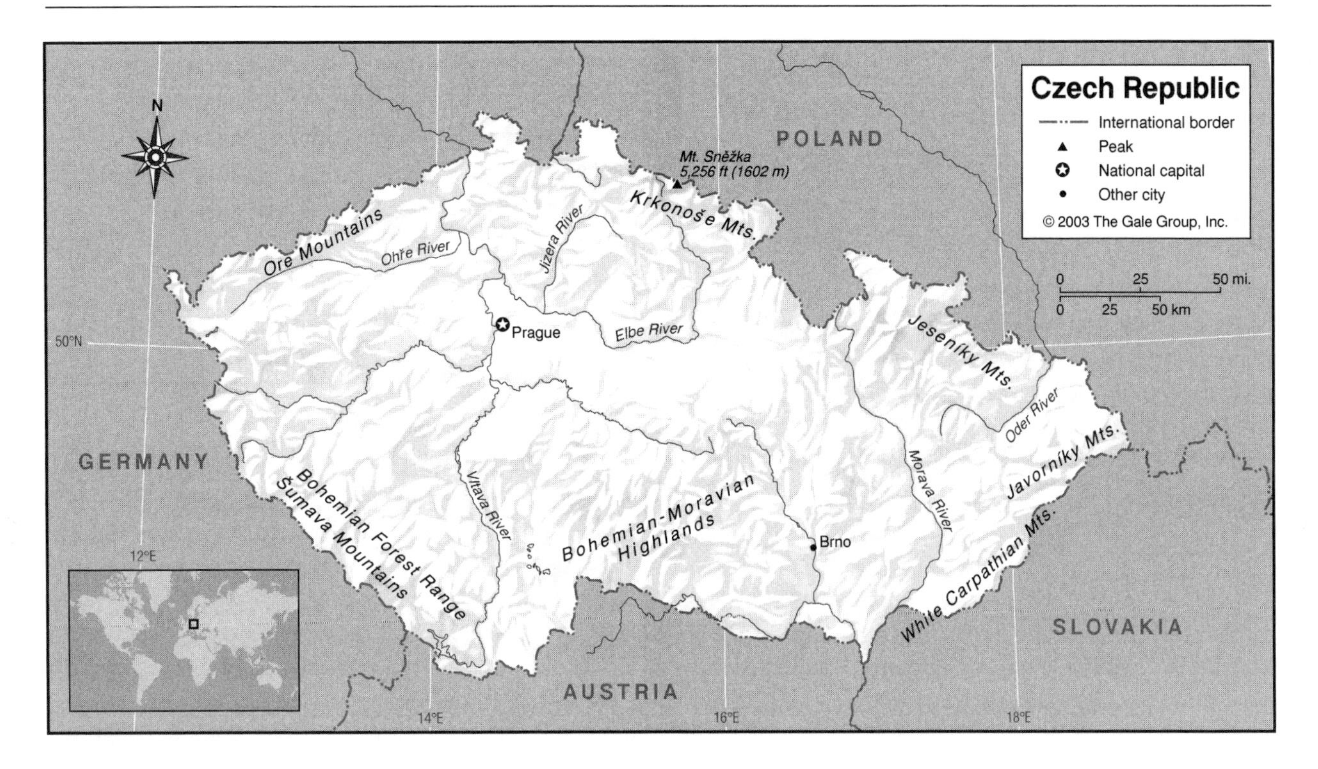

separate the Czech portion of Silesia from Moravia to the south. The eastern border—both of Moravia and the Czech Republic itself—is demarcated by the Javorníky Mountains and the Bíele Karpaty Mountains (White Carpathian Mountains).

In the center of the country, the Bohemian-Moravian Highlands separate Bohemia from Moravia. The šumava Mountains to the southwest, which include the Bohemian Forest Range, mark the borders with Austria and Germany in the south. The northeastern border with Germany is formed by the Ore Mountains (Erzgebirge).

Plateaus

The mountain ranges of Bohemia encircle a plateau that averages about 1,640 ft (500 m) above sea level and is shaped roughly like an oval. The capital city of Prague is located near the center of the plateau.

Canyons

One of the Czech Republic's most famous topographical features is the Moravian Karst, a highland area in southern Moravia where the erosion of limestone hills over time has created a dramatic landscape of caves and canyons.

Hills and Badlands

There are hills among the mountains surrounding the Bohemian plateau and in the Bohemian Forest to the south. Much of Moravia is hilly as well. There are sandstone hills with deep gorges in the northwestern part of the region, and there is a well-known winery district in its southern hills.

INLAND WATERWAYS

Lakes

Most lakes in the Czech Republic are manmade. In the southern part of Bohemia, near Ceské Budìjovice, is a region of artificial lakes and fish ponds where carp are bred. This part of the country is home to the Czech Republic's largest (artificial) lake, Lake Rozmberk, which covers some 1,235 acres (500 hectares), and the Lipno Dam, located near the southernmost part of the country, just north of the border with Austria. The Czech Republic also has many mineral springs.

Rivers

The Czech Republic's many rivers belong to three major systems. In the northwest, the Elbe (Labe) River, flows northward into Germany, ultimately draining into the North Sea. It is the longest river that flows through the Czech Republic. Among its tributaries are the Jizera,

GEO-FACT

In 1997 the Oder River experienced the worst flooding in the nearly two centuries that weather records had been kept to that date. The floods, which covered over one-third of the country, killed forty people, injured over two thousand, and left ten thousand homeless.

the Ohře, and the Vltava (or Moldau), which at 267 mi (430 km) is the longest river found entirely within Czech territory. In the northeast, the Oder (Odra) River flows north to Poland, later draining into the Baltic. The Morava River, the principal river of Moravia, flows southward through the eastern part of the country, to Austria and the Danube River.

Population Centers – Czech Republic

(1994 POPULATION ESTIMATES)

Name	Population	Name	Population
Prague (capital)	1,216,568	Olomouc	105,998
Brno	390,073	Liberec	100,934
Ostrava	326,049	Hradec Kralove	100,839
Pizen	172,055		

SOURCE: "Population of Capital Cities and Cities of 100,000 and More Inhabitants." United Nations Statistics Division.

THE COAST, ISLANDS, AND THE OCEAN

The Czech Republic is landlocked.

CLIMATE AND VEGETATION

Temperature

The Czech Republic essentially has a continental climate, although in Bohemia it is moderated somewhat by ocean influences from the Baltic Sea, so that there is less variation in temperature between different times of day. Nevertheless, the country as a whole is known for its changeable weather. Winters are cold, with average January temperatures between 25°F and 28°F (-4°C and -2°C). Both the Moravian lowlands and the Bohemian highlands can experience bitter cold, with temperatures below 0°F (-18°C). Summers are hot and wet, with frequent storms and temperatures ranging between 55°F and 73°F (13°C to 23°C). Summer temperatures above 86°F (30°C) are common in Moravia.

Rainfall

Rainfall is heaviest in the spring and summer, with the greatest rainfall occurring in July. Average annual rainfall ranges from 20 to 30 in (50 to 76 cm) in low-lying areas to over 50 in (127 cm) in the Krkonoše Mountains. Fog is common in the lowlands.

Grasslands

The central and southern Moravian lowlands are part of the Danube Basin and are similar to the lowlands they join in southern Slovakia. Cultivated varieties of vegetation are predominant, but some original steppe grassland remains. To the north, the lowlands of the Oder River form a narrow strip along the border with Poland. Parts of the Bohemia Plateau, especially in the west, can also be considered grasslands.

Forests and Jungles

About one-third of the Czech Republic is forested. Most forests are deciduous or mixed and vary according to elevation, with deciduous trees like oak at lower altitudes, spruce and beech higher up, and dwarf pines at the greatest heights.

HUMAN POPULATION

Urban dwellers account for more than two-thirds of the country's population. The population is densest in north and central Bohemia and in Moravia, with the fewest people living in the šumava Mountains to the south. Prague is both the capital and the major population center. The most important city in Moravia is Brno.

NATURAL RESOURCES

The coal mines of northern Bohemia have played a dominant role in the economy of the Czech Republic and the former Czechslovakia. However, sulfur emissions from the burning of brown coal by industry have created serious pollution problems. Aside from coal, other natural resources include timber, iron, tin, tungsten, lead, zinc, uranium. kaolin, clay, and graphite.

FURTHER READINGS

Beattie, Andrew, and Timothy Pepper. *Off the Beaten Track: Czech and Slovak Republics.* Old Saybrook, Conn.: Globe Pequot Press, 1995.

Czech Centers. "Czech Republic." www.Czech.cz (accessed Mar. 11, 2002).

Holtslag, Astrid. *The Czech Republic.* New York: Hippocrene Books, 1994.

Ivory, Michael. *Essential Czech Republic.* Lincolnwood, Ill.: Passport Books, 1994.

Klaus, Vaclav. *Renaissance: The Rebirth of Liberty in the Heart of Europe.* Washington, D.C.: Cato Institute, 1997.

Sayer, Derek: *The Coasts of Bohemia: A Czech History.* Translated by Alena Saye. Princeton: Princeton University Press, 1998.

Denmark

- **Area:** 16,638 sq mi (43,094 sq km; includes the island of Bornholm in the Baltic Sea, but excludes the Faeroe Islands and Greenland) / World Rank: 133

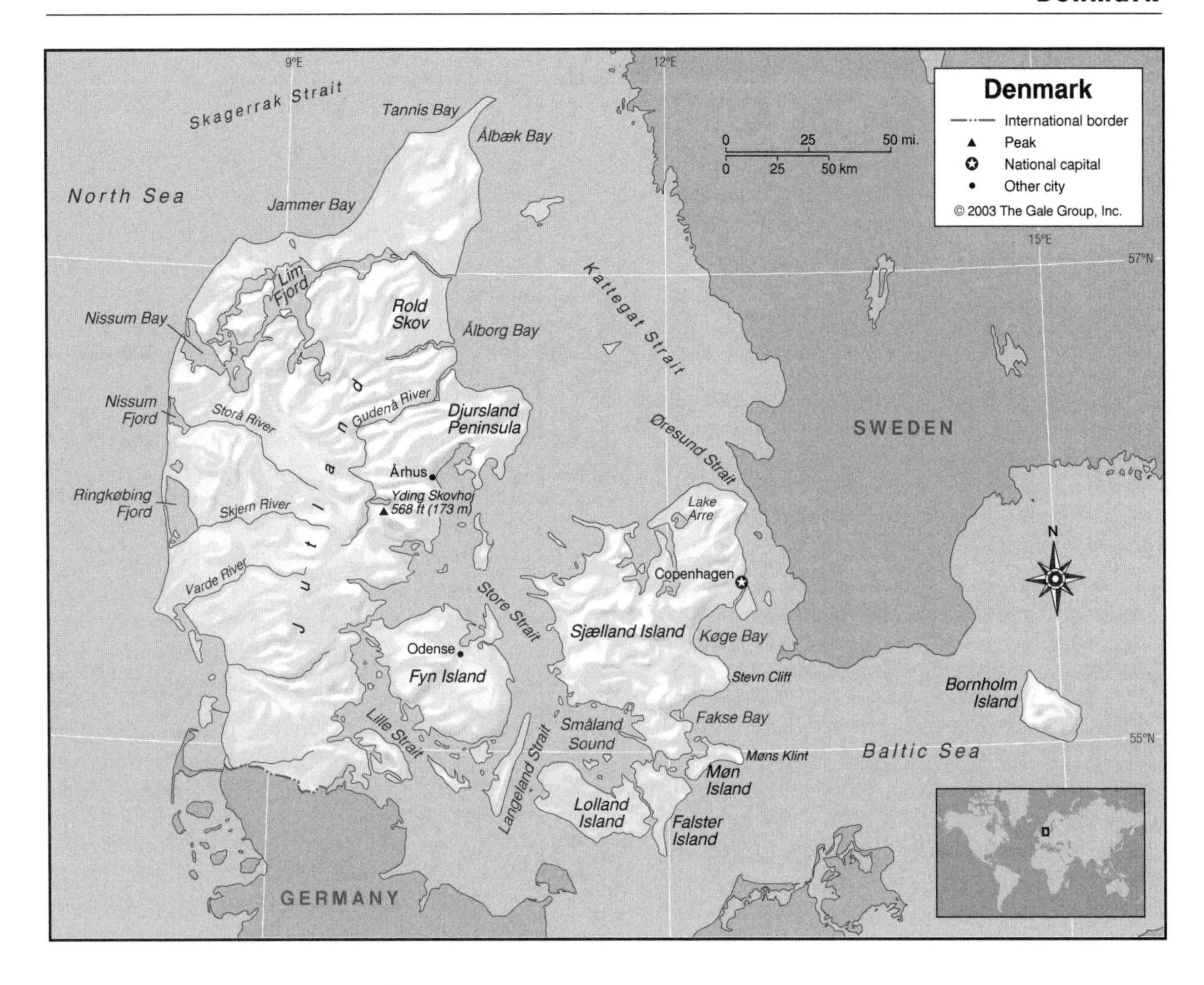

- **Location:** Northern and Eastern Hemispheres, on the Jutland peninsula north of Germany and nearby islands, between the Baltic and North Seas. / Faeroe Islands: In the North Atlantic Ocean northwest of Britain. / Greenland: Between the North Atlantic and Arctic Oceans northeast of Canada.
- **Coordinates:** 56°00′N, 10°00′E
- **Borders:** 42 mi (68 km), all with Germany
- **Coastline:** 4,545 mi (7,314 km) / Faeroe Islands: 614 mi (1,117 km)
- **Territorial Seas:** 12 NM
- **Highest Point:** Yding Skovhoj, 568 ft (173 m) / Faeroe Islands: Slaettaratindur, 2,894 ft (882 m)
- **Lowest Point:** Lammefjord, 23 ft (7 m) below sea level
- **Longest Distances:** 250 mi (402 km) N-S / 220 mi (354 km) E-W
- **Longest River:** Gudenå River, 100 mi (160 km)
- **Largest Lake:** Arre, 15.7 sq mi (40.6 sq km)
- **Natural Hazards:** Coastal flooding
- **Population:** 5,252,815 (July 2001 est.) / World Rank: 104
- **Capital City:** Copenhagen, on the eastern shore of Sjælland
- **Largest City:** Copenhagen, 1,326,000 (2000 est.)

OVERVIEW

The small nation of Denmark occupies a number of large islands and most of the Jutland (Jylland) peninsula that separate the North Sea from the Baltic Sea. It is a low-lying country, with its surface relief characterized by glacial moraine deposits, which form undulating plains with gently rolling hills interspersed with lakes. The moraines consist of a mixture of clay, sand, gravel, and boulders, carried by glaciers from the mountains of Scandinavia and raised from the bed of the Baltic Sea, with an admixture of limestone and other rocks. Between the hills are extensive level plains, created when the meltwater washing away from the glaciers deposited stratified sand and gravel outside the ice limit. These heathland plains are the site of the country's densest settlements.

GEO-FACT

Early on the morning of August 14, 1999 the 10-mi (16-km) Øresund Bridge (Øresundbron) opened. The bridge connects Malmo, Sweden, and Copenhagen, Denmark across the Øresund.

The boundary line between the sandy West Jutland and the loam plains of East and North Denmark is the most important geographical division of the country. West of the line is a region of scattered farms; to the east, villages with high population density. Valleys furrow the moraine landscape.

The coastlines of eastern Jutland and many of the islands are heavily indented with fjords, bays, and other inlets, forming numerous natural harbors. Many of the islands are separated only by narrow straits.

MOUNTAINS AND HILLS

Denmark is a low-lying country. While there are many hills and ridges, the highest point, Yding Forest Hill (Yding Skovhoj) in eastern Jutland, is only 173 m (568 ft) above sea level.

INLAND WATERWAYS

Lakes

Dozens of lakes dot the middle interior region of Jutland known as the Lakeland region. The largest lake in the country is Arre, (40.6 sq km; 15.7 sq mi); it lies between Helsingør and Hillerød on Sjælland island.

Rivers

The Gudenå River, the longest river at about 100 mi (160 km) follows intersecting valley systems as it flows from the interior of Jutland north to the Kattegat strait. Other smaller rivers include the Storå, the Skjern, and the Varde, all of which flow from the interior Jutland into the North Sea. Many of the country's rivers have been artificially rerouted.

THE COAST, ISLANDS, AND THE OCEAN

Major Islands

There are 406 islands in Denmark (of which only 97 are inhabited), accounting for over one-third of its land area. The largest are Sjælland (2,709 sq mi / 7,015 sq km); Fyn (1,152 sq mi / 2,984 sq km), Lolland (480 sq mi / 1,234 sq km), Bornholm (588 sq km / 227 sq mi), and Falster (198 sq mi / 514 sq km). All of these islands except for Bornholm lie between Jutland and Sweden. Bornholm, Denmark's easternmost island, is southeast of Sweden in the Baltic Sea. It is a nature reserve that is accessible only by boat. There are no cars, modern buildings, or domesticated animals (cats or dogs) on the island.

FAEROE ISLANDS. Denmark administers the Faeroe Islands, an archipelago of 17 inhabited islands and one uninhabited island in the Atlantic Ocean, northwest of Britain. Among the larger islands are Stromp (174 sq mi / 374 sq km), Ostero (110 sq mi / 266 sq km), Vago (69 sq mi / 178 sq km), Sydero (59 sq mi / 153 sq km), and Sando (44 sq mi / 114 sq km). The Faeroe landscape is rugged, characterized by a stratified series of basalt sheets with intervening thinner layers of solidified volcanic ash (tufa). Glacial action has carved the valleys into trough-shaped hollows and formed steep peaks; the highest point is Slaettaratindur, at 2,894 ft (882 m) on Ostero.

GREENLAND. The world's largest island, Greenland is located off the coast of North America in the Arctic Ocean. Although a part of Denmark, Greenland has limited home rule. For details on its geography see the Greenland entry.

Surrounding Waters

Denmark is almost completely surrounded by water, only the southern part of the Jutland peninsula is connected to the European mainland. The Skagerrak separates Jutland from Norway in the northwest. The Kattegat lies between Jutland and Sweden to the east. The narrow Lille Strait separates the island of Fyn from the mainland. The Store and Langeland Straits lie between Fyn and the easternmost islands. The Øresund separates Sjælland from Sweden, and the smaller islands of Falster, Lolland, and Møn lie to the south across the Småland Sound. The Baltic Sea lies to the southeast.

The Coasts and Beaches

The coastline of Jutland and the nearby islands are highly indented. On the west side of the peninsula there are two great fjords, Ringkøbing and Nissum. Further north is Nissum Bay. The northern coast is more regular, with the broad Jammer and Tannis Bays. In the east are Ålbæk and Ålborg Bays. These are punctuated by a number of fjords, most notably Lim Fjord, which stretches all the way across Jutland from Ålborg Bay to Nissum Bay in the west. The southern coast of Ålborg Bay juts east to form the Djursland Peninsula, south of which is Arhus Bay and many smaller fjords. On Sjælland, the capital of Copenhagen is situated on Køge Bay, with Stevn Cliff and Fakse Bay further to the south.

Along sections of the coast of Jutland and southern Lolland, dikes have been constructed to protect the low-lying coastline from seawater. White chalk cliffs are found along the coastline of the small island of Møn, lying south of Sjælland. The cliffs rise from the beach

Population Centers – Denmark

(2000 POPULATION)

Name	Population	Name	Population
Copenhagen (metropolitan area, capital)	1,075,851	Frederiksberg	91,076
Århus	217,260	Esbjerg	73,341
Odense	145,062	Randers	55,761
		Kolding	53,447

SOURCE: Statistics Denmark.

Counties – Denmark

2001 POPULATION ESTIMATES

Name	Population	Area (sq mi)	Area (sq km)	Capital
Århus	640,637	1,761	4,561	Århus
Bornholm	44,126	227	588	Bornholm
Frederiksborg	368,116	520	1,347	Frederiksborg
Fyn	472,064	1,346	3,486	Fyn
Copenhagen	615,115	203	526	Copenhagen (Kobenhagen)
North Jutland	494,833	2,383	6,173	Nordjylland
Ribe	224,446	1,209	3,131	Ribe
Ringkobing	273,517	1,874	4,853	Ringkobing
Roskilde	233,212	344	891	Roskilde
South Jutland	253,249	1,520	3,938	Sonderjylland
Storstrom	259,691	1,312	3,398	Nykobing
Vejle	349,186	1,157	2,997	Vejle
West Zealand	296,875	1,152	2,984	Vestja'lland
Viborg	233,921	1,592	4,122	Viborg

SOURCE: Statistics Denmark.

about 422 ft (128 m) in an area known as Møn Cliff (Møns Klint).

CLIMATE AND VEGETATION

Temperature

Climate in Denmark is temperate. Days are typically humid and overcast; winters are mild and windy, and summers are cool. The mean temperature in February, the coldest month, is 32°F (0°C), and in July, the warmest, 63°F (17°C).

Rainfall

Rainfall falls fairly evenly throughout the year, the annual average amounting to approximately 24 in (61 cm).

Forests and Jungles

Over 10 percent of Denmark's land area is covered with trees, but almost none of this is primary forest. The woodlands are predominated by beech and oak trees, with other species including elm, hazel, maple, pine, birch, aspen, linden, and chestnut. Denmark's largest contiguous area of woodland is Rold Forest (Rold Skov), a public forest (30 sq mi / 77 sq km) that contains Denmark's only national park, Rebild Bakker. Located near the city of Ålborg, It is the last section of natural forest that once covered the eastern part of Jutland.

HUMAN POPULATION

The population of Denmark proper is 5,352,815 (2001 estimate), giving the country an overall population density of 124 persons per sq km (322 per sq mi). Almost 90 percent of the population live in urban areas, with one quarter of the population living in Copenhagen.

NATURAL RESOURCES

The economy relies primarily on commerce and industry, but some 56 percent of the total land area of Denmark is cultivated and it is an exporter of food. Minerals are limited, and comprise, in large part, the clays, peats, and other deposits common to boggy country. The surrounding waters yield a good supply of fish, and there are offshore deposits of petroleum and natural gas in the North Sea.

FURTHER READING

Bendure, Glenda. *Denmark.* London: Lonely Planet, 1999.

Holbraad, Carsten. *Danish Neutrality: A Study in the Foreign Policy of a Small State.* New York: Clarendon Press, 1991.

Keillor, Garrison. "Civilized Denmark." *National Geographic,* July 1998, Vol. 194, No. 1, p. 50+.

Kostyal, K. H. "Danish Light (Danish Jutland Peninsula)." *National Geographic Traveler.* July-August 1998, Vol. 15, No. 4, p. 96+.

Graham-Campbell, James. *The Viking World.* New Haven: Ticknor & Fields, 1980.

Miller, Kenneth. *Denmark: A Troubled Welfare State.* Boulder, CO: Westview, 1990.

Symington, Martin. *Passport's Illustrated Travel Guide to Denmark.* Lincolnwood, IL: Passport Books, 1996.

Woodward, Christopher. *Copenhagen.* Manchester: Manchester University Press, 1998.

Djibouti

- **Area:** 8,494 sq mi (22,000 sq km) / World Rank: 150
- **Location:** Northern and Eastern Hemispheres, Eastern Africa, bordering Eritrea to the north, the Gulf of Aden to the east, Somalia to the southeast, and Ethiopia to the south and west
- **Coordinates:** 11°30′N, 43°E

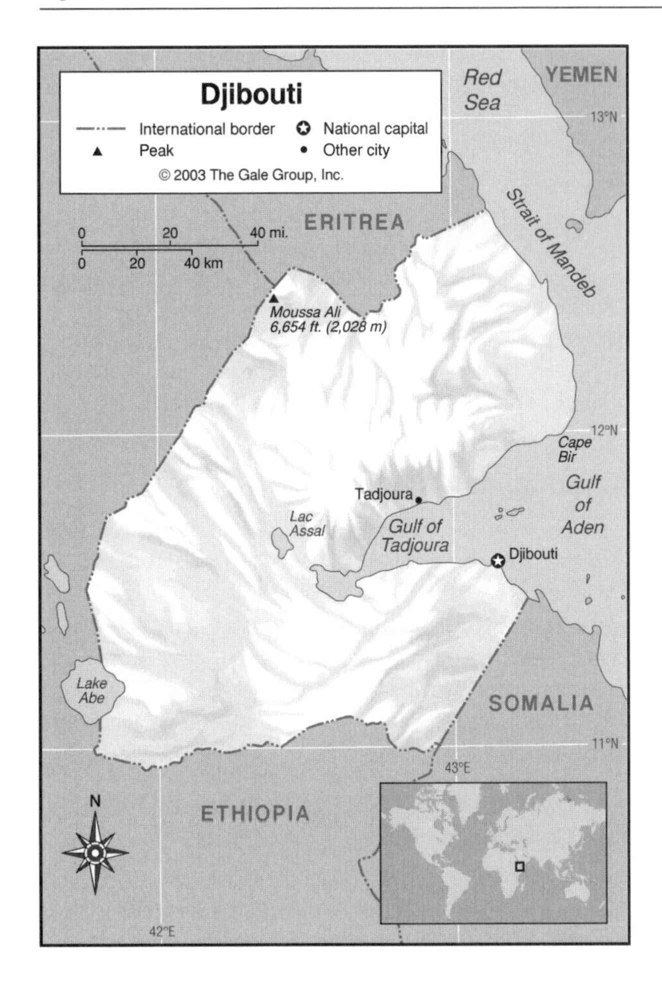

- **Borders:** 316 mi (508 km) / Eritrea, 70 mi (113 km); Ethiopia, 209 mi (337 km); Somalia, 36 mi (58 km).
- **Coastline:** 195 mi (314 km)
- **Territorial Seas:** 12 NM (22 km)
- **Highest Point:** Moussa Ali, 6,654 ft (2,028 m)
- **Lowest Point:** Lac Assal, 509 ft (155 m) below sea level
- **Longest Distances:** approx. 132 mi (213 km) NE-SW / 96 mi (155 km) SE-NW
- **Longest River:** None
- **Natural Hazards:** Earthquakes, drought
- **Population:** 460,700 (July 2001 est.) / World Rank: 162
- **Capital City:** Djibouti, on the Gulf of Tadjoura
- **Largest City:** Djibouti, 355,000 (2001)

OVERVIEW

Djibouti is a small, desert country on the coast of the Horn of Africa. Its eastern coast borders the Strait of Mandeb (Bab el Mandeb), which connects the Red Sea to the Gulf of Aden. Djibouti is part of the Afar Triangle, a three-sided structural depression that forms part of the East African Rift Valley. The country can be divided into three major geographic regions: a coastal plain at elevations of less than 650 ft (200 m), mountains backing the plain, and a desert plateau behind the mountains.

Djibouti is on seismically active terrain, at the meeting point of the Arabian and African tectonic plates, with frequent tremors and thick layers of lava flow from past volcanic activity.

MOUNTAINS AND HILLS

The rugged volcanic mountains in the northern part of the country average 3,300 ft (1,000 m) in elevation. The highest peak is Moussa Ali (6,654 ft / 2,028 m). Low mountains separate the coast from Djibouti's central plateau region, which rises from 1,000 to 5,000 ft (300 to 1,500 m).

INLAND WATERWAYS

The desert terrain of Djibouti is broken in places by salt lakes fed by saline underground aquifers. The largest is Lac Assal. At 509 ft (155 m) below sea level is the lowest point in Africa and the second-lowest in the world. It is also the world's saltiest body of water and reaches temperatures of up to 135°F (57°C) in the summer. There are no permanent inland water courses and very little fresh groundwater of any kind. In the far west is Lake Abe, formed by Ethiopia's Awash River.

THE COAST, ISLANDS, AND THE OCEAN

Djibouti's eastern shore forms most of the west bank of the Strait of Mandeb, the connecting point between the Gulf of Aden to the south and the Red Sea to the north. The coastline is deeply indented south of Cape Bir to form the Gulf of Tadjoura, which is 28 mi (45 km) across at its entrance and penetrates 36 mi (58 km) inland.

Much of the coastline consists of white, sandy beaches and it is fringed by picturesque coral reefs. The capital city of Djibouti, which is the site of a deepwater port, is built on these reefs.

CLIMATE AND VEGETATION

Djibouti's climate is extremely hot and dry. The average high temperature in the capital is 87°F (31°C) in the cool season (October to April) and 99°F (37°C) in summer. The dry hamsin wind increases the already hot summer temperatures, which can rise as high as 113°F (45°C). Rainfall is infrequent, averaging under 5 in (13 cm) annually.

About 90 percent of Djibouti's terrain, essentially all of the interior of the country, is flat, barren, desert land made up of volcanic rock. Vegetation is minimal. The coastal plain is more fertile, but still requires irrigation.

Districts – Djibouti

Name	Area (sq mi)	Area (sq km)	Capital
Ali Sabih (Ali-Sabieh)	925	2,400	Ali Sabib
Dikhil	2,775	7,200	Dikhil
Djibouti	225	600	Djibouti
Obock	2,200	5,700	Obock
Tadjouri (Tadjourah)	2,825	7,300	Tadjoura

SOURCE: *Geo-Data: The World Geographical Encyclopedia,* 2nd ed. Detroit: Gale Research, 1989.

HUMAN POPULATION

Because the vast majority (about two-thirds) of Djibouti's population is concentrated in or around the capital, and much of the rest of the country is uninhabitable desert, Djibouti has been called a city-state or mini-state.

NATURAL RESOURCES

Djibouti is a poor country with few natural resources. Its economy relies heavily on the trade that passes through the Strait of Mandeb. The country's scant mineral resources include salt and limestone.

FURTHER READINGS

Aboubaker Alwan, Daoud, and Yohanis Mibrathu. *Historical Dictionary of Djibouti.* Lanham, Md.: Scarecrow Press, 2000.

"Drought And Economic Refugees Overburden Capital." *Africa News Service,* September 6, 2001.

Gordon, Frances Linzee. *Ethiopia, Eritrea, and Djibouti.* Oakland, Calif.: Lonely Planet Publications, 2000.

Saint Viran, Robert. *Djibouti, Pawn of the Horn of Africa.* Metuchen, N.J.: Scarecrow Press, 1981.

"Tiny Djibouti's Port is Thriving as Neighbors' Problems Continue." *The Wall Street Journal,* October 16, 2000.

University of Pennsylvania African studies Web site. *Djibouti Page.* http://www.sas.upenn.edu/African_Studies/Country_Specific/Djibouti.html (accessed March 12, 2002).

Dominica

- **Area:** 291 sq mi (754 sq km) / World Rank: 177
- **Location:** Northern and Western Hemispheres, Caribbean Sea, between Guadeloupe and Martinique
- **Coordinates:** 15°25′N, 61°20′E
- **Borders:** None
- **Coastline:** 92 mi (148 km)
- **Territorial Seas:** 12 NM
- **Highest Point:** Morne Diablotin 4,747 ft (1,447 m)
- **Lowest Point:** Sea level
- **Longest Distances:** 29 mi (47 km) N-S / 16 mi (26 km) E-W
- **Longest River:** None of significant size
- **Natural Hazards:** Hurricanes, flash floods
- **Population:** 70,786 (July 2001 est.) / World Rank: 189
- **Capital City:** Roseau, southwestern Dominica
- **Largest City:** Roseau, 21,000 (2000 est.)

OVERVIEW

Dominica, an island in the eastern Caribbean Sea, lies at the midpoint of the Lesser Antilles, between the French possessions of Guadeloupe to the north and Martinique to the south.

The rugged, mountainous terrain that covers much of the interior is the island's outstanding physical feature. The two mountainous regions, in the north and south, are separated by the Layou River plain at the center of the island. Lush vegetation and abundant wildlife of the rain forests cover the country's elevated lands.

Signs of Dominica's volcanic origins, and its relative geological newness, include hot springs, sulfur springs bubbling from volcanic vents, and Boiling Lake, one of the country's best-known features. The island is located on the Caribbean Tectonic Plate.

MOUNTAINS AND HILLS

Mountains

Deep ridges, ravines, and valleys are etched in the densely wooded mountains. The island's highest peak, Morne Diablotin, is located in the mountains to the north, while its second-highest, Morne Trois Pitons—which at 4,667 ft (1,387 m) is nearly as high as Diablotin—is situated in the south. Other high peaks include Morne au Diable, Morne Brule, Morne Couronne, Morne Anglais, and Morne Plat Pays.

INLAND WATERWAYS

Dominica has a number of thermally active lakes, of which the best known is Boiling Lake, in the southeastern part of the island. There are many streams and rivers, but none are navigable. The main rivers are Indian, Espagnol, Layou, Roseau, and Queens running west to the Caribbean Sea, and Toulaman, Hodges, Tweed, Clyde,

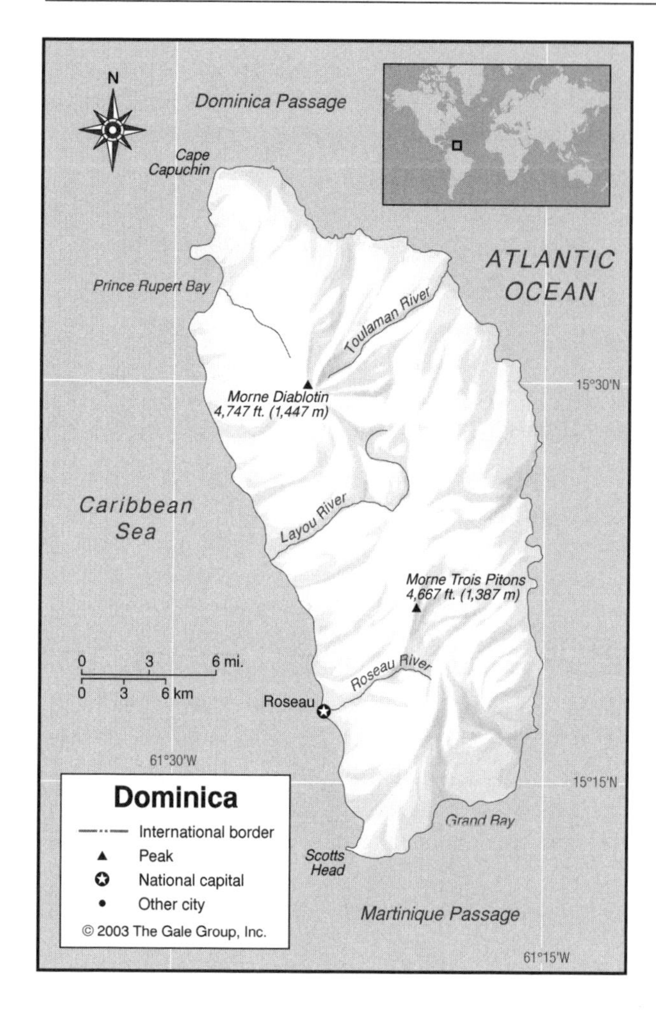

Maclaralin, Grand Bay, Rosalie, and Wanerie running east to the Atlantic.

THE COAST, ISLANDS, AND THE OCEAN

Dominica is located between the Caribbean Sea and the North Atlantic Ocean, at the midpoint of the Leeward Islands. Guadeloupe is to the north, across the Dominica Passage; Martinique is south, across the passage of the same name.

A thin coastal strip lies between the sea and the mountains. The coast, which is heavily indented on the eastern side of the island, is fringed with coral reefs.

GEO-FACT

Boiling Lake is the world's second-largest thermally active lake. The pressure of gases escaping from the volcanic vent underneath regularly raises the lake's water level by as much as 3 ft (1 m).

Population Centers – Dominica

(1991 POPULATION ESTIMATES)

Name	Population
Roseau	16,243
Portsmouth	4,000
Marigot	3,000
Atkinson	2,500

SOURCE: "Population of Capital Cities and Cities of 100,000 and More Inhabitants." United Nations Statistics Division.

Black, gray, and white volcanic sand is found on the beaches. Cape Capuchin marks the northern end of the island, with Prince Rupert Bay not far south. Scotts Head and Grand Bay are at the southern end of the island.

CLIMATE AND VEGETATION

Temperature

Tempered by sea breezes, Dominica's tropical climate is generally mild and pleasant. Summer temperatures average 82°F (28°C) and may rise as high as 90°F (32°C). Winter temperatures average 77°F (25°C).

Rainfall

Dominica has a dry season in the spring and a rainy season in summer, with rainfall especially heavy during the hurricane season in late summer. Average annual rainfall ranges from about 75 in (191 cm) near the coast to over 200 in (508 cm) in the mountains.

Forests and Jungles

Dominica's mountains are covered with dense forest growth. The most heavily wooded island in the Lesser Antilles, it is known for the rich and varied vegetation of its rainforests, which are protected by a park system. The government has created forest reserves in the north (21,770 acres / 8,708 hectares) and east (1,013 acres / 405 hectares). The numerous tree species include breadfruit, white cedar, coconut, cocoa, and many more, including many species of palm tree. Flowering plants include bougainvillea, frangipani, hibiscus, and poinsettia.

HUMAN POPULATION

It is estimated that over two-thirds of the population is urban. Most of the island's estimated 3,000 Caribs live in a special reserve in eastern Dominica.

NATURAL RESOURCES

In spite of its largely mountainous terrain, over one-fifth of Dominica's land is arable. Other important resources include its forests and the hydropower potential of its many streams.

FURTHER READINGS

Baker, Patrick L. Centering the Periphery: *Chaos, Order, and the Ethnohistory of Dominica.* Montreal: McGill-Queen's University Press, 1994.

Commonwealth of Dominica Web site. http://www.ndc.dominica.dm/ (accessed Mar. 14, 2002)

Kincaid, Jamaica. *Autobiography of My Mother.* New York: Farrar, Straus, & Giroux, 1996.

Philpott, Don. *Dominica.* Lincolnwood, Ill.: Passport Books, 1996.

Sullivan, Lynne M. *Dominica & St Lucia Alive!* Edison, N.J.: Hunter, 2002.

Dominican Republic

- **Area:** 18,810 sq mi (48,730 sq km) / World Rank: 130
- **Location:** Northern and Western Hemispheres, eastern two-thirds of the island of Hispaniola, south of the Atlantic Ocean, north of the Caribbean Sea, bordering on Haiti in the east
- **Coordinates:** 19°00′N, 70°40′W
- **Borders:** 177 mi (275 km)
- **Coastline:** 800 mi (1,288 km)
- **Territorial Seas:** 6 NM
- **Highest Point:** Pico Duarte, 10,417 ft (3,175 m)
- **Lowest Point:** Lake Enriquillo, 151 ft (46 m) below sea level
- **Longest Distances:** 240 mi (386 km) E-W / 162 mi (261 km) N-W
- **Longest River:** Yaque del Norte, 170 mi (280 km)
- **Largest Lake:** Lake Enriquillo, 190 sq mi (500 sq km)
- **Natural Hazards:** Hurricanes
- **Population:** 8,581,477 (July 2001 est.) / World Rank: 83
- **Capital City:** Santo Domingo, located on the southeastern coast of the country
- **Largest City:** Santo Domingo, 3,601,000 (2000 est.)

OVERVIEW

The Dominican Republic shares the Caribbean island of Hispaniola with Haiti. The Republic has a rugged and mountainous terrain, with fertile valleys in the central and eastern areas. The highest mountain, Pico Duarte, is the highest point in the West Indies, and Lake Enriquillo (Lago Enriquillo) is the lowest-lying lake in the West Indies.

MOUNTAINS AND HILLS

Mountains

The principal mountain systems are four parallel ranges extending in a northwesterly direction in the western part of the country and a single minor chain—the Cordillera Oriental—in the east. The core of the system is the Cordillera Central, which rises in the east near Santo Domingo and veers northwestward into Haiti, where it becomes the Massif du Nord. The Cordillera Central divides the country into two parts; its ridges crest between 5,000 and 8,000 ft (1,524 and 2,438 m), but there are individual peaks with considerably greater heights. The highest peak in the West Indies, Pico Duarte, has an elevation of 10,414 ft (3,174 m) and is found in the Cordillera Central.

The two ranges that lie to the south of the Cordillera Central, the Sierra de Neiba and the Sierra de Baoruco, begin as escarpments flanking Neiba Bay (Bahía de Neiba) in the southwest and continue northwestward to join corresponding ranges in Haiti. Both crest generally at elevations of between 3,000 and 4,000 ft (914 and 1,219 m) but have peaks as high as 6,000 ft (1,828 m). The eastern part of the Sierra de Neiba is separated from the remainder of the range by the Yaque del Sur River and is known as the Sierra de Martin Garcfa. The Sierra de Baoruco forms an extension of the southern mountain ranges of Haiti. North of the Cordillera Central lies the Cordillera Septentrional, a mountain range characterized by precipitous slopes and deeply etched valleys.

Hills and Badlands

The Cordillera Oriental is less a range of mountains than a narrow band of hills representing an eastward terminal spur of the Cordillera Central. It extends westward some 85 mi (137 km) from the Atlantic coast along the southern shore of the Bay of Samaná (Bahía de Samaná) to the foothills of the Cordillera Central about 30 mi (48 km) north of Santo Domingo. The western third of the range is rolling and permits fairly easy access from the capital city to the interior lowlands. The remainder is more rugged. Elevations are generally less than 1,000 ft (305 m) except in the extreme east where a few isolated promontories rise to over 2,000 ft (610 m).

INLAND WATERWAYS

Lakes

The largest of the country's natural lakes is Lake Enriquillo in the Neiba Valley. A remnant of the strait that once occupied the area, its waters are 140 ft (43 m) below sea level. Although it is fed by many streams from the surrounding mountains and has no outlet, the high rate of evaporation in the valley is causing its waters gradually to recede. On Isla Cabritos, a small island in the center of Lake Enriquillo, there is a national park that supports and preserves the habitat of the crocodile. A dam on the Yaque del Norte River at Tavera creates a reservoir and provides irrigation for central Cibao.

Rivers

The rivers of the Dominican Republic for the most part are shallow, subject to wide seasonal change in flow and consequently of little use for transportation. Flowing out of the several highlands in varying directions, they form a variety of drainage systems.

The Cibao Valley has two systems. On its western flank it is drained into the Atlantic near Monte Cristi by the Yaque del Norte, the country's longest river; east of Santiago de los Caballeros in the Vega Real, the Yuna drains eastward into the Bay of Samaná (Bahía de Samaná).

South of the Cordillera Central, the San Juan Valley is also divided between two hydrographic systems with opposite watersheds. Near the frontier it is drained by a tributary of the Artibonito River (Río Artibonito). This stream continues westward across the border and becomes the principal watercourse of Haiti. To the southeast, it is drained by the Yaque del Sur, which flows into the Caribbean at the Bay of Neiba (Bahía de Neiba).

Wetlands

The most extensive marshland extends inland from the delta of the Yuna River on the Bay of Samaná (Bahía de Samaná). At the opposite end of the Cibao, there are salt marshes south of Monte Cristi Bay (Bahía de Monte Cristi).

THE COAST, ISLANDS, AND THE OCEAN

Oceans and Seas

The Dominican Republic lies between the Atlantic Ocean and the Caribbean Sea. The Mona Passage lies to the east of the country, separating it from Puerto Rico. On the Atlantic coast of the Dominican Republic is an offshore rocky ledge. This platform is highly developed in the shallow waters of the Bay of Samaná (Bahía de Samaná), and continues in a westerly direction along the northern coasts of the Dominican Republic and Haiti. The platform extends seaward from a few hundred yards to more than 30 mi (48 km) at a maximum depth of 200 ft (61 m). At irregular intervals the shelf rises to form tiny islands and jagged coral reefs that lie close to the surface, and represent hazards to navigation in waters to the east of Monte Cristi.

Major Islands

Among the numerous islands scattered off the Dominican coastline, only three are permanently inhabited, and none are of significant economic importance. The largest, Saona Island (Isla Saona), has maximum dimensions of 15 mi by 4 mi (24 km by 6 km) and is located at the southeastern tip of Hispaniola. The 20 sq mi (52 sq km) of Beata Island (Isla Beata) lie off the Pedernales Peninsula in the extreme west.

The Coast and Beaches

Sandy beaches and rocky escarpments mark the northern coast. There are a few sheltered coves, but access to the interior is difficult and there are few major ports. The Bay of Monte Cristi (Bahía de Monte Cristi) marks the westernmost part of the north coast. Further east Cape Francés Viejo projects north into the Atlantic. South and east of Cape Francés Viejo, the Samaná Peninsula projects eastward, forming the narrow bay of the same name.

The Caribbean coast in the south is better suited to port development. Reefs and islets are relatively few, and access from ports to the interior is easier. The best of the natural harbors are located on the broad estuaries of rivers that meet the Caribbean at Santo Domingo, San Pedro de Macorís, and La Romana. The Pedernales Peninsula juts into the Caribbean at the west end of this coastline, with the Bay of Neiba (Bahía de Neiba) on its eastern side. Otherwise the coast is fairly even, meeting with the north coast to form Cape Engaño at the eastern end of the island.

CLIMATE AND VEGETATION

Temperature

The Dominican Republic has a semitropical climate, tempered by the prevailing easterly winds. Temperatures

Population Centers – Dominican Republic

(1993 POPULATION ESTIMATES)

Name	Population
Santo Domingo (capital)	2,134,779
Santiago (de los Caballeros)	690,548
La Romana	150,000
San Francisco de Macoris	130,000
San Pedro de Macoris	125,000

SOURCE: "Population of Capital Cities and Cities of 100,000 and More Inhabitants." United Nations Statistics Division.

Provinces – Dominican Republic

2000 POPULATION ESTIMATES

Name	Population	Area (sq mi)	Area (sq km)	Capital
Azua	243,157	938	2,430	Azua
Bahoruco	124,592	531	1,376	Neiba
Barahona	179,945	976	2,528	Barahona
Capital District	2,677,056	570	1,477	Santo Domingo
Dajabón	78,045	344	890	Dajabon
Duarte	318,151	499	1,292	San Francisco De Macoris
Elías Piña	66,267	550	1,424	
El Seibo	105,447	641	1,659	El Seibo
Espaillat	228,173	386	1,000	Moca
Hato Mayor	87,595	514	1,330	Hato Mayor
Independencia	41,778	719	1,861	Jimaní
La Altagracia	128,627	1,191	3,084	Higuey
La Romana	213,628	209	541	La Romana
La Vega	390,314	916	2,373	La Vega
María Trinidad Sánchez	142,030	506	1,310	Nagua
Monseñor Nouel	174,923	388	1,004	Bonao
Monte Cristi	103,711	768	1,989	Monte Cristi
Monte Plata	174,126	841	2,179	Monte Plata
Pedernales	19,698	373	967	Pedernales
Peravia	223,273	626	1,622	Baní
Puerto Plata	302,799	726	1,881	Puerto Plata
Salcedo	106,450	206	533	Salcedo
Samaná	82,135	382	989	Samaná
Sánchez Ramirez	194,282	453	1,174	Cotuí
San Cristóbal	519,906	604	1,564	San Cristóbal
San Juan	265,562	1,375	3,561	San Juan
San Pedro de Marcorís	260,629	450	1,166	San Pedro de Marcorís
Santiago	836,614	1,205	3,122	Santiago de los Caballeros
Santiago Rodriguez	65,853	394	1,020	Sabaneta
Valverde	198,979	220	570	Mao

SOURCE: National Statistics Office, Dominican Republic, as cited by Johan van der Heyden, Geohive, http://www.geohive.com (accessed June 2002).

ranging from 64° to 84°F (18° to 29°C) are registered during the winter, and 73° to 95°F (23° to 35°C) in the summer. Temperatures are highest along the coast and are much cooler in the mountains.

Rainfall

Annual precipitation averages about 60 in (152.5 cm), but varies considerably from 17 in (43 cm) in the arid west to 53 in (135 cm) in the east, to 82 in (208 cm) in the mountainous areas of the north. The wet season is from June to November, with the dry season from December to May. Tropical hurricanes every few years, and can cause great damage.

Grasslands

The largest of the lowland regions is the Caribbean Coastal Plain; the plain covers more than 1,100 sq mi (2,849 sq km). It is composed principally of a limestone platform formed by corals and alluvial deposition. Inland, there are calcareous soils of high fertility, and to the west of Santo Domingo there are infertile soils derived from acid clays. The region is the center of the country's cattle-raising and sugar industries.

The country's other lowlands consist for the most part of long valleys that, like the mountains that define them, extend in a northwesterly direction from origins close to the Caribbean Sea to corresponding lowlands in Haiti. In these, the fertile alluvial soils of their flood plains and terraces are suitable for intensive agriculture, and shallower soils provide good pasture. The most extensive of the valleys, the Cibao, is the breadbasket of the country.

Forests and Jungles

In 1997 about 12.4 percent of the total land area consisted of forests. Dense rain forests can be found in the wetter portions of the country, and scrub woodland flourishes along the mountain slopes.

HUMAN POPULATION

The southern coastal plains around Santo Domingo and the Cibao Valley are the two most densely populated areas of the country. About 60 percent of the population lives in urban areas Almost 75 percent of the population of mixed Spanish and African descent and 95 percent report their religious affiliation as Roman Catholic.

NATURAL RESOURCES

The minerals nickel, bauxite, gold, and silver have long been an essential part of the republic's economy. Salt Mountain, located west of Barahona, is one of the largest known salt deposits in the world. Off the coasts lie some 3,000 sq mi (7,770 sq km) of fishing banks.

FURTHER READINGS

Bell, Ian. *The Dominican Republic.* Boulder, CO: Westview Press, 1980.

Furlong, Kate A. *Dominican Republic.* Edina, MN: Abdo Publications Co., 2000.

Lannom, Gloria. "The Jewel of the Dominican Republic." *Faces: People, Places, and Cultures,* February 1999, Vol. 15, Issue 6, p. 14.

"New Lizard Ties for 'World's Smallest.'" *Science News,* December 8, 2001, Vol. 1, Issue 23, p. 356.

East Timor

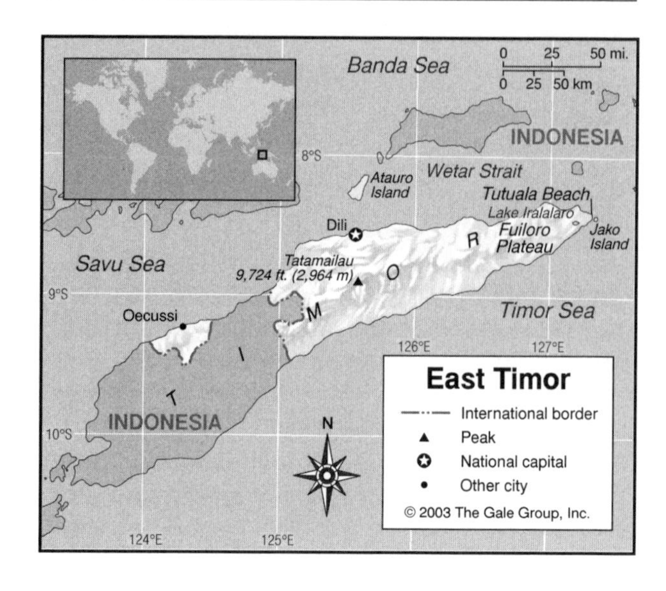

- **Area:** 5,641 sq mi (14,609 sq km) / World Rank: 157
- **Location:** The eastern half of Timor Island in the Indonesian Archipelago, Southern and Eastern Hemispheres, east of Indonesia.
- **Coordinates:** 8°00′S, 123°00′E
- **Borders:** 107 mi (172 km), all with Indonesia
- **Coastline:** 385 mi (620 km)
- **Territorial Seas:** Not established
- **Highest Point:** Tatamailau, 9,724 ft (2,964 m)
- **Lowest Point:** Sea level
- **Longest Distances:** 57 mi (92 km) N-S; 165 mi (265 km) E-W
- **Longest River:** Lois River, 50 mi (80 km)
- **Largest Lake:** Iralalaro, 4 mi by 2 mi (6.5 km by 3 km)
- **Natural Hazards:** Flooding; droughts; earthquakes; forest fires; tsunamis
- **Population:** 779,567 (2000 est.) / World Rank: 154
- **Capital City:** Dili, on the north coast
- **Largest City:** Dili, 67,000 (1999 estimate)

OVERVIEW

East Timor (Timor Lorosa'e) had its beginnings as a Portuguese colony. After the Portuguese relinquished control, East Timor was taken over by Indonesia in 1975. In 1999, Indonesia allowed the inhabitants of East Timor to have a referendum on independence, which was passed by an 80 percent vote. Many Indonesians opposed this decision and there was fighting in East Timor, with significant loss of life and property, until an international peacekeeping force was put in place. The new nation of East Timor became officially independent on May 20, 2002. Many aspects of the new country, such as it territorial waters, remained to be determined as of mid-2002.

East Timor consists of the eastern half of Timor Island, one of the Lesser Sunda Islands, plus the enclave of Oecussi (30 sq mi / 78 sq km) on the north coast of Indonesian half of the island (West Timor). The Banda Sea is to the north, the Timor Sea to the south. The country is primarily mountainous, with many short streams, and narrow coastal plains and wetlands.

MOUNTAINS AND HILLS

The Ramalau, the central mountain range of East Timor, is characterized by deep valleys and looming cliffs. Tatamailau (9,724 ft /2,964 m) is the highest peak in the country. Six other peaks are above 6,566 ft (2,000 m): Sabiria, Usululi, Harupai, Cablake, Laklo, and Matebian. Coffee, the most important cash crop in East Timor, is grown in the foothills that surround this range. River gorges and deep streambeds cut through the center of the country.

Fuiloro, a 1,640 to 2,297 ft (500 to 700 m) plateau in the east, is the remnant of a fossil atoll. Nari, Lospalos and Rere are other eastern plateaus. Baucau and Laga are coral-rock plateaus along the north coast, and the Maliana Plateau rises along the West Timor border.

INLAND WATERWAYS

Lakes

The largest lake in East Timor is Lake Iralalaro, 4 by 2 mi (6.5 by 3 km) in the far east of the island, surrounded by much of the country's remaining rainforest, a Protected Wild Area. Smaller lakes include Be Malae, Maubara, and Tibar.

Rivers

East Timor has 25 rivers or streams, originating in the central mountains. They are strong torrents during rainy periods, but their water levels drop severely in the dry months. Soil erosion on deforested slopes in the watershed has led to siltation and flooding. Significant rivers are the country's longest, the Lois (50 mi /80 km), the Laklo, Karau Ulun, and Tafara, all in the south. The Tono River runs through Oecussi. There are hot springs along the Marobo River, in the north border region; and waterfalls throughout the country.

Wetlands

The wetlands of East Timor are mostly marshes in estuaries along the south coast, and small mangrove swamps. The transitional government outlawed mangrove cutting and damage to wetlands.

THE COAST, ISLANDS AND THE OCEAN

Oceans and Seas

The rough waters of the Timor Sea of the Indian Ocean on the south, and the calmer Banda Sea of the Pacific Ocean on the north, enclose East Timor. The deep Wetar Strait separates East Timor from Indonesia's Wetar Island to the north. Australia is about 311 mi (500 km) to the south across the Timor Gap. The enclave of Oecussi is on the Savu Sea. Timor Lorosa'e has extensive coral reefs but they are damaged by dynamite fishing

Major Islands

Atauro Island (54 sq mi / 141 sq km), to the north of Dili. Jaco Island (4 sq mi / 11 sq km) off the easternmost point, is a Protected Wild Area.

The Coast and Beaches

East Timor's coastline has little indentation, with steep slopes along the north coast, and river outlets meeting the sea. The easternmost point is Tutuala Beach, which is a Protected Wild Area, as is Christo Rei Beach. Dili, the capital, is located on a bay on the north coast.

CLIMATE AND VEGETATION

Temperature

East Timor has an equatorial climate with two basic seasons: the hot northwest monsoon of November-May, and the cooler southeast monsoon of April-December. The average annual temperature is 70° F (21° C), with a range of 64° to 90° F (18° to 32° C) and humidity averaging 73 percent.

Rainfall

A yearly average of 47 to 59 in (120 to 150 cm) of rain falls on East Timor. Precipitation varies greatly according to coast and terrain. Due to its proximity to Australia, the south receives more rain than the north.

Grasslands

East Timor has extensive grasslands on the coastal plains and hillsides. Invasive imperata grass is rampant where the forests have been cut and burned.

Districts – East Timor

1999 POPULATION ESTIMATES

Name	Population	Area (sq mi)	Area (sq km)	Capital
Aileu	32,500	281	729	Aileu
Ainaro	44,100	308	797	Ainaro
Ambeno	54,500	315	815	Pante Makasar
Baucau	97,600	577	1,494	Baucau
Bobonaro	90,700	528	1,368	Maliana
Cova-Lima	63,900	473	1,226	Suai
Dili	179,600	144	372	Dili
Ermera	89,500	288	746	Ermera
Lautem	52,100	657	1,702	Los Palos
Liquica	54,800	320	543	Liquica
Manatuto	34,900	659	1,706	Manatuto
Manufahi	37,200	512	1,325	Same
Viqueque	59,600	688	1,781	Viqueque

SOURCE: Indonesian National Electoral Board, as cited by Johan van der Heyden, Geohive, http://www.geohive.com (accessed June 2002).

Deserts

An area between Venilale and Los Palos in the far east of the island has been desertified to the point that it's known as "dead earth" where very little will grow.

Forests and Jungles

The forests of East Timor exist only in patches, including a few last groves of natural sandalwood (which the island was once famous for) and Eucalyptus urophylla. The country's deciduous and evergreen forests have decreased from 50 percent cover in 1975 to less than 10 percent in 2002, and primary forest cover is less than 1 percent. The forests were deliberately destroyed in Indonesian military operations, cut for firewood and timber, or burned for agricultural clearing. Some primary forest is still found at the eastern end of the island and in the Oecussi enclave.

HUMAN POPULATION

East Timor's population has undergone tremendous upheaval. During the Indonesian occupation, an estimated 200,000 East Timorese died, nearly a third of the population. After the 1999 independence referendum thousands more were killed and much of the rest of the population was displaced. Many of the country's citizens were still refugees or only recently resettled in 2002. In 2000, the population density of East Timor was estimated at 132 people per sq mi (51 people per sq km), with just eight percent living in urban areas.

NATURAL RESOURCES

East Timor has petroleum reserves offshore in the Timor Sea between the island and Australia. Other mineral resources include some marble, gold, and manganese. None of these resources, including the offshore petroleum, have been developed. Subsistence agriculture and fishing are the primary economic activities.

FURTHER READINGS

Cardoso, Luis. *Crossing: A Story of East Timor.* London: Granta, 2002.

Nunes, Mario N. *The Natural Resources of East Timor.* http://www.pcug.org.au/~wildwood/01jannaturalresources.html (Accessed June 13, 2002).

Periplus Adventure Guides. *East of Bali: From Lombok to Timor.* Boston: Tuttle Publishing, 2001.

Tanter, Richard, Mark Selden, and Stephen R. Shalom, eds. *Bitter Flowers: East Timor, Indonesia and the World Community.* Lanham, Md.: Rowman & Littlefield, 2001.

University of Coimbra. *Timor Net.* http://www.uc.pt/timor (Accessed June 13, 2002).

Ecuador

- **Area:** 106,888 sq mi (276,840 sq km) including Galápagos Islands/ World Rank: 73
- **Location:** Northern, Southern, and Western Hemispheres, on the equator in South America, south of Colombia, west and north of Peru, east of the Pacific Ocean
- **Coordinates:** 2°00′S, 77°30′W
- **Borders:** 1,158 mi (2,010 km) total / Colombia, 366 mi (590 km); Peru, 880 mi (1,420 km)
- **Coastline:** 1,398 mi (2,237 km)
- **Territorial Seas:** 200 NM of territorial seas, plus waters over the continental shelf that lies between the mainland and the Galápagos Islands
- **Highest Point:** Chimborazo, 20,681 ft (6,267 m)
- **Lowest Point:** Sea level
- **Longest Distances:** 444 mi (714 km) N-S / 409 mi (658 km) E-W
- **Longest River:** Putumayo, 980 mi (1,575 km)
- **Natural Hazards:** Subject to frequent earthquakes, landslides, and volcanic activity; periodic droughts and flooding
- **Population:** 13,183,978 (July 2001 est.) / World Rank: 62
- **Capital City:** Quito, northern end of the country in an Andean valley 14 mi (22km) south of the equator
- **Largest City:** Guyaquil, on the west coast, 2,127,000 (2000 est.)

OVERVIEW

Ecuador is a small South American country that takes its name from the equator, which passes through the north of the country. The dominant topographical features are two parallel ranges of the Andes Mountains, which separate a fertile coastal lowland on the west and the more extensive lowlands of the Amazon Basin on the east. Streams that rise in the Andes flow westward to the Pacific or eastward toward the Amazon to form the drainage systems.

The country's mainland divides naturally into a coastal lowland known as the Costa, a central mass made up of the Andean highlands called the Sierra, and an interior lowland that forms part of the Amazon Basin called the Eastern Region (Oriente). A fourth region is made up of the Galápagos Islands. Most geographers consider the two lowland regions to extend up the approaches of the Andes to an elevation of about 1,600 ft (487 m). On the basis of this definition, the Eastern Region would include about half of the territory, and the remainder would be divided equally between the other two regions.

Ecuador is very geologically active, with many volcanoes and frequent earthquakes. It is situated on the South American Tectonic Plate, with the Nazca Plate off the coast to the west.

MOUNTAINS AND HILLS

Mountains

The principal features of the Sierra region are two parallel ranges of the Andes Mountains. On the west, the Cordillera Occidental is a compact high range extending roughly north to south the full length of the country. To its east, the Cordillera Central is less a true mountain range than a series of lofty peaks. Both ranges are of volcanic origin. Between them lies a trench with elevations from 7,000 to 10,000 ft (2,133 to 3,048 m).

East of the crests of the Cordillera Central, the downward slope to the floor of the Amazon Basin is interrupted by lower mountains; aerial mapping undertaken during the 1960s indicated that these lower mountains actually form a third parallel range with elevations of as high as 13,000 ft (3,962 m). The system is identified by Ecuadorian geographers as the Cordillera Oriental. The range is broken at the midpoint by the wide valley of the Pastaza River.

In all there are twenty-two peaks with elevations in excess of 14,000 ft (4,267 m). Many are active or dormant volcanoes. The highest, Chimborazo at 20,702 ft (6,310 m), is a snow-capped volcano located in the central portion of the country. Cotopaxi, at 19,344 ft (5,896 m) is one of the loftiest active volcanoes in the world, and the twin peaks of dormant Pichincha overlook Quito. South of Azuay Province the volcanoes disappear, the mountain chains are lower and less symmetrical, and the path of the Inter-Andean Lane becomes obscured by a more complex mountain pattern.

Plateaus

There is no consensus with respect to the exact elevation on the convoluted eastern slopes of the Andes at which the Eastern Region begins. It encompasses 50 percent or more of Ecuador and consists principally of an alternately flat and gently undulating expanse of tropical rain forest. The region is watered by a multitude of rivers and streams, but the low gradient of the terrain after the rivers pour into the Amazon Basin results in generally poor drainage.

Hills and Badlands

The trench between the Cordillera Occidental and Cordillera Central was named the Avenue of the Volcanoes by the nineteenth-century naturalist Baron Alexander von Humboldt, and is now often referred to as the Inter-Andean Lane (Callejón Interandino). It makes up about three-eighths of the Sierra region. Hill systems run between the mountain ranges, breaking the Lane into a series of *hoyos* (intermount basins) in which most of the region's population live. In all, there are about a dozen, 25–40 mi (40 to 64 km) across from north to south, descending in altitude from north to south. In the southern quarter of the Sierra the terrain is increasingly broken, the soil is poorer, and the hoyos are valleys spilling into the Costa or the Eastern Region rather than true basins.

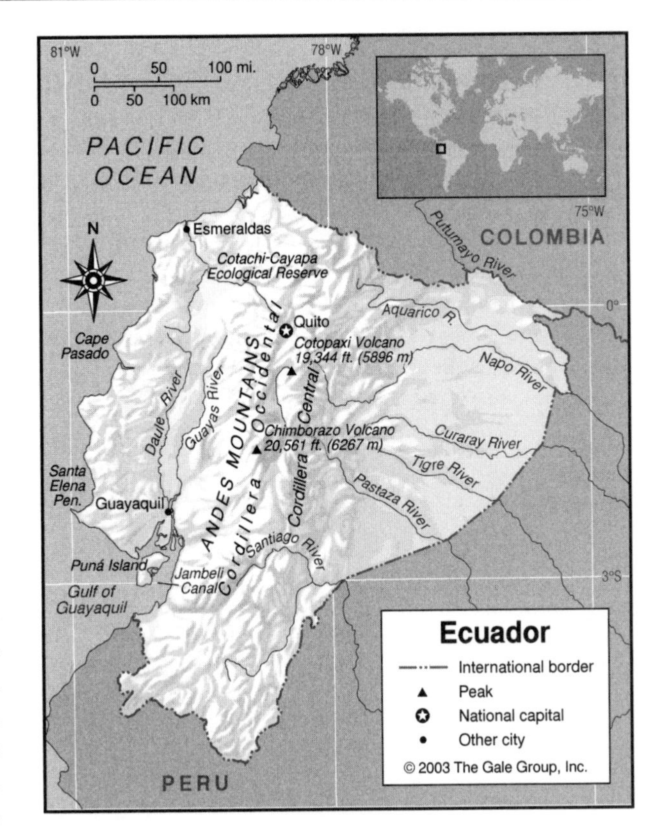

INLAND WATERWAYS

Lakes

There are more than 275 lakes in the Sierra region, including many volcanic crater lakes. Among the most famous is the Cuicocha Volcano lagoon, in the Cotachi-Cayapas Ecological Reserve. Situated in a collapsed volcanic crater, the lake is 600 ft (200 m) deep and almost 2 mi (3 km) in diameter.

Rivers

All of Ecuador's major rivers arise in the Andes. The most important river system of the Costa region is that of the Guayas River and its tributaries, especially the Daule. They flow from the north-central part of the Sierra south and west into the Gulf of Guayaquil. This drainage system creates the country's richest agricultural zone.

Many rivers flow east out of the Andes into the Eastern Region. Among the most significant are the Pastaza, Napo, Santiago (or Zamora), Paute, Curaray, Tigre, Morona (Macuma) and Aguarico. These rivers have carved deep trenches that interfere with land transportation and limit the amount of land suitable for cultivation. However some of these river valleys also provide access into the Inter-Andean Lane from the east, particularly the Paute and the Pastaza. The longest river in Ecuador is the Putumayo (980 mi / 1,575 km), which flows east along the border with Colombia. All Eastern Region waters eventually find their way to the Atlantic through the Amazon River.

THE COAST, ISLANDS, AND THE OCEAN

Oceans and Seas

Ecuador's western boundary is the Pacific Ocean. The continental shelf extends to the Galápagos Islands roughly 600 mi (965 km) to the west. The cold Peruvian Current keeps the climate of the coast and the Galápagos Islands moderate.

The Galápagos Islands

The Galápagos Islands, a province of Ecuador, lie far off the western coast of the country at 89° to 92°W, right on the equator. The largest islands are Isabela Island, Santa Cruz Island, San Salvador Island (or Isla Santiago), Fernandina Island, Santa María Island, Pinta Island, San Cristóbal Island, Marchena Island, and Española Island. Only five of the islands have permanent populations, and over half of the people live on San Cristóbal Island, which also serves as the seat of the administrative government. The highest elevation on the island is a 5,540 ft (1,689 m) volcanic peak, Mt. Azul. The islands are famous for their unique plant and animal life, which inspired Charles Darwin to develop the theory of evolution.

The Mainland Coastline

Like the rest of the Pacific coast of South America, that of Ecuador has few good natural harbors. The major port city of Guayaquil lies 33 mi (53 km) up the Guayas River from the Gulf of Guayaquil (Golfo de Guayaquil). The Gulf is an indentation at the southwestern end of Ecuador's coast, separated from the open ocean by the Santa Elena Peninsula. The large inhabited Puná Island lies in the Gulf, separated from the mainland in to the west by the Jambelí Canal. Esmeraldas, near the Colombian frontier, is the country's second seaport.

Population Centers – Ecuador

(1997 POPULATION ESTIMATES)

Name	Population
Guayaquil	1,973,880
Quito (capital)	1,487,513
Cuenca	255,028
Machala	197,350

SOURCE: Projected from "Population of Capital Cities and Cities of 100,000 and More Inhabitants." United Nations Statistics Division.

Provinces – Ecuador

2001 CENSUS OF POPULATION

Name	Population	Area (sq mi)	(Area (sq km)	Capital
Azuay	598,504	3,124	8,092	Cuenca
Bolívar	168,874	1,599	4,142	Guaranda
Cañar	206,953	1,344	3,481	Azogues
Carchi	152,304	1,446	3,744	Tulcan
Chimborazo	403,185	2,338	6,056	Riobamba
Cotopaxi	350,450	2,007	5,198	Latacunga
El Oro	515,664	2,281	5,908	Machala
Esmeraldas	386,032	5,854	15,162	Esmeraldas
Galápagos Islands	13,555	3,086	7,994	Puerto Baquerizo Moreno
Guayas	3,256,763	8,256	21,382	Guayaquil
Imbabura	345,781	1,921	4,976	Ibarra
Loja	404,085	4,429	11,472	Loja
Los Rios	650,709	2,459	6,370	Babahoyo
Manabi	1,180,375	6,990	18,105	Portoviejo
Morona-Santiago	113,300	10,200	26,418	Macas
Napo	79,610	20,200	52,318	Tena
Pastaza	61,412	11,687	30,269	Puyo
Pichincha	2,392,409	6,404	16,587	Quito
Tungurahua	441,389	1,201	3,110	Ambato
Zamora-Chinchipe	76,414	7,102	18,394	Zamora

SOURCE: Republic of Ecuador, 6th Census of Population, November 2001.

CLIMATE AND VEGETATION

Temperature

Ecuador has a tropical climate overall and it varies little over the course of the year, but does differ greatly from one region to another. The cold Peruvian Current off the coast keeps the Costa cool for a tropical region, with temperatures ranging between 76° to 90°F (25° to 31°C). In the Sierra, temperatures depend on altitude and can vary greatly over the course of the day. The highest mountains are snow-covered year-round. The Eastern Region normally has a warm, humid, and rainy climate. The average temperature varies from 72°C to 80°F (23°C to 26°C). The Galápagos Islands enjoy warm and dry weather, with an average yearly temperature of 85°F (28°C).

The Sierra dry season is June to August, which coincides with the wettest months in the Eastern Region. The climate in the highlands varies according to the altitude. During the year, a subtropical climate prevails on the Andean valleys; at higher altitudes it is cool during the day and colder at night. In Quito, the average temperature is 55°F (13°C), with 50 in (127 cm) of rain. At the highest elevations (above 17,000 ft / 5,200 m), there is snow year-round.

Rainfall

Due to the effects of the Peruvian Current, very little rain falls along the southern coast of Ecuador. Rainfall increases in the north, and the region around Esmeraldas can see 97 in (250 cm) annually. The lower part of the Sierra generally has heavy rainfall, with precipitation decreasing with altitude. Both the Sierra and the Costa get most of their rain between December and June. The Eastern Region is rainy year-round, however, with some areas receiving nearly 200 in (500 cm) of rain annually. The Galápagos receive little rainfall; most of it comes between January and April.

Grasslands

The Costa is sometimes identified in English as the Coastal Lowlands and in Spanish as the Litoral (Littoral). It includes the basin made up of the Guayas River drainage system, and is the country's richest agricultural zone.

The Costa is widest in a central belt between Cape Pasado and the Santa Elena Peninsula. Near both the northern and southern extremities of the region, the Sierra Highlands intrude close to tidewater. At intervals, subtropical river valleys that are physical extensions of the Costa penetrate far into the Sierra.

The diversity of natural features of the Costa is so great that it can be considered to be a single geographic region only because the terrain rises abruptly from it to the Andean Sierra. Multiple climatic conditions, soils, forms of vegetation, and settlement patterns set it apart from the more homogeneous Sierra and Eastern Region.

Forests and Jungles

The Eastern Region is part of the Amazon River basin, which contains the world's largest tropical rainforest. Roughly 55 percent of the country, mostly in the Eastern Region, remains covered by native forest, including Yasuni National Park.

HUMAN POPULATION

The population is made up of descendants of Spanish colonialists, native people, and African slaves. A population of over 13 million, of which 47 percent live in urban areas, indicated a growth rate of 2 percent.

NATURAL RESOURCES

Known mineral resources are copper, gold, silver, clays, zinc, kaolin, limestone, marble, tin, lead, bismuth, and sulfur. Extensive petroleum fields lie off-shore and in the Eastern Region. Ecuador also has some of the richest fishing in the world, especially in tuna supplies. Ecuador's rainforests also represent a huge supply of timber.

FURTHER READINGS

Dyott, George Miller. *On the Trail of the Unknown in the Wilds of Ecuador and the Amazon.* London: Butterworth, 1926.

Handelsman, Michael. *Culture and Customs of Ecuador.* Westport, CT: Greenwood Press, 2000.

Murray, Gell-Mann. *The Quark and the Juguar: Adventures in the Simple and the Complex.* New York: W. H. Freeman and Co., 1994.

Simmons, Beth A. *Territorial Disputes and Their Resolution: The Case of Ecuador and Peru.* Washington, D.C.: U.S. Institute of Peace, 1999.

Egypt

- **Area:** 386,599 sq mi (1,001,450 sq km) / World Rank: 31
- **Location:** Northern and Eastern Hemispheres, northwestern most part of Africa, east of Libya, north of Sudan, west of the Red Sea, south of the Mediterranean Sea. Extends into the Sinai Peninsula in Asia, where it borders in the east on Israel
- **Coordinates:** 27°00′N, 30°00′E
- **Borders:** 1,667 mi (2,689 km) / Israel, 165 mi (266 km; includes Gaza Strip, 7 mi / 11 km); Libya, 713 mi (1,150 km); Sudan 789 mi (1,273 km)
- **Coastline:** 1,442 mi (2,325 km)
- **Territorial Seas:** 12 NM
- **Highest Point:** Mt. Catherine, 8,625 ft (2,629 m)
- **Lowest Point:** Qattara Depression, 439 ft (133 m) below sea level
- **Longest Distances:** 997 mi (1,572 km) SE-NW / 743 mi (1,196 km) NE-SW
- **Longest River:** Nile, 4,160 mi (6,693 km)
- **Largest Lake:** Lake Nasser, 1,522 sq miles (3,942 sq km)
- **Natural Hazards:** Droughts; hot, driving windstorms (*khamsin*); earthquakes and volcanic activity
- **Population:** 69,536,644 (July 2001 est.) / World Rank: 15
- **Capital City:** Cairo, in the northeast portion of the country, on the banks of the lower Nile
- **Largest City:** Cairo, 6,542,000 (1990 est.)

OVERVIEW

Most of Egypt is hot, dry desert, which covers about 96 percent of the country's surface. Over 96 percent of the population finds shelter and food in the remaining territory—the long, narrow, Nile Valley and its delta, which has a total land area of only about 15,000 sq mi (38,850 sq km). Two dominant characteristics of life in Egypt: overpopulation and the preeminence of the Nile River—overshadow all others.

The entire country lies within the wide band of the Sahara Desert that stretches from the Atlantic coast across North Africa to the Red Sea. The topographic channel through which the Nile flows across the Sahara causes an interruption in the desert, and the contrast between the Nile Valley and the rest of the country is abrupt and dramatic.

The majority of Egypt lies on the African Tectonic Plate, although the Sinai Peninsula lies on the Arabian Plate. Unlike most of North Africa and Arabia, the country has had an extremely disturbed geological history that has produced four major regions: the Nile Valley and Delta; the Western Desert; the Arabian Desert (Eastern Desert) and Red Sea Highlands; and the Sinai Peninsula.

MOUNTAINS AND HILLS

Mountains

The Red Sea Highlands run along the coast of the sea that they are named for. It is a region of hills and rugged mountains, and is extremely arid. Notable peaks include Mt. Shāīyb al-Banāt (7,173 ft / 2,186 m) and Mt. Hamātah (6,485 ft / 1,977 m). The Al-'Ajmah Mountains on the Sinai Peninsula are geologically an extension of the Red Sea Highlands. They run through the southern part of the peninsula. Egypt's highest peak, Mt. Catherine (Gebel Katherina; 8,625 ft / 2,629 m) is located here.

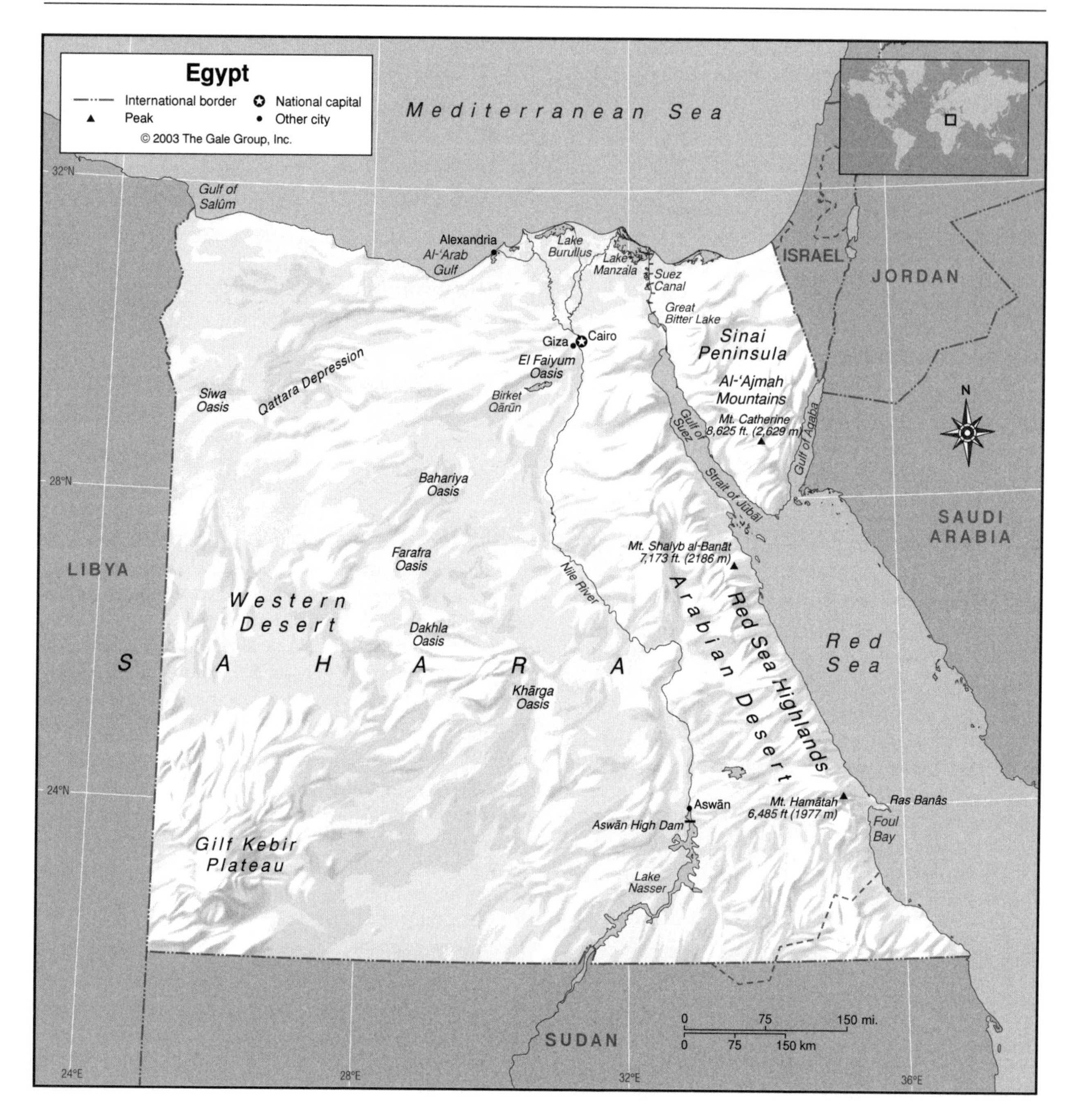

Plateaus

The Gilf Kebir rises out of the desert near the southwest boundary with Libya. It has an altitude of over 3,000 ft (914 m), an exception to the otherwise flat terrain of western Egypt.

The Arabian Desert rises abruptly from the Nile Valley, sloping upwards in a plateau of sand, before giving way to the rocky hills and mountains of the Red Sea Highlands.

Depressions

There are seven important depressions in the Western Desert. The largest is Qattara in the northwest. Halfway between the Nile and the Libyan border and 50 mi (80 km) from the Mediterranean coast, the Qattara Depression is a desolate area of badlands, salt marshes, and brackish lakes, lying mostly below sea level. The other depressions are smaller but more hospitable, especially the El Faiyum and Khargah depressions.

INLAND WATERWAYS

Lakes

Lake Nasser was formed by the damming of the River Nile with the Aswān High Dam. The lake extends south from the dam some 200 mi (322 km) to the border and an additional 99 mi (159 km) into Sudan. Only 6 to 11 mi (9 to 18 km) wide, Lake Nasser's waters fill the narrow groove between the cliffs of sandstone and granite created by the flow of the river over many centuries. The creation of Lake Nasser regulates the flow of the Nile. It ended the annual floods of the river, but also prevented fertile silt from being carried further downstream.

In the north near the coast, the Nile delta surrounds a series of lakes; most notable among them are lakes Maryut, Idku, Burullus, and Manzala. The Great Bitter Lake forms a part of the Suez Canal. Birket Qārūn is a salt lake in the El Faium depression.

The Nile River

The Nile River (Al-Bahr) extends across Egypt from south to north for roughly 992 mi (1,600 km). With a total length of a 4,160 mi (6,693 km), the Nile is the longest river in the world, although other rivers carry more water. The Egyptian Nile is a combination of the White Nile, originating in Lake Victoria in Uganda and Tanzania; the Blue Nile; and the Atbara; both of which originate in Ethiopia. All of these rivers meet in Sudan; throughout its length in Egypt no tributary streams enter the Nile. It enters Egypt in the form of Lake Nasser. From the Sudanese border to the Mediterranean Sea there is an average gradient of one foot to 13,000 feet (3,962 m) and an average flow of two to four miles per hour. The river's tendency to hug its east bank has produced a wider cultivable area on the west bank.

North of the capital city of Cairo, the Nile branches out into a delta. Historically there were as many as seven channels to the delta, but now only two remain, the Rosetta in the west and the Damietta in the east. Between and around these channels are many small streams, irrigation canals, ponds, lakes, and marshes, growing saltier as one approaches the sea.

THE COAST, ISLANDS, AND THE OCEAN

Egypt is bordered by the Mediterranean Sea on the north. The Mediterranean coast is marked by the Gulf of Salûm near the Libyan border and Al-'Arab Gulf west of the Nile delta. Although undeveloped and relatively unpopulated, miles of white sand beaches populate the Egyptian coast along the Mediterranean Sea. The azure water is warm in summer and cold in winter.

In the east lies the Red Sea. The Sinai Peninsula projects into the northern end of the Red Sea, forming two narrow gulfs. The Gulf of Aqaba is east of the peninsula; the Gulf of Suez is to the west and is separated from the open sea by the Strait of Jūbāl. The coastline is regular, with the exception of Ras Banâs in the south and the associated Foul Bay.

The Suez Canal connects the Mediterranean Sea with the Gulf of Suez. The canal is 101 mi (163 km) long, and at least 179 ft (55 m) wide and 40 ft (12 m) deep throughout its length. It has been one of the world's most important waterways since its completion in 1869.

GEO-FACT

The Nile was once famous for its floods. These floods were due to heavy seasonal rainfall in Ethiopia, which caused the flow of the Blue Nile and Atbara to fluctuate. The floods were unpredictable and could be destructive, but also provided vast amounts of fresh, fertile, soil. The damming of the river and the formation of Lake Nasser have ended the great Nile floods, but in the past the flood of the main Nile generally occurred in Egypt during the months of August, September, and October. It sometimes began as early as June at Aswān and often did not completely wane until January. There were rare years when the flood hardly occurred.

CLIMATE AND VEGETATION

Temperature

Egypt experiences mild winters (November to April) and hot summers (May to October). In Alexandria, located in the north on the Mediterranean coast, the average temperature ranges from 56°F (13°C) in December and January to 79°F (26°C) in July and August. Cairo, located further south, posts average lows of 57°F (14°C) in January and average highs of 82°F (28°C) in July. Aswān, located in the southern region, is considerably warmer with average temperatures of 60°F (16°C) in January and 93°F (34°C) in July, although highs exceeding 120°F (50°C) are not uncommon.

Rainfall

Except for the areas along the Mediterranean coast where, winter rains are common, rainfall in Egypt's harsh desert climate is scarce to nonexistent. During the summer months even the coast receives little or no rain.

The Nile Valley and Delta

The Nile Valley and its delta is a long narrow strip of fertile land created by the Nile's never-ending supply of fresh water and sediment. It is in effect the world's largest oasis, and makes up virtually all of Egypt's fertile land. The delta is roughly 155 mi (250 km) wide at the seaward base and about 100 mi (160 km) from north to south.

Population Centers – Egypt

(1996 POPULATION ESTIMATES)

Name	Population	Name	Population
Cairo (El-Qahira, capital)	6,789,500	Shubra el-Khaymah	870,700
Alexandria (El-Iskandariyah	3,328,200	Port Said	469,500
Giza (El-Giza or El Jizah)	2,221,900	Suez	417,600

SOURCE: Central Authority for Population, Mobilization and Statistics (CAPMAS) CAPMAS, National Information Center (NIC), Cairo.

Once a broad estuary, it was gradually filled by the Nile's sediment to become rich farmland.

The Western Desert

The Western Desert accounts for almost three-fourths of the total land area of Egypt. To the west of the Nile this immense desert spans the area from the Mediterranean south to the Sudanese border. It is a barren region of rock and sand, with occasional ridges or depressions but very little vegetation.

There are seven important depressions in the Western Desert, and all are considered oases except the largest, Qattara, which contains only salt water. Limited agriculture, some natural resources, and permanent habitation characterize the remaining six depressions. As oases these depressions have fresh water in sufficient quantities, provided either by the Nile waters or from local groundwater sources.

The Siwa Oasis, close to the Libyan border and west of Qattara, is isolated from the rest of the country, but has sustained life since ancient times.

The El Faiyum Oasis, sometimes called the Faiyum Depression, is 40 mi (64 km) southwest of Cairo. Around 3,600 years ago a canal was constructed from the Nile to the El Faiyum Oasis, probably to divert excessive floodwaters there. Over time this has produced an irrigated area of over 700 sq mi (1,813 sq km).

On the floors of the remaining depressions, artesian water is available to support limited populations. The Bahariya Oasis lies 210 mi (338 km) southwest of Cairo, and the Farafra Oasis, larger but sparsely populated, lies directly south. The Dakhla and Khārga oases complete the chain to the south.

The Arabian Desert

The desert east of the Nile is quite dissimilar from the Western Desert. While equally arid, it is more elevated and rugged, with the Red Sea Highlands along the shoreline.

HUMAN POPULATION

Almost all Egyptians live in the Nile River valley. The population is densest around Cairo and in the delta region, with more than 84,000 per sq mi (32,500 per sq km). Cairo itself is one of the largest cities in Africa. Outside of the Nile Valley population is restricted to the coasts, the oases, and the Suez Canal area.

NATURAL RESOURCES

Egypt is rich in mineral resources: petroleum, natural gas, iron ore, phosphates, manganese, limestone, gypsum, talc, asbestos, lead, and zinc are all present. Phosphates are located along the Nile, the Red Sea, and the Western Desert. Coal is found in the Sinai. At Aswān, iron ore deposits have been mined. Many regions of the country have been excavated since ancient times, although there are many areas that remain unexplored. Gold and copper deposits also exist; the cost of extraction is not warranted because of the low-grade quality of the minerals. The Nile Valley also supports rich farmland, and the Suez Canal sees busy shipping traffic.

FURTHER READINGS

Boraas, Tracey. *Egypt.* Mankato, MN: Bridgestone Books, 2001.

Bridges, Marilyn. *Egypt: Antiquities from Above.* Boston: Little, Brown, 1996.

Carpenter, Allan. *Egypt.* Chicago: Children's Press, 1972.

Deady, Kathleen W. *Egypt.* Mankato, MN: Bridgestone Books, 2001.

Feinstein, Stephen. *Egypt in Pictures.* Minneapolis, MN: Lerner Publications Co., 1988.

Manley, Bill. *The Penguin Historical Atlas of Ancient Egypt.* New York: Penguin Books, 1996.

Manley, Deborah, ed. *The Nile: A Traveler's Anthology.* London: Cassell, 1996.

Roberts, Paul William. *River in the Desert: Modern Travels in Ancient Egypt.* New York: Random House, 1993.

El Salvador

- **Area:** 8,124 sq mi (21,040 sq km) / World Rank: 151
- **Location:** Western and Southern Hemispheres, southern Central America isthmus on the Pacific Ocean, bordered by Honduras to the northeast and Guatemala to the west.
- **Coordinates:** 13°50′N, 88°55′W
- **Borders:** 339 mi (545 km) / Guatemala, 126 mi (203 km); Honduras, 213 mi (342 km)

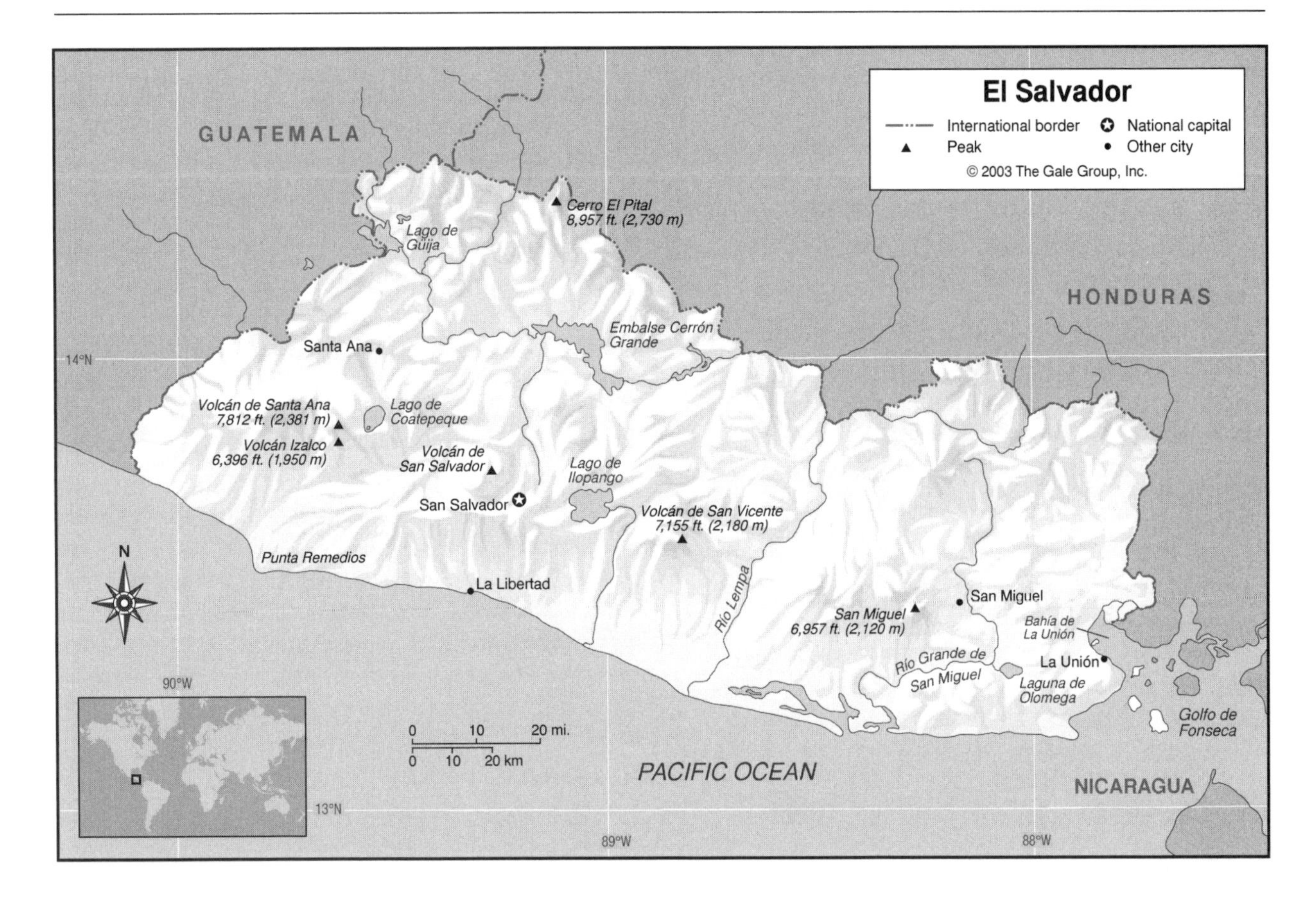

- **Coastline:** 191 mi (307 km)
- **Territorial Seas:** 200 NM
- **Highest Point:** Cerro El Pital, 8,957 ft (2,730 m)
- **Lowest Point:** Sea level
- **Longest Distances:** 88 mi (142 km) N-S; 168 mi (270 km) WNW-ESE
- **Longest River:** Río Lempa, 200 mi (320 km)
- **Largest Lake:** Ilopango, 25 sq mi (65 sq km)
- **Natural Hazards:** Relatively frequent earthquakes, active volcanoes
- **Population:** 6,237,662 (July 2001 est.) / World Rank: 97
- **Capital City:** San Salvador, located in the west-central part of the country
- **Largest City:** San Salvador, population 1,415,000 (2000 est.)

OVERVIEW

El Salvador is the smallest, most densely populated country in Middle America. It borders the Pacific Ocean on the southern Central American isthmus, and is the only Central American country that does not border the Caribbean Sea. This tiny "Land of Volcanoes" contains more "Ring of Fire" volcanoes than any other Central American country.

El Salvador is one of the most seismically active, hence earthquake-vulnerable areas in the Western Hemisphere. The capital, San Salvador, has been completely destroyed twice by major earthquakes, once in 1756 and again in 1854. More recent earthquakes in central El Salvador killed at least 1,500 people in 1986 (and left 500,000 homeless); in January and February of 2001, two earthquakes caused more than 1,000 fatalities.

The land is buffeted between two areas of active tectonic plate movement. In southern El Salvador on the Pacific Ocean side, the Cocos Plate forces material under the relatively motionless Caribbean Plate (a process termed "subduction"), accounting for frequent earthquakes near the coast. As the ocean floor is forced down, rocks melt, and the molten material pours up through fissures, producing volcanoes and geysers.

North of El Salvador, the North American Plate abuts one edge of the same stationary Caribbean Plate, creating a major fault that runs the length of Río Motagua Valley in Guatemala. Motion along this fault generates earthquakes in both Guatemala and northernmost El Salvador.

MOUNTAINS AND HILLS

Mountains

Two volcanic-formed mountain ranges run roughly northwest to southeast across northern and southern El

GEO-FACT

In Morozan, in northeast El Salvador, two caves—Espíritu Santo and Cabeza de Duende—have well-preserved pre-Columbian paintings on the walls.

Salvador, with a broad high plateau between. The northern Sierra Madre range is a continuous chain, with elevations from 5,200 to 7,210 ft (1,580 to 2,200 m). The southern coastal range is a discontinuous chain composed of more than twenty volcanoes in five clusters. Near the western end is Volcán de Santa Ana, the highest volcano in the country at 7,812 ft (2,381 m). Also at the western end is Volcán Izalco (6,396 ft / 1,950 m), known as "Lighthouse of the Pacific," which last erupted in 1966, making it El Salvador's most recently active volcano. Other volcanoes in the chain are Volcán de San Salvador northwest of the city of San Salvador, Volcán de San Vicente (7,155 ft / 2,180 m) south of the city of San Vicente, and San Miguel (6,957 ft / 2,120 m) southwest of the city of San Miguel. The highest mountain in El Salvador is not a volcano—Cerro El Pital sits on the Honduras-El Salvador border and towers to a height of 8,957 ft (2,730 m).

Plateaus

The central valley running east and west between the two mountain ranges is actually a rolling plateau peppered with lava fields, escarpments, and geysers. Comprising most of the land in the country, this high plain averages 30 miles (50 km) wide with an average elevation of 2,000 ft (600 m).

INLAND WATERWAYS

Lakes

El Salvador contains hundreds of tiny lakes, and a few larger ones. The largest lake, scenic Lago de Ilopango, lies just east of San Salvador and contains emerald-blue water in the caldera of an inactive volcano. In the late 1800s, an island, Islas Quemadas or Burnt Island, appeared in the middle of the lake, either due to receding water levels or seismic activity.

A second volcanic lake, Lago de Coatepeque, is smaller in surface area but so deep its lowest point is unknown. It is part of Cerro Verde National Park, located due north of Lago Ilopango. A third lake, Lago de Guija, lies in the northwest on the border with Guatemala.

There are two significant manmade lakes, both above dams on the Río Lempa. The largest reservoir is Embalse Cerrón Grande. Slightly further downstream is a second dam and reservoir.

Rivers

The Río Lempa is the longest—and only navigable—river in El Salvador. Río Lempa originates in Guatemala, flowing for a short distance through Honduras before entering El Salvador, where approximately 160 miles (257 km) of its 200 mi (320 km) flows. It turns east near Lago de Guija (Lake Guija), where it is fed by a tributary from the lake. From there the Río Lempa continues easterly about halfway across the country, then south to empty into the Pacific Ocean. The area around the mouth of the Río Lempa is known as Isla Montecristo; it is undeveloped with lush stands of mangrove. Hundreds of smaller rivers and streams drain the highlands directly into the Pacific Ocean or are tributaries of Río Lempa.

The Río Grande de San Miguel flows in the eastern part of country, originating north of San Francisco and flowing southward past San Miguel. It is joined by a tributary that flows from Laguna de Olomega, meandering westward for about 25 mi (40 km) before turning south to its Pacific Ocean mouth. Another river, the Río Jiboa, flows from Lago de Ilopango to the Pacific, where its mouth marks the country's approximate midpoint.

Wetlands

Most of El Salvador's fragile wetlands are threatened by pollution of various kinds, as are wetlands throughout the world. Most estuaries are polluted by shrimp farming, logging, agricultural runoff, and the effects of overpopulation. One exception may prove to be the 3,880-acre Laguna del Jocotal. In May 1999, the marshland was registered by the Ramsar Convention on Wetlands as an internationally significant wetland. Laguna del Jocotal is at the southeastern tip of the country, north of the Río Grande de San Miguel, near San Miguel Volcano (13°15′N, 88°16′W). Laguna del Jocotal is a permanent freshwater lake 10 ft (3 m) deep during the wet season; it recedes to less than 4 ft (1.1 m) deep during the dry season. The lake is eutrophic (especially supportive of plant life) and much of the surface is covered with floating vegetation. In 1978 a wildlife sanctuary was created at the site.

THE COAST, ISLANDS, AND THE OCEAN

Oceans and Seas

El Salvador's southern border is the Pacific Ocean. Off the coast is a deep ocean trench, the Middle America Trench, which was created by movement of the Cocos Tectonic Plate.

The Coast and Beaches

The area between the coastal range and the coast beach is relatively narrow with its widest point, about 20 miles (32 km), at the eastern end of the country and disappearing at the western end. The beaches are black volcanic sand with frequent marshes. Near the small port of

La Libertad, volcanoes fall steeply to the sea, leaving virtually no beach. At its southeastern tip, El Salvador faces Nicaragua across the Golfo de Fonseca (Gulf of Fonseca), with La Unión Bay lying between El Salvador and Honduras, just off the town of La Unión in the northwestern Gulf.

Further west is Jiquilsco Bay, a narrow inlet that forms a long westward-reaching finger of water. Still further west of La Libertad is the popular 45 mi (75 km) stretch of beach known as Costa de Bálsamo (Balsam Coast). Punta Remedios (Remedios Point) is near the westernmost end of the country.

Population Centers – El Salvador

(2000 ESTIMATES)

Name	Population
San Salvador	1,415,000
Soyapango	252,000
Santa Ana	202,000
San Miguel	183,000

SOURCE: Direccion General de Estadística y Censos (Directory of Statistics and Censuses), Ministerio de Economía (Ministry of Economy), El Salvador.

Departments – El Salvador

POPULATIONS FROM 1992 CENSUS

Name	Population	Area (sq mi)	Area (sq km)	Capital
Ahuachapán	261,188	479	1,240	Ahuachapán
Cabañas	138,426	426	1,104	Sensuntepeque
Chalatenango	177,320	779	2,017	Chalatenango
Cuscatlán	178,502	292	756	Cojutepeque
La Libertad	513,866	638	1,653	Nueva San Salvador
La Paz	245,915	473	1,224	Zacatecoluca
La Unión	255,565	801	2,074	La Unión
Morazán	160,146	559	1,447	San Francisco (Gotera)
San Miguel	403,411	802	2,077	San Miguel
San Salvador	1,512,125	342	886	San Salvador
Santa Ana	458,587	781	2,023	Santa Ana
San Vicente	143,003	457	1,184	San Vicente
Sonsonate	360,183	473	1,226	Sonsonate
Usulután	310,362	822	2,130	Usulután

SOURCE: Direccion General de Estadística y Censos (Directory of Statistics and Censuses), Ministerio de Economía (Ministry of Economy), El Salvador.

CLIMATE AND VEGETATION

Temperature

Temperatures in tropical El Salvador vary more with altitude than with season. The average temperature on the central highlands is 74°F (28°C) year round. Along the coast and at lower altitudes it is hotter, and in the northern mountains cooler. Even at the highest elevations the climate is temperate, rarely approaching freezing even in the winter.

Rainfall

Most rainfall occurs during the winter/wet season, which is from May to October. The heaviest rains are along the coast. Rainfall along the coast during the wet season averages 85 in (216 cm). During the same season the northwest averages 60 in (150 cm). Summer is the dry season, lasting from November to April.

Mostly due to deforestation, heavy rains have become a hazard. In 1998 Hurricane Mitch caused massive landslides, loss of life, and destruction.

Grasslands

Starting in the early 1900s, forests in the central high plateau were cleared and farmed, creating large areas of grasslands across much of the country.

Deserts

Although there are no true deserts in El Salvador, an estimated half of the land is severely eroded from deforestation, farming, and development. Much of this land is on the way to becoming desert. (This phenomenon, known as desertification, is a worldwide problem.)

Forests and Jungles

As recently as the 1950s, El Salvador was predominantly forest and jungle. However, most areas have since been logged, cleared, settled, and farmed. Today, only about 5 percent of the land is true jungle or unspoiled forest. Of that, some is now protected in two national parks.

High in the northeast corner, where El Salvador, Guatemala, and Honduras meet, the three countries have agreed to protect an area called El Trifinio International Biosphere Reserve. The El Salvador portion is named Montecristo National Park (the reserve is centered on the mountain). Montecristo National Park is nearly perpetually covered in clouds and mist. It is a spectacular true rain forest, an increasingly rare type of ecosystem. Within the boundary of the park are giant ferns, air plants, and ground-level areas never sunlit supporting flora that can grow only in such an environment. The park protects a few species of mammals including endangered jaguars, jungle foxes, tree-dwelling spider monkeys, and opossums.

Near the southwest coast, near the border of Guatemala, is the Bosque El Imposible (Impossible Forest) National Park. It is named for a famously dangerous pass that is part of a traditional mule trail employed to bring coffee to the coast. An El Salvador environmental organization, SalvNatura, has managed the park since 1990. Covering 12,000 acres (5,000 hectares), the park is home to four hundred species of trees and nearly three hundred species of birds, as well as unique animals such as the Tamandua anteater (antbear), pumas, and hundreds of species of butterflies. Eight rivers flow through the park's sometimes steep terrain. (The lowest point is about 1,000 ft (300 m) and the highest about 4,600 ft (1,400 m). Three extinct volcanoes are within the park boundaries.

HUMAN POPULATION

Overpopulation is a critical problem in El Salvador. The country has the most highly dense population in Central America: 777 persons per sq mi (about 300 per sq km). Nearly 80 percent of the population is concentrated in the central plateau.

As with many Central American nations, the majority (about 95 percent) of the population is mestizo, that is, of mixed indigenous and European ancestry.

NATURAL RESOURCES

The major natural resource of El Salvador is coffee, which is grown extensively across the plateau region. Petroleum and a few metals are minor resources. About half the power is hydroelectric; geothermal resources are used with potential for more use. The surviving natural forest areas have research potential.

FURTHER READINGS

Boland, Roy C. *Culture and Customs of El Salvador.* Westport, Conn.: Greenwood, 2000.

Brauer, Jeff, and Bea Weiss. *On Your Own in El Salvador.* Charlotsville, Va.: On Your Own Publications, 2001.

Kelly, Joyce. *An Archaeological Guide to Northern Central America: Belize, Guatemala, Honduras, and El Salvador.* Norman, Olka.: University of Oklahoma Press, 1996

Lonely Planet. *El Salvador* http://www.lonelyplanet.com/dest/cam/els.htm (Accessed May 17, 2002).

Towell, Larry. *El Salvador.* New York: W.W. Norton, 1998.

Equatorial Guinea

- **Area:** 10,831 mi (28,051 sq km) / World Rank: 144
- **Location:** Western Africa, bordering Cameroon to the north, Gabon to the south and east, and the Bight of Biafra in the Atlantic Ocean to the west. Crossed by the equator, all of the country is in the Eastern Hemisphere, parts are in the Southern and the Northern Hemispheres.
- **Coordinates:** 2°00′N, 10°00′E

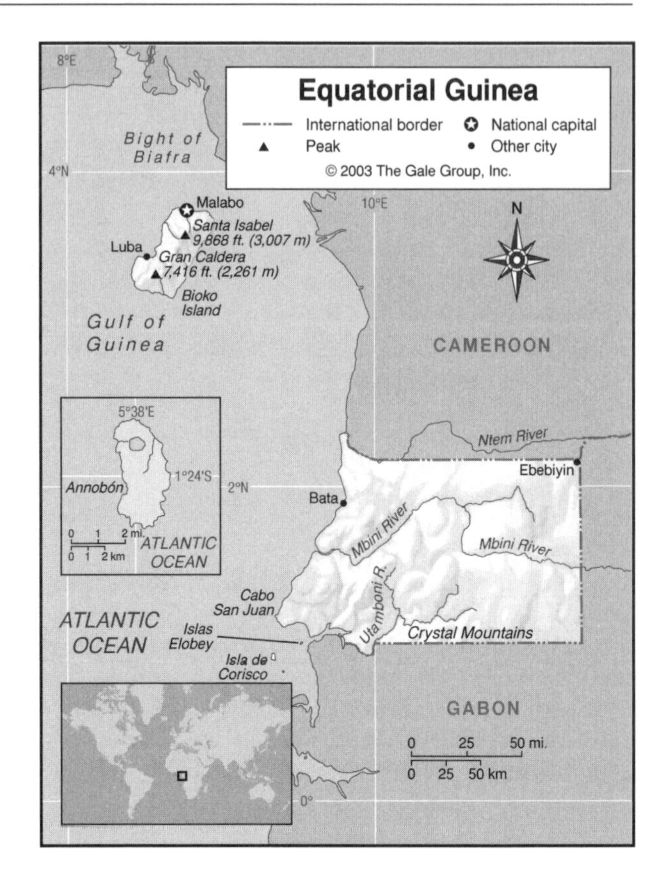

- **Borders:** 334 mi (539 km) total / Cameroon, 117 mi (189 km); Gabon, 217 mi (350 km)
- **Coastline:** 183 mi (296 km)
- **Territorial Seas:** 12 NM
- **Highest Point:** Santa Isabel, 9,865 ft (3,007 m)
- **Lowest Point:** Sea level
- **Longest Distances:** Mbini: 154 mi (248 km) ENE-WSW / 104 mi (167 km) SSE-NNW; Bioko 46 mi (74 km) NE-SW / 23 mi (37 km) SE-NW
- **Longest River:** Mbini, 155 mi (248 km)
- **Natural Hazards:** violent windstorms, flash floods
- **Population:** 486,060 (July 2001 est.) / World Rank: 160
- **Capital City:** Malabo, located on north coast of Bioko Island
- **Largest City:** Malabo, 112,800 (2002 est.)

OVERVIEW

Equatorial Guinea consists of a mainland province, Mbini (Río Muni), and five inhabited islands: Isla de Bioko (formerly Macias Nguema Biyogo, and prior to that Fernando Po), Annabón (formerly Pagalu), Ilas Elobey, made up of Elobey Grande and Elobey Chico, and Isla de Corisco. Bioko is located 20 mi (32 km) from the coast of Cameroon, and Annabón is located 220 mi

(350 km) from mainland Gabon. Corisco and the Elobey Islands are off the southwest coast of Mbini close to the Gabonese coast. The total land area is slightly smaller than the state of Maryland.

Mbini on the African mainland is a jungle enclave with a coastal plain rising steeply toward the west. In the interior the plain gives way to a succession of valleys separated by low hills and spurs of the Crystal Mountains. The islands are all volcanic in origin. In 2002 Equatorial Guinea was in the midst of maritime boundary and border disputes with Nigeria, Cameroon, and Gabon.

MOUNTAINS AND HILLS

Bioko has two large volcanic formations separated by a valley that bisects the island. In the north of the island is Santa Isabel (9,865 ft / 3,007 m). In the south is Gran Caldera (7,416 ft / 2,261 m). All of the other islands are also volcanic, but of much lower elevation. On the mainland the coastal plains rise to interior ridges of the Crystal Mountains, which separates the coast from the inland plateau. The highest peaks are Monte Chocolate (3,609 ft / 1,100 m) and Monte Chime (3,937 ft / 1,200 m).

INLAND WATERWAYS

The main rivers are the Mbini (formerly Río Benito), the Ntem (formerly Río Campo), and Río Muni. The Mbini, which divides the mainland province of the same name into two, is not navigable except for a 12 mi (20 km) stretch. The Ntem flows along part of the northern border with Cameroon. The Río Muni is not properly a river at all but an estuary of several rivers among which the Utamboni is the most notable. The islands have only storm arroyos and small cascading rivers.

THE COAST, ISLANDS, AND THE OCEAN

Oceans and Seas

The islands and mainland are separated by the Bight of Biafra. The Bight is part of the broad Gulf of Guinea, from which the country takes its name. The Gulf is part of the Atlantic Ocean.

Major Islands

Bioko is a volcanic island roughly 779 sq mi (2,018 sq km) in size, making it the largest island in the Gulf of Guinea. The other islands are also volcanic, but are much smaller than Bioko. Annabón is 7 sq mi (18 sq km) in size, Corisco covers 6 sq mi (15 sq km), and the Great and Little Elobeys are each about 1 sq mi (2.5 sq km). Bioko and Annabón are part of the volcanic chain starting with the Cameroon Highlands and outcropping into the Atlantic as far as St. Helena.

Population Centers – Equatorial Guinea

(2002 POPULATION ESTIMATES)

Name	Population
Malabo (capital)	112,800
Bata	40,000

SOURCE: Projected from 1983 data, United Nations Statistics Division, and 1999 data, World Bank.

Provinces – Equatorial Guinea

2000 POPULATION ESTIMATES

Name	Population	Area (sq mi)	Area (sq km)	Capital
Annobón	4,400	6.6	17	Palé
Bioko Norte	112.800	300	776	Malabo
Bioko Sur	12,200	479.2	1,241	Luba
Centro Sur	69,000	3,384	9,931	Evinayong
Kié-Ntem	87,200	1,522	3,943	Ebebiyin
Litoral	118,000	2,573	6,665	Bata
Wele-Nzas	62,000	2,115	5,478	Mongomo

SOURCE: *Geo-Data*, 1989 ed., and United States Agency for International Development, Africa Division.

The Coast and Beaches

Sandy shores and estuaries make up the coastal mainland. Near Mbini's southern tip, Cabo San Juan protrudes into the sea. On Bioko the coastline is high and rugged in the south but lower and more accessible in the north.

CLIMATE AND VEGETATION

Temperature

Equatorial Guinea lies near the equator and as a result has a warm climate that varies mainly by altitude. At Malabo, temperatures range from 61°F (16°C) to 91°F (33°C). In Mbini, the average temperature is about 80°F (27°C).

Rainfall

Annual rainfall varies from 76 in (193 cm) at Malabo to 430 in (1,092 cm) at Ureka, and also on Bioko Island.

Forests and Jungles

Most of the country, including the islands, is tropical rain forest.

HUMAN POPULATION

The population growth rate is 2.46 percent and life expectancy at birth is 54 years. Population density was about 40 per sq mi (15 per sq km) in 1996, and approximately 48 percent of the population was urban. Bioko Island is the most densely settled part of the country.

NATURAL RESOURCES

Equatorial Guinea is rich in petroleum, timber, and unexploited deposits of gold, manganese, uranium, titanium, and iron ore. Although little of the land is considered arable, most of the population is engaged in subsistence farming. Offshore oil deposits are at the heart of Equatorial Guinea's maritime boundary disputes.

FURTHER READINGS

"A Corner of Africa Where Dreams Gush Like Oil." *New York Times*, July 28, 2000, p. A4.

Hecht, David. "Gushers of Wealth, But Little Trickles Down: Oil in Equatorial Guinea." *Christian Science Monitor*, July 21, 1999, p. 7.

Onishi, Norimitsu. "Oil Riches, and Risks, in Tiny African Nation." *New York Times*, July 23, 2000, pp. A1, A4.

World Resources Institute and World Bank. *Tropical Forests: A Call for Action*, Washington D.C.: World Resources Institute and World Bank, 1985.

Eritrea

- **Area:** 46,842 sq mi (121,320 sq km) / World Rank: 98
- **Location:** Northern and Eastern Hemispheres, eastern Africa, bordering the Red Sea, north of Djibouti and Ethiopia and east of Sudan.
- **Coordinates:** 15°00′ N, 39°00′ E
- **Borders:**1,013 mi (1,630 km) / Djibouti 70 mi (113 km); Ethiopia, 567 mi (912 km); Sudan, 376 mi (605 km)
- **Coastline:** 1,388 mi (2,234 km)
- **Territorial Seas:** 12 NM
- **Highest Point:** Soira, 9,902 ft (3,018 m)
- **Lowest Point:** Denakil Depression, near Kulul, 246 ft (75 m) below sea level
- **Longest Distances:** 520 mi (830 km) NW-SE; 250 mi (400 km) N-S
- **Longest River:** Tekeze, 470 mi (755 km)
- **Natural Hazards:** Droughts; locust swarms
- **Population:** 4,298,269 (July 2001 est.) / World Rank: 118
- **Capital City:** Asmara, centrally located 50 mi (85 km) west of the Red Sea
- **Largest City:** Asmara, 431,000 (2000 est.)

GEO-FACT

Fishing was largely suspended during the decades of war for independence from Ethiopia. As a result, the relatively unpolluted waters of the Red Sea are full of a wide variety of fish and coral species, making them popular with adventurous divers. The largest island, Dahlak Kebir, has ancient ruins of both indigenous and Arabic communities.

OVERVIEW

Slightly larger than Pennsylvania, Eritrea resembles a funnel lying on its side tilted southeastwardly. It occupies the northern portion of a high, mountainous plateau reaching north from Ethiopia to the Red Sea. The mountains descend to a network of high hills on the northeast and to a low, arid coastal strip along the Red Sea. A corridor of low rolling plains marks the southwestern perimeter with Sudan. Bordering Ethiopia in the southeast, the Danakil Depression at its deepest point lies 423 ft (130 m) below sea level. The hottest temperatures in the world have been reported there. Only 3 percent of the land is arable.

Eritrea lies along the boundary between the African and Arabian Tectonic Plates. The Great Rift Valley, which extends from Mozambique in southern Africa all the way north into the Middle East, passes near Eritrea's eastern border.

As a result of a December 12, 2000, peace agreement ending a two-year war with Ethiopia, the United Nations (UN) is administering a 15 mi (25 km) wide security zone within Eritrea until a joint boundary commission delimits a final boundary between the two countries. Eritrea gained its independence from Ethiopia in 1993.

MOUNTAINS AND HILLS

Mountains

Ethiopia's northwestern highlands extend into Eritrea, reaching elevations of more than 6,500 ft (2,000 m) above sea level. A line of seismic belts extends along the length of Eritrea and the Danakil Depression, but serious earthquakes have not been recorded in the area during the twentieth century.

Plateaus

Eritrea shares the northeast section of the Ethiopian high plateau, which in appearance looks more like a set of rugged uneven mountains. The plateau, also known as

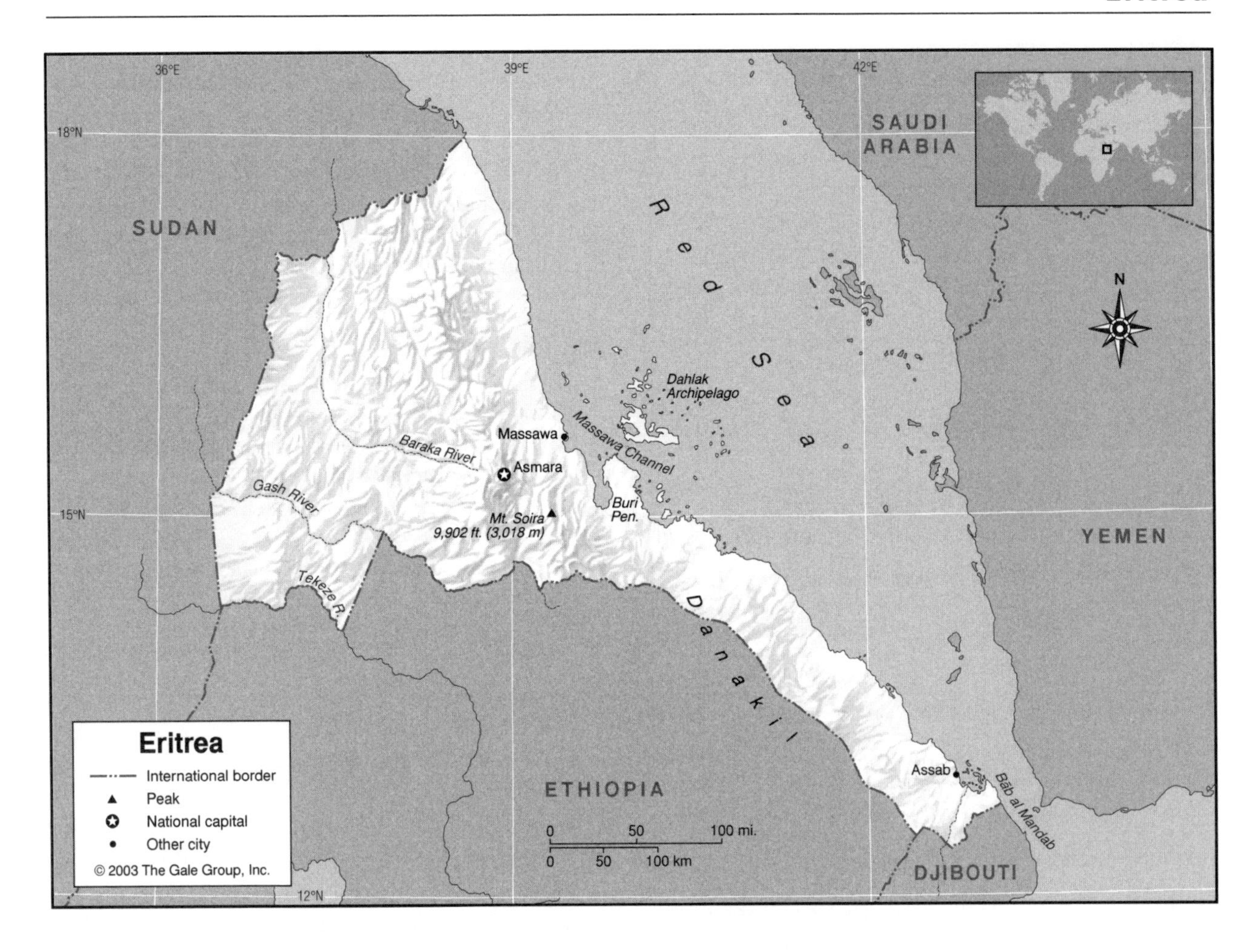

the Northwestern Highlands, rises up on the western scarp of the Great Rift Valley and projects northward from Addis Ababa in Ethiopia to the Red Sea coastline in Eritrea. It descends to the Red Sea coast in a series of hills.

INLAND WATERWAYS

Rivers

The Tekeze and the Mereb form sections of the southern border with Ethiopia. The Gash drains westward to Kassala in Sudan, and the Baraka flows northward to Sudan from its source near Asmara. Volume in these rivers is highly seasonal, they are sometimes completely dry.

Wetlands

Coastal hills drain inland into saline lakes and sinks from which commercial salt is extracted.

THE COAST, ISLANDS, AND THE OCEAN

Oceans and Seas

Eritrea borders on the Red Sea, a busy shipping channel, potentially rich in oil and natural gas.

Major Islands

The Dahlak Archipelago, a collection of coralline islands, lies opposite the Buri Peninsula. They are separated from the mainland by the Massawa Channel. The many islands are mostly small and sparsely inhabited.

The Coast and Beaches

Although subject to torrid temperatures much of the year, Eritrea's coastal beaches and Red Sea islands hold significant tourism potential. The hot, arid, and treeless coastal lowlands range from 10 to 50 mi (16 to 80 km) wide.

CLIMATE AND VEGETATION

Temperature

Along the Red Sea temperatures average between 81° F (27° C) to 86° F (30° C) in the daytime, but at midyear in the Danakil Depression in the southeast temperatures may hit 140° F (60° C). The highlands are moderate with temperatures that average about 63° F (17° C). The coast enjoys a Mediterranean-like climate when the northeast trade winds blow in January.

Rainfall

Rainfall varies according to season, elevation, and location. The semi-arid western hills and lowlands along the Sudanese border receive up to 20 in (50 cm) of rain with the heaviest rainfall in June through August. In Jan-

Population Centers – Eritrea

(2000 POPULATION ESTIMATES)

Name	Population
Asmara (capital)	431,000
Assab	55,000

SOURCE: Projected from United Nations Statistics Division data.

Provinces – Eritrea

POPULATIONS FROM 1995 ESTIMATE

Name	Population	Area (sq mi)	Area (sq km)	Capital
Anseba	400,846	8,541	22,120	Keren
Debub	702,502	3,749	9,709	Mendeferas
Debubawi Keyih Bahri	189,627	not available	not available	Assab
Gash Barka	515,667	38,356	99,341	Barentu
Maekel	502,300	2,155	5,581	Asmara
Semenawi Keyih Bahir	392,653	12,425	32,180	Massawa

SOURCE: Eritrean Ministry of Local Government.

uary monsoons originating in Asia cross the Red Sea bringing rain to the coastal plains and the eastern escarpment. Eastern lowlands receive less than 20 in (50 cm) of rainfall annually, while the much cooler and somewhat wetter highlands receive up to twice that amount.

Deserts

Eritrea has semi-arid western hills and a very dry and hot coastal strip along the eastern seaboard. The desert-like coast is home to acacia, cactus, aloe vera, prickly pear, and olive trees. The Danakil Depression is a desert region.

HUMAN POPULATION

In July 2001, the population was estimated at 4,298,269, with a high annual growth rate of 3.84 percent. Life expectancy was 56.18 years. Eritreans are fairly evenly divided between Tigrinya-speaking Christians, who have traditionally lived in the highlands, and Muslims, who inhabit the western lowlands, northern highlands, and the east coast. Coptic Christians, Roman Catholics, and Protestants comprise the Christian faiths. The principal languages spoken are Afar, Amharic, Arabic, Tigre and Kunama, Tigrinya, and other Cushitic languages.

NATURAL RESOURCES

The country's natural wealth is still largely unmapped. Gold, zinc, copper, and iron ore are found in the Eritrean part of the Ethiopian plateau. The Dallol Depression south of Massawa commands significant potash deposits, while salt is extracted from saline pools and sinks within the depression. The Red Sea and its coral islands offer rich fishing grounds and hold promise as a tourist destination. Pending further exploration and evaluation, the Red Sea could prove to hold significant deposits of offshore oil and natural gas.

FURTHER READINGS

Africa South of the Sahara 2002. "Ethiopia." London: Europa Publishers, 2001.

Ellingson, L. *The Emergence of Eritrea, 1958-1992.* London: James Currey Publishers, 1993.

Killion, Tom. *Historical Dictionary of Eritrea.* Lanham, Md.: Scarecrow Press, 1998.

Papstein, Robert J. *Eritrea: A Tourist Guide.* Lawrenceville, N.J.: Red Sea Press, 1995.

Estonia

- **Area:** 17,462 sq mi (45,226 sq km) / World Rank: 132
- **Location:** Northern and Eastern Hemispheres, in Eastern Europe, bordering the Baltic Sea and Gulf of Finland, between Latvia and Russia
- **Coordinates:** 59°00′N, 26°00′E
- **Borders:** 392 mi (633 km) / Latvia, 210 mi (339 km); Russia, 182 mi (294 km)
- **Coastline:** 2,352 mi (3794 km)
- **Territorial Seas:** 12 NM
- **Highest Point:** Suur Munamägi, 1,043 ft (318 m)
- **Lowest Point:** Sea level
- **Longest Distances:** Not available
- **Longest River:** Pärnu, 89 m (144 km)
- **Largest Lake:** Lake Peipus, 1,386 sq mi (3,555 sq km)
- **Natural Hazards:** Subject to springtime flooding
- **Population:** 1,423,316 (July 2001 est.) / World Rank: 146
- **Capital City:** Tallinn, located on the northern coast
- **Largest City:** Tallinn, 499,000 (2000 est.)

OVERVIEW

Smallest of the Baltic states, Estonia is a low, flat country with a hilly region in the southeast. It has a long, shallow coastline on the Baltic Sea, with many islands off the coast. Over a third of the country is forest. The country is dotted with more than 1,000 natural and artificial lakes. The capital city Tallinn is the largest city and chief port. The Pärnu, Narva, and Ema are the country's chief rivers. Estonia is on the Eurasian Tectonic Plate.

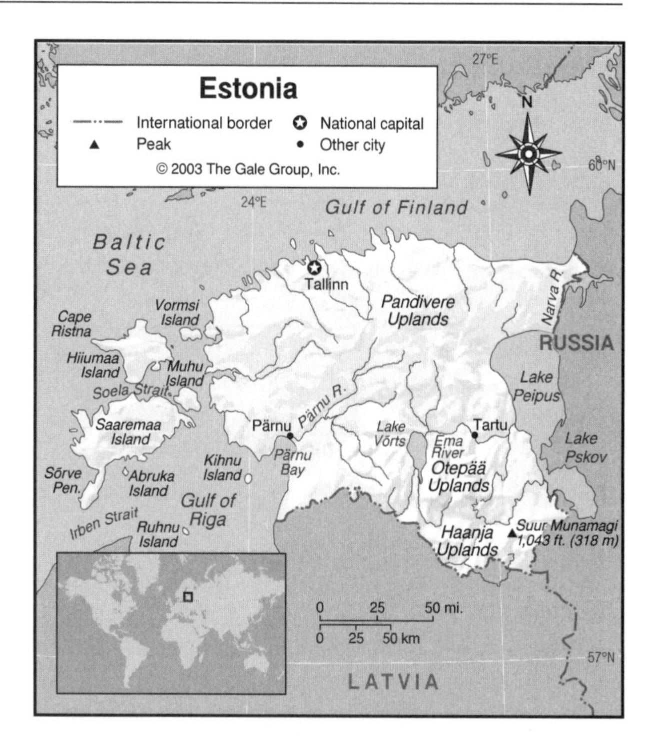

MOUNTAINS AND HILLS

Hills

Estonia is mostly a low-lying plain, but there are some modest hills in the central and southern regions. These are the Pandivere, Otepää, and Haanja Uplands. The country's highest point, Suur Munamägi, 1043 ft (318 m), is in the extreme southeast corner of the country near the Russian border.

Along the north coast there is an area of slightly elevated limestone known as the Glint. There, waterfalls as high as 185 ft (56 m) tumble down the exposed limestone cliffs.

INLAND WATERWAYS

Lakes

Lakes and reservoirs cover fully five percent of Estonia's territory. The two largest lakes, Lake Peipus on the eastern border with Russia and Lake Võrts (Võrtsjarv) in south central Estonia, account for nearly four-fifths of the total lake area. Lake Peipus covers 1360 sq mi (3520 sq km). A long, narrow channel connects it on the south with the smaller Lake Pskov. Lake Peipus is drained on the north by the Narva River, which flows into the Gulf of Finland. Fishing is the chief industry, and Lake Peipus is navigable for about eight months of the year. Lake Võrts's area is 105 sq mi (270 sq km).

Rivers

The Pärnu is the longest river in Estonia at 89 mi (144 km). It flows southwest, emptying into the Gulf of Riga at Pärnu Bay. Other important rivers include the Ema in the southeast and the Narva, which forms the northeastern border with Russia.

Wetlands

More than 20 percent of Estonia is considered wetlands.

THE COAST, ISLANDS, AND THE OCEAN

The Baltic Sea

The Baltic Sea lies north and west of Estonia. The coastal waters are shallow, and dotted with over 1,500 islands. The western part of the country borders on the open sea. The rest of Estonia's coastline is on two major inlets of the Baltic.

The Gulf of Finland reaches east about 250 mi (400 km) between Finland on the north and Estonia and Russia on the south. Its width varies from 12 to 80 mi (19 to 129 km), narrowest at the eastern end. The Narva River empties into Narva Bay at the northeastern end of the country, and links the Gulf of Finland with Lake Peipus to the south.

The Gulf of Riga is to the southwest of mainland Estonia, directly south of Estonia's major islands, with Latvia on the far shore. It is about 90 mi (145 km) long from north to south and ranges from 45 to 80 mi (72 to 129 km) wide from east to west. Pärnu Bay extends northeastward off of this Gulf into Estonia.

Major Islands

There are thousands of islands along Estonia's coastline. The largest islands lie west of the mainland. Saaremaa is the largest island at 1048 sq mi (2714 sq km). It lies between the Baltic Sea and the Gulf of Riga. The Sõrve Peninsula extends off the southern end of the island, and is separated from Latvia by the Irben Strait. Raising livestock and tourism are the principal economic activities of this low-lying island.

Hiiumaa, the next largest of Estonia's islands, measures 371 sq mi (961 sq km) in area. It is located in the Baltic Sea, southwest of the entrance to the Gulf of Finland. The Soela Strait separates it from Saaremaa to the south. Its most distinctive feature is Cape Ristna, which projects off the west coast into the Baltic. Fishing and tourism are the island's chief industries. Many of its inhabitants are of Swedish descent.

Population Centers – Estonia

(1993 POPULATION ESTIMATES)

Name	Population
Tallinn (capital)	447,672
Tartu	107,303

SOURCE: "Population of Capital Cities and Cities of 100,000 and More Inhabitants." United Nations Statistics Division.

The other islands are all much smaller. Vormsi and Muhu Islands lie between the larger islands and the Estonian mainland. Arbuka, Kihnu, and Ruhnu Islands are in the Gulf of Riga.

CLIMATE AND VEGETATION

Temperature

Estonia's marine location keeps the coastal climate moderate, but in the interior temperatures are typically more extreme. Summers in Estonia are generally cool, with temperatures rarely exceeding 64°F (18°C). Winters are cold; the temperature usually remains below freezing from mid-December to late February.

Rainfall

July and August are the wettest months. Annual precipitation is moderate, ranging from 19 to 27 in (500 to 700 mm), averaging about 23 in (568 mm). Rain and melting snow cause some flooding of rivers in the spring.

Grasslands

While Estonia is a flat country, much of its area is forested or marshy. Approximately 25 percent of the land (926,000 hectares) is considered arable, but with no permanent crops. Permanent pastures (181,000 hectares) comprise 11 percent of land use. Sixty-eight sq mi (110 sq km) of land is irrigated for crop production (1996 estimate).

Forests

Forty-four percent of Estonia consists of forests and woodlands (about 1.8 million hectares). Meadows cover about 252,000 hectares. Trees are chiefly pine, birch, aspen, and fir. Wildlife includes elk, deer, and wild boar. Beaver, red deer, and willow grouse have been protected by legislation because of their dwindling numbers.

HUMAN POPULATION

Estonia is the smallest of any republic of the former U.S.S.R. with a population of 1,423,316 (July 2001 estimate). This equates to 83 persons per sq mi (32 per sq km). The northern part is the most densely inhabited. A highly urbanized country, 73 percent of the residents live in cities or towns. Almost a third live in the capital, Tallinn, on the north coast. Among other important cities are Tartu, an industrial and cultural center; and Pärnu, a major seaside resort.

NATURAL RESOURCES

Estonia's natural resources include timber, oil shale, peat (a carbon-rich material used as fuel and mulch), phosphorite, amber, cambrian blue clay, limestone, and dolomite. Manufacturing and commerce are the most important part of the economy, however.

FURTHER READINGS

Fitzmaurice, John. *The Baltics: A Regional Future?* New York: St. Martin's Press, 1992.

Hiden, John, and Patrick Salmon. *The Baltic Nations and Europe: Estonia, Latvia, and Lithuania in the Twentieth Century.* New York: Longman, 1991.

Lieven, Anatol. *The Baltic Revolution: Estonia, Latvia, Lithuania, and the Path to Independence.* New Haven: Yale University Press, 1993.

Pettai, Vello A. "Estonia: Old Maps and New Roads." *Journal of Democracy,* 4, No. 1, January 1993, pp. 117-25.

Raun, Toivo V. *Estonia and the Estonians. 2d ed.* Stanford, California: Hoover Institution Press, 1991.

Ethiopia

- **Area:** 435,186 sq mi (1,127,127 sq km) / World Rank: 28
- **Location:** Northern and Eastern Hemispheres; eastern Africa, west of Somalia, north of Kenya, east of Sudan, south of Eritrea and Djibouti.
- **Coordinates:** 8°00′ N; 38°00′ E
- **Borders:** 3,300 mi (5,311 km) / Djibouti, 209 mi (337 km); Eritrea, 567 mi (912 km); Kenya 516 mi (830 km); Somalia, 1,010 mi (1,626 km); Sudan, 998 mi (1,606 km)
- **Coastline:** None
- **Territorial Seas:** None
- **Highest Point:** Ras Deshen, 15,157 ft (4,620 m)
- **Lowest Point:** Danakil Depression, 410 ft (125 m) below sea level
- **Longest Distances:** 1,018 mi (1,639 km) E-W; 980 mi (1,577 km) N-S
- **Longest River:** Shabeelle, 1,250 mi (2,011 km)
- **Largest Lake:** Laka Turkana, 2,473 sq mi (6,405 sq km)
- **Natural Hazards:** Earthquakes, volcanoes, drought

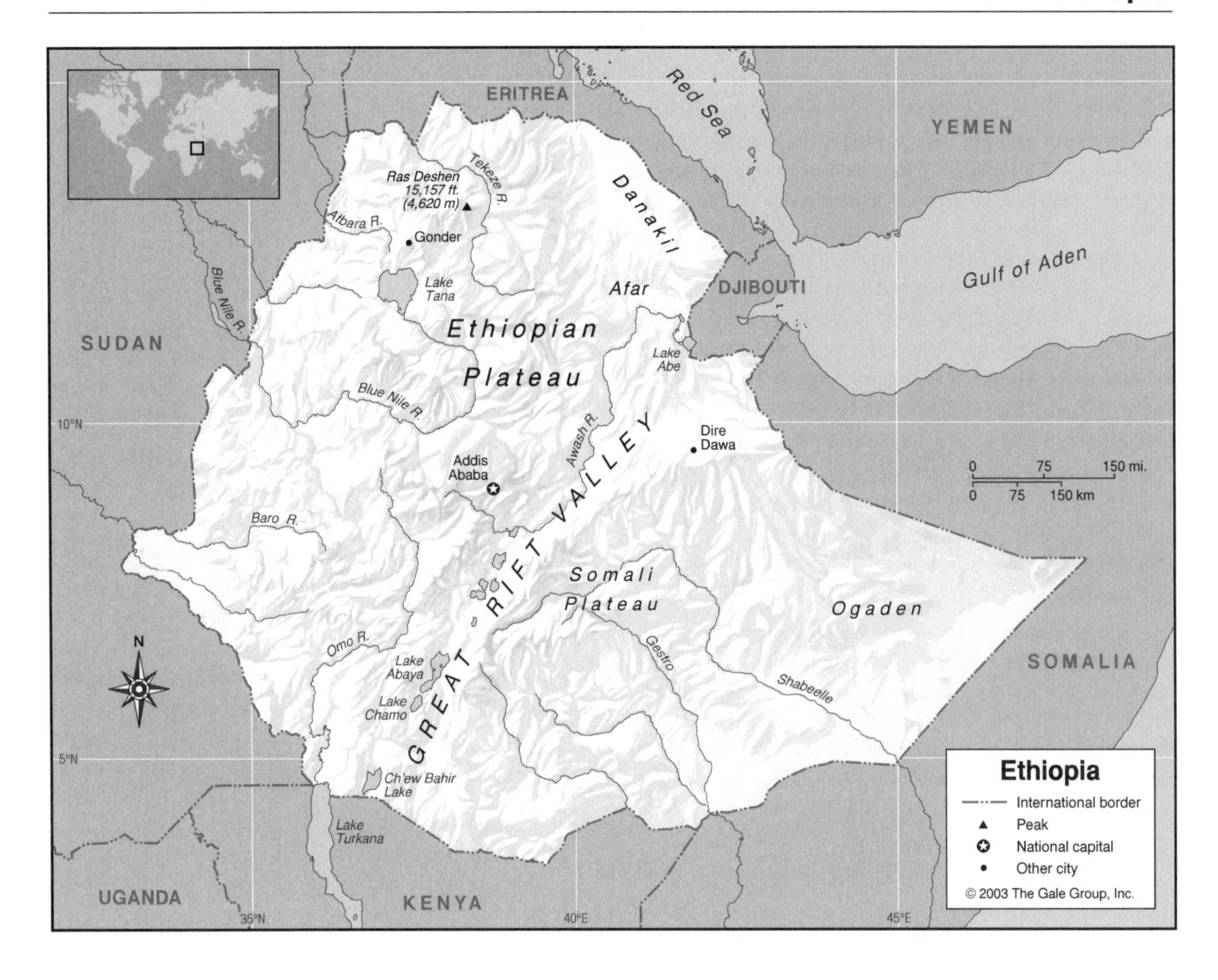

- **Population:** 65,891,874 (July 2001 est.) / World Rank: 18
- **Capital City:** Addis Ababa, centrally located on the western rim of the Great Rift Valley
- **Largest City:** Addis Ababa, 3,112,00 (2000 est.)

OVERVIEW

Ethiopia is slightly less than twice the size of Texas, and occupies the bulk of the Horn of Africa, the easternmost extension of that continent. It has some of the most spectacular scenery in Africa, highlighted by a massive highland complex of mountains and plateaus divided by the deep Great Rift Valley and a series of lowlands along the periphery of the higher elevations. The wide diversity of terrain is fundamental to regional variations in climate, natural vegetation, soil composition, and settlement patterns.

Most of Ethiopia is seismically active. It is located on the African Tectonic Plate, with the Arabian Tectonic Plate somewhat further north, beyond Eritrea. The Great Rift Valley extends across the country from the southwest to the northeast.

The Ogaden border region in the southeast, subject to attacks since 1961, is claimed by Somalia and has never been demarcated. Ethiopia's border with Eritrea, which achieved independence from Ethiopia in 1993, is also disputed.

MOUNTAINS AND HILLS

Mountains and Plateaus

The highland that comprises much of the country consists of two regions: the Ethiopian Plateau in the west—bisected by the Great Rift Valley—and, merging with it in the east, the Somali Plateau. The Ethiopian Plateau is higher and is rugged and mountainous, while the Somali Plateau is a sparsely populated, flat, arid, and rocky semi-desert.

Northward from Addis Ababa the Ethiopian Plateau inclines slightly toward the west and northwest, then abruptly descends near the boundary with Sudan. Given the rugged nature of these massifs and the surrounding tableland, its name is somewhat misleading. Little of the Ethiopian Plateau is flat, except for a scattering of level-topped mountains known to Ethiopians as ambas. The highest point is a volcanic cone in the northeast, Ras Deshen (Mount Rasdajan). Ras Deshen is Africa's fourth-highest mountain. In contrast to the steep scarps of the plateau along the Great Rift Valley and in the north, the western and southwestern slopes of the Northwestern Highlands descend somewhat less abruptly and are broken more often by the river exits. Between the high plateau and the Sudanese border in the west lies a narrow strip of sparsely populated, tropical lowland that belongs politically to Ethiopia but whose people are related to those in Sudan.

South of Addis Ababa the plateau is also rugged, but its elevation is slightly lower than in its northwestern section. The eastern segment beyond the Great Rift Valley exhibits characteristics almost identical to its western counterpart.

The existence of small volcanoes, hot springs, and many deep gorges indicates that large segments of the land mass are still geologically unstable. A number of volcanoes occur in the Danakil area, and hot springs and steaming fissures are found in other northern areas of the Great Rift Valley. A line of seismic belts extends along the length of Eritrean border and the Danakil Depression; however, serious earthquakes have not been recorded in the area during the twentieth century.

The Great Rift Valley

Some geographers, especially Ethiopians, consider the Great Rift Valley a distinct region. This most extensive fault on the earth's surface extends from the Middle East's Jordan Valley to the Shire tributary of the Zambezi River in Mozambique. The vast segment that runs through the center of Ethiopia is marked in the north by the Danakil Depression, a large triangular-shaped basin that in places it 410 ft (125 m) below sea level and said to be one of the hottest places on earth. To the south, at approximately 9° north latitude, the rift becomes a deep trench slicing through the high plateau from north to south, its varying width averaging 30 mi (48 km). The Awash River courses through the northern section of the trench.

Canyons

While the Great Rift Valley is by far the most impressive of Ethiopia canyons, millennia of erosion have produced steep-sided valleys throughout the country, in places 1 mi (1.6 km) deep and several kilometers wide. In these valleys flow rapid waters unsuitable for navigation, but adequate as potential sources of hydroelectric power and irrigation. The Blue Nile winds in a great arc starting at Lake Tana and courses in an arc through canyons more than 4,000 ft (1,200 m) in depth before making its exit into Sudan to merge with the White Nile.

INLAND WATERWAYS

Lakes

The southern half of the Ethiopian segment of the Great Rift Valley is dotted by a chain of large lakes. Some are freshwater, fed by small streams from the east; others contain various salts and minerals. Lake Turkana (Rudolf), fed by the Omo River, is the largest. Most of Lake Turkana is in Kenya, with only the northermost portion extending into Ethiopia. Other lakes in the southern Rift Valley are Ch'ew Bahir, Chamo, and Abaya. Lake Abe is located in the northern part of the Rift Valley on the border with Djibouti. It is fed by the Awash River.

Lake Tana is located in the northwest, on the Ethiopian Plateau. It is the largest lake located entirely within Ethiopia (approx. 1,110 sq mi / 2,849 sq km), and is the source of the Blue Nile.

Rivers

All of the country's rivers originate in the highlands and flow outward in many directions through deep gorges. Most of the northern and western rivers are a part of the vast Nile River system. Most notable of these is the Blue Nile (Abay), which flows out of Lake Tana towards the center of the country before curving northwest into Sudan. It meets the White Nile in the center of that country, forming the Nile River. The Atbara feeds into the Nile further north in Sudan. This last tributary river to the Nile has its source in Ethiopia, as does its own tributary, the Tekeze. Together, the Blue Nile and the Atbara provide about 70 percent of the water in the Nile River. The Baro River in southwestern Ethiopia is another Nile tributary. Altogether these four Nile tributaries account for about half of the outflow of water from the country.

In the northern half of the Great Rift Valley the Awash River flows between steep cliffs. Originating some 50 mi (80 km) west of Addis Ababa, it courses northward and descends several thousand feet to the valley floor. It is joined by several tributaries until it becomes a river of major importance, only to disappear into the saline lakes of the Danakil Depression, most notably Lake Abe. The Omo River rises near the source of the Awash, but flows south into Lake Turkana at the other end of the Ethiopia's portion of the Great Rift Valley.

In the southeast portions of the Somali Plateau, seasonally torrential rivers provide drainage toward the southeast. Chief of these is the Shabeelle, which has its source in several smaller rivers in the south and flows into Somalia. While it does not carry as much water as the Blue Nile, the Shabeelle is longest river to flow through Ethiopia. It is a tributary of the Gestro (Jubba), which also has its source in Ethiopia and flows into Somalia. The Gestro generally flows year round into the Indian Ocean, thanks in part to its northern tributary, the Dawa. In contrast the Shabeelle can dry up in the deserts of Somalia before ever reaching the Gestro.

Wetlands

Sections of marshy lowlands exist along the Sudanese border in the west and southwest.

THE COAST, ISLANDS, AND THE OCEAN

When Eritrea became independent in 1993, Ethiopia lost its entire coastline on the Red Sea. It has since been landlocked.

Population Centers – Ethiopia

(1994 POPULATION ESTIMATES)

Name	Population	Name	Population
Addis Ababa (capital)	3,112,000*	Nazret	147,088
Dire Dawa	194,587	Harar	122,932
Gonder	166,593	Mekele	119,779

* 2000 estimate

SOURCE: United Nations Statistics Division.

CLIMATE AND VEGETATION

Temperature

Ethiopia has three zones: cool, temperate, and hot, known as the dega, the weina dega, and the kolla. In the highlands above 7,800 ft (2,400 m) in elevation, daily highs range from near freezing to 61° F (16° C), with March, April, and May the warmest months. Nights are usually cold, and it is not uncommon to greet the day with light frost. Snow is found at the highest elevations. Daily highs at lower elevations from 4,875 ft to 7,800 ft (1,500 m to 2,400 m) range from 61° F (16° C) to 86° F (30° C). Below 4,875 ft (1,500 m) is the hot zone, with daytime temperatures averaging 81° F (27° C), but soaring to 104° F (40° C) in the Ogaden in midyear.

Rainfall

Ethiopia is affected by the seasonal monsoon trade winds from the Atlantic Ocean that cross the African continent. The country receives most of its rain from mid-June to mid-September with the high plateau experiencing a second, and light, rainy season from December to February. Converging winds in April and May bring lighter rains known as the balg. Annual precipitation is heaviest in the southwest, reaching up to 80 in (200 cm). Up to 48 in (122 cm) of rain falls annually in the highlands. The Ogaden in the east receives as little as four inches (10 cm), and precipitation in the Great Rift Valley and Danakil Depression is negligible.

Grasslands

High mountain elevations above the tree line colonized by grasslands are under intensive cultivation. Even steep slopes and marginal areas are cultivated. The Borena and Ogaden plains in the south are characterized by grassy range lands, and are highly vulnerable to drought and erosion, especially from overgrazing.

Deserts

The Somali Plateau in the east is semi-arid. The northern end of the Great Rift Valley in Ethiopia is desert-like, especially the Danakil Depression.

Forests and Jungles

Highlands in remote areas above 5,850 ft (1,800 m) are covered with a varied assortment of evergreens and conifers, especially zigba and tid. However, owing to population pressures, forests have retreated into the most inaccessible areas. Broadleaf semi-tropical forests at lower and wetter elevations in the southwest have also been subjected to extensive cutting and commercial exploitation.

HUMAN POPULATION

More than 70 percent of the population is concentrated on the Ethiopian Plateau. Overall, the population is growing by 2.7 percent annually (2001 est.), with life expectancy at 44.68 years. Ethiopia has one of the highest populations of any country in Africa, and its resources are often strained as a result.

NATURAL RESOURCES

By the mid-1990s, only one quarter of the country had been geologically mapped, rendering estimates of Ethiopia's natural resources at best inexact. The west and southwest regions are thought to be rich in alluvial gold and platinum deposits. Potash has been found in the Dallol Depression, but exploitation will require joint efforts between Ethiopia and Eritrea. The Afar plain has geothermal potential and fossil fuel deposits have been identified in the southeast. In addition to existing power plants on the Awash River, rivers within the Blue Nile river basin offer the country significant hydroelectric power and irrigation potential.

FURTHER READINGS

Africa South of the Sahara, 2002. "Ethiopia." London: Europa Publishers, 2001.

CyberEthiopia.com. http://www.cyberethiopia.com (accessed June 20, 2002).

Ethiopia: The Politics of Famine. New York: Freedom House, 1990.

Ofcansky, Thomas P., and LaVerle Berry, eds. *Ethiopia: A Country Study.* Fourth edition. Area Handbook Series. Washington, D.C.: Federal Research Division, Library of Congress.

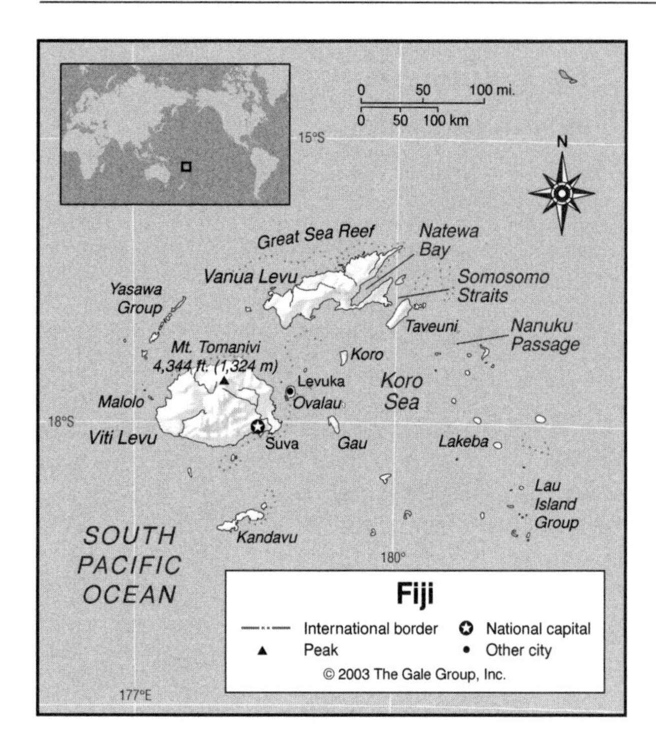

> **GEO-FACT**
>
> The tagimaucia, a beautiful red-and-white flowering plant that resembles the hibiscus, blooms only on the banks of the Tagimaucia River in the mountains of Taveuni island.

Fiji

- **Area:** 7,054 sq mi (18,270 sq km) / World Rank: 154
- **Location:** Southern, Eastern, and Western Hemispheres, in the South Pacific Ocean, north of New Zealand
- **Coordinates:** 18°00′S, 175°00′E
- **Borders:** None
- **Coastline:** 702 mi (1,129 km)
- **Territorial Seas:** 12 NM
- **Highest Point:** Tomaniivi, 4,344 ft (1,324 m)
- **Lowest Point:** Sea level
- **Longest Distances:** 370 mi (595 km) SE-NW / 282 mi (454 km) NE-SW
- **Longest River:** Rewa, 95 mi (150 km)
- **Natural Hazards:** Cyclones
- **Population:** 844,330 (July 2001 est.) / World Rank: 153
- **Capital City:** Suva, on the island of Viti Levu
- **Largest City:** Suva, 166,000 (2000 est.)

OVERVIEW

Fiji is an island nation in the South Pacific, about 1,700 mi (2,735 km) northeast of Sydney, 1,100 mi (1,769 km) north of Auckland, and 2,776 mi (4,466 km) southwest of Honolulu. It consists of around 300 islands—about one-third of which are inhabited—and some 500 islets. Fiji is roughly one-third of the way from New Zealand to Hawaii. The 180° meridian of longitude passes through the Fijian archipelago; however, the International Date Line is located farther east in this region of the South Pacific so that it will avoid passing through populated areas. The part of the Pacific surrounded by the archipelago is called the Koro Sea.

Fiji lies on the Australian Tectonic Plate. Most of its islands were volcanically created, although there are no currently active volcanoes. The volcanic peaks of many of the smaller islands are capped with coral limestone, especially in the eastern portion of the archipelago. The larger islands are generally mountainous, with flatter land along their river deltas and fertile coastal plains.

MOUNTAINS AND HILLS

Fiji's largest island, Viti Levu, has a central mountain range dividing it down the middle with some peaks rising higher than 3,000 ft (914 m), including Fiji's highest mountain, Mt. Tomaniivi (4,344 ft / 1,324 m). The mountain system includes the picturesque Nausori Highlands. The next-largest island, Vanua Levu, also has a central range, which spans its length, with peaks of roughly equal height. Fiji's other large islands are also mountainous, with slopes that often rise dramatically near the shoreline. Some of the higher mountain peaks on Fiji's large islands give way to plateaus before descending to the lowlands near the coast.

INLAND WATERWAYS

Rivers

The Rewa is the major river on Viti Levu, the largest island, and is navigable for 70 mi (113 km). The island also has other river systems, including those of the Nadi, Ba, and Sigatoka. All of these rivers rise in the island's central mountains. The main river on Vanua Levu is the Ndreketí.

Wetlands

Mangrove swamps are found on the eastern coastlines of many of Fiji's islands.

THE COAST, ISLANDS, AND THE OCEAN

Oceans and Seas

Fiji is located in the South Pacific Ocean and surrounds the Koro Sea. In addition to the coral reefs that fringe the islands, there are also circular or U-shaped coral atolls and coral barrier reefs that encircle large coastal lagoons. Some of Fiji's best-known reefs are the Great Sea Reef, the Rainbow Reef, and the Great Astrolabe Reef. The reefs, rocks, and shoals in the waters off Fiji make navigation on the Koro Sea dangerous.

The Somosomo Straits separate the islands of Vanua Levu and Taveuni, which is in turn separated from the Lau island group by the Nanuku Passage. Several other passages separate the various islands and island groups.

Major Islands

By far the two largest islands in the archipelago are Viti Levu (4,010 sq mi / 10,386 sq km) near its western end and Vanua Levu (2,137 sq mi / 5,535 sq km), which reaches almost to the northernmost point in the archipelago. Viti Levu accounts for more than half the country's total land area. It is also the third-largest island in the Pacific. The next-largest islands in the group—Kadavu, Taveuni, Gau, Koro, and Ovalau—have areas between 40 and 200 sq mi (100 and 500 sq km).

The islands in the central part of the archipelago, east of Viti Levu, make up the area called Lomaiviti, or Central Fiji. There are seven larger islands and several smaller ones. The closest is Ovalau, home to Levuka, Fiji's one-time capital.

The easternmost islands, fifty-seven in number, are collectively known as the Lau Group. With a mere land area of 62 sq mi (160 sq km), they stretch over an expanse of ocean covering 43,232 sq mi (112,000 sq km). The island of Lakeba is centrally situated in this group.

At the northwest end of Fiji lies a string of islands called the Yasawa Group, including the island of Malolo. The Polynesian island of Rotuma, located 440 mi (708 km) north of Suva, also belongs to Fiji although it is not a part of the Fijian archipelago.

The Coast and Beaches

Fiji is known for its sandy beaches, which support a thriving tourist industry, as well as the beauty of the coral reefs that fringe its islands. The coastline of Vanua Levu is much more deeply indented than that of Viti Levu and includes the long, narrow Natewa Bay.

Population Centers – Fiji

(1996 POPULATION ESTIMATES)

Name	Population
Suva (capital)	77,366
Lautoka	36,083
Nadi	9,170
Labasa	6,491
Ba	6,314

SOURCE: Fiji Island Statistics Bureau.

CLIMATE AND VEGETATION

Temperature

Fiji's tropical climate is modified by easterly trade winds. Temperature variation between seasons is modest. High temperatures in the summer (October to March) reach 85°F (29°C); winter lows generally go down only as far as 68°F (20°C). Cooler temperatures are recorded at higher elevations.

Rainfall

Annual rainfall ranges from an average of 70 in (178 cm) on the dry leeward sides of the islands to 120 in (305 cm) on the windward sides. The leeward sides have a dry season from April to October, while rainfall is distributed throughout the year on the windward sides.

The cyclone season lasts from November to April. Truly disastrous storms are rare, but cyclones have caused numerous deaths and extensive property and crop losses during Fiji's history.

Grasslands

There are grasslands on the flat, dry western halves of Fiji's larger islands.

Forests and Jungles

About half of Fiji is forested, with forests concentrated in the eastern, windward sides of the islands, which receive the bulk of the rainfall. In particular, the southeastern portions of some islands have dense tropical vegetation. Thousands of plant species have been found in the rainforests of Fiji. The national flower is the hibiscus. Coconut palms and casuarinas are among the plants found in the drier coastal areas.

HUMAN POPULATION

Some 90 percent of Fiji's population live on the two largest islands of Viti Levu and Vanua Levu: 70 percent on Viti Levu and nearly 20 percent on Vanua Levi.

Nearly half the population lives in urban areas.

NATURAL RESOURCES

Dense forestland is found on portions of some islands. Gold is Fiji's most important mineral resource. There is a growing fishing industry. The country's mountains and rivers are a potential source of hydroelectric power. Tourism and subsistence agriculture are the most important parts of the economy, however.

FURTHER READINGS

Fiji: A Lonely Planet Travel Survival Kit. Berkeley, Calif.: Lonely Planet Publications, 2000.

Fiji Online. *Ni Sa Bula, Namaste & Welcome To Fiji Online.* http://www.fiji-online.com.fj/ (accessed March 23, 2002).

Lal, Brij V. *Broken Waves: A History of the Fiji Islands in the Twentieth Century.* Honolulu: University of Hawaii Press, 1992.

Siers, James. *Fiji Celebration.* New York: St. Martin's Press, 1985.

Wibberley, Leonard. *Fiji: Islands of the Dawn.* New York: Washburn, 1964.

Wright, Ronald. *On Fiji Islands.* New York: Viking, 1986.

Finland

- **Area:** 117,942 sq mi (305,470 sq km) / World Rank: 65
- **Location:** Northern and Eastern Hemispheres, Northern Europe, bordering Norway to the north, Russia to the east, the Gulf of Finland to the south, Sweden and the Gulf of Bothnia to the west
- **Coordinates:** 64°N, 26°E
- **Borders:** 1,629 mi (2,628 km) total / Norway, 452 mi (729 km); Sweden, 363 mi (586 km); Russia, 814 mi (1,313 km)
- **Coastline:** 698 mi (1,126 km)
- **Territorial Seas:** 12 NM
- **Highest Point:** Mount Haltia 4,343 ft (1,328 m)
- **Lowest Point:** Sea level
- **Longest Distances:** 719 mi (1,160 km) N-S / 335 mi (540 km) E-W
- **Longest River:** Kemi, 343 mi (552 km)
- **Largest Lake:** Saimaa, 1,700 sq mi (4,403 sq km)
- **Natural Hazards:** None of significance
- **Population:** 5,175,783 (July 2001 est.) / World Rank: 106
- **Capital City:** Helsinki, on the southern coast
- **Largest City:** Helsinki, 1,163,000 (2000 est.)

OVERVIEW

Located in northeastern Europe, Finland is one of the world's northernmost countries—roughly one-third of the country lies north of the Arctic Circle. It is a generally low-lying country. The terrain is close to sea level in the southern half of the country, rising in the north and northeast. Finland's outstanding physical feature is the multitude of lakes that were formed when the glaciers retreated at the close of the Ice Age. The same phenomenon created the marshes that gave Finland its native name—*Suomi,* which means "swamp." Other dramatic geographical features are its vast and plentiful forests and the thousands of islands that dot the coastline.

Nearly the entire northern half of Finland, including its most elevated terrain, belongs to the larger region known as Lapland. Lapland stretches across Norway, Sweden, Finland, and Russia from the Norwegian Sea in the west to the White Sea in the east, lying largely within the Arctic Circle. It is one of the coldest zones in Europe, with the timberline passing through it. The easternmost part of Finland is Karelia, part of which was ceded to the Soviet Union at the close of World War II. It is dominated by the Saimaa Canal, one of the most impressive structures in Finland.

In geological terms, Finland is part of the Fenno-Scandian Shield, and it is situated on the Eurasian Tectonic Plate. The relief of Finland has been considerably affected by the continental glacier, which on retreating left the bedrock littered almost everywhere with the deposits of earth and stones known as moraine. The resulting formations can be seen most clearly in the shape of complex features such as the Salpausselkä Ridge and the numerous ridges known as eskers. Another reminder of the Ice Age is the fact that Finland is still emerging from the sea: its area grows by 2.7 sq mi (7 sq km) annually.

In the region bordering the Gulf of Bothnia, the land rises by 3 ft (90 cm) every century; in the Helsinki area it rises by 1 ft (30 cm). The surface of the land has been scoured and gouged in recent geological times by glaciers that have left thin deposits of gravel, sand, and clay. Finland exercises rights over the continental shelf to a depth of 656 ft (200 m) or to the depth of exploitation.

MOUNTAINS AND HILLS

Mountains

Finland's small mountain region is found at the extreme northwest, bordering Sweden and Norway, where peaks rise to an average height of 3,281 ft (1,000 m). The highest point is Mount Haltia (Haltiatunturi).

Hills and Badlands

Most of the densely forested upland in the north and east of Finland consists of landforms with rounded ridgetops averaging between 1,500 and 2,500 ft (457 and 762 m) above sea level, but there is a major interruption around Lake Inari (Lake Enar), which occupies a plain at elevations of 300 to 600 ft (91 to 183 m). More than half of eastern Finland is hilly, with the land gently sloping toward the southwest.

GEO-FACT

The two bodies of water bordering Finland—the Gulf of Finland and the Gulf of Bothnia—can freeze over entirely for months at a time due to cold winter temperatures.

INLAND WATERWAYS

Lakes

A network of interconnected lakes and rivers covers the greater part of southern Finland—altogether, about 10 percent of Finland's area consists of inland water. In relation to its size, Finland has more lakes than any other country—their total number has been estimated at close to 200,000. There are 55,000 lakes that are at least 656 ft (200 m) in breadth and 19 large lakes, including the artificial reservoirs of Lokan and Porttipahdan, that are more than 77 sq mi (200 sq km) in area. The largest, Lake Saimaa, is the fifth-largest lake in Europe. Other large lakes include Inari to the north, Oulujärvi in the central part of the country, and Päijanne and Pielinen in the south.

Most of Finland's lakes are quite shallow, the average depth being only 23 ft (7 m). Only three lakes have depths greater than around 300 ft (91 m)—the greatest is just over 328 ft (100 m). The lakes are dominated by long, sinuous eskers (sand-and-gravel ridges) rising scores of feet above their surfaces and generally covered with lofty pines and flanked by sandy beaches.

Rivers

Drainage patterns are directly related to Finland's surface features. The north is drained by long rivers, such as the Muonio, the Tornio (Torneå), and the Kemi. In the central part of the country the streams become shorter, except for the Oulu. They also are more sluggish and flow across land that must be ditched before it can be used for cultivation. In the lake district in the southeast, rivers are long and narrow and dammed by the great east-to-west double ridge called the Salpausselkä, which runs eastward from Helsinki parallel to the Gulf of Finland coast. The area south of the lake district and westward along the coast is drained mostly by a series of short streams.

Some of the northern rivers, such as the Kemi, empty into the freshwater Bothnian Gulf, but others, including the Paats and the Tenu (Tano), drain into the Arctic, and some have carved dramatic gorges through to Russian Karelia. These torrents are among the most unspoiled in the country. Farther south is a series of parallel rivers that originate at the high point of Suomenselka and flow northwest to the broad coastal plain of the Gulf of Bothnia. Among these are the Oulu, Pyha, and Lapuan Rivers.

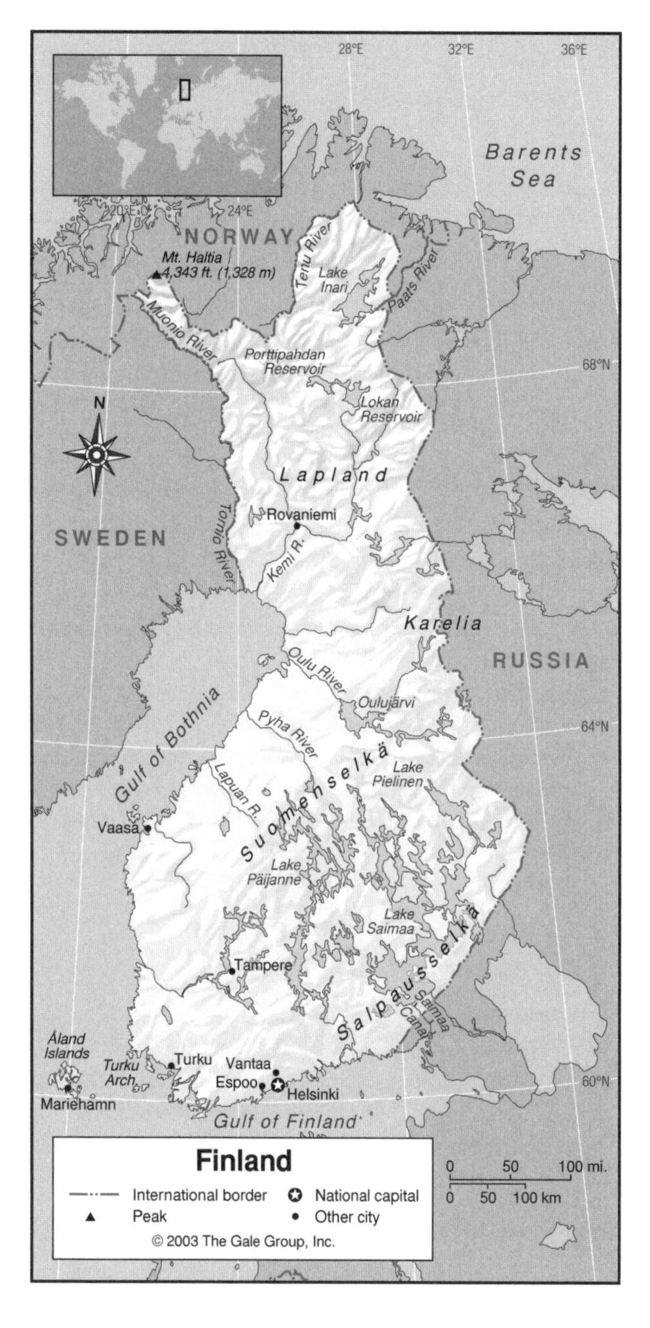

Wetlands

Both above and below the tree line, the north country has extensive swamps, and about a third of the area is covered with bogland. The vast expanses of swamp are the least attractive elements in the northern landscape.

THE COAST, ISLANDS, AND THE OCEAN

Oceans and Seas

Finland is bordered on the west by the Gulf of Bothnia and on the south by the Gulf of Finland. At some points, only a narrow strip of land in Norway separates it from the Barents Sea to the north, and some of its rivers drain northward to that direction.

Major Islands

The Åland Islands in the Gulf of Bothnia off the southwest coast are an autonomous region of Finland. They have an area of 600 sq mi (1,552 sq km) and encompass over 6,500 islands and islets, only about 80 of which are inhabited. They are farther from shore than any of Finland's other islands, although as they extend toward land they merge with Finland's other major island group, the Turku (or Turun) Archipelago, which is closer to shore.

Another major group of islands lies of the western coast near Vaasa.

The Coast and Beaches

Finland's heavily indented coastal zone, which has been called the "golden horseshoe," is dominated by the cities of Helsinki and Turku, the former capital of the country. The entire coast is paralleled by an island zone. It reaches its greatest breadth and complexity in the southwest, with the Turku Archipelago, which encompasses over 15,000 islands and islets reaching all the way to the Åland Islands. The islands in the Gulf of Finland are low-lying, while those in the southwest rise to elevations of 400 ft (122 m) and higher.

CLIMATE AND VEGETATION

Temperature

In spite of its proximity to the Arctic, Finland has a relatively mild climate, thanks to the warming influence of the Gulf Stream. However, temperatures are colder in the north, with winter lows down to -22°F (-30°C) and permanent snowcaps on the northern slopes of its highest peaks. Temperatures for the country as a whole average 7°F to 27°F (-14°C to -3 °C) in winter and about 55°F to 65°F (13°C to 18°C) in summer. Summer temperatures average about 68°F / 20°C in the southern part of the country, with daytime summer rising as high as 86°F (30°C).

The north of Finland is famous for its "midnight sun"—for about 70 days beginning in mid-May, the sun never sets and is visible even at night.

Even the southern part of the country can have as many as 19 hours of sunlight on summer days. Another climate-related phenomenon experienced in the north is *kaamos,* the sunless winter, when it is dark even at the height of day and spectacular displays of northern light are visible in the sky.

Rainfall

Average annual precipitation varies from about 17 in (43 cm) in the north to 28 in (71 cm) the south.

Population Centers – Finland

(2001 POPULATION ESTIMATES)

Name	Population	Name	Population
Helsinki (capital)	559,718	Vantaa	179,856
Espoo	216,836	Turku	173,686
Tampere	197,774	Oulu	123,274

SOURCE: "Largest Municipalities." Finland in Figures, Statistics Finland, Demographic Statistics.

Grasslands and Tundra

Low-lying plains make up much of the coast. South of the Salpausselkä ridge, the plain is narrow, where it borders the Gulf of Finland. It widens in the southwest and west, where it borders the Gulf of Bothnia. Finland's farmland is concentrated in this region. Cloudberries grow in the sphagnum swamps of the northern boglands.

Forests and Jungles

Aside from its lakes, Finland's forests are its best-known natural feature. More than two-thirds of the country is forested, and its forests range from temperate to arctic. Species found at lower elevations include maple, ash, elm, birch, aspen, and elder. At lower latitudes conifers, especially pine and spruce, predominate, and dwarf species of willow and birch are found in the extreme north. Finland has over 1,000 species of flowering plants. The Åland Islands and Turku Archipelago are particularly rich in both flora and fauna.

HUMAN POPULATION

Population density becomes progressively higher from north to south, with the greatest concentration of people, and the largest number of major cities, in the coastal region to the south and west. The southern third of the country is home to its largest cities and towns, which cluster along the coast and are home to nearly two-thirds of the population. The only major town in the north is Rovaniemi, the capital of the Lapp region, where scattered Saami (Lapps) preserve their ancestors' nomadic way of life.

NATURAL RESOURCES

Finland's forests support its timber, wood pulp, and paper industries. Mineral resources include silver, iron ore, copper, and zinc.

FURTHER READINGS:

Engman, Max, and David Kirby, eds. *Finland: People, Nation, State.* Bloomington: Indiana University Press, 1989.

Lange, Hannes. *The Visito's Guide to Finland.* Translated by Andrew Shackleton. Edison, N.J.: Hunter, 1987.

Mead, W. R., and Helmer Smeds. *Winter in Finland: A Study in Human Geography.* New York: Praeger, 1967.

Ministry for Foreign Affairs for Finland. *Virtual Finland.* http://virtual.finland.fi/ (accessed March 16, 2002).

Singleton, Frederick Bernard. *A Short History of Finland.* Cambridge: Cambridge University Press, 1989.

Trotter, William R. *Winter Fire.* New York: Signet, 1994.

Ward, Philip. *Finnish Cities: Travels in Helsinki, Turku, Tampere, and Lapland.* Cambridge: Oleander, 1987.

France

- **Area:** 211,208 sq mi (547,030 sq km) / World Rank: 49
- **Location:** Northern, Western, and Eastern Hemispheres; crossed by the Prime Meridian; in Western Europe bordering Belgium, Luxembourg, and Germany on the northeast; Germany, Switzerland, and Italy on the east; the Mediterranean Sea and Spain on the south; the Atlantic Ocean on the west; the English Channel on the northwest. / Has numerous dependencies around the world including many islands in the Atlantic Ocean, Caribbean Sea, Indian Ocean, and Pacific Ocean, as well as French Guiana in northern South America.
- **Coordinates:** 46°N, 2°E
- **Borders:** 1,795 mi (2,889 km) / Andorra, 35 mi (56.6 km); Belgium, 385 mi (620 km); Germany, 280 mi (451 km); Italy, 303 mi (488 km); Luxembourg, 45 mi (73 km); Monaco, 2.8 mi (4.4 km); Spain, 387 mi (623 km); Switzerland, 356 mi (573 km)
- **Coastline:** 2,130 mi (3,427 km)
- **Territorial Seas:** 12 NM
- **Highest Point:** Mont Blanc, 15,772 ft (4,807 m)
- **Lowest Point:** Rhône River delta, 7 ft (2 m) below sea level
- **Longest Distances:** 598 mi (962 km) N-S / 590 mi (950 km) E-W
- **Longest River:** Rhine, 820 mi (1,320 km)
- **Largest Lake:** Bourget, 11,120 acres (4,500 hectares)
- **Natural Hazards:** Floods and avalanches
- **Population:** 59,551,227 (July 2001 est.) / World Rank: 21
- **Capital City:** Paris, north-central France
- **Largest City:** Paris, 9.6 million (2000 metropolitan est.)

OVERVIEW

France is the largest country in Western Europe and the third-largest in Europe, surpassed only by Russia and Ukraine. Roughly hexagonal in shape, it is bordered by three different bodies of water (the Atlantic Ocean, the English Channel, and the Mediterranean) and three mountain chains (the Pyrenees to the south and the Jura and Alps to the east and southeast). The tiny principality of Monaco, a self-contained enclave, lies entirely within French borders, at the far southeastern tip of the country.

Although France's topography is varied, it can be broken down into three major types of terrain. The plateaus of the four Hercynian Massifs form a V shape that covers much of the center of the country, with the Massif Central at its midpoint. The higher mountain peaks of the Pyrenees, the Jura, and the Alps rise in the south and east, forming natural borders with the neighboring countries of Spain, Switzerland, and Italy. Between these geographical features are the low-lying plains of the Paris Basin and the regions to the west.

France shows a rare combination of national unity and regional variety. The following ten regions have been identified based on geographical and cultural factors: the Nord; the Paris Basin; the East; Burgundy and the Upper Rhine; the Alps; Mediterranean France; Aquitaine and the Pyrenees; the Massif Central; the Loire Valley and Atlantic France; and Armorica. An additional area of France is the large Mediterranean island of Corsica, which lies 100 mi (160 km) to the south.

France also has a number of overseas departments and territories throughout the world, which are remnants of the global empire it forged in the eighteenth and nineteenth centuries. All of these are islands except for French Guiana, on the northeast coast of South America (the last remaining foreign possession on that continent), an underdeveloped territory without significant resources whose interior is still partly unexplored.

Geologically, mainland France's major divisions are between the older, worn-down massif formations, the higher, younger mountain peaks, and the plains. France lies on the Eurasian Tectonic Plate.

MOUNTAINS AND HILLS

Mountains

France has three major mountain systems: the Pyrenees, the Alps, and the Jura mountains.

The Pyrenees extend for over 280 mi (450 km), from the Atlantic Ocean to the Mediterranean Sea and along the southwestern coast of France, forming a barrier that rises above 10,000 ft (3,048 m) and more or less seals off the border with Spain. The peaks do not vary dramati-

cally in height. The central part of the system consists of a series of parallel ridges with few passes, and those that do exist are high and difficult to traverse.

The French Alps in the southeastern part of the country represent only a small part of the whole Alpine chain, but even so they occupy 15,000 sq mi (38,849 sq km) of French territory and include Europe's highest peak, Mont Blanc, as well as its greatest expanses of glacier and permanent snowcaps. The French section, representing the broad outer slope of the great chain at its western extremity, extends as far north as Lake Geneva and as far westward as the Rhône river, forming a natural barrier with Italy and Switzerland in the southeast. They include the Maritime Alps, the Provence Alps, and the Dauphiné Alps. The passes through the Alps are easier to cross than those of the Pyrenees to the south.

The limestone ridges of the Jura Mountains rise to 5,000 ft (1,524 m), forming France's eastern border with Switzerland north of Lake Geneva. Geologically the same age as the French Alps, the Jura cover an area of some 5,000 sq mi (12,950 sq km), with hills in the south and high plateaus in the north, and extend into Switzerland in the northeast. Their highest peak is Mt. Neige, at 5,653 ft (1,723 m).

Plateaus

The four Hercynian Massifs are variously composed of granite, sandstone, or shale. The Ardennes Plateau in the northeast, occupying 500 sq mi (1,554 sq km), is the western tip of the Middle Rhine uplands, which extend into France from Belgium at elevations of less than 1,500 ft (457 m). Open valleys lie between its ridges, traversed by the Meuse and Sambre rivers.

Southeast of the Ardennes, the Vosges Massif rises to rounded granite summits over 4,000 ft (1,219 m) high. To its east are the plains of Baden and Alsace, where the Rhine flows; to the west, the land slopes down to Lorraine. The highest points in the Vosges, called *ballons*, are located near the Alps; the most elevated is the Ballon de Guebwiller, at 4,669 ft (1,423 m). Together the Ardennes and the Vosges enclose the Paris Basin on its eastern side, separating it from the Plain of Alsace.

The Armorican Massif, which protects the Paris Basin in the west, has a much greater expanse. It covers 25,000 sq mi (64,750 sq km), thrusting out into the Atlantic and the English Channel in two rocky promontories, Brittany and the Côtentin Peninsula. But its hills, trending east to west in a series of ridges, seldom exceed 1,200 ft (365 m) in height.

Finally there is the Massif Central, which covers roughly one-sixth of the country's total area and rises to over 5,000 ft (1,524 m) at its highest elevations. This granite plateau separates northern from southern France. The Paris Basin lies to its north, the Rhône-Saône valley to its east, the Languedoc region to its south, and Aquitaine to the southwest. Its geological history as an area of volcanic activity has left rocky formations and craters. A medieval cathedral in the city of Clermont-Ferrand was constructed from black lava rock.

Hills and Badlands

There are hills in many parts of France, but those especially noted for their hilly terrain are the northwest region of Lower Normandy and Brittany, the Champagne region northeast of the Paris Basin, which is one of France's most famous wine-growing areas, and the southern region of Provence.

INLAND WATERWAYS

Lakes

France's inland waterways include a number of both natural and artificial lakes. The largest natural lake, Lake Bourget, lies at the western edge of the Alps, as does Lake Annecy. There are also lakes in the Vosges Massif and in the valleys of the Jura Mountains. Past volcanic activity created the lakes of the Massif Central. Ponds and lagoons lie along the Atlantic coast in the Landes region and the Mediterranean coast in Languedoc. Major reservoirs are found in the Massif Central (Sarrans and Bort-les-Orgues) and in the Alps (Serre-Ponçon).

GEO-FACT

Scientists named the Jurassic Period (145 to 208 million years ago) for France's Jura Mountains, because fossils discovered there date back to this period.

Rivers

The drainage system of France is based on five major rivers. In the north, the Seine, the most gentle, regular, and navigable of French rivers, flows across the Paris Basin for 485 mi (780 km). Before draining into the English Channel at Le Havre, it is joined by its tributaries, the Yonne, the Marne, and the Oise. It has a number of islands, of which the most famous is the Île de la Cité in Paris.

The Loire, whose river basin occupies the central part of France, is the longest river located entirely in France (634 mi / 1,020 km) and covers the largest area (44,400 sq mi / 115,000 sq km). It rises in the Massif Central and flows northwestward to Orléans, joined by its tributary, the Allier, then westward to the Atlantic. The terrain of the Loire's basin is uneven, giving the river an irregular and unpredictable flow. Sudden floods generated in its upper course and in that of the Allier pose a constant threat to the basin lands; to counter it, levees have been constructed along its banks. A series of canals link the Loire to the Seine.

At 357 mi (575 km) in length, the Garonne is the shortest of France's major rivers. It rises in the Pyrenees, across the border with Spain, and empties into the Bay of Biscay at Bordeaux, draining an area of 22,000 sq mi (57,000 sq km). It is reinforced by its tributaries, including the Tarn, the Aveyron, and the Dordogne, which flow from the Massif Central. The Garonne reaches its maximum volume in spring and is capable of flooding catastrophically, with sudden rises in level over 30 ft (9 m).

The Rhône is the largest and most complex of French rivers, with very rapid currents and a volume three times that of the Seine. Rising in Switzerland, it gathers its major tributary, the Saône, at Lyon and flows southward through France for 324 mi (521 km) out of its total length of 505 mi (813 km), emptying into the Mediterranean. Fed principally by Alpine tributaries, it also receives waters from the Saône in winter and the Doubs, flowing from the Jura, in spring, giving it a constant flow throughout the years, although floods sometimes occur.

Lastly, there is the Rhine, which is considered more a European river than a specifically French one. It flows along the eastern border for about 118 mi (190 km), fed by Alpine streams. The Moselle and the Meuse, which drain the Paris Basin, are both tributaries of the Rhine that join it in neighboring countries.

There are some smaller rivers in the northeast; the best known is the Somme, which flows into the English Channel. Many of France's cities and towns are connected by a system of canals, including the Canal du Midi, which links Toulouse in the southwest with the Languedoc coast, the Canal du Nivernais in Burgundy, and the Nantes-Brest Canal in Brittany.

Wetlands

The marshes of the Camargue region, located on the Mediterranean coast between the two mouths of the Rhône River, are formed by alluvial deposits from the Rhône and its tributaries. Covering 304 sq mi (787 sq km), the Camargue consists largely of salt flats, which flood during the winter, support vegetation in the spring, and dry out in summer. The Camargue is known for its unusual fauna, which include the pink flamingo.

THE COAST, ISLANDS, AND THE OCEAN

Oceans and Seas

France has coastlines on the English Channel, the Bay of Biscay (on the North Atlantic Ocean), and the Mediterranean Sea. In addition, a small portion of its northern coast borders the North Sea.

The Coast and Beaches

Indentations are found in all sections of France's coastline. In the north, the Seine River empties into the English Channel in a bay of the same name to the north of Normandy, and the Saint-Malo Gulf lies between the Côtentin Peninsula and Brittany. This coastline is marked by dramatic chalk cliffs that drop abruptly to sandy beaches bordering the English Channel. The coast of Brittany is particularly rugged; its southern shore, which faces the Bay of Biscay, part of the Atlantic Ocean, is called the Côte Sauvage ("wild coast").

Farther south, along the Bay of Biscay, port cities include St.-Nazaire and La Rochelle, and there is a deep coastal indentation north of Bordeaux where the Garonne River and its tributary, the Dordogne, empty into the Bay of Biscay. Fine sand lines the beaches along this coast, and it is characterized by dunes in the southern area known as the Landes.

The western part of France's Mediterranean coast borders the Golfe de Lion (Gulf of Lion). The shoreline on the western edge of this bay consists of sandbars and lagoons. Farther east it has a shoreline of headlands and bays that includes the port city of Marseille and the marshland area called the Camargue. The eastern part of the Mediterranean coast is the Côte d'Azur, the famous resort area known as the French Riviera that lies between the hills of Provence and the sea. The tiny country of Monaco lies at the easternmost end of this region. Pebbles predominate on the beaches of the Mediterranean coastline.

Nearby Islands

France's largest island, and the fourth-largest in the Mediterranean, is Corsica. Separated by over 100 mi (160 km) of sea from the mainland, the mountainous island exhibits structurally the same features as Provence. The island rises to over 5,500 ft (1,676 m) and has a coastal plain only on its eastern side.

France also has a number of islands off its coast in the Atlantic Ocean. The largest of these are Ouessant Island, off the tip of Brittany; Belle-Île-En-Mer to Brittany's south; and Île de Ré and Oléron Island near La Rochelle.

Island Dependencies

The northernmost of France's overseas island dependencies, located in the Atlantic Ocean off the coast of Newfoundland, is the archipelago of St. Pierre and Miquelon, consisting of two islands and a number of rocky islets. (One of the major islands, Miquelon, is actually two land forms separated by an isthmus.) Farther south, in the Lesser Antilles between the Caribbean Ocean and the Atlantic, lie the tropical islands of Guadeloupe and Martinique.

Guadeloupe, the more northerly of the two departments, is a group of islands, while Martinique is a single volcanic island.

Several French islands are located in the Indian Ocean. Mayotte, part of the Comoros archipelago that lies east of Mozambique, is the only island in that group that declined to join the others when they declared independence in 1975. Farther south, to the east of Madagascar, is the volcanic island of Réunion, where a narrow plain encircles a mountainous central massif. Réunion administers two other French dependencies that lie in the Mozambique Channel between Mozambique and Madagascar: the wooded island of Europa and the atoll of Bassas da India.

France also has an overseas territory farther south in the Indian Ocean, collectively called the Southern and Antarctic Lands. The Southern Lands, located southeast of the African continent, comprise two individual volcanic islands (St. Paul Island and Amsterdam Island) roughly halfway between Africa and Australia (heading east) and two archipelagos (the Kerguélin and Crozet islands) farther south, with the latter about halfway between Africa and Antarctica (heading southeast). "Antarctic Lands" refers to a section of Antarctica called "Adelie Land," which has been claimed by France since 1840, although this claim is not universally recognized.

France has three dependencies in the Pacific. New Caledonia, located east of Australia, consists of a long reef-fringed main island, Grande Terre (or New Caledonia Island), an archipelago (the Loyalty Islands), and a number of smaller islands and atolls. Farther east, located northeast of Fiji and west of Samoa, is the overseas territory of Wallis and Futuna, which consists of two island groups located 144 mi (230 km) apart. Wallis comprises a main island and several smaller ones; Futuna consists of two islands. The dependency that spans the widest area is French Polynesia, a group of five archipelagos with a total of some 130 islands, located about halfway between South America and Australia. Its island groups include the Society, Marquesas, Tuamotu, Gambier, and Tubuai (or Austral) islands. The Society Islands is the largest group, accounting for about 40 percent of the dependency's territory and more than four-fifths of its population. The largest island in the group is Tahiti, which is also the largest island in the dependency.

CLIMATE AND VEGETATION

Temperature

France various regions have three major types of climate: oceanic, continental, and Mediterranean. Temperatures generally increase from north to south. The western part of the country, which borders the Atlantic Ocean, has a temperate, humid oceanic climate, characterized by relatively small annual temperature variations, heavy precipitation, and overcast skies, with cool summers and winters. The oceanic climate is purest in the northwest and becomes modified farther south in the Aquitaine basin, which has warmer summers and milder winters. Average temperatures in Brest, at the tip of Brittany, are 43°F (6°C) in January and 61°F (16°C) in July.

Much of eastern and central France has a continental climate, with a wider range of temperatures, especially in the northeast, and greater variations between seasons. Winters are cold and snowy, and storms are frequent in June and July. The continental climate is tempered somewhat in the Paris Basin. Paris has an average annual temperature of 53°F (11°C). The eastern part of the country has the most severe winters.

The Mediterranean climate predominates in the south and southeast, stretching inland from the coast to the lower Rhône valley. Winters are mild and humid, with only short periods of frost, and summers are hot and dry. Temperatures above 90°F (32°C) are common. Annual temperatures in Nice, on the Côte d'Azur, average 59°F (15°C). Southern France occasionally experiences a cold northern wind called the *mistral,* which can reach speeds of 65 mph (105 kph).

Rainfall

Rainfall ranges from as low as 17 in (43 cm) on the Languedoc coast to 50 in (130 cm) at high elevations in the mountains of the Alps, Pyrenees, and Jura, on the Massif Central, and in the northwest. Annual rainfall averages 27 in (68 cm) in Paris and 39 in (100 cm) in Bordeaux.

Grasslands

France's plains are mostly located in the Paris Basin to the north and in a series of lowland regions in the western part of the country. The Paris Basin is the cradle of France, occupying one-third of the nation's territory. It is encircled by France's four major massifs: the Ardennes, the Vosges, the Massif Central, and the Armorican Massif. At the center of the basin lies Paris itself. Southwest of

Population Centers – France

(1999 CENSUS)

Name	Population
Paris (capital)	11,174,743
Lyon (Lyons)	1,648,216
Marseille-Aix-en-Provence	1,516,340
Lille	1,143,125

SOURCE: *Recensement de la Population, Mars 1999.* Institut National d'Etudes Démographiques (INSEE)

Regions – France

1999 CENSUS OF POPULATION

Name	Population	Area (sq mi)	Area (sq km)	Capital
Alsace	1,734,145	3,197	8,280	Strasbourg
Aquitaine	2,908,359	15,949	41,309	Bordeaux
Auvergne	1,308,878	10,044	26,013	Clermont-Ferrand
Basse Normandie	1,422,193	6,791	17,589	Caen
Bretagne	2,906,197	10,505	27,209	Rennes
Bourgogne	1,610,067	12,194	31,582	Dijon
Centre	2,440,329	15,116	39,151	Orléans
Champagne-Ardenne	1,342,363	9,887	25,606	Châlons sur Marne
Corse	260,196	3,351	8,680	Ajaccio
Franche-Comté	1,117,059	6,256	16,202	Besançon
Haute-Normandie	1,780,192	4,756	12,318	Rouen
Île-de-France	10,952,011	4,637	12,011	Paris
Languedoc-Roussillon	2,295,648	10,570	27,376	Montpellier
Limousin	710,939	6,541	16,942	Limoges
Lorraine	2,310,376	9,092	23,547	Metz
Midi-Pyrérées	2,551,687	17,509	45,349	Toulouse
Nord-Pas-de-Calais	2,555,020	4,793	12,413	Lille
Pays de la Loire	3,222,061	12,387	32,082	Nantes
Picardie	1,857,834	7,490	19,399	Amiens
Poitou-Charentes	1,640,068	9,965	25,809	Poitiers
Provence-Côte d'Azur	4,506,151	12,124	31,400	Marseille
Rhône-Alpes	5,645,407	16,872	43,698	Lyon

SOURCE: "Population aux dernier recensements." *Recensement de la Population, Mars 1999.* Institut National d'Etudes Démographiques (INSEE).

the Paris Basin, along the valley of the Loire River, lie the plains of Anjou and, to their south, Poitou. Still farther south are the lowlands of Aquitaine, including the basins of the Garonne and the Adour rivers and the plain of Landes, which borders the Bay of Biscay.

Forests and Jungles

France's vegetation matches its topography in variety. More than one-quarter of France is forested. Oak, beech, chestnut, and pine predominate in the Paris Basin. Larch, Norway maple, and beech are among the trees that grow in the eastern part of the country. Beech grows on the lower slopes of the Alpine regions, with fir, larch, and mountain pine higher up. Chestnut and beech trees are found in the Massif Central, oak, cypress, poplar, and willow in the Aquitaine Basin, and extensive pine forests along the western border. Typical Mediterranean flora thrive in the south of the country, including olive, fig, and mulberry trees, grapevines, laurel bushes, a characteristic type of scrub called *maquis,* and an abundance of wild herbs. Colorful expanses of lavender, sunflowers, and other plants grace the fields of Provence. In springtime a variety of lilies, orchids, and other flowers can be found in the Alpine meadows of the southeast.

HUMAN POPULATION

Much of France's population is concentrated in the north and southeast, while the central part of the country is more sparsely populated. Roughly three-quarters of France's population is concentrated in urban areas. Nearly one-sixth live in Paris and its environs, which constitute the country's major urban center, as well as its political, cultural, and economic capital. Next in size are Marseille, the nation's major port, and Lyon, an important commercial and industrial center. Declining farm incomes in the last part of the twentieth century were accompanied by widespread depopulation of rural areas.

NATURAL RESOURCES

France has a strong, modern, economy, based largely on commerce and industry. Although an increasingly small percent of the population is engaged in agriculture, France's arable land (one-third of the total land) remains an important natural resource; France is one of the world's major exporters of food. France also has the highest percentage of forested land among the countries of the European Union. However, it is not rich in mineral resources. After the oil shortages of the 1970s, an extensive nuclear power program was developed to compensate for the country's small reserves of oil and natural gas. By the mid-1990s the country was able for the first time to generate enough power to meet over half its electricity needs.

FURTHER READINGS

Buchwald, Art. *Don'tt Forget to Write.* Illustrated by Laszlo Matulay. Cleveland: World Publishing, 1960.

Busselle, Michael. *France: The Four Seasons.* London: Pavilion Books, 1994.

Cobb, Richard. *Paris and Elsewhere.* London: John Murray, 1998.

Edmondson, John. *France.* Lincolnwood, Ill.: Passport Books, 1997.

Franceway.com. *Franceway.* http://www.franceway.com/ (accessed Mar. 21, 2002).

Haine, W. Scott. *The History of France.* Westport, Conn.: Greenwood Press, 2000.

Mayle, Peter. *A Year in Provence.* London: Hamish Hamilton, 1989.

Michelin France: Landscape, Architecture, Tradition. Boston: Little, Brown, 1995.

French Guiana

Overseas Department of France

- **Area:** 35,135 sq mi (91,000 sq km) / World Rank: 113
- **Location:** Northern and Western Hemispheres, in northern South America, bordered by Brazil on the east and south, Suriname on the west, and the Atlantic Ocean on the north.
- **Coordinates:** 4°00′ N, 53°00′ W
- **Borders:** 735 mi (1,183 km) total / Brazil, 418 mi (673 km); Suriname, 317 mi (510 km)
- **Coastline:** 235 mi (378 km)
- **Territorial Seas:** 12 NM
- **Highest Point:** Bellevue de l'Inini, 2,791 ft (851 m)
- **Lowest Point:** Sea level
- **Longest Distances:** 250 mi (400 km) N-S / 190 mi (300 km) E-W
- **Longest River:** Maroni River, 420 mi (680 km)
- **Natural Hazards:** Frequent heavy rain, flooding
- **Population:** 177,562 (July 2001 est.) / World Rank: 178
- **Capital City:** Cayenne, on the northern coast
- **Largest City:** Cayenne, 90,000 (1999 est.)

OVERVIEW

Located on the South American Tectonic Plate, French Guiana is an overseas department of France. If it were independent, French Guiana would be the smallest country in South America. French Guiana's landscape consists of a swampy coast backed by a forested plateau and low mountains.

MOUNTAINS AND HILLS

Most of French Guiana's interior is a low-lying plateau, slowly rising to the Tumuc-Humuc Mountains along the southern border with Brazil. The tallest peak in French Guiana, Bellevue de l'Inini (2,791 ft / 851 m), is located here. The mountains and plateau are part of the Guiana Highlands that run throughout northeastern South America.

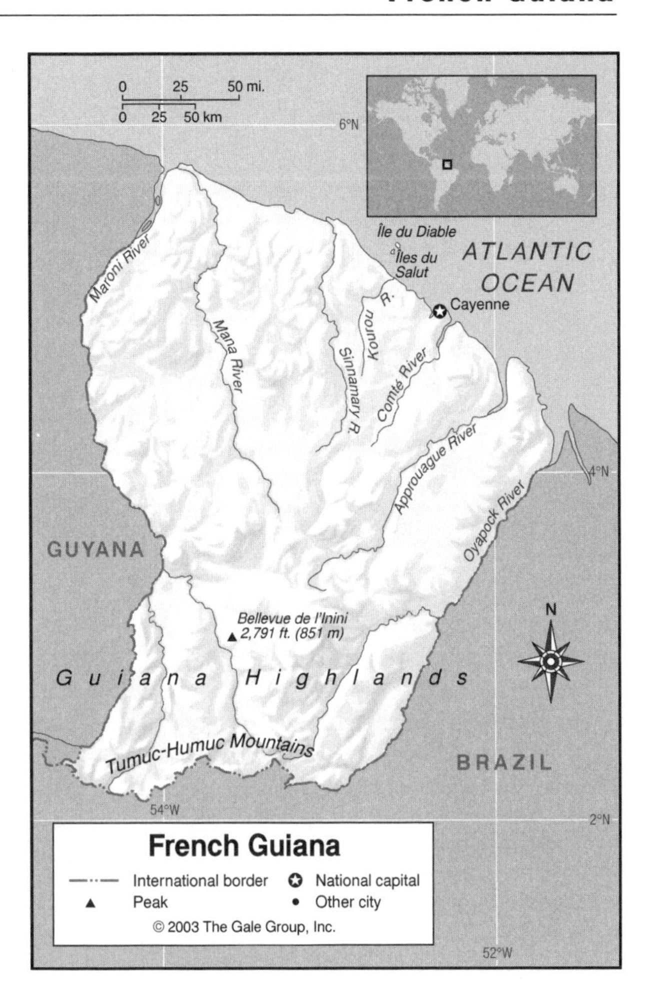

INLAND WATERWAYS

Rivers

There are seven river basins in French Guiana. The Maroni River is the largest; it marks the border with Suriname. The Oyapock flows along the eastern border with Brazil. The Mana, Sinnamary, Kourou, Comté, and Approuague are the others. Each of these rivers has a network of tributaries or basins.

Wetlands

Mangrove trees cover much of the French Guianese coast. These short trees create swampy mudflats with their extensive roots. The swamps extend from 6-19 mi (10-30 km) inland.

THE COAST, ISLANDS, AND THE OCEAN

Oceans and Seas

French Guiana's Atlantic shoreline is dominated by mangrove swamps. There are a few sandy beaches, and these and cleared areas of the swamps are where most of the population lives.

Major Islands

A number of islands are part of French Guiana. The Îles du Salut include Île du Diable (Devil's Island), Royale, and Saint-Joseph. Devil's Island was once the site of infamous French penal colonies (prison camps). Other islands include the Pere and Mere Islands, the Malingre and Remire Islands, and the two Connetables.

CLIMATE AND VEGETATION

Temperature

French Guiana is located near the equator and sees little variation in its temperatures throughout the year. The average temperature along its coast is 80°F (26°C).

Rainfall

From January to June, there is a heavy rainy season (with May receiving the most rain), and rainfall ranges from 140-160 in (350-400 cm). Humidity averages 85 percent.

Forests and Jungles

Tropical rain forests cover 90 percent of the land area in the country's vast interior known as the *terres hautes.* They are sparsely populated and are largely untouched by mankind.

HUMAN POPULATION

A little over half of the country's population lives in Cayenne, the capital city; another forty-five percent lives in the coastal lowlands. The remaining five percent is scattered across the sparsely populated interior.

NATURAL RESOURCES

The country's natural resources include mineral resources such as gold and bauxite, and exports include fish and timber. French Guiana trades mainly with France, which subsidizes its economy. There is a major French space launch facility at Kourou.

FURTHER READINGS

Halle, Francis, and Raphael Gaillarde. "A Raft Atop the Rain Forest." *National Geographic,* October 1990, Vol. 178, No. 4, p. 128.

French Guiana. Parsippany, N.J.: Dun & Bradstreet, 1999.

Lonely Planet. *French Guiana.* http://www.lonelyplanet.com/destinations/south_america/french_guiana/ (accessed March 7, 2002).

Luxner, Larry. "Ariane Program Proves Boon to French Guiana: Launchpad Vies with NASA." *Journal of Commerce and Commercial*, July 26, 1990, Vol. 385, No. 27278, p. 5A.

WorldTravelGuide.Net. *French Guiana.* http://www.wtg-online.com/data/guf/guf.asp (accessed March 7, 2002).

GEO-FACT

Because of its limited population and booming economy, Gabon relies heavily on laborers from other African nations, including Benin, Cameroon, Equatorial Guinea, Mali and Senegal. Foreigners make up at least 20 percent of the population in Gabon.

Gabon

- **Area:** 103,347 sq mi (267,667 sq km) / World Rank: 76
- **Location:** Northern, Southern, and Eastern Hemispheres, on the west coast of Africa on the Equator, bordered on the north by Cameroon, on the east and south by the Republic of the Congo, on the west by the Atlantic Ocean, and on the northwest by Equatorial Guinea.
- **Coordinates:** 1°00′S, 11°45′E
- **Borders:** 1,585 mi (2,551 km) / Cameroon, 185 mi (298 km); Republic of the Congo, 1,182 mi (1,903 km); Equatorial Guinea, 217 mi (350 km)
- **Coastline:** 550 mi (885 km).
- **Territorial Seas:** 12 NM
- **Highest Point:** Mt. Iboundji, 5,167 ft (1,575 m)
- **Lowest Point:** Sea level
- **Longest Distances:** 446 mi (717 km) NNE-SSW; 400 mi (644 km) ESE-WNW.
- **Longest River:** Ogooué River, 690 mi (1,100 km).
- **Natural Hazards:** None
- **Population:** 1,221,175 (July 2001 est.) / World Rank: 149
- **Capital City:** Libreville, located on the northwest coast
- **Largest City:** Libreville, population 419,000 (2000 est.)

OVERVIEW

The low-lying coastal plain of Gabon is narrow in the north—approximately 18 mi (29 km)—and broader in the estuary regions of the Ogooué River. The interior relief is more complex, though nowhere is it dramatic. In the north, mountains enclose the valleys of the Woleu and Ntem Rivers, and the Ivindo Basin. In southern Gabon, the coastal plain is dominated by granitic hills. Between the Ngounié and the Ogooué Rivers the Chaillu Massif rises to 3,000 ft (915 m). Almost the entire country is contained in the basin of the Ogooué River and its two major tributaries.

MOUNTAINS AND HILLS

Mountains

Rivers descending from the interior have carved deep channels in the face of the escarpment, dividing it into distinct blocks and separating the Crystal Mountains from the Chaillu Massif. The Crystal Mountains run roughly north to south across the country, just west of the center. The highest point in Gabon is the peak of Mt. Iboundji, which reaches an altitude of 5,167 ft (1,575 m). It is located in the northern Crystal Mountains.

Plateaus

The plateaus cover the north and east and most of the south of the country. They rise from the coastal lowlands, which range in width from 20 to 125 mi (30 to 200 km), and form a band more than 60 mi (96 km) wide of a rocky escarpment, which ranges in height from 1,480 to 1,970 ft (450 to 600 m).

INLANDS WATERWAYS

Rivers

Virtually the entire territory is contained in the basin of the Ogooué River, which is about 690 mi (1,100 km) long and navigable for about 250 mi (400 km). It flows from the southeastern point of Gabon and winds its way up through the center of the country, turning west and cutting through the Crystal Mountains to reach its mouth on the Atlantic Ocean at Port-Gentil. Its two major tributaries are the Ivindo and the Ngounié, which are navigable for 50 to 100 mi (80 to160 km) into the interior. The Ivindo drains the northeastern part of Gabon, and the Ngounié runs parallel to the Crystal Mountains along their western face. The Ogooué is also fed in the east by the Sébé River. The relatively short Gabon River rises just inside Equatorial Guinea and flows southwest into Gabon, over the Kinguélé Falls, then dumps into the Atlantic at Kango.

Wetlands

South of the Ogooué River mouth, there are coastal areas bordered by lagoons and mangrove swamps.

THE COAST, ISLANDS AND THE OCEAN

Gabon faces the South Atlantic Ocean south of the Bight of Biafra and Gulf of Guinea. The northern coastline is deeply indented with bays, estuaries, and deltas as far south as the mouth of the Ogooué River, featuring Cape Santa Clara in the north, and Cape Lopez, the westernmost point, just north of the Ogooué River mouth. These bays and estuaries form excellent natural shelters providing good ports and harbors. South of the Ogooué River are numerous lagoons, such as Ndogo, M'Goze and Nkomi, along the coastline. Continuing even farther south, the coast becomes more precipitous.

CLIMATE AND VEGETATION

Temperature

Gabon has the moist, hot and humid climate typical of tropical regions. The hottest month is January, with an average high at Libreville of 88°F (31°C) and an average low of 73°F (23°C). Average July temperatures in the capital range between 68° and 82°F (20 and 28°C).

Rainfall

From June to September there is virtually no rain but high humidity; there is occasional rain in December and January. During the remaining months, rainfall is heavy. The excessive rainfall is caused by the condensation of moist air resulting from the meeting, directly off the coast, of the cold Benguela Current from the south and the warm Guinea Current from the north. At Libreville, the average annual rainfall is more than 100 in (254 cm). Farther north on the coast, it is 150 in (381 cm).

Forests

About 85 percent of the country is covered by heavy rain forest. The dense green of the vegetation never changes, since more than 3,000 species of plants flower and lose their leaves at different points throughout the year according to species. Tree growth is especially rapid; in the more sparsely forested areas, the trees tower as high as 200 ft (60 m), and the trunks are thickly entwined with vines. In the coastal regions, marine plants abound, and wide expanses are covered with tall papyrus grass.

The World Wildlife Fund has launched two projects in Gabon to support the management and protection of native wildlife. In the northeast Minkebe region is an 11,000 sq mi (30,000 sq km) native forest, where most of the native wildlife has lived, undisturbed. In the southwest, the Gamba Protected Area Complex provides lowland rain forest habitat for chimpanzees.

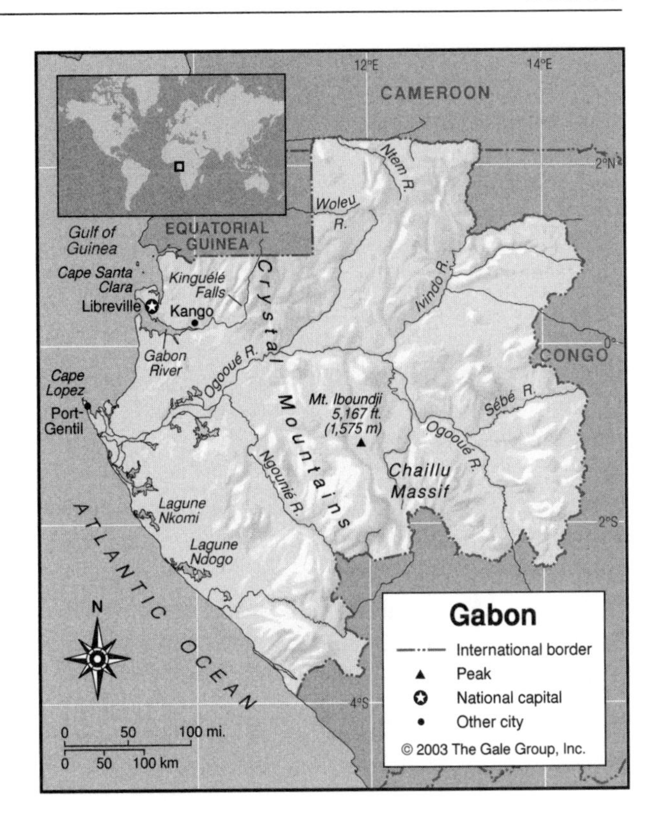

HUMAN POPULATION

Most of the people live on the coast or are concentrated along rivers and roads; large areas of the interior are sparsely inhabited. It was estimated that 55 percent of the population lived in urban areas in 2000. The capital city, Libreville, had a 2000 population of 419,000, or roughly one-third the entire Gabon population. Another major center is Port-Gentil, with about 164,000 inhabitants.

NATURAL RESOURCES

Gabon is one of the richest countries in Africa. Its forests and offshore petroleum reserves are its chief sources of wealth. Forests have provided ebony and mahogany, and lumber is the country's second leading export behind oil. While minerals such as iron, manganese, and uranium were once more important in Gabon's export trade, the oil sector now accounts for 75 percent of exports. Gabon is member of the Organization of Petroleum Exporting Countries (OPEC), but continues to face fluctuating prices for its oil, as well as timber, manganese, and uranium exports. On the fertile land of the northwestern region, cocoa and coffee are the most prominent products for export.

FURTHER READINGS

Gardinier, David. *Historical Dictionary of Gabon. 2nd edition.* Scarecrow Press Inc, 1994.

Gray, Christopher. *Colonial Rule and Crisis in Equatorial Africa: Southern Gabon*. Rochester: University of Rochester Press, 2002.

Gunnar Anzinger. *Government on the WWW*: Gabon. http://www.gksoft.com/govt/en/ga.html (Accessed May 2002).

Yates, Douglas Andrew. *The Rentier State in Africa: Oil Rent Dependency Neocolonialism in the Republic of Gabon*. Ternton, N.J.: Africa World Press, 1996.

Gambia

- **Area:** 4,363 sq mi (11,300 sq km) / World Rank: 161
- **Location:** Northern and Western Hemispheres, in western Africa, bordering the North Atlantic Ocean and surrounded by Senegal.
- **Coordinates:** 13°28′N, 16°34′W
- **Borders:** Senegal, 460 mi (740 km)
- **Coastline:** 50 mi (80 km)
- **Territorial Seas:** 12 NM
- **Highest Point:** Unnamed location, 173 ft (53 m)
- **Lowest Point:** Sea level
- **Longest Distances:** 210 mi (338 km) E-W; 29 mi (47 km) N-S
- **Longest River:** Gambia River, 700 mi (1,126 km)
- **Largest Lake:** None of significant size
- **Natural Hazards:** Drought
- **Population:** 1,411,205 (July 2001 est.) / World Rank: 147
- **Capital City:** Banjul, located on a peninsula facing the Atlantic coast on the south-westerly bank of the mouth of the Gambia River
- **Largest City:** Serekunda, near Banjul, population 207,000 (2002 est.)

OVERVIEW

A product of colonial trading history, Anglophone (English-speaking) Gambia is an enclave of Francophone (French-speaking) Senegal. Its serpentine borders surround the Gambia River and border the Atlantic Ocean. Between 1982 and 1989, Gambia established a short-lived political federation with Senegal called the Confederation of Senegambia. In area, the Gambia is roughly twice the size of the state of Delaware, making it the African continent's smallest country.

Most of Gambia is low-lying but it is generally divided into three regions on the basis of topographical features: the valley floor built up of alluvium with areas known as Bango Faros; a dissected plateau edge, consisting of sandy and often precipitous hills alternating with broad valleys; and a sandstone plateau which extends, in parts, across the border into Senegal.

Gambia is situated on the African Tectonic Plate, not near any major faults or plate boundaries, so seismic activity is rare. The Gambia River, the country's major waterway, is its most distinguishing geographic feature.

MOUNTAINS AND HILLS

Gambia is primarily the area surrounding the Gambia River. Thus, it occupies a fairly flat fluvial plateau dissected by streams, a few steep hills of insignificant height, and broad valleys.

INLAND WATERWAYS

The Gambia River rises in Guinea and follows a twisting path for about 1,000 mi (1,600 km) to the sea. In its last 292 mi (470 km), the river flows through Gambia, narrowing to a width of 3 mi (5 km) at Banjul. During the dry season, tidal saltwater intrudes as far as 155 mi (250 km) upstream. The Gambia River is one of the finest waterways in West Africa and is navigable as far as Kuntaur, 150 mi (240 km) upstream, by seagoing vessels and as far as Koina by vessels of shallow draft. Brown mangrove swamps line the lower reaches on both sides of the river for the first 90 mi (145 km) from the sea. Behind these mangroves are the "flats", which are submerged completely during the wet season. The land then gives way to more open country and in places to red ironstone cliffs. The land on either side away from the river is generally open savanna with wooded areas along the drainage channels. In addition to the lower reaches of the Gambia River, the coast is marked with capes and lagoons. Banjul is located on a lowland peninsula separated from the mainland by swamps.

THE COAST, ISLANDS, AND THE OCEAN

On its west side, Gambia borders the Atlantic Ocean. Most of its border is composed of the peninsula on which the cities of Banjul and Serekunda sit. Sandy white beaches cover most of Gambia's 44 mi (71 km) coast (also called the "smile coast"). A strip of beach hotels and resorts is located to the west of Banjul near Cape St. Mary. Sand dunes line the coast at Gambia's southern border with Senegal. Though it has no islands in the ocean, several islands are found in the Gambia River. These include the historic James Island, which the French had named St. Andrews, and historic McCarthy Island where Georgetown is located.

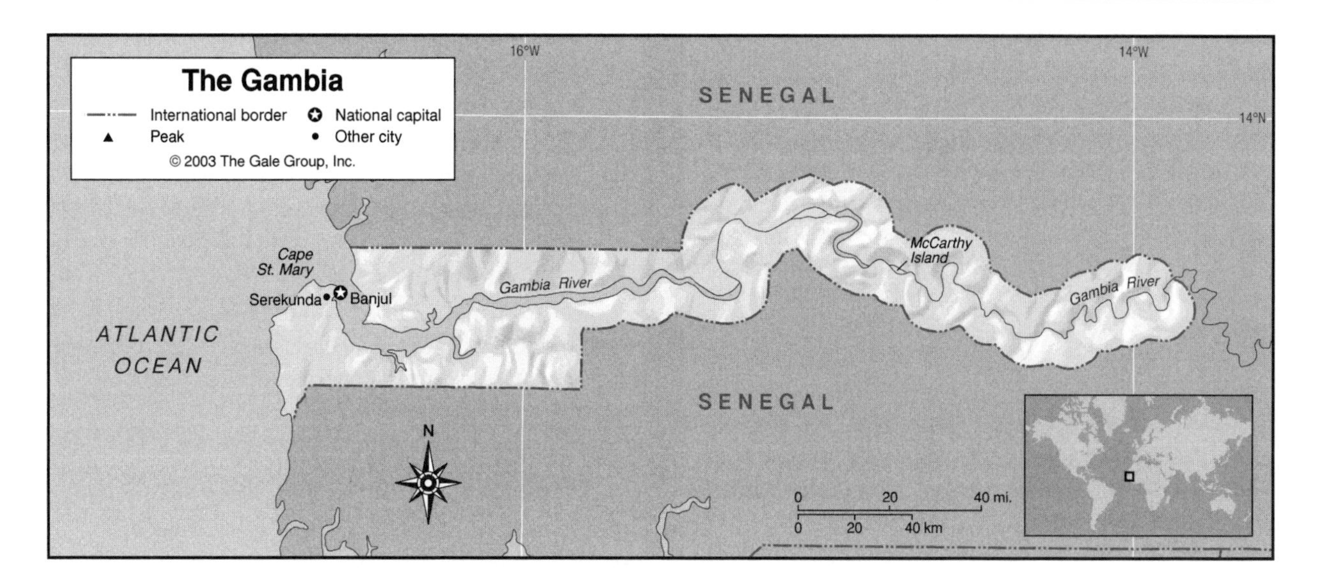

CLIMATE AND VEGETATION

Temperature

Gambia has a subtropical climate with distinct cool and hot seasons. From November to mid-May there is uninterrupted dry weather, with temperatures as low as 61°F (16°C) in Banjul and surrounding areas. Hot, humid weather predominates the rest of the year, with a rainy season from June to October. During this period, temperatures may rise as high as 109°F (43°C) but are usually lower near the sea. Mean temperatures range from 73°F (23°C) in January to 81°F (27°C) in June along the coast, and from 75°F (24°C) in January to 90°F (32°C) in May inland.

Rainfall

The average annual rainfall ranges from 36 in (92 cm) in the interior of the country to 57 in (145 cm) along the coast. Rainfall has decreased by 30 percent in the last thirty years, leading to problems with agricultural yield. Most of the rain falls from June to October.

Grasslands

Slash-and-burn agriculture and the use of trees for charcoal has resulted in grassy expanses mixed with scrub and fire-resistant trees in the open savanna away from the river.

Forests and Jungles

Along the river Gambia is mainly sub-tropical savanna with tropical forest, mangroves, and bamboo found along the lower Gambia River. Mahogany, rosewood, oil palm, and rubber are found along the river banks giving Gambia a park-like appearance. Away from the river savanna takes over. Due to agriculture and wood fuel, only 9 percent of the forests have survived deforestation. The decrease in rainfall over the last 30 years has encouraged desertification.

HUMAN POPULATION

Gambia has a population density of more than 344 people per sq mi (133 people per sq km), mostly due to its small size. The majority of the population lives in the peninsular region near the Atlantic coast with the interior on the country sparsely populated. Almost 15 percent of the population lives in the Serekunda metropolitan area, but this is the only city with more than 60,000 inhabitants. Most of the country has not seen urbanization and still lives dependent on agriculture and fishing.

NATURAL RESOURCES

Gambia is poor in mineral resources, with no real deposits of any kind. Its greatest natural resources are its fish and unspoiled natural areas including its beaches and fertile land. Sixteen percent of the land is arable; 9 percent is meadows and pastures; 20 percent is forest and woodland; and 55 percent is used for other purposes. The leading exports are fish and peanuts and their related products.

Population Centers – The Gambia

(1993 CENSUS)

Name	Population
Serekunda	151,450
Banjul (capital)	42,407
Brikama	42,480
Bakau	38,062
Farafenni	21,142

SOURCE: Census of Population, The Gambia, April 15, 1993. As cited on Thomas Brinkhoff: City Population. http://www.citypopulation.de (accessed July 2002).

FURTHER READINGS

Africa South of the Sahara 2002. "The Gambia." London: Europa Publications Ltd, 2002.

AllAfrica Global Media. *Equatorial Guinea and the Bakassi Dispute.* http://allafrica.com/stories/200203250065.html (Accessed March 30, 2002).

Hughes, Arnold, and Harry A. Gailey. "Historical Dictionary of The Gambia." *African Historical Dictionaries, No. 79.* Lanham, Md., and London: The Scarecrow Press, Inc, 1999.

Park, Mungo. *Travels in the Interior Districts of Africa.* N.p., London, 1899.

PrimaNET Communications Inc. *Gambia.* http://www.gambia.com (Accessed April 1, 2002).

Vollmer, Jurgen. *Black Genesis, African Roots: A Voyage from Juffure, The Gambia, to Mandingo Country to the Slave port of Dakar, Senegal.* New York: St. Martin's Press, 1980.

Webb, Patrick. "Guests of the Crown: Convicts and Liberated Slaves on McCarthy Island, The Gambia." *The Geographical Journal* 160 (2): 136-7, July 1994.

Georgia

- **Area:** 26,807 sq mi (69,700 sq km) / World Rank: 121
- **Location:** Located in southwestern Asia, east of the Black Sea, in the Northern and Eastern Hemispheres, bordering Turkey and Armenia to the south, Azerbaijan to the southeast, Russia to the north
- **Coordinates:** 42°00′N, 43°30′E
- **Borders:** 906 mi (1,461 km) / Armenia, 102 mi (164 km); Azerbaijan, 200 mi (322 km); Russia, 448 mi (723 km); Turkey 156 mi (252 km)
- **Coastline:** 192 mi (310 km)
- **Highest Point:** Mount Shkhara, 17,064 ft (5,201 m)
- **Lowest Point:** Black Sea 0 ft (m)
- **Longest Distances:** Not available
- **Longest River:** Kura River, 941 mi (1,514 km)
- **Natural Hazards:** Earthquakes, rockslides
- **Population:** 4,989,285 (July 2001 est.) / World Rank: 110
- **Capital City:** Tbilisi, located east-central part of the country
- **Largest City:** Tbilisi, 1.4 million (2000 est.)

OVERVIEW

Although a small country, Georgia features extremely diverse terrain, with both high mountain ranges and fertile coastal lowlands. Most of the country is mountainous, however, with the Greater Caucasus Mountains in the north and the Lesser Caucasus in the south. In the mountains, earthquakes and landslides frequently destroy life and property. In the west the Kolkhida Lowland borders the Black Sea, while in the east are the plains of the Kura River Basin. The country is situated in the isthmus between the Caspian and Black seas.

Included within Georgia's boundaries are two autonomous republics: Ajaria in Georgia's southwestern corner, and Abkhazia in the northwest. Another autonomous region is South Ossetia, in the north-central part of Georgia. Separatists have sought to detach these areas from Georgia, especially in Abkhazia and South Ossetia.

MOUNTAINS AND HILLS

Mountains

About 85 percent of the total land area of Georgia consists of rugged mountains. The Greater Caucasus along the northern border with Russia are the tallest, and mark the boundary between Asia and Europe. Mount Shkhara (17,064 ft / 5,201 m), on the Georgian-Russian border, is the highest peak. Mount Kazbek (16,526 ft / 5,037 m), also in this chain, is the tallest mountain fully within Georgia's borders. Many other peaks exceed 15,000 ft (4,500 m). In the south, Lesser Caucasus peaks rarely exceed 10,000 ft (3,000 m). In the center of the country, the Suram Mountains follow a northeast-southwest path across the center of the country, connecting the Greater and Lesser Caucasus Ranges. West of the Suram range, relief drops off. Along the river valleys and the coast of the Black Sea, elevations are generally less than 300 ft (100 m).

Plateaus

A high plateau known as the Kartaliniya Plain follows the eastern side of the Suram Range, along the Kura River to the border with Azerbaijan. Further east, a semiarid region called the Iori Plateau borders the Iori River.

INLAND WATERWAYS

The Kura is the largest river in Georgia. It flows 941 mi (1,514 km) from its source in Turkey across the plains of eastern Georgia, through the capital, Tbilisi, and on into Azerbaijan, before entering the Caspian Sea. The largest river in western Georgia, the Rioni, flows from the Greater Caucasus into the Black Sea at the port of Pot'i. In Georgia's Soviet period, engineers turned the Rioni River lowlands into prime subtropical agricultural land, straightened and banked much of the river, and built an extensive network of canals. The country's other rivers include the Iori, Khrami, and Inguri.

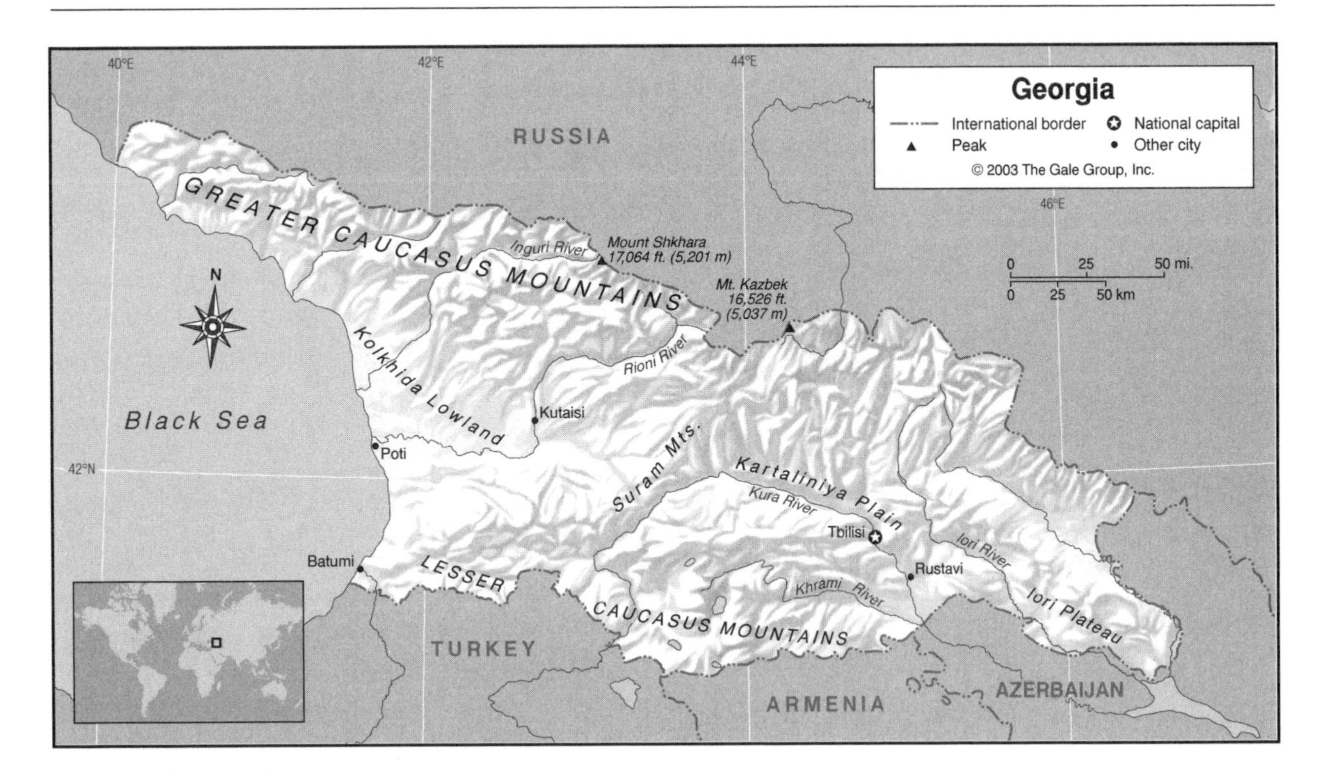

THE COAST, ISLANDS, AND THE OCEAN

Georgia's coast marks the easternmost extent of the Black Sea, the body of water that lies between southeastern Europe and Asia Minor. Through this sea, Georgia is connected by water with the Mediterranean and, eventually, the Atlantic Ocean. Pot'i and Batumi are the principal Black Sea ports in Georgia.

CLIMATE AND VEGETATION

Temperature

The Georgian climate is notably humid, warm, and pleasant on the Black Sea coast—almost subtropical. It is protected from truly cold weather by the Greater Caucasus Mountains to the north, and even in midwinter the average temperature is 41°F (5°C). The average summer temperature along the coast is 72°F (22°C). The plains region to the east, blocked from the sea by the Suram Mountains, is more continental in climate with hot summers and cold winters. Summer temperatures there range from 68°F (20°C) to 75°F (24°C), while in winter the range is from 36°F (2°C) to 39°F (4°C). The climate grows much cooler in the mountains, with alpine conditions starting at about 6,800 ft (2,100 m); above 12,000 ft (3,600 m) the mountains have snow and ice year-round.

Rainfall

The areas along the Black Sea coast and inland through the Kolkhida Lowlands experience high humidity and heavy precipitation of 40 to 80 in (1,000 to 2,000 mm) per year. The Black Sea port of Batumi receives 100 in (2,500 mm) of rain per year. At higher elevations humidity is lower and rainfall averages 18 to 32 in (500 to 800 mm) per year.

HUMAN POPULATION

The population of Georgia is 4,989,285 (July 2001 estimate), with an average population density of 192 persons per sq mi (74 per sq km). Nearly 60 percent of that is urban. For the most part people live along the coast of the Black Sea and in the river valleys, in particular that of the Kura River, location of the capital, Tbilisi, which has a population of 1.3 million. Migration from rural areas caused its population to grow by 18 percent between 1979 and 1989. The second largest city is Kutaisi, on the upper Rioni River, with a population of about 235,000.

NATURAL RESOURCES

Georgia's natural resources include manganese deposits, iron ore, copper, lead, gold, marble, alabaster,

Population Centers – Georgia

(1990 POPULATION ESTIMATES)

Name	Population
Tbilisi (capital)	1,268,000
Kutaisi	236,000
Rustavi	160,000
Batumi	137,000
Sokhumi	122,000

SOURCE: "Population of Capital Cities and Cities of 100,000 and More Inhabitants." United Nations Statistics Division.

and minor coal and oil deposits. Significant oil deposits are located in the Black Sea shelf near the port cities of Batumi and Pot'i, but these remain undeveloped. Georgian mineral waters are well known and draw visitors from surrounding countries. The country's forests and woodlands also contribute pulp and timber, and its rivers are an important source of hydroelectric power. Agriculture remains one of Georgia's most productive economic sectors. Georgia's western coastal climate and soils are conducive to tea and citrus growth. Approximately 1,544 sq mi (4,000 sq km) of land is irrigated for crop production (1993 estimate).

FURTHER READINGS

Jones, Stephen. "The Caucasian Mountain Railway Project: A Victory for Glasnost?" *Central Asian Survey*, Vol. 8, No. 2, 1989, pp. 47-59.

Parsons, Robert. "Georgians." *The Nationalities Question in the Soviet Union*. Edited by Graham Smith. New York: Longman, 1990, pp. 180-96.

Rosen, Roger. *The Georgian Republic*. Lincolnwood, Illinois: Passport Books, 1992.

Suny, Ronald G. *The Making of the Georgian Nation*. Bloomington: Indiana University Press, 1988.

Germany

- **Area:** 137,847 sq mi (357,021 sq km) / World Rank: 63
- **Location:** Northern and Eastern Hemispheres, central Europe, bordering the Baltic Sea and the North Sea, south of Denmark, east of Belgium, Luxembourg, and the Netherlands, northeast of France, north of Switzerland, north and west of Austria, and west of the Czech Republic and Poland.
- **Coordinates:** 51°00′N, 9°00′E
- **Borders:** 2,248 mi (3,618 km) / Austria, 487 mi (784 km); Belgium, 104 mi (167 km); Czech Republic, 401 mi (646 km); Denmark, 42 mi (68 km); France, 280 mi (451 km); Luxembourg, 84 mi (135 km); Netherlands, 359 mi (577 km); Poland, 302 mi (456 km); Switzerland, 208 mi (334 km)
- **Coastline:** 1,484 mi (2,389 km)
- **Territorial Seas:** 12 NM
- **Highest Point:** Zugspitze, 9,721 ft (2,963 m)
- **Lowest Point:** Freepsum Lake, 6.6 ft (2 m) below sea level
- **Longest Distances:** 530 mi (853 km) N-S; 404 mi (650 km) E-W
- **Longest River:** Danube, 1771 mi (2850 km)
- **Largest Lake:** Lake Constance, 220.7 sq mi (571.5 sq km)
- **Natural Hazards:** Subject to periodic flooding
- **Population:** 83,029,536 (July 2001 est.) / World Rank: 12
- **Capital City:** Berlin, located in northeastern Germany
- **Largest City:** Berlin, population 3,337,000 (2000 est.)

OVERVIEW

Germany is roughly the size of Montana, and is the sixth largest country in Europe. The unification of East Germany and West Germany in 1990 enlarged the territory of the former West German republic by 30 percent and increased its population by 20 percent. Germany lies on the Eurasian Tectonic Plate.

Topographically, the new Germany is composed of northern lowlands, central uplands, Alpine foothills, and Bavarian Alps. The northern plain covers the upper one-third of the country and contains the coastal area in the far north. Inland, the plain becomes hilly and is crisscrossed by rivers and valleys. These hills open to the Alpine Foreland where a series of north-south ranges interspersed with deep valleys climbs to the wooded slopes and craggy peaks of the German-Austrian Alps.

MOUNTAINS AND HILLS

Mountains

The Bavarian Alps represent a small fringe of high mountains that extend in a narrow strip along the country's southern boundary, and are vital to the country's tourism industry. They range eastward from Lake Constance on the Swiss border to just west of Salzburg on the Austrian border. They are divided into three sections: The Allgäuer portion extends from the Lake Constance to the Lech River and contains attractions such as Neuschwanstein, site of whimsical King Ludwig's mountain castle at Königsee. The Bavarian Alps comprise the central section between the Lech and the Inn rivers and contain the highest point in Germany: the Zugspitze, reaching 9,721 ft (2,963 m); from the Inn to the Salzburg Alps lies the third and eastern-most section, which includes the resort of Berchtesgaden where Hitler's retreat, the Eagle's Nest, was located. Several other peaks with altitudes of more than 8,202 ft (2,500 m) such as Watzmann 8,901 ft (2,713 m), Hochfrottspitze 8,691 ft (2,649 m), and Madelgabel 8,678 ft (2,645 m) rise majestically over the Bavarian fore Alps.

Also in the southern highlands, the Haardt Mountains stretch into southwestern Germany from France, following the Rhine River. Moving northwest along the

Rhine the elevations gradually diminish, reaching the Taunus Mountains, then finally lowering to the Seven Hills and Rothaar Hills in west Germany.

Plateaus

The Central German Uplands consist of a massive rectangular block of slate and shale covered by gently rolling plateau averaging 1,312 ft (400 m) in elevation with peaks from 2,625–2,952 ft (800–900 m). The plateau extends from the Rheinish Uplands on the French border to the Erzegebirge Range—part of the Bohemian Massif—on the Czech border. The plateau is cut into sections by several crisscrossing river valleys from the Rhine and the Moselle in the west, the Weser in the center, to the Elbe on the east. The southern edge is demarcated by the Main River, which flows westerly to the Rhine.

Hills and Uplands

Germany's Central Uplands are part of the Central European Uplands extending from the Massif Central in France into Poland and the Czech Republic. The landscape consists of hills and high ridges and contrasts broad, tilted blocks of sedimentary rocks with deep, trough-like valleys and lowlands.

In the center of Germany are the Rhon Mountains, whose highest point is Mount Grosser Beer at 3,221 ft (982 m). Just north and slightly to the east of this range are the Kyffhäuser Mountains. The Harz forms the northwest section, and its highest point, Brocken Peak, reaches a height of 3,743 ft (1,141 m). To the southeast along the Czech border are the Erzegebirge with elevations reaching 3,980 ft (1,213 m) at Fichtelberg. Many major industrial centers are situated along the base of the Erzegebirge. Traditional passages into the Central Uplands include the 'Hessian Corridor,' between Hanover and Frankfurt passing through Kassel on the Weser River. Good agricultural land is found at the base of the Thüringer Wald surrounding Erfurt, but soils in the southernmost districts are poor and are not favorable for cultivation.

The Central Uplands are bordered on the south by the south German scarplands, a succession of escarpments and intervening valleys stretching across the country from southern Baden-Württemberg to the northeastern corner of Bavaria. Sections of these uplands are formed by the extension of the Jura ranges from France and Switzerland. One of these ranges constitutes the Black Forest and a second forms the Swabian Alb and its extension, the Franconian Alb. In the Black Forest, the Feldberg reaches an elevation of 4,908 ft (1,496 m). The two albs are about 25 mile (40 km) wide, and in several places exceed altitudes of 2,953 ft (900 m). They form an arc some 248 mi (400 km) long extending to the Central Uplands near Bayreuth.

The deep incision of the valleys and their spectacular landscapes make the south German scarplands most distinctive. They give way to the gentle Alpine Foothills comprising all of Bavaria and the eastern portion of Baden-Württemberg. Most of this region is in the upper Danube River basin and is crossed by the Danube's main Alpine tributaries—the Iller, the Lech, the Isar and the Inn.

INLAND WATERWAYS

Lakes

The northern lowlands contain numerous lakes, varying in size and shape particularly in northeastern Germany and around Berlin. Lakes in this region include Lake Müritz, Lake Kummerow, Lake Plau, and Lake Schwerin. In general these lakes are of little commercial value because of their shallow depth.

The Alpine Foothills are speckled with many lakes of clear, clean water and steep, wooded banks; in high elevations are found glacial lakes adding to the spectacular charm of Alpine meadows. Several lakes dot the landscape in this area of southern Germany including Ammer Lake, Lake Chiem, and Starnberger See. Lake Constance (Bodensee) through which passes the upper Rhine River, is shared with Switzerland and Austria. It is Germany's largest lake, with a surface area of 220.7 sq mi (571.5 sq km), of which 118 sq mi (305 sq km) lie within Germany. It is 46 mi (74 km) long and reaches a maximum depth of 827 ft (252 m).

GEO-FACT

The Schlossberg caves near Hamburg are Europe's largest sandstone caves. Their vast multi-tiered colored sandstone rooms are connected by long corridors, about 3,000 ft (1000 m) of which may be hiked by visitors. Part of a fortress that was destroyed in 1714, the caves were not rediscovered until around 1930.

Rivers

The greater part of the country drains into the North Sea via the Rhine, the Ems, the Weser, the Saale, and the Elbe Rivers. A small area north and northeast of Hamburg drains into the Baltic Sea via the Oder River on the Polish border. The divide between the watersheds of the Danube and Rhine basins winds round Baden-Württemberg and Bavaria, most of which, excepting a small area north of Lake Constance, drains into the Danube.

Germany's two longest rivers are the Rhine and the Danube (Donau). The Rhine, which rises in Switzerland, flows into Lake Constance, then west along Germany's southern border with Switzerland, before turning north into Germany itself. The western part of Germany is called the Rhineland, after the river. The Rhine receives a steady flow from melting snow in the winter and in the summer from the Neckar, the Main, and the Moselle, it principal tributaries in Germany (the Moselle has its headwaters in France). The Rhine curves west again and branches into a delta shortly after exiting Germany for the Netherlands, after which it flows into the North Sea. The Rhine is one of the largest and most commercially important rivers in Western Europe.

The Danube rises in the southwestern part of the country, not far from the Rhine, but flows northeast until it reaches the Bavarian Forest, where it curves southeast and exits into Austria at Passau. It then follows a winding, generally eastern, course, finally emptying into the Black Sea 1771 mi (2850 km) from its source. It flows for 402 mi (647 km) within Germany itself. The second longest river in Europe, the Danube is a vital commercial and transportation route.

Germany has an extensive system of canals that effectively link all of its major rivers together. A series of canals runs across the middle of the country, including the Dortmund-Ems Canal, connecting the Rhine with the Ems; the Mittelland Canal, connecting the Ems with the Wesser and Elbe; and other shorter canals. The Main-Danube Canal crosses through the Franconian Alb to connect those two rivers. In northern Germany the Nord-Ostsee Canal (Kiel Canal) connects the estuary of the Elbe River on the North Sea with the Baltic Sea at Kiel. It is one of the world's busiest canals.

Wetlands

The North Sea coast has wide expanses of sand, marsh, and mudflats (watten). On the Baltic side, the northern sections of Schwerin and Neubrandenburg districts, which are also coastal, are dotted with marshes and numerous lakes.

THE COAST, ISLANDS, AND THE OCEAN

Oceans and Seas

Germany faces the North Sea and Baltic Sea to the north. A narrow strip of land on which Germany borders Denmark separates the two seas. On the Baltic Sea side the German coast is indented by Mecklenburg Bay.

Major Islands

In the North Sea, a former line of inshore sand dunes became the East Frisian Islands when the shoreline sank during the thirteenth century. These islands have maximum elevations of less than 115 ft (35 m) and have been subject to eroding forces that have washed away whole sections during storms. In 1854, for example, the only village on Wangerooge, the easternmost of the main East Frisian Islands, was washed away. While the East Frisian Islands are strung along the coast in a nearly straight line, having long axes roughly parallel to the coast, those in the North Frisian Islands are irregularly shaped and haphazardly positioned.

At 358 sq mi (927 sq km) in area, Germany's largest island is Rügen. It lies in the Baltic Sea off Stralsund. Another large island, Fehmarn, is located at the northern edge of Mecklenburg Bay in the Baltic.

The Coast and Beaches

On the North Sea side, the coastal mud flats between the Frisian islands and the shore are exposed at very low tides and are crossed by innumerable channels. The mud and sand are constantly shifting making navigation treacherous.

The Schleswig-Holstein coast on the Baltic Sea differs markedly from that of the North Sea. It is indented by a number of small fjords with steep banks, and the deep water and shelter of the fjords provide safe sailing conditions. Fishing villages are common on this coast, which is flat and sandy. Farther east, the coastline is uneven but also generally flat and sandy. The continuous action of wind and waves has created sand dunes and ridges, and sandbars connect the mainland with some of its offshore islands.

The Jasmund National Park lies along the northeast shore of the Baltic sea, and is characterized by dramatic chalk cliffs. The Königsstuhl is the highest point of this coastline, which reaches 386 ft (117 m).

CLIMATE AND VEGETATION

Temperature

Gulf Stream westerly winds from the North Sea moderate temperatures throughout the year. In the lowlands, mid-winter temperatures average more than 35°F (1.6°C), while summer temperatures average between 61° and 64°F (16° and 18°C). In the south, temperatures are somewhat more extreme, averaging about 28°F (-2°C) in winter and 67°F (19.4°C) or higher in summer. The yearly mean for the entire country is 48°F (9°C).

Rainfall

Rainfall varies from 79 in (200 cm) in the Alps to 16 in (40 cm) in the vicinity of Mainz. In the maritime region, precipitation varies between 24 and 25 in (61 and 64 cm), which approximates the national yearly average of between 24 and 31.5 in (60 and 80 cm).

Grasslands

Grasslands, pastures, and cultivated areas cover significant portions of the lowland plains, the Bavarian foothills, and the valleys and lower slopes of the Alps. Alpine meadows provide rich summer pastures. Barren moors cover the tops of the Harz mountains in the Central Uplands.

At least a third of the country lies in an area of northern plains known as the central lowlands. These lowlands are part of a great plain that extends across north-central Europe, broadening from Belgium and the Netherlands until it reaches the Ural Mountains. The terrain is gentle, and the landscape is marked by few sharp contrasts. Landform areas merge into one another; no significant natural boundaries bar communications or distinguish one section of the country from another. Elevation in this region rarely exceeds 492 ft (150 m) and that in the central and western part. The land slopes imperceptibly toward the sea.

Population Centers – Germany

(2000 POPULATION ESTIMATES)

Name	Population
Rhein-Ruhr North (Duisburg, Essen, Krefeld, Mülheim, Oberhausen, Bottrop, Gelsenkirchen, Bochum, Dortmeund, Hagen, Hamm, and Herne)	6,531,000
Rhein-Main (Darmstadt, Frankfurt am Main, Offenbach, and Wiesbaden)	3,681,000
Berlin (capital)	3,319,000
Rhein-Ruhr Middle (Düsseldorf, Mönchengladbach, Remscheid, Solingen, and Wuppertal)	3,233,000
Rhein-Ruhr South (Bonn, Cologne, Leverkusen)	3,050,000
Stuttgart	2,672,000
Hamburg	2,664,000
Munich	2,291,000

SOURCE: "Table A.12. Population of Urban Agglomerations with 750,000 Inhabitants or More in 2000, by Country 1950–2015. Estimates and Projections: 1950–2015." United Nations Population Division, World Urbanization Prospects: The 2001 Revision.

States – Germany

1999 POPULATION ESTIMATES

Name	Population	Area (sq mi)	Area (sq km)	Capital
Baden-Württemberg	10,475,932	13,804	35,751	Stuttgart
Bayern	12,154,967	27,241	70,553	München
Berlin	3,386,667	344	891	Berlin
Brandenburg	2,601,207	11,381	29,477	Potsdam
Bremen	663,065	156	404	Bremen
Hamburg	1,704,735	292	755	Hamburg
Hessen	6,051,350	8,152	21,114	Wiesbaden
Mecklenburg-Vorpommern	1,789,322	8,947	23,172	Schwerin
Niedersachsen	7,898,760	18,320	47,450	Hanover
Nordhein-Westfalen	17,999,800	13,153	34,068	Düsseldorf
Rheinland-Pfalz	4,030,773	7,663	19,847	Mainz
Saarland	1,071,501	992	2,568	Saarbrücken
Sachsen	4,459,686	7,110	18,413	Dresden
Sachsen-Anhalt	2,648,737	7,895	20,447	Magdeburg
Schleswig-Holstein	2,777,275	6,072	15,727	Kiel
Thüringen	2,449,082	6,244	16,172	Erfurt

SOURCE: Federal Statistical Office of Germany, as cited on Geohive, http://www.geohive.com (accessed 24 June 2002); and Geo-Data, 1989 ed.

Forests

Germany is dotted with patches of forest. A mixture of deciduous and conifer forests is found in the Central Uplands and southern scarplands such as the Thüringer Wald, the Bavarian and Bohemian Forests on the eastern frontiers, and the Black Forest in the southwest. In addition, the upper elevations in the Uplands surrounding the Rhine River are heavily forested, as are the Harz at lower levels. Conifers cover Alpine slopes.

HUMAN POPULATION

After more than four decades (1945–89) of separate development, ideology, and living standards, Germans are seeking a new identity to harmonize and integrate two essentially different peoples. The fall of the Berlin Wall in December 1989 opened the floodgates for some 700,000 East Germans to migrate to the West in search of opportunity and prosperity, leading to an influx in the urban population. East Germany was relatively sparsely populated compared with West Germany, and the country's population densities still vary greatly. There are many densely populated pockets interspersed with much less populated countryside regions. The greater Berlin region is the most dense area with a population density of 9,565 people per sq mi (3,693 people per sq km). The surrounding countryside north to the Baltic Sea and south to the Czech Republic, by contrast, has an average density of only 216 per sq mi (83 per sq km). Central Germany is even less densely populated. The southwest has a population density of 762 per sq mi (294 per sq km) while the southeast has 446 people per sq mi (172 people per sq km). The central western region is the most populous, but covers more area than Berlin and has a density of 1,368 people per sq mi (528 people per sq km). There are much fewer people in the lands to the northwest and southwest.

NATURAL RESOURCES

Germany's natural resources include iron ore, coal, potash, timber, lignite, uranium, copper, natural gas, salt, and nickel. The country also has fertile agricultural belts composed of loess, a fine silt that provides a thick soil cover that is favorable for intensive cultivation of crops such as wheat, barley, and sugar beets. Much of the country's mineral wealth, including sizable reserves of brown coal and potash, is found along the Elbe and Saale rivers. Also, the Saar basin, which has changed hands between France and Germany on several occasions, is noted for its rich coalfields of about 25 mi (40 km) long and 8 mi (13 km) wide.

FURTHER READINGS

German Government. http://www.bundesregierung.de (Accessed June 2002).

Godberg, Louis. "Where Have All the Eastern Germans Gone?" *Guardian Weekly*, October 31, 1993, 15.

Jones, Alun. *The New Germany: A Human Geography.* New York: John Wiley and Sons, 1994.

Mellor, Roy E.H. *The Two Germanies: A Modern Geography.* London: Harper and Row, 1978.

Solsten, Eric, ed. *Germany: A Country Study.* 3rd ed. Area Handbook Series. Federal Research Division, Library of Congress. Washington, D.C.: Department of the Army, 1996.

Velotours. *Welcome to Lake Constance.* http://www.bodensee-info.com/ (Accessed June 2002).

Ghana

- **Area:** 92,100 mi (238,540 sq km) / World Rank: 79
- **Location:** Western Africa, bordering the Gulf of Guinea, between Côte d'Ivoire and Togo
- **Coordinates:** 8° N; 2° W
- **Borders:** 1,298 mi (2,093 km) / Togo 544 mi (877 km); Côte d'Ivoire 414 mi (668 km); Burkina Faso 340 mi (548 km)
- **Coastline:** 334 mi (539 km)
- **Territorial Seas:** 12 NM
- **Highest Point:** Mt. Afadjato 2,887 ft (880 m)
- **Lowest Point:** Sea level
- **Longest Distances:** 284 mi (458 km) NNE-SSW; 184 mi (297 km) ESE-WNW
- **Longest River:** Volta 992 mi (1600 km)
- **Largest Lake:** Lake Volta, 3,276 sq mi (8,485 sq km)
- **Natural Hazards:** Droughts, and dry, dusty harmattan winds that occur from January to March
- **Population:** 19,361,100 (2002 est.) / World Rank: 50
- **Capital City:** Accra, located on the Gulf of Guinea
- **Largest City:** Accra 1,605,400 (2002 est.)

OVERVIEW

Ghana faces the Gulf of Guinea in the great bulge of West Africa. Situated on the African Plate, between Togo on the east, Côte d'Ivoire (Ivory Coast) on the west, and Burkina Faso on the north and northwest. The coast is characterized by strong surfs, which make landing difficult except at artificially constructed harbors. Average elevation is relatively low, mostly between sea level and about 1,000 ft (305 m).

Five major geographical regions can be distinguished. In the southern part of the country are the low plains, part of the belt that extends along the entire coastal area of the Gulf of Guinea. To the north of these plains lie three distinct regions—the Ashanti Uplands, the Volta Basin, and the Akwapim-Togo Ranges. The fifth region, the high plains, occupies the northern and northwestern parts of the country. These plains also are part of a belt stretching generally eastward and westward through West Africa.

MOUNTAINS AND HILLS

Mountains

The Akwapim-Togo Ranges in the eastern part of the country consist of a generally rugged complex of folded strata, with many prominent heights composed of volcanic rocks. The ranges begin west of Accra and continue in a northeasterly direction through the Volta Region and finally cross the frontier in the upper part of that region completely into the Republic of Togo.

In their southeastern part the ranges are bisected by a deep, narrow gorge that has been cut by the Volta River. The head of this gorge is the site of the Akosombo Dam, which impounds the water of the river to form Lake Volta. The ranges south of the gorge form the Akwapim section of the mountains. The average elevation in this section is about 1,500 ft (475 m) and the valleys are generally deep and relatively narrow. North of the gorge for about 50 mi (80 km), the Togo section has broader valleys and generally low ridges. Beyond this point, the folding becomes more complex, and heights increase greatly, with several peaks rising above 2,500 ft (762 m). The country's highest point, Mount Afadjato at 2887 ft (880 m), is located in this area.

Plateaus

The general terrain in the northern and northwestern part of the country outside the Volta Basin region consists of a dissected plateau, which averages between 500 and 1,000 ft. (152 and 304 m) in elevation and in some places is even higher. Soils in the high plains have generally greater fertility than in the Volta Basin, and the population density is considerably higher. Grains are a major crop, and livestock raising is a major occupation.

The Kwahu Plateau, forming the northeastern and eastern part of the uplands, has a quite different geologic structure from the uplands and consists largely of relatively horizontal sandstone. Elevation averages 1,500 ft (457 m) and high points rise to over 2,500 ft (762 m). The greater height of the plateau gives it a comparatively cooler climate.

The Volta Basin

The Volta Basin region occupies the central part of the country and covers about 45 percent of the country's total area. Much of the southern and southwestern part of this basin is under 500 ft (152 m) in elevation, whereas in the northern section, lying above the upper part of Lake Volta and the Black Volta, elevations are from about 500 to 750 ft (152 to 228 m). The edges of the basin are characterized by high scarps. The Kwahu Plateau marks the

GEO-FACT

Until the latter half of the 1960s, the Black Volta and the White Volta came together near the middle of the country to form the Volta River, which from this confluence flowed first southeastward, then south, for about 310 mi (499 km) to the Gulf of Guinea. In 1964 the closing of a dam across the Volta at Akosombo, roughly some 50 mi (80 km) upstream from its mouth, created the world's largest artificial lake.

southern end of the basin, although it forms a natural part of the Ashanti Uplands. The Konkori Scarp, on the western edge of the basin, and the Gambaga Scarp, in the north, have elevations from about 1,000 to 1,500 ft (304 to 457 m).

Hills and Uplands

The Ashanti Uplands lie just to the north of the Akan Lowlands area and extend from the Ivory Coast border, through the western and part of the northern Brong-Ahafo Region and the Ashanti Region (excluding its eastern section), to the eastern end of the Kwahu Plateau. With the exception of the Kwahu Plateau, the uplands slope gently toward the south, gradually decreasing in elevation from about 1,000 to 500 ft (304 m to 152 m). Erosion of the crystalline rocks that underlie this area has left a number of hills and ranges, trending generally southwest to northeast, which in places reach heights between 1,500 and 2,500 ft (457 and 762 m). In the southernmost part, their valleys become more open, and the region merges imperceptibly into the Akan Lowlands with an elevation between sea level and 500 ft (152 m). These lowlands make up the greater part of the low plains and are broken by hill ranges with a few peaks exceeding 2,000 ft (609 m), although most high points rarely rise above 1,000 ft (304 m).

INLAND WATERWAYS

Lakes

Lake Volta is the world's largest artificial lake (3,276 sq mi/8,485 sq km), formed by the impoundment of the Volta River behind Akosombo Dam. Ghana's one large natural lake, Lake Bosumtwi (18 sq mi/46 sq km), is located about twenty miles southeast of Kumasi. It occupies the steep-sided caldera of a former volcano. Several small streams flow into the lake, but because there is no drainage, its level is gradually rising.

Rivers

The entire country is interlaced by streams and rivers. The stream pattern is closest in the moister south and southwest. North of the Kwahu Plateau, in the eastern part of the Ashanti Region, and in the western part of the Northern Region, the pattern is much more open and makes access to water more difficult. Stream flow is not regular throughout the year, and during the dry seasons the smaller streams and rivers dry up or have greatly reduced flow, even in the wetter areas of the country.

A major drainage divide runs from the southwestern part of the Akwapim-Togo Ranges northwestward through the Kwahu Plateau and then irregularly westward to the Ivory Coast border. Almost all streams and rivers north and east of this divide are part of the vast Volta drainage system, which covers some 61,000 sq mi (157,989 sq km) or more than two-thirds of the country. To the south and southwest of the plateau several smaller independent river systems flow directly into the Gulf of Guinea. The most important are the Pra, the Ankobra, and the Tano. Only the Volta, Ankobra, and Tano rivers are navigable by launches or lighters, and this is possible only in their lower sections. Apart from the Volta, only the Pra and the Ankobra rivers permanently pierce the sand dunes; most of the other rivers terminate in brackish lagoons.

The largest river, the Volta, has three branches, all of which originate in Burkina Faso. The Black Volta forms the northwest border, then flows southeastward into Ghana to the east. The White Volta and the Red Volta both enter the country in the northeast. About twenty-five miles inside the border, the Red Volta joins the westward-flowing White Volta, which eventually turns and flows southward through approximately the central part of the country into Lake Volta behind the Akosombo Dam.

Wetlands

The Afram Plains in the vicinity of the Afram River are swampy or flooded during the rainy season, and were largely submerged in the formation of Lake Volta. Near the coast the intermittent drainage from the Accra plains empties into the gulf through a series of river valleys. The valleys are often swampy during the rainy seasons, and their outlets are periodically blocked by sandbars to form lagoons.

THE COAST, ISLANDS, AND THE OCEAN

Oceans and Seas

Ghana's coast stretches for 334 mi (539 km) along the Gulf of Guinea of the Atlantic Ocean.

The Coast and Beaches

The coast consists mostly of a low sandy shore, behind which stretches the coastal plain. Except in the

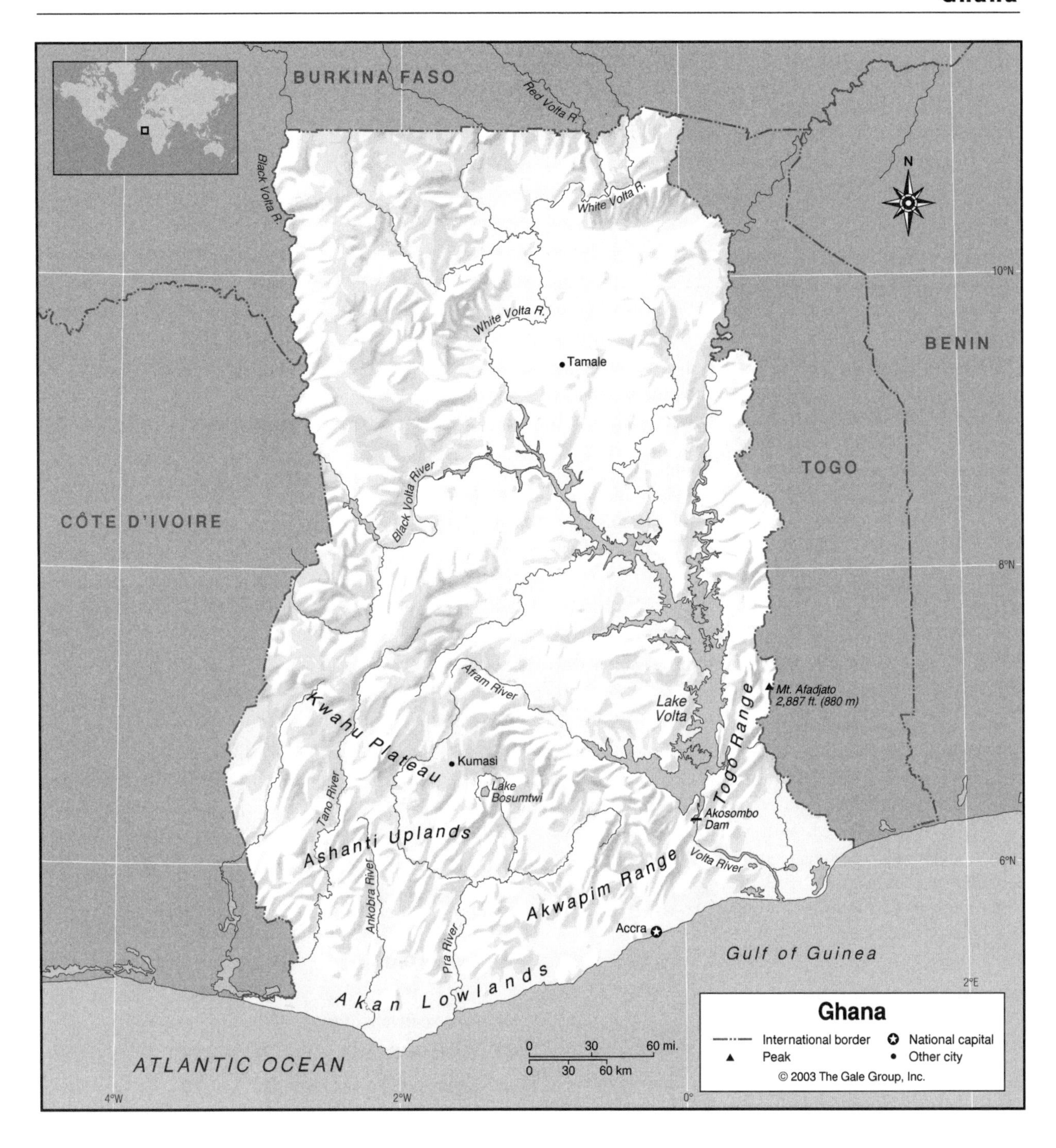

west where the forest comes down to the sea, the plain is mostly flat and generally covered with grass and scattered fan palms. Most of Ghana's rivers terminate in brackish lagoons along the coast, but there are no natural harbors.

The Volta Delta, which forms a distinct subregion of the low plains, projects out into the Gulf of Guinea in the extreme southeast. As this delta grew outward over the centuries, sandbars developed across the mouth of the Volta River and also of some smaller rivers that empty into the gulf in the same area, resulting in the formation of numerous lagoons, some of large size. Dense groves of coconut palms also are found, and at places inland in the drier, older section of the delta, oil palms grow in profusion. The main occupation in the delta is fishing and this industry supplies dried and salted fish to other sections of the country.

CLIMATE AND VEGETATION

Temperature

Average temperature ranges between 70-90° F (21-31° C) with relative humidity between 50% and 80%.

Population Centers – Ghana

(2002 POPULATION ESTIMATES)

Name	Population
Accra (capital)	1,605,400
Kumasi	700,000
Tamale	300,000
Tema	250,000

SOURCE: Projected from United Nations Statistics Division data.

Rainfall

Rainfall is affected by the seasonal position of the inter-tropical convergence zone, the boundary between the moist southwesterly winds and the dry northeasterly winds. Except in the north, there are two rainy seasons, from April through June and from September to November. Squalls occur in the north during March and April, followed by occasional rain until August and September, when the rainfall reaches its peak. Rainfall ranges from 33-87 in (83-220 cm) per year.

Grasslands

Grasslands dominate the south comingled with mixed coastal scrub, and in the northern savanna with deciduous trees.

Deserts, Forests, and Jungles

Ghana's forest belt extends northward from the western coast on the Gulf of Guinea about 200 mi (320 km) and eastward for a maximum of about 170 mi (270 km) and is broken up into heavily wooded hills and steep ridges. Over-cultivation, overgrazing, heavy logging, over-cutting of firewood, and mining have taken their toll on forests and woodland with deforestation proceeding at an annual rate of 278 sq mi.

HUMAN POPULATION

In 2001, the population growth rate was 1.79%. Life expectancy at birth was 57 years; the total fertility rate was 3.82%. It was estimated that 38% of the population lived in urban areas in 2000, up from 31% in 1980. The major ethnic group is the Akan family with some 44% of the population, which includes the Twi or Ashanti in central Ghana, and the Fanti in the coastal areas.

NATURAL RESOURCES

Ghana is well-endowed with natural resources including gold, timber, bauxite and aluminum, manganese ore, diamonds, and tuna.

FURTHER READINGS

Africa South of the Sahara 2002. "Ghana." London: Europa Publications Ltd, 2001.

Boateng, E.A. *Geography of Ghana.* 2nd ed. Cambridge: Cambridge University Press, 1966.

Dickson, Kwamina B. *A Historical Geography of Ghana.* Cambridge: Cambridge University Press, 1971.

Ghana Home Page. http://www.ghanaweb.com/ (accessed April 10, 2002).

Kropp, Dagubu, M.E. *The Languages of Ghana.* London: Kegan Paul for the International African Institute, 1988.

Owusu-Ansah, David and Daniel M. McFarland. *Historical Dictionary of Ghana.* Metuchen: Scarecrow Press, 1995.

Greece

- **Area:** 50,942 sq mi (131,940 sq km) / World Rank: 96
- **Location:** Northern and Eastern Hemispheres, Southern Europe, bordering the Former Yugoslav Republic of Macedonia and Bulgaria on the north, Turkey and the Aegean Sea on the east, the Mediterranean Sea on the south, the Ionian Sea on the southwest, and Albania on the northwest
- **Coordinates:** 39°N, 22°E
- **Borders:** 752 mi (1,210 km) / Albania, 175 mi (282 km); Bulgaria, 307 mi (494 km); Turkey, 128 mi (206 km); The Former Yugoslav Republic of Macedonia, 142 mi (228 km)
- **Coastline:** 8,498 mi (13,676 km)
- **Territorial Seas:** 6 NM
- **Highest Point:** Mt. Olympus, 9,571 ft (2,917 m)
- **Lowest Point:** Sea level
- **Longest Distances:** 584 mi (940 km) N-S / 80 mi (772 km) E-W
- **Longest River:** Maritsa, 300 mi (480 km)
- **Natural Hazards:** Earthquakes
- **Population:** 10,623,835 (July 2001 est.) / World Rank: 70
- **Capital City:** Athens, southern mainland Greece
- **Largest City:** Athens, 3.1 million (2000 est.)

OVERVIEW

Greece is located at the southern tip of the Balkan Peninsula in Southern Europe. It is a country of peninsulas and islands, surrounded by the sea. Mainland Greece lies between the Aegean, Mediterranean, and Ionian Seas, and one-fifth of the country is made up of the hundreds of islands, many of them uninhabited, that lie in these bodies of water.

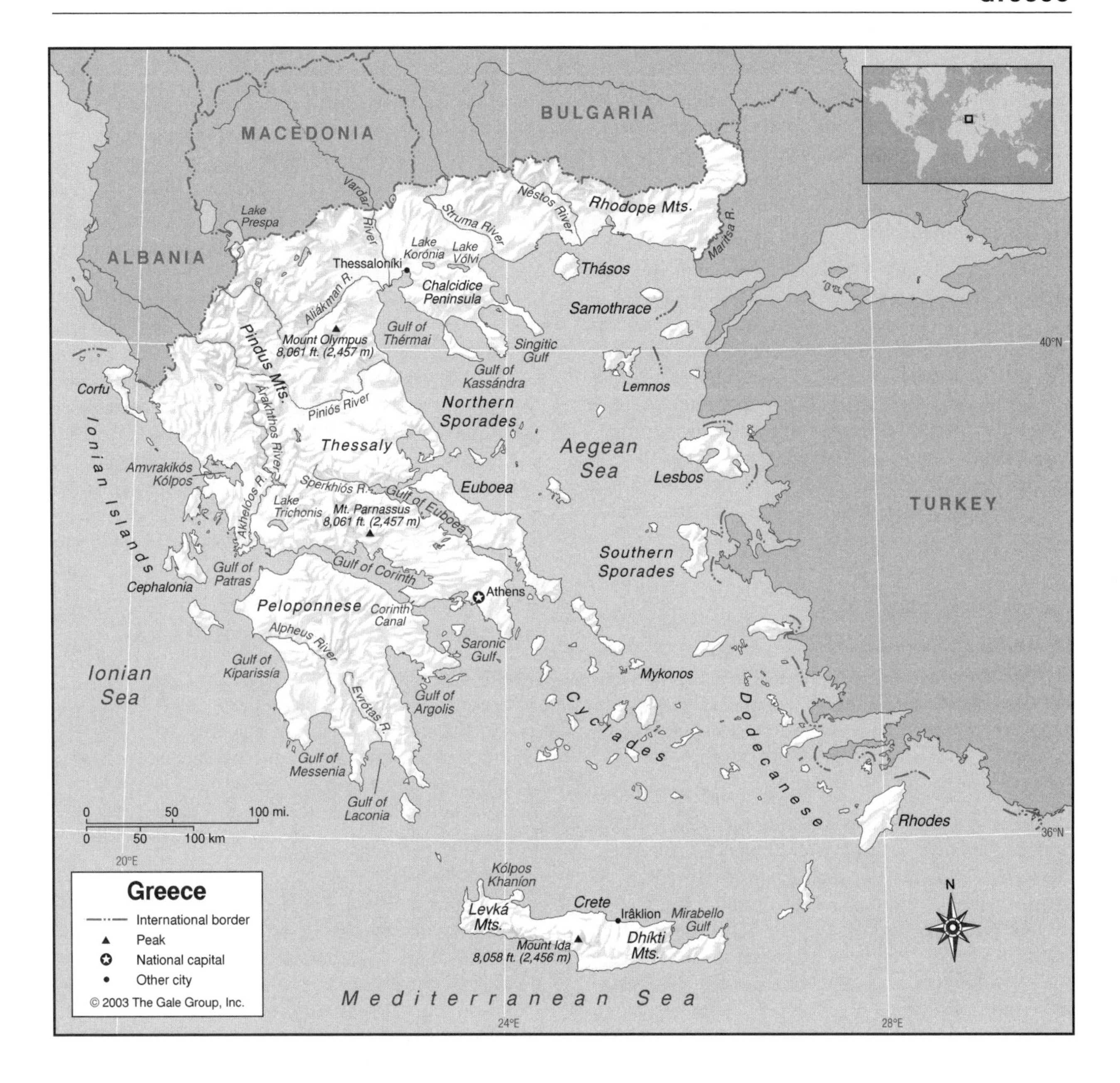

Greece's terrain is generally rugged, with mountain ranges and their spurs running northwest to southeast through much of the mainland. Altogether, four-fifths of Greece is covered by mountains. The remaining, low-lying land consists of basins that lie between the mountain chains, river plains, and narrow intermittent coastal strips. The mainland includes a variety of topographical regions that can be divided in a number of different ways. The simplest is the division into northern, central, and southern areas.

The northern part of mainland Greece consists of a long strip of land between the northern shore of the Aegean Sea and the southern borders of Bulgaria and the Former Yugoslav Republic of Macedonia. It includes Thrace and the eastern and central portions of Greek Macedonia. Most of this region is occupied by the Rhodope Mountains, extending southward from Bulgaria, and the plains of Greece's northern rivers. The region is bounded on the east by the Maritsa River, which forms the boundary with Turkey.

The central part of the mainland corresponds to the bulk of the Greek peninsula, extending south from the borders with Albania and the Former Yugoslav Republic of Macedonia. It is dominated by the Pindus Mountains, Greece's most extensive mountain range. To the east, between mountain spurs, lie the plains of Thessaly and, farther to the southeast, Boeotia and Attica. To the west lie Epirus and, farther south, Arkananía.

South of the Gulf of Corinth, forming the southern part of the mainland, lies the Peloponnese, a peninsula of 8,278 sq mi (21,446 sq km) connected to Attica by an isthmus that is only 4 mi (6.4 km) across at its narrowest point. (The Corinth Canal cuts through the isthmus to connect the Gulf of Corinth with the Saronic Gulf, so the Peloponnese is technically an island rather than a peninsula.) A series of ridges extending southward give the Peloponnese its distinctive "four-fingered" shape. Although mountainous, it has a narrow coastal plain around its entire periphery.

Greece is situated at the convergence of the Eurasian and African Tectonic Plates, making it geologically unstable and prone to frequent earthquakes and tremors, which are sometimes associated with volcanic eruptions. The famous statue known as the Colossus of Rhodes was toppled by an earthquake in 227 B.C. In the ninth century Corinth was destroyed by an earthquake that reportedly killed 45,000 people. In the 1950s the country was again ravaged by earthquakes, particularly along the Ionian Islands, along the western coast, and on some of the Aegean Islands.

MOUNTAINS AND HILLS

Mountains

The Rhodope Mountains in northern Greece, which lie mostly beyond the border in Bulgaria, rise to over 7,000 ft (1,800 m) in many places. Their highest point in Greece is Mt. Órvilos, at 7,287 ft (2,212 m).

The Pindus Mountains, Greece's major mountain range, belong to the same system as the Dinaric Mountains to the north, which extend through Croatia, Yugoslavia, and Albania. In central Greece (between the northern border and the Gulf of Corinth), the range is divided into three segments by the Métsovon pass and, farther south, Mt. Timfristós. Mountain spurs extend into the eastern part of central Greece, separated by structural depressions, the largest of which is Thessaly. The mountain spur north of Thessaly is home to Greece's highest peak, the legendary Mt. Olympus, mythic home of the Greek gods. The Pindus extends southeastward through the mainland peninsula to the Gulf of Corinth, where Mt. Parnassus (8,061 ft / 2,457 m) is located. Geologically, the mountains of the Peloponnese, which rise to heights of 7,800 ft (2,377 m), are a continuation of the Pindus chain.

Many of the Greek islands are also mountainous. On Euboea, Mt. Dhírfis reaches 5,717 ft (1,743 m). Crete's Levká and Dhíkti Mountains are the most significant of the island mountain chains, and contain Mt. Ida (8,058 ft / 2,456 m).

INLAND WATERWAYS

Lakes

Lake Korónia and Lake Vólvi mark the northern end of the Chalcidice Peninsula. Lake Vistonis in western Thrace, although called a lake, is actually a lagoon. Another major lake is Lake Trichonis near the southern end of the Pindus Mountains. Greece shares Lake Prespa, along its northern border, with Albania and the Former Yugoslav Republic of Macedonia.

Rivers

Greece has relatively few rivers. Those it does have are short, and although they flow swiftly down its mountains none are navigable for commercial purposes. In the north, the Maritsa (Evros), Néstos, Struma, and Vardar flow across the plains of Thrace and Macedonia and into the northern Aegean Sea. The rivers of central Greece, which rise in the Pindus Mountains, are the Aliákman, Árakhthos, Akhelóos, Pinriós, and Sperkhiós. These drain either into the Aegean or Ionian seas. The Aliákman (200 mi / 320 km) is the longest river located entirely in Greece. The major rivers of the Peloponnese are the Alpheus and Evrótas. The rivers of northern Greece flow year round, while those farther south tend to dry up in the summer months.

THE COAST, ISLANDS, AND THE OCEAN

Oceans and Seas

Greece is bounded on the south by the Mediterranean Sea. To the west is the Ionian Sea. On the east is the Aegean Sea, an arm of the Mediterranean demarcated by Crete to the south and connected to the Black Sea through the Dardanelles, the Sea of Marmara, and the Bosporus.

Major Islands

Greece's major island regions are the Ionian Islands, which hug the western coast from Albania to the Peloponnese; the Aegean Islands, scattered about the sea of the same name; and Crete, which separates the Aegean and Mediterranean seas. Euboea, a long and mostly narrow section of central Greece that is generally considered part of the mainland, is in fact also an island, separated from the mainland by the Gulf of Euboea, and by a strait that in some places is only 200 ft (61 m) wide.

The seven major islands of the Ionian group are Corfu, Leukas (Levkas), Cephalonia (Kefallinía), Paxos, Ithaca, Zákinthos, and Skorpios. The northernmost and second-largest of the islands is Corfu, part of which lies off the southern coast of Albania. Cephalonia, at 302 sq mi (782 sq km) the largest of the Ionian Islands, is directly west of the Gulf of Patras. Ithaca, east of Cephalonia, is famous as the home of Homer's Odysseus.

The northernmost Greek islands in the Aegian are Thásos, Samothrace (Samothráki), and Lemnos (Límnos). Continuing southward near the Turkish coast are the islands of Lesbos (Lésvos), Chios, and Sámos. Lesbos is split nearly in two by the Gulf of Kalloní. Across the Aegean, roughly opposite Lesbos, are the small islands of the Northern Sporades, north of Euboea. ("Sporades" means "scattered.") A few miles southeast of Attica lie the twenty-nine islands of the Cyclades, extending from Andros in the north to Thíra and Anáfi in the south. They encircle the tiny but historically sacred island of Delos, near Mykonos.

East of the Cyclades, just off the southwestern coast of Turkey, are the 18 inhabited islands of the Southern Sporades, which include the twelve islands of the Dodecanese Archipelago, whose largest and most famous island is Rhodes (540 sq mi; 1,399 sq km). Other important islands of this group are Kos, Kálimnos, and Pátmos, the latter noted as the place where St. John, in exile, wrote the Book of Revelation.

Crete, the site of the earliest European civilization, is the largest of Greek islands and the fifth-largest Mediterranean island, with an area of 3,207 sq mi (8,308 sq km). It is long, narrow, and, like all of Greece, mountainous. The mountains that cover the island are separated into four distinct masses by structural depressions. The island ranges from about 7 mi (11.2 km) to 36 mi (58 km) wide and is about 152 mi (245 km) long.

The Coast and Beaches

The coast of the Greek peninsula is mostly rocky, although there are some strips of lowland along the shore. The most distinctive formation along the coast of the northern Greek mainland is the Chalcidice (Chalkidhiki) Peninsula, from which three narrow, smaller peninsulas jut into the Aegean. These peninsulas form the Gulf of Kassándra and the Singitic Gulf. The port city of Thessaloníki is located on a natural harbor at the western end of Chalcidice, on the Gulf of Thérmai. It is second only to Athens both as a port and as a city. Farther to the east, the Thracian coastline is generally smooth and uniform.

The coast of central Greece is relatively even, except where it is indented by the Gulf of Euboea in the Aegean and the Amvrakikós Kólpos on the Ionian shore. Further south, the coast is very deeply indented on both sides of the country. Together, the Saronic Gulf in the east and the Gulf of Patras and Gulf of Corinth in the west nearly separate the Peloponnese from the rest of mainland Greece.

The coast of the Peloponnese has an uneven, four-fingered shape, forming the Gulf of Argolis, Gulf of Laconia, Gulf of Messenia, and Gulf of Kiparissía. It has several good harbors and includes some plains areas, notably around the cities of Sparta, Árgos, and Ilia. At its southern end, cliffs meet the sea on the capes of Akirítas, Matapan, and Maléa.

The northern coast of Crete is heavily indented, most notably by the Kólpos Khaníon in the west and Mirabello Gulf in the east, but has some fine natural harbors.

GEO-FACT

To explain their barren, rocky landscape, the Greeks adopted the legend that God poured all the world's soil through a sieve, and created Greece from the rocks that were left over.

CLIMATE AND VEGETATION

Temperature

Greece has a temperate Mediterranean climate moderated by both sea and mountain breezes. Summers are hot and dry, while winters are generally cool and rainy, but the weather at higher elevations is colder and wetter. Average January temperatures range from 43°F (6°C) in the northern city of Thessaloníki, to 50°F (10°C) in Athens, toward the southern end of the mainland peninsula, to 54°F (12°C) at Irâklion on Crete. The average July temperature at sea level is near 80°F (27°C), with the thermometer topping 100°F (38°C) on the hottest days.

Rainfall

Rainfall increases from south to north, ranging from 16 in (41 cm) in Athens to about 50 in (127 cm) on the island of Corfu. In addition to being lighter in the south, the rain is also less evenly distributed, with almost no rainfall during the winter months. In the north, rain is also lighter in the summer, but not as light as in the south.

Grasslands

The most extensive plains in Greece are found at the mouths of the Struma and Néstos rivers in the northern part of the country and in Thessaly, whose lowlands constitute the country's most fertile farmland.

Attica is mountainous in the north but levels off to plains that extend from Athens to the end of the peninsula. Though less fertile than those of Thessaly, these plains are arable enough to support the production of grains, olives, grapes, figs, and cotton, while the foothills make good pastures. Fertile lowlands are also found in the alluvial plains of the Peloponnese.

Forests

Vegetation varies with elevation. Mediterranean vegetation including olive, date, and fig, is found up to about 1,500 ft (460 m), oak, pine, and chestnut to 3,500 ft (1,070 m), and beech and fir at higher elevations. The Mediterranean scrub called maquis is abundant on the lower slopes of central and southern Greece.

Population Centers – Greece

(1991 POPULATION ESTIMATES)

Name	Population	Name	Population
Athens	772,072	Pátrai	153,344
Thessaloníki	406,413	Péristéri	137,288
Volos	383,967	Iráklion	116,178
Piraeus	182,671	Calithèa	114,233

SOURCE: United Nations Statistics Division.

Regions – Greece

POPULATIONS FROM 1991 CENSUS

Name	Population	Area (sq mi)	Area (sq km)	Capital
Thráki (Thrace)	604,254	5,466	14,157	Comotini
Attikí (Attica)	3,764,348	1,470	3,808	Athens
Dhytikí Ellás (West Greece)	742,419	4,382	11,350	Patras
Dhytikí Makedhonía (West Macedonia)	302,750	3,649	9,451	Kozani
Iónioi Nísoi (Ionian Islands)	214,274	891	2,307	Corfu
Ípiros (Epirus)	352,420	131,168	339,728	Ioannina
Kedrikí Makedhonía (Central Macedonia)	1,862,833	7,393	19,147	Thessaloníki
Kríti (Crete)	601,159	3,219	8,336	Iráklion
Nótion Aiyaíon (South Aegean)	298,745	2,041	5,286	Hermoupolis
Pelopónniisos (Peloponesus)	632,955	5,981	15,490	Tripolis
Stereá Ellás (Central Greece)	608,655	6,003	15,549	Lamia
Tessalía	754,893	5,420	14,037	Larissa
Vóreion Aiyaíon (North Aegean)	200,066	1,481	3,836	Mytilene

SOURCE: National Statistical Service of Greece as cited on Geo-Hive, 2002, http://www.geohive.com (accessed July 2002).

HUMAN POPULATION

Nearly two-thirds of the population lives in cities—nearly one quarter in the Athens metropolitan area alone. Of Greece's islands, 170 are inhabited. Other urban centers, including Thessaloníki, Pátri, and Irâklion, have also grown rapidly in recent decades.

NATURAL RESOURCES

Greece is rich in mineral resources, including petroleum, lignite, bauxite, marble, and the rare mineral chromium. The country has a modern, mixed economy, with industry, services, and tourism playing large roles.

FURTHER READINGS

Boatswain, Tim, and Colin Nicolson. *A Traveler's History of Greece.* 4th edition. New York: Interlink Books, 2001.

Dicks, Brian. *Greece: The Traveler's Guide to History and Mythology.* North Pomfret, Vt.: David & Charles, 1980.

Greece, Athens and the Mainland. New York: DK Publishing, 1997.

GrecceNow Project. *Greece Now.* http://www.greece.gr/ (accessed March 28, 2002).

The Greek Islands. New York: DK Publishing, 1997.

The Thomas Cook Guide to Greek Island Hopping. Peterborough, United Kingdom: Thomas Cook, 1998.

Van Dyck, Karen, ed. *Greece.* Insight Guides, APA Publications. Boston: Houghton Mifflin, 1995.

Greenland

Dependency of Denmark

- **Area:** 840,00 sq mi (2,175,600 sq km) / World size ranking: 14
- **Location:** An island, north of Canada on the North American continent, in the Northern and Western Hemispheres, bounded on the north by the Arctic Ocean, on the east by the Greenland Sea, on the southeast by Denmark Strait, on the south by the Atlantic Ocean and on the west by Baffin Bay and Davis Strait.
- **Coordinates:** 72° 00′ N, 40° 00′ W
- **Borders:** No international boundaries.
- **Coastline:** 27,333 mi (44,087 km)
- **Territorial Seas:** 3 NM
- **Highest Point:** Gunnbjorn, 12,136 ft (3,700 m)
- **Lowest Point:** Sea level
- **Longest Distances:** 1,660 mi (2,670 km) N-S; 800 mi (1,290 km) E-W
- **Longest River:** None of significant size.
- **Natural Hazards:** Continuous permafrost covers two-thirds of the island in the north
- **Population:** 56,352 (July 2001 est.) / World Rank: 194
- **Capital City:** Nuuk, located on the southwestern coast
- **Largest City:** Nuuk, 14,000 (2000)

OVERVIEW

Greenland is the largest island in the world, and is compossed of possibly the oldest rocks on Earth. Located on the North American Tectonic Plate, the island once was part of the North American continent, and is believed to have been separated from it by continental drift. More than half the country is within the Arctic Cir-

GEO-FACT

Above the Arctic Circle between late May and mid-July, the sun never entirely sets. This phenomenon is known as the midnight sun. Conversely, for much of the winter the sun does not come above the horizon.

cle, and of Greenland's total land area, 84 percent lies under its ice cap (or ice sheet). Of the ice-free area, which is along the middle west and south coasts, some 38,000 sq mi (150,000 sq km) are inhabited. Mountains run along the east and west coasts, and large glaciers and deep fjords line the coasts as well. In the southern region, there are lowland areas.

The climate is arctic, and, because of the island's size, temperatures vary between the various regions. Winters are severely cold, while summers are relatively mild, especially in sheltered areas. Snow falls in any month.

MOUNTAINS AND HILLS

Mountains

Mountains are situated along the eastern and western coasts. The highest range runs along the east coast, and houses the highest point on the island, Gunnbjorn Mountain (12,136 ft / 3,700 m).

Glaciers

Greenland's interior, and much of its coast, is entirely covered with ice. This is the second largest ice cap in the world; only that of Antarctica is larger. In some places it is estimated to be 11,000 ft (3,355 m) thick. Due to the enormous weight of the ice, the interior of the island has sunk into a vast concave basin that reaches a depth of 1,180 ft (360 m) below sea level.

Greenland's coasts are home to numerous glaciers, extensions of the interior ice cap. In the 1990s, Jakobshavn Glacier, located on the west coast, was found to be moving down from the ice cap at a rate of 100 ft (30 m) per day. Large icebergs break off from the glaciers into the sea and float mostly southward.

THE COAST, ISLANDS, AND THE OCEAN

Oceans and Seas

An island, Greenland is surrounded by the Arctic Ocean, Greenland Sea (to the northeast), Denmark Strait (to the southeast), and the Atlantic Ocean. Baffin Bay and Davis Strait lie between Greenland and the northeast coast of Canada. Many smaller islands lie off the coast of Greenland, the most important of which is Qeqertarsuaq.

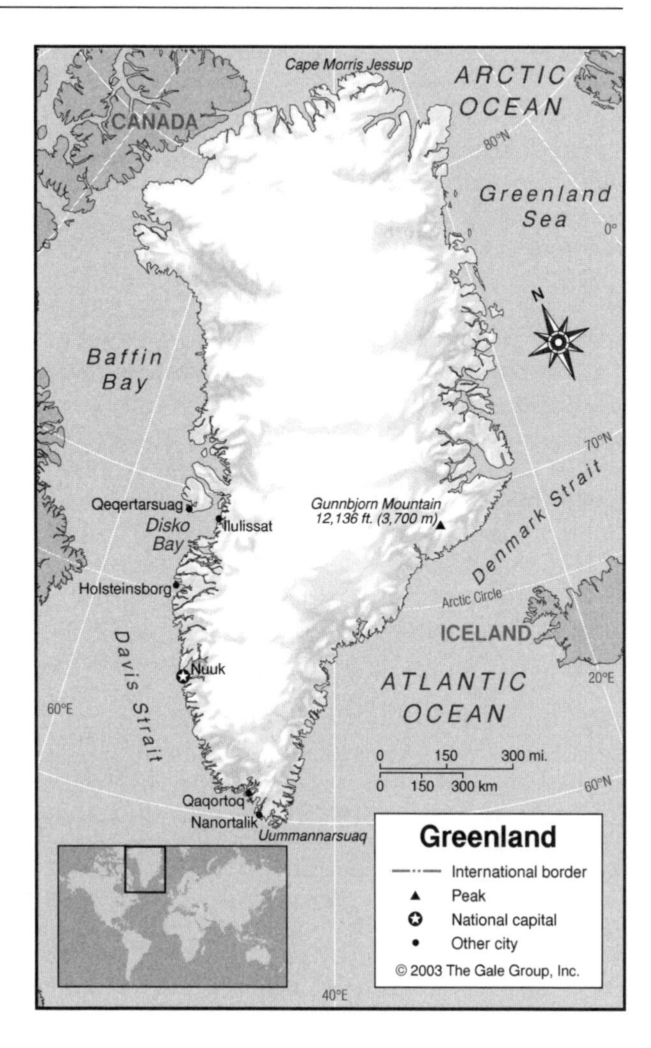

The Coast and Beaches

Deep fjords and glaciers line the coasts; in many areas the great ice sheet runs directly into the sea. Disko Bay is located on the western coast, and Ilulissat is the largest town in this area. A few harbors include Nanortalik, Qaqortoq, and Nuuk (Godthåb), the capital city. Uummannarsuaq (Cape Farewell) is located on the southernmost point of the island, and Cape Morris Jessup is the northernmost point. The Tunulliarfik (Eriksfjiord) flows in the southwest, and is the site of early settlements by Nordic settlers.

CLIMATE AND VEGETATION

Temperature

Greenland's climate is arctic, but is relatively mild along the coasts, particularly in the west and south. The northern region of the country and much of the interior rarely sees temperatures above freezing, never for prolonged periods. Temperature changes can be sudden in any one locality.

Winters are generally severe throughout the country. Even in the far south, the temperature often reaches –4°F (–20°C) or even lower. In the northern regions and on the ice, winter temperatures can plummet to –40°F (–40°C) for weeks. Maximum daytime temperatures in the summer, in the south, average between 50 and 64°F (10 and 18°C), and in the north, between 41 and 50°F (5 and 10°C).

Rainfall and Snow

Precipitation, mostly snow, is moderately heavy around the coast. It increases from north to south, and ranges between 10 and 45 in (25-114 cm). The northern part of the ice sheet receives less snow than the western and southern regions. Snow can, and usually does, fall during any month, throughout the country.

Grasslands

Lowland areas exist in the south, and in late summer, wild berries and wild flowers, including Arctic poppies, dandelion, harebell, and chamomile, grow in abundance. Not many trees exist on Greenland, but deciduous trees, such as alder, dwarf arctic birch, and willow, grow in the south in sheltered fjords.

HUMAN POPULATION

Greenland's population is grouped in a number of scattered settlements, which vary in size. Most of the population lives on the west coast. The eastern, northeastern, and extreme northern areas are almost completely uninhabitable.

NATURAL RESOURCES

Greenland is known to possess reserves of zinc, lead, iron ore, coal, molybdenum, gold, platinum, uranium, and cryolite. Oil and gas are suspected to be present. The extreme climate has prevented exploitation of most of these resources. The economy depends on the harvesting of fish, seals, and whales from the surrounding ocean, as well as subsidies from the Danish government.

FURTHER READINGS

Dupre, Lonnie. *Greenland Expedition: Where Ice is Born.* Minnetonka, Minn.: North Word Press, 2000.

Greenland Tourism a/s. http://www.greenland-guide.dk/gt/visit/green-10.htm#top (Accessed March 18, 2002).

Lepthien, Emilie. *Greenland.* Chicago: Children's Press, 1989.

Swaney, Deanna. *Iceland, Greenland & the Faroe Islands – a travel survival kit.* 2nd ed. Berkeley: Lonely Planet Publications, 1994.

Grenada

- **Area:** 131 sq mi (340 sq km) / World Rank: 192
- **Location:** Northern and Western hemispheres, in the Caribbean Sea, about 68 mi (109 km) south-southwest of St. Vincent and 90 mi (145 km) northwest of Trinidad
- **Coordinates:** 12°07′N, 61°40′W
- **Borders:** No international boundaries
- **Coastline:** 75 mi (121 km)
- **Territorial Seas:** 12 NM
- **Highest Point:** Mt. St. Catherine, 2,756 ft (840 m)
- **Lowest Point:** Sea level
- **Longest Distances:** 21 mi (34 km) NE-SW / 12 mi (19km) SE-NW
- **Longest River:** None of significant size
- **Natural Hazards:** Hurricanes
- **Population:** 89,227 (July 2001 est.) / World Rank: 186
- **Capital City:** St. George's, on the western coast of the main island
- **Largest City:** St. George's, 33,000 (2000 est.)

OVERVIEW

The country of Grenada consists of the main island of Grenada and a number of smaller islands and islets. Grenada is the most southerly of the Windward Islands; the other islands in the country are part of the Grenadines chain that extends north to the country of St. Vincent. The islands are almost wholly volcanic, with a mountain mass at the center of the main island. Grenada lies on the Caribbean Tectonic Plate.

MOUNTAINS AND HILLS

Volcanic in origin, the terrain of Grenada is very rugged. The mountain mass in the center of the main island consists of a number of ridges fanning out across the island. Mount Saint Catherine, the country's highest point (2,756 ft / 840 m), is located in these central highlands.

INLAND WATERWAYS

Lakes have formed in some of the extinct volcanic craters on Grenada. Grand Etang, at the center of the main island, is the largest of the crater lakes. Lake Antoine and Levera Pond are close by. While many short, fast-running streams cross the terrain of the main island, there are no rivers of note in the country. Mangrove swamps can be found along the coast.

THE COAST, ISLANDS, AND THE OCEAN

Oceans and Seas

Grenada is in the southeastern corner of the Caribbean Sea. Lying beneath the sea off the coasts of both Grenada and Carriacou is some of the Caribbean's most dramatic underwater scenery. Coral reefs abound, and the wreck of the S.S. Bianca C, the largest shipwreck in the Caribbean, is also near Grenada.

Major Islands

The country consists of the island of Grenada, the most southerly of the Windward Islands; the islands of Carriacou, Ronde, and Petit Martinique to the north; and a number of smaller islets of the Grenadines. The small islets include Diamond, Green, Sandy, Caille, Les Tantes, Frigate, Large, and Saline Islands.

The Coast and Beaches

The coastline of Grenada is dotted with many beaches and small bays. The best-known beach and principal tourist area is Grand Anse Bay, near St. George's, a broad beach with white sand. The bay is formed by Point Salines, which juts westward at the southern end of Grenada.

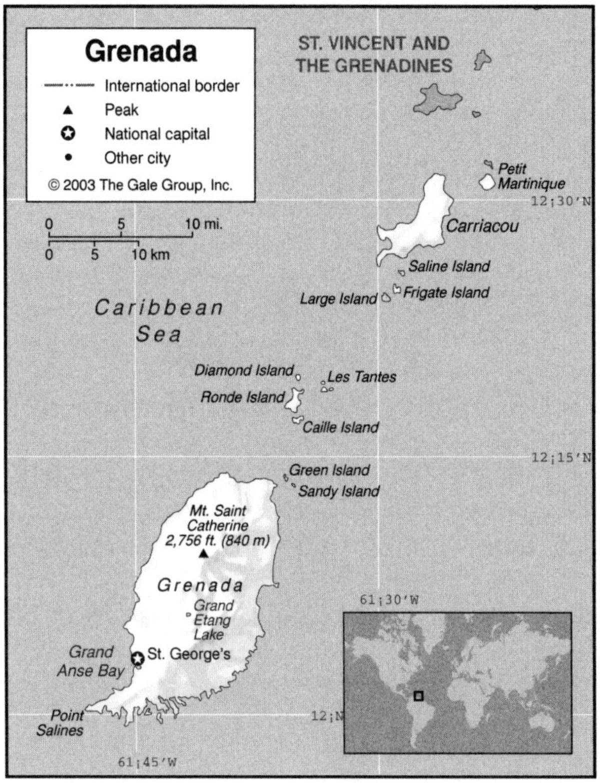

CLIMATE AND VEGETATION

Temperature

Grenada has a tropical climate moderated by cooling trade winds, with temperatures ranging from 75°F (24°C) to 87°F (30°C) year round. The lowest temperatures occur between November and February.

Rainfall

Annual rainfall is roughly 60 in (150 cm) along the coast, although it can be double that in the central highlands. The driest season is between January and May. Even during the rainy season, from June to December, it rarely rains for more than an hour at a time and generally not every day. Hurricanes are a danger between June and November.

Parishes – Grenada

Name	Area (sq mi)	Area (sq km)
Carriacou	13	34
St. Andrew	35	91
St. David	18	47
St. George's	26	67
St. John	15	39
St. Mark	9	23
St. Patrick	17	44

SOURCE: *Geo-Data: The World Geographical Encyclopedia,* 2nd ed. Detroit: Gale Research, 1989.

Forests and Jungles

Rainforests cover the middle altitudes, and give way to the dry forests of the lowlands. Deforestation is occurring due to farming activities and the use of wood products for fuel.

HUMAN POPULATION

Grenada's population is made up of citizens of African, East-Indian, and European descent. The largest proportion of the population, about 75 percent, is descended from former African slaves. Most people live on Grenada itself, with more than one-third living in or near St. George's.

NATURAL RESOURCES

While tourism is the major economic activity, Grenada has long been a major source of spices, and its major crops for export include nutmeg, mace, bananas, and cocoa beans. Timber, tropical fruit, and deepwater harbors are an integral part of this country's natural resources.

FURTHER READINGS

Brizan, George. *Grenada, Island of Conflict: from Amerindians to People's Revolution, 1498–1979.* Totowa, NJ: Biblio Distribution Center, 1984.

Philpot, Don. *Grenada.* Lincolnwood, IL: Passport Books, 1996.

Schoenhals, Kai P., and Richard A. Melanson. *Revolution and Intervention in Grenada: The New Jewel Movement, the United States, and the Caribbean.* Boulder, CO: Westview Press, 1984.

Thorndike, Tony. *Grenada: Politics, Economics, and Society.* Boulder, CO: L. Rienner Publishers, 1985.

Guadeloupe

Overseas Department of France

- **Area:** 687 sq mi (1,780 sq km) / World Rank: 173
- **Location:** Northern and Western Hemispheres, in the Lesser Antilles in the Caribbean Sea, most northern of the Windward Islands
- **Coordinates:** 16°15′N, 61°35′W.
- **Borders:** 6.4 mi (10.2 km), all with Netherlands Antilles (on St. Maarten)
- **Coastline:** 191 miles (306 km)
- **Territorial Seas:** 12 NM
- **Highest Point:** Soufrière, 4,813 ft (1,467 m)
- **Lowest Point:** Sea level
- **Longest Distances:** 42 mi (67 km) E-W / 37 mi (60 km) N-S
- **Longest River:** None of significant size
- **Natural Hazards:** Hurricanes, volcanic activity
- **Population:** 431,170 (July 2001 est.) / World Rank: 165
- **Capital City:** Basse-Terre, on the southwestern coast of the island of Basse-Terre
- **Largest City:** Les Abymes, 63,054 (March 1999), on Grand-Terre

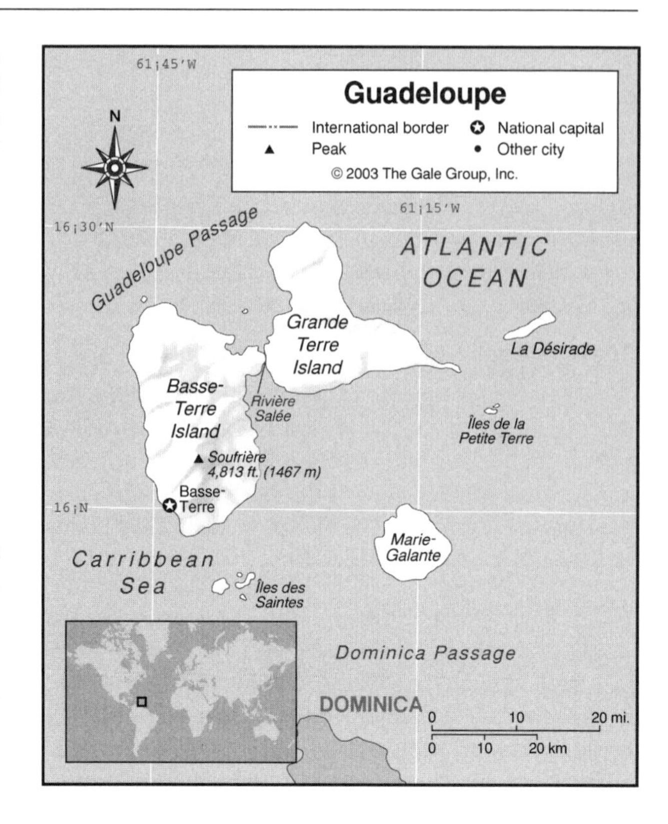

OVERVIEW

Guadeloupe is an archipelago of nine inhabited islands in the eastern Caribbean Sea. The main islands are between Antigua and Barbuda to the north and Dominica to the south, but there are also some islands further north near the Netherlands Antilles. It is a possession of France, which terms it an Overseas Department. The islands are mostly hilly or mountainous; some are volcanic in origin. Guadeloupe is on the Caribbean Tectonic Plate.

MOUNTAINS AND HILLS

Soufrière, at 4,813 ft (1,467 m), is the highest point in Guadeloupe. Located on Basse-Terre island, Soufrière is an active volcano that last erupted in the 1800s. The islands of Les Saintes and Saint-Barthélemy are volcanic in formation, with high, rugged terrain. Grande-Terre features rolling hills and limestone plains. The island of La Désirade has hills with elevations reaching nearly 900 ft (270m).

INLAND WATERWAYS

Basse-Terre's mountains receive much rainfall and this feeds numerous small rivers on the island. The Rivière Salée, actually a narrow channel of flowing seawater, divides Basse-Terre and Grande-Terre. Mangrove swamps can be found on the islands, near the coastal regions.

THE COAST, ISLANDS, AND THE OCEAN

The main islands in Guadeloupe are Basse-Terre and Grande-Terre. They are separated by the Rivière Salée, a narrow seawater channel. The islands of Marie-Galante, La Désirade, Îles des Saintes, and Îles de la Petite Terre lie nearby the main islands. Much further north are Saint-Barthélemy and Saint-Martin, the last of which is divided between Guadeloupe and the Netherlands Antilles.

All of the islands are part of the Lesser Antilles, with the main islands in the Windward Islands chain and the northern islands in the Leeward Islands chain. The Guadeloupe Passage runs between the Caribbean Sea and the Atlantic Ocean to the north of the main islands; the Dominica Passage connects these bodies of water to the south.

Arrondissements – Guadeloupe

POPULATIONS FROM 1992 CENSUS

Name	Area (sq mi)	Area (sq km)	Capital
Basse-Terre	332	861	Basse-Terre
Pointe-a-Pitre	297	769	Pointe-a-Pitre
Saint-Martin-Saint-Barthelemy	29	75	Marigot

SOURCE: *Geo-Data: The World Geographical Encyclopedia*, 2nd ed. Detroit: Gale Research, 1989.

CLIMATE AND VEGETATION

Temperature

Guadeloupe is warm year-round, tempered by trade winds or *alizés,* but the evenings are coolest in winter (December to February). The temperature hovers around 75°F (23°C) in winter and 90°F (32°C) in summer; the average humidity is 77 percent.

Rainfall

Annual rainfall ranges from 39 in (99 cm) on the outlying island of La Désirade to between 200–400 in (500–1000 cm) on the mountains of Basse-Terre island. February to April are the driest months, with rain falling an average of seven days a month and the humidity keeping in the realm of the tolerable. The wettest months are July to November, which is also hurricane season.

Forests and Jungles

Forty percent of the land area of Guadeloupe is forest and woodland. Bamboo, mangrove, and tropical hardwoods are abundant.

HUMAN POPULATION

Guadeloupe's population is concentrated on the two main islands. The vast majority of the inhabitants are descendants of Africans brought to the islands as slaves; they are predominantly Roman Catholic in faith.

NATURAL RESOURCES

Guadeloupe has no significant mineral resources. The economy is subsidized by France. The primary natural resources are tropical crops, especially bananas and sugar. The islands also have good beaches and a climate that fosters tourism.

FURTHER READINGS

The Civilized Explorer. *Guadeloupe-The Civilized Island.* http://www.cieux.com/gdlp.html (accessed April 3, 2002).

Guadeloupe. Cincinnati, OH: Seven Hills Book Distributors, 1994.

Pineau, Gisèle. *The Drifting of Spirits.* Translated by Michael Dash. London: Quartet Books, 1999.

Sullivan, Lynne M. *Martinique and Guadeloupe Alive!* Edison, NJ: Hunter, 2002.

Guatemala

- **Area:** 42,042 sq mi (108,890 sq km) / World Rank: 106
- **Location:** Northern and Western Hemispheres in Central America, south of Mexico, west of Belize, Honduras, and El Salvador, coastlines on the Pacific Ocean in the south and the Caribbean Sea in the east
- **Coordinates:** 15°30′N, 90°15′W
- **Borders:** 1,046 mi (1,687km) / Belize, 165 mi (266 km); El Salvador, 126 mi (203 km); Honduras, 159 mi (256 km); Mexico, 597 mi (962 km).
- **Coastline:** 205 mi (330 km)
- **Territorial Seas:** 12 NM
- **Highest Point:** Tajumulco Volcano, 13,830 ft (4,211 m)
- **Lowest Point:** Sea level
- **Longest Distances:** 284 mi (457 km) NNW-SSE / 266 mi (428 km) ENE-WSW
- **Longest River:** Usumacinta River, 690 mi (1110 km).
- **Largest Lake:** Lake Izabal, 324 sq mi (817sq km)
- **Natural Hazards:** earthquakes, volcanoes, Caribbean hurricanes
- **Population:** 12,974,361 (July 2001 est.) / World Rank: 63
- **Capital City:** Guatemala City, located in a highland valley south of the Sierra de Las Minas Mountains in the southeastern portion of the country
- **Largest City:** Guatemala City, 2,907,500 (2002 est.)

OVERVIEW

While Guatemala is slightly smaller in area than the state of Tennessee, it features great variety in climate and landforms. Much of the country is comprised of highlands. There are plateaus and hills where the great majority of people live, and also many high mountains and volcanoes. The mountain systems are more related to those of the West Indies than to those of North and South America, running east-west rather than north-south. In the north is the lowland region called El Petén.

Drainage to the Caribbean predominates, although there are many short and unnavigable rivers that flow to the Pacific from the southern highlands. Guatemala is on

the Caribbean Tectonic Plate, near the boundaries with the Cocos and the Caribbean Plates, and as a consequence earthquakes and volcanism are frequent and sometimes destructive.

MOUNTAINS AND HILLS

Mountains

The Sierra Madre is a system of mountains and high plateaus extending from Mexico through Guatemala to El Salvador and Honduras with more than 30 volcanoes, some still active, dotting the southern escarpment. The highest four are Tajumulco, the highest peak in the country at 13,830 ft (4,211 m) above sea level, Tacana (13,300 ft / 4,053 m), Fuego (12,579 ft / 3,839 m), and Agua (12,307 ft / 3,751 m).

Many smaller ranges make up the Sierra Madre or branch off from it. The Sierra de Chuacús branches due east from the Sierra Madre in the central part of the country. East-north-east of these mountains lie the Sierra de las Minas and the Mico Mountains. Together, these three chains extend across the entire country, making north to south travel difficult.

The Sierra de los Cuchumatanes, a great limestone massif, enters Guatemala from Mexico in the northwest. The height of the Cuchumatanes ranges between 9,000 and 11,000 ft (2,743 and 3,352 m). To the east, but separated from the Cuchumatanes by the valley of the Salinas River, lies the Sierra de Chama. Still farther east and extending nearly to Amatique Bay lies the Sierra de Santa Cruz, just north of the Polochic River-Lake Izabal lowland.

Plateaus

The vast area of El Petén, comprising about one-third of the national territory, extends to the north of the mountain ranges into the Yucatan Peninsula. It is a rolling limestone plateau, between 500 and 700 ft (152 and 213 m) above sea level, covered with dense tropical rainforest, occasionally interspersed with savannas. The soils are relatively poor for agriculture.

There are other, smaller, plateaus among the mountains in the south. Many are lava plateaus, and can be as high as 8,000 ft (2,438 m) above sea level in the western section of the Sierra Madre.

Canyons

The lava plateaus and ash-filled basins of the mountains are often separated by deep ravines difficult to cross even on foot. Rivers falling abruptly from the mountains have cut these canyons out of the soft volcanic soil. Pockets of dense population are often isolated from one another by these ravines.

INLAND WATERWAYS

Lakes

There are two important lakes of volcanic origin in the Sierra Madre highlands. Lake Atitlán is said to be one of the most beautiful lakes in the world. The volcanoes Atitlán, San Pedro, and Toliman line its shores. The lake, over 1,000 ft (304 m) deep in places, receives a number of rivers, but its drainage is underground. Lake Amatitlán, just south of Guatemala City, is smaller and less spectacular. Steam rises from this warm-water lake, and medicinal sulfur springs are found along the banks. The lake has its outlet in the Michatoya River.

In the east is Lake Izabal, the largest lake in the country (27 mi / 43 km long, 12 mi / 19 km wide). It empties into the Caribbean through the Dulce River. Lake Petén Itzá is in the north. It is 15 mi long, 2 mi wide, and about 165 ft (50 m) deep, and has no visible outlet because its drainage is underground.

Rivers

Eighteen principal, though relatively short, rivers flow from the mountains to the Pacific Ocean. They are navigable only for short distances in small boats, but they have great potential for the production of hydroelectric power and, in fact, serve to supply the major portion of electric power available in the country.

The Motagua River rises in the Sierra de Chuacús and flows east for about 250 mi (402 km) until it empties into the Gulf of Honduras. On the last few miles of its course it serves as the boundary between Guatemala and Honduras. It is navigable for the last 120 mi (193 km) of its length, and receives many tributaries.

The Polochic River rises in Alta Verapaz and flows west, emptying into Lake Izabal, the largest lake in the country. The outlet of Lake Izabal is the Dulce River, which flows into Amatique Bay. Not far to the north is the Sarstún River. It serves, in the latter part of its course, as the boundary between Belize and Guatemala and links the El Petén region with the coast.

Further to the north is the Usumacinta River basin, which covers most of the El Petén region and some of the surrounding area. The Usumacinta flows northeast along the Mexican border before continuing into that country. Major tributaries include the Salinas River, the Pasión River, and the San Pedro River. The Belize River and the Azul River both rise in El Petén and empty into the Caribbean.

Wetlands

Wet lagoons filled with mangrove swamps lie just inland from the sandy Pacific shore. The Chiquimulilla Canal, which runs 70 mi (112 km) from the port of San Jose to the Salvadorean border, is part of the coastal lagoon but has been dredged to allow river traffic.

The swampy Polochic River-Lake Izabal lowland lies north of the Sierra de las Minas and the Mico Mountains and is separated from the Motagua River valley by them.

GEO-FACT

In the heart of El Petén's jungles is Tikal National Park. Within the park is one of the major centers of the native Mayan civilization, inhabited from the 6th century B.C. to the 10th century A.D. Its ceremonial center contains superb temples and palaces, and ramps leading to public squares. Remains of dwellings are scattered throughout the surrounding countryside.

THE COAST, ISLANDS, AND THE OCEAN

The Pacific Coast

Guatemala's Pacific coast is straight and open, with no natural harbors and relatively shallow offshore waters. Long stretches of black sand line the coast, with mangrove swamps behind them, and a coastal plain behind that.

The Caribbean Coast

In the east, Guatemala borders on the Caribbean Sea, in the form of the Gulf of Honduras. The coast along the Gulf itself is flat and open to Caribbean storms. Amatique Bay, however, which is 16 km (10 mi) wide and 40 km (25 mi) long, is sheltered, and the country's major port, Puerto Barrios, is located on its shores. Three valley corridors extend inland from the Caribbean coast, linking it to the interior.

CLIMATE AND VEGETATION

Temperature

The climate ranges from hot and humid in parts of the lowlands to very cold in the highlands, where frosts are common in some months and where snow falls occasionally. Average annual temperatures at the coast range from 77–86°F (25–30°C), 68°F (20°C) in the central highlands, and 59°F (11°C) in the higher mountain areas.

Rainfall

The rainy season lasts from May through October inland, and into December along the coast; the dry season from November (or January) to April in the same regions.

The annual rainfall is heavy in the El Petén, the largest geographic region, averaging 80 in (203 cm) in the north and 150 in (441 cm) in the south.

Grasslands

The Pacific coastal plain is predominantly savanna, interspersed with semi-deciduous forests that line the streams originating in the highlands. Most of the savanna is given over to cattle ranching.

Forests and Jungles

In 1994, forests covered nearly 55 percent of the total land area of Guatemala, an increase of almost 12 percent from 1979. This increase masks the fact that over half of Guatemala's indigenous forests were destroyed in the 20th century. The type of forest varies by location. The highlands above 5,000 ft (1,524 m) are covered by the remnants of a once extensive pine and oak forest, which was cleared for the highland subsistence agriculture that now prevails. The forest cover disappears above 10,000 ft (3,048 m). In the north, much of El Petén remains covered by rain forest.

HUMAN POPULATION

Guatemala has a population exceeding 13.7 million people and a population growth rate of 2.6 percent. Most of the population lives in the southern highlands; the coasts and El Petén are sparsely settled. The indigenous

Population Centers – Guatemala

(2002 POPULATION ESTIMATES)

Name	Population
Guatemala City (capital)	2,907,500
Mixco	209,791
Esquintla	180,000
Quezal Tenango	171,700

SOURCE: Projected from United Nations Statistics Division data.

Departments – Guatemala

POPULATIONS FROM 1994 CENSUS

Name	Population	Area (sq mi)	Area (sq km)	Capital
Alta Verapaz	543,777	3,354	8,686	Cobin
Baja Verapaz	155,480	1,206	3,124	Salami
Chimaltenango	314,813	764	1,979	Chimaltenango
Chiquimula	230,767	917	2,376	Chiquimula
El Progreso	108,400	742	1,922	Progreso
Escuintla	386,534	1,693	4,384	Escintla
Guatemala	1,813,825	821	2,126	Guatemala City
Huehuetenango	634,374	2,857	7,400	Huehuetenango
Izabal	253,153	3,490	9,038	Puerto Barrios
Jalapa	196,940	797	2,063	Jalapa
Jutiapa	307,491	1,243	3,219	Jutiapa
Peten	224,884	13,843	35,854	Ciudad Flores
Quezaltenango	503,857	753	1,951	Quezal Tenango
Quiche	437,669	3,235	8,378	Santa Cruz
Retalhuleu	188,764	717	1,856	Retalhuleu
Sacatepequez	180,647	180	465	Antigua Guatemala
San Marcos	645,418	1,464	3,791	San Marcos
Santa Rosa	246,698	1,141	2,955	Cuilapa
Solola	222,094	410	1,061	Solola
Suchitepequez	307,187	969	2,510	Mazatenango
Totonicapan	272,094	410	1,061	Totonicapin
Zacapa	157,008	1,039	2,690	Zacapa

SOURCE: 1994 Census of Population, National Statistics Institute of Guatemala.

Maya, who continue to speak Mayan languages and follow Mayan traditions, make up about half of the total populace. Most of the rest are assimilated Mestizo (assimilated Mayans or mixed Amerindian-Spanish.

NATURAL RESOURCES

Petroleum and nickel are present in Guatemala. The forests yield timber and rare tropical woods. Guatemala's rivers are an excellent source of hydropower, and the waters off the coast waterways are rich in fish. Agriculture is a major part of the economy, with coffee, sugar, and bananas among the most valuable crops.

FURTHER READINGS

Barry, Tom. *Guatemala: A Country Guide.* Albuquerque, NM: Inter-Hemispheric Education Resource Center, 1989.

Cummins, Ronnie. *Guatemala.* Milwaukee, WI: G. Stevens Children's Books, 1990.

Kelsey, Vera, and Lilly de Jongh Osborne. *Four Keys to Guatemala.* New York: Funk & Wagnells Company, 1939.

Perl, Lila. *Guatemala, Central America's Living Past.* New York: Morrow, 1982.

Guinea

- **Area:** 94,926 sq mi (245,857 sq km) / World Rank: 77
- **Location:** Northern and Western Hemispheres, on coastal West Africa, south of Senegal, southwest of Mali, west of Côte d'Ivoire, north of Liberia, north of Sierra Leone, and east of Guinea-Bissau.
- **Coordinates:** 11°00′N, 10°00′W
- **Borders:** 2,112 mi (3,399 km) / Senegal, 205 mi (330 km); Mali, 533 mi (858 km); Cote d'Ivoire, 379 mi (610 km); Liberia, 350 mi (563 km); Sierra Leone, 405 mi (652 km); Guinea-Bissau, 240 mi (386 km)
- **Coastline:** 199 mi (320 km)
- **Territorial Seas:** 12 NM
- **Highest Point:** Mount Nimba, 5,748 ft (1,752 m)
- **Lowest Point:** Sea level
- **Longest Distances:** 516 mi (831 km) SE-NW; 306 mi (493 km) NE-SW
- **Longest River:** Niger River, 2,460 mi (4,100 km)
- **Natural Hazards:** Flooding, harmattan haze may reduce visibility during dry season
- **Population:** 7,613,870 (July 2001 est.) / World Rank: 88
- **Capital City:** Conakry, built upon a peninsula facing the Atlantic Ocean
- **Largest City:** Conakry, population 1,896,000 (2000 est.)

OVERVIEW

Guinea is slightly smaller than Oregon and is situated on the southwestern edge of the great bulge of West Africa, between roughly 7° and 12.5° north of the equator. It has four geographic regions, each with its own morphological features. Lower Guinea, or Maritime Guinea, consists mainly of a coastal plain that rises steeply to high central plateaus known as the Fouta Djallon, or 'The Fouta' in Middle Guinea. To the northeast are broad savannas in Upper Guinea, and to the southeast a combination of mountains and uplands in the Forest Region.

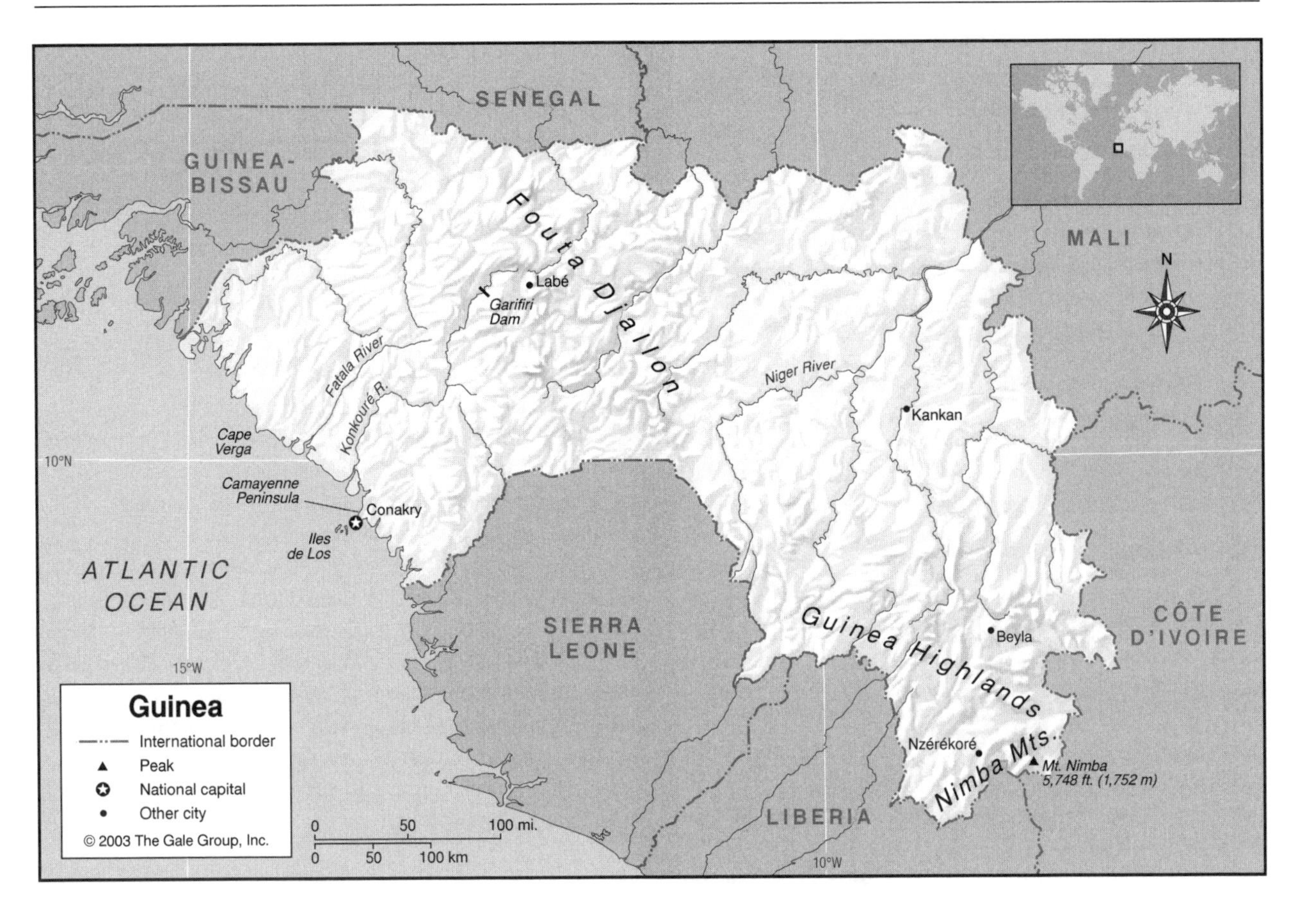

MOUNTAINS AND HILLS

Mountains and Uplands

The Guinea Highlands in the Forest Region have general elevations ranging from about 1,500 ft (457 m) above sea level in the west to over 3,000 ft (914 m) in the east; peaks at several points attain 4,000 ft (1,219 m) and higher. Southeast of Nzérékoré are the Nimba Mountains on the Liberian and Ivory Coast frontiers. Located in this range is Mount Nimba, Guinea's highest point at 5,748 ft (1,752 m).

Plateaus

The Fouta Djallon occupies most of Middle Guinea and consists of a complex, elevated, relatively level plateau. About 5,000 sq mi (12,950 sq km) of this area are over 3,000 ft (914 m) above sea level. The plateaus are deeply cut in many places by narrow valleys, many of which run at roughly right angles, giving the region a checkerboard appearance. A number of major valleys extend for long distances, providing important lines of communication; the railroad from Conakry to Kankan runs in part through one of these valleys. In the south, foothills occur in steep steps having escarpments from several hundred to well over 1,000 ft (304 m) high.

> **GEO-FACT**
>
> The Garifiri hydroelectric dam on the Konkouré River, inaugurated by French President Jacques Chirac, opened ahead of schedule in 1999. It features a 75-megawatt power plant, a reservoir of 7.51 billion cubic feet (2 billion m^3), and a spillway that evacuates 70,580 cubic feet (2,000 m^3) per second.

INLAND WATERWAYS

Rivers

Guinea is the "water tower" of West Africa. Over one-half of West Africa's principal rivers rise either in the Fouta Djallon or the Guinea Highlands of the Forest Region. These highlands divide the upper Niger River basin from rivers flowing westward through Guinea, Sierra Leone, and Liberia to the Atlantic Ocean, and from those flowing northward into the Gambia and Sénégal river watersheds.

The Niger River system drains more than one-third of the country's total area, while many short rivers originating either in the Fouta Djallon or in its foothills cascade onto, and then wind through, the coastal plain to estuaries along the Atlantic. Among the more important for navigation purposes are the Rio Nunez and the Fatala. The Konkouré River, north of Conakry, provides hydroelectric power for the capital.

Wetlands

Flooding occurs during the rainy season along the banks of sluggish rivers in the Niger River basin, including stretches of the Niger. Tidal marshes and swampy flats surround Atlantic coast estuaries.

THE COAST, ISLANDS, AND THE OCEAN

Oceans and Seas

Guinea's irregular coast is broken up by a number of bays and estuaries facing the Atlantic Ocean.

Major Islands

The Îles de Los, a cluster of small volcanic islands off Conakry, are inhabited and draw tourists during the dry season when seas are calm.

The Coast and Beaches

Mangroves line Guinea's coast through which oxbow rivers open onto the sea. The coast is broken at only two points where spurs of resistant rock formations jut into the ocean; one is found at Cape Verga in the north, and the other is the Camayenne (or Kaloum) Peninsula on which Conakry is situated. Tides are high along the entire coast, reaching fifteen or more feet, which results in brackish water in estuaries many miles inland. Behind the coastal swamps lies an alluvial plain averaging about thirty miles wide but considerably narrower in its central section.

CLIMATE AND VEGETATION

Temperature

Temperature varies according to region and season in Guinea. Conakry is humid nearly all year round with fairly uniform temperatures from 73°F (23°C) to 84°F (29°C). Temperatures in the Fouta Djallon and Forest highlands are more moderate, and in the dry season may vary daily by 25°F (14°C).

Rainfall

Conakry and the maritime region receive as much as 169 in (430 cm) of monsoon rains annually with half of the rainfall in July-August. The Fouta receives about 60 to 80 in (150 to 200 cm), while the Forest highlands receive 110 in (280 cm) annually.

Grasslands

Tall grasses interspersed with lightly wooded savanna is the principal feature of Upper Guinea; grasses also have colonized deforested areas of the Forest Region plains and highlands.

Forests and Jungles

Dense rain forest—now largely secondary growth—characterizes the Forest Region in areas below 2,000 ft (609 m). Higher areas are more lightly forested. The area around Beyla and Nzérékoré consists of rolling plains at one time probably covered by rain forest.

HUMAN POPULATION

Much of Guinea's population is highly concentrated in one area—roughly a quarter of the people live in the captial city. Other than Kankan, none of the other cities boasts more than 100,000 inhabitants. The other cities and towns, however, are well distributed across the country. Unlike many African nations, with so many rivers Guinea can support settlement throughout its land. The overall population density for the country is just over 79.4 people per sq mi (30 per sq km). The Peuhl constitute 35 percent of the population, the Malinke 26 percent, the Soussou 20 percent, and the Foresters 15 percent, with smaller groups making up the remainder. French is the official language along with eight Guinean languages. Until recently Guinea hosted almost half a million refugees from Sierra Leone and Liberia, although many had returned home by 2002.

NATURAL RESOURCES

Guinea is rich in iron ore, diamonds, gold, and uranium. Located on the Atlantic Ocean as well as having numerous rivers, it is has a near endless supply of fish. Its rivers also afford Guinea the resource of hydropower. Additionally, the country possesses more than 30 percent of the world's bauxite reserves; Guinea is the second largest bauxite producer in the world.

FURTHER READINGS

Boubah.com. http://www.boubah.com/English.htm (Accessed May 2002).

Laye, Camara. *The Dark Child.* Trans. Eva Thoby-Marcelin. New York: Farrar, Straus and Giroux, 1954.

Nelson, Harold D., et al, eds. "Area Handbook for Guinea." *Foreign Area Studies.* Washington, D.C.: American University, 1975.

Niane, Djibril Tamsir. *Sundiata: An Epic of Old Mali.* Trans. G.D. Pickett. Essex: Longman, 1965.

O'Toole, Thomas. *Historical Dictionary of Guinea.* Third Edition. Lanham, Md., and London: The Scarecrow Press, 1994.

Guinea-Bissau

- **Area:** 13,948 sq mi (36,125 sq km) / World Rank: 136
- **Location:** Northern and Western Hemispheres, on the coast of West Africa, west of Guinea and south of Senegal.
- **Coordinates:** 12°00′N, 15°00′W
- **Borders:** 450 mi (724 km) / Senegal, 210 mi (338 km); Guinea, 240 mi (386 km)
- **Coastline:** 217 mi (350 km)
- **Territorial Seas:** 12 NM
- **Highest Point:** Point on the Planalto de Gabú, 984 ft (300 m)
- **Lowest Point:** Sea level
- **Longest Distances:** 209 mi (336 km) N-S; 126 mi (203 km) E-W
- **Natural Hazards:** Drought, harmattan-induced dusty hazes, brush fires
- **Population:** 1,315,822 (July 2001 est.) / World Rank: 148
- **Capital City:** Bissau, centrally located on a coastal estuary facing the Atlantic Ocean
- **Largest City:** Bissau, population 233,000 (2000 est.)

OVERVIEW

Guinea-Bissau is located on the coast of West Africa where a large cluster of islands is found on the extensive continental shelf. The country is made up of a mainland, the Arquipélago dos Bijagós, and various coastal islands. The mainland relief consists of a coastal plain and a transition plateau forming the Planalto de Bafatá in the center and the Planalto de Gabú abutting on the Fouta Djallon. Guinea-Bissau is slightly less than three times the size of the state of Connecticut.

MOUNTAINS AND HILLS

Plateaus

Aside from the low-lying coastal plain and islands, Guinea-Bissau's most defining characteristic is the transitional plateau rising gradually from the plain to a few hundred feet in elevation. In the center of the country this plateau is called the Planalto de Bafatá, and to the eastern frontier with Guinea's Fouta Djallon highland, the Planalto de Gabú. The Planalto de Gabú reaches a maximum height of almost 1,000 ft (310 m) near the city of Buruntuma on the Guinea border.

INLAND WATERWAYS

Rivers

The country is drained by a number of meandering rivers flowing into the Atlantic through wide estuaries. There are six main rivers in Guinea-Bissau. The first, the Cacheu, flows near the northern border with Senegal and is also known as Farim for part of its course. The Mansôa flows from the center of the country and dumps into the Atlantic near the city of Bissau. The Gêba originates in Senegal and bisects the country. The Corubal originates in Guinea and meanders close to the southern border. On the southern border with Guinea is the Cacine. The last of the major rivers is the Rio Grande. These rivers provide the principal means of transportation. Ocean-going vessels of shallow draught can reach most of the main towns, and flat-bottomed tugs and barges can reach smaller settlements except those in the northeast. The water resources in this small country are abundant, but they are badly distributed in space and in time: 90 percent of the flow occurs in six months.

Wetlands

The low-lying coastal plain is characterized by wetlands that are submerged at high tide. Owing to excessive monsoon rains during the rainy season, swamps and marshes occur further inland as well.

THE COAST, ISLANDS, AND THE OCEAN

Oceans and Seas

Guinea-Bissau faces the Atlantic Ocean to the west. Coral reefs and islands dominate the coastal region.

Major Islands

Guinea-Bissau contains many islands. Located to the southwest of the capital city of Bissau, the Arquipélago dos Bijagós consists of over 18 islands, among them are Caravéla, Caraxe, Formosa, Uno, Orango, Orangozinho, Bubaque, and Roxa. The country also includes various other coastal islands such as Jeta, Bolama, Melo, Pecixe, Bissau, Areicas, and Como.

The Coast and Beaches

The coast of Guinea-Bissau is very irregular and deeply indented by swampy estuaries called "rias." Rias

GEO-FACT

About 10 percent of Guinea-Bissau's land is submerged at high tide. In the coastal area, problems of salt intrusion exist in the dry season and many "anti-salt" dams have been constructed.

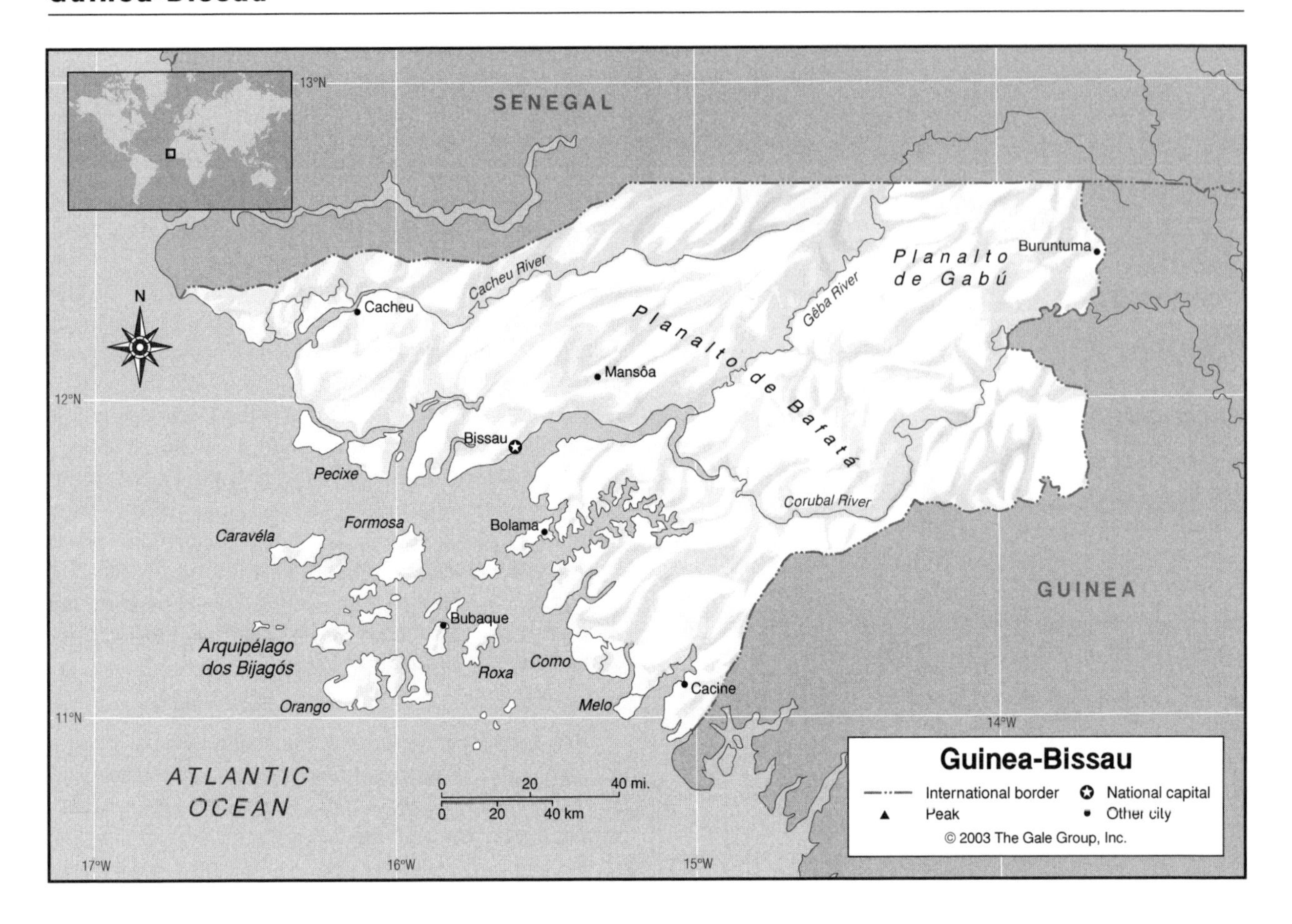

are fed by serpentine, mangrove-lined tidal rivers. The capital, Bissau, is located on the largest of these estuaries that snakes nearly into the center of the country.

CLIMATE AND VEGETATION

Temperature

Guinea-Bissau has a very moderate, tropical climate. The average temperature does not vary significantly all year—in the cooler rainy season, temperatures average from 79° to 82° F (26° to 28° C), and during the dry harmattan season, temperatures do not exceed 75° F (24° C) on average.

Rainfall

The rainy season lasts from mid-May to mid-November with rainfall exceeding 78 in (198 cm). Because of monsoon winds blowing off the ocean, the bulk of the rain falls during July and August. The harmattan season reverses the wind direction, blowing dry, dusty air from the Sahel across the country from mid-December to mid-April. This brings cooler temperatures and almost no precipitation.

Grasslands

Forty-six percent of the land in Guinea-Bissau is meadows and pastures. Savanna predominates in the east and northeast providing a mixture of lightly wooded forest interspersed with grasses.

Forests and Jungles

About 38 percent of the land is covered in forests and woodlands. Mangroves dominate the coastal region while tangled forests and jungle are found in the interior plains. Thick forests give way to less dense savanna cover and grasses on the planaltos.

HUMAN POPULATION

In 1999, the population density in Guinea-Bissau was 114 people per square mile (44 people per sq km), with most of the population living in small fishing or farming towns of less than 50,000 people. Nearly the entire population is of African tribal descent; Guinea-Bissau's largest ethnic group is the Balanta accounting for 32 percent of the population; followed by the Fula (Fulani or Peuhl), 22 percent; the Manjaca, 14.5 percent; the Mandinga (Malinke), 13 percent; Papel, 7 percent; and other about 10 percent. Europeans and mulattos make up 2 percent. Nearly half of the population holds indigenous beliefs, while 45 percent profess Islam, and 5 percent Christian faiths. Portuguese is the official language, but Crioulo and African traditional languages are widely spoken.

NATURAL RESOURCES

Somewhat challenged in the way of natural resources, Guinea-Bissau's economy depends heavily on fishing and agriculture. However, the land is rich in timber, phosphates, and bauxite. Unexploited deposits of petroleum are potentially a source of great wealth. Only nine percent of the land is arable with a mere 1 percent given to permanent farming.

FURTHER READINGS

Africa South of the Sahara 2002. "Guinea-Bissau." London: Europa Publishers, 2001.

Freire, Paulo. *Pedagogy in Process: The Letter to Guinea-Bissau.* N.p., 1978.

Lobban, R.A., and P.K. Mendy. *Historical Dictionary of Guinea-Bissau. 2nd ed.* Lanham, Md.: Scarecrow Press, 1997.

Lopes, Carlos. *Guinea-Bissau: From Liberation Struggle to Independent Statehood.* Boulder: Westview Press, 1987.

Guyana

- **Area:** 83,000 sq mi (214,970 sq km) / World Rank: 83
- **Location:** Northern and Western Hemispheres, Northern South America, bordering the Atlantic Ocean to the north, Suriname to the east, Brazil to the south and southwest, and Venezuela to the west
- **Coordinates:** 5°N, 59°W
- **Borders:** 1,526 mi (2,462 km) / Brazil, 694 mi (1,119 km); Suriname, 372 mi (600 km); Venezuela, 460 mi (743 km)
- **Coastline:** 285 mi (459 km)
- **Territorial Seas:** 12 NM
- **Highest Point:** Mt. Roraima, 9,301 ft (2,835 m)
- **Lowest Point:** Sea level
- **Longest Distances:** 500 mi (807 km) N-S; 270 mi (436 km) E-W
- **Longest River:** Essequibo, 600 mi (966 km)
- **Natural Hazards:** Flash floods
- **Population:** 697,181 (July 2001 est.) / World Rank: 157
- **Capital City:** Georgetown, northeast Guyana
- **Largest City:** Georgetown, 254,000 (2000 est.)

OVERVIEW

Guyana is a small independent republic located on the northeastern coast of South America, between Suriname and Venezuela. A former British colony, it is the only member of the British Commonwealth—and the only English-speaking country—in South America. Guyana is situated on the South American Tectonic Plate.

Guyana has four major types of terrain. A narrow but densely populated strip of plains extends the full length of the coast, varying in width from 2 to 30 mi (3.2 to 48 km). Beyond the coastal plain, and covering most of the country, lies a hilly, forested interior whose terrain is distinguished by the white-sand-and-clay composition of its soil. The interior also includes two major savanna regions, and highlands that rise in the south and west.

MOUNTAINS AND HILLS

Mountains

The Pakaraima Mountains rise from the Kaieteurian Plateau in the western part of the country, with peaks of over 9,000 ft (2,743 m) near Venezuela and Brazil, including the country's highest point, Mt. Roraima. This inaccessible mountain range separates Guyana from the Orinoco River watershed, and is the source of several of the main tributaries of the Essequibo River. It is also the site of the dramatic Kaieteur Falls, the world's seventh most forceful waterfall. Its falls plunge straight down 721 ft (220 m) at a rate of 175,500 gal (664,338 l) per second. Kaieteur is only one of many waterfalls in Guyana, including several other large ones.

Farther south the Kanuku Mountains extend east-west in the southwestern part of Guyana. Covering 1,750 sq mi (4,531 sq km) and reaching heights of 3,000 ft (914 m), they cut the Rupununi savanna region into two sections. The Acarai Mountains rise to elevations of over 2,000 ft (610 m) in the southeast.

Plateaus

The Kaieteurian Plateau, which together with the Pakaraima Range dominates west-central Guyana, is generally less than 2,000 ft (610 m) in elevation. Sedimentary sandstone and shale cover this ancient crystalline plateau, which was once below sea level.

Hills and Badlands

Beginning about 40 mi (64 km) inland, Guyana's hilly *zanderij* ("white-sand") area extends down the center of the country in a band whose width varies from 80 to100 mi (129 to161 km), widening in the southeast and covering over three-fourths of the country. The hills, whose elevations range from 50 ft (15 m) to 400 ft (120 m), gradually rise from west to east.

INLAND WATERWAYS

Rivers

Guyana has four major rivers—the Courantyne, Berbice, Demerara, and Essequibo—which flow northward

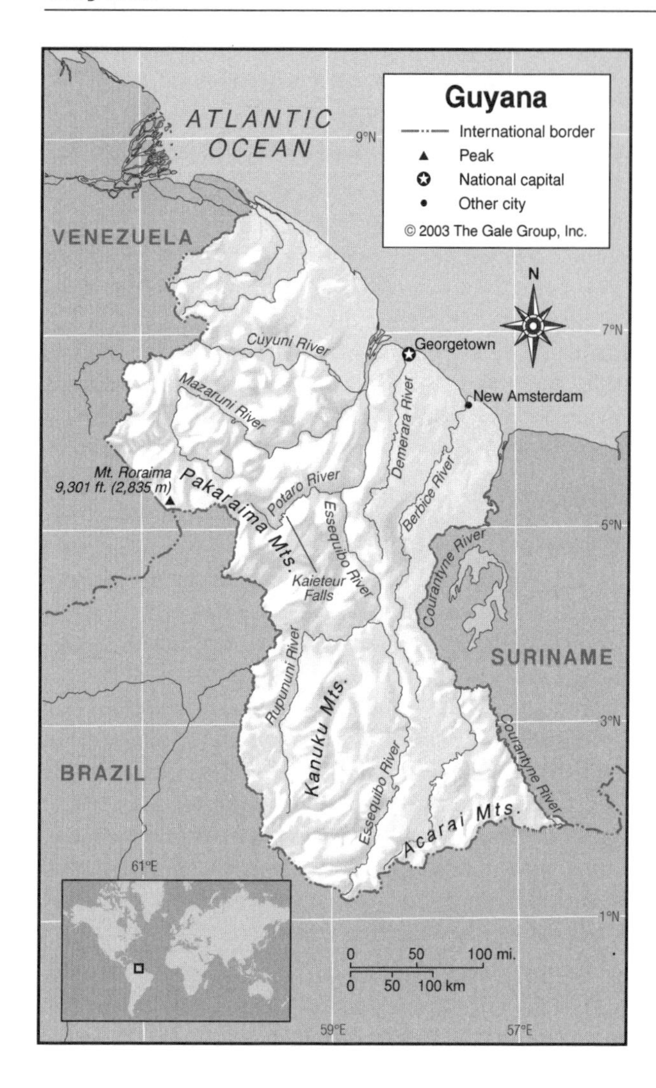

and empty into the Atlantic. Their courses become parallel in the northeastern part of the country as they near the coast.

The longest and widest is the Essequibo, which has its source in Brazil, as does the Courantyne, whose course forms Guyana's border with Suriname. The Potaro, Mazaruni, and Cuyuni Rivers, all tributaries of the Essequibo, drain the northwestern part of the country. The Rupununi River flows through the savanna land in the southwest that bears its name.

Guyana's many waterways are largely unnavigable, with passage impeded by swamps, rapids, sandbars, and other hazards.

Wetlands

Poor drainage in the mountains and savanna areas causes flooding that leads to the formation of swampland along Guyana's rivers. In addition, the coastal plain is cut off from the forested interior zone by a barrier of swamps that are prevented from intruding into the croplands of the plain by a series of "back-dams." These swamps serve a useful purpose as natural reservoirs during periods of drought.

THE COAST, ISLANDS, AND THE OCEAN

Oceans and Seas

The northern coast of Guyana borders the southeastern North Atlantic Ocean.

The Coast and Beaches

The deep indentation at the mouth of the Essequibo River divides Guyana's coast into two nearly equal sections. The one to the west is smooth, while the one to the east, which includes the mouths of the country's four major rivers, is more indented. The two major indentations occur at the mouth of the Essequibo and the mouth of the Courantyne, at the eastern end of the coast on the border with Suriname.

The coast is protected by 140 mi (225 km) of seawall and an extensive system of drainage canals to keep it from flooding at high tide, as much of the coastal plain lies below sea level. Guyana has no beaches in the usual sense. Silt carried to the coast by the rivers that drain into the Atlantic keeps the water off Guyana a brown churning mass of sandbars, semiliquid mud, and clouded water. Shifting sandbars off the coast at Georgetown and New Amsterdam require special shipping provisions at these ports. Mud flats continue for up to 15 mi (24 km) offshore before navigation is considered free.

CLIMATE AND VEGETATION

Temperature

Guyana has a hot, humid subtropical climate moderated by trade winds off the Atlantic. There is little temperature variation between seasons. Temperatures rarely rise above 90°F (32°C) or fall below 70 °F (21°C). The average annual temperature in the capital city of Georgetown is 81°F (27°C).

Rainfall

Average annual rainfall ranges from about 65 in (165 cm) in the savanna regions to 90 in (229 cm) on the coast and in elevated parts of the interior. The coastal areas have two rainy seasons—one between November and January and the other between May and July—while the savanna has only one, between April and August.

Grasslands

The narrow coastal plain varies in width from 10 to 40 mi (16 to 65 km). Except for palm trees and other tropical vegetation, the coastal plain looks much like the coast of the Netherlands due to the extensive system of sea defenses. The first mile behind the seawall is generally used for either pasturage or rice, which can survive the higher water table and higher salinity of the soil. The coastal plain is made up largely of highly fertile alluvial mud that overlays the white sand and clay carried seaward by the rivers of Guyana.

Guyana has two savanna regions. The largest is the Rupununi in the extreme southwestern part of the country, with elevations of approximately 250 ft (76 m). Divided into northern and southern sections by the Kanuku Mountains, it covers approximately 6,000 sq mi (15,540 sq km). A second area of savanna, the "intermediate savanna," lies about 60 mi (96 km) inland from the mouth of the Berbice River. Enclosed by the forests of the white sands belt, it covers nearly 2,000 sq mi (5,180 sq km). Both savannas support only sparse grasses for pasturage.

Forests and Jungles

The tropical rain forests that cover at least three-fourths of Guyana are thought to contain as many as 1,000 tree species. The wallaba and greenheart thrive in the areas with sandy soils, while the species found in the swampier regions include crabwood as well as various rubber trees, and hardwoods including the hubaballi. Coconut palms and other palm species grow in the savannas.

HUMAN POPULATION

The narrow coastal strip, which accounts for 5 percent of Guyana's territory, is home to 90 percent of its population.

NATURAL RESOURCES

The white-sand hills of Guyana's interior are a rich source of bauxite, the country's most important mineral resource. Also mined are gold, diamonds, and crushed stone. Forests cover most of the interior, but limited accessibility reduces their potential as a source of timber, although greenheart and other hardwoods are processed for commercial purposes.

FURTHER READINGS

Adamson, Alan H. *Sugar Without Slaves: The Political Economy of British Guiana*. New Haven, Conn.: Yale University Press, 1972.

Burnett, D. Graham. *Masters of All They Surveyed: Exploration, Geography, and a British El Dorado*. Chicago: University of Chicago Press, 2000.

Guyana News and Information. http://www.guyana.org/ (accessed April 4, 2002).

Spinner, Thomas J. *A Political and Social History of Guyana, 1945-1983*. Boulder, Colo: Westview Press, 1984.

Waugh, Evelyn. *Ninety-Two Days*. London: Duckworth, 1986.

Williams, Brackette F. *Stains on My Name, War in My Veins: Guyana and the Politics of Struggle*. Durham: Duke University Press, 1991.

Haiti

- **Area:** 10,714 sq mi (27,750 sq km) / World Rank: 146
- **Location:** Western third of the island of Hispaniola, in the Northern and Western Hemispheres, south of the Atlantic Ocean, west of the Dominican Republic, north and east of the Caribbean Sea
- **Coordinates:** 19°00′N, 72°25′W
- **Borders:** 170.7 mi (275km), all with the Dominican Republic
- **Coastline:** 1098 mi (1,771 km).
- **Territorial Seas:** 12 NM
- **Highest Point:** Mt. La Selle, 8,793 ft (2,680 m)
- **Lowest Point:** Sea level
- **Longest Distances:** 300 mi (485 km) ENE-WSW / 240 mi (385 km) SSE-NNW
- **Longest River:** Artibonite, 170 mi (280 km)
- **Largest Lake:** Lake Saumâtre, 65 sq mi (168 sq km)
- **Natural Hazards:** Hurricanes, other severe storms, earthquakes, floods, droughts
- **Population:** 6,964,549 (July 2001) / World Rank: 93
- **Capital City:** Port-au-Prince, on the western coast of Haiti, off the Bay of Port-au-Prince
- **Largest City:** Port-au-Prince, 1,791,000 (2000)

OVERVIEW

Haiti occupies the western one-third of Hispaniola, the second-largest island in the Caribbean; the eastern two-thirds is occupied by the Dominican Republic. Intricately convoluted mountains and hills cover most of the Haitian countryside. Less than 20 percent of the land lies at elevations below 600 ft (183 m), and about 40 percent is at elevations in excess of 1,500 ft (457 m). Haiti's coast is deeply indented in the west by the Gulf of Gonâve (Golfe de la Gonâve), with the long, narrow, Tiburon Peninsula in the south.

MOUNTAINS AND HILLS

Mountains

The border between Haiti and the Dominican Republic follows an irregular line extending from north to south, but the relief features of Hispaniola follow an east-

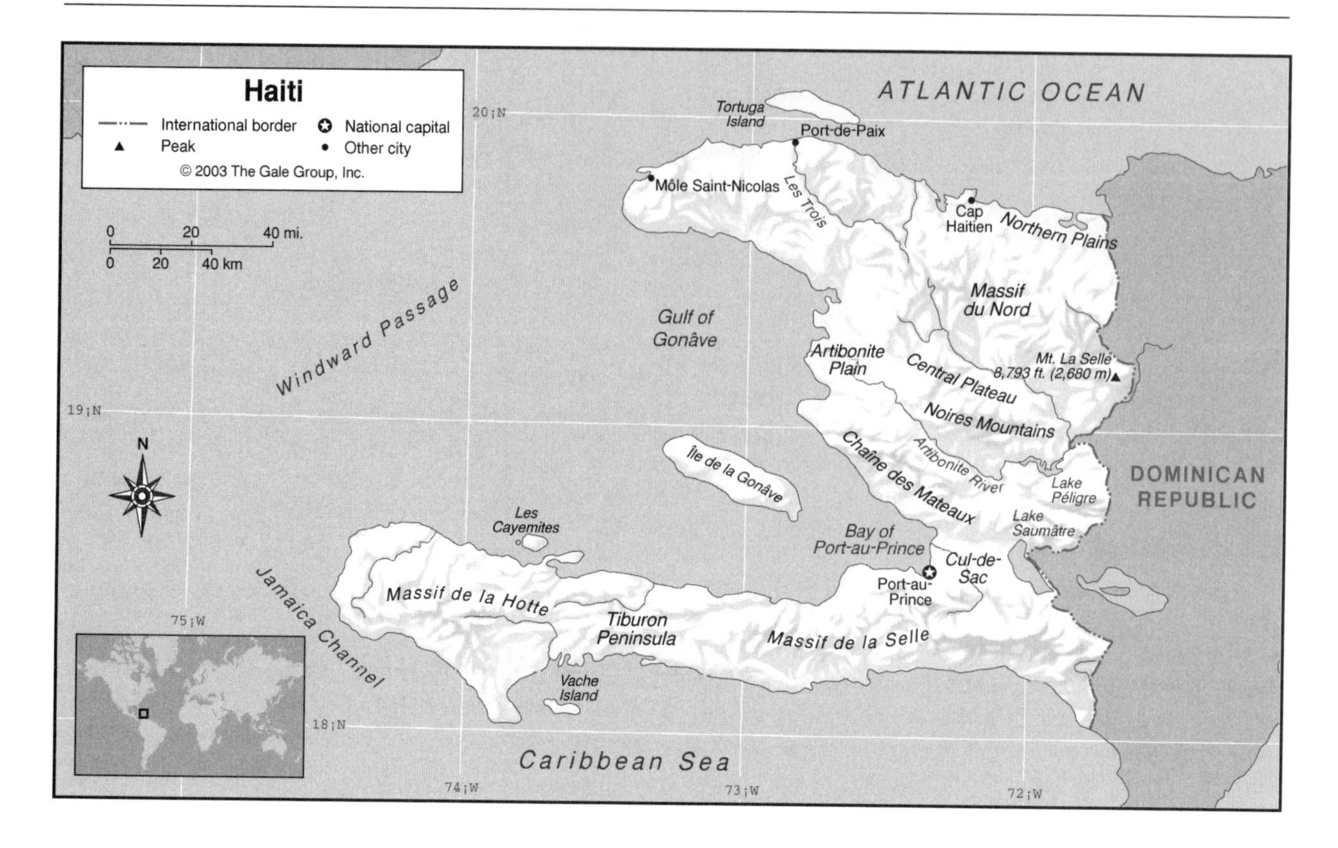

west axis. As a consequence, the principal mountain ranges and intermountain valleys are shared by the two countries.

The intricate highland pattern that covers more than three-fourths of Haiti is characterized by narrow-crested east-west ranges and spurs extending in random directions. Although there are at least five major systems and numerous spurs, the ranges meet one another to form a highland conglomerate that is discontinuous only in the south where the Cul-de-Sac lowland extends eastward from the Gulf of Gonâve at Port-au-Prince to the Dominican frontier.

In the north, the most extensive of the mountain systems is the Massif du Nord, which slants southeastward from the Atlantic Ocean near Port-de-Paix across the Dominican border. It is rugged and intricately dissected. Its complex geology includes sedimentary, magmatic, and plutonic rock, and limestone cliffs scar its slopes. To its west at the extremity of the island, satellite ranges extend to Môle-St.-Nicolas. To the southwest, the range called the Noires Mountains has altitudes up to 2,000 ft (610 m) and extends laterally across the country to a point where its approaches are separated by the Artibonite River from the Chaine de Mateaux, a range with a southwesterly axis that extends from the Gulf of Gonâve into the Dominican Republic as the Sierra de Neiba.

Separated from the northern mountains by the Cul-de-Sac is another mountain system that extends the full length of the long southern peninsula of Haiti to the frontier and into the Dominican Republic as the Sierra de Bahoruco. In the west it is the Massif de la Hotte, and in the east it is the Massif de la Selle. The latter range has several peaks with elevations of over 7,000 ft (2,133 m), and the country's highest peak, Mount La Selle (8,844 ft / 2,680 m).

Plateaus

Southward from the Massif du Nord, the Central Plateau extends eastward from the Noires Mountains to the Dominican frontier, where it joins the San Juan Valley. Its more than 840 m (1,351 km) of rolling terrain make it the largest of the country's flatlands. Slightly dissected and composed of consolidated and unconsolidated sediments, the plateau has an average elevation of about 1,000 ft (305 m) and its relatively thin soils are useful principally for pasturage.

INLAND WATERWAYS

Lakes

Lake Saumâtre (Etang Saumâtre) is located in the Cul-de-Sac close to the frontier and is the habitat of many exotic species of tropical wildlife. It is the largest lake in the country. There are also several smaller natural lakes and a reservoir known as Lake Péligre (Lac de Péligre), formed by the damming of the upper Artibonite River.

Rivers

Although over a hundred streams flow throughout Haiti, the only large river is the Artibonite. It is shallow but long, and its flow averages ten times that of any of the others. Second in length is the Les Trois, which spills into the Atlantic at the town of Port-de-Paix.

Wetlands

Wetlands in Haiti include the area around lake Lake Saumâtre (Etang Saumâtre, Saumatre Lagood or Lago Azuei), which used to be a channel separating Haiti from the Dominican Republic during the last ice age.

THE COAST, ISLANDS, AND THE OCEAN

Oceans and Seas

Haiti lies between the Atlantic Ocean and the Caribbean Sea. These are connected by the Windward Passage and the Jamaica Channel, which lie between Haiti and Cuba and Haiti and Jamaica, respectively.

Much of the Haitian coastline is rimmed by an underwater sedimentary platform that extends around the island of Hispaniola. There are many protected anchorages, but waters close to the shoreline tend to be shallow. Coral reefs are common, especially around Vache Island and the Cayemites.

Major Islands

Haiti includes the islands of Tortuga, Gonâve, Les Cayemites, and Vache. Haiti also claims the uninhabited island of Navassa, presently a U.S. possession, about 31 mi (50 km) west of Hispaniola.

The largest of the islands is Gonâve, located in the gulf of the same name off Port-au-Prince. Its area of approximately 80 sq mi (207 sq km) is made up of rugged terrain, and its highest point, Morne la Pierre, rises to more than 2,500 ft (762 m). Second in size is Tortuga, with an area of 70 sq mi (181 sq km). It lies in the Atlantic Ocean off Port-de-Paix.

The Coast and Beaches

Haiti's coastline is irregular and forms both a long southern peninsula, the Tiburon, and a shorter northern one. The peninsulas flank the large Gulf of Gonâve (Golfe de la Gonâve). At its eastern end the Gulf forms the Bay of Port-au-Prince.

CLIMATE AND VEGETATION

Temperature

Haiti's climate is tropical, with slight variations dependant upon altitudes and season; humidity is high in the coastal regions. The average annual temperature ranges from 70-86°F (22–30°C), with a slightly lower temperature in the interior highlands.

Population Centers – Haiti

(2000 POPULATION ESTIMATES)

Name	Population
Port-au-Prince (capital)	1,791,000
Carrefour	564,000
Delmas	465,000
Cap-Haitien	100,638

SOURCE: Projected from United Nations Statistics Division data.

Departments – Haiti

Name	Area (sq mi)	Area (sq km)	Capital
Artibonite	1,750	4,532	Gonaïves
Centre	1,429	3,700	Hinche
Grande Anse	1,268	3,284	Jérémie
Nord	790	2,045	Cap-Haitien
Nord-Est	676	1,752	Fort-Liberté
Nord-Ouest	899	2,330	Port-de-Paix
Ouest	1,795	4,649	Port-au-Prince
Sud	1,117	2,894	Les Cayes
Sud-Est	855	2,215	Jacmel

SOURCE: *Geo-Data: The World Geographical Encyclopedia,* 2nd ed. Detroit: Gale Research, 1989.

Rainfall

Haiti has two rainy seasons: April through June and October through November. Annual rainfall on the western coast near Port-au-Prince averages 54 in (137 cm). Haiti's dry season is from November to January. Hurricanes and tropical storms occur periodically between June and October. Rainfall tends to be higher on the mountain slopes and lower on the sheltered leeward slopes and in the valleys. It is semiarid where the mountains in the east cut off trade winds that bring moisture and cooler temperatures.

Grasslands

The most important of the lowland regions of the country are the Northern Plain, the Artibonite Plain, and the Cul-de-Sac. There are also scattered stretches of narrow coastal plain and small coastal basins, as well as pockets of level land tucked into the mountains.

The Northern Plain, which has an area of about 150 sq mi (362 sq km) located between the Atlantic Ocean and the Massif du Nord, extends eastward from near Cap Haitien to the Dominican border. Its rich soils are formed in part by erosion and in part by alluvial deposition. The plain is a geographical extension of the Cibao Valley in the Dominican Republic.

Separated from the Central Plateau by the Noires Mountains and located to the north of the Chaine de Mateaux, the funnel-shaped Artibonite Plain has an area of about 300 sq mi (777 sq km). Drained by the Artibonite River that crosses the central part of the country after rising in the Dominican Republic, the region is generally fertile, but near the coast its soils are too alkaline for intensive agriculture.

In the far south, the 150 sq mi (388 sq km) that make up the Cul-de-Sac lie between the Chaine de Mateaux and the Massif de la Selle. Extending eastward from Port-au-Prince to the border, the Cul-de-Sac becomes the Neiba Valley in the Dominican Republic. It is a down-faulted depression once filled by the waters of an ocean channel that separated the mountain ridges to the south from the mainland.

HUMAN POPULATION

Haiti has a population of nearly 7 million with an annual growth rate of 1.4 percent (July 2001). With a density of 717 persons per sq mi (277 persons per sq km), Haiti is the most densely populated country in the Western Hemisphere. It is estimated that 35 percent of the population lives in urban areas.

NATURAL RESOURCES

Haiti's natural resources include bauxite, copper, calcium carbonate, gold, marble, and hydropower. Few of these resources are exploited. The economy is extremely weak, with most workers devoted to subsistence agriculture.

FURTHER READINGS

Dell'Oro, Suzanne. *Haiti.* Mankato, MN: Bridgestone Books, 2002.

Griffiths, John. *Take a Trip to Haiti.* New York: F. Watts, 1989.

Hanmer, Trudy J. *Haiti.* New York: F. Watts, 1988.

Leyburn, James G. *The Haitian People.* Westport, CN: Greenwood Press, 1980.

Honduras

- **Area:** 43,267 sq mi (112,090 sq km) / World Rank: 102
- **Location:** Central America, Northern and Western hemispheres, bordered by the Caribbean Sea on the north and east, Nicaragua and the Pacific in the south, El Salvador in the southwest, Guatemala in the west
- **Coordinates:** 15°00′N, 86°30′W
- **Borders:** 1,454 mi (2,340 km) / El Salvador, 212 mi (342 km); Guatemala, 159 mi (256 km); Nicaragua, 572.6 mi (922 km)
- **Coastline:** 440 mi (710 km)
- **Territorial Seas:** 12 NM
- **Highest Point:** Cerro Las Minas, 9,417 ft (2,870 m)
- **Lowest Point:** Sea level
- **Longest Distances:** 412 mi (663 km) ENE-WSW / 197 mi (317 km) NNW-SSE
- **Longest River:** Coco River, 470 mi (750 km).
- **Largest Lake:** Lake Yojoa, roughly 150 sq mi (400 sq km)
- **Natural Hazards:** Earthquakes, hurricanes along Caribbean coast
- **Population:** 6,406,052/ World Rank: 96
- **Capital City:** Tegucigalpa, in the southwestern highlands
- **Largest City:** Tegucigalpa, 1.2 million (2002 est.)

OVERVIEW

Honduras, part of the isthmus of Central America, is the second largest Central American republic, with coasts on both the Pacific and the Caribbean Sea. It is heavily mountainous, but has lowland areas along both coasts and a deep, flat, basin that winds between the mountains from the plains around San Pedro Sula to the Gulf of Fonseca. Honduras also has many rivers, some of which have extensive valleys.

Honduras is located on the Caribbean Tectonic Plate, near its boundaries with the Cocos and North American Plates. Earthquakes are frequent as a result, although they are generally mild. Over 80 percent of the land is mountainous, thereby limiting the area suitable for cultivation and pastures. Much of the small amount of cultivated area is located in the flatlands, lofty plateaus, and river valleys that are between, and parallel to, the mountains. These temperate valleys and flatlands are also the primary areas of settlement except for the north coast banana district, which was retrieved from tropical forests in the twentieth century.

MOUNTAINS AND HILLS

Mountains

Honduras is the most mountainous country in Central America, and two distinct series of mountain ranges divide the country roughly into two halves; the north and the south.

In the north, mountain ranges extend from the Guatemala border on the west to the Platano River on the east. These northern ranges are all extensions of the Central American Cordillera, a mountain chain that traverses Central America from Mexico to Nicaragua. The chains of the Central American Cordillera trend east-northeast and west-southwest. They run largely parallel to the coast and to each other. The northern mountain ranges were formed by changes in the earth's surface several million years ago. Underneath the surface cover of limestone and sandstone, the mountains are composed of granite, mica, slate, and other materials.

The Volcanic Highlands extend from the border with El Salvador in the southwest and across the southern part of the country to the border with Nicaragua in the east. Unlike the mountains of the north, these southern ranges are newer, consisting of lava formed by volcanic eruption some 12 million years ago. Volcanic material has both eroded and been ejected from these highlands and forms fertile soil.

The Volcanic Highlands are higher overall than the Central American Cordillera chains. The highest peaks in the country, Cerro Las Minas (9,417 ft / 2,870 m) and Mount Celaque (9,345 ft / 2,848 m) are found here.

Plateaus

In the areas between one mountain range and the other, in both the Central American Cordillera ranges and the Volcanic Highlands, are plateaus, river valleys, and savannas. These intermountain flatlands average two to seven miles in width and are flanked by mountains usually 3,000 to 7,000 ft (914 to 2,133 m) in height. Historically, these level lands have been the most highly populated regions.

Canyons

A series of interconnected valleys at various elevations extends across the entire north-south distance of the country. These valleys were formed by rivers, most notably the Ulúa and its tributary the Humuya in the north and the Nacaome and Choluteca in the south.

INLAND WATERWAYS

Lakes

Lake Yojoa is the only large natural lake in Honduras. Surrounded by massive mountains, the lake itself sits at an altitude of approximately 2,200 ft (669 m) above sea level. It is drained on the south by Río Tepemechín and on the north by Río Blanco.

Rivers

There are many large river systems in Honduras. They have formed the valleys in which many of the people live, and their alluvial deposits have contributed to the fertility of the soil.

In the north, from west to east, are the Chamelecón, the Ulúa, the Aguán, the Sico, the Paulaya, the Platano,

the Sicre, the Patuca, and the Coco rivers. All the rivers in the north flow into the Caribbean Sea. The Ulúa and its tributaries drain one-third of the country. The Coco actually rises in the south, then flow north along the border with Nicaragua. It is the longest river in Honduras.

Other than the Coco, all the rivers that arise in the south flow toward the Pacific Ocean. The Lempa, Sumpul, and the Goascoran Rivers run nearly the entire length of Honduras's borders with El Salvador. Further east are the Nacaome and the Choluteca, which drains into the Gulf of Fonseca.

Wetlands

Along the northeast coastline and extending across the border into Nicaragua is the Mosquito Coast region. It is low-lying and swampy, largely covered by mangroves. It includes the large Laguna de Caratasca.

THE COAST, ISLANDS, AND THE OCEAN

Oceans and Seas

Honduras has a large northern coastline along the Caribbean Sea and a shorter one along the Pacific Ocean to the south. There are many large coral reefs in the Caribbean off Honduras's northern coast.

Major Islands

The small Cajones Cays (Swan Islands) lie some 110 mi (177 km) north-northeast of Punta Patuca in the Caribbean Sea. Also in the Caribbean are the Islas de la Bahía (Bay Islands) which include Guanaja, Utila, and, the largest, Roatán. Honduras also controls some small islands in the Gulf of Fonseca.

The Coast and Beaches

Honduras's northern coast is long and even, running east from the Gulf of Honduras for most of its length before curving south as it approaches the Nicaraguan border at Cape Gracias a Dios. A major inlet on the Caribbean coast, the Laguna de Caratasca provides a natural harbor for Puerto Lempira. The Pacific coast is much shorter and uneven. It is all on the sheltered waters of the Gulf of Fonseca.

Population Centers – Honduras

(1991 POPULATION ESTIMATES)

Name	Population
Tegucigalpa (capital)	670,100
San Pedro Sula	437,000

SOURCE: Projected from United Nations Statistics Division data.

Departments – Honduras

Name	Area (sq mi)	Area (sq km)	Capital
Atlántida	1,641	4,251	La Ceiba
Choluteca	1,626	4,211	Choluteca
Colón	3,427	8,875	Trujillo
Comayagua	2,006	5,196	Comayagua
Copán	1,237	3,203	Santa Rosa de Copán
Cortés	1,527	3,954	San Pedro Sula
El Paraíso	2,787	7,218	Yuscarán
Francisco Morazán	3,068	7,946	Tegucigalpa
Gracias a Dios	6,421	16,630	Puerto Lempira
Intibucá	1,186	3,072	La Esperanza
Islas de la Bahía	100	261	Roatán
La Paz	900	2,331	La Paz
Lempira	1,656	4,290	Gracias
Ocotepeque	649	1,680	Nueva Ocotepeque
Olancho	9,402	24,351	Juticalpa
Santa Barbara	1,975	5,115	Santa Bárbara
Valle	604	1,565	Nacaome
Yoro	3,065	7,939	Yoro

SOURCE: *Geo-Data: The World Geographical Encyclopedia,* 2nd ed. Detroit: Gale Research, 1989.

CLIMATE AND VEGETATION

Temperature

Temperature in Honduras varies primarily by elevation, not season. Along the coasts the temperatures average 88°F (31°C). As one proceeds up the higher elevations, cooler temperatures prevail, moving from 84°F (29°C) at 1,000 to 2,500 ft (300 to 760 m) to 73°F (23°C) at the highest elevations (above 2,500 ft / 760 m).

Rainfall

Honduras has two seasons; the dry season lasts from November through April, and the wet season from May through October. Although temperatures in Honduras are warm the entire year, rainfall and humidity vary, depending on the altitude, winds, location relative to the coast, and the mountains' blockage of cloud formations. On the northern Caribbean coast average rainfall is over 95 in (240 cm), and drops to only 33 in (84 cm) in the south near the capital. The Caribbean coast is subject to hurricanes.

Grasslands

Tropical lowland areas are found on both coasts, but are much larger in the north. The plains extend particularly far inland along the Ulúa River valley, about 75 mi (121 km). The southern coastal plains are much shorter, the lowlands extend only about 25 mi (40 km inland).

Forests and Jungles

Inland from the northern coast, great stands of Caribbean pines cover large portions of land. Other trees include hardwoods such as walnut, mahogany, cedar, and ebony. It is estimated that about 54 percent of Honduras is covered with forest and woodland.

HUMAN POPULATION

Honduras has a population of over 6.4 million with an annual growth rate of 2.43 percent. Most people live in the river valleys and plateaus in the west of the country, particularly in the Ulúa River valley and in Tegucigalpa, the capital. About 90 percent of the population is of mestizo or mixed descent. Native Indians are about 7 percent, and are concentrated near the Guatemalan border.

NATURAL RESOURCES

Commercial fishing exists in both the Gulf of Fonseca and the Gulf of Honduras in the Caribbean. Shrimp accounts for nearly 45 percent of the catch, and export of fish is an important part of the Honduran economy. Agriculture is also very important; bananas and coffee are the most valuable crops.

Honduras is known to have deposits of gold, silver, copper, lead, zinc, iron ore, antimony, and coal, but transportation difficulties limit their exploitation. Hydropower supplied almost 88 percent of electrical generation in Honduras in 1998.

FURTHER READINGS

Haynes, Tricia. *Let's Visit Honduras.* London: Burke Publishing Co., 1985.

Norsworthy, Kent. *Inside Honduras.* Albuquerque, NM: Inter-Hemispheric Education Resource Center, 1993.

Weddle, Ken. *Honduras in Pictures.* New York: Sterling Publishing Co., 1977.

Hungary

- **Area:** 35,919 sq mi (93,030 sq km) / World Rank: 107
- **Location:** Northern and Eastern Hemispheres, bordering Slovakia to the north, Ukraine to the northeast, Romania to the east, Yugoslavia to the south, Croatia to the southwest, and Slovenia and Austria to the west.
- **Coordinates:** 47°00′N, 20°00′E
- **Borders:** 1,248 mi (2,009 km) / Austria, 227 mi (366 km); Croatia, 204 mi (329 km); Romania, 275 mi (443 km); Slovakia, 320 mi (515 km); Slovenia, 63 mi (102 km); Ukraine 64 mi (103 km); Yugoslavia, 94 mi (151 km)
- **Coastline:** None
- **Territorial Seas:** None
- **Highest Point:** Kékes, 3,327 ft (1,014 m)
- **Lowest Point:** Tisza River, 256 ft (78 m)
- **Longest Distances:** 167 mi (268 km) N-S; 328 mi (528 km) E-W
- **Longest River:** Danube, 1,725 mi (2,776 km)
- **Largest Lake:** Balaton, 231 sq mi (598 sq km)
- **Natural Hazards:** Flooding
- **Population:** 10,106,017 (July 2001 est.) / World Rank: 78
- **Capital City:** Budapest, north-central Hungary
- **Largest City:** Budapest, population 2,017,000 (2000 metropolitan est.)

OVERVIEW

Hungary is a country located in the Carpathian Basin, in the heart of Central Europe. Before World War I, as part of the great Austro-Hungarian Empire, it stretched from the Carpathian Mountains in the north to the Adriatic Sea in the south, occupying parts of present-day Slovakia, Slovenia, Ukraine, Romania, Croatia, and Yugoslavia. Reduced to one-third of its former size by the treaty that ended the war, Hungary became a landlocked, predominantly flat country, with more than four-fifths of its terrain at altitudes below 656 ft (200 m).

The country can be divided into four major regions. To the north a long system of low mountains and hills stretches across the country for 250 mi (400 km) from the southwest to northeast, beginning at the Slovenian border and forming most of the border with Slovakia to the north. East of the Danube River and south of this mountain system is the Great Alföld, a large fertile plain that is Hungary's largest region and agricultural heartland. The northern mountains divide the land west of the Danube into two regions. In the northwest corner of the country is the Little Alföld. To the south is the hilly region known as Transdanubia, between the mountains and the Danube.

Hungary is located on the Eurasian Tectonic Plate.

MOUNTAINS AND HILLS

Mountains

Rising to elevations of 1,300 to 2,300 ft (400 to 700 m), the dolomite and limestone uplands and volcanic peaks of the Bakony Mountains constitute the major geographical feature in the part of Hungary's mountain system that lies to the west of the Danube River. The highest point in this range is the 2,309-ft high Mount Kőris, located just northwest of Veszprém. To the west of the Bakony Range, the uplands between Lake Balaton and the Austrian and Slovenian borders are foothills of the Alpine system. Most elevations here are below 1,000 ft (304 m), but a few isolated spots on the Austrian border rise to nearly 3,000 ft (914 m). Farther east, the Pilis Mountains rise between the Bakony range and the Danube.

The mountains and related hills east of the Danube account for 3,100 sq mi (4,988 sq km) of the country's area. They constitute the lower volcanic fringe of the Carpathian Mountains, the only uplands in the country that are part of the Carpathian system. The individual ranges in the group extend northeastward from the gorge of the Danube River near Esztergom for about 140 mi (225 km). Their highest point—and the highest point in Hungary—is Mount Kékes in the Mátra range, pushing to a height of 3,327 ft (1,014 m).

Hills and Badlands

The hills of Hungary's band of northern uplands rise to elevations of 800 to 1,000 ft (244 to 305 m). Farther south, the Transdanubia region is composed of rolling, hilly land rising to elevations of 2,000 ft (610 m). The Mecsek Mountains to the south are part of this region.

INLAND WATERWAYS

Lakes

Lake Balaton, 75 mi (120 km) southwest of Budapest, is Hungary's largest lake, and also the largest freshwater lake in Central Europe. It is about 45 mi (72 km) long, and its width varies, never reaching more than eight miles. It averages a little more than ten feet in depth and has a surface area of 231 sq mi (598 sq km). There are few other lakes in Hungary. Lake Fertő (Neusiedler See in German) on the northwestern border is shared with Austria, and Hungary's portion is only about one-fourth of the total. A shallow lake, with depths averaging only about three feet, it frequently freezes over entirely. Lake

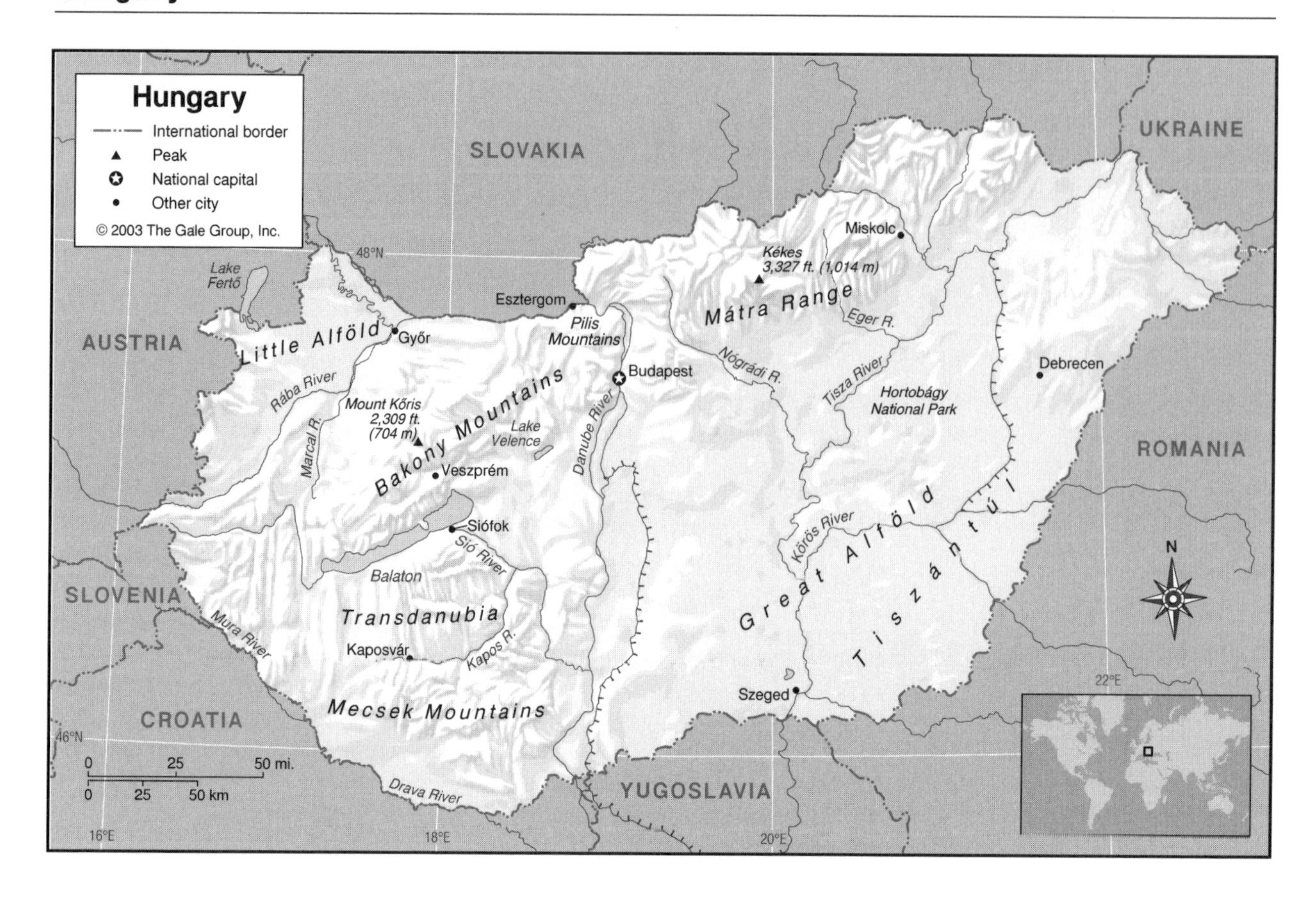

Velence, between Lake Balaton and Budapest, is adjusted artificially to maintain water depths between three and six feet to keep it suitable for swimming, even though it does not have good natural beaches. Hungary also has many mineral springs, which are used for both health and recreational purposes.

Rivers

Hungary's longest and most important river is the Danube. The Danube enters the country in the northwest where it forms the western portion of the border with Slovakia. It flows eastward until it bends north of Budapest and heads south across the width of the country, roughly at its center, until it crosses the border with Yugoslavia. Altogether, about 240 mi (386 km) of the Danube's total length of 1,725 mi (2,776km) borders or flows through Hungary. As it flows through Hungary, the Danube falls very little. It is about 440 ft (134 m) above sea level when it reaches Hungary, and about 280 ft (85 m) above sea level as it leaves the country for Yugoslavia. By far the greater portion of the fall is along the Slovakian border in the gorge of the river north of Budapest. Both the Rába River, which flows into the Danube on the Slovakian border, and the Drava, which joins it much later in the south, rise in the Alpine foothills at the western end of Hungary. Hungary's capital city of Budapest is located on both banks of the Danube, and eight bridges across the river link the two parts of the city.

The next most important river, draining much of eastern Hungary, is the Tisza, which is yet another tributary of the Danube. It rises in the Carpathian Mountains in Ukraine, enters Hungary in the northeast, and flows southward in a meandering course through the eastern part of the Great Alföld, joining the Danube farther south in Yugoslavia. Its streambed is flat, and it has little or no valley. Its flow is highly irregular, and during early and late spring floods it may carry fifty times as much water as it does during the summer. The Tisza has three major tributaries of its own, the Eger River, the Nógrádi River, and the Körös River.

Other notable rivers in Hungary include the Mura River, the Kapos River, the Sió River, and the Marcal River. The Mura feeds into the Drava and along with it helps to form the border with Croatia. The Kapos rises in the southwestern Mecsek Mountains and also near the city of Kaposvár and flows northeast to the Sió. The Sió flows out of Lake Balaton at Siófok and courses southeast until it joins with the Danube. The Marcal River originates in the Bakony Mountains and flows north until it joins with the Rába at Győr.

Wetlands

An extensive series of levees have been built in Hungary's plains to prevent the formerly disastrous flooding of the Tisza and Danube rivers. At various times in the nineteenth century the floods came close to destroying

GEO-FACT

The Carpathian Basin is rich in geothermal resources, and geothermal aquifers underlie nearly all of Hungary, sending large volumes of water between 104°F (40°C) and 158°F (70°C) to the earth's surface. Much of this water is used to heat greenhouses. In southwest Hungary, the water in geothermal aquifers reaches temperatures of 284°F (140°C), and four-fifths of all greenhouses are heated with it. It is also cooled and used for drinking water.

the cities that later combined to make up Budapest and virtually leveled the city of Szeged.

THE COAST, ISLANDS, AND THE OCEAN

Hungary is a landlocked country.

CLIMATE AND VEGETATION

Temperature

Hungary has a continental climate, with Atlantic and Mediterranean influences, characterized by cold winters, warm summers, and abrupt seasonal transitions. The mean temperature ranges from 25° to 32°F (-4° to 0°C) in January, and 64° to 73°F (18° to 23°C) in July. However, temperatures as high as 109°F (43°C) have been recorded, while the record low is -29°F (-34°C).

Rainfall

Rainfall decreases from west to east; the plains around the Tisza River depend on irrigation to prevent crop failure from summer drought. Average annual rainfall ranges from around 20 in (51 cm) in the east to approximately 30 in (76 cm) in the west.

Grasslands

Hungary has two distinct plains regions. The larger and more important one is the Great Alföld, which spreads across central and eastern Hungary, occupying all of the land south of the northern mountain system, from the Danube River to the Romanian and Ukrainian borders in the east. Composed of floodplains, loess plains, and sand dunes, it is a fertile basin with average elevations of slightly more than 300 ft (91 m). Its entire western boundary is formed by the Danube, and it is traversed from north to south by the Tisza River farther to the east. The area east of the Tisza is called Tiszántúl.

Population Centers – Hungary

(2002 POPULATION ESTIMATES)

Name	Population	Name	Population
Budapest (capital)	1,858,000	Szeged	172,000
Debrecen	209,000	Pécs	163,000
Miskolc	181,000	Györ	133,000

SOURCE: Hungarian Central Statistical Office.

Counties – Hungary

2001 POPULATION ESTIMATES

Name	Population	Area (sq mi)	Area (sq km)	Capital
Baranya	406,330	1,732	4,487	Pés
Bacs-Kiskun	545,989	3,229	8,362	Kecskemdt
Békés	399,061	2,175	5,632	Békéscaba
Borsod-Abaúj-Zemplén	749,104	2,798	7,248	Miskolc
Csongrád	428,144	1,646	4,263	Hódmezövászárhely
Fejér	428,922	1,689	4,374	Székesfehévár
Györ-Moson-Sopron	435,256	1,549	4,012	Györ
Hajdú-Bihar	552,478	2,398	6,212	Debrecen
Heves	326,800	1,404	3,637	Eger
Jász-Nagykum-Szolnok	418,601	2,165	5,608	Szolnok
Komárom	316,780	869	2,250	Tatábanya
Nógrád	220,600	982	2,544	Salgótarján
Pest	1,089,478	2,469	6,394	Budapest
Somogy	336,799	2,331	6,036	Kaposvár
Szabolcs-Szatmár-Bereg	587,994	2,293	5,938	Nyíregyháza
Tolna	250,337	1,430	3,704	Szekszárd
Vas	268,591	1,288	3,337	Szombathely
Veszprem	388,000	1,810	4,689	Veszprem
Zala	299,112	1,461	3,784	Zalaegerszeg
Capital City				
Budapest	1,739,569	203	525	

SOURCE: "Area, Population, Population Density, 2002." Hungarian Central Statistical Office.

North of Hungary's mountain system, in the northwest corner of the country, is the Little Alföld, whose composition and elevation are similar to those of the larger plain to the south. Bordered mostly by Austria and by the southwestern edge of Slovakia, it is drained by the Rába and other, smaller rivers that rise in the Alpine foothills of western Hungary.

Forests and Jungles

Most of Hungary's natural forest has given way to cultivated land. However, there is still wooded land in the mountain regions, and the Bakony Mountains are sometimes referred to as the Bakony Forest. Beech is the predominant tree at the highest elevations, with oak and other deciduous species in the woodlands at lower levels. After the destruction of its original forests, much of the Great Alföld was formerly dry steppe country inhabited by herders traditionally known as the *puszta,* but this area has largely been reclaimed for cultivation through irrigation. Steppe areas remain in the northeastern region known as the Hortobágy.

HUMAN POPULATION

Hungary experienced a major rural-to-urban migration in the second half of the twentieth century, by the end of which roughly two-thirds of the population lived in cities and towns, with about one-third of urban dwellers concentrated in the Budapest metropolitan area. As for the rural areas, the villages in the Little Alföld to the northwest tend to be smaller and more widely dispersed than those in the central Great Alföld region.

NATURAL RESOURCES

Hungary's greatest natural resource is its fertile land. It is well suited for the growth of many crops, many of which it exports, including wheat, corn, barley, grapes, sugar beets, and potatoes. Its land is also useful for raising livestock. Hungary also has its share of mineral resources; bauxite, manganese, and brown coal are mined in the Transdanubia region west of the Danube.

FURTHER READINGS

Dent, Bob. *Hungary. 2nd ed. Blue Guides.* New York: W.W. Norton, 1998.

Department of Control Engineering and Information Technology. Hungarian Home Page. *Technical University of Budapest.* http://www.fsz.bme.hu/hungary/homepage.html (Accessed Apr. 3, 2002).

Ivory, Michael. *Essential Hungary.* Lincolnwood, Ill.: Passport Books, 1998.

Parsons, Nicholas T. *Hippocrene Inside's Guide to Hungary.* New York : Hippocrene Books, 1992.

Radkai, Marton. "Hungary." *Insight Pocket Guides.* APA Publications. Boston: Houghton Mifflin, 1994.

Richardson, Dan, and Charles Hebbert. *Hungary: The Rough Guide.* 3rd ed. London: Rough Guides, 1995.

Iceland

- **Area:** 39,769 sq mi (103,000 sq km) / World Rank: 107
- **Location:** Northern and Western Hemispheres, Northern Europe, island between the Greenland Sea and the North Atlantic Ocean, northwest of the United Kingdom and southeast of Greenland.
- **Coordinates:** 65°00′N, 18°00′W
- **Borders:** None
- **Coastline:** 3,099 mi (4,988 km)
- **Territorial Seas:** 12 NM
- **Highest Point:** Öraefajökull, 6,952 ft (2,119 m)
- **Lowest Point:** Sea level
- **Longest Distances:** 304 mi (490 km) E-W; 194 mi (490 km)
- **Longest River:** Thjórsá River, 147 mi (237 km)
- **Largest Lake:** Thingvallavatn, 32 sq mi (84 sq km)
- **Natural Hazards:** Earthquakes; volcanic activity
- **Population:** 277,906 (July 2001 est.) / World Rank: 172
- **Capital City:** Reykjavík, located on southwestern coast
- **Largest City:** Reykjavík, 156,000 (2000 est.)

OVERVIEW

The westernmost European country, Iceland is an island in the North Atlantic Ocean just below the Artic Circle, 200 mi (322 km) east of Greenland, 645 mi (1,038 km) west of Norway and 520 mi (837 km) northwest of Scotland. Slightly smaller than Kentucky, the mainland comprises 39,769 sq mi (103,000 sq km), including 157 sq mi (408 sq km) of lakes. The islands and skerries comprise 58 sq mi (150 sq km). Of the many islands, the most notable are the Westman Islands off the southern coast.

Iceland consists mainly of a central volcanic plateau with elevations raging from 2,297 to 2,625 ft (700 to 800 m) ringed by mountains. Lava fields cover about one-ninth of the country, and glaciers about one-eighth. Geologically, the country is still very young and bears signs of still being in the making. It appears abrupt and jagged without the softness of outline that characterizes more mature landscapes. The average height 1,640 ft (500 m) is above sea level, and one-quarter of the country lies below the 656 ft (200 m) contour line.

The largest lowland areas include Árnessýsla, Rangárvallasýsla, and Vestur-Skaftafellssýsla in the South and Myar in the West. In the plateaus, land is broken into more or less tilted blocks, with most leaning toward interior of the country. Glacial erosion has played an important role in giving the valleys their present shape. In some areas, such as between Eyjafjördhur and Skagafjördhur, the landscape possesses alpine characteristics. There are numerous and striking gaping fissures within the post glacially active volcanic belts.

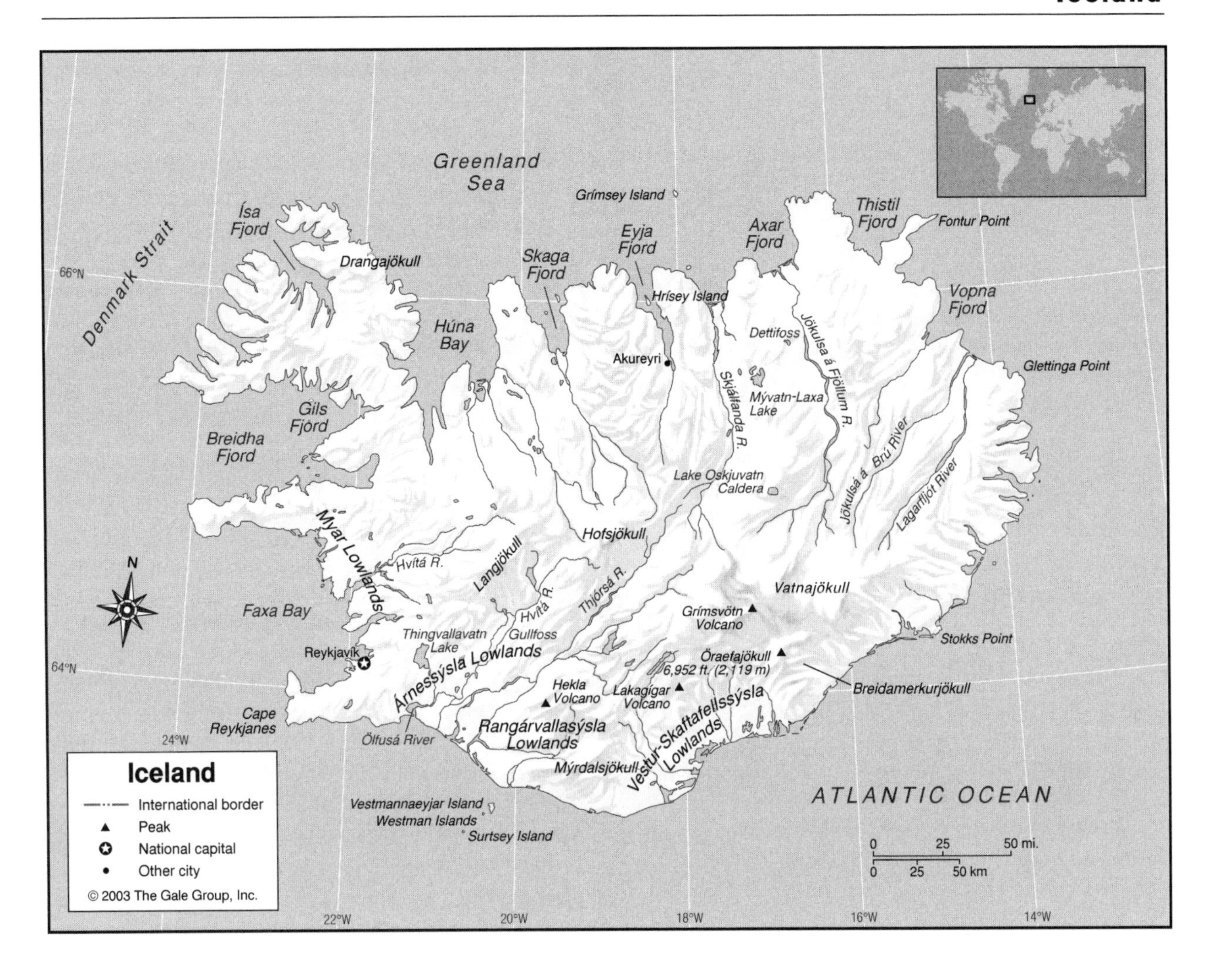

MOUNTAINS AND HILLS

Mountains

Iceland has mountain peaks that ring a central plateau. Most peaks are affected by the underlying thermal activity that characterizes most of the country, where nearly every type of volcanic activity is found. Lava-producing fissures forming so-called center rows are the most common. The most memorable one is the Lakagígar, which poured out the most extensive lava flow in history covering 218 sq mi (565 sq km). The crater rows follow the same direction as the Pleistocene ridges. Shield volcanoes such as the Skjaldbreidhur have also produced a great amount of lava but have not been active for the past 1,000 years.

Iceland also has active volcanoes of the central type fed by the magma chambers. Many of them are blanketed by perpetual ice, including two that have erupted most frequently in recent years, Grímsvötn and Kada. The latter has erupted about 20 times in the past 1,000 years. Each eruption of volcanoes is accompanied by a water flood. These floods occur every five to ten years, sometimes without volcanic eruptions. The eruption of Öraefajökull in 1362 devastated the settlement at the foot of the volcano. The most famous Icelandic volcano is Hekla, which was renowned throughout the Roman Catholic world during the Middle Ages for being the "Abode of the Damned."

The highest mountain on Iceland is Öraefajökull, which reaches a height of 6,952 ft (2,119 m) on the southeastern coast of the island. It is covered by the Vatnajökull glacier.

Plateaus

The inland plateau is a rugged, barren area above sea level. A fault line runs across Iceland. It makes the plateau the land of violent natural wonders, including volcanoes, hot springs, steaming geysers, glaciers, and glistening lava fields. Earthquakes are frequent in Iceland, but rarely dangerous. The most disastrous ones occurred in the southern lowlands in 1784 and 1896, leaving many farms in ruins.

Glaciers

Glaciers cover an area of 4,323 sq mi (11,200 sq km), or 11 percent of the total land area. Nearly all types of glaciers, from small cirque glaciers to extensive plateau icecaps, are represented. The biggest of these icecaps, Vatnajökull, with an area of 3,204 sq mi (8,300 sq km) and a maximum thickness of 3,281 ft (1,000 m) is larger than all the glaciers in continental Europe put together. One of its southern outlets, Breidamerkurjökull, reaches more than 394 ft (120 m) below sea level. Other large icecaps are Langjökull 396 sq mi (1,025 sq km) and Hofsjökull 368 sq mi (953 sq km) in the Central Highlands, Mýrdalsjökull 270 sq mi (700 sq km) in the south and Drangajökull 62 sq mi (160 sq km) in the Northwest. The altitude of the glaciation limit is the lowest, about 1,961 ft (600 m), in the Northwest; and the highest, over 4,922 ft (1,500 m), in the highlands north of Vatnajökull. Since about 1890, the glaciers have greatly thinned and retreated, and some of the smaller ones have almost disappeared. During the 1960s the retreat began to slow down, and some of the glaciers are now advancing again.

GEO-FACT

Low-temperature areas with hot springs are found all over Iceland. Some of the hot springs are spouting springs or geysers, the most famous of which is the Great Geysir in Haukadalur in South Iceland, from which the international word *geyser* is in fact derived. It has been known to eject a column of hot water to a height of about 200 ft (60 m). Another renowned geyser in the vicinity of the Great Geysir is Strokkur.

INLAND WATERWAYS

Lakes

Iceland possesses numerous lakes, mostly of tectonic origin. Others resulted from the deepening of valleys by glacial erosion or damming by lava flows, glacial deposits and rockslides. Small crater lakes are common, especially in the Landmannalaugar-Veidivötn area, where the Lake Oskjuvatn Caldera has an area 4.2 sq mi (11 sq km) and a depth 712 ft (217 m). On the sandy shores, lagoon lakes are common. The largest lake is Thingvallavatn, 32 sq mi (84 sq km), in the southwest. However, Lake Mývatn, in the northeast, is well-known for the large variety of birds that inhabits its shores, and for its excellent fishing.

Rivers

Due to the heavy rainfall, Icelandic rivers are numerous and relatively large. Thjórsá, the longest river, has a length of 147 mi (237 km) and has an average discharge of 13,770 cu ft per second (390 cu m per second). Jökulsá á Fjöllum, the second longest river, is 128 mi (206 km) long. Other major rivers include Hvítá and Ölfusá in the south, Skjálfanda in the north, and Lagarfljót and Jökulsá á Brú in the east.

Icelandic rivers are mainly of two types: glacial and clear-water rivers. The former usually divide into numerous more or less intertwined tributaries that constantly change their courses and swing over the outwashed plains lying below the glaciers. This is especially true of the rivers running south from Vatnajökull. In that area, it was extremely difficult to build a permanent road, since the bridges and parts of the roads were constantly being washed away when the glacial rivers reached their maximum discharge, usually in July and August.

Clear-water rivers are of two kinds. One drains the old basalt areas and has a variable discharge with maximum flow in late spring. The other kind drains regions covered with post-glacial lava and usually has small variations in discharge, which makes them ideally suited for hydroelectric power production. Swift currents make Icelandic rivers for the most part unnavigable.

An impressive characteristic of the youthful Icelandic landscape is its waterfalls, and among the most famous are Gullfoss in Hvítá, Dettifoss in Jökulsá á Fjöllum, Aldeyjarfoss and Godhafoss in Skjálfandafljót, Hraunfossar in Hvítá in Borgarfjördhur, and Skógafoss in Skógá.

Wetlands

Iceland has presently three sites designated as wetlands of international importance, with surface area of 228 sq mi (590 sq km). The Grudnarfjördur wetland is an estuary and sea-bay consisting of mudflats rich in invertebrates, supporting musselbanks, and saltmarsh vegetation. Part of the region of Mývatn-Laxa Lake is a marsh complex fed by both cold and thermal springs. The site supports freshwater marshes, a rich submerged flora, algal communities, woodland, bog and moorland. The abundant invertebrate fauna provide food for large numbers of water-birds. The site is especially important for two duck species—that in Europe nest only in Iceland—and for large number of molting Anatidae. The last site, Thjórsárver, includes abundant pools and lakes and extensive marshland dominated by sedges.

THE COAST, ISLANDS, AND THE OCEAN

Oceans and Seas

Iceland is bordered on the north by the Greenland Sea, on the south and southeast by the North Atlantic Ocean and on the west by Denmark Strait.

Major Islands

There are numerous islands around the coast, some of them inhabited. The largest ones are the Westman

Islands in the south, Hrísey in the north, and Grímsey at the Arctic Circle. Explosion pits are found throughout the country and, in the south, submarine eruptions often occur off the coast. The last one, which began in 1963 and ended in 1967, built up the island of Surtsey, which now covers an area of 1.1 sq mi (2.8 sq km). Several small islands are formed due to submarine eruptions like Surtsey, while others are destroyed—such as Vestmannaeyjar crater, which erupted in 1973 and buried one-third of the island of the same name.

The Coast and Beaches

Icelandic coasts can be divided into two main types. In regions not drained by the debris-laden glacial rivers, the coasts are just irregular, incised with numerous fjords and smaller inlets. They offer many good natural harbors where the fjords have been deepened by glacial erosion. The other type of coast is sand, with smooth outlines featuring extensive offshore bars with lagoons behind them. The beaches from Djúpivogur in the southeast to Ölfusá in the southwest belong to this category.

The peninsula on which Reykjavík sits encloses the Faxa Bay. Moving from Cape Reykjanes, which heads this peninsula, and moving around the coast to the east one can travel halfway around the island without meeting with any notable features other than Stokks Point. However, after turning around Glettinga Point the first of the many fjords on the north shore appears. These fjords are, from east to west: Vopna Fjord, separated from Thistil Fjord by Fontur Point; Axar Fjord; Eyja Fjord; and Skaga Fjord. West of Skaga Fjord the coast sweeps in forming Húna Bay, then turns north toward Denmark Strait. The west coast is also cut by many fjords, among them are Ísa Fjord, Gils Fjord, and Breidha Fjord.

CLIMATE AND VEGETATION

Temperature

Iceland has a relatively mild and equable climate, despite its high altitude and its proximity to the Arctic. Because of oceanic influences, notably the North Atlantic Drift (a continuation of the Gulf Stream), climatic conditions are moderate in all sections of the island. The mean annual temperature at Reykjavík is about 41°F (about 5°C), with a range from 31°F (-1°C) in January to 52°F (11°C) in July. In the northwestern, northern, and eastern coastal regions—which are subject to the effects of polar currents and drifting ice—temperatures are generally lower. Windstorms of considerable violence are characteristic during much of the winter season.

Rainfall

Annual precipitation ranges between about 50 and 80 in (about 127 and 203 cm) along the southern coast, and is only about 20 in (about 51 cm) along the northern coast. The southern slopes of some of Iceland's interior mountains receive up to about 180 in (about 457 cm) of precipitation per year.

HUMAN POPULATION

The population of Iceland is extremely homogeneous, being almost entirely of Scandinavian and Celtic origin. Beginning in the 1940s a large-scale movement to the coastal towns and villages occurred. Some 92 percent of the people now live in cities and towns. Iceland is one of the least densely populated countries in the world with a population density of only 7 people per sq mi (2.7 per sq km). However, the interior of the island is uninhabited so the actual population density is much higher. The inhabitants are distributed among coastal cities and towns with more than half of the population living in the capital of Reykjavíc.

NATURAL RESOURCES

Until the close of the nineteenth century, agriculture was the chief occupation, with fishing as a supplementary source of income. By the middle of the twentieth century, however, fishing and fish processing had become the major industries. Iceland is very rich in natural heat flow and within the neo-volcanic areas averages 2 or 3 times the global average. Hydroelectric power potential is abundant and is being developed to further industrialization.

FURTHER READINGS

Baxter, Colin. *Iceland.* Moray, Scotland: Baxter, Colin Photography, Ltd, 2001.

IcelandUSA.net. *Iceland: A Quick Glance.* http://www.icelandusa.net/iceland.htm (Accessed June 13, 2002).

McBride, Francis. *Iceland, Vol. 37.* Santa Barbara, Calif.: Clio Press, 1996.

Swaney, Deanna. *Lonely Planet: Iceland, Greenland and the Faroe Islands 2001.* London: Lonely Planet Publications, 2001.

India

- **Area:** 1,269,345 sq mi (3,287,590 sq km) including India-administered Kashmir / World Rank: 8
- **Location:** Northern and Eastern hemispheres, on the South Asian subcontinent, from the Himalayas to the Indian Ocean
- **Coordinates:** 20°00′N, 77°00′E

- **Borders:** 8,744 mi (14,103 km) / Bangladesh, 2,513 mi (4,053 km); Bhutan, 375 mi (605 km); China, 2,096 mi (3,380 km); Myanmar, 907 mi (1,463 km); Nepal, 1,048 mi (1,690 km); Pakistan, 1,805 mi (2,912 km)
- **Coastline:** 4,340 mi (7,000 km)
- **Territorial Seas:** 12 NM
- **Highest Point:** Kanchenjunga, 28,209 ft (8,598 m)
- **Lowest Point:** Sea level
- **Longest Distances:** 1,997 mi (3,214 km) N-S / 1,822 mi (2,933 km) E-W
- **Longest River:** Brahmaputra, 1,800 mi (2,900 km)
- **Largest Lake:** Chilka Lake, 425 sq mi (1,100 sq km)
- **Natural Hazards:** earthquakes, cyclones, floods, droughts, landslides, thunderstorms, hailstorms, dust storms
- **Population:** 1,029,991,145 (July 2001 est.) / World Rank: 2
- **Capital City:** New Delhi, in the center of northern India
- **Largest City:** Mumbai, 12,147,100 (2002 est.)

OVERVIEW

One of the largest countries on earth, as well as one of the most heavily populated, India is a nation of great geographic diversity. The extraordinary geographic variety of India can be divided into three main regions: the Himalayan mountain range of the north; the broad and flat alluvial plain of the Ganges River to the south of the mountains; and, even further south, the vast peninsula that juts into the Indian Ocean, creating the Arabian Sea and the Bay of Bengal, with small island chains offshore. India's mountainous northeastern region is nearly separated from the rest of the country by Bangladesh and Nepal. Mostly due to population pressure, India's wonderfully diverse geographical features, encompassing everything from snowy peaks to desert to rainforest, are at risk from environmental damage. Many local groups have organized to fight pollution and protect wildlife.

India was originally attached to a land mass that included southern Africa and Antarctica some 100 million years ago, but then broke off to form the separate Indian tectonic plate. This plate eventually collided with Asia. The Himalayas and Ganges plain were formed by this collision, and there continues to be much seismic activity in the northern part of the country.

A discussion of India's geography is complicated by the fact that it has ongoing border disputes with many of its neighbors. Part of the southern border with Bangladesh is undefined, as is part of the border with China in the northeast (the McMahon Line). Since their creation as independent countries in 1947, India and Pakistan have disputed ownership of the northern region of Jammu and Kashmir, a simmering conflict which has broken into open warfare between the neighbors in 1948, 1965, and 1971, and continues to be a source of sporadic fighting. China also occupies portions of northeastern Jammu and Kashmir that are claimed by India, which caused fighting in 1962. A de-facto line of control divides Jammu and Kashmir, with the eastern sector along the Siachen Glacier undemarcated. Of the 85,806 sq mi (222,236 sq km) of Jammu and Kashmir, 16,500 sq mi (42,735 sq km) were controlled by India in 2002, and are dealt with in this entry.

The names of several well-known Indian locations were changed in the 1990s by local political parties. Most noteworthy are the cities of Mumbai (formerly Bombay), Chennai (formerly Madras) and Kolkata (formerly Calcutta) and Bangla State (formerly West Bengal).

MOUNTAINS AND HILLS

The Himalayas

The name Himalaya, which means "abode of snow" in Sanskrit, is given to the tremendous system of mountain ranges, the loftiest in the world, that extends along the northern frontiers of Pakistan, India, Nepal, and Bhutan. The Himalayas are made up of three more or less parallel ranges. The northernmost and highest are the Greater Himalayas. The world's tallest mountains are found in this range, with most peaks over 20,000 ft (6,096 m). India's highest mountain is in this range, the five-peaked Kanchenjunga (28,208 ft / 8,597 m) on the border between Nepal and India. An enormous 19 mi (31 km) glacier sprawls down its west side. Other great peaks include Kamet (25,447 ft / 7,756 m) and Nanda Devi (25,645 ft / 7,817 m), which lie north of New Delhi and west of Nepal.

South of the Greater Himalayas is the Lesser Himalayas range. Their peaks are mostly between 5,000 and 12,000 ft (1,524 and 3,657 m) in height; although some exceed 15,000 ft (4,572 m). The Outer Himalayas are the southernmost and lowest of the three ranges, with peaks between 3,000 and 4,000 ft (914 and 1,219 m) in height.

In each of the main Himalayan ranges the southward slopes are too steep either to accumulate snow or to support more than sparse tree growth. The northward slopes, which are much gentler, are generally forest clad below the snow line. Because these mountain ranges are geologically young, earthquakes are not infrequent among them.

A number of passes through India's Himalayan barrier have been known from ancient times. Of these, the Karakoram Pass (18,290 ft / 5,575 m high) is one of the best known and most important. It is situated about 200

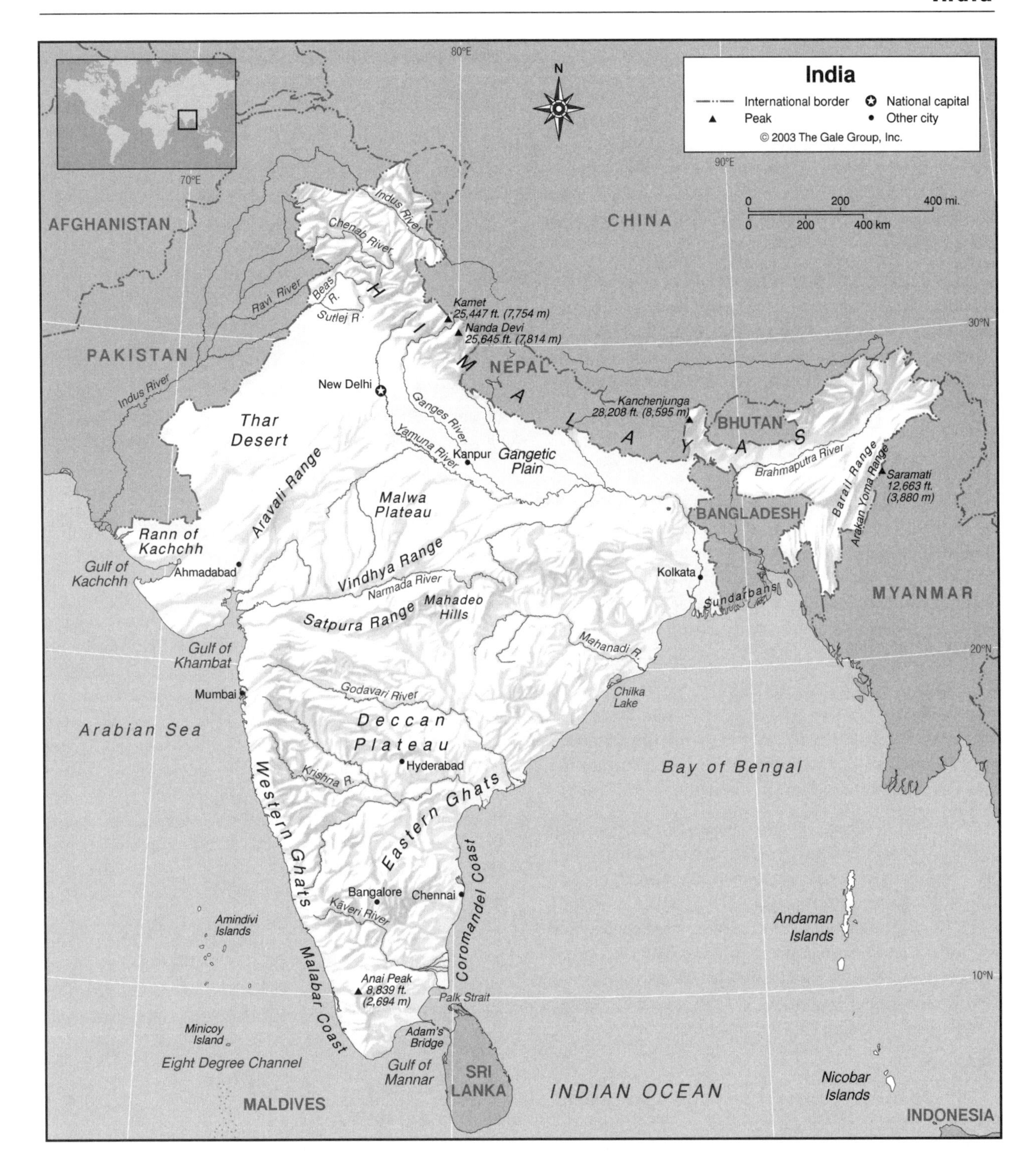

mi (322 km) northeast of Srinagar, in the disputed region of Jammu and Kashmir. Other important passes include the Jelep La and Natu La, northeast of Darjeeling in the east of India.

Other Mountains

There are many other mountain ranges in India, although none nearly so large and high as the Himalayas. At the southern end of the country are the two mountain ranges called the Ghats. The Western and Eastern Ghats parallel the coasts of the peninsula and separate the interior plateau from the coastal plains. The mountains called the Western Ghats have an average elevation of 3,500 ft (1,066 m) although in the Anaimalis part of the range Anai Peak reaches 8,839 ft (2,694 m). The range runs fairly close to the ocean, facing it like a wall. Near the tip of the peninsula it converges with the Eastern Ghats. The Eastern Ghats are disconnected and much lower than the Western Ghats, averaging only about 2,000 ft (610 m) in elevation.

A complex of hills and low mountain ranges, varying in elevation from about 1,500 to 4,000 ft (457 to 1,219 m) occupies the northern part of the peninsula, separating the Gangetic Plain to the north from the plateaus of the peninsula. The main ranges in that area are the Aravali, the Vindhya, and the Satpura in the west; the Mahadeo, Maikala, and Bhanrer hills in the center; and the Kaimur, Hazaribagh, and the jungle-clad hills of the Chota Nagpur to the east.

The easternmost part of India, nearly separated from the rest of the country by Bangladesh, is very mountainous. The chief ranges here are the Barail Range and the Arakan Yoma Range along the border with Myanmar, whose highest peak is Saramati (12,663 ft / 3,860 m). These ranges are sometimes considered a southern extension of the Himalayas.

Plateaus

The peninsula is an area of great complexity, but can be described generally as a plateau region roughly bounded by ranges of low mountains and hills that geographically block the peninsula off not only from the plain to the north but also from the deserts and semiarid regions of Rajasthan to the northwest and the coastal flats of Gujarat to the west.

The largest of the identifiable plateaus are the central Malwa Plateau between the Aravali and Vindhya Ranges; the Chota Nagpur in the northeast of the peninsula, which contains the most important mineral concentrations in India; and the Deccan plateau.

The name Deccan, which means "south," is often applied loosely to all the elevated land of southern India. More properly, however, it refers to the western portions of the irregular central plateau. The Deccan is actually not a single plateau but a series of plateaus topped by rolling hills and intersected by many rivers. The Deccan plateau system averages about 2,500 ft (762 m) in elevation in the west and about 1,000 ft (305 m) in the eastern parts.

Canyons

The Himalayan regions of Ladakh, Zanskar, and Sikkim possess many deep canyons, as do hill regions such as Madhya Pradesh at India's center. The Brahmaputra River cuts a deep gorge through the mountains of northeast India, as does the Ganges at its source in the Himalayas.

INLAND WATERWAYS

Lakes

India's landscape contains a variety of lakes; salt water and fresh water as well as natural and human-made. Water levels in most lakes are jeopardized by siltation and sedimentation, and many are affected by pollution from industry and agriculture.

Chilka Lake, the largest brackish water lake in Asia, is also the largest lake in India. A 425 sq mi (1,100 sq km) lagoon, Chilka's bird and fish habitat and its rich algae content are threatened by silt, sedimentation, and commercial interests. Wular Lake, in India-controlled Jammu and Kashmir, is India's largest freshwater lake (78 sq mi / 202 sq km). It holds large quantities of floating vegetation and is an important source of fish and of irrigation water. In Tamil Nadu state an extensive system of shallow irrigation reservoirs known as "tanks" has been maintained since the 8th century C.E.

GEO-FACT

The neem tree, a tropical evergreen native to India, is known as "the village pharmacy" because every part of it is used, for purposes ranging from: insect repellent, to medicines against fever and infection, skin lotions, and even toothpaste. Efforts by a foreign company to patent a neem-based pesticide were denied in India's courts. Neem tree planting is encouraged in India as well as other tropical regions including Africa for shade, organic pest control, and its many other uses.

The Great Rivers of India

The Indus River, rising in the Tibetan Himalayas of China, flows through Indian-controlled Jammu and Kashmir before entering Pakistan. The Indus has five principal tributaries, also of Himalayan origin, that are of importance to India: the Sutlej, Beas, Ravi, Chenab, and Jhelum. These rivers drain part of the Indian state of Punjab, whose name derived from *panch ab,* meaning five waters or rivers.

South of Punjab and east of the desert region of western India is the main feature of the Gangetic Plain and the most revered and mightiest of India's rivers—the Ganges. The origin of the Ganges is identified in an ice cave about 30 mi (48 km) north of Nanda Devi, almost on the line of the India-China frontier and 3,139 m (10,300 ft) above sea level. The river is about 1,560 mi (2,510 km) in length.

The Ganges cuts a gorge south through the Himalayas, then enters the Gangetic Plain at the city of Hardwar. Proceeding south and then eastward, it winds through

Uttar Pradesh in north central India, and is joined by its major tributary, the Yamuna, at Allahabad. The Yamuna, companion to the Ganges, also rises in the Himalayas and through much of its length flows only 80 to 120 km (50 to 75 mi) to the west, and later to the south, of the Ganges.

After merging with the Yamuna, the Ganges then continues in a southeasterly direction and has numerous tributaries. The river has shifted its course many times over the years, and as it approaches the border with Bangladesh there are many distributary streams and rivers that branch of to the south. They are the beginnings of the enormous Ganges Delta, most of which is found within Bangladesh. After entering that country, the Ganges merges with the Brahmaputra River before emptying into the Bay of Bengal.

The 1,700 mi (2,900 km) long Brahmaputra River rises in southwestern Tibet, not far from the separate sources of both the Indus and the Ganges but on the northern side of the Himalayas. Most of its course is in China, but more than a thousand miles to the east it turns south and enters India in its northeastern corner. The Brahmaputra then curves west and flows through northeast India in the Assam Valley along a narrow plain, before entering Bangladesh where it merges with the Ganges.

Other Rivers

South of the Gangetic Plain there are six major rivers. Four of these peninsular rivers, the Mahanadi, the Godavari, the Krishna, and the Kâveri, flow into the Bay of Bengal. All of these rivers rise in the Western Ghats, flow eastward, and deposit silt in fertile deltas. Most of the rivers are fast flowing during the rainy season but are reduced to a fraction of their size or may even be dry during the hot season. All have reached their base level of erosion and thus have broad shallow valleys that tend to flood during the rainy season. Several of the rivers have waterfalls and cascades in their upper courses.

The Mahanadi, which rises in Madhya Pradesh, is about 560 mi (900 km) long and is an important source of irrigation water in Orissa state. The Godavari has its source north of Mumbai and follows a general southeasterly course for 900 mi (1,488 km) to its fertile delta on the coast of Andhra Pradesh. The Kâveri rises in southern Karnataka and flows irregularly southeastward for approximately 475 mi (764 km). Harnessed for irrigation since ancient times, an estimated 95 percent of the water of the Kâveri is put to use before the river empties into the Bay of Bengal. The Krishna River rises in the west of the peninsula, not far from Mumbai and only 40 mi (64 km) from the Arabian Sea, but then flows east for roughly 800 mi (1,285 km) before reaching the Bay of Bengal

Only two major rivers of the peninsula flow into the Arabian Sea—the Narmada and the Tāpi. The Narmada rises in eastern Madhya Pradesh, flows through Gujarat State, then forms a thirteen-mile wide estuary at the Gulf of Khambhat. The shorter Tāpi River follows a companion course south of the Narmada, and is a tidal river for the last 32 mi (51 km) of its 450 mi (724 km) length.

Rivers are regarded with love and reverence in Indian culture. Several, especially the Ganges, the Yamuna, and the Godavari, are regarded as sacred by Hindus and are visited and bathed in by millions of pilgrims annually.

Wetlands

The Ganges Delta on the Bay of Bengal includes an ecologically significant region of mangrove swamp known as the Sundarbans, which stretches from Bangla State of India into neighboring Bangladesh. As one of the last remaining natural habitats of endangered Bengal tigers, much of the Sundarbans is designated a wildlife preserve. On the other side of India, in Gujarat state, the Ranns of Kachchh are massive salt marshes that flood during the monsoon. They hold India's largest mangrove forests.

Eight Indian locations are designated as Wetlands of International Importance under the Ramsar Convention on Wetlands. Chilka Lake's wetland is a breeding area for 33 waterbird species, and the lake holds 118 species of fish. Harike Lake in Punjab State is a shallow reservoir that is an important bird sanctuary. The Kanjli site is a stream converted into an irrigation reservoir, with diverse flora and fauna. Keoladeo National Park, in Rajasthan State, is designated a World Heritage Site. A complex of ten lagoons, Keoladeo is a bird and wildlife sanctuary but is threatened by water shortages, invasive plant species, and grazing. The Loktak Lake site, in Manipur State, northeast India, is a freshwater lake with adjacent swamplands. Ropar, a National Wetland, in Punjab, is a habitat for many mammal, bird and fish species. It is threatened by siltation, invasive weeds, and pollution. The saline Sambar Lake, in Rajasthan, is fed by four streams and surrounded by sand flats. The Wular Lake wetland area in Jammu and Kashmir has extensive marshland.

THE COAST, ISLANDS AND THE OCEAN

Oceans and Seas

India's peninsula projects into the Indian Ocean, between the Arabian Sea and Bay of Bengal, placing it on vital maritime trade routes between the Middle East, Africa, and East Asia. Major ocean ports along the peninsula include Mumbai, Cochin, Kandla, and Marmagao on the Arabian Sea and Vishakapatnam and Chennai on the Bay of Bengal.

Major Islands

Two groups of islands belonging to India—each group classed and governed as union territory—lie one to either side of the southern tip of the country. The areas and populations of these island chains are very small.

The eastern group, the Andaman and Nicobar Islands, formed of an undersea mountain range, are in the Bay of Bengal. The Andamans form a generally north-south chain some 800 mi (1,287 km) east of the city of Chennai. The Nicobar chain of fewer and smaller islands stretches south-southeast from the Andamans and is separated from them by the Ten Degree Channel. The overall length of the Andaman-Nicobar group is about 500 mi (804 km). The total land area of these low lying, lightly populated islands is roughly 3,200 sq mi (8,287 sq km).

In the Arabian Sea is the second group, composed of the Laccadive, Minicoy, and Amindivi Islands, are collectively named Lakshadweep. This irregular scattering of small coral islands stretches about 200 mi (320 km) from the northwest to the southeast. The total area is only about 18.5 sq mi (50 sq km). Most, although not all, of these low-lying small islands are occupied, and population density is high on the inhabited islands. Directly south of Lakshadweep, separated from them by the Eight Degree Channel, is the small island nation of the Republic of Maldives.

The Coast and Beaches

Not far south of where the India-Pakistan border meets the ocean, the broad, short Kathiwar Peninsula projects into the Arabian Sea. The Gulf of Kachchh lies to the north of this peninsula, and the Gulf of Khambhat to the south and east. The Gulf of Kachchh includes a Marine National Park, which is an effort to protect coral reefs and wetland wildlife habitat. South of the Gulf of Kachchh, the coast continues with few inlets and a flat sandy shore to its southernmost point, Cape Comorin. The southern section of this coastline is known as the Malabar Coast.

The eastern coast of India, on the Bay of Bengal, begins in the northeast at the fragmented Ganges River delta and continues generally southwest before curving more to the south, at which point it becomes known as the Coromandel Coast. The channel between India and the island nation of Sri Lanka is called the Palk Strait, and a series of small islands nearly forming a 30 mile (48 km) long natural causeway almost linking the two countries is known as Adam's Bridge. Just south of that, the Gulf of Mannar indents India's southern tip, where Cape Comorin joins the two coasts of the immense Indian peninsula.

CLIMATE AND VEGETATION

Temperature

India experiences an array of different climate conditions due to its great size and varying terrain. The Greater Himalayan region has a dry, subarctic climate, but the valleys and outer ranges are temperate or subtropical. The inland of the peninsula and Ganges-Brahmaputra drainage areas range from subtropical to temperate. The coasts of the peninsula are humid and tropical.

India's four seasons are determined by the monsoons, a pattern of winds sweeping across southern Asia. There is a dry, cool season (winter) of the northeast monsoon in December-March; a hot season (spring) in April-May; the rainy season (summer) of the southwest monsoon in June-September; and a less-rainy season (autumn) in October-November. India's north has frost in the cool season and temperatures as high as 120°F (49°C) in the hot season. As an example of South India's climate, the city of Chennai has an average temperature of 83°F (28°C). Temperatures for the entire nation reach an average high of 100° to 104°F (38° to 40°C) and dip to an average low of 50°F (10°C).

Rainfall

India's weather is characterized by intense, sudden changes, such as the onset of the monsoon, flash floods, or violent thunderstorms. The monsoons bring dry air from the northeast in the winter and spring, then southwestern rains from the Indian Ocean in the summer. The arrival of the southwestern monsoon is unpredictable and its duration fluctuates. Cyclones from the Indian Ocean often affect coastal areas in April-June and September-December. Rainfall varies extremely in India, from the Thar Desert with less than 5 in (13 cm) yearly, to Cherrapunji in the northeastern mountains, known as the world's rainiest place, with an average of 450 in (105 cm) per year. Rainfall for the entire nation of India averages 41 in (105 cm). The great majority of that precipitation deluges the subcontinent during the southwestern monsoon period. Snow falls in the Himalayan area, which also produces hailstorms, which sweep down over the peninsula. Dust storms affect many regions of India.

The Gangetic Plain

The Gangetic (or Indo-Gangetic) Plain lies at the foot of the Himalayan mountain barrier, extending from Assam and the Bay of Bengal on the east into Pakistan and to the Arabian Sea on the west. Covering some 300,000 sq mi (776,996 sq km), it extends roughly 1,500 mi (2,414 km) from east to west. Once a gulf between the peninsula and the Himalayas, the Gangetic plain is the product of the continual deposits of alluvium borne by the Indus and Ganges river systems. The trough has been so filled over the ages that it looks to the eye like a level plain. Between Delhi and the Bay of Bengal, about 950 mi (1,528 km) to the east, there is a drop of only about 660 ft (201 m) in elevation. The entire region is very fertile and very densely populated.

Other grasslands in India include the Terai region in the low mountains along the border of Nepal, which

Population Centers – India

(2001 CENSUS)

Name	Population	Name	Population
Greater Mumbai (Bombay)	11,914,398	Ahmadabad	3,515,361
Delhi Municipal Corporation	9,817,439	Hyderabad	3,449,878
Kolkata (Calcutta)	4,580,544	Pune (Poona)	2,540,069
Bangalore	4,292,223	Kanpur (Cawnpore)	2,532,138
Chennai (Madras)	4,216,268	Surat	2,433,787

SOURCE: Office of the Registrar General, India. "Cities with More Than One Million Population, Census of India 2001.

includes savanna and alpine grassland types. Bamboo grasslands occur across the Himalayan foothills, especially in northeast India.

Deserts

Below the state of Punjab and extending southwest along the Pakistani border is the sparsely populated Thar Desert, which covers the large part of the state of Rajasthan. 3,000 sq km (1,158 sq mi) of its terrain of sand dunes and flat thorn scrub is protected as the Thar Desert National Park. No rivers of any significance flow through the desert to compensate for its scarcity of rain, but the Indira Ghandi Canal carries some water in from the Punjab region to the north. The Thar contains a large supply of white marble that has been used in many of India's notable buildings. There also are sizable deposits of limestone, gypsum, lignite, fuller's earth, and salt.

Forests and Jungles

India's forest cover is estimated at approximately 19 percent of the country. Madhya Pradesh in the center of the country and Arunachal Pradesh in the extreme northeast are the states with the most forest cover. There is a great range of types of forest, including alpine scrub in the Himalayan regions; temperate evergreen in Jammu and Kashmir and other hill areas; tropical rainforest in the Western Ghats, northeastern states, and islands; and mangroves in the Sundarbans on the Bay of Bengal and in Gujarat.

Indian forests are under pressure from livestock grazing, firewood and charcoal collection, logging, shifting cultivation, and clearing for settlement. Firewood and charcoal are estimated to comprise 90 percent of India's wood usage. The Indian government has promoted efforts towards reforestation using species such as eucalyptus, teak, and rubberwood. Attempts have been made to control and decrease logging. India has 441 wildlife sanctuaries and 80 national parks, but these protected areas are often poorly secured and most are routinely encroached upon. Local environmental movements work to protect the remaining forests.

States and Union Territories – India

2001 CENSUS OF POPULATION

Name	Population	Area (sq mi)	Area (sq km)	Capital
States				
Andhra Pradesh	75,727,541	106,204	275,068	Hyderabad
Arunachal Pradesh	1,091,117	32,333	83,743	Itanagar
Assam	26,638,407	30,285	78,438	Dispur
Bihar	82,878,796	67,134	173,877	Patna
Chhatisgarh	20,795,956	52,197	135,191	Raipur
Goa	1,343,998	1,430	3,702	Panaji
Gujarat	50,596,992	75,685	196,024	Gandhingar
Jharkhand	26,909,428	30,778	79,714	Ranchi
Haryana	21,082,989	17,070	44,212	Chandigarh
Himachal Pradesh	6,077,248	21,495	55,673	Simla
Jammu and Kashmir	10,069,917	39,145	101,387	Srinagar
Karnataka	52,733,958	74,051	191,791	Bangalore
Kerala	31,838,619	15,005	38,863	Trivandrum
Madhya Pradesh	60,385,118	171,215	443,446	Bhopal
Maharashtra	96,752,247	118,800	307,690	Mumbai (Bombay)
Manipur	2,388,634	8,621	22,327	Imphal
Meghalaya	2,306,069	8,660	22,429	Shillong
Mizoram	891,058	8,140	21,081	Aizawl
Nagaland	1,988,636	6,401	16,579	Kohima
Orissa	36,706,920	60,119	155,707	Bhubaneswar
Punjab	24,289,296	19,445	50,362	Chandigarh
Rajasthan	56,473,122	132,140	342,239	Jaipur
Sikkim	540,493	2,740	7,096	Gangtok
Tamil Nadu	62,110,839	50,216	130,058	Chennai (Madras)
Tripura	3,191,168	4,049	10,486	Agartala
Uttaranchal	8,479,562	20,650	53,483	Dehra Dun
Uttar Pradesh	166,052,859	113,673	294,411	Lucknow
West Bengal	80,221,171	34,267	88,752	Kolkata (Calcutta)
Union Territories				
Andaman and Nicobar Islands	356,265	3,185	8,249	Port Blair
Chandigarh	900,914	44	114	Chandigarh
Dadra and Nagar Haveli	220,451	190	491	Silvassa
Daman and Diu	158,059	43	112	Daman
Delhi	13,782,976	572	1,483	Delhi
Lakshadweep	60,595	12	32	Kavaratti
Pondicherry	973,829	190	492	Pondicherry

SOURCE: "Provisional Population Totals: India." Census of India 2001.

HUMAN POPULATION

Historical migrations, and an abundance of river-nourished, fertile land have filled India with its multiethnic population over the centuries. At the beginning of the 21st century, one sixth of all human beings were believed to live in India. This huge population strains the country's resources and continues to expand. As of 2001 India had a 1.55 percent estimated annual growth rate, and is expected to surpass China as the world's most populous nation by the middle of the 21st century.

Population density is extremely high, with an estimated km 733 per sq mi (283 people per sq km). One quarter of Indians live in cities, which grow at unsustainable rates, causing breakdowns in civil infrastructure. Twenty-three Indian cities have over a million inhabitants and three—Mumbai, Kolkata, and Delhi—have over ten million. Both urban dwellers and rural villagers face many challenges, as some 35 percent of the people of India subsist below the poverty line (1994 estimate).

NATURAL RESOURCES

The subcontinent of India contains many mineral resources including iron ore, bauxite, manganese, mica, titanium, chromite, diamonds, limestone, gypsum, lignite, and salt. The land also provides agriculture, fishing, palm tree products, and bamboo. The remaining forests, wetlands, and other environmentally protected areas are significant for scientific study, botanical medicines, and ecotourism. In addition to reserves of coal, oil, and natural gas, India has great potential for energy from hydropower, solar, wind, and geothermal sources. Although India is in every sense rich in natural resources, this wealth is stretched thin by population size, soil erosion, desertification, pollution, substandard construction, and industrial accidents.

FURTHER READINGS

Alter, Stephen. *Sacred Waters: A Pilgrimage up the Ganges River.* New York: Harcourt Brace, 2001.

Bradnock, Robert, and Roma Bradnock. *Footprint India Handbook 2001.* Bath, UK: Footprint Handbooks, 2001.

Frater, Alexander. *Chasing the Monsoon: A Modern Pilgrimage through India.* New York: Henry Holt, 1992.

Gadgil, Madhav, and Guha, Ramachandra. *This Fissured Land: An Ecological History of India.* Berkeley, CA: University of California Press, 1993.

Keay, John. *The Great Arc: The Dramatic Tale of How India was Mapped.* New York: Harper Perennials, 2001.

Maps of India web site. http://www.mapsofindia.com (accessed March 18, 2002).

Neem Foundation. *The Neem Foundation web site.* http://www.neemfoundation.org (accessed March 18, 2002).

Roy, Arundhati. *The Cost of Living.* New York: Modern Library, 1999.

Shiva, Vandana. *Tomorrow's Biodiversity: Prospects for Tomorrow.* London: Thames and Hudson, 2001.

WWF India. *World Wide Fund for Nature—India.* http://www.wwfindia.org (accessed March 18, 2002).

Indonesia

- **Area:** 741,096 sq mi (1,919,440 sq km) / World Rank: 17
- **Location:** An archipelago in Southeast Asia, between the Indian Ocean on the west and south and the Pacific Ocean on the east and north. Bordering Malaysia in northern Borneo, East Timor in eastern Timor, and Papua New Guinea in eastern New Guinea. Crossed by the equator, in the Northern, Southern, and Eastern Hemispheres.
- **Coordinates:** 5° 00′ S, 120° 00′ E
- **Borders:** 1,719 mi (2,774 km) / East Timor, 106 mi (172 km); Malaysia, 1,104 mi (1,782 km); Papua New Guinea, 508 mi (820 km)
- **Coastline:** 33,999 mi (54,716 km)
- **Territorial Seas:** 12 NM
- **Highest Point:** Puncak Jaya, 16,503 ft (5,030 m)
- **Lowest Point:** Sea level
- **Longest Distances:** 3,275 mi (5,271 km) E-W; 1,373 mi (2,210 km) N-S
- **Longest River:** Kapuas River, 710 mi (1,143 km)
- **Largest Lake:** Lake Toba, 502 sq mi (1,300 sq km)
- **Natural Hazards:** Floods, droughts, tsunamis, earthquakes, volcanoes, and landslides, as well as environmental hazards of deforestation, pollution, and haze from burning forests and peat swamps.
- **Population:** 228,437,870 (July 2001 est.) / World Rank: 4
- **Capital City:** Jakarta, in northwest Java
- **Largest City:** Jakarta, 8,621,000 (2000 estimate)

OVERVIEW

Indonesia consists of more than 13,000 islands scattered for about 3,200 mi (5,149 km) above and below the equator between the Indian and Pacific Oceans, in the largest archipelago in the world. Five major islands make up 90 percent of Indonesia's land area. These are Sumatra, Java, Sulawesi, plus parts of Borneo (which it shares with Malaysia and Brunei), and New Guinea, which the Indonesian province of Papua, known until January 2002 as Irian Jaya, shares with the nation of Papua New Guinea. Indonesia also contains about 30 smaller island groups, the largest of which is Nusa Tenggara, which includes the islands of Lombok, Sumba, Sumbawa, Flores, and Timor. In 1999, East Timor gained independence from Indonesia. Indonesia disputes ownership of Sipadan and Ligitan islands with Malaysia.

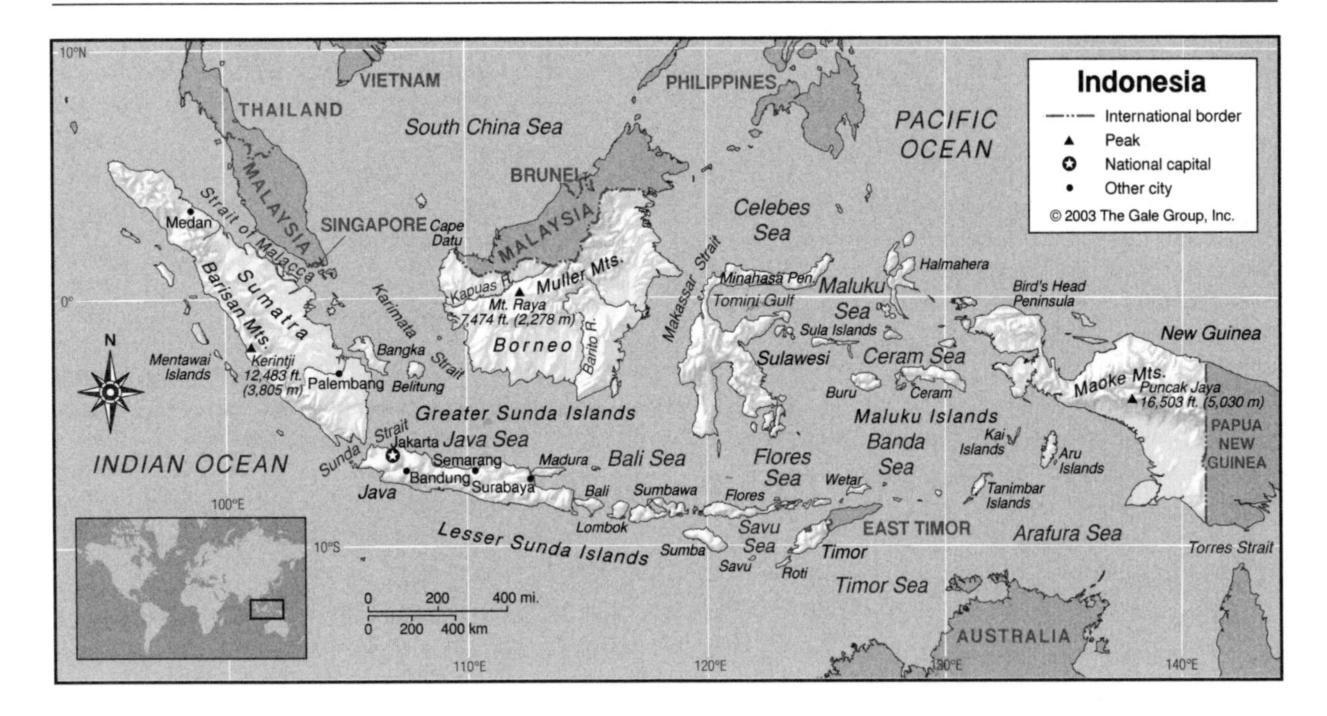

Along the length of Indonesia's island chain the landscape is highly varied, and volcanic mountains stand out in sharp relief on most of the larger islands. Once thickly forested, Indonesia now suffers from one of the world's worst rates of forest loss. Indonesia is both a contributor to global warming through its rampant deforestation and air pollution, and a potential victim as sea levels rise to threaten its coastline. The nation has a very large and growing population.

MOUNTAINS AND HILLS

Mountains

The mountains of Indonesia are chains that run underneath the sea and show their peaks and ridges above it in the form of islands. In earlier geological periods the eastern and southeastern regions of Asia went through several stages of folding, which can be discerned in two long mountain systems that intersect each other in the eastern Indonesian islands of Sulawesi (Celebes) and Halmahera. The first, consisting of two parallel ridges that are a continuation of the western Burmese chain, runs through Sumatra, Java, Bali, and Timor, and curves sharply in a great hook through the southeastern islands to Ceram and Buru. The second runs southwest through the Philippines into eastern Indonesia; hence the complexity of the mountain structure and the peculiar shape of Sulawesi and Halmahera, which have ranges running north-south and east-west. Sulawesi is extremely mountainous, with peaks rising in places to well over 8,000 ft (2,438 m).

The Barisan Mountains of Sumatra follow the island's west coast and are intersected by short but rapid streams. The highest peaks reach more than 12,000 ft (3,600 m), with Kerintji (12,483 ft / 3,805 m) being the tallest. On Java the mountains also lie close to the shoreline of the Indian Ocean but, being less continuous and rugged than those of Sumatra, they allow frequent access from north to south. The highest peaks are in the Tengger Mountains in the east. Many of the islands of Nusa Tenggara and the Maluku (the islands between Sulawesi and New Guinea) are mountainous. On Bali, Lombok, and Ceram there are peaks of over 10,000 ft (3,048 m).

Papua in New Guinea has towering non-volcanic mountains, the highest in Indonesia. The Maoke Mountains extend almost the entire length of the province in a general northwest to southeast direction, a massive barrier between the northern and southern parts of Papua. Some peaks are covered with snow throughout the year, including the 16,503 ft (5,030 m) tall Puncak Jaya, the country's loftiest peak. Also known as Jayawijaya or the Carstenz Pyramid, it is the highest mountain in the Pacific Basin. Puncak Jaya is counted (for Australia/Oceania) as one of the "Seven Summits" sought by mountaineers who attempt to climb the highest peak on every continent.

The Muller Mountains of Borneo run mainly along Indonesia's northern border with Malaysia. Mount Raya (7,474 ft / 2,278 m) is the highest peak.

Volcanoes

Lying along the borders of the Eurasian, Australian, and Philippine Tectonic Plates, Indonesia is the most highly volcanic region in the world. More than 100 peaks either are active or were active until recently. The volcanic ash enriches the soil, and the greatest population density is to be found in the regions where volcanoes have been active. Thus Java, with the most volcanoes, is by far the most densely populated of the islands. The greatly superior productivity of a volcanic region apparently justifies the risk of living there. However, at times large crop areas are burned out or badly scorched, and villages threatened by the lava streams are deserted. Indonesians also coined the word "lahar," which is now used as a scientific term for mudslides in volcanic areas. Despite these dangers, villagers return to volcanic slopes.

Far greater catastrophes can also occur. In 1883 the Krakatau (Krakatoa; 2,667 ft / 812 m) volcano in the Sunda Strait erupted in the largest known natural explosion, killing 30,000 people in the greatest volcanic disaster in history. Smoke and dust from its eruption darkened skies around the world for months, with significant effects on climate and temperature. Krakatau then lay dormant, and renewed growth appeared on its slopes until 1927, when it threatened to erupt once more. In 1928 a new cone arose, forming a crater island. The most recent eruption was in 1952. In 1963 the Gunung Agung (10,308 ft / 3,141 m) volcano on Bali erupted, killing thousands.

In addition to volcanic disasters, Indonesia has often suffered from earthquakes. A 1992 earthquake killed more than 2,500 people on the island of Flores, a 1994 earthquake and tsunami in eastern Java killed more than 200 people, and in June 2000 an earthquake measuring 7.9 on the Richter Scale caused more than 100 deaths in southern Sumatra.

Plateaus

The island of Sumatra has significant plateau areas, including Tanah Karo with approximately 1,930 sq mi (5,000 sq km) of fertile volcanic soil; the Agam Plateau; and the Maninjau Plateau which rises 2,296 ft (700 m) above Maninjau Lake. The landscape of Java is elevated in the Gunung Sewu ("Thousand Hills") Plateau, which is riddled with caves; and the Dieng Plateau, an extinct volcanic caldera, 6,562 ft (2,000 m) above sea level, an area famous for its mineral lakes and ancient Hindu temple ruins.

Canyons

Rivers have carved dramatic canyons in some regions of Sumatra and Java. Sumatra's terrain is cut through by Sianok Canyon, a 492 ft (150 m) deep limestone gorge that is 9 mi (15 km) long; the Harau Valley nature reserve, which is 492 to 1,312 ft (150 to 400 m) wide, with walls 262 to 984 ft (80 to 300 m) deep; and the Anai Valley gorge. The Green Canyon, a nature reserve, is in western Java close to the coast.

Hills and Badlands

Many hill areas on Bali and Java are covered with rice terraces, which help to prevent soil erosion. On Java, tea plantations occupy numerous hillsides as well. The area of volcanic foothills of the Bandung district is the best-known hill region of Java. The islands of Nusa Tenggara, including Lombok and Timor, have grass-covered hills. Much of Sulawesi is highland, including the region called Torojaland in the south of the island. Kalimantan's north-central region is distinguished by hilly terrain.

INLAND WATERWAYS

Lakes

More than 500 lakes are scattered across Indonesia. Their types include tectonic basin, volcanic crater, ancient glacial, floodplain, salt lagoon, and artificial lakes. Indonesian lakes are often heavily used for irrigation, and industry or agricultural chemicals have polluted many lakes and ponds.

By far the largest Indonesian lake is Lake Toba in northern Sumatra, covering more than 502 sq mi (1,300 sq km) between towering cliffs that once were the rim of a volcanic crater. Toba is one of the deepest lakes in the world, plunging over 1,476 ft (450 m). It is also one of the highest, at 2,953 ft (900 m) above sea level. Samosir Island, 630 sq km (243 sq mi) occupies the center of the lake. In addition to Toba, notable Sumatran lakes include Manindjau and Singkarak.

The central region of Sulawesi has a pair of deep lakes: Lake Towuti, which is 30 mi (48 km) wide, and Lake Matana. Lake Poso is in north-central Sulawesi. In northern Sulawesi, lakes include Limboto and Tandano. In Papua, Panai Lake, 15 mi (24 km) long and 11 mi (18 km) wide, lies in the central highlands, with two smaller lakes nearby. Kalimantan's lakes include the three Mahakam lakes, located in a geological depression. The Mahakam River basin, an important bird habitat, contains 96 lakes altogether. The island of Flores is famous for a trio of lakes at the top of volcanic Mount Keli Mulu, each with different-colored water (green, maroon and black) due to variation in mineral content.

Rivers

Rivers and rivulets are found in every part of the islands. Although most rivers are short, they are often important for irrigation and are occasionally destructive in periods of exceptionally heavy rainfall. Indonesia's rivers are extremely vulnerable to riverbed siltation from soil erosion, agricultural chemical pollution, and industrial pollution.

Major rivers can be found on Kalimantan, Java, Papua, and Sumatra. Some are navigable for part of their length and were associated with the development of old trading centers that served as river ports and later as provincial capitals, such as Djambi, Palembang, and Banjarmasin. Indonesia's longest river, the Kapuas, which is 710 mi (1,143 km) long, is in Kalimantan, flowing from the north-central mountains to the South China Sea. Much of its length is to some degree navigable. Other major rivers in Kalimantan are the Barito, Mahakham, and Rajang (most of which is in Malaysian Sarawak, where a controversial dam project is underway). Southern Kalimantan is crisscrossed with a network of hundreds of smaller rivers.

Sumatra's rivers include the Batanghari and Musi in the south, and the Indragiri and Kampar in the center of the island. Java's rivers are used for irrigation and include the Solo which is Java's longest at 348 mi (560 km), Tarum, and Brantas. Many rivers wind through Papua, including the Mamberamo, which runs into the Pacific Ocean. A 20 year "mega-project" to dam the Mamberamo, to provide power for an aluminum smelter and industrialize its valley, has been opposed by environmental groups and Papuan activists.

Wetlands

Coastal Sumatra has large swamp and mangrove forest wetland regions. Much of southern Kalimantan is covered with extensive peat swamps, their soft soil composed of decaying plant matter. The peat swamps have been logged-out for valuable trees and have burned and smoldered out of control during the dry season.

Indonesia has two areas, Berbak and Sentarium, designated as Wetlands of International Importance under the Ramsar Convention on Wetlands. Berbak is a National Park consisting of 444 sq mi (1,150 sq km) of peat swamp and 173 sq mi (450 sq km) of freshwater swamp forest in the Djambi region of southern Sumatra. The forests of Berbak are flooded nearly year round. More than 150 tree species are found in this wetland, which is the habitat of endangered mammals (including the extremely rare Sumatran rhinoceros and Sumatran tiger) and reptiles as well as waterbirds. Danau Sentarium, a National Park in Kalimantan, is a complex of freshwater lakes, rivers, peat swamp and freshwater swamp forest. It is the last large primary freshwater swamp forest remaining in Kalimantan. The Danau Sentarium park is the habitat of more than 185 fish species and 200 bird species.

THE COAST, ISLANDS AND THE OCEAN

Oceans and Seas

Citizens of Indonesia often refer to their country as "Tanah Air Kitah," "Our Land and Water," which illustrates the importance of the seas surrounding the archipelago. Indonesia forms a natural barrier between the Indian Ocean to the south and west, the open Pacific Ocean to the northeast, and the South China Sea to the north. South of the island of Java is the lowest point in the Indian Ocean, the Java Trench, some 24,000 ft (7,300 m) deep. The Trench was formed by the subduction of the Australian Tectonic Plate beneath the Eurasian Plate, and its instability is a major source of earthquakes. Between Timor and Australia is the Timor Trough, which is approximately 9,842 ft (3,000 m) deep. A series of agreements was concluded in the early 1970s relating to territorial waters and continental shelf and seabed boundaries between Indonesia and neighboring countries. In the waters directly off the islands of Indonesia are at least 10 percent of the world's coral reefs. Fishing practices and land erosion increasingly endangers these important marine ecosystems.

GEO-FACT

The Wallace Line is an ecological boundary that separates the Asian and Australian habitat regions of the Indonesian archipelago. Animals, plants, and even ethnic groups differ greatly between the two regions, separated by the channel between the island of Bali and the Lombok Islands and the sea to the west of Sulawesi.

There are a vast number of seas, straits, and passages found around the islands of Indonesia. The South China Sea is north of Borneo. The Karimata Strait connects it to the Java Sea. The Strait of Malacca, running between Sumatra and mainland Malaysia and connecting the South China Sea to the Bay of Bengal, is one of the busiest waterways in the world. Most ships heading to the east coast of Asia from the west pass through this strait, as does most traffic from East Asia heading west. In 1971 the Indonesian and Malaysian governments implemented their respective 12 NM territorial water limits, which overlap in the narrowest part of the Strait, meaning that they do not regard the strait as an international waterway. The Sunda Strait between Java and Sumatra is also heavily traveled. The northernmost tip of Sumatra is separated from India's Nicobar Islands by the Great Channel.

Further east among the islands is the Makassar Strait between Borneo and Sulawesi. It connects the Sulawesi (Celebes) Sea in the north with the Java, Bali, and Flores Seas in the south. The Bali Sea is north of Bali and Lombok, while the Flores Sea is north of Sumbawa and Flores. There are several small seas amongst the Maluku Islands, including the Maluku (Molucca) Sea, Ceram Sea, and Banda Sea. The seas off these islands and Sulawesi are rich in oil and are the site of much offshore drilling, with attendant pollution. The Timor Sea and Arafura Sea lie between Indonesia and Australia.

Major Islands

The islands of Indonesia are part of the Malay Archipelago, which also includes the Philippines. The Indonesian part of the archipelago includes more than 13,000 islands, many of them only a few acres in size. Not all of these islands have been officially named, and only about 1,000 are inhabited.

Most of the islands rise from the submerged Sunda shelf, considered a continuation of the Asian continent. The western and central islands are known as the Sunda Islands. Sumatra, Java, Borneo (the Indonesian part of which is called Kalimantan), and Sulawesi, along with the surrounding islands, are known as the Greater Sunda Islands. Borneo is the largest; at 290,320 sq mi (751,929 sq km) it is the third-largest island on earth. Smaller islands in this region include Bangka, Belitung, and the Mentawi Islands. The island of Madura, near Java's north-east coast, is particularly heavily populated, and "trans-migration" programs have sent many Madurese to settle on other islands.

Further east are the Lesser Sunda Islands. They begin with Bali and extend to Timor. Lombok, Sumbawa, Flores, and Sumba are the other large islands in this chain, which is also known as Nusa Tenggara. Along with Savu and Roti Islands, they enclose the Savu Sea. Even further to the east are the Maluku Islands, formerly called the Moluccas. Most of the Maluku are found in groups of small and medium-sized islands, such as the Tanimbar Islands, Aru Islands, Kai Islands, and Sula Islands. Halmahera, Wetar, Buru, and Ceram are the largest individual islands. The Maluku were called the Spice Islands by 15th and 16th century European explorers, who found them to be rich in the spices valuable in Europe at that time. Some smaller islands that housed important European colonies include Ambon, Tindore, and Ternate. The Lesser Sunda Islands and Maluku are spread over a sea area of more than 1 million sq mi (2,589,988 sq km), although the islands themselves are only 60,000 sq mi (155,399 sq km) in land area.

New Guinea, the island of which Indonesia's Papua state is the western half, is the second largest island in the world (341,631 sq mi / 884,824 sq km). It and a few associated island groups are exposed parts of a different submerged platform, this one part of the Australian continental shelf.

The Coast and Beaches

Indonesia has one of the world's longest coastlines. The southwestern islands are similar in that their Indian Ocean shores tend to be steep, with few sandy beaches, while their northern and eastern coasts are mostly flat in terrain, with the shorelines expanding relentlessly due to sedimentation. River deltas dominate Sumatra's eastern side in particular. In the late 20th century increased siltation from deforestation throughout Indonesia blocked many harbors, making dredging necessary for shipping access.

Sulawesi is formed from four conjoined peninsulas, with the long, northernmost, Minahasa Peninsula curved around the Tomini Gulf, while the two southern arms enfold the Bone Gulf. Between the two is the Gulf of Todo. Kalimantan has a jagged coastline with numerous river deltas that empty into the South China Sea, Java Sea, Makassar Strait, and Celebes Sea. Cape Datu marks the northwestern edge of Kalimantan. At the far side of the archipelago, the northwest Arfak region of Papua is known as the Bird's Head Peninsula. It is indented by the Sarera Gulf on its northern coast, which borders the Pacific Ocean. On the south of Papua, river deltas empty onto the Arafura Sea.

CLIMATE AND VEGETATION

Temperature

Indonesia has a tropical climate, with high humidity (an average of 82 percent) and high temperatures. Two basic seasons are experienced: a rainy season from November to March, and a hot, drier season from April through October. Temperatures in Indonesia's capital, Jakarta, generally range from 73° F (23° C) to 91° F (33° C).

Rainfall

Average yearly rainfall for Indonesia as a whole is approximately 78 in (200 cm). In lowland areas, the average rainfall is 70-125 in (180-320 cm); while in the mountains it can reach as much as 238 in (610 cm) annually. The fearsome typhoons of the South China Sea spend themselves before reaching Indonesian waters, and the gales that blow from time to time through the Torres Strait, between Australia and New Guinea, seldom move farther than the extreme southeastern islands of the archipelago, so the seas of Indonesia are generally calm.

Grasslands

Many of the Lesser Sunda Islands, including Sumba, Lombok, Sumbawa, and Timor, have extensive grassland areas, as do parts of Sumatra, Kalimantan, Sulawesi, and Papua. Most of these grasslands are tracts where forests have been cut or burned. Invasive grasses such as imperata are encroaching on more and more terrain on Indonesia's islands, as grassland, which burns easily in the dry season, replaces forest. On Sulawesi, this former-forest grassland becomes very sparse, almost desert-like, in the dry season. Bamboo, both wild and cultivated, grows in many parts of Indonesia, although wild bamboo is also being cleared.

Forests and Jungles

Indonesia has for centuries been rich with a great variety of forest types: dipterocarp rainforests in Sumatra, Kalimantan and Papua; monsoon forests in the Lesser Sunda Islands; coastal mangrove forests; and alpine forests in the mountains of Papua. Indonesia has been estimated to be the habitat of 12 percent of world's mammal species and 16 percent of bird species as well as 11 percent of plant species. The ancient rainforests of Kalimantan have survived as an especially diverse ecosystem.

The pressures of the expanding Indonesian population in the 1970s and 1980s led to forest cutting for agricultural expansion and firewood, increasing soil erosion and depletion of soil. Government programs promoting "transmigration" of islanders from heavily populated regions to forest areas increased forest loss. Commercial logging of Indonesia's rainforests for the international trade in wood products took place on a massive scale during the 1980s and 1990s. The commercial timber operations often clear-cut trees and left logged areas bare of vegetation without replanting, causing desertification. In many areas, natural forests were replaced by monoculture plantations including oil palms, coffee, rubber, and pulpwood trees such as eucalyptus.

During 1983 an enormous fire, apparently spread by dead wood left after logging, destroyed approximately 30,000 sq km (11,583 sq mi) of forests in Kalimantan. Then, in 1997-98, a series of long-burning fire outbreaks destroyed more than 17,374 sq mi (45,000 sq km) of forests and peat swamps, mainly in Sumatra and Kalimantan. Mostly started by timber and agri-business companies (particularly oil palm plantation companies) to clear land, these fires rampaged quickly through forests degraded by logging. Haze, a form of thick air pollution from the fires' smoke, spread over other islands of Indonesia and neighboring Malaysia and Brunei. In addition to rainforest destruction, coastal mangrove forests, found in Papua, Sumatra and Kalimantan, are endangered by shrimp-farm pollution and land clearing.

Satellite map-based reports by environmental organizations reveal that more than 20,000 sq km (7,722 sq mi) of Indonesia's remaining forest is being destroyed by logging, fires and agricultural clearing annually, and that Indonesia's forest cover declined from approximately 1,619,957 sq km (625,468 sq mi) in 1950 to an estimated 980,148 sq km (378,437 sq mi) in 2000. The World Bank predicts the extinction of all lowland dipterocarp rainforests in Sumatra by 2005 and in Kalimantan by 2010, and extinction of coastal mangrove forests by 2015. The World Bank warns that at the present rate of destruction, only scrub mountain forests will survive in Indonesia by the year 2020. Lowland rainforest is already nearly gone on the island of Sulawesi.

Population Centers – Indonesia

(1995 POPULATION ESTIMATES)

Name	Population	Name	Population
Jakarta (capital)	9,160,500	Ujung Pandang (Makassar)	1,091,800
Surabaya	2,701,300	Malang	763,400
Bandung	2,368,200	Padang	721,500
Medan	1,909,700	Bandjarmasin	534,600
Semarang	1,366,500	Surakarta	516,500
Palembang	1,352,300		

SOURCE: Badan Pusat Statistik, Republik Indonesia (Statistics Indonesia, Republic of Indonesia)

Provinces – Indonesia

POPULATIONS FROM 2000 CENSUS

Name	Population	Area (sq mi)	Area (sq km)	Capital
Bali	3,151,162	2,147	5,561	Denpasar
Bengkulu	1,567,432	8,173	21,168	Bengkulu
DKI Jakarta	8,389,443	228	590	Jakarta
Jambi	2,413,846	7,345	44,924	Jambi
Jawa Barat	35,729,537	7,877	46,300	Bandung
Jawa Tengah	31,228,940	3,207	34,206	Semarang
Jawa Timur	34,783,640	8,503	47,922	Surabaya
Kalimantan Barat	4,034,198	6,664	146,760	Pontianak
Kalimantan Selatan	2,985,240	4,541	37,660	Banjarmasin
Kalimantan Tengah	1,857,000	8,919	152,600	Palangkaraya
Kalimantan Timur	2,455,120	8,162	202,440	Samarinda
Lampung	6,741,439	12,860	33,307	Tanjung Karang
Maluku	1,205,539	28,767	74,505	Ambon
Nanggroe Aceh Darussalam	3,930,905	21,387	55,392	Banda Aceh
Nusa Tenggara Barat	4,009,261	7,790	20,177	Mataram
Nusa Tenggara Timur	3,952,279	18,485	47,876	Kupang
Riau	4,957,627	36,511	94,562	Pakanbaru
Sulawesi Selatan	8,059,627	28,101	72,781	Ujung Pandang
Sulawesi Tengah	2,218,435	26,921	69,726	Palu
Sulawesi Tenggara	1,821,284	10,690	27,686	Kendari
Sulawesi Utara	2,012,098	7,345	19,023	Menado
Sumatera Barat	4,248,931	19,219	49,778	Padang
Sumatera Selatan	6,899,675	40,034	103,688	Palembang
Sumatera Utara	11,649,655	27,331	70,787	Medan
DI Yogyakarta	3,122,268	1,224	3,169	Yogyakarta

SOURCE: "Population of Indonesia by Province 1971, 1980, 1990, 1995, and 2000." 2000 Population Census, BPS Statistics Indonesia.

HUMAN POPULATION

Indonesia's overall population density was estimated at 119 people per sq km (308 per sq mi) in 2001. Only 40 percent of Indonesians live in urban areas (1999 estimate), but the urban population density tends to be extremely high, with more than 44,030 per sq mi (17,000 people per sq km) in the cities of Jakarta and Surabaya (1995 estimate). The population is also concentrated on a few islands: Roughly 60 percent of Indonesians live on the islands of Java, Madura and Bali, which combined have just 7 percent of the nation's land. In contrast, Papua contains 1 percent of the population with 22 percent of Indonesia's land. These disparities have brought on severe ethnic conflict when Indonesian government programs called "transmigration" resettled people from densely populated islands to less populated provinces, particularly Kalimantan and Papua.

Effective family planning programs reduced Indonesia's population growth rate from 2.3 percent in 1972 to 1.6 percent in 2001 but there is still an ongoing population explosion, which causes environmental pressures such as deforestation, pollution, and land erosion, as well as contributing to chronic poverty, civil unrest, and depletion of resources. Some 20 percent of Indonesians subsist below the poverty line, according to a 1998 estimate.

NATURAL RESOURCES

Indonesia's mineral resources are rich, and include tin, nickel, bauxite, copper, gold, and silver as well as major reserves of coal, oil, and natural gas. There is also great potential for generating geothermal, solar and wind power. Other Indonesian resources include farm products, fishing, and rubber. The provincial location of major resources such as mining in Papua, and petroleum in Sumatra and Kalimantan, has brought about conflicts with Indonesia's Java-based central government and demands for localized control of resource extraction and income.

FURTHER READINGS

Barber, Charles, and James Schweithelm. *Trial By Fire: Forest Fires and Forestry Policy in Indonesia's Era of Crisis and Reform.* Washington, D.C.: World Resources Institute, 2000.

Biodiversity Support Program. *Kemala.* http://www.bsp-kemala.or.id (Accessed April 29, 2002).

Blair, Lawrence, and Loren Blair. *Ring of Fire: An Indonesia Odyssey.* Rochester, Vt.: Inner Traditions International, 1991.

Dalton, Bill. *Moon Handbooks: Indonesia.* Emeryville, Calif: Avalon Travel Publishing, 1995.

Daws, Gavan, and Marty Fujita. *Archipelago: Islands of Indonesia.* Berkeley: University of California Press, 1999.

Environmental Investigation Agency. *The Final Cut: Illegal Logging in Indonesia's Orangutan Parks.* London: Environmental Investigation Agency, 1999.

Forest Watch Indonesia, Global Forest Watch, World Resources Institute. "State of the Forests Indonesia." Washington, DC: World Resources Institute, 2002.

Van Oosterzee, Penny. *Where Worlds Collide: The Wallace Line.* Ithaca, N.Y.: Cornell University Press, 1997.

Iran

- **Area:** 636,296 sq mi (1,648,000 sq km) / World Rank: 19
- **Location:** Northern and Eastern Hemispheres; southwestern Asia, between the Caspian Sea and Persian Gulf, bordered on the north by Armenia, Azerbaijan, and Turkmenistan, on the east by Afghanistan and Pakistan, and on the west by Iraq and Turkey.
- **Coordinates:** 32°00′N, 53°00′E
- **Borders:** 3,380 mi (5,440 km) / Afghanistan, 582 mi (936 km); Armenia, 22 mi (35 km); Azerbaijan proper, 268 mi (432 km); Azerbaijan-Naxcivan enclave, 111 mi (179 km); Iraq, 906 mi (1,458 km); Pakistan, 565 mi (909 km); Turkey, 310 mi (499 km); Turkmenistan, 616 mi (992 km)
- **Coastline:** 1,556 mi (2,510 km) / Caspian Sea 392 mi (630 km); Gulf of Oman and Persian Gulf, 1,168 mi (1,880 km)
- **Territorial Seas:** 12 NM
- **Highest Point:** Mount Damāvand, 18,606 ft (5,671 m)
- **Lowest Point:** Caspian Sea, 92 ft (28 m) below sea level
- **Longest Distances:** 1,398 mi (2,250 km) SE-NW 870 mi (1,400 km) NE-SW
- **Longest River:** Kārūn, 553 mi (890 km)
- **Largest Lake:** Lake Urmia, 1,879 sq mi (4,868 sq km)
- **Natural Hazards:** Earthquakes; floods; droughts; sand and dust storms; hailstorms
- **Population:** 66,128,965 (July 2001 est.) / World Rank: 17
- **Capital City:** Tehran, in the central north
- **Largest City:** Tehran, population 7,723,000 (2002 est.)

OVERVIEW

The topography of Iran (known as Persia prior to 1935) consists of two main mountain ranges wrapped around a basin holding deserts and salt marshes. In the north Iran meets the Caspian Sea, and in the south, the Persian Gulf and Gulf of Oman. Iran is rich in resources and trade routes. Settlement is mainly in the mountain regions, the coasts, and some oases. In the areas where agriculture is viable, crops thrive as long as there is adequate water. However, Iran has a delicate environmental balance as forests and farmland decrease and desert increases.

Caspian Sea boundaries are still undefined between Iran, Azerbaijan, Russia, Kazakstan, and Turkmenistan. Two islands in the Persian Gulf are occupied by Iran but also claimed by the United Arab Emirates (UAE): Lesser Tunb and Greater Tunb; and one island is jointly administered by Iran and the UAE: Abu Musa.

Iran lies on the Eurasian Tectonic Plate on some of the world's most active fault lines. Its western border sits right where this plate bumps into the Arabian Plate and the effect of the Arabian Plate impacting the Eurasian Plate has caused the bent and rippled layers of rock in the Zagros Mountains. In the southeast, the Eurasian Plate collides with the Indian Plate not too far outside Iran's borders. Subterranean shifts produced numerous faults in the earth's crust, where frequent and devastating earthquakes occur, with the western region being hit the hardest. As many as 140,000 people were killed in Iranian earthquakes during the twentieth century.

GEO-FACT

Iran has a huge network of undergound water canals called *qanats*, with about 50,000 *qanats* covering an estimated 248,548 mi (400,000 km). In the absence of major rivers, the *qanats* were Iran's traditional irrigation source, constructed with underground storage structures. Water-use analysts have called for a return to the *qanat* system and smaller scale irrigation projects as the best ways to combat ongoing water shortages throughout Iran.

MOUNTAINS AND HILLS

Mountains

The broken and irregular ranges of Iran's mountains, extending from Armenia and Azarbaijan in the north to Pakistan in the south, are barren, but the valleys that intersperse them are fertile. In the north of Iran, where the mountains reach 7,000 to 9,000 ft (2,133 to 2,743 m), livestock grazing and settlements can be found above 4,000 ft (1,219 m).

The narrow Elburz Range curves west to east along the Capsian Sea coast. Iran's capital, the sprawling city of Tehran, is located on the south side of the Elburz range. The highest of Iran's mountains, Mt. Damāvand 18,606 ft (5,671 m), a symmetrical volcanic cone, is in the Elburz Range just northeast of Tehran.

The forbidding Zagros Range, a group of parallel mountain chains, runs northwest to southeast through Iran. Much of the Zagros Range towers above 9,842 ft (3,000 m), until it declines in height in the southeast to an average of less than 4,921 ft (1,500 m). The Zagros Range extends down to the Persian Gulf and Gulf of Oman coasts in rocky cliffs.

Plateaus

Iran is located on the Plateau of Iran, a high triangular plateau with average elevations of 3,000 to 5,000 ft (914 to 1524 m) that includes parts of Afghanistan and Pakistan. Great salt deserts such as Dasht-e-Lūt and Dasht-e-Kavīr occupy the eastern section of the Plateau of Iran, and mountains cut through the center and west of it. The plateau has an area of approximately one million square miles (2,590,000 sq km), of which about 600,000 sq mi (1,554,000 sq km) is in Iran. The region was formed and shaped by the uplifting and folding effect of three giant plates pressing against each other: the Arabian, Eurasian, and Indian Plates.

Canyons

The Zagros Mountains have deep folds and eroded valleys, where streams and small rivers have created deep gorges. In the Zagros region are found the Kārūn River Canyon, Sezar River Gorges, Bactiara River Canyon, and deep canyons in the vicinity of the Gahar Lakes.

Hills and Badlands

The foothills of Iran's mountain ranges are terraced for farming and towns. The Elburz foothills follow the Caspian Sea coast. In the Zagros foothills, salt domes cover Iran's major oil fields. The Kandovan hills in northwest Iran are a group of rock formations with inhabited cave-dwellings.

INLAND WATERWAYS

Lakes

The lakes of Iran are few, and mostly small. Many lakes and most shallow wetlands of Iran dried up during the catastrophic drought of 1998–2001, the worst in three decades.

Lake Urmia (Orumiyé) is Iran's largest intact lake, with an average surface area of 1,879 sq mi (4,868 sq km). It can vary from 1,158 to 2,317 sq mi (3,000 to 6,000 sq km) depending on seasonal conditions. A salt lake, Urmia is in the northwest near the Turkish border, at 4,255 ft (1,297 m) above sea level.

Lake Helmand, a lake/wetland system extending into Afghanistan, has been a freshwater lake, used for irrigation and fishing. The lake system decreased from about 57,915 sq mi (150,000 sq km) to 12,355 sq mi (32,000 sq km) during the twentieth century, dwindling to about 1,235 sq mi (3,200 sq km) in the dry seasons. Lake Helmand dried up almost completely during the 1998–2001 drought.

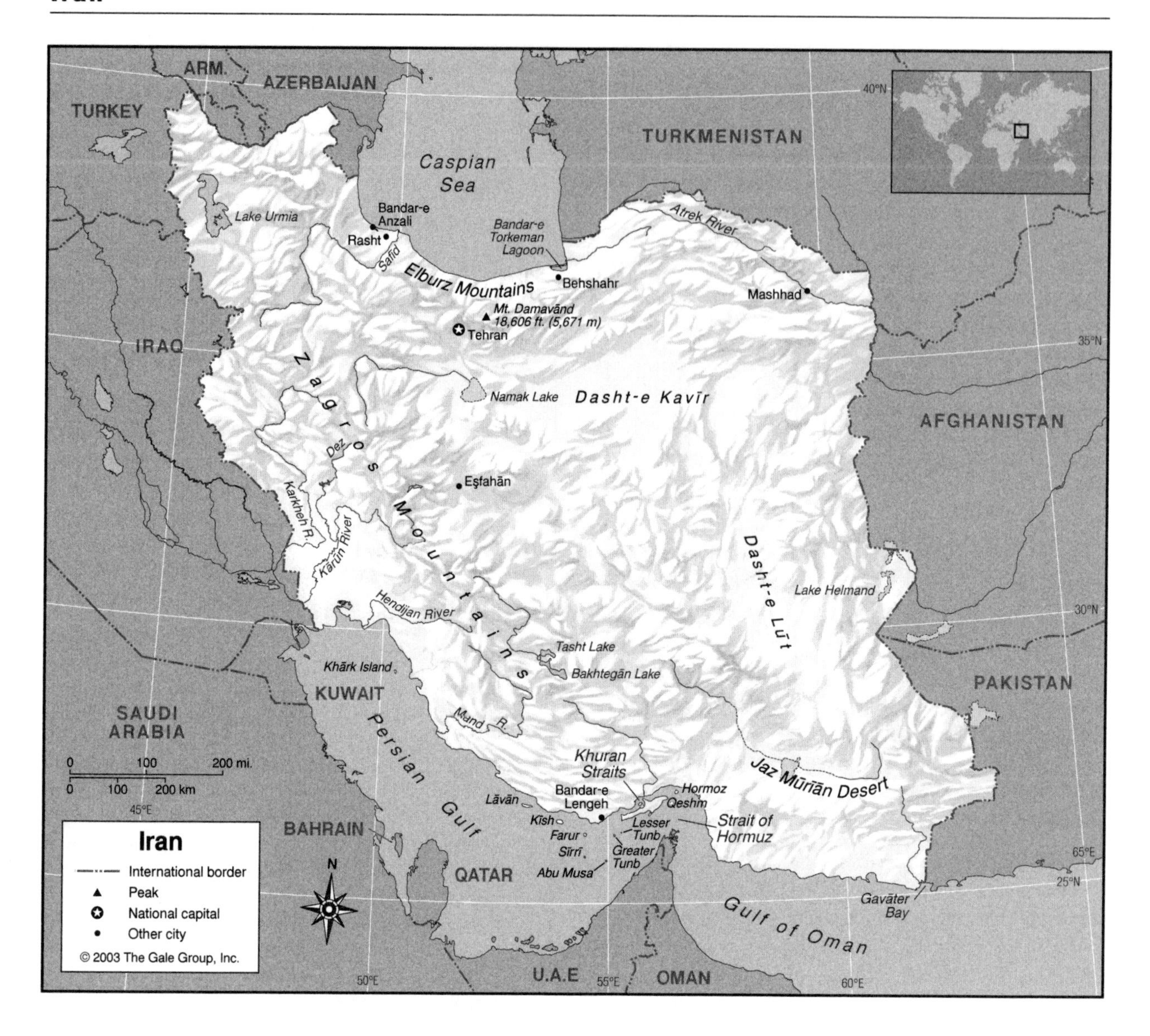

The lakes in Fars Province (southwest Iran) were hit particularly hard by the drought and most dried up almost completely. Notable lakes in the southwest include (with pre-drought sizes): Bakhtegān Lake, 290 sq mi (750 sq km); Tasht Lake, 171 sq mi (442 sq km); and Moharloo Lake, 80 sq mi (208 sq km).

Important lakes of central Iran include: Namak Lake, 697 sq mi (1,806 sq km); and Howz Soltan Lake, 41 sq mi (106 sq km). Snowmelt feeds the 8,366-ft (2,550 m) high twin Gahar Lakes in the Zagros Mountains.

Rivers

Iran's rivers drain into the Persian Gulf and Gulf of Oman in the south, the Central Plateau, Lake Urmia in the northwest, and the Caspian Sea. Some rivers are dry most of the time but fill from snowmelt in the spring.

The Kārūn, at 553 mi (890 km), is Iran's longest river and its only navigable one. Still, it is only navigable for just 112 mi (180 km), and by shallow draft vessels only. The Kārūn runs from the Zagros Mountains to the Persian Gulf delta region in western Iran, where these related rivers are also located: the Karkheh (469 mi / 755 km); the Dez (320 mi / 515 km); the Hendijan (303 mi / 488 km) and the Jarahi (272 mi / 438 km).

Other notable rivers of Iran include: the Sefidrood (475 mi / 765 km), Atrek (332 mi / 535 km), and Safīd draining into the Caspian Sea; the Mand (426 mi / 685 km) in in the southwest; and the Zayande (251 mi / 405 km) flowing through the city of Isfahan in the Zagros foothills.

Wetlands

The low basins of central Iran have extremely shallow lakes that dry up and leave thick, broken salt crusts known as *kavirs* with mud marshes beneath them. Iran also has major areas of coastal wetlands, including those bordering the Caspian Sea.

There are 18 sites in Iran that are designated Wetlands of International Importance under the Ramsar International Convention on Wetlands, which was inaugurated at Iran's Caspian coastal resort of Ramsar in 1971. Caspian wetlands sites include the Anzali Mordab marsh complex, a bird migration area; Bandar-e Torkeman Lagoon; and other lagoons.

In western Iran, the Ramsar sites include the Shadegan wetland, delta mudflats on the Iraq border which were badly damaged in the Iran-Iraq war of 1980–88; the Parishan and Dasht-e Arjan marshes in southwestern Iran; and the Neyriz Lakes and Kamjan Marshes in a wildlife refuge in the southwest. Lake Urmia, with its brackish marshes, birds and fish species, in the northwest, is a Ramsar site, as is the dying Helmand Lake in the east.

Offshore wetlands sites include the Khuran Straits between the mainland and Qeshm island; and estuaries on the Strait of Hormuz, featuring mangroves and salt marshes that are significant bird wintering sites.

Many of Iran's wetlands dried up during the three-year drought at the turn of the twenty-first century. They have also been threatened by invasive plant species, pollution, water diversion for agriculture, agricultural land expansion, road building, and shrimp farming.

THE COAST, ISLANDS AND THE OCEAN

Oceans and Seas

The Caspian Sea, at 163,800 sq mi (424,240 sq km) the world's largest landlocked body of water (although shrinking), borders northern Iran. A shallow sea of salt water, the Caspian is 92 ft (28 m) below sea level. Because of massive reserves of natural gas, demarcation of rights to the Caspian Sea's waters has become a contentious issue among all of the countries bordering it.

Iran's southwest embraces the Persian Gulf, a 615-mi (990-km) extension of the Indian Ocean. Iran shares the Persian Gulf with Iraq, Kuwait, Saudi Arabia, Bahrain, Qatar, the United Arab Emirates, and Oman. From the Persian Gulf, the 34-mi (55-km) Strait of Hormuz, one of petroleum shipping's most strategic points, leads into the Gulf of Oman, which then widens to the Indian Ocean's Arabian Sea. Pollution from oil tankers and military ships, overfishing, destructive fishing methods, agricultural chemical runoff, sewage, and industrial waste are problems in the Persian Gulf and Gulf of Oman.

Major Islands

Iran occupies 16 islands in the Persian Gulf, although three are contested by the United Arab Emirates; 11 of the islands are inhabited. The Persian Gulf islands are seabird nesting sites in late spring, and also are nesting sites for endangered sea turtles. The coral reefs around islands in the Persian Gulf are barely surviving temperature fluctuations, algae, oil spills and other pollution, tourist damage, and construction. Qeshm, 515 sq mi (1,335 sq km), a mountainous oblong in the Strait of Hormuz, is the largest island in the Persian Gulf. Other, much smaller islands in and near the Strait of Hormuz include Kīsh (Qeys), Hormoz, Hendurabi, Farur, Sīrrī, Abu Musa, and Lāvān. Khārk Island is close to the northern end of the Persian Gulf.

The Coast and Beaches

Iran's Caspian Sea coast, with much of the shore formed from former seabed as the water recedes, begins in the west at the border of Azerbaijan, sweeps southeast to the lagoon port of Bandar-e Anzali, and continues east to the Bandar-e Torkeman lagoon above Behshahr town. The coast then turns straight north to the Turkmenistan border.

Southwest Iran meets the northwest end of the Persian Gulf at the border with Iraq. At this end of the Gulf the coastal plain is wide, containing the delta of the Kārūn River, which adjoins neighboring Iraq's Tigris and Euphrates River deltas. Estuaries with mudflats and salt marshes are found in this region, and there are hundreds of seasonal creek outlets in non-drought years, many emptying into Moosa Bay.

The Persian Gulf coast continues southeast to Būshehr Bay and Naayband Bay in Būshehr province. This stretch where the mountains come right down to the sea is marked by rocky shores and cliffs. This rugged coast, especially around Naayband Bay and the harbor of Bandar-e Lengeh, is considered particularly vulnerable to oil spills. At Hormozgān Province, the coastline curves inward sharply, sheltering Qeshm island, with seasonal creek outlets in Khamir Harbor across from Qeshm. The Strait of Hormuz is formed here, with Oman's Masandam peninsula projecting across the strait. This section of coast has sandy beaches on a narrow coastal strip, including the white sand beach at Koohestak. After the Strait of Hormuz the coastline veers south and then east on the Gulf of Oman, indenting at Chaabahar Bay and again at Gavāter Bay on the Pakistan border.

CLIMATE AND VEGETATION

Temperature

Iran has an arid and semi-arid climate with subtropical areas along the coasts. There are four seasons: a warm spring, hot summer, brief autumn, and cold winter. The central deserts and Persian Gulf coast are especially hot in summer, with some of the world's highest recorded temperatures occurring in the desert. The average annual temperature in northern Iran is 50°F (10°C); the average annual temperature in southern Iran is 77° to 86°F (25° to 30°C). Iran's climate is dry except for belts of high humidity along the Caspian Sea and Persian Gulf. There are strong seasonal winds, often causing dust and sandstorms.

Rainfall

Iran's average annual precipitation is 11 in (27 cm) during non-drought years. Less than 14 percent of the land receives more than 52 percent of the precipitation. The most rainfall occurs along the Caspian Sea coast, past the Elburz range. The rains arrive mostly in the winter, when snow also affects the mountainous regions. In some areas precipitation is entirely lacking for long periods of time. Sudden storms with heavy rains a few times per year may provide the entire annual rainfall. Extremely low rainfall brought on the drought crisis at the turn of the twenty-first century, while normal rain and snowfall finally returned in 2002.

Grasslands

Iran lacks substantial pasture lands; marsh reedbeds and shrubbery are used for livestock grazing. There are some grasslands in upland areas such as the hills around Isfahan, and foothills in the southeast.

Deserts

More than 115,831 sq mi (300,000 sq km) of Iran is covered with deserts. That coverage is increasing through the process of desertification as farmland, grassland, and forests are losing vegetation and then soil. The drought of 1998–2001 increased desert area when lakes and wetlands dried up.

Iran's immense Lūt Desert, which includes the Dasht-e-Kavīr and Dasht-e-Lūt, and adjacent Namakzār-e Shahdād, covers some 30,888 sq mi (80,000 sq km). It is one of the hottest places on earth with temperatures reaching as high as 135°F (57°C). The Lūt Desert goes without rain for years at a time. The results of sudden large rainfalls are preserved for years and sculpted by winds, creating illusions of rivers and cities. "Sand mountains" rise up to 1,558 ft (475 m) in the desert's eastern sector, and there are also sand dunes moved by wind. The region contains an interior area lacking in all life forms, even bacteria. The similar Jaz Mūrīān Desert lies to the south of the Lūt Desert.

The outer deserts are scrubland, habitats for rare Asiatic cheetahs and koulans (Asian zebras). Inner desert areas are covered with hard layers of stones, gravel and pebbles. Salt lakes and marshes create salt flats when they dry out; there are also saltwater springs and salt mines in the Iranian deserts. Scattered oases, linked by roads, are shaded by groves of date palms, poplars and other trees.

Forests and Jungles

An estimated 11 percent of Iran is forested, but only 5 percent of these remaining forests are protected and there is a very low level of reforestation. Deforestation in Iran, mostly caused by logging and grazing, causes sedimentation, flooding, and soil erosion.

Population Centers – Iran

(1996 CENSUS)

Name	Population
Tehran (Teheran)	6,758,845
Mashhad (Meshed)	1,887,405
Eşfahān (Isfahan)	1,266,072
Tabriz	1,191,043
Shiraz	1,053,025

SOURCE: "1996 National Census of Population and Housing." Statistical Centre of Iran.

The Caspian Sea region has the largest forests, which have mostly deciduous tree species, including oak, elm, beech, and linden. Golestān National Park in the Caspian region, near the Turkmenistan border, is highly biodiverse, with deciduous and conifer tree species. Sisangan National Park, near the Azerbaijan border, is another Caspian forest.

There are forests of oaks and other deciduous trees in the Zagros Mountains, and wild pistachio forests are still found in the foothills of southeast Iran. The Persian Gulf and Gulf of Oman coasts have thick stands of palms and mangrove forests.

HUMAN POPULATION

Iran's population growth rate is 1.6 percent (2001), with an overall population density of 104 per sq mi (40 people per sq km) (2000). However, the population distribution is very uneven. The eastern half of the country is only home to about one-fifth of the population, and just as many live within a 50-mile radius of Tehran. Sixty-four percent of Iranians live in cities (2002), with urban infrastructure stressed by growth, particularly in ever-expanding Tehran.

NATURAL RESOURCES

Iran has one of the world's leading petroleum (oil and natural gas) reserves, as well as coal. Its major oil fields are located in the Zagros foothills of Iran's southwest. Despite these fuel resources, severe urban air pollution from fossil fuels may necessitate a search for alternative power generation, and there is excellent solar and wind energy potential. Other resources include chromium, copper, iron, lead, manganese, zinc, sulphur and bauxite, as well as agriculture.

FURTHER READINGS

Bartol'd, V. V. (Vasilii Vladimirovich). *An Historical Geography of Iran,* tr. Svat Soucek; ed. with an introduction by C. E. Bosworth. Princeton, NJ: Princeton University Press, 1984.

Green Party of Iran. http://www.iran-e-sabz.org/ (Accessed June 3, 2002).

Metz, Helen Chapman. *Iran: A Country Study.* Washington, D.C.: Government Printing Office, 1990.

Net Iran. http://www.netiran.com (Accessed June 3, 2002).

Wearing, Alison. *Honeymoon in Purdah: An Iranian Journey.* New York: Picador USA, 2000.

Iraq

- **Area:** 168,754 sq mi (437,072 sq km) / World Rank: 59
- **Location:** Northern and Eastern Hemispheres, Middle East, bordering the Persian Gulf, south of Turkey, west of Iran, north of Kuwait and Saudi Arabia, and east of Jordan and Syria.
- **Coordinates:** 33°00′N, 44°00′E
- **Borders:** 2,256 mi (3,631 km) / Iran, 906 mi (1,458 km); Jordan, 112 mi (181 km); Kuwait, 150 mi (242 km); Saudi Arabia, 506 mi (814 km); Syria, 376 mi (605 km); Turkey, 206 mi (331 km)
- **Coastline:** 36 mi (58 km)
- **Territorial Seas:** 12 NM
- **Highest Point:** Mount Ebrāhīm, 11,811 ft (3,600 m)
- **Lowest Point:** Sea level
- **Longest Distances:** 611 mi (984 km) SSE-NNW; 454 mi (730 km) ENE-WSW
- **Longest River:** Euphrates, 2,235mi (3,596 km)
- **Largest Lake:** Ath-Tharthār, 579 sq mi (1,500 sq km, approx.)
- **Natural Hazards:** Dust storms; soil salinity; flooding
- **Population:** 23,331,985 (July 2001 est.) / World Rank: 44
- **Capital City:** Baghdad, centrally located on the Tigris River
- **Largest City:** Baghdad, 4.3 million (2000 est.)

OVERVIEW

Iraq can be divided into four main regions: the Syrian Desert in the west and southwest; mountains in the north and northeast; a rolling upland between the upper Tigris and Euphrates rivers; and the alluvial plain through which the Tigris and Euphrates flow in the southeast.

The area between Iraq's two great rivers is the heartland of the country. It has been known since ancient times as Mesopotamia. The northernmost part of Mesopotamia is also sometimes called Al Jazīrah. In the north the terrain surrounding the rivers is somewhat elevated, but as they approach the middle of the country they enter a vast alluvial plain that extends to the Persian Gulf. Here the Tigris and Euphrates rivers lie in natural embankments above the level of the plain in many places. The rivers have multiple channels and feed many lakes and marshes. The mountains of the northern highlands are rugged and inhospitable, but the are the source of several tributary rivers to the Tigris, and their foothills have good soil and adequate rainfall. The western deserts are almost completely uninhabited.

MOUNTAINS AND HILLS

Mountains

The northeastern highlands begin just southwest of a line drawn from Mosul to Kirkūk and extend north to the borders with Turkey and Iran. High ground, separated by broad, undulating steppes, gives way to mountains ranging from 3,280 to 13,123 ft (1,000 to nearly 4,000 m) near the Iranian and Turkish borders. The high mountains are an extension of the Zagros Mountains of Iran and include Iraq's highest peak, Mount Ebrāhīm (11,811 ft / 3,600 m).

Plateaus

Iraq derives its name from the term "cliff." West of the central river plain rises a plateau that extends into Syria, Jordan, and Saudi Arabia, reaching heights of about 3,281 ft (1,000 m). Some of this plateau is revealed in exposed cliff rock, but the boundaries between Iraq and its western neighbors are physically indistinguishable.

Hills and Uplands

Al Jazīrah is a hilly region extending into Iraq from Syria and Turkey, with the Tigris and Euphrates flowing through it. Water in the area flows in deeply cut valleys, and irrigation is much more difficult than it is in the lower plain. Much of this zone may be classified as desert.

INLAND WATERWAYS

Lakes

There are many lakes in central Iraq. They are fed largely by the floods of the Tigris and the Euphrates, and by distributory streams and canals from these rivers. As a result their size varies considerably depending on the flow of the rivers, but the largest are Ath-Tharthār, Ar-Razzāzah, and Hawr al-Habbānīyah. South of Baghdad the lakes tend to be increasingly saline, reflecting the heavy silt content of the two great rivers and the poor drainage in this region. Lake Al-Qādisīyah is a sizeable reservoir on the Euphrates in the northwest.

Rivers

The Euphrates is the longest river in the country. Originating in Turkey, it flows through Syria before

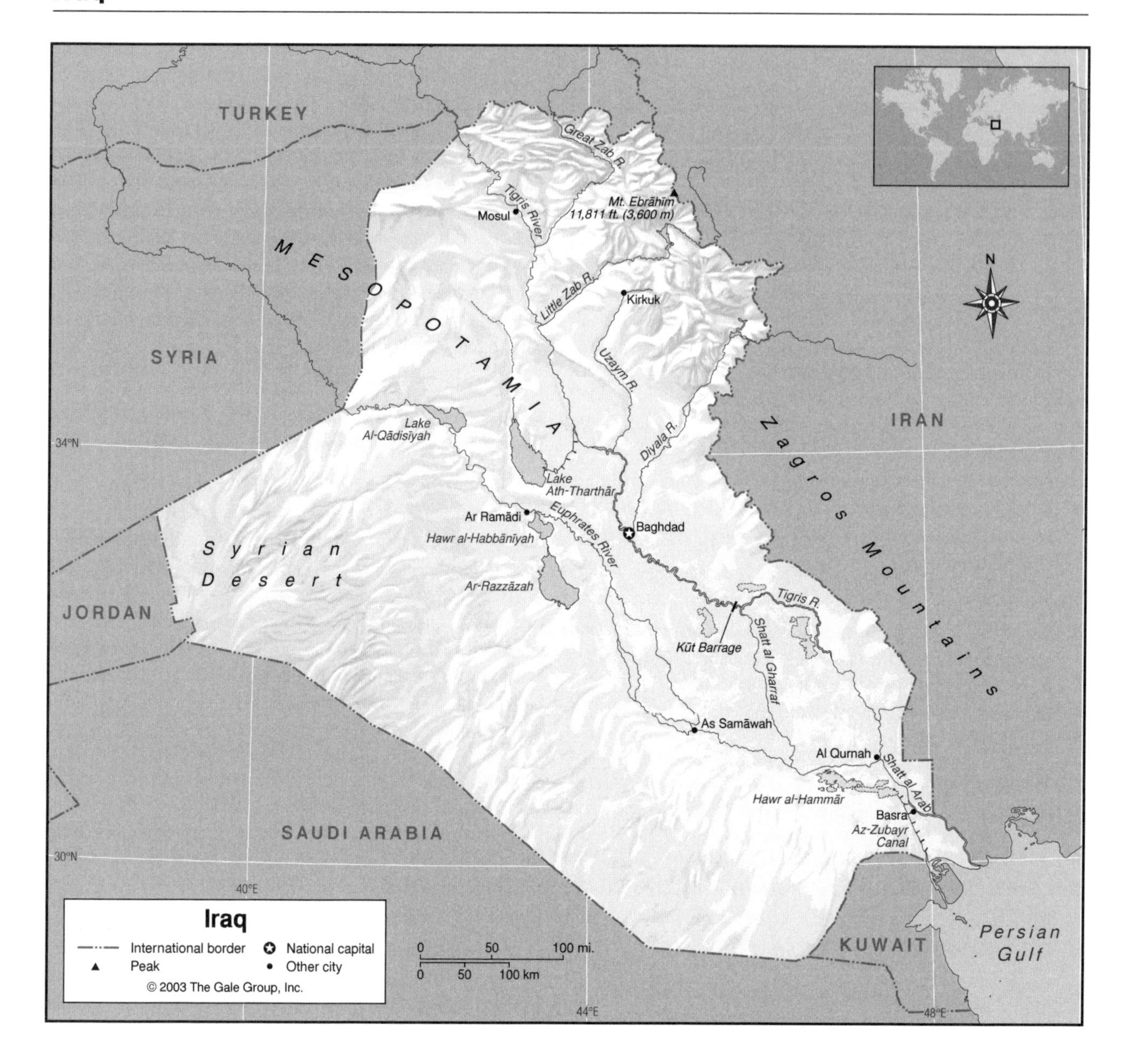

entering Iraq and receives several tributaries in that country. Once within Iraq it has no permanent tributaries, but is fed by the wadis of the western desert during the winter rains. The Euphrates winds through a gorge 1.2 to 10 mi (2 to 16 km) wide in the Al Jazīrah before reaching the plains at Ar Ramādi. Once on the plain, the river flows through many channels—some natural, some artificial—and feeds numerous lakes and marshlands. Near As Samāwah, it reforms as a single channel and flows southeast, through the Hawr al-Hammār to join the Tigris at Al Qurnah.

The Tigris also rises in Turkey and flows through a brief section of Syria before entering Iraq. It has many tributaries in Iraq, all of which enter it from the northeast. The most important are the Great Zab, Little Zab, Uzaym, and Diyala. All of these join the Tigris above Baghdad except for the Diyala, which joins it about 22 mi (36 km) below the city. Once into the plains the Tigris has many of the same qualities as the Euphrates. At the Kūt Barrage much of the water is diverted into the Shatt al Gharraf, which was once the main channel of the Tigris. Water from the Tigris enters the Euphrates through the Shatt al Gharraf well above the actual confluence of the two main channels at Al Qurnah.

After the Euphrates and Tigris rivers converge, they are known as the Shatt al Arab. Shatt al Arab flows for roughly 120 mi (193 km) southeast from Al Qurnah into the Persian Gulf. The river forms the border between Iran and Iraq for about half its length, and receives the Kārūn River of Iran here, carrying large quantities of silt. Siltation makes travel through the lower part of the river difficult; dredging is necessary to maintain a channel for oceangoing vessels to reach Iraq's major port at Basra. The Az-Zubayr Canal also provides a connection from Basra to the sea.

The waters of the Tigris and Euphrates are essential to the life of the country, but they may also threaten it. The rivers are at their lowest level in September and October and at flood in March, April, and May. They may carry forty times as much water during these floods than at low mark.

Wetlands

Hawr al-Hammār is a large area swampy area near the confluence of the Tigris and Euphrates rivers. It has formed as the result of centuries of flooding from these rivers and inadequate drainage. Much of it is a permanent marsh, but some parts dry out in the early winter, and other parts become marshland only in years of great flood.

THE COAST, ISLANDS, AND THE OCEAN

Iraq has a short coastline on the Persian (Arabian) Gulf between Iran and Kuwait. The coastline is made up entirely of the Shatt al-Arab river delta.

CLIMATE AND VEGETATION

Temperature

Summer temperatures range from 72°F to 84°F (22°C to 29°C) minimum and from 100.4°F to 109.4°F (38°C to 43°C) maximum—in the shade. Temperatures higher than 118.4°F (48°C) have been reported, with June through August the hottest months. Winter temperatures range from 27°F to about 61°F (-3°C to about 16°C), but have been recorded below 7°F (-14°C) in the western desert. Severe winter frost is frequent in the north.

Rainfall

Ninety per cent of the precipitation falls between November and April, mostly from December through March. May through October are dry. Mean annual rainfall is between four and seven inches (10 and 17 cm). Rainfall is higher in the foothills southwest of the mountains (between 12.5 and 22.5 in / 32 and 57 cm) and in the mountains themselves rainfall reaches 39.4 in (100 cm) annually.

Plains

The alluvial plain of Mesopotamia begins north of Baghdad and extends to the Persian Gulf. The Tigris and Euphrates rivers lie above the level of the plain in many places, held within natural embankments, and the whole area is in a sense a delta, as it is interlaced by the river's many natural and manmade channels. The Tigris and Euphrates both carry a heavy content of silt. During the frequent floods of these rivers, they deposit the silt over a wide area, forming fertile farmland. These regular silt deposits create much fertile land, It has been estimated that the delta plains are built up at the rate of nearly twenty centimeters per century. In some areas major floods lead to the deposit, in temporary lakes, of as much as thirty centimeters of mud. However, the rivers also carry large amounts of salts. These, too, are spread on the land by flooding and irrigation, and can make the soil infertile, especially in the south as the concentration grows heavier.

Population Centers – Iraq

(1995 POPULATION ESTIMATES)

Name	Population	Name	Population
Baghdad (capital)	4,478,000	Al-Basra	410,000
Mosul	700,000	As-Sulaymāniyah	375,000
Irbīl	500,000	An-Najaf	315,000
Kirkuk	430,000		

SOURCE: Projected from United Nations Statistics Division, 1987.

Deserts

The area west and southwest of the Euphrates River is a part of the Syrian Desert, which also covers sections of Syria, Jordan, and Saudi Arabia. The region, sparsely inhabited by pastoral nomads, consists of a wide, stony plain interspersed with rare sandy stretches. A complicated pattern of wadis, watercourses that are dry most of the year, runs from the border to the Euphrates. Some are more than 248 mi (400 km) long and carry brief but torrential floods during the winter rains.

HUMAN POPULATION

In 2001, Iraq had a population density of 138 people per sq mi (53 people per sq km). Mesopotamia is the most densely populated region, followed by the northern uplands. In the semi-arid northwest most settlements are on rivers. The desert region is nearly uninhabited. Flooding and warfare dislocated inhabitants of the delta area during the 1980s and 1990s. About 76 percent of the population is Arab. Most of the rest (19 percent) are Kurds, who live predominantly in the north.

NATURAL RESOURCES

Iraq has substantial reserves of oil. Mosul and Kirkuk in the northeast are the sites of Iraq's most important oil fields. Iraq also produces natural gas, phosphates and sulfur.

FURTHER READINGS

ArabNet. *Iraq.* http://www.arab.net/iraq/iraq_contents.html. (Accessed July 7, 2002).

The Middle East and North Africa 2002. 48th Ed. "Iraq." London: Europa Publishers, 2001.

Lloyd, Seton F.H. *Twin Rivers: A Brief History of Iraq from the Earliest Times to the Present Day*. Oxford, 1943.

Tripp, Charles. *A History of Iraq*. Cambridge: Cambridge University Press, 2000.

Ireland

- **Area:** 27,135 sq mi (70,280 sq km) / World Rank: 120
- **Location:** Northern and Western Hemispheres, western Europe, on the island of Ireland, in the North Atlantic Ocean to the west of Great Britain, bordering Northern Ireland (United Kingdom).
- **Coordinates:** 53°00′N, 8°00′W
- **Borders:** United Kingdom, 224 mi (360 km)
- **Coastline:** 900 mi (1,448 km)
- **Territorial Seas:** 12 NM
- **Highest Point:** Carrantuohil, 3,415 ft (1,041 m)
- **Lowest Point:** Sea level
- **Longest Distances:** 302 mi (486 km) N-S; 171 mi (275 km) E-W
- **Longest River:** Shannon, 240 mi (386 km)
- **Largest Lake:** Lough Corrib, 65.6 sq mi (170 sq km)
- **Natural Hazards:** None
- **Population:** 3,840,838 (July 2001 est.) / World Rank: 122
- **Capital City:** Dublin, eastern coast of Ireland
- **Largest City:** Dublin, population 993,300 (2002 est.)

OVERVIEW

Ireland occupies five-sixths of the island of Ireland in the eastern part of the North Atlantic. Situated on the European continental shelf, it lies at the westernmost edge of Europe, to the west of Great Britain. The northeastern corner of the island is occupied by Northern Ireland, which belongs to Britain and is separated from the independent republic to its south by a winding border.

Ireland lies on the Eurasian Tectonic Plate. The island's topography is the opposite of that found on many of the world's islands and land masses, which have an elevated interior surrounded by a coastal plain. In Ireland, the lowlands are in the center, and the uplands on the perimeter. The country's low, central limestone plateau rimmed by coastal highlands has been compared to a gigantic saucer. In spite of these coastal highlands, Ireland is generally a low country—only about 20 percent of its terrain is more than 500 ft (150 m) above sea level, and even its mountains rarely exceed altitudes of 3,000 ft (900 m). Ireland's peat bogs, although rapidly diminishing in number, are still the country's most distinctive physical feature.

MOUNTAINS AND HILLS

Mountains

Ireland has a number of mountain systems, of which the highest rise to elevations of about 3,000 ft (914 m), while the lower ranges have peak elevations between 2,000 and 3,000 ft (610 and 914 m). Among the higher ranges are the granite peaks of the Wicklow Mountains, which extend roughly north-south for 40 mi (64 km) in the southeastern part of the country, between Dublin and Wexford. Farther south are the Knockmealdown, Galty, and Comeragh mountains of Waterford. The sandstone mountains of Cork and Kerry counties in the southwest include the Slieve Mish, Caha, and Slieve Mishkish mountains, as well as Macgillycuddy's Reeks on the Ivreagh peninsula where the country's highest peak, Mt. Carrantuohil, is found. Carrantuohil is a dome-shaped peak rising to a height of 3,415 ft (1,041 m) that is popular with hikers and climbers. Lower mountain ranges include the Twelve Bens of Bennebeola in the western county of Galway next to Lough Corrib, and the Derryveagh Range and Blue Stack Mountains in the far north near Donegal. The Silvermine Mountains are located in the center of the country near Lough Derg.

INLAND WATERWAYS

Lakes

Ireland's slow-moving rivers widen into loughs (lakes) at many points in the central lowlands, where limestone fields and bogs overlie limestone rock, before moving on to their winding estuaries and the sea. The largest lake in Ireland is Lough Corrib. The lough is 42,000 acres (17,000 ha) in size, stretching 33 miles (53 km) north from near the city of Galway into County Mayo. It is known for its wild brown trout and salmon fishing. Two of the other largest loughs in Ireland are Lough Mask (which is connected to Lough Corrib by an underground stream) and Lough Conn. All three are located in the western counties of Galway and Mayo. The largest lough on the Irish island—and in all of the British Isles—is Lough Neagh in Northern Ireland. Lough Hyne, located in the southwestern-most part of the island, has sunk below sea level over time and is one of Europe's only saltwater lakes (or inland seas). In the center of the island is the large Lough Derg. North of this lake is Lough Ree. Both are relatively large lakes on the island. North of Lough Ree is Lough Allen. All three lakes are located at points where the Shannon River widens and slows. Located at Killarney National Park among Macgillycuddy's Reeks in the "Ring of Kerry" is Lough Leane, offering stunning views of wildlife and natural landscape.

Rivers

The rivers of Ireland have courses of considerable variety and are among the most attractive features of the landscape. Beginning as mountain streams, they cross the central lowlands as wide, slow-moving, alkaline streams, frequently surrounded by bogs and marshes, and in many cases reach their marshy estuaries through valleys before draining into the sea. A few shorter and more rapid waterways flow directly from the mountains into the ocean. The Shannon, the most important and the longest, rises in the plateau near Sligo Bay, flows sluggishly over the western part of the lowland, receiving slow-moving tributaries, and then fills Lough Derg before beginning its final spurt through rapids to its estuary. Altogether, it drains over 4,000 sq mi (10,360 sq km) of the central lowlands.

Other rivers of the lowlands include the Boyne and the Barrow. The Clare and Moy rivers flow through the west, the Finn in the north, and the Bandon, Suir, and Blackwater are among the southern rivers.

Wetlands

Ireland has both coastal and interior wetlands, where a variety of waterfowl and other birds, including the peregrine falcon, find refuge in winter.

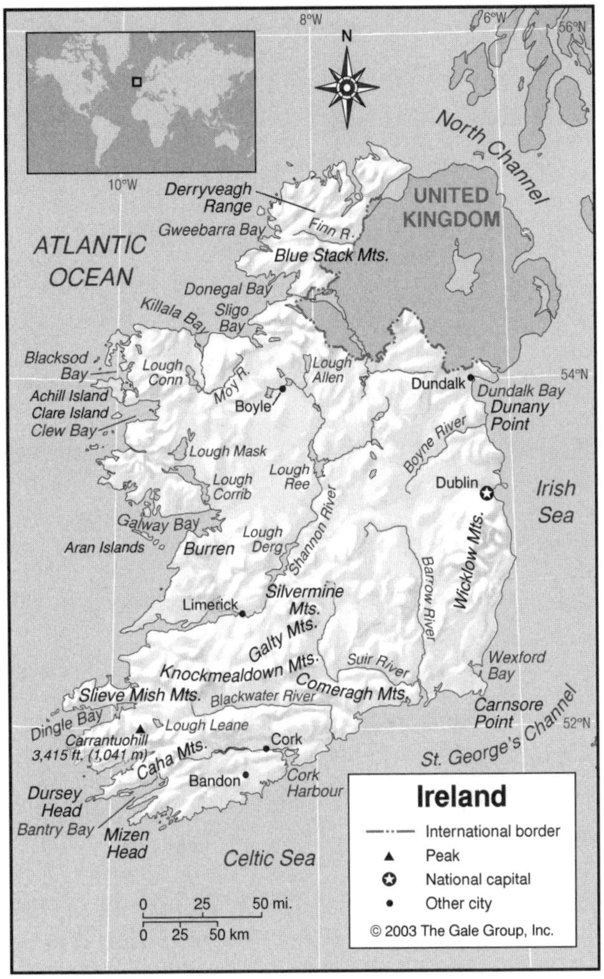

THE COAST, ISLANDS, AND THE OCEAN

Oceans and Seas

Ireland is bounded on the east by the Irish Sea, on the southeast by St. George's Channel, on the south by the Celtic Sea, and on the north and west by the Atlantic Ocean. The North Channel separates the island from Scotland. Most of it borders Northern Ireland, but it does touch the northeastern part of Ireland.

Major Islands

There are a number of small islands off the western coast. At the mouth of Clew Bay sit Achill Island and Clare Island. The former is the largest island off the coast of Ireland (176 sq mi / 456 sq km), and features the highest sea cliffs in Europe as well as five EU Blue Flag beaches. Always receiving a fair number of tourists, in recent years it has also received record numbers of visitors from all over the world on pilgrimages to Our Lady's House of Prayer. Though not the largest, the best known of Ireland's islands are the three Aran Islands at the mouth of Galway Bay. They attract many tourists every year, being home to primitive stone forts and dolmens, very old Christian churches, and other remains that survive from early Christian times as well as medieval castles.

The Coast and Beaches

There is a sharp contrast between Ireland's smooth eastern coast, which faces England and Wales, and the deeply indented coasts to the west and northwest. The latter have numerous bays and inlets, of which the largest are Donegal Bay and Galway Bay, in whose mouth the Aran Islands are located. The deepest coastal indentation, however, is the mouth of the mouth of the Shannon River in the southwest, separating the counties of Clare and Limerick. The southwestern corner of Ireland has deep, parallel, fjordlike indentations between a series of capes where the mountains of Kerry and Cork jut out into the sea. These fingerlike capes—two of which terminate in Dursey Head and Mizen Head—encompass Bantry Bay and Dingle Bay. Other bays on the western coast are Clew Bay, Blacksod Bay, Killala Bay, and Gweebarra Bay. The eastern coast features Wexford Bay, which is cradled by Carnsore Point, and Dundalk Bay cutting between Dunany Point and Northern Ireland. Though Ireland has many good ports, Cork, situated in Cork Harbour on the Celtic Sea coast, is the most important.

GEO-FACT

The Wexford Slobs, located within Wexford Harbor on the southeastern coast of Ireland, serve as a haven for about half the world's population of the Greenland White-fronted Goose, which migrate from Greenland every winter and feed there.

Since the Irish coastline consists mostly of highlands, much of it is rocky and it often features sheer, rocky cliffs diving into the ocean. However, there are also long stretches of sandy beach known as strands. Many are lined with dunes, formed by soil that the rivers have carried seaward. More than 450 mi (724 km) of Ireland's coast are covered by these dunes.

CLIMATE AND VEGETATION

Temperature

Ireland has a mild maritime climate governed by its proximity to the Atlantic Ocean. Average temperatures range from 39° to 45°F (4° to 7°C) in January and February, and from 57° to 61°F (14° to 16°C) in July and August. Southwesterly and westerly winds generated by depressions blowing in from the ocean combine to keep Ireland's weather humid and highly changeable. A commonplace about Irish weather is, "If you don't like it, wait a couple of minutes!" A character in James Joyce's masterwork *Ulysses* refers to the weather on a June day in Dublin as being "uncertain as a child's bottom." In addition to rapid changes in the weather the winds can bring violent storms in winter, although snow is common only in certain areas in the center of the country.

Rainfall

Due to the prevailing westerly winds bringing the humid air from the Atlantic, the western coast of Ireland receives much more rain than the rest of the island. Average annual rainfall ranges from roughly 30 in (76 cm) in the eastern part of the country to more than 100 in (250 cm) in the western highlands.

Grasslands

The central lowlands may be roughly demarcated by an east-west line running from Dundalk to Boyle and beyond in the north, and a similar axis from Dublin to Galway in the south. Their average elevation is about 200 ft (60 m), although the terrain is broken in many places by hills, ridges, and loughs. The well-watered pasturelands are perfect for raising livestock, although the peat bogs can be used only for fuel, not grazing. Grass grows throughout Ireland, up to the higher mountain elevations.

The Irish peat bog, a patchwork of deep pools, lawnlike expanses, and low hills, created by heavy rainfall and poorly drained soil, constitutes a distinctive environment that favors the growth of moss, heather, royal fern, and other flora.

Forests and Jungles

Ireland has lost almost all of its original forest cover, which included birch, elm, yew, and ash; the 5 percent of its land covered by forestland is the result of forestation programs. Introduced tree species include beech and lime.

Heather, ferns, bracken, and woody shrubs are found on Ireland's mountain slopes.

The many miles of hedgerows that crisscross the lowlands consist principally of tall grasses and hawthorn bushes. Common Irish wildflowers include gorse, hogweed, and ragged robin. Ferns and moss grow in the crevices of the stone walls and replace the hedgerows near the western coast.

The Burren

Grass and an unusual variety of both Alpine and Mediterranean plants grow out of bare limestone rock in a distinctive area known as the Burren, a rocky expanse covering more than 50 sq mi (135 sq km) on the coast of County Clare in west-central Ireland. Its plant species include the alpine bearberry, burnet rose, spring gentian, bloody cranesbill, saxifrage, and more than twenty varieties of orchids.

HUMAN POPULATION

Ireland has an overall population density of about 137 people per square mile (53 people per sq km); however more than 40 percent of the population lives within 60 miles (97 km) of Dublin. Its urban centers, which are home to about 60 percent of its population, are relatively small compared to those elsewhere in Europe—other than Cork and Limerick, none have a population greater than 60,000. Most of the land has remained undeveloped, and the Irish countryside is dotted with small towns and farms.

NATURAL RESOURCES

Ireland's grasslands, which are primarily used as pastureland, are among its most important natural resources. Peat from the country's bogs is widely used for fuel. The country's natural gas reserves, whose commercial exploitation began in the 1970s, have largely been depleted. Other mineral resources include lead, zinc, coal, gypsum, and limestone. Ireland is also highly dependent on tour-

Population Centers – Ireland

(1991 POPULATION ESTIMATES)

Name	Population
Dublin (metropolitan area, capital)	915,516
Cork	174,400

SOURCE: "Population of Capital Cities and Cities of 100,000 and More Inhabitants," United Nations Statistics Division.

Administrative Districts – Ireland

POPULATIONS FROM 2002 CENSUS

Name	Population	Area (sq mi)	Area (sq km)
Carlow	45,845	346	896
Cavan	56,416	730	1,891
Clare	103,333	1,231	3,188
Connacht	466,050	6,611	17,122
Cork	448,181	2,880	7,460
Donegal	137,383	1,865	4,830
Dublin	1,122,600	356	922
Galway	208,826	2,293	5,940
Kerry	132,424	1,815	4,701
Kildare	163,995	654	1,694
Kilkenny	80,421	796	2,062
Laoighis	58,732	664	1,719
Leinster	2,105,449	7,580	19,633
Leitrim	25,815	581	1,525
Limerick	175,529	1,037	2,686
Longford	31,127	403	1,044
Louth	101,802	318	823
Mayo	117,428	2,084	5,398
Meath	133,936	902	2,336
Monaghan	52,772	498	1,291
Munster	1,101,266	9,315	24,127
Offaly	63,702	771	1,998
Roscommon	53,803	951	2,463
Silgo	58,178	693	1,796
Tipperary North Riding	61,068	771	1,996
Tipperary South Riding	79,213	872	2,258
Ulster	246,571	3,093	8,012
Waterford	101,518	710	1,838
Westmeath	72,027	681	1,763
Wexford	116,543	908	2,351
Wicklow	114,719	782	2,025

SOURCE: Central Statistics Office, Ireland. As cited by Johan van der Heyden. Geohive. http://www.geohive.com (accessed August 2002).

ism, and is famous for its fine glass and crystal, clothing, and beer, among other products. It is estimated that at any one given time, one out of every three people in Ireland is a tourist.

FURTHER READINGS

De Breffny, Brian. *In the Steps of St. Patrick.* New York: Thames and Hudson, 1982.

Department of Arts, Heritage, Gaeltacht and the Islands. *Heritage of Ireland.* http://www.heritageireland.ie/ (Accessed May 4, 2002).

Hawks, Tony. *Round Ireland with a Fridge.* New York : Thomas Dunne Books, 2000.

Keneally, Thomas. *Now and in Time to Be: Ireland and the Irish.* Photos by Patrick Prendergast. New York: W.W. Norton and Co., 1992.

Morris, Jan. *Ireland: Your Only Place.* New York : C.N. Potter, 1990.

Powers, Alice Leccese, ed. *Ireland in Mind: An Anthology.* New York : Vintage Books, 2000.

Wilson, David A. *Ireland, a Bicycle and a Tin Whistle.* Montreal: McGill-Queens University Press, 1995.

Israel

- **Area:** 8,019 sq mi (20,770 sq km) / World Rank: 152
- **Location:** Northern and Eastern Hemispheres, in the Middle East, bordering Lebanon to the north, Syria to the northeast, Jordan and the West Bank to the east, the Gulf of Aqaba to the south, Egypt and the Gaza Strip to the southwest, the Mediterranean Sea to the west
- **Coordinates:** 31°30′N, 34°45′E
- **Borders:** 625 mi (1,006 km) / Egypt, 158 mi (255 km); Gaza Strip, 32 mi (51 km); Jordan, 148 mi (238 km); Lebanon, 49 mi (79 km); Syria, 47 mi (76 km); West Bank, 191 mi (307 km)
- **Coastline:** 170 mi (273 km)
- **Territorial Seas:** 12 NM
- **Highest Point:** Mt. Meron, 3,963 ft (1,208 m)
- **Lowest Point:** Dead Sea, 1,339 ft (408 m) below sea level
- **Longest Distances:** 200 mi (320 km) N-S / 70 mi (110 km) E-W
- **Longest River:** Jordan, 200 mi (322 km)
- **Largest Lake:** Dead Sea, 370 sq mi (962 sq km)
- **Natural Hazards:** drought and sandstorms
- **Population:** 5,938,093 (July 2001 est.) / World Rank: 99
- **Capital City:** Jerusalem, in east-central Israel
- **Largest City:** Tel Aviv-Jaffa, central Israel on the coast, 2.2 million (2000 est.)

OVERVIEW

Israel is a small country in the Middle East, on the eastern shore of the Mediterranean Sea. It occupies most of the region historically known as Palestine. A dramatic variety of physical landscapes are concentrated within

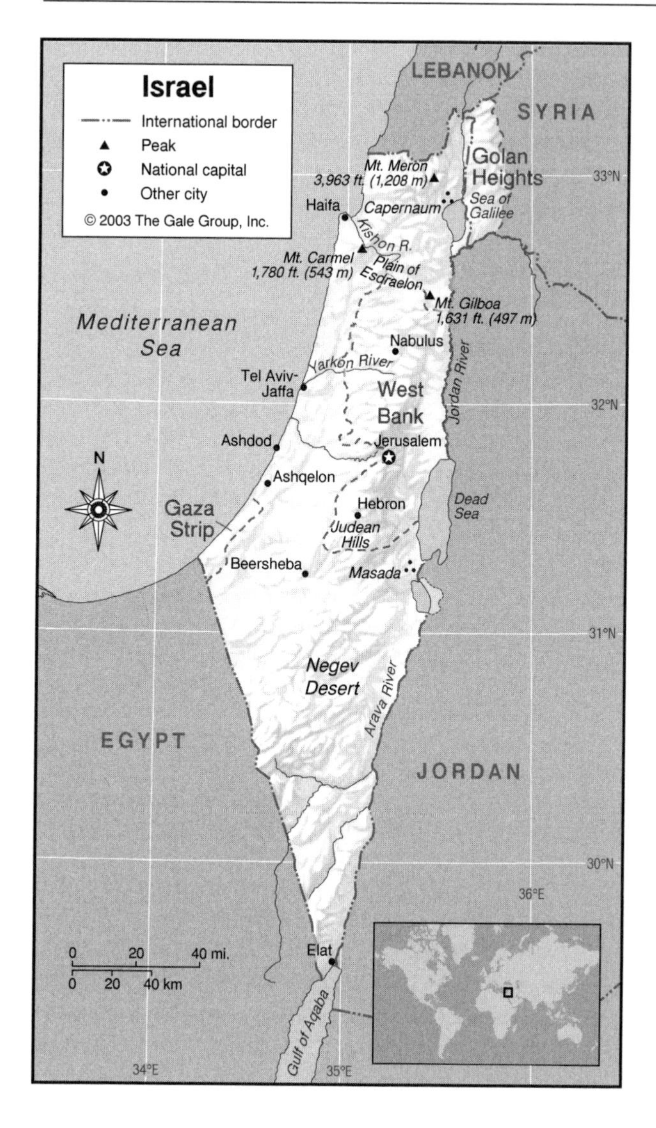

Israel's narrow borders, including the hills and mountains of the Galilee, the Mediterranean coastal plains, the parched expanse of the Negev Desert, and the Dead Sea—the lowest point on the surface of the earth. The country can be divided into four major regions: the coastal plain to the west, the upland areas in the central and northern parts of the country, the Great Rift Valley along its eastern border, and the Negev Desert to the south.

In addition to the land within the country's original 1948 borders, Israel also controls the areas known as the West Bank, the Golan Heights, and the Gaza Strip. These are collectively called the occupied territories. They were captured from Jordan, Syria, and Egypt during wars in 1967 and 1973, and have been occupied by Israel since that time. Their dimensions are not included in the overall totals for Israel, but their geography will be discussed in this essay. It should be understood, however, that these occupied territories are not recognized as an integral part of Israel. The Golan Heights are claimed by Syria; the Gaza Strip and the West Bank have long sought to become independent of Israel. Even outside of the occupied territories, Israel's borders are not necessarily fully accepted by its neighbors, and are the result of cease-fires rather than permanent agreements.

The Golan Heights, located to the northeast along the Israeli-Syrian border, is an upland region covering 660 sq mi (1,710 sq km). Its major topographical sections are the Hermon Range to the north and the Golan plateau to the south. The West Bank (2,270 sq mi / 5,878 sq km) is the former Jordanian section of Palestine, whose major physical features are the hills of Judaea and Samaria and the Jordan River Valley. The Gaza Strip is a narrow strip of land at the southern end of Israel's Mediterranean coast, adjacent to Egypt, with an area of only 140 sq mi (363 sq km) and a maximum width of only 8 mi (13 km).

Israel is situated along the border between the African Tectonic Plate and the Arabian Tectonic Plate. The border between these two plates forms part of the Great Rift Valley, the world's most extensive geological fault, which extends southward through eastern Africa as far south as Mozambique.

MOUNTAINS AND HILLS

Mountains

The highest mountain in Israel (not counting the occupied territories) is Mt. Meron in the mountains of Upper Galilee, which rises to 3,963 ft (1,208 m); the peaks of Lower Galilee rise to only about half this height. To the southwest, at the edge of the coastal city of Haifa, Mount Carmel, rising to 1,790 ft (546 m), consists of two ranges with a fertile lowland in between. Topping all these elevations is that of Mitzpeh Shlagim in the Golan Heights, which rises to over 9,297 ft (2,224 m). It is the second-highest peak of the Hermon Range, whose highest point, Mt. Hermon, is in Syria. The mountains of the Golan are composed of softstone or dolomite, pierced by natural caves, and intersected by wadis (mostly dry river beds).

Valleys

Fringing Israel's mountains are several valleys: the Hula, between the mountains of Upper Galilee and the Golan Heights; the Capernaum, near the Sea of Galilee; and the Jezreel Valley, on the Plain of Esdraelon, between Mount Gilboa and the Hill of Moreh. The Great Rift Valley runs from north to south along the border between Israel and Jordan. In this region it is known as the Jordan Valley. It is the northernmost extension of an enormous geological fault, forming the Red Sea further south and then extending across much of Eastern Africa.

Plateaus

The Golan Plateau in the Golan Heights extends south of the Hermon range. Its basalt surfaces were formed by volcanic activity, and it is marked by craters and deep canyons.

Hills and Badlands

The hills of Galilee, from which Mt. Meron rises, are located in the northern part of the country. Farther south, two hilly regions—Judaea and Samaria—make up most of the West Bank and also extend into Israel proper. Nestled in these hills lie the cities of Jerusalem, Nābulus, and Hebron. Parts of Samaria are 3,149 ft (960 m) above sea level. The Judaean hills are a compact range 50 mi (80 km) long and 9 to 12 mi (14 to 19 km) across, with an average altitude of 2,460 ft (750 m).

INLAND WATERWAYS

Lakes

Israel's two major lakes (both called "seas") are the Sea of Galilee (also called Lake Tiberias, or the Kinneret) in the northeast, bordering the Golan Heights, and the Dead Sea to its south along the border with Jordan. The two bodies of water are connected by the Jordan River. The Sea of Galilee is actually fresh water, the largest such body in the country. The Dead Sea is a large saltwater lake. It is the lowest spot on the surface of the earth. Its high salt and mineral content gives it a bright green hue and makes it extremely easy to float in—it is possible to "sit" on the surface of the lake and remain afloat. Lake Hula, once found north of the Sea of Galilee, has been drained and survives in part as a nature reserve.

Rivers

Most of Israel's other rivers are seasonal, drying up in the summer. All except the Jordan flow into the Mediterranean. Israel's major rivers are the Jordan; the Yarqon, which drains into the Mediterranean near Tel Aviv; and the Kishon, which enters the Mediterranean farther north, near Haifa.

The Jordan River, which makes up part of the border between Israel and Jordan, is the country's largest and best-known river, and its main source of water. Three of its sources—the Banyas, the Dan, and the Hasban—rise on Mount Hermon in the Golan Heights. Along its 200 mi (322 km) course, the Jordan descends over 2,300 ft (701 m) to the lowest point on earth, the Dead Sea. While the river swells during the rainy season, it is for most of the year a small, muddy stream that can easily be forded at several points.

THE COAST, ISLANDS, AND THE OCEAN

Oceans and Seas

Israel lies on the eastern shore of the Mediterranean Sea. The Mediterranean coastline is almost entirely smooth. The only indentation is at the mouth of the Kishon River, at the port city of Haifa. In most places the shoreline is sandy and bordered by agricultural land. In the extreme south, Israel also has a short shoreline on the Gulf of Aqaba, an extension of the Red Sea. The soft pink and red coral that line the shores of the Gulf of Aqaba give the Red Sea its name. Both the coral and the plentiful marine life in these waters make Elat a popular diving and snorkeling center, and home to an extensive underwater observatory and aquarium.

CLIMATE AND VEGETATION

Temperature

Israel has a Mediterranean climate, with variations influenced by elevation and by proximity to the coast. Summers are hot and dry; winters are short, warm, and wet. Average summertime temperatures range from 75°F (24°C) at Safed, in the hills of Galilee, to 93°F (34°C) in Elat, the southernmost point of the Negev Desert, where high temperatures in August can reach 114°F (46°C). The hot, dry desert wind called the *hamsin* can raise the already high summer temperatures even higher, as well as filling the air with sand and dust. Temperatures in January, the coldest month, average 56°F (13°C) on the coastal plain and 60°F (16°C) in the southern desert.

Rainfall

Rainfall is lightest in the south, ranging from 1 in (3 cm) per year south of the Dead Sea to 44 in (118 cm) in the hills of Galilee. Most rain falls between October and April. In the south, dew adds several inches to annual precipitation.

Grasslands

Israel's narrow coastal plain includes the Plain of Judaea, south of Tel Aviv, and the Plain of Sharon, between Tel Aviv and Haifa.

Israel's largest plains area is the fertile Esdraelon Plain, which extends southeast from the Mediterranean coast near Mt. Carmel to the Jordan River valley, separating Galilee from Samaria. It is drained by the Kishon River in the west and the Harod River in the east. The Valley of Jezreel lies at its eastern end, between Mount Gilboa and the Hill of Moreh, and the entire plain itself is often referred to as Jezreel.

Deserts

Descending eastward to the Dead Sea, the Judaean Hills turn into the Judaean Desert, a scenic wilderness. To the south is Masada, the site of a heroic first-century resistance to a protracted Roman siege. At the junction of the desert and the Dead Sea, the Dead Sea Scrolls were discovered in the late 1940s.

Where the Judaean Hills end in the south, the Negev Desert begins. Although it comprises two-thirds of Israel's land area, it contains only a small percentage of the population. The northern Negev is a region of low

Population Centers – Israel

(1997 POPULATION ESTIMATES)

Name	Population	Name	Population
Jerusalem (capital)	622,091	Beersheba (Be'er Sheva)	160,364
Tel Aviv-Jaffa (Yafo)	348,570	Petah Tiqwa	155,169
Haifa	264,301	Bene Beraq	133,943
Holon	162,896	Ramat Gan	127,434

SOURCE: Central Bureau of Statistics, Israel.

Districts – Israel

1997 POPULATION ESTIMATES

Name	Population	Area (sq mi)	Area (sq km)	Capital
Central (Ha Merkaz)	1,307,830	479	1,242	Ramla
Haifa (Hefa)	774,914	330	854	Haifa
Jerusalem (Yerushalayim)	701,734	215	557	Jerusalem
Northern (Ha Zafon)	1,001,849	1,347	3,490	Nazareth
Southern (Ha Darom)	813,445	5,555	14,387	Beersheba
Tel Aviv	1,139,980	66	170	Tel Aviv-Jaffa

SOURCE: Central Bureau of Statistics, Israel.

sandstone hills, steppes, and fertile plains abounding in canyons and wadis. The central Negev, to the south, is a tract of bare, rocky peaks and craters. The Arava, an extremely parched stretch of desert between the Dead Sea and the Red Sea, has an average annual rainfall of less than 1 in (2.5 cm), and its summer temperatures are very high. The Elat Mountains form the southern tip of the Negev triangle, with sharp pinnacles of gray and red granite broken by dry gorges and sheer cliffs. Elat is the region's major city and port, as well as a popular vacation resort.

Forests and Jungles

Much of Israel's original forestland was destroyed over time, cleared for cultivation or grazing land, or for timber. However, an estimated 200 million new trees were planted in the twentieth century, and tree planting became an important symbol of the reclamation of the land for productive purposes and beautification. Species planted include citrus, eucalyptus, and conifers. Scrub grows in the Negev desert, and maquis in the country's hilly regions.

HUMAN POPULATION

Although the original resettlement of modern Israel was largely rural, industrialization and mechanization led to a major population shift, and today some 90 percent of Israelis are urban dwellers, with nearly half living in the major urban centers of Jerusalem, Tel Aviv-Jaffa, and Haifa. The latter two are located along Israel's narrow coastal strip, which claims nearly two-thirds of the country's inhabitants in these two metropolitan areas, Ashqelon-Ashdod, and the Gaza Strip. Another densely populated region is the Esdraelon Plain. A high proportion of the country's collective settlements (kibbutzim) and cooperative farms (moshavim) are located in this region, much of which consists of former marshland that was drained and transformed into productive land.

Israel was established in 1948 to be a homeland for the Jewish people of the world. Roughly 80 percent of the population of Israel (not including the occupied territories) was Jewish in 2001, many of them immigrants. Since the 1970s Israel has engaged in a highly controversial program of establishing Jewish settlements in the occupied territories, especially in the West Bank. There were roughly 370,000 Jewish settlers in the occupied territories in 2000; they are included in Israel's overall population.

NATURAL RESOURCES

The Dead Sea is rich in minerals, including potash and bromine. Phosphate, granite, low-grade copper deposits, sand, clay, and gypsum are found in the Negev Desert, and there are oil deposits in the northern Negev and north of Beersheba. Israel also has a strong technology and manufacturing sector. Nevertheless, the economy depends in no small part on money sent by other nations and people, especially Jews, living abroad.

FURTHER READINGS

Bellow, Saul. *To Jerusalem and Back: A Personal Account.* Boston: G. K. Hall, 1977.

Ciment, James. *Palestine/Israel: The Long Conflict.* New York: Facts on File, 1997.

Grossman, David. *The Yellow Wind.* Translated by Haim Watzman. New York: Farrar, Straus and Giroux, 1988.

Kochav, Sarah. *Israel: Splendors of the Holy Land.* London: Thames and Hudson, 1995.

Peres, Shimon. *The Imaginary Voyage: With Theodor Herzl in Israel.* New York: Arcade, 1999.

Richler, Mordecai. *This Year in Jerusalem.* New York: Knopf, 1994.

Italy

- **Area:** 116,305 sq mi (301,230 sq km) / World Rank: 71
- **Location:** Northern and Eastern Hemispheres, bordering Switzerland and Austria to the north, Slovenia to

the northeast, the Adriatic Sea to the east, the Ionian Sea to the southeast, the Mediterranean Sea to the south, the Tyrrhenian Sea and the Ligurian Sea to the west, and France to the northwest

- **Coordinates:** 42°50′N, 12°50′E
- **Borders:** 1,199 mi (1,932 km) / Austria, 267 mi (430 km); France, 303 mi (488 km); Vatican City, 2 mi (3.2 km); San Marino, 24 mi (39 km); Slovenia, 144 mi (232 km); Switzerland, 459 mi (740 km)
- **Coastline:** 4,712 mi (7,600 km)
- **Territorial Seas:** 12 NM
- **Highest Point:** Mt. Blanc, 15,771 ft (4,807 m)
- **Lowest Point:** Sea level
- **Longest Distances:** 236 mi (381 km) NE-SW; 735 mi (1,185 km) SE-NW
- **Longest River:** Po, 417 mi (673 km)
- **Largest Lake:** Garda, 143 sq mi (370 sq km)
- **Natural Hazards:** Active volcanoes, earthquakes, mudslides, floods
- **Population:** 57,679,825 (July 2001 est.) /World Rank: 22
- **Capital City:** Rome, south-central Italy
- **Largest City:** Milan, north-central Italy, 4.3 million (2000 est.)

GEO-FACT

The violent blasts and powerful vortexes produced by the currents that blow across the Strait of Messina, between the Ionian and Tyrrhenean Seas, were personified as the monsters Scylla and Charybdis in Homer's *Odyssey.* Scylla was on the Calabrian coast, Charybdis on the coast of Sicily.

OVERVIEW

The southern European nation of Italy occupies a long, slender peninsula shaped like a high-heeled boot that extends southeastward into the Mediterranean Sea between the islands of Corsica and Sardinia to the west and the Balkan Peninsula to the east; the country also fans out in all directions onto the European continent, toward the neighboring countries of France, Switzerland, Austria, and Slovenia. The major islands of Sicily and Sardinia, as well as many smaller islands and archipelagos, also form part of Italy's territory. The tiny independent republic of San Marino is a self-contained enclave about two-thirds of the way up the eastern coast of Italy. Vatican City in Rome is another independent entity within Italian territory.

Italy is predominantly mountainous, with the Alps accounting for nearly all the terrain of the land boundaries that separate it from its neighbors. The same mountain system, in the form of the older and less elevated Apennines, extends southward down the entire length of the peninsula. Italy's only major lowland region is the Po River valley, encircled by the northern Alpine ranges.

In terms of both political administration and local tradition, Italy is divided into many different subregions. However, these subregions can be viewed as belonging to larger areas. One such overall demarcation is the fourfold division into the regions north of the peninsula; the peninsular regions as far south as Campania and Apulia; the southernmost part of the peninsula (commonly called the Mezzogiorno); and the islands.

The development of volcanic phenomena bears witness to Italy's geologic youth. Vesuvius near Naples, Etna in Sicily, and Stromboli and Vulcano in the Lipari Islands are active volcanoes. There are also numerous extinct volcanoes, including those whose craters house the volcanic lakes of central Italy. Seismic activity is widespread and strong throughout the entire Apennine region and in southern Italy. Periodically Italy has suffered from major earthquakes, such as the Calabrian-Sicilian earthquake of 1908, which leveled the cities of Reggio di Calabria and Messina, claiming 100,000 lives. Italy is located on the Eurasian Tectonic Plate.

MOUNTAINS AND HILLS

Mountains

The two principal mountain ranges are the Alps and the Apennines. The Alps are made up of a series of massifs and chains running almost parallel to each other. The average elevations are higher toward the west, and the distances between them increase as the range becomes fan-shaped in the east.

The Alps are commonly divided into three ranges. The Western Alps begin a short distance west of Genoa (Genova) at Cadibona Pass near Savona and sweep in a great arc to Lake Maggiore. They include more than 50 peaks over 10,000 ft (3,048 m) high, including the highest peak of both Italy and France, Mont Blanc (Monte Blanco), as well as the highest peak entirely within Italy, Gran Paradiso (13,323 ft/4,061 m). The Central Alps, extending from Lake Maggiore to the Adige River, also have more than 50 peaks higher than 10,000 ft (3,048 m), but in contrast to the Western Alps, there are valleys between the mountain ranges, making communication

easier. They also cover a larger area than the Western Alps, collect more precipitation, and have large glaciers.

The Eastern Alps run from the Adige River to the Tarvis Pass on the Slovenia border. They have wider valleys and less steep terrain than the Central and Western Alps. Also called the Venetian Alps, the Eastern Alps are divided into the Dolomites, the Carnic Alps, and the Julian Alps. The Dolomites have eighteen peaks over 10,000 ft (3,048 m) high.

The Apennine system is formed not by consecutive chains, like the Alps, but by staggered sections joined by negotiable passes. They are more rounded and less elevated than the Alps and formed of softer rock—sandstone and clay in the north, limestone and dolomite in the center, and granite and limestone in the south. Their highest elevation, at Monte Corno in the Gran Sasso Range, is only 9,500 ft (2,895 m). The Peloritani, Nebrodi, and Madonie Ranges in northern Sicily are a continuation of the Apennines, completing the overall arc of the chain, which—from Liguria to Palermo—resembles a giant, narrow, inverted letter *C*. The Sicilian peaks are geologically similar to their more northerly counterparts. Except for Mt. Etna, they are relatively low.

In the eastern part of Sardinia, mountains rise steeply from the sea, divided into northern and southern systems by the Plain of Campidano. The highest point is Gennargentu (6,017 ft/1,834 m).

Hills and Badlands

Italy is a hilly country—hills cover roughly as much of its terrain as do mountains (about 40% in each case). The majority of Italy's hills are in the peninsula, to either side of the Apennine chain, where the mountains are flanked by more modest uplands that combine plateaus, hills, and isolated groups of mountains.

To the west, this terrain, which is called the Anti-Apennines, or sub-Apennines, forms a broad band across Tuscany, descending to the Tyrrhenian Sea and continuing southeastward past Rome. To the east, in the regions of Emilia-Romagna and Marche, the hills, composed of clay and sand, are densely cultivated. On Sicily, the mountain chains that continue the Apennine system also descend to hills in the eastern part of the island.

INLAND WATERWAYS

Lakes

Italy has some 1,500 lakes—more than it has rivers. Most are found in the north of the country, in the Alpine foothills at the edge of the Po Valley. They include many small, elevated lakes as well as the large, well-known ones that lie in the valleys. These are—in descending order by size—Garda, Maggiore, Como, Iseo, and Lugano.

Volcanic lakes, which fill the craters of extinct volcanoes, are found in the peninsula. The best known of these is Lake Bolsena, which has two islands. Other volcanic lakes include Bracciano, Trasimerto, Vico, Albano, and Nemi. The third type of lake found in Italy is the coastal lake. This category includes Lakes Comacchio, Orbetello, Massaciuccoli, Fondi, Lesina, Varano, and Salpi.

Rivers

Since most of Italy's many rivers flow across the narrow Italian peninsula from the Apennines to one of the surrounding seas, most of them are short. The longest rivers are in the northern, continental part of the country. The longest and most important is the Po, which has the largest basin. It traverses the northern regions nearly all the way from the French border in the west to the Gulf of Venice in the east, with most of its tributaries flowing from the Alpine lakes of the north. Italy's second-longest river, also found in the northern part of the country, is the Adige, which rises in the northernmost reaches of the Italian Alps and flows south to empty into the Gulf of Venice north of the mouth of the Po.

In the peninsula, a number of rivers flow in a parallel series crossing the Marche, Abruzzi, and Molise regions, including the Reno, the most important river flowing into the Adriatic. On the western side of the peninsula, emptying into the Tyrrhenian Sea, the Arno and its tributaries flow through Tuscany; the Tiber is among the rivers that flow through Latium and Campania. The principal river draining the southern end of the peninsula is the Bradano. The Simeto and the Salso are Sicily's most important rivers, and the Tirso is the principal one in Sardinia.

Wetlands

Italy's largest wetland area is Venice, a city surrounded by a shallow lagoon in the Adriatic Sea north of the Po delta, with a network of canals crossed by bridges.

Much of the original marshland that made up the coast of the Tyrrhenian Sea has been reclaimed and developed. However, marshy lowlands created by river flooding are still found along the coast of Tuscany, in an area called the Maremma, which literally means "swamp." A national park has been created here, at the mouth of the Ombrone River, to preserve the distinctive fauna and flora of the region.

THE COAST, ISLANDS, AND THE OCEAN

Oceans and Seas

The Italian peninsula is surrounded by four seas: the Adriatic, Ionian, Ligurian, and Tyrrhenian Seas. There is almost no spot in Italy that lies farther than 75 mi (120 km) from a coastline.

The long Adriatic Sea, which lies between the Italian and the Balkan peninsulas, has an area of 51,000 sq mi (132,100 sq km) and is known for the wealth of high-quality fish in its waters. At its northern end is the Gulf of Venice. The Gulf of Manfredonia is found about two-thirds down the coast to the south. At its the southern end the Strait of Otranto connects it with the Ionian Sea, a triangular body of water bordered on the west by Sicily and the southeastern coast of the Italian peninsula, on the east by the Balkan Peninsula, and on the south by the Mediterranean. At its deepest point, the Ionian Sea reaches a depth of 2.75 mi (4.4 km), the greatest depth recorded in Mediterranean waters. The large Gulf of Taranto in the Ionian is located between the "toe" and "heel" of the Italian "boot."

The narrow, funnel-shaped Strait of Messina connects the Ionian and Tyrrhenian seas, separating Sicily from the Calabria Peninsula, at the tip of the Italian Peninsula. Also known as the Italian Sea, the Tyrrhenian Sea occupies a triangular area between the islands of Sicily, Sardinia, and Corsica and the western coast of the peninsula. To its north is the smaller Ligurian Sea, with the Gulf of Genoa on its northern shore.

The parts of the Mediterranean that lie to the south and southwest of Sicily are the Malta Channel, which separates it from Malta, and the Sicilian Channel, which lies between Sicily and the Tunisian coast.

Major Islands

With an area of 9,926 sq mi (25,708 sq km), Sicily, located just west of the "toe" of the Italian "boot," is both Italy's largest island and the largest island in the Mediterranean. The second-largest island, Sardinia, located northwest of Sicily, just south of the French island of Corsica, is close to Sicily in size, with an area of 9,300 sq mi (24,090 sq km).

Among Italy's smaller islands are those belonging to the Tuscan Archipelago, the largest island named Elba. Other islands in the Tyrrhenian Sea include the Ponza group and the islands of Ischia and Capri off the coast of Naples, which are popular vacation spots. At the southern edge of the Tyrrhenian Sea, north of Sicily, are the volcanic Lipari Islands, the site of Stromboli, one of the country's three active volcanoes. The strategically located island of Pantelleria lies in the Sicilian Channel between Sicily and the coast of Africa.

The Coast and Beaches

Italy's major harbors are the Gulfs of Genoa and Spezia on the Ligurian coast; Civitavecchia, an artificial harbor at the central part of the Tyrrhenian coast; the Gulfs of Gaeta and Salerno farther south, as well as the city of Naples; the deeply indented Gulf of Taranto on the Ionian Sea; Brindisi, a natural port on the Strait of Otranto; and, on the Adriatic Coast, the Gulf of Manfredonia, and the lagoons of Venice. The Venetian lagoon is Italy's largest, covering 136,000 acres (55,039 hectares). It contains grassy islets, bog land that emerges at low tide, and small islands whose shape shifts with the tide.

The shoreline in Liguria to the north includes both rocky areas and level stretches of gravel. Farther south, between Tuscany and Campania, promontories separate expanses covered of sandy beach and dunes. The coast of Calabria is mostly elevated. Most of the Adriatic coast is flat, with a complex system of lagoons shaping the shoreline in the area around the Po delta and the Gulf of Venice.

CLIMATE AND VEGETATION

Temperature

Due to Italy's varied topography and its latitudinal extension, it has considerable climatic variation, from subtropical conditions in Sicily to year-round snowcaps in parts of the Alpine region. The northern part of the country has a continental climate, with cold winters and warm summers. Farther south, the climate becomes Mediterranean, with cool winters, hot, dry summers, and less differentiation between seasons. Average January temperatures range from 35°F (2°C) in Milan (northern Po basin), to 45°F (7°C) in Rome (central part of the peninsula), to 52°F (11°C) in the Sicilian city of Taormina. Average July readings for the same cities are Milan, 75°F (24°C), Rome, 77°F (25°C), Taormina, 79°F (26°C).

Rainfall

Rainfall is lower in the south and higher in the north, with average annual rainfall ranging from about 20 in (50 cm) in Sicily, Sardinia, and the southeast coast of the Italian peninsula to 80 in (200 cm) in the Alpine regions. In addition to latitude, rainfall also varies with elevation, and is highest in the upper regions of the Alps and Apennines. Rainfall in the north is not only heavier but also more evenly distributed throughout the year.

Grasslands

Plains account for around 20 percent of Italy's terrain. The most extensive plains region is the Po Basin, which has an area of over 17,000 sq mi (44,030 sq km) and an average elevation of under 330 ft (101 m), although it rises to 1,800 ft (549 m) at its highest point. It is drained by the Po River and its tributaries, as well as the Adige and other rivers. At the opposite end of northern Italy, on the Ligurian coast, is the narrow coastal plain of the Gulf of Genoa, whose scenic beauty and pleasant climate have made it the Italian Riviera.

Plains regions of the Italian peninsula include the Tuscan plains, consisting of the Arno River basin and, farther to the south, the marshlands of the Maremma; the Roman countryside, or Compagna, on both banks of the Tiber, and its coastal extension in the form of the reclaimed Pontine Marshes; the fertile plains of the Campania region; and the lowlands of Apulia, including the Salentina Peninsula (the "heel" of the Italian "boot"), at the southeastern end of the peninsula.

Forests and Jungles

Much of Italy's native forestland has been cleared to make way for cultivation. In the higher Alpine regions, however, the native fir, larch, and pine still stand. Reforestation has covered other upland regions with evergreens and eucalyptus. Oak, beech, and chestnut grow on the lower Alpine slopes and in the Apennines, and poplar and willow are found in the Po River Valley. Mediterranean vegetation predominates in the lower reaches of the peninsula and on the islands. Along the coast, cypress and olive trees, as well as myrtle, laurel, thyme, and other plants have replaced the original cork and oak trees.

Human Population

Cities have played an important role in Italian life throughout the country's history, and more than two-thirds of the inhabitants of modern Italy are urban dwell-

Population Centers – Italy

(2000 POPULATION ESTIMATES)

Name	Population	Name	Population
Milan	4,251,000	Turin	1,294,000
Naples	3,012,000	Genoa	890,000
Rome (capital)	2,649,000	Florence	778,000

SOURCE: "Table A.12. Population of Urban Agglomerations with 750,000 Inhabitants or More in 2000, by Country 1950–2015. Estimates and Projections: 1950–2015." United Nations Population Division, World Urbanization Prospects: The 2001 Revision.

Provinces – Italy

2001 POPULATION ESTIMATES

Name	Population	Area (sq mi)	Area (sq km)	Capital
Abruzzo	1,281,283	4,168	10,794	L'Aquila
Basilicata	604,807	3,858	9,992	Potenza
Calábria	2,043,288	5,823	15,080	Catanzaro
Campania	5,782,244	5,249	13,595	Naples (Nápoli)
Emilia-Romagna	4,008,663	8,542	22,123	Bologna
Friuli-Venezia Giulia	1,188,594	3,030	7,846	Trieste
Lazio	5,302,302	6,642	17,203	Rome
Liguria	1,621,016	708	5,416	Genoa (Génova)
Lombardia	9,121,714	9,211	23,857	Milan (Milano)
Marche	1,469,195	3,743	9,694	Ancona
Molise	327,177	1,713	4,438	Campobasso
Piemonte	4,289,731	9,807	25,399	Turin (Turino)
Puglia	4,086,608	7,470	19,348	Bari
Sardegna	1,648,044	9,301	24,090	Cágliari
Sicily (Sicilia)	5,076,700	9,926	25,708	Palermo
Toscana	3,547,604	8,877	22,992	Florence
Trentino-Alto Adige	943,123	5,259	13,620	Bolzano
Umbria	840,482	3,265	8,456	Perúgia
Valle d'Aosta	120,589	1,259	3,262	Aosta
Veneto	4,540,853	7,091	18,365	Venice (Venézia)

SOURCE: "Population Projections Years 2001–2005." ISTAT. Instituto Nazionale di Statistica.

ers. Since Italy's regions were independent of each other for many years, they developed their own urban centers, and cities with populations of more than 100,000 are still found throughout the country, although a disproportionate number (roughly half) are located in the more economically developed north.

The Po Valley is one of the most densely populated parts of Italy.

NATURAL RESOURCES

Mineral resources include mercury, marble, and oil and gas reserves.

The rivers of the Anti-Apennines in central Italy, including the Arno and the Tiber, are sources of hydroelectric power.

FURTHER READINGS

Altman, Jack, and Jason Best. *Discover Italy.* Oxford, England: Berlitz, 1993.

Casserly, Jack. *Once Upon a Time in Italy: The Vita Italiana of an American Journalist.* Niwot, Colo.: Roberts Rinehart Publishers, 1995.

Cahill, Susan, ed. *Desiring Italy.* New York: Fawcett Columbia, 1997.

Cummings, Martha T. *Straddling the Borders: The Year I Grew Up in Italy.* Boston: Branden, 1999.

Harrison, Barbara Grizzuti. *Italian Days.* New York: Weidenfeld & Nicolson, 1989.

McGinniss, Joe. *The Miracle of Castel di Sangro.* Boston: Little, Brown, 1999.

Simon, Kate. *Italy: The Places in Between.* New York: Harper & Row, 1970.

Windows on Italy. http://www.mi.cnr.it/WOI/woiindex.html (accessed Mar. 31, 2002).

Jamaica

- **Area:** 4,243 sq mi (10,990 sq km) / World Rank: 162
- **Location:** Northern and Western Hemispheres, Caribbean Sea, south of Cuba and southwest of Haiti
- **Coordinates:** 18°15′ N, 77°30′ W
- **Borders:** No international borders.
- **Coastline:** 634 mi (1,022 km)
- **Territorial Seas:** 12 NM
- **Highest Point:** Blue Mountain Peak, 7,402 ft (2,256 m)
- **Lowest Point:** Sea level
- **Longest Distances:** 146 mi (235 km) N-S; 51 mi (82 km) E-W
- **Longest River:** Black River, 44 mi (71 km)
- **Natural Hazards:** Hurricanes
- **Population:** 2,665,636 (July 2001 est.) / World Rank: 134
- **Capital City:** Kingston, southeastern coast
- **Largest City:** Kingston, 621,000 (2000 est.)

OVERVIEW

Jamaica is an island nation in the Greater Antilles and a member of the British Commonwealth. Located 90 mi (145 km) south of Cuba and 100 mi (161 km) west of Haiti, it is the third-largest Caribbean island.

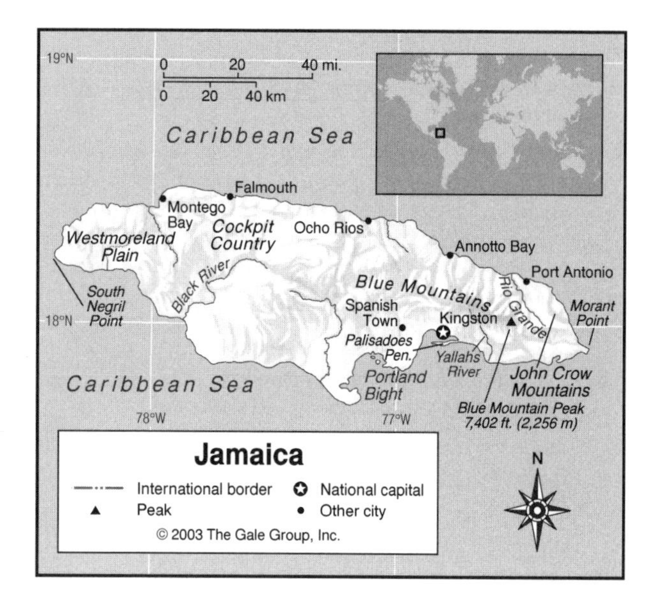

Jamaica is not easy to separate into geographic regions. Its overall topographical arrangement consists of coastal plains and valleys fringing an interior plateau of white limestone that covers most of the island, extending from east to west along its length. The uneven plateau surface is broken by twisting valleys, ranges of limestone hills and mountains, broad basins, and by two mountain ranges of contrasting composition and appearance.

Vestiges of volcanic activity occur in Jamaica in the form of lava cones and hot springs, and there have been occasional serious earthquakes. A 1907 earthquake followed by a tidal wave wrecked much of the Kingston area and took 800 lives. Jamaica lies on the Caribbean Tectonic Plate.

MOUNTAINS AND HILLS

Mountains

The crests of the Blue Mountains, the country's principal mountain system, form part of the boundary between the parishes of Portland and Saint Thomas near the eastern end of the island, where they follow a southeast to northwest axis for a distance of about 50 mi (80 km). The system comprises two ranges, the first of which is the more northerly and extensive of the two, and in it Blue Mountain Peak rises to 7,402 ft (2,256 m), the country's highest elevation. From this central peak, narrow ravines and sharp ridges descend like spokes of a wheel. The second range, also known as the Port Royal Mountains, extends southeastward from the principal range. It rises from the Liguanea Plain north of Kingston and reaches elevations of up to about 4,000 ft (1,219 m).

The John Crow Mountains rise in the extreme northeast of the island, between the Rio Grande and the sea. Their rugged terrain remains only partially explored.

GEO-FACT

Jamaica has several radioactive hot springs, and one—the Milk River Bath—is said to have the highest level of radioactivity of any hot spring in the world.

Plateaus

Elevations on Jamaica's central plateau range from near sea level to a maximum of about 3,000 ft (914 m) on the crests of the limestone uplands. Along much of the coastline, especially in the north, the plateau extends almost to tidewater, and in parts of the parishes [counties] of Saint Ann and Saint Elizabeth it rises in steep coastal cliffs that reach as high as 2,000 ft (609 m). The plateau is ruggedly irregular, and its most characteristic landscape is known to geologists as karst—an irregular limestone terrain with sinkholes, underground caverns and streams, steep hills, and caves. The karst landscape is most distinctive in the Cockpit Country, an area of about 200 sq mi (518 sq km) located largely in the western parish of Trelawney.

INLAND WATERWAYS

Rivers

Jamaica's rivers flow either northward or southward to the sea from springs in the interior highlands. These springs are so numerous that Jamaica is sometimes called the Isle of Springs. Jamaica's major rivers include the Yallahs in the southeast, the Rio Grande in the south-central part of the country, and the Black River (44 mi / 71 km) in the west—Jamaica's longest river and the only one that is navigable for a significant distance.

Wetlands

Swamps, partially drained, are located along the lower course of the Black River and in the vicinity of Morant Point at the eastern tip of the island, and South Negril Point at its western tip.

THE COAST, ISLANDS, AND THE OCEAN

Oceans and Seas

At the eastern end of the island and along the northern coast the ocean plunges to great depths not far from the shoreline. Near the resort town of Port Antonio the water drops to a depth of 600 ft (182 m) no more than one-half mile offshore, and the Bartlett Trough that separates Jamaica from Cuba reaches a depth of 23,000 ft (7,010 m). West of Kingston, however, relatively shallow water off the southern coast covers a sunken highland that extends southwestward to the coast of Central America.

Islands

There are cays (small coral and sand islands) in the Portland Bight, or bay, on the south coast, and a few scattered coral formations occur elsewhere as well, particularly at the eastern end of the island. Jamaica's only offshore territories are the Morant Cays, about 40 mi (60 km) southeast of Morant Point; and the more extensive Pedro Cays, about 60 mi (96 km) south of the southwestern coast.

The Coast and Beaches

The shoreline is indented by numerous harbors, of which the harbor at Kingston is the largest. It is sheltered on its southern flank by the Palisadoes Peninsula, an eight-mile-long sandspit that connects several coral islands. There are extensive coral reefs near the southeast coast. The northern coastal plain is known for its white-sand beaches, which draw three-fourths of Jamaica's tourist trade. In Manchester and much of Saint Elizabeth parishes, the limestone plateau drops directly to the sea in steep cliffs.

Population Centers – Jamaica

(1991 POPULATION ESTIMATES)

Name	Population
Kingston (capital)	103,771
Spanish Town	92,383
Portmore	90,138
Montego Bay	83,446
May Pen	46,785

SOURCE: Statistical Institute of Jamaica.

Parishes – Jamaica

POPULATIONS FROM 1998 ESTIMATE

Name	Population	Area (sq mi)	Area (sq km)	Capital
Clarendon	227,000	462	1,196	May Pen
Hanover	67,900	174	450	Lucea
Kingston	110,152	8	22	Kingston
Manchester	182,900	321	830	Mandeville
Portland	79,200	314	814	Port Antonio
Saint Andrew	595,948	166	431	Half-Way-Tree
Saint Ann	161,900	468	1,213	Saint Ann's Bay
Saint Catherine	409,200	460	1,192	Spanish Town
Saint Elizabeth	149,000	468	1,212	Black River
Saint James	175,900	230	595	Montego Bay
Saint Mary	112,900	236	611	Port Maria
Saint Thomas	91,300	287	743	Morant Bay
Trelawny	72,700	338	875	Falmouth
Westmoreland	137,700	312	807	Savanna-la-Mar

SOURCE: Statistical Institute of Jamaica.

CLIMATE AND VEGETATION

Temperature

Jamaica has a tropical climate moderated by northeast trade winds, with some variation in temperature due to altitude but little seasonal variation. The average annual temperature varies from 81°F (27°C) on the coast to 55°F (13°C) in the Blue Mountains.

Rainfall

Rainfall varies both by region and altitude, ranging from as little as 30 in (75 cm) in some places on the south coast to 130 in (330 cm) in Port Antonio on the northeast coast, to 200 in (500 cm) or more in the Blue Mountains.

Grasslands

The northern coastal plain is narrow but extends almost continuously from the vicinity of Annotto Bay in the east to Montego Bay in the west. It is broadest to the south of Falmouth. On the southern coast the plains are discontinuous but much more extensive than on the north. They are relatively dry, and savanna landscape is characteristic. The city of Kingston lies on the broad Liguanea Plain, an expanse of 130 sq mi (337 sq km) that extends inland and westward. West of the Liguanea Plain a second coastal lowland stretches inland from the coast in Saint Catherine and Clarendon parishes. The Westmoreland Plain occupies much of the western extremity of the island.

Forests and Jungles

Although most of Jamaica's native trees have been cut, about one-fifth of the island is still forested. In addition to isolated stands of the original hardwoods (ebony, rosewood, mahogany), tree species include substantial amounts of bamboo as well as introduced species of pine and eucalyptus. Almost the entire coastline east of Falmouth is thickly fringed with coconut palms. Typical savanna vegetation, with grasses and scattered trees, is found in the west and southwest. There are thousands of flowering plant species in Jamaica; orchids alone account for about two hundred.

HUMAN POPULATION

Slightly more than half the population lives in cities and towns, and most of these are located on the coastal plains. The white-sand beaches and scenic mountain views on the northern coast have made Montego Bay, Ocho Rios, and Port Antonio the principal tourist centers.

NATURAL RESOURCES

Jamaica is a leading producer of bauxite. Other mineral resources include limestone and gypsum.

FURTHER READINGS

Boot, Adrian, and Michael Thomas. *Jamaica: Babylon on a Thin Wire*. New York: Schocken Books, 1977.

Baker, Christopher P. *Jamaica*. 2nd ed. Oakland, Calif.: Lonely Planet Publications, 2000.

Hurston, Zora Neale. *Tell My Horse: Voodoo and Life in Haiti and Jamaica*. San Bernardino, Calif.: Borgo Press, 1992.

Mordecai, Martin, and Pamela Mordecai. *Culture and Customs of Jamaica*. Westport, Conn.: Greenwood Press, 2001.

Statistical Institute of Jamaica. http://www.statinja.com/ (accessed Apr. 5, 2002).

Wilson, Annie. *Essential Jamaica*. Lincolnwood, Ill.: Passport Books, 1996.

Japan

- **Area:** 145,883 sq mi (377,835 sq km) / World Rank: 62
- **Location:** Eastern and Northern Hemispheres, Northeast Asia, bordered by the Pacific Ocean, the Sea of Japan, the East China Sea, and the Philippine Sea.
- **Coordinates:** 36°00′ N, 138°00′ E
- **Borders:** None
- **Coastline:** 18,486 mi (29,751 km)
- **Territorial Seas:** 12 NM (but as little as 3 NM in some straits and narrows)
- **Highest Point:** Mt. Fuji, 12,388 ft (3,776 m)
- **Lowest Point:** Hachiro-gata, 13.1 ft (4 m) below sea level
- **Longest Distances:** 1,869 mi (3,008 km) NE-SW; 1,022 mi (1,645 km) SE-NW
- **Longest River:** Shinano, 228 mi (367 km)
- **Largest Lake:** Biwa, 260 sq mi (673 sq km)
- **Natural Hazards:** Earthquakes, volcanoes, landslides, land subsidence, typhoons, tsunamis, flooding
- **Population:** 126,771,662 (July 2001 est.) / World Rank: 9
- **Capital City:** Tokyo, on the Pacific side of central Honshū island
- **Largest City:** Tokyo, 12,188,038 (2001)

OVERVIEW

The Japanese archipelago forms a crescent off the eastern coast of Asia. The Korean Peninsula is the nearest point on the Asian continent. Japan's location has made it a maritime trading and ocean-fishing nation. Japan has four principal islands; from north to south they are: Hokkaidō, Honshū, Shikoku, and Kyūshū. The four major islands are separated only by narrow straits and form a natural entity. The nation also has more than 3,000 smaller islands, including the Ryukyu archipelago extending far to the southwest of the main islands. A territorial dispute with Russia concerns small islands north of Hokkaidō: Etorofu, Kunashir and the Shikotan and Habomai island groups.

The terrain on all of the major islands is primarily mountainous. The lowland areas that exist are mainly along the shore and are densely populated. The mountains remain largely covered by forest land. Japan lies along the boundary between the Eurasian, North American, and Pacific Tectonic Plates. As a result, earthquakes are common throughout the islands, as are volcanoes.

MOUNTAINS AND HILLS

Mountains

The Japanese islands essentially are the summits of mountain ridges that have been uplifted near the outer edge of the Asian continental shelf. Consequently, mountains take up some 75% of the land. A long spine of mountain ranges runs down the middle of the archipelago dividing it into two halves: one fronting the Pacific Ocean, the other facing the Sea of Japan.

Although the mountains are steep, most of them are not very high. Central Honshū island, however, has a convergence of three mountain chains, the Akaishi, Kiso and Hida, forming the Japanese Alps, which include many peaks that exceed 10,000 ft (3,048 m). Other ranges include the Ōu, Chūgoku, Daisetsu, and the Kitami Mountains. Snow lingers late into spring on the Japanese Alps, but there are no true glaciers in Japan.

The highest point in the country is the renowned Mount Fuji (Fujiyama), a symmetrical dormant volcano that rises to 12,388 ft (3,776 m) in central Honshū, outside of the Alps. The second highest peak is Kitadake (10,472 ft/3,192 m) and the third highest is Hotakadake (10,466 ft/3,190 m); both are in central Honshū.

A tenth of the world's volcanoes are found in Japan. Of Japan's 265 known volcanoes, 20 have been active since the beginning of the 20th century. They are particularly numerous in Hokkaidō, the Fossa Magna region of central Honshū, and Kyūshū. The mountainous areas of Japan contain wide craters and cones of every form ranging from the ash cone of Mount Fuji on Honshū to the volcanic dome of Daisetsu on Hokkaidō. Recent eruptions have included Mount Unzen, on Kyūshū island, in 1991-93; Mount Usu on Hokkaidō in March 2000; and Mount Oyama on Miyako island, south of Tokyo, in September-October 2000.

Landslides that shake loose entire mountainsides are generally composed of clay, and may be from 20 to 75 ft

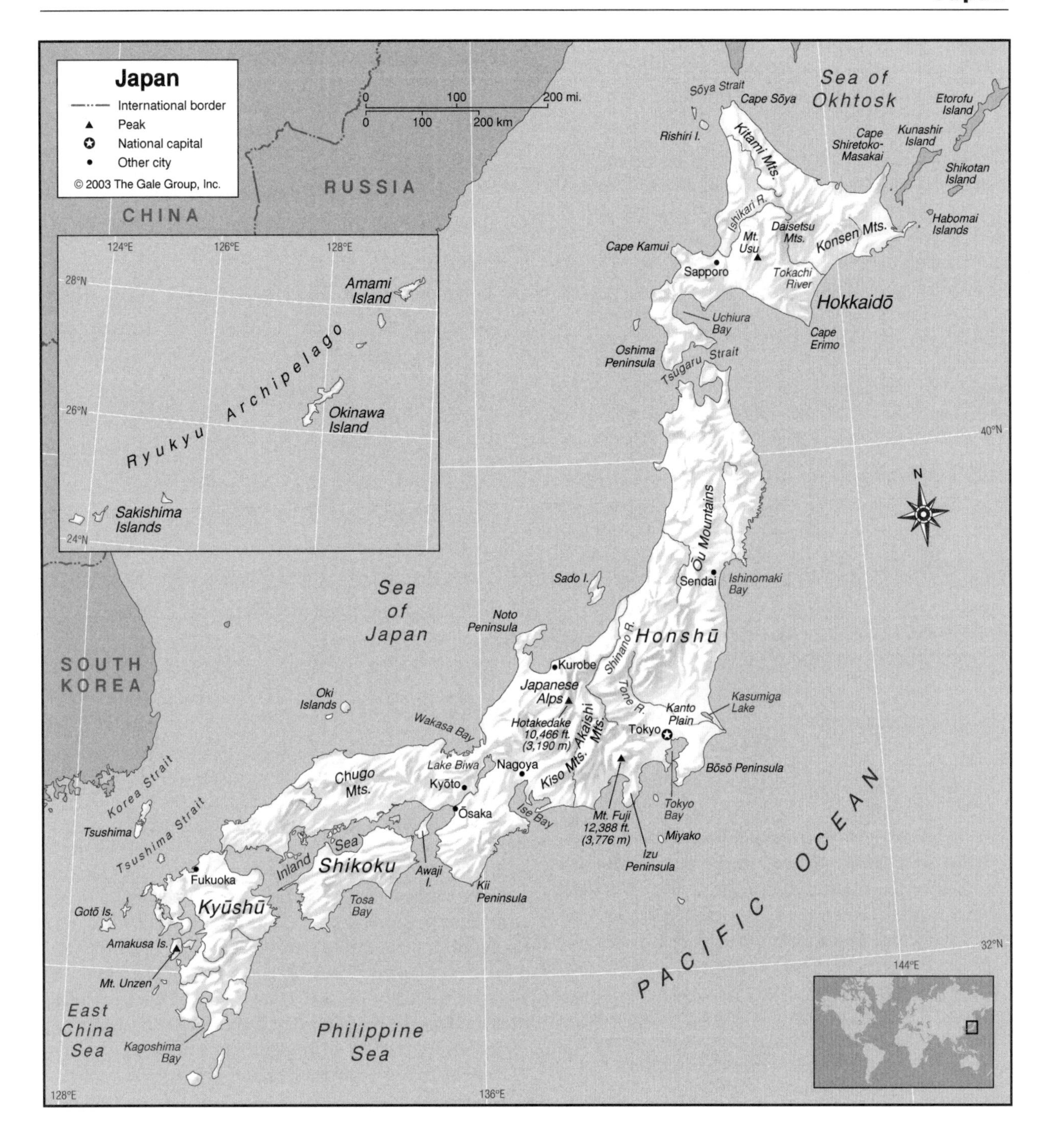

(6 to 23 m) deep, several hundred feet wide, and up to 2.5 mi (4 km) long. Such landslides are especially frequent on the Sea of Japan side of Honshū.

Plateaus

Volcanic activity has shaped many of Japan's plateaus, while others consist of ancient limestone. The Shiga Highlands, in Jo-Shin-Etsu National Park, central Honshū, is a lava plateau 4593 to 5,577 ft (1,400 to 1,700 m) in height. The Hachimantai Plateau, volcanic in origin, in northern Honshū, is 4,593 to 5,249 ft (1,400 to 1,600 m) above sea level. The Akiyoshi-dai Plateau of western Honshū is a limestone platform that is riddled with 420 caves. The Atetsu Plateau, in the same region, is also limestone-based. Northern Honshū's Bandai Plateau contains lakes and marshes. Other plateaus on Honshū include Nihon Daira near Mount Fuji; Midagahara in the Japanese Alps; and the Musashino Plateau, near Tokyo.

The Ebino Plateau, 3,937 ft (1,200 m) above sea level, stands within Japan's first national park, Kirishima Yaku, on Kyūshū island. The Takachihokyo Plateau, near

Kyūshū's Mount Aso, is lava-based with a river-eroded valley and rock formations. The Shiroito Plateau of Kyūshū has a stone aqueduct that irrigates rice farms

Canyons

Japan's rivers have cut deep gorges through the mountain ranges. Suwa, Minakami and Momiji Canyons on the Tone River in the Japanese Alps are known for their whitewater rapids. Kurobe Gorge, in central Honshū is Japan's deepest, plunging 4,921 to 6,562 ft (1,500 to 2,000 m); it has a dam at its south end. Dakigaeri Gorge is a national park in northern Honshū. The Oobako and Kobako Canyons on Hokkaidō feature rocky terrain and waterfalls, as does Soun-kyo Gorge. Noteworthy river gorges on the other islands include Oboke Gorge, Shikoku island; Takachiho Gorge, Kyūshū island; and Yabakei Gorge, also on Kyūshū.

Hills and Badlands

Foothills border the coastal plains of Japan. Away from the coasts ascending terraces mark the foothills, which provide a transition from these plains to the mountain ranges. On the approaches to the mountains, the slopes become steeper and are laced by numerous watercourses, isolating groups of hills. The Hakone hills, in central Honshū, are typical of this type of terrain.

GEO-FACT

Japan is very prone to earthquakes, with more than 1,500 recorded annually. Most are minor tremors, but major earthquakes, although rare, can result in thousands of deaths. The Great Kanto Earthquake of 1923 was one of the most destructive of all time, causing powerful tremors and resulting fires that destroyed most of Tokyo and Yokohama, with a loss of more than 100,000 lives. More recently the 7.2 Richter Scale earthquake at Kobe on January 17, 1995, killed 5,100 people and destroyed 102,000 buildings. Due to the danger they pose, Japan has become a world leader in research on the causes and prediction of earthquakes and the construction of earthquake-proof buildings.

INLAND WATERWAYS

Lakes

The landscape of Japan contains numerous and varied lakes. The biggest is Lake Biwa, 260 sq mi (673 sq km) in area, which fills a fault basin on Honshū. Lake Biwa is affected by pollution as well as the demand for fresh water from the cities of Osaka and Kyoto. The second largest lake is Kasumiga (65 sq mi/168 sq km) near Tokyo; the third largest is Saroma (58 sq mi/150 sq km) on Hokkaidō; and the fourth largest is Inawashiro (40 sq mi/103 sq km) in Bandai-Asahi National Park, northern Honshū. The fifth largest lake, Nakaumi and adjacent Shinjiko Lake, on the Sea of Japan coast of western Honshū, are threatened by land reclamation projects.

Rivers

Although the country is exceptionally well watered, the absence of large plains has prevented the formation of a major river system. The longest river, the Shinano, is only 228 mi (367 km) long, and the second longest is the Tone, 200 mi (322 km); both are in central Honshū. The third longest is Hokkaidō's Ishikari (166 mi / 268 km). Japan's rivers tend to be swift-flowing and unsuitable for navigation. The mountainous terrain and the absence of glaciers make the flow highly irregular. Early summer rains account for a large part of the annual precipitation and turn slow streams into raging torrents. In winter, the riverbeds are transformed into wide stretches of gravel furrowed by thin trickles of water. Rivers are used mostly for hydroelectric production and for irrigation. Extensive dams have been built for flood control, hydropower, and irrigation diversion, disrupting natural river ecosystems.

Wetlands

Eleven areas in Japan are designated Wetlands of International Importance under the Ramsar International Convention on Wetlands. Those on Honshū include Lake Biwa and its surrounding marshes; Izu-numa and Uchi-numa lakes and peat swamps; Katano-kamoike pond/marsh, a major bird habitat; and Yatsu-higata, a mudflat shorebird habitat near Tokyo. The Hokkaidō sites are Akkeshi-ko and Bekambeushi-shitsugen, a lake and saltmarsh complex; Kiritappu-shitsugen, a peat bog; Kushiro-shitsugen, a wildlife habitat containing reedbeds; Kutcharo-ko, a reed swamp; and Utonai-ko, a lake with surrounding swamps. There are also wetland sites on Okinawa and Niigata islands. Japan's wetlands are threatened by water pollution, reclamation of land for development, and extraction of water.

THE COAST, ISLANDS AND THE OCEAN

Oceans and Seas

No point in Japan is more than 93 mi (150 km) from sea waters. Japan is bound on the north by the Sea of Okhotsk, on the east and south by the Pacific Ocean, on the southwest by the East China Sea, and on the west by the Sea of Japan. The islands of Honshū, Shikoku, and Kyūshū enclose Japan's narrow Inland Sea. The Korean

Strait, approximately 124 mi (200 km) across, separates southwest Japan from South Korea. The Sōya Strait (La Perouse Strait) runs between northern Japan and Russia's Sakhalin Island. Tsugaru Strait separates Hokkaidō and Honshū islands. Warm and cold ocean currents blend in the waters surrounding Japan.

Undersea earthquakes often expose the Japanese coastline to dangerous tidal waves, known as tsunamis. Japan's coral reefs have been severely damaged by sedimentation from construction and agriculture, and by over-fishing. Environmentalists try to protect the remaining intact reefs around southern islands such as Okinawa, where tourist facility construction poses a threat.

Major Islands

The northern island of Hokkaidō, (30,394 sq mi / 78,719 sq km) was long looked upon as a remote frontier area because of its forests and rugged climate. Hokkaidō is divided along a line extending from Cape Sōya to Cape Erimo. The eastern half includes the Daisetsu Mountains, at the foot of which lie the plains of Tokachi and Konsen. The western half is milder and less mountainous.

Honshū, Japan's largest island (87,182 sq mi/225,800 sq km) curves south/southwest between Hokkaidō and Kyūshū. Tohoku, the northern region of Honshū, has flat, well drained alluvial plains. In the center of Honshū is the Kanto region, which includes the Tokyo-Yokohama metropolis.

The Chubu region, lying west of Kanto, has three distinct districts: Hokuriku, a "snow country" coastal strip on the Sea of Japan with stormy winters; Tosan, the central highlands, including the Japanese Alps; and Tokai, a narrow corridor lying along the Pacific coast.

The Kinki region of Honshū lies to the southwest and consists of a narrow area stretching from the Sea of Japan on the north to the Pacific Ocean on the south. It includes Japan's second largest commercial industrial complex centered on Osaka and Kobe, and the two former imperial cities of Nara and Kyoto.

The Chugoku region occupies the western end of Honshū and is divided into two distinct districts by mountains running through it. The northern, somewhat narrower, part is called "San'in" (shady side), and the southern part, "San'yo" (sunny side.)

Shikoku island (7,160 sq mi / 18,545 sq km), separated from western Honshū by the Inland Sea, is divided by mountains into a northern sub-region on the Inland Sea and a southern part on the Pacific Ocean. Most of the population lives in the northern zone. The southern part is mostly mountainous and sparsely populated.

Kyūshū (14,454 sq mi / 37,437 sq km) the southernmost of the main islands, is divided by the Kyūshū Mountains, which run diagonally across the middle of the island. The northern part is one of Japan's most industrialized regions.

Other Islands

There are thousands of small islands in Japan's possession. Some of the largest located near the main islands are Tsushima, Sado, Rishiri, and Awaji Island, as well as the Gotō, Oki, and Amakusa Islands.

Japan also has many islands located further out in the Pacific Ocean. These include the Nanpo Chain, the Bonin (Ogasawara) islands, Iwo Jima, and the Volcano Islands, which are located in the Pacific Ocean some 683 mi (1,100 km) south of central Honshū. The Ryukyu Archipelago includes over 200 islands and islets of which fewer than half are populated. They extend in a chain from southeast of Kyūshū to within 120 mi (193 km) of Taiwan. Okinawa (485 sq mi/1,256 sq km) is the largest and most populated of the Ryukyu islands.

In addition to the thousands of small natural islands, Japan also has artificial landfill islands, such as Kobe's Port Island, and Osaka's Kansai International Airport.

The Coast and Beaches

Japan's coastline has been highly modified by projects such as land reclamation, port construction, and sea walls. At the head of most of the bays where Japan's major cities are located the land is subsiding (sinking), causing buildings to sink up to 1.5 in (4.5 cm) annually. Since 1935 the port area of Osaka has subsided as much as 10 ft (3 m). The effects of global warming, a general increase in the average temperature worldwide, threaten the beaches of Japan, an estimated 90 percent of which would disappear with a 3.4 ft (1 m) rise in the sea level.

The coastline of Hokkaidō island has a rough diamond shape, with the capes of Sōya in the north, Shiretoko-Masakai in the east, Erimo in the south, and Kamui in the west, forming its corners. A southwestern peninsula of Hokkaidō, Oshima, curves around Uchira Bay and ends in the promontories of Shiragami and Esan.

Honshū has large indentations along its Pacific coast, such as the Bōsō, Izu, and Kii peninsulas, and the bays of Ishinomaki, Tokyo and Ise (Nagoya). On the Pacific side, flat shores are found at the head of the principal bays where major cities are situated. North of Tokyo Bay is a type of landscape called *suigo* ("land of water") where the plain is exactly at sea level, protected by levees and locks and by a system of pumps. In contrast to the Pacific coast, Honshū's Sea of Japan shoreline is less indented, with the central Noto Peninsula and Wakasa Bay serving as exceptions to long curves of flat shoreline.

Shikoku island has a violin shape, with the Inland Sea on the north and Tosa Bay curving into the south. The southern and western coasts of Kyūshū island, including Kagoshima Bay, are deeply fragmented and fractured.

CLIMATE AND VEGETATION

Temperature

Japan is in the temperate zone, with the exception of the subtropical southern island chains. There are four distinct seasons: Winter (December-February), Spring (March-May), Summer (June-August) and Autumn (September-November.) The average annual temperature is 59° F (15° C) with a winter range of 15° to 61° F (-9° to 16° C) and a summer range of 68° to 82° F (20° to 28° C). Humidity is high, ranging from 50 percent to 75 percent.

Rainfall

The peak rainy season is from May to October with regional variations. Yearly rainfall averages 39 to 98 in (100 to 250 cm). Southern Shikoku island is particularly vulnerable to typhoons, violent cyclonic storms from the Pacific. In regions bordering the Sea of Japan the winter monsoon, laden with snow, can be destructive. Snowfall is generally heavy along the western coast where it covers the ground for almost four months.

Floods are common, especially in the Pacific coastal areas where the subsidence of land makes it necessary to raise large embankments and dikes against rivers that flow at a level well above the surrounding plains. During periods of heavy rains, waters bearing great quantities of alluvium can break through the embankments, inundating adjacent fields and covering them with a thick carpet of gravel and sand. Sometimes typhoons bringing fresh torrents of water to the rivers convert whole plains into vast lakes and sweep away roads and railroads.

Plains

Japan has few regions of level, open, land. Most of those that exist are areas in which masses of river-borne soil have accumulated. Accordingly, most of the plains are located along the coasts, including the largest, Kanto, where Tokyo is located; the Nobi plain surrounding Nagoya; the Kinki plain in the Osaka-Kyoto area; the Sendai plain in northeastern Honshū; and the Ishikarai and Tokachi Plains on Hokkaidō. Japan's plains are almost completely urbanized; little of the natural ground cover remains.

Forests and Jungles

About 67 percent of Japan's land is forested. This percentage includes plantations of cedar and cypress species that replaced natural forests during the 20th century, as well as secondary forest and stands of old-growth trees. Most of Japan's forest is of temperate zone types, including conifer, deciduous, and alpine species. There are also subtropical forests on the Ryukyu Islands. Nearly all of Japan's remaining forests are in mountainous areas. Many are under official protection as national parks and Forest Ecosystem Reserves. Continuing threats to the forests include dams, road building, and recreation construction. Japan is an enormous importer of logs and wood pulp, but relies on overseas forests to supply its large scale timber and paper consumption.

Prefectures – Japan

Name	Area (sq mi)	Area (sq km)	Capital
Aichi	1,984	5,138	Nagoya
Akita	4,483	11,612	Akita
Aomori	3,713	9,617	Aomori
Chiba	1,988	5,150	Chiba
Ehime	2,190	5,672	Matsuyama
Fukui	1,618	4,191	Fukui
Fukuoka	1,915	4,960	Fukuoka
Fukushima	5,322	13,794	Fukushima
Gifu	4,091	10,596	Gifu
Gumma	2,454	6,356	Maebashi
Hiroshima	3,269	8,466	Hiroshima
Hokkaido	32,247	83,519	Sapporo
Hyōgo	3,235	8,378	Kōbe
Ibaraki	2,353	6,094	Mito
Ishikawa	1,620	4,197	Kanazawa
Iwate	5,899	15,279	Morioka
Kagawa	727	1,882	Takamatsu
Kagoshima	3,539	9,165	Kogoshima
Kanagawa	927	2,402	Yokohama
Kōchi	2,744	7,107	Kōchi
Kumamoto	2,860	7,408	Kumamoto
Kyōto	1,781	4,613	Kyōto
Mie	2,231	5,778	Tsu
Miyagi	2,815	7,292	Sendai
Miyazaki	2,986	7,735	Miyazaki
Nagano	5,245	13,585	Nagano
Nagasaki	1,588	4,112	Nagasaki
Nara	1,425	3,692	Nara
Niigata	4,857	12,579	Niigata
Oita	2,447	6,337	Oita
Okayama	2,737	7,090	Okayama
Okinawa	870	2,254	Naha
Osaka	721	1,868	Osaka
Saga	939	2,433	Saga
Saitama	1,467	3,799	Urawa
Shiga	1,551	4,016	Otsu
Shimane	2,559	6,628	Matsue
Shizuoka	3,001	7,773	Shizuoka
Tochigi	2,476	6,414	Utsunomiya
Tokushima	1,600	4,145	Tokushima
Tokyo	835	2,162	Tokyo
Tottori	1,349	3,493	Tottori
Toyama	1,642	4,252	Toyama
Wakayama	1,824	4,725	Wakayama
Yamagata	3,601	9,327	Yamagata
Yamaguchi	2,358	6,106	Yamaguchi
Yamanashi	1,723	4,463	Kofu

SOURCE: *Geo-Data: The World Geographical Encyclopedia,* 2nd ed. Detroit: Gale Research, 1989.

HUMAN POPULATION

Japan has a very high population density of 880 per sq mi (340 people per sq km), putting it among the most densely populated countries on Earth. This is despite a low population growth rate of 0.17% (2001). An estimated 78 percent of Japan's people lived in urban areas in 2000. These cities have population densities as high as

Population Centers – Japan

(2000 POPULATION ESTIMATES)

Name	Population	Name	Population
Tokyo (capital)	8,130,000	Kyōto	1,468,000
Yokohama	3,427,000	Fukuoka	1,341,000
Osaka	2,599,000	Kawasaki	1,250,000
Nagoya	2,171,000	Hiroshima	1,126,000
Sapporo	1,822,000	Sendai	1,08,000
Kobe	1,494,000	Kitakyushu	1,011,000

SOURCE: "Table 2.10. Population of Major Cities." *Statistical Handbook of Japan 2001.* Statistics Bureau (Japan), Ministry of Public Management, Home Affairs, Posts and Telecommunications.

14,245 per sq mi (5,500 people per sq km). The capital, Tokyo, is the world's most populous city.

NATURAL RESOURCES

Japan has few natural resources. The most important are fishing, agriculture, and hydropower. There is also considerable geothermal and wind power generation potential. The Japanese economy is nevertheless very advanced, with its large industrial and commercial sectors giving it one of the world's highest Gross National Products (GNPs).

FURTHER READINGS

Booth, Alan. *The Roads to Sata: A 2,000 Mile Walk through Japan.* Tokyo: Kodansha International, 1997.

Bornoff, Nicholas. *The National Geographic Traveler. Japan.* Washington, D.C.: National Geographic Society, 2000.

Nature Conservation Society of Japan. http://www.nacsj.or.jp (accessed June 20, 2002).

Public Broadcasting System. *NOVA Online: Japan's Secret Garden, Lake Biwa.* http://www.pbs.org/wgbh/nova/satoyama (accessed June 20, 2002).

Sutherland, Mary. *National Parks of Japan.* Tokyo: Kodansha International, 1995.

Jordan

- **Area:** 35,637 sq mi (92,300 sq km) / World Rank: 112
- **Location:** Northern and Eastern Hemispheres, in the Middle East, bordering Syria to the north, Iraq to the northeast, Saudi Arabia to the east and south, the Red Sea to the southwest, and Israel and the Israeli-occupied West Bank to the west, the Israeli-occupied Golan Heights to the northwest.
- **Coordinates:** 31°00′N, 36°00′E
- **Borders:** 1,006 mi (1,619 km) / Iraq, 112 mi (181 km); Israel, 148 mi (238 km); Saudi Arabia, 452 mi (728 km); Syria, 233 mi (375 km); West Bank, 60 mi (97 km)
- **Coastline:** 16 mi (26 km)
- **Territorial Seas:** 3 NM
- **Highest Point:** Jabal Ramm, 5,689 ft (1,734 m)
- **Lowest Point:** Dead Sea, 1,339 ft (408 m) below sea level
- **Longest Distances:** 349 mi (562 km) NE-SW; 217 mi (349 km) SE-NW
- **Longest River:** Jordan, 200 mi (322 km)
- **Largest Lake:** Dead Sea, 370 sq mi (962 sq km)
- **Natural Hazards:** Drought
- **Population:** 5,153,378 (July 2001 est.) / World Rank: 107
- **Capital City:** Amman, northwestern Jordan
- **Largest City:** Amman, 1,183,000 (2000 estimate)

OVERVIEW

Jordan is a Middle Eastern country located to the northwest of the Arabian Peninsula. It is landlocked except for its southernmost edge, where some 16 mi (26 km) of shoreline along the Gulf of Aqaba provide access to the Red Sea. Jordan annexed territory west of the Jordan River after a 1948-49 war with Israel, and it became known as the West Bank. The West Bank was occupied by Israel after the 1967 war between these countries. Jordan surrendered its claim to the region in 1988.

The eastern four-fifths of the country is desert. Jordan's western border is formed by a structural depression occupied by the Jordan River Valley, the Dead Sea (the lowest point on earth), and, farther to the south, the Wadi al Arabah (Wadi al Jayb), which rises gradually through a barren desert until it reaches sea level about halfway to the Gulf of Aqaba. The depression to the west is separated from the desert along its entire length by an upland area known as the Eastern Heights or Mountain Heights Plateau.

The Jordan River Valley forms the northern portion of the Great Rift Valley, an enormous north–south geological rift that continues southward along the Red Sea and into eastern Africa as far as Mozambique. With a total length of 3,000 mi (4,827 km), it is the largest geological fault system in the world. Jordan's location at the edge of this rift places it near the dividing line between the Arabian and African Tectonic Plates.

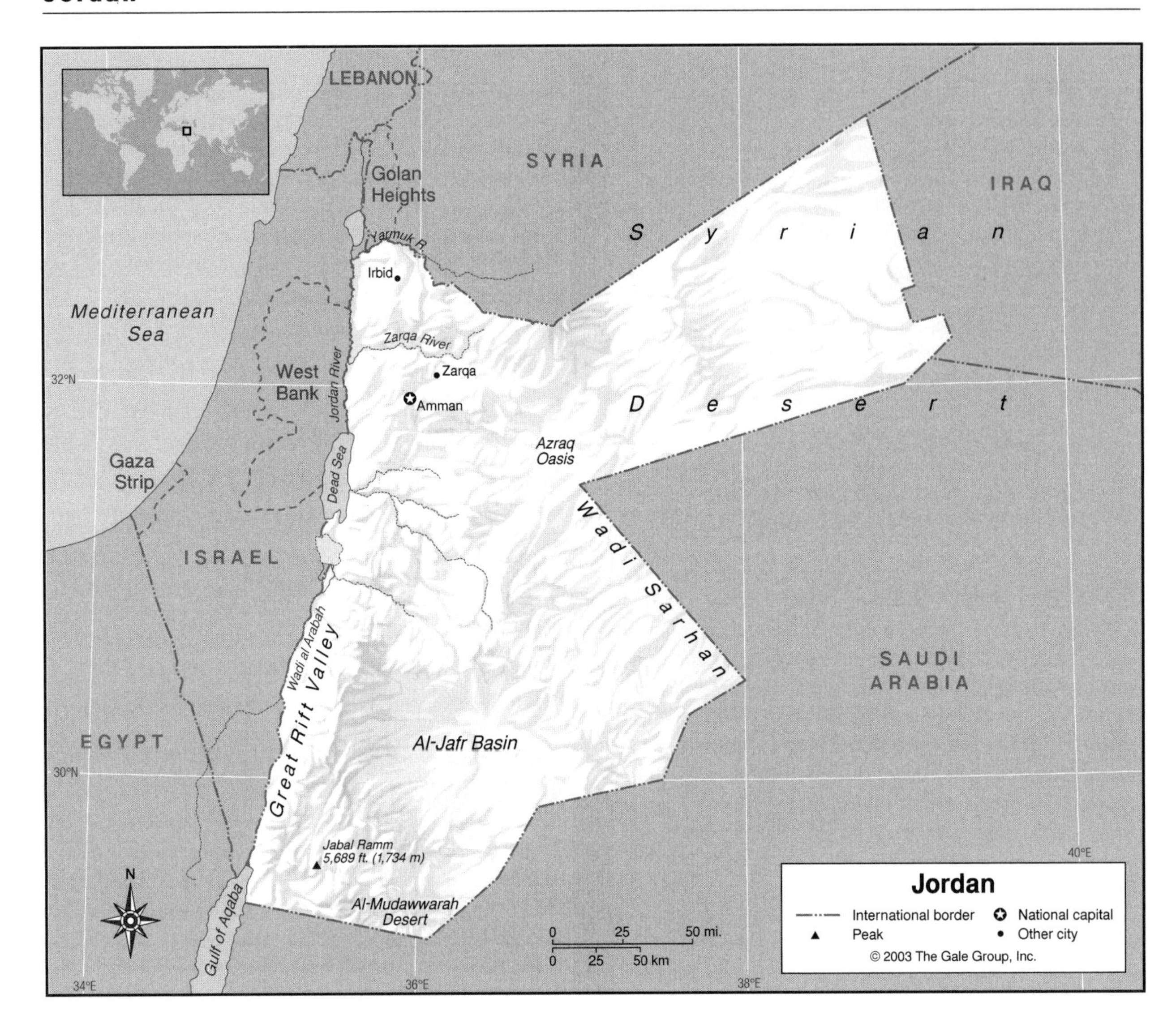

MOUNTAINS AND HILLS

Mountains

The high sandstone and granite formations in the southwestern part of Jordan rise to over 5,000 ft (1,524 m) and include the country's highest point, Jabal Ramm.

Plateaus

Separating the country's western rift from the desert is a chain of high, deeply cut limestone plateaus with average elevations between 3,000 and 4,000 ft (about 900 to 1,200 m), and summits reaching over 5,000 ft (1,524 m) in the south. The western edge of this plateau region forms an escarpment along the eastern edge of the Jordan River–Dead Sea depression. The plateau is traversed by the seasonally or completely dry river beds known in Jordan as wadis.

The northernmost part of the plateau, extending from the northern border to slightly north of Amman and known as the Northern Highlands, is the country's most fertile area, where fruits and vegetables are grown. It corresponds to the biblical Land of Gilead.

INLAND WATERWAYS

Lakes

Jordan shares the Dead Sea with Israel and with the occupied West Bank. The world's lowest body of water (and lowest point), this saltwater lake (or inland sea) has a high concentration of minerals that makes it at seven or eight times as salty as the ocean, and also makes it impossible for any form of life to survive within its waters, hence its name.

The large Azraq Oasis in the northern part of the country is the most important source of water in the Jordanian desert.

Rivers

The primary river of the country is the Jordan, which rises near the conjunction of the Israeli, Syrian, and Lebanese borders before flowing south to become Jordan's northwestern border. The river's principal tributary is the Yarmūk River, which forms parts of the Jordanian, Syrian, and Israeli borders. The Zarqa River, the second main tributary of the Jordan, is the only sizable river located entirely within Jordan.

THE COAST, ISLANDS, AND THE OCEAN

The southwestern edge of the country has a short border on the Gulf of Aqaba, which provides an opening to the Red Sea, which gets its name from the soft coral that lines its shores. The short coastline at Aqaba has sandy beaches that, together with a scenic mountain backdrop, make it popular with tourists.

CLIMATE AND VEGETATION

Temperature

Jordan has a Mediterranean climate, with cool winters and hot, dry summers. Average temperatures in Amman are 39°F to 54°F (4°C to 12°C) in January and 64°F to 90°F (18°C to 32°C) in August. The summertime heat is exacerbated by the khamsin, a hot, dry desert wind from the Arabian peninsula that can last for several days. Temperatures are especially high in the region surrounding the Dead Sea, where summer highs of around 100°F (38°C) are common and the highest temperature ever recorded was 124°F (51°C).

Rainfall

Average annual rainfall ranges from less than 4 in (10 cm) in the south to around 20 in (58 cm) in the northwest. Most rain falls between November and April.

Grasslands

There are some steppe grasslands in Jordan's desert regions. In these areas sagebrush is the most common form of plant life, and there are occasional shrubs and trees. The plateaus in the northwestern part of the country are Jordan's most fertile land and are given over to agriculture.

Deserts

The eastern four-fifths of Jordan are part of the Syrian Desert (Northern Arabian Desert), which also extends over portions of Syria, Iraq, and Saudi Arabia. Elevations in Jordan's desert region range from about 2,000 to 3,000 ft (roughly 600 to 900 m). A forbidding landscape called the Black Desert or Basalt Desert makes up the northern and northeastern parts of the Jordanian desert, extending into Syria and Iraq. With a characteristic terrain of rough black rocks and boulders, it is formed of volcanic lava and basalt.

There are several distinct areas in the desert of central and southern Jordan, some of which are demarcated by concentrations of the dry water beds called wadis. To the east, stretching southeast along the lower part of the angle formed by the border with Saudi Arabia, is the Wadi Sarhan, whose waters drain northward to the Azraq Oasis east of Amman. In the southeast part of the country lies the Al-Jafr Basin, where a number of wadis intersect. In the southwestern corner of the country is the Wadi Rum, where elevated sandstone and granite formations and deep canyons create one of the most dramatic landscapes in Jordan. To the east, the land descends to the scattered hills, low mountains, and broad wadis of the Al-Mudawwarah Desert.

Forests and Jungles

Some forest land can be found in the Jordan River Valley and the highlands that border it. The vegetation here is typical Mediterranean, including olive, eucalyptus, and cedar trees. Other tree species include evergreen oak and pine, as well as pistachio and cinnabar. Flowering plants include poppies, roses, cyclamen, and irises.

HUMAN POPULATION

About three-fourths of Jordan's population consists of urban dwellers. Most of Jordan's main urban centers are located on the uplands that separate the Jordan River Valley from the desert to the east, and population density is especially high in the northern part of this region. Some of Jordan's Bedouin population is still nomadic or seminomadic.

NATURAL RESOURCES

The Dead Sea is rich in minerals, especially potash and phosphates. Limestone, gypsum, marble, and salt are among the other raw materials that are mined in Jordan.

FURTHER READINGS

Caulfield, Annie. *Kingdom of the Film Stars: Journey into Jordan.* Oakland, Calif.: Lonely Planet Publications, 1997.

Hamilton, Masha. *Staircase of a Thousand Steps: A Novel.* New York : BlueHen Books, 2001.

Layne, Linda L. *Home and Homeland: The Dialogics of Tribal and National Identities in Jordan.* Princeton, N.J.: Princeton University Press, 1994.

Rollin, Sue, and Jane Streetly. *Jordan.* Blue Guides. New York: W. W. Norton, 1996.

Sicker, Martin. *Between Hashemites and Zionists: The Struggle for Palestine, 1908-1980.* New York: Holmes & Meier, 1989.

Kazakhstan

- **Area:** 1,049,149 sq mi (2,717,300 sq km) / World Rank: 10
- **Location:** Northern and Eastern Hemispheres; Central Asia; south of Russia; northwest of China; north of Kyrgyzstan, Turkmenistan, and Uzbekistan; east of the Caspian Sea
- **Borders:** 7,447 mi (12,012 km) / China, 950 mi (1,533 km); Kyrgyzstan, 652 mi (1,051 km); Russia, 4,245 mi (6,846 km); Turkmenistan, 235 mi (379 km); Uzbekistan, 1,366 mi (2,203 km)
- **Coordinates:** 48°00′N, 68°00′E
- **Coastline:** Landlocked with no ocean coasts, borders Aral Sea (663 mi / 1,070 km) and Caspian Sea (1,174 mi / 1,894 km)
- **Territorial Seas:** None.
- **Highest Point:** Khan Tangiri Shyngy, 20,991 ft (6,398 m)
- **Lowest Point:** Vpadina Kaundy, 433 ft (132 m) below sea level
- **Longest River:** Irtysh River, 2,760 mi (4,441 km)
- **Largest Lake:** Lake Balkhash, 7,030 sq mi (18,200 sq km)
- **Population:** 16,731,303 (July 2001 est.) / World Rank: 55
- **Capital City:** Astana, north-central Kazakhstan
- **Largest City:** Almaty, in the southeast near the Kyrgyzstan border, 1.3 million (2000 est.)

OVERVIEW

Kazakhstan lies in the center of western Asia, with a small part of the northwestern corner of the country in Europe. At 1,030,810 sq mi (2,669,800 sq km), it is the largest country in Central Asia, and of the former republics of the Soviet Union, only Russia is larger.

Topography varies greatly across this vast country. There are three mountainous regions, one in the northeast where the Altay Shan are found, one in the southeast among the Tian Shan, and one in the northwest where the southernmost of the Ural Mountains stretch into the country. Between these widely separated mountains are vast stretches of desert and steppe, a harsh terrain of bare rock and sand dunes. Roughly three-quarters of the country is considered arid or semi-arid.

Despite its desert character and the fact it is a landlocked country, Kazakhstan has considerable water resources. Both the Caspian and the Aral Seas are partially within the republic. In the east lies Lake Balkhash, an inland lake that is partially fresh and partially saline from the salts that leech into its waters from the land. Melting and runoff from mountain snows and glaciers feed the country's many rivers, although some of these are seasonal and dry up in summer.

Kazakhstan lies on the Eurasian Tectonic Plate. The Tian Shan mountains were uplifted and folded during the Paleozoic era. Deep faulting of their primarily sedimentary rock has been responsible for frequent, severe earthquakes along the rim of this range.

MOUNTAINS AND HILLS

Mountains

In the southeast of Kazakhstan, extending across the borders with Kyrgyzstan and China, are the rugged Tian Shan. These are one of Central Asia's major mountain systems. The Tian Shan cover an area of roughly 400,000 sq mi (1,036,000 sq km), which makes them comparable in size to the American Rocky Mountains. The chain is some 1500 mi (2414 km) in length and 200 to 300 mi (320 to 480 km) wide. There are many high peaks, and Kazakhstan's tallest mountain, Khan Tangiri Shyngy (Mount Tengri; 20,991 ft / 6,398 m) can be found here.

Alpine glaciers, some as long as 12 mi (19 km), are common on the steep eastern slopes, and these feed may of Kazakhstan's river systems. The southern slopes of the range are predominantly arid, except where the Ili River flows between two branches of the system. Highland meadows create natural pastures on the northern slopes of the Tian Shan at around elevations of 3,000 ft (900 m).

The Altay Mountains enter the country in its northeastern corner. This is another high mountain range, with peaks over 15,000 ft (4,572 m), but most of the range lies in Russia and China.

The Urals are a large mountain chain stretching all the way across Russia from the Arctic Ocean and into northwestern Kazakhstan, approximately 1,500 mi (2,400

GEO-FACT

The Aqsu-Zhabaghly Nature Reserve lies south of Qyzylorda and about 60 mi (100 km) west of Zhambyl, just to the north of Uzbekistan. It provides a protected habitat for the animals and plants native to Kazakhstan, including the snow leopard, several species of bear, and the ibex.

km) altogether. These mountains, along with the Ural River, are the landform that are considered to physically separate the continents of Europe and Asia. The Urals are the weathered remnants of an old mountain range that rose 250 million years ago near the end of the Paleozoic era, and are relatively low in height. In Kazakhstan, they run in three parallel chains. The easternmost range is particularly low, with peaks reaching about 2,200 to 2,800 ft (670 to 850 m). Moving west, the other two chains are higher, reaching up to 5,230 ft (1594 m). These two ranges feature many rivers; and all three of the range have heavy deciduous forests and mountain pastures.

Plateaus and Depressions

There are many elevated but relatively flat areas in central and western Kazakhstan, as well as places where the terrain dips down to form great basins and depressions. Some of these depressions are filled with water, forming Kazakhstan's lakes and seas. Others are dry. The Caspian Depression is a vast lowland extending between Kazakhstan and Russia. Located in both Europe and Asia, it features some of the lowest elevations to be found in either continent. Lying north of the Caspian Sea, the depression covers roughly 75 sq mi (200 sq km). Its lowest elevation (in Russia) is some 92 ft (28 m) below sea level; at its highest it rises to about 479 ft (149 m) above sea level. Many major rivers—as well as numerous lesser rivers and streams—cross the Caspian Depression, including the Volga in Russia and the Ural in Kazakhstan.

Located entirely within Kazakhstan, the Karagiye Depression lies in the extreme southeast, east of the Caspian Sea. Here is located Kazakhstan's lowest elevation, 433 ft (132 m) below sea level. Livestock raising is the primary economic activity in the region. South and east of Karagiye is the Ustyurt (Ust Urt) Plateau, an elevated region separating the Caspian and Aral Seas. Further east, beyond the Aral Sea, is the Turan Steppe, a vast region of plateaus and desert that extends south into Uzbekistan and Turkmenistan.

INLAND WATERWAYS

Lakes

In southeastern Kazakhstan is Lake Balkhash, whose shape forms a long, narrow arc, a third of which runs northeast before turning directly east toward China. The lake actually consists of two parts separated by the narrow Uzun-Aral Strait. The eastern waters are deeper and more saline than those of the western part. Total length of Lake Balkhash is 376 mi (605 km); its maximum width is 44 mi (71 km). In area it covers some 7030 sq mi (18,200 sq km), making it the largest lake in the country, although the so-called Caspian and Aral Seas are both larger. The lake lies at about 1,120 ft (341 m) above sea level.

Lake Balkhash is fed principally by the Ili River, which enters near the southern tip, but other rivers also flow into it from the northeast and from the southeast. Many islands, peninsulas, and strips of shallow water can be found along the lake's southern shores.

Kazakhstan has three other significant lakes. Lakes Alakol' and Tengiz are both salt lakes. In the far northeast near the border with China lies Lake Zaysan, which is a freshwater lake.

Rivers

A good many of Kazakhstan's rivers and streams, as well as its lakes, evaporate in summer. The Emba, in northwestern Kazakhstan, often dries up completely before reaching the Caspian Sea. Nevertheless, Kazakhstan does have some permanent rivers of major economic and geographic significance. The Tobol and Ishim Rivers originate in north-central Kazakhstan and flow northward into Russia, where they join other rivers and eventually reach the ocean.

The Irtysh River enters the country from China and flows west through Lake Zaysan, then curves northwest into Russia. It is the longest river to pass through Kazakhstan, and among the largest rivers in Asia, flowing for a total of 2,760 mi (4,441 km) before emptying into the Ob' River in Russia, which eventually leads to the Arctic Ocean. It is navigable for most of its length in Kazakhstan, and many cities are located near it.

Other than these three rivers of northeastern Kazakhstan, all of the country's rivers and streams are landlocked. In southeastern Kazakhstan, the Ili River flows westerly about 800 mi (1,287 km) from its headwaters in China through the city of Qapshaghay and northwest into Lake Balkhash. The river forms a rich delta that extends 115 mi (185 km) from the lake. The Ili River Valley is noted for its fertility, and it supports a significant agrarian and livestock population all along its length. During the rainy season the normally shallow river is navigable for about 280 miles (450 km) east of Lake Balkhash. A dam located near Qapshaghay has restricted the flow of water from the Ili to Lake Balkhash by about a third. As a result the lake's level has dropped by more than 6 ft (2 m) since the dam was completed in 1970.

The Syr Darya is one of the major rivers of Central Asia. With origins in Uzbekistan the Syr Darya flows northwest through Kazakhstan into the Aral Sea. It is 1,370 mi (2,200 km) in total length. During the last half of the twentieth century the Syr Darya was used heavily for irrigation, to the extent that from 1974 to 1986 almost none of its waters reached the Aral Sea. As a result of this, and similar drawing off of the waters from the Aral's major tributary in Uzbekistan, the volume of water in the Aral Sea has fallen by half.

The Ural River flows from the Ural Mountains in southern Russia into northwestern Kazakhstan. It runs south through the town Oral into the Caspian Sea. In total it flows some 1,509 mi (2,428 km). Along with the mountains of the same name, the Ural is traditionally considered the boundary between Asia and Europe.

This Irtysh-Qaraghandy Canal, located in the uplands of central Kazakhstan, was the largest water-diversion project (by volume) in the former Soviet Union. Only one other canal is longer, the Garagum Canal in Turkmenistan. The canal supplies water for recreational, agricultural, industrial, and other uses. Copper, coal, and manganese mining are a few of the many large-scale extractive industries supported by the canal.

THE COAST, ISLANDS, AND THE OCEAN

The Aral Sea

The Aral Sea is really a very large saltwater lake that lies across the border of southwestern Kazakhstan and northwestern Uzbekistan. Located east of the much larger Caspian Sea, the Aral is surrounded by deserts and has no outlets to other bodies of water.

The size of this inland lake has been steadily shrinking over the last several decades. In 1960 it ranked fourth among the world's largest lakes. Its area then was approximately 25,660 sq mi (66,458 sq km); its north-south dimension was 266 mi (428 km) and east-west it measured 181 mi (292 km). The average depth of the Aral was 53 ft (16 m) with a maximum depth of 226 ft (69 m). Since 1960 however, the Aral Sea has been reduced to nearly half its former size. This comes about from crop irrigation projects that have heavily diverted the waters of the two principle rivers that feed the Aral, the Amu Darya in the south (in Uzbekistan) and the Syr Darya in the east, in Kazakhstan. Many negative environmental changes seriously threaten the local populace as well as the ecology of the region in general. Since 1988 the drop in sea level has caused the Aral Sea to divide into two distinct seas.

The Caspian Sea

Nearly half of Kazakhstan's eastern border is on the Caspian Sea. Like the Aral Sea to the east, the Caspian is landlocked; it has no outlet to other seas, lakes, or oceans. While this means that it could technically be considered a lake, it is rarely treated as such because of its salty waters and vast size. The Caspian Sea is the world's largest landlocked body of water. Its size varies from 130 to 271 mi (210 to 436 km) in its east-west direction, and it measures some 750 mi (1,210 km) running north and south. Although deepest in the southern portions, it has a mean depth of about 550 ft (170 m). It covers approximately 143,000 sq mi (371,000 sq km). Other countries that border it besides Kazakhstan include Azerbaijan on the west, Russia on the northeast, Turkmenistan on the east, and Iran on the south.

For unknown reasons, water levels have been rising steadily in the Caspian Sea since the late 1970s (in contrast to the situation in the Aral Sea). Millions of acres of land have been flooded in Atyraū Province in the sea's northern section. By the year 2020 the city of Atyraū itself faces inundation, as do some ninety other populated areas if the Caspian continues to rise at its present rate. Also threatened are many of Kazakhstan's oil fields, located near the northern shores.

Kazakhstan's coastline on the Caspian Sea runs for 1,174 mi (1,894 km). Irregular in shape, the coast juts deeply into the country at its northern end. South of this the Mangyshlak Peninsula juts northwest into the water.

Population Centers – Kazakhstan

(1999 CENSUS OF POPULATION)

Name	Population	Name	Population
Almaty	1,129,400	Astana (capital)	313,000
Qaraghandy	436,900	Ust-Kamenogorsk	311,000
Shymkent	360,100	Pavlodar	300,500
Taraz	330,100		

SOURCE: National Statistical Agency of Kazakhstan, cited on GeoHive. http://www.geohive.com (accessed 20 May 2002).

CLIMATE AND VEGETATION

Temperature

Located thousands of miles from the ocean, Kazakhstan's climate is extremely continental, with cold winters and hot summers. Temperatures also vary greatly by region. Average January temperatures are –2°F (-3°C) in the north and 25°F (18°C) in the south; July temperatures average 66°F (19°C) in the north and range from 66° to 79°F (28° to 30°C) in the south. Temperature extremes can reach much higher or lower than these averages, however. In the winter they may fall below -49°F (-45°C) and in summer they can reach 113°F (45°C). Strong, cold northern winds make winters in the steppes especially harsh.

Rainfall

Generally little precipitation falls in Kazakhstan, ranging from less than 4 in (10 cm) in the south-central desert regions to between 10 and 14 in (25 and 35 cm) on the steppes, where flash floods are common with summer thunderstorms. In the mountains, precipitation (largely snow) averages 60 in (150 cm) per year. The sun shines a great deal in Kazakstan; on average, 260 days in the south and 120 days in the north.

Oblasts – Kazakhstan

1999 POPULATION ESTIMATES

Name	Population	Area (sq mi)	Area (sq km)	Capital
Almaty	1,614,800	86,448	223,900	Almaty
Aqmola	583,300	46,873	121,400	Astana
Aqtöbe (Aktyubinsk)	718,900	116,062	300,600	Aqtöbe
Atyraū	458,700	45,791	118,600	Atyraū
Batys Qazaqstan	641,800	58,417	151,300	Oral
Mangghystaū	350,000	63,938	165,600	Aqtau
Ongtüstik Qazaqstan	2,017,900	45,290	117,300	Shymkent
Pavlodar	854,200	48,185	124,800	Pavlodar
Qaraghandy (Karagande)	1,507,400	165,251	428,000	Qaraghandy
Qostanay	1,083,400	75,676	196,000	Qostanay
Qyzylorda	621,300	87,259	226,000	Qyzylorda
Shyghys Qazaqstan	1,612,300	109,382	283,300	Öskemen
Soltüstik Qazaqstan	1,082,400	47,568	123,200	Petropavl
Zhambyl	999,600	55,714	144,300	Taraz

SOURCE: National Statistical Agency of Kazakhstan, cited on GeoHive. http://www.geohive.com (accessed May 20, 2002).

Grasslands

Roughly 10 percent of Kazakstan is treeless prairie or prairie with mixed forest. These grassland areas are located in the Ural River basin in the west or in northern portion of the country. In the 1950s and 1960s, the Soviets introduced wide-scale wheat farming in the prairies of the northern and central parts of Kazakhstan. Much of the soil has subsequently been lost to wind erosion. Some estimates claim that 60 percent of the nation's original pastureland is now desert.

Deserts and Steppes

Most of Kazakhstan (about 75 percent) is desert, semi-desert, or steppe (arid grassy plains). The largest deserts are located in the south, the Kyzyl Kum and the Betpaqdala. Only a few scrub plants grow in these areas. The Greater Barsuki Desert lies northwest of the Aral Sea.

On the dry steppes, plants such as wormwood, tamarisk (salt cedar), and feather grass, all capable of resisting drought, are found in abundance. In the northern steppes, non-native grain crops dominate the native vegetation.

Forests

Forests and woodlands make up only 4 percent of Kazakhstan's territory. Most trees are conifers, found on the mountains of the east and southeast, where spruce, larch, cedar, and juniper grow heavily. The steppe and desert areas have virtually no trees.

HUMAN POPULATION

Kazakhstan had an estimated population of 16,731,303 (2001), which works out to an average population density of 16 persons per sq mi (6 per sq km). About 60 percent of that number is urban, which means that Kazakhstan is the most urbanized republic in Central Asia. Almaty, the former capital, is the largest city.

Astana, the new capital (1997), is still relatively small. A contributing factor to population distribution is the arid climate that makes much of the land uninhabitable. Most people live in the more hospitable east and north. Some areas in the harsh south and west have less than one person per square mile.

Kazakhstan is rich in mineral resources, including coal, iron ore, manganese, chrome ore, nickel, cobalt, copper, titanium, nickel, wolfram, tungsten, molybdenum, lead, zinc, bauxite, gold, silver, and uranium. The country also enjoys large deposits of petroleum and natural gas in the western Caspian Sea area. In the southern city of Zhambyl, fertilizer plants are well supplied by significant phosphate mines. Some 8,494 sq mi (22,000 sq km) of land is irrigated for crop production (1996 estimate).

FURTHER READINGS

Dorian, James P. "The Kazakh Oil Industry: A Potential Critical Role in Central Asia," *Energy Policy*, 22, August 1994, pp. 685-98.

Ferdinand, Peter. *The New States of Central Asia and Their Neighbors*. New York: Council on Foreign Relations Press, 1994.

Mandelbaum, Michael, ed. *Central Asia and the World: Kazakhstan, Uzbekistan, Tajikistan, Kyrgyzstan, and Turkmenistan*. New York: Council on Foreign Relations Press, 1994.

United States. Central Intelligence Agency. *Kazakhstan: An Economic Profile*. Springfield, Virginia: National Technical Information Service, 1993

Kenya

- **Area:** 224,962 sq mi (582,650 sq km) / World Rank: 48
- **Location:** Eastern Hemisphere on the equator, eastern Africa, south of Sudan and Ethiopia, west of Somalia, northwest of the Indian Ocean, north of Tanzania, and east of Uganda.
- **Coordinates:** 1°00′N, 38°00′E
- **Borders:** 2,141 mi (3,446 km) / Ethiopia, 516 mi (830 km); Somalia, 424 mi (682 km); Sudan, 144 mi (232 km); Tanzania, 478 mi (769 km); Uganda, 580 mi (933 km)
- **Coastline:** 333 mi (536 km)
- **Territorial Seas:** 12 NM
- **Highest Point:** Mt. Kenya, 17,057 ft (5,199 m)
- **Lowest Point:** Sea level
- **Longest Distances:** 703 mi (1,131 km) ENE-WSW; 637 mi (1,025 km) ENE-WSW
- **Longest River:** Tana River, 450 mi (724 km)
- **Largest Lake:** Lake Victoria, 26,828 sq mi (69,484 sq km)
- **Natural Hazards:** Drought in northern and eastern region; flooding during rainy season
- **Population:** 30,765,916 (July 2001 est.) / World Rank: 36
- **Capital City:** Nairobi, located in south-central Kenya
- **Largest City:** Nairobi, population 2,510,800 (2000 est.)

OVERVIEW

Slightly more than twice the size of the state of Nevada, Kenya has a great diversity of terrain, ranging from barrier reefs off the Indian Ocean coast to sandy desert, forested uplands, and perpetually snow-covered Mt. Kenya. A particularly prominent feature is the section of the Great Rift Valley of East Africa that runs through Kenya. The most striking physiographic distinction, however, is between the large area of higher land encompassing roughly the southwestern one-third of the country and the remaining two-thirds consisting of low plateaus and plains. Geographically, the country may conveniently be divided into seven major regions: a coastal belt; a region of plains behind the coastal strip; a low eastern plateau region; a region of plainlands covering approximately the northern one-fifth of the country; the fertile Kenya Highlands; the north-south Rift Valley Region bisecting the Kenya Highlands; and an area of western plateaus that forms part of the Lake Victoria basin.

MOUNTAINS AND HILLS

Mountains

The Kenya Highlands region consists of two major divisions lying respectively east and west of the great north-south Rift Valley. Tectonic activity played a major part in the formation of the highlands. This includes the early upwarping that resulted in what is known in geologic terminology as the Kenya Dome, and the faulting and displacement, both major and minor, across this dome that produced the Great Rift Valley and many of the region's numerous escarpments. Great outpourings of lava have added thousands of feet to the elevation over broad areas.

A striking feature on the eastern edge of the highlands is Mount Kenya, an extinct volcano and the coun-

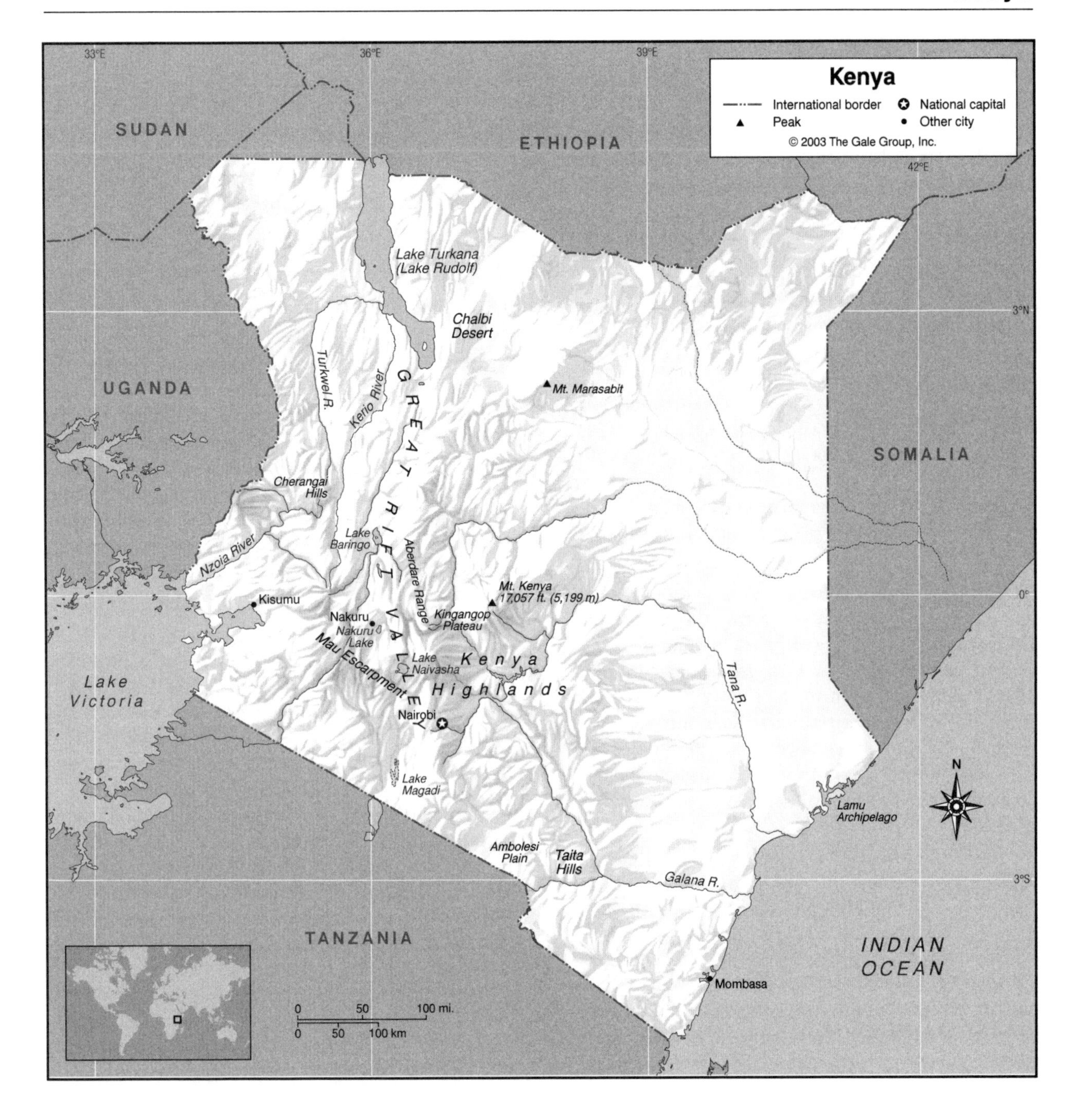

try's highest point, which rises to 17,058 ft (5,199 m). Another important subdivision of the eastern highlands is the area east of the Aberdane Range, which is populated by the Kikuyu, the country's largest ethnic group.

The Great Rift Valley of eastern Africa, formed by a long series of faulting and differential rock movements, extends in Kenya from the Lake Turkana area in the north generally southward through the Kenya Highlands and into Tanzania. In the vicinity of Lake Rudolph the elevation of the valley floor is less than 1,500 ft (457 m) above sea level, but southward it rises steadily until in its central section in the area of Lake Naivasha the elevation is close to 6,200 ft (1,889 m). From that point southward it drops off to about 2,000 ft (610 m) at the Kenya-Tanzania border.

The central section of the valley, about 40 mi (64 km) wide, is rimmed by high escarpments. On the east is the Kinangop Plateau, some 2,000 ft (610 m) above the valley floor, and east of that plateau lies the Aberdare Range, which has elevations above 13,000 ft (3,962 m). On the valley's western side is the Mau Escarpment, rising to nearly 10,000 ft (3,352 m) and farther north the Elgeyo Escarpment and the Cherangai Hills, the latter having elevations of over 11,000 ft (3,352 m). The valley floor has been subjected to extensive volcanic activity, and several cones rise high above it; the area remains one of latent volcanism, hot springs, and steam emerging at numerous spots.

Volcanoes and lava heaps divide the central section into compartments in which lie a series of lakes that are remnants of an earlier larger lake or lakes. The northern and southern parts of the valley receive a yearly rainfall averaging between 10 to 20 in (25 to 50 cm).

Plateaus

The Eastern Plateau Region consists of a belt of plain lands extending north and south to the east of Kenya Highlands. Elevations run mainly between 1,000 and 3,000 ft (300 and 900 m) except for the Chyulu Range and the Taita Hills, which rise to over 7,000 ft (2,134 m). The region appears monotonous except for the isolated hills and pinnacles that were left during the erosional development of the plains. The southern part of the region includes Ambolcsi plains known as the site of the Ambolesi and Tsavo national parks.

The Western Plateau Region forms part of the extensive down-warpped basin in which Lake Victoria lies. The region consists mainly of faulted plateaus marked by escarpments that descend in a gentle slope from the Kenya Highlands region to the shore of the lake. The region is divided by the secondary Kano Rift Valley into northern and southern components having different features. This faulted valley lies at a right angle to the main rift running through the highlands and is separated from that valley by a great lava mass.

To the southwest of Mount Kenya, the Kingangop Plateau, a relatively small, 38-mile (60-km) long plateau with some of Kenya's densest forest cover, is home to Aberdare National Park. The park is home to elephant, rhinoceros, and antelope.

INLAND WATERWAYS

Lakes

Kenya has two significant lakes: Lake Victoria and Lake Turkana (Rudolf). Lake Victoria is bordered by Uganda, Kenya and Tanzania, has an area of 26,830 sq mi (69,490 sq km)—one-third of which is within Kenyan borders—and lies 3,720 ft (1,130 m) above sea level. The lake is 209 mi (337 km) long at its greatest length, and stretches about 150 mi (about 240 km) at its greatest width. It is the world's second largest freshwater lake, after Lake Superior.

Lake Turkana (Rudolph), approximately 155 mi (250 km) long and having a maximum width of some 35 mi (56 km), has no present outlet. An earlier connection with the Nile River has been postulated based in part on the contemporary presence of the giant Nile perch in its water. The area west of the lake is quite arid, and rainfall is under 10 in (25 cm); in some years it is almost negligible. Rivers and streams flowing through the area—including the large Turkwel and Kerio Rivers, which originate in the Kenya Highlands and in the rainy season empty into Lake Rudolf—dry up at certain times of the year. Water holes remain, however, and at various other points water lies only a short distance below riverbeds.

Lakes of less significance—Baringo, Nakuru, and Naivahsa, and Magadi—lie in or near the Eastern Rift.

Rivers

The principal drainage pattern centers in the Kenya Highlands Region, from which streams and rivers radiate in a generally eastward direction toward the Indian Ocean, westward to Lake Victoria, and northward to Lake Turkana. A secondary drainage system is formed by rivers in the southern highlands of Ethiopia, which extends into Kenya along the eastern section of their mutual boundary. These rivers are all seasonal, and those that receive sufficient water at flood times to reach the sea do so through Somalia.

The two largest perennial rivers—and the only navigable ones—are the Tana and the Galana Rivers, both of which empty into the Indian Ocean. The Tana River, approximately 450 mi (724 km) long, rises in the southeastern part of the Kenya Highlands. From there it flows in a great arc northeastward along the highlands, then enters the sea at Kipini. The Galana River rises in the southern part of the Kenya Highlands and with its tributaries flows into the Indian Ocean north of Malindi. Several smaller rivers that originate in the eastern Kenya Highlands area usually disappear in the semiarid region east of the highlands. On the western slope of the Kenya Highlands, generally parallel rivers empty into Lake Victoria. The largest river in that area, the Nzoia (about 160 mi / 257 mi), eventually reaches Lake Victoria after flowing through Lake Kanyaboli and the Yala Swamp.

THE COAST, ISLANDS AND THE OCEAN

Oceans and Seas

Kenya faces the Indian Ocean. A coral reef running for more than 300 mi (480 km) lies just off the Kenyan coast and protects its coastal beaches from destructive waves. There are three marine parks along the coast: Kisite, Watumu, and Milindi.

The Coast and Beaches

Extending some 250 mi (402 mi) from the Tanzanian border in the south and the Somalia border in the north, the coastal region exhibits somewhat different features in its southern and northern parts. The shoreline in the larger southern part (below the Tana River delta) is formed largely of coral rock and sand and is broken by bays, inlets, and branched creeks.

Mangrove swamps line these indentations, but along the ocean are many stretches of coral sand that form attractive beaches. At certain places, islands were formed, the most notable of which is Mombasa, used for centuries as a port. The principal physiographic feature of the northern section is the Lamu Archipelago, formed by the drowning of the coastline as a result of a rise in the ocean level.

The coastal hinterland, forming the southern part of this region, is a relatively low erosional plain broken only in a few places by small, somewhat higher, hill groups. The plain rises very gradually westward from an elevation of about 500 ft (152 m) at the so-called coastal ranges on its eastern edge to about 1,000 ft (304 m) where it meets the Eastern Plateau Region. The Tana Plains section of the region is mainly a depositional plain—equally featureless and deficient in rainfall—extending northward from the upper part of the Coastal Region to the northern plainlands. The perennial Tana River flows across the plain on its course from the Kenya Highlands to the Indian Ocean.

GEO-FACT

Conservation of wildlife and efforts to restore the endangered African elephant and black rhino populations within reserves are a high priority in Kenya. Five biosphere reserves have been recognized under the United Nations Educational, Scientific, and Cultural Organization (UNESCO) Man and the Biosphere Program.

CLIMATE AND VEGETATION

Temperature

The climate of Kenya is as varied as its topography. Climatic conditions range from the tropical humidity of the coast through the dry heat of the hinterland and northern plains to the coolness of the plateau and mountains. The coastal temperature averages 81°F (27°C), and the temperature decreases slightly less than about 3°F (2°C) with each 1,000-ft (300-m) increase in altitude. The annual average temperature in Nairobi is 66°F (19°C) whereas in the arid northern plains it ranges from 70° to 81°F (21° to 27°C).

Rainfall

Seasonal variations are distinguished by the duration of rainfall rather than by changes of temperature. Most regions of the country have two rainy seasons: the long rainy season between April and June, and the short one between October and December. The average annual rainfall varies from 5 in (13 cm) in the most arid regions to 76 in (193 cm) near Lake Victoria. The coast and highland areas receive an annual average of 40 in (102 cm).

Grasslands

The vast Northern Plainlands Region stretches from the Uganda border on the west to Somalia. It consists of a series of plains of differing origins—mainly erosional or formed by great outpourings of lava—and includes within its limits Lake Turkana and the Chalbi Desert. The entire area east of the Chalbi Desert supports vegetation only of the semidesert type. Certain spots are more favored climatically; however, including Mount Marasabit, which at higher elevations may receive 30 in or more (76 cm or more) of rain annually and has an upper forest cover.

South-central Kenya features savannah grassland, and in the south near the Tanzania border the Amboseli National Park protects grassy plains that are home to elephant herds.

Deserts

The Chalbi Desert has Kenya's only environment classifiable as true desert. This extensive area was once covered by a lake that resulted from the damming of natural drainage by volcanic activity in the Mount Marasabit area. The plains themselves around Mount Marasabit consist of a vast lava plateau. Those farther eastward have developed on the continental basement rock; erosion here, as in the Eastern Plateau Region, has dotted the landscape with inselbergs of varying shapes and sizes. The plains closest to the Somalia border are underlain by sedimentary deposits. At the center of the desert is Lake Turkana.

Forests

Much of the original forest has been cut down, and the land is now used intensively for crops, both for subsistence and for cash. Forest still covers large areas of the northern part of the western highlands. In western Kenya, the Kakamega Forest Reserve, an area of tropical rain forest, is found in the midst of agricultural lands. The forest supports diverse plant and animal life, especially a number of primate species.

HUMAN POPULATION

Kenya's population has increased with remarkable rapidity in recent decades. About 75 percent of the population lives on only 10 percent of the land and 33 percent of the population lives in urban areas. The most populous area is the Great Rift Valley, followed by the southeastern region. The northeastern and coastal regions are the least populated, the northeast only sparsely.

Population Centers – Kenya

(1991 POPULATION ESTIMATES)

Name	Population
Nairobi (capital)	1,600,000
Mombasa	500,000
Kisumu	190,000
Nakuru	175,000

SOURCE: Projected from "Populations of Capital Cities and Cities of 100,000 and More Inhabitants." United Nations Statistical Division.

Provinces – Kenya

POPULATIONS FROM 1999 CENSUS

Name	Population	Area (sq mi)	Area (sq km)	Provincial Headquarters
Central	3,724,159	5,087	13,176	Nyeri
Coast	2,487,264	32,279	83,603	Mombasa
Eastern	4,631,779	61,734	159,891	Embu
Nairobi	2,143,254	264	684	Nairobi
North Eastern	962,143	48,997	126,902	Garissa
Nyanza	4,392,196	6,240	16,162	Kisumu
Rift Valley	6,987,103	67,131	173,868	Nakuru
Western	3,368,776	3,228	8,360	Kakamega

SOURCE: Central Bureau of Statistics, Ministry of Finance and Planning, Kenya.

NATURAL RESOURCES

Kenya has soda ash, salt barites, rubies, flavor spar, and minor deposits of gold. Tourism related to wildlife safaris is a mainstay for the economy, and the national parks of Kenya are among the richest in natural beauty in the world. Kenya's chief agricultural exports are coffee, cotton, and wheat.

FURTHER READINGS

Maxon, Robert M., and Thomas P. Ofcansky. *Historical Dictionary of Kenya.* Metuchen, N.J.: Scarecrow Press, 1999.

Ojany, Francis F., and Reuben B. Ogendo. *Kenya: A Study in Physical and Human Geography.* Boston: Longman Publishing Group, 1975.

Otoole, Thomas. *Kenya in Pictures.* Minneapolis: Lerner Publishing Company, 1997.

Kiribati

- **Area:** 277 sq mi (717 sq km) / World Rank: 179
- **Location:** Group of islands in the Pacific Ocean, between Hawaii and Australia, lying where the equator and international date line intersect.
- **Coordinates:** 1°25′N, 173°00′E
- **Borders:** No international borders
- **Coastline:** 709 mi (1,143 km)
- **Territorial Seas:** 12 NM
- **Highest Point:** Unnamed location on Banaba, 266 ft (81 m)
- **Lowest Point:** Sea level
- **Longest River:** None of significance
- **Natural Hazards:** Typhoons, occasional tornadoes, changes in sea level to low level islands
- **Population:** 94,149 (July 2001 est.) / World Rank: 185
- **Capital City:** Tarawa, located on Tarawa Island
- **Largest City:** Tarawa, population 28,000 (2000 est.)

OVERVIEW

Kiribati comprises three island groups of 33 low atolls. The three island groups are dispersed over an area of 1.1 million sq mi (3 million sq km) in the mid Pacific: the Gilbert Islands on the equator; the Phoenix Islands to the east; and the Line Islands to the north of the equator.

The Gilbert group consists of Abaiang, Abemama, Aranuka, Arorae, Banaba (formerly Ocean Island), Beru, Butaritari, Kuria, Maiana, Makin, Marakei, Nikunau, Nonouti, Onotoa, Tabiteuea, Tamana, and Tarawa.

The Phoenix group is composed of Birnie, Abariringa (Kanton), Enderbury, Nikumaroro (Gardner), Orona (Hull), McKean, Rawaki (Phoenix), and Manra (Sydney).

The Line Group encompasses Kiritimati (Christmas), Tabuaeran (Fanning), Malden, Starbuck, Vostock, Teraina (Washington), Caroline, and Flint; the last two are leased to commercial interests on Tahiti. Only some of the islands are inhabited. With an area of 186 sq mi (481 sq km), Kiritimati (Christmas Island) is the largest atoll in the world.

MOUNTAINS AND HILLS

The islands of Kiribati are low-lying, with little variation in elevation.

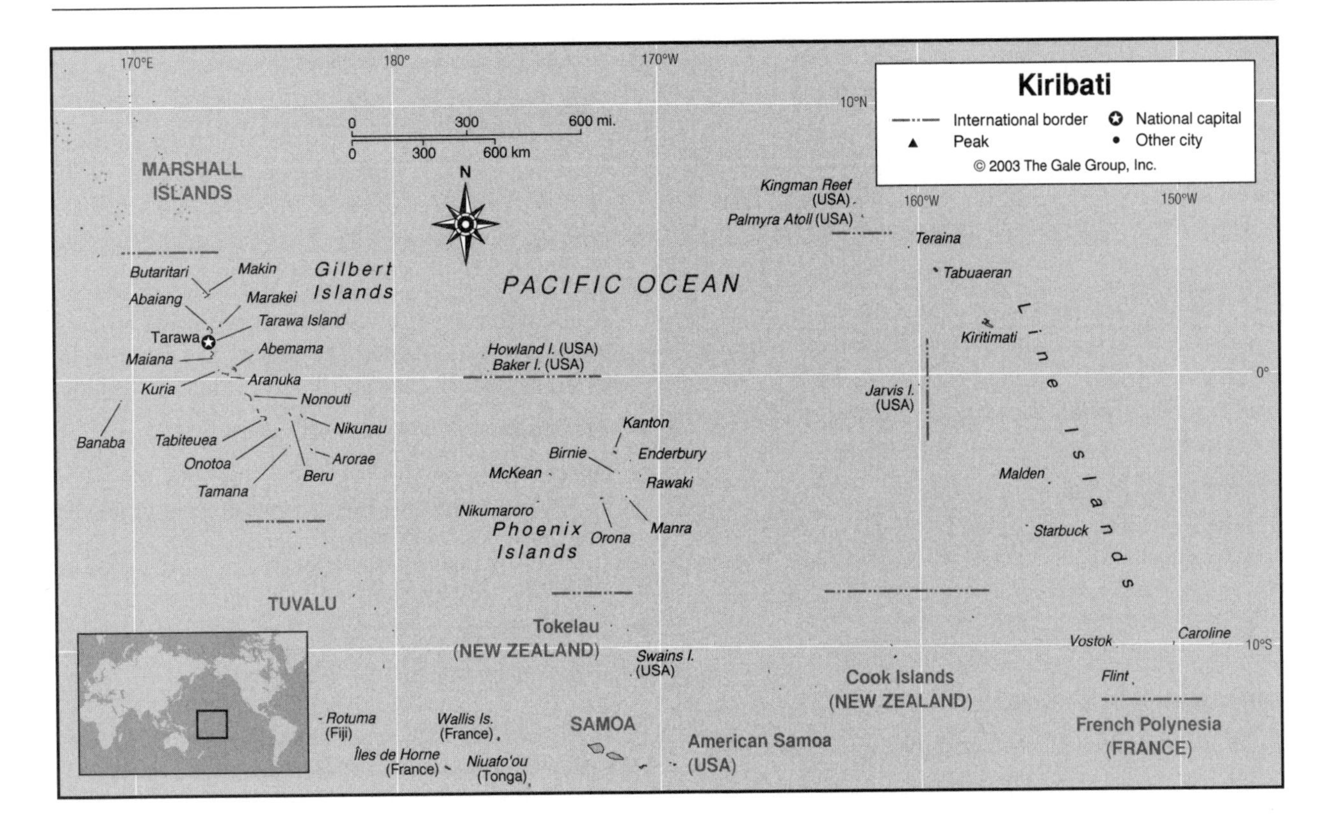

INLAND WATERWAYS

Lakes

There are dozens of lakes and ponds sprinkled across the interiors of the islands. Kiritimati has several large lagoons lying in its interior, including Manulu Lagoon in the north, Isles Lagoon in the center, and Fresh Water Lagoons on the south.

THE COAST, ISLANDS, AND THE OCEAN

Oceans and Seas

As almost all of the islands are coral atolls (except for Banaba), they are built on a submerged volcanic chain and are low-lying. A reef encloses a lagoon in most of the atolls.

Major Islands

Kiritimati (Christmas Island), representing about half the total land area of Kiribati, is the largest of the world's coral atolls. The other Line Islands—Tabuaeran, Malden, Starbuck, Vostok, Teraina, and Flint—are either sparsely inhabited or uninhabited, although Tabuaeran Island has become a stop for cruise ships. Most have been explored for phosphate mining potential at some time in the past.

Banaba (Ocean) Island is among the most westerly islands in Kiribati, and was once a rich source of phosphate. But of all the islands making up the country, Banaba has suffered the most negative environmental effects from phosphate mining, which include air pollution, water pollution, loss of green cover, and diminished aesthetic appeal of the natural surroundings. The land quality and phosphate resources have both deteriorated to the point that Banaba is no longer mined or habitable.

The Coast and Beaches

Because of Kiribati's low-lying land, it is sensitive to changes in sea level; a rise of even 2 ft (60 cm) in sea level would leave Kiribati uninhabitable. On the east coast of Kiritimati is a large, sweeping bay, named the Bay of Wrecks because of the many ships that have struck the coral reefs that lie just off the shore there. The western coast of Kiritimati forms a large, reverse C shape, enclosing a lagoon.

CLIMATE AND VEGETATION

Temperature

Located in the equatorial region, Kiribati's climate is tempered by the easterly trade winds, and humidity is high during the November to April rainy season. Occasional gales and tornadoes occur on the islands, even though they lie outside the tropical hurricane belt. The average temperature is 81°F (27°C). Between the cool and hot months, on average, there is less than 1 percent variation. Daily temperatures, however, range between 77 and 90°F (25 and 32°C).

Rainfall

Near the equator, rainfall averages 40 in (102 cm), and in the extreme north and south, it averages 120 in

Island Groups – Kiribati

Name	Area (sq mi)	Area (sq km)	Capital
Gilberts Group	110	285	Bairiki Islet
Line Group	207	535	Kiritimati
Phoenix Group	11	29	Kanton

SOURCE: *Geo-Data: The World Geographical Encyclopedia,* 2nd ed. Detroit: Gale Research, 1989.

(305 cm). The islands also face the possibility of severe droughts.

Forests and Jungles

The soil is poor and rainfall is variable on the islands, making cultivation of most crops impossible. Coconut palms and pandanus trees, however, grow without difficulty on most of the islands.

HUMAN POPULATION

While the overall population density is about 269 people per square mile (104 people per sq km), Kiribati's population is unevenly distributed. Forty percent of the total population lives in Tarawa, the capital, with 37 percent of the population living in urban areas. Some islands of the Phoenix and Line groups are uninhabited.

NATURAL RESOURCES

Kiribati has few natural resources; the economy relies on the exports of fish and copra, which is dried coconut meat and the source of coconut oil. The islands also had some of the purest phosphate lime found in nature, but mining was discontinued in 1979.

FURTHER READINGS

"Ashore on Fanning Island." *Travel Weekly,* February 4, 2002, 61 (5): C4.

Columbus Publishing. *WorldTravelGuide.Net.* "Kiribati." http://www.wtg-online.com/data/kir/kir.asp (Accessed March 12, 2002).

Troost, J. Maarten. "Taking Atoll." *The Washington Post,* June 28, 1998, p. F1.

Willis-Richards, Jonathan. *Kiribati.* http://www.collectors.co.nz/kiribati/geography.html (Accessed March18, 2002).

Korea, Democratic People's Republic of

North Korea

- **Area:** 46,540 sq mi (120,540 sq km) / World Rank: 99
- **Location:** Northern and Eastern Hemispheres, eastern Asia, northern half of the Korean Peninsula bordering the Korea Bay and the Sea of Japan, between China and South Korea.
- **Coordinates:** 40°00′N, 127°00′E
- **Borders:** 1,040 mi (1,673 km) / China, 880 mi (1,416 km); South Korea, 148 mi (238 km); Russia, 12 mi (19 km)
- **Coastline:** 1,550 mi (2,495 km)
- **Territorial Seas:** 12 NM
- **Highest Point:** Paektu-san 9,003 ft (2,744 m)
- **Lowest Point:** Sea level
- **Longest Distances:** 447 mi (719 km) NNE-SSW; 231 mi (371 km) ESE-WNW
- **Longest River:** Yalu, 500 mi (800 km)
- **Largest Lake:** Kwangpo (salt lagoon), 5 sq mi (13 sq km)
- **Natural Hazards:** Subject to drought in late spring, often followed by severe flooding; occasionally subject to typhoons in early fall
- **Population:** 21,968,228 (July 2001 est.) / World Rank: 49
- **Capital City:** P'yŏngyang, in the southwest
- **Largest City:** P'yŏngyang, 2,500,000 (2000 est.)

OVERVIEW

The Democratic People's Republic of Korea (DPRK), or North Korea, comprises a little more than half of the Korean peninsula. To the west lies the Korea Bay and to the east the Sea of Japan. The northern frontier shares an 880 mi (1,416 km) border with the People's Republic of China (PRC), and a 12 mi (19 km) border with Russia in the extreme northeastern corner about 75 mi (120 km) southwest of the Russian city of Vladivostok. The terrain is mountainous, with Paektu-san (Mount Paektu), an extinct volcano, as the highest point. A demilitarized zone (DMZ) separates North Korea from the Republic of Korea, North Korea's neighbor to the south on the Korean Peninsula. A series of plains extends along the coasts on either side of the country. North Korea is situated on the Eurasian Tectonic Plate.

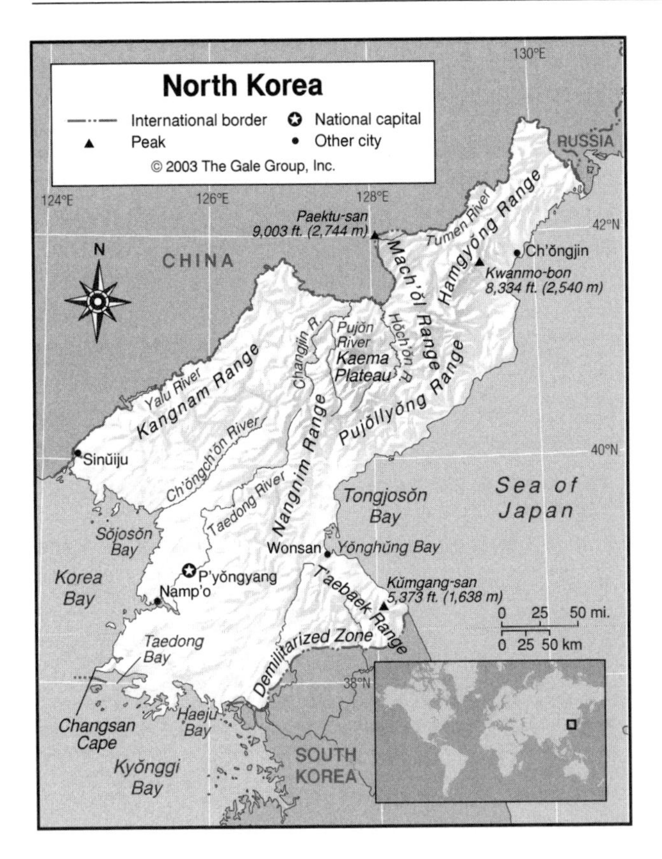

MOUNTAINS AND HILLS

Mountains and uplands cover 80 percent of the territory; the proportion is as high as 90 percent in the northern region. The major mountain ranges form a crisscross pattern extending from northwest to southeast and northeast to southwest. The Mach'ŏl Range extends from the vicinity of Paektu-san on the Chinese border in a southeasterly direction toward the east coast. This range has peaks of over 6,500 ft (1,981 m) in altitude. At the summit of Paektu-san, the country's highest peak, is a crater lake, Cho'onji (Heavenly Lake).

Running northeasterly from the center of the Mach'ŏl Range toward the Tumen River valley is the Hamgyŏng Range, which also has a number of peaks of over 6,500 ft (1,981 m), including Kwanmo-bon (Mount Kwanmo), 8,334 ft (2,540 m). The southwest extension of the Hamgyŏng Range is known as the Pujŏllyŏng Range. Running from north to south and marking the drainage divide for the eastern and western halves of the country is the Nangnim Range, averaging 1,499 m (4,920 ft). To the west of the Nangnim Range are two less prominent ranges, the Myohyang and (in the center of the country) the Puktae, both of which are from 500 to 1,000 m (1,640 to 3,280 ft) in height. Running in a southwestern direction from the Nangnim Range along the Yalu River (which forms the border with China) is the Kangnam Range, the name of which means 'south of the river.'

Korea's other major mountain chain, the T'aebaek Range, rises south of Wŏnsan and extends down the eastern side of the peninsula; it is often called the "backbone of Korea." Only a short portion of its length is in North Korea, but this section includes the scenic Kŭmgang-san ("Diamond Mountains") comprising the heart of North Korea's largest national park. The granite mountains that rise near the shore of the sea of Japan feature nearly vertical sheer walls, deep canyons, and spectacular waterfalls.

The terrain east of the Hamgyŏng and Pujŏllyŏng consists of short, parallel ridges that extend from these mountains to the Sea of Japan, creating in effect a series of isolated valleys accessible only by rail lines branching off from the main coastal track. West of the T'aebaek Range, the terrain of central Korea is characterized by a series of lesser ranges and hills that gradually level off into plains along the west coast.

In some areas where the rock formations are limestone, caves abound. One of the best-known caves is located near Yongbyon on the southern side of the Ch'ŏngch'ŏn River. Known as T'ongnyonggul, it is about 3 mi (5 km) long, with many chambers, some of which reach widths of 500 ft (150 m) and with ceilings as high as 150 ft (50 m).

Plateaus

To the west of the Hamgyŏng and Pujŏllyŏng ranges lies Kaema Plateau, sometimes referred to as the "roof of Korea". The Kaema Plateau is a heavily forested basaltic tableland with relatively low elevation averaging 3,280 to 4,950 ft (1000 to1,500 m).

INLAND WATERWAYS

Rivers

The major rivers flow in a westerly direction to Korea Bay, the northern extent of the Yellow Sea. The major river is the Yalu, which flows from Paektu-san to Korea Bay, a distance of almost 500 mi (800 km). Because its course cuts through rocky gorges for much of its length, its alluvial plains are less extensive than its size would suggest. Oceangoing vessels can dock at Sinŭiju, and small craft can travel upstream as far as Hyesan. Although it is important for transportation and irrigation, the Yalu's main value lies in its hydroelectric power potential. Dams have been built on the Yalu and four of its tributaries, the Changjin, Hŏch'ŏn, Pujŏn, and Tongno rivers.

The Ch'ŏngch'ŏn River flows in the valley between the Kangnam and the Myohyang Mountain Ranges.

THE COAST, ISLANDS, AND THE OCEAN

Oceans and Seas

Korea Bay off the west coast is shallow and has an unusually great tidal range—from 20 to 40 ft (6 to 12 m). To the east, the coastal waters of the Sea of Japan are very deep, averaging about 5,500 ft (1,676 m).

Major Islands

There are hundreds of small islands off the west coast of North and South Korea. None of the islands under North Korea's control are notable.

The Coast and Beaches

The west coast along the Korea Bay is highly indented and irregular, and it is studded with a multitude of small offshore islands. A considerable portion of the tidelands have potential value as agricultural land, reed fields, and salt evaporation facilities. Some reclamation has been undertaken in North and South P'yongan provinces. The main port on the west coast is Namp'o, which is located at the mouth of the Taedong River south of Sŏjosŏn Bay and is a center for both international and domestic trade. Further south the coast carves out two more bays, Taedong Bay, which cuts into the coast south of Changsan Cape, and Haeju Bay, which is tucked in away from the larger Kyŏnggi Bay.

In the east, where steep mountains lie close to the coastline along the Sea of Japan, the coastline is relatively smooth, with few islands and only two bays: the large Tongjosŏn Bay, and the smaller Yŏnghŭng Bay. The coast is washed by both warm and cold currents, contributing to a wide variety of marine life, and causing the coastal region to be frequently shrouded in dense fog.

Population Centers – Democratic People's Republic of Korea (North Korea)

(1993 POPULATION ESTIMATES)

Name	Population
P'yŏngyang (capital)	2,741,260
Hamhung-Hungnam	709,730
Chongjin	582,480
Kaesong	334,433

SOURCE: "1993 Census: Demographic Yearbook." United Nations.

Provinces – Democratic People's Republic of Korea (North Korea)

Name	Area (sq mi)	Area (sq km)	Capital
Chagang-do	6,300	16,200	Kanggye
Hamgyong-bukto	6,100	15,900	Chongjin
Hamgyong-namdo	7,400	19,200	Hamhung
Hwanghae-bukto	3,300	8,600	Sariwŏn
Hwanghae-namdo	2,900	7,600	Haeju
Kangwŏn-do	4,100	10,700	Wŏnsan
P'yŏngan bukto	4,600	12,000	Sinŭiju
P'yŏngan namdo	4,700	12,300	P'yŏngsan
Yanggang-do	5,400	14,100	Hyesan
Special Cities			
Kaesong-si	485	1,255	
Namp'o-si	291	753	
P'yŏngyang -si	695	1,800	P'yŏngyang

SOURCE: *Geo-Data: The World Geographical Encyclopedia,* 2nd ed. Detroit: Gale Research, 1989.

CLIMATE AND VEGETATION

Temperature

The temperature varies from north to south during winter, with the average January temperature 1°F (-17°C) along the northern border and 18°F (-8°C) at P'yŏngyang, the capital. Summer temperatures have less variation from north to south, averaging 70°F (21°C) in the north, and 75°F (24°C) at P'yŏngyang.

Rainfall

Approximately 60 percent of the rainfall, 30–40 in (75–100 cm) annually, occurs in June through September. The northernmost regions receive less rainfall, averaging 20 in (50 cm).

Grasslands

The plains regions are important to the nation's economy, although they constitute only one-fifth of the total area. Most of the plains are alluvial, built up from silt deposited on their banks by rivers in their middle and lower courses. Other plains, such as the P'yŏngyang peneplain, were formed by thousands of years of erosion from surrounding hills. A number of plains areas exist on the west coast, including the P'yŏngyang peneplain and the Unjon, Anju, Chaeryŏng, and Yonbaek plains. Of these, the Chaeryŏng and the P'yŏngyang are the most extensive, each covering an area of about 200 sq mi (618 sq km). They are followed in size by the Yonbaek Plain, which comprises about 120 sq mi (315 sq km); the rest are about 80 sq km (207 sq km) each. The plains support most of the country's farmlands, and their small size indicates the severe physical limitations placed on agriculture.

Forests and Jungles

North Korea has an extensive coniferous forest located in its mountainous interior, especially in the north. Species include pine, spruce, fir, and cedar. The lowlands have been deforested for purpose of agricultural cultivation.

HUMAN POPULATION

Like its neighboring country to the south, most of the land in North Korea is mountainous, leaving the population in scattered settlements in the valleys and lowlands. Also like their neighbors in South Korea, North Koreans began migrating to urban areas following World War II. Most of the population resides in the region south of the Kangnam Mountains and west of the Nangnim Mountains, extending as far south as the P'yŏngyang municipal district. The least populated region is the mountainous northern region near the border with China. North Korea is not nearly as densely populated as South Korea, with an average population density of about 466 people per sq mi (180 people per sq km). Only seven of its cities have more than 300,000 people, and P'yŏngyang alone has more than a million.

NATURAL RESOURCES

Because of the mountainous terrain, much of the country's natural resources have not yet been fully explored or exploited. Natural resources include coal (along the Taedong River basin in the center of the country), lead, tungsten, zinc, graphite, magnesite, iron ore (centered in the southwestern peninsula region), copper, gold, pyrites, salt, and fluorspar. There is significant hydroelectric power potential, and wind-power generating equipment is installed in the P'yŏngyang region.

FURTHER READINGS

Breen, Michael. *The Koreans: Who They Are, What They Want, Where Their Future Lies.* New York: St. Martin's, 1999.

Oberdorfer, Don. *The Two Koreas: A Contemporary History.* New York: Basic Books, 2001.

Oh, Kongdan, and Ralph C. Hassig. *North Korea Through the Looking Glass.* Washington, D.C.: Brookings Institution Press, 2000.

Oliver, Robert Tarbell. *A History of the Korean People in Modern Times: 1800 to the Present.* Newark: University of Delaware Press, 1993.

Korea, Republic of

South Korea

- **Area:** 38,023 sq mi (98,480 sq km) / World Rank: 109
- **Location:** Northern and Eastern Hemispheres, Eastern Asia, southern half of the Korean Peninsula bordering the Sea of Japan and the Yellow Sea.
- **Coordinates:** 37°00′N, 127°30′E
- **Borders:** North Korea, 148 mi (238 km)
- **Coastline:** 1,508 mi (2,413 km)
- **Territorial Seas:** 12 NM
- **Highest Point:** Halla San, 6,398 ft (1,950 m)
- **Lowest Point:** Sea level
- **Longest Distances:** 399 mi (642 km) NNE-SSW; 271 mi (436 km) ESE-WNW
- **Longest River:** Naktong, 324 mi (521 km)
- **Natural Hazards:** Subject to typhoons, resulting in wind and flood damage; occasional mild earthquakes in southwest
- **Population:** 47,904,370 (July 2001 est.) / World Rank: 25
- **Capital City:** Seoul, located in the northwest
- **Largest City:** Seoul, 12,200,000 (2000 metropolitan est.)

OVERVIEW

The Republic of Korea (South Korea) occupies the southern part of the Korean Peninsula that projects to within 120 mi (193 km) of the principal Japanese islands of Honshū and Kyūshū on the southeast. Elongated and irregular in shape, the peninsula separates the Sea of Japan from the Yellow Sea; the seas are known in Korea as the Eastern Sea and the Western Sea, respectively. The country was divided into North and South Korea along a line that follows just north of the 38th parallel, the peninsula's narrowest point (about 135 mi/217 km). South Korea is situated on the Eurasian Tectonic Plate.

As a result of the 1953 armistice agreement concluding the Korean War, about 45 percent of the Korean peninsula, or 37,910 sq mi (98,190 sq km), falls below the demarcation line. The demarcation line divides the 2.5 mi- (4 km-) wide Demilitarized Zone (DMZ), which is largely uninhabited.

MOUNTAINS AND HILLS

Mountains

While the Korean peninsula is very rugged and mountainous, the elevations in South Korea are generally less than those in the north. The T'aebaek Mountain Range in South Korea runs northeast to southwest along the Sea of Japan. Dividing the country into east and west is the Sobaek Mountain Range, running in a generally northeast-to-southwest direction; these mountains have prevented easy travel and interaction between the regions throughout history. The highest peak on the South Korean mainland is Chiri-san, at 6,283 ft (1,915 m), located in the south-central part of the country at the southern end of the Sobaek mountain range. The country's highest peak, Halla San, which rises to a height of 6,398 ft (1,950 m), lies on Cheju Do (Cheju Island) off the

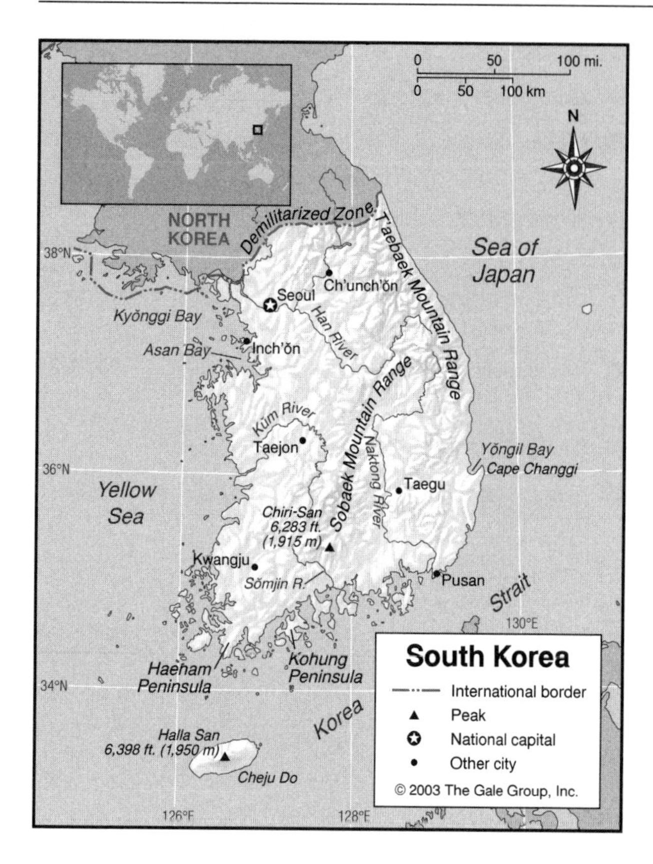

southern tip of the country and has a small crater lake at its summit.

West of Ch'ŏngju lies Maisan (Horse Ears Mountain), a two-peaked mountain that resembles the ears of a horse. In the central and south, limestone caves, with dramatic stalagmites and stalactites, may be found. One of the most famous is Kosudonggul, known as the "Underground Palace." Hills separate the Sobaek mountain range from the coastal plains in the south.

INLAND WATERWAYS

Of a comparatively large number of rivers and streams, four are of major importance: the Han River and the Kŭm River flow west to the Yellow Sea; the Naktong River and the Sŏmjin River flow south to the Korea Strait. In addition, the Yŏngsan and Tongjin rivers water South Korea's main rice growing areas. Because of their very low gradients, the rivers to the west of the T'aebaek Mountains watershed were used for transportation. These west-flowing rivers have built up extensive plains at the points where they flow into the sea. River navigation declined in importance in modern times with the introduction of new means of transport, the diversion of water for irrigation, and the construction of dams. River flow is highly seasonal, with the heaviest flows occurring in the summer months. Floods are common in the basins associated with the major river systems, particularly in estuary areas along the west coast. During much of the year, however, the rivers are shallow, exposing very wide, gravelly riverbeds. Near Ch'unch'ŏn in the north are three artificial lakes, giving the city the nickname, "City of Lakes." The Naktong River Basin in the southeast is a complex of structural basins and river floodplains separated from one another by low hills. The Naktong River forms an extensive delta where it reaches the sea a few miles west of Pusan, South Korea's major port.

THE COAST, ISLANDS, AND THE OCEAN

Oceans and Seas

The Yellow Sea lies to the west of South Korea. It is relatively shallow, and has an extremely large tidal range. At low tide, large mud flats are exposed. The Korea Strait separates South Korea from Japan. The Japanese island, Tsushima, lies in the Korea Strait, with the Western Channel and Eastern Channel separating it from South Korea and Japan's Kyūshū, respectively. The Sea of Japan forms the open body of water to the north east of South Korea; the waters of the Sea of Japan are deep, and the tidal range is small.

Major Islands

Cheju Do, formed from a volcanic eruption, features unusual lava formations on the coast near the city of Cheju. Directly east of South Korea in the Sea of Japan is Ullung-do (Ullung Island).

The Coast and Beaches

The southeast coastline may be divided at the Naktong River mouth near Pusan. To the north of this point the coast is relatively smooth, consisting of alternating headlands and bays; the latter have small lowlands at their heads, but they are not as isolated from the interior as their counterparts farther north. From the watershed divide close to the east coast, the land slopes sharply and abruptly to the narrow and discontinuous lowland of the coast. This coast is fairly regular, with few islands and bays, the major one being Yŏngil Bay enclosed in Cape Changgi.

To the west of the Naktong River mouth the coast becomes much more complex. The central and western regions of the southern coastline, where the various arms of the Sobaek Mountains reach the sea, feature a number of basins that create an intricate coastline of extensive, highly irregular peninsulas—including Kohung and Haenam Peninsulas—flanked by abruptly rising islands; offshore, the basins contain deep water. At times the peninsulas almost enclose equally irregular bays deeply penetrating the land. Around the western coast near Seoul, the tiny Asan Bay reaches into the mainland. This part of the coastline is part of the larger Kyŏnggi Bay shared with North Korea.

CLIMATE AND VEGETATION

Temperature

South Korea has a continental climate, with hot, rainy summers and cold winters. Temperatures range from 71–83°F (22–29°C) in the summers, and 19–33°F (-7–1°C) in the winter months, with warmer winter temperatures along the southern coast, and lower temperatures in the interior.

Rainfall

Rainfall averages between 40 and 50 in (100 and 150 cm) annually, but many areas experience less rainfall. Rainfall is greatest in the south, and in inland mountainous regions. The coastal areas receive the least rainfall, on average.

Grasslands

In the southern coastal regions, inland from the coast, the plains are fertile and agriculturally productive, although small in some areas. The center of bamboo cultivation is in the west-central region, near Chinan.

Forests and Jungles

Coniferous forests, comprised of pine, maple, elm, fir, poplar and aspen, cover approximately three-quarters of the land. The mild southern coastal area's forests are populated with bamboo, evergreen oak, and laurel. Deforestation, due to rapid urbanization and population growth, have diminished these forests as natural habitats for South Korea's native animal population.

Population Centers – South Korea

(1995 POPULATION ESTIMATES)

Name	Population	Name	Population
Seoul (capital)	10,229,262	Inch'ŏn	2,307,618
Pusan	3,813,814	Taejon	1,272,143
Taegu	2,449,139	Kwangju (Gwangju)	1,257,504

SOURCE: National Statistical Office, Republic of Korea.

Provinces – Republic of Korea

2000 POPULATION ESTIMATES

Name	Population	Area (sq mi)	Area (sq km)	Capital
Cheju-do	513,000	705	1,825	Cheju
Chŏlla-bukto	1,890,000	3,108	8,050	Chonju
Chŏlla-namdo	1,996,000	4,729	12,249	Kwangju
Ch'ungch'ŏng-bukto	1,466,000	2,870	7,433	Chongju
Ch'ungch'ŏng -namdo	1,845,000	3,411	8,835	Taejon
Kangwŏn-do	1,487,000	6,523	16,894	Ch'unch'ŏn
Kyŏnggi-do	8,979,000	4,193	10,859	Suwon
Kyŏngsang-bukto	2,725,000	7,506	19,441	Taegu
Kyŏngsang-namdo	2,978,000	4,577	11,855	Changwon
Special Cities				
Inch'ŏn	2,307,618	121	313	
Kwangju	1,257,504	193	500	
Pusan	3,813,814	203	526	
Seoul	10,229,262	237	614	
Taegu	2,449,139	176	456	
Taejon	1,272,143	207	536	

SOURCE: National Statistical Office, Republic of Korea.

HUMAN POPULATION

South Korea is one of the most densely populated countries, with an overall population density of about 1,243 people per sq mi (480 people per sq km). However, 70 percent of the land in South Korea is mountainous and most of the population is concentrated in the valleys. Additionally, continuing migration to urban areas over the last few decades has further concentrated the population, so actual population densities are much higher. South Korea has twenty cities with populations of 300,000 or more, and three-quarters of the entire population lives in one of these metropolitan areas. Moreover, about one-quarter of the population lives in the capital city of Seoul alone.

NATURAL RESOURCES

Coal is the chief energy resource; most of the country's coal is a grade of anthracite that is not high enough for industrial uses. Other mineral resources include iron ore, lead, copper, zinc, tungsten, and limited gold and silver. Agriculture is widespread; crops include rice, barley, fruits, and vegetables. Livestock such as cattle, pigs, and chicken are raised.

FURTHER READINGS

Amsden, Alice H. *Asia's Next Giant: South Korea and Late Industrialization.* New York: Oxford University Press, 1992.

Breen, Michael. *The Koreans: Who They Are, What They Want, Where Their Future Lies.* New York: St. Martin's, 1999.

Clifford, Mark L. *Troubled Tiger: Businessmen, Bureaucrats and Generals in South Korea.* Armonk, N.Y.: Sharpe, 1998.

Lie, John. *Han Unbound: The Political Economy of South Korea.* Stanford, Calif.: Stanford University Press, 2000.

Oberdorfer, Don. *The Two Koreas: A Contemporary History.* New York: Basic Books, 2001.

Savada, Andrea Matles, and William Shaw, eds. *South Korea: A Country Study.* Washington, D.C.: Library of Congress, 1992.

Song, Byung-Nak. *The Rise of the Korean Economy.* New York: Oxford University Press, 1997.

World Gazetteer. *Korea (South).* http://www.world-gazetteer.com/fr/fr_kr.htm (Accessed June 2002).

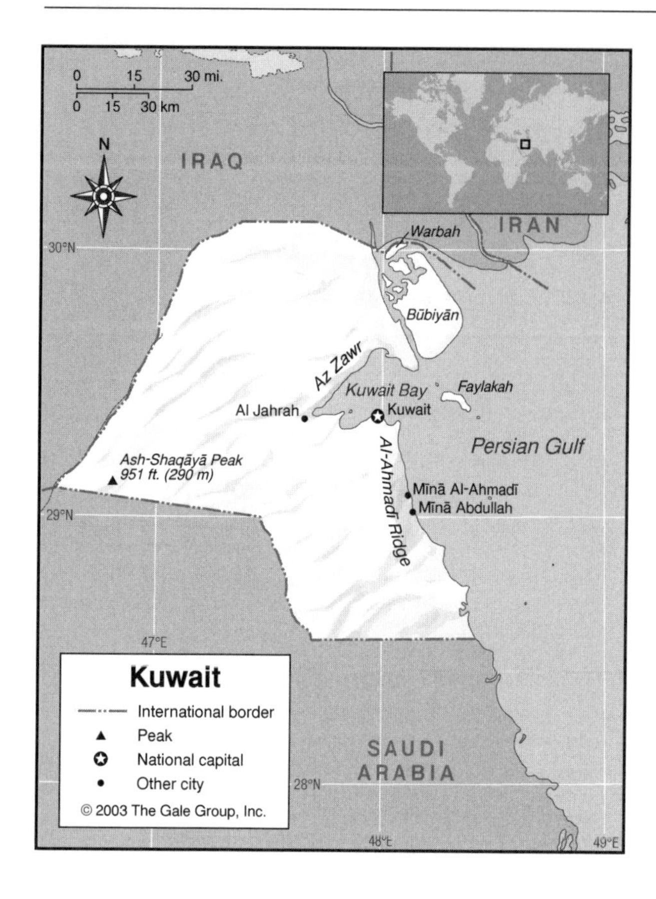

Kuwait

- **Area:** 6,880 sq mi (17,820 sq km) / World Rank: 155
- **Location:** Northern and Eastern Hemispheres, in the Middle East, bordering Iraq to the north and west, the Persian Gulf to the east, and Saudi Arabia to the south and southwest.
- **Coordinates:** 29°30′N, 45°45′E
- **Borders:** 288 mi (464 km); Iraq, 150 mi (242 km); Saudi Arabia, 138 mi (222 km)
- **Coastline:** 310 mi (499 km)
- **Territorial Seas:** 12 NM
- **Highest Point:** Unnamed location, 1,004 ft (306 m)
- **Lowest Point:** Sea level
- **Longest Distances:** 109 mi (176 km) NE-SW; 127 mi (205 km) SE-NW
- **Longest River:** None of significant length
- **Natural Hazards:** Damaging rainstorms and flooding, sandstorms
- **Population:** 2,041,961 (July 2001 est.) / World Rank: 142
- **Capital City:** Kuwait city, on Kuwait Bay
- **Largest City:** Kuwait, population 1,187,000 (2000 metropolitan est.)

OVERVIEW

Kuwait is a small Middle Eastern country located at the head of the Persian Gulf and surrounded by the much larger neighboring states of Saudi Arabia, Iraq, and Iran. Located on the coastal plain that rings much of the gulf, Kuwait has one deeply indented bay, on which its capital and major population center is located, and another coastal indentation that is the site of several uninhabited islands. Otherwise, its terrain is largely composed of flat or rolling desert land, with maximum elevations reached at its western and southwestern corners. Kuwait is located on the Arabian Tectonic Plate.

MOUNTAINS AND HILLS

Kuwait has no actual mountain ranges or distinct plateaus. Its terrain gradually rises from near sea level at the coast to elevations of about 650 ft (198 m) in the northwest and to about 1,000 ft (305 m) at its westernmost edge. The country's two other noticeable points of elevation are the Az Zawr escarpment on the northern shore of Kuwait Bay, which rises to 475 ft (145 m), and the Al-Ahmadī ridge south of the bay (450 ft / 137 km), where the coastal town of Mīnā Al-Ahmadī is located. Also of note is Ash-Shaqāyā Peak in the western corner of the country, rising to a height of 951 ft (290 m).

INLAND WATERWAYS

Some of Kuwait's wadis, or desert basins, fill with water during the winter rains, but the country has no permanent rivers or lakes. There is an oasis at Al Jahrah, at the western end of Kuwait Bay.

THE COAST, ISLANDS, AND THE OCEAN

Oceans and Seas

Kuwait is located at the northwestern edge of the Persian (or Arabian) Gulf, which empties into the Arabian Sea by way of the Strait of Hormuz and the Gulf of Oman.

GEO-FACT

Distilled water is the main source of drinking water in Kuwait, which has some of the world's most sophisticated desalination facilities.

Major Islands

There are nine islands off the coast of Kuwait, of which the largest are Būbiyān and Warbah, both uninhabited. The only one that is inhabited is Faylakah, at the edge of Kuwait Bay.

The Coast and Beaches

Kuwait Bay is the only major coastal indentation entirely within Kuwait. The city of Kuwait, the national capital, lies near the mouth of the bay, which is the sole deepwater harbor on the western coast of the Persian Gulf. Kuwait's low-lying coast is characterized by areas of marshland, as well as mud flats, sand bars, and islets.

Population Centers – Kuwait

(1995 POPULATION ESTIMATES)

Name	Population
Hawalli	150,000
Salmiya	130,000
Kuwait City (capital)	85,000

SOURCE: Projected from United Nations Statistical Division data.

Governates – Kuwait

POPULATIONS FROM 1995 CENSUS

Name	Population
Ahmadi	263,960
Asima	305,694
Jahra	224,515
Farwaniya	427,018
Hawalli	467,103

SOURCE: Ministry of Planning, Kuwait.

CLIMATE AND VEGETATION

Temperature

Kuwait has a desert climate, with elevated humidity in the coastal region due to the proximity to the Persian Gulf. Summer temperatures average about 90°F (32°C), with daytime highs commonly reaching 110°F (43°C) or higher. Readings as high as 130°F (54°C) have been recorded. Winter temperatures average between 50°F and 60°F (10°C and 15°C). The prevailing northwesterly wind, which exerts a cooling influence in summer, is called the *shamal.*

Rainfall

Average annual rainfall is less than 10 in (25 cm), and less than 5 in (13 cm) in the southern part of the country. The rainy season occurs between October and April and is characterized by cloudbursts—sudden, violent storms bringing tremendous amounts of rain that can cause severe property damage.

Deserts

The Kuwaiti desert is undulating and gravelly, with few hills or ridges. The only trees apart from those in the date gardens at the oasis of Al Jahrah are tamarisks found in Kuwait City and a few villages, and the most characteristic vegetation consists of low, hardy bushes and scrub. The most common desert shrub is the arfaj, which grows to a maximum height of 2.5 ft (0.7 m). For a brief period in the spring, if winter rains have been adequate, the wadi beds (desert basins) are covered with grass and flowering desert shrubs. The spring green period, however, is very brief, for the May sun dries the grass and withers the shrubs and the June winds soon bury everything in driving sand.

HUMAN POPULATION

Almost the entire population lives in urban areas, with more than 90 percent occupying the cities of the coastal strip between Al-Jahrah and Mīnā'Abd Allah. This region, which includes Kuwait City, is collectively called the Metropolitan Area. Most of the desert is uninhabited.

NATURAL RESOURCES

Kuwait's primary natural resource is petroleum, which has been the primary basis of its economy since the 1930s. The country also has natural gas reserves and a fishing and shrimping industry.

FURTHER READINGS

Anscombe, Frederick F. *The Ottoman Gulf: The Creation of Kuwait, Saudi Arabia, and Qatar.* New York: Columbia University Press, 1997.

Facey, William, and Gillian Grant. *Kuwait by the First Photographers.* London: I. B. Tauris, 1999.

KuwaitOnLine. *Kuwait.* http://www.kuwaitonline.com/ (Accessed April 8, 2002).

Longva, Anh Nga. *Walls Built on Sand: Migration, Exclusion, and Society in Kuwait.* Boulder, Colo.: Westview Press, 1997.

Rahman, H. *The Making of the Gulf War: Origins of Kuwait's Long-standing Territorial Dispute with Iraq.* Reading, England: Ithaca, 1997.

Robison, Gordon, and Paul Greenway. *Bahrain, Kuwait & Qatar.* London: Lonely Planet, 2000.

Vine, Peter, and Paula Casey. *Kuwait: A Nation's Story.* London: Immel, 1992.

Kyrgyzstan

- **Area:** 76,641 sq mi (198,500 sq km) / World Rank: 86
- **Location:** Northern and Eastern Hemispheres, central Asia, west of China, south of Kazakhstan, east of Uzbekistan, northeast of Tajikistan.
- **Coordinates:** 41°00′N, 75°00′E
- **Borders:** 2,410 mi (3,878 km) / China, 533 mi (858 km); Kazakhstan, 652 (1,051 km); Tajikistan, 539 mi (870 km); Uzbekistan, 681 mi (1,099 km)
- **Coastline:** Landlocked
- **Highest Point:** Jengish Chokusu, 24,406 ft (7,439 m)
- **Lowest Point:** Kara-Daryya, 433 ft (132 m)
- **Longest Distances:** Not available
- **Longest River:** Naryn River, 504 mi (807 km)
- **Largest Lake:** Lake Issyk-Kul, 2,360 sq mi (6,100 sq km).
- **Natural Hazards:** Earthquakes, landslides
- **Population:** 4,753,003 (July 2001 est.) / World Rank: 112
- **Capital City:** Bishkek, in north-central Kyrgyzstan
- **Largest City:** Bishkek, 662,000 (2000 est.)

OVERVIEW

Landlocked into east Central Asia, Kyrgyzstan (the Kyrgyz Republic) is the smallest of the Central Asian countries that became independent after the breakup of the Soviet Union in the 1990s, covering just 76,641 sq mi (198,500 sq km). There are a number of small enclaves within southwestern Kyrgyzstan that belong to neighboring Uzbekistan or Tajikistan.

The character of Kyrgyzstan is predominantly mountainous. Only about 10 percent of the terrain is below 4,900 ft (1,500 m); and more than 50 percent surpasses 8,200 ft (2,500 m). Some 3 percent of the country is covered by permanent snowfields and glaciers. Indeed, studies estimate that Kyrgyzstan's 6,500 glaciers contain a staggering 850 billion cubic yards (650 billion cubic meters) of water. This abundance of mountain moisture is the source of Kyrgyzstan's many lakes and fast-flowing rivers.

The primary mountain range in Kyrgyzstan is the great Tian Shan, whose peaks, valleys, and basins essentially define the whole republic. In addition the Trans Alai mountains in the south, part of the Pamirs, are also significant. The only land flat enough to be suitable for large-scale agriculture is in the Chu, Talas, and Fergana valleys of the north and east.

Kyrgyzstan's mountains are geologically young, so the peaks are sharply uplifted, separated by deep valleys. An active seismic belt still lifting these mountains upward causes frequent and sometimes devastating earthquakes. The country is located on the Eurasian Tectonic Plate.

GEO-FACT

Kyrgyzstan is frequently subject to many types of natural and man-made disasters. A powerful earthquake in August 1992 left several thousand people homeless in the southwestern city of Jalal-Abad; others over the years have destroyed towns and villages, killing thousands. On the steep mountain slopes, deforestation and overgrazing have set conditions for great mudslides and avalanches, some of which have even inundated entire villages.

MOUNTAINS AND HILLS

Mountains

Kyrgyzstan is a land of high peaks and steep valleys. It lies where two great Central Asian mountain systems, the Tian Shan and the Pamirs, come together. The Tian Shan Mountains run northeast to define the country's eastern border with China; Kyrgyzstan's southern border with Tajikistan follows the Trans Alai Range along the northernmost part of the Pamirs.

The Tian Shan is the largest system of mountains in Asia outside of the Himalayas, and its highest point, Jengish Chokusu, (Pik Pobedy, Victory Peak; 24,406 ft / 7,439 m) is the highest peak in Kyrgyzstan. A series of secondary mountain ranges are considered part of the Tian Shan system. In Kyrgyzstan these include the Ala Tau, running generally east to west across northern Kyrgyzstan. Another chain, the central Fergana Mountains, runs southeast to northwest. The Tian Shan Mountains were uplifted and folded during the Paleozoic era. Deep faulting of their primarily sedimentary rock has been responsible for frequent, severe earthquakes along the rim of this range.

In southern Kyrgyzstan the Trans Alai Range of the Pamirs extends into the country. Mountains in the Trans Alai system are generally about 16,000 ft (4,880 m) in

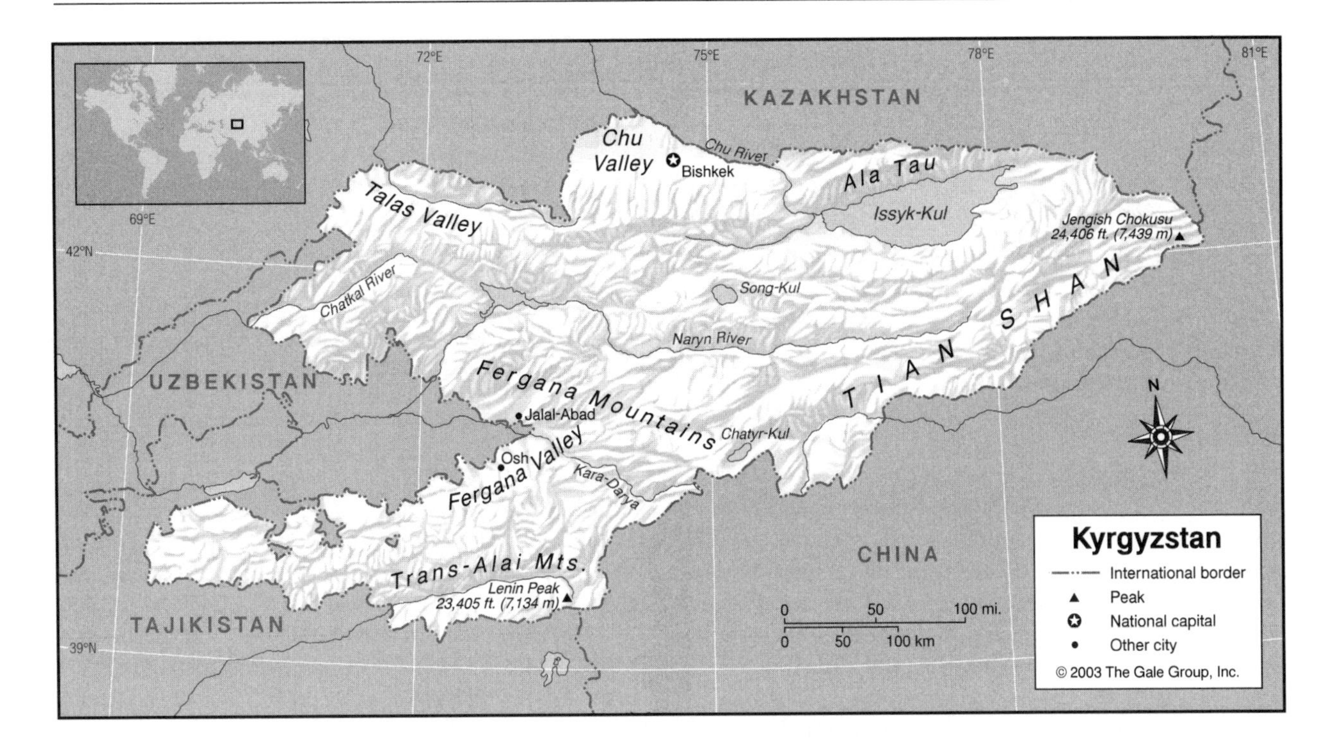

height, but Lenin Peak (Kaufmann Peak) towers at a height of 23,405 ft (7,134 m) on the border with Tajikistan.

INLAND WATERWAYS

Lakes

There are nearly 2,000 lakes in Kyrgyzstan, located in the higher elevations of 9,840 to 13,120 ft (3,000 to 4,000 km). Most are small, but together they have a combined surface area of some 2,703 sq mi (7,000 sq km). Lake Issyk-Kul, however, is nearly this large in itself; at 2,360 sq mi (6,100 sq km) and with a depth of 2,298 ft (700 m), it is Kyrgyzstan's largest lake. Among Central Asian lakes, Issyk-Kul is second in size only to the Aral Sea and, like the Aral Sea, it is saline. A mountain lake, Issyk-Kul is located at 5,273 ft (1,607 m), in the northeastern Tian Shan range. Both climate and human intervention (irrigation) have caused the lake's level to fall in recent years. Some commercial fishing interests operate on the lake year-round, as it never freezes. Two other large lakes, Song-Kul and Chatyr-Kul (which is also saline), lie in the Naryn Basin.

Rivers

The majority of Kyrgyzstan's many rivers are small, fast-flowing runoff streams with origins in the melting snows of the high eastern mountains. Not one of these is navigable, however, not even the country's largest river, the Naryn, which flows west from its origins in the mountains of the northeast, across central Kyrgyzstan, then southward through the Fergana Valley and into Uzbekistan. There it converges with other rivers, to become the great Central Asian Syr Darya. In the north, the Chu River flows northwestward, eventually drying up in the desert country of southern Kazakhstan.

THE COAST, ISLANDS, AND THE OCEAN

A landlocked country, Kyrgyzstan has no seacoasts.

CLIMATE AND VEGETATION

Temperature

Average temperatures vary significantly by region in Kyrgyzstan. The coldest January temperatures are in the mountain valleys, where readings have been known to fall below -22° F (-30°C). (The lowest temperature on record is -65° F /-54°C.) The warmest January average is 25° F (-4°C), near the southern city of Osh and around Lake Issyk-Kul, which never freezes. In July the average temperature varies from 81° F (27°C) in the Fergana Valley (record high of 111° F / 44°C), to 14° F (-10°C) on the high mountain peaks.

Kyrgyzstan's chiefly continental climate is highly determined by its mountains, as well as its location, free of any large body of water, in the center of the Eurasian landmass. Nevertheless, significant local variations do occur. The Fergana Valley is subtropical while the northern foothills are temperate. A dry continental climate, warmed by desert winds from the south, predominates in the lower mountain slopes; a polar climate is the norm for the high mountain elevations.

Rainfall

Like temperatures, precipitation rates, which include snow as well as rainfall, are largely a product of Kyrgyzstan's mountains. Precipitation is greater in the western mountains and less in the flatter, lower regions of north-central Kyrgyzstan.

Average precipitation levels range from 4 to 20 in (100 to 500 mm) in the valleys and 7 to 40 in (180 to 1000 mm) in the mountains. Extremes vary from less than 4 in (100 mm) per year on the west bank of Issyk-Kul to 79 in (2,000 mm) per year in the mountains above the Fergana Valley.

Forests

Only 4 percent of Kyrgyzstan is forest. Conifers predominate in the lower valleys and northern mountain slopes. Kyrgyzstan can boast the world's largest natural growth walnut forest. Deer, mountain goats, and mountain sheep are abundant, but the country's forests also support many rare, protected species like the Tian Shan bear, the red wolf, and the snow leopard. Thousands of migrating birds find refuge in Kyrgyzstan's high mountain lakes.

Population Centers – Kyrgyzstan

(1995 POPULATION ESTIMATES)

Name	Population
Bishkek (capital)	583,900
Osh	218,300
Jalal-Abad	80,000
Tokmak	72,000

SOURCE: "Population of Capital Cities and Cities of 100,000 and More Inhabitants." United Nations Statistics Division.

Oblasts – Kyrgyzstan

1999 CENSUS OF POPULATION

Name	Population	Area (sq mi)	Area (sq km)	Capital
Batken	382,426	6,564	17,000	Batken
Chuy	770,811	7,799	20,200	Bishkek
Jalal-Abad	869,259	13,011	33,700	Jalal-Abad
Naryn	249,115	17,452	45,200	Naryn
Osh	1,175,998	11,274	29,200	Osh
Talas	199,872	4,402	11,400	Talas
Ysyk-Kol	413,149	16,641	43,100	Karakol

SOURCE: National Statistical Committee, Kyrgyz Republic.

HUMAN POPULATION

With a population of 4,753,003 (July 2001 estimate), Kyrgyzstan has an average density of 59 persons per sq mi (23 per sq km). However, two areas comprise the principal population centers: the Chu Valley in the north and the Fergana Valley in the southwest. Not as urbanized as other Central Asian states, Kyrgyzstan has a population of which only 39 percent lives in cities and towns. The capital and largest city is Bishkek (population 662,000), situated on the Chu River in the far north; Osh, in the Fergana Valley, is second in size.

NATURAL RESOURCES

Kyrgyzstan's many fast-flowing rivers are considered one of the country's chief natural resources because they provide an abundance of hydroelectric power. Mineral resources include rare earth metals, small but locally exploitable oil and natural gas, coal, nepheline, uranium (although no longer mined), antimony, mercury, bismuth, lead, and zinc. Kyrgyzstan also has major gold deposits; the largest gold mine of the former Soviet Union was located at Makmal. Other significant deposits, valued in the billions of dollars, have been discovered at Kumtor and Jerui. Most of the country's gold is located in north-central Kyrgyzstan and in the Chatkal River area of the northwest. Although only 4 percent of the total land area is cultivated, with 3,475 sq mi (9,000 sq km) under irrigation, agriculture is the primary occupation of Kyrgyzstan's population

FURTHER READINGS

Anderson, John. *Kyrgyzstan, Central Asia's Island of Democracy.* Amsterdam, Netherlands: Harwood Academic Publishers, 1999.

Kyrgyzstan. Parsippany, N.J.: Dun & Bradstreet, 1999.

Pilkington, John. "Kyrgyzstan: A Tale of Two Journeys," *Geographical Magazine,* 65 (April 1993): 8-12.

Sparks, John. *Realms of the Russian Bear: A Natural History of Russia and the Central Asian Republics.* Waltham, Massachusetts: Little Brown, 1992.

Thomas, Paul. *The Central Asian States: Tajikistan, Uzbekistan, Kyrgyzstan, Turkmenistan.* Brookfield, Conn.: Millbrook Press, 1992.

Laos

- **Area:** 236,800 sq km (91,400 sq mi) / World Rank: 81
- **Location:** Northern and Eastern Hemispheres in Southeast Asia; bordering Burma, China, and Vietnam in the north; Vietnam in the east; Thailand in the west and southwest; Cambodia in the south
- **Coordinates:** 18°00′N, 105°00′E

- **Borders:** 3,151 mi (5,083 km) / Burma, 146 mi (235 km); Cambodia, 335 mi (541 km); China, 262 mi (423 km); Thailand, 1087 mi (1,754 km); Vietnam, 1321 mi (2,130 km)
- **Coastline:** Laos is completely landlocked
- **Territorial Seas:** None
- **Highest Point:** Mt. Bia, 9,252 ft (2,820 m)
- **Lowest Point:** 230 ft (70 m)
- **Longest Distances:** 1,162 km (722 mi) SSE-NNW / 478 km (297 mi) ENE-WSW
- **Longest River:** Mekong River, 2,700 mi (4,350 km)
- **Largest Lake:** Ngum Reservoir, 96 sq mi (250 sq km)
- **Natural Hazards:** floods, droughts
- **Population:** 5,635,967 (2001 est.) World Rank: 101
- **Capital City:** Vientiane, on the Mekong River near the border with Thailand
- **Largest City:** Vientiane, 555,100 (1997 est.)

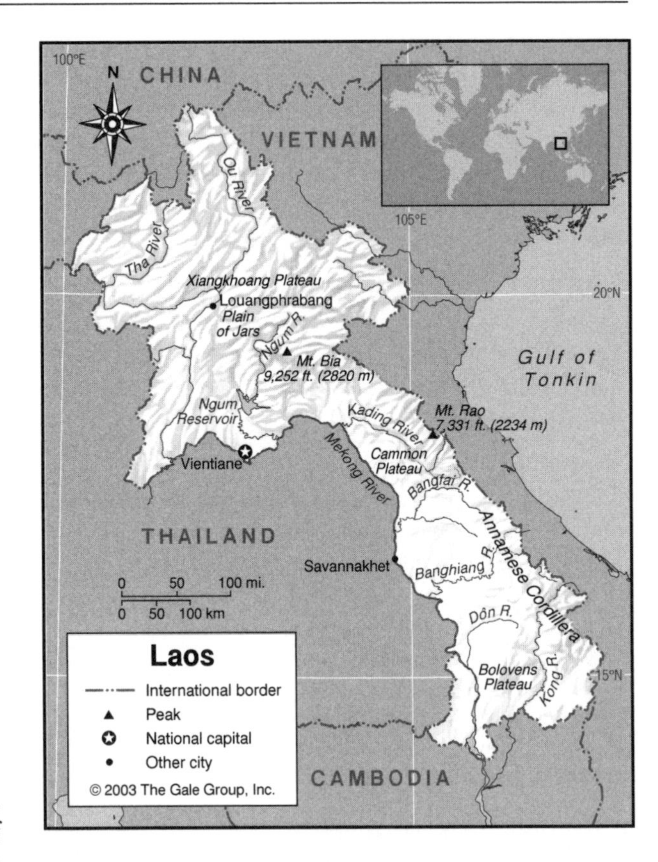

© 2003 The Gale Group, Inc.

OVERVIEW

Laos, the only landlocked Southeast Asian country, lies at the heart of the Indochina Peninsula. It consists of a northern region, centered on the valley of the Mekong River, with a narrower panhandle extending off to the southeast, with the Mekong along the western border. Less than three-fifths of the national territory is contained in the northern section of the country and over two-fifths in the country's southern panhandle. Away from the Mekong, the high mountains of the Annamese Cordillera extend across the country. They have long acted as barriers to communication with the countries lying to the northwest and east. The Mekong River on the west, however, has served as a link with southeastern neighbors. Laos is situated on the Eurasian Tectonic Plate.

MOUNTAINS AND HILLS

Mountains

The northern and northeastern section of the country, north of the Laotian panhandle and away from the Mekong River, is characterized by rugged mountain terrain. The main ranges run northeast to southwest, a continuation of the folded mountains that sweep southward from Tibet, and are sharp crested and steep sloped. Several ranges are around 5,000 ft (1,524 m) in height, and many peaks are well over 6,000 ft (1,829 m). The country's highest mountain, Mount Bia (Phou Bia), rising 9,252 ft (2,820 m) above sea level, is situated here, near the beginning of the panhandle.

The chief topographic feature of the Laotian panhandle is the Annamese Cordillera, which runs along the entire eastern side of this section of the country. The chain parallels the flow of the Mekong River. In its upper portion, mountains resemble those in northern Laos, having rugged peaks and deep valleys; peaks are over 5,000 ft (1,524 m) in elevation, including Mt. Rao (7,331 ft / 2,234 m). Elevations then moderate somewhat to form the Cammon Plateau. South of this the chain enters a limestone region characterized by steep ridges and peaks and sinkholes. This is followed in turn by another a comparatively flat area, the Bolovens Plateau. From this point to the southern end of Laos the chain again becomes very rugged, and its elevations rise to 6,500 ft (1,981 m); the high point is over 7,500 ft (2,286 m).

Plateaus

A prominent feature of the northern part of the country is the Plain of Jars (Thong Hai Hin), located on the Xiangkhoang Plateau. The plateau, lying mostly between about 3,330 and 4,000 ft (1,015 and 1,219 m) above sea level, has relatively infertile soils. The plain takes its name from large stone jars of ancient origin that have been found there. The Phouane Plateau is another significant region of plateau land in northern Laos.

At the neck of the panhandle section the Annamese Cordillera is buttressed by several plateaus, including the Cammon and Nakai Plateaus. About one quarter of the Nakai Plateau is scheduled to be flooded by the Nam Theun II dam project. A mixture of forest, grassland and wetland, the Nakai Plateau is a habitat for endangered elephants, tigers, and clouded leopards as well as rare bird species. From the plateaus the land slopes more gently westward to the alluvial plains along the Mekong. Prominent in the southern part of the country is the fertile Bolovens Plateau. Almost encircled by a high escarpment, this plateau has an elevation of about 3,500 ft (1,067 m).

Canyons

When the Mekong River enters Laos, it runs through steep limestone gorges north of the city of Louangphrabang. The Hin Boon River in central Laos cuts through narrow limestone canyons.

Hills and Badlands

Laos has several areas of karst limestone hill formations. Vangvieng is a region in northwest Laos with karst foothills pockmarked by caves and tunnels. Nam Phoun is a hill district National Biodiversity Area, in northeast Laos near the border of Vietnam. Some 444 sq mi (1,150 sq km) in area, Nam Phoun is one of the last habitats for the Sumatran rhinoceros and wild elephant herds.

INLAND WATERWAYS

Lakes

Laos boasts few lakes. The largest by far is Ngum reservoir, 96 sq mi (250 sq km) which was created by the Nam Ngum hydropower dam, which flooded a forest area. Underwater logging takes place in the reservoir, which is dotted with many islands. Many bomb craters from the United States' massive aerial bombardment of Laos during the 1960s and 1970s have filled with water, becoming ponds.

Rivers

The Mekong River and its tributaries drain almost all of Laos. Only a few small rivers in the east flow into Vietnam and the Pacific Ocean. The Mekong flows through Laos for 1,122 mi (1,805 km) and is the center of its economic life. Its flood plains create the major wet-rice lands, its waters provide fish, and it transports the country's freight.

In the northern part of the country the course of the Mekong is characterized by narrow valleys and deep gorges. This is the only part of the country where the river is entirely within Laos's borders. Downstream from Vientiane the river forms the border with Thailand; it is also here that the valley widens and alluvial plains are found on each side. The river is navigable from Louangphrabang to Savannakhet. Downstream from Savannakhet there are many rapids and navigation is only possible at high water. The Kong falls at the extreme south are completely impassable.

The Mekong's tributaries in the north include the Tha and the Ou Rivers, which flow in deep, narrow valleys, as well as the Ngum River, which enters the Mekong east of Vientiane through an alluvial plain. In the south the main tributaries are the Kading, Bangfai, Banghiang, and Dôn Rivers. All of these rivers flow west out of the mountains, forming fairly wide and fertile valleys. Another large tributary, the Kong, flows south from Laos into Cambodia before joining the Mekong. During September-October 2000 disastrous floods affected Laos, Cambodia, and Vietnam. United Nations agencies determined that the main cause of the flooding was deforestation in the upper Mekong region.

THE COAST, ISLANDS AND THE OCEAN

Laos is a land-locked nation. The closest sea is the Gulf of Tonkin of the Pacific Ocean, more than 31 mi (50 km) to the east across Vietnam.

CLIMATE AND VEGETATION

Temperature

Laos has a tropical monsoon climate with three seasons: a cool dry season in November-February, a hot dry season in March-April, and a rainy season in May-October. Temperatures average 82°F (28°C), ranging from highs of 104°F (40°C) along the Mekong in March-April to lows of 41°F (5°C) in the mountains in January. Humidity averages 70 to 80 percent.

Rainfall

Annual rainfall in Laos averages 69 in (175 cm). Most of this rain occurs during the southwest monsoon between May and October. Rainfall can be anywhere from 50 to 90 in (127 to 229 cm) during this period.

Grasslands

Laos's most fertile flatlands are found in the valleys and flood plains of the Mekong and its tributaries. These are most extensive near the capital of Vientiane, and all are extensively cultivated. Away from the river there are the Xiangkhoang, Cammon, Nakai, and Bolovens plateaus. Of these, only the last three are particularly fertile. The Nam Ha National Protected Area, in the highlands of northern Laos, combines savanna and mountain forest, home to elephants, tigers, and many bird species.

Forests and Jungles

Much of northern Laos is still covered with evergreen rainforest, although accessible areas are increasingly being reduced to secondary growth or completely clear cut. In southern Laos there is a mixture of secondary evergreen rainforest and deciduous monsoon forest. Deforestation is relentless throughout the country, brought about by legal and illegal logging, slash and burn agriculture, firewood and charcoal use, and forest fires. Efforts to control the loss of Laotian forests include government promotion of irrigated lowland rice growing in

Population Centers – Laos

(1996 POPULATION ESTIMATES)

Name	Population
Vientiane (capital)	180,000
Savannakhét	80,000
Pakxé	50,000
Louangphrabang	30,000

SOURCE: Projected from United States Agency for International Development data.

Provinces – Laos

Name	Area (sq mi)	Area (sq km)	Capital
Attapu	3,985	10,320	Attapu
Bokèo	2,392	6,196	Ban Houayxay
Bolikhamxai	5,739	14,863	Muang Pakxan
Champasak	5,952	15,415	Pakxé
Houaphan	6,371	16,500	Sam Neua
Khammouan	6,299	16,315	Thakhek
Louangnamtha	3,600	9,325	Louangnamtha
Louangphabang	6,515	16,875	Louangphabang
Oudômxai	5,934	15,370	Muang Xay
Phôngsali	6,282	16,270	Phôngsali
Salavan	4,128	10,691	Salavan
Savannakhét	8,407	21,774	Savannakhét
Vientiane	6,149	15,927	Muang Phôn-Hông
Xaignabouli	6,328	16,389	Muang Xayabury
Xékong	2,959	7,665	Ban Phone
Xiangkhoang	6,131	15,880	Xiangkhoang

SOURCE: *Geo-Data: The World Geographical Encyclopedia,* 2nd ed. Detroit: Gale Research, 1989.

place of shifting slash and burn agriculture; sometimes this is attempted through forced resettlement of highland farming communities.

HUMAN POPULATION

Roughly 80 percent of Laotians live in rural areas. Although the population density of Laos was a very low 54 per sq mi (21 people per sq km) in 2002, the living space is severely limited by terrain, since much of the country is too mountainous for farming. The Laotian government has warned that with the high growth rate of 2.8 percent (2001 estimate), the country's population could double to around ten million by 2025, causing an extreme strain on natural resources and the land available for cultivation.

NATURAL RESOURCES

The resources of Laos include hydropower, tin, gypsum, coal, sapphires, copper, gold, iron, agriculture, river fish, timber, and other forest products (botanical medicines, rattan, resins). Laos is a very poorly developed country, however, with no railroads and little electricity. Most of the population is engaged in subsistence agriculture.

FURTHER READINGS

Eliot, Joshua, and Bickersteth, Jane. *Footprint Laos Handbook.* Bath, UK: Footprint Handbooks, 2000.

International Rivers Network. *IRN's Mekong Campaign.* http://www.irn.org/programs/mekong/ (accessed April 11, 2002).

Lao Embassy. *Discovering Laos.* http://www.laoembassy.com/discover/ (accessed April 11, 2002).

Savada, Andrea Matles, ed. *Laos: A Country Study.* Washington DC: Library of Congress, 1996.

Stuart-Fox, Martin. *A History of Laos.* Cambridge, UK: Cambridge University Press, 1997.

Latvia

- **Area:** 24,938 sq mi (64,589 sq km) / World Rank: 124
- **Location:** Northern and Eastern Hemispheres, eastern Europe, east of the Baltic Sea, south of Estonia, north of Lithuania, west of Russia
- **Coordinates:** 57°00′N, 25°00′E
- **Borders:** 713 mi (1,150 km) / Belarus, 88 mi (141 km); Estonia, 211 mi (339 km); Lithuania, 281 mi (453 km); Russia, 135 mi (217 km)
- **Coastline:** 330 mi (531 km)
- **Territorial Seas:** 12 NM
- **Highest Point:** Gaizinkalns, 1,024 ft (312 m)
- **Lowest Point:** Sea level
- **Longest Distances:** 131 mi (210 km) N-S / 281 mi (450 km) E-W
- **Longest River:** Daugava River, 632 mi (1,020 km)
- **Natural Hazards:** None
- **Population:** 2,385,231 (July 2001 est.) / World Rank: 138
- **Capital City:** Riga, north-central Latvia
- **Largest City:** Riga, 921,000 (2000 est.)

OVERVIEW

Along with Estonia and Lithuania, Latvia is one of the Baltic states of northeastern Europe. Its capital, chief seaport, and largest city is Riga, which is found on the shores of the Gulf of Riga, a deep indentation in the country's northern coast. Approximately three-fourths of Latvia is an undulating plain cultivated for farming, part of the

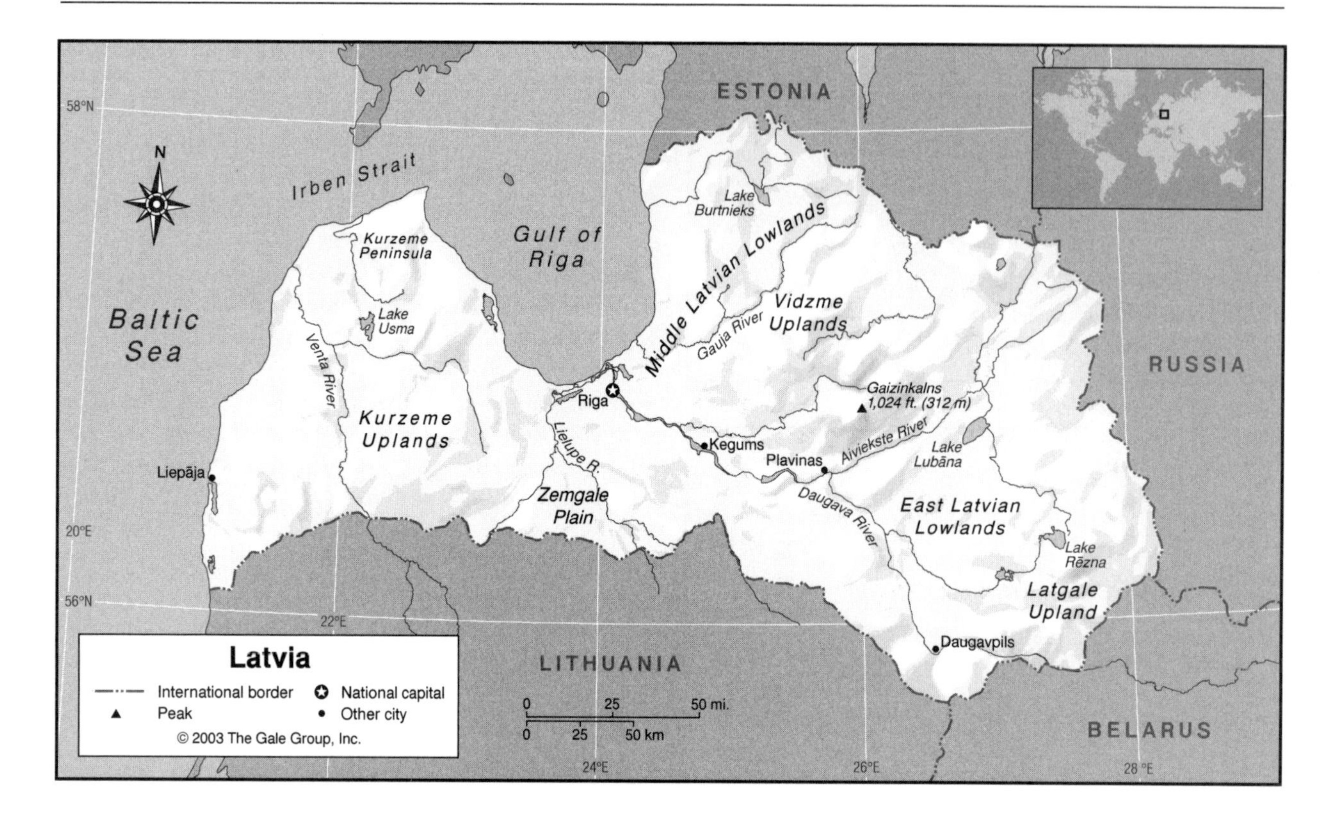

vast European Plain. An estimated 10 percent of Latvia lies below sea level. The remaining fourth of the country consists of uplands with moderate-sized hills, also used for farming.

The Latvian landscape, with its mounded hills (moraines) and lakes, streams, marshes, and peat bogs, was formed by continental glaciers during the Quaternary period and the Pleistocene ice age. The nation is located on the Eurasian Tectonic Plate.

MOUNTAINS AND HILLS

The Latvian terrain is primarily a low-lying plain, and there are no true mountains. The only relief are three upland regions consisting of hills formed by glacial activity. The Kurzeme Uplands lie in the west, and are split into eastern and western portions by the Venta River. The highest elevation in the country, Gaizinkalns (1,023 ft / 312 m), is found in Vidzeme Uplands, east of the Gulf of Riga. This upland is the single largest area more than 660 ft (200 m) above sea level in the Baltic region. Further south and east is the Latgale Upland.

INLAND WATERWAYS

Lakes

Latvia has a wet climate and contains many lakes both large and small, particularly in the southeast. Major lakes include Usma, in the west; Burtnieks, in the north-central area; and Lakes Lubāna and Rāzna in the east.

Rivers

Latvia's largest river, the Daugava (called the Dvina in neighboring Belarus), is one of the most important rivers of the Baltic region. With headwaters in Russia, the Daugava flows into Belarus and continues northwest through Latvia, finally emptying into the Gulf of Riga. Its total length is 632 mi (1,020 km). At its mouth the river is nearly a mile wide (1.5 km).

With a total descent of 321 ft (98 m), the Daugava is an excellent source of hydroelectric power. Dams have formed reservoirs at Kegums, Plavinas, and near Riga.

Lesser but still significant Latvian rivers include the Venta, in the west; the Lielupe, in central Latvia; the Gauja, in the northeast; and the Aiviekste, in the east. Although some dams have been constructed on these rivers, they account for very little of the country's total hydroelectric power production.

Wetlands

Peat bogs, swamps, and marshes are found throughout the country, and cover about 10 percent of the Latvian terrain. Stunted forest growth is the primary feature.

THE COAST, ISLANDS, AND THE OCEAN

Along the Baltic Sea, the Latvian coastline runs uninterrupted until the Gulf of Riga juts into it on the north, where it forms the Kurzeme Peninsula on the western side. Altogether the Latvian coast runs 329 mi (531 km).

It is known as a beautiful coastline, with many sandy beaches.

The Gulf of Riga is shared by Latvia and Estonia. Its north-south measurement is about 90 mi (145 km), and east-west it ranges from 45 to 80 mi (72 to 129 km). The capital city and port of Riga is located near the mouth of the Daugava River, which empties into the Gulf. The western entrance to the Gulf is the Irben Strait, between the Kuzeme Peninsula and Estonia's Saaremaa Island.

Population Centers – Latvia

(1994 POPULATION ESTIMATES)

Name	Population
Riga (capital)	837,976
Daugavpils	121,063
Liepāja	102,450

SOURCE: "Population of Capital Cities and Cities of 100,000 and More Inhabitants." United Nations Statistics Division.

CLIMATE AND VEGETATION

Temperature

Summers in Latvia are generally cool, but winters are mild. Atlantic Ocean air masses influence the country's moderate, maritime climate, as well as its annual precipitation, which is fairly high. January temperatures range from 31°F (-2.8°C) in Liepaja, on the western coast, to 44°F (6.6°C) in Daugavpils in the southeast. In July they range from 62°F (16.7°C) in Liepāja to 64°F (17.6°C) in Daugavpils.

Rainfall

Latvia's coastal climate means cloudiness, high humidity, and precipitation most of the year. Only 72 days see sun, 44 are foggy, and it rains or snows 180 days on average. Measured in Riga, annual precipitation is ranges between 22 and 31 in (56 and 79 cm).

Grasslands

Most of Latvia is low, level terrain, part of the European Plain. It is largely suitable for farming, but the heavy annual precipitation means that much of Latvia's agricultural land requires drainage. The most fertile and profitable area is the central Zemgale Plain south of Riga. Other lowlands include the Middle and the East Latvian Lowlands, and coastal lowlands. Large parts of all of these lowlands are covered by forest.

Forests

Forty-six percent of Latvia consists of forests and woodlands of pine, spruce, aspen, and birch; lumber and wood products are important Latvian exports. Blueberries, mushrooms, and cranberries grow in abundance on the forest floors. However, few forests are mature, due to unmanaged cutting and violent storms that have destroyed Latvia's trees by the millions. Nevertheless, the country supports many thriving species of wildlife, including elk, deer, moose, wild boar, and fox; also wolves, lynx, beaver, otter, black storks, and eagles. The coast has a significant population of seals.

HUMAN POPULATION

With a population of 2,385,231 (July 2001 estimate), Latvia has a population density of 96 persons per sq mi (37 per sq km). Nearly 73 percent of Latvian residents live in cities or towns. More than a third live in the capital, Riga, near the Gulf of Riga. Among other important cities are Daugavpils in the southeast, an industrial center; and Liepaja on the western Baltic seacoast, a major seaport. Many other urban population centers lie along the country's rivers.

NATURAL RESOURCES

Latvia is poor in mineral resources. The most significant deposits in the country are of peat (a carbon-rich material used as fuel and mulch); limestone, gypsum, sand, gravel, and clay (all used in construction); amber; and dolomite. Large amounts of timber are available in Latvia's forests, however, and it also has a great deal of arable land.

FURTHER READINGS

Dreifelds, Juris. "Two Latvian Dams: Two Confrontations," *Baltic Forum*, 6, No. 1, Spring 1989, p. 11-24.

Lieven, Anatol. *The Baltic Revolution: Estonia, Latvia, Lithuania, and the Path to Independence.* New Haven: Yale University Press, 1993.

Plakans, Andrejs. *The Latvians: A Short History.* Stanford, California: Stanford University, 1995.

United States. National Technical Information Service. *Latvia: An Economic Profile.* Washington: August 1992.

Lebanon

- **Area:** 4,015 sq mi (10,400 sq km) / World Rank: 163
- **Location:** Northern and Eastern Hemispheres, in the Middle East, bordering Syria to the north, east, and southeast, Israel to the south, and the Mediterranean Sea to the west.
- **Coordinates:** 33°50′N, 35°50′E

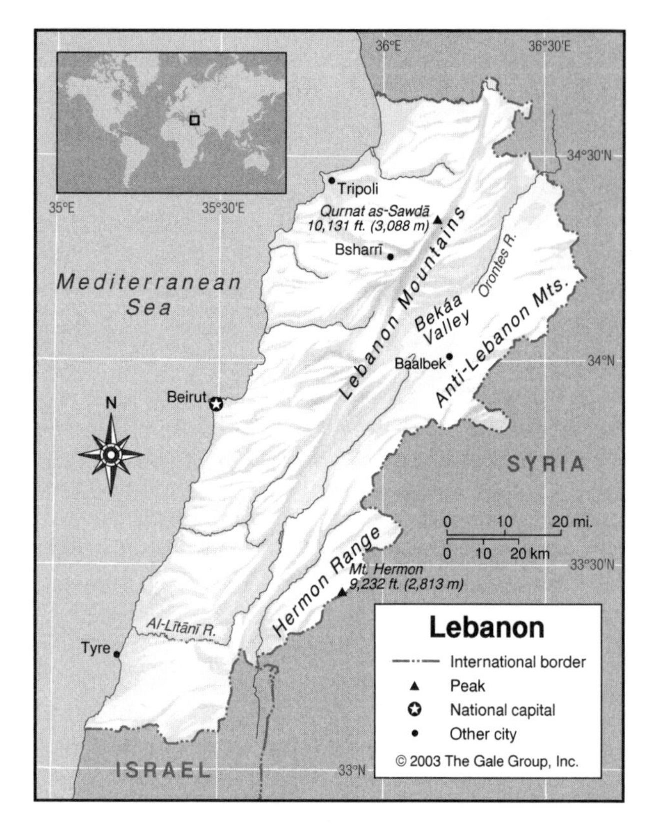

- **Borders:** 282 mi (454 km) / Israel, 49 mi (79 km); Syria, 233 mi (375 km)
- **Coastline:** 140 mi (225 km)
- **Territorial Seas:** 12 NM
- **Highest Point:** Qurnat as-Sawdā, 10,131 ft (3,088 m)
- **Lowest Point:** Sea level
- **Longest Distances:** 135 mi (217 km) NE-SW; 35 mi (56 km) SE-NW
- **Longest River:** Al-Līṭānī, 90 mi (145 km)
- **Natural Hazards:** Sandstorms and duststorms
- **Population:** 3,627,774 (July 2001 est.) / World Rank: 124
- **Capital City:** Beirut, on central Mediterranean Coast
- **Largest City:** Beirut, population 2,058,000 (2000 est.)

OVERVIEW

Lebanon is a small Middle Eastern country on the eastern coast of the Mediterranean Sea. It is mostly mountainous, and the dominant topographical feature is a central mountain range spanning most of the country's length and reaching almost to the coast. In addition to this range there are three other distinct geographical regions: a narrow coastal plain; a second mountain system in the east, on the border with Syria; and the Bekáa Valley, which separates the coastal and interior mountains. Despite the small size of the country, these regions are well demarcated and are characterized by differences in climate and, to a certain extent, religious communities and ethnicity. The Bekáa Valley belongs to the same geological rift that continues southward into the Jordan River Valley and the Great Rift Valley of eastern Africa. Lebanon itself is situated near the dividing line between the African and Arabian Tectonic Plates.

MOUNTAINS AND HILLS

Mountains

The major mountain range, called the Lebanon Mountains, extends about 100 mi (161 km) northeast to southwest, or nearly the entire length of the country. Its width ranges from 6 to 35 mi (10 to 56 km) as it runs down the center of the country. The peaks rise rapidly from the coast, reaching their highest elevations in the northern part of the country and gradually decreasing in elevation as they extend southward. Lebanon's highest point, Qurnat as-Sawdā (10,131 ft / 3,088 m), is situated in the northern part of this range.

To the east, Lebanon's border with Syria is demarcated by a second mountain system composed of two different ranges, the Anti-Lebanon Mountains to the north and the Hermon Range to the south. The mountains in these ranges are generally lower than those of the Lebanon Range to the west, although Mt. Hermon, which rises to 9,232 ft (2,813 m), is the country's second-highest peak.

Plateaus

The fertile Bekáa Valley separates Lebanon's two parallel mountain systems, reaching maximum elevations of around 3,000 ft (914 m). At 110 mi (180 km) long, it extends the entire length of the Lebanon Mountains, with widths varying from 6 to 16 mi (10 to 26 km), and constitutes the country's greatest expanse of essentially level terrain. The depression it occupies, which is a northern extension of the Great Rift Valley, continues farther south as the Jordan River Valley.

INLAND WATERWAYS

Rivers

Lebanon has few year-round rivers. Its most important, and longest, river is the Al-Līṭānī, which rises at the watershed of the Bekáa Valley, near Baalbek, and flows southwest and then westward, draining into the Mediterranean near the city of Tyre. Another major river is the Orontes, which flows through the northern Bekáa Valley into Syria and then Turkey before emptying into the Mediterranean. There are many smaller rivers originating in the Lebanon Mountains and running west to the Mediterranean Sea.

THE COAST, ISLANDS, AND THE OCEAN

Lebanon is located on the eastern shore of the Mediterranean Sea. The coastline is relatively smooth with no major indentations. It has few good natural harbors but has, instead, many shallow, curved bays. The cities of Beirut, Tripoli, and Tyre are all located at spots where the coastline juts out into the sea, with Beirut occupying the largest one. The northern part of the coast is mostly rocky; south of Beirut, it becomes sandy in places.

CLIMATE AND VEGETATION

Temperature

Lebanon has a subtropical, temperate Mediterranean climate, with hot, dry summers and cool, humid winters. Temperatures rarely exceed 90°F (32°C). Average temperatures in Beirut are 82°F (28°C) in the summer and 55°F (13°C) in the winter. Temperatures are expectedly cooler in the mountains.

Rainfall

Average annual rainfall ranges from about 15 in (38 cm) in the Bekáa Valley, to 35 in (89 cm) on the coast, to more than 50 in (127 cm) the mountains. Four-fifths of the annual rainfall occurs in the winter months, between November and March.

The peaks of the Lebanon Mountains experience heavy winter snows and are snow covered from winter to spring.

Forests and Jungles

Much of the country's original forestland has been lost over time, including most of the famous cedar trees referred to in the Bible. The oldest surviving cedars are found near Bsharrī in the northern part of the Lebanon Mountains. Other trees found in Lebanon include olive and fig trees at lower elevations and oak, fir, juniper, and cypress on the higher mountain slopes. Tamarisks grow on the coastal plains.

HUMAN POPULATION

Most of Lebanon's population is concentrated along the coastal plain, where the major cities are located. Since the displacements caused by the civil war of the 1970s and 1980s, Christians and Muslims are more segregated by region that they were formerly, with Christians concentrated in the north and Muslims in the south. The overall population density is 901 people per sq mi (348 people per sq km), but more than half of the population lives in Beirut.

GEO-FACT

The name *Lebanon* comes from the Arabic name for the Lebanon Mountains, *Djebel Libnan,* which means "milky-white mountains" (a reference to its snow-covered peaks).

NATURAL RESOURCES

Lebanon's mineral resources include salt, limestone, silica, and gypsum. The mountains are a good source of iron ore. Along with its Mediterranean climate, Lebanon's rainfall, which is above average for a Middle Eastern country, provides important water resources for irrigation and the growth of olives, tobacco, and various fruits and vegetables; most are grown in the coastal plain region. The unusually high rainfall also provides Lebanon with water resources for itself and other countries in the region that are deficient.

FURTHER READINGS

Friedman, Thomas L. *From Beirut to Jerusalem.* New York: Farrar, Straus, Giroux, 1989.

Haag, Michael. *Syria and Lebanon.* Cadogan Guides. Old Saybrook, Conn.: Globe Pequot Press, 1995.

Malik, Habib C. *Between Damascus and Jerusalem: Lebanon and Middle East Peace.* Washington, D.C.: Washington Institute for Near East Policy, 2000.

Reid, Carlton, Kathryn Leigh, and Jamie Kennedy. *Lebanon, A Travel Guide.* Newcastle upon Tyne, England: Kindlife, 1995.

Salibi, Kamal S. *A House of Many Mansions: The History of Lebanon Reconsidered.* Berkeley: University of California Press, 1988

Timerman, Jacobo. *The Longest War: Israel in Lebanon.* Trans. Miguel Acoca. New York : Knopf, 1982.

ArabNet. http://www.arab.net/lebanon/lebanon_contents.html (Accessed April 9, 2002).

Lesotho

- **Area:** 11,720 sq mi (30,355 sq km) / World Rank: 140
- **Location:** Eastern and Southern Hemispheres, surrounded by South Africa.
- **Coordinates:** 29°30′S, 28°30′E
- **Borders:** 565 mi (909 km), all with South Africa

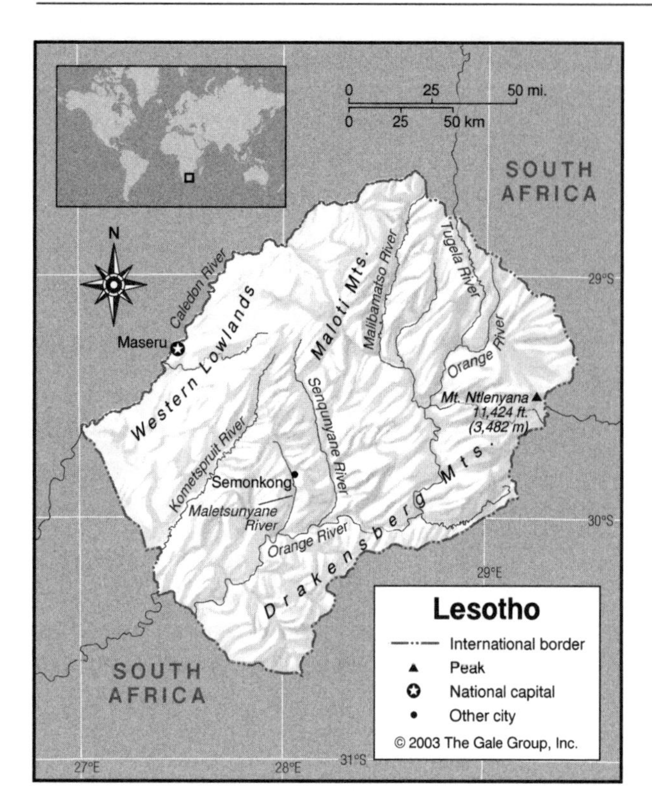

- **Coastline:** None
- **Territorial Seas:** None
- **Highest Point:** Mount Ntlenyana, 11,424 ft (3,482 m)
- **Lowest Point:** Junction of the Orange and Kometspruit rivers, 4,593 ft (1,400 m)
- **Longest Distances:** 154 mi (248 km) NNE-SW; 112 mi (181 km) ESE-WNW
- **Longest River:** Orange River, 1,400 mi (2,250 km)
- **Natural Hazards:** Drought
- **Population:** 2,177,062 (July 2001 est.) / World Rank: 139
- **Capital City:** Maseru, western Lesotho
- **Largest City:** Maseru, population 397,000 (2000 est.)

OVERVIEW

Lesotho, situated on the African Tectonic Plate, is one of the smallest countries in Africa, and one of only three sovereign nations in the world to be completely surrounded by another country (the other two similar enclaves are San Marino and the Vatican). It borders the South African provinces of KwaZulu/Natal to the east, Eastern Cape to the south, and Orange Free State to the north and west. Located on the Drakensburg Escarpment, which forms its eastern border with the KwaZulu province in South Africa, Lesotho is mostly mountainous. Even its "lowlands," a strip of land lying lengthwise along its NE-SW border, have an average elevation of 5,000 to 6,000 ft (1,524 to 1,829 m).

MOUNTAINS AND HILLS

Mountains

Mountains cover two-thirds of Lesotho, giving it the nickname "the Switzerland of Africa." Lesotho's highlands are part of the Drakensburg Mountains, which rise in the east and then drop abruptly at the border with South Africa. The Maloti Mountains, in the center of the country, constitute a spur of the Drakensburg system, joining it in the north. The average elevation of the highlands is over 8,000 ft (2,438 m). They rise to heights of over 10,000 ft (3,048 m) in the east and northeast, reaching their highest point at Mount Ntlenyana on the eastern border.

Plateaus

The northern area where the Maloti Mountains join the Drakensburg system consists of a high plateau with average elevations between 8,900 and 10,500 ft (2,700 and 3,200 m).

Hills and Badlands

The Cave Sandstone Foothills constitute an intermediate region between the highlands and the lowlands, with an average altitude of between 6,000 and 7,000 ft (1,829 and 2,134 m). These foothills form a narrow band bordering the Eastern Highlands of the Drakensburg and Maloti Mountains, which cover two-thirds of Lesotho.

INLAND WATERWAYS

Most of Lesotho is drained by the Orange River and the Caledon, which forms the country's western border. The Orange and Tugela Rivers, as well as the tributaries of the Caledon, rise in the northern plateau region where the Maloti Mountains merge with the main Drakensburg Range. Three other important rivers flow in a north to south direction and feed into the Orange. The Kometspruit (Makhaleng) is in western Lesotho, the Senqunyane flows through the center of the country, and the Malibamatso run through northeastern Lesotho. The Maletsunyane River is notable for the Maletsunyane Falls, located in Semonkong. The drop is some 630 ft (192 m), making it the tallest single-drop waterfall in southern Africa.

THE COAST, ISLANDS, AND THE OCEAN

Lesotho is landlocked.

GEO-FACT

Lesotho is the only country in the world whose lowest elevation is more than 3,281 ft (1,000 m) above sea level.

CLIMATE AND VEGETATION

Temperature

Lesotho has a dry, temperate climate, with mean temperatures of 70°F (21°C) in summer (November to January) and 45°F (7°C) in winter (May to July). Extremes range from 90°F (32°C) to 20°F (–7°C) in the lowlands, with winter temperatures in the highlands sometimes plummeting below 0°F (–18°C). On average, there are more than 300 sunny days per year.

Rainfall

Annual rainfall ranges from 24 in (60 cm) in the lowlands to 75 in (191 cm) in the mountains. Most rain falls between October and April. Lesotho is prone to damaging hail in the summer and periodic disastrous drought. The Maloti Mountains are generally snowcapped in winter.

Grasslands

Occupying roughly a quarter of the country's land area, the Western Lowlands extend eastward from the Caledon River to the Cave Sandstone Foothills. They consist of undulating basins and plains ranging in width from 6 to 40 mi (10 to 64 km), with altitudes averaging between 5,000 and 6,000 ft (1,524 and 1,829 m).

Vegetation

Due to erosion, overgrazing, and overcultivation, Lesotho has sparse vegetation other than grassland, and very few trees. Native tree species include the Cape willow and the wild olive. Subalpine grasses are found on the high northern plateau, and red oat grass in the Cave Sandstone Foothills.

HUMAN POPULATION

More than two-thirds of the population lives in the Western Lowlands, which is the country's major agricultural region. Most live in villages or small towns. An estimated 28 percent of the population lives in urban areas.

NATURAL RESOURCES

Lesotho is poor in natural resources and highly dependent on South Africa economically. Its limited mineral resources include alluvial diamonds, and the hydroelectric potential provided by the rivers that rise in its mountains. Lesotho's primary natural resource is water, of which it has plenty. Since 1998, Lesotho has been exporting water to South Africa, and using it within the country to promote subsistence farming and livestock grazing in the non-mountainous regions. Besides livestock and leather, corn, wheat, and barley are among Lesotho's exports.

Districts – Lesotho

1995 POPULATION ESTIMATES

Name	Population	Area (sq mi)	Area (sq km)	Capital
Berea	206,200	858	2,222	Teyateyaneng
Butha-Buthe	135,400	682	1,767	Butha-Buthe
Leribe	349,500	1,092	2,828	Hlotse
Mafeteng	259,000	818	2,119	Mafeteng
Maseru	400,200	1,652	4,279	Maseru
Mohales Hoek	231,300	1,363	3,530	Mohales Hoek
Mokhotlong	100,300	1,573	4,075	Mokhotlong
Qachas Nek	86,800	907	2,349	Qachas Nek
Quthing	151,900	1,126	2,916	Quthing
Thaba-Tseka	136,200	1,649	4,270	Thaba-Tseka

SOURCE: *Geo-Data,* 1989 ed., and Johan van der Heyden, Geohive, http://www.geohive.com (accessed June 2002).

FURTHER READINGS

Baedeker South Africa. New York: Macmillan Travel, 1996.

Gay, John, Debby Gill, and David Hall, eds. *Lesotho's Long Journey: Hard Choices at the Crossroads.* Maseru, Lesotho: Sechaba Consultants, 1995.

Mbendi Information Services. *Lesotho.* http://www.mbendi.co.za/land/af/le/p0005.htm (Accessed May 30, 2002).

Murray, Jon, and Jeff Williams. *South Africa, Lesotho & Swaziland.* London: Lonely Planet, 2000.

Tonsing-Carter, Betty. *Lesotho.* New York: Chelsea House, 1988.

Turco, Marco. *Visitors' Guide to Lesotho: How to Get There, What to See, Where to Stay.* Johannesburg: Southern Book Publishers, 1994.

Liberia

- **Area:** 43,000 sq mi (111,370 sq km) / World Rank: 103
- **Location:** Northern and Western Hemispheres, in Western Africa, bordering Guinea to the north, Côte d'Ivoire to the east, the Atlantic Ocean to the south and southwest, and Sierra Leone to the northwest.
- **Coordinates:** 6°30′N, 9°30′W
- **Borders:** 985 mi (1,585 km) / Guinea, 350 mi (563 km); Côte d'Ivoire, 445 mi (716 km); Sierra Leone, 190 mi (306 km)

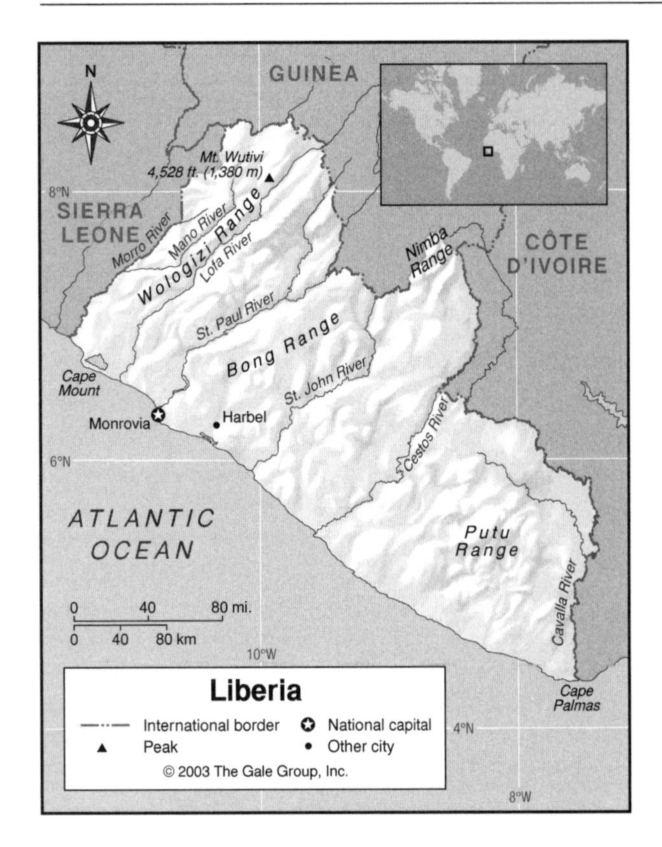

- **Coastline:** 360 mi (579 km)
- **Territorial Seas:** 200 NM
- **Highest Point:** Mt. Wutivi, 4,528 ft (1,380 m)
- **Lowest Point:** Sea level
- **Longest Distances:** 170 mi (274 km) NNE-SSW; 341 mi (548 km) ESE-WNW
- **Longest River:** Cavalla, 320 mi (520 km)
- **Natural Hazards:** Dust storms due to harmattan winds blowing from the Sahara
- **Population:** 3,225,837 (July 2001 est.) / World Rank: 131
- **Capital City:** Monrovia, northwestern Liberia
- **Largest City:** Monrovia, 1,413,000 (2000 est.)

OVERVIEW

Liberia, Africa's oldest republic, is located at the western edge of the continent, on the Atlantic coast between Sierra Leone and Côte d'Ivoire. Starting from a coastal plain that is 25 mi (40 km) wide, the terrain gradually rises through two more major geographical regions: a belt of forested hills, and beyond it an upland region of plateaus and low mountains that reach their highest elevations at several places hear the northern border as they approach the Guinea Highlands.

GEO-FACT

Liberia's coast was traditionally referred to as the Grain Coast, a reference to the "Grains of Paradise," or malagueta peppers, that initially attracted European traders.

MOUNTAINS AND HILLS

Mountains

There are scattered mountain ranges in Liberia's upland plateau region, including the Putu Range in the southeast, the Bong Range near the center of the country, and the Wologizi and Nimba ranges in the north. The highest point in the country, Mt. Wutivi in the Wologizi Range, rises to 4,528 ft (1,380 m).

Plateaus

Beyond Liberia's coastal plain and forested hills lies a rolling plateau broken abruptly by low mountains constituting spurs of the Guinea Highlands. This belt of high ground, which extends over the Liberian border into Guinea, forms part of a major watershed between streams that flow across Liberia to the Atlantic Ocean and the great Niger River basin on the northeast. Ranging in elevation from 1,000 ft (305 m) to over 4,000 ft (1,219 m) in the high northern uplands, Liberia's inland plateau region is the country's largest geographical division.

Hills and Badlands

Between the coastal plain and the interior plateau is a band of heavily wooded, hilly country about 20 mi (32 km) wide, with elevations of between 200 and 500 ft (60 to 150 m).

INLAND WATERWAYS

Rivers

Most of Liberia's rivers flow in roughly parallel courses from the interior plateau to the ocean. Several, including the Lofa, the St. Paul, and the St. John, rise in the Guinea Highlands north of the border with Guinea. The Mano and Morro rivers to the west form parts of the border with Sierra Leone, while to the east the Cavalla River, rising in the Nimba Mountains and flowing south and southeast, forms more than half of the border with Côte d'Ivoire. The St. Paul River forms part of the border with Guinea. The Cestos River flows out of Côte d'Ivoire and nearly bisects Liberia. However, even with so many rivers, navigation is severely limited by rapids, waterfalls, and other barriers.

THE COAST, ISLANDS, AND THE OCEAN

Oceans and Seas

Liberia is bordered on the west and southwest by the Atlantic Ocean. Since the country is only a few degrees north of the equator, it is also near the dividing point between the North Atlantic and South Atlantic oceans. The surf is normally heavy all along the coast but is worst at the height of the rainy season. The tidal range is moderate.

The Coast and Beaches

The coastal region, a belt of gently rolling low plains extending 20 to 30 mi (32 to 48 km) inland, is broken along the shore by river estuaries, tidal creeks, and swamps, as well as a few prominent, rocky capes and promontories. In the northwest, not far from the border with Sierra Leone, Cape Mount rises steeply from the sea to an elevation of over 1,000 ft (305 m), overlooking a broad tidal lagoon. Cape Mesurado, the site of Monrovia, and the lagoon that lies behind it are similar features on a smaller scale. Farther to the southeast, several other fairly prominent headlands break the monotony of the low shoreline. Its southeast point on the border with Côte d'Ivoire is marked by Cape Palmas. The rather straight, sandy shoreline is only slightly indented by the mouths of rivers that are so obstructed by shifting bars, submerged rocks, and sandpits that they provide no natural harbors.

CLIMATE AND VEGETATION

Temperature

Liberia has a hot, humid tropical climate with little seasonal variation, although temperatures are cooler in the interior highlands than along the coast. The mean temperature is 81°F (27°C). The tropical heat is tempered by ocean breezes, and also by the dry desert wind called the harmattan, which blows in December, often bringing sandstorms with it.

Rainfall

Most rain falls during the rainy season between April and November. Rainfall varies from about 70 in (178 cm) in the northern uplands to 200 in (510 cm) on the coast.

Forests and Jungles

Extensive mangrove, pandanus, and palm thickets are found near the coastal estuaries of Liberia's rivers. Other coastal vegetation includes rattan, oil, and coconut. Raffia palm and ferns thrive in the drier parts of the coast.

Both evergreen and deciduous rainforest covers most of the interior with more than 200 tree species. Hardwoods include teak, mahogany, camwood, and red ironwood. Of the remaining tree species, rubber and coffee are especially important to the country's economy.

Counties – Liberia

1999 POPULATION ESTIMATES

Name	Population	Area (sq mi)	Area (sq km)	Capital
Bong	299,825	3,127	8,099	Gbarnga
Grand Bassa	215,338	3,382	8,759	Buchanan
Grand Cape Mount	120,141	2,250	5,827	Robertsport
Grand Gedeh	94,497	6,575	17,029	Zwedru
Grand Kru*	39,062			Barclayville
Lofa	351,492	7,475	19,360	Voinjama
Margibi	219,417	1,260	3,263	Kakata
Maryland*	71,977			Harper
Montserrado	843,783	1,058	2,740	Bensonville
Nimba	338,887	4,650	12,043	Saniquillie
Sinoe	79,241	3,959	10,254	Greenville
Territories				
Bomi	114,316	755	1,955	Tubmanburg
Rivercess	38,167	1,693	4,385	Rivercess City

*Grand Kru was formerly part of Maryland; total Maryland territory prior to the split was 2,066 sq mi/5,351 sq km.

SOURCE: United Nations Humanitarian Assistance Coordinating Office in Liberia. As cited by Johan van der Heyden. Geohive. http://www.geohive.com (accessed June 2002).

HUMAN POPULATION

The capital city of Monrovia is the country's political, economic, and cultural hub. The city and its environs—especially the area between Monrovia and Harbel—constitute by far the most densely populated part of Liberia. In fact, it is the only large city in the entire country and is home to almost half of Liberia's population. Some of the upland regions are nearly uninhabited. The overall population density of the country is 78.6 people per square mile (30 people per sq km).

NATURAL RESOURCES

Liberia's plentiful rainfall and warm, tropical climate make its agricultural production its number one resource. Bananas, cocoa, rubber, rice, coffee, and sugarcane are among its many products grown and exported. Its mountains are an important source of valuable natural resources including iron ore, gold, and diamonds. The hilly forested country between the coastal plain and the interior plateau provides an excellent source of timber.

FURTHER READINGS

Daniels, Anthony. *Monrovia Mon Amour: A Visit to Liberia.* London: John Murray, 1992.

Huband, Mark. *The Liberian Civil War.* Portland, Ore.: F. Cass, 1998.

Zemser, Amy Bronwen. *Beyond the Mango Tree.* New York: HarperCollins, 2000.

Libya

- **Area:** 679,362 sq mi (1,759,540 sq km) / World Rank: 18
- **Location:** Northern and Western Hemispheres, on the northern part of the African continent, bordered by the Mediterranean Sea on the north, Egypt on the east, the Sudan on the southeast, Chad and Niger on the south, Algeria on the west, and Tunisia on the northwest.
- **Coordinates:** 25°00′N, 17°00′E
- **Borders:** 2,723 mi (4,383 km) / Algeria, 610 mi (982 km); Chad, 656 mi (1,055 km); Egypt, 715 mi (1,150 km); Niger, 220 mi (354 km); Sudan, 238 mi (383 km); Tunisia, 285 mi (459 km)
- **Coastline:** 1,100 mi (1,770 km)
- **Territorial Seas:** 12 NM
- **Highest Point:** Bīkkū Bīttī, 7,438 ft (2,267 m)
- **Lowest Point:** Sabkhat Ghuzayyil, 154 ft (47 m) below sea level
- **Longest Distances:** 1,236 mi (1,989 km) SE-NW; 933 mi (1,502 km) NE-SW
- **Longest River:** No perennial rivers
- **Natural Hazards:** Subject to hot, dry, dust-laden ghibli (a southern wind lasting one to four days in spring and fall); dust storms; sandstorms
- **Population:** 5,240,599 (July 2001 est.) / World Rank: 105
- **Capital City:** Tripoli, in the northern region on the Mediterranean coast
- **Largest City:** Tripoli, 2,404,000 (2000 metropolitan est.)

OVERVIEW

More than 600 millions years ago, an enormous mountain range once covered Libya, which lies on the African Tectonic Plate, formed during the Precambrian period 570 million years ago. Over the centuries, the sea advanced, then retreated, over the region, and the water, wind, and temperature changes eroded the mountains, leaving the sands and plateaus that comprise Libya's landscape.

The fourth-largest country in Africa, Libya is sectioned into three main geographical areas: Tripolitania, Cyrenaica, and Fezzan. Tripolitania covers the northwestern corner of the country, Cyrenaica covers the eastern half (and as the country's largest geographic region, it covers almost half of Libya), and the Fezzan covers the land south of Tripolitania. Tripolitania and Cyrenaica are made up of low-lying land and plateaus, with Tripolitania containing the Nafūsah Plateau, and Cyrenaica houses the Jabal al-Akhdar (Green Mountains). Fezzan is home to desert lands, including the Sahara.

Libya's climate is influenced by the desert that comprises most of Libya and the Mediterranean Sea. Extreme temperatures dominate the summer months, and winters are mild. At night in the desert, winter temperatures, however, fall below zero. No permanent rivers exist in Libya, only watercourses that frequently flood from runoff during the occasional rainfall.

MOUNTAINS AND HILLS

Mountains

The Tibesti Massif, a rugged mountain range, runs along the southern border near Chad, and it houses Libya's highest point, Bīkkū Bīttī (Bette Peak) at 7,436 ft (2,267 m). The Al-Akhdar Mountains run along the northeastern Mediterranean Coast. In the center of the country are the lower Al-Harūj Al-Aswad Hills.

Plateaus

In the northwest region of the country, Tripolitania is home to a series of terraces that rise slowly from sea level along the coastal plain of Al-Jifarahh until they reach Nafūsah Plateau. This upland plateau, made of limestone, contains sand, shrubs, and scattered masses of stone, and elevations reach 3,300 ft (1,000 m). Southward from the Nafūsah Plateau is the Al Hamādah Al Hamrā' (the Red Desert), a rocky plateau comprised of red sandstone. Its flat landscape stretches hundreds of miles to the southwest Fezzan Desert region; the rocky plateaus of the Fezzan Desert have been shaped by wind and extreme temperature changes.

INLAND WATERWAYS

Rivers

In Libya there are no permanent rivers, only *wadis* (watercourses). They catch the infrequent runoff from rainfall during the rainy season, commonly causing flash floods, then run dry during the hot weather of the summer months.

THE COAST, ISLANDS, AND THE OCEAN

The Mediterranean Sea comprises Libya's northern border. The coastal plain is often marshy, yet beaches stretch more than 1,000 mi (1,600 km) along the Mediterranean Sea. Along the shore of the western region surrounding Tripoli, coastal oases alternate with sandy beaches and lagoons for more than 180 miles (300 km). The Gulf of Sidra is nestled between the Tripolitania and Cyrenaica regions. Important ports are located along the coast, including Benghazi, Tobruk, and Darnah.

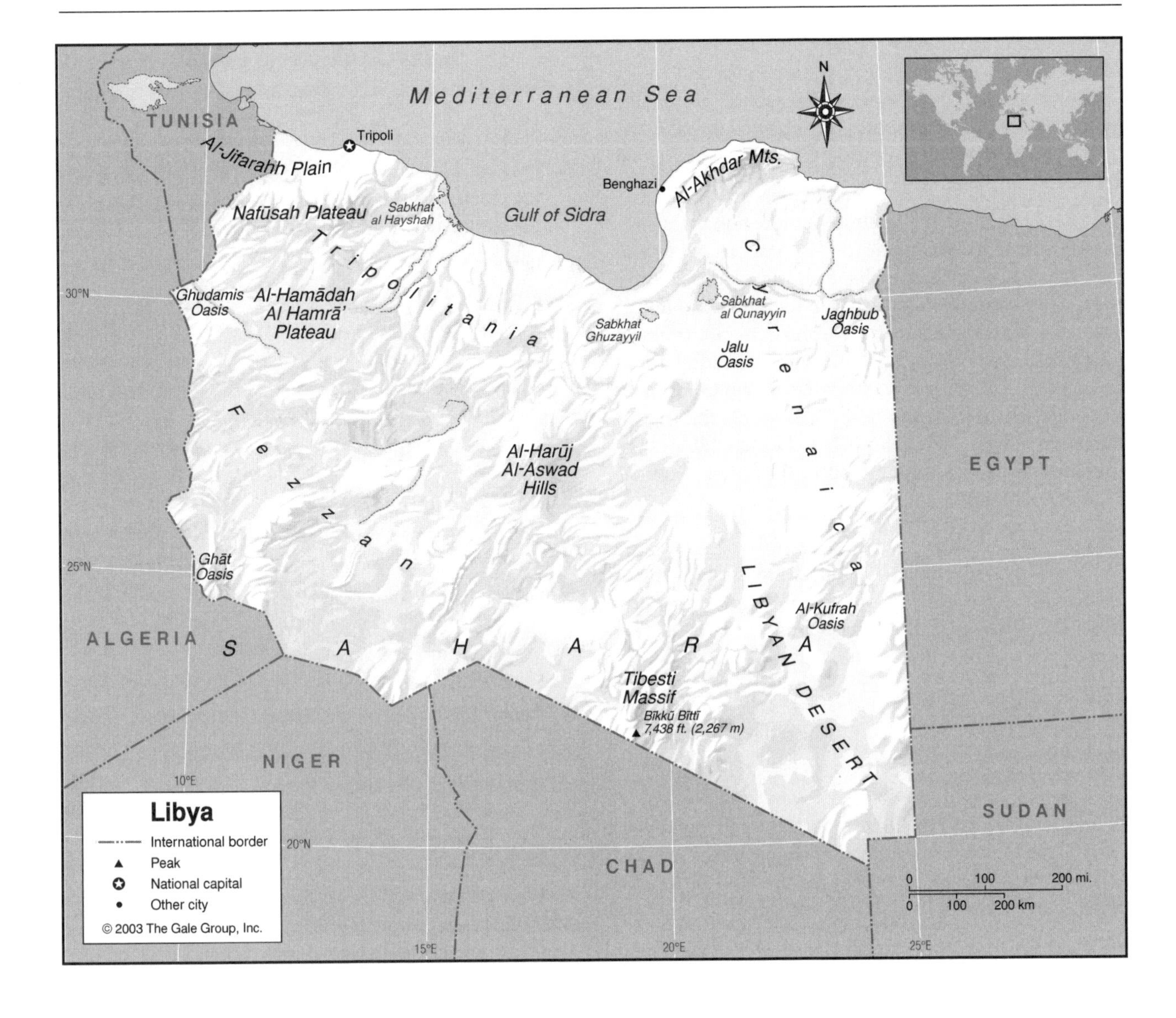

CLIMATE AND VEGETATION

Temperature

Libya's climate is influenced by the Mediterranean Sea and Sahara Desert. The ghibli (a hot, dry desert wind that lasts one to four days in spring and fall) varies temperatures by as much as 30° to 40°F (17° to 22°C) in both the summer (June-September) and winter (October-May) months. Summer highs along the northwestern coast are between 104° and 115°F (40° and 46°C), and temperatures in the farther south reach even higher degrees. In the northeastern region, summer temperatures range from 81° to 90°F (27° to 32°C). In January, temperatures average 55°F (13°C) in the northern region.

During the summer months in southern Libya, there is virtually no rainfall and temperatures quickly climb to over 122°F (50°C). Daytime winter temperatures range between 59° and 68°F (15° and 20°C), and fall below 32°F (0°C) at night.

Rainfall

Rainfall varies between the different regions. The northeastern region receives 16 to 24 in (40 to 60 cm) of rain yearly, while other regions receive less than 8 in (20 cm) and the Sahara has less than 2 in (5 cm). A short winter period brings rain, which usually causes floods. Evaporation is high, making severe droughts common.

Grasslands

In the northeastern area of Cyrenaica (the region that covers almost half of Libya), the land rises from a coastal plain to the Jabal al-Akhdar (Green Mountains) to a height of just under 3,000 ft (915 m). The lower slopes are covered with flowers, and at the higher elevations there are shrubs and juniper. In the southern region, a pastoral zone of sparse grassland gives way to the vast Sahara Desert.

Deserts

The Sahara Desert, the largest desert in the world, is located in the southern part of the country. It is a vast barren wasteland of rocky plateaus and sand. The part of the Sahara in eastern Libya and western Egypt and Sudan is known as the Libyan Desert. Agriculture is possible only in a few scattered oases, which include Jalu and Jaghbub. The three largest oases in Libya's desert region are Al-Kufrah, Ghāt, and Ghadāmis.

The Fezzan, in the southwestern region, is also a desert, with ergs (vast sand dunes) that reach several hundred feet and change shape slowly in the shifting wind. They cover about one-fifth of the land. Also in this area there is a series of sabkhas (depressions on the desert floor) that contain underground water, creating occasional oases. Most of the Fezzan is flat, except for the area along the southern border near Chad, where the rugged mountain range, Tibesti Massif, is located. The range contains Libya's highest point, Bīkkū Bīttī (Bette Peak) at 7,436 ft (2,267 m).

Forests and Jungles

Forest areas of shrubby junipers are located in the Jabal al-Akhdar (Green Mountains) in the northeastern area of Cyrenaica. In isolated regions, there are a few conifer trees, and in inaccessible regions in Tripolitania, there are some forests remnants.

HUMAN POPULATION

Ninety percent of the country's population lives near Tripoli, the capital city in the northeastern region of Libya. This coastal area receives enough rainfall to support agriculture. Small oasis communities dot the largely uninhabited desert in the southern regions.

NATURAL RESOURCES

Libya's natural resources are petroleum, natural gas, and gypsum. The discovery of the Zaltan Oil Field in 1959 in the Sahara Desert launched Libya's economic growth, and the country is one of the world's leading oil producers. Natural gas is located in the northern region, and gypsum is located on the northern coast. Reserves of iron ore, potassium, sulfur, and magnesium also exist in the country.

FURTHER READINGS

ArabNet. *Libya: Geography.* http://www.arab.net/libya/geography/libya_geography.html (Accessed March 4, 2002).

Malcolm, Peter. *Libya.* New York: Marshall Cavendish, 1999.

Miftah Shamali. *Libya.* http://www.i-cias.com/m.s/libya/index.htm (Accessed March 4, 2002).

Virtual Dimensions Inc. *Libya.* http://www.libyaonline.com/libya/index.html (Accessed March 14, 2002).

Liechtenstein

- **Area:** 62 sq mi (160 sq km) / World Rank: 199
- **Location:** Northern and Eastern Hemispheres, bordering Austria to the north and east and Switzerland to the west and south.
- **Coordinates:** 47°10′N, 9°32′E
- **Borders:** 47 mi (76 km) / Austria, 22 mi (35 km); Switzerland, 25 mi (41 km)
- **Coastline:** None
- **Territorial Seas:** None
- **Highest Point:** Grauspitz, 8,527 ft (2,599 m)
- **Lowest Point:** Ruggeller Riet, 1,411 ft (430 m)
- **Longest Distances:** 15.2 mi (24.5 km) N-S; 5.8 mi (9.4 km) E-W
- **Longest River:** Rhine, 820 mi (1,320 km)
- **Natural Hazards:** None
- **Population:** 32,528 (July 2001 est.) / World Rank: 197
- **Capital City:** Vaduz, west-central Liechtenstein
- **Largest City:** Vaduz, population 6,000 (2000 est.)

OVERVIEW

With an area slightly smaller than that of Washington, D.C., Liechtenstein is one of the smallest countries in the world, and the fourth smallest in Europe. Shaped like an elongated triangle, the principality is sandwiched between the Swiss cantons of Graubünden and St. Gall to the south and west, and the Austrian province of Vorarlberg to the north and east. The western third of the country lies on flat land in the floodplain of the Rhine River, which forms its western boundary. The eastern region consists of Alpine highlands.

Liechtenstein, which is situated on the Eurasian Tectonic Plate, sits directly on the divide between the Eastern and Western Alps, which passes through its capital city of Vaduz.

MOUNTAINS AND HILLS

Mountains

Liechtenstein's Alpine foothills and peaks are located on a spur of the Rhaetian Alps called the Rhätikon Massif. Its mountains are traversed by three main valleys, and their highest point is the Grauspitz, located on the southeastern border with Switzerland at a height of 8,527 ft (2,599 m).

GEO-FACT

Along with Uzbekistan, Liechtenstein is one of only two countries in the world that are doubly landlocked—that is, surrounded only by countries that are themselves landlocked.

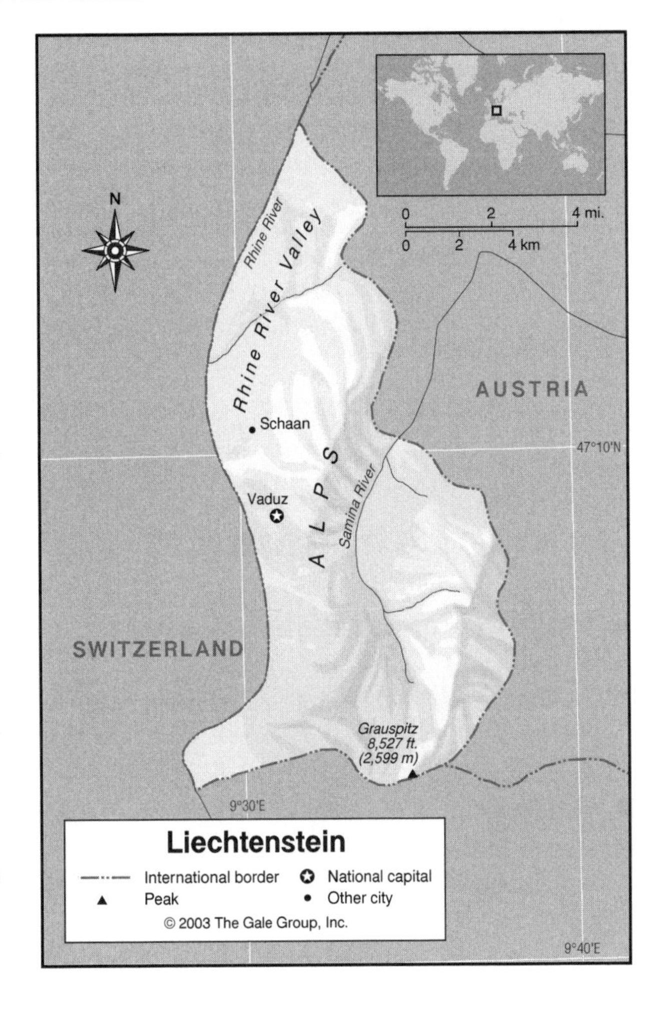

Hills and Badlands

An unusual hill formation, called the Eschnerberg, rises to heights of 2,395 ft (730 m) on the flat terrain of Liechtenstein's western plains area.

INLAND WATERWAYS

Rivers

Most of Liechtenstein is drained by the Rhine River and its tributaries. The mountain valleys to the east are drained by the Samina River, which rises in the southeast and flows northward through Liechtenstein's mountains into Austria.

Wetlands

Former marshland on the banks of the Rhine was reclaimed for agricultural use in the first half of the twentieth century.

THE COAST, ISLANDS, AND THE OCEAN

The principality of Liechtenstein is landlocked.

CLIMATE AND VEGETATION

Temperature

Liechtenstein has a continental climate tempered by a warm south wind called the *fohn*. Even at the upper Alpine elevations, winter temperatures rarely drop below 5°F (-15°C), and lowland temperatures average 24°F (-5°C) in January. Summer highs are generally between 68 and 82°F (20 and 28°C).

Rainfall

Annual precipitation ranges from 36 to 45 in (91 to 114 cm). The higher Alpine peaks are snowcapped year round.

Grasslands

Meadows and pastureland make up about 40 percent of the total land area; the remainder is either forest or rugged mountainous terrain.

Forests and Jungles

About one-fifth of Liechtenstein is forested. Forests on the lower mountain slopes include maple, alder, sycamore, red beech, larch, and various evergreens, as well as Alpine flowers such as gentian, Alpine rose, and edelweiss. At the highest elevations, the mountains are snowcapped.

HUMAN POPULATION

Most of Liechtenstein's residents live in the Rhine River Valley. Roughly one-third of the population is concentrated in the two largest communes, Vaduz and Schaan.

NATURAL RESOURCES

Liechtenstein has no commercially exploitable natural resources except for arable land and hydroelectric power. Instead, Liechtenstein's economy relies on its formidable industry and financial service sector. Low business taxes and easy incorporation rules have drawn many international companies to establish at least nominal offices in Liechtenstein.

FURTHER READINGS

Cussans, Thomas, ed. *Fodor's Switzerland.* New York: Fodor's Travel Publications, 1988.

Duursma, Jorri. *Self-determination, Statehood, and International Relations of Micro-states: The Cases of Liechtenstein, San Marino, Monaco, Andorra, and the Vatican City.* Cambridge: Cambridge University Press, 1996.

Frommer's Switzerland and Liechtenstein. New York: Prentice Hall Travel, 1994.

Greene, Barbara. *Valley of Peace: The Story of Liechtenstein.* Vaduz: Liechtenstein Verlag, 1947.

Lonely Planet. *Destination Liechtenstein.* http://www.lonelyplanet.com/destinations/europe/liechtenstein/ (Accessed May 14, 2002).

Lithuania

- **Area:** 25,174 sq mi (65,200 sq km) / World Rank: 123
- **Location:** Northern and Eastern Hemispheres, in Eastern Europe, east of the Baltic Sea, west and north of Belarus, south of Latvia, and northeast of Poland and Russia (Kaliningrad).
- **Coordinates:** 56°00′N, 24°00′E
- **Borders:** 791 mi (1,273 km) / Belarus, 312 mi (502 km); Latvia, 281 mi (453 km); Poland, 57 mi (91 km); Russia, 141 mi (227 km)
- **Coastline:** 62 mi (99 km)
- **Territorial Seas:** 12 NM
- **Highest Point:** Juozapinė, 958 ft (292 m)
- **Lowest Point:** Sea level
- **Longest Distances:** 172 mi (276 km) N–S; 233 mi (373 km) E–W
- **Longest River:** Neman, 582 mi (936 km)
- **Largest Lake:** Lake Druksiai, 17.2 sq mi (44.5 sq km)
- **Natural Hazards:** None
- **Population:** 3,610,535 (July 2001 est.) / World Rank: 125
- **Capital City:** Vilnius, located in the southeast
- **Largest City:** Vilnius, 553,000 (estimated 2001 population)

OVERVIEW

Along with Latvia and Estonia, Lithuania is one of the so-called Baltic states. It is located on southeastern shore of the Baltic Sea in northeastern Europe. Once part of the Soviet Union, Lithuania gained its independence in 1991. Its capital and largest city, Vilnius, lies near the western border on the Neris River.

With an area of about 25,200 sq mi (65,300 sq km), Lithuania is the largest of the Baltic states. Its topography is characterized by alternating regions of highlands and lowlands, but the primary feature is a low-lying central plain. Like that of other states in the region, the Lithuanian landscape was formed by continental glaciers during the Pleistocene ice age. No elevation is greater than 1,000 feet (305 meters). Highlands lie to the east and southeast of the central plain, and to the west the land is hilly but becomes low again along the coast. The plains of the southwestern and central regions are noted for their fertile soil.

Lithuania is a land of rivers and lakes (nearly 3,000 lakes, mostly in the east). The Neman (Nemunas) is the longest river, which flows north from Belarus into central Lithuania, then west into the Baltic Sea. Lithuania is on the Eurasian Tectonic Plate.

MOUNTAINS AND HILLS

There are hills and uplands to either side of Lithuania's central plain. In the west is the Žemaičai Upland. To the southeast are the Baltic Highlands, including the Ašmena Highland. None of these hills are very tall; the highest elevation is at Juozapinė, 958 ft (292 m), in the southeast region on the border of Belarus.

INLAND WATERWAYS

Lakes

Lithuania has 2,833 lakes larger than one hectare; in addition there are 1,600 ponds smaller than one hectare. Most are located in eastern Lithuania. Lake Druksiai is the largest lake, covering about 17.2 sq mi (44.5 sq km), while the deepest is Lake Tauragnas (200 ft / 61 m), and the longest Asveja Lake (14 mi / 22 km). A large reservoir on the Neman near Kaunas is known as Kaunas Sea.

Rivers

Lithuania is notable for its many rivers as well as it lakes; some 758 rivers are longer than ten kilometers. However, few are navigable; only 372 mi (600 km) of the country's rivers can be traveled. The Neman is longest, entering the country from Belarus in the south and flowing for roughly 295 mi (475 km) before entering the Baltic Sea. It forms the border with Russian Kaliningrad along its lower course. Other significant rivers include the Neris, 316 mi (510 km); the Venta, 215 mi (346 km); and the Šešupė,185 mi (298 km).

Wetlands

Like its northern neighbor Latvia, Lithuania has many marshes and swamps. However, most of the country's original wetlands have been drained for agriculture. Remaining wetlands are found mostly in the north and west.

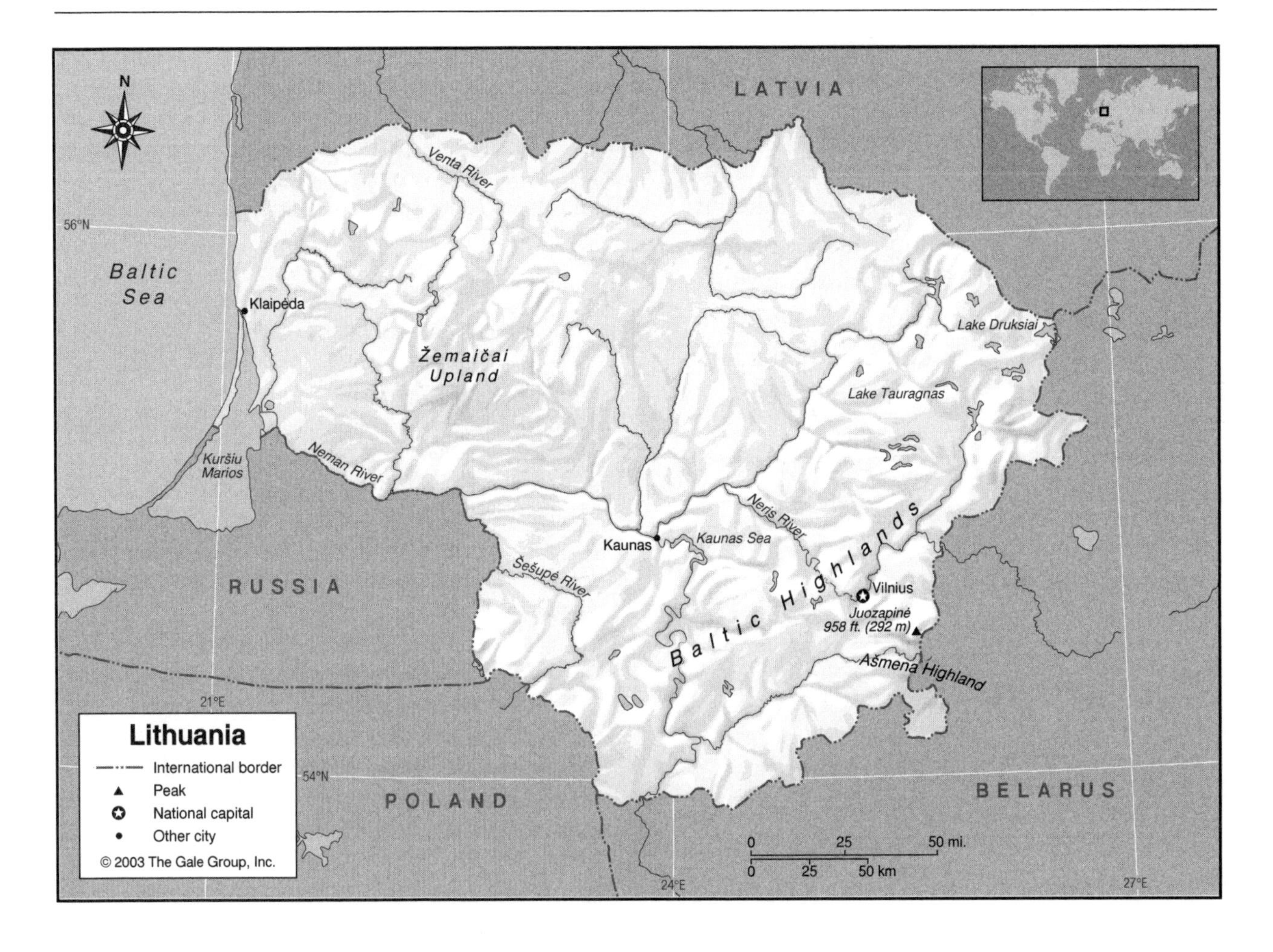

THE COAST, ISLANDS, AND THE OCEAN

Lithuania's Baltic Sea coast is short, only 67 mi (108 km). A long, narrow sandbar forms an offshore lagoon along the southern half of the coastline called Kuršiu Marios (Courland Lagoon).

CLIMATE AND VEGETATION

Temperature

Although its climate is continental, Lithuania's proximity to the Baltic Sea provides a moderating maritime influence with mild winters and cool summers. In the east, however, conditions may vary from this pattern. The west has a growing season of 202 days; in the east it lasts 169 days. Overall the climate is mild. In January temperatures average 35° F (1.6°C) on the western coast and 36° F (2.1°C) in Vilnius in the east. In summer, temperatures average 64° F (17.8°C) on the coast and 65° F (18.1°C) in Vilnius.

Rainfall

Western Lithuania is more rainy than the rest of the country, with an average annual precipitation of 33 in (85 cm), compared to 24 in (49 cm) in the central plains and 28 in (72 cm) on the east coast.

Forests

About 28 percent of Lithuania's land was forested in 2002, with patches of forest scattered throughout the country. The coastal region and the south favor pines, while oak trees predominate the central region, although they are relatively scarce. Mushrooms and berries are abundant. Lithuania has set aside large forested areas as nature reserves, which support many species of wildlife. Mammals include elk, deer, wolves, foxes, and wild boar; bird species include white storks, herons, geese, ducks, and hawks.

HUMAN POPULATION

Lithuania's population is 3,610,535 (July 2001 estimate), which gives it a population density of 55 persons per sq km (143 persons per sq mi). The country's population has declined in the recent past as a result of low birth rates and emigration. Although noted for its beautiful rural and woodland areas, the nation is quite urbanized; some 73 percent of the population lives in the cities. However, Lithuania does not have a single dominating urban center like most other countries of the former Soviet Union. The largest city is the capital, Vilnius (population 639,000); second largest is Kaunas (population

GEO-FACT

Like many of the former republics of the Soviet Union, Lithuania has had to struggle with the Soviet legacy of exploited resources and environmental disregard. For example, the Kuršiu Marios, a beautiful natural area separated from the Baltic Sea by a strip of high dunes and pine forests, is about 85 percent contaminated.

Population Centers – Lithuania

(1993 POPULATION ESTIMATES)

Name	Population	Name	Population
Vilnius (capital)	581,500	Shauliai	147,900
Kaunas	421,600	Panevezhis	131,800
Klaipėda	204,300	Panevezys	122,000

SOURCE: "Population of Capital Cities and Cities of 100,000 and More Inhabitants." United Nations Statistics Division.

Counties – Lithuania

2000 POPULATION ESTIMATES

Name	Population	Area (sq mi)	Area (sq km)
Alytus (Alytaus)	187,769	2,095	5,425
Kaunas (Kauno)	701,529	3,154	8,170
Klaipeda (Klaipėdos)	385,768	2,219	5,746
Mairjampole	188,634	1,723	4,463
Panevezys	299,990	3,043	7,881
Siauliai	379,096	3,379	8,751
Taurage	134,275	1,496	3,874
Telsiai	179,885	1,598	4,139
Utena	185,962	2,780	7,201
Vilnius	850,0064	3,726	9,650

SOURCE: Population Census Division, Statistics Lithuania.

418,087), an industrial and commercial center; followed by Klaipėda (population 202,929), an important seaport.

NATURAL RESOURCES

Lithuania has relatively few natural resources. The nation does have an abundant quantity of the materials used for making high-quality cement, glass, and ceramics: limestone, clay, quartz sand, gypsum sand, and dolomite. Some iron ore deposits have been found in the southern Lithuania. The country's marsh areas produce peat (a compact, high-carbon material used for fuel and mulch), and the fossil tree resin, amber, is found along the Baltic shore. Oil and natural gas resources are limited. Although oil was discovered in the 1950s, only a few wells in the western part of the country are productive; and there is no environmentally approved method that would allow greater commercial exploitation..

FURTHER READINGS

Bite, Vita. "Lithuania: Basic Facts," *CRS Report for Congress.* Washington. D.C.: Library of Congress, Congressional Research Service, August 6, 1992.

Economist Intelligence Unit. *Country Profile: Lithuania.* London: The Economist, 1995.

Lithuania: An Economic Profile. Washington, D.C.: United States National Technical Information Service, August 1992.

Lithuanian Folk Culture Center. *The Lithuanians.* http://www.lfcc.lt/publ/thelt/node4.html (Accessed June 18, 2002).

Pakalnis, Romas. "The Future of Lithuanian Nature Is the Future of Lithuania," *Science, Arts, and Lithuania,* No. 1(1991): 16-21.

World Bank. *Lithuania: The Transition to a Market Economy.* Washington, D.C.: World Bank, 1993.

Luxembourg

- **Area:** 998 sq mi (2,586 sq km) / World Rank: 170
- **Location:** Northern and Eastern Hemispheres, in Western Europe, bordering Germany to the east, France to the south, and Belgium to the west and north.
- **Coordinates:** 49°45′N, 6°10′E
- **Borders:** 221 mi (356 km) / Belgium, 92 mi (148 km); France, 45 mi (73 km); Germany, 84 mi (135 km)
- **Coastline:** None
- **Territorial Seas:** None
- **Highest Point:** Buurgplaatz, 1,834 ft (559 m)
- **Lowest Point:** Moselle River, 436 ft (133 m)
- **Longest Distances:** 51 mi (82 km) N-S; 35 mi (57 km) E-W
- **Longest River:** Moselle, 320 mi (515 km)
- **Natural Hazards:** None of significance
- **Population:** 442,972 (July 2001 est.) / World Rank: 163
- **Capital City:** Luxembourg, southern Luxembourg
- **Largest City:** Luxembourg, 81,800 (2001 est.)

OVERVIEW

Luxembourg, one of the world's smallest countries and the only grand duchy, is a landlocked nation located at the heart of Western Europe situated on the Eurasian Tectonic Plate. Together with Belgium and the Netherlands, it is part of a group known as the Benelux countries (formerly the Low Countries). In spite of its small size, there is considerable variety in Luxembourg's terrain, which includes parts of three different topographical areas: the Lorraine plateau of northern France, the foothills of Belgium's Ardennes Mountains, and Germany's Moselle Valley. The intersection of these features carves Luxembourg into two major geographic regions. The northern third of the country, known as the Oesling, is a plateau region belonging to the Ardennes system of southeastern Belgium. The southern two-thirds, known as Gutland, or the Bon Pays, is a less elevated region consisting of hills and broad valleys.

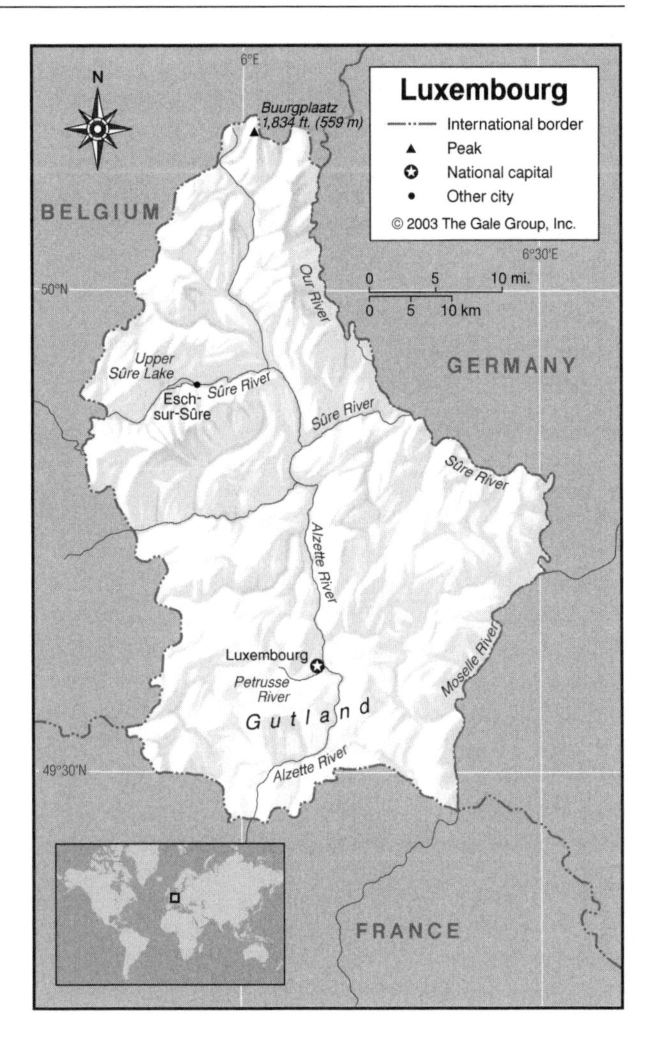

MOUNTAINS AND HILLS

Plateaus

The Ardennes region that forms Luxembourg's northern uplands consists of a schist and sandstone plateau rising to an average elevation of 1,500 ft (450 m) and deeply carved by the valleys of the Sûre River and its tributaries. The highest point is Buurgplaatz in the north (1,834 ft / 559 m), also the highest point in all of Luxembourg.

Hills and Badlands

The southern two-thirds of Luxembourg consists of fertile, gently rolling land with an average elevation of about 750 ft (229 m). The Moselle River Valley in the east is known for its vineyards, and there is a mining region to the southwest, near the border with France. The city of Luxembourg sits on a rocky height with steep cliffs on three sides dropping to the Alzette River.

INLAND WATERWAYS

Lakes

The most important lake is the Upper Sûre lake, on the course of the Sûre River as it winds across the upper portion of the country. Esch-sur-Sûre is located at its eastern end of the lake, which is the site of both a nature reserve and a hydroelectric dam.

Rivers

Luxembourg's major rivers are the Moselle, the Sûre, and the Our, which together form its border with Germany. The Moselle, which originates in France and has a total length of 320 mi (515 km), demarcates Luxembourg's eastern border for 19 mi (31 km). The Sûre, which rises in Belgium, flows eastward across Luxembourg in a meandering course for 107 mi (172 km), gathering tributaries from both the northern and southern parts of the country, before reaching the German border and then flowing southward into the Moselle. In the south, the Alzette River flows northward through the center of the country until it reaches the Sûre.

THE COAST, ISLANDS, AND THE OCEAN

Luxembourg is landlocked.

CLIMATE AND VEGETATION

Temperature

Luxembourg has a temperate climate, with cool summers, mild winters, and plentiful precipitation. The city of Luxembourg, in the south-central part of the country, has average temperatures of 33°F (0.6°C) in January and 63°F (17°C) in July. In the Oesling region to the north, temperature averages for both seasons are somewhat lower. The Moselle River Valley in the east has an espe-

Population Centers – Luxembourg

(2000 POPULATION ESTIMATES)

Name	Population	Name	Population
Luxembourg City (capital)	79,893	Differdange	17,050
Esch-sur-Alzette	24,873	Petange	13,479
Dudelange	17,300	Sanem	12,770

SOURCE: L'Informatique au Service des Communes et de Leurs Citoyens, Luxembourg.

Cantons – Luxembourg

2000 POPULATION ESTIMATES

Name	Population	Area (sq mi)	Area (sq km)
Capellen	36,533	77	199
Clerveaux	11,827	128	332
Diekirch	26,417	92	239
Echternach	13,005	72	186
Esch-sur-Alzette	129,173	94	243
Grevenmacher	20,917	82	211
Luxembourg (Ville et Campagne)	127,808	92	238
Mersch	20,844	86	224
Redange	13,245	103	267
Remich	15,526	49	128
Vianden	2,805	21	54
Wiltz	11,119	102	265

SOURCE: L'Informatique au Service des Communes et de Leurs Citoyens, Luxembourg.

cially pleasant climate, leading to its nickname of the "Little Riviera."

Rainfall

Rainfall, which varies from about 30 in (76 cm) to 50 in (127 cm) annually, is generally heavier in the north. However, the southwest gets more rain than the rest of the Bon Pays region.

Forests

Forests of oak and beech grow on the low mountains of the northern Oesling region, interspersed with pastures and open fields. Other tree species include willow, black alder, linden, pine, and spruce. The major forested area in the south is the Müllerthal, or "Little Switzerland," area in the east-central part of this region, known for its beech forests. The Moselle River Valley is famous as a wine-growing region.

HUMAN POPULATION

Most of Luxembourg's population is concentrated in the southern two-thirds of the country (the Gutland, or Bon Pays, region). The capital, the city of Luxembourg, is located on the Alzette and Petrusse rivers in the heart of this region. Roughly 90 percent of Luxembourg's residents live in urban areas, mostly in towns with populations of under 30,000.

GEO-FACT

Human settlement of the area where Luxembourg's second-largest city, Esch-sur-Alzette, stands today in southern Luxembourg can be traced back 5,000 years.

NATURAL RESOURCES

Luxembourg's major natural resource has been the iron deposits located in the southwestern part of the country, but these deposits are no longer mined due to depletion. Other resources include slate, and the hardwood from the large oak trees that grow in the northern part of the country. Luxembourg depends largely on services, especially banking, for its economy. There are also factories that produce rubber, steel, and various chemicals.

FURTHER READINGS

Belgium and Luxembourg. Oakland, Calif.: Lonely Planet Publications, 2001.

Belgium and Luxembourg: The Rough Guide. London: Penguin, 1997.

De Vries, Andri. *Live and Work in Belgium, the Netherlands and Luxembourg.* 2nd ed. Oxford: Vacation Work, 1998.

Egan, E. W. *Belgium and Luxembourg in Pictures.* New York: Sterling, 1966.

Little, Gary. *Luxembourg Central.* http://www.luxcentral.com/index.shtml. (Accessed 29 April 2002).

Widing, Katherine, and Jerry Widing. *Cycling the Netherlands, Belgium, and Luxembourg.* Osceola, Wis.: MBI Publishing, 1998.

Macedonia, Former Yugoslav Republic Of

- **Area:** 9,781 sq mi (25,333 sq km) / World Rank: 148
- **Location:** Northern and Western Hemispheres, southeastern Europe, south of Yugoslavia, west of Bulgaria, north of Greece, and east of Albania.

- **Coordinates:** 41°50′N, 22°00′E
- **Borders:** 465 mi (748 km) / Albania, 94 mi (151 km); Bulgaria, 92 mi (148 km); Greece, 142 mi (228 km); Yugoslavia, 137 mi (221 km)
- **Coastline:** None
- **Territorial Seas:** None
- **Highest Point:** Golem Korab, 9,032 ft (2,753 m)
- **Lowest Point:** Vardar River, 164 ft (50 m)
- **Longest Distances:** Approximately 109 mi (175 km) N-S; approximately 134 mi (216 km) E-W
- **Longest River:** Vardar River, 241 mi (388 km).
- **Largest Lake:** Lake Ohrid, 134 sq mi (348 sq km)
- **Natural Hazards:** Earthquakes
- **Population:** 2,046,209 (July 2001 est.) / World Rank: 141
- **Capital City:** Skopje, in north central Macedonia.
- **Largest City:** Skopje, 448,600 (2002 estimate)

OVERVIEW

The Former Yugoslav Republic of Macedonia (FYROM; Macedonia) lies inland in the middle of the Balkan Peninsula. The nearest open water is the Adriatic Sea, on the far side of Albania to the west, and the Aegean Sea beyond Greece to the southeast. About 80 percent of its territory is mountainous, with large and high massifs giving way to some extensive flat valleys and plains. The valleys are interconnected by low passes or deep ravines. There are some interior highlands in the north central region and southwest corner of Macedonia.

Macedonia is on the Eurasian Tectonic Plate. A thrust fault line extends in a north to south direction in east central Macedonia. The structural seam in the Earth's crust periodically shifts, causing earth tremors and occasional destructive earthquakes. In 1963, an earthquake destroyed much of Skopje and killed 1,066 people.

The name Macedonia has historically been used to describe a region that includes parts of modern Greece, Bulgaria, and the FYROM. The ancient kingdom that was based there ruled Greece for centuries, and produced its most famous conqueror, Alexander the Great. When the FYROM declared independence from Yugoslavia in 1991 and took Republic of Macedonia for its name, Greece objected. To them, Macedonia is a Greek name and an important part of Greek history and culture, which the new country could not rightfully claim. Due to the ongoing controversy, many countries refer to the Republic of Macedonia as the Former Yugoslav Republic of Macedonia, or other names.

MOUNTAINS AND HILLS

Mountains

The FYROM is mostly covered by mountains; the average altitude of the country is about 2,800 ft (850 m). The mountains are a complicated mass, with ridges running in many different directions, and no truly dominant range. Some of the highest ranges are the Jakupica, in central FYROM; Korab in the west; Plačkovica in the east; and Kožuf and Nidže in the south. There are 34 mountain peaks exceeding 6,560 ft (2,000 m) above sea level, ranging from Mount Belasica (6,657 ft / 2,029 m) to Golem Korab (9,032 ft / 2,753 m), the highest peak in Macedonia. Along the northern border with the Kosovo region of Yugoslavia, Šar Planina, at 50 mi (80 km) long and 6-12 mi (10-20 km) wide, is the largest natural massif in Macedonia, reaching a peak of 9,012 ft (2,747 m).

There are dozens of glacial caves within the mountains, some of which feature water. One of these is Djonovica (located between Gostivar and Kičevo), which extends about 2,000 ft (600 m) underground.

Valleys and Canyons

Macedonia has 19 separate lowland areas, covering about 2,970 sq mi (7,690 sq km), with around 1,900 sq mi (4,900 sq km) of valley basin lowlands. Macedonia's canyons link the lowlands. There are 114 separate canyons in Macedonia totaling 185 mi (297 km), ranging from the 1.4-mi (2.3-km) Boshavica River to the 26.4-mi (42.5-km) Radika canyon. The Derven, Taor, and Demir Kapija canyons are on the Vardar River. Demir Kapija has nearly vertical sides and several small caves.

Population Centers – Former Yugoslav Republic of Macedonia

(1994 CENSUS OF POPULATION)

Name	Population	Name	Population
Skopje (capital)	545,228	Bitola	108,203
Tetovo	172,171	Gostivar	108,181
Kisela Voda	146,746	Gazi Baba	100,259
Kumanovo	127,814	Prilep	94,183
Karpoch	127,462	Strumica	91,047

SOURCE: State Statistical Office, Macedonia.

INLAND WATERWAYS

Lakes

Macedonia has 53 natural and artificial lakes. The three largest lakes are of tectonic origin: Ohrid, Prespa, and Dojran. Lake Ohrid is in southwestern corner of Macedonia, covering 134 sq mi (348 sq km), of which 89 sq mi (230 sq km) are within Macedonia's borders; the rest is within Albania. Lake Ohrid is some 18.9 mi (30.4 km) long and 9 mi (14.5 km) wide, with its surface 2,280 ft (695 m) above sea level. The clarity of the water extends some 70 ft (21.5 m) down, and the lake's maximum depth is 942 ft (287 m). Lake Prespa is the second largest lake in Macedonia, with a surface area of 106 sq mi (274 sq km), but only 68 sq mi (177 sq km) are within Macedonian territory; Greece and Albania share the rest. At 2,799 ft (853 m) above sea level, the water in Lake Prespa gradually percolates through the porous limestone and ends up in the lower Lake Ohrid, not far to the northwest.

Macedonia also has 25 glacial mountain lakes, known as *oci*, or mountain "eyes." Additionally, there are numerous mineral springs. The Katlanovo Spa outside Skopje is fed by several springs and has been utilized for its therapeutic 115°F (46°C) waters since the Roman era. There are 15 artificial lakes, the largest of which is Mavrovo. Formed in 1953, Lake Mavrovo covers about 5.3 sq mi (13.7 sq km).

Rivers

Macedonia's rivers flow into one of three basins: the Aegean, Adriatic, or Black Sea. The Vardar River enters from Yugoslavia in the north and flows southeast across the country for 187 mi (301km) before crossing into Greece, eventually emptying into the Aegean. The Vardar is the most important river in the country, draining 80 percent of its territory. Within Macedonia, the Vardar has 37 tributaries, including the Bregalnica and the Crna. The Strumica, in southeast FYROM, is the only other river of note flowing into the Aegean.

The Crni Drim River drains the westernmost 13 percent of Macedonia. It flows north out of Lake Ohrid and into Albania before turning west and draining into the Adriatic Sea. Less than 0.2% of the country is drained by the Binacka Morava River, which has its source in Macedonia. The Binacka Morava only flows a few miles through FYROM before crossing into Yugoslavia and eventually emptying into the Danube River and the Black Sea.

THE COAST, ISLANDS, AND THE OCEAN

The FYROM is landlocked.

CLIMATE AND VEGETATION

Temperature

FYROM's climate is a blend of continental and Mediterranean, with very cold winters and hot summers. The average annual temperature for the country is 52.7°F (11.5°C). Maximum summer temperatures in the lowlands can reach 104°F (40°C), and the coldest winter temperatures can drop to around -22°F (-30°C).

Rainfall

Due to the influence of the Mediterranean, rainfall is moderate in the Vardar River valley. Annual rainfall is scattered throughout the year and only averages about 20–28 in (500–700 mm).

Forests and Jungles

Macedonia has four national parks (Galičica, Mavrovo, Pelister, and Jasen), which have a total area of about 272,000 acres (110,000 hectares). The high mountains are covered mostly with pine trees. Lower mountains have a canopy of beech and oak trees. The Macedonian Pine is an ancient native species found in the forests on Mount Pelister near Lake Prespa.

HUMAN POPULATION

The population of the FYROM is concentrated in the lowlands. About 42 percent of the country's population lives in the Vardar River valley of central Macedonia. Another 39 percent lives in western Macedonia. Only 19 percent lives in the east. The capital of Skopje is by far the largest city; most of the towns have fewer than 15,000 inhabitants.

NATURAL RESOURCES

Iron ore is mined near Kičevo; zinc, lead, and copper are also mined. There are deposits of iron and nickel in the Vardar River valley area, and well as discoveries of molybdenum, tungsten, mercury, and gold. Macedonia also has marble and stone quarries near Prilep, Gostivar, and Tetovo. Wood and lumber production are centered in Bitola.

FURTHER READINGS

Brân, Zoë. *After Yugoslavia*. Oakland, Calif.: Lonely Planet, 2001.

Georgieva, Valentina, and Sasha Konechni. *Historical Dictionary of the Republic of Macedonia*. Lanham, Md.: Scarecrow Press, 1998.

Government of the Republic of Macedonia. http://www.gov.mk/English/index.htm (Accessed July 10, 2002).

Pettifer, James, ed. *The New Macedonia Question*. New York: St. Martin's Press, 1999.

Scekic, Jovan, ed. *This Was Skopje*. Beograd: Federal Secretariat for Information, 1963.

Madagascar

- **Area:** 226,657 sq mi (587,040 sq km) / World Rank: 47
- **Location:** An island in the Indian Ocean, Southern and Eastern Hemispheres, southeast of Africa, east of Mozambique.
- **Coordinates:** 20°00′S, 47°00′E
- **Borders:** None
- **Coastline:** 3,000 mi (4,828 km)
- **Territorial Seas:** 12 NM
- **Highest Point:** Maromokotro, 9,449 ft (2,880 m)
- **Lowest Point:** Sea level
- **Longest Distances:** 995 mi (1,601 km) NNE-SSW; 360 mi (579 km) ESE-WNW
- **Longest River:** Mangoky, 350 mi (560 km)
- **Largest Lake:** Alaotra, 69 sq mi (180 sq km)
- **Natural Hazards:** Cyclones
- **Population:** 15,982,563 (July 2001 est.) / World Rank: 58
- **Capital City:** Antananarivo, central Madagascar
- **Largest City:** Antananarivo, population 1,128,000 (2000 est.)

GEO-FACT

The lemur, Madagascar's most distinctive wildlife species, evolved from primates thought to have reached the island on logs millions of years ago and evolved independently of monkeys and other primate species. The roughly thirty species on the island vary considerably in size and appearance. The ring-tailed lemur has a striped, raccoon-like tail. The aye-aye lemur, now a protected species, was traditionally thought to bring bad luck and killed on sight. The name *lemur* itself comes from the Roman word for the ghosts of the dead.

OVERVIEW

Madagascar is an island nation sitting on the African Tectonic Plate in the Indian Ocean, off the coast of Mozambique. It is the world's fourth-largest island (excluding Australia), and one of its southernmost countries—the most southerly part of the island lies below the Tropic of Capricorn. The country is sometimes called the Great Red Island because of the bright red color of its laterite soil, which erosion washes into the rivers coloring them red as well. In some places along the coast, the ocean itself is stained red by silt from the rivers that drain into it.

The island can be broadly divided into three major regions: 1) a narrow coastal plain to the east, about 30 mi (48 km) wide, with alluvial soil; 2) a large central region covered by plateaus and mountains; and 3) a sloping, hillier, and less clearly defined coastal region to the west, 60 to 125 mi (97 to 201 km) wide. Madagascar is famous for its unique wildlife and vegetation, which developed and diversified in isolation from the fauna and flora of mainland Africa. Three-quarters of its plant and animal species are found nowhere else on earth, which makes the island a popular site for both nature study and ecotourism.

MOUNTAINS AND HILLS

Mountains

Most of Madagascar's mountains were apparently thrust upward by long-term lateral shifting in the foundation rock, but some of the highest mountains are of volcanic origin, including those of the Tsaratanana and Ankaratra Massifs. In the north, the Tsaratanana Massif, which separates the northernmost region from the rest of the country, includes the country's highest point, Mt. Maromokotro at 9,436 ft (2,876 m). The Ankaratra Massif, which occupies the center of the island, forms a watershed between three river basins; its highest point is Mt. Tsiafajavona (8,668 ft / 2,642 m). To the south, the granite expanse of the Andringitra Massif north of Tôlanaro rises to 8,720 ft (2,658 m) at Boby Peak, its highest point. The low Ambohitra Mountains at the northern-

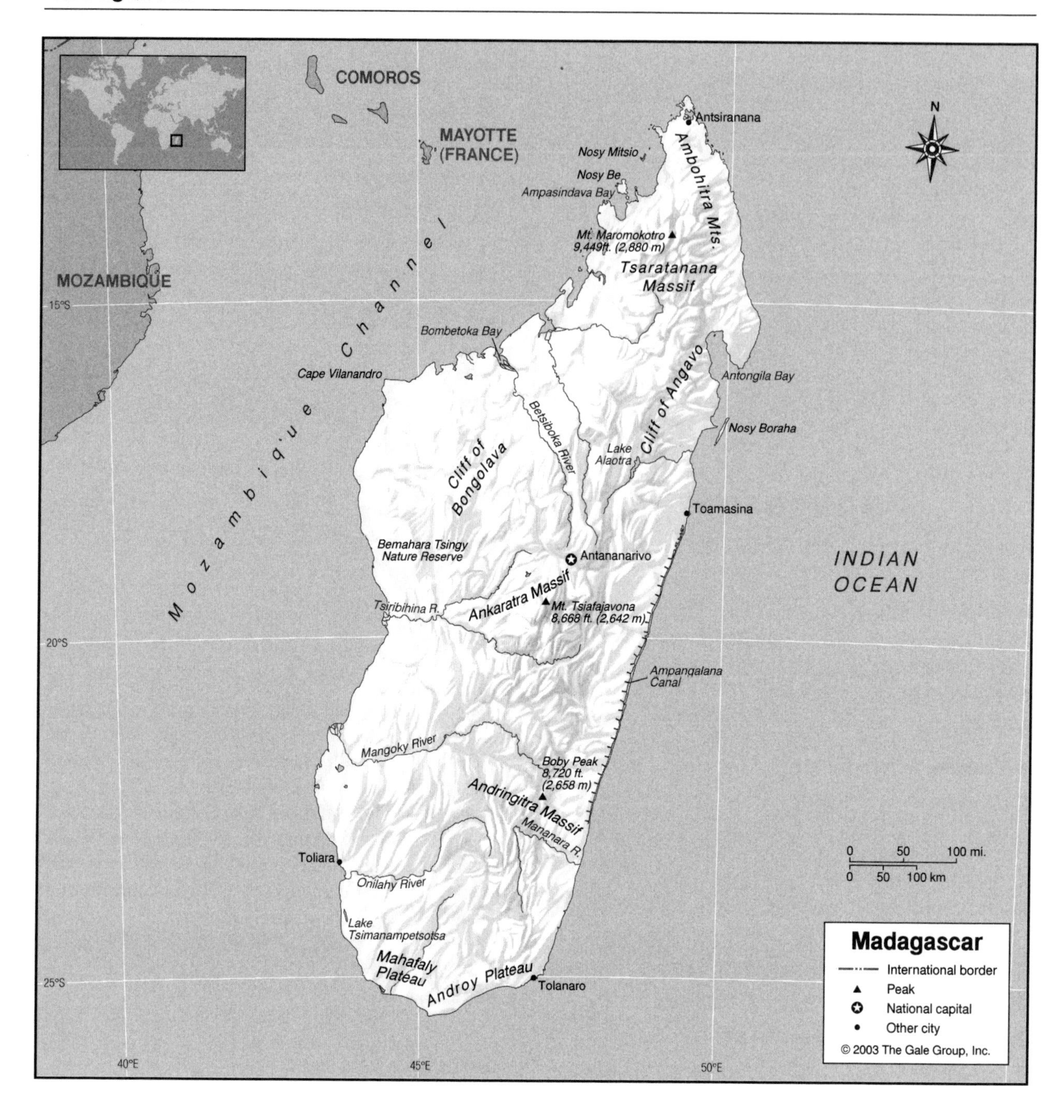

most part of the island contain a number of volcanic craters.

Plateaus

The central plateau region has average elevations of 2,500 to 4,500 ft (800 to 1,400 m) but rises to heights of over 8,000 ft (2,438 m) in several places.

Although wide areas of the region are covered by rounded and sometimes baretopped hills of nearly uniform height—especially in the west—there is still considerable diversity in the landforms, elevations, and geologic origins of these highlands, which include terraced valleys where rice is grown and rolling pastureland. Elevation is gradually steeper in the east, with Ankaratra Plateau being bordered by the sheer Cliff of Angavo (the Great Cliff). The descent is more gradual in the southwest and the south, terminating at the ocean with Mahafaly Plateau and Androy Plateau. The northwest is bounded by the Cliff of Bongolava.

Jagged needlelike limestone pinnacles created over the centuries by rainfall are a distinctive feature found in parts of the plateau region, including the Tsing'i Bemahara nature reserve in the west. (*Tsing'i* is the native word for these limestone "needles.") Much of the surface layer in the middle elevations and many prominent topographic features developed from the ancient granite structure underlying the island, which is similar to the substructure of the African continent. Lava flows occurring in geologically recent times covered the older granite landforms in a few highland areas, providing the basis for some of the country's best soils. The Tsaratanana Massif is primarily granite but is partially covered by lava from eruptions in the immediate area. Rock and soil from other volcanic activity are associated with the Ankaratra Massif.

INLAND WATERWAYS

Lakes

Madagascar has a number of volcanic lakes, of which only a few are of significant size. The largest is Lake Alaotra in the northeast, on the Ankaratra Plateau. There is a large saltwater lake, Lake Tsimanampetsotsa, at the southwestern end of the island, near Toliara.

Rivers

The short rivers on the eastern part of the island are torrents rushing down the steep slopes of the escarpment that borders the coastal plain and either draining into the coastal lagoons or cascading into the ocean in the form of rapids and waterfalls. These rivers include the Mananara, Faraony, Ivondro, and Maningory. Despite heavy forest growths, rainwater drains quickly down the eastern escarpment, into the main river channels, and back to the ocean. On the western part of the island, rivers flow sluggishly westward across a broad coastal zone, and alluvial soil material carried down from the highlands is deposited behind the coast or in the deltas along the Mozambique Channel. The mouths of these rivers, which are longer and larger than those in the east, are frequently blocked by sandbars. The major western rivers include the Mangoky, Tsiribihina, Betsiboka, Onilahy, and Manambajo.

THE COAST, ISLANDS, AND THE OCEAN

Oceans and Seas

Madagascar is located in the southwestern part of the Indian Ocean, opposite Mozambique. It is separated from the African continent by the Mozambique Channel, whose width is 250 mi (400 km). In the Channel, coral reefs fringe the northwestern coast of the island.

Major Islands

Small volcanic islands, including Nosy Mitsio and Nosy Be, border the northwestern coast. The only such island to the east is the 64-sq mi (165.8 sq km) Nosy Boraha, south of Antongil Bay.

The Coast and Beaches

The narrow eastern coastal plain is covered mostly by sandy beaches. The shoreline is almost perfectly straight for more than half its length. Running parallel to it for some 400 mi (644 km) is a narrow, artificial waterway called the Ampangalana Canal that links a series of lagoons. Toamasina, near the northern end of the canal, is the country's major port. North of the canal area is Madagascar's deepest coastal indentation, Antongil Bay, and the only island off its eastern coast, Nosy Boraha. North of the bay, the coast becomes relatively smooth once again, terminating in a sharp point in the extreme north, beyond the smaller Antsiranana Bay. The eastern coast south of the canal is rocky.

The western coast is more irregular and indented than that to the east. Most of these indentations occur in the northwest section, which is fringed with coral reefs, bordered by small islands, and broken up by a number of estuaries and bays, including the Bombetoka and Ampasindava bays. Cape Vilanandro sits at the point where the western coastline straightens before turning due south. Farther south, the coastline, although curved, is smoother, with mangrove trees and small dunes at its edges.

CLIMATE AND VEGETATION

Temperature

Madagascar's climate is strongly influenced by southeasterly trade winds, and temperatures are also moderated by altitude. The coastal areas are hottest, and the highest elevations of the plateau regions are the coolest. Temperatures range from 50°F (10°C) to 78°F (26°C) in July (the coolest month) and from 61°F (16°C) to 84°F (29°C) in December, the hottest month.

Rainfall

The hot season between November and April is also the rainy season, while drier weather prevails the rest of the year. Rainfall on the eastern, or windward, side of the island is heaviest, with an annual average of almost 150 in (380 cm) at Antongil Bay. Monsoons bring precipitation to the northwestern coast, which averages 83 in (211 km) annually, compared with the arid southwest where the average drops to a mere 14 in (36 cm). Annual rainfall on the plateau falls between these extremes, averaging about 53 in (135 cm).

Grasslands

The destruction of Madagascar's original forestland has left savanna grasslands over most of the island, with a characteristic landscape of prairie grasses broken by thickets of bamboo and small, scattered trees. These steppes are green in the rainy season, but during the dry season they turn to brown and to the characteristic reddish clay color associated with the earth and rivers of the island. A type of bush called savoka is found on the eastern coastal plain.

Deserts

Arid conditions produce a desert environment in the southernmost part of the island, with spiny desert vegetation resembling that which can be found at the same latitude on the African continent, including dwarf trees, cacti, thorn scrub, aloe and other succulents, and other types of xerophytes (hardy plants that resist drought).

Forests and Jungles

About three-fourths of Madagascar's original evergreen and deciduous forests have been cut down to make way for rice cultivation or grazing land, leaving only about one-tenth of the island forested. Most of this forestland is concentrated on the steep eastern slope of the central plateau region, where evergreens and hardwoods including ebony, rosewood, and raffia palm can be found.

Mangroves, screw pines, palm trees, and reeds grow in the fringe of wooded land that encircles the coast, except in the arid southwest. Also growing in this area is the distinctive baobab tree, with its long, bare, bulbous trunk and small, rootlike bunch of branches at the top. This tree, which is actually a succulent, grows only in Africa, Australia, and Madagascar, and most of its species are found only in Madagascar.

About four-fifths of the flowering plants found in Madagascar's forests are endemic, including some 1,000 species of orchid. One of the island's most famous species is the pitcher plant, whose open "lid" attracts and then traps insects that the plant then feeds on. Madagascar is also known for its profusion of medicinal plants, including the rosy periwinkle, the source of chemical compounds used to treat leukemia.

HUMAN POPULATION

About one-third of the population in Madagascar lives in urban areas. The western part of the country is the least densely populated. On the central plateau, the greatest population concentration is between Fianarantsoa and the capital city of Antanarivo. Antanarivo is the only urban center with a population of more than 1 million.

NATURAL RESOURCES

With very little arable land and only modest mineral reserves, Madagascar is poor in natural resources, a poverty that is mirrored in the dire state of its economy. The country does grow and export modest amounts of rice, bananas, tobacco, coffee, sugar, vanilla, and other crops. Cotton is grown and not only exported, but also used in the textile industry. Chromite, graphite, and mica are commercially exploited. Other mineral deposits include semiprecious stones, and low-grade coal and iron ore.

FURTHER READINGS

Attenborough, David. *Journeys to the Past: Travels in New Guinea, Madagascar, and the Northern Territory of Australia.* Guildford, England: Lutterworth Press, 1981.

Eveleigh, Mark. *Maverick in Madagascar.* London: Lonely Planet, 2001.

Kottak, Conrad Phillip. *The Past in the Present: History, Ecology, and Cultural Variation in Highland Madagascar.* Ann Arbor: University of Michigan Press, 1980.

Lanting, Frans. *Madagascar: A World Out of Time.* New York: Aperture, 1990.

PBS. *The Living Edens.* http://www.pbs.org/edens/madagascar/ (Accessed April 12, 2002).

Stratton, Arthur. *The Great Red Island.* New York: Scribner, 1964.

Malawi

- **Area:** 45,745 sq mi (118,480 sq km) / World Rank: 100
- **Location:** Southern and Eastern Hemispheres, in southeastern Africa, bounded on the north and east by Tanzania, on the east, south and southwest by Mozambique, and on the west by Zambia.
- **Coordinates:** 13°30′S, 34°00′E
- **Borders:** 1,790 mi (2,881 km) / Mozambique, 975 mi (1,569 km); Tanzania, 295 mi (475 km); Zambia, 520 mi (837 km)
- **Coastline:** None
- **Territorial Seas:** None
- **Highest Point:** Mt. Mulanje, 9,849 ft (3,002 m)
- **Lowest Point:** Shire River at Mozambique border, 121 ft (37 m)
- **Longest Distances:** 530 mi (853 km) N-S; 160 mi (257 km) E-W
- **Longest River:** Shire River, 250 mi (400 km)
- **Largest Lake:** Lake Malawi, 11,400 sq mi (29,600 sq km)
- **Natural Hazards:** None
- **Population:** 10,548,250 (July 2001 est.) / World Rank: 71
- **Capital City:** Lilongwe, located in the central part of the country
- **Largest City:** Blantyre, in the south, population 502,053 (1998 est.)

OVERVIEW

Malawi, an inland nation in southeastern Africa, is well within the southern tropics. Its territory extends north-south for 560 mi (901 km) at an average width of less than 100 mi (161 km) in a southern segment of the East African Rift Valley. Comparatively, the area occupied by Malawi is slightly larger than the state of Pennsylvania.

A complex geologic history has contributed to the formation of a landscape of great diversity in elevations and relief features. Within its borders are 9,425 sq mi (24,410 sq km) of water area, mostly in Lake Malawi and two smaller lakes. Floodplains, marshes, hills, plateaus, escarpments, and mountains range from a few hundred feet above sea level in the lower valley of the Shire River to rugged peaks of more than 8,500 ft (2,590 m) in elevation in several widely separated sections of the country.

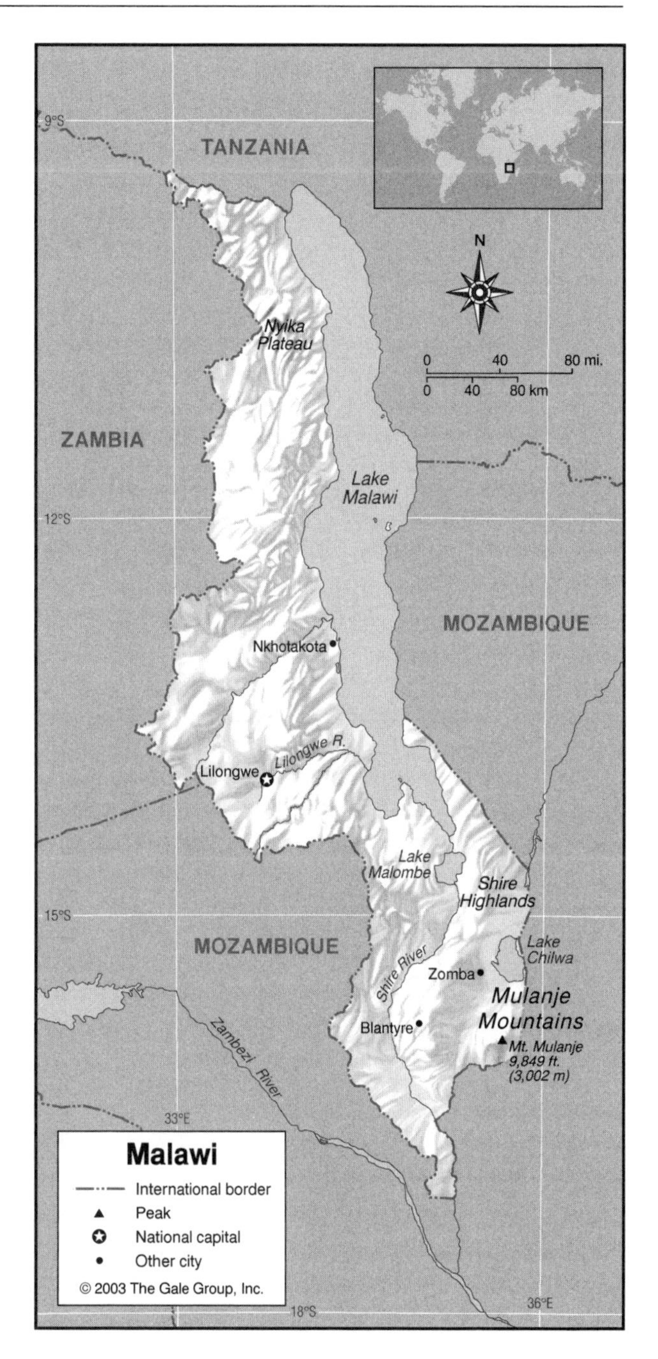

MOUNTAINS AND HILLS

Mountains

A few mountains ranges rise above the level of the highest plateaus. In the north, several peaks on the Nyika Plateau reach 8,500 ft (2,590 m). None of the mountains in the central region exceeds Dedza Mountain, which is just over 7,400 ft (2,255 m) above sea level. In the southern region, Zomba Mountain, standing north of the town of Zomba, has an elevation of more than 6,800 ft (2,072 m). The Mulanje Mountains (also called the Mulanje Plateau, or the Mulanje Massif) near the southeastern border is the highest mountain complex in the country. Various sections of several square miles each are above 9,000 ft (2,743 m) and the highest pinnacle, Mt. Mulanje, rises to 9,849 ft (3,002 m).

Plateaus

Malawi's most important geographic features are the plateaus, which form three-fourths of the land area. The surface of most of these elevated plains has been eroded into low rolling hills and open, shallow valleys.

The Shire Plateau in the southern region, best known by European settlers as the Shire Highlands, covers about 2,800 sq mi (7,251 sq km). It is heavily cultivated, producing both commercial and subsistence crops. Altitudes vary from 2,500 to 4,000 ft (762 to 1,219 m). Blantyre, Malawi's largest town, and Zomba stand in the western sections of this plateau, which slopes gently eastward toward Lake Chilwa. A much broader plateau in the central region, covering about 9,000 sq mi (23,309 sq km) is commonly known as the Lilongwe Plain. It has numerous broad valleys and dambos (areas of moist soils on impermeable subsurface layers) separated by low rounded hills.

The Nyika Plateau in the northern region is the highest formation of this kind in Malawi. It covers some 9,000 sq mi (23,309 sq km) at elevations between 7,000 and 8,000 ft (2,133 and 2,438 m). It is sparsely populated and is less productive than the plateaus farther south.

Hills and badlands

The country has about six other plateaus, some of which are known locally as hills or plains. Most of them are extensive flat or rolling surfaces between 2,500 to 4,500 ft (762 to 1,371 m) above sea level, close to the median altitude for all of Malawi.

INLANDS WATERWAYS

Lakes

Lake Malawi (also called Lake Nyasa), one of the largest and deepest lakes in the world, extends from north to south for more than 350 mi (563 km) occupying the floor of a major southern segment of the East African Rift Valley system. The part of the rift just south of the present-day shores of Lake Malawi was also formerly a part of the lake, and Lake Malombe is a shallow remnant of this larger ancient lakebed. The only internal drainage basin that does not flow into Shire River drains into Lake Chilwa, a complex of lakes and marshes in the southwest that has no outlet to the sea.

Rivers

The Shire River drains the overrun from Lake Malawi, flowing southward across the old lakebed, through Lake Malombe, and then continuing southward toward the Zambezi River. The Lilongwe River is dry for nearly one month each year, and its average flow throughout the dry season is less than 5 percent of its average during the rainy season. West of Zomba, numerous rapids and cataracts have restricted the use of the river for transportation during more recent times.

Wetlands

The shallow and saline Lake Chilwa is subject to seasonal variations in water level. Surrounded by dense swamps, neutral to acid marshes, and seasonally inundated grassland floodplains, the lake consists of numerous islands, two of which are permanently inhabited.

Population Centers – Malawi

(1998 POPULATION)

Name	Population
Blantyre	502,053
Lilongwe (capital)	440,471
Mzuzu	86,980
Zomba	65,915

SOURCE: "1998 Malawi Population and Housing Census," National Statistical Office, Malawi.

Districts – Malawi

POPULATIONS FROM 1998 CENSUS

Name	Population	Area (sq mi)	Area (sq km)	Capital
Blantyre	809,397	777	2,012	Blantyre
Chikwawa	356,682	1,836	4,755	Chikwawa
Chiradzulu	236,050	296	767	Chiradzulu
Chitipa	126,799	1,353	3,504	Chitipa
Dedza	486,682	1,399	3,624	Dedza
Dowa	411,387	1,174	3,041	Dowa
Karonga	194,572	1,141	2,955	Karonga
Kasungu	480,659	3,042	7,878	Kasungu
Lilongwe	1,346,360	2,378	6,159	Lilongwe
Machinga	369,614	2,303	5,964	Machinga
Mangochi	610,239	2,422	6,272	Mangochi
Mchinji	324,941	1,296	3,356	Mchinji
Mulanje	428,322	1,332	3,450	Mulanje
Mwanza	138,015	886	2,295	Mwanza
Mzimba	610,994	4,027	10,430	Mzimba
Nkhata Bay	164,761	1,579	4,090	Nkhata Bay
Nkhotakota	229,460	1,644	4,259	Nkhotakota
Nsanje	194,924	750	1,942	Nsanje
Ntcheu	370,757	1,322	3,424	Ntcheu
Ntchisi	167,880	639	1,655	Ntchisi
Rumphi	128,360	2,298	5,952	Rumphi
Salima	248,214	848	2,196	Salima
Thyolo	458,976	662	1,715	Thyolo
Zomba	546,661	996	2,580	Zomba

SOURCE: "1998 Malawi Population and Housing Census," National Statistical Office, Malawi.

THE COAST, ISLANDS, AND THE OCEAN

Malawi is landlocked.

CLIMATE AND VEGETATION

Temperature

Variations in altitude in Malawi lead to wide differences in climate. The vast water surface of Lake Malawi has a cooling effect, but because of the low elevation, the margins of the lake have long hot seasons and high humidity, with a mean annual temperature of 75°F (24°C). Lilongwe, in Central Malawi, at an elevation of 3,415 ft (1,041 m), has a moderately warm climate with adequate rainfall. The average daily minimum and maximum temperatures in November, the hottest month, are 63°F (17°C) and 84°F (29°C) respectively; those in July, the coolest month, are 45°F (7°C) and 73°F (23°C).

Rainfall

In general, the four seasons may be divided into the cool (May to mid-August); the hot (mid-August to November); the rainy (November to April), with rains continuing longer in the northern and eastern mountains; and the post-rainy (April to May), with temperatures falling in May.

Precipitation is heaviest along the northern coast of Lake Malawi, where the average is more than 64 in (163 cm) per year. About 70 percent of the country averages about 30 to 40 in (75 to 100 cm) annually.

Grasslands

Although about half of the country is classified as forest, some of this includes grasslands, thicket, and scrub.

Forests

There are indigenous softwoods in better-watered areas, with bamboo and cedars on Mt. Mulanje; evergreen conifers grow on the Shire Highlands and in the highlands. Mopane, baobab, acacia, and mahogany trees are among those found at lower elevations.

HUMAN POPULATION

The population density in 1998 was 290 people per sq mi (112 people per sq km), one of the highest in Africa.

GEO-FACT

Lake Malawi National Park, centered in the south of Lake Malawi, was the world's first park established to protect marine life of a tropical, deep water, Rift Valley lake. The waters protected by the park are teeming with the spectacular tropical fish known as the mbuna.

The country is predominantly rural, with only about one-fifth of the population living in urban areas. Most of the population is concentrated in the southeastern region within an arc formed by Lake Chilwa, Zomba, and Blantyre. Another area of population concentration surrounds Lilongwe west of the southernmost part of Lake Malawi. The remainder of the country is sparsely populated.

NATURAL RESOURCES

Malawi's natural resources include limestone and minor deposits of gemstones such as amethyst, ruby, and opal. Deposits of coal, bauxite, and uranium have been identified, but with the exception of coal, had not been mined significantly as of 2002. The economy is predominantly agricultural with about 80 percent of the population living in rural areas. Agriculture accounts for 37 percent of the Gross Domestic Product (GDP) and 85 percent of export revenues. Fishing, mostly in Lake Malawi, is also an important industry. There is abundant water and hydroelectric power potential.

FURTHER READINGS

Briggs, Philip. *Bradt: Guide to Malawi.* 2nd ed. Chalfont St Peter, UK: Bradt Publications, 1999.

O'Toole, Thomas. *Malawi in Pictures.* Minneapolis: Lerner Publishing Group, Feb., 1989.

Young, Anthony. *A Geography of Malawi*, limited ed. London: Evans Bros., 1991.

Malaysia

- **Area:** 127,317 sq mi (329,750 sq km) / World Rank: 66
- **Location:** Southeastern Asia, southern portion of Malaya Peninsula and northern one-third of the island of Borneo, bordering Indonesia to the south and the South China Sea and Thailand to the north
- **Coordinates:** 2°30′N, 112°30′E
- **Borders:** 1,658 mi (2,669 km) / Brunei, 237 mi (381 km); Indonesia, 1,107 mi (1,782 km); Thailand, 314 mi (506 km)
- **Coastline:** 2,905 mi (4,675 km)
- **Territorial Seas:** 12 NM
- **Highest Point:** Mount Kinabalu, 13,451 ft (4,100 m)
- **Lowest Point:** Sea level
- **Longest Distances:** Peninsular Malaysia extends 465 mi (748 km) SSE-NNW and 200 mi (322 km) ENE-WSW. On Borneo, Sarawak extends 422 mi (679 km) NNE-SSW and 158 mi (254 km) ESE-WNW, while Sabah is 256 mi (412 km) E-W; 204 mi (328 km) N-S.
- **Longest River:** Rajang River, 350 mi (565 km)
- **Largest Lake:** Kenyir Reservoir, 143 sq mi (369 sq km)
- **Natural Hazards:** Flooding, landslides
- **Population:** 22,229,040 (July 2001 est.) / World Rank: 48
- **Capital City:** Kuala Lumpur, southwest peninsular Malaysia
- **Largest City:** Kuala Lumpur, 1,297,526 (2000 est.)

OVERVIEW

Malaysia consists of two noncontiguous geographic regions. Peninsular Malaysia (50,806 sq mi / 131,587 sq km), formerly West Malaysia, occupies the southern third of the Malay Peninsula of the Asia mainland. East Malaysia occupies the northern quarter of the island of Borneo. East Malaysia is divided into two parts: Sabah (28,725 sq mi / 74,398 sq km) in the north and Sarawak (48,050 sq mi / 124,449 sq km) in the southwest. These two regions are almost, but not quite, separated by Brunei and Indonesia, which are the other two countries on Borneo. Borneo and the Malay Peninsula are separated by the South China Sea.

About four-fifths of Malaysia's terrain is covered by rain forests, jungle, and swamp. Peninsular Malaysia's terrain consists of a range of steep forest covered mountains with coastal plains to the east and west. Sarawak encompasses an alluvial swampy coastal plain, an area of rolling country interspersed with mountain ranges, and a mountainous interior, the greater part of which is covered with rain forest. Sabah is split in two by the Crocker Mountains, which extend north and south some 30 mi (48 km) inland from the west coast.

Malaysia's history and its strategic location along the Strait of Malacca and southern South China Sea have led to international disputes over rights and ownership.

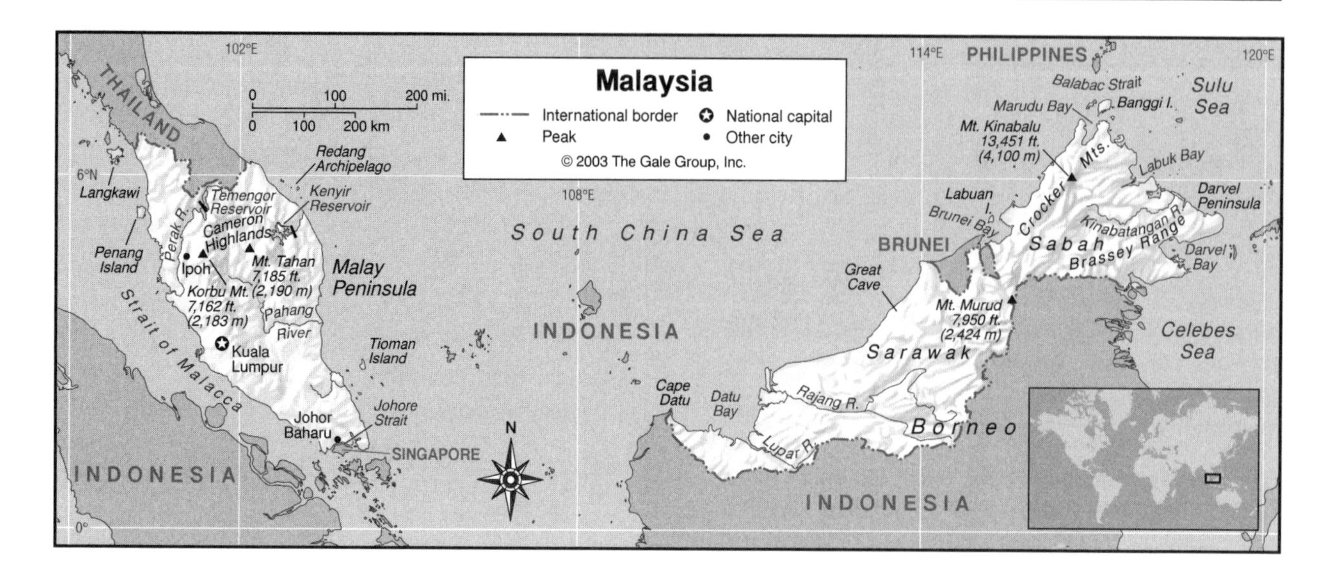

Malaysia is one of several countries that has claims on the Spratly Islands in the South China Sea. The Philippines dispute Malaysia's ownership of Sabah.

MOUNTAINS AND HILLS

Mountains

In Peninsular Malaysia (PM) the northern regions are divided by a series of mountain ranges known as the Cameron Highlands that rise abruptly from the wide, flat coastal plains. The main range, running along the backbone of the peninsula, is the Titiwangsa, stretching for 310 mi (500 km) southward from the border of Thailand. Its highest peak is Korbu, 7,162 ft (2,183 m). A secondary chain lies to the east. It is generally lower in altitude, but does hold the highest mountain in Peninsula Malaysia, Mt. Tahan (7,185 ft / 2,190 m).

The interior of Sarawak is an irregular mountainous mass of unconnected ranges with a mean elevation of about 5,000 ft (1,525 m). Mount Murud is Sarawak's highest peak at 7,950 ft (2,424 m). Mulu 7,793 ft (2,376 m) is its second highest peak, and is famous for its caves.

The interior ranges of Sabah bordering Indonesia are the same complex masses as those of Sarawak. The only continuous mountain range of East Malaysia, the Crocker Range, stretches from 30 mi (48 km) inland from the west coast rising to Malaysia's highest peak, Mount Kinabalu 13,450 ft (4,100 m). Mt. Kinabalu is the highest peak between the Himalayas and New Guinea. The Brassey Range is parallel to but lower than the Crocker Range.

Hills

Hills dominate the terrain in the central Cameron Highlands, between the two mountain chains. The average elevation in this area is 5,999 ft (1,829 m) above sea level. It is regarded as the "Green Bowl" of the country, supplying its produce of cabbages, tomatoes, lettuces, and green peppers to major cities in Malaysia, as well as to Singapore.

Caves

The Malaysian climate, with its combination of heavy rainfall and high temperatures, sets ideal conditions for the formation of limestone caves. Spectacular cave complexes can be found throughout the country. Gua Kelam (Dark Caves), located near the Thai border traverse approximately 1,214 ft (370 m) of limestone hills. Tempurung Cave, near the city of Ipoh, is a white marble and limestone formation made up of five huge domes whose ceilings resemble coconut shells running from east to west with a stream throughout its 0.9 mi (1.5 km) length.

Gunung Mulu National Park is a World Heritage site, within which is one of the most extensive and spectacular limestone cave systems on Earth. Mulu's Sarawak Chamber is the largest natural chamber in the world: 1,968 ft (600 m) in length, 1,361 ft (415 m) in width, and 984 ft (300 m) high. Nearby Deer Cave has two huge entrances at either end of the mountain it penetrates. It is the largest known cave passage, 7,085 ft (2,160 m) long and 728 ft (222 m) deep. Nearly a million bats exodus its colony every evening.

The Great Cave, 7,085 ft (2,160 m) long and 722 ft (220 m) deep, located in Sarawak's Niah National Park, is one of the largest in the world. Evidence of human existence in Borneo dating as far back as 40,000 years ago was discovered in the Niah Caves. The skull of a young Homo sapiens; some tools made out of stone, bone and iron; and cave drawings were found.

INLAND WATERWAYS

Lakes

In the northeast of Peninsular Malaysia is Kenyir Reservoir (143 sq mi / 369 sq km), the largest artificial lake in

Southeast Asia. It is dotted with about 340 islands, which were once hilltops and highlands, and more than 14 waterfalls and numerous rapids. Temengor is another large reservoir, near the Thai border. Tasik Bera, located in southwest Pahang, is the largest natural freshwater lake on the Malay Peninsula. It is situated in the saddle of the main and eastern mountain ranges of the Peninsula, with an area of approximately 270 sq mi (700 sq km).

Rivers

Peninsular Malaysia's main watershed follows the Titiwangsa mountain range about 50 mi (80 km) inland, roughly parallel to the west coast. The rivers flowing to the east, south, and west of this range are swift and have cut some deep gorges, but on reaching the coastal plains they become sluggish. Almost all the states in Malaysia have adopted the names of the principal rivers flowing through their respective territories. Peninsular Malaysia's longest river is the Pahang (285 mi / 458 km). It has its source in the central Cameron Highlands, then flows south and east into the South China Sea. The second longest river on the mainland, the Perak, flows south out of the Temengor Reservoir for 200 mi (322km), parallel with the west coast, before entering the Strait of Malacca. The Kelantin (150 mi / 242 km), which flows north out of the Cameron Highlands, has spectacular waterfalls at Mount Strong and Lata Beringin.

The Rajang flows westward across Sarawak for 350 mi (565 km), making it the longest river in the country. Sarawak's other major river is the Lupar. These rivers and their tributaries are the primary means of inland travel in Sarawak. The same can be said of the Kinabatangan in Sabah, which is almost as long as the Rajang at 349 miles (563 km). The Libang River valley in Sarawak separates the two halves of Brunei from each other.

Wetlands

Malaysia has the world's fifth most extensive mangrove area. The mangrove forests serve as a delicately balanced ecosystem, more than 50 percent of which is found in Sabah. In Sarawak, the wetlands cover an extensive area mainly classified as peat swamp forests covering some 1.2 million hectares. Significant as freshwater habitats, this area is under protection laws.

THE COAST, ISLANDS, AND THE OCEAN

Oceans and Seas

The South China Sea borders Peninsular Malaysia on the east and both Sarawak and Sabah on the west. The South China Sea region is the world's second busiest international sea lane. More than half of the world's supertanker traffic passes through the region's waters. The Strait of Johore is a narrow channel, about 0.8 miles (1 km) wide, separating Malaysia from Singapore at Malaysia's southern tip. To Malaysia's west is the Strait of Malacca, separating Malaysia and Indonesia. It is the shortest route for ships traveling between the northern Indian Ocean and the Pacific, making it a vital shipping route. The Andaman Sea on Malaysia's northwestern coast is part of the Indian Ocean. Sabah is bounded to the north by the Balabac Strait, which connects the South China Sea to the Sulu Sea to the east. The Celebes Sea is southeast of Sabah.

Major Islands

There are islands in all the waters surrounding Malaysia. Langkwai (140 sq mi / 363 sq km in area) is off the northwest coast in the Andaman Sea. Langkwai is actually made up of 99 islands, the largest of which is Palua Senga Besar. Penang (110 sq mi / 285 sq km) is also in the Andaman Sea. A mountainous island with heights of up to 2,719 ft (829 m), it was the site of one of the earliest British colonies in the region and remains densely populated. Off Malaysia's east coast in the South China Sea is Tioman Island, the largest of a group of 64 volcanic islands. The Redang Archipelago comprises nine islands in the South China Sea.

Off Sarawak's coast is the large, swampy, island of Betruit (161 sq mi / 417 sq km). Labuan is an island off the coast of Sabah at the mouth of the Brunei Bay. It encompasses one main island and six smaller ones, thus covering an area of 36 sq mi (92 sq km). Banggi (170 sq mi / 440 sq km) is the largest of the islands off Sabah's northern coast. Sarawak and Sabah are themselves located on northern Borneo, the third largest island on Earth (290,320 sq mi / 751,929 sq km). Malaysia shares Borneo with Brunei and Indonesia. Borneo is part of the Malay Archipelago, most of which is part of Indonesia.

Malaysia, along with the Philippines, China, Taiwan, Vietnam, and Brunei, claims several atolls of the Spratly Islands in the South China Sea. A huge oil reserve is suspected in this region. Palau Batu Putih (Pedra Branca Island) is disputed with Singapore, which is itself a small island nation south of the Malay Peninsula. Sidipan and Ligitan Islands are in dispute with Indonesia.

The Coast and Beaches

The west coast of Peninsular Malaysia is characterized by muddy beaches and wide river plains. Mangrove swamps are common. On the east coast there are many sandy beaches, some quite narrow. The two coasts together form a diamond shape, narrow in the north, broadening near the middle of the peninsula, then narrowing again until they meet in the south. There are no major inlets or capes. Sarawak also has a regular coastline, with the exception of Datu Bay in the west. Both the Rajang and Lupar empty into Datu Bay in wide, swampy, deltas.

Sabah has a more rugged coastline than the other parts of Malaysia, with the mountain ranges often reach-

Population Centers – Malaysia

(1991 POPULATION ESTIMATES)

Name	Population	Name	Population
Kuala Lumpur (capital)	1,145,074	Kelang	368,228
Ipoh	468,765	Petaling Jaya	351,719
Johore Baharu	442,250	Kota Baharu	234,603

SOURCE: "Populations of Capital Cities and Cities of 100,000 and More Inhabitants." United Nations Statistics Division.

Departments – Malaysia

POPULATIONS FROM 2000 CENSUS

Name	Population	Area (sq mi)	Area (sq km)	Capital
Johor	2,740,625	7,330	18,985	Johore Baharu
Kedah	1,649,756	3,639	9,425	Alor Setar
Kelantan	1,313,014	5,765	14,931	Kota Baharu
Labuan	76,067	36	92	Labuan
Melaka	635,791	640	1,658	Melaka
Negeri Sembilan	859,924	2,565	6,646	Seremban
Pahang	1,288,376	13,884	35,960	Kuantan
Perak	2,051,236	8,110	21,005	Ipoh
Perlis	204,450	307	795	Kangar
Pulau Pinang	1,313,449	398	1,031	Pinang (Georgetown)
Sabah	2,603,485	29,460	73,711	Kota Kinabalu
Sarawak	2,071,506	48,050	124,449	Kuching
Selangor	4,188,876	3,072	7,956	Shah Alarn
Terengganu	898,825	5,002	12,955	Kuala Terengganu
Wilayah Persekutuan	1,379,310	94	243	Kuala Lumpur

SOURCE: "Key Summary Statistics by State, Malaysia, 2000." Department of Statistics, Malaysia.

ing very near the shore. Brunei Bay is in the west, Marudu, the north. Labuk and Darvel Bays are separated by the Darvel Peninsula in eastern Sabah. A number of offshore islands around Sabah support extensive and diverse coral reefs. In general, Malaysia's coasts have optimal conditions for reef development.

CLIMATE AND VEGETATION

Temperature

Malaysia is characterized by fairly high but uniform temperatures, ranging from 73° to 88° F (23° to 31° C), with high humidity. Lying very close to the equator, Malaysia's seasons are based primarily on rainfall patterns.

Rainfall

Peninsular Malaysia experiences copious rainfall, averaging about 100 in (250 cm) annually in two monsoon seasons. The heaviest rains are from October to December or January (the northwest monsoon season). Except for a few mountain areas, the most abundant rainfall is in the east coast region, where it averages more than 120 in (300 cm) per year. Elsewhere the annual average is 80 to120 in (200 to 300 cm), with the northwestern and southwestern regions having the least rainfall. The nights are usually cool because of the nearby seas. The southwest monsoon season (April to October) is characterized by squalls and thunderstorms.

Forests and Jungles

The jungles of Malaysia are said to be the oldest in the world. Covering more than two thirds of the country they stretch from the mangrove swamps of the west coast, through freshwater swamps, to lowland hardwood forests, heath forests, and mountain forest. There are believed to be around 8,500 species of flowering plants and ferns, and 2,500 species of trees, in Malaysia's forests. About 59 percent of Malaysia's total land area is tropical rainforest. The Titiwangsa Range has the largest remaining continuous forest tract in Peninsular Malaysia. Most of the Sabah's interior is covered with tropical forest, and rainforests cover the greater part of Sarawak. Malaysia's forests play a vital role in both its economic life and its climate.

HUMAN POPULATION

In 2000 it was estimated that 57 percent of the population lived in urban areas. The estimated 2001 population growth rate was 1.96 percent. The population is predominantly ethnic Malay, but there are substantial Chinese and Indian minorities.

NATURAL RESOURCES

Malaysia has large deposits of tin on the Malay Peninsula. Other mineral resources include copper, iron ore, and bauxite. Petroleum exports are a major part of the economy, and natural gas is also available. The rainforests provide abundant timber and other forest products.

FURTHER READING

Aiken, Robert S., et al. *Development and Environment in Peninsular Malaysia.* New York: McGraw-Hill International Book Co., 1982.

Fascinating Malaysia. http://www.fascinatingmalaysia.com/naad/index.html (Accessed June 21, 2002).

Hutton, Wendy. *Discovering Sabah.* Kota Kinabalu, Sabah, Malaysia: Natural History Publications (Borneo), 2001.

Komoguchi, Yoshimi, ed. *Human Ecology in Rural Malaysia.* Tokyo: Institute for Applied Geography, Komazawa University, 1995.

Major, John S. *The Land and People of Malaysia and Brunei.* New York: HarperCollins, 1991.

Rain, Nick. *Enchanting Islands and Coastal Havens : Malaysia, Thailand, Singapore. Kuala Lumpur, Malaysia*: S. Abdul Majeed, 1995.

United Nations System-Wide Earthwatch. *Malaysia's Islands.* http://www.unep.ch/islands/IHC.htm (Accessed June 21, 2002).

Voon, Phin Keong, and Tunku Shamsul Bahrin, eds. "The View from Within: Geographical Essays on Malaysia and Southeast Asia." *Malaysian Journal of Tropical Geography.* Kuala Lumpur: Dept. Geography, University of Malaysia, 1992.

Maldives

- **Area:** 116 sq mi (300 sq km) / World Rank: 194
- **Location:** Straddling the equator, in the Eastern, Northern, and Southern Hemispheres, in the Indian Ocean, south of India, west of Sri Lanka
- **Coordinates:** 3°15′N, 73°00′E
- **Borders:** None
- **Coastline:** 400 mi (644 km)
- **Territorial Seas:** 12 NM
- **Highest Point:** Unnamed location on Wilingili island in the Addu Atoll, 7.9 ft (2.4 m)
- **Lowest Point:** Sea level
- **Longest Distances:** 510 mi (823 km) N-S / 82 mi (133 km) E-W
- **Longest River:** None of significant size
- **Natural Hazards:** Sensitivity to sea level rise because of the Maldives's level, low-lying land
- **Population:** 310,764 (July 2001 est.) / World Rank: 170
- **Capital City:** Malé, in the Malé Atoll, on the largest island in the chain
- **Largest City:** Malé, 68,000 (2000 est.)

OVERVIEW

Consisting of an archipelago of almost 1,200 coral islands and sand banks in the Indian Ocean, Maldives is the smallest country in Asia. Some of the islands are level and low-lying and are gradually washing away into the ocean; others are in the process of formation and are constantly growing in size. Most islands have freshwater lagoons, and all have coastal reefs. The largest atoll group is the Malé Atoll, where the capital city is located.

Because of its equatorial position, the Maldives have a humid and hot climate, with two monsoon seasons. More violent storms are possible in the northern atolls than those in the south.

GEO-FACT

Because of concerns that rising sea levels may eventually submerge the low-lying islands of Maldives, the government has undertaken programs to prepare for the worst. A six-foot concrete retaining wall, known as the Great Wall of Malé, has been constructed to protect the capital. Even more drastic is the construction of 4 sq mi (10 sq km) artificial island, Hulhumale off the coast of Malé. With an elevation of 6 ft (2 m), Hulhumale is about 2 ft (.6 m) higher than Malé itself. The government plans to use Hulhumale to support the country's population if the other islands become inhabitable due to flooding. People will be encouraged to settle on Hulhumale when it is completed, even before flooding threatens their homes. Trees and rabbits are already thriving there, and light industry will be established on the island as soon as possible.

MOUNTAINS AND HILLS

Consisting of coral islands, the Maldives are almost completely flat.

INLAND WATERWAYS

The islands of the Maldives are too small to support inland waterways of any significant size.

THE COAST, ISLANDS, AND THE OCEAN

Oceans and Seas

The Maldives are located in the Indian Ocean, about 400 mi (645 km) southwest of Sri Lanka. Four ocean channels cross through the archipelago from east to west. These are the Kardiva Channel, Veimandu Channel, One and a Half Degree Channel, and Equatorial Channel.

A protective, fringing coral reef surrounds each island, and many islands have freshwater lagoons. Small

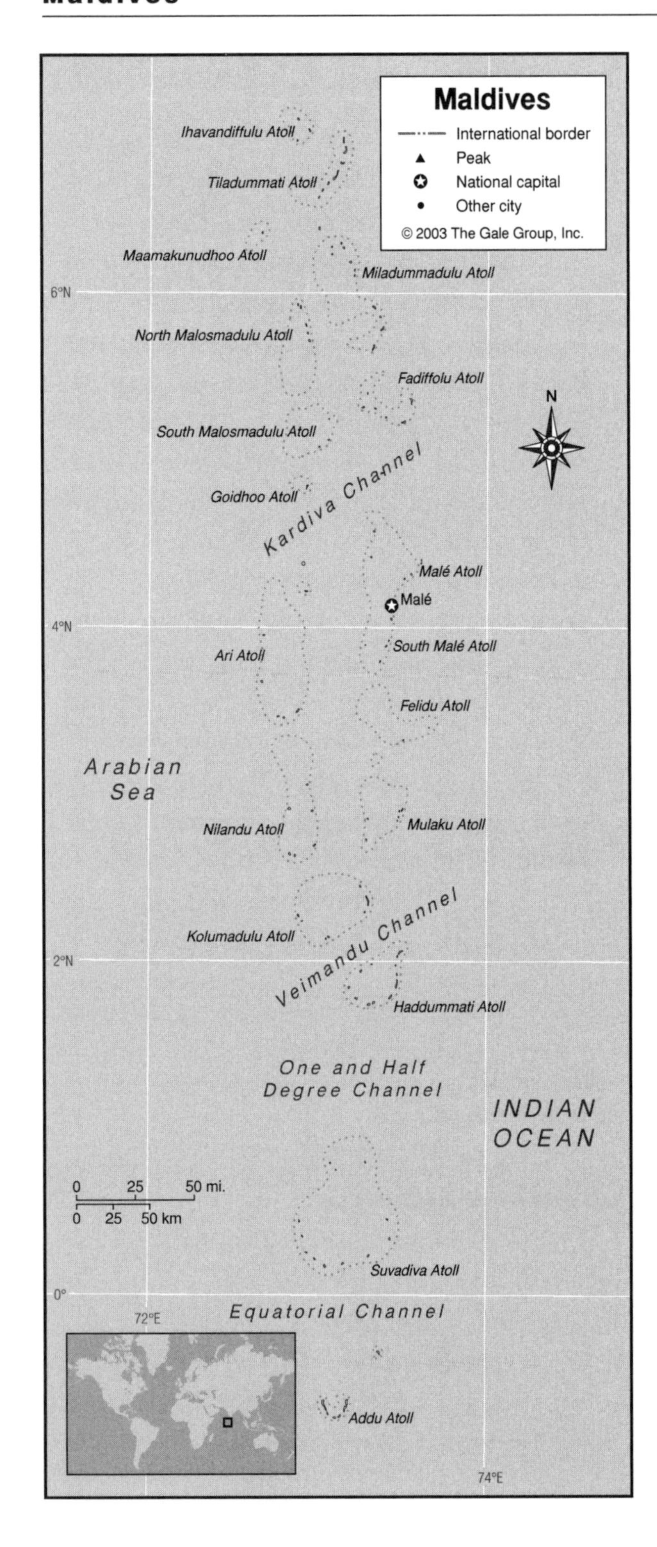

patch reefs and faroes (unusual ring-shaped reefs) are located in Malé Atoll's lagoon.

Major Islands

The atoll groups, from north to south, are: Ihavandiffulu Atoll, Tiladummati Atoll, Miladummadulu Atoll, North Malosmadulu and South Malosmadulu Atolls, and Fadiffolu Atoll; separated from these atolls by the Kardiva Channel are Malé Atoll, South Malé Atoll, Ari Atoll, Felidu Atoll, Nilandu Atoll, Mulaku Atoll, and Kolumadulu Atoll; further south, separated by the Veimandu Channel is Haddummati Atoll; the One and a Half Degree Channel separates the Suvadiva Atoll; and the Equatorial Channel separates the most southerly atoll, Addu Atoll.

All the islands of the Maldives are small, but the island of Malé, location of the capital city, is the most densely populated and developed. It is 1.2 mi (2 km) long and just over one-half mile (1 km) wide. Land reclamation projects have increased the size of Malé, but travelers must use an airport on a neighboring island and then be transported to Malé by boat, since there is not enough room for an airport runway. Sea walls surrounded the island on all sides.

To the far south in the Maldives lies Addu Atoll, where the town of Seenu is located. The British used these south-lying atolls as an air base during World War II, and undertook some engineering projects during the 1950s following the end of the war. Among these was a causeway connecting Fua Mulaku, Midu, and Hitadu, and Gan islands in the Addu Atoll.

The Coast and Beaches

White coral sand covers Maldives's flat beaches, and there is no trace of yellow or black coloring in the sand, as there is on other beaches in the world.

CLIMATE AND VEGETATION

Temperature

The Maldives' climate is equatorial—usually hot and humid, with an average temperature of about 81°F (27°C). During the northeast monsoon (from November to March), the weather is mild and comfortable, while the southwest monsoon season (from June to August) is extremely rainy and violent.

Rainfall

In the south, rainfall averages approximately 150 in (380 cm), and in the north it averages 100 in (250 cm).

Grasslands

Dense scrub covers the islands. The central islands are less fertile than the northern and southern groups, and the western islands are less fertile than the eastern ones.

Forests and Jungles

There are no thick jungles on the islands because of the poor soil, but small areas of rainforest exist on the larger islands that have more precipition. Coconut, mango, plantain, banyan, and mango trees thrive in the tropical climate, as well as flowers and shrubs.

Departments – Maldives

POPULATIONS FROM 1995 CENSUS

Name	Population	Capital
Alifu Alifu	4,852	Rasdhoo
Alifu Dhaal	6,764	Mahibadhoo
Baa	10,147	Eydhafushi
Dhaalu	5,587	Kudahuvadhoo
Faafu	3,573	Magoodhoo
Gaafu Alifu	9,453	Villingili
Gaafu Dhaalu	15,067	Thinadhoo
Ghaviyani	8,034	Fuvahmulah
Haa Alifu	15,893	Dhidhdhoo
Haa Dhaalu	17,755	Kulhudhuffushi
Kaafu	7,855	Male
Laamu	11,320	Fondahoo
Lhaviyani	9,250	Naifaru
Malé	34,583	Malé
Meemu	5,543	Muli
Noonu	11,457	Manadhoo
Raa	15,129	Ugoofaaru
Seenu	23,835	Hithadhoo
Shaviyani	11,792	Funadhoo
Thaa	11,620	Veymandhoo
Vaavu	1,823	Felidhoo

SOURCE: Census of Population by Locality, 1995, Ministry of Planning and National Development, Maldives.

HUMAN POPULATION

The capital, Malé, the only urban settlement, had an estimated population of 68,000 in 2000, and the size of the island is roughly 0.75 mi (1.9 sq km). About 200 of the 1,200 islands are inhabited.

NATURAL RESOURCES

Fish is Maldives's only natural resource of note, especially yellowfin tuna and skipjack, with almost half of the harvest dried, frozen, or canned for export. Also important to the economy is the tourism industry.

FURTHER READINGS

Balla, Mark, and Willox, Robert. *Maldives & Islands of the East Indian Ocean.* 2nd Edition. Berkeley, CA: Lonely Planet, 1993.

Heyerdahl, Thor. *The Maldives Mystery.* Bethesda, Md.: Adler & Adler, 1986.

Maldive Holidays. *Maldives... the Last Paradise.* http://www.maldive.com/geog/mgeog.html (accessed March 13, 2002).

Reynolds, C.H.B. *Maldives.* Santa Barbara, CA: Clio Press, 1993.

TheMaldives.com. *Maldives Geography.* http://www.themaldives.com/Maldives/Maldives_geography.htm (accessed March 13, 2002).

Mali

- **Area:** 478,767 sq mi (1,240,000 sq km) / World Rank: 25
- **Location:** Northern and Western hemispheres, West Africa, with Algeria to the north, Niger to the east, Burkina Faso and Cote d'Ivoire to the south, Guinea to the southwest, and Senegal and Mauritania to the west.
- **Coordinates:** 17°00′N, 4°00′W
- **Borders:** 4,661 mi (7,243 km) / Algeria, 855 mi (1,376 km); Burkina Faso, 621 mi (1,000 km); Côte d'Ivoire, 331 mi (532 km); Guinea, 533 mi (858 km); Mauritania, 1,390 mi (2,237 km); Niger, 510 mi (821 km); Senegal, 260 mi (419 km)
- **Coastline:** None
- **Territorial Seas:** None
- **Highest Point:** Mount Hombori Tondo, 3,789 ft (1,155 m)
- **Lowest Point:** Sénégal River, 75 ft (23 m)
- **Longest Distances:** 1,852 kin (1,151 mi) ENE-WSW; 1,258 km (782 mi) NNW-SSE
- **Longest River:** Niger, 2,600 mi (4,185 km)
- **Largest Lake:** Lake Faguibine, 228 sq mi (590 sq km)
- **Natural Hazards:** Subject to harmattan dust storms and frequent drought
- **Population:** 11,008,518 (July 2001 est.) / World Rank: 68
- **Capital City:** Bamako, in southwest Mali
- **Largest City:** Bamako, population 1,160,000 (2000 est.)

OVERVIEW

Mali, a landlocked nation, is located in western Africa and is crossed by the Niger River. The country's terrain is mostly flat, arid, and sandy. Mali can be roughly divided into three geographic regions: the southern region, where rainfall is greatest; the Sahel, the semi-desert region in the center of the country; and the Sahara Desert region of the far north.

The Sénégal River flows in the western section of the county. The Niger, one of Africa's major rivers, forms a semicircle in the south-central region, separating the semi-arid Sahel from the highlands. The desert region of the north is dotted by oases that were stopovers for caravans traveling the Sahara Desert in ancient times. Most of the population lives in the south, in the cities and town along the Niger and Baoulé and Bani.

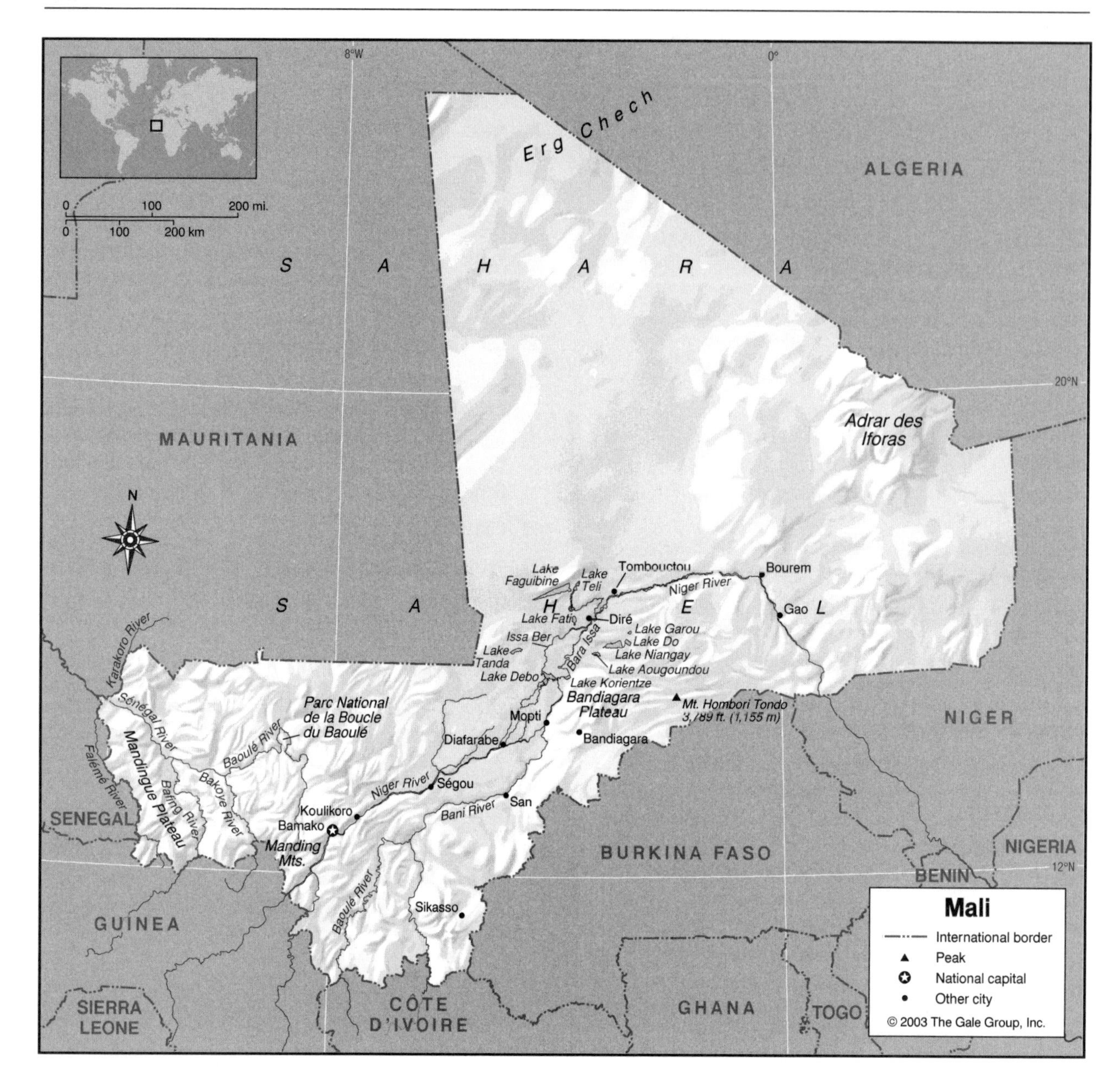

MOUNTAINS AND HILLS

Mountains

In the south the Fouta Djallon Highlands and the Manding Mountains provide a barrier separating Mali from Guinea. These mountains are relatively low, with deep valleys formed by the rivers and their tributaries.

The eastern region contains two spectacular mountain ranges: the Bandiagara Plateau and the Hombori Mountains, the highest points of which are the holy mountain called the Hand of Fatima, and Mount Hombori Tondo. Mount Hombori Tondo is the highest point in Mali, with an elevation of 3,789 ft (1,155 m).

In the south-central area, dramatic sandstone cliffs (2,000 ft / 600 m) in the area of Bandiagara run from southwest to northeast. The local people, the Dogon, have built villages into sheer faces of escarpments in the area of the cliffs. As of the twenty-first century, these ancient, abandoned or sparsely populated villages, despite their remote location, have become a tourist attraction. The pyramidal or rectangular structures are built of mud, with the wood supports protruding at regular intervals. Rock paths connect the structures. Dogon people sleep on the flat roofs of their dwellings.

Plateaus

Other than the Bandiagara Plateau there are two plateau regions in Mali. The Adrar des Iforas is an eroded massif (sandstone plateau) that rises to 2,640 ft (800 m) in northeastern Mali near the Niger and Algeria borders. It is part of the Hoggar Mountain System that extends into Algeria. In the opposite corner of the country the Mandingue Plateau runs along the border with Senegal turning south and extending into Guinea.

GEO-FACT

Tombouctou (Timbuktu) has been a center of Islamic learning since the seventeenth century. The city's Sankore Mosque, of golden clay with its wooden support structure protruding, is a well-know landmark and center for Islamic study in Africa.

INLAND WATERWAYS

Lakes

After the September–December rainy season, the delta region of the Niger—about 19,000 sq mi (30,000 sq km) in total area—is flooded. Grasslands become green, and the seasonal lakes—Debo, Fati, Teli, Korientze, Tanda, Niangay, Do, Garou, Aougoundou, and others—are filled with water. Following the rainy season, there is fishing on the lakes, especially on Lake Debo.

The only two perennial lakes of any real size are located in the center of the country on either side of the Niger River. To the east of the river sits Lake Niangay, and northwest of this lake is the larger Lake Faguibine. Lake Faguibine is the largest lake in Mali, with a rainy season surface area of 228 sq mi (590 sq km).

Rivers

Two main rivers cut through Mali: the Niger and the Sénégal. The Niger River traverses Mali for 1,060 mi (1,700 km), more than one-third of its total length of 2,600 mi (4,185 km). Beyond the town of Ségou, the Niger forms a vast inland delta and then joins with its main tributary, the Bani, at Mopti. The Bani River itself is navigable for about 140 mi (225 km) between San and Mopti. Beyond Mopti the Niger breaks up into two channels, the Bara Issa and the Issa Ber, that spread out in a broad flood plain covering 40,000 sq mi (103,600 sq km). A number of shallow seasonal lakes lie in this area. Just above Diré between Lakes Niangay and Fagubine, the two main branches join again, changing to an eastern direction beyond Tombouctou (Timbuktu) and making a great bend toward the southeast at Bourem. Depending on rainfall, the Niger is often navigable in Mali for large craft from Koulikoro to Gao during the high-water period (August to January). However, between Koulikoro and Bamako there are rapids. Fishing on the Niger and in the lakes of the flood plain is an important economic activity.

In western Mali, the Sénégal River is formed at the small town of Bafoulabé through the confluence of the Bafing and Bakoye Rivers. The Falémé River lies along the border with Senegal. It joins other tributaries—including the Karakoro River that originates in Mauritania near the Mauritania, Mali, and Senegal borders—and becomes the Sénégal. The Gorgol River, which also originates in Mauritania, joins it about 125 mi (200 km) downstream.

Wetlands

After the rainy season floods the Niger River Delta area, the swamps between Mopti and Diafarabe flood, providing habitat for wading birds such as the white heron.

THE COAST, ISLANDS, AND THE OCEAN

Mali is a landlocked nation.

CLIMATE AND VEGETATION

Temperature

Temperatures range by season and region. In Bamako in the southwest, temperatures in June-September average 68°F (20°C). In the February-May hot, dry season temperatures average 95°F (35°C). In the Sahelian region, the average annual temperature is 86°F (30°C).

Rainfall

The rainy season is from June to September although this really only applies to the south; northern regions rarely receive rainfall. Average annual rainfall in the south is approximately 55 in (140 cm), and in the north, rainfall averages only 8 in (20 cm). However, precipitation varies considerably from year to year. In is not uncommon for less than 3 in (8 cm) of rain to fall annually in the far northern Sahara Desert area.

Grasslands

The center part of Mali, lying between Mauritania and Niger, is the semi-arid Sahel. The Sahel is the region that lies between the Sahara Desert and the forests the lie closer to the Atlantic coast of Africa. Historically, the Sahel was dedicated to grazing, but years of drought have caused much of the central land to begin the transition to desert. In the upper southern region, the Niger and Bani rivers join to form a rich inland delta with green grasses during the wet season.

Periodically through history, most recently in the 1970s (when an estimated 200,000 people died due to drought and the resulting famine) and the early 1980s, the Sahel experiences severe drought. The recovery is more difficult following each severe drought, since soil erosion, lack of available water for irrigation, deforestation, and the pressure of a growing population all combine to prevent the ecosystem from returning to its normal healthy state. The terrain in places like Tombouctou, which was once in the heart of the Sahel, is now more desert-like. International organizations in the early twenty-first century were sponsoring tree planting and

Population Centers – Mali

(POPULATIONS FROM 2000 CENSUS)

Name	Population
Bamako (capital)	1,016,167
Sikasso	125,000

SOURCE: Projected from United Nations Statistics Division data.

Regions – Mali

POPULATIONS FROM 1998 CENSUS

Name	Population	Area (sq mi)	Area (sq km)	Capital
Bamako	1,178,977	103	267	Bamako
Gao	495,178	124,323	321,996	Gao
Kayes	1,424,657	76,356	197,760	Kayes
Koulikoro	1,620,811	34,685	89,833	Koulikoro
Mopti	1,405,370	34,257	88,752	Mopti
Ségou	1,652,594	21,671	56,127	Ségou
Sikasso	1,839,747	29,529	76,480	Sikasso
Tombouctou	496,312	157,907	408,977	Tombouctou

SOURCE: Ministry for the Administration of Territories and Locales of the Republic of Mali. As cited by Johan van der Heyden. Geohive. http://www.geohive.com (accessed June 2002).

other environmental restoration programs to halt the desertification of the Sahel.

Deserts

The Niger River Valley forms the southernmost extent of the Sahara Desert. Northern Mali lies completely within the Sahara Desert's territory. In the extreme north, the Erg Chech, a region that straddles Mali and Algeria, is characterized by ergs—deep and shifting parallel dunes in the sand. This region also contains vast plains known as the Tanezrouft, where reddish sandstone formations are the extension of the Ahaggar Mountains of Algeria, and Taoudenni, where salt has been mined for centuries.

In the oases (low-lying places where water allows some vegetation to grow) of the Sahara, small stands of trees such as a desert form of acadia and date palm may be found.

Forests and Jungles

Parc National de la Boucle du Baoulé, a nature reserve and national park, lies south of the Baoulé River in southern Mali just north of the border with Côte d'Ivoire. The territory contains mostly wooded savannah, forest areas, and provides protected habitat for a number of species of African birds and animals, including buffalo, antelope, warthog, elephant, hippopotamus, hyena, chimpanzee, and baboon. Illegal hunting has drastically reduced the populations of both giraffe and lion.

HUMAN POPULATION

Mali's average population density is 23 people per sq mi (9 people per sq km), but it is much higher along the Niger River where roughly three-fourths of the county's population live. As the environment has suffered due to cyclical severe drought, more and more people are moving to the cities; in 1980, barely 10 percent of the population lived in cities in Mali, but by 1990, the percentage had grown to exceed 30 percent. As a result, the cities in Mali and its neighbors are struggling to provide services—housing, power, water, sewer, garbage collection, health care—to the growing population.

NATURAL RESOURCES

Gold is mined in the south along the border with Côte d'Ivoire. There are also phosphate mines, limestone and marble quarries, and a small salt mine in the north near Taoudenni. Deposits of other minerals, including bauxite, iron ore, manganese, tin, and copper have been discovered, but the development of mining requires improved infrastructure. Mali also uses its rivers, which are swiftly flowing at points, to provide hydropower.

FURTHER READINGS

Bingen, R. James, David Robinson, and John M. Staatz, eds. *Democracy and Development in Mali.* East Lansing: Michigan State University Press, 2000.

Celati, Gianni. *Adventures in Africa.* Chicago: University of Chicago Press, 2000.

Durou, Jean-Marc. *Sahara.* New York: Harry N. Abrams, 2000.

Embassy of Mali in Washington D.C. http://www.maliembassy-usa.org/index.html (Accessed June 13, 2002).

Imperato, Pascal James. *Historical Dictionary of Mali.* Lanham, Md.: Scarecrow Press, 1996.

Keenan, Jeremy. *Sahara Man: Travelling with the Tuareg.* London: J. Murray, 2001.

PBS Online. *The Sahara.* http://www.pbs.org/wnet/africa/explore/sahara/sahara_overview_lo.html (Accessed June 13, 2002).

Scott, Chris. *Sahara Overland: A Route and Planning Guide.* Surrey, Eng.: Trailblazer Publications, 2000.

Malta

- **Area:** 122 sq mi (316 sq km) / World Rank: 193
- **Location:** Part of a chain of five islands in the central Mediterranean Sea, 58 mi (93 km) south of Sicily in the Northern and Eastern Hemispheres
- **Coordinates:** 35°50′N, 14°35′E

GEO-FACT

Malta is the site of the world's most ancient temple complexes, about 6,000 years old. The islands' limestone megaliths are centuries older than Stonehenge or the Egyptian pyramids.

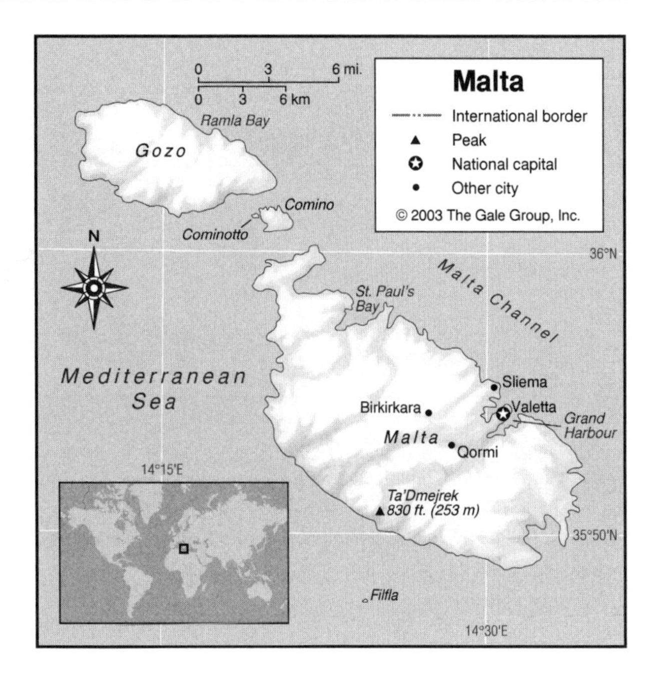

- **Borders:** None
- **Coastline:** 157 mi (252.81 km)
- **Territorial Seas:** 12 NM
- **Highest Point:** Ta'Dmejrek, 830 ft (253 m)
- **Lowest Point:** Sea level
- **Longest Distances:** 28 mi (45 km) SE-NW / 8 mi (13 km) NE-SW
- **Longest River:** None of significant size
- **Natural Hazards:** None
- **Population:** 394,583 (July 2001 est.) / World Rank: 168
- **Capital City:** Valletta, located on the east coast of Malta
- **Largest City:** Valletta, 99,000 (2000 est.)

OVERVIEW

Malta is an archipelago of five islands in the central Mediterranean Sea in the African Plate. Three of the islands (Malta, Gozo, and Comino) are inhabited, and two (Cominotto and Filfla) are uninhabited. The islands are almost treeless, with rocky terrain that has openings that form deep harbors, bays, creeks, and rocky coves. Summers are fairly hot and dry, and winters are rainy and mild. There are at least six hours of daylight year-round, with almost 12 hours in summer.

MOUNTAINS AND HILLS

The islands' terrain consists of low hills (mostly limestone formations) running east to northwest to a height of 786 ft (239 m). There is little vegetation, and no forests. One of the islands, Gozo, is greener and hillier than Malta, and its coast has high uneven cliffs.

INLAND WATERWAYS

Due to their small size, the islands of Malta do not support any significant waterways.

THE COAST, ISLANDS, AND THE OCEAN

Malta is surrounded by the Mediterranean Sea. The island of Malta is the largest in the country, accounting for 94.9 mi (245.8 km) of the total area. Gozo (25.9 mi / 67.1 km) and Comino (about 1 mi / 2.8 km) are much smaller.

There are about 20 beaches on Malta, ranging from rocky to sandy textures. Gozo also has some popular beaches, one of which is at Ramla Bay of the northern shore, where the beach features reddish sand. The main attraction on Comino is Santa Maria Bay, known for its clear waters and the coastal lagoon, Blue Lagoon.

CLIMATE AND VEGETATION

Temperature

The average winter temperature is 48°F (9°C), with January being the coldest month. The average summer temperature is 88°F (31°C), the highest temperature which occurs during midsummer (July to August).

Rainfall

Most rainfall occurs between November and January, and average rainfall is approximately 22 in (56 cm) per year.

HUMAN POPULATION

It is estimated that 91 percent of the population lives in urban areas. The capital, Valletta, which is also the chief port, houses most of the population. Other larger cities include Birkirkara, Qormi, and Sliema. The countryside to the north is rugged and sparsely populated.

NATURAL RESOURCES

Limestone is one of Malta's few natural resources. Agriculture is limited because of the rocky land of the

Regions – Malta

Name	Area (sq mi)	Area (sq km)
Gozo and Comino	27	70
Inner Harbour	6	15
Northern	30	78
Outer Harbour	12	32
South Eastern	20	53
Western	27	69

SOURCE: *Geo-Data: The World Geographical Encyclopedia,* 2nd ed. Detroit: Gale Research, 1989.

islands; most food has to be imported. Industrial raw materials need to be imported as well.

Wheat, barley, and grapes are grown and consumed, while fruit, seeds, potatoes, onions, wine, and cut flowers are major exports.

FURTHER READINGS

Berg, Warren G. *Historical Dictionary of Malta.* Lanham, Md.: Scarecrow Press, 1995.

Clews, Hilary A., ed. *The Year Book, 1987.* Sliema: De La Salle Brothers, 1987.

Ellis, William S. "Malta: The Passion of Freedom." *National Geographic,* June 1989, pp. 700-17.

LonelyPlanet.com. *Malta—Lonely Plant World Guide.* http://www.lonelyplanet.com/destinations/europe/malta/ (accessed March 1, 2002).

VisitMalta.com. *Malta—Welcome to the Heart of the Mediterranean.* http://www.visitmalta.com/ (accessed March 1, 2002).

Marshall Islands

- **Area:** 70 sq mi (181 sq km) / World Rank: 198
- **Location:** Northern and Eastern Hemispheres, in the central Pacific Ocean, between Hawaii and Papau New Guinea, near Kiribati
- **Coordinates:** 9°00′N, 168°00′E
- **Borders:** none
- **Coastline:** 230 mi (370 km)
- **Territorial Seas:** 12 NM
- **Highest Point:** Unnamed location on Likiep, 33 ft (10 m)
- **Lowest Point:** Sea level
- **Longest River:** None of significant size.
- **Natural Hazards:** Typhoons
- **Population:** 70,822 (July 2001 est.) / World Rank: 188
- **Capital City:** Majuro, located on the Majuro island
- **Largest City:** Majuro, 28,000 (2000 est.)

OVERVIEW

Lying in the central-western part of the Pacific Ocean, the Marshall Islands are comprised of 1,152 islands (five of which are major islands) and 29 atolls, which form two almost parallel chainlike formations known as the Ratak (Sunrise), or Eastern, group and the Ralik (Sunset), or Western, group. Most of the islands have an atoll formation of narrow strips of low-lying land enclosing a lagoon.

MOUNTAINS AND HILLS

There are no elevations of note in the Marshall Islands; the average elevation for the country is 7 ft (2 m) above sea level.

INLAND WATERWAYS

The islands of the Marshall Islands are too small to support any bodies of water larger than small lagoons and ponds.

THE COAST, ISLANDS, AND THE OCEAN

Located in the central Pacific Ocean, Marshall Islands has 870 reef systems, and about 160 coral species. This area in the Pacific is also home to numerous Japanese and American World War II wrecks.

Atolls, narrow strips of low land that enclose a lagoon, make up the majority of Marshall Islands. The Ratak Group includes Mili, Majuro, Maloelap, Wotje, Likiep, Rongelap, Ailinginae, Bikini, Enewetok, Ujelang Atolls. The Ralik Group includes Namorik, Ebon, Jaluit, Ailinglaplap, and Kwajalein Atolls. There are also low coral limestone and sand islands and islets. Calalien Pass, the main channel in Majuro, is deep and wide-open, and it allows large container ships to pass between the ocean and the lagoon.

CLIMATE AND VEGETATION

Temperature

Since the Marshall Islands are located near the equator, the climate is hot and humid and there is little change between seasonal temperatures. Daily temperatures generally vary between 70°F and 93°F (21°C and 34°C). The high temperatures are cooled from December through March by trade winds that blow in from the northeast.

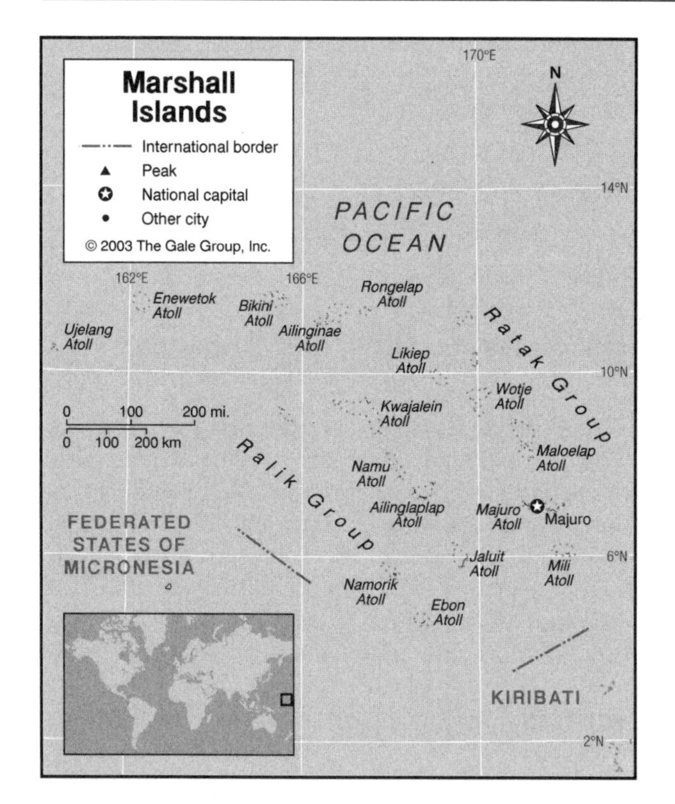

Rainfall

Per month, rainfall averages approximately 12-15 in (30-38 cm). The wettest months are October and November, and the driest are December through April. Because average rainfall increases from north to south, the northern atolls average 70 in (178 cm) annually, while the southern atolls average 170 in (432 cm).

Forests and Jungles

Coconut palms, breadfruit, pandanus, and citrus trees are the dominant tree species, and about 22,000 acres (8,900 ha) of land is planted with coconut palms.

HUMAN POPULATION

In 2000, about 72 percent of the population lived in urban areas. Approximately 60 percent of the total population reside on Majuro and Ebeye, which are atolls. Twenty-four of the 29 atolls and five major islands are inhabited. The outer atolls are sparsely populated.

GEO-FACT

Bikini Atoll in the northern Marshall Islands was the site of U.S. nuclear weapons tests between 1947 and 1962.

NATURAL RESOURCES

Marshall Islands' natural resources are deep seabed minerals, phosphate deposits, and marine products. Copra, dried coconut meat, is produced on almost all islands and atolls. Hawaii, located 2,100 mi (3,381 km) away, is Marshall Islands' closest major trading partner.

FURTHER READINGS

Dibblin, Jane. *Day of Two Suns: U.S. Nuclear Testing and the Pacific Islanders.* New York: New Amsterdam, 1990.

LonelyPlanet.com. *Marshall Islands.* http://www.lonelyplanet.com/destinations/pacific/marshall_islands/environment.htm (accessed March 18, 2002).

Republic of the Marshall Islands. http://www.rmiembassyus.org/about/geography.html (accessed March 18, 2002).

U.S. Department of State. *Trust Territory of the Pacific Islands, 1986.* Washington, D.C.: U.S. Department of State, 1986.

Weisgall, Jonathan M. *Operation Crossroads: The Atomic Tests at Bikini Atoll.* Annapolis, Md.: Naval Institute Press, 1994.

Martinique

An overseas department of France

- **Area:** 425 sq mi (1,100 sq km) / World Rank: 174
- **Location:** Northern and Western Hemispheres, bordered by the Caribbean Sea and Atlantic Ocean, near the South American continent, part of the Lesser Antilles chain in the Caribbean Sea, between the islands of Dominica and St. Lucia
- **Coordinates:** 14°40′N, 61°00′W
- **Borders:** None
- **Coastline:** 217 mi (350 km)
- **Territorial Seas:** 12 NM
- **Highest Point:** Mount Pelée, 4,583 ft (1,397 m)
- **Lowest Point:** Sea level
- **Longest Distances:** 47 mi (75 km) SE-NW / 21 mi (34 km) NE-SW
- **Longest River:** None of significant size
- **Natural Hazards:** Hurricanes, flooding, volcanic activity, earthquakes
- **Population:** 418,454 (July 2001 est.) / World Rank: 166
- **Capital City:** Fort-de-France, located on the western coast of the island
- **Largest City:** Fort-de-France, 100,000 (1996 est.)

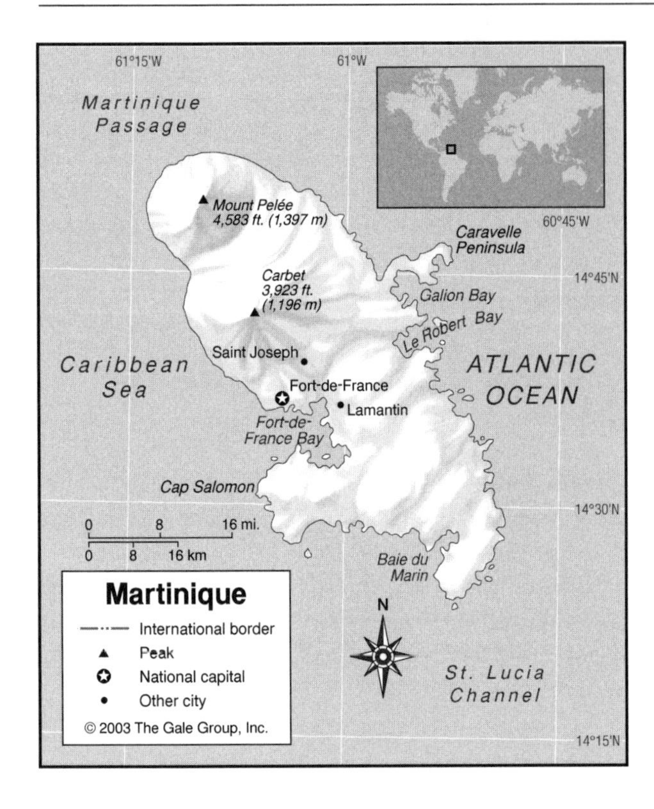

GEO-FACT

Saint-Pierre was the capital of Martinique until the eruption of Mount Pelée in 1902 destroyed the city and killed 30,000 inhabitants. The capital was moved to Fort-de-France after this destruction, where it remains today.

OVERVIEW

Martinique is part of the Lesser Antilles, a line (or arc) of volcanic islands in the Caribbean. The line is formed by the subduction of the North American Plate under the Caribbean Plate (which formed during the Cenozoic-Mesozoic period 240 million years ago). The two highest peaks on the island are volcanoes, and the rest of the landscape is composed of mountains, hills, and plateaus. Reefs of different origins are located along the island's coasts. The coasts are also home to several bays. Martinique is a possession of France.

MOUNTAINS AND HILLS

Most of Martinique is mountainous, with the two highest peaks, Pelée 4,582 ft (1,397 m) and Carbet 3,923 ft (1,196 m), being volcanoes. Mount Pelée lies in the northwestern area, while Carbet is in the central area. Steep cliffs line the northern coastline, while the island's other southern regions are characterized by hills, plateaus, and forests.

INLAND WATERWAYS

Martinique's only inland waters are small streams and ponds.

THE COAST, ISLANDS, AND THE OCEAN

Oceans and Seas

An island, Martinique is surrounded by the Caribbean Sea on the south and west, and the Atlantic Ocean on the north and east. A barrier reef of algal origin lies off the east coast of the island. South of the Caravelle Peninsula in the north-central region of the island, and parallel to the coast, a barrier reef made of coral and algae extends for about 15 mi (25 km). The Martinique Passage is to the north of the island, the St. Lucia Channel to the south.

The Coast and Beaches

White sandy beaches are located in the southern region of Martinique. In the north, especially on the western coast, the beaches are shorter, and the sand is dark gray or black and of volcanic origin. Fort-de-France, Le Robert, Galion, and Baie du Marin, to name a few, are large and deeply indented bays with extensive mangrove forests and seagrass beds. At the southernmost tip, Cap Salomon juts out into the ocean.

Along the northern and northwest coasts, the beach sand is black. The northwest shoreline is ringed by lush palm trees, while at the northernmost tip of the island, steep cliffs surround a narrow strip of beach.

CLIMATE AND VEGETATION

Temperature

Temperatures are mild year-round, and a rainy season occurs between June and October. Humidity is high year-round as well. The average temperature for the island is about 80°F (26°C). The mountainous northern area is cooler and rainier than the coastal region.

Rainfall

Annual rainfall averages about 75 in (190 cm), and a rainy season occurs between June and October. In April (the driest month), measurable rain falls an average of 13 days, and about twice as often in September (the rainiest month). Average humidity ranges from 80 percent in March and April to 87 percent in October and November.

Forests and Jungles

Approximately 25 percent of the land is wooded, housing European and tropical trees. Rainforests cover the slopes of the mountains in the northern interior, with lush underbrush including ferns, vines, and bamboo groves. Stands of mahogany, locust, rosewood, and other species of hardwood are also located in this area.

HUMAN POPULATION

Half of Martinique's population lives in Fort-de-France (the capital, located along the western coast), Saint Joseph, and Lamantin, all of which rest in the central area of the island. The remaining population is unevenly distributed throughout Martinique.

NATURAL RESOURCES

The island's coastal scenery, beaches, and cultivable land are its natural resources. Bananas and sugarcane are the main crops, and pineapples, mangoes, citrus fruits, avocados, cacao, and coffee are also grown. As an overseas department of France, trade is mainly with that country.

FURTHER READINGS

The Civilized Explorer. *Martinique—The Cosmopolitan Island.* http://www.cieux.com/mrtnq.html (accessed March 11, 2002).

French Coral Reef Initiative. *Martinique.* http://www.environnement.gouv.fr/ifrecor/domtom/mainta.htm (accessed March 11, 2002).

Martinique. Montréal: Ulysses Travel Publications, 1994.

Rosette-Rose, Robert. *Martinique, French Indies.* New York: Vilo, 1982.

Mauritania

- **Area:** 397,955 sq mi (1,030,807 sq km) / World Rank: 30
- **Location:** Northern and Western Hemispheres, in western Africa, bordering Algeria to the northeast, Mali to the east and southeast, Senegal to the southwest, the Atlantic Ocean to the west, and Western Sahara to the northwest.
- **Coordinates:** 20°00′N, 12°00′W
- **Borders:** 3,153 mi (5,074 km) / Algeria, 288 mi (463 km); Mali, 1,390 mi (2,237 km); Senegal, 505 mi (813 km); Western Sahara, 970 mi (1,561 km)
- **Coastline:** 469 mi (754 km)
- **Territorial Seas:** 12 NM
- **Highest Point:** Mount Ijill, 3,002 ft (915 m)

GEO-FACT

The nineteenth-century shipwreck of the frigate *Meduse,* immortalized in a famous painting by Théodore Géricault, occurred off the coast of Mauritania. Many of those who did not die aboard the ill-fated raft crafted by the passengers perished ashore in a futile trek across the desert.

- **Lowest Point:** Sebkha de Ndrhamcha, 10 ft (3 m) below sea level
- **Longest Distances:** 941 mi (1,515 km) NE-SW; 816 mi (1,314 km) SE-NW
- **Longest River:** Senegal River, 1,015 mi (1,663 km)
- **Natural Hazards:** Sirocco winds and drought
- **Population:** 2,747,312 (July 2001 est.) / World Rank: 133
- **Capital City:** Nouakchott, western Mauritania on the Atlantic Ocean
- **Largest City:** Nouakchott, population 694,000 (2000 est.)

OVERVIEW

Mauritania is an arid country on the African Tectonic Plate in western Africa. It forms a transitional zone between the Islamic, Arabic-speaking countries of North Africa's Maghreb region and the sub-Saharan countries to the south. Mauritania's terrain is generally a flat plain with occasional ridges and cliff-like outcroppings. The country can be roughly divided into two regions: the Saharan region, which is desert and covers the northern two-thirds, and the southern third of the country, which is mostly semidesert and coastal plain.

MOUNTAINS AND HILLS

Mauritania is largely flat, but in places its rocky plateau lands attain heights of over 1,500 ft (457 m). Its highest point is an enormous block of hematite, Mount Ijill in the northwest, topping out at 3,002 ft (915 m). Mauritania is nearly bisected by the sandstone plateaus that extend down the center of the country on a north-south axis, rising to elevations of over 1,000 ft (300 m). The Affollé Hills mark the south-central region of Mauritania on the border with Mali.

INLAND WATERWAYS

Lakes

Lake D'Aleg, Lake Rkiz, and a few other saltwater lakes are scattered throughout Mauritania. None are of considerable size, and due to recurrent droughts in recent decades they are even smaller than they once were.

Rivers

Most of Mauritania has little or no drainage to the sea. The Senegal River, which forms the boundary between Mauritania and Senegal, is the only permanent river between southern Morocco and central Senegal. Rising in Guinea, it flows north and west to the sea at Saint Louis in Senegal. Tributaries of the Senegal River drain the fertile southwestern corner of Mauritania.

Wetlands

The Banc d'Arguin National Park, Mauritania's only national park, is a nature reserve on the coastline bordering the Baie de Lévrier. It consists of a narrow strip of coastal wetlands, some small islands, and the sand dunes of the adjacent desert area. The reserve is known for the wide array of migratory birds that winter there.

THE COAST, ISLANDS, AND THE OCEAN

Oceans and Seas

Mauritania borders the North Atlantic Ocean. The waters off the coast of Mauritania are among the richest fishing areas in the world. Mauritania's claim to the continental shelf extends to 200 nautical miles or to the edge of the continental margin.

MAJOR ISLANDS

The largest island belonging to Mauritania is Île Tidra, which lies close to shore in the Baie de Lévrier, between the cities of Tanoudert and Nouamrhar.

THE COAST AND BEACHES

Mauritania's Atlantic coast is sandy, flat, and dotted with the saltwater pools known as *sebkhas*. Battering surf and shifting sand banks mark the entire length of the shoreline. From its southernmost point at the marshy Senegal River delta, the coastline remains smooth for somewhat more than half its length, marked only by an occasional high dune until reaching Cap Timiris, the only significant promontory. North of this point it is indented to form the Baie de Lévrier, which lies between Cap Timiris and the long peninsula of Cap Blanc, the northernmost point on the coast. This bay, one of the largest natural harbors on the west coast of Africa, measures 27 mi (43 km) long by 20 mi (32 km) wide at its broadest point. Jutting southward, Cap Blanc is 30 mi (48 km) long and 8 mi (13 km) wide and is divided between Western Sahara and Mauritania, with the Mauritanian city, only port, and rail terminus of Nouadhibou located on its eastern shore.

CLIMATE AND VEGETATION

Temperature

An area encompassing roughly the northern two-thirds of the country has an extremely hot, arid, Saharan climate, with afternoon high temperatures in the hottest months averaging 100°F (38°C) and often exceeding 115°F (46°C) in the interior areas. The southern part of the country has a semidesert, Sahelian climate. Summer is generally from June to September, and average summer temperatures at Kiffa, in the southern region, are around 79°F (26°C). The coastal region, although still arid, has the most moderate temperatures due to trade winds off the Atlantic. The average temperature in the coastal city of Nouakchott is around 75°F (24°C) during September, which is the hottest month in this region.

Rainfall

Northeasterly winds and the harmattan wind from the east keep Mauritania's climate dry, especially in the north where rain often does not fall for years or more at a time. Rainfall increases gradually from north to south as the rainy season becomes longer. Average annual rainfall at Nouadhibou is between 1 and 2 in (3 and 5 cm) and falls only between September and November. Farther north and east, rainfall is too rare and sparse to be measured. At the opposite end of the scale, Sélibaby in the far southern Senegal Valley region averages about 25 in (64 cm) of rainfall annually, within a rainy season that lasts from June to October. Annual variations in rainfall are also an important factor. The centrally located town of Atar went five years without any rain in the 1980s, but received almost 10 in (25 cm) in 1927.

Grasslands

Thin grass, scrubland, and savannah characterize the semidesert basins in the southern third of the country, which include the low-lying plains of Brakna and Trarza as well as the Hodh el Gharbi basin in the southeast. Among tree species of this region are the baobob and the acacia, which is harvested for gum.

Variously known as the Chemama or the Pre-Sahel, the Senegal River Valley zone on the country's southwestern border consists of a narrow, fertile belt of land 250 mi (400 km) long and extending 10 to 20 mi (16 to 32 km) north of the Senegal River. Completely dominated by the seasonal cycle of the river, the valley supplies more than 80 percent of the country's agricultural production. In addition to acacias, willow and jujube trees grow in this region.

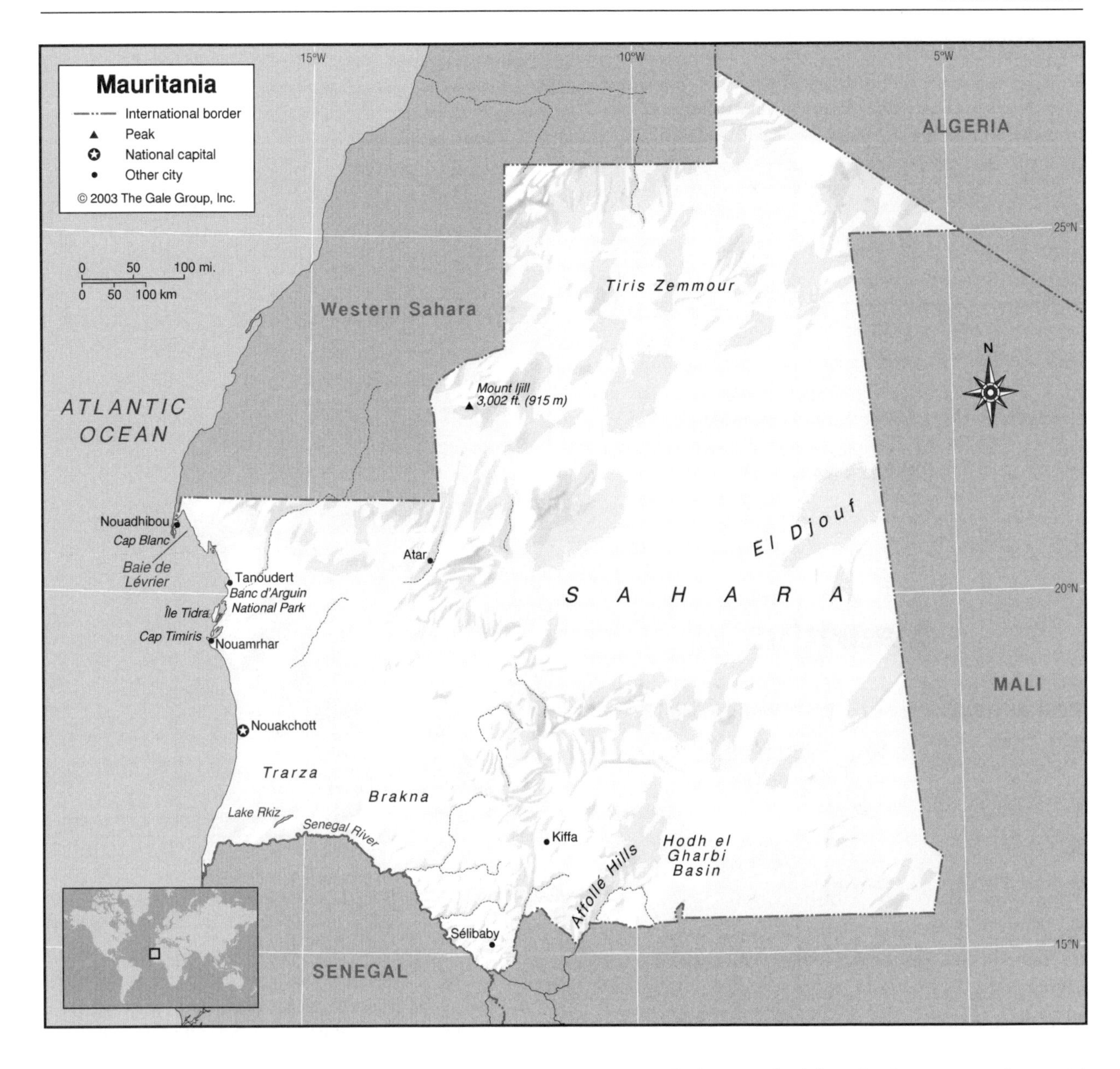

Deserts

The northern two-thirds of Mauritania is true Saharan desert, with vegetation other than cacti—such as date palms—found only in oases. Sand dunes cover about half of Mauritania, often arranged in long ridges with a northeast-southwest orientation and heights of up to 300 ft (91 m). In the far eastern part of the country, known as El Djouf, the terrain encompasses both rocky and sandy desert.

HUMAN POPULATION

Population in the arid northern part of the country is very sparse, with more than 90 percent of Mauritania's population concentrated in the semidesert region that comprises the southern one-third of the country. The overall population density is about 5 inhabitants per square mile (2 per sq km), but this figure varies from 0.26 per sq mi (0.1 per sq km) in Tiris Zemrnour to 47.9 per sq mi (18.5 per sq km) in the Gorgol region, a roughly triangular area between Nouakchott and the Senegal River in the southwestern corner of the country. Forty-six percent of the total population lives in rural areas.

NATURAL RESOURCES

Mauritania's coastal waters are an abundant source of fish, one of the best in the world. Fish and fish products are among the most important exports, however supplies are dwindling due to overexploitation. The country's rich deposits of iron ore, copper, and gypsum are also integral to the economy. Iron ore alone comprises half of Mauritania's total exports. Other mineral resources beginning to be mined in Mauritania include gold, diamonds, and phosphate.

FURTHER READINGS

Arepo Solutions Ltd. *Geography—Mauritania, Worldsuface.com.* http://www.worldsurface.com/browse/static.asp?staticpageid=156 (Accessed May 31, 2002).

Celati, Gianni. *Adventures in Africa.* Chicago: University of Chicago Press, 2000.

Edwards, Ted. *Beyond the Last Oasis: A Solo Walk in the Western Sahara.* Salem, N.H.: Salem House, 1985.

Handloff, Robert E. *Mauritania, A Country Study.* Federal Research Division, Library of Congress. Washington, D.C.: U.S. Government Printing Office, 1990.

Hudson, Peter. *Travels in Mauritania.* London: Virgin, 1990.

Lonely Planet World Guide. *Mauritania.* www.lonelyplanet.com/destinations/africa/mauritania/ (Accessed May 7, 2002).

"Morocco Handbook with Mauritania." *Footprint Handbooks.* Lincolnwood, Ill.: Passport Books, 1997.

Mauritius

- **Area:** 718 sq mi (1,860 sq km) / World Rank: 172
- **Location:** Southern and Eastern hemispheres, island in the Indian Ocean east of Madagascar.
- **Coordinates:** 20°17′S, 57°33′E
- **Borders:** None
- **Coastline:** 110 mi (177 km)
- **Territorial Seas:** 12 NM
- **Highest Point:** Piton de la Rivière Noire, 2,717 ft (828 m)
- **Lowest Point:** Sea level
- **Longest Distances:** 38 mi (61 km) N-S; 29 mi (47 km) E-W
- **Longest River:** Grand River South East, 25 mi (40 km)
- **Natural Hazards:** Cyclones from November to April; surrounded by coral reefs that are potentially hazardous to ships
- **Population:** 1,189,825 (July 2001 est.) / World Rank: 150
- **Capital City:** Port Louis, on the northwest coast
- **Largest City:** Port Louis, 165,000 (2000 est.)

OVERVIEW

Mauritius is a picturesque island nation, with rugged volcanic features and a large fertile plain. The compact island is the worn and eroded base of an extinct volcano.

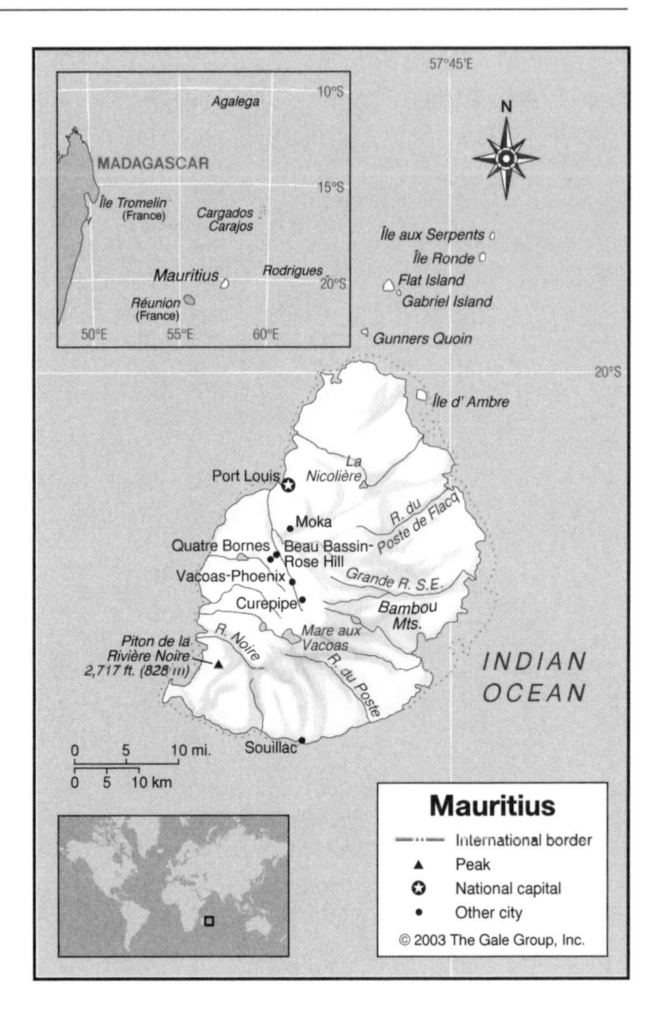

It stands on a mostly undersea feature called the Mascarene Plateau (a ridge that for much of its length is now underwater in the Indian Ocean and runs in a north-south direction), formerly a land bridge between Asia and Africa. The island's surface consists of a broad plateau that slopes toward a northern coastal plain from elevations of approximately 2,200 ft (670 m) near the southern coastline. Several low mountain groups and isolated peaks rise above the level of the plateau to give the appearance of a more rugged landscape. A coral reef nearly encircles the island. Mauritius sits on the African Tectonic Plate, but not near enough to any plate boundaries or fault lines to experience any major earthquakes or tectonic activity.

MOUNTAINS AND HILLS

Mountains

The entire island of Mauritius is of volcanic origin, having risen from the sea floor roughly 10 million years ago. Three mountain ranges border the central plateau of Mauritius: Moka in the northwest, Grand Port in the east, and Black River in the southwest. The highest peak on the island, Piton de la Rivière Noire, whose name means mountain of the Black River, is in the southwest region of the country in the Black River Mountain Range.

Plateaus

From elevations of approximately 2,200 ft (670 m) near the southern coastline, a broad central plateau slopes toward a northern coastal plain.

INLAND WATERWAYS

Lakes

Grand Bassin and Bassin Blanc, both of which lie in craters of extinct volcanoes, are the country's two natural lakes. Grand Bassin, about 4 mi (6 km) southeast of Mare aux Vacoas in the southwest, is believed to be sacred by Hindus. Several reservoirs are also located on the island, including La Nicolière in the north, Piton du Milieu in the central area, and Mare aux Vacoas, the largest reservoir, in the south.

Rivers

Numerous rivers flow through Mauritius. The Grand River South East, the country's longest river, is located in the central-eastern region. The other main rivers are Rivière Noire (Black River), Rivière du Poste, Grand River North West, Rivière La Chaux, and Rivière des Créoles. Several waterfalls exist, with the highest being the Tamarin Falls, in the west, at 961 ft (293 m) high.

THE COAST, ISLANDS, AND THE OCEAN

Oceans and Seas

A large coral reef entirely surrounds Mauritius except for a few small breaks along the coast. A large break in the reef occurs on the southern coast between Souillac and Le Bouchon, and a smaller break occurs on the western coast at Flic-en-Flac.

Major Islands

The inhabited island Rodrigues, which is a dependency of Mauritius, lies about 350 mi (560 km) to the northeast of Mauritius. It has an area of about 42.5 sq mi (110 sq km) and a population of slightly more than 34,000. Mauritius' other dependency, Agalega, is a group of two islands—North and South—that lie 697 mi (1,122 km) north of Mauritius. Agalega has a combined area of 27 sq mi (70 sq km).

Coral atolls surround Mauritius, including the Cargados Carajos Shoals (St. Brandon Group). Nature preserves protect the natural habitat on Île Ronde (Round Island) and Île aux Serpents (Serpent Island), among others.

The Coast and Beaches

A few long stretches of white sand beaches line the country on the north and east, and a lagoon exists at Flic-en-Flac on the western coast. Part of the southern coast is rocky and steep, with black cliffs.

Districts – Mauritius

2000 POPULATION ESTIMATES

Name	Population	Area (sq mi)	Area (sq km)
Black River	55,100	100	259
Flacq	125,600	115	298
Grand Port	108,100	101	262
Moka	75,600	89	230
Pamplemousses	118,900	69	179
Plaines Wilhems	362,300	78	202
Port Louis	138,700	17	44
Rivière du Rampart	99,600	57	148
Savanne	66,300	94	243
Rodrigues	35,700	40	104

SOURCE: Central Statistics Office, Mauritius.

CLIMATE AND VEGETATION

Temperature

Mauritius has a maritime climate, and temperatures vary within the differing altitudes. Being in the tropics, Mauritius' climate is humid, and there are prevailing southeast winds. At sea level, temperatures range from 64° to 86°F (18° to 30°C), and at an elevation of 1,500 ft (460 m), they range from 55° to 79°F (13° to 26°C). The warmest months are October through April (the summer months), and the coolest are June through September (winter).

Rainfall

The prevailing southeast winds affect the average rainfall in the central plateau, while the coasts experience little rainfall averages in comparison. Due to the tradewinds, the central plateau and windward slopes experience heavy rains from October to March. These areas have an annual average rainfall of more than 200 in (500 cm). On the coast, yearly rainfall averages about 40 in (100 cm). From April to September, daily showers occur, and between December and April, occasional tropical cyclones strike Mauritius.

Forests and Jungles

Approximately 22 percent of Mauritius' total land area is forest.

HUMAN POPULATION

An estimated 41 percent of the population lives in urban areas. The capital, Port Louis, is the largest city, with other large cities being Beau Bassin/Rose Hill, Quatre Bornes, Curepipe, and Vacoas-Phoenix. About 2 percent of the population lives on the island of Rodrigues.

NATURAL RESOURCES

Fish and arable land are the natural resources of Mauritius, and sugarcane is the major crop. The economy is based on sugar, tourism, and on export-oriented manufacturing, including clothing.

FURTHER READINGS

Columbus Publishing. "Mauritius." *World Travel Guide.* http://www.wtgonline.com/data/mus/mus.asp (Accessed March 18, 2002).

COMPNet USA-Mauritius. *Mauritius Island-Online.* http://www.maurinet.com (Accessed July 8, 2002).

Government of Mauritius. *Geography & Climate.* http://ncb.intnet.mu/govt/geograph.htm (Accessed March 19, 2002).

Lonely Planet. *Mauritius.* http://www.lonelyplanet.com/destinations/africa/mauritius/environment.htm (Accessed March 18, 2002).

Mauritius, Réunion, & Seychelles. New York: Langenscheidt Publishers, 2000.

Mediatool Ltd. *Mauritius-Info.* http://www.mauritius-info.com/travel/guidetou/geo.shtml (Accessed March 19, 2002).

NgCheong-Lum, Roseline. *Culture Shock! Mauritius.* Singapore: Time Books International, 1997.

Singh, Sarine, Robert Strauss, and Deanna Swaney. *Mauritius, Réunion, & Seychelles: A Travel Survival Kit.* 3rd ed. Oakland, Calif.: Lonely Planet Publications, 1998.

Mexico

- **Area:** 761,606 sq mi (1,972,550 sq km) / World Rank: 15
- **Location:** Northern and Western Hemispheres, in North America, bordering the United States to the north and northeast, the Gulf of Mexico and the Caribbean Sea to the east, Guatemala and Belize to the southeast, and the North Pacific Ocean to the south and west.
- **Coordinates:** 23°00′N, 102°00′W
- **Borders:** 2,820 mi (4,538 km) / Belize, 155 mi (250 km); Guatemala, 598 mi (962 km); United States, 2,067 mi (3,326 km)
- **Coastline:** 5,797 mi (9,330 km)
- **Territorial Seas:** 12 NM
- **Highest Point:** Pico de Orizaba, 18,701 ft (5,700 m)
- **Lowest Point:** Laguna Salada, 33 ft (10 m) below sea level
- **Longest Distances:** 660 mi (1,060 km) ENE-WSW; 2,000 mi (3,200 km) SSE-NNW
- **Longest River:** Rio Grande, 1,885 mi (3,034 km)
- **Largest Lake:** Lake Chapala, 405 sq mi (1,050 sq km)
- **Natural Hazards:** Earthquakes, tsunamis, hurricanes, and volcanoes
- **Population:** 101,879,171 (July 2001 est.) / World Rank: 11
- **Capital City:** Mexico City, southern Mexico
- **Largest City:** Mexico City, 8.6 million (2001 est.)

OVERVIEW

Mexico is the northernmost—and by far the largest—country on the isthmus that connects North and South America. It is considered part of North America, while the much smaller countries to its south make up Central America. Most of Mexico's mountain ranges, plateaus, and lowlands are continuations of landforms of the southwestern United States. Extending southeastward from its border with the U.S., Mexico forms a generally narrowing cone, broken in the northwest by the long, narrow peninsula of Baja California and in the extreme southeast by the blunt peninsula of Yucatán. It can be divided into five major geographical regions: 1) the Pacific Northwest, consisting of the northwestern edge of the main land mass plus the Baja California Peninsula; 2) the large Central Plateau, extending southward down the center of the country and including the Sierra Madre ranges; 3) the lowlands of the Gulf Coast and Yucatán Peninsula; 4) Central Mexico, occupying the transverse volcanic range at the southern end of the Central Plateau; and 5) the highlands to the south of that region.

A fairly broad coastal plain borders the Gulf of Mexico, and a narrower and broken strip of lowland borders the Pacific Ocean. Great mountain ranges extend roughly parallel to the two coastal lowlands, and between them lies a great interior plateau region. The plateau narrows to the south and terminates in a transverse mountain range consisting of a series of volcanic cones, some of them still active. This highland region of volcanoes is the heartland of Mexico. Most of the large cities and the densest rural population are located in its basins and valleys. Southward, the range of volcanoes drops away to the basin of the Balsas River flowing into the Pacific. The south, however, is mountainous except in a few river valleys, along the coasts, on the peninsula of Yucatán, and on the low-lying Isthmus of Tehuantepec.

Situated at the edge of the North American Tectonic Plate, the Cocos Plate, and the Pacific Plate, Mexico is also in the seismically active "Ring of Fire" that encircles the Pacific and is associated with frequent earthquakes and other seismic events. The series of volcanic peaks

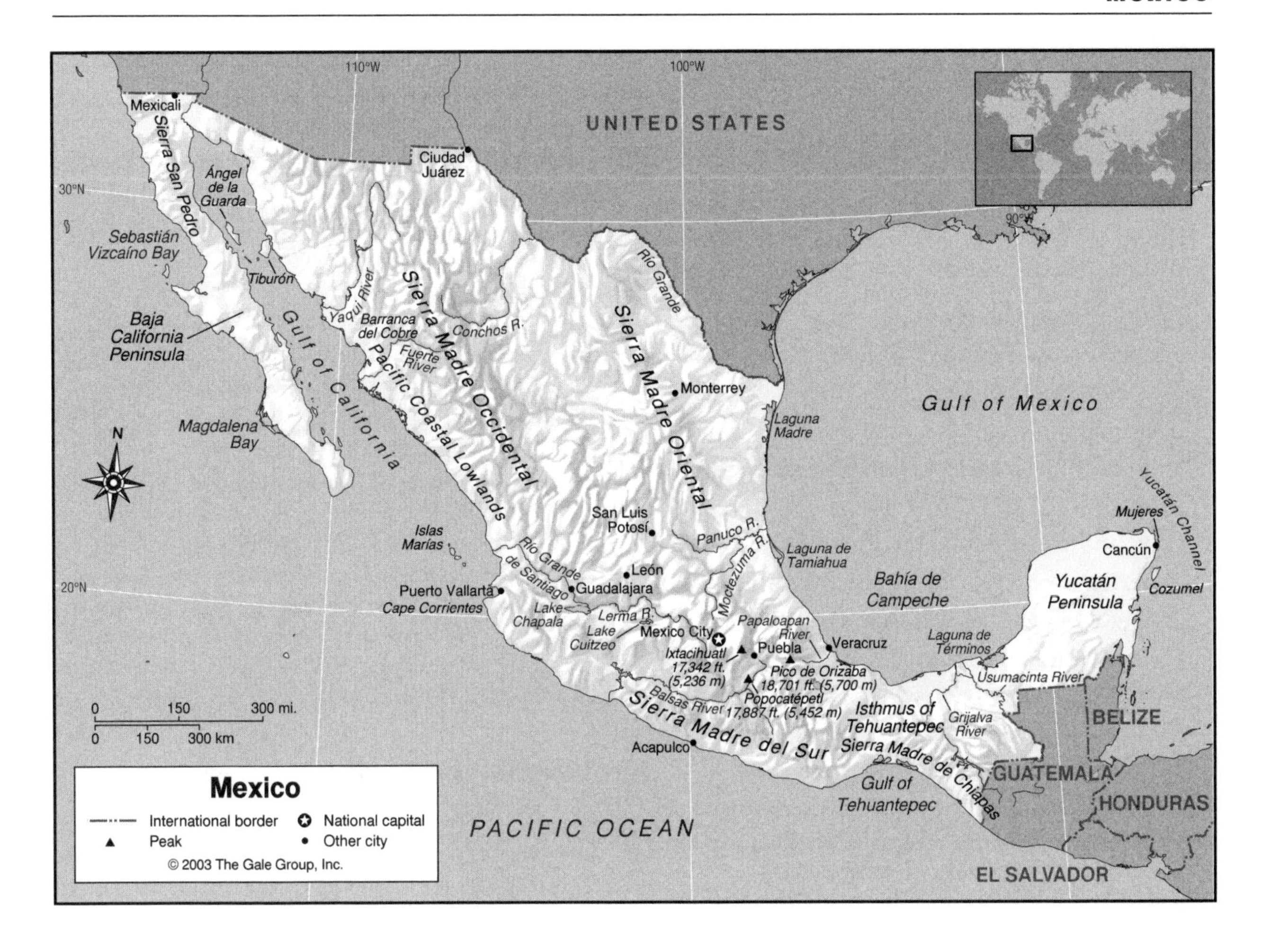

lying between 18° and 20° north latitude in Mexico's heartland constitute a fracture zone. Some geographers regard this volcanic fracture zone, a region of frequent and sometimes serious earthquakes, as the true termination of the North American continent. Nearly all of the volcanoes of the fracture zone are extinct, but some volcanism continues. Ixtacihuatl last erupted in the eighteenth century, but its crest is still sometimes topped by a plume of smoke.

MOUNTAINS AND HILLS

Mountains

The massive Sierra Madre Occidental that forms the western edge of the Central Plateau is the country's most extensive mountain system and an extension of the Sierra Nevada range of the United States. Its peaks average 8,000 to 9,000 ft (2,438 to 2,743 m). At the eastern edge of the Central Plateau, the shale and limestone peaks of the Sierra Madre Oriental range, rising to maximum heights of over 12,000 ft (3,658), form an extension of the Rocky Mountain range.

The loftiest peaks of the volcanic fracture zone at the southern edge of the Central Plateau form a kind of mountain range that extends laterally from the Pacific almost to the Gulf of Mexico. It is known variously as the Cordillera Neovolcanica, the Sierra Volcanica Transversal, and the Transverse Volcanic Range. Most of the country's highest peaks are in this chain. Although in the south, it is often referred to as Central Mexico because it is the country's heartland, containing almost half of Mexico's population. This region was an ancient center of civilization and is one of the world's loftiest areas of concentrated population after those of Tibet and the Andean countries. The terrain includes the dissected cones of thousands of old volcanoes interspersed by broad basins and valleys. The range is anchored on the east, not far from the Gulf of Mexico, by the perfect volcanic cone of Orizaba. With an elevation of 18,702 ft (5,700 m), it is the country's highest mountain. The second and third in height, Popocatépetl (17,887 ft/5,452 m) and Iztaccíhuatl (17,342 ft/5,286 m) respectively, are visible from Mexico City.

South of the fracture zone, between the broad Balsas River basin and the Pacific coast, lies the Sierra Madre del Sur range with altitudes of up to 10,000 ft (3,048 m). This range is not related to the mountain systems of northern Mexico but resembles the submerged chain of mountains whose peaks appear as the islands of the Greater Antilles.

GEO-FACT

Tenochtitlán, on the site of present day Mexico City, was the capital of the Aztec Empire. It had a population of roughly 200,000 when it was first visited by Europeans in 1519, making it one of the world's largest cities at the time. Located on an island in the middle of a lake, Tenochtitlán's boundaries had already been expanded by the Aztecs using dams and landfills when it was conquered by the Spanish in 1521. They renamed it Mexico City, and it became the center of Spain's American empire. The Spanish continued the Aztec's work of draining the lake, and now present day Mexico City covers the former lakebed.

Beyond the Isthmus of Tehuantepec, the Sierra Madre de Chiapas extends to the Guatemalan border, separated from the Pacific by a fairly broad coastal plain. The remaining major mountain system, the Chiapas Highlands, occupies most of the interior east of the Isthmus of Tehuantepec and south of the peninsula of Yucatán.

On the narrow finger of the Baja California Peninsula the Sierra San Pedro Mountains, with elevations of up to 9,000 ft (2,743 m), drop abruptly into the Gulf of California.

Plateaus

Mexico's large Central Plateau runs between the eastern and western Sierra Madre mountain ranges, extending southward down the center of the country from the U.S. border to the volcanic transverse mountains at Isthmus of Tehuantepec, narrowing somewhat from north to south. Average elevations range from 4,000 ft (1,219 m) in the north to over 8,000 ft (2,438 m) as the plateau of Central Mexico is neared. At approximately its midpoint, near the city of San Luis Potosí, the plateau is interrupted by a series of cross ranges between the two Sierra Madre systems. The drier, low-lying part of the plateau north of this point encompasses parts of the Sonoran Desert to the north and is called the Mesa del Norte. It is a dry pastoral countryside of great desert basins known as bolsons. These depressions drain internally to salty lakes or salt flats bordered by gradually sloping shallow alluvial fans and rimmed by steeply sloped hill ranges. To the south lies the Mesa Central, which is higher, wetter, and flatter than the area to the north. It contains rich farmland and is more densely populated.

Canyons

The Sierra Madre Occidental has a number of steep canyons called *barrancas.* The most dramatic is the Barranca del Cobre; it is the Mexican counterpart to the Grand Canyon in the United States.

Hills and Badlands

The northern section of the Central Plateau (the Mesa del Norte) includes extensive flat areas, but it is broken by numerous hill ranges, most of them longitudinal. Central Mexico's terrain includes rolling hills in addition to its volcanic peaks and basins. The southern highlands, consisting of the states of Guerrero, Oaxaca, and Chiapas, are located to the south of both Central Mexico and the Gulf Coastal Plain. Cut off from the rest of the country by the plateau of Central Mexico, this is the most rural and slowest growing of Mexico's regions, and the one in which Indian life most closely adheres to its traditional patterns. The natural features of this region resemble those of Central America.

INLAND WATERWAYS

Lakes

Mexico has only a small number of natural lakes. Lake Chapala on the outskirts of Guadalajara is the country's largest natural lake and a popular recreation and retirement center. Covering an area of 405 sq mi (1,050 sq km), it is approximately 50 mi (75 km) long and has a maximum width of around 13 mi (20 km). The westward-flowing Lerma River drains into it, and the lake in turn empties into the Río Grande de Santiago River, which flows northwestward to the Pacific. Another relatively large lake is Lake Cuitzeo in Central Mexico.

Rivers

Mexico is drained by few major rivers. The longest is the Río Grande (known in Mexico as the Río Bravo del Norte), which flows southeast for some 1,300 mi (2,092 km) in Mexico before draining into the Gulf of Mexico. It forms more than half of Mexico's border with the United States. Its tributary, the Conchos, drains a large part of the Mesa del Norte. The Moctezuma-Panuco River flows eastward through the Sierra Madre Oriental into the Gulf of Mexico, draining the eastern part of the Mesa Central. Farther south, two larger rivers flow into the Gulf: the Papaloapan River, whose mouth is near Veracruz, and the Grijalva-Usumacinta river, which flows through the Chiapas Highlands. The countless streams that form their delta have created Mexico's most extensive coastal swamplands.

The Lerma River rises in the volcanic highlands, near Mexico City, then flows westward to Lake Chapala, from where the Río Grande de Santiago completes the journey to the Pacific. Flowing westward farther south is the Balsas River, which provides hydroelectric power through a dam at the Sierra Madre del Sur. Rivers that have been dammed for irrigation purposes include the Yaqui, the Fuerte, and the Culiacán Rivers, all of which flow through the narrow Pacific coastal plain. The Morelos Dam on the Colorado River at the head of the Gulf of California has converted the desert land of the Mexicali Valley into an important agricultural area devoted primarily to cotton farming.

Wetlands

The Gulf Coast and the Yucatán Peninsula together make up a lowland region that stretches from the northern border to the Caribbean coast in an arc that is unbroken except for spurs of the Sierra Madre Oriental that extend eastward to the coast north of Veracruz. While the coast in the northern state of Tamaulipas is relatively arid, the fairly broad Veracruz coastal plain is fertile, hot, and marshy, and formerly so insect-infested that Veracruz was known as the "City of Death." Drainage of swamps and other health and sanitation programs, however, have made the coastal plain of both Veracruz and Tabasco an area of rapid population growth.

THE COAST, ISLANDS, AND THE OCEAN

Oceans and Seas

Most of Mexico's eastern coast borders the Gulf of Mexico, but the eastern shore of the Yucatán Peninsula borders the Caribbean Sea. The Yucatán Channel, between the northeast tip of the peninsula and southern Cuba, divides the Caribbean from the Gulf of Mexico. The narrow Isthmus of Tehuantepec that connects the Mexican mainland to the Yucatán Peninsula and Central America is bordered by two gulfs, the Bahía de Campeche (to the north) in the Gulf of Mexico and the much smaller Gulf of Tehuantepec (to the south) in the Pacific.

The long western coast of the Mexican mainland borders the Pacific Ocean to the south and, to the north, the long, narrow Gulf of California (formerly the Sea of Cortez), which is enclosed between the main isthmus and the Baja California Peninsula. Near the midpoint of the coast, near Puerto Vallarta, the otherwise relatively straight shore hooks out at Cape Corrientes. Even though its coastline is less than half that of the mainland, the Baja California Peninsula is more distinguishing, featuring two bays. The Sebastián Vizcaíno is a large bay cutting into the coast in the north. Toward the south, the much smaller Magdalena Bay is enclosed by offshore islands. The waters of the Pacific off the western coast of the peninsula are known for the array of marine life they harbor, and are especially famous as the only place in the world where the gray whale calves.

The waters of Mexico's both eastern and western coasts feature extensive coral reefs, concentrated mostly around the southern half of the country. Located just off the coast of Cozumel, Palancar Reef, one of the largest coral reefs in the world, became famous among divers after marine explorer Jacques Cousteau "discovered" the area in the 1950s. The second-largest barrier reef in the world, the Great Maya Reef, is also located off the coast of the Yucatán Peninsula. Also off the Yucatán Coast lies the Middle American Trench. The northern edges of this trench rise sharply to form the Cayman Ridge, the tops of which comprise the Cayman Islands, while the southern edge drops steeply to the Cayman Trench, which runs in an east-west direction and reaches a depth of 24,720 ft (7,535 m), the deepest point in the Caribbean Sea.

Major Islands

Mexico has several islands off the Pacific coast of the Baja peninsula and dotting the Gulf of California, the two largest being Tiburón and Ángel de la Guarda. The Islas Marías are a few islands in the Pacific opposite the southern end of the western coastal plain north of Puerto Vallerta. Several more large islands exist near the northeastern tip of the Yucatán Peninsula, including the islands of Cozumel and Mujeres.

The Coast and Beaches

Mexico's coastline includes sandy beaches, which draw visitors to coastal resort areas like Acapulco and Cancún (known for its white-sand beaches and deep blue waters), but in parts of Baja California and the southern Pacific coastline the mountains come right down to the sea. Still other parts of the coast are bordered with mangrove-lined lagoons.

The northern part of all three coastlines in the Pacific region—both coasts of the Baja peninsula and the western coast of the mainland—are heavily indented, with multiple bays and inlets. The shoreline in the southern reaches of the mainland coastal plain (and the southern part of Baja California's western coast) becomes much smoother. The coast is still smooth but more uneven as it curves around the southern highlands, and then becomes almost perfectly smooth at the Gulf of Tehuantepec.

The northernmost section of Mexico's Gulf Coast is the site of a distinctive inland waterway called the Laguna Madre, which has an almost identical counterpart—called by the same name—just north of the U.S. border along the Texas Gulf Coast. These narrow strips of water are two of only three coastal lagoons in the world that are hypersaline (i.e. saltier than the ocean). (The third is in the Crimea.) They are ecologically valuable for their beds of seagrass, which nurture marine life in the area. A short distance south of the Laguna Madre is another good-sized lagoon area, the Laguna de Tamiahua.

The Yucatán Peninsula has a broad indentation at its southwest corner, in the state of Campeche, called the Laguna de Términos. There are also two more major inlets on its Caribbean coast.

CLIMATE AND VEGETATION

Temperature

About half of Mexico lies to the south of the tropic of Cancer but, due to altitude and other geographical factors, temperatures in the north can be higher than in the south. The various temperature zones in most of Mexico are generally categorized by altitude rather than latitude. Areas at elevations of up to 3,000 ft (914 m)—the coastal lowlands and the Yucatán peninsula—are *terra caliente* ("hot land") and have a tropical climate. The plateau (3,000 to 6,000 ft / 914 to 1,829 m) is *terra templada* ("temperate land"), with a moderate climate, and the mountains (above 6,000 ft / 1,829 m) are *terra fria* ("cold land"). The average annual temperatures for cities in these three regions are, respectively, Veracruz, 77°F (25°C); Jalapa, 66°F (19°C); and Pachuca, 59°F (15°C).

In the south temperatures may vary by as little as 10°F (5°C) between seasons, but temperature extremes are much greater in the north. Baja California and the Sonoran desert can have readings as high as 110°F (43°C) in summer (July) and as low as 32°F (0°C) in winter (January).

Rainfall

Rainfall varies greatly by region, ranging from under 10 in (25 cm) per year in Baja California to 200 in (500 cm) in the rainforests of Tabasco. The north generally gets less rainfall than the south, but the entire Gulf Coastal Plain is a wet area. Most of the rain in the central part of the country falls between May and August.

Grasslands

The plateau lands south of the tropic Cancer are classified as either tropical or highland savanna and are covered with grass and shrubs with a few scattered trees. The Pacific Coastal Lowlands on the Gulf of California is a narrow plain extending inland for a distance of only 10 to 15 mi (16 to 24 km). This area is in contrast to the desert and steppe land that surrounds it.

Deserts

Much of the area north of the tropic of Cancer is considered tropical desert or steppe land. Tropical desert land dominated by scrub, brush, and cacti extends over much of the Baja California Peninsula and the northwestern quarter of the mainland extending southward into Sinaloa. The semiarid Balsas Depression south of the transverse volcanic highlands also has a desert environment.

Population Centers – Mexico

(POPULATIONS FROM 2000 CENSUS)

Name	Population	Name	Population
Mexico City (Cuidad de México; capital)	8,591,309	León	1,133,576
Guadalajara	1,647,720	Monterrey	1,108,499
Puebla (de Zaragoza)	1,346,176	Mexicali	764,902
Nezahualcóyotl	1,224,924	Culiacán	744,859
Ciudad Juárez	1,217,818	Acapulco (de Juárez)	721,011
Tijuana	1,212,232	Mérida	703,324
		Chihuahua	670,208
		San Luis Potosí	669,353

SOURCE: INEGI (Instituto Nacional de Estadística, Geografía e Informá, Mexico). *Estados Unidos Mexicanos. XII Censo General de Población y Vivienda, 2000.*

Unusual plant life found in the deserts of the Baja peninsula includes the boojum and elephant trees and the world's tallest cactus, called the cardón. Desert flowers that bloom in the spring or summer include the yucca, agave, and ocotillo.

Forests and Jungles

Much of the original deciduous and evergreen forest of the Mesa Central and Sierra Madre Occidental has been destroyed, although there remain some substantial forested areas on the upper slopes of the Occidental range.

The Gulf Coastal Plain, the Chiapas Highlands, and the Yucatán Peninsula are dominated by dense tropical rain forests. In some areas, more than a hundred different tree species can be found within an area of 2 acres (1 hectare). The central and northeastern parts of the Yucatán are covered with relatively low deciduous trees, while the forest becomes even more luxuriant in the wettest regions to the northwest and south (including parts of Chiapas state) and changes to broadleaf evergreens at elevations of around 6,500 ft (1,981 m).

The layers of vegetation in these tropical forests include the woody climbing plants called lianas, as well as orchids, bromeliads, and other epiphytes.

HUMAN POPULATION

Two-thirds of the country is mountainous, and Mexicans tend to think of directions as up and down rather than in terms of the four quarters of the compass. The mountains have tended to perpetuate regionalism, and access to many areas is so difficult they have remained economically undeveloped and culturally isolated.

Urbanization increased rapidly in the last two decades of the twentieth century, with three-fourths of the population—up from two-thirds—now living in

States – Mexico

POPULATIONS FROM 2000 CENSUS

Name	Population	Area (sq mi)	Area (sq km)	Capitals
Aguascalientes	944,285	2,112	5,471	Aguascalientes
Baja California Norte	2,487,367	26,997	69,921	Mexicali
Baja California Sur	424,041	28,369	73,475	La Paz
Campeche	690,689	19,619	50,812	Campeche
Chiapas	3,920,892	28,653	74,211	Tuxtla Gutiérrez
Chihuahua	3,052,907	94,571	244,938	Chihuahua
Coahuila de Zaragoza	2,298,070	57,908	149,982	Saltillo
Colima	542,627	2,004	5,191	Colima
Distrito Federal (Mexico City)	8,605,239	571	1,479	-
Durango	1,448,661	47,560	123,181	Durango
Guanajuato	4,663,032	11,773	30,491	Guanajuato
Guerrero	3,079,649	24,819	64,281	Chilpancingo
Hidalgo	2,235,591	8,036	20,813	Pachuca de Soto
Jalisco	6,322,002	31,211	80,836	Guadalajara
México	13,096,686	8,245	21,355	Toluca de Lerdo
Michoacán de Ocampo	3,985,667	23,138	59,928	Morelia
Morelos	1,555,296	1,911	4,950	Cuernavaca
Nayarit	920,185	10,417	26,979	Tepic
Nuevo Léon	3,834,141	25,067	64,924	Monterrey
Oaxaca	3,438,765	36,275	93,952	Oaxaca
Puebla	5,076,686	13,090	33,902	Puebla
Querétaro de Arteaga	1,404,306	4,420	11,449	Querétaro
Quintana Roo	874,963	19,387	50,212	Chetumal
San Luis Potosí	2,299,360	24,351	63,068	San Luis Potosí
Sinaloa	2,536,844	22,521	58,328	Culiacán
Sonora	2,216,969	70,291	182,052	Hermosillo
Tabasco	1,891,829	9,756	25,267	Villahermosa
Tamaulipas	2,753,222	30,650	79,384	Ciudad Victoria
Tlaxcala	962,646	1,551	4,016	Tlaxcala
Veracruz-Llave	6,908,975	27,683	71,699	Jalapa
Yucatán	1,658,210	14,827	38,402	Mérida
Zacatecas	1,353,610	28,283	73,252	Zacatecas

SOURCE: INEGI (Instituto Nacional de Estadística Geografía e Informática), *Estados Unidos Mexicanos. XII Censo General de Población y Vivienda, 2000. Tabulados Básicos y por Entidad Federativa. Bases de Datos y Tabulados de la Muestra Censal.*

urban areas. The major concentration of urban centers is an area reaching across the southern edge of the Central Plateau from Puebla to Guadalajara and including Mexico City, the country's capital and largest city as well as the world's second-largest city, surpassed only by Tokyo. More than half the nation's population lives in this region. The northern and southernmost parts of the country are more sparsely settled.

NATURAL RESOURCES

Mexico is very rich in natural resources due to its size and varied climate. It ranks among leading countries in mining, producing large amounts of silver, gold, mercury, and oil and petroleum products. It also has significant reserves of copper, lead, zinc, and manganese. Mexico also has around 120 million acres (49 million hectares) of forestland for sources of wood and lumber. Agriculture is another important source of income for Mexico. The climate supports the growth of many crops; among the leaders are corn, wheat, beans, barley, and citrus fruits of all kinds. The northwest region on the Gulf of California includes a narrow western coastal plain extending inland for a distance of only 10 to 15 mi (16 to 24 km) that is one of the country's most productive commercial farming areas. Mexico also relies on a large tourism industry. People come to Mexico for its history, its sandy beaches and lavish Caribbean resorts, and for its world-renowned cuisine.

FURTHER READINGS

Annerino, John. *The Wild Country of Mexico: La Tierra Salvaje de México.* San Francisco: Sierra Club Books, 1994.

Butler, Ron. *Dancing Alone in Mexico: From the Border to Baja and Beyond.* Tucson: University of Arizona Press, 2000.

Embassy of Mexico. http://www.mexican-embassy.dk (Accessed June 4, 2002).

1UpTravel.com. http://www.1uptravel.com/geography/mexico.html (Accessed April 14, 2002).

Route of the Mayas. Knopf Guides. New York: Knopf, 2002.

Ryan, Alan, ed. *The Reader's Companion to Mexico.* San Diego: Harcourt Brace Jovanovich, 1995.

Tree, Isabella. *Sliced Iguana: Travels in Unknown Mexico.* London: Hamish Hamilton, 2001.

Wauer, Roland H. *Naturalis's Mexico.* College Station: Texas A&M University Press, 1992.

Micronesia, Federated States of

- **Area:** 271 sq mi (702 sq km) / World Rank: 180
- **Location:** Northern and Eastern Hemispheres, Oceania, island group in the North Pacific Ocean, about three-quarters of the way from Hawaii to Indonesia
- **Coordinates:** 6°55′N, 158°15′E
- **Borders:** None
- **Coastline:** 3,798 mi (6,112 km)
- **Territorial Seas:** 12 NM
- **Highest Point:** Totolom, 2,595 ft (791 m)
- **Lowest Point:** Sea level
- **Longest Distances:** 1,800 mi (2,898 km) E-W from Kosrae to Yap

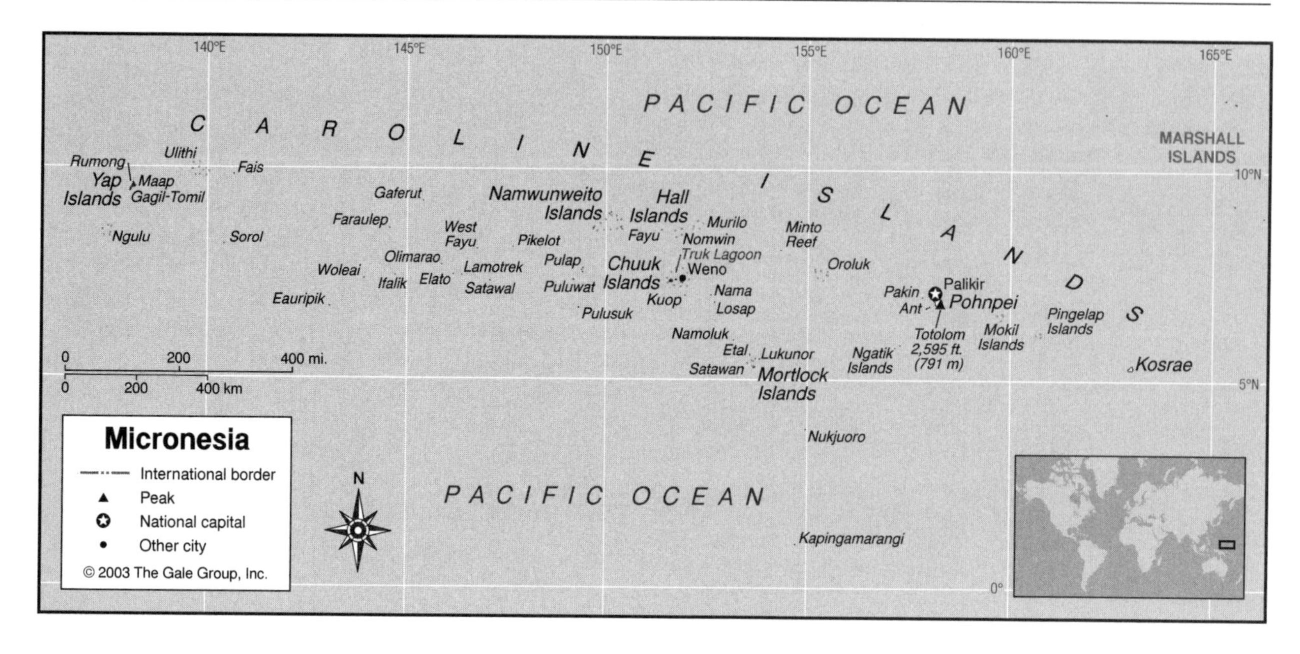

- **Longest River:** None of significant length
- **Natural Hazards:** Tropical typhoons
- **Population:** 134,597 (July 2001 est.) / World Rank: 181
- **Capital City:** Palikir, on Pohnpei in the eastern end of the archipelago
- **Largest City:** Weno, on Weno Island in the center of the island groups, population 24,900 (2002 est,)

OVERVIEW

The Federated States of Micronesia (FSM) is located in the western Pacific Ocean within the Caroline Islands archipelago and encompasses more than one million square miles of ocean (2.6 million sq km). It is the second largest land and sea area in Micronesia. Its four states consist of four major island groups. From east to west they are Kosrae, Pohnpei (Ponape), Chuuk Islands (formerly Truk), and Yap Islands. The territory is made up of 607 islands, including mountainous islands of volcanic origin and coral atolls, of which 40 are a significant size; 65 of the islands are inhabited. The outer islands of all states are mainly coral atolls. Subsistence farming and fishing are the primary economic activities. However, licensing to commercial fishing fleets for rights to operate in FSM territorial waters provides nearly 30 percent of domestic budgetary revenue. Tourism, with its associated demands, is a developing industry, catering mostly to sport divers. Geographical isolation and a lack of adequate facilities hinder development.

MOUNTAINS AND HILLS

Mountains

The island of Kosrae is largely mountainous with two main peaks: Fenkol (Mount Crozer), 2,080 ft (634m), and Matanti, 1,913 ft (583m). Pohnpei contains a large volcanic island, with the highest elevation that of Mount Totolom at 2,595 ft (791 m), which is the highest point in the FSM. Chuuk (Truk) consists of 14 islands that are mountainous and of volcanic origin.Yap has four large, high islands, with the peak elevation that of Mount Tabiwol at 584 ft (178 m). Yap is at the southern end of a great submarine ridge and volcanic outcropping has occurred in the five largest island clusters.

INLAND WATERWAYS

The four states of Micronesia have a total of 2,766 sq mi (7,164 sq km) of lagoons within their coastal borders. Pohnpei, the largest and tallest island in the FSM, has peaks that get much rainfall annually, creating more than 40 rivers that feed the upper rain forest and create spectacular waterfalls.

THE COAST, ISLANDS, AND THE OCEAN

Major Islands

Within Micronesia the four states center around one or more "high islands." Kosrae, the smallest and easternmost state, consists of five closely situated islands, but is essentially one high island of 42 sq mi (119 sq km). Pohnpei state (133 sq mi / 344 sq km) consists of the single large island of Pohnpei (130 sq mi / 137 sq km) and 25 smaller islands within a barrier reef, in addition to 137 outer islands, of which the major atolls are Mokil, Pingelap, Kapingamarangi, Nukuoro, and Ngatik. Chuuk (Truk) state (49 sq mi / 127 sq km) includes the large Truk lagoon, enclosing 98 islands, and major outer island groups including the Mortlocks, Halls, Western, and Namwunweito islands. Yap (46 sq mi / 118 sq km), the westernmost state, consists of four large islands and seven smaller islands surrounded by barrier reefs, in addition to 134 outer islands, of which the largest are Ulithi and Woleai. Roads connect Yap, Gagil-Tomil, and Maap. Rumong is accessible only by boat.

The Coast and Beaches

Low sheltered coastal areas of the FSM islands are covered in mangrove forests. The Chuuk islands are an "almost atoll," encircled by a barrier reef. The "Truk Lagoon" is one of the largest enclosed lagoons in the world, circled by a 140-mi (225-km) long barrier reef, and covering an area of 822 sq mi (2,129 sq km). Of the 80 countries that have coral reefs the FSM ranks thirteenth by area with 1.53 percent in the worlds reefs spanning 4,340 sq mi (11,241 sq km). During the past century, FSM coral reefs suffered from soil erosion resulting from logging, agriculture, major coastal construction (dredging and filling), military occupation, and World War II battles, along with poaching of giant clams, sharks, trochus (marine gastropod), and other commercial species from remote reefs. Ports and harbors are located at Colonia (Yap), Kolonia (Pohnpei), Lele (Kosrae), Moen (Chuuk).

GEO-FACT

Land ownership in the FSM is vested in family held trusts and land use rights are passed down intergenerationally within the extended family system. Families or clans may have different factions, all of whom may assert interest in the land. Small holdings characterize ownership and subsurface rights are synonymous with surface rights.

CLIMATE AND VEGETATION

Temperature

The climate is maritime tropical, with little seasonal or diurnal variation in temperature, which averages 80°F (27°C) year round. The humidities average over 80 percent.

Rainfall

The northeast trade winds that prevail from November to December and April to May frequently bring heavy rainfall. The short and torrential nature of the rainfall, which decreases from east to west, results in an annual average of 200 in (508 cm) in Pohnpei and 120 in (305 am) in Yap. Reputedly, Pohnpei is one of the wettest places on earth. The eastern islands are located on the southern edge of the typhoon belt and occasionally suffer severe damage from typhoons, which are a threat from June to December.

Forests and Jungles

There is moderately heavy tropical vegetation, with tree species including tropical hardwoods on the slopes of higher volcanic islands and coconut palms on the coral atolls. Pohnpei and Kosrae have the only remaining patches of montane cloud forest in Micronesia. In the Yap islands, forest covers 40 percent of total land area but is largely secondary growth.

HUMAN POPULATION

The FSM has the largest population in Micronesia. The majority of the Micronesian population lives in coastal areas of the high islands, leaving the mountainous interiors largely uninhabited. It was estimated that 30 percent of the population lived in urban areas in 2000. No significant permanent emigration has occurred. However, there is internal FSM migration. Population growth and urban pollution such as sewage and garbage disposal are increasingly becoming problematic.

NATURAL RESOURCES

The ocean is the FSM's most important resource, providing marine products and deep-seabed minerals. The FSM's exclusive economic zone covers more than 1 million sq mi (2.6 million sq km) of ocean, which contains the worlds most productive tuna fishing grounds. Each state in the FSM has extensive forest cover that is used for construction, firewood, and handicrafts. Copra remains the main cash crop throughout the FSM.

FURTHER READINGS

Action Atlas. *Coral Reefs.* http://www.motherjones.com/coral_reef/micronesia.html (Accessed May 7, 2002).

Earthwatch. *Island Directory.* http://www.unep.ch/islands/isldir.htm (Accessed May 19, 2002).

Federated States of Micronesia. www.fsmgov.org/info (Accessed May 7, 2002).

Gillett, Robert D. *Traditional Tuna Fishing: A Study at Satawal, Central Caroline Islands.* Honolulu: Bishop Museum Press, 1987.

International Centre for Island Studies. http://www.islandstudies.org// (Accessed May 19, 2002).

Myers, Robert F. *Micronesian Reef Fishes: A Comprehensive Guide to the Coral Reef Fishes of Micronesia.* Barrigada, Guam: Coral Graphics, 1999.

Pacific Island Region. http://www.spc.org.nc/En/region.htm (Accessed May 19, 2002).

Pollard, Stephen J. *Micronesia's Success in Business: Cooperation or Competition. Pacific Islands Development Program,*

Reports on the Role of the Private Sector Project. Honolulu: East-West Center, 1995.

Seach, John. *Volcanoes of Micronesia.* http://www.volcanolive.com/mariana.html (Accessed May 19, 2002).

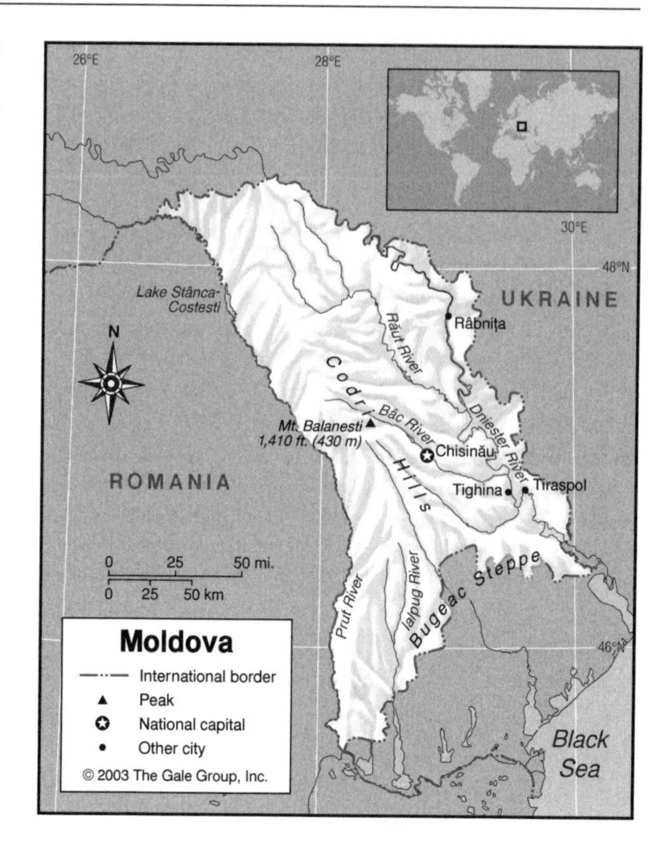

Moldova

- **Area:** 13,067 sq mi (33,843 sq km) / World Rank: 138
- **Location:** Northern and Eastern Hemispheres, in eastern Europe, east of Romania, south and west of Ukraine.
- **Coordinates:** 47°00′N, 29°00′E
- **Borders:** 864 mi (1,389 km) / Romania, 280 mi (450 km); Ukraine, 583 mi (939 km)
- **Coastline:** None
- **Territorial Seas:** None
- **Highest Point:** Mount Balănesti, 1,410 ft (430 m)
- **Lowest Point:** Dniester River, 6.6 ft (2 m)
- **Longest Distances:** 90 mi (150 km) E–W; 210 mi (340 km) N-S
- **Longest River:** Dniester, 870 mi (1400 km)
- **Natural Hazards:** Landslides
- **Population:** 4,431,570 (July 2001 est.) / World Rank: 115
- **Capital City:** Chisinău, in the center of the country
- **Largest City:** Chisinău, 765,000 (2000 estimate)

OVERVIEW

Moldova is a completely landlocked country of about 13,000 sq mi (33,700 sg km), after Armenia the second smallest republic of the former U.S.S.R. It is located in southeastern Europe, east of Romania and north, west, and northeast of Ukraine. The entire border with Romania lies along the Prut River in the west; on the east, the Dniester (Nistru) River follows some of the northern border with Ukraine but flows mostly within the nation's eastern region. Moldova's capital and largest city, Chisinău, is situated nearly in the center of the country.

The Moldovan terrain is mostly a hilly plain cut by many deep river and stream valleys. In general the terrain slopes gradually south toward the Black Sea, although Moldova is separated from the sea by a narrow arm of Ukraine. Its average elevation is only 482 ft (147 m) above sea level.

MOUNTAINS AND HILLS

Moldova's hills are more accurately described as rolling, hilly plains that rise in elevation to the north as they approach the foothills of the Carpathian Mountains. The hill country is cut by deep ravines and gullies from the country's many rivers and streams. The highest terrain are the Codri Hills of west-central Moldova, where Mount Balănesti rises to 1,409 ft (430 m).

Most of Moldova lies on deep layers of sedimentary rock. In the higher elevations of the north, near the foothills of the Carpathian Mountains, crystalline igneous outcroppings can be found.

INLAND WATERWAYS

Rivers

Moldova has more than 3,000 rivers and streams, but only eight are more than 60 mi (100 km) long. The two largest rivers are the Dniester (called the Nistru in Moldova) and the Prut, which both originate in the Carpathian Mountains north of Moldova in Ukraine. The longer Nistru flows south through eastern Moldova. It forms a short section of the Moldova/Ukraine border in the northeast, flows into Moldova, then borders Ukraine again in the southeast. It finally reenters Ukraine in the south shortly before emptying into the Black Sea. Its width ranges on average from 500 to 750 ft (152 to 229 m), with a maximum of 1,400 ft (427 m). The Dniester can be navigated through most of its length in Moldova, and only freezes over in severe winters.

The second longest river is the Prut, a major tributary of the Danube River. The Prut River forms the entire border of Moldova and Romania, with Lake Stânca-Costesti part of its course, before flowing south into the Danube. The Danube then goes eastward until it empties into the Black Sea. Like the Nistru, the Prut originates in the Carpathian Mountains in southwestern Ukraine; it flows a total distance of 564 mi (909 km). It is navigable well into Moldova, about 200 mi (320 km). Other smaller Moldovan rivers include the Ialpug, the Bâc, and the Răut.

Wetlands

Saline marshes can be found along the lower reaches of the Prut and other river valleys of southern Moldova.

THE COAST, ISLANDS, AND THE OCEAN

Moldova is narrowly separated fom the Black Sea by Ukraine, and is landlocked.

CLIMATE AND VEGETATION

Temperature

The Moldovan climate is continental, with conditions kept somewhat moderate by the influence of the Black Sea. Winters are generally dry and mild, with average daily temperatures in January ranging from 23° to 27° F (-5° to -3° C). The long summers are warm; average daily temperatures in July are over 68° F (20° C), and daily highs may even reach 104° F (40° C).

Rainfall

Precipitation in Moldova is typically light, and sometimes irregular, even characterized by dry spells. Rainfall is least in the south, on average 14 in (35 cm) per year. In higher elevations it can exceed 20 in (60 cm). Early summer and October are the rainy seasons, with heavy showers and thunderstorms common, often causing erosion and river silting. Overall Moldova's climate is excellent for agriculture, grape growing in particular.

Grasslands

Southern Moldova lies in an area called the Bugeac Steppe. However, in Moldova essentially the entire steppe zone has been cultivated.

Forests

About 13 percent of Moldova is forest and woodland, with the central hill country the most densely forested. Oak and hornbeam trees predominate, but linden, maple, beech, and wild fruit trees are also found. Many wildlife species inhabit these forests, including roe and spotted deer. Badgers, martens, ermines, and polecats are abundant, as are wild boars, foxes, and hares. Bird species include larks, jays, blackbirds, and migratory geese.

Population Centers – Moldova

(1992 POPULATION ESTIMATES)

Name	Population
Chisinău (Kishinev, capital)	667,100
Tiraspol	186,200
Balti (Beltsy)	159,000
Bendery	132,700

SOURCE: "Population of Capital Cities and Cities of 100,000 and More Inhabitants." United Nations Statistics Division.

HUMAN POPULATION

Moldova's population is 4,431,570 (July 2001 estimate), for an average density of 343 persons per sq mi (132 per sq km). Most of the inhabitants are concentrated in north and central Moldova. The country had the highest population density of any republic in the Soviet period; however, it was one of the least urbanized. About 53 percent of the people live in cities. Most country dwellers live in large villages.

NATURAL RESOURCES

Moldova has an abundant quantity of the sedimentary rock materials used for making high-quality cement and other construction products: sand, gravel, gypsum, lignite, phosphorite, and limestone. More than three-quarters of the country is covered in an exceptionally fertile type of soil called chernozem, which is ideal for agriculture. Some 14 percent of the nation's arable land is

GEO-FACT

Like other countries now independent of the former U.S.S.R., Moldova struggles to recover from Soviet ecological mismanagement. The country suffered extreme degradation in the name of industrial and agricultural output. Pesticides were used with utter disregard, polluting the topsoil almost beyond its ability to recover. Furthermore, agricultural methods that leveled forests to make way for vineyards compounded Moldova's erosion problems. Industrial emission controls were practically unknown.

used for permanent crops and 13 percent for pastures; approximately 3,110 sq km are irrigated.

FURTHER READINGS

Moldova. Minneapolis: Lerner, 1993.

Moldova: The Republic of Moldova. Chisinău, Moldova: Moldpres, 1995.

Verona, Sergiu. "Moldova Republic: Basic Facts." *CRS Report for Congress.* (92-182F.) Washington, D.C.: Library of Congress, Congressional Research Service, 1992.

The Republic of Moldova Site. http://www.moldova.org/ (accessed June 18, 2002).

Monaco

- **Area:** 0.7 sq mi (1.9 sq km) / World Rank: 206
- **Location:** Northern and Western Hemispheres in southern Europe; bordered by France on all sides except the Mediterranean coast to the south and southeast
- **Coordinates:** 43°44′N, 7°24′E
- **Borders:** 2.7 mi (4.4 km) total, all with France
- **Coastline:** 2.5 mi (4.1 km)
- **Territorial Seas:** 12 NM
- **Highest Point:** Mont Agel , 459 ft (140 m)
- **Lowest Point:** Sea level
- **Longest Distances:** 1.98 mi (3.18 km) E-W / .68 mi (1.1 km) N-S
- **Longest River:** None of significant length
- **Natural Hazards:** None
- **Population:** 31,842 (July 2001 est.) / World Rank: 198
- **Capital City:** Monaco-Ville, located in the southern area of the country
- **Largest City:** Monte Carlo, 12,300 (2000 est.)

OVERVIEW

Monaco is the the second-smallet country in Europe and the world. The entire country is urbanized, there are no forests or agricultural lands. There is little geographic variation in this tiny country but it is often divided into four regions based on economic activities: La Condamine is the central business district around the port. Monte Carlo is the northern entertainment district, site of the famous casino. Monaco-Ville is the location of the palace, it is situated on a rocky projection about 200 ft (60 m) above sea level. Fontvieille is an industrial area south of La Condamine.

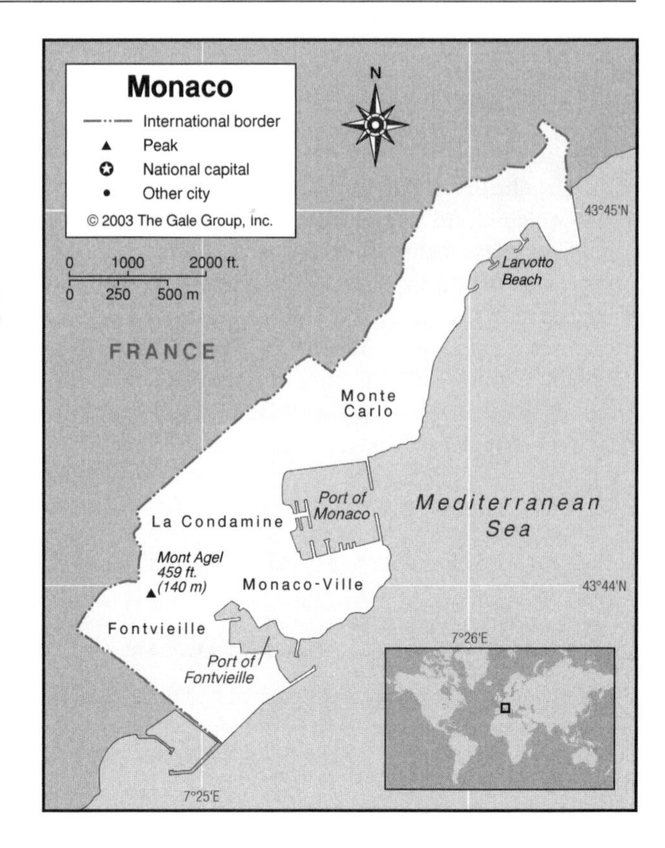

Having a sunny climate, Monaco only receives rain about 60 days out of the year. Winters are mild, and the summer heat is comfortable because of the cooling breezes from the bordering Mediterranean Sea.

MOUNTAINS AND HILLS

Most of the country is hilly, rugged and rocky, with various sea cliffs. Elevations rise away from the coast; the surrounding area of France is mountainous.

INLAND WATERWAYS

Monaco has no inland waterways of note, due to its small size.

THE COAST, ISLANDS, AND THE OCEAN

The Mediterranean Sea lies to the east and south of Monaco. Monaco's coastline contains several cliffs as well as the Monte Carlo and Larvotto beaches in Monte Carlo. The Port of Monaco is located off the central coast, and the Port of Fontvieille is in the south. They are separated by the small peninsula where Monaco-Ville is located.

CLIMATE AND VEGETATION

Temperature

Monaco's winters are mild, with temperatures rarely falling below freezing; January's average temperature is

46°F (8°C). The average temperature high in July and August (the summer months) is 79°F (26°C). Sea breezes moderate the summer heat. On average, the sun shines for 7 hours a day.

Rainfall

About 300 days of the year have no precipitation. Rainfall averages about 30 in (77 cm) per year.

HUMAN POPULATION

Owing in part to its small size, Monaco is the most densely populated nation in the world. Most of the people in Monaco are resident foreigners; native Monégasques make up only 16 percent of the population. One hundred percent of Monaco's population lives in urban areas.

NATURAL RESOURCES

Monaco does not have any natural resources or agriculture. It imports all of its energy from France. Its economy is based primarily on tourism, and Monaco's casino and resorts are world famous. Real estate, light industry, and financial services are also important parts of the economy.

FURTHER READINGS

Campbell, Siri. *Inside Monaco.* Glen Ellyn, IL: MCI, 1996.

Edwards, Anne. *The Grimaldis of Monaco.* New York: Morrow, 1992.

Hopkins, Adams. *Essential French Riviera.* Lincolnwood, Ill.: Passport Books, 1994.

LonelyPlanet.com. *Monaco.* http://www.lonelyplanet.com/destinations/europe/monaco/ (Accessed March 1, 2002).

Monte-Carlo Online. *Monte-Carlo.* http://www.monte-carlo.mc/principalitymonaco/index.html (Accessed March 1, 2002).

Mongolia

- **Area:** 604,247 sq mi (1,565,000 sq km) / World Rank: 20
- **Location:** North Asia, north of China and south of Russia.
- **Coordinates:** 46° 00′ N, 105° 00′ E
- **Borders:** 5072 mi (8,161.9 km) / Russia 1,867 mi (3,005 km); China 2,904 mi (4,673 km); also touches Kazakstan at westernmost point
- **Coastline:** None
- **Territorial Seas:** None
- **Highest Point:** Nayramadlïn Orgil (also Huyten Orgil or Mount Huyten), 14,350 ft (4,374 m)
- **Lowest Point:** Hoh Nuur depression, 1,709 ft (518 m)
- **Longest Distances:** 1,471 mi (2,368 km) E-W; 783 mi (1,260 km) N-S
- **Longest River:** Orhon, 698 mi (1,124 km)
- **Largest Lake:** Uvs Lake, 1,300 sq mi (3,366 sq km)
- **Natural Hazards:** Dust storms, snowstorms, floods, droughts, fires, earthquakes
- **Population:** 2,655,000 (2001) / World Rank: 135
- **Capital City:** Ulaanbaatar, in north-central Mongolia
- **Largest City:** Ulaanbaatar, population 773,700 (2000)

OVERVIEW

Mongolia's mountainous, arid terrain, with less than 1 percent of the land considered arable, has led to its inhabitants' dependence on livestock grazing, and the establishment of crucial trade routes to elsewhere in Asia. While the country is undeniably remote, it is never really isolated. In the past it has sometimes been known as Outer Mongolia to differentiate it from neighboring Chinese-administered Inner Mongolia.

In Mongolia there are five topographic regions: the mountainous Altai (the largest mountain system); the Great Lakes depression (lakes and plains); the mountainous Hangayn-Hentiyn (medium altitude, old mountains, gentle slopes and valleys); the uplifted eastern plains (smooth and rolling, excellent pastures, forests and rivers); and the Gobi (hilly in the west with salt lakes and marshes in flat lowlands, sand desert, semi-arid or arid.)

The country has greatly varied and distinctive ecosystems: alpine, upland and taiga forests; steppe grassland; desert grazing land; and truly barren desert. An extraordinary array of plant and animal life is found in these different habitats, including the very rare snow leopard and Gobi brown bear. Efforts are being made to preserve Mongolia's wildlife habitats and natural terrain, with more than a dozen nature reserves established, even while the nation becomes increasingly developed and more resources are extracted.

MOUNTAINS AND HILLS

Mountains

The high mountains of Mongolia rise mostly in the west. Some of the peaks are long extinct volcanoes. The lofty Altai Shan Range is part of a chain that continues over the border into China, Russia, and Kazakstan; it runs northwest to southeast in Mongolia. Some 200 glaciers cascade through the Altai Range. Mongolia's highest mountain is Nayramadlïn Orgil (Huyten Orgil or Mount

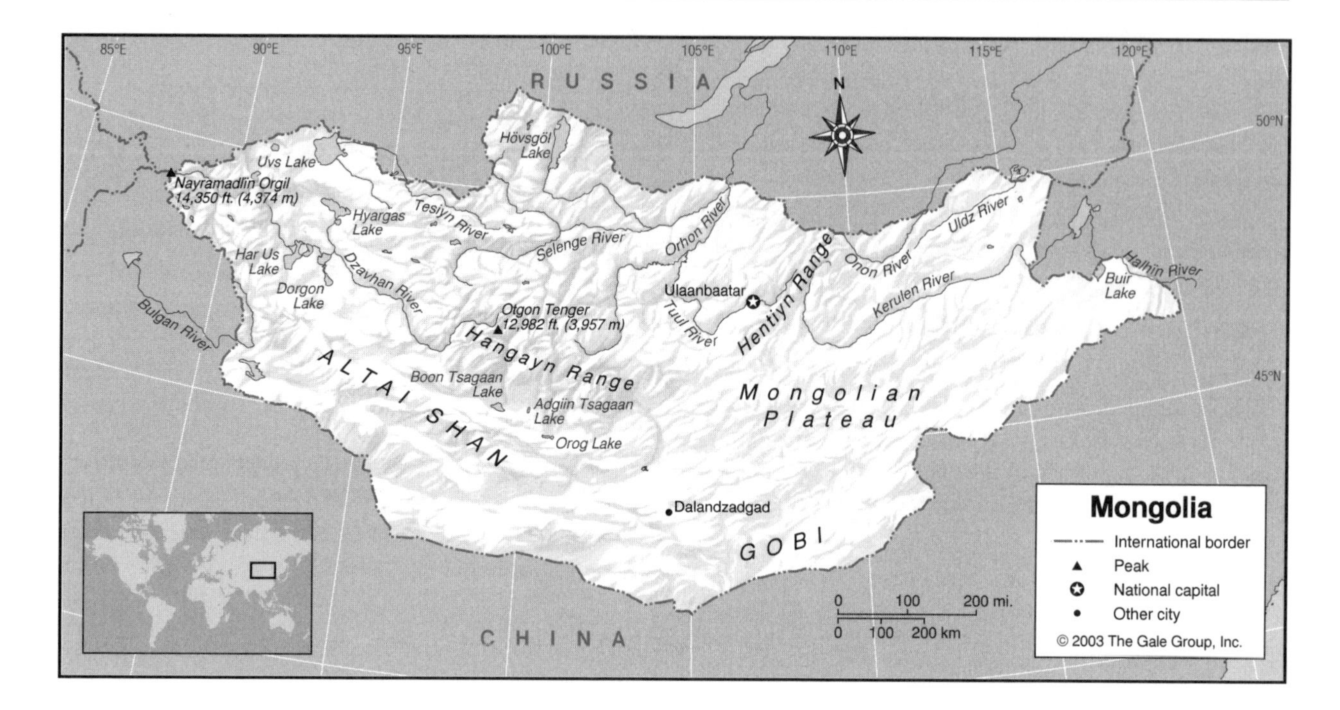

Huyten, also called Mount Nayramadlïn) rising 14,350 ft (4,374 m) in the Tawan Bogdo group of the Altai Range at Mongolia's westernmost extension, the four-corners meeting point with Russia, Kazakstan, and China. The second highest mountain in Mongolia, Mt. Chajrchan Uul, 14,311 ft (4,362 m), is in the central Altai Range.

The Hangayn (Khangai) Range in central Mongolia has generally lower mountains. The Hangayn Range's highest peak is Otgon Tenger, 12,982 ft (3,957 m) Another chain of low mountains is the Hentiyn (Khentei) Range in north central Mongolia, sprawling along and across the Russian border. Both the Hangayn and Hentiyn Ranges feature considerable forests and alpine meadows, and are important habitats for wildlife including ibex and snow leopards.

Plateaus

Mongolia is an immense plateau with an average height of 3,000 to 5,000 ft (914 to 1,524 m) above sea level. Nowhere is the country lower than 1,500 ft (457 m) above sea level.

Passageways across the Mongolian Plateau, between the mountain ranges, vary from 1,200 to 2,000 mi (1,931 to 3,218 km) in length. Since the seventeenth century, the main trade route across Mongolia proceeded from Russia to China, via the city of Ulaanbaatar and the Gobi desert.

Canyons

Within the Gobi desert, in the Dalanzadgad region, Gobi Gurvansaikhan National Park contains winding canyons of colorfully streaked sandstone. Yolym Am (Yol canyon), containing a permanently frozen stream, is in this area. Also in the National Park are the Flaming Cliffs (including the Nemegt, Khermiin Tsav, and Bayanzag canyons). The canyons of the Flaming Cliffs contain the archeological sites originally excavated by American fossil-hunter Roy Chapman Andrews in the early twentieth century. The first examples of dinosaur eggs were found there, as well as many significant dinosaur skeletons from the late Cretaceous period. The Gobi desert's dry climate has also preserved later mammal and plant fossils of great significance.

Hills and Badlands

Mongolia's elevations decline from northwest to southeast, decreasing gradually from alpine snow peaks to rolling contours, mesas, ridges and low hills, and eventually to completely flat plains. The foothills of the Altai Range stretch south and east into the Gobi, forming a terrain of bare desert hills.

INLAND WATERWAYS

Lakes

More than 4,000 lakes, mostly of glacial or volcanic origin, relieve the dry landscape of Mongolia. For the most part the lakes are located high above sea level and they freeze over every winter; those with outlets usually have fresh water. Most of the 16 biggest lakes are found in the northwest. The country has more than 200 developed sites of hot and cold natural mineral water springs.

In the Great Lakes depression of northwest Mongolia there are at least 300 lakes, as well as high waterfalls and springs. Uvs Lake, a saltwater lake at 2,490 ft (759 m) above sea level in this region, is Mongolia's largest with a surface area of 1,300 sq mi (3,366 sq km). Fed by some

200 rivers and rivulets, Uvs Lake is surrounded by a spectrum of ecosystems, from snowy mountains to desert. The lake basin experiences extremes of cold and warm weather—it is one of ten worldwide locations being studied in the International Geosphere-Biosphere Program on global climate change. Uvs Lake and adjacent areas (including sand dunes, marshes, and mountains) are designated a "Strictly Protected Area" by the Mongolian government. Also in the Great Lakes area, Har Us, Hyargas, and Dörgön Lakes are a trio of connected, large, shallow lakes within Har Us Nuur National Park.

The freshwater alpine Hövsgöl Lake, 1,069 sq mi (2,770 sq km) in north central Mongolia near the Russian border, is connected by the Selenge River to Russian Siberia's immense Lake Baikal. As many as 45 rivers and rivulets empty into Hövsgöl Lake. It is the deepest lake in Central Asia, with depths of up to 905 ft (276 m). Coniferous taiga forest, a Siberian ecosystem, surrounds the lake. Measures have been taken to protect Hövsgöl Lake from the pollution suffered by Lake Baikal, including a ban on oil shipping by barges.

In the Gobi foothills of eastern Mongolia, the Valley of the Lakes holds four salt lakes: Boon Tsagaan, Taatsiin Tsagaan, Adgiin Tsagaan, and Orog. The fresh-water Buir Lake, at 235 sq mi (668 sq km), is the largest lake in the east of Mongolia. It connects to Hulun Lake, in China.

Rivers

With more than 1,200 rivers, Mongolia has three drainage systems: to the Arctic Ocean, to the Pacific Ocean, and to the desert or salt lakes. Rivers draining north to the Arctic Ocean include the Selenge River (via Lake Baikal, which drains by the Angara and Yenisey Rivers to the Arctic Ocean), as well as the Shishkhed and Bulgan Rivers. The Selenge River arises in the Hangayn uplands of northern Mongolia, and flows north into Russia's Lake Baikal. The Selenge has a total length of 616 mi (992 km), about 370 mi (595 km) of which is within Mongolia. The Selenge's watershed area, approximately 109,000 sq mi (28,231 sq km), is the greatest of the river basins inside Mongolia. Among the Selenge's numerous tributaries are the Orhon which, with a length of 698 mi (1,126 km), is the longest river in Mongolia, and the Tuul (437 mi / 703 km) where the capital, Ulaanbaatar, is located.

The Kerulen, Onon, Uldz, and Halhïn rivers of northeast Mongolia flow into the Amur River of Russia, which continues east to the Pacific Ocean. The longest river of the Amur/Pacific group of Mongolia is the Kerulen, which is 675 mi (1,086 km) long. Mongolia's other river systems are found in the depression of the Great Lakes and in the Central Asian basin, including the Dzavhan (500 mi / 804 km), Tesiyn (350 mi / 563 km), and Khobdo (310 mi / 499 km) rivers. The river system in the Gobi region is negligible; the few small rivers of the northern portion of the desert zone rise in the Hangayn Range but vanish into salt lakes.

The water supply of Mongolia's rivers is not plentiful. In a land of low precipitation, rain must account for 80 to 85 percent of the river water. During the country's usual six months of winter, the rivers are covered with very thick ice and smaller waterways freeze solid. Meltwater from the mountains provides 15 to 20 percent of the yearly runoff. The mountain districts of the north contain the greatest potential for hydroelectric power generation. Most waterways—with the exception of the Selenge—are not suitable for navigation.

Wetlands

Mongolia contains many salt marshes and a variety of lake-centered wetland environments. Valuable examples of these are the six sites in Mongolia that are designated Wetlands of International Importance under the Ramsar International Convention on Wetlands. They are: Ayrag Nuur, a shallow lake which is a bird habitat; Har Us Nuur National Park, a complex of three shallow lakes; Mongol Daguur, a Nature Reserve in a volcanic basin, including marshes and lakes; Ogii Nuur, a freshwater lake with surrounding marshes; Terhiyn Tsagaan, a volcanic lake with marshes; and the Valley of Lakes, a chain of four salt lakes and extensive reed marshes with important bird populations. There are also extensive marshes around Uvs Lake, where many bird species nest.

GEO-FACT

Przewalski horses, which Mongolians call "takh," are the last existing ancestors of the modern domesticated horse, but the species became extinct in the wild by 1970. With only 300 of the small, heavy-maned horses surviving in captivity, a project was begun to breed groups of them for good health and genetic differences and then reintroduce them to the old wild habitat of the Mongolian steppes. Starting in 1992, more than 1,500 of those Przewalski horses have been set free in central Mongolia's Hustain Nuruu nature reserve.

THE COAST, ISLANDS AND THE OCEAN

Mongolia is a landlocked nation. The closest ocean is the Pacific's Yellow Sea, which is 435 mi (700 km) to the east across northeast China.

CLIMATE AND VEGETATION

Temperature

Mongolia has two climatic zones: the continental zone in the north, and the desert in the south. The country's high altitude gives it inhospitably cold, dry, and harsh weather.

Temperatures can fluctuate radically each day, dropping to much colder levels at night, and they differ greatly season to season. There is an especially long winter, with freezing temperatures the norm from October to April. The temperature can plunge to as low as -62°F (-52°C) in January. Mongolia's average winter temperature is -13°F (-24°C), with an average range of -5° to -22°F (-21° to -30° C). Spring is a brief windy and stormy transition period of five to six weeks around May. Summer lasts from June to August, with an average temperature of 65°F (20°C), ranging from 50° to 80°F (10° to 27°C). Autumn is a five- to six-week transition period around September. Mongolia's average humidity is 65 percent in summer and 75 percent in winter.

Rainfall

Most of Mongolia's rainfall occurs from May to September. The country usually has at least 250 sunny days each year. Rainfall is considerably heavier in the north, and nearly nonexistent in the southern Gobi desert. Mongolia's annual average rainfall is a low 8 to 9 in (20 to 22 cm), with an average of 14 in (36 cm) in the north, and less than 4 in (10 cm) in the south. The country experienced devastating heavy snowstorms in the winters at the start of the twenty-first century.

Less than 70 percent of Mongolia's land has a consistent supply of water. Winter freezes cut off access to even this supply from surface waters and the many wells. Melted snow and ice become the sources of winter water for household and workplace purposes. The best relative water situation is in the north, because it is the region of major rivers and heavier precipitation. Where surface streams are lacking, as in the south, below-surface water becomes particularly important. Well-drilling and related construction of irrigation systems have been promoted by Mongolian governments in recent decades.

Grasslands

Mongolia is famous for its useful and beautiful grasslands, with at least 79 percent of the country covered with natural meadows and pastures. The southeast is an area of particularly extensive grasslands, which are known as steppes in Central Asia. The steppe regions, historically occupied by nomadic herders, are very vulnerable to even small variations in annual rainfalls, which can have a considerable effect on the growth of grass and therefore on the raising of livestock. The steppe hills and plains are covered with many varieties of grasses, and are grazed by domestic animals including sheep, goats (Mongolian goats are the primary source of cashmere), horses, cattle, yaks, and camels; as well as wild antelopes including enormous migratory herds of gazelles.

Population Centers – Mongolia

(1999 POPULATION ESTIMATES)

Name	Population
Ulaanbaatar	691,000
Darhan	72,600
Erdenet	65,700
Choybalsan	38,500

SOURCE: "Table 3.4 Urban Population." National Statistical Office of Mongolia.

Provinces – Mongolia

2000 POPULATION ESTIMATES

Name	Population	Area (sq mi)	Area (sq km)	Capitals
Arhangay	95,500	21,000	55,000	Tsetserleg
Bayanhongor	84,100	45,000	116,000	Bayanhongor
Bayan-Olgiy	91,000	18,000	46,000	Olgiy
Bulgan	61,000	19,000	49,000	Bulgan
Darhan-Uul	84,800	100	200	Darhan
Dornod	76,100	47,000	122,000	Choybalsan
Dornogovi	51,500	43,000	111,000	Saynshand
Dundgovi	51,100	30,000	78,000	Mandalgovi
Dzavhan	88,000	32,000	82,000	Uliastay
Erdenet	73,900	300	800	Erdenet
Govi-Altay	62,800	55,000	142,000	Altay
Hentiy	70,800	32,000	82,00	Ondorhaan
Hovd	86,300	29,000	76,000	Hovd
Hövsgöl	118,300	39,000	101,000	Moron
Ömnögovi	46,900	64,000	165,000	Dalanzadgad
Övörhangay	110,800	24,000	63,000	Arvayheer
Selenge	100,900	16,000	42,000	Suhbaatar
Sühbaatar	55,900	32,000	82,000	Baruun-Urt
Töv	97,500	31,000	81,000	Zuunmod
Ulaanbaatar	773,700	800	2,000	Ulaanbaatar
Uvs	89,700	27,000	69,000	Ulaangom

SOURCE: National Statistical Office of Mongolia. http://statis.pmis.gov.mn (accessed 28 May 2002).

Deserts

The great desert known as the Gobi occupies one third of Mongolia, and extends far south into China's Inner Mongolia. It is the world's largest cold-climate desert. Less than 2 in (5 cm) of rain falls in the Gobi in a year, with no rainfall at all in parts of the desert.

There are two types of desert within the Gobi. One is a scrubland with coarse, stunted bunchgrass and hardy bushes, which is dry but can be grazed by camels. It contains numerous species of plants, many of which bloom in summer if they can receive enough moisture during the year. The other type of desert in the Gobi is a landscape of sand dunes mixed with stone or gravel, with little to no vegetation. Alarming increases of this eroded and barren form of desert wasteland have occurred due to over-grazing in the scrub desert and in transitional grasslands.

Forests and Jungles

As much as 10 percent of Mongolia has forest and woodland cover. Deforestation has taken place as trees and shrubs have been cut for fuelwood and timber, or destroyed by livestock grazing. There are attempts at reforestation in some areas.

Mongolia's forests are mostly in the mountain regions, especially the Hentiyn and Hangayn ranges of the north, where tree species include deciduous birch and larch, plus coniferous cedar, fir and pine. The conifer forests of the northern mountains form an ecosystem known as "taiga," which is the same as the vast forests of neighboring Siberia. Far to the south, in the scrublands of the Gobi desert, there are forest-like groves of tall saxual shrubs, but these are diminishing due to climate changes and livestock grazing.

HUMAN POPULATION

The population density of Mongolia, at fewer than 5 people per sq mi (2 people per sq km), is one of the lowest in world. This lack of human population pressure has helped to conserve the country's resources and to preserve its environment, although increasing development has brought new challenges. The population growth rate is 1.47 percent (2001). The proportion of urban dwellers is 59 percent (2000), with nearly a third of Mongolia's people living in the capital, Ulaanbaatar.

NATURAL RESOURCES

Agriculture and livestock raising accounts for most of Mongolia's economy, with the major agricultural products being wheat, potatoes, oats, and barley. There are considerable mineral resources as well: copper, gold, coal, molybendenum, tungsten, and tin account for a large part of Mongolia's industrial production. Mongolia also has deposits of phosphates, oil, feldspar, nickel, silver, iron, zinc, and wolfram. The country contains considerable potential for energy generation from hydropower, solar, and wind sources.

FURTHER READINGS

Foundation for the Preservation and Protection of Przewalski's Horse. http://www.treemail.nl/takh (Accessed May 2002).

Kawaoka, Michie. *Mongolia World.* http://plaza.harmonix.ne.jp/~michie/mongolia.html (Accessed May 2002).

Lawless, Jill. *Wild East: The New Mongolia.* Toronto: ECW Press, 2000.

Man, John. *Gobi: Tracking the Desert.* New Haven, Conn.: Yale University Press, 1999.

Novacek, Michael. *Dinosaurs of the Flaming Cliffs.* New York: Anchor Books, 1997

Morocco

- **Area:** 172,414 sq mi (446,550 sq km; of which the Western Sahara comprises 97,344 sq mi / 252,120 sq km) / World Rank: 58
- **Location:** Northern and Western hemispheres, Northern Africa, bordered on the north by the Mediterranean Sea and the two Spanish enclaves of Ceuta and Melilla, on the south and southeast by Algeria, on the south by the Western Sahara, and on the west by the North Atlantic Ocean. Moroccan-controlled Western Sahara also borders on Mauritania.
- **Coordinates:** 32°00′N, 5°00′W
- **Borders:** 1,254 mi (2,081 km) / Algeria, 969 mi (1,559 km); Spain (Ceuta), 3.9 mi (6.3 km); Spain (Melilla), 6.0 mi (9.6 km); Western Sahara, 275 mi (443 km)
- **Coastline:** 1,140 mi (1,835 km)
- **Territorial Seas:** 12 NM
- **Highest Point:** Mt. Toubkal, 13,665 ft (4,165 m)
- **Lowest Point:** Sebkha Tah, 180 ft (55 m) below sea level
- **Longest Distances:** 1,124 mi (1,809 km) NE-SW; 326 mi (525 km) SE-NW
- **Longest River:** Oum er Drâa 744 mi (1200 km; intermittent)
- **Natural Hazards:** Earthquakes; periodic droughts
- **Population:** 30,645,305 (July 2001 est.) / World Rank: 37
- **Capital City:** Rabat, on the northwestern coast
- **Largest City:** Casablanca, south of Rabat on the northwestern coast, 3,344,300 (2002)

OVERVIEW

Situated at the northwestern corner of Africa, Morocco is strategically located along the Strait of Gibraltar. Morocco has four distinct geographic regions. In the north there is a fertile coastal plain along the Mediterranean as well as the rugged Er Rif Mountains. The Atlas Mountains, extending across the country from the southwest to northeast and into Algeria, are another region. Between these two regions is a third, an arc of wide

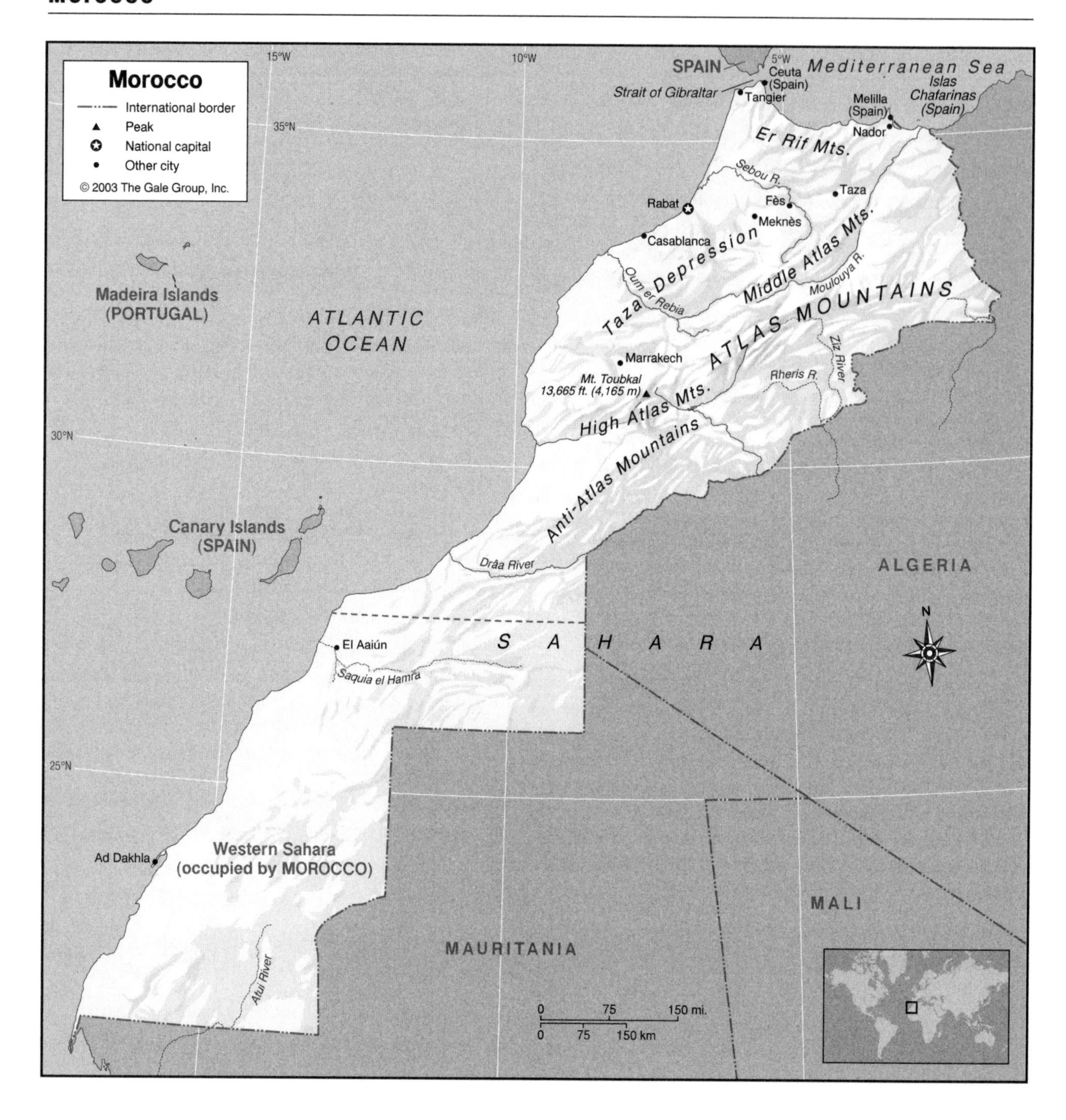

coastal plains along the country's western seaboard, bounded by the Er Rif and Atlas Mountain Range. Finally, south of the Atlas there are semiarid grasslands that merge with the Sahara Desert along the southeastern borders of the country.

In 1976, Morocco occupied the northern two-thirds of the former Spanish colony now called Western Sahara (Spanish Sahara), and in 1979 it occupied the remainder of that territory. Morocco claims and administers the territory as its own, but guerilla forces challenge these claims and the area's sovereignty is unresolved. Many statistics produced by or about Morocco include Western Sahara, and the region is therefore described in this essay.

Morocco contests Spain's control of five small regions near Morocco. Three are islands off Morocco's coast: Peñón de Alhucemas, Peñón de Vélez de la Gomera, and Islas Chafarinas. The other two are small Spanish enclaves on Morocco's Mediterranean coast: Ceuta and Melilla.

MOUNTAINS AND HILLS

Mountains

The Atlas Mountains are the largest and most important mountain range in North Africa, extending from Morocco to Tunisia (about 1,488 mi / 2,400 km) in a series of creased mountain chains. Morocco's portion of the Atlas Mountains includes the Middle Atlas, High Atlas, and Anti-Atlas.

The High Atlas (Western Atlas, Great Atlas) is the highest, stretching for more than 400 mi (644 km) with ten peaks of over 13,000 ft (3,965 m). Mount Toubkal, south of Marrakech, reaches to13,665 ft (4,165 m), the highest point in the country. The Middle Atlas stretches for 156 mi (251 km) east of the High Atlas, extending into Algeria. Mount Bounaceur is the highest point in the Middle Atlas 10,909 ft (3,326 m). West and south of the High Atlas is the Anti-Atlas range. Although not as tall as the High Atlas, the terrain in the Anti-Atlas is very rugged. It is about 250 mi (403 km) long. South of the Atlas is the Sirwa, a volcanic outcropping and a ridge of black lava that connects the High Atlas and Anti-Atlas. The Sirwa reaches a maximum height of 9,254 ft (2,822 m).

The Er Rif near the northern coast are not part of the Atlas ranges. They are made up of steep cliffs. The highest peak in the Er Rif is Tidghine (8,085 ft / 2,465 m), south of Ketama. Also in Er Rif is Toghobeit Cave. At 3,918 ft (722 m) deep it is one of the most fantastic open caverns in the world.

INLAND WATERWAYS

Lakes

Lake Semara is in the Western Sahara. Lake Chiker, near Taza, is usually dry during the summer months. Deep water mountain lakes are in the Middle and High Atlas: Tigalmamine, 53 ft (16 m) deep; Sidi Ali, 213 ft (65 m) deep; and Isti, 311 ft (95 m) deep, are examples.

Rivers

Morocco has the most extensive river system in North Africa. Principal rivers flowing south or westward into the Atlantic are the Oumer Rebia (344 mi / 555 km long), Sebou (Sebu; 310 mi / 500 km long), Bouregreg (155 mi / 250 km long), Tensift (167 mi / 270 km long), and Drâa (744 mi / 1,200 km). The Drâa marks part of the border with Algeria and, running through a desert area, is sometimes dry. The Ziz and Rheris both flow south out of the Atlas Mountains into the heart of the Sahara. The Moulouya (Muluya) flows 347 mi (560 km) northeast from the Atlas to the Mediterranean, and is the longest river in the country that consistently reaches the sea.

THE COAST, ISLANDS, AND THE OCEAN

The Mediterranean Sea is north of Morocco, and is connected to the Atlantic Ocean by the Strait of Gibraltar north of Tangier. Spain lies on the far side of the strait. The Mediterranean coast between Tangier and Nador has a string of creeks, bays, sheltered beaches, and cliffs along the Mediterranean shore, ideal for recreational uses. The Atlantic coast is often rocky, but has some long stretches of fine sand and calm bays, including the harbors at Rabat and Casablanca.

CLIMATE AND VEGETATION

Temperature

Morocco has two climatic zones, coastal and interior. Temperature variations are relatively small along the Atlantic coast, while the interior is characterized by extreme variations. The north and central areas are a Mediterranean climate, moderate and subtropical, cooled by the Mediterranean Sea and Atlantic Ocean. These areas characteristically have warm wet winters and hot dry summers. The average temperature hovers at around 68°F (20°C). In the northern part of the interior the climate is predominantly semiarid. Winters can be quite cold and the summers very hot. In the mountain ranges temperatures can drop to 0°F (-18°C). Mountain peaks in both the Atlas and Rif mountain ranges are snow capped throughout most of the year.

Rainfall

The western slopes of the Atlas Mountains are well watered, at the expense of the interior, which they block from the Atlantic or Medieterranean. The rainy seasons are from April to May and October to November. Maximum annual rainfall (30-40 in / 75-100 cm) occurs in the northwest. Other parts of the country receive much less; half of all arable land receives no more than 14 in (35 cm) per year.

Plains

With the exception of the Er Rif, all of Morocco north of the Atlas Mountains is a fertile plain. This area is also known as the Taza Depression. There are some semiarid grasslands in the south beyond the Atlas Mountains; these eventually give way to the Sahara Desert. Semiarid plains can also be found in northern Western Sahara.

Deserts

The Sahara Desert, the largest in the world, intrudes into the southeastern fringes of Morocco and covers much of Western Sahara. Except along the coast rainfall is erratic and temperatures extremely high.

Forests

The mountainous regions of Morocco contain extensive areas of forest, including large stands of cork oak, evergreen oak, juniper, cedar, fir, and pine.

HUMAN POPULATION

Morocco had a population density of 178 people per sq mi (69 people per sq km) in 2001. It was estimated that 55 percent of the population lived in urban areas in 2000.

Population Centers – Morocco

(1993 POPULATION ESTIMATES)

Name	Population
Casablanca	2,943,000
Rabat (capital)	1,220,000
Fès	564,000
Marrakech	602,000

SOURCE: "Population of Capital Cities and Cities of 100,000 and More Inhabitants, United Nations Statistics Division.

These figures would be higher if not for the inclusion of Western Sahara, which is very sparsely settled with a total population of perhaps 250,000 people. Most of Morocco's population lives in the fertile plains between the mountain ranges and along the Mediterranean coast.

NATURAL RESOURCES

Morocco is still chiefly an agrarian society but mineral resources, including phosphates, iron ore, lead, manganese, silver, copper, tin, zinc, coal and petroleum are also mined. Traditional sustenance resources are fish and salt.

FURTHER READINGS

Demeude, Hugues. *Morocco.* Köln: Evergreen, 1998.

Italia, Bob. *Morocco.* Minneapolis, Minn.: Abdo Pub., 2000.

Morocco. http://bubl.ac.uk/link/m/morocco.htm (Accessed July 6, 2002)

Solyst, Annette. *Morocco.* New York: Friedman/Fairfax, 2000.

Jacobshagen, H. Volker, ed. *The Atlas System of Morocco: Studies on Its Geodynamic Evolution.* New York: Springer-Verlag, 1988.

Western Sahara Geography. http://www.arso.org/05-2.htm (accessed July 6, 2002).

Wilkins, Frances. *Morocco.* Philadelphia: Chelsea House Publishers, 2001.

Mozambique

- **Area:** 309,496 sq mi (801,590 sq km) / World Rank: 37
- **Location:** Southern and Eastern hemispheres, on the southern part of the African continent, south of Tanzania, northeast of South Africa, east of Swaziland, southeast of Zimbabwe, Zambia, and Malawi, with the Indian Ocean to the east
- **Coordinates:** 18°15′S, 35°00′ E
- **Borders:** 2,840 mi (4,571 km) / Malawi, 975 mi (1,569 km); South Africa, 305 mi (491 km); Swaziland, 65 mi (105 km); Tanzania, 470 mi (756 km); Zambia, 260 mi (419 km); Zimbabwe, 765 mi (1,231 km)
- **Coastline:** 1,535 mi (2,470 km)
- **Territorial Seas:** 12 NM
- **Highest Point:** Monte Binga, 7,992 ft (2,436 m)
- **Lowest Point:** Sea level
- **Longest River:** Zambezi, 1,650 mi (2,650 km)
- **Largest Lake:** Lake Malawi, 11,400 sq mi (29,600 km)
- **Natural Hazards:** Droughts and floods in central and southern provinces; cyclones
- **Population:** 19,371,057 (July 2001 est.) / World Rank: 52
- **Capital City:** Maputo, in the southernmost part of Mozambique
- **Largest City:** Maputo, population 3,017,000 (2000 est.)

OVERVIEW

Mozambique, at a size approximately twice that of California, is an expansive and geographically diverse nation in the southeast of Africa bordering the Indian Ocean. The Zambezi River divides the country into distinct northern and southern halves. The north is known for its mountainous regions and plateaus, notably the Livingstone-Nyasa Highlands, the Shire or Namuli Highlands, and the Angonia Highlands in the northeast. The westernmost regions are particularly mountainous, giving way to plateaus and uplands as one travels eastward. South of the Zambezi are the more fertile plains, most notably in the area surrounding the river, as well as uplands in the center of the country, marshes, and coastal lowlands. Inland areas are dry and not conducive to vegetation. The country is approximately 44 percent coastal lowlands, 26 percent elevated hills and plateaus, 17 percent low plateaus and hills, and 13 percent mountainous.

Mozambique is located on the African Tectonic Plate and experiences little or no tectonic activity.

MOUNTAINS AND HILLS

Mountains

Mountainous regions in Mozambique are found throughout the western end of the country. Most mountain peaks rise from plateau regions, although many mountains are isolated in the landscape. The Great Rift Valley, which starts in Jordan near Syria, terminates in Mozambique near Beira at Sofala Bay. A wide variety of animal species, including lions, reside in this area.

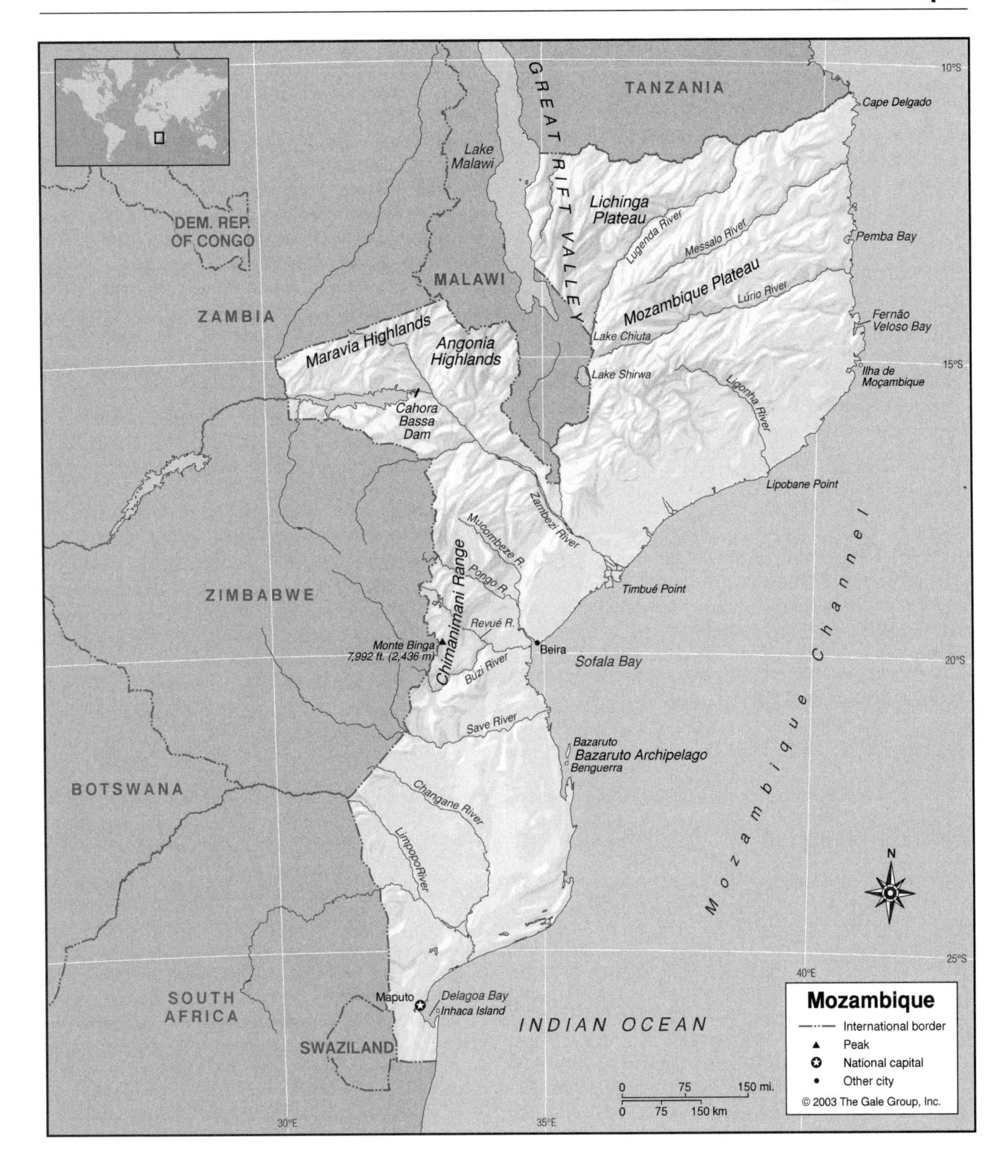

The country shares the Chimanimani Mountain Range with Zimbabwe, where Mozambique's highest peak, Mt. Binga, is located. Alluvial gold has been extracted from these mountains.

Plateaus

There are many plateaus of varying elevations throughout the northwestern portion of Mozambique, which generally increase in elevation as one travels westward. These plateaus help support the country's many farmers, supporting both cash crops and feeding livestock.

The province of Niassa, which borders Lake Malawi in northern Mozambique, is the largest and tallest in the country with 25 percent of its area covered by the Lichinga Plateau, which reaches elevations of up to 4,920 ft (1,500 m). The entire province itself has an average eleva-

GEO-FACT

The Limpopo is a sluggish river other than at flood tide and was given the euphonious epithet of "the great gray-green, greasy Limpopo River, all set about with fever-trees" by Rudyard Kipling.

tion of 2,296 ft (700 m). The plateau is a heavily wooded savanna, with dry and open woodland areas covered in acacia trees. On the other side of the Lugenda River is the Mozambique Plateau. This plateau is similar to the Lichinga, though lower in elevation. It reaches from the center of the country all the way to the Indian Ocean.

The Angonia and Maravia Highlands in northwest Mozambique on the Zambia border are some of the most fertile lands in all of Mozambique. Crops such as peaches, apples, and potatoes are grown in this area.

Hills and Badlands

The area in northeastern Mozambique between the Lúrio and Ligonha Rivers is the location of some of the most magnificent vertical granite rock faces in all of Africa, and is a favorite rock-climbing destination. Rolling hills are commonly found east of areas with particularly mountainous terrain. Vegetation is sparse in these savanna regions and the land does not support many crops.

INLAND WATERWAYS

Lakes

Three lakes in northern Mozambique—Lake Malawi, Lake Chiuta, and Lake Shirwa—form part of the border with Malawi but have remained largely unexploited.

Navigable Lake Malawi (also known as Lake Nyasa) borders Mozambique, Tanzania, and Malawi. The lake is an incredible 11,400 sq mi (29,600 km) in surface area, about one-third of which is within Mozambique's territory. Its deepest waters—which reach a maximum depth of 2,316 ft (706 m)—are found in this third.

Rivers

Mozambique is rich in rivers, with 25 throughout the country. Many of these rivers flow out from the western highlands to the Indian Ocean or the Mozambique Channel in the east. Five major basins and several smaller ones drain the country, although several have their major catchment areas as far away as eastern Angola. Flow tends to fluctuate, owing to the alternation of rainy and dry seasons. The greatest flow takes place between January and March, the least between June and August.

The largest and most important basin is that of the Zambezi River, which flows southeast across the heart of Mozambique into the Indian Ocean and has historically been the principal means of communication between inland central Africa and the coast. Its waters make the soil in the land surrounding it some of the most fertile land in the country. From the Maravia Highlands downstream the valley is low lying and has a very gentle slope, since the land has an elevation of less than 500 ft (152 m). Upstream the river enters a narrow gorge, which prompted the construction of the Cahora Bassa Dam. The Cahora Bassa Dam, the largest hydroelectric power dam in Africa, powers the capital city of Maputo and provides electricity for parts of South Africa and Zimbabwe as well.

The Limpopo River in the south flows through Botswana, Zimbabwe, Mozambique, and South Africa. It is fed mainly by the Changane River and drains the Limpopo Basin. It is susceptible to serious flooding, the effects of which are compounded when cyclones occur in the wet months. Also particularly notable is the Save (or Sabi) River in the center of the country, which, along with the Búzi and Revué Rivers, drains the southern Mozambique Plain. In the northeast draining the Mozambique Plateau are the Lugenda River, the Messalo River, the Lúrio River, and the Ligonha River.

Wetlands

Much of the area around the mouth of the Zambezi and a belt running along the lower reaches of the Pongo River and its tributary the Mucombeze north to the Zambezi is marshy, hindering north-south communications and promoting the spread of disease. Mangrove swamps are common near the coast of the Sophala and Zambezia provinces. These wetlands provide excellent conditions for many marine species, most notably prawns.

THE COAST, ISLANDS, AND THE OCEAN

Oceans and Seas

Bordering Mozambique to the east is the Mozambique Channel, a strait in the Indian Ocean that separates Africa from Madagascar. It is approximately 1,000 miles (1,600 km) long and at its widest point stretches more than 600 miles (950 km). This area is particularly susceptible to cyclones. Many coral reefs line the channel, attracting large numbers of divers from around the world. Coral islands also exist in the channel.

Major Islands

Mozambique is dotted with many small islands along its coastline. Ilha de Moçambique, located 2 mi (3 km) off the coast of the Nampula province, is a small but culturally significant island. Formerly a Portuguese colonial capital, this 1.5-mi (2.5-km) long and 0.4-mi (0.6-km) wide island is assessable via a mainland bridge.

Inhaca Island, located 18.6 mi (30 km) from Maputo, is a 7.8-mi (12.5-km) long and 4.7-mi (7.5-km) wide island known for its sandy beaches and ideal diving and fishing locations.

The Bazaruto Archipelago, also known as the Paradise Islands, is located 6 mi (10 km) off the country's coast and was formed from sands deposited from the Lipopo river thousands of years ago. Santa Carolina, Bazaruto, Ibo, Benguerra, and Magaruque are the most popular islands, boasting clear blue waters, sandy beaches, palm trees, coral reefs, crocodiles, many species of tropical fish, and other tropical wildlife such as the samango monkey. The whole area became declared a national park in 1970.

The Coast and Beaches

The expansive coastlines of Mozambique are jagged and provide for numerous bays and beaches. The coastal areas are ideal for the cultivation of rice, maize, sugar cane, and cashews. The coastal waters are rich in prawns, one of the country's leading exports. Fishermen often frequent the coastlines, as small and large catch fish are abundant.

Located in the southeast of Mozambique, Tofo (sometimes Tofu) and Barra beaches are known for their sand dunes, mangroves, and palm groves, as well as their tropical wildlife including parrots and monkeys. Wimbi Beach is particularly notable for its coral reefs, a favorite among snorkelers. Its white coral beaches, lined by palm trees, provide an ideal tropical setting. The beaches of Mozambique are well preserved, and wildlife—including humpback whales, turtles, flamingoes, dolphins, and manta rays—thrives.

Several points, capes, and bays dot the coastline, including Delagoa Bay, Indigo Bay, Fernão Veloso Bay, Sofala Bay, and Pemba Bay. Starting in the south, the coast moves northeasterly from Delagoa bay then curves smoothly to the north. At Ponta São Sebastião it cuts back to the south before turning back to the north. Here it cuts back slightly to the west until Sofala Bay, where the coastline heads northeast, continuing almost straight past Timbué Point and Lipobane Point. It then curves and heads due north where it ends at Cape Delgado.

CLIMATE AND VEGETATION

Temperature

Between the months of November and March, temperatures are usually between 81°F and 84°F (27°C and 29°C) throughout most of the country, though they are lower in the interior uplands. Between April and October, temperatures are cooler, averaging between 64°F and 68°F (18°C and 20°C).

Rainfall

The wet season runs from November to March, when 80 percent of all rainfall occurs. Rainfall is lowest in the southwest portion of the country, which receives an annual average of 12 in (30 cm). It is highest near the western hills and the central areas near the Zambezi, as well as along the central coast where annual averages are 53-59 in (135-150 cm).

Grasslands

Areas close to the major rivers in Mozambique are particularly fertile and support a variety of plants and trees, including lemon, orange, lychee, and mango.

Deserts

Much of southern and central Mozambique that is inland from the coastline suffers from poor, sandy, infertile soil. Little vegetation other than dry scrubs can be supported on this land.

Forests and Jungles

Approximately two-thirds of the land supports woodland vegetation. The majority of Mozambique's forested areas are located along plateaus and contain the miombo forest type, categorized by having dry, deciduous trees of varying heights. The northernmost regions as well as those surrounding the mouth of the Zambezi River are the richest in woodland. Tropical forests are also prevalent, with lush vegetation and African game species such as zebras, wildebeests, and even elephants, but mangroves are relatively rare and are found near coastal regions.

HUMAN POPULATION

Mozambique has an overall population density of 63.2 people per sq mi (24.4 people per sq km), with the northeastern and southern regions being slightly above that average, and the central region well under, at 44.3 per sq mi (17.1 per sq km). However, the population is concentrated in three general areas: along the southern and southeastern coastal region, along the northeastern coastal region, and in the plateau region southeast of Lake Malawi. The only area with significant settlement in the central region is along the Revué River. There are also significant settlements in the northwestern highlands and along the Zambezi River.

The ten most significant tribal groups in Mozambique inhabit separate areas. The largest of the groups, the Makua, resides north of the Zambezi River. In the south of the Zambezi, the Tsonga is the most prevalent ethnic group, accounting for approximately 23 percent of the entire country's population. The Shona or Karanga people live in the central region, and the Chopi live along the southeastern coast. The tribal groups make up 99.66 percent of the country's population; Euro-Africans, Europeans, and Indians account for the remaining 0.34 percent.

Population Centers – Mozambique

(1997 POPULATION ESTIMATES)

Name	Population	Name	Population
Maputo (capital)	989,386	Nacala-Porto	164,309
Matola	440,927	Quelimane	153,187
Beira	412,588	Tete	104,832
Nampula	314,965	Xai-Xai	103,251
Chimoio	177,608	Gurue	101,367

SOURCE: Instituto Nacional de Estatística (INE), Mozambique.

Provinces – Mozambique

2002 POPULATION ESTIMATES

Name	Population	Area (sq mi)	Area (sq km)	Capital
Cabo Delgado	1,525,634	31,902	82,625	Pemba
Gaza	1,266,431	29,231	75,709	Xai-Xai
Inhambane	1,326,848	26,492	68,615	Inhambane
Manica	1,207,332	23,807	61,661	Chimoio
Maputo Cidade	1,044,618	116	300	Maputo
Maputo Provincia	1,003,992	9,944	25,756	Matola
Nampula	3,410,141	31,508	81,606	Nampula
Niassa	916,672	49,829	129,056	Lichinga
Sofala	1,516,166	26,262	68,018	Beira
Tete	1,388,205	38,890	100,724	Tete
Zambézia	3,476,484	40,544	105,008	Quelimane

SOURCE: Instituto Nacional de Estatística (INE), Mozambique.

NATURAL RESOURCES

Mozambique boasts a variety of natural resources, but industry has remained largely undeveloped. Large deposits of gold, graphite, and marble exist throughout the mountainous regions, and north-central regions are rich in columbite, beryl, feldspar, kaolin, and semiprecious stones. Hydropower from the Cahora Bassa Dam is also an important resource, giving power to sections of Mozambique, South Africa, and Zimbabwe. Oil prospecting has begun, but the country's major domestic fuel source is coal, of which the country has an estimated two billion tons.

FURTHER READINGS

Darch, Colin. *Mozambique.* Santa Barbara, Calif.: Clio Press, 1987.

James, R.S. *Mozambique.* New York: Chelsea House, 1988.

Lord, Graham. *Ghosts of King Solomon's Mines Mozambique and Zimbabwe: A Quest.* London: Sinclair-Stevenson, 1991.

Slater, Mike. *Mozambique.* London: New Holland, 1997.

Waterhouse, Rachel. *Mozambique: Rising from the Ashes.* Oxford: Oxfam, 1996.

Myanmar

- **Area:** 261,969 sq mi (678,500 sq km) / World Rank: 41
- **Location:** Northern and Eastern Hemispheres, in Southeast Asia, bordered by India and Bangladesh in the northwest, China in the northeast, Laos in the east, Thailand in the east and southeast, and the Indian Ocean to the south and the west.
- **Coordinates:** 22° 00′ N, 98° 00′ E
- **Borders:** 3,643 mi (5,876 km) / Bangladesh, 120 mi (193 km); China, 1,355 mi (2,185 km); India, 907 mi (1,463 km); Laos, 146 mi (235 km); Thailand, 1116 mi (1,800 km)
- **Coastline:** 1197 mi (1930 km)
- **Territorial Seas:** 12 NM
- **Highest Point:** Hkakabo Razi, 19,295 ft (5,881 m)
- **Lowest Point:** Sea level
- **Longest Distances:** 1,200 mi (1,931 km) N-S; 575 mi (925 km) E-W
- **Longest River:** Mekong River, 2,600 mi (4,200 km)
- **Largest Lake:** Indawgyi Lake, 45 sq mi (116 sq km)
- **Natural Hazards:** Earthquakes, flooding, drought, cyclones
- **Population:** 41,994,678 (2001 estimate) / World Rank: 27
- **Capital City:** Yangon, on the Yangon River in the delta region
- **Largest City:** Yangon, 4 million (2001 estimate)

OVERVIEW

Myanmar, the largest nation of mainland Southeast Asia, has an extraordinary variety of terrain, from glaciers in the north to coral reefs in the south. In outline, Myanmar frequently is compared to a diamond-shaped kite with a long tail sharing the Malay Peninsula with Thailand. The country has mountainous frontiers, a high plateau in the northeast, and a fertile central plain ending in the deltas of the Irrawaddy (Ayeyarwady) and Sittang (Sittoung) Rivers. Myanmar is located on the Eurasian Tectonic Plate and is seismically active territory, with frequent earthquakes. Northern and central Myanmar have many extinct volcanoes.

In the late 1980s the military government changed the name of the country from Burma to Myanmar and changed the names or spellings of many geographic features. In this essay the most widely used version of these names are used.

MOUNTAINS AND HILLS

Mountains

There are many mountain ranges throughout the country. Myanmar's northern mountains, including the Patkai and Kumon Ranges, are among the southernmost extensions of the Himalayas. These mountains are very high and rugged and include Hkakabo Razi at the northernmost tip of the country. At 19,295 ft (5,881 m), it is the highest peak in the nation.

The mountains run south along the western border with India and Bangladesh. This belt is composed of many ranges, including the Patkai, Mangin and the Chin Hills, which continue southward to the extreme southwestern corner of the country. The Arakan (Rakhine) Mountains extend southeastwards along the coast from there before giving way to the Irrawaddy flood plains. Notable peaks in the west include Saramati (12,663 ft / 3,860 m) and Mt. Victoria (10,016 ft / 3,053 m).

In central Myanmar, the north-south Pegu Yoma (Bago) Mountains break up the flatness of the alluvial plains between the Irrawaddy and Sittang Rivers. In the southeast, the Dawna and Bilauktaung Ranges mark the border with Thailand on the Malay Peninsula.

Plateaus

In northeast Myanmar, the Shan Plateau, 57,816 sq mi (149,743 sq km) in area, rises to an average elevation of about 3,000 ft (914 m) above sea level. Its western edge is clearly marked by a north-south cliff that often rises 2,000 ft (610 m) in a single step.

Canyons

The Shan Plateau features deep limestone river gorges. The most notable are the gorge of the Salween (Thanlwin) River and Gokteik Gorge, cut by the Namtu River.

Hills and Badlands

Steep craggy limestone hills with many caves are found in the Shan Plateau and in the southeastern part of the country. Elsewhere in the country there are foothill areas leading up to the mountain chains.

INLAND WATERWAYS

Lakes

Myanmar's largest lake, Indawgyi, 45 sq mi (116 sq km), is thought to have been formed by an earthquake. The second largest is the shallow Inle Lake, which covers about 26 sq mi (67 sq km) on the Shan Plateau. It is the residue of an inland sea that is still shrinking. The lower Chindwin River basin has several crater lakes. Other lakes and ponds are for the most part either closed bodies

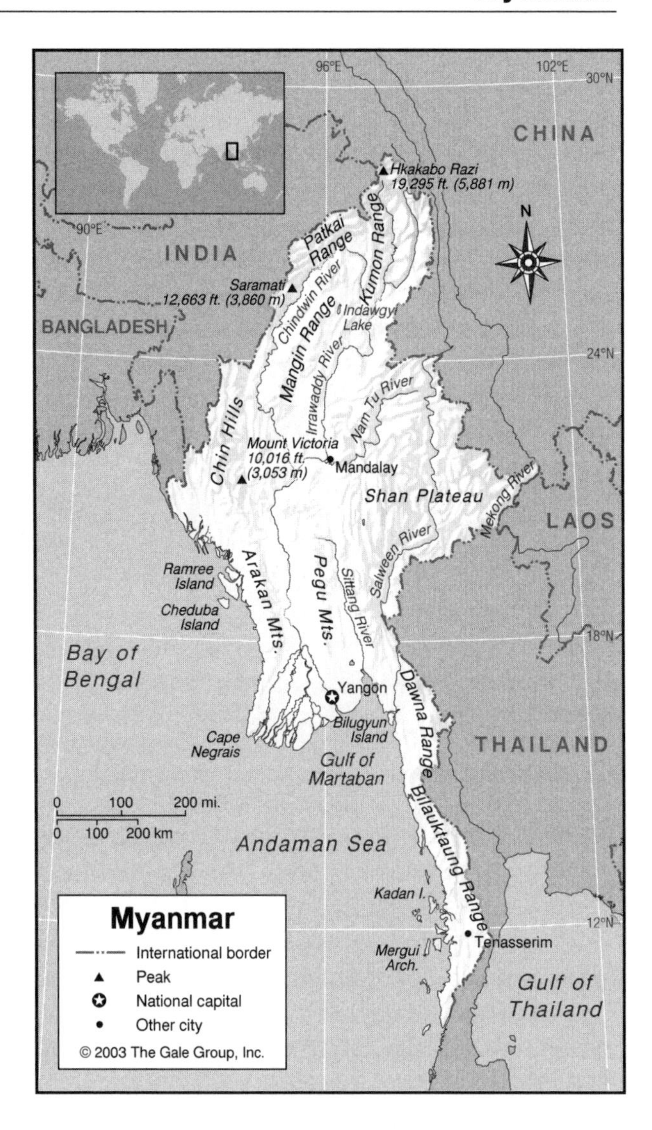

in the courses of former rivers, or are artificial ones formed by reclaiming marshes.

Rivers

The Irrawaddy (Ayeyarwady) River, 1,350 mi (2,170 km) is Myanmar's primary drainage system. Rising in the far north of Myanmar, the Irrawaddy flows south across the entire country before entering the sea through a nine-channel delta. It is the longest river found entirely within Myanmar. The Irrawaddy's most important tributary is the Chindwin River (600 mi / 960 km). It drains the northwest and is fed by tributary streams from the mountains of the Indian frontier. The Sittang (Sittoung) River (300 mi / 483 km) rises just south of Mandalay and parallels the Irrawaddy on its eastern flank. The lower valleys of the Irrawaddy and Sittang rivers form a vast, low-lying delta area of about 10,000 sq mi (25,900 sq km) that continually expands into the sea due to silting.

Myanmar's other large river, the Salween (Thanlwin), rises in China and flows south across the Shan Plateau in eastern Myanmar. The Salween covers 823 mi (1,325 km) within Myanmar, in a series of rapids and waterfalls through steep, narrow valleys, with enormous changes in level. Plans to dam the Salween have caused international controversy, and mining operations have polluted northern rivers.

In the eastern Shan State the mighty Mekong River (2,600 mi / 4,200 km) forms Myanmar's 146 mi (235 km) long border with Laos. In the southeast, many short streams run westward to the Andaman Sea, most notably the Tenasserim. There are also a number of small rivers in the southwest, flowing south out of the mountains into the Bay of Bengal.

Wetlands

The Irrawaddy delta region has salt and fresh water swamps with mud flats and abundant bird life. The coast of the southeastern region features numerous lagoons adjoined by salt marshes.

THE COAST, ISLANDS AND THE OCEAN

Oceans and Seas

Myanmar's western shores curve along the Bay of Bengal, coming to a point at Cape Negrais. The Irrawaddy delta and southeastern region's coasts together frame the upper corner of the Andaman Sea, joining at the Gulf of Martaban. All of these are parts of the Indian Ocean.

Major Islands

Offshore there are many large islands and hundreds of smaller ones. The islands of Myanmar's west coast and delta have been formed by erosion of the shoreline. Just off the northwest coast, the large Ramree (520 sq mi/ 1,350 sq km) and Cheduba (202 sq mi/523 sq km) islands are part of the Ramri Group. Bilugyun is a large island on the southwest coast. Also in the southwest is an undersea ridgeline forming the Mergui Archipelago of some 900 islands ranging in size from Kadan Island (170 sq mi / 440 sq km) to small rocky outcroppings.

The Coast and Beaches

In the northwest of Myanmar, the coast has rocky ridges with deep channels. Mud flats are covered with mangroves. After Cape Negrais Myanmar's southern delta coast is formed by silt from the Irrawaddy and other rivers. From the mouth of the Sittang River, the coast stretches south with inlets, rocky cliffs, and coral reefs.

CLIMATE AND VEGETATION

Temperature

The average annual temperature is 82°F (28°C). Temperatures can dip below 32°F (0°C) in mountainous areas, and soar as high as 113°F (45°C) on the central plains. Humidity ranges from 66 percent to 83 percent. Three seasons are experienced: a cool winter from November to February, a hot season in March-April, and a rainy season when the southwest monsoon arrives, from May through October.

Rainfall

Most of the country's rainfall occurs during the monsoon. Annual average rainfall is 200 in (508 cm) along the coast and 30 in (76 cm) for central regions. Frost and snow occur in the high mountains of the north.

Grasslands

The vast deltas and flood plains of the Irrawaddy and Sittang Rivers form the heart of Myanmar and are its most productive farmland. Some grassland is mixed with scrub growth in the large tracts of denuded former forest on the Shan Plateau and in the "dry zone" of central Myanmar. Bamboo grows extensively in many parts of the country.

Deserts

The "dry zone" of upper central Myanmar has seven rainless months each year, during which its rivers go dry and windstorms are frequent. The terrain there is characterized by erosion and scrub vegetation, including cactus and acacia.

Forests and Jungles

Until recent decades, Myanmar has been the repository of much of the last large temperate and tropical rainforests in mainland Asia, as well as deciduous monsoon forests and coastal mangrove forests. All of these are dwindling, primarily due to unchecked timber operations. Traditional sustainable logging practices were largely abandoned for clearcutting of valuable hardwoods, particularly teak, in the central regions during the 1970s and 1980s. By the early 1990s, Myanmar's deforestation rate, estimated by satellite mapping, was the third highest in the world. The loss of much of the eastern and western forest cover occured when the military government expanded timber exports in the 1990s. Myanmar's rate of deforestation has more than doubled since 1988, according to a World Resources Institute study. Logging concessions have also been extended to pristine areas of the far north. Deforestation, estimated at nearly 5,000 sq km (1,930 sq mi) per year, is also caused by land-clearing for farms, roads, and military bases. The country's remaining forest cover, now less than 30 percent, is mostly found in the relatively inaccessible mountain areas of the north and northeast. The loss of forest cover in Myanmar has threatened animal and plant populations, and caused landslides, river siltation, flooding, and drought.

HUMAN POPULATION

Myanmar has a moderate growth rate of 1.3 percent and density of 176 people per sq mi (68 people per sq

Population Centers – Myanmar

(2000 POPULATION ESTIMATES)

Name	Population	Name	Population
Yangon (captial)	2,520,000	Bassein	148,000
Mandalay	535,000	Taunggyi	112,000
Moulmein	225,000	Akyab	110,000
Bago	155,000	Monywa	109,000

SOURCE: Projected from United Nations Statistics Division data, 1983.

States – Myanmar

2000 POPULATION ESTIMATES

Name	Population	Area (sq mi)	Area (sq km)	Capital
Ayeyarwady	6,779,000	13,567	35,138	Bassein
Bago	5,099,000	15,214	39,404	Bago
Chin	481,000	13,907	36,019	Hakha
Kachin	1,272,000	34,379	89,041	Myitkyina
Kayah	266,000	4,530	11,733	Loikaw
Kayin	1,489,000	11,731	30,383	Pa-an
Magway	4,548,000	17,305	44,820	Magway
Mandalay	6,574,000	14,364	37,024	Mandalay
Mon	2,502,000	4,748	12,297	Moulmein (Mawlamyine)
Rakhine	2,744,000	14,200	36,778	Sittwe (Akyab)
Sagaing	5,488,000	36,535	94,625	Sagaing
Shan	4,851,000	60,155	155,801	Taunggyi
Tanintharyi	1,356,000	16,735	43,343	Tavoy
Yangon	5,560,000	3,927	10,171	Yangon (Rangoon)

SOURCE: Myanmar Data on Internet, Table 2.03, Estimated Population of Myanmar by State and Division, and by Sex.

km) by 1999 estimates. Some 27 percent of the population live in urban areas; most people make their living through agriculture. In 1997 it was estimated that 23 percent of Myanmar's people live in poverty.

NATURAL RESOURCES

Myanmar is rich in natural resources but remains one of world's poorest countries, with an economy based primarily on agriculture. Fishing, timber, and other forest resources (including botanical medicines) are important. Minerals found in Myanmar include jade, rubies and other gemstones, gold, silver, copper, tin, zinc, antimony, tungsten, lead, nickel, cobalt, coal, limestone, marble, chromium, oil (onshore), and natural gas (offshore).

FURTHER READINGS

AsianInfo.Org. *Myanmar.* http://www.asianinfo.org/asianinfo/myanmar/myanmar.htm (Accessed April 11, 2002).

Beck, Jan. *Historical Dictionary of Myanmar.* Lanham, Md.: Scarecrow Press, 1995.

Brunner, Jake, Chantal Elink, and Kirk Talbot. *Logging Burma's Frontier Forests: Resources and the Regime.* Washington, D.C.: World Resources Institute, 1998.

Kyi, Aung San Suu. *Letters from Burma.* London: Penguin Books, 1997.

Namibia

- **Area:** 318,696 sq mi (825,418 sq km) / World Rank: 35
- **Location:** Southern and Eastern Hemispheres, on the west coast of Africa, bordering Angola and Zambia to the north, Botswana to the east, and South Africa to the southeast.
- **Coordinates:** 22°00′S, 17°00′E
- **Borders:** 2,376 mi (3,824 km) / Angola, 855 mi (1,376 km); Botswana, 845 mi (1,360 km); South Africa, 531 mi (855 km); Zambia 145 mi (233 km)
- **Coastline:** 977 mi (1,572 km)
- **Territorial Seas:** 12 NM
- **Highest Point:** Konigstein, 8,550 ft (2,606 m)
- **Lowest Point:** Sea level
- **Longest Distances:** 931 mi (1,498 km) SSE-NNW; 547 mi (880 km) ENE-WSW (excluding the Caprivi Strip)
- **Longest River:** Zambezi, 1,650 mi (2,650 km)
- **Natural Hazards:** Subject to severe drought
- **Population:** 1,797,677 (July 2001 est.) / World Rank: 144
- **Capital City:** Windhoek, in the center of the country
- **Largest City:** Windhoek, 190,000 (2000 est.)

OVERVIEW

Namibia, on the southwest coast of Africa, is primarily a large desert and semi-desert plateau with an average elevation of 3,543 ft (1,080 m). There are four distinct geographical regions in Namibia: the Namib Desert, a strip of desert that lies along the Atlantic Coast; the central plateau; the southeastern Kalahari Desert; and the northeastern woodland savannah. Extending from the northeast corner of the country is the Caprivi Strip, a narrow panhandle extending between Angola and Zambia on the north and Botswana on the south. Namibia lies on the African Tectonic Plate.

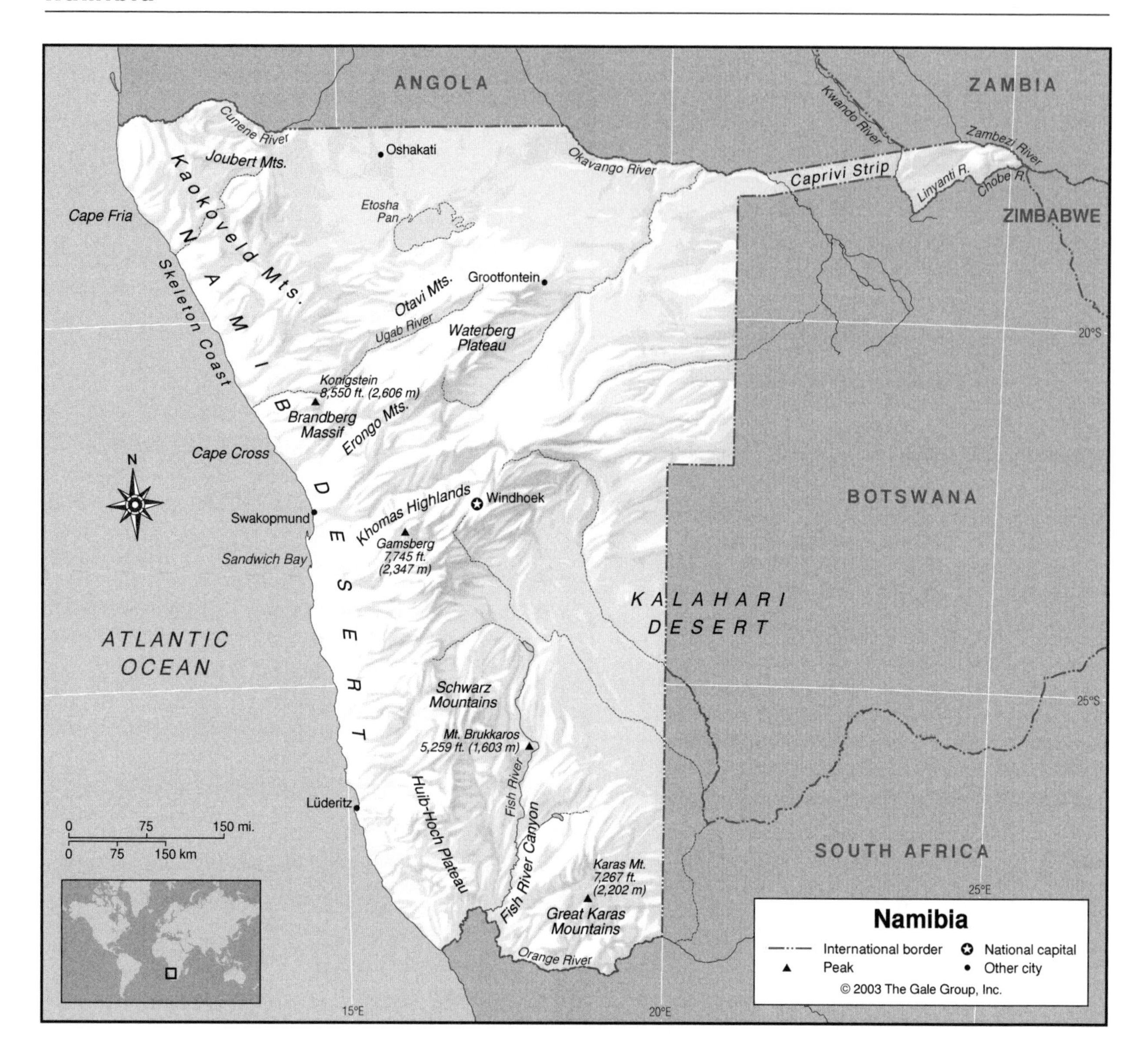

MOUNTAINS AND HILLS

Mountains

The highest mountain in Namibia is a massif rising from the gravelly plains known as the Brandberg. Its peak, Konigstein, reaches 8,550 ft (2,606 m). In 1917 the White Lady rock painting was discovered in a ravine, Maack's Shelter, at the base of the mountain. West of the Brandberg lie the Gobobose Mountains, where the crater of an extinct volcano, the Messum Crater, is found. Just south of the Brandberg Massif, in the region northeast of Swakopmund, are the sharp peaks, Groot Spitzkoppe (5,702 ft / 1,728 m) and Klein Spitzkoppe (5,227 ft / 1,584 m). About 40 mi (60 km) north of the Brandberg in the Kaokoveld Mountains, Twyfelfontein is a west-facing mountain slope covered with more than 2,000 rock engravings (where the designs have been chipped into the rock), some dating to 3300 B.C. The Kaokoveld Mountains run along the Namib Desert parallel to the coast. At their northern extent they run into the Joubert Mountains.

The Khomas Highlands run east to west from Windhoek toward the sea, and include the flat-topped Gamsberg (7,745 ft / 2,347 m). In the north-central region there are two mountain ranges: the Erongo Mountains, about 94 mi (150 km) from the Brandberg with maximum elevations about 7,653 ft (2,319 m), and the Otavi Mountains even further north. Northeast of Windhoek are the Eros Mountains, which reach a maximum elevation of 6,270 ft (1,900 m).

In the south, there are two main mountain ranges. The first, the Schwarz Mountains, runs north to south along the western bank of the Fish River. Its highest peak is Mount Brukkaros at 5,259 ft (1,603 m). The second range, the Great Karas Mountains, runs southwest to northeast across the southeastern corner of the country beginning to the east of Fish River Canyon. The highest point in this range is Karas Mountain, which reaches an altitude of 7,267 ft (2,202 m). The country's second-highest peak, Von Moltkeblick (8,184 ft / 2,480m), is found in the Auas Mountains in southeastern Namibia.

Plateaus

The central plateau has elevations between 3,300 and 6,600 feet (1000 and 2000 m). The terrain features mountain peaks, rock formations, and broad sweeping plains or savannah. In the northwest, the plateau runs into the Kaokoveld, a remote and desolate area of high elevation, home to many rare species of African animals. Further east toward the center of the country, just south of Grootfontein in an area known as the Kaukauveld, the red sandstone Waterberg Plateau rises about 660 ft (200 m) above the savannah and extends for more than 30 mi (50 km). It is the centerpiece of a large area that was designated parkland to protect the habitat of rare and endangered species in 1972. The southwestern corner of the country sits on the Huib-Hoch Plateau.

Canyons

Fish River Canyon lies in the dry, stone-covered plain in south-central Namibia. With an estimated length of 100 mi (160 km), maximum width 17 mi (27 km), and depth of 1,815 ft (550 m), it is the second-largest natural gorge in Africa.

Hills and Badlands

North of the Ugab River are two interesting geological features: Burnt Mountain, a hill displaying outcroppings of purple, black, and deep gray rock; and a dramatic mass of perpendicular volcanic rock called the Organ Pipes.

INLAND WATERWAYS

Lakes

The Etosha Pan in northwestern Namibia is known both as the "Great White Place" because of the appearance of its dry, saline, clay soil, and also as "Land of Dry Water" because it is a dry lake for much of the year. It has been protected as a nature preserve since 1907; the intermittent Ekuma and Oshigambo Rivers feed the Etosha Pan, periodically creating a large shallow lake where flamingoes congregate.

Rivers

The only permanent rivers lie on or near the country's borders. The Cunene River forms the northwestern border with Angola, and the Okavango forms the northeastern border. The Zambezi River, though one of the longest rivers in Africa with a total length of 1,650 mi (2,650 km), only touches Namibia where it forms the far eastern border of the Caprivi Strip with Zambia. The system of the Kwando, Linyanti, and Chobe Rivers forms the easternmost border between the Caprivi Strip and Botswana. The Orange River forms the southern border with South Africa.

GEO-FACT

The herds of elephants that roam northwest Namibia have adapted to desert conditions, with larger feet and smaller bodies than other elephants. Elephants live in desert conditions in only two places in the world—Namibia and Mali.

Along the northern border with Angola, the Cunene River courses to the Atlantic Ocean. Two dramatic waterfalls lie on the Cunene: Epupa Falls is actually a series of cascades created by the river dropping almost 200 ft (60 m) over the short distance of just one mile (1.5 km). The Ruacana Falls are 400 ft (120 m) high and 2,300 ft (700 m) wide at full flood.

During the rainy season (generally from November to March), the intermittent rivers may be filled with water and may even pose flash flood hazards; at other times they are dry riverbeds, sometimes dotted with pools filled with fish. Intermittent rivers that flow west to the Atlantic Ocean include the Kuiseb, Swakop, Omaruru, Hoarusib, Hoanib, Ugab, and Khumib. The Nossob, a tributary of the Orange River, flows along the Kalahari Desert into Botswana. Another Orange River tributary, the Fish, flows in south-central Namibia. Intermittent rivers that flow north include the Marienfluss, Omatako, and Cuvelai, which flows from its source in Angola to the Etosha Pan.

Wetlands

Sandwich Harbor, the coastal area around Sandwich Bay, is a wetland fed both by salt water flowing with the tides and by freshwater seeping up from aquifers. It attracts a wide variety of wading birds and serves as a breeding ground for marine life.

THE COAST, ISLANDS, AND THE OCEAN

Oceans and Seas

The cold Benguela Ocean Current contributes to the overall climate of Namibia and causes the dense fog that almost always hangs over much of the coast, especially in the north.

Major Islands

Namibia has only twelve few small, rocky islands off of its coast. The islands are uninhabited except for colonies of penguins and scientists performing research on them.

The Coast and Beaches

The 300-mi (500-km) stretch of Atlantic Coast, from roughly the Cunene River on the Angola border to the Ugab River, is known as the Skeleton Coast. This remote, foggy stretch of coastline that is characterized by dramatic sand dunes, deep canyons, and mountains, marks the extreme western edge of the Namib Desert. It got its name from the many shipwrecks that occurred there. A park covering about 6,200 sq mi (16,000 sq km) is maintained in the area south of Cape Fria.

Just north of Swakopmund is Cape Cross, where Africa's largest colony of cape fur seals (numbering 100,000 to 200,000) may be found. Cape Cross is so named because Portuguese explorer (and first European to visit Namibia) Diogo Cao erected a cross in honor of the Portuguese king when he landed there in 1486.

Access to the coast south of Lüderitz to the South African border, an area rich in diamonds, is restricted.

Population Centers – Namibia

(1991 POPULATION ESTIMATES)

Name	Population	Name	Population
Windhoek (capital)	147,056	Swakopmund	17,681
Oshakati	21,603	Tsumeb	16,211
Rehoboth	21,439	Otjiwarongo	15,921
Rundu	19,366	Keetmanshoop	15,032

SOURCE: Government of Namibia.

Magisterial Districts – Namibia

Name	Area (sq mi)	Area (sq km)	Capital
Caprivi	7,154	18,530	Katima Mulilo
Erongo	25,129	65,086	Swakopmund
Hardap	42,618	110,382	Mariental
Karas	62,696	162,384	Keetmanshoop
Kavango	16,153	42,771	Rundu
Khomas	14,513	37,590	Windhoek
Kunene	52,721	136,549	Outjo
Ohangwena	3,872	10,029	none
Omaheke	33,668	87,202	Gobabis
Omusati	4,846	12,552	Oshakati
Oshana	1,999	5,180	Etosha
Oshikoto	10,333	26,765	Tsumeb
Otjozondjupa	40,666	105,327	Okahandja

SOURCE: *Geo-Data: The World Geographical Encyclopedia,* 2nd ed. Detroit: Gale Research, 1989.

CLIMATE AND VEGETATION

Temperature

The average temperature along the coast ranges from 73°F (23°C) in summer to 55°F (13°C) in winter. Inland, the temperatures may be somewhat higher, except at higher elevations, where they are lower.

Rainfall

There is little rainfall in Namibia. The rainy season is November to March, with most of the rainfall occurring from January to March. Rain typically occurs during widely scattered, brief thunderstorms. Average annual rainfall along the Atlantic Coast is less than 2 in (5 cm); 14 in (35 cm) fall in the central highlands; and 28 in (70 cm) in the northeast. Because of the erratic rainfall, droughts are frequent and areas of the country may go years without any rain.

Grasslands

African savannah (grassland) dotted by solitary shrubby trees are common in vast areas of the country, everywhere except for the desert on the western coast.

Deserts

The Namib Desert follows the full length of the Atlantic Coastline, varying in width from 30 to 88 mi (50 to 140 km). The terrain features dramatic stretches of dunes, dry riverbeds, and deep canyons, sometimes lined with dramatic rock formations. From Swakopmund to Lüderitz some of the highest sand dunes found anywhere in the world extend about 44 mi (70 km) inland. Remains of shipwrecks also dot the beach.

For most of the year the Etosha Pan in the north-central part of the country is a vast area of dry, salty white sand. However, in the wet season it fills with water as temporary rivers drain into it.

In the southeast, the elevation decreases and the terrain becomes a sandy strip that eventually merges into the Kalahari Desert, which straddles the border with Botswana. The Kalahari features relatively flat expanses of red sand covered in some areas with sparse vegetation.

Forests and Jungles

The country's highest rainfall occurs in the northeast, where there is woodland savannah featuring dense vegetation covering the plains.

HUMAN POPULATION

Namibia is sparsely populated, with an overall population density of only 5 persons per sq mi (2 persons per sq km). About one-third of the population lives in metropolitan areas, and more than half inhabits the regions north of Windhoek. This region surrounding Windhoek is the second most densely populated region, along with the Caprivi Strip. The region around the Etosha Pan in the far north is the most densely populated at 30 inhabitants per sq mi (12 inhabitants per sq km). The rest of the country has less than 2.5 people per sq mi (1 person per sq km).

NATURAL RESOURCES

Namibia is one of the world's leading producers of gem-quality diamonds. The most significant diamond mine area is in the southwest. One of the world's largest open uranium mines is on the west coast near Swakopmund. While wildlife viewing and tourism is one of the country's main economic activities, its other natural resources include gold, lead, tin, lithium, cadmium, zinc, salt, vanadium, natural gas, hydropower, and fish.

FURTHER READINGS

Allen, Benedict. *The Skeleton Coast: A Journey through the Namib Desert.* London, Eng.: BBC Books, 1997.

Ballard, Sebastian. *Namibia Handbook.* Lincolnwood, Ill.: Passport Books, 1999.

Bannister, Anthony. *Namibia: Africa's Harsh Paradise.* London, Eng.: New Holland, 1990.

Cardboard Box Travel Shop. *Namibian Geography.* http://www.namibian.org/travel/namibia/geography.htm (Accessed June 12, 2002).

Dressler, Thomas. *Scenic Namibia.* Cape Town, South Africa: Sunbird, 2000.

E-Tourism Namibia. http://www.e-tourism.com.na/ (Accessed June 12, 2002).

Grotpeter, John J. *Historical Dictionary of Namibia.* Metuchen, N.J.: Scarecrow, 1994.

Nauru

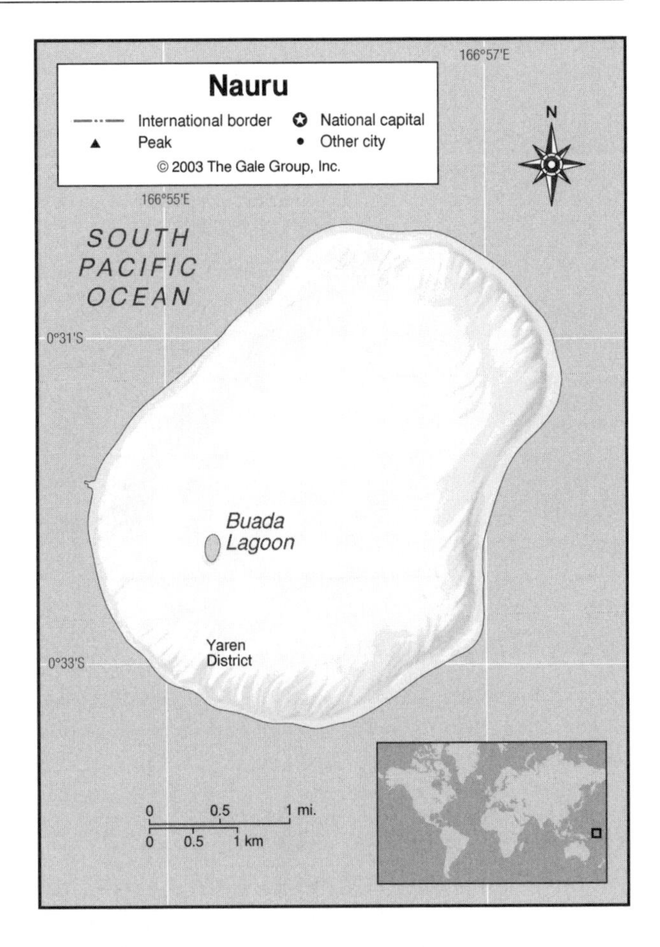

- **Area:** 8.1 sq mi (21 sq km) / World Rank: 205
- **Location:** Southern and Eastern Hemispheres, in the West-Central Pacific; 33 mi (53 km) from the Equator; 2,200 mi (3,539 km) northeast of Sydney, Australia; 2,445 mi (3,934 km) southwest of Honolulu; nearest neighbor is Banaba, about 190 mi (305 km) to the east
- **Coordinates:** 0°32′S, 166°55′E
- **Borders:** None
- **Coastline:** 18.6 mi (30 km)
- **Territorial Seas:** 12 NM
- **Highest Point:** At 202 ft (61 m) lies an unnamed central plateau that makes up the highest land mass of the island; no single elevation of note
- **Lowest Point:** Sea level
- **Longest Distances:** 3.5 mi (5.6 km) NNE-SSW / 2.5 mi (4 km) ESE-WNW
- **Longest River:** None of significant size
- **Largest Lake:** Buada Lagoon , 300 acres (120 ha)
- **Natural Hazards:** Periodic droughts
- **Population:** 12,088 (July 2001 est.) / World Rank: 204
- **Capital City:** None
- **Largest City:** The Yaren district, which functions as the capital, 10,000 (2000 est.)

OVERVIEW

Nauru is an oval-shaped island in the West-Central Pacific Ocean. It is the smallest nation in Asia and is located in the Pacific Plate. The island is encircled by a sandy beach, which gradually rises, forming a fertile section no wider than 300 yards (275 m). A coral cliff rises from this belt to a central plateau.

MOUNTAINS AND HILLS

There are no mountains on the island. A central plateau of phosphate-bearing rock comprises the majority of Nauru's land mass, making it one of the largest phosphate-rock islands in the Pacific.

Districts – Nauru

Name	Area (sq mi)	Area (sq km)
Alwo	0.4	1.1
Anabar	0.6	1.5
Anetan	0.4	1.0
Anibare	1.2	3.1
Baitsi	0.5	1.2
Boe	0.2	0.5
Buada	1.0	2.6
Denigomodu	0.3	0.9
Ewa	0.5	1.2
Ojuw	0.4	1.1
Meneng	1.2	3.1
Nibok	0.6	1.6
Uaboe	0.3	0.8
Yaren	0.6	1.5

SOURCE: *Geo-Data: The World Geographical Encyclopedia,* 2nd ed. Detroit: Gale Research, 1989.

GEO-FACT

Nauru is a member of the Pacific Islands Forum, a group of 16 countries organized in the 1970s in response to France's nuclear testing in French Polynesia. At the turn of the century, the group is facing another environmental crisis—rising sea levels due to global warming. For these low lying islands, the consequences of climate change include destruction of freshwater sources, more intense storms, loss of crops to seawater, and coastal erosion.

INLAND WATERWAYS

The permanent, often brackish, Buada Lagoon (Lake Buada, Blue Lagoon) is the only lake of significance on the island.

THE COAST, ISLANDS, AND THE OCEAN

The island is surrounded by a coral reef, exposed at low tide and dotted with pinnacles. The reef is bounded seaward by deep water, and inside by a sandy beach.

CLIMATE AND VEGETATION

Temperature

Nauru has a tropical climate that is tempered by sea breezes. From November to February is the westerly monsoon season. Temperatures range from 75 to 91°F (23 to 32°C).

Rainfall

Nauru experiences widely variable rainfall: 12 in (30.5 cm) to as much as 180 in (457.2 cm). Rainfall provides the majority of the nation's water supply.

Grasslands

There are large areas of scrub brush on the plateau, with the occasional coconut and tamanu tree. Tropical flowers grow on Nauru, but do not flourish as on other Pacific islands.

HUMAN POPULATION

Nauru has a population of 12,000 with an annual growth rate of 2 percent. There are no cities. The major ethnic groups are: Nauruan, 58 percent; other Pacific Islander, 26 percent; Chinese, 8 percent; European, 8 percent. The major religion is Christianity (two-thirds Protestant, one-third Roman Catholic). In both size and population it is one of the smallest republics in the world.

NATURAL RESOURCES

Phosphates are the only resource native to the island. Fish are abundant off the coast of Nauru, supplying tuna and bonita catches.

FURTHER READINGS

Pacific Island Travel. *Nauru.* http://www.pacificislandtravel.com/nauru/introduction.html (accessed May 23, 2002).

U.S. Department of State. *Background Notes, Nauru.* Washington, D.C.: Bureau of Public Affairs, Office of Public Communication, Editorial Division, 1988.

Nepal

- **Area:** 54,363 sq mi (140,800 sq km) / World Rank: 95
- **Location:** Northern and Eastern Hemispheres, in South Asia's Himalayan region, between China and India.
- **Coordinates:** 28°00′N, 84°00′E
- **Borders:** 1,818 mi (2,926 km) / China, 768 mi (1,236 km); India, 1,050 mi (1,690 km)
- **Coastline:** None
- **Territorial Seas:** None
- **Highest Point:** Mount Everest, 29,030 ft (8,850 m, estimated)
- **Lowest Point:** Kanchan Kalan, 230 ft (70 m)

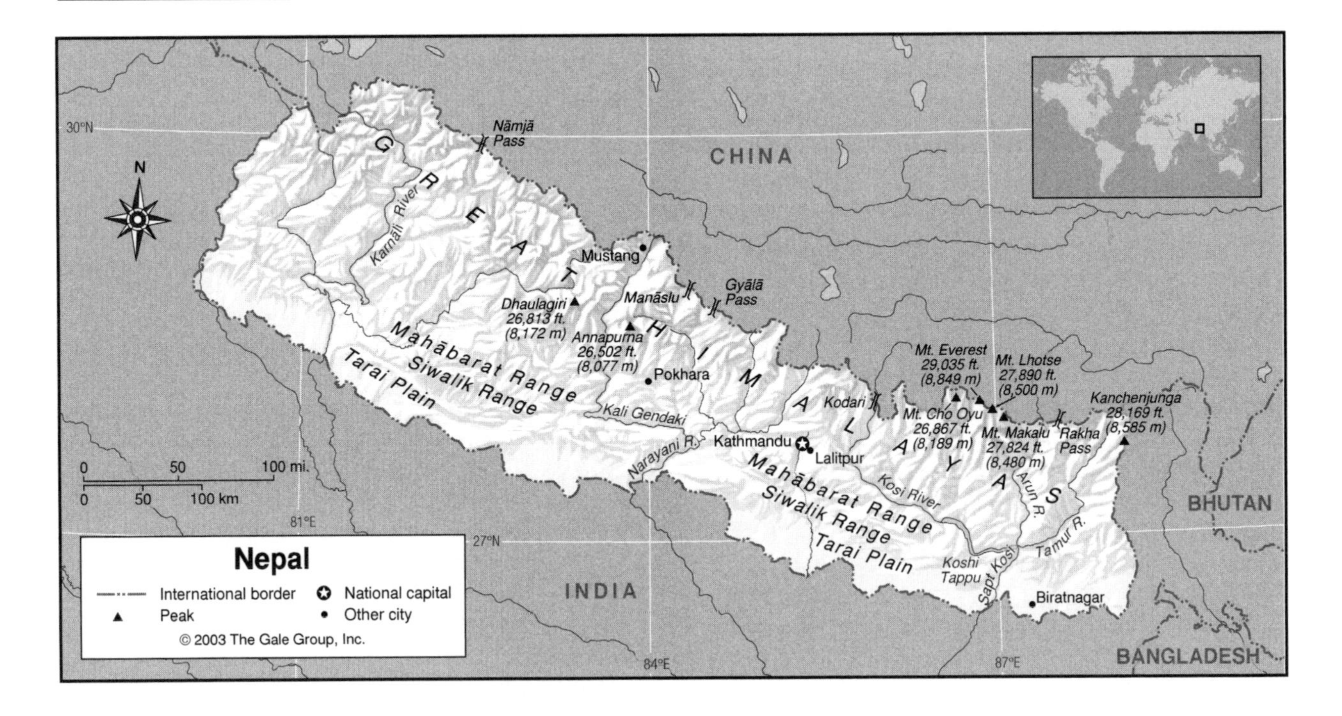

- **Longest Distances:** 550 mi (885 km) SE-NW; 125 mi (201 km) NE-SW
- **Longest River:** Karnali River, approximately 570 mi (915 km)
- **Largest Lake:** Rara Lake, approximately 4 sq mi (11 sq km)
- **Natural Hazards:** Floods, landslides, earthquakes, drought, severe storms
- **Population:** 25,284,463 (July 2001 est.) / World Rank: 40
- **Capital City:** Kathmandu, in Nepal's central valley
- **Largest City:** Kathmandu, 553,000 (2000 est.)

OVERVIEW

Nepal is a mountainous land, situated in the southern slopes of the Himalayas between China's Tibet region to the north and India to the south. The complex mountain mass within Nepal's borders contains seven of the world's ten highest peaks. Six are more than 26,000 ft (7,924 m) above sea level. Nepal can be divided into three distinct geographic regions, forming horizontal bands across the rectangle-shaped country: the Mountain Region, which constitutes almost three fourths of the total area; the central hill area, which includes the Katmandu Valley; and the Tarai, a narrow flat belt that extends along the boundary with India in the northern part of the Gangetic Plain. The country lies along the boundary between the Indian Tectonic Plate and the Eurasian Plate, and sees much seismic activity as a result.

MOUNTAINS AND HILLS

The Himalayas

Nepal's Mountain Region is part of the Himalayas, formed by the collision of the Indian subcontinent with the Asian land mass around 25 million years ago. The Great Himalayas are in the north. Their major heights in northeastern Nepal generally define the boundary with Tibet, while in the northwest they lie just to the south of the boundary.

South of the Great Himalayas are the Lesser Himalayas, which in Nepal form the Mahābharat Range. South of these is the Siwalik Range, part of the Outer Himalayas. Both of these are much lower than the Great Himalayas and are part of Nepal's Hill Region, although in most other countries they would be considered mountains.

The whole Mountain Region is marked by a series of parallel north-south ridges flanking deep, narrow, southward-sloping valleys. The rivers in the principal valleys rise some 50 to 100 mi (80 to 160 km) inside Tibet on the high plateau north of the boundary. These streams are older than the mountain mass through which they flow, having created their valleys by erosion as the mountain barrier lifted around them. Thus, the actual watershed is not generally the line of high peaks in the region itself, but the Tibetan plateau farther north.

Based mainly on differences in physical features and climate, the Mountain Region may be subdivided into three general areas by two lines, one running generally northward from Kathmandu and the other about 150 mi (241 km) to the west, extending northward from the foothills near the boundary with India. From east to west, these subdivisions are designated the Eastern Mountains, the Western Mountains and the Far Western Mountains.

The Eastern Mountain has an area of 10,114 sq mi (26,195 sq km). This region has five of the seven highest peaks in the world. The most famous is the world's very highest, Mount Everest (Sagarmatha in Nepalese, Chomlungma in Tibet), at 29,028 ft (8,847 m). It is located on the border with China. Then there are Mount Lhotse, 27,890 ft (8,500 m); Mount Makalu, 27,824 ft (8,480 m); and Mount Cho Oyu, 26,867 ft (8,189 m). All are situated on a 30 mi (77 km) segment of the boundary with Tibet. Forming a massive, white, saw-toothed range, they are visible on clear days from points throughout eastern Nepal. The icy, stormy summit of Mount Everest was first climbed by Tenzing Norgay and Edmund Hillary in 1953, ascending by the southern (Nepalese) face. Trails leading up the principal valleys among the Eastern Mountains cross the Tibetan frontier over four difficult but well-known passes: Khangla Deorali and Rakha to the east of the peaks, and Nangpa and Kodari (Kuti) to the west.

The world's third-tallest mountain, Kanchenjunga (28,169 ft / 8,585 m), towers along the eastern border with India. The mountain features five peaks, and a vast glacier on its western side. It was first climbed in 1955.

The 11,076 sq mi (28,686 sq km) of the Western Mountain area holds a jumble of ridges and deep valleys projecting at various angles from the main Himalayan range. Two mountains dominate the area: Dhaulagiri, 26,813 ft (8,172 km) in height, and Annapurna, which is 26,502 ft (8,077 m) high. Both are within 50 mi (80 km) of the town of Pokhara, which is less than 5,000 ft (1,524 m) above sea level. Another great peak lies further east, Manāslu (26,775 ft / 8,155 m). The principal pass from this part of the region into Tibet is at Rasua Garhi, commonly called the Girange Dzong Pass after a nearby Tibetan town, about 50 mi (80 km) north of Kathmandu. Other passes to the west include the Gyālā, the Kore La, and the Yansang Bhanjyang.

The Far Western Mountain area, 18,879 sq mi (48,896 sq km) in area, is the driest and most sparsely inhabited section of the Mountain Region. Its scattered settlements are generally confined to the river valleys. The southward drainage pattern is interrupted in many places by east-west ranges around which the streams zigzag on their way to the Ganges. Three passes in this area lead into Tibet: the Nāmjā, the Takhu La and the Nara Lagna (on the Karnali River), all at elevations of approximately 16,000 ft (4,876 m).

Plateaus

Dolpo is a 2,100 sq mi (5,439 sq km) plateau bordering Tibet in Nepal's northwest. It includes the 1,373 sq mi (3,555 sq km) Shey-Phoksumdo National Park, which is a habitat for the rare snow leopard. In the far west, the Khaptad Plateau at 9,842 ft (3,000 m) elevation is a National Park with grassland and forest.

GEO-FACT

The wild yak, still found in the mountains of Nepal, is the mammal that lives at the highest altitude. These endangered yaks, with large lung capacity, can exist at altitudes of up to 20,000 ft (6,096 m) but have difficulty surviving below 10,000 ft (3,048 m). Yaks are domesticated throughout the Himalayas as pack animals, and for their meat, milk, and warm wool.

Canyons

Nepal has numerous deep canyons and river gorges. The world's deepest river gorge is said to be Kali Gandak at 22,860 ft (6,967 m) deep, between the peaks of Dhaulagiri and Annapurna in north-central Nepal. The high altitude valley of Mustang, north of the Himalayas, contains many dry, eroded canyons. Nepal's rivers carve mazes of canyons, especially along the courses of the Bhote Koshi and Karnali.

Hills and Badlands

Nepal's central hill region is north of the Tarai, south of the Great Himalayas; its hills are called the Pahar complex. These hills (which might be considered mountains in other countries) are from 1,968 to 13,123 ft (600 to 4,000 m) in height. The two ranges are the Siwalik and the Mahabharat. On the northern edge of the Tarai is the Siwalik Range, sometimes called the Churia Hills or Churia Range, which rises to heights of almost 5,000 ft (1,524 m). This range is paralleled some 20 mi (32 km) to the north by the narrow Mahabharat, with elevations above 10,000 ft (3,048 m). The valleys are settled and farmed, and the hills are sculpted with terraced fields or grazed by livestock. The hill region includes the populous Kathmandu Valley, just south of the junction between the Eastern and Western Mountains. It is a circular basin of only 218 sq mi (565 sq km), thought to be a dried-up lakebed, and it contains some of Nepal's largest cities, including the capital, Kathmandu.

INLAND WATERWAYS

Lakes

Rara Lake is Nepal's largest, approximately 4 sq mi (11 sq km), located at an elevation of 9,600 ft (2,990 m) in a National Park in the remote northwest of the country. The world's highest lake, Tilicho, is located in eastern

Nepal, at an elevation of 16,140 ft (4,919 m). A 1999 survey found 2,323 glacial lakes in Nepal, twenty of which are considered particularly dangerous as monsoon rains cause them to burst their banks in violent floods. Glacial melting due to global warming is thought to have brought about the formation of glacial lakes, most of which appeared in the last half of the twentieth century. The overflow of Sabai Tsho, a glacial lake, created a massive flood of the Kosi River in 1998.

Rivers

Numerous streams and rivers flow generally southward out of Nepal's northern mountains, then meander across the Tarai Plain and finally join the Ganges in northern India. The presence of fertile alluvial soil at stream junctions and at other places in the valley bottoms is a major determinant in the settlement pattern, so most of the largest population concentrations exist along the rivers and their principal tributaries.

Three separate river systems, each having its headwaters on the Tibetan plateau, drain almost all of Nepal. The Kosi River drains the Eastern Mountains; the Narayani, the Western Mountains; and the Karnali, the Far Western Mountains. After plunging through deep gorges, the waters of these streams deposit their heavy sediment and debris on the plains of the Tarai.

The Narayani cuts through the Western Mountains. Its Kali Gandak tributary, with a streambed at the elevation of 3,630 ft (1,106 m), flows between the region's highest peaks, Dhaulagiri and Annapurna, which are only 22 mi (35 km) apart. The Kosi River has seven major tributaries. The principal one, the Arun, rises almost 100 mi (160 km) inside the Tibetan plateau. Two other major tributaries, the Sun Kosi and the Tamur, join the Arun to form the southward-flowing Sapt Kosi. The Karnali River is noted for its deep gorges and rapid current. The Kosi and Karnali have become tourist destinations for river rafters and kayakers. Large hydropower dams, intended to provide energy to India as well as Nepal, have been built and proposed, causing environmental controversies.

Wetlands

Wetlands are estimated to cover about 5 percent of Nepal, with 242 wetland sites, 163 of which are in the densely populated Tarai. Nepal has four wetlands of particular importance. Koshi Tappu, in the Tarai, covering 68 sq mi (175 sq km), is a nature reserve on the flood plain of the Sapta Kosi River. A mixture of marshes, mud flats, and reed beds, Koshi Tappu is the habitat of water birds and the last herds of wild water buffalo in Nepal. It is designated a Wetland of International Importance under the Ramsar Convention on Wetlands. In 2002 Nepal's government declared three other wetlands to be of special significance: Ghodaghodi Tal, 10 sq mi (25 sq km); Beeshazar Tal, over 12 sq mi (32 sq km); and Jagdishpur Reservoir, approximately 1 sq mi (2 sq km). All three are biodiverse habitats for birds, fish, and reptiles.

THE COAST, ISLANDS AND THE OCEAN

Nepal is a landlocked country. The nearest sea access is 400 mi (644 km) to the southeast on the Indian Ocean's Bay of Bengal.

CLIMATE AND VEGETATION

Temperature

Nepal has four seasons: Winter in December-February, which is cold and clear, with some snow; Spring in March-May, which is warm, with some rain showers; Summer in June-August, when the monsoon rains appear; and Autumn in September-November, cool and clear. Nepal's climate varies by elevation. Above 16,000 ft (4,877 m) the temperature stays below freezing and there is permanent snow and ice. The average temperature in the Kathmandu Valley ranges from 36° to 64°F (2° to 18°C) in January, to 68° to 84°F (20° to 29°C) in July. In the Tarai the annual range is from 44° to 104°F (7° to 40°C).

Rainfall

Roughly 80 percent of Nepal's precipitation happens during the summer monsoon season. Annual rainfall in the Kathmandu Valley averages 51 in (130 cm), with a range from less than 10 in (25 cm) to more than 236 in (600 cm).

Grasslands

The Tarai Region, with a total area of 8,969 sq mi (23,220 sq km), consists mainly of an alluvial plain along the boundary with India. A northern extension of the Gangetic Plain, the Tarai varies between 150 and 600 ft (46 and 183 m) in altitudes and between 5 and 55 mi (8 and 88 km) in width. The Tarai is crossed by numerous streams which, particularly in the east, carry down tons of silt, sand, gravel, and huge boulders from the mountains in the north, during the annual monsoon floods. Most rivers in the Tarai overflow their banks onto wide floodplains during the rainy season.

Natural grasslands cover approximately 14 percent of Nepal, but that percentage is decreasing. The alluvial grasslands of Nepal are, along with some in northeast India, the last remnants of a formerly widespread South Asian ecosystem of tall saccharum grass ("elephant grass") that provided a habitat for diverse wildlife. These natural flood plains grasslands now only remain in protected areas of the Tarai, and those are frequently encroached on for settlement and grazing. At the higher altitudes, grasslands such as Ramaroshan and the Khaptad National Park, in the far west, are alpine and Trans-

Population Centers – Nepal

(2000 POPULATION ESTIMATES)

Name	Population
Kathmandu	500,000
Lalitpur	175,000
Biratnagar	125,000
Bhaktapur	120,000
Patan	110,000

SOURCE: Compiled from United Nations and United States Agency for International Development estimates.

Himalayan meadows that function as grazing areas for wild sheep, antelopes, deer, and livestock (including yaks).

Forests and Jungles

Nepal has a mixture of deciduous and evergreen conifer forests, as well as bamboo and rhododendron forests. An estimated 42 percent of Nepal is covered with forest and scrub woodlands, but the country is undergoing widespread deforestation. Forest products are used for fuel, timber (for use in Nepal and sale to India), and livestock fodder. Agricultural clearing, forest fires, and undergrowth cutting (for fuel and fodder) have severely accelerated deforestation. In the hill regions, deforested areas are terraced or become scrubland; in the Tarai, deforested areas are usually converted to farmland. Forests provide 75 percent of energy needs and 40 percent of fodder in Nepal. Enormous amounts of wood are consumed for fuel, but sustainable development projects are trying to decrease wood usage by promoting alternative technologies such as solar ovens and micro-hydro power generation.

HUMAN POPULATION

Nepal has a growth rate of 2.27 percent (2001 estimate), which analysts consider excessive, as the country's population could double in three decades. Already the population, which is centered in the Kathmandu Valley and the Tarai, has created a shortage of agricultural land, with soil depletion and erosion very common. The Tarai, with 23 percent of the land and 48.5 percent of the population, is deforested and over-farmed. Dangerous levels of air pollution from fuel burning, automobile emissions, and industry are frequent in the Kathmandu valley. Almost half of Nepal's people survive below the poverty line, according to most estimates. Nepal's population density has been estimated at 151 people per sq mi (58 people per sq km). The 2001 national census found that 15 percent of the population live in cities and 85 percent in rural areas. Only 7 percent live in the Mountain Region.

NATURAL RESOURCES

Nepal's timber and agricultural resources are stressed to the point of unsustainability. Potential exists for large- and small-scale hydropower and solar energy generation, and there are deposits of lignite, copper, cobalt, and iron. The rugged landscape of Nepal, with its famous peaks and wild rivers, is a resource for tourism and adventure expeditions.

FURTHER READINGS

Kelly, Thomas, and V. Carroll Dunham. *The Hidden Himalayas.* New York: Abbeville Press, 2001.

Krakauer, Jon. *Into Thin Air: A Personal Account of the Everest Disaster.* New York: Anchor Books, 1998.

Matthiessen, Peter. *The Snow Leopard.* New York: Penguin USA, 1996.

Moran, Kerry. *Nepal Handbook.* Emeryville, Calif.: Avalon Travel Publishing, 1999.

Nepal Internet Users Group. *Nepal Net.* http://www.panasia.org.sg/nepalnet (accessed April 5, 2002).

The Netherlands

- **Area:** 16,033 sq m (41,526 sq km) / World Rank: 134
- **Location:** Northern and Eastern Hemispheres, in Western Europe, bordering the North Sea, between Belgium and Germany
- **Coordinates:** 52°30′N, 5°45′E
- **Borders:** 638 mi (1,027 km) / Germany, 359 mi (577 km); Belgium, 280 mi (450 km)
- **Coastline:** 280 mi (451 km)
- **Territorial Seas:** 12 NM
- **Highest Point:** Vaalserberg, 1,053 ft (321 m)
- **Lowest Point:** Prins Alexanderpolder, 23 ft (7 m) below sea level
- **Longest Distances:** 194 mi (312 km) N-S / 164 mi (264 km) E-W
- **Longest River:** Rhine, 820 mi (1320 km)
- **Largest Lake:** IJsselmeer, 467 sq mi (1,210 sq km)
- **Natural Hazards:** Flooding
- **Population:** 15,981,472 (July 2001 est.) / World Rank: 59
- **Capital City:** Amsterdam, on the north-central coast (note that The Hague, on the west coast, is the seat of government)
- **Largest City:** Amsterdam, 1,150,000 (2000 est.)

OVERVIEW

The Netherlands (also known as Holland) is a low-lying country on the shores of the North Sea in Europe. It is famous for its dams and dikes, some of which date back many centuries, constructed to reclaim large swaths of land from the sea and stabilize its coastlines. Nearly a quarter of The Netherlands is below sea level. The country's many large rivers, including the Rhine and its distributaries, have also been dammed and canalized to control flooding.

The Netherlands may be divided into two main regions, one comprising areas below sea level (Low Netherlands) and the other those above sea level (High Netherlands). Although primarily based on elevation, this classification coincides with the broad division of the country according to its geological formation. The High Netherlands was formed mainly in the Pleistocene Age (which began about 2 million years ago and ended about 10,000 years ago) and is composed chiefly of sand and gravel. The Low Netherlands is relatively younger, having been formed in the Holocene Age (less than 10,000 years ago) and consists mainly of clay and peat. There are other differences: the High Netherlands is undulating and even hilly in places, with farms alternating with woodland and heath. The Low Netherlands is predominantly flat, and is intersected by natural and man-made waterways. Dunes and dikes protect the Low Netherlands against flooding. The western and northern regions of the country consist of about 5,000 polders (plots of land reclaimed from the sea), representing over 950 sq mi (2,500 sq km).

MOUNTAINS AND HILLS

Plateaus

The South Limburg Plateau is the only part of the country not classed as lowland. The hills, which rise to over 1,000 ft (300 m), are the foothills of the Central European Plateau. This is also virtually the only area of the country where rocks can be found at or near surface levels.

Hills and Badlands

The highest point is Vaalserberg (1,053 ft / 321 m) in the hills of the South Limburg Plateau on the German border. Low hills (up to 328 ft / 100 m) can be found in the eastern part of the country, the result of ancient glacial activity.

INLAND WATERWAYS

Rivers and Canals

The Rhine River and the Meuse (Maas) River dominate the western and central part of the country. The Rhine enters The Netherlands from Germany around the middle of their border. It soon branches out into two major arms, the Neder Rijn (called Lek in its lower course) and the Waal. They flow west, roughly parallel to each other, and never more than about 19 mi (30 km) apart. Both branches have many tributaries entering them and distributaries branching off from them before they reach the North Sea, some of which connect the two branches to each other.

The Meuse River is the largest tributary of the Rhine in the Netherlands. It enters the country in the far southeast and flows north to the middle of the country before curving to the west. In this part of its course it is only a few miles south of the Waal, and eventually the two rivers meet, then flow into the North Sea.

The IJssel River is a major distributary of the Neder Rijn, which it branches off from shortly after that river's beginning. The IJssel flows north, receives a number of small tributaries, and then empties into the IJsselmeer.

The Schelde (Scheldt; Escaut) River enters The Netherlands from Belgium in the southwest. It almost immediately widens into a broad estuary and flows into the North Sea.

The Netherlands has a very extensive system of canals that run throughout almost the entire country. The North Sea Canal connects Amsterdam and the Markermeer (Marker Lake) to that body of water. The Amsterdam-Rhine River Canal is just one of several that connect the city and that river. A network of canals including the Wilhelmina, Zuid-Willems, and Juliana Canals connects the southern part of the country to the Rhine River and to other canals in Belgium. In addition, many of The Netherlands' natural rivers, including all of its largest rivers, have had their shores reinforced (canalized) to prevent them from flooding or shifting their courses.

Lakes

There are many small lakes located in the northern and western portions of The Netherlands. In the northeast there is a network of more than 30 lakes that are all interconnected by canals. Some of the largest of these are Lake Fluessen, Lake Sloter, and Sneek Lake. Southwest of these is the IJsselmeer, a freshwater lake that was formed by the construction of the Barrier Dam (completed in 1932). Prior to construction of the dam, this body of water was a shallow, salty, arm of the North Sea known as the Zuider Zee. South of the IJsselmeer is the Markermeer (Marker Lake), another freshwater body enclosed by a dam.

Wetlands

In the transitional region between the High and Low Netherlands marshy conditions conducive to peat formation exist, especially in the center of the country.

THE COAST, ISLANDS, AND THE OCEAN

Oceans and Seas

To the west, the North Sea borders the coast of The Netherlands. Lying along the coast as it curves eastward at the northernmost end of the country is a shallow protected body of water, the Waddenzee, a popular nesting area for birds.

Major Islands

The West Frisian Islands were formed when the North Sea broke through a series of dunes along The Netherlands' ancient northern coastline. The area behind the dunes became the Waddenzee, while the tallest of the dunes remained intact as islands. From west to east, the largest of the islands are Texel, Vlieland, Terschelling, Ameland, and Schiermonnikoog. Vlieland is the site of a national park.

The Coast and Beaches

The North Sea coastline of The Netherlands consists mostly of dunes. North of about the midpoint of the country, the northwestern coastline is characterized by low-lying sandy dunes. The dunes, about one-half mile (several kilometers) wide in places, were created by the action of wind and water; in some areas, the dunes may reach nearly 100 ft (30 m) high.

Further south, the delta region is formed where the major rivers flow into the North Sea. The delta region is characterized by a series of islands, some connected by dikes or dams, and waterways, many of which have been connected by canals. The largest inlets are the Westerschelde and Oosterschelde in the south of the country;

the most of the Oosterschelde is blocked by a dam of the same name.

CLIMATE AND VEGETATION

Climate

The Netherlands shares the temperate maritime climate common to much of northern and western Europe. The average temperature ranges from 34°F to 41°F (1°C to 5°C) in January and from 55°F to 72°F (13°C to 22°C) in July. Because The Netherlands has few natural barriers, such as high mountains, the climate varies little from region to region.

Rainfall

Annual precipitation averages 30 in (765 mm).

Grasslands

The western and northern regions of the country consist of polders (land reclaimed from the sea) where the water level is mechanically controlled at about 3 ft (1 m) below ground level, thus permitting cultivation. However, there are polders that were reclaimed by earthen dikes in the late nineteenth century. These soils of these polders is marshy, and is too wet to be used for cultivation, but may be used for grazing livestock. Polders do not necessarily lie below sea level, although this is the case with the IJsselmeer polders, which are 11.5 ft (3.5 m) below sea level and with polders created by draining lakes, which can lie below 22.0 ft (6.7 m). In areas of young marine clay and along the rivers, many polders lie above the average sea level, which means that it is not always necessary to pump the water out. Almost half of the land area of The Netherlands is made up of polders.

Vegetation

The country is known for its flowers, especially cultivated varieties such as Dutch tulips, and wild flowers such as daisies, buttercups, and heather blooms. While the mild climate is ideal for these varieties, the lack of sunshine restricts the growing of crop foods.

HUMAN POPULATION

The majority of inhabitants of The Netherlands are Dutch descended from Franks, Frisians, and Saxons. Most residents of Friesland Province are Frisian, a distinct cultural group with its own language.

NATURAL RESOURCES

The Slochteren reserve in the northeast is one of the world's largest fields of natural gas. There are also significant offshore deposits of oil in the North Sea in The Netherlands' waters. Besides this, the country has a diverse modern economy, with significant agriculture, industry, commerce, and shipping.

FURTHER READINGS

Blom, J. C. H., and E. Lambert. *History of the Low Countries.* Translated by James C. Kennedy. Providence, RI: Berghahn Books, 1999.

Dash, Mike. *Tulipomania: The Story of the World's Most Coveted Flower and the Extraordinary Passions It Aroused.* New York: Crown, 2000.

North, Michael. *Art and Commerce in the Dutch Golden Age.* Translated by Catherine Hill. New Haven, CT: Yale University Press, 1997.

Schama, Simon. *The Embarrassment of Riches: An Interpretation of Dutch Culture in the Golden Age.* Vintage, 1997.

———. *Patriots and Liberators: Revolution in the Netherlands, 1780-1813.* Vintage, 1992.

Westermann, Mariet. *A Worldly Art: The Dutch Republic, 1585-1718.* New York: Abrams, 1996.

Netherlands Antilles

A dependency of the Netherlands

- **Area:** 370.6 sq mi (960 sq km) / World Rank: 176
- **Location:** Divided geographically into two groups. The Windward group, Curaçao and Bonaire, are in the Caribbean Sea north of Venezuela. The Leeward Islands group—St. Maarten, Saba, and St. Eustatius—is more than 500 mi (800 km) northeast of the second group, between the Caribbean Sea and Atlantic Ocean.
- **Coordinates:** 12° 15′ N, 68° 45′ W.
- **Borders:** 6.33 mi 10.2 (km), all with Guadeloupe (France).
- **Coastline:** 226 mi (364 km).
- **Territorial Seas:** 12 NM
- **Highest Point:** Mt. Scenery, 2,828 ft (862 m).
- **Lowest Point:** Sea level.
- **Longest Distances:** The Leeward group of the Netherlands Antilles extends 17.5 mi (27.5 km) from north to south and 10 mi (15 km) from east to west; the Windward group extends 60 mi (90 km) from east to west and 20 mi (30 km) from north to south.
- **Longest River:** None of significant size.
- **Natural Hazards:** St. Maarten, Saba, and St. Eustatius are subject to hurricanes from July to October.
- **Population:** 212,226 (July 2001 est.) / World Rank: 175
- **Capital City:** Willemstad, on Curaçao.
- **Largest City:** Willemstad, 58,000 (2001 est.)

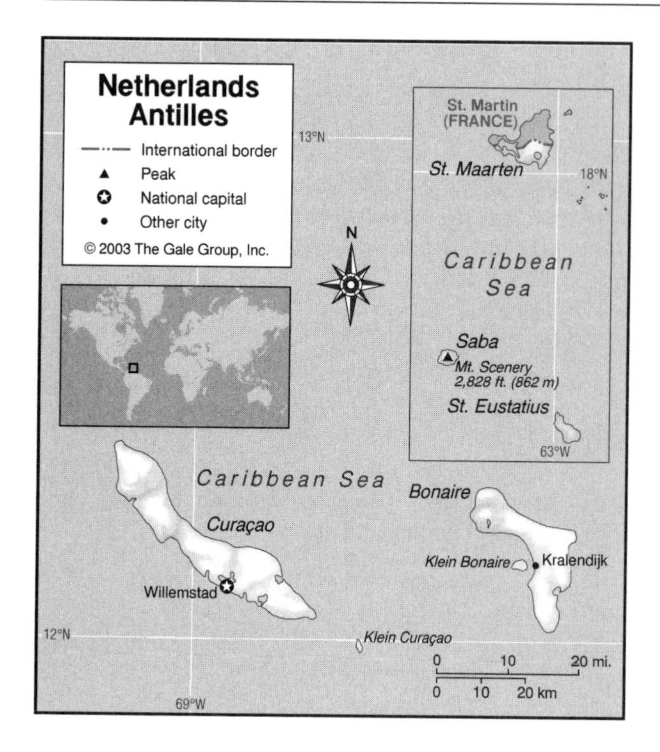

OVERVIEW

The Netherlands Antilles consists of five islands: Bonaire, Curaçao, Saba, St. Eustatius (Statia) and half of Saint Martin (Sint Maarten). These islands are divided into two groups about 500 mi (800 km) apart in the Caribbean Sea. The main group, off the coast of Venezuela, consists of Bonaire and Curaçao. They are part of the Windward Islands. Aruba, located nearby, seceded from the Netherlands Antilles in 1986. The other three islands are far to the north, in the Leeward Island chain that forms the boundary between the Atlantic Ocean and the Caribbean Sea.

MOUNTAINS AND HILLS

The Windward Island group is generally flat, but the smaller Leeward Islands are mountainous. Mt. Scenery on the island of Saba at 2,828 ft (862 m) is the highest point and an extinct volcano.

INLAND WATERWAYS

The islands of the Netherlands Antilles are too small to support waterways of any significance. The northern (Leeward) islands are generally wetter than those in the Windward Islands, which are semi-arid.

THE COAST, ISLANDS, AND THE OCEAN

The two groups of the Netherlands Antilles are located on either side of the Caribbean Sea, with the Windward Islands near the Venezuelan coast and the Leeward Islands between the Caribbean and the Atlantic Ocean.

Curaçao (171 sq mi / 444 sq km) and Bonaire (111 sq mi / 288 sq km) are by far the largest islands in the Netherlands Antilles. The Leeward Islands are all much smaller. The largest, St. Martin (13 sq mi / 34 sq km), is actually just the southern part of that island, the northern half of which belongs to France in the form of Guadeloupe. St. Eustatius (8 sq mi/21 sq km) and Saba (5 sq mi/ 13 sq km) are smaller still.

CLIMATE AND VEGETATION

Temperatures average between 77°F (25°C) and 88°F (31°C). Rainfall on an annual basis averages 42 in (107 cm) on the Leeward Islands in the north and 20 in (51 cm) on the Windward Islands in the south.

HUMAN POPULATION

With a population of more than 200,000, the Netherlands Antilles has a .97 percent annual growth rate. Curaçao is the most populous island.

NATURAL RESOURCES

Phosphates are mined on Curaçao, salt on Bonaire.

FURTHER READINGS

Gastmann, Albert. *Historical Dictionary of the French and Netherlands Antilles.* Metuchen, N.J.: Scarecrow Press, 1978.

Gravette, A. Gerald. *The Netherlands Antilles.* New York: Hippocrene Books, 1989.

Hiss, Philip Hanson. *Netherlands America: The Dutch Territories in the West.* New York: Duell, Sloan, and Pearce, 1943.

Lutz, William. *Netherlands Antilles.* New York: Chelsea House, 1987.

New Zealand

- **Area:** 103,738 sq mi (268,680 sq km) / World Rank: 75
- **Location:** Islands in the South Pacific Ocean, Southern and Eastern Hemispheres, southeast of Australia and the Tasman Sea.
- **Coordinates:** 41°00′S, 174°00′E
- **Borders:** None
- **Coastline:** 9,404 mi (15,134 km)

- **Territorial Seas:** 12 NM
- **Highest Point:** Mt. Cook, 12,349 ft (3,764 m)
- **Lowest Point:** Sea level
- **Longest Distances:** 994 mi (1,600 km) NNE-SSW; 280 mi (450 km) ESE-WNW
- **Longest River:** Waikato, 264 mi (425 km)
- **Largest Lake:** Lake Taupo, 234 sq mi (606 sq km)
- **Natural Hazards:** Earthquakes; volcanoes
- **Population:** 3,864,129 (July 2001 est.) / World Rank: 121
- **Capital City:** Wellington, southeast tip of the North Island
- **Largest City:** Auckland, north-central part of the North Island, 1,079,304 (2001 census population)

OVERVIEW

New Zealand lies in the southwestern Pacific Ocean and consists of two main islands and a number of smaller ones. The main North and South Islands, separated by the Cook Strait, lie on an axis running from northeast to southwest, except for the low-lying Northland Peninsula on the North Island. New Zealand also has jurisdiction over foreign affairs and defense matters for the self-governing entities of Tokelau and the Cook Islands and claims land in Antarctica in and near the Ross Sea.

New Zealand is very mountainous, with less than a quarter of its land below 656 ft (200 m). The South Island is by far the more mountainous of the two main islands. A massive mountain chain, the Southern Alps, runs the entire length of the island, and outlying ranges extend to the north and the southwest. The natural divisions of South Island are the Canterbury Plains to the east; the central mountain highlands, which covers much of the island; and a narrow western coast. The Canterbury Plains rise steadily but imperceptibly inland until they abut abruptly against the foothills at a height of from 1,110 to 1,400 ft (335 to 427 m).

The North Island is characterized by hill country. The mountain highland here is narrow and lies in the east, but the rest of the country is composed of hills and a central Volcanic Plateau. There is little coastal lowland and even in the west, where it is widest, the Egmont Peak (also called Mt. Taranaki) rises well over 8,000 ft (2,438 m). The narrow northern peninsular section of North Island is mostly low-lying, though its surface is often broken and irregular.

North Island lies at or near the boundary between the Australian and Pacific Tectonic Plates. The eastern border of the North Island constitutes a subduction zone between the two plates. The plate boundaries for South Island are uncertain. Compared with some other parts of the almost continuous belt of earthquake activity around the rim of the Pacific, such as Japan, Chile and the Philippines, the level of seismic activity in New Zealand is moderate, although earthquakes are common. It may be compared roughly with that prevailing in California. There are a number of volcanoes, especially on North Island. The movement of the plates sometimes causes avalanches in the Southern Alps.

MOUNTAINS AND HILLS

Mountains

Three-fourths of New Zealand is mountainous. The Southern Alps extend for some 300 mi (483 km) on South Island and include New Zealand's highest peak, Mt. Cook, as well as about 350 glaciers, of which the largest is the Tasman, extending for 18 mi (29 km). There are at least 223 named peaks higher than 7,546 ft (2,300 m), and 15 that are higher than 10,000 ft (3,764 m). In contrast, the highest peak on the North Island, Ruapehu, is only 9,177 ft (2,797 m) high.

The southernmost section of the South Island mountain system is Fiordland at the island's southwestern edge, appropriately named for its deep, canyon-like valleys filled at the coast by saltwater fjords and inland by the arms of freshwater lakes. In contrast to Fiordland, the wettest part of New Zealand, is the mountain interior of Otago, whose mountains are separated by terraced, gravel-floored basins, some of them occupied by lakes.

The mountain highlands of North Island are a continuation of the Southern Alps. The Tararua, Ruahine, Kaimanawa, and Huiarau Ranges extend across the island on the same southwest to northeast axis as the higher mountains to the south. Their highest peak is Makorako, at 5,700 ft (1,737 m).

Nearby but separate from North Island's highlands are the peaks of three extinct volcanoes, rising to heights of between 6,457 and 9,177 ft (1,968 to 2,797 m) on the Volcanic Plateau. The tallest of these is Ruapehu, the highest point on North Island. Further west the land is lower overall, but is dominated by the extinct volcano of Mt. Egmont (or Taranaki), whose elevation is 8,260 ft (2,518 m). In the area around Rotorua on North Island, hot springs and geysers provide evidence of the volcanic system underlying the area's geology.

Plateaus

The wide Volcanic Plateau, with a terrain of lava, pumice stone, and volcanic ash, lies north and west of the Kaimanawa range on North Island. Deep-cut hill country with short but steep slopes occupies most of its rim. The elevation of the plateau decreases and its slopes become gentler toward the coast in the west.

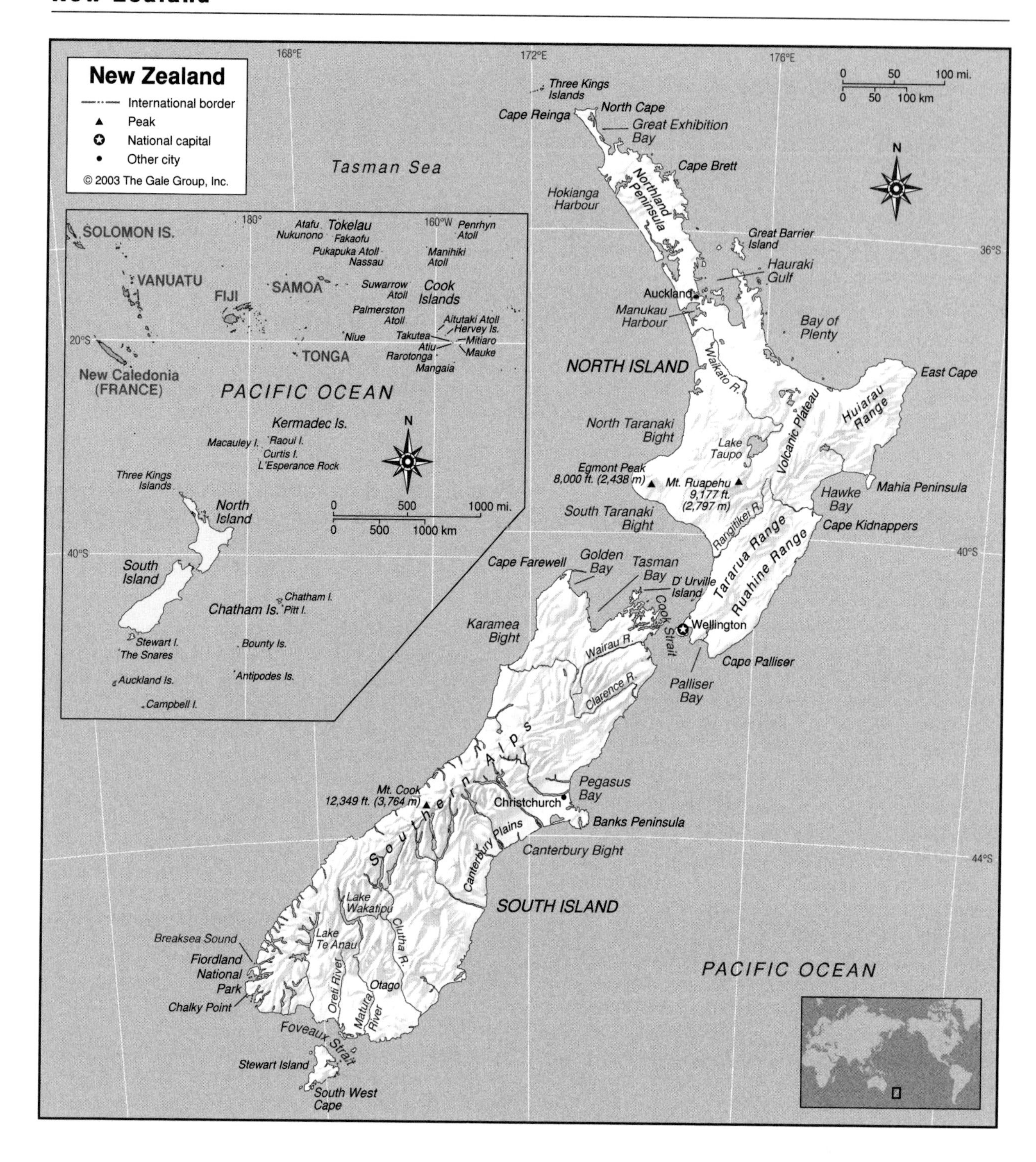

Hills and Badlands

North Island is characterized by undulating terrain. Much of the land surrounding the mountain ranges and the Volcanic Plateau on all sides is hill country. North of Hawke Bay in the east, the mountain ranges are flanked by deeply corrugated hill country. On South Island, the central section of the narrow coastal strip, the only area where the mountains are set back from the coast, is made up of broken hill country.

INLAND WATERWAYS

Lakes

New Zealand has many lakes; the rivers that flow through its mountains often rise from or drain into them. The lakes on South Island in particular are noted for their magnificent scenery. The country's largest natural lake is Lake Taupo on the North Island, followed by Lakes Te Anau and Wakatipu on the South Island. Lake Benmore is New Zealand's largest artificial lake.

Rivers

The rivers are swift flowing and shallow, and only a few are navigable. They are, however, suitable for hydroelectric power generation, with a high rate of flow and reliable volume of ice-free water. The longest river is the Waikato (264 mi / 425 km), which flows northwestward across the North Island and empties into the Tasman Sea, as do the Wanganui and Rangitaiki. Rivers that flow into the Pacific from the South Island include the Clutha, the Taieri, and the Clarence, while the Mataura, Wairau, and Oreti flow from the South Island into the Foveaux Strait. The Clutha is the South Island's longest river, and its volume is the greatest of any river in the country.

Wetlands

The terrain of Northland, the northernmost part of the North Island, includes peat bogs and swamplands with mangrove trees, interspersed with hilly areas.

GEO-FACT

New Zealand has several species of flightless birds, of which the most famous is the kiwi, the national emblem. These flightless species were able to evolve and survive because the uniquely isolated environment on the islands lacked the predators that would have required them to have more highly developed defense mechanisms. Many of New Zealand's plant species were similarly able to evolve without natural defenses. These plant and animal species are now protected to ensure their continued survival.

THE COAST, ISLANDS, AND THE OCEAN

Oceans and Seas

New Zealand lies in the South Pacific Ocean southeast of Australia, from which it is separated by the Tasman Sea. At the Tamaki Isthmus on the North Island, the two bodies of water are separated by only 1 or 2 mi (2 or 3 km) of land. The North and South Islands are separated by the Cook Strait, which is 16 to 90 mi (26 to 145 km) wide. The Foveaux Strait lies between South Island and Stewart Island to the southeast.

Major Islands

South Island is the largest in New Zealand; it has an area of 57,870 sq mi (149,883 sq km). North Island, which has an area of 44,274 sq mi (114,669 sq km), is by far the second largest. Stewart Island, to the southeast of South Island, is next largest at only 674 sq mi (1,746 sq km). Great Barrier Island is northeast of North Island. Other islands include the Chatham Islands (372 sq mi / 963 sq km), which lie 536 mi (862 km) to the east of South Island, and many small islands lying near North or South Island, including D'Urville, Three Kings Islands, and the Snares Islands.

The Coast and Beaches

North Island has the more heavily indented coastline of the two main islands. Its bays include North and South Taranaki Bights to the west and Palliser Bay to the south. Cape Palliser is the southernmost point on the island. The wide Hawke Bay is in the east, with Cape Kidnappers at its southern end and the Mahia Peninsula in the northeast. The even wider Bay of Plenty is around East Cape, on the north side of the island.

Northland, the long arm of land that juts out to the northwest on North Island, is deeply indented on both sides near its midpoint. At Auckland the land mass narrows to a width of only 1 to 2 mi (2 to 3 km), with the almost totally enclosed Manukau Harbor to its west, and the complex pattern of islands, channels, and bays of the Hauraki Gulf to the east. The east coast and northern tip of Northland have multiple bays and harbors, including Cape Brett, Great Exhibition Bay, and North Cape. The west coast is almost completely smooth except for Hokianga Harbour, coming to a point at Cape Reinga.

The northern and southern ends of South Island have numerous bays, of which the largest are Golden Bay and Tasman Bay in the north. The long eastern and western coastlines are smoother, with the major indentations consisting of Karamea Bight at the northern end of the west coast and Canterbury Bight and Pegasus Bay to the east, with the Banks Peninsula jutting out between them somewhat less than halfway down the coast. The coast of Fiordland to the southeast is broken up into numerous sounds and inlets, including Breaksea Sound and Chalky Inlet. Southwest Cape, on Stewart Island, is the southernmost point in New Zealand (excluding small islands). At the northwestern corner of South Island is Cape Farewell.

Island Dependencies

There are three island groups in the Pacific Ocean that are dependant on New Zealand to varying degrees without being an integral part of that country.

The Cook Islands are located roughly halfway between New Zealand and Hawaii, in the middle of the South Pacific Ocean. The islands have local self-government but rely on New Zealand for foreign affairs and defense in a free association. There are two chains of islands, a series of seven low-lying coral atolls in the north, and a group of eight larger and more elevated volcanic islands in the south. The later are the home of most of the population and the capital of Avarua. The islands have a land area of 92 sq mi (240 sq km).

Niue Island, which extends more than 102 sq mi (263 sq km), is one of the world's largest coral islands. It is located east of the Cook Islands, and like those islands is self-governing in local affairs. The capital is Alofi.

Tokelau is another island chain in the middle of the South Pacific Ocean, northwest of the Cook Islands. They consist of three small coral atolls and surrounding islets, with a land area of only 5 sq mi (12 sq km). Tokelau is a territory of New Zealand.

Population Centers – New Zealand

(PRELIMINARY 2001 CENSUS DATA)

Name	Population	Name	Population
Auckland	1,079,304	Hamilton	165,576
Wellington (capital)	340,719	Napier-Hastings	116,292
Christchurch	340,053	Dunedin	109,563

SOURCE: Statistics New Zealand.

CLIMATE AND VEGETATION

Temperature

New Zealand has a mild oceanic climate with little seasonal variation. The climate varies primarily with altitude and by location, as temperatures grow cooler towards the south. Mean annual temperatures range from about 52°F (11°C) in the southern part of South Island to 59°F (15°C) in Northland, the northernmost part of the North Island. Daytime high temperatures in summer generally vary from 70° to 81°F (21° to 27° C); winter highs are usually at least 50°F (10°C). Temperatures rarely extend beyond the extremes of 14°F (-10°C) and 95°F (35°C).

Rainfall

Due to prevailing westerly and northwesterly winds, the westerly mountain slopes of both islands receive the heaviest rainfall. Average rainfall for the country as a whole ranges from 25 to 60 in (64 to 152 cm), but the total for specific locations varies from as little as 12 in (30 cm) in Central Otago on South Island to as much as 315 in (800 cm) in Fjordland, at the southwestern edge of the same island.

Grasslands

The Canterbury Plains on the east coast of South Island are New Zealand's largest plains area, stretching for about 200 mi (320 km) and reaching as much as 40 mi (64 km) inland. North Island has coastal plains bordering the Bay of Plenty, Hawke Bay, and on Northland. There are also plains areas south of the Volcanic Plateau and along the Waikato.

Forests and Jungles

New Zealand is home to many unique plant (and animal) species that have evolved and diversified in isolation for millions of years. At least three-fourths of its plants species are endemic (found only in New Zealand), and New Zealand's forests are considered to be among the most "ancient" in character in the world. They are primarily temperate rainforest in nature with spectacularly tall broadleaf evergreens and a dense undergrowth of ferns and vines. Forest once covered most of the islands but now extend over less than one-fourth of New Zealand. Forests can be found throughout the islands, particularly in mountainous regions or national parks. Forests of Kauri pine, a distinctive hardwood species that was once widespread, are now found only in special forest reserves. Four unique softwoods that provide timber are the rimu, matai, totara, and kahikatea (white pine) trees. Introduced species include poplar, birch, elm, and radiata pine. Some 145 species of fern constitute most of the undergrowth in the damp forests on the western slopes of South Island. Alpine wildflowers bloom on the upper slopes of the tall peaks on the island, including the daisy-like Mount Cook lily, which blooms in December.

HUMAN POPULATION

Three-quarters of New Zealand's population lives on the North Island. Within that island, most people live in the north, or near the capital of Wellington. About half of the country's population is found in these areas. Around 90 percent of New Zealand's population is urban. Auckland and Wellington are the major cities of the North Island; the major cities on the South Island are Christchurch and Dunedin.

NATURAL RESOURCES

Planted forests provide New Zealand with abundant timber resources; the most important and productive imported tree species in these forests is the radiata pine. Hydropower provides most of the country's electricity, and its geothermal areas are an increasingly important power resource. Mineral resources include gold, iron sands, coal, and natural gas.

FURTHER READINGS

Barber, Laurie. *New Zealand: A Short History.* London: Hutchinson, 1989.

Eyewitness Travel Guides. "New Zealand." New York: Dorling Kindersley, 2001.

Hanbury-Tenison, Robin. *Fragile Eden: A Ride through New Zealand.* Topsfield, Mass.: Salem House, 1989.

McDermott, John W. *How to Get Lost and Found in New Zealand.* Wellington: A. H. & A. W. Reed, 1976.

Sinclair, Keith, ed. *The Oxford Illustrated History of New Zealand.* New York: Oxford University Press, 1997.

Tourism New Zealand. *100% Pure New Zealand.* http://www.purenz.com/ (Accessed June 13, 2002).

Nicaragua

- **Area:** 49,998 sq mi (129,494 sq km) / World Rank: 97
- **Location:** Northern and Western Hemispheres, on the Central American Isthmus, north of Costa Rica and south of Honduras.
- **Coordinates:** 13°00′N, 85°00′W
- **Borders:** 765 mi (1,231 km) / Costa Rica, 192 mi (309 km); Honduras, 573 mi (922 km)
- **Coastline:** 565 mi (910 km)
- **Territorial Seas:** 200 NM
- **Highest Point:** Pico Mogotón, 7,999 ft (2,438 m)
- **Lowest Point:** Sea level
- **Longest Distances:** 293 mi (472 km) N-S; 297 mi (478 km) E-W
- **Longest River:** Río Coco, 423 mi (680 km)
- **Largest Lake:** Lago de Nicaragua, 3,089 sq mi (8,000 sq km)
- **Natural Hazards:** Earthquakes; volcanoes; hurricanes
- **Population:** 4,918,393 (July 2001 est.) / World Rank: 111
- **Capital City:** Managua, far west center of the country
- **Largest City:** Managua, population 1,319,000 (2000 metropolitan est.)

OVERVIEW

Nicaragua is the largest country in Central America, and boasts the largest freshwater lake in the Americas after the Great Lakes of North America. The country is roughly an equilateral triangle: southwest/northeast along the Honduran border, north/south on the Caribbean, and southeast/northwest along the Costa Rican border and Pacific Ocean.

The land naturally divides into three topographic zones: the Pacific Lowlands, the Central Highlands, and the Atlantic Lowlands. The Pacific Lowlands is a band about 47 mi (75 km) wide along the Pacific Ocean between Honduras and Costa Rica. The plain is punctuated by clusters of volcanoes, immediately to the east of which is a Great Rift ("crustal fracture")—a long, narrow depression passing along the isthmus from the Golfo de Fonseca in the north to the Río San Juan at the bottom of the country. To the northeast are the Central Highlands, including the highest mountains and coolest temperatures, where the majority of Nicaragua's coffee is grown. The sparsely populated Atlantic Lowlands comprise more than half the area of Nicaragua. These lowlands and Mosquito Coast are the traditional home of the Miskito peoples. The area is tropical rain forest and savannas crossed by scores of rivers flowing to Caribbean.

Nicaragua is situated on the Caribbean Tectonic Plate, but just off the Pacific Coast is the Cocos Plate. Frequent earthquakes and volcanic eruptions result from action of the Caribbean and Cocos Tectonic Plates. Nicaragua has hundreds of minor earthquakes and shocks each year, and occasionally a serious quake. In 1931 and in 1972 earthquakes virtually destroyed the capital city of Managua. Central Managua has yet to be rebuilt.

MOUNTAINS AND HILLS

Mountains

Nicaragua has three inland mountain ranges and a chain of volcanoes. Cordillera Isabella runs southwest to northeast, toward the Honduran border. Cordillera Dariense runs nearly west to east, defining the southern edge of the triangular central highlands. The rugged mountain terrain between is composed of ridges from 1,968 to 5,905 ft (900 to 1,800 m) high. River valleys drain mostly to the Caribbean. Cordillera Los Maribios is the chain of volcanoes starting in the northwest. Cutting across the Atlantic Lowlands in the southeast are three smaller mountain ranges. From north to south they are the Huapí Mountains, the Amerrique Mountains, and the Yolaina Mountains. The highest peak in Nicaragua, Pico Mogotón, sits on the Honduran border about one hundred miles inland from the Pacific Ocean. The peak rises to a height of 7,999 ft (2,438 m).

Volcanoes

A chain of seventeen volcanoes runs along the Pacific Coast. Six have erupted in the last hundred years. The most significant active volcanoes are: Concepción, San Cristóbal, Telica and Masaya. Volcán Concepción, Nicaragua's second highest volcano, is one of its most active volcanoes. This symmetrical volcano, on the north end of Isla de Ometepe in the middle of Lago de Nicaragua, has had frequent moderate eruptions in the twentieth century; it threw ash over the countryside in December 2000.

A complex of five volcanoes northwest of Managua is named for the "oldest," San Cristóbal (El Viejo), which is the highest peak of the Maribios Range. Volcán Casita, immediately east of San Cristóbal, was the site of a catastrophic landslide in 1998.

Volcán Telica, located northwest of the city of León, has erupted frequently in the since the 1800s. Telica's steep cone is topped by a double crater 2,300 ft (700 m) wide.

Volcán Masaya, near Managua, is one of only four volcanoes on earth with a constant pool of lava that neither increases nor recedes. It is the focal point of one of Nicaragua's oldest national parks.

Canyons

Nicaragua has more than 90 principal rivers running through canyons of various depths. In comparison to mountain ranges in North and South America, and even to adjacent Honduras, Nicaragua's highest mountains are modest, and few canyons are notably deep.

Hills and Badlands

The fumaroles (steam vents), hot springs, and boiling mudpots of Hervideros de San Jacinto (The Swarms of San Jacinto), southeast of Telica, are an extensive geothermal area frequented by tourists.

INLAND WATERWAYS

Lakes

Lago de Nicaragua is the largest freshwater lake in Central or South America, and one of the most spectacular bodies of water in the Americas. It fills the southern portion of the Great Rift running parallel to the Pacific Ocean. The lake is 99 mi (160 km) long, 40 mi (65 km) at its widest, and is 105 ft (32 m) above sea level. It is relatively shallow, with an average depth of 66 ft (20 m) and maximum of 197 ft (60 m). With a total surface area of 3,089 sq mi (8,000 sq km), the lake is sprinkled with many islands including Ometepe Island.

Lago de Managua connects to Lago de Nicaragua by the Tipitapa River. The lake is 32 mi (52 km) long and up to 16 mi (25 km) wide with an area of 396 sq mi (1,025 sq km). It is only 98 ft (30 m) at its deepest. On the lake's southwest side is a peninsula, Chiltepe, that holds two small crater lakes: Xiloá and Apoyeque.

Rivers

Nicaragua has nearly a hundred principal rivers; most drain the Central Highlands, through the Atlantic Lowlands and empty along the Mosquito Coast. The majority of them are relatively short rivers with a few longer ones, such as Río Grande Matagalpa. A few rivers feed Lakes Managua and Nicaragua. Río Coco, Nicaragua's longest river, flows 423 mi (680 km) from the northwest highlands to the Caribbean Sea, forming Nicaragua's border with Honduras.

The river carrying the largest volume of water is Río San Juan, which is only 110 mi (180 km) long. It flows from the southeast corner of Lake Nicaragua east to the Caribbean Sea. This deep, navigable river is the boundary between Nicaragua and Costa Rica.

Wetlands

With many rivers, Nicaragua has many wetlands. Besides the entire Caribbean Coast, which is mostly swampy and marshy land, there are three areas of particular note. Deltas del Estero Real (315 sq mi / 816 sq km) in the Golfo de Fonseca is a natural reserve, part of the large mangrove systems of the gulf shared with El Salvador and Honduras. Humedales de San Miguelito is near where Río San Juan exits Lago de Nicaragua. It is home to diverse species of birds, fish, reptiles and mammals. Lagunar de Tisma is a small area of lake, marsh, and river ecosystems on the northwest shores of Lago de Nicaragua.

THE COAST, ISLANDS, AND THE OCEAN

Oceans and Seas

Nicaragua has coasts on the Pacific Ocean and on the Caribbean Sea. There are coral reef systems off the eastern coast—including the largest hard carbonate bank in the Caribbean—yet most are not near the mainland due to sediment runoff from the many rivers. Closer to the shore, reef systems form four groups of islands: the Moskitos Cays , Man-of-War Cays, (Guerrero Cays), Pearl Cays, and the Corn Islands. The last three groups are inhabited.

Major Islands

Scores of islands dot the huge Lago de Nicaragua; the most prominent is Isla de Ometepe. The dumbbell-shaped island—once two islands—was formed by a volcano at each end. Its total area is 106 sq mi (276 sq km), including the Isthmus of Istián that connects the two sections of the island. At the south end of Lago de Nicaragua is a group of 36 small islands collectively named Archipiélago Solentiname. Some of the larger islands in the group are Venada, San Fernando, Mancarroncito, and Mancarrón.

Besides islands in the freshwater lakes, there are a few islands off the Caribbean shore and none on the Pacific side. The two Corn Islands are 43 mi (70 km) off the southern coast, 5 mi (8 km) apart. Great Corn Island is about 3 sq mi (8 sq km), and Little Corn Island is half that size.

The Moskitos Cays is an offshore island group with coral reefs 7.5 mi (12 km) from the north shore. The area is home to several endangered species including the Hawksbill Turtle, Caribbean manatee, Tucuxi freshwater dolphin, and caiman crocodile.

The two other coralline island groups, the Pearl Cays and the Man-of-War Cays, also sit not far from the main-

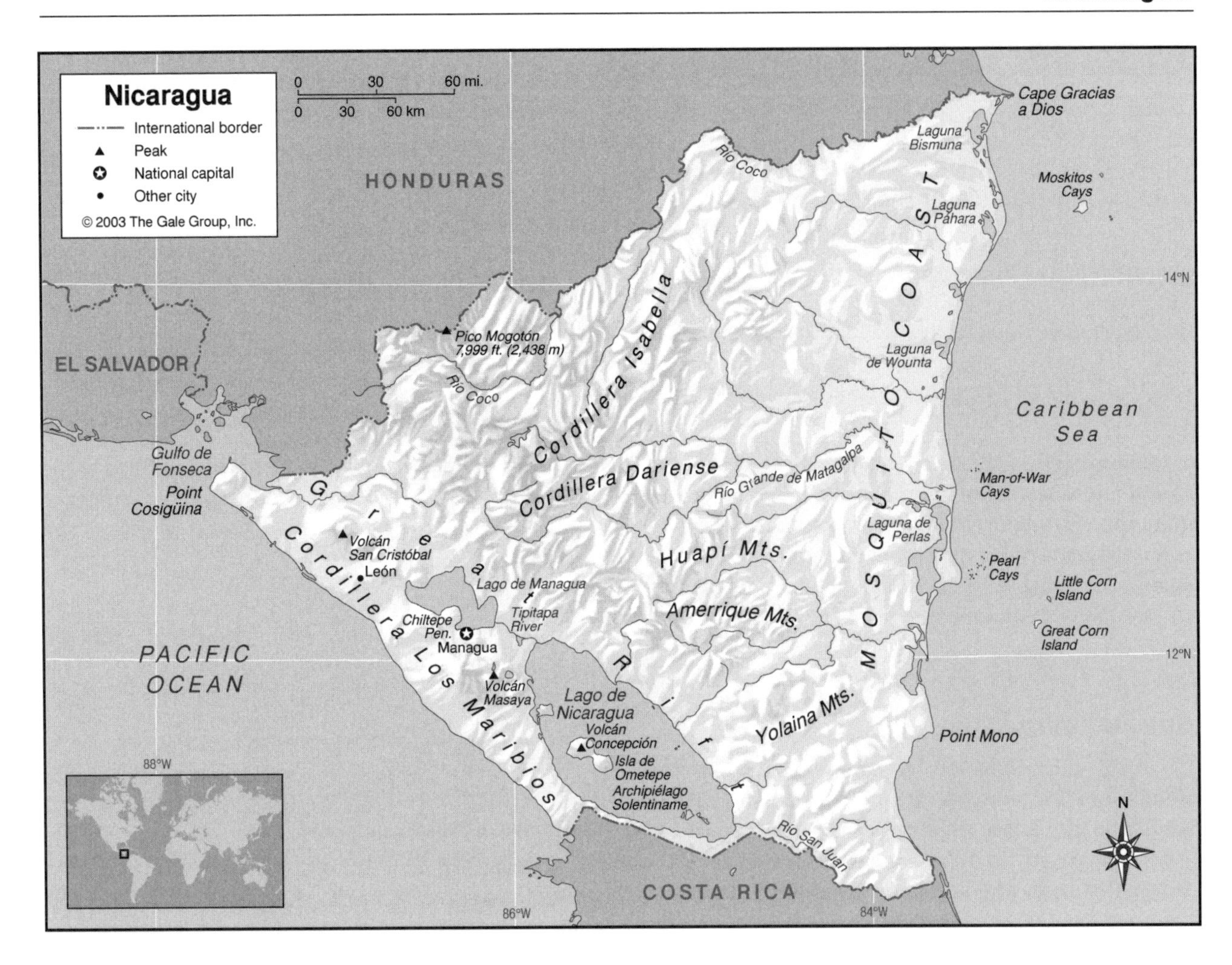

land. They are sparsely populated with villages mainly of fishermen.

The Coast and Beaches

The most hospitable, populated coast is the Pacific. There are no significant lagoons along the Pacific side, and the only significant features are at the northernmost point where the coast turns inland at Point Cosigüina to carve out the Gulf of Fonseca. The relatively remote and sparsely populated Atlantic Lowlands and Mosquito Coast are periodically broken by lagoons and estuaries where major rivers end. The largest of these lagoons, from north to south, are: Laguna Bismuna, Laguna Páhara, Laguna Karatá, Laguna de Wounta, and Laguna de Perlas. The northern end of the coast is marked by Cape Gracias a Dios, and near its southern extremity Point Mono sticks out into the sea.

CLIMATE AND VEGETATION

Temperature

In Nicaragua temperature is more affected by elevation than by season. On the flat lands (east and west) daytime temperatures average 85°F (29°C), and night temperatures drop to 70°F (21°C) or below. In the central highlands, temperatures are lower—about 70°F (21°C) in the daytime, and about 60°F (15°C) at night. In the very high mountains, temperatures can approach freezing at night.

Rainfall

The rainy season (winter, invierno) is from May to November; the dry season (summer, verano) is from June to October. The Mosquito Coast gets the greatest rainfall, from 90 to 200 in (76 to 508 cm) per year. Less rain—30 to 90 in (76 to 229 cm)—falls on the Central Highlands and falls over a longer period of the year. On the Pacific Coast, annual rainfall ranges from 40 to 60 in (102 to 152 cm) per year.

Hurricanes have periodically exacted severe damage on Nicaragua. The most devastating in recent years were Hurricanes Mitch (October 1998) and Joan (1988).

Forests and Jungles

Nicaragua has numerous rainforests, some protected as reserves. Ecologically, two exceptional reserves are Reserva Natural Miraflor and Reserva Biológic Indo-Maiz. Miraflor (80 sq mi / 206 sq km) is remarkably pristine, and has tropical savannah at lower altitude, pine forest higher up, and cloud forest at its highest. Miraflor

GEO-FACT

The Great Rift along the western part of Nicaragua is so near sea level that, until construction of the Panama Canal, it was considered the most likely site for joining the Atlantic and Pacific Oceans.

contains a tiny lake at 4,528 ft (1,380 m) altitude, and a 196-ft (60-m) waterfall.

Biológic Indo-Maiz covers 1,400 sq mi (3,626 sq km). Only a few square miles within the preserve is habitat for more species of birds, trees, and insects than are on the entire continent of Europe. Indo-Maiz protects the largest contiguous extent of primary rainforest in Central America, the Bosawás Biosphere Reserve (2,820 sq mi / 7,300 sq km).

HUMAN POPULATION

About three-fifths of Nicaragua's population live along the Pacific Lowlands. Most of the remaining population is in the Central Highlands. Only an estimated 2 or 3 percent live in the Atlantic Lowlands and along the Mosquito Coast. The average population densities for these regions are as follows: Atlantic Lowlands, a little more than 26 people per sq mi (10 people per sq km); Central Highlands, 115 per sq mi (44 per sq km); Pacific Lowlands, 444 per sq mi (171 per sq km). The largest ethnic group in Nicaragua is mestizo (of mixed indigenous and European ancestry).

NATURAL RESOURCES

Nicaragua is very dependent on its agricultural production, which includes coffee, cotton, sugar, bananas, beans, and rice. The Central Highlands, including the highest mountains and coolest temperatures, has the soil and climate best suited for the growth of coffee, which along with cotton comprises Nicaragua's largest cash crops. The country also has mineral resources, the most significant being copper, lead, timber, and fish; however, these resources are not yet exploited to their potential. Geothermal resources and hydropower from dams on the rivers provide electricity.

FURTHER READINGS

Glassman, Paul. *Nicaragua Guide: Spectacular and Unspoiled.* Champlain, N.Y.: Travel Line, 1996.

Haverstock, Nathan A. "Nicaragua in Pictures." *Visual Geography Series.* Minneapolis: Lerner Publications, 1993.

Meehan, John F., et al. *Managua, Nicaragua Earthquake of December 23, 1972; Reconnaissance Report.* Oakland, Calif.: Earthquake Engineering Research Institute, 1973.

Niger

- **Area:** 489,191 sq mi (1,267,000 sq km) / World Rank: 23
- **Location:** Northern and Eastern Hemispheres in Africa, bordering Libya to the northeast, Chad to the east, Nigeria to the south, Benin and Burkina Faso to the southwest, Mali to the west, and Algeria to the northwest.
- **Coordinates:** 16°00′N, 8°00′E
- **Borders:** 3,540 mi (5,697 km) / Algeria, 594 mi (956 km); Benin, 165 mi (266 km); Burkina Faso, 390 mi (628 km); Chad, 730 mi (1,175 km); Libya, 220 mi (354 km); Mali, 510 mi (821 km); Nigeria, 930 mi (1,497 km)
- **Coastline:** None
- **Territorial Seas:** Landlocked
- **Highest Point:** Mt. Gréboun, 6,378 ft (1,944 m)
- **Lowest Point:** Niger River, 656 ft (200 m)
- **Longest Distances:** 1,146 mi (1,845 km) ENE-WSW; 637 mi (1,025 km) SSE-NNW
- **Longest River:** Niger, 2,600 mi (4,184 km)
- **Largest Lake:** Lake Chad, 9,950 sq mi (28,457 sq km) in October
- **Natural Hazards:** Drought
- **Population:** 10,355,156 (July 2001 est.) / World Rank: 73
- **Capital City:** Niamey, southwestern Niger
- **Largest City:** Niamey, population 587,000 (2000 est.)

OVERVIEW

Landlocked Niger is the second largest country in West Africa (surpassed only by Algeria) and the tenth largest on the continent. Niger is a dry country, with four-fifths of its land covered by desert, but there is diversity in its topography, which includes plains, plateau regions, and mountains. The country can be divided into three major regions: the arid, inhospitable desert to the north and northeast, a transitional Sahelian region in the center, and a small fertile area in the south, between the Niger River basin in the southwest and the Lake Chad basin in the southeast.

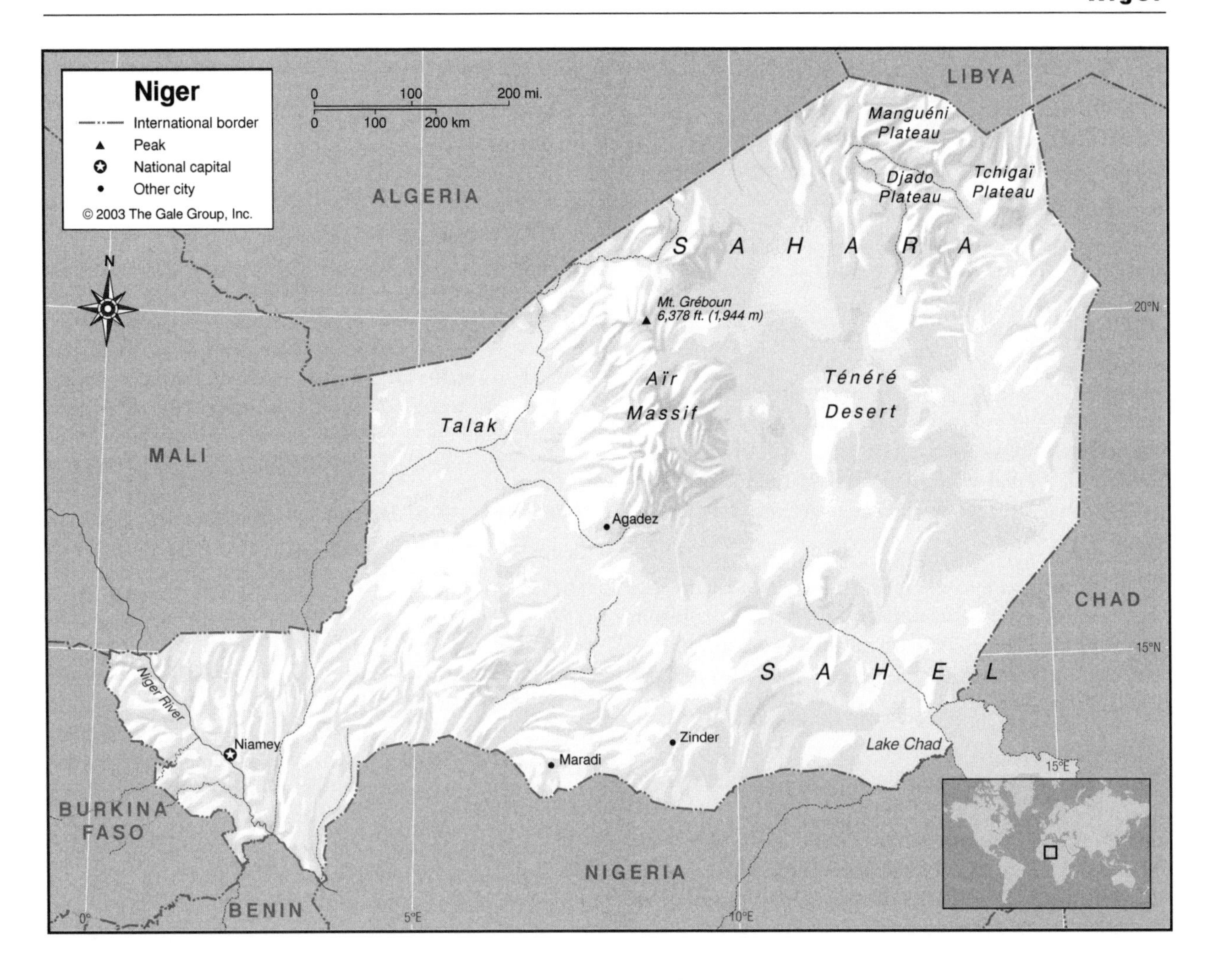

MOUNTAINS AND HILLS

Mountains

In the north-central region are the mountains of the volcanic Aïr Massif, which belong to the same system as Algeria's Ahaggar Mountains and extend southward more than 248 mi (400 km) from Niger's border with Algeria. The mountains cover an area of approximately 30,880 sq mi (80,000 sq km), and their average elevation is between 2,000 and 3,000 ft (600 to 900 m). Their highest point is Mt. Gréboun in north-central Niger, at 6,378 ft (1,944 m).

Plateaus

The Manguéni, Djado, and Tchigaï plateaus are clustered together in the northeastern corner of the country near the border with Libya. Their average elevation is about 2,600 ft (800 m). The mountains of the Manguéni Plateau are a continuation of Chad's Tibesti Mountains.

INLAND WATERWAYS

Lakes

About 1,000 sq mi (2,590 sq km) of Lake Chad lies within the southeastern tip of Niger. The size of the lake, which Niger shares with Chad and Nigeria, varies greatly depending on the time of year, decreasing in the dry season to nearly one-third of its size. In October, its surface area can exceed 9,950 square miles (28,457 sq. km), but by late April or early May, it is usually roughly 3,000 square miles (8,580 sq. km).

Rivers

The lifeline of the country is the Niger River, which flows year round across the southwestern corner of the country for about 350 mi (563 km) in a northwest-to-southeast course, while its tributaries flow during the rainy season only. The Kamadougou Yobé River in the southeast drains into Lake Chad, forming part of Niger's border with Nigeria.

THE COAST, ISLANDS, AND THE OCEAN

Niger is landlocked.

CLIMATE AND VEGETATION

Temperature

Niger's climate is one of the hottest on earth. Between February and July high temperatures on the plateaus in the northeast can hit 122°F (50°C). In January, readings can drop to a low of 46°F (8°C) in the desert regions, which have both the hottest temperatures and the greatest contrast between highs and lows. The harmattan wind blows across the eastern desert for much of the year.

Rainfall

Rainfall varies markedly between Niger's Saharan and Sahel regions. The arid north receives only very small amounts of rain, while the central Sahel region has a short rainy season. The southernmost strip has a longer one that can last as long as four months. Most of Niger receives less than 14 in (36 cm) of rain annually, and less than 4 in (10 cm) of rain falls in almost half the country. The capital city of Niamey, in the Niger River valley, receives an annual average of about 22 in (56 cm), most of it during a rainy season between May and October.

Grasslands

There are some steppe grasslands in the Sahel region that are mostly used as pastureland. The southwest region drained by the Niger River is a savannah with low bushes and trees including the baobab, kapok, bastard mahogany, and tamarind.

Deserts

Dense vegetation, including acacias and palm trees, is found in the deep valleys of the Aïr Mountains, which are called koris. The Ténéré Desert that lies to the east of the mountains and the Talak to the west have vast expanses of shifting sand dunes (called ergs) where no vegetation grows, as well as dunes anchored by sparse, scrubby vegetation. They are extensions of the vast Sahara Desert that dominates the interior of northern Africa. Plant life is supported by oases in these regions and in the northern Sahel, which includes the city of Agadez.

HUMAN POPULATION

Niger is a mostly rural country—only about 20 percent of its people live in cities or towns. The overall population density for the country is 20.5 people per square mile (7.9 people per sq km), but much of Niger's population is concentrated in the fertile southern region of the country. In the deserts, only the areas with oases can support human life, and nomadic herders are found there. Most of the land is otherwise uninhabited.

NATURAL RESOURCES

With its scarcity of water and arable land, Niger's most important natural resources are its minerals, of which the most abundant is uranium, which alone constituted 65 percent of Niger's exported commodities in 1998. Salt, sulfate, and phosphates are also mined, and the country has reserves of manganese, copper, zinc, lead, silver, and other minerals. In the southern savanna, the climate is suitable for the growth of cowpeas, onions, rice, cotton, and peanuts. Some pastureland in the savanna region is also the site of livestock farms.

GEO-FACT

The name of Niamey, the capital of Niger, comes from the phrase "settle and acquire," an order given by a long-ago tribal leader in connection with the prime riverbank land on which the city is located.

FURTHER READINGS

Brent, Peter Ludwig. *Black Nile: Mungo Park and the Search for the Niger.* London : Gordon & Cremonesi, 1977.

Charlick, Robert B. *Niger: Personal Rule and Survival in the Sahel.* Boulder: Westview Press, 1991.

Chilson, Peter. *Riding the Demon: On the Road in West Africa.* Athens: University of Georgia Press, 1999.

Hollett, Dave. *The Conquest of the Niger by Land and Sea: From the Early Explorers and Pioneer Steamships to Elder Dempster and Company.* Abergavenny, Gwent, Wales: P. M. Heaton, 1995.

Mbendi Information Services. *Niger.* http://www.mbendi.co.za/land/af/ni/p0005.htm (Accessed May 11, 2002).

Miles, William. *Hausaland Divided: Colonialism and Independence in Nigeria and Niger.* Ithaca, NY: Cornell University Press, 1994.

ThinkQuest. *The Living Africa.* http://library.thinkquest.org/16645/contents.html (Accessed May 29, 2002).

Watson, Jane Werner. *The Niger: Africa's River of Mystery.* Champaign, Ill.: Garrard, 1971.

Nigeria

- **Area:** 356,669 sq mi (923,768 sq km) / World Rank: 33
- **Location:** Northern and Eastern Hemispheres, in western Africa, bordering the Gulf of Guinea, southwest of Chad, west of Cameroon, east of Benin, south of Niger.
- **Coordinates:** 10°00′N, 8°00′E
- **Borders:** 2,514 mi (4,047 km) / Chad, 54 mi (87 km); Cameroon, 1,050 mi (1,690 km); Benin, 480 mi (773 km); Niger, 930 mi (1,497 km)
- **Coastline:** 530 mi (853 km)
- **Territorial Seas:** 12 NM

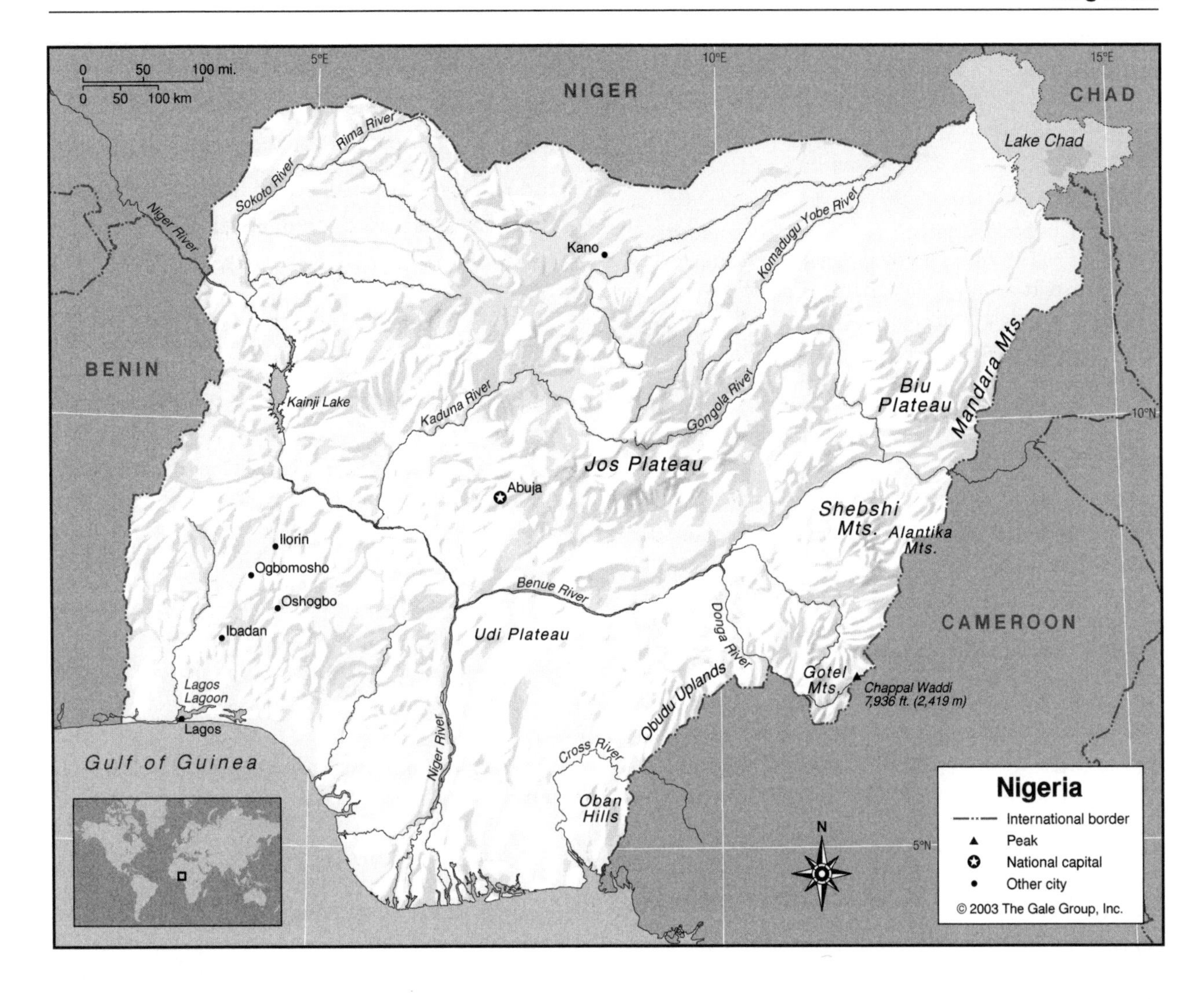

- **Highest Point:** Chappal Waddi, 7,936 ft (2,419 m)
- **Lowest Point:** Sea level
- **Longest Distances:** 700 mi (1,127 km) E-W; 650 mi (1,046 km) N-S
- **Longest River:** Niger River, 2,600 mi (4,184 km); inside Nigeria, 730 mi (1,175 km)
- **Largest Lake:** Lake Chad, size varies seasonably from 3,000 to10,000 sq mi (8,000 to 26,000 sq km)
- **Natural Hazards:** Periodic droughts, flooding
- **Population:** 126,635,626 (July 2001 est.) / World Rank: 10
- **Capital City:** Abuja, in central Nigeria
- **Largest City:** Lagos, on the Gulf of Guinea coast, population 8,000,000 (2000 est.)

OVERVIEW

Nigeria is the easternmost of the countries that face the Gulf of Guinea in the West African bulge. It sits on the center of the African Tectonic Plate and lies entirely in the tropics, its southern edge being only a few degrees above the equator and its northern border well below the tropic of Cancer. It is slightly more than twice the size of California.

The outstanding geographic feature of the country is the basin of the Niger and Benue rivers, running east and west through the center of the country. South of the basin the elevation is generally under 1,000 ft (304 m) except for some plateau surfaces. To the north of it is a broad plateau region that occupies the country to its northern border with elevations of 1,000 to over 4,000 ft (304 m to 1,219 m). On the east the country is bordered by mountainous regions, in which the highest point is located. At least 12 major geographic divisions are distinguishable, stretching in generally east-west zones across the country.

Nigeria is at odds with Cameroon, Niger, and Chad over the delimitation of international boundaries in the vicinity of Lake Chad. A proposed settlement is currently awaiting ratification by each nation. It is also in dispute with Cameroon and Equatorial Guinea over demarcation of their tri-state maritime boundary in the oil-rich Gulf of Guinea.

MOUNTAINS AND HILLS

Mountains

Nigeria's boundary with Cameroon is characterized for about 500 mi (804 km) by mountainous country. The northern part of the highlands consists of several hill groups, with high points around 3,600 ft (1,097 m). To the south of these are the Mandara Mountains, a dissected plateau with a general elevation of about 4,000 ft (1,219 m) encompassing an area of some 300 mi (482 km) in length with an average width of about 20 mi (32 km).

The central part of the border region consists of the Adamawa Highlands, a discontinuous series of mountain ranges and high plateau surfaces situated between the Benue River valley and the Donga River valley. They include the Alantika Mountains along the border and, separated in the west by a lower plains area, the Shebshi Mountains. The Shebshi Mountains, generally at an elevation of 3,500 ft (1,066 m), are a dissected plateau with highly eroded lower slopes. The highest surveyed point in the country, Chappal Waddi, at a height of 7,936 ft (2,419 m), is located in these hills. To the southwest of the Adamawa Highlands lies the Nigerian section of the Bamenda Highlands, 4,000 ft (1,219 m) in elevation. The Gotel Mountains rise up along the southeastern border with Cameroon.

Plateaus

With the exception of the coastal plains and the Niger-Benue valley, Nigeria consists mostly of high plains and plateaus. Directly to the east of the Lower Niger Valley are the Udi and Igala plateaus, and the Akwa-Orlu Uplands. The general elevation of these plateaus is about 1,000 ft (304 m) above sea level, with escarpments rising considerably higher. Between the low western coastal plains and the Niger-Benue River Valley lie the Western High Plains, or Plateau of Yorubaland, part of the belt of high plains that extends through West Africa. Plateau surfaces vary in elevation from about 750 ft to 1,200 ft (228 m to 372 m), with some dome-shaped hills attaining a height of 2,000 ft (609 m) above sea level.

North of the Niger-Benue valley lies a broad plateau, the Northern High Plains or the High Plains of Hausaland. The central section of the plateau extends for about 300 mi (482 km) from east to west with stepped plains ranging from about 600 ft (183 m) above sea level at the outer edge to roughly 3,000 ft (914 m) in the area surrounding the Jos Plateau. The Jos Plateau covers an area of about 3,000 sq mi (7,770 sq km), separated from the surrounding plains area by pronounced escarpments. The area's general elevation is over 4,000 ft (1,219 m) above sea level, and hills in its eastern part attain heights of over 5,800 ft (1,767 m).

The Biu Plateau to the east of the Gongola River basin covers about 2,000 sq mi (5,180 sq km). The upper level of the plateau, from 2,000 to 3,000 ft (609 to 914 m), is separated from the Northern High Plains by a pronounced scarp. Inactive volcanic cones are found in the northern part of the area.

GEO-FACT

Although little archeological research has been done to this point in Nigeria, the earliest fossil skeleteon having negroid features was discovered at Iii Ileru. This fossil dates back 10,000 years, testifying to the antiquity of civilization in the area.

INLAND WATERWAYS

Lakes

In the northeast corner, a portion of Lake Chad lies within Nigeria's borders. Lake Chad alternately advances and recedes over considerable distances in the flat plains area on the Nigerian side. Between December and January, the lake may cover up to 10,000 sq mi (25,900 sq km). However, during the ensuing months it may diminish to less than half that size with depths of only 4 to 16 ft (1.2 to 5 m). Little water is supplied to the lake from rivers in Nigeria; its principal source is the Chari River in the Republic of Chad.

In the far western part of the country is Kainji Lake, formed in 1968 by the damming of the Niger River. Kainji Lake was developed as a combined hydroelectric power and river navigation project. The lake itself extends for about 85 mi (137 km) in a section of the Niger River valley from Kainji to a point beyond Yelwa and has a width of from 9 to 15 mi (14 to 24 km). At maximum level it covers an area of about 480 sq mi (1,243 sq km).

Rivers

The valleys of the Niger and Benue rivers, which account for most of the country's drainage, form a great east-west arc approximately across the middle of the country. The Niger River valley extends from the border with Benin on the west, and the Benue River valley extends from the eastern border with Cameroon. Near Lokoja in the center of the country the two rivers join and change course to flow southward to the Gulf of Guinea. The Niger travels some 1,800 mi (2,896 km) from its source in Guinea through Mali, Niger, and Benin, and traverses Nigeria for approximately 730 mi

(1,175 km). The Benue, which rises in Cameroon, flows about 495 mi (796 km) inside Nigeria to its confluence with the Niger River.

The most important river outside this system is the Cross River in the southeast. The Cross originates in southern Cameroon and enters the country through the Eastern Highlands. It was a major route for the slave trade.

South of the Western High Plains several rivers flow directly into the Gulf of Guinea or its fringe lagoons. In the north from the Jos Plateau radiate rivers that flow in the direction of Lake Chad, or into the Niger-Benue system including the Sokoto, Kaduna, Rima, Komadugu Yobe, and Gongola rivers.

Wetlands

Wetlands are found along the coastal lagoons and Niger Delta, the Niger River flood plains, and in the Lake Chad basin. On the northern edges of the coastal lagoons, many smaller rivers lose themselves in freshwater swamps. Open flood plains extend between Yelwa and Jebba in the Niger valley, and eastward from Jebba to the confluence of the Niger and Benue rivers extensive swampy plains are found up to 50 mi (80 km) wide. Extremely low gradients in the Lake Chad basin impede river-flow, and during much of the year the flood plains are swampy.

THE COAST, ISLANDS, AND THE OCEAN

Oceans and Seas

Nigeria faces the Gulf of Guinea in the Atlantic Ocean with the Bight of Benin to the west and the Bight of Biafra to the southeast.

Major Islands

Islands of solid ground within the Niger delta are occupied by populated settlements, and Lagos is located on a group of islands at the western end of Lagos Lagoon.

The Coast and Beaches

Low swampy land is part of the coastal belt extending along the entire Gulf of Guinea coast of West Africa, which varies up to 20 mi (32 km) or more in width. The outer edge of the coastal area consists of sand spits and changes to mud as the coast nears the Niger Delta. Behind the outer spits and lagoons creeks of varying size generally parallel the coast and form a continuous waterway from the border with Benin on the west to the tributaries of the Niger Delta in the east. Unlike the westerly coastal areas, longitudinal lagoons are completely absent to the east of the Niger Delta.

Niger Delta

One of the major features of the West African coastline is the Niger Delta, which projects into the Gulf of Guinea from the southern coast of Nigeria. This great bulge of sedimentary material, deposited by the Niger River, stretches some 75 to 80 mi (120 to 128 km) from its apex below the town of Aba to the sea; it covers an area of about 10,000 sq mi (25,900 sq km). The water of the Niger flows through this delta in a series of radial tributaries. For navigational purposes, the two most important are the Forcados and Nun rivers.

The outer edge of the delta is fringed by sand spits and ridges, varying in width from a fraction of a mile to 10 or more miles (16 or more km). Behind these ridges are mangrove swamps covering about 4,000 sq mi (10,360 sq km), and farther inland an extensive area of freshwater swamps is found. The delta is the site of large natural gas and oil deposits.

CLIMATE AND VEGETATION

Temperature

The climate varies from equatorial in the south, tropical in the center, to arid in the north, with midday temperatures that surpass 100°F (38°C), but relatively cool nights, dropping as low as 54°F (12°C). On the Jos Plateau, tempreatures are more moderate. Near the coast temperatures rarely exceed 90°F (32°C), but humidity is high and nights are hot.

Rainfall

Inland there are two distinct seasons: a wet season from April to October with generally lower temperatures, and a dry season from November to March with hotter temperatures. Along the coast annual rainfall varies from about 70 in (180 cm) in the west to about 170 in (420 cm) in certain parts of the east. Inland it decreases to around 50 in (130 cm) over most of central Nigeria and only 20 in (50 cm) in the extreme north.

Grasslands

The uppermost levels of the Obudu Uplands and the Oban Hills, westward extensions of the Bamenda highlands, are covered grasslands. The Western High Plains are covered largely with savanna parkland and grass.

Forests and Jungles

Tropical rain forests form a belt roughly 80 mi (130 km) wide across the southern zone with trees such as African mahogany, irokol. African walnut, and obeche reaching heights of 200 ft (60 m). These forests are found in the Obudu Uplands and the Oban Hills in the east and the plains in Western State, whose physical features are often masked by vegetation. The central and western sections in Mid-Western State with gentle slopes and elevations mostly under 400 ft (122m) contain extensive and luxuriant forest areas in protected reserves. Mangroves dominate the coast, while freshwater swamp forests with palms, abura, and mahogany predominate the area immediately inland. Beyond the tropical belt to the north

Population Centers – Nigeria

(2000 POPULATION ESTIMATES)

Name	Population
Lagos	8,665,000
Ibadan	1,549,000

SOURCE: "World Urbanization Prospects: The 2001 Revision," United Nations Population Division, and projections from United Nations Statistics Division.

are found tall grasses and deciduous trees of small stature, characteristic of the savanna.

HUMAN POPULATION

Nigeria is by far the most populated country in Africa, having nearly double the population of Egypt, the second most populated African nation. The overall population density is 324 people per sq mi (125 people per sq km); however a large percentage of the population lives within a couple hundred miles of the coast, leaving the majority of the country sparsely populated. Population densities in the south near Lagos and the rich agricultural lands around Enugu and Owerri far exceed 1,036 inhabitants per sq mi (400 inhabitants per sq km). Migration to urban areas has increased over the past decades, leading to much overcrowding. The city of Lagos itself has a population density of 7,459 per sq mi (2,880 per sq km).

NATURAL RESOURCES

Nigeria is rich in tin, columbite, iron ore, coal, limestone, lead, and zinc. The Jos Plateau is the site of tin and other metals that have made the region economically important. The delta region is rich in natural gas and petroleum, and this makes up Nigeria's chief product. Important coal-bearing formations are exposed in the scarps of the Akwa-Orlu Uplands and the Udi and Igala plateaus east of the Lower Niger Valley.

FURTHER READINGS

Achebe, Chinua. *Anthills of the Savannah.* New York: Doubleday, 1987.

Africa South of the Sahara, 2002. "Nigeria." London: Europa, 2001.

Floyd, Barry. *Eastern Nigeria: A Geographical Review.* London: Macmillan, 1969.

Grove, Alfred Thomas. *The Changing Geography of Africa.* Oxford: Oxford University Press, 1989.

Soyinka, Wole. *The Open Sore of a Continent: A Personal Narrative of the Nigerian Crisis.* New York: Oxford University Press, 1996.

Udo, Reuben K. *Geographical Regions of Nigeria.* London: Heinemann, 1970.

US Energy Information Administration. *West African Gas Pipeline (WAGP) Project.* http://www.eia.doe.gov/emeu/cabs/wagp.html (Accessed May 2002).

Norway

- **Area:** 125,182 sq mi (324,220 sq km) / World Rank: 68
- **Location:** Northern and Eastern Hemispheres in Northern Europe, on the Scandinavian peninsula, west of Sweden and Russia, northwest of Finland, east of the Norwegian Sea and the North Sea, north of the Skagerrak Strait, south of the Barents Sea and the Arctic Ocean / Has island dependencies in the Arctic and Southern Oceans.
- **Coordinates:** 62°00′N, 10°00′E
- **Borders:** 1,562 mi (2,515 km) total / Finland, 453 mi (729 km); Sweden, 1,006 mi (1,619 km); Russia, 104 mi (167 km)
- **Coastline:** 13,594 mi (21,925 km, estimated)
- **Territorial Seas:** 4 NM
- **Highest Point:** Galdhøpiggen, 8,100 ft (2,469 m)
- **Lowest Point:** Sea level
- **Longest Distances:** 1,089 mi (1,752 km) NNE-SSW / (267 mi (430 km) ESE-WNW
- **Longest River:** Glåma River, 372 mi (598 km)
- **Largest Lake:** Lake Mjøsa, 140 sq mi (362 sq km)
- **Natural Hazards:** Avalanches and rockslides, extreme cold
- **Population:** 4,503,440 (July 2001 est.) / World Rank: 114
- **Capital City:** Oslo, located in the southeast region of the country
- **Largest City:** Oslo, 507,467 (2000 est.).

OVERVIEW

A mountainous country dominated by glaciers, Norway contains some of the oldest rocks on earth. But it is Norway's coastline, cut with countless fjords, for which the country is most famous. The fjords, bays deeply indented in the mainland and often bounded by high plateaus, form breathtaking natural harbors. Some 1,700 glaciers totaling 3,400 sq km cover Norway's interior. Average elevations are in excess of 1,500 ft (457 m) and only one-fifth of the country's total area sits lower than 500 ft (150 m) above sea level. Almost one-third of the country sits north of the Arctic Circle.

Norway consists of five geographic regions. West Country (Vestlandet) is an area carved by glaciers and features majestic fjords and the abrupt slope of the western Scandinavian mountains toward the North Sea. Connected to the West Country by numerous valleys, the East Country (Ostlandet), along the more moderate eastern slopes of these southern mountain ranges, contains rolling hills and valleys in which is found some of the country's richest agricultural soil. The Trondheim (Trøndelag) Depression forms a natural boundary between the northern and southern halves of the country. It is a region of hills, valleys, and fjords north of the high mountain ranges. Further north is North Norway (Nord Norge), which is marked by fjords, mountains, vast snowfields, and some of Europe's largest glaciers. In the far south is an area of agricultural lowlands known as South Country (Sorlandet).

MOUNTAINS AND HILLS

Mountains

Norwegian mountain ranges are roughly divided into three groups. The Kjølen in the north are the greatest of the three groups. They form a natural barrier between Norway and Sweden and extend northward toward the border with Finland. Further south are the Dovrefjell, which abut the Trondheim Depression in the south and along with it splits the northern and southern areas of the country.

Norway's highest mountain ranges, the Langfjell, lie south of the Trondheim Depression and the Dovrefjell in the broadest area of the country. The ranges, comprised of sharp peaks called fjells and high plateaus called vidder, run southwest to northeast, and divide the West and East countries. Among them are the Rondane Mountains and the Jotunheimen. Within the latter is Galdhøpiggen, Scandinavia's highest mountain at 8,100 ft (2,469 m).

Plateaus

Ice Age glacial movement carved countless plateaus into the Norwegian landscape, some of them very large. The Norwegian Plateau includes the western and eastern vidders in the high mountains of the central region. Other major plateaus are the Finnmark Plateau in the far north, and the Hardangervidda in the south, a desert plateau with an elevation of 6,004 ft (1,830 m) and an area of 2,500 sq mi (6,474 sq km), with steep sides scarred and grooved by waterfalls and valleys.

Glaciers

Most of the northern end of the Norwegian Plateau in the country's central region is covered by icecaps. Among the glaciers in this area are Jostedalsbreen, the largest glacier in Europe at 580 sq mi (1,502 sq km) in area and possibly 1,500 ft (457 m) thick, and Folgefonn, the top of which is over 5,000 ft (1,524 m) above sea level. Norway's northern extremes, including the Finnmark Plateau, are also heavily glaciated. Other large snowfields including Hallinskarvet in the Hardangervidda, Snohetta in the Dovrefjell, Seiland near Hammerfest, and Oksfjordjokel near Kvanangen.

INLAND WATERWAYS

Lakes

Glacial lakes abound in Norway. Nearly one-twelfth of the country is under fresh water, sometimes so deep that the waterbed is far below the level of the sea. The Mjøsa, at 140 sq mi (363 sq km) in area and 1,982 ft (452 m) in depth, is by far the largest of these lakes. The levels of the lakes vary by as much as 1,300 ft (316 m) in altitude. Most of the larger lakes are 400 ft (122 m) above the sea and were perhaps heads of fjords that have since been sealed off from the ocean.

Rivers

Norway has numerous glacier-fed rivers. Most are swift and turbulent, rushing through steep valleys and rocky gorges. The only navigable rivers are the Glåma and the Dramselva. The Glåma is the longest river in Scandinavia at 350 mi (563 km). It rises more than 2,000 ft (610 m) above sea level at Aursunden Lake and flows south into the Skagerrak. Many lakes widen the stream, and the river is famous for its waterfalls. The Dramselva rises in the central part of the country and also flows south entering Oslo Fjord at Drammen. Other major rivers in the south are the Otra, Sira, and the two Lågen Rivers. In the extreme north are the Reisa and the Tana.

THE COAST, ISLANDS, AND THE OCEAN

Oceans and Seas

Norway's long coastline borders on many bodies of water. In the south is the Skagerrak, separating the country from Denmark. The Atlantic Ocean lies to the west, in the form of the North Sea in the southwest between Norway and England, and the Norwegian Sea along the northwestern coast. In the north is the Barents Sea, an extension of the Arctic Ocean.

Major Islands

Except in the southwest and the far north, the Norwegian coast has a girdle of islands called the skjaergard. Containing roughly 50,000 islands total, the island zone reaches its broadest width of over 37 mi (60 km) at the southern approaches to the Trondheim Fjord. The outer islands, protruding from relatively shallow waters, rarely exceed 100 ft (30 m) in height, while the inner islands

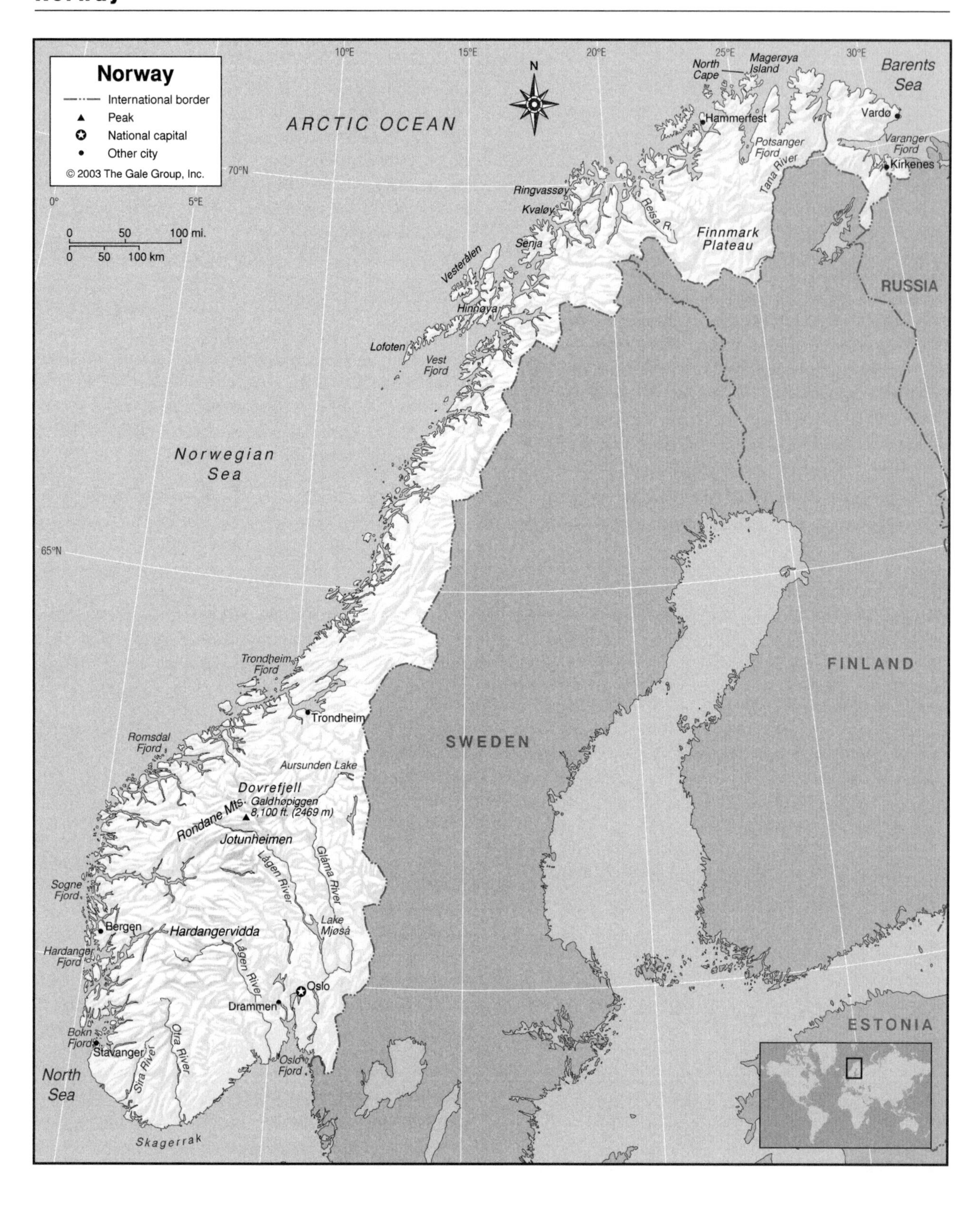

may rise to 1,000 ft (305 m). These islands are characterized by a series of rock terraces known as strandflats.

The Lofoten and Vesterålen islands off the northwestern coast are the country's most extensive island chains. They are formed from glaciers that covered the tops of partially submerged ancient volcanic ranges. The larger islands of Hinnøya, Kvaløy, Senja, and Ringvassøy also lie off the northwest coast.

Island Dependencies

The Svalbard archipelago, which includes the Spitzbergen archipelago, North-East Island, Edge Island, and Barents Island, is located north of Norway in the

Arctic Ocean. They have an area of 24,208 sq mi (62,700 sq km). Ice sheets and permafrost cover most of the islands, but they are also the sites of the most northerly permanent settlements in Europe. A dependency of Norway since 1925, Svalbard is at the center of a maritime border dispute between Norway and Russia.

The Beerenberg volcano (7,470 ft / 2,277 m), the world's northernmost active volcano, created uninhabited Jan Mayen Island in the Arctic Ocean northeast of Iceland. Jan Mayen has an area of 144 sq mi (373 sq km). Long dormant, Beerenberg erupted anew in 1970.

Bouvet Island was claimed for Norway in 1927. It is located far to the south, in the South Atlantic Ocean, between Africa and Antarctica. It is uninhabited and is almost completely covered by ice. Norway has additionally claimed Peter I Island off the Antarctic coast, and is one of many countries that have put its claims on the continent itself in abeyance under the Antarctic Treaty.

The Coast and Beaches

Except in the southernmost part of the country, Norway's coastline is extremely irregular. Deep fjords extend far into the interior of the country in many places. These troughs were incised by glaciers into the interior plateau. Their central stretches are of great depth, but their mouths are relatively shallow and characterized by deltaic flats. The fjord zone reaches its greatest breadth along the west coast, where the longest and deepest fjord, Sogne Fjord (Sognafjorden), is located. Sogne Fjord is approximately 127 mi (204 km) long. Its walls rise sharply from the coast, reaching elevations of 5,000 ft (1,500 m) in some places. Other major fjords in the south include Oslo Fjord, Hardanger Fjord, Bokn Fjord, and Romsdals Fjord. In the middle of the country the Trondheim Fjord extends 78 mi (126 km) into the interior. Arctic fjords such as Tana, Porsangen, and Varanger tend to be broader and somewhat shorter.

In the far north of Norway, on Magerøya Island, is North Cape (Nordkap). This is the northernmost point in all of Europe.

Population Centers – Norway

(2001 POPULATION ESTIMATES)

Name	Population
Oslo (capital)	773,498
Bergen	205,759
Stavanger/Sandnes	162,083
Trondheim	140,631

SOURCE: Statistics Norway.

Counties – Norway

POPULATIONS FROM 1992 CENSUS

Name	Population	Area (sq mi)	Area (sq km)	Capital
Akershus	467,052	1,898	4,917	Oslo
Aust-Agder	102,178	3,557	9,212	Arendal
Buskerud	236,811	5,763	14,927	Drammen
Finnmark	74,059	18,779	48,637	Vardø
Hedmark	187,103	10,575	27,388	Hamar
Hordaland	435,219	6,036	15,634	Bergen
Møre og Romsdal	243,158	5,832	15,104	Molde
Nord-Trøndelag	127,108	8,673	22,463	Steinkjer
Nordland	239,109	14,798	38,327	Bodø
Oppland	182,701	9,753	25,260	Lillehammer
Oslo	507,467	175	454	Oslo
Østfold	248,217	1,615	4,183	Moss
Rogaland	373,210	3,529	9,141	Stavanger
Sogn og Fjordane	107,589	7,195	18,634	Leikanger
Sør-Trøndelag	262,852	7,271	18,831	Trondheim
Telemark	165,038	5,913	15,315	Skien
Troms	151,160	10,021	25,954	Tromsø
Vest-Agder	155,691	2,811	7,281	Kristiansand
Vestfold	212,775	856	2,216	Tønsberg

SOURCE: Statistics Norway.

CLIMATE AND VEGETATION

Temperature

The warm waters of the Gulf Stream coupled with prevailing westerly winds moderate temperatures along the west and southwest coast, where high temperatures average 38°F (3°C) in January and 66°F (19°C) in July. The climate is more extreme and temperature ranges are broader in Norway's interior. The arctic north is much colder than the south, but even here the Gulf Stream keeps temperatures relatively warm and the coast ice-free. Oslo, in the southern interior, has an average high temperature of 82°F (28°C) in July and 41°F (5°C) in January.

Known as the land of the midnight sun, the far north has 24-hour daylight from May through July. Oslo and the rest of the southern region have summer daylight from about 4 a.m. to 11 p.m. Conversely, from November to the end of January the sun never rises above the horizon in the north.

Rainfall

The coastal areas of the west, affected most by Atlantic weather disturbances, receive almost year-round rainfall. Some areas average 130 in (330 cm). Precipitation is not as great in the interior, but it is increasing enough that its works with the cooler summers to increase the size of the glaciers. Oslo, in the southern interior, averages 30 in (760 mm) of precipitation a year.

Grasslands

Arable land is sparse, located primarily in the far south and southwest. Only 3 percent of the country's total area is used for farming. Potatoes and grains, particularly wheat, oats, and barley, are grown in the East Country and Trøndelag. Livestock and dairy farming are found in the west and north.

Tundra

Treeless heath covers the far northern region and much of the high plateaus. Temperatures here are so cold that the ground is frozen most or all of the year. Hardy dwarf shrubs and wildflowers are the predominant vegetation in the tundra areas.

Forests

Forests cover 27 percent of the country. Deciduous forests of maple, ash, oak, elm, and hazel are located in the southern and southwestern coastal regions. Boreal coniferous forests of Scotch pine and Norway spruce cover the central and eastern coastal areas.

HUMAN POPULATION

Norway's population of 4,503,440 (2001 est.) is virtually homogeneous. Norwegians are almost all of Germanic (Nordic, Alpine, and Baltic) descent. Some 20,000 Lapps (Saami) and 7,000 descendants of Finnish immigrants constitute the only notable minority groups. Most Norwegians live in the cities of the south.

NATURAL RESOURCES

Norway is rich in natural resources, particularly oil and natural gas. Oil platforms off Norway's coast in the North Sea have made it one of the world's largest oil exporters. Other mineral resources include iron ore, zinc, lead, coal, copper, pyrites, and nickel. In addition, Norway's waters provide a rich harvest of fish, and much good timber is available in its forests. With heavy rainfall and rapid glacier-fed rivers, hydropower is extremely abundant.

FURTHER READINGS

Galenson, Walter. *A Welfare State Strikes Oil: The Norwegian Experience.* Lanham, Mass.: University Press of America, 1986.

Jones, Gwyn. *A History of the Vikings.* New York: Oxford University Press, 1984.

Selbyg, Arne. *Norway Today: An Introduction to Modern Norwegian Society.* New York: Oxford University Press, 1987.

Stewart, Janice S. *The Folk Arts of Norway.* 3rd edition. New York: Nordhus, 1999.

Zickgraf, Ralph. *Norway.* Broomall, Penn.: Chelsea House, 1999.

Oman

- **Area:** 82,031 sq mi (212,460 sq km) / World Rank: 84
- **Location:** Northern and Eastern Hemispheres, in the Middle East, on the southeast edge of the Arabian Peninsula bordering the Arabian Sea, Persian Gulf, and Gulf of Oman, northeast of Yemen, east of Saudi Arabia, and southeast of the United Arab Emirates.
- **Coordinates:** 21°00′N, 57°00′E
- **Borders:** 854 mi (1,374 km) / Yemen, 179 mi (288 km); Saudi Arabia, 420 mi (676 km); United Arab Emirates, 255 mi (410 km)
- **Coastline:** 1,300 mi (2,092 km)
- **Territorial Seas:** 12 NM
- **Highest Point:** Jabal Sham, 9,957 ft (3,035m)
- **Lowest Point:** Sea level
- **Longest Distances:** 604 mi (972 km) NE-SW; 319 mi (513 km) SE-NW
- **Longest River:** There are no perennial rivers
- **Largest Lake:** There are no perennial lakes
- **Natural Hazards:** Sandstorms and dust storms, drought
- **Population:** 2,622,198 (July 2001 est.) / World Rank: 136
- **Capital City:** Masqat, on the northeastern coast of the Gulf of Oman
- **Largest City:** Masqat, population 635,000 (2000 est.)

OVERVIEW

The sultanate of Oman is located in the extreme southeastern corner of the Arabian Peninsula. It is bordered by the United Arab Emirates to the northwest, Saudi Arabia to the north and west, Yemen to the southwest, the Gulf of Oman to the northeast, and the Arabian Sea to the southeast and east. Oman is mostly desert, but 15 percent is mountainous. Oman consists of four major regions: Musandam Peninsula, the Al Batinah coastal plain, Oman interior (mountain range/plateau), and Dhofar region. Furthest north is the tip of the Musandam Peninsula and the Ras Al-Jabal, a low mountain range. The fertile coastal plain, Al-Bātinah, slopes to the foothills of the Western Hajar. The Al-Hajar Mountain range is the highest in eastern Arabia. The Dhofar (Zufar) region is a lush vegetated coastal plain giving rise to the Al-Qarā Mountains. Owing to its climate and geography, Oman's most pressing issue is the maintenance of an adequate supply of water for domestic and agricultural use.

MOUNTAINS AND HILLS

Mountains

The Hajar (the Rock) mountains form two ranges: the Hajar al-Gharbi, or Western Hajar, and the Hajar al-Shargi, or Eastern Hajar. They are divided by the Wadi Sanā'il, a valley that forms the traditional route between Masqat and the interior. The general elevation is about 4,000 ft (1,219 m) but the peaks of a high limestone massif known as the Jabal Akhdar (Green Mountain) lie between the western and eastern portions of the range rising to nearly 10,000 ft (3,048 m) above sea level in some places. Jabal Akhdar has a breadth from north to south varying from 6 to 25 mi (10 to 40 km) and a maximum height of 9,957 ft (3,035m) in Jabal Sham. The mountains of Al Hajar Ash Sharqi have a maximum elevation of 7,059 ft (2,152m). This is the only habitat in eastern Arabia above 6,658 ft (2,030m) in elevation. While these spectacular mountains form a breathtaking landscape of dramatic peaks and precipices, they also provide important protection for endemic and relict species of plants and animals, mostly of Indo-Iranian origin. Prior to 1961 little was known of the flora and fauna of Al Hajar's montane woodlands as inhospitable tribal politics and terrain limited scientific research.

Plateaus

The coastal plain blends into an area of hills, which in turn gives way to a plateau with an average height of about 300 m (1,000 ft). It is mostly stony and waterless, arable only at oases, extending to the sands of the Rub'al Khālī Desert. Inland, in the Al-Wusta region is the desert plateau, Jiddat al-Harāsīs, encompassing the Al-Huqf Depression. It is here that the last sightings of the Arabian Oryx in the wild were recorded, whose numbers were depleted drastically by hunters. In 1974 the Arabian Oryx Reserve was established there.

INLAND WATERWAYS

Rivers

In the Hajar and its foothills a small number of wadis (gullies/watercourses) are found. However, there are no perennial rivers in Oman. Much of the water that flows in these temporary runoffs only flows during the wet season, and what does not make it to the sea is quickly evaporated.

THE COAST, ISLANDS, AND THE OCEAN

Oceans and Seas

Oman borders the Arabian Sea and the Gulf of Oman, the latter of which separates Arabia from the Middle East. Evidence based on comparison of the plant associations in the Al-Hajar range to those from southeastern Iran suggest migrations that in the past occurred when sea levels fell to 394 ft (120 m) below present levels, resulting in a land bridge between Asia and Arabia.

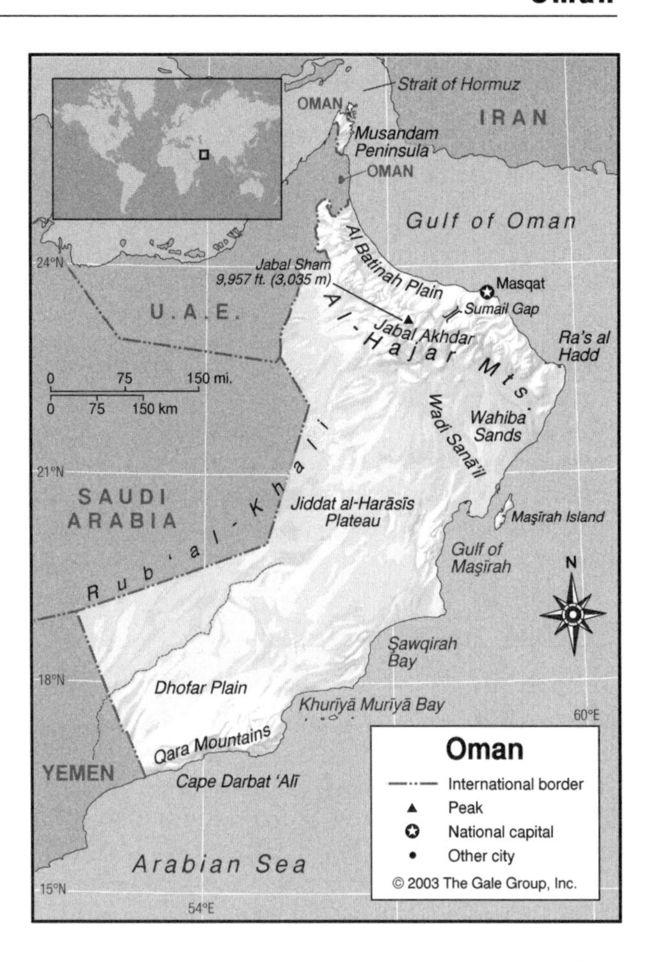

Major Islands

Along the Arabian Sea length of coast and separated from it by about 10 mi (16 km) is the barren island of Maşīrah. It is 40 mi (64 km) long, virtually uninhabited. More importantly, it contains the world's largest nesting population of loggerhead turtles, with between 23,000 and 30,000 females nesting each year.

The Coast and Beaches

The Bātinah Plain runs 167 mi (270 km) in length between the foothills of the Hajar Mountains and the Gulf of Oman, varying in width from 2 to 19 mi (3 to 30 km). It is scored along its length by wadis (watercourses/gullies) from the mountains. The Bātinah plain has inlets of water (khors) not infrequently containing mangrove stands creating significant habitats for birds. An extremely rugged area exists where two inlets, the Elphinstone and Malcom, cut into the coastline south of the Strait of Hormuz. Towering cliffs of 3,000 to 4,000 ft (914 to 1,219 km) enclose the Elphinstone in height. The coastline on the Arabian Sea is rather jagged and indented. Ra's al Hadd is the point sticking out into the ocean that separates the Gulf of Oman from the Arabian

GEO-FACT

There are many caverns in Oman. One of the largest in the world, Teyq Cave, is 820 feet (250m) deep and 10,595 cubic feet (300 m^3) in size.

Sea. From this point the coast is rather straight going south until it cuts sharply in near Maşīrah Island to the Gulf of Maşīrah. The coastline then sweeps in and out forming two more bays, Şawqirah Bay and Khurīyā Murīyā Bay, before terminating at Cape Darbat 'Alī. The coast southward to Dhofar is particularly barren and forbidding.

CLIMATE AND VEGETATION

Temperature

Oman's climate is arid subtropical. However, the climate differs somewhat from one region to another. The interior is generally very hot, with temperatures reaching 129°F (54°C) in the hot season from May to October. The coastal areas are hot and humid from April to October. The prevailing summer wind, the Gharbi, makes the heat more oppressive. In the south, the Dhofar region has a more moderate climate.

Rainfall

Average annual precipitation is 2 to 4 in (5 to 10 cm), depending on the region and the prevailing summer wind. Annual rainfall in Masqat averages 4 in (10 cm), falling mostly in January. Dhofar is subject to the southwest monsoon, and rainfall up to 25 in (64 cm) has been recorded in the rainy season from late June to October. While mountain areas receive more plentiful rainfall, some parts of the coast, particularly near the island of Maşīrah, sometimes receive no rain at all within the course of a year. An unusual feature of Oman weather is an eastern coastal region characterized by dense fog. This region starts close to the Bar al-Hikman, opposite Maşīrah Island, and extends southward and inland 74 mi (120 km) over the stony plateau of the Jiddat Al-Harāsīs, and continuing southward over the escarpment Mahrāt Mountains of Dhofar and into Yemen. This is an area of little rainfall, but the dense fogs that can limit visibility to 33 ft (10m) are influenced by the southwest monsoon. The fog-affected escarpments of Dhofar are some of the most species-rich habitats in Oman.

Deserts

At the Sumail Gap the Sharqiya region begins. To the south are the isolated Wahiba Sands. Southeast lies the Jalaan, a vast sandy plain that stretches from the Arabian coast inland to meet the Wahiba Sands. Described as "perfect specimens of a sand of sea" this small 5,792 sq mi (15,000 sq km) desert stretches for 112 mi (80 km) north to south and 50 mi (80 km) east to west. The Wahiba Sands are the largest areas of lithified sand dunes in the world. Surface dunes that can reach an impressive height of 328 ft (100 m) cover the region. Their variety and formation are considered representative of the desert's evolution over millions of years. The Dhahira area and the southern region (Dhofar) border the Rub'al Khālī (Empty Quarter). Dhahira is a semi-desert plain, sloping from the southern flanks of the Western Hajar into the Rub'al Khālī. Dhofar's border with the Kingdom of Saudi Arabia runs through the Empty Quarter. Situated mainly in Saudi Arabia, the Rub'al Khālī is one of the largest sand deserts in the world. It covers an area of more than 250,000 sq mi (650,000 sq km) and extends to 744 mi by 310 mi (1,200 km by 500 km). It is one of the driest places on earth, receiving almost no rain at all.

Population Centers – Oman

(2000 POPULATION ESTIMATES)

Name	Population
Masqat (Muscat)	635,000
Nizwa	100,000

SOURCE: "World Urbanization Prospects: The 2001 Revision," United Nations Population Division, and projections from United Nations Statistics Division.

HUMAN POPULATION

Traditional Omani society consisted of four categories: seafarers who fished and traded; agriculturists of the Bātinah coast and the south, and those of the Interior who employ the aflaj (aqueduct) system of irrigation; the mountain people of Dhofar and the Musandam; and the Bedouin of the desert areas. In 2000 it was estimated that 84 percent of the population lived in urban areas. The area around Masqat and the Bātinah Coast have more than half the population.

NATURAL RESOURCES

Oman does not have the vast oil reserves of some of its neighbors; although the outlook for further reserves is promising, Oman's complex geology makes exploration and production a challenge. Natural resources that are found in Oman and exploited are petroleum, natural gas, and copper. Much of the country's industry and trade is based on these substances. Situated on the Arabian Sea, Oman also has a sizable fishing economy. Asbestos, limestone, chromium, and gypsum can be found in Oman's mountains. In addition, Oman's location is a resource: it is strategically located on the Musandam Peninsula adjacent to the Strait of Hormuz, a vital transit point for world crude oil, and thus derives income from controlling this trade route.

FURTHER READINGS

Advancing Women in Leadership. "Advancing Women." *International Business and Career Community.* http://www.advancingwomen.com/awl/spring99/awl_spring99.html (Accessed May 9, 2002).

Chatty, Dawn. *Mobile Pastoralists: Development Planning and Social Change in Oman.* New York: Columbia University Press, 1996.

Evans, M.I. *Important Bird Areas of the Middle East.* Cambridge, U.K.: Birdlife International, 1994.

Jiddat Al Harasis 026. http://www.meteorites21.com/j_a_h026.htm (Accessed May 11, 2002).

Kay, Shirley. *Enchanting Oman.* Dubai, United Arab Emirates: Motivate Publishing, 1988.

Middle East & Islamic Studies Collection. http://www.library.cornell.edu/colldev/mideast/universi.htm (Accessed May 10, 2002).

Middle East and Jewish Studies. *Women in the Middle East.* http://www.columbia.edu/cu/lweb/indiv/mideast/cuvlm/women.html (Accessed May 9, 2002).

Ministry of Information. *Sultanate of Oman.* http://www.omanet.com/ (Accessed May 10, 2002).

NASA. *Gemini Earth Photographs, Ar-Rub′ al-Kahli.* http://www.hq.nasa.gov/office/pao/History/SP-4203/phot01.htm (Accessed May 11, 2002).

Natural History of Oman & Arabia. http://www.oman.org/nath00.htm (Accessed May 10, 2002).

Newcombe, Ozzie. *The Heritage of Oman: a Celebration in Photographs.* Reading, Berkshire, U.K.: Garnet Publishing, 1995.

Oman Daily Observer. "Oman: People & Heritage." Oman: Oman Daily Observer, 1994.

Pakistan

- **Area:** 310,403 sq mi (803,940 sq km) / World Rank: 36
- **Location:** Northern and Eastern Hemispheres, in South Asia, between the Himalayan Mountains and the Arabian Sea, west of India, east of Iran and Afghanistan, and south of China.
- **Coordinates:** 30°00′N, 70°00′E
- **Borders:** 4,209 mi (6,774 km) / China, 325 mi (523 km); India, 1,809 mi (2,912 km); Iran, 565 mi (909 km); Afghanistan, 1,510 mi (2,430 km)
- **Coastline:** 650 mi (1,046 km)
- **Territorial Seas:** 12 NM
- **Highest Point:** K2, 28,251 ft (8,611m)
- **Lowest Point:** Sea level
- **Longest Distances:** 1,165 mi (1,875 km) NE-SW; 625 mi (1,006 km) SE-NW
- **Longest River:** Indus River, 1,988 mi (3,200 km)
- **Natural Hazards:** Earthquakes, flooding, drought, cyclones
- **Population:** 144,616,639 (July 2001 est.) / World Rank: 7
- **Capital City:** Islamabad, in the northeast
- **Largest City:** Karachi, in the southeast on the Arabian Sea, population 9.6 million (2001 estimate)

OVERVIEW

Pakistan can be divided into three major geographic areas: the northern highlands, the Indus River plain, and the Baluchistan Plateau. About one-third of the Pakistan-India frontier is the cease-fire line in the Jammu and Kashmir region, disputed between the two countries since their independence. Jurisdiction over Kashmir has been a matter of longstanding conflict between Pakistan and India, and has resulted in open warfare.

Pakistan lies at the border of three tectonic plates: the Arabian, Indian, and Eurasian. The Arabian Plate converges with the Eurasian Plate at the coastline in southeastern Pakistan. On Pakistan's eastern and northeastern border the Eurasian Plate collides with the Indian Plate. Seismic activity is high along this border, and the region surrounding Quetta is also prone to frequent and devastating earthquakes.

MOUNTAINS AND HILLS

Mountains

The northern highlands are a convergence of some of the most rugged, formidable mountains in the world. The Himalayas stretch from northeast India to the northeast corner of Pakistan, where they merge into the Karakoram and Pamirs mountain ranges. West of the Pamirs are the heights and steep valleys of the Hindu Kush.

In the northern mountains, virtually all elevations are higher than 8,000 ft (2,438 m) above sea level. More than fifty peaks are above 22,000 ft. (6,705 m). The soaring summits of K2 (Mount Godwin Austen) in the Karakoram Range, the world's second highest mountain (28,251 ft/8,611 m), and Nanga Parbat (26,660 ft/8,126 m) in the Himalayan Range, have been often deadly challenges to climbing expeditions. Enormous glaciers sprawl across this region, including Baltoro and Pasu, each more than 31 mi (50 km) long.

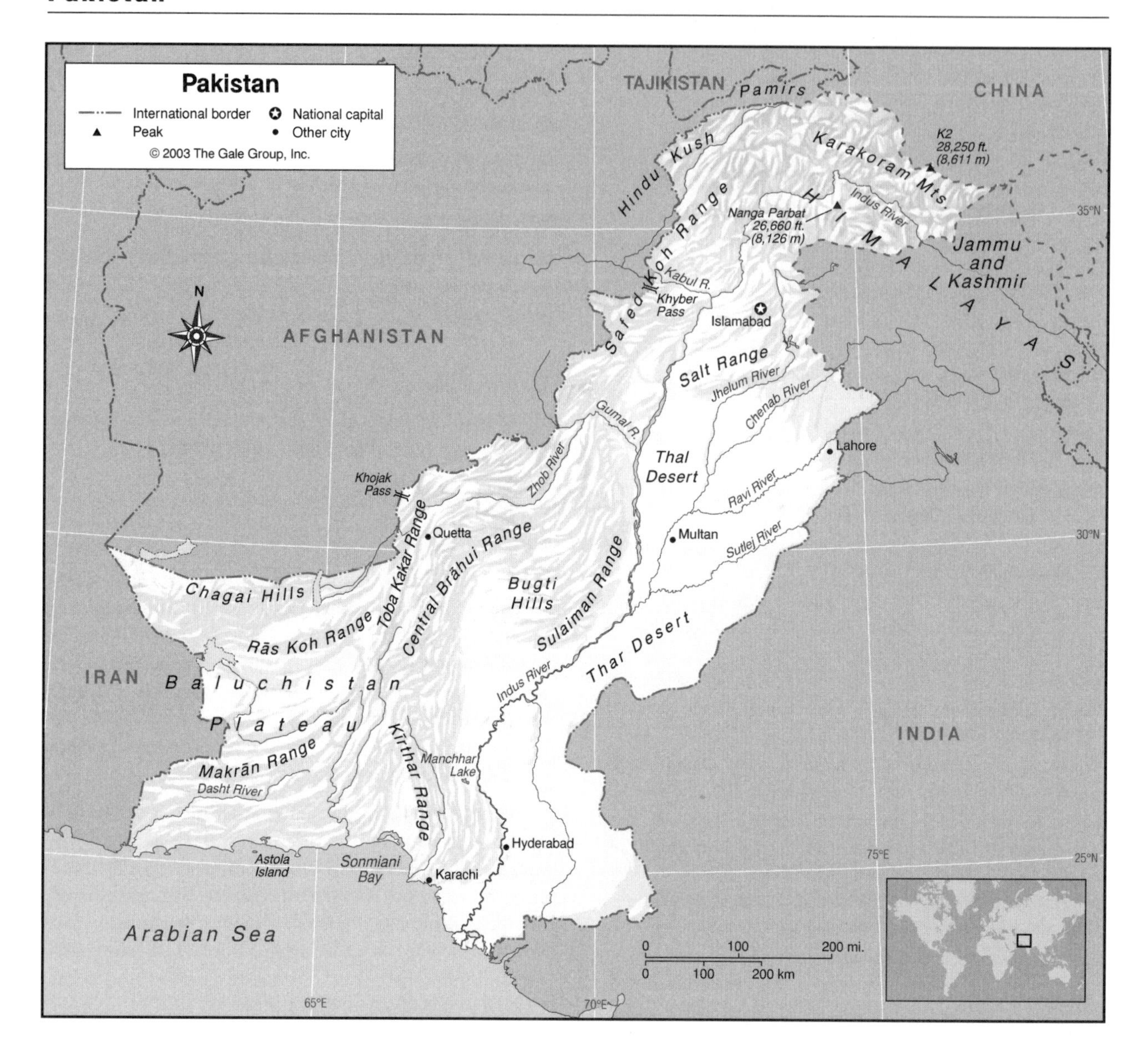

The Safed Koh Range south of the northern highlands and west of the Indus River plain reaches 15,620 ft (4,761 m) in its extension to the Afghanistan border. This arid scrubland includes the strategic Khyber Pass, which connects the Peshawar Valley to Afghanistan. South of the Safed Koh and clustered near the border are the mountains of Waziristan. Beyond them, the Toba Kakar Range of about 9,000 ft (2,743 m) average crest elevation extends from northern Baluchistan to the Khojak Pass. The Rās Koh Range west of the city of Quetta, and the Chagai Hills extending further west to the Pakistan-Iran-Afghanistan tri-point, complete the western highlands.

Hills

The Margalla Hills, 2,000 to 3,000 ft (610 to 914 m) high, are foothills of the northern mountains, overlooking Islamabad, the capital. The Swat and Chitral Hills in the northwest have heights of 5,000 to 6,000 ft. (1,524 to 1,829 m).

The Central Brāhui Range extends for 175 mi (282 km) south from Quetta and then divides into the Kīrthar Range that extends southeast, and the Makrān Range that reaches to the west as far as the Iranian border. Southeast from Quetta, the Bugti Hills merge into the Sulaiman Range, separating the country's east and west, with summits of 6,000 to 7,000 ft (1,828 to 2,133 m).

The Salt Range (containing coal and salt mines) rises between the Potwar Plateau and the fertile, irrigated plains of the upper Indus Valley. This roughly east-west range has some peaks of nearly 5,000 ft (1,524 m), but most do not exceed 2,500 ft (762 m).

Plateaus

The Baluchistan Plateau, at an elevation of 3,000 to 4,000 ft (914 to 1,219 m), is defined and enclosed by the western mountain ranges along the Afghan border and by those extending southwards from Quetta. The plateau is an arid tableland of approximately 135,000 sq mi (350,945 sq km) with interior drainage and dry lake beds. Large natural gas reserves lie beneath it. The Potwar Plateau in the region at the foot of the mountains south of Islamabad is a dry, eroded area where most of Pakistan's oil is located. Adjacent to Indian-administered Kashmir, the Deosai Plateau is a 1,337 sq mi (3,464 sq km) upland National Park that is a major bear habitat.

Canyons

Northern Pakistan has many narrow, twisting canyons, particularly in Hunza. The Indus River rushes through the steep Attock Gorge near the Khyber Pass.

INLAND WATERWAYS

Lakes

In Pakistan's southeast is Manchhar Lake. It was once a large body of fresh water (roughly 100 sq mi / 259 sq km) and a major habitat for birds and fish, but pollution and water diversion have shrunk the lake dramatically and made its waters increasingly saline. Other lakes in the lower Indus region face extinction, including Kerjhar Lake and Hammal Lake. Kinjhar (Kalri) Lake, Haleji Lake, and Drigh Lake are wildlife sanctuaries in this region. Further north the Khabbaki, Uchali and Jahlar Lakes complex is a Wildlife Sanctuary, which is a wintering area for numerous bird species. The far northern basin known as Snow Lake is a massive snowbed comprising the Sim Gang glacier and a frozen glacial lake with ice more than 9 mi (15 km) thick.

Rivers

The Indus River is an irrigation lifeline for much of the country. The Indus rises in the Tibetan Himalayas. After crossing the Indian administered portion of Jammu and Kashmir, it enters Pakistan and flows southwest for 1,000 mi (1,609 km) to the Arabian Sea. At Attock, the Indus receives the waters of the Kabul River from the west. After being joined by the Gumal River, the Indus continues south to Mithanhot, where it is joined by its major tributary, the Panjnad. The short Panjnad River, about 75 mi (121 km) long, is actually the combined input of the "five rivers of the Punjab": the Jhelum, Chenab, Ravi, Beas, and Sutlej. The principal river of Baluchistan is the Zhob, running along the southern slopes of the Toba Kakar Range and north into the Gumal River. In southern Baluchistan several minor rivers flow into the Arabian Sea; these include the Dasht, Mashkai, Nal, and Porali.

Pakistan has two major river dams. In northern Punjab, near Kashmir, the Mangla Dam sits on the Jhelum river. The second is the Tarbela Dam on the Indus near Taxila. Dams on the Indus River for hydropower or agricultural water diversion have been extremely controversial. Provincial governments of Sind and Baluchistan believe that Punjab Province is diverting too much water from the Indus. Intensive irrigation has led to a crisis of waterlogging and salinity throughout the farmlands of the Indus Basin. In this syndrome, salty water seeps from canals into surrounding soil, which the salt renders useless for farming as the water evaporates.

Wetlands

Pakistan has sixteen areas designated as Wetlands of International Importance under the Ramsar Convention on Wetlands. Most are in the southern half of the country, including the Indus Dolphin Reserve, which holds the last 500 of the blind dolphin species; the Jiwani Coastal Wetland, a mangrove forest belt which stretches to Iran; and the Miani Hor, a shallow bay with mud flats and mangrove forests. Waste disposal, commercial fishing, oil drilling, and pollution pose threats to these southern wetlands. In other regions, Ramsar sites include the Thanedar Wala, a floodplain on the Kurram River in the northwest, and the Taunsa and Chashma Barrages, reservoirs in Punjab that are major bird habitats.

THE COAST, ISLANDS AND THE OCEAN

The coastline of Pakistan meets the Arabian Sea of the northern Indian Ocean. On the western coast, Baluchistan's Ormara Turtle Beaches, about 6 mi (10 km) long, are a habitat for endangered sea turtles; and mud volcanoes sputter along the shore. The central coast is indented by Sonmiani Bay. The coast has few settlements, except for Pakistan's largest city, the port of Karachi. The city's beaches are badly polluted by oil spills, sewage, and industrial toxic waste, which pours directly into the ocean. To the southeast of Karachi, the Indus River delta is approximately 130 mi (210 km) wide. Pakistan's only major offshore island is Astola (Haft Talar), about 15 mi (25 km) south of Baluchistan in the Arabian Sea, with an area of 19 sq mi (50 sq km). Astola is a turtle nesting area and a bird and reptile habitat.

CLIMATE AND VEGETATION

Temperature

Pakistan is in the temperate zone and varies greatly in weather conditions, from the humid coast, to the dry, hot desert interior, to the icy mountains in the north. Four seasons are experienced: winter in December-February; a hot, dry spring in March-May; the arrival of the southwest monsoon in June-September; and the northeast monsoon in October-November. In the north and west, the rainy season occurs during the winter.

In the north the capital, Islamabad, has average temperatures ranging from a low of 35°F (2°C) in January to a high of 104°F (40°C) in June. The southern port of Karachi has average temperatures varying from a low of

55°F (13°C) in winter to a high of 93°F (34°C) in summer.

Rainfall

Arid conditions prevail in most of Pakistan, which misses the full force of the monsoons. Punjab has had major fluctuations in monsoon rainfall in recent decades, with droughts in some years and floods in others. On Pakistan's plains the average annual rainfall is a mere 5 in (13 cm) while in the highlands it is 35 in (89 cm). Hailstorms are common, and snow falls in the north in winter. The lofty mountains of the north are permanently cloaked in snow and ice.

Grasslands

Khunjerab National Park in Hunza has high altitude grasslands, which have been the subject of conflicts over grazing rights; they are the habitat of endangered Marco Polo sheep and snow leopards. Also in the north, Shandur-Hundrup National Park contains upland meadows.

Indus River Plain

The upper Indus River plain, in Punjab, varies from about 500 to 1,000 ft (152 to 304 m) in elevation and consists of fertile alluvium deposited by the rivers. The lower Indus Plain, corresponding to generally the province of Sind, is lower in altitude. On the Indus plain, grasslands called "doabs" provide grazing on the strips of land between rivers.

Deserts

Pakistan's Thal Desert is south of the Salt Range, between the Indus and Jhelum rivers. The Thar Desert (Cholistan Desert) lies south of the Sutlej River along the Pakistan-India border. Both are extensions of India's Thar Desert.

The Baluchistan Plateau is largely a desert area with erosion, sand dunes, and sandstorms. There is also a dry region in the northern Chilas-Gilgit area, which is in the Himalayan rain shadow. In addition to existing deserts, the environmental change called desertification is occurring across Pakistan, with more than one third of the country considered at risk. Deforestation, depletion of soil, and water shortages are causing desertification as vegetation is cut and stripped away.

Population Centers – Pakistan

(2000 POPULATION ESTIMATES)

Name	Population	Name	Population
Karachi	10,032,000	Gujranwala	1,325,000
Lahore	5,452,000	Multan	1,263,000
Faisalabad	2,142,000	Hyderabad	1,221,000
Rawalpindi	1,521,000	Peshawar	1,066,000

SOURCE: Population Division, Department of Social and Economic Affairs, United Nations Secretariat. World Urbanization Prospects, The 2001 Revision. http://www.un.org/esa/population/publications/wup2001/wup2001dh.pdf (accessed September 3, 2002).

Forests and Jungles

Coniferous and deciduous forests, scrub woods, mangrove forests, and tree plantations grow in Pakistan. Some 40 percent of the forests are conifer or scrub woods, found mainly in mountain watershed areas. Pakistan's forest cover has been reduced to just 4 percent or lower. Deforestation in northern Pakistan has caused severe erosion when exposed mountain soil is washed away, resulting in excessive sediment in the Indus River. Grazing and fuelwood collection have decreased the coastal mangrove forests as well.

GEO-FACT

The Karakoram Highway, completed in 1986 as a joint project between Pakistan and China, winds from north of Islamabad to the city of Kashgar in China, at elevations reaching 15,072 ft. (4,594 m). Cutting 808 mi (1,300 km) through rocky terrain, the highway follows ancient Silk Road trade routes.

HUMAN POPULATION

In 2001, the nation's population density was estimated to be 466 people per sq mi (178 people per sq km). A high growth rate of 2.11 percent (2001 estimate) is affecting resources, the environment, and civil society in Pakistan. Thirty-seven percent of Pakistanis live in cities (2000 estimate) where infrastructure is under stress from growing populations. In particular, the southern port of Karachi has become a mega-city with a metro area holding more than 12 million people.

NATURAL RESOURCES

Pakistan has many power and energy resources. Buried beneath the Baluchistan Plateau lie vast natural gas reserves. Oil reservoirs have been found in the Potwar Plateau, but it is estimated that not even 2 percent of the oil reserves in the country has been discovered. Due to the many rivers and large amounts of sunshine in Pakistan, hydropower and solar energy are resources increasingly being tapped. Additionally, Pakistan produces and exports a low-grade coal. The country is also home to various mineral resources, including iron, copper, limestone, salt, antimony, bauxite, gypsum, and lignite.

Though mostly dry, hot desert, 25 percent of the land is arable and wheat, cotton, sugarcane, rice, and corn are grown successfully. The worst resource problem is the increasing shortage of water due to salination, pollution, and the population's use of water at double what would be sustainable rates.

FURTHER READINGS

Pakistan High Commission, Brunei. http://www.brunet.bn/gov/emb/pakistan (accessed June 6, 2002).

Reeves, Richard. *Passage to Peshawar: Pakistan, Between the Hindu Kush and the Arabian Sea.* New York: Simon and Schuster, 1984.

Shaw, Isobel. *Pakistan Handbook.* Emeryville, Calif.: Avalon Travel Publishing, 1998.

Shirahata, Shiro. *Karakoram: Mountains of Pakistan.* Seattle, Wash.: Mountaineers Books, 1998.

Sustainable Development Networking Programme. http://www.edu.sdnpk.org (accessed June 6, 2002).

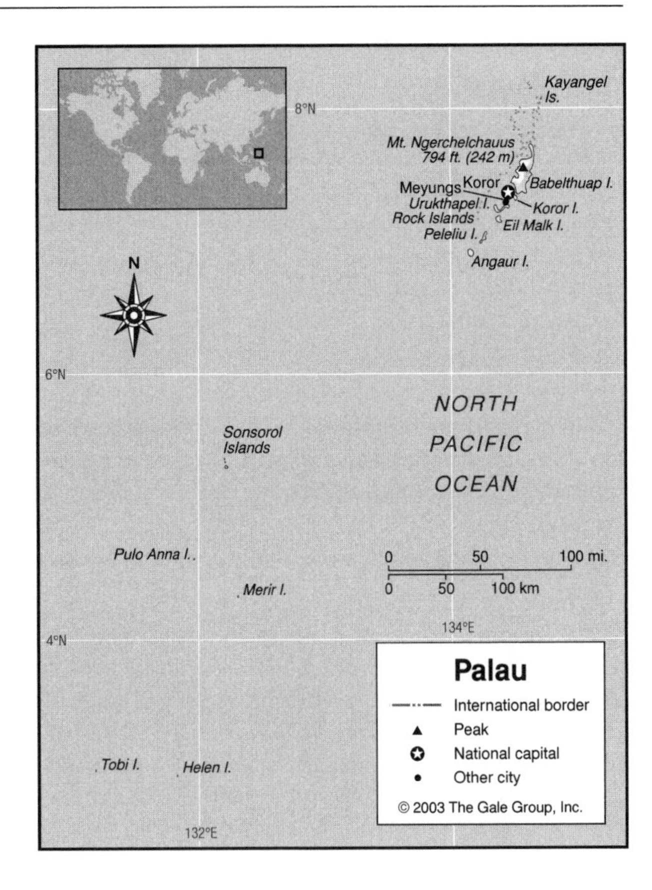

Palau

- **Area:** 177 sq mi (458 sq km) / World Rank: 185
- **Location:** A group of islands in the western extremities of the North Pacific Ocean, in Oceania, in the Northern and Eastern Hemispheres, southeast of the Philippines.
- **Coordinates:** 7°30′N, 134°30′E
- **Borders:** None
- **Coastline:** 944 mi (1,519 km)
- **Territorial Seas:** 3 NM
- **Highest Point:** Mount Ngerchelchauus, 794 ft (242 m)
- **Lowest Point:** Sea level
- **Longest River:** None of significant length
- **Natural Hazards:** Typhoons
- **Population:** 19,092 (July 2001 est.) / World Rank: 201
- **Capital City:** Koror, located on Koror Island, approximately in the center of the island group (Note: a new capital is being built about 12 mi [20 km] northeast of Koror)
- **Largest City:** Koror, population 12,000 (2000 est.)

OVERVIEW

Palau, or Belau, the westernmost archipelago in the Caroline chain, consists of six island groups totaling more than 200 islands roughly oriented north-south. The islands include four types of topographical formation: volcanic, high limestone, low platform, and coral atoll. Palau's volcanic and limestone islands sustain distinctly different vegetation communities. Limited natural resources and few skilled personnel constrict economic growth. The burgeoning tourist sector is dependent on regional financial stability. In 1999 foreigners made up 46 percent of the total work force. About one-fifth of Paluans live abroad, mainly on Guam.

MOUNTAINS AND HILLS

Mountains

The highest point in Palau, Mount Ngerchelchauus, is 794 ft (242 m) above sea level. The peak is located on the main island of Babelthuap, which, compared to the rest of the islands comprising Palau, is high and mountainous. Many of the other islands are low, coral islands.

INLAND WATERWAYS

Wetlands

Mangrove forests exist in coastal areas and the lower portions of rivers. Swamp forests are found in low-lying areas, just inland of mangroves and above tidal areas.

THE COAST, ISLANDS, AND THE OCEAN

Major Islands

Babelthuap is the largest island with an area of 153 sq mi (397 sq km). It is also the second-largest island in Micronesia after Guam. The second largest island in Palau is Urukthapel. Koror Island, containing the capital and most of the population, has an area of 7.1 sq mi (18 sq km). Other islands include Eli Malk, the islands of Peleliu and Angaur, which are low platform reefs, and Sonsorol and Hatohobei, the two smallest islands. Kayangel is a coral atoll.

Palau is also home to the world-famous Rock Islands. The Rock Islands are a cluster of more than 200 rounded knobs of forest-capped limestone that plunge steeply into the sea.

The Coast and Beaches

The Palau barrier reef encircles the Palau group, except Angaur Island and the Kayangel atoll. The reef encloses a lagoon (489 sq mi / 1,267 sq km) on the western side, containing a large number of small elevated limestone islets known as the Rock Islands. The dramatic marine environment of extensive coral rock formations, caves, and reefs—including the famous Great Barrier Reef—and abundance of sea life surrounding Palau make it a prime spot for snorkeling as well as scientific research. The waters are warm year round, and many of the islands have beautiful white sandy beaches that attract tourists and SCUBA divers from around the world.

States – Palau

POPULATIONS FROM 1995 CENSUS

Name	Population	Area (sq mi)	Area (sq km)	Capital
Aimeliik	419	20	52	Ulimang
Airai	1,481	17	44	Airai
Angaur	193	3	8	Ngaramasch
Hatohobei	51	1	3	Hatohobei
Kayangel	124	1	3	Kayangel
Koror	12,299	7	18	Koror
Melekeok	261	11	28	Melekeok
Ngaraard	421	14	36	Ngermechau
Ngarchelong	253	4	10	Imeong
Ngardmau	162	18	47	Chol
Ngatpang	221	18	47	Oikuul
Ngchesar	228	16	41	Ngerkeai
Ngeremlengui	281	25	65	Ollei
Ngiwal	176	10	26	Ngetkip
Peleliu	575	5	13	Kloulkubed
Rock Islands	0	18	47	
Sonsorol	80	1	3	Dongosaru

SOURCE: 1995 Census of Population and Housing, Republic of Palau.

CLIMATE AND VEGETATION

Temperature

Located near the equator, Palau's climate is maritime tropical, characterized by little seasonal and diurnal (day/night) variation. The annual mean temperature is 82°F (28°C) in the coolest months.

Rainfall

Annually there is high precipitation and a relatively high humidity of 82 percent with heavy rainfall from May to November. Short torrential rainfall produces up to 150 in (381 cm) of precipitation annually. Although outside of the main typhoon path, in the months from June through November damaging storms occur.

Grasslands

Most of the islands are either rock or tropical forest, however grasslands cover large areas of Babelthuap where forests have been cleared.

Forests and Jungles

Palau accommodates Micronesia's greatest diversity of terrestrial flora and fauna. Palau's tropical moist forests are of 7 main types: upland, swamp, mangrove, atoll, limestone, plantation, and palm. Dense tropical broadleaf forests cover most of the volcanic and all of the limestone islands.

HUMAN POPULATION

Almost two-thirds of the population lives in the capital city of Koror. The nearby city of Meyungs has a little more than a thousand people, making it the second largest. For 2000 it was estimated that 73 percent of the population lived in urban areas, mostly in these two cities. The population density on the island of Koror is roughly 1,850 people per sq mi (730 per sq km). However, the rest of the population is fairly spread out in little towns or villages on seven of the other islands—other than Koror and Meyungs there are no cities of more than 1,000 people on the islands.

NATURAL RESOURCES

Palau's number one natural resource is its tourism. It is also rich in fish and marine products, as well as craft items made from shells, pearls, wood, and limestone. Its forests provide wood, and there are some mineral deposits (especially gold) on some of the islands, as well as deep-seabed minerals that can be excavated.

FURTHER READINGS

Action Atlas Coral Reefs. *The Pacific: Palau.* http://www.motherjones.com/coral_reef/palau.html (Accessed May 2, 2002).

Brower, Kenneth. *1944-With Their Islands around Them.* New York: Holt, Rinehart and Winston, 1974.

BUBL Information Service. *Palau.* http://bubl.ac.uk/link/p/palau.htm (Accessed May 2, 2002).

Dahl, Arthur L. *Review of the Protected Areas System in Oceania.* Gland, Switzerland: International Union for Conservation of Nature and Natural Resources, Commission on National Parks and Protected Areas, in collaboration with the United Nations Environment Programme, 1986.

Faulkner, Douglas. *This Living Reef.* New York: Quadrangle-New York Times Book Co., 1974.

Palau National Communications Corporation. www.palaunet.com (Accessed June 20, 2002).

World Gazetteer. *Palau.* http://www.world-gazetteer.com/fr/fr_pw.htm (Accessed June 2002).

Panama

- **Area:** 30,193 sq mi (78,200 sq km) / World Rank: 118
- **Location:** Northern and Western Hemispheres, an isthmus in Central America linking North and South America, west of Colombia and east of Costa Rica.
- **Coordinates:** 9°00′N, 80°00′W
- **Borders:** 345 mi (555 km) / Colombia, 140 mi (225 km); Costa Rica, 205 mi (330 km)
- **Coastline:** 1547 mi (2490 km)
- **Territorial Seas:** 12 NM
- **Highest Point:** Volcán Barú, 11,401 ft (3,475 m)
- **Lowest Point:** Sea level
- **Longest Distances:** 480 mi (772 km) E-W; 115 mi (185 km) N-S
- **Longest Rivers:** Chucunaque River and Chepo River, 134 mi (215 km)
- **Largest Lake:** Gatun Lake, 161 sq mi (418 sq km)
- **Natural Hazards:** Floods; drought; earthquakes
- **Population:** 2,845,647 (July 2001 est.) / World Rank: 132
- **Capital City:** Panama City, at the center of the country on the Pacific coast
- **Largest City:** Panama City, population 704,117 (2000 est.)

OVERVIEW

The Republic of Panama occupies the Isthmus of Panama, a narrow land bridge between North and South America. Panama, an S-shaped isthmus, divides the Pacific and Atlantic Oceans and, with a great canal at its midsection, links them. The country's narrowest point is just 30 mi (48 km) across, and its widest is 115 mi (185 km).

Panama is traversed by two parallel mountain ranges, between which are valleys and plains. The highest lands are toward the Costa Rican border, and the interior of the country where the Panama Canal is found has the lowest elevation. Panama is seated on the Caribbean Tectonic Plate, but just offshore three other plates bump into the Caribbean: to the west the Cocos Plate, to the south the Nazca Plate, and to the southeast the South American Plate. The action of these plates on each other in the Miocene Epoch caused the Isthmus of Panama to rise out of the ocean. As they kept pushing on each other, the mountain ranges and volcanoes of Panama rose. The continued interaction of the Cocos, Nazca, and Caribbean Plates on each other today cause frequent earthquakes in Panama. However, its volcanoes have not erupted in hundreds of years.

MOUNTAINS AND HILLS

Mountains

A spine of mountains formed by an undersea volcanic chain divides Panama into Pacific and Caribbean (Atlantic) regions. These main ranges are the Serrianía de Tabasará in Panama's west and the Cordillera de San Blas in the east. A gap between them in the center of the country is where the Panama Canal was built. In the Cordillera Talamanca, right on the Costa Rican border, is Volcán Barú (formerly known as Volcán Chiriquí), the highest point in Panama at 11,401 ft (3,475 m). The peak of Barú, a long-extinct volcano, has views of both the Pacific and Caribbean on clear days. In the east there are three other smaller mountain ranges. The Majé Mountains run parallel to the Gulf of Panama shore. Running into Panama from Colombia along the Pacific and Caribbean coasts, respectively, are the Sapo Mountains and the Darien Mountains.

Plateaus

The El Santuario Plateau rises 1,212 ft (4,000 m) in the Boquete district of Chiriquí Province, near the border with Costa Rica.

Canyons

There are narrow river canyons in the rugged terrain of western Panama. Erosion has carved gorges in Darién, the thickly forested region in the east.

Hills and Badlands

Hills dominate Chiriquí Province, in the west, particularly the Boquete district, where coffee is grown on the hillsides. The Peninsula de Azuero and much of the country's center are hilly and are occupied by farming communities.

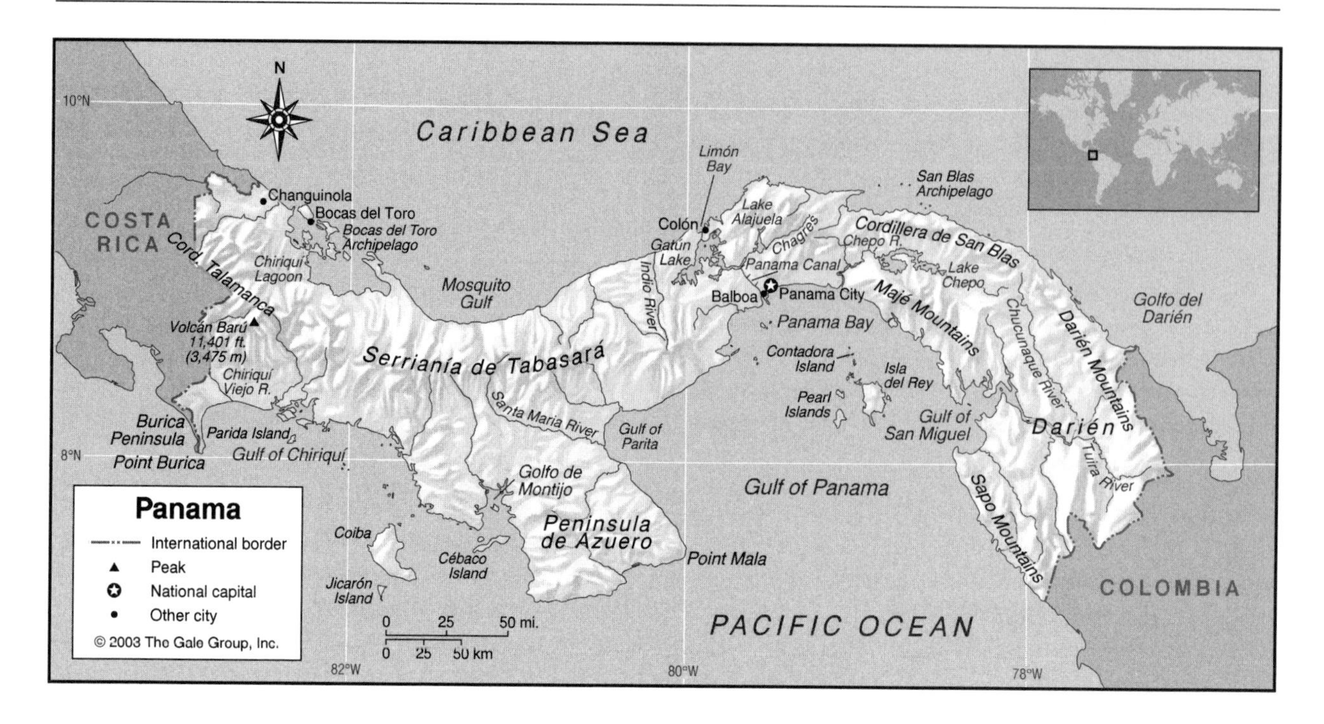

INLAND WATERWAYS

Lakes

Gatun Lake, formed by damming the Rio Chagres, is Panama's largest lake, with an area of 161 sq mi (418 sq km). Located in the center of Panama at 85 ft (26 m) above sea level, Gatun Lake is an important bird habitat and includes the Barro Colorado wildlife refuge. Gatun Lake and Lake Alajuela (Madden) are supplied by rains and provide the water of the Panama Canal and the drinking water for Panama City. Lake Chepo is another large reservoir in central Panama.

Rivers

Panama has more than 500 rivers, most of which are quite short. Rivers flowing into the Pacific include two that are tied for Panama's longest: the Chucunaque (134 mi / 215 km) and the Chepo (134 mi / 215 km). The Chepo has been dammed for hydroelectric power. Other rivers with Pacific outlets are the Santa Maria (104 mi / 168 km), Chiriquí Viejo (100 mi / 161 km), and the Tuira (79 mi / 127 km).

The more than 150 rivers draining into the Caribbean include the Chagres (78 mi / 125 km), Changuinola (68 mi / 110 km), Indio (57 mi / 92 km), and Cricamola (38 mi / 62 km). There is a hydroelectric dam on the Chagres, which has its source in mountain cloud forest. The Chagres waters run into Lakes Gatun and Arejuela, which form the 40 mi (65 km) Panama Canal.

The highly strategic Panama Canal was completed in 1914, with the Canal and adjacent areas (the Canal Zone) under United States administration. In 1977 a treaty mandated the transfer of the Canal from US jurisdiction to Panama's government. On December 31, 1999, the Panama Canal, Canal Zone, and former US military bases were handed over to Panama.

Wetlands

In Panama there are three sites designated as Wetlands of International Importance under the Ramsar Convention on Wetlands. Punta Patino in Darién (where there are extensive swamps) is a Private Nature Reserve on a coastal plain with mangroves, salt flats, and reefs, and is a seabird habitat. The Golfo de Montijo on the Pacific coast is a complex of coastal marshes, mangrove forests, and seasonally flooded grassland. San San-Pond Sak, in Bocas del Toro on the Costa Rican border, is a river basin complex of shallow lakes, mangrove forests, and peat bogs. It is an important bird habitat.

THE COAST, ISLANDS AND THE OCEAN

Oceans and Seas

The Pacific Ocean is to the south of Panama, and the Caribbean Sea of the Atlantic Ocean is to the north. Coral reefs are found along the coastlines, notably the protected coral reef at Isla Bastimentos National Park of Bocas del Toro, which is located off the northwestern coast and is also a sea turtle nesting site. Panama claims the seabed of the continental shelf, which has been defined by the country to extend to the 500-meter submarine contour.

The waters of Panama's Pacific coast, especially within the Gulf of Panama and the Gulf of Chiriquí, are extremely shallow (less than 590 ft / 180 m), with extensive mud flats. Because of this, the tidal range in this area is extreme. The range between high and low tide on the

Pacific coast exceeds 275 in (700 cm) while on the Caribbean coast it is only 27 in (70 cm).

The Caribbean coastline is marked by several good natural harbors; however Cristóbal, near the Panama Canal, is the only one of significant traffic. The major port on the Pacific is Balboa.

Major Islands

On the Caribbean side, the 366-island San Blas Archipelago stretches for more than 100 mi (160 km) down the eastern Panama coast. The Bocas del Toro Archipelago extends along the west of Panama to the border of Costa Rica.

The Pacific coast has many more islands than the Caribbean. Within the Gulf of Panama are the Pearl Islands, Isla Del Ray, and Contadora Island. Coiba, Panama's largest island at 104 sq mi (271 sq km), sits in the Gulf of Chiriquí along with Jicarón Island, Cébaco Island, Parida Island, and hundreds of much smaller islands and islets.

The Coast and Beaches

From the Costa Rica border in the west, Panama's Caribbean coastline is indented by the Chiriquí Lagoon, and then by the broad Mosquito Gulf, before curving north to the city of Colón and the port of Cristobal on Limon Bay, which is the Caribbean entrance to the Panama Canal. After the Canal, this coast sweeps south to the Golfo del Darién and the border of Colombia.

Panama's Pacific coast curves south from Point Burica on the Burica Peninsula, with the Gulf of Chiriquí extending down to the Azuero Peninsula. Turning around Point Mala, on the northeast of the Azuero Peninsula is the Gulf of Panama, with Panama Bay at its apex. There, the capital, Panama City, marks the Pacific entrance to the Panama Canal. The Gulf of Panama is indented on the west by the Gulf of Parita, and on the east by the Gulf of San Miguel where rivers flow down from the highlands of Darién. Then the coast continues to the border of Colombia.

CLIMATE AND VEGETATION

Temperature

Panama has a tropical climate with temperate areas at the higher elevations of 2,297 to 4,921 ft (700 to 1,500 m). There are two seasons: a rainy "winter" from May to December, when humidity is 90 to 100 percent, and a drier "summer" from January to April, when the northeast trade winds arrive. Panama's average temperature is 84°F (29°C) on the coasts and 64°F (18°C) in the highlands.

Rainfall

Rainfall patterns are different on Panama's Caribbean and Pacific Coast regions. The Caribbean Coast and mountain slopes get rain throughout the year, receiving from 59 to 140 in (150 to 355 cm) annually. The Pacific Coast experiences a more distinct dry season, and has annual rainfall of 45 to 90 in (114 to 229 cm).

From year to year Panama has considerable variation in the amount of rainfall, as the country is affected by El Niño and La Niña weather patterns. Panama is not in the path of Caribbean hurricanes.

Grasslands

Some regions of natural savannahs exist on Panama's Pacific Coast. There are also cattle ranches on the country's central plains; some 20 percent of Panama is considered pastureland. Invasive grass species take hold in areas deforested by burning and in abandoned pastures.

Forests and Jungles

With species from North and South America, Panama has an extremely high level of biodiversity, especially in the tropical rainforests, which are found on Caribbean-region mountain slopes particularly in the Darién region near Colombia. In addition to the rainforests, there are dwarf forests and cloud forests in the mountains, and mangrove forests on the coasts.

Panama was 91 percent forest-covered in 1850, which has decreased to about 44 percent. Deforestation has slowed but is still a major problem, especially for the Canal watershed. At least 290 sq mi (750 sq km) of Panama is deforested every year. The forests of Darién are threatened by fires, such as those meant to burn grasslands for agriculture, which spread into the forests in 1998. Other causes of forest loss in Panama include logging, fuelwood cutting, road construction, plantations, and mining. As much as 30 percent of Panama's land is under some degree of official protection, as forest reserves, national parks, or wildlife refuges. Darién National Park is a UNESCO World Heritage Site and Biosphere Reserve extending along most of Panama's border with Colombia. The park's 2,305 sq mi (5,970 sq km) of mountains and river basins are covered with primary and secondary tropical rainforests, dwarf and cloud forests, and wetlands. Darién National Park is home to jaguars, ocelots, giant anteaters, tapirs, howler monkeys, and many other wildlife species.

Population Centers – Panama

(2000 POPULATION ESTIMATES)

Name	Population
Panama City (capital)	704,117
San Miguelito	304,000

SOURCE: "World Urbanization Prospects: The 2001 Revision," United Nations Population Division, and projections from United Nations Statistics Division.

HUMAN POPULATION

Panama has a population growth rate of 1.3 percent and density of 96 people per sq mi (37 people per sq km). However, the population is unequally distributed. The eastern third of the country, which is mostly rain forest and swampland, is only sparsely populated. Also falling into this category is most of the Atlantic Coast region, except for an area of banana plantations near the Costa Rican border. The Bay of Panama region is the most populated, with a density of 386 per sq mi (149 per sq km). The second most densely populated region is that surrounding the Gulf of Chiriquí, which has a density of 153 per sq mi (59 per sq km). The Central lands are home to most of the rural population, including farmers and fishermen. An estimated 58 percent of the population lives in cities and towns.

NATURAL RESOURCES

Panama's natural resources include copper, limestone, gold, agriculture, fishing, and hydropower. However, its economy relies heavily on services including banking and transportation—the country's greatest asset is its location on a vital shipping lane, with ownership of the Panama Canal. Though not exploited to its potential, the copper mine in Cerro Colorado, in the southwestern corner, has the potential to be one of the largest copper mines in the world.

FURTHER READINGS

Ared Networks. http://www.ared.com (Accessed June 13, 2002).

Espino, Ovidio Diaz. *How Wall Street Created a Nation: J.P. Morgan, Teddy Roosevelt and the Panama Canal.* New York: Four Walls Eight Windows, 2001.

Friar, William. *Adventures in Nature: Panama.* Emeryville, Calif.: Avalon Travel Publishing, 2001.

Smithsonian Tropical Research Institute. http://www.stri.org (Accessed June 13, 2002).

Ventocilla, Jorge, et al. *Plants and Animals in the Life of the Kuna.* Austin: University of Texas Press, 1995.

Papua New Guinea

- **Area:** 178,704 sq mi (462,840 sq km) / World Rank: 55
- **Location:** Group of islands in the southwest Pacific including the eastern half of the island of New Guinea, Southern and Eastern Hemispheres, between the Coral Sea and the South Pacific Ocean, east of Indonesia
- **Coordinates:** 6°00′S, 147°00′E
- **Borders:** 510 mi (820 km), all with Indonesia
- **Coastline:** 3,201 mi (5,152 km)
- **Territorial Seas:** 12 NM
- **Highest Point:** Mount Wilhelm, 14,793 ft (4,509 m)
- **Lowest Point:** Sea level
- **Longest Distances:** 1,294 mi (2,082 km) NNE-SSW / 718 mi (1,156 km) ESE-WNW
- **Longest River:** Fly River system, over 746 mi (1,200 km)
- **Largest Lake:** Lake Murray, 250 sq mi (647 sq km)
- **Natural Hazards:** Volcanic activity, earthquakes, tsunamis, drought
- **Population:** 5,049,055 (July 2001 est.) / World Rank: 109
- **Capital City:** Port Moresby, on the southeast coast
- **Largest City:** Port Moresby, 252,000 (2000 census)

OVERVIEW

The island of New Guinea, the second largest in the world (316,605 sq mi / 820,003 sq km), is divided in half between Papua New Guinea and the Indonesian province of Papua (formerly Irian Jaya). The border between the two is a nearly straight north-south line. Mountainous New Guinea was pushed up by the colliding Australian and Pacific Tectonic Plates. New Guinea's mountains have been isolators, producing diversity in languages, customs, and wildlife. They form chains crossing the island, with riverine plains interspersed. Hundreds of smaller volcanic and coral islands lie off the eastern shore to complete the nation of Papua New Guinea, but 85 percent of the total land area is on the island of New Guinea itself.

MOUNTAINS AND HILLS

Mountains

The island of New Guinea is rugged, with many high peaks. At the center of the island are the Highlands, consisting of the Bismarck Range and its subsidiaries. Papua New Guinea's highest peak, Mt. Wilhelm, (14,793 ft / 4,509 m), is in the Bismarck Range. The second highest, Mt. Giluwe, an extinct volcano 14,327 ft (4,367 m) high, is also in the central mountain complex. Another major mountain range, the Owen Stanley, is found in the southeast. The highest peak here is Mt. Victoria (13,363 ft / 4,073 m). The Finisterre, Sarawat and Rawlinson Ranges line the northeast coast. They are made of coral limestone and extend onto New Guinea from the sea. In the west of New Guinea, near the Indonesian Papua border, the Star, Hindenburg, and Victor Emmanuel ranges are made of limestone and contain deep caves.

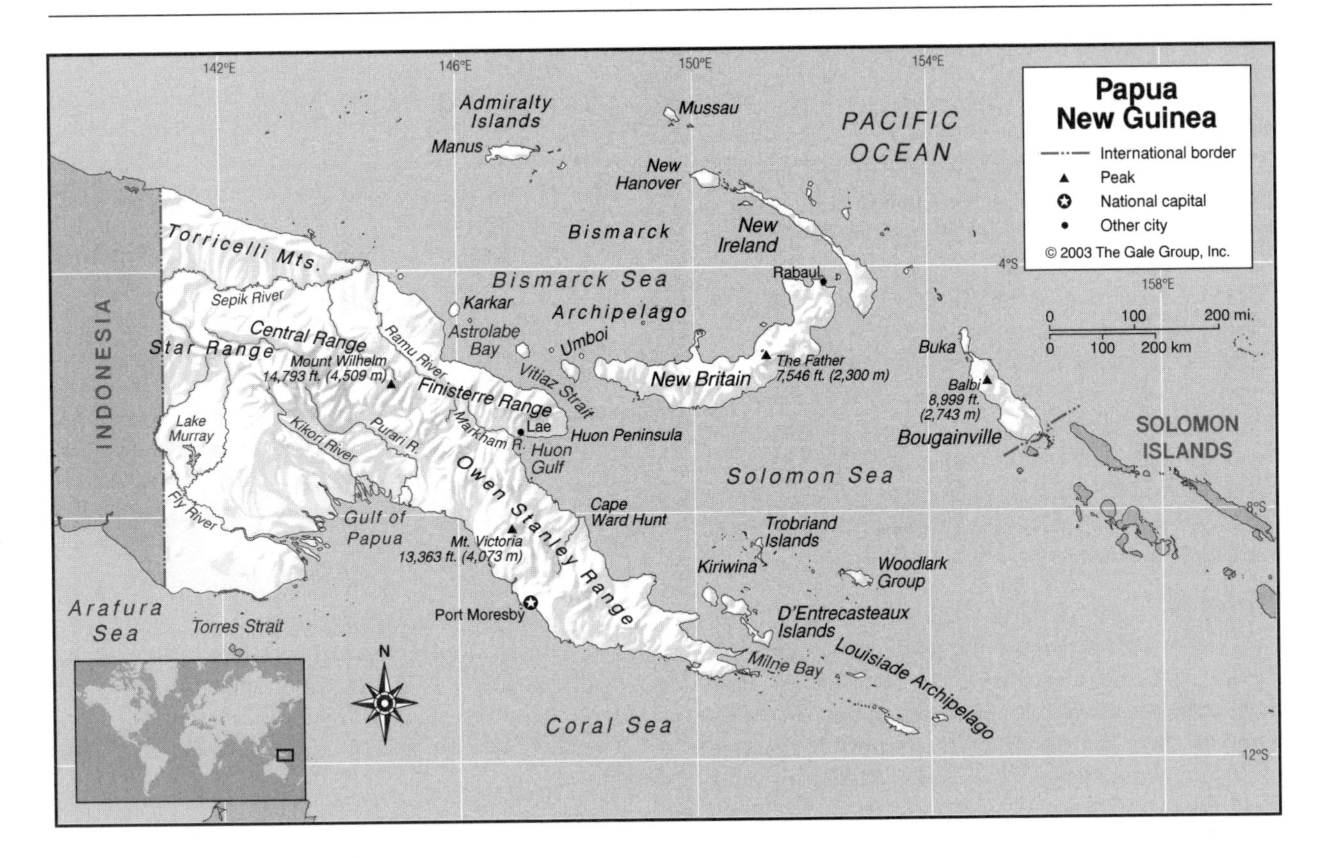

Many of the other islands in Papua New Guinea are volcanic in origin, with rugged terrain and peaks above 5,000 ft (1,524 m). The Father, on New Britain, rises to 7,546 ft (2,300 m), but the tallest peak on the outer islands is Balbi (8,999 ft / 2,743 m), on Bougainville.

Volcanoes

Papua New Guinea is rife with active volcanoes, over a dozen of which have erupted during the last one and a half centuries. Most of the volcanoes are on the southeastern arm of New Guinea, the large island of New Britain, and other islands. In September 1994 the eruption of two volcanic cones destroyed half the town of Rabaul on New Britain island, but due to timely evacuation there were few casualties.

Plateaus

The Great Papuan Plateau is in the central mountains, rising 4,921 to 6,562 ft (1,500 to 2,000 m) above sea level. It is a limestone formation with many caves, and petroleum deposits. The Oriomo Plateau rises in the west. Nine rivers run through it east-west into Indonesian Papua. Tabubil Plateau, also in west, is the site of the enormous Ok Tedi copper and gold mine. Sogeri Plateau, 29 mi (46 km) from the capital, Port Moresby, is 2,625 ft (800 m) in elevation, and home to many bird species.

Canyons

The rugged mountain terrain of New Guinea is cut through by numerous canyons, gorges, and ravines. The upper Fly River is a region of many deep gorges. The region between the northeastern coastal mountains and the central Highlands, where the Sepik and Markham Rivers and their tributaries flow, is known as the Central Depression.

Hills and Badlands

The steep mountains of the Highlands have very few foothills. Hill areas of the island of New Guinea include the upper Sepik region and the countryside surrounding Port Moresby.

INLAND WATERWAYS

Lakes

Lake Murray is a 250 sq mi (647 sq km) freshwater lake that connects with the Strickland River, in Western Province. The lake and river are badly polluted with chemicals from nearby mining operations. Lake Kutubu, in the southern Highlands, is designated a Wetland of International Importance under the Ramsar Convention on Wetlands. The 19 sq mi (49 sq km) site includes the pristine lake, which contains ten unique fish species, plus adjacent swamp forest.

The Muruk Lakes are a group of freshwater lakes and saltwater lagoons in the Sepik River region, with mangrove forests around them. Also in the Sepik region are the Chambri Lakes, a 83 sq mi (216 sq km) system of linked shallow lakes and swamps; and the Blackwater Lakes.

Rivers

Papua New Guinea's largest rivers are the Fly, Purari, and Kikori, which flow southward into the Gulf of Papua; and the Sepik, Markham, and Ramu, flowing northward into the Pacific. The Fly River and Sepik River are crucial transportation routes. Rising in the Star Mountains, the twisting Fly River is navigable for 500 mi (805 km). It is 50 mi (80 km) wide at its entry to the Gulf of Papua. The Fly forms a 746 mi (1,200 km) river system with the Ok Tedi and Strickland Rivers, and all three have been polluted by the Ok Tedi mining project. The Sepik River, 698 mi (1,126 km) long, has its source in the Victor Emmanuel Mountains. It is wide and navigable its whole length and has no real delta.

Wetlands

In the lowlands along the southern frontier with Indonesian Papua there are vast wetlands fed by the Fly and Sepik River systems. Mangrove forests fringe the coasts in these areas and sago palms provide the staple food for human settlements. Tonda Wildlife Management Area is a Ramsar Wetland of International Importance. Near the Indonesian Papua border, Tonda's 2,278 sq mi (5,900 sq km) site includes coastal plains that flood seasonally as well as grasslands and mangroves, all of which are waterbird habitat. The northwest of the island of New Britain is another significant wetland region.

THE COAST, ISLANDS AND THE OCEAN

Oceans and Seas

The seas surrounding New Guinea all belong to the Pacific Ocean. The Bismarck Sea is to the north of the main island of New Guinea and encircled by the Bismarck Archipelago. To the east of New Guinea is the Solomon Sea, enclosed by New Britain, Bougainville, and the Solomon and Trobriand Islands. Milne Bay at the southeastern end of New Guinea is enclosed by the D'Entrecasteaux Islands. The Coral Sea is south of New Guinea and north of Australia, with a major indentation on Papua New Guinea's southern coast, the Gulf of Papua. The Torres Strait separates New Guinea and the northern tip of Australia, with just 99 mi (160 km) between them, leading to the Arafura Sea of Indonesia.

15,444 sq mi (40,000 sq km) of coral reefs, rich in marine life, lie close to the shore of New Guinea. The Ontong Java Plateau, one of the world's largest ocean lava platforms, is to the northeast of New Guinea; and the Eastern and Papuan Plateaus lie beneath the Coral Sea. In July 1998 a tsunami produced by a submarine earthquake slammed Papua New Guinea's north coast with waves as high as 46 ft (14 m), killing over 2,500 people.

Major Islands

Papua New Guinea includes more than 1,400 islands besides New Guinea itself. Off the north coast of New Guinea are the volcanic Manam, Karkar, Long, and Umboi islands. To the east of New Guinea are the islands of the Bismarck Sea and Solomon Sea. In the western Bismarck Archipelago, New Britain is 370 mi (595 km) long and 50 mi (80 km) wide, formed of a chain of volcanoes. New Ireland, also in the Western Bismarck Archipelago, is 200 mi (354 km) long and only 7 mi (11 km) wide. Limestone mountains form its spine. New Hanover and Mussau are smaller islands in the western Bismarck Archipelago. Further west in the Bismarck Archipelago is Manus, which 50 mi (80 km) long and 20 mi (32 km) wide. It plus surrounding coral atolls form the Admiralty Islands.

The two largest islands in Papua New Guinea's part of the Solomon Islands (not to be confused with the country of the same name) are mountainous, mineral-rich Bougainville which is 127 mi (204 km) long and 50 mi (80 km) wide, and Buka, 35 mi (56 km) long and 9 mi (14 km) wide. There are also many small atolls near these islands.

Many of the 22 small islands comprising the Trobriand Group in the Solomon Sea are low coral types; they include Kaileuna, Kiriwina, Kitava, and Vakuta. Other island groups in the Solomon Sea include the D'Entrecasteaux Group, the Louisiade Archipelago, and the Woodlark Group.

The Coast and Beaches

The north coast of Papua New Guinea's side of the island of New Guinea slopes to the southeast from the Indonesian border, along Cape Moem and the outlets of the Sepik and Ramu Rivers. The coast indents at Astrolabe Bay by the town of Madang. Further southeast, the Huon Peninsula protrudes above the Huon Gulf, an indentation of the Solomon Sea. The Vitiaz Strait flows between the Huon Peninsula and New Britain. Cape Ward Hunt extends southeast from the Huon Gulf, leading into the long extending arm of southeast New Guinea, formed by the Owen Stanley Range. Oro Bay indents the coastline and Cape Nelson protrudes from it with lava rock fiords. On the south side of the arm is the natural harbor of Port Moresby, the capital. The south coast of New Guinea curves around the Gulf of Papua, broken up by the outlets of the Purari, Fly, and numerous other rivers, before it meets the southern point of the Indonesian border. Much of this region is marshland.

CLIMATE AND VEGETATION

Temperature

Papua New Guinea is a tropical country, but has two main seasons and two transition periods: December-

March brings the northwest monsoon; April is a transition; May-October brings the southeast monsoon; and November is a transition. Average lowland temperatures range from 70°F to 90°F (21°C to 32°C) while the Highlands have temperatures as cold as 37°F (3°C.) Lowland humidity averages 75 to 90 percent, and Highland humidity averages 65 to 80 percent.

Rainfall

Most of Papua New Guinea gets its rain from the northwest monsoon in December-March, but some areas, such as Lae and the Trobriand Islands, get their main rainfall in May-October. The Solomon Islands and Louisiade Archipelago are out of the monsoon pattern and rainfall occurs there year-round.

New Guinea has much regional weather variation, with micro-climates differing from valley to valley. Port Moresby has a rain shadow micro-climate, receiving less than 50 in (127 cm) of rain per year. Rainfall is heaviest in New Guinea's western river basin region, averaging up to 230 in (584 cm) a year. The average annual rainfall for the whole country is 80 to 100 in (203 to 254 cm). Snow and ice cover the highest mountain peaks.

Grasslands

Savannahs, mixed with scrub melaleuca woods and swamps, stretch from the Fly and Sepik river systems into Indonesian Papua. These coastal plains grasslands flood during the rainy season. The valleys in the central Highlands have grasslands produced by burning forests to clear land for agriculture. Alpine grasslands exist at elevations above 11,000 ft (3,353 m) where there is a moist, cool climate.

Forests and Jungles

Tropical rainforest covers as much as 77 percent of Papua New Guinea. These forests are a wealth of biodiversity. Papua New Guinea has an estimated 11,000 plant species, 250 mammal species, and 700 bird species. Unchecked logging for export during the 1980s significantly decreased the forest cover. Government moratoriums have been imposed on new logging. An eco-forestry movement promotes the use of small-scale sawmills and logging for local use only.

In addition to tropical rainforest, Papua New Guinea has coastal mangrove, monsoon, conifer, and alpine forests. Following clear cutting for timber, many natural forests are replaced with oil palm and rubber tree plantations.

HUMAN POPULATION

With a low density of 26 people per sq mi (10 people per sq km, 1999 estimate), Papua New Guinea would appear not to have population problems. However there is a high growth rate of 2.43 percent (2001 estimate) which may cause shortages of farmland and water. Papua New Guinea is a rural country, with only 18 percent of the population living in urban areas (2001 estimate).

Provinces – Papua New Guinea

Name	Area (sq mi)	Area (sq km)	Administrative Center
Central	11,400	29,500	Port Moresby
Chimbu	2,350	6,100	Kundiawa
Eastern Highlands	4,300	11,200	Goroka
East New Britain	6,000	15,500	Rabaul
East Sepik	16,550	42,800	Wewak
Enga	4,950	12,800	Wabag
Gulf	13,300	34,500	Kerema
Madang	11,200	29,000	Madang
Manus	800	2,100	Lorengau
Milne Bay	5,400	14,000	Alotau
Morobe	13,300	34,500	Lae
National Capital District	100	240	Port Moresby
New Ireland	3,700	9,600	Kavieng
Northern	8,800	22,800	Popondetta
North Solomons	3,600	9,300	Kieta
Southern Highlands	9,200	23,800	Mendi
Western	38,350	99,300	Daru
Western Highlands	3,300	8,500	Mount Hagen
West New Britain	8,100	21,000	Kimbe
West Sepik	14,000	36,300	Vanimo

SOURCE: *Geo-Data: The World Geographical Encyclopedia,* 2nd ed. Detroit: Gale Research, 1989.

NATURAL RESOURCES

Papua New Guinea's rich mineral resources include gold, copper, silver, chromite, cobalt and nickel. Other resources include natural gas and oil, hydropower, geothermal potential, timber, and fishing. There has been violent conflict over copper-mining and logging on the island of Bougainville.

FURTHER READINGS

Feld, Stephen. *Bosavi.* Washington D.C.: Smithsonian Folkways, 2001.

Papua New Guinea Eco-Forestry Forum. http:\\www.ecoforestry.org.pg (accessed May 17, 2002).

Salak, Kira. *Four Corners: Into the Heart of New Guinea.* New York: Counterpoint Press, 2001.

Sillitoe, Paul. *A Place Against Time: Land and Environment in the Papua New Guinea Highlands.* New York: Routledge, 1996.

Wantoks Communications, Limited. *Papua New Guinea Online.* http:\\www.niugini.com (accessed May 17, 2002).

Paraguay

- **Area:** 157,047 sq mi (406,750 sq km) / World Rank: 60
- **Location:** Southern and Western Hemispheres, South America, bordering Brazil to the northeast and east, Argentina to the south and southwest, and Bolivia to the west and north.
- **Coordinates:** 23°00′S, 58°00′W
- **Borders:** 2,436 mi (3,920 km) / Argentina, 1,168 mi (1,880 km); Bolivia, 466 mi (750 km); Brazil, 802 mi (1,290 km)
- **Coastline:** None
- **Territorial Seas:** None
- **Highest Point:** Cerro Pero, 2,762 ft (842 m)
- **Lowest Point:** Junction of Paraguay River and Paraná River, 151 ft (46 m)
- **Longest Distances:** 305 mi (491 km) ENE-WSW; 616 mi (992 km) SSE-NNW
- **Longest River:** Paraguay River, 1,500 mi (2,414 km)
- **Largest Lake:** Itaipu Reservoir, 521 sq mi (1,350 sq km)
- **Natural Hazards:** Flooding
- **Population:** 5,734,139 (July 2001 est.) / World Rank: 100
- **Capital City:** Asunción, southwest Paraguay
- **Largest City:** Asunción, population 1,262,000 (2000 est.)

OVERVIEW

Located in the south-central interior of the South American continent and bisected laterally by the tropic of Capricorn, Paraguay is separated from Argentina on the west by the Pilcomayo and Paraguay Rivers and on the south by the Paraná River. On the east it is separated from Argentina and Brazil by the higher reaches of the Paraná, and on the north and northwest its border with Bolivia is marked by small streams and by surveyed boundary lines. Paraguay lies on the South American Tectonic Plate.

Paraguay is seventh in size among the South American nations and one of only two landlocked countries on the continent (the other is Bolivia). Flowing almost directly southward out of Brazil, the river from which Paraguay takes its name divides the country into two regions with strikingly different characteristics. The three-fifths of Paraguay that lies to the north and west of the river is the Chaco, a hot, flat, semiarid plain with little vegetation and few inhabitants. The two-fifths to the south and east is Eastern Paraguay, which has a lush and diverse landscape, a wealth of fertile land, and is home to nearly the entire population of the country. It is sometimes referred to as Paraguay Proper. The easternmost part of this region forms the western end of the Paraná Plateau, which covers parts of Brazil and Argentina as well as Paraguay. Two westward spurs of the plateau extend toward the Paraguay River, one in the north and one in the center of the country, at Asunción. One lowland plain lies between these two ridges and east of the river; another lies to the south. A 1945 decree officially labels Paraguay's two major regions as Occidental (Western) and Oriental (Eastern) Paraguay.

MOUNTAINS AND HILLS

Mountains

The mountains of the Paraná Plateau include the Cordillera de Amambay, which extends southward from Brazil along the Paraguay/Brazil border. To the southeast is the Cordillera de San Rafael, which contains the country's highest peak, Cerro Pero, reaching a height of 2,762 ft (842 m). Both of these ranges are relatively low, rolling mountains. Also in this region are the Cordillera de Caaguazú and the Sierra de San Joaquin. These two ranges enclose the Paraná Plateau; the Cordillera de Caaguazú is a small range running northwest to serve as the western border of the plateau, and the Sierra de San Joaquin form the northern border, extending east to the Aracanguy Hills at the Brazilian border.

Plateaus

The heavily wooded Paraná Plateau occupies one-third of Eastern Paraguay and extends its full length from north to south and up to 90 mi (145 km) westward from the Brazilian and Argentine frontiers. Its western edge is demarcated by an abrupt escarpment that descends from an altitude of about 1,500 ft (457m) at Pedro Juan Caballero in the north to about 600 ft (183 m) at its southern extremity. The plateau slopes moderately to the east and south, and its surface is very uniform.

Eroded extensions of the Paraná Plateau that reach almost to the Paraguay River further divide Eastern Paraguay into subregions. The first of these extensions—the Northern Upland—lies within the portion northward from the Aquidaban River to the Apa River on the Brazilian frontier. For the most part it consists of a rolling plateau about 600 ft (183 m) above sea level and 250 to 300 ft (76 to 91 m) above the plain farther to the south. The second extension of the Paraná Plateau occurs in the vicinity of Asunción and is known as the Central Hill Belt. Its rolling terrain is extremely uneven, and small, isolated peaks are numerous.

Hills and Badlands

The most conspicuous features of the Central Lowland are flat-topped hills covered with thick forest that project 20 to 30 ft (6 to 9 m) above the grassy plain to

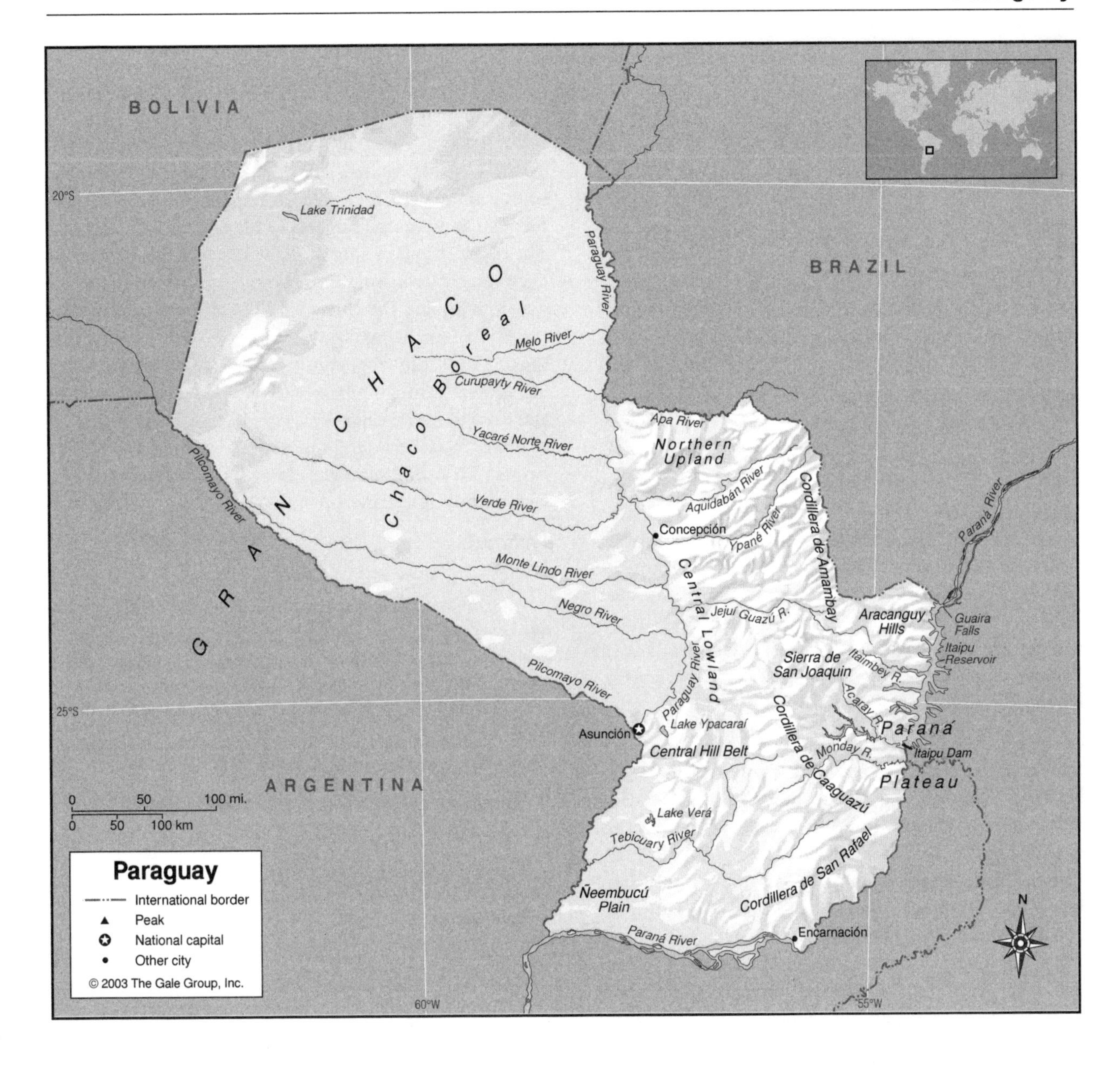

cover areas ranging from a few acres to several square miles. Weathered remnants of rock that are geologically related to the plateau in the east are called islas de monte (mountain islands), and their margins are known as costas (coasts).

INLAND WATERWAYS

Lakes

The largest freshwater lakes are the shallow Lake Ypacaraí in the Central Hill Belt and Lake Ypoá in the Ñeembucú Plain near Lake Paraguay. Lake Ypacaraí is a pleasure resort, and Lake Ypoá is the largest natural lake in the country. Lake Ypoá has a surface area of about 100 sq mi (260 sq km) and is navigable for small boats. Also of note in southern Paraguay is Lake Verá.

Due to poor drainage, the water of the numerous ponds and marshes of the Chaco region is too salty to be used for drinking or irrigation. The major lake in this region is Lake Trinidad in the northeastern corner of the country.

Dwarfing all of these natural lakes in size is Itaipu Reservoir, on the Paraná River. It has a surface area of roughly 521 sq mi (1,350 sq km). The reservoir was created by the Itaipu Dam.

Rivers

The Paraguay and Paraná Rivers and their tributaries define most of the country's frontiers, provide all of its drainage, and serve as indispensable transportation routes. Most of the larger towns of the interior, as well as Asunción, are river ports.

GEO-FACT

The prairies and swamps of Paraguay's Chaco region, while nearly uninhabited by humans, provide a habitat for a diverse array of wildlife, including such unusual species as anteaters, armadillos, tapirs, peccaries, and the capybara, the world's largest rodent, which can grow to a length of more than 4 ft (1.2 m).

Rising in the Pantanal Wetlands of Brazil, the Paraguay River forms part of the northeastern border with Brazil, then flows across the country on a southward course, then marks the southwestern border with Argentina, for a total length in Paraguay of about 1,128 km (700 mi). It is shallow and flows sluggishly sometimes overflowing its low banks to form temporary swamps, damaging crops and destroying villages. It empties into the Paraná River at the extreme southeast of the country, having flowed for a total of 1,500 mi (2,414 km).

The Paraná River also rises in Brazil and flows south. It marks the most eastern and southeastern borders between Paraguay and Brazil and Argentina, flowing some 500 mi (804 km) from the Brazilian frontier at the Guaira Falls to its juncture with the Paraguay. At the Guaira Falls, it tumbles 300 ft (91 m) in a series of steps into a narrow gorge. From this point it flows southward between gorge walls diminishing from 300 ft (91 m) in the north to about 100 ft (30 m) at Encarnación, where the river curves westward. The northern part of this gorge where the cliffs are highest has been filled by the Itaipu Reservoir, which was formed by the massive Itaipu Dam. After Encarnación, the gorge disappears and the river widens to a maximum of 18 mi (29 km) as the course becomes braided and large river islands appear. After merging with the Paraguay the Paraná curves south and flows across Argentina, then empties into the Atlantic Ocean as part of the Rio de la Plata. With a total length of 3,030 mi (4,875 km), it is the second longest river in South America, and among the very longest on Earth.

The third-largest river, the Pilcomayo, is a tributary of the Paraguay and enters the parent river near Asunción after following the entire length of the frontier between the Chaco and Argentina. During most of its course the river is sluggish and marshy, although small craft can navigate the lower reaches. The Verde and Monte Lindo rivers also enter the Paraguay from the Chaco. Other Chaco streams, including the Negro, Yacaré Norte, Curupayty, and Melo, overflow their badly defined channels in summer but are reduced to trickles or vanish altogether in winter to form salt marshes.

Major tributaries of the Paraguay River entering it from Eastern Paraguay include the Apa, Aquidaban, Ypané, and Jejuí Guazú. They descend rapidly from their source in the Paraná Plateau to the lower lands; here they broaden and become sluggish as they meander westward. The major tributary south of the plateau is the Tebicuary, beginning its course near Caazapá and flowing almost due west to the Paraguay. The rivers flowing eastward across Eastern Paraguay as tributaries of the Paraná are shorter, faster flowing, and narrower. They follow winding courses through gorges carved in the Paraná Plateau and drop to the parent river in steplike falls. Some sixteen of these rivers, including the Acaray, Monday, and Itaimbey, as well as numerous smaller streams, enter the Paraná above Encarnación.

Wetlands

There is a great deal of swampland in the southern portion of the Chaco region. The Ñeembucú Plain also has considerable swampland, and the Central Lowland becomes swampy during the rainy season when rivers overflow their poorly defined banks.

THE COAST, ISLANDS, AND THE OCEAN

Paraguay is landlocked and thus has no coastline.

CLIMATE AND VEGETATION

Temperature

Most of Eastern Paraguay lies south of the tropic of Capricorn and thus has a subtropical climate, while the Chaco region to the west, which lies mostly between the two tropics, has a tropical climate. In Paraguay there are basically two seasons: summer (October through March) and winter (May through August), with April and September serving as transitional months. Average summer temperatures range from about 77°F (25°C) to 100°F (38°C), although summer highs in the east usually do not rise much above 90°F (32°C), whereas highs in the west can top 109°F (43°C). Average winter temperatures are usually between about 60°F (16°C) and 70°F (21°C).

Rainfall

Rainfall is heaviest on the Paraná Plateau in the east, where it averages more than 60 in (152 cm) annually, decreasing to about 50 in (127 cm) in the lowlands east of the Paraguay River. Average rainfall is about 30 in (76 cm) in the Chaco region west of the river. Most of the rain falls in the summer months, but rainfall is generally irregular, posing the danger of both flooding and drought.

Population Centers – Paraguay

(1994 POPULATION ESTIMATES)

Name	Population
Asunción (capital)	862,800
Ciudad del Este	135,000

SOURCE: Projected from United Nations Statistics Division data.

Grasslands

Geologically, the Chaco region of Paraguay is a part of the South American Gran Chaco, which extends from Argentina in the south to the fringes of Bolivia and Brazil in the north. Because it is located in the northern part of the Chaco, the complete but seldom-used name of the part in Paraguay is Chaco Boreal. This region is bounded on the east by the Paraguay River and on the southwest by the Pilcomayo River. Except for low hills in the northeast, the featureless landscape is virtually flat plains, broken by intermittent rivers and streams and by extensive swamps in the south. Vegetation ranges from some deciduous woodland near the Paraguay River to desert-type scrub farther west. Much of the land is low-wooded plains or swampland. From an altitude of over 1,000 ft (305 m) in parts of the northeast, the plain slopes southward imperceptibly to an altitude of some 300 ft (91 m) at the confluence of the Pilcomayo and the Paraguay rivers, the lowest point in the country.

Between the two westward extensions of the Paraná Plateau lies the Central Lowland, an area of low elevation and relief, sloping gently upward toward the plateau and covered largely with savannah grasslands. Valleys of its westward-flowing rivers are broad and shallow, and their courses are subject to flooding that creates seasonal swamps. In the southwestern part of Paraguay's eastern region lies the Ñeembucú Plain. This alluvial flatland has a slightly west and southwest incline obscured by gentle undulations marking its surface. The Tebicuary river, a major tributary of the Paraguay River, bisects this swampy lowland, which is broken in its central portion by rounded swells of land up to 100 ft (30 m) in height.

Forests and Jungles

Extensive deforestation reduced Paraguay's forest cover from 50 percent in the 1940s to less than 33 percent by the end of the century. The majority of forestland (and vegetation generally) is found in the Paraná Plateau in the east, due to the higher rainfall in that region. The area still abounds in many species of hardwood, including urunday, lapacho, cedron, curupay, and numerous types of palm trees. Other distinctive native trees include the yerba, from which yerba maté tea is made, and the quebracho, found in northern areas of the Chaco. Paraguay is also known for its great variety of medicinal plants.

HUMAN POPULATION

More than 95 percent of the population lives in Eastern Paraguay, with the heaviest concentrations on the westward spurs of the Paraná Plateau known as the Northern Uplands (north of Concepción) and Central Hill Belt (in the vicinity of Asunción). More than half the population is urban.

NATURAL RESOURCES

The most important natural resources are timber from the many hardwood species found on the forested Paraná Plateau. The Itaipu Dam on the Paraná River is the world's largest power plant. Jointly owned by Paraguay and Brazil, it provides almost all of Paraguay's electricity and a major portion of Brazil's, as well as attracting many tourists. Mineral reserves include limestone, iron ore, manganese, sandstone, and copper.

FURTHER READINGS

Argentina, Uruguay and Paraguay. Berkeley: Lonely Planet Publications, 1992.

Miranda, Carlos R. *The Stroessner Era: Authoritarian Rule in Paraguay.* Boulder, Colo.: Westview Press, 1990.

"Public Saves Park." *American Forests* 106 (Winter 2001): 12.

Roett, Riordan, and Richard Scott Sacks. *Paraguay: The Personalist Legacy.* Boulder, Colo.: Westview Press, 1991.

Whigham, Thomas. *The Paraguayan War.* Lincoln: University of Nebraska Press, 2002.

World News Network. *Paraguay.com.* http://www.paraguay.com/ (Accessed May 12, 2002).

Peru

- **Area:** 496,226 sq mi (1,285,220 sq km) / World Rank: 21
- **Location:** Southern and Western Hemispheres, on the west coast of South America, west of Bolivia and Brazil, north of Chile, and south of Colombia and Ecuador.
- **Coordinates:** 10°00′S, 76°00′W
- **Borders:** 3,440 mi (5,536 km) / Bolivia, 559 mi (900 km); Brazil, 969 mi (1,560 km); Chile, 99 mi (160 km); Colombia, 930 mi (1,496 km); Ecuador, 882 mi (1,420 km)
- **Coastline:** 1,500 mi (2,414 km)
- **Territorial Seas:** 200 NM
- **Highest Point:** Nevado Huascarán, 22,205 ft (6,768 m)

- **Lowest Point:** Sea level
- **Longest Distances:** 800 miles (1,287 km) SE-NW ; 350 miles (563 km) NE-SW
- **Longest River:** Amazon, 3,900 mi (6,275 km)
- **Largest Lake:** Lake Titicaca, 3,212 sq mi (8,320 sq km)
- **Natural Hazards:** Earthquakes; tsunamis; flooding; landslides; mild volcanic activity
- **Population:** 27,483,864 (July 2001 est.) / World Rank: 38
- **Capital City:** Lima, at the midpoint of the Pacific Coast
- **Largest City:** Lima, population 7,443,000 (2000 est.)

OVERVIEW

Peru is the third largest country in South America; only Brazil and Argentina are larger. It stretches from its northern tip at the equator south along the Pacific Ocean, and inland to the Amazon Basin. Peru is a country of geographic extremes: Consider, for example, that two canyons in Peru are twice as deep as the Grand Canyon in the United States. Peru has the highest navigable lake in the world, and has some of the highest and most spectacular mountains. Off the Pacific Ocean shoreline is a trench as deep as the Andes Mountains are high. The driest desert on earth is in Peru.

Peru has three major topographic regions running north and south: La Costa, La Sierra, and La Selva. La Costa, bordering the Pacific Ocean, is a 1,500-mile (2,414-km) long desert, only 10 miles (16 km) wide at one point, widening to about 100 miles (160 km) in the north and south. La Sierra is the Peruvian portion of the Cordillera de los Andes, a vast mountain range crossing Peru and parts of Bolivia, Chile, and Ecuador. La Selva covers roughly 60 percent of Peru; it is the rainforest region of the Amazon Basin between the mountains of La Sierra and the eastern foothills.

Peru has occasional volcanic activity and earthquakes from the effect of the offshore Nazca Tectonic Plate moving under the South American Plate on which Peru sits.

MOUNTAINS AND HILLS

Mountains

Covering the greater part of the country, Peru's spectacular Andes Mountains are subdivided into three main parallel ranges. They are, from west to east: the Cordillera Occidental, the Cordillera Central, and the Cordillera Oriental. The Cordillera Occidental is further divided into the adjacent Cordillera Blanca and Cordillera Huayhuash. Nevado Huascarán, Peru's highest mountain towering to 22,204 ft (6,768 m), is in the Cordillera Blanca about 60 miles inland from the coastal city of Chimbote. The Cordillera Huayhuash is on average lower but includes Nevada Yerupajá at 21,765 ft (6,634 m) and Cerro Jyamy at 17,050 ft (5,197 m). In the south are two of the highest volcanoes in the world: Volcán Misti, which rises to 19,031 ft (5,801 m) at the edge of Arequipa, and the slightly shorter Volcán Yucamani, which reaches 17,860 ft (5,444 m).

Canyons

Cañon del Colca, in southwest Peru, is twice as deep as the United States' Grand Canyon: 10,607 feet (3,182 m). However, unlike the Grand Canyon, parts of Colca Canyon are inhabited. The canyon attracts numbers of visitors who not only want to view the magnificent canyon itself, but also wish to see the Andean condors, which hunt and nest in the canyon. It is only recently that the canyon has been fully traversed. Nearby Cañon del Cotahuasi is less explored. Some observers think that Cotahuasi is deeper than Colca, and may prove to be the world's deepest canyon.

INLAND WATERWAYS

Lakes

The Peruvian Andes are speckled with dozens of small lakes filled by milky-blue glacial water and one notably large lake, Lake Titicaca, by far the largest lake in the country. At 12,650 feet (3,856 m) above sea level, Titicaca is the world's highest navigable lake. It is in the mountains in Peru's southeastern corner on the border with Bolivia; the lake is nearly equally shared between the two countries. Titicaca is 136 miles (220 km) long and 37 miles (60 km) at its widest. Its surface covers a total of 3,212 sq miles (8,320 sq km), and its maximum depth is 1,181 ft (360 m).

Peru's newest lake was created by the especially severe 1998 El Niño. In the northern desert district of Piura, a lake formed from rainfall and drainage off the western mountains. It became the second largest lake in Peru. Water is receding, but a significant lake remains.

Rivers

Peru's mountains have two distinct drainage systems. About 60 rivers flow generally westward through the coastal plains to empty into the Pacific. They are relatively short and low-volume. The rivers swell for the few rainy months, then diminish or even dry up during the arid season. Rio Santa is an exception; it is larger in volume than the other rivers flowing into the Pacific, and flows mostly from north to south for 100 miles (160 km). Other rivers that empty into the Pacific include the Chicama, the Huaura, the Pisco, and the Ica Rivers.

Scores of rivers flow eastward into the Amazon Basin. Because of heavy rainfall, these rivers carry a tremendous volume of water. Many of these rivers are tributaries that create the Amazon, one of the world's great rivers.

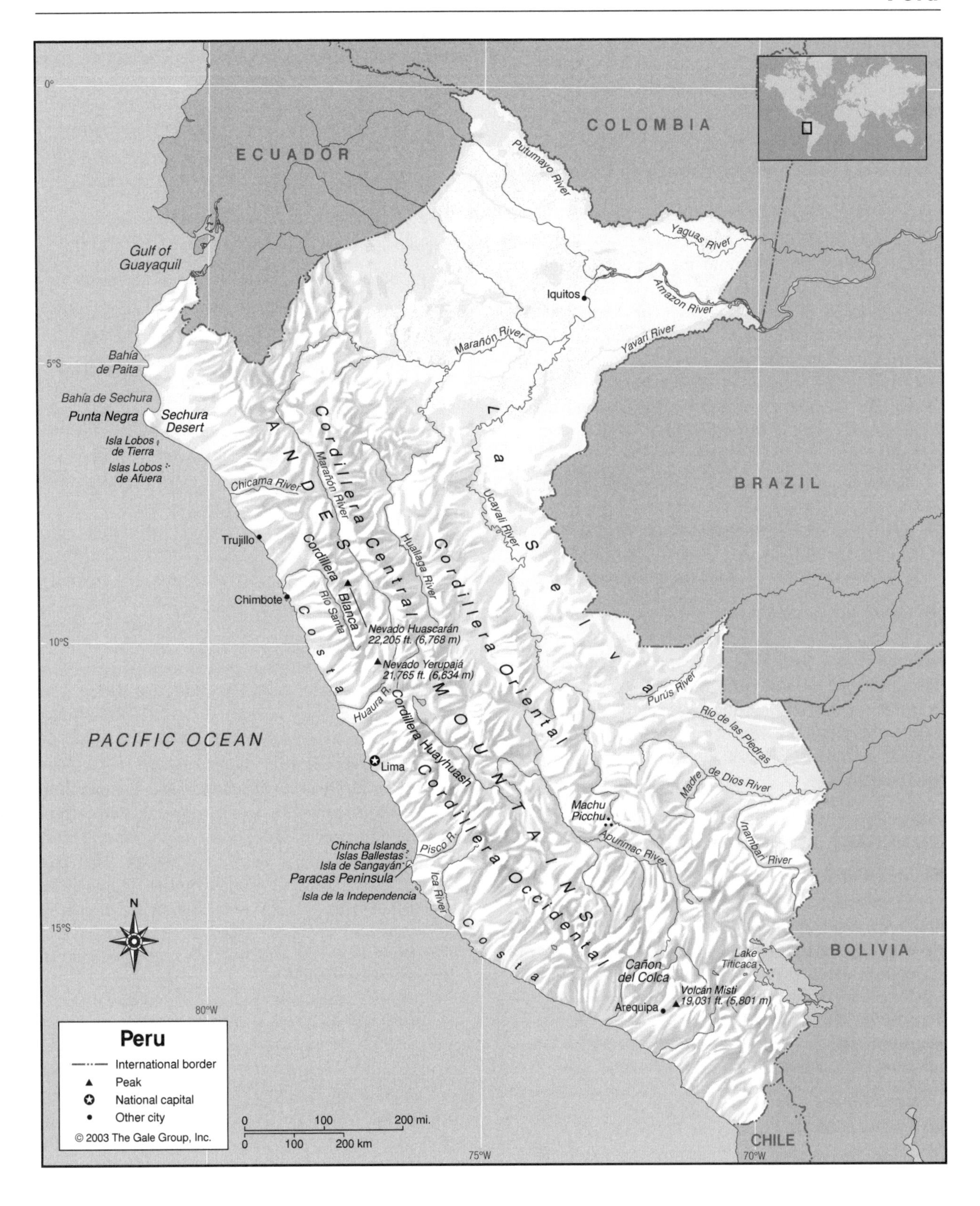

Though second in length to the Nile, the massive Amazon carries an estimated 60 times more water. Ships of 16 feet (5 m) draft navigate unobstructed from the Atlantic Ocean to Iquitos in Peru, nearly the entire river's length. Less than one-tenth, or a total of 368 mi (592 km) of the Amazon flows through Peru. The source of the Amazon is usually described as a point where two major Peruvian rivers, the Marañón and the Ucayali, converge. In 2000, a National Geographic expedition established the precise source of the Amazon to be a stream running from

Nevado Mismi, an 18,363-foot (5,597-m) mountain in the central-south Andes. It is the furthest point from which water flows year-round into the Amazon.

Major northeastern Peruvian rivers that contribute to the Amazon include the Marañón, the Ucayali, and the Yavarí. The Marañón flows northeast from the Andes and the Ucayali flows north from central Peru; both join the Amazon in the northeast. The Marañón has many tributaries, including the Napo, Mantaro, Huallaga, Tigre, and Pastaza. Rivers that feed into the Ucayali include the Urubamba and the Apurímac. The Urubamba River flows through El Valle Sagrado (The Sacred Valley) beside and below the ancient city of Machu Picchu. The Yavarí River flows somewhat parallel and to the east of the Urubamba and makes up most of Peru's border with Brazil. The Putumayo and Yaguas later join the Amazon in Brazil, the former is important in Peru because it forms the border with Columbia.

In southeastern Peru there are several important rivers. The Purús, Río de las Piedras, Madre de Dios, and Inambari drain the region north of Lake Titicaca. They all flow northeast and join the Amazon thousands of miles later, in Brazil.

Wetlands

Besides various estuaries, one remote large wetland in the northeast Selva region is especially interesting. Reserva Pacaya Samiria (8,031 sq mi / 20,800 sq km) is a complex expanse of alluvial terraces and floodplains covered by tropical rainforest. It contains two river basins, permanent freshwater lakes, and seasonally flooded, forested wetlands.

THE COAST, ISLANDS, AND THE OCEAN

Oceans and Seas

The western border of Peru is the Pacific Ocean. Offshore, the ocean floor drops quickly into the Peru-Chile Trench, a trench that is 1,100 miles (1,770 km) long and averages a depth of 16,400 ft (5,000 m)—as deep the Andes Mountains are tall. That is, one can view the bottom of the Pacific off Peru as an inverted mountain range. Cold water rising in the underwater trench generates the chill costal winds named the Peru Current.

Major Islands

The Islas de los Uros, in the Peruvian part of Lake Titicaca, may be the most unique inhabited islands in South America. The Uros are actually floating islands made of reed, and are also called Islas Flotantes. The largest islands in the group are Toranipata, Huaca Huacani, and Santa Maria. Lake Titicaca also has more than 30 normal islands on each side of the Peru/Bolivia border. On the Peruvian side, two important islands, both to the east of Islas de los Uros are Isla Taquile and Isla Amantani. The latter contains Inca ruins. A third, Isla Esteves, is connected to the mainland town of Puno by a causeway.

GEO-FACT

The Nazca lines, made on plains near the southern coast of Peru, are an enigmatic ancient phenomenon. These graceful lines, thought to be formed by the pre-Incan Nazca peoples who removed large stones and set them next to the depressions thus made in the ground, create a series of drawings, some of which are 1,000 feet (305 m) across. A particularly mysterious aspect of the Nazca lines is that they can be viewed whole only from the air. From there they resemble animals such as a spider, a monkey, and a whale. There is little agreement among archeologists regarding why they were made.

Because the ocean floor is so steep few islands appear off the Pacific coast of Peru, and those that do are relatively small. Starting from the north, a few kilometers from the shore of the Sechura Desert are Isla Lobos de Tierra and Islas Lobos de Afuera, the latter being actually two tiny islands. Much further south near the mouth of the Pisco River, within Reserva Nacional de Paracas, are several islands notable for the rare sea animals and birds that live there including the most northern habitat of penguins. From north to south they are: the Chincha Islands, Islas Ballestas, Islas de Sangayán, and Isla de la Independencia.

The Coast and Beaches

The Andes Mountains drop quickly to the coastline down the length of the country, and as a result the coastline is somewhat featureless, with few ports, bays, or dramatic points. The Pacific coast begins at the border with Ecuador in the Gulf of Guayaquil running southwest. A section of this north coast near Ecuador does have two inlets, Bahía de Paita (Paita Bay) and the larger Bahía de Sechura (Sechura Bay) on either side of Punta Negra. After completing Sechura Bay, the coastline quickly turns and runs southeast. Below Lima the coast is also less monotonous at the mouth of the Pisco River, jutting out to form Paracas Peninsula near the town of Pisco complete with a small inlet the offshore islands. After forming the Paracas Peninsula the coastline heads more to the east, ending uneventfully at Chile.

CLIMATE AND VEGETATION

Temperature

Peru has two seasons corresponding to rainfall rather than temperature: summer, from January to March, and winter during the remainder of the year. Because of extremes in topography, average temperatures vary greatly between regions.

In the La Sierra region, temperatures average 47°F (8°C) all year. To the east in the *montaña* forests the temperature is warmer but still fairly moderate. Dropping to La Selva and the jungles of the Amazon Basin, temperatures average 68°F (20°C) and as reach as high as 95°F (35°C) during the hottest months. The Coast (La Costa) is also warm all year, averaging 68°F (20°C). Despite being a desert area, the relatively moderate temperatures are credited to nearly constant cold air movement. The Peru (Humboldt) Current is a wind blowing from the Pacific Ocean toward the equator, the source of which is the very cold waters in Peru-Chile Trench.

In addition to the chilly Peru Current, Peru is affected by a second weather phenomenon: El Niño. Every several years (four to ten), El Niño presents the strongest climate-changing phenomenon on earth. El Niño is a warm current originating from the Central Pacific along the coasts of Peru and Ecuador that, among other effects, brings flooding rains and unusually warm temperatures to Peru. The name, referring to the infant Christ, was coined by Peruvian fishermen because the weather system begins near Christmas. El Niño has strong worldwide effects on climate, including repercussions on fishing, agriculture, and animal and plant life.

Rainfall

Most rain and moisture originates from trade winds to the east, blowing across the Amazon Basin. Because the mountains trap nearly all the rains, the coastal plain is relatively dry year-round, averaging less than one in (2.5 cm) in Lima. However, during the winter season, a nearly constant mist, the *garua*, shrouds the coast. In extreme contrast, the eastern jungles receive an average annual rainfall of 100 in (245 cm), in some years getting as much as 140 in (350 cm).

Deserts

The west side of Peru, bordering the Pacific Ocean, is desert. One particularly inaccessible area in the far northwest is the Sechura Desert. This desert consists of shifting sand dunes and borax lakes. It is a national reserve area.

The driest area of anyplace on earth is at Peru's far south near the Chilean border. This region marks the beginning of the Atacama Desert, an area that virtually never receives rain and is measurably drier than the Sahara Desert.

Population Centers – Peru

(1993 CENSUS OF POPULATION)

Name	Population	Name	Population
Lima (metropolitan area, capital)	6,321,173	Piura	277,964
Arequipa	619,156	Iquitos	274,759
Trujillo	509,312	Chimbote	268,979
Chiclayo	411,536	Huancayo	258,209
		Cuzco	174,336

SOURCE: INEI (Instituto Nacional de Estadistica and Informatica), Peru.

Departments – Peru

POPULATIONS FROM 1993 CENSUS

Name	Population	Area (sq mi)	Area (sq km)	Capital
Amazonas	428,095	15,945	41,297	Chachapoyas
Ancash	1,107,828	14,158	36,669	Huaráz
Apurimac	463,131	7,934	20,550	Abancay
Arequipa	1,101,005	24,528	63,528	Arequipa
Ayacucho	550,751	17,058	44,181	Ayacucho
Cajamarca	1,498,567	13,486	34,930	Cajamarca
Constitutional Province of Callao	787,154	57	148	Callao
Cusco	1,208,689	29,471	76,329	Cuzco
Huancavelica	443,213	8,139	21,079	Huancavelica
Huánuco	811,865	13,088	33,897	Huánuco
Ica	687,334	8,205	21,251	Ica
Junín	1,246,663	15,944	41,296	Huancayo
La Libertad	1,506,122	8,973	23,241	Trujillo
Lambayeque	1,121,358	5,304	13,737	Chiclayo
Lima	7,748,528	13,058	33,821	Lima
Loreto	907,341	146,342	379,025	Iquitos
Madre de Dios	99,452	30,271	78,403	Puerto Maldonado
Moquegua	156,750	6,065	15,709	Moquegua
Pasco	264,702	9,356	24,233	Cerro de Pasco
Piura	1,636,047	14,055	36,403	Piura
Puno	1,263,995	27,947	72,382	Puno
San Martín	757,740	20,197	52,309	Moyobamba
Tacna	294,214	5,881	15,232	Tacna
Tumbes	202,088	1,827	4,732	Tumbes
Ucayali	456,340	38,931	100,831	Pucallpa

SOURCE: INEI (Instituto Nacional de Estadistica and Informatica), Peru.

Forests and Jungles

Some of the world's most spectacular forest and jungles are in Peru. An enormous band of tropical cloud forests (*montaña*) form a natural border between the Andes and the Amazon Basin to the east. Starting at about 8,200 feet (2,500 m) and below, the mountain vegetation changes from grasses to bushes, shrubs, and then trees. This transition in vegetation is sharply noticeable, hence the Spanish name for it: *ceja de la montaña* (eyebrow of the forest). Further east and south, toward Brazil, is La Selva, the lowland jungle and rainforest region of Peru. Over some areas of this region the forest is so dense that access to it exists only along the rivers.

HUMAN POPULATION

Peru is unique among American countries in the proportion and absolute numbers of indigenous people. Nearly one half of the population is Amerindian, more than any other South American country, and about one-fifth the total for all the Americas. The next largest population is mestizo (mixed Amerindian and Caucasians) followed by Caucasians, and then other ethnic groups. The ethnic composition is reflected in the fact that Peru has two official languages: Spanish and Quechua.

The densest population areas are along the coast, especially around Lima and to the north. Next most populous are the Andean foothills and rainforests. The large eastern jungles—and most of the country—are sparsely populated.

NATURAL RESOURCES

Peru's most valuable natural resources are copper, silver, phosphates and zinc. In recent years, zinc and gold mining have led the country's mineral exports. In the southwestern Andes there are considerable iron mines. In the northwest, coal deposits are abundant. There are also reserves of liquid petroleum gas in the basin of the Ucayali River. Peru's extensive coastline on the Pacific Ocean also serves to give it abundant resources of fish, of which anchovies, mackerels, and sardines are most important. Its rivers and rainforests provide the country with valuable hydropower and timber resources, respectively.

FURTHER READINGS

Barrett, Pam. *Insight Guide: Peru.* London: Geocenter International, 2002.

Boehm, David A. *Peru in Pictures.* Minneapolis: Lerner Publications, 1997.

Pearson, David, and Les D. Beletsky. *Peru: The Ecotravellers' Wildlife Guide.* Burlington, Mass.: Academic Press, 2000.

Lutz, Richard L. *The Hidden Amazon: The Greatest Voyage in Natural History.* Salem, Oreg.: Dimi Press, 1998.

Lyle, Gary. *Peru.* Broomall, Pa.: Chelsea House Publications, 1998.

Mennen-Vela, Mónica. *Virtual Peru.* http://www.virtualperu.net/ (Accessed June 3, 2002).

Wright, Ruth M., and Alfredo Valencia Zegarra. *The MacHu Picchu Guidebook: A Self-Guided Tour.* Boulder, Colo.: Johnson Books, 2001.

The Philippines

- **Area:** 115,800 sq mi (300,000 sq km) / World Rank: 72
- **Location:** An archipelago in the Northern and Eastern Hemispheres, between the South China Sea and the Pacific Ocean
- **Coordinates:** 13°00′N, 122°00′E
- **Borders:** No international boundaries
- **Coastline:** 22,499 mi (36,289 km)
- **Territorial Seas:** Determined by treaty and irregular in shape, extending up to 100 NM from shore in some locations
- **Highest Point:** Mount Apo, 9,692 ft (2,954 m)
- **Lowest Point:** Sea level
- **Longest Distances:** 1,150 mi (1,851 km) SSE-NNW / 660 mi (1,062 km) ENE-WSW
- **Longest River:** Agusan River, 240 mi (386 km)
- **Largest Lake:** Laguna de Bay, 356 sq mi (922 sq km)
- **Natural Hazards:** Earthquakes, volcanoes, landslides, typhoons, tsunamis, drought, floods
- **Population:** 82,841,518 (2001 est.) / World Rank: 13
- **Capital City:** Manila, on southeastern Luzon
- **Largest City:** Manila, 1,673,000 (2000 est.)

OVERVIEW

The Philippine Archipelago contains about 7,100 islands (some no more than rocks), and extends over 1,000 mi (1,609 km) from north to south. Only 154 of these islands exceed 5 sq mi (13 sq km) in area. The two largest islands, Luzon in the north and Mindanao in the south, comprise about 65 percent of the total land area of the archipelago. Possession of the Spratly Islands, to the southwest of the archipelago, is contested by the Philippines, China, Malaysia, Taiwan, and Vietnam.

The very complex and volcanic origin of most of the islands is visible in their varied and rugged terrain. Mountain ranges divide the island surfaces into narrow coastal strips and shallow interior plains or valleys. The sea and mountain barriers have been constant factors shaping the historical, cultural, and demographic development of the nation.

MOUNTAINS AND HILLS

Mountains

All of the Philippine Islands are volcanic in origin, and as a result the country is very mountainous. The northern part of Luzon Island is extremely rugged. Luzon's highest peak, Mount Pulog, rises to 9,626 ft

(2,934 m). The island has three mountain ranges that run roughly parallel in a north-south direction. A range in the east, the Sierra Madre, runs so close to the island's eastern shore that there is hardly any coastal lowland. The valley of the Cagayan River separates this eastern range from a large mountain complex to the west, the Cordillera Central. On the west, the Zambales Mountains extend southward and terminate at Manila Bay. Southeastern Luzon consists of a large convoluted peninsula which is a mountainous and volcanic area, containing the active 7,941 ft (2,420 m) Mount Mayon volcano.

The large island of Mindanao has five major mountain systems, some formed by volcanic action. The eastern edge of Mindanao is highly mountainous, including the Diuata Mountains, with several elevations above 6,000 ft (1,828 m), and the southeastern ranges, which reach a high point of 9,200 ft (2,804 m). In central Mindanao there is a broad mass of rugged mountain ranges, one of which bisects the island from north to south. This range contains 9,692 ft (2,954 m) Mount Apo, the highest peak in the country, overlooking Davao Gulf. The broad Cotabato Lowland separates these highlands and a southwestern coastal range.

Volcanoes

The Philippine Islands are situated in a region of considerable geological instability. Most of the islands are located on the Eurasian Tectonic Plate, but a major fault line extends along the eastern part of the archipelago, the boundary with the Philippine Plate. The Philippines are part of the volcanic western Pacific "Ring of Fire," with 37 volcanoes, of which 18 are active. In 1991-92 Mount Pinatubo erupted, with 900 people killed. Its lahar mudslides were particularly destructive. Mount Mayon erupted in February 2000.

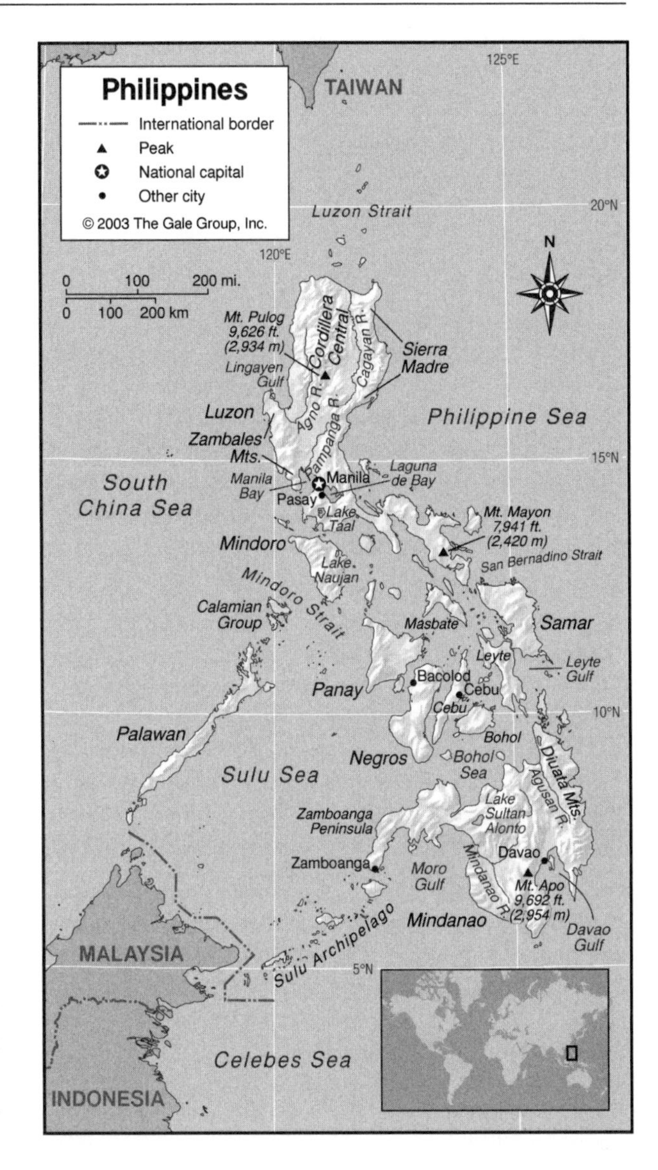

Plateaus

The central mountain complex of Mindanao extends into the northwest corner of the island, terminating in the Bukidnon-Lanao Plateau. At approximately 2,000 ft (609 m) in elevation, the plateau is interspersed with extinct volcanic peaks. On southeast Negros, the volcanic rock Tablas plateau rises 500 to 1,000 ft (152 to 305 m.)

Canyons

Many of the Philippines's rivers have dug canyons through the mountains. Particularly deep canyons fissure the Bukidnon-Lanao Plateau of Mindanao. Pagsanjan Gorge National Park, 39 mi (63 km) southeast of Manila on Luzon, is a river gorge with 300 ft (91 m) high sides.

Hills and Badlands

The low Ragay Hills overlook Ragay Gulf on the Bicol Peninsula of southeastern Luzon. To the south on Samar the terrain is broken up by rocky hills, 500 to 1,000 ft (152 to 305) high. In central Bohol there is a 20 sq mi area known as the Chocolate Hills, with 1,268 mounds, each 164 to 656 ft (50 to 200 m) high, covered in grass which turns brown in the dry season. Their origin has not been determined but they are thought to be eroded coral limestone. There are also hill areas on Panay and nearby Guimaras, and on Masbate, Tablas, and Romblon.

INLAND WATERWAYS

Lakes

On Luzon, southeast of Manila Bay and joined to it by the short Pasig River, is the largest lake in the Philippines, the freshwater Laguna de Bay, which at one time was probably an extension of Manila Bay. It has a water surface of 356 sq mi (922 sq km.) Surrounded by an urbanized region, the lake's water is contaminated by sewage and toxic waste. A few miles to the southwest of Laguna de Bay lies Lake Taal, which has an active volcano in its center that erupted in 1965. Other crater lakes are Lake Danao and Lake Balinsasayan in southeast Negros.

On Mindanao, atop the Bukidnon-Lanao Plateau, is Lake Sultan Alonto (formerly Lake Lanao), 134 sq mi (347 sq km) in area, the second largest lake in the country. The shallow Lake Buluan is in Mindanao south of the Plateau. The lowland of Mindoro contains Lake Naujan, one of the country's larger lakes, which has many fish and bird species.

Rivers

For the most part the country's mountainous terrain causes drainage systems characterized by short, turbulent streams. The larger rivers are not navigable except for short distances. Most main streams and their tributaries are subject to extensive and damaging floods during the heavy rainfall of the monsoon seasons and typhoons. Urban sewage, livestock waste, mining runoff, and industrial chemicals pollute many rivers.

The Cagayan River drains the fertile Cagayan Valley of northern Luzon. It flows northward and empties into the sea at Aparri. This wild river is threatened with contamination as mining operations take over land from agriculture in the valley. The low-lying Central Luzon Plain is interlaced by a network of rivers and streams. Two of the plain's more important rivers are the Agno, which flows northward into Lingayen Gulf, and the Pampanga, which empties into Manila Bay. The short Pasig River flows through the city of Manila.

Two large rivers are found on Mindanao. The Agusan River, the longest in the country, flows northward through the Agusan Valley into the Bohol Sea. The Mindanao River and its tributaries drain the Cotabato Lowland, emptying into Moro Gulf.

Wetlands

The Central Luzon Plain is not much above sea level, and has extensive swamps along the north of Manila Bay and the Candaba Swamp. Mindanao's Davao-Agusan Trough, a lowland in the east, contains seasonally flooded areas. Other Mindanao wetlands include the Libungan and Libuasan marshes, in the central-southern part of the island where tributaries of the Rio Grande come together.

There are four "Wetlands of International Importance" in the Philippines, as designated under the Ramsar Convention on Wetlands. These are: Agusan Marsh Wildlife Sanctuary which includes rare swamp forest and peat forest; Olango Island Wildlife Sanctuary, a shorebird habitat; Naujan Lake National Park; and Tubbataha Reefs National Marine Park.

THE COAST, ISLANDS AND THE OCEAN

Oceans and Seas

Branches of the Pacific Ocean surround the Philippines, and its islands enclose between them many more. The east coast of the Philippines faces the open Pacific Ocean, where the Philippine Trough (Emden Deep) plunges to 34,219 ft (10,430 m). The northwest faces the South China Sea. The southwest surrounds the Sulu Sea on three sides. The Celebes (Sulawesi) Sea is in the south, between the island of Mindanao and the Indonesian island of Sulawesi. The Bohol Sea is to the north of Mindanao. The Visayan Sea is encircled by Panay, Masbate, Cebu, Negros, and other islands. The Sibuyan Sea meets southern Luzon and eastern Mindoro. The Camotes Sea lies between Cebu, Leyte, and Bohol. The Samar Sea is between Samar and Masbate. Leyte Gulf separates Leyte and Samar. The Philippines's biodiverse coral reefs are endangered by fishing (using dynamite and poisons), and sedimentation from deforestation.

There are countless narrows between the Philippine islands. Principal among them are the San Bernadino Strait and Verde Island Passage, which permit ocean travel across the northern part of the archipelago. The Surigao Strait allows travel between the Pacific and the Bohol Sea in the south. The Mindoro Strait lies between Mindoro and the Calamian Group of islands. A number of channels north of the country make up the Luzon Strait, separating that island from Taiwan.

Major Islands

The northernmost part of the Philippines, the Luzon region consists of Luzon Island, many much smaller adjacent islands, and the small island groups of Batan and Babuyan to the north. The largest of the Philippine islands, Luzon has an area of 40,420 sq mi (104,687 sq km), which is more than one third of the country's total area. In shape it resembles an upright rectangle with an irregular southeastern peninsula. The main part of the island is roughly 250 mi (402 km) in length and has a width generally between 75 and 100 mi (120 and 160 km).

Just south of Luzon lies Mindoro, with an area of 3,758 sq mi (9,733 sq km.) The island is largely mountainous and has high peaks rising above 8,000 ft (2,438 m). A coastal lowland lies to the east and northeast of the mountain zone.

Southwest of Mindoro is the Calamian Group of islands, with the long, narrow island of Palawan beyond them. It has a length of over 275 mi (442 km), a width varying from 5 to 30 mi (8 to 48 km), and an area of 4,500 sq mi (11,655 sq km). Narrow coastal strips border a ridge of rugged mountains that run its entire length. Over 1,100 smaller islands and islets surround Palawan.

The Visayan Islands are grouped in a roughly circular pattern around the Visayan Sea. They include seven large, populated islands that range in size from Masbate, which is 1,262 sq mi (3,268 sq km) in area, to Samar, 5,050 sq mi (13,079 sq km). The others are Bohol (1,492 sq mi, 3,865 sq km), Cebu (1,707 sq mi, 4,421 sq km), Leyte (2,785 sq mi, 7,213 sq km), Panay (4,446 sq mi, 11,515 sq km), and

Negros (4,905 sq mi, 12,703 sq km). Including numerous islets, the Visayan group includes over half of the islands that make up the country. Samar and Leyte, the easternmost islands, act as a buffer for the other islands against the full force of typhoons originating from the Pacific Ocean. Samar's interior has rugged mountains, but elevations are generally below 1,000 ft (305 m). Leyte is separated from Samar by a narrow strait. A central mountain range divides Leyte. The long narrow island of Cebu is the site of the country's largest copper mine and also produces low-grade coal and limestone for cement. West of Cebu, Negros, and Panay are roughly similar in area, with lowland plains that permit intensive agriculture.

The Mindanao region consists of the island of Mindanao and numerous small offshore islands, including a mineral-rich group off the northeastern coast. Mindanao, the second largest of the Philippine Islands, has an area of 36,537 sq mi (94,630 sq km). In the east of the island, two mountain ranges are separated by the Agusan River. To the southwest of those ridges several rivers meet in the Cotabato Basin, and mountain peaks lead to the Bukidnon-Lanao Plateau. West of the Plateau, the island narrows to an isthmus ten miles wide, from which the long Zamboanga Peninsula protrudes to the southwest for some 170 mi (273 km.) The peninsula is covered largely with mountains and possesses limited coastal lowlands.

Southwest of the Zamboanga Peninsula of Mindanao is the Sulu Archipelago, a string of smaller islands of volcanic and coral origin protruding from a submarine ridge that joins Mindanao to the Malaysian state of Sabah, on the island of Borneo. A chain 200 mi (322 km) long, the Sulu Archipelago has over 800 islands. Its total area is about 1,600 sq mi (4,144 sq km.) Its three principal islands are Basilan, directly offshore from Zamboanga; Jolo; and Tawi-Tawi, near Sabah.

The Coast and Beaches

Among the Visayan Islands there are two large gulfs: Leyte in the southeast and Panay in the west. Luzon's western coast is indented by Lingayen Gulf; south of it the Bataan Peninsula extends around Manila Bay. The capital city of Manila is located on the bay's eastern shore. Tayabas Bay indents Luzon's southern coastline, with the Bondoc Peninsula between it and the Ragay Gulf. The southeastern extension of Luzon ends in the Sorsogon Peninsula. North of the peninsula on the east coast is Lamon Bay, then, moving further north, Dingalan Bay, and Escarpada Point. The mountains of the east, an extension of the Sierra Madre, drop off sharply to the Pacific Ocean and leave little accessible level terrain and no protective harbors against the heavy Pacific surf.

Mindanao's very irregular shape is characterized by a number of sizable gulfs and bays and several large peninsulas that give it an extremely long coastline. Mindanao's northernmost point is the Surigao Peninsula, with Butuan Bay to its west. Further southwest along the jagged coast, Iligan Bay makes a deep indentation, making a narrow isthmus, which connects the large Zamboanga Peninsula to the rest of Mindanao. Sibuguey and Baganian Peninsulas protrude from the south coast of the Zamboanga Peninsula on Moro Gulf, with Pagadian Bay on the south of the isthmus and Illana Bay continuing the southwest coast. Sarangani Bay indents the coast just above its southernmost part, Tinaca Point. North of that point is Davao Gulf, defined by Cape San Agustin. The jagged east coast continues north, with several river outlets.

GEO-FACT

The mountain rice terraces of northern Luzon's Cordillera are a UNESCO World Heritage site. Built by the indigenous Ifugao people over the last two millennia, the terraces follow mountain contours over 3,281 ft (1,000 m) high, creating an agricultural landscape that is both productive and harmonious with nature.

CLIMATE AND VEGETATION

Temperature

The Philippines has a tropical maritime climate, with two seasons: November-April, when the northeast monsoon brings rain, and May-October, when the southwest monsoon brings cool, dry weather. The average temperature is 80°F (27°C) with a range between 73°-90°F (23° and 32°C). Humidity averages 77 percent.

Rainfall

The annual average rainfall varies from 38 to 106 in (96 to 406 cm). The northern islands are often heavily affected by seasonal typhoons. which cause destructive winds and flooding rains.

Grasslands

Wild imperata and saccharum grasslands are common throughout the islands. These areas of tall, thick grass may have originated in repeated burning of forests to clear land for agriculture and pastures, as well as from volcanic burning. A 1997 survey found 17 percent of the Philippines covered with grasslands. Efforts are being made to convert grasslands to plantations, or to reforest them. Savannahs, mixing grasslands and scrub woods, are found in Luzon's Cagayan Valley, and amid the hills of Mindoro, Negros, and Masbate, as well as on Panay, and on Mindanao's Bukidnon-Lanao Plateau.

Population Centers – Philippines

(2000 POPULATION ESTIMATES)

Name	Population	Name	Population
Quezon City	2,173,831	Davao City	1,147,116
Manila (capital)	1,581,082	Cebu City	718,821
Caloocan City	1,177,604	Zamboanga City	601,794

SOURCE: "Total Population, Number of Households, Average Household Size, Population Growth Rate and Population Density by Region, Province, and Highly Urbanized City: as of May 1, 2000." *Census 2000 Final Counts.* National Statistics Office, Philippines.

Provinces – Philippines

MID-2000 CENSUS OF POPULATION

Name	Population	Area (sq mi)	Area (sq km)
Autonomous Region in Muslim Mindanao	2,412,159	--	--
Bicol	4,674,855	6,808	17,633
Cagayan Valley	2,813,159	14,055	36,403
Caraga	2,095,367	--	--
Central Luzon	8,030,945	7,039	18,231
Central Mindanao	2,598,210	8,994	23,293
Central Visayas	5,701,064	5,773	14,951
Cordillera Administrative Region	1,365,220	--	--
Eastern Visayas	3,610,355	8,275	21,432
Ilocos	4,200,478	8,328	21,568
National Capital Region	9,932,560	246	636
Northern Mindanao	2,747,585	10,937	28,328
Southern Mindanao	5,189,335	12,237	31,693
Southern Tagalog	11,793,655	18,117	46,924
Western Mindanao	3,091,208	7,214	18,685
Western Visayas	6,208,733	7,808	20,223

SOURCE: "Total Population, Number of Households, Average Household Size, Population Growth Rate and Population Density by Region, Province, and Highly Urbanized City: as of May 1, 2000." *Census 2000 Final Counts.* National Statistics Office, Philippines.

Forests and Jungles

The forests of The Philippines have been reduced to just 3 percent cover, with approximately 2,702 sq mi (7,000 sq km) remaining. Most of that is secondary growth, and hardly any is pristine. Primary forest persists only in the mountains of Palawan, Mindoro, and Mindanao, and in the Sierra Madres of northeast Luzon. These last Philippine rainforests have an extraordinary level of biodiversity. Forests throughout the islands have been destroyed by illegal and legal logging, as well as by agricultural clearing.

HUMAN POPULATION

The Philippines has high population density, with 660 people per sq mi (255 per sq km) and a high growth rate of 2.36 percent (2001). Migration to cities from rural areas has been an ongoing trend, with 59 percent of the population living in urban areas (2001 estimate), where overcrowding and stresses on infrastructure are constant problems.

NATURAL RESOURCES

The Philippines's greatest resources are agriculture in its rich volcanic soils and fish from the surrounding waters. It also has significant deposits of many minerals, including nickel, manganese, chromite, cobalt, silver, salt, gold, copper, limestone, and coal. Petroleum is present, with offshore deposits a source of maritime boundary disputes. Geothermal energy and hydropower are readily available due to the Philippines's geological activity and many fast-flowing streams.

FURTHER READINGS

Broad, Robin, and John Cavanagh. *Plundering Paradise: the Struggle for the Environment in the Philippines.* Berkeley CA: University of California Press, 1994.

Franca, Luis H. *Eye of the Fish.* New York: Kaya Press, 2001.

VolcanoWorld. *Tectonics and Volcanoes of the Philippines.* http://volcano.und.nodak.edu/vwdocs/volc_images/southeast_asia/philippines/tectonics.html (Accessed April 27, 2002).

Wenzel, Eberhard. *Philippines.* http://www.ldb.org/philippi.htm (accessed April 27, 2002).

Wernstedt, Frederick. *The Philippine Island World.* Berkeley CA: University of California Press, 1967.

Poland

- **Area:** 120,728 sq mi (312,685 sq km) / World Rank: 70
- **Location:** Located in Central Europe in the Northern and Eastern Hemispheres; south of the Baltic Sea; south and west of Russia, Lithuania, Belarus, and the Ukraine; north of Slovakia and the Czech Republic; east of Germany
- **Coordinates:** 52°00′N, 20°00′E
- **Borders:** 1,794 mi (2,888 km) total / Russia, 128 mi (206 km); Lithuania, 56.5 mi (91 km); Belarus, 376 mi (605 km); Ukraine, 266 mi (428 km); Slovakia, 276 mi (444 km); Czech Republic, 409 mi (658 km); Germany, 283 mi (456 km)

- **Coastline:** 305 mi (491 km)
- **Territorial Seas:** 12 NM
- **Highest Point:** Mt. Rysy, 8,199 ft (2,499 m)
- **Lowest Point:** Raczki Elblaskie, 6.6 ft (2 m) below sea level
- **Longest Distances:** 428 mi (689 km) E-W / 403 mi (649 km) N-S
- **Longest River:** Vistula, 661 mi (1064 km)
- **Largest Lake:** Lake Śniardwy, surface area of 42.3 sq mi (109.7 sq km), depth up to 51 ft (17 m)
- **Natural Hazards:** Floods
- **Population:** 38,633,912 (July 2001 est.) / World Rank: 30
- **Capital City:** Warsaw, located on the Vistula River in Eastern Poland
- **Largest City:** Warsaw, 1,609,000 (2002 est.)

OVERVIEW

Poland is an unbroken plain extending from the Baltic shore to the Carpathian Mountains in the south. Differences in climate and terrain occur, in bands that extend east to west, accounting for the wide variations in land utilization and population density. The coastal area lacks natural harbors except those at Gdańsk-Gdynia and Szczecin. The coast and the adjoining lake district have fewer natural resources, fertile soils, and people than areas to the south. The vast plains south of the lake district have more fertile soils, a longer growing season, and a denser population than the northern regions. The southern foothills and mountains contain most of the country's mineral wealth and much of the most fertile soils and have attracted the greatest concentration of industry and people. Poland is situated on the Eurasian Plate.

MOUNTAINS AND HILLS

Mountains

Mt. Rysy (Mount Rysy), the highest peak at 8,199 ft (2,499 m), is on the Czech and Slovak borders about 59 mi (95 km) south of Krakow (Kraków) in the Tatra (Tatry) range of the Carpathians. Six other peaks on the Polish side of the Tatra Mountains reach 6,233 ft (1,900 m) or more. The Sudety Mountains are lower, only one peak exceeding 5,249 ft (1,600 m). In both ranges a total of about 115 sq mi (300 sq km) rises about 3,280 ft (1,000 m). Neither range is rugged enough over large areas to limit habitation, and in only a few isolated places is the population density below the country average. Most of the more rugged slopes are in the Tatra Mountains; many slopes in the Sudety range are gentle and can be cultivated or used as meadows and pastures on dairy farms. Poland's average elevation is 567 ft (173 m); more than 90 percent of its area lies below 984 ft (300 m).

Hills and Badlands

Only 3 percent of the nation of Poland—chiefly in the far south and southwest and extending across the country parallel to the southern border in a belt roughly 55 to 74 mi (90 to 120 km) wide—rises above 1,640 ft (500 m). These are small highland areas in the Carpathian and Sedeten (Sudety) Mountains, shared with Czech Republic and Slovakia.

This area of foothills of the two mountain ranges to the south of the central lowlands blend into the mountains in the extreme south and in the southwestern corner of the country.

North of the central lowlands are the hills, forests, and lakes created by the recession of the most recent glacier a millennia ago. The effects of glaciation are the most prominent features of the terrain for 124 mi (200 km) or more inland from the Baltic Sea in the western part of the country but for a much shorter distance in the east. The earth and stone carried by glaciers embossed what would have been a nearly flat area. Glacial action led to the formation of many lakes and low hills.

INLAND WATERWAYS

Lakes

Northeast Poland is rich in lakes that were formed by the retreating Scandinavian glacier. Much of the inland area that comprises the lake district, which extends across the entire country (from the Vistula valley in the West to the Russian border in the North and East, and south to Podlasie Plain and Mazovian Plain), is poorly drained. The many lakes add beauty and value to the region, but its swampland has been difficult to reclaim. Most of the lakes are small and shallow; yet nearly a dozen, including some very small ones, have depths of 164 ft (50 m) or more.

The lake district is subdivided into smaller regions. In the Pomeranian Lake District there are 4,192 lakes, which occupy over 290,000 acres (115,000 ha), while in the Masurian Lake District there are 2,561 lakes occupying altogether almost 355,000 acres (142,000 ha).

In central Poland (Wielkopolska-Kujawy Lowland and Polesie Lubelskie) there are 1,711 lakes, which cover 132,500 acres (53,000 ha). In the rest of the country there are only 895 lakes with the total surface area of 17,000 acres (6,800 ha).

Rivers

By far the greatest portion of the country drains northwestward to the Baltic Sea by way of the Vistula (Wisla) and Oder (Odra) Rivers. Most other rivers join

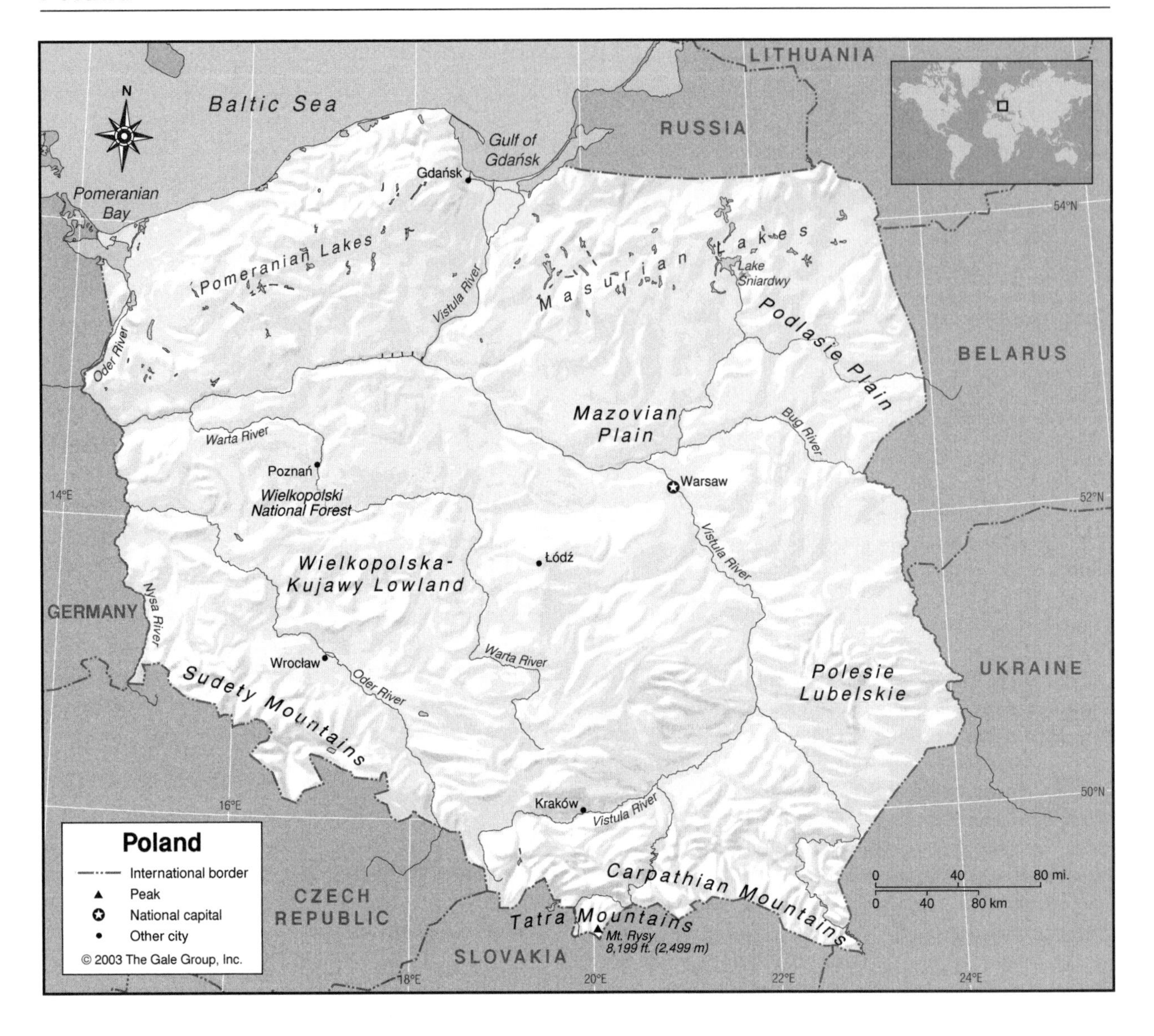

the Vistula and Oder systems, but a few streams in the northeastern region reach the sea through Russian territory.

The Vistula is the greatest river in Poland. The Vistula and its tributaries drain the country's largest basin, an area almost double the Polish portion of the Oder basin. The Vistula basin includes practically all of the southeastern and east-central regions and much of the northeast. The Vistula rises in the Tatra range of the Carpathians in Upper Silesia, near the boundaries of Poland, the Czech Republic, and Slovakia, and exits at the Gulf of Gdańsk (Danzig). Its catchment area includes 56 percent of the country. The Vistula links the old Polish capital of Krakow, with the new capital Warsaw, and completes its flow to the Baltic Sea at Gdańsk, the water exit for Poland's economy. Since Viking times it has been part of an important trade route from the east to northern and western Europe. This navigational use has hardly altered the character of much of the river. Most of its tributaries flow to it from the east, rising in Russia or near the Russian border. One of them, the Bug, forms about 174 mi (280 km) of the Polish-Russian border.

The Oder, which with the Nysa (Neisse) River forms most of the border between Poland and Germany, is fed by several other rivers and streams, including the Warta, which drains a large section of central and western Poland. The Oder reaches the Baltic Sea through the harbors and bays north of Szczecin.

Wetlands

There are large areas of swampland in the northern lake district because of poor drainage, and land has been hard to reclaim.

THE COAST, ISLANDS, AND THE OCEAN

At both Gdańsk-Gdynia and Szczecin on the Baltic Sea are natural harbors. The coastline is a narrow lowland dotted with bays (including Pomeranian Bay), lakes,

GEO-FACT

The name of Poland's longest river, the Vistula, was recorded by the Roman historian Tacitus in 98 A.D. in his *De origine et situ Germanorum*, an account of the origin and location of the German tribes. The German tribes included the Goths, who gave the river its name.

Population Centers – Poland

(1993 POPULATION ESTIMATES)

Name	Population	Name	Population
Warsaw (Warszawa, capital)	1,643,203	Poznań	582,839
Lódź	835,807	Gdańsk	462,239
Kraków (Cracow)	744,203	Szczecin	417,115
Wroclaw	641,386	Bydgoszcz	384,101
		Katowice	359,776
		Lublin	350,731

SOURCE: "Population of Capital Cities and Cities of 100,000 and More Inhabitants." United Nations Statistics Division.

Provinces – Poland

1999 POPULATION ESTIMATES

Name	Population	Area (sq mi)	Area (sq km)	Capital
Dolnośląskie	2,977,611	7,702	19,947	Wroclaw
Kujawsko-Pomorskie	2,100,771	6,938	17,969	Bydgoszcz
Lódzkie	2,652,999	7,034	18,219	Lódz
Lubelskie	2,234,937	9,697	25,114	Lublin
Lubuskie	1,023,483	5,399	13,984	Gorzów Wielkopolski
Malopolskie	3,222,525	5,847	15,144	Kraków
Mazowieckie	5,069,977	13,744	35,597	Warsaw
Opolskie	1,088,272	3,634	9,412	Opole
Podkarpackie	2,126,001	6,921	17,926	Rzeszów
Podlaskie	1,222,709	7,791	20,179	Bialystok
Pomorskie	2,192,268	7,063	18,292	Gdańsk
Slaskie	4,865,512	4,747	12,294	Katowice
Swietokrzyskie	1,322,747	4,507	11,672	Kielce
Warmińsko-Mazurskie	1,465,577	9,344	24,202	Olsztyn
Wielkopolskie	3,355,332	11,515	29,825	Poznań
Zachodnio-pomorskie	1,732,838	8,842	22,901	Szczecin

SOURCE: Central Statistical Office, Poland. Cited by Johan van der Heyden, Geohive, http://www.geohive.com (accessed June 2002).

and promontories (high-points of land or rock projecting into a body of water).

CLIMATE AND VEGETATION

Temperature

Poland's continental climate is affected by westerly winds. Summers are generally cool, with only the southern portions of the country experiencing notable humidity. Winter can be cold to frigid. Average temperatures range from 21–30°F (-6 to -1°C) in January and 55–75°F (13–24°C) in July.

Rainfall

Annual average precipitation ranges from 20 in (50 cm) in the lowlands to 53 in (135 cm) in the mountains. For the country as a whole, the average annual precipitation rate is 25 in (64 cm).

Grasslands

Most of Poland lies in the North European Plain that extends from the North Sea coast of the Netherlands to the Ural Mountains in the Russia. The lower land is found just south of the Gulf of Gdańsk, where approximately 23 sq mi (60 sq km) lie below sea level.

Geographers usually divide the country into five topographic zones, each extending from west to east. The largest, accounting for three-fourths of Poland's territory, is the great central lowlands area. It is narrow in the west but expands to both the north and the south as it extends eastward. At the eastern border it includes everything from near the northeastern tip of the country to about 124 mi (200 km) from the southeastern corner.

Forests

It was estimated in 2001 that twenty-nine percent of Poland's land mass is forested. The most important varieties of trees are pine, larch, spruce, and fir, which account for 70 percent of the forests; deciduous species include elm, beech, and birch. Difficult growing conditions—degraded, dry sandy soils; wide temperature fluctuations in winter; hurricane-strength winds; drought conditions in summer—have contributed to the forests suffering natural adversity. Loss of forest land through territorial redistribution and wartime destruction has been offset by a governmental program of reforestation. Despite these multiple difficulties, forestry has continually developed as an important sector of the domestic economy; although Poland no longer produces enough wood for its own purposes, much less any for export.

The Wielkopolski National Forest, located in a reservation in the Rogalin province, is well known for its thousand-year-old oak trees.

HUMAN POPULATION

Poland's population exceeds 38,600,000, although the growth rate is in decline (-0.03 percent in 2000). Poles are the dominant ethnic group, accounting for almost 98 percent of the population in the 1990s. The country was more diverse before WWII with Poles accounting for only 70 percent of the population. However, the changes in national boundaries and populations shifts that occurred at the close of the war resulted in a ethnically homogeneous population. As of 2000 Germans accounted for only 1.3 percent of the population, Ukrainians 0.6 percent, and Byelorussians 0.5 percent.

Religions represented are Roman Catholic, 95 percent (about 75 percent practicing); Eastern Orthodox, Protestant, and others comprise 5 percent. At the beginning of World War II, Jews constituted over 3.3 million of the Polish population—the largest of any country—nearly 10 percent of the total. The majority of the Jews were murdered during the German occupation in World War II, and many others emigrated in the succeeding years. As of 1990 only 3,800 Jews lived in Poland.

NATURAL RESOURCES

The mineral deposits—coal, sulfur, copper, natural gas, silver, uranium, lead, zinc, glass sand, and salt—in Poland are its greatest natural resource. The most important raw material is hard coal. Arable land, which accounts for 47 percent of total land mass, is an important resource to the Polish domestic economy.

FURTHER READINGS

Barnett, Clifford R. *Poland, Its People, Its Society, Its Culture.* New Haven, CT: HRAF Press, 1958.

Brandys, Marian. *Poland.* Garden City, NY: Doubleday, 1974.

Corona, Laurel. *Poland.* San Diego, CA: Lucent Books, 2000.

Furlong, Kate A. *Poland.* Edina, MN: Abdo Publishing, 2001.

McCollum, Sean. *Poland.* Minneapolis, MN: Carolrhoda Books, 1999.

Portugal

- **Area:** 35,672 sq mi (92,391 sq km) / World Rank: 111
- **Location:** Northern and Western Hemispheres, Western Europe, bordering Spain to the northeast, east, and southeast and the Atlantic Ocean to the southwest and west
- **Coordinates:** 39°30′N, 8°00′W
- **Borders:** 754 mi (1,214 km), all with Spain
- **Coastline:** 1,114 mi (1,793 km)
- **Territorial Seas:** 12 NM

GEO-FACT

The name *Madeira*, taken from the Portuguese word for wood, was given to these islands due to the dense forests their discoverers found there.

- **Highest Point:** Estrela, 6,532 ft (1,991 m; on the mainland) / Ponta do Pico, 7,713 ft (2,351 m; in the Azores Islands)
- **Lowest Point:** Sea level
- **Longest Distances:** 349 mi (561 km) N-S; 135 mi (218 km) E-W
- **Longest River:** Tagus, 626 mi (1,007 km)
- **Natural Hazards:** Earthquakes
- **Population:** 10,066,253 (July 2001 est.) / World Rank: 79
- **Capital City:** Lisbon, on the west coast
- **Largest City:** Lisbon, 1,971,000 (2000 est.)

OVERVIEW

Portugal, located at the westernmost edge of continental Europe, occupies approximately one-sixth of the Iberian peninsula, which it shares with Spain.

Although Portugal's boundary with Spain was fixed before those of any other European countries, there are few natural frontiers between the two nations; many of Portugal's geographical features are continuations of those in Spain.

In addition to its continental territory, Portugal also has jurisdiction over two autonomous island groups in the Atlantic—the Azores and Madeira—remnants of a once far-flung empire.

Portugal's major topographical dividing lines are the Douro and Tagus (Tejo) rivers and the Serra da Estrela, all of which run across the country from east to west. The area north of the Douro has low but rugged mountains. The Serra da Estrela, in roughly the center of the country, is the highest mountain range. The Tagus River forms a dividing line between the hilly to mountainous regions of the north and the rolling plains of the south. There is another region of low mountains in the far south of the country, with the plains of Algarve beyond them.

Portugal is situated on the Eurasian Tectonic Plate, not far north of its boundary with the African Tectonic Plate. There are several zones of intense seismic activity as well as major geological faults. The largest zones are concentrated in the Algarve, the greater Lisbon area, and

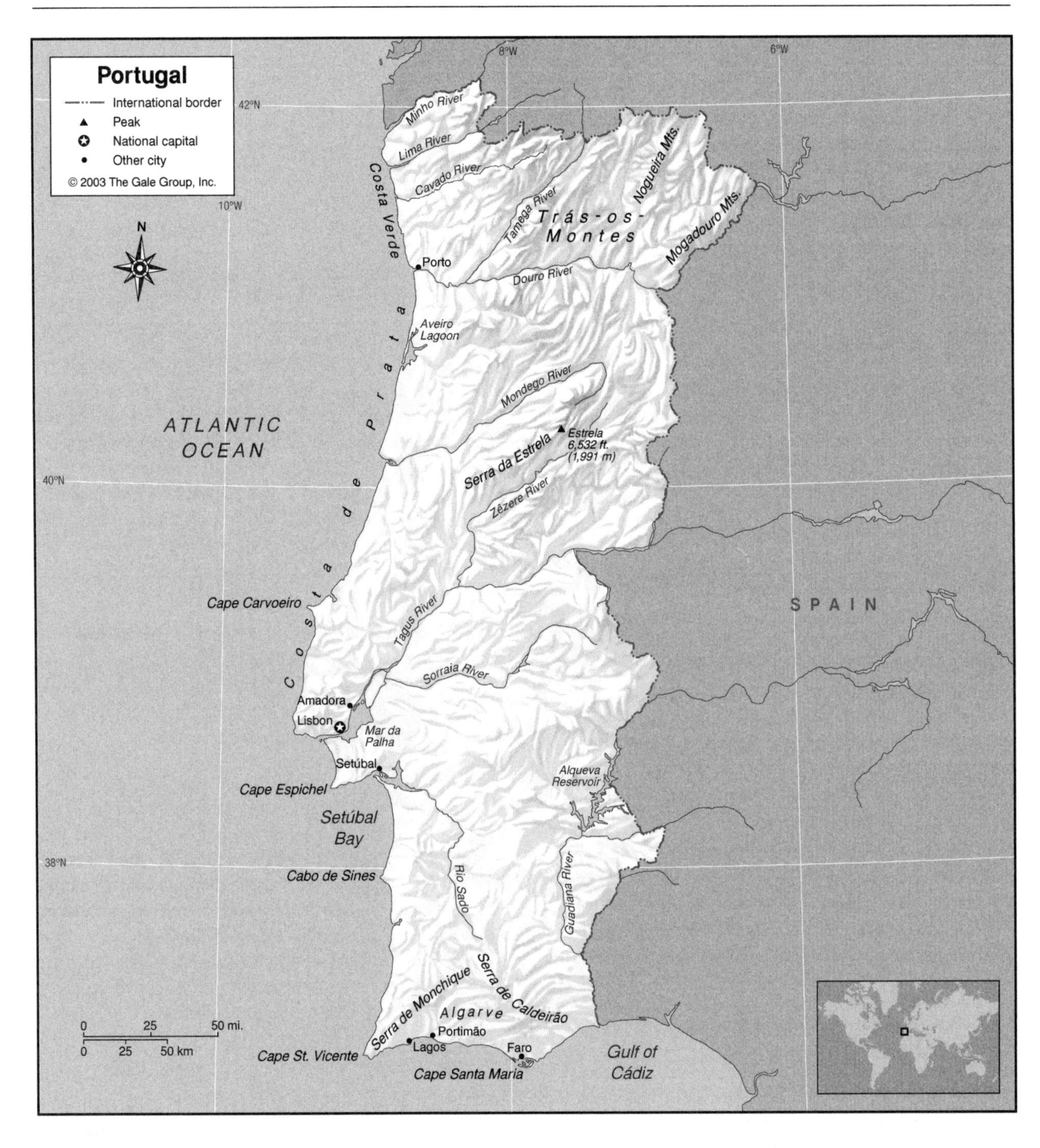

the Tagus River estuary. A disastrous earthquake on November 1, 1755, killed an estimated 20,000 people and caused extensive damage in Lisbon, Setúbal, Lagos, Portimão, and Faro. Seismic activity in the twentieth century has been centered in the north.

MOUNTAINS AND HILLS

Mountains

The most mountainous part of Portugal is north of the Douro River, in the region known as Trás-os-Montes, or "beyond the mountains." Its ranges are part of the same system as the Cantabrian Mountains of Spain, and include the Nogueira and Mogadouro Mountains. However, Portugal's tallest mountain range is the Serra da Estrela in the central part of the country, which includes the highest peak, also called Estrela (6,532 ft / 1,991 m). The southern end of the country is marked by two low mountain ranges: the Serra de Monchique in the west and the Serra de Caldeirão in the east.

Hills and Badlands

The region south of the Tagus, Alentejo (literally, the land across the Tagus, or Tejo) is an area of gently rolling hills that generally rise to about 600 ft (183 m) but occasionally reach between 900 and 1,500 ft (274 and 457 m). The northwestern part of the country, Minho, is also hilly. The beautiful river valleys of the Lima, Cavado, and Tamega cross it from northeast to southwest, providing some of the finest scenery in Europe. The terraced hills along the Douro in the Trás-os-Montes region east of Minho are Portugal's famous port wine region.

Population Centers – Portugal

(1991 POPULATION ESTIMATES)

Name	Population
Lisbon (Lisboa, capital)	663,394
Porto (Oporto)	302,467
Amadora	181,774
Funchal	115,403
Setúbal	103,634

SOURCE: "Population of Capital Cities and Cities of 100,000 and More Inhabitants." United Nations Statistics Division.

INLAND WATERWAYS

Rivers

Of the ten largest rivers in Portugal, five have their origins in Spain and form part of the Spanish-Portuguese boundary at one or more points in their courses toward the Atlantic. The remaining five are entirely within Portugal and are short, the longest being the 108-mile-long (174-kilometer-long) Rio Sado.

The Douro, on whose estuary Porto is located, is the largest river in northern Portugal. The total length of this river is 584 mi (940 km), of which 460 mi (740 km) are in Spain and 124 mi (200 km) in Portugal. The Douro drains 7,200 sq mi (18,653 sq km) in Portugal. The Minho River along the northern border is the other major river of the region.

The longest river in Portugal as well as in the Iberian Peninsula is the Tagus, with a total length of 621 mi (999 km), and a 142-mi (228-km) course in Portugal. The Tagus basin, one of the most fertile regions in Portugal, extends more than 9,620 sq mi (24,922 sq km). Its estuary, the Mar da Palha, is one of the world's great natural harbors and the site of the capital city of Lisbon. The Sorraia in the south and the Zêzere in the north are major Portuguese tributaries of the Tagus. The Mondego River drains the area north of the Zêzere and south of the Douro.

The major river of southern Portugal is the Guadiana. Like Portugal's other large rivers it has its source in Spain. The Guadiana flows along the border with Spain, curves into Portugal, then curves back to the east to form the most southeastern part of the Spanish border before emptying into the Gulf of Cádiz. The large Alqueva Reservoir can be found on the section of the Guadiana within Portugal's borders.

THE COAST, ISLANDS, AND THE OCEAN

The Coast and Beaches

Mainland Portugal has an Atlantic coastline of more than 500 mi (805 km). Most of the coastline is smooth, but there are indentations at the mouths of the major rivers. The major harbors are at the mouths of the Tagus, Mar da Palha, and the Sado, Setúbal Bay.

The forested northern part of the coast, famous for the vineyards around Porto (where the grapes for port wine are grown), is called the Costa Verde, or "green coast." The midsection of the coast, between Porto and Lisbon, is called the Costa de Prata, or "silver coast." Its most prominent feature is the saltwater Aveiro Lagoon. Aveiro's network of canals has led to comparisons with Venice, Italy, while the fertile land reclaimed from the sea in this region has given rise to the nickname "the Portuguese Holland."

Several capes jut out into the Atlantic in the southern half of the coast, including Cape Carvoeiro, Cape Espichel south of Lisbon, Cabo de Sines, Cape St. Vicente at the southwestern end of the country, and Cape Santa Maria along the Gulf of Cádiz in the south.

Major Islands

Portugal has no islands of note near its coastline, but possesses two archipelagos located well out into the Atlantic Ocean. The Madeiran archipelago, about 600 mi (960 km) southwest of the mainland, consists of the island of Madeira, which is 34 mi (55 km) long and 14 mi (23 km) wide; the smaller island of Porto Santo; and the uninhabited Desertas and Selvagens islets. On the main island the land rises sharply from the coast to a height of 6,106 ft (1,861 m) at Mt. Ruivo de Santana.

The Azorean archipelago, about 800 mi (1,300 km) due west of Portugal, is a volcanic mountain chain of nine islands divided into three groups: São Miguel and Santa Maria to the east; Terceira, Pico, Faial, São Jorge, and Graciosa in the center; and Flores and Corvo to the northwest. Thermal springs are features on the largest island, San Miguel. The tallest mountain in all of Portugal is Ponta do Pico (7,713 ft / 2,351 m) on Pico.

CLIMATE AND VEGETATION

Temperature

Proximity to the Atlantic keeps Portugal's climate generally temperate. The northwest has a maritime climate, with short, cool summers and mild winters. In the northeast the climate is more continental, with sharper

Districts and Regions – Portugal

Name	Area (sq mi)	Area (sq km)	Capital
Aveiro	1,084	2,808	Aveiro
Azores (Autonomous Region)	868	2,247	Ponta Delgada
Beja	3,948	10,225	Beja
Braga	1,032	2,673	Braga
Bragança	2,551	6,608	Bragança
Castelo Branco	2,577	6,675	Castelo Branco
Coimbra	1,524	3,947	Coimbra
Évora	2,854	,393	Évora
Faro	1,915	4,960	Faro
Guarda	2,131	5,518	Guarda
Leiria	1,357	3,515	Leiria
Lisboa	1,066	2,761	Lisbon (Lisboa)
Madeira (Autonomous Region)	306	794	Funchal
Portalegre	2,342	6,065	Portalegre
Porto	925	2,395	Porto
Santarém	2,605	6,747	Santarém
Setúbal	1,955	5,064	Setúbal
Viana do Castelo	871	2,255	Viana do Castelo
Vila Real	1,671	4,328	Vila Real
Viseu	1,933	5,007	Viseu

SOURCE: *Geo-Data: The World Geographical Encyclopedia,* 2nd ed. Detroit: Gale Research, 1989.

contrasts between the seasons. The central part of the country has hot summers and mild, rainy winters, and the south has a dry climate with long, hot summers. Average temperatures in Lisbon are about 75°F (24°C) in July and about 40°F (4°C) in January.

Rainfall

Average annual rainfall ranges from more than 120 in (305 cm) in the northwestern grape-growing region to 20 in (51 cm) on the southern coast. Average annual rainfall in Lisbon is 27 in (69 cm).

Grasslands

The plains of the Alentejo region, between the Tagus and the southern mountains, account for one-third of the country's total area and are Portugal's agricultural heartland.

The coastal plain between the Tagus and the Douro, which extends inland up to 30 mi (48 km) from the ocean, contains salt marshes and alluvial deposits and stretches of sand dunes, in some places 2 to 5 mi (3 to 8 km) wide.

Forests and Jungles

Between one-fourth and one-fifth of Portugal is forested. Pine and other evergreens are found in the northern and central coastal areas. Cork and olive trees grow in the central grasslands. (Portugal is the world's number one cork producer, accounting for half the total produced.) Evergreen brush is found in the dry southern region.

HUMAN POPULATION

In 2001, Portugal had a population density of 282 people per sq mi (109 people per sq km). The population is mostly rural, with only 36 percent living in cities. Lisbon and Porto account for most of the urban dwellers. Approximately two-thirds of the population lives in coastal areas, and especially near the mouths of the Douro and Tagus Rivers.

NATURAL RESOURCES

Portugal's most important natural resources are its forests and arable land, and the fish in its coastal waters. Mineral resources include tin, iron ore, tungsten, uranium ore, and coal.

FURTHER READINGS

Kempner, Mary Jean. *Invitation to Portugal.* New York: Atheneum, 1969.

Proper, Datus C. *The Last Old Place: A Search Through Portugal.* New York: Simon and Schuster, 1992.

Saramago, Josi. *Journey to Portugal: In Pursuit of Portugal's History and Culture.* Translated from the Portuguese and with notes by Amanda Hopkinson and Nick Caistor. New York: Harcourt, 2001.

Stoop, Anne de. *Living in Portugal.* Paris: Flammarion, 1995.

Symington, Martin. *Essential Portugal.* Lincolnwood, Ill.: Passport Books, 1994.

Puerto Rico

A Commonwealth of the United States of America

- **Area:** 3,515 sq mi (9,104 sq km)/ World Rank: 165
- **Location:** Northern and Western Hemispheres, between the Caribbean Sea and the Atlantic Ocean, east of the Dominican Republic.
- **Coordinates:** 18°15′N, 66°30′W
- **Borders:** None
- **Coastline:** 311 mi (501 km)
- **Territorial Seas:** 12 NM
- **Highest Point:** Cerro de Punta, 4,390 ft (1,338 m)
- **Lowest Point:** Sea level
- **Longest Distances:** 111 mi (179 km) E-W; 36 mi (58 km) N-S
- **Longest River:** Río La Plata, 46 mi (74 km)
- **Natural Hazards:** Hurricanes, drought

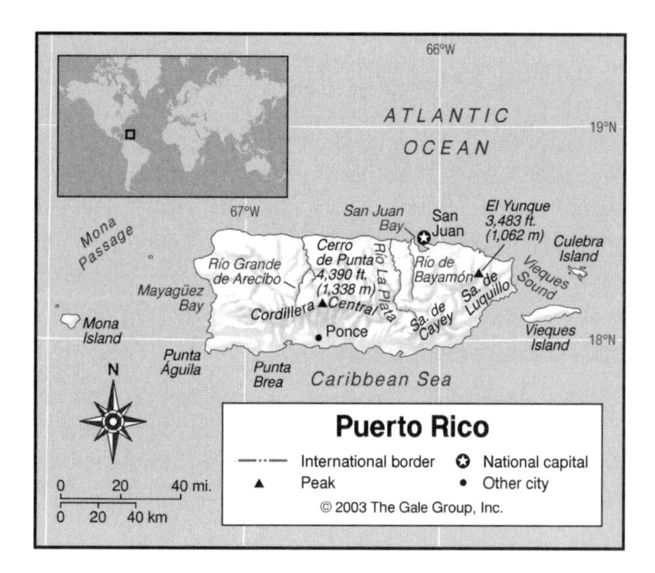

- **Population:** 3,937,316 (July 2001 est.) /World Rank: 119
- **Capital City:** San Juan, on the northern coast
- **Largest City:** San Juan, 421,958 (2000)

OVERVIEW

Puerto Rico is located at the eastern end of the Greater Antilles, between the Atlantic Ocean and the Caribbean Sea. It is a commonwealth of the United States of America, 1,000 mi (1,609 km) southeast of the U.S. mainland, between the island of Hispaniola to the west and the Virgin Islands to the east. In addition to its main island, Puerto Rico also includes three smaller ones, Vieques and Culebra to the east and Mona to the west.

Puerto Rico's main island, rectangular in shape, has a hilly and mountainous interior ringed by a narrow coastal plain. The major mountain system is the Cordillera Central, which bisects the western and central parts of the island. Other than the coastal plain, the major lowland area is the Turabo Valley, a largely agricultural area that lies between three mountain chains in the eastern part of the island.

Located near the division between the Caribbean and North American Tectonic Plates, Puerto Rico is a seismically active region that has sporadic earthquakes.

MOUNTAINS AND HILLS

Mountains

Steep mountain slopes cover nearly one-fourth of the island. The highest and longest mountain range is the Cordillera Central, which extends 60 mi (97 km) across the center of the island and reaches elevations of over 3,000 ft (914 m). Puerto Rico's highest peak, Cerro de Punta, is part of this system, which rises rapidly from the southern coast and more gradually in the north. The other major mountain system is the Sierra de Luquillo in the east, where the country's most famous peak, El Yunque (3,483 ft / 1,062 m) is located. A third mountain range—the Sierra da Cayey—is in the southeast.

Hills and Badlands

Puerto Rico's steep mountains descend to foothills before giving way to the coastal plains. While the original forests have been preserved on many of the mountains, which are too steep for cultivation, farming takes place on the foothills.

In the northwest, an area of low hills with elevations between 200 and 300 ft (60 and 90 m) marks the karst region, an area where rain has carved limestone rocks into hills, caves, and tunnels.

INLAND WATERWAYS

The major rivers flow northward over the mountains to the coast. These include the Río La Plata (the longest), the Río Grande de Loiza (the widest), the Río de Bayamon, and the Río Grande de Arecibo. Many of these rivers feature dams and small reservoirs. The rivers in the south are fewer, shorter, and smaller in volume.

THE COAST, ISLANDS, AND THE OCEAN

Oceans and Seas

Like the other islands of the Greater Antilles, Puerto Rico is bordered on the north by the rough, cold waters of the Atlantic Ocean and on the south by the warmer, calmer Caribbean Sea. It is separated from the island of Hispaniola to the west by the Mona Passage, and from the Virgin Islands to the east by the Vieques Sound and the Virgin Passage.

The waters just off the coast are shallow, but the bottom drops to 6,000 ft (1,829 m) 2 mi (3 km) to the north. Some 40 mi (64 km) farther north lies the Puerto Rico Trench. At the western end of the trench is the Milwaukee Depth, where the ocean floor plunges to a depth of more than 27,493 feet (8,380 meters)—among the greatest ocean depths in the world, and the greatest known depth in the Atlantic.

Major Islands

Vieques, Puerto Rico's largest island aside from the main island, has an area of about 52 sq mi (135 sq km), much of it occupied by a U.S. naval training facility. Its mountains retain some of their original rainforest. Culebra, which also lies to the east of the main island, is an archipelago consisting of a largely flat main island surrounded by 20 islets. Mona Island, to the west, has an area of 20 sq mi (52 sq km).

The Coast and Beaches

Puerto Rico's coastline is moderately indented at most. San Juan Bay is in the northeast, the site of the commonwealth's capital. Mayagüez Bay marks the western end of the island; Punta Águila and Punta Brea are in the southwest. The shore has both rocky and sandy beaches.

CLIMATE AND VEGETATION

Temperature

Trade winds from the northeast moderate Puerto Rico's tropical climate. Temperatures year round generally stay in between 70°F and °80°F (21°C and 27°C), although more extreme temperatures are possible in lower inland areas and on the southern coast. The mean temperature in San Juan is 75°F (24°C) in January and 81°F (27°C) in July. Hurricanes are a hazard between August and October; in the history of Puerto Rico, dozens of these storms have caused property damage and loss of life.

Rainfall

Average annual rainfall varies from 36 in (91 cm) in the south to 60 in (152 cm) at San Juan to as much as 180 in (457 cm) in the mountains. Rainfall is distributed fairly evenly throughout the year.

Grasslands

Coastal plains ring the main island. They have a maximum width of 15 mi (24 km), and the strip of plain on the north is only 5 mi (8 km) wide. This industrialized and urbanized area is the island's major population and commercial center. Agricultural land is still found in the plains of the south and west and the Turabo Valley in the east.

Forests and Jungles

Although little of Puerto Rico's original forest remains, the northern part of the country is rich in planted tree species, and rainforest covers much of the mountainous area. There are a number of wildlife reserves. The most extensive is the El Yunque National Forest in the Sierra de Luquillo, which has more than 200 tree species and about the same number of fern species. One of Puerto Rico's most distinctive sights is the royal poinciana tree, also called the flamboyant, which has brilliant red flowers. The drier south is home to scrub vegetation, cactus, and other thorny plants.

HUMAN POPULATION

Most of Puerto Rico's population lives in the northern coastal plain. Puerto Ricans are citizens of the United States of America, and large numbers immigrated to the mainland during the 20th century. Nevertheless, at the end of the twentieth century the population was more than twice what it had been in 1930.

Population Centers – Puerto Rico

(1990 CENSUS OF POPULATION)

Name	Population
San Juan (capital)	437,745
Bayamón	220,262
Ponce	190,900
Carolina	177,806

SOURCE: 1990 Census of Population and Housing, U.S. Department of Commerce, 1990 CPH-2-53.

NATURAL RESOURCES

Natural resources include forest and agricultural land, hydropower potential, and onshore and offshore oil reserves.

FURTHER READINGS

Luxner, Larry. *Puerto Rico.* Boston: Houghton Mifflin, 1995.

Marino, John. *Puerto Rico: Off the Beaten Path.* Guilford, Conn.: Globe Pequot Press, 2000.

Pariser, Harry S. *The Adventure Guide to Puerto Rico.* Edison, N.J.: Hunter Publications, 1996.

Porter, Darwin, and Danforth Prince. *Frommer's Portable Puerto Rico.* New York: Hungry Minds, 2001.

Scott, David Logan, and Kay W. Scott. *Guide to the National Park Areas. Eastern States.* Old Saybrook, Conn.: Globe Pequot Press, 1999.

Welcome to Puerto Rico! http://welcometopuertorico.org/ (accessed June 22, 2002).

Qatar

- **Area:** 4,416 sq mi (11,437 sq km) / World Rank: 160
- **Location:** Northern and Eastern Hemispheres, in the Middle East, on a peninsula surrounded by the Persian Gulf, north of Saudi Arabia.
- **Coordinates:** 25°30′N, 51°15′E
- **Borders:** Saudi Arabia, 37 mi (60 km)
- **Coastline:** 350 mi (563 km)
- **Territorial Seas:** 12 NM
- **Highest Point:** Qurayn Abu al Bawl, 338 ft (103 m)
- **Lowest Point:** Sea level

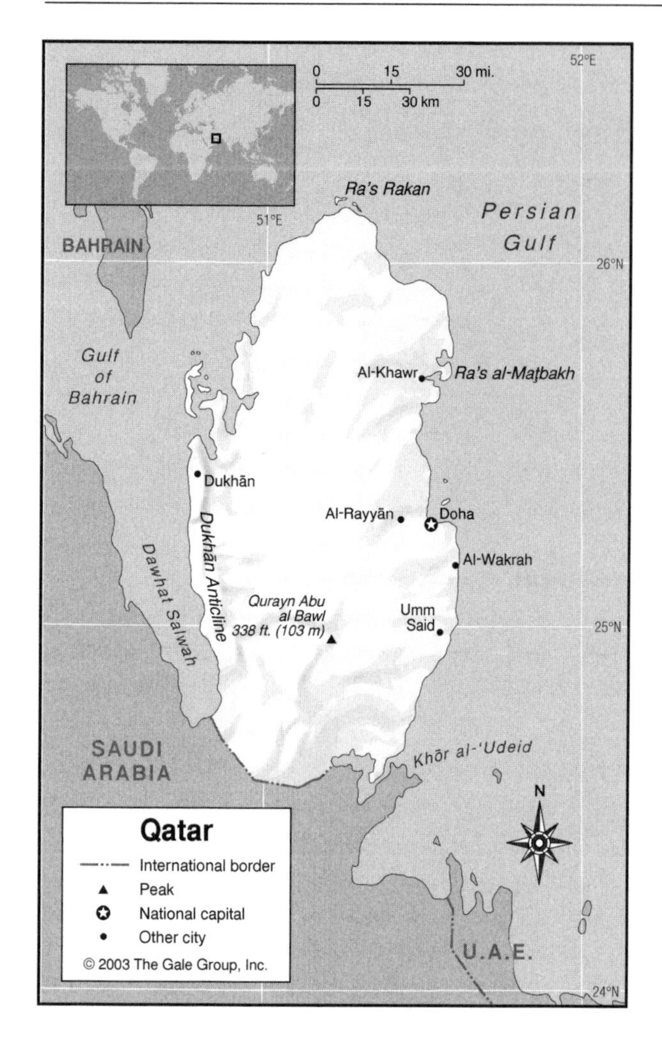

- **Longest Distances:** 100 mi (160 km) N-S; 55 mi (90km) E-W
- **Longest River:** Perennial rivers do not exist
- **Natural Hazards:** Haze, dust storms, sandstorms
- **Population:** 769,152 (July 2001 est.) / World Rank: 155
- **Capital City:** Doha, on the central eastern coast
- **Largest City:** Doha, population 355,000 (2000 est.)

OVERVIEW

Qatar consists of a tiny peninsula projecting northward into the Persian Gulf from the larger Arabian Peninsula. Sitting on the Arabian Tectonic Plate, away from any major faults or plate boundaries, the terrain is mostly flat and barren desert covered with loose sand and gravel, with some low hills. Limited natural, fresh water resources are increasing dependence on large-scale desalination facilities. Until recent decades Qatar was an undeveloped, impoverished area, with a scant living provided by the traditional occupations of pearl diving, fishing, and nomadic herding. Oil, first discovered in 1940, dominates the Qatari economy. Oil revenues provide Qataris with a per capita income comparable to industrialized nations in the West.

MOUNTAINS AND HILLS

Plateaus

A low central limestone plateau, which contains a number of shallow wadis, rises from the east and north.

Hills and Badlands

An elevated limestone formation, the Dukhān anticline (under which lies the Dukhān oil field) lies along the West Coast. Hills and sand dunes reach an altitude of 131 ft (40m) above sea level in the western and northern parts of the country.

INLAND WATERWAYS

Rivers

Though it has no perennial rivers, Qatar is characterized by a number of geographical features that are peculiar to the western side of the Arabian Gulf. These include rainwater-draining basins found mainly in the north and central areas of the country.

THE COAST, ISLANDS, AND THE OCEAN

Oceans and Seas

Qatar borders the Persian Gulf on the north and east, and on the western side of the peninsula the Gulf of Bahrain cuts into Qatar to form Dawhat Salwah. In the southeast an inlet of the Persian Gulf is known to local English speakers as the Inland Sea, or Khōr al-'Udeid in Arabic.

Major Islands

Qatar has a few islands under its possession, the most important of which is Halul. Halul, which has an area of only about 0.58 sq mi (1.5 sq km), lies about 60 mi (90 km) east of Doha and is used for storing oil found in offshore wells and loading it onto ships for trade.

The Coast and Beaches

The coastline of Qatar is part of a regional low desert plain. A notable feature of the coastal area is the prevalence of salt pans, shallow depressions of salt flats (sabkhas). Coral reefs and shallow waters retard navigation. However, there are a couple good ports. The capital of Doha resides on a sizable, though shallow, harbor. Umm Said also provides a commercial harbor. Other ports have been made by digging channels to deepen the shallow waters. These include Al-Khawr and Al-Wakrah. Qatar also has two important capes, one of the northernmost point, Ra's Rakan, and one jutting into the Persian Gulf just north of Al-Khawr, Ra's Al-Maṭbakh.

GEO-FACT

In 2001, Qatar resolved its longstanding border disputes with both Bahrain and Saudi Arabia. The International Court of Justice (ICJ) awarded the Hawar Islands to Bahrain and adjusted its maritime boundary with Qatar; a final border resolution was agreed to with Saudi Arabia.

CLIMATE AND VEGETATION

Temperature

Qatar has a desert climate that is characterized by extremely hot and dry conditions in the summer from May to October, and milder in winter. Mean temperatures in June are 108°F (42°C), dropping to 59°F (15°C) in winter.

Rainfall

Average annual precipitation is below 3 inches (7.6 cm). Most of the rainfall occurs during the winter months, sometimes only in localized heavy downpours. Humidity along the coast frequently reaches 90 percent during summer.

Deserts

Qatar is a flat desert with scanty vegetation. An extension of the Rub'al-Khali (Empty Quarter) reaches northward from Saudi Arabia and the United Arab Emirates with massive sand dunes surrounding Khōr al-'Udeid in the south of Qatar.

HUMAN POPULATION

The majority of the inhabitants of Qatar live in the capital, Doha, or in the neighboring cities of Al-Wakrah and Ar Rayyān. Dukhān and Umm Said have grown considerably due to the oil industry. Much of the country is uninhabited desert, with most of the population concentrated in cities along the coast.

NATURAL RESOURCES

Qatar's natural resources are petroleum, natural gas, and fish. Qatar contains the third-largest natural gas reserves and the largest non-associated gas field in the world.

FURTHER READINGS

ArabNet. *Qatar.* http://www.arab.net/qatar/qatar_contents.html (Accessed May 7, 2002).

Energy information. *Qatar.* http://www.eia.doe.gov/cabs/qatar2.html (Accessed May 7, 2002).

Ferdinand, Klaus. *Bedouins of Qatar.* New York: Thames and Hudson, 1993.

Library of Congress Country Studies. *Qatar.* http://lcweb2.loc.gov/frd/cs/qatoc.html (Accessed May 7, 2002).

Vine, Peter. *The Heritage of Qatar.* London: IMMEL Pub., 1992.

Winckler, Onn. *Population Growth, Migration and Socio-Demographic Policies in Qatar.* Tel Aviv: Moshe Dayan Center for Middle Eastern and African Studies, 2000.

Romania

- **Area:** 91,699 sq mi (237,500 sq km) / World Rank: 80
- **Location:** Northern and Eastern Hemispheres, bordering Ukraine to the north, Moldova and Ukraine to the northeast, the Black Sea to the east, Bulgaria to the south, Yugoslavia to the southwest, and Hungary to the northwest.
- **Coordinates:** 46°00′N, 25°00′E
- **Borders:** 1,558 mi (2,508 km) / Bulgaria, 378 mi (608 km); Hungary, 275 mi (443 km); Moldova, 279 mi (450 km); Ukraine (east), 105 mi (169 km); Ukraine (north), 225 mi (362 km); Yugoslavia, 296 mi (476 km)
- **Coastline:** 140 mi (225 km)
- **Territorial Seas:** 12 NM
- **Highest Point:** Moldoveanu, 8,346 ft (2,544 m)
- **Lowest Point:** Sea level
- **Longest Distances:** 490 mi (789 km) E-W; 295 mi (475 km) N-S
- **Longest River:** Danube, 1,775 mi (2,857 km)
- **Largest Lake:** Razelm, 150 sq mi (390 sq km)
- **Natural Hazards:** Earthquakes; landslides
- **Population:** 22,364,022 (July 2001 est.) / World Rank: 47
- **Capital City:** Bucharest, south-central Romania
- **Largest City:** Bucharest, population 2,130,000 (2000 metropolitan est.)

OVERVIEW

Romania, located in southeastern Europe on the Eurasian Tectonic Plate, is the largest country on the Balkan Peninsula. It has a short coast on the Black Sea to the east and is surrounded on all other sides by five neighboring countries. The Carpathian Mountains, Romania's major

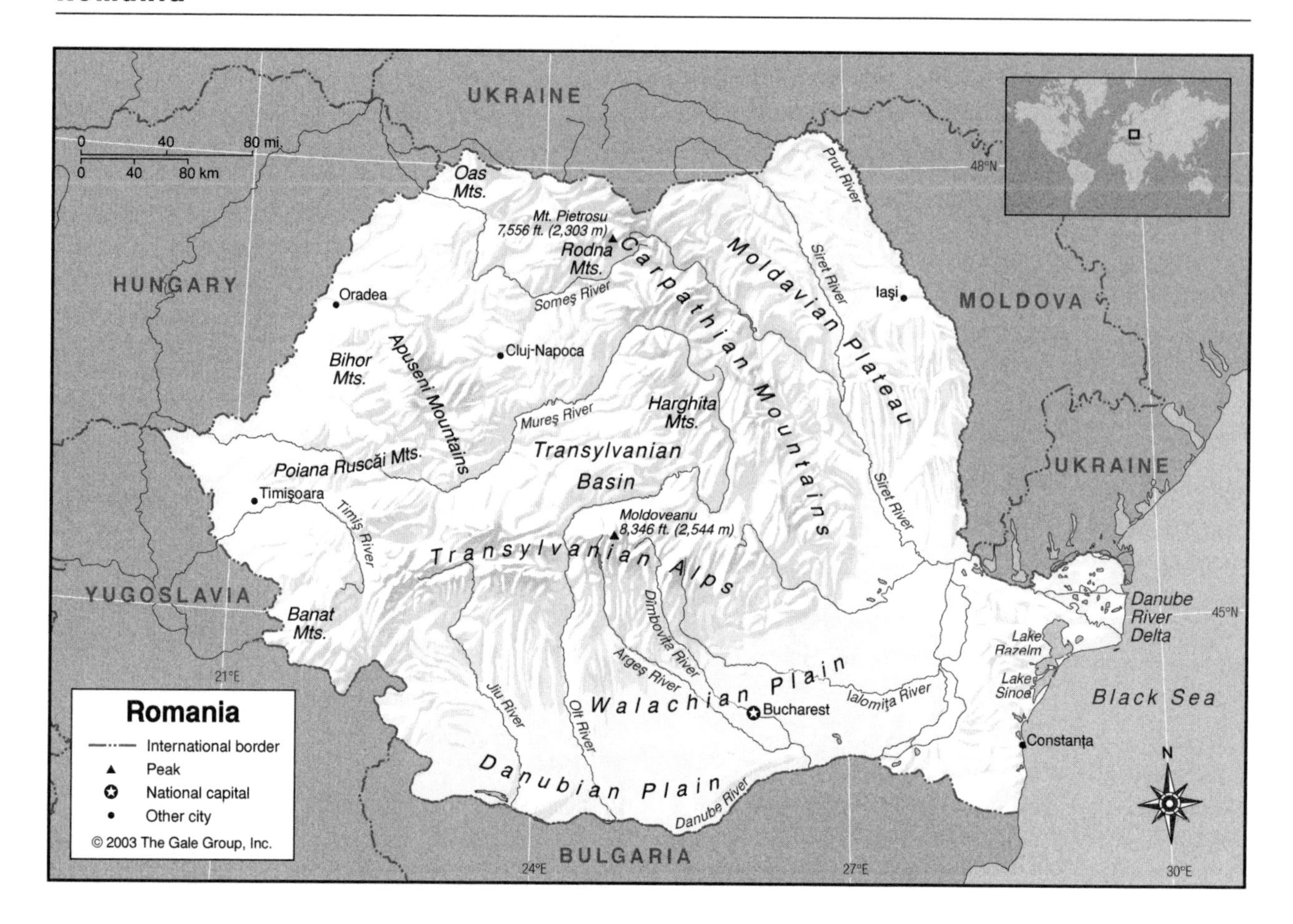

physical feature, define its overall topographical pattern. Roughly forming an arc in the center of the country, their various branches separate the plateau of the Transylvanian Basin in the center from a wide band of lowlands on the periphery, extending to the country's eastern, southern, and western borders.

Romania is traditionally divided into several distinct regions. Transylvania, which forms a large wedge in the north and northwest and makes up one-third of Romania, is by far the largest region. It encompasses the central plateau of the Transylvanian Basin, all of the Carpathian Mountains except for the most southeastern section, and the hilly terrain in the northwestern part of the country. Walachia, which curves around Transylvania in the south and southeast, is the country's major lowland region, encompassing the plains of the Danube River to the south of the Transylvanian Alps. The part of Walachia west of the Olt River is a subregion known as Oltenia. Dobruja occupies the southeastern corner of Romania, bounded by the path of the Danube where the river flows northward for about 100 mi (160 km) before it again turns to the east for its final passage to the sea. Moldavia, in the northeast, constitutes about one-fourth of the country's area. Much of this region is hilly or mountainous, and heavily forested. To the southwest, in the opposite corner of the country from Dobruja, is the Banat region.

MOUNTAINS AND HILLS

Mountains

The ranges in the eastern part of the country are referred to as the Moldavian Carpathians. They have maximum elevations of about 7,500 ft (2,286 m) and are the most extensively forested part of the country. Their highest peak, Mr. Pietrosu (7,556 ft / 2,303 m), rises in the Rodna Mountains in the far north at the border with Ukraine. Two volcanic ranges, the Oas and Harghita Mountains, extend for about 250 mi (400 km) along the western edge of the Moldavian Carpathians. They contain Romania's only crater lake, the St. Ana Lake, as well as roughly 2,000 mineral-water springs.

The slightly higher southern ranges, called the Transylvanian Alps, form the southern border of Transylvania and have the highest peaks and the steepest slopes in the country. Romania's highest point, Mt. Moldoveanu, rises to a height of 8,346 ft (2,544 m) about 100 mi (161 km) northwest of Bucharest. Among the alpine features of the Transylvanian Range are glacial lakes, upland meadows and pastures, and bare rock along the higher ridges. Portions of the mountains are predominantly limestone, with caves, waterfalls, and underground streams.

The ranges in the west are generally lower, and unlike those in the east and south, they are not an unbroken ridge of mountains. The northernmost group is the Bihor

Mountains, originating south of the city of Oradea. The southernmost is the Banat Mountains, in the extreme southwestern corner of the country. In between these two ranges are the perpendicular ranges of the Poiana Ruscăi Mountains and the Apuseni Mountains. These four ranges are not as rugged as those to the south and east, and average elevations run considerably lower: only a few points in the Bihor Mountains approach 6,000 ft (1,828 m), as compared with maximum elevations of nearly 7,500 ft (2,286 m) in the Moldavian Carpathians and over 8,000 ft (2,438 m) in the Transylvanian Alps. The various mountain groups of the western Carpathians are separated by a series of structural depressions, called "gates" because they provide gateways through the mountains. The best known is the Iron Gate on the Danube, in the southeastern corner of Romania. A large dam built on the river in this area in the 1970s is a major source of hydroelectric power.

On the outer fringes of the eastern and southern Carpathian Mountains is a band of lower, but still elevated, terrain called the Subcarpathians, which rises to elevations between 1,300 and 3,300 ft (400 to 1,000 m).

Plateaus

The Transylvanian Plateau, at elevations averaging 1,200 ft (365 m), lies in the center of Romania, ringed by the three branches of the Carpathian Mountains.

Its terrain includes valleys and rounded hills, and it is bordered on the west by an area of the eroded limestone known as karst.

The Moldavian Plateau is marked by hills and narrow valleys and extends across the eastern region of Moldavia between the Subcarpathians and the Prut River, rising to between 1,600 and 2,000 ft (488 and 610 m). Farther south, in the northern, inland part of the Dobruja region, is a plateau that rises to a maximum height of 1,532 ft (467 m).

Hills and Badlands

Hills cover much of Romania, as part of both the mountain and plateau regions as well as the transitional regions between the mountain ranges. The hills are mostly rolling plains with well-watered and fertile soil.

INLAND WATERWAYS

Lakes

Romania is said to have 2,500 lakes, but most are small, and lakes occupy only about 1 percent of the country's total surface area. The largest lakes are along the Danube River and the Black Sea Coast. Some of those along the coast, including the largest, the 150 square mile (390 sq km) Lake Razelm, are salt lakes that are open to the sea. These and a few of the freshwater lakes are commercially important for their fish. The many smaller ones scattered throughout the mountains are usually glacial in origin and add much to the beauty of the resort areas.

GEO-FACT

Bran Castle in the Transylvanian Alps was the home of the fifteenth-century Romanian prince Vlad Tepes ("The Impaler"), on whom the British author Bram Stoker based the vampire in his 1897 Gothic novel *Dracula.*

Rivers

All of Romania's rivers and streams drain to the Black Sea. Except for the minor streams that rise on the eastern slopes of the hills near the sea and flow directly into it, all join the Danube River. Those flowing southward and southeastward from the Transylvanian Alps drain to the Danube directly. Those flowing northward and eastward from Moldavia and Bukovina reach the Danube by way of the Prut River. Most of the Transylvanian streams draining to the north and west—including the Mureş and Someş Rivers—flow to the Tisza River, which joins the Danube in Yugoslavia, north of Belgrade.

The Danube is customarily divided into three sections; most of the portion in Romania—from the Iron Gate to the Black Sea—is known as its lower course. As it approaches its delta, it divides into a number of channels. It also forms several lakes, some of them quite large. At the delta it divides into three major and several minor branches. The delta has an area of about 1,000 sq mi (2,590 sq km) and grows steadily, as the river deposits billions of cubic feet of sediment into the sea annually. Its main tributaries flowing through Romania include the Siret, Ialomiţa, Argeş, Olt, Jiu, and Timiş. The Argeş has an important tributary of its own: the Dîmboviţa River.

Wetlands

The Dobruja region provides Romania's access to the Black Sea and contains most of the Danube River Delta. Much of the Danube River Delta and a belt of land up to 20 mi (32 km) wide along most of the river's length, from the delta westward to the Iron Gate rapids, is marshland. The majority of this land is not easily exploited for agricultural purposes, although some of the reeds and natural vegetation have limited commercial value. The delta is a natural wildlife preserve, particularly for waterfowl, and is large enough so that many species can be protected. Willows flourish in parts of the delta, and there are a few deciduous forests in the north-central section.

THE COAST, ISLANDS, AND THE OCEAN

Oceans and Seas

Romania borders the western end of the Black Sea. The floor of the Black Sea is composed of a shallow shelf that extends about 6 to 7 miles (10 to 11 km) from the coast of Romania. On this shelf the average sea depth is 330 to 360 ft (100 to 110 m). This shelf then drops steeply to the sea floor, which is unusually flat and reaches depths of 7,200 ft (2,195 m). Romania claims the continental shelf off its coast to a depth of 656 ft (200 m).

The Coast and Beaches

The marshy delta of the Danube River makes up the northern third of the coast, while two large saltwater lagoons—Lake Razelm and Lake Sinoe—open onto the sea at the central part of the coastline. To the south, steep cliffs extend to the sea, fringed by white sandy beaches whose popularity with tourists has given this area a reputation as thc "Romanian Riviera."

Population Centers – Romania

(1994 POPULATION ESTIMATES)

Name	Population	Name	Population
Bucharest (capital)	2,060,551	Galati	326,728
Constanţa	348,575	Cluj-Napoca	326,017
Iaşi	339,889	Brasov	324,210
Timişoara	327,830		

SOURCE: "Population of Capital Cities and Cities of 100,000 and More Inhabitants." United Nations Statistics Division.

Districts – Romania

1998 POPULATION ESTIMATES

Name	Population	Area (sq mi)	Area (sq km)	Capital
Alba	400,563	2,406	6,231	Alba Iulia
Arad	478,092	2,954	7,652	Arad
Argeş	674,809	2,626	6,801	Piteşti
Bacău	748,894	2,551	6,606	Bacău
Bihor	624,157	2,909	7,535	Oradea
Bistriţa-Năsăud	326,217	2,048	5,305	Bistriţa
Botoşani	461,889	1,917	4,965	Botoşani
Brăila	388,606	1,824	4,724	Brăila
Braşov	633,140	2,066	5,351	Braşov
Buzău	507,607	2,344	6,072	Buzău
Călăraşi	332,309	1,959	5,074	Călăraşi
Caraş-Severin	359,230	3,283	8,503	Resita
Cluj	723,915	2,568	6,650	Cluj-Napoka
Constanţa	747,218	2,724	7,055	Constanţa
Covasna	230,992	1,431	3,705	Stîntu Gheorghe
Dâmboviţa	553,953	1,559	4,036	Tîrgovişte
Dolj	747,840	2,862	7,413	Craiova
Galaţi	642,759	1,708	4,425	Galaţi
Giurgiu	297,368	1,404	3,636	Giurgiu
Gorj	396,549	2,178	5,641	Tîrgu Tiu
Harghita	342,892	2,552	6,610	Miercurea-Ciuc
Hunedoara	531,238	2,709	7,016	Deva
Ialomiţa	304,873	1,718	4,449	Slobozia
Iaşi	828,476	2,112	5,469	Iaşi
Maramureş	533,088	2,400	6,215	Baia Mare
Mehedinţi	325,167	1,892	4,900	Drobeta-Turnu-Severin
Mureş	602,721	2,585	6,696	Tirgu Mureş
Neamţ	584,801	2,274	5,890	Piatra Neamţ
Olt	512,597	2,126	5,507	Slatina
Prahova	860,481	1,812	4,694	Ploieşti
Sălaj	258,109	1,486	3,850	Zalău
Satu Mare	392,084	1,701	4,405	Satu Mare
Sibiu	444,522	2,093	5,422	Sibiu
Suceava	713,782	3,303	8,555	Suceava
Teleorman	463,307	2,224	5,760	Alexandria
Timiş	682,409	3,356	8,692	Timişoara
Tulcea	265,173	3,255	8,430	Tulcea
Vâlcea	432,375	2,203	5,705	Rimnicu Vîlcea
Vaslui	464,032	2,045	5,297	Vaslui
Vrancea	391,628	1,878	4,863	Focşani

SOURCE: Romania INS (Institutul National de Statistica).

CLIMATE AND VEGETATION

Temperature

Romania has a transitional continental climate with moderating influences from the Black Sea and variations due to altitude. In general, winters are cold and summers are warm. Temperatures are lower in the more elevated Transylvania Plateau in the northwest, and temperature extremes are greater in the plains of the east and south, where the continental influence is strongest. Average temperatures in Bucharest, the capital, are 27°F (-3°C) in January and 73°F (23°C) in July.

Rainfall

Average annual rainfall ranges from about 15 in (38 cm) in the eastern lowland region of Dobruja to 50 in (125 cm) or more in the Carpathian Mountains.

Grasslands

Much of the original grassland vegetation of the steppe-like lowland area in the eastern and southern parts of the country has given way to human settlement and cultivation. Nearly all of the Walachian Plain and Danubian Plain to the south—except for the marshes along the Danube River and the seriously eroded foothills of the mountains—is cultivated. Where the original vegetation remains, short grasses grow in the drier areas, and taller grasses closer to the rivers.

Forests and Jungles

Forested areas, mostly in the Carpathian Mountains, cover about one-fourth of Romania. Oak and other broadleaf deciduous tree varieties are common at lower elevations, giving way to beech and then conifers, with alpine pastureland at the highest altitudes. Bukovina, a small, forested region in the north of Moldavia, is part of a larger area that was mostly annexed to Ukraine by the former Soviet Union following World War II. The approximately 1,300 sq mi (3,366 sq km) of Bukovina remaining in Romania is particularly well-forested and picturesque.

HUMAN POPULATION

The overall population density of Romania is about 251 people per sq mi (97 people per sq km), though the population is unequally distributed. The largest towns and cities are located in the Walachian Plain and Danubian Plain regions in the south, as well as the other plains at the periphery of the country. The density in the plains regions is closer to 415 inhabitants per sq mi (160 inhabitants per sq km). However, urban dwellers make up less than 60 percent of the population, and the rural settlements in hills and valleys of the Subcarpathian belt are also densely populated. The mountainous regions are the least densely populated areas.

NATURAL RESOURCES

Hydropower from the rivers flowing down the Carpathian Mountains provides an important energy source. Romania was formerly one of Europe's major petroleum producers, but its reserves have declined substantially since the 1970s. Other mineral resources include coal and iron ore. Romania's arable land and forests are also among its most important natural resources. Fishing also contributes to Romania's economy, with 80 percent of the catch taken from the lower Danube and its delta, and the rest coming from the lakes of the Dobruja Region and the Black Sea Coast.

FURTHER READINGS

Burford, Tim. *Hiking Guide to Romania.* Old Saybrook, Conn.: Globe Pequote Press, 1996.

Codrescu, Andrei. *The Hole in the Flag: A Romanian Exile's Story of Return and Revolution.* New York: W. Morrow, 1991.

Harold, Dennis-Jones. *Where to Go in Romania.* London: Settle Press, 1994.

Richardson, Dan. *Romania: The Rough Guide.* New York: Penguin, 1995.

Williams, Nicola. *Romania and Moldova.* Hawthorn, Victoria: Lonely Planet, 1998.

Russia

- **Area:** 6,592,771 sq mi (17,075,200 sq km) / World Rank: 1
- **Location:** Northern, Eastern, and Western Hemispheres; Eastern Europe but primarily in north Asia; south of the Arctic Ocean and east of the North Pacific Ocean; north of Azerbaijan, Georgia, Kazakhstan, North Korea, Latvia, and Mongolia; east of Belarus, Estonia, Finland, and Norway; and northeast of China. Kaliningrad Oblast is a separate region in Europe, north of Poland, and southeast of Lithuania.
- **Coordinates:** 60°00′N, 100°00′E
- **Borders:** 12,403 mi (19,961 km) / Azerbaijan, 176 mi (284 km); Belarus, 596 mi (959 km); China, 2,265 mi (3,605 km); Estonia, 183 mi (294 km); Finland, 816 mi (1,313 km); Georgia, 449 mi (723 km); Kazakhstan, 4,254 mi (6,846 km); Latvia, 135 mi (217 km); Lithuania, 141 mi (227 km); Mongolia, 2,165 mi (3,485 km); North Korea, 12 mi (19 km); Norway, 104 mi (167 km); Poland, 128 mi (206 km); and Ukraine, 979 mi (1,576 km)
- **Coastline:** 23,396 mi (37,653 km)
- **Territorial Seas:** 12 NM
- **Highest Point:** Mount El'brus, 18,481 ft (5,633 m)
- **Lowest Point:** Caspian Sea, 92 ft (28 m) below sea level
- **Longest Distances:** Approximately 2,400 mi (4,000 km) N-S; 6,200 mi (10,000 km) E-W.
- **Longest River:** Ob', 3,362 mi (5,410 km; includes Irtysh tributary).
- **Largest Lake:** Lake Baikal, 11,870 sq mi (30,510 sq km)
- **Natural Hazards:** Flooding; extreme cold; earthquakes; volcanoes in the east
- **Population:** 145,470,197 (July 2001 est.) / World Rank: 6
- **Capital City:** Moscow, in the northwest
- **Largest City:** Moscow, 9,299,000 (2000 estimate)

OVERVIEW

Russia is by far the largest country in the world, containing one-ninth of the Earth's land area. Russia is also the most expansive country. Starting from its islands in the Arctic Sea one can travel to the Caucasus Mountains in the south, nearly halfway to the equator, without ever entering another country. Russia is even more extensive from east to west. From the Baltic Sea on the east coast to Big Diomede Island in the Bering Strait, the country's east-west measurement approaches 6,200 mi (10,000 km), almost half the circumference of the earth.

Considering the nation's geography from west to east, Russia can be categorized into several large regions. They include the Great European Plain; the Ural Mountains; the mountain systems and ranges along much of Russia's southern border; and Siberia, which includes the West Siberian Plain, the Central Siberian Plateau, and the mountain ranges of northeastern Siberia and the Kamchatka Peninsula.

Russia
International border
Peak
National capital
Other city
© 2003 The Gale Group, Inc.
0°
20°E
40°E
80°E
120°E
160°E
180°
80°N
60°N
40°N
N
ARCTIC OCEAN
Svalbard (NORWAY)
Franz Josef Land
Barents Sea
Novaya Zemlya
Kara Sea
Severnaya Zemlya
Komsomolets Island
October Revolution Island
Bol'shevik Island
Cape Chelyuskin
Laptev Sea
New Siberian Islands
Proliv Dmitriya Lapteva
East Siberian Sea
Chukchi Sea
Wrangel Island
Long Strait
Bering Strait
Chukchi Peninsula
Gulf of Anadyr
UNITED STATES
Bering Sea
Koryak Mts.
Sredinnyy Range
Kamchatka Peninsula
Klyuchevskaya Sopka 15,584 ft. (4,750 m)
Shelikhova Bay
Kolyma Range
Kolyma Lowlands
Cherskiy Range
Verkhoyansk Range
Lena River
Sea of Okhotsk
Kuril Islands
Sakhalin Island
Tatar Strait
Sikhote-Alin' Mts.
Amur River
Ussuri River
Lake Khanka
Vladivostok
Peter the Great Bay
JAPAN
Sea of Japan
NORTH KOREA
NORWAY
SWEDEN
FINLAND
Baltic Sea
Gulf of Finland
EST.
LAT.
LITH.
Kaliningrad Oblast'
BELARUS
UKRAINE
Murmansk
Kola Pen.
White Sea
Kolguyev Island
Vaygach Island
Arkhangel'sk
Northern Dvina R.
St. Petersburg
Lake Peipus
Lake Ladoga
Lake Onega
Valdai Hills
Great European Plain
Moscow
Nizhniy Novgorod
Kazan'
Samara
Don R.
Dnieper R.
Sea of Azov
Rostov-na-Donu
Volgograd
Volga R.
Kama R.
Ural R.
Black Sea
Mt. El'brus 18,481 ft. (5,633 m)
CAUCASUS MTS.
GEORGIA
ARM.
AZER.
Caspian Sea
Mt. Narodnaya 6,212 ft. (1,894 m)
URAL MOUNTAINS
Yekaterinburg
Gulf of Ob'
Gydan Pen.
Taymyr Peninsula
Lake Taymyr
North Siberian Lowlands
West Siberian Plain
Vasyugan'ye Swamp
Ob' River
Irtysh River
Omsk
Lake Chany
Novosibirsk
Novosibirsk Res.
Yenisey River
Central Siberian Plateau
SIBERIA
Lower Tunguska R.
Stony Tunguska R.
Angara River
Vilyui R.
Aldan River
Bratsk Res.
Irkutsk
Sayan Mts.
Lake Baikal
Vitim R.
Yablonovyy Range
Shilka R.
Argun R.
Stanovoy Mts.
Zeya Res.
Mt. Pelukha 15,157 ft. (4,619 m)
ALTAY SHAN
KAZAKHSTAN
UZBEKISTAN
TURKMENISTAN
KYRGYZSTAN
TAJIK.
IRAN
MONGOLIA
CHINA
0 250 500 mi.
0 250 500 km

Most of Russia is located on the Eurasian Tectonic Plate but eastern Russia is on the North American Plate. Their exact boundary is uncertain. The Pacific Plate is located off of Russia's eastern coastline. The movement of these three plates against each other is a cause of significant earthquakes and volcanoes in this region, especially on Kamchatka. Seismic activity is also common in the Caucasus Mountains in the southwest.

Russia was even larger in the past than it is today. Finland, Alaska, and parts of modern day Poland were owned by Russia at various times in its history. After World War I, Russia technically ceased to be an independent country, becoming part of the Union of Soviet Socialist Republics (U.S.S.R., Soviet Union). Russia was by far the largest of the republics that made up the Soviet Union, however, and was considered to be the ruling power of that nation. The Soviet Union started to dissolve in 1991. Eventually the nations of Armenia, Azerbaijan, Belarus, Estonia, Kazakhstan, Kyzgystan, Latvia, Lithuania, Moldova, Tajikistan, Turkmenistan, Ukraine, and Uzbekistan all became independent of the Soviet Union. Russia's borders with many of these new countries had not been completely settled by 2001. The end of the Soviet Union also left a small portion of Russia, the Kaliningrad Oblast, separated from the rest of the country. Russia also has long-standing disputes with China over their border and with Japan over control of the southernmost of the Kuril Islands, which the Soviet Union occupied in 1945.

MOUNTAINS AND HILLS

Mountains

With nine major mountain ranges, Russia can be considered among the most mountainous countries in the world. Eastern Russia is by far more mountainous than the west, while the center section of the country is primarily low plains.

The Urals are perhaps the best known of Russia's mountain ranges, as they define the boundary between Asia to the east and Europe to the west. A lengthy range, the Urals extend 1,300 mi (2,100 km) from the northern border of Kazakhstan all the way to the Arctic Ocean. However, the highest peak, Mount Narodnaya, is only 6,212 ft (1,894 m) in elevation. The Urals have never offered any significant barrier to travel.

Located between the Black and Caspian seas, the Caucasus Mountains consist of two major chains separated by lowlands. The northern Greater Caucasus form most of the border between Russia, Azerbaijan, and Georgia, as well as marking the boundary between Asia to the south and Europe to the north. These mountain systems are made up of granite, other crystalline rocks, and some volcanic formations. Elevations in the Greater Caucasus reach a maximum of 18,476 ft (5,633 m) at the extinct volcano Mount El'brus, the highest peak in both Russia and Europe.

Russia's other mountains are far to the east. The Altay Shan and Sayan Mountains are found in the area north of Mongolia, west of Lake Baikal. Further east are the Yablonovyy Range and Stanovoy Mountains. They follow much of the southern border of central and eastern Siberia on toward the Pacific Ocean, where they join the ranges of the far east. Collectively, these ranges form the watershed between the Arctic Sea to the north and the Pacific Ocean and Central Asia to the south and southeast. The Altay Shan are the tallest and include Mt. Pelukha (15,157 ft / 4,619 m). The other ranges average less than 10,000 ft (3,048 m) in height.

The topography east of the Lena River is predominantly mountainous, with the elevations becoming become higher and more rugged moving east. Major ranges in this region are Verkhoyanskiy, Cherskiy, Kolyma, Koryak, and Sredinnyy. The easternmost ranges feature live volcanoes. As many as 120 volcanoes dot the Kamchatka Peninsula, and no less than 23 are active. Klyuchevskaya Sopka, the highest of these, is 15,584 ft (4,750 m) tall. Moving offshore, these same mountains form the Kuril Islands, where 30 volcanoes are active out of some 100 present. Across the Sea of Okhotsk, in Russia's southeasternmost area, there are several low mountain ranges, including the Sikhote-Alin' Mountains and the mountains of Sakhalin Island.

Plateaus

The Central Siberian Plateau is an enormous stretch of rolling land between the Yenisey and the Lena Rivers. Heights of this vast plateau range from 1,600 to 2,300 ft (500 to 700 m) on average. Its surface is eroded by the many rivers, some forming deep canyons. Layers of sedimentary rock, subsequently intruded by volcanic lava, were deposited long ago on top of igneous and metaphoric rock. Within the layers of sedimentary rock are rich deposits of coal.

Hills

There are many regions of hills and uplands in Russia. The Valdai Hills are the most noteworthy. Although not particularly tall (600-1000 ft / 182-304 m), they are among the highest elevations in the Great European Plain of western Russia. Many important rivers have their source there, including the Volga.

INLAND WATERWAYS

The Caspian Sea

On Russia's southern border between Europe and Asia is the Caspian Sea. Although it is rarely thought of as such, it not a true sea but rather the world's largest lake (143,550 sq mi / 371,795 sq km). This is because the Caspian is held in a vast land depression, with no outlet to any ocean. Although many rivers drain into it, the water escapes only through evaporation; the Caspian's salinity results from accumulated salts left from evaporation.

Lakes

Most Russian lakes were formed by glaciation. The largest such lakes in European Russia are Ladoga (6,835 sq mi / 17,703 sq km) and Onega (3,701 sq mi / 9,609 sq km), northeast of St. Petersburg. They are also the two largest lakes in all of Europe (as the Caspian is generally not counted). Other large lakes in western Russia include Lake Peipus on the Estonian border and the reservoirs of the Volga River.

Lake Baikal in southern Siberia is the largest lake in Russia and in Asia (again excluding the Caspian, as well as the so-called Aral Sea). It is 392 mi (632 km) long and 32 mi (59 km) wide, with a surface area of 11,870 sq mi (30,510 sq km). It has a maximum depth of 5,715 ft (1,742 meters), making it the deepest body of freshwater on Earth. Due to its great depth, Lake Baikal also has the greatest volume of any freshwater lake. It is said to contain one-fifth of the Earth's fresh surface water. Other large Siberian lakes include Lakes Taymyr, Chany, Khanka and the Novosibirsk, Bratsk, and Zeya Reservoirs. There are countless smaller lakes.

Rivers

Most of Russia's urban population lives along the banks of the nation's many rivers. The most important commercial river in Russia is the Volga, which is also the longest river in Europe. The Volga begins in the hills west of Moscow and flows southeastward for 2,293 mi (3,689 km) to the Caspian Sea. The Volga River system drains nearly 550,000 sq mi (1.4 million square km), and four of Russia's largest cities are located on its banks: Nizhniy Novgorod, Samara, Kazan', and Volgograd. The Kama River flows west out of the southern Urals and into the Volga; this too is a major waterway for Russia and for Europe.

Also located in European Russia, the Dnieper and the Don drain large portions of the region. Although the Dnieper flows mainly through Belarus and Ukraine, it has headwaters in the hills west of Moscow. The Don flows from its origins in the Central Russian Upland south of Moscow for 1,153 mi (1,860 km) before emptying into the the Sea of Azov at Rostov-na-Donu. Further east is the Ural River, which flows south from the Ural Mountains into Kazakhstan before reaching the Caspian Sea. The Ural River is traditionally considered part of the boundary between Europe and Asia. Several canals connect most of European Russia's rivers, and altogether the rivers have provided a vital transportation system, carrying fully two-thirds of the nation's inland water traffic.

GEO-FACT

Russia is such an enormous country that it can be difficult to grasp. Not only is it the largest country on Earth, but it is almost twice the size of any of the next largest countries—Canada, China, and the United States. In fact, it is almost equal in area to the entire continent of South America, meaning that it is larger than Antarctica, Australia, or Europe.

A number of major rivers drain into the Pacific and Arctic Oceans from the Siberian plateau and mountain areas in the east. Three of these river systems are larger than the Volga. The Irtysh-Ob' system flows through the West Siberian Plain, emptying into the Arctic at the Gulf of Ob'. The Irtysh is the longer of the two rivers, but is tributary to the Ob'. Together they have a length of 3,335 mi (5,380 km) making them the longest river system in Russia.

On the far side of the Central Siberian Plateau is the Lena, the longest individual river in Russia at 2,700 mi (4,400 km). It too empties into the Arctic, and has many large tributaries including the Aldan, Vitim, and Vilyui. The third great Arctic river is the Yenisey (2,480 mi/4,000 km), which flows across the Central Siberian Plateau. Its largest tributary, the Lower Tunguska, is itself roughly 2,000 mi (3,226 km) long. Other major tributaries include the Stony Tunguska and Angara. Together, the Lena and Yenisey drain some 3,088,800 sq mi (8 million sq km), sending nearly 165,000 cubic ft (50,000 cubic m) of water into the Arctic Ocean every second.

The Amur River (1,768 mi / 2,874 km) is the most important Siberian river flowing into the Pacific. Its major tributaries the Argun, Ussuri, and Shilka, and together they drain most of southeastern Siberia and northeastern China. The Amur River and its primary tributary the Ussuri River comprise a significant section of the boundary between Russia and China.

As a resource, the number and size of Russia's inland waterways can be misleading. For example, nearly 85 percent of the nation's rivers flow through sparsely populated areas and empty into the Arctic and North Pacific oceans. In contrast, the most highly populated areas, which also require the most water, are located in the warmest climates with the highest evaporation rates. Water supplies in the densely populated European river basins, such as that of the Don, often fail to meet demand.

Wetlands

The same river systems that account for such an enormous flow of water into the Arctic Ocean are also responsible for creating vast swamps in the West Siberian Plain. Snow and ice in the warmer regions where the rivers have their sources thaw well before the northern regions, causing great flooding to the north. The Vasyugan'ye Swamp in the center of the West Siberian Plain, for example, covers 18,500 sq ft (48,000 sq km). The same effect can be observed with other Siberian river systems.

In all nearly 10 percent of Russian territory can be classified as swampland. Much of this is concentrated in the West Siberian Plain, which lies between the Ural Mountains and the Yenisey River. This plain is a vast area of lowlands, probably the largest expanse of flat land anywhere in the world. It stretches from the steppes of Central Asia in the south to the Arctic Ocean in the north, covering a region nearly 1,100 mi (1,800 km) wide. Flat and poorly drained, these lowlands feature many swamps, marshes, and peat bogs, with significant oil and natural gas deposits in their central and northern parts.

THE COAST, ISLANDS, AND THE OCEAN

The Coast and Beaches

No country in the world can surpass Russia's 23,396 mi (37,653 km) of coastline. Yet most of this coastline is so far north that it is frozen for much of the year. Despite the fact that frozen harbors mean Russia has very few outlets to the ocean that remain open all year, Russian shipping and fishing thrives on all its seas.

The Arctic Ocean

The majority of Russia's coastline is on the Arctic Ocean and its seas. Located almost entirely north of the Arctic Circle, much of the water here remains frozen for the better part of the year. One exception is the area in the far west, where the Gulf Stream current warms the waters of the Barents Sea near the Kola Peninsula, allowing the port of Murmansk to function year round. South of the Kola Peninsula is the deep inlet of the Arctic called the White Sea. Further east is the Kara Sea. The Gulf of Ob' and the estuary of the Yenisey River punctuate the coastline here, with the Gydan Peninsula between them. Continuing to the east, the Taymyr Peninsula extends north, reaching mainland Russia's northernmost point at Cape Chelyuskin. The Laptev Sea and the East Siberian Sea are connected by the Proliv Dmitrya Lapteva, with Long Strait even further east connecting the East Siberian Sea to the Chukchi Sea.

The Pacific Ocean

The Chukchi Peninsula stretches out to become Russia's easternmost point, with the Chukchi Sea of the Arctic Ocean to the north and the Bering Sea of the Pacific to the south. The two bodies of water are connected by the Bering Strait, which separates Siberia and Alaska by a mere 53 mi (86 km). To the south of the Chukchi Peninsula is the Gulf of Anadyr, and further south is the large Kamchatka Peninsula. Kamchatka encloses the Sea of Okhotsk to the west. Most of the rest of Russia's Pacific coastline is on this sea, but in the southeast is the Sea of Japan, connected to Okhotsk by the Tatar Strait. Russia's principal Pacific port, Vladivostok, is found on Peter the Great Bay in this sea.

The Black Sea and the Baltic Sea

Western Russia is connected to the Atlantic Ocean by its coasts on the Baltic and Black Seas. The Baltic coastline is very short. Most of it is on the Gulf of Finland, where St. Petersburg is located. The Kaliningrad Oblast also has a Baltic shoreline. In the southwest Russia touches the Black Sea and its major inlet, the Sea of Azov. Even here, the harbor of Rostov-na-Donu is frozen for several months of the year.

Major Islands

Many islands lie within the Arctic and Pacific Oceans off the shores of Russia. Franz Josef Land is comprised of about 100 small islands in the Arctic Ocean; it is the northernmost part of Russia and is among the northernmost land on Earth. Other large Arctic islands are Novaya Zemlya, Vaygach Island, Wrangel Island, and the Severnaya Zemlya and New Siberian Islands groups. Many small islands and island chains are scattered among the large groups.

In the Pacific, the Kuril Islands curve southwest from the Kamchatka Peninsula to Japan. Although the Kuril Islands are under Russian administration, Japan and Russia dispute ownership of the four southernmost islands. Also lying in the Pacific is Sakhalin, a large island that separates the Seas of Okhotsk and Japan.

CLIMATE AND VEGETATION

It is said that Russia has only two seasons: summer and winter. Only a slight exaggeration, this statement accurately characterizes the country's harsh climate with its long, cold winters and short, cool summers. These conditions are owing to Russia's location in the high latitudes; more than half the country lies above 60° north latitude with only relatively small areas below 50° north. Furthermore, the high mountains that form Russia's southern border effectively block out warm air masses. Any influence the warm waters of the Pacific Ocean might have is essentially nullified by the predominant movement of the country's weather systems from east to west. In winter, Siberia lies under a vast high pressure cell centered in Mongolia, which keeps the region enveloped in air of surpassing frigidity. The magnitude of this cold is not easy to grasp. Soil in the far northern permafrost can be frozen several hundred meters deep, and even into southern Siberia the land is covered by snow for more than six months. The annual average temperature for most of Siberia is below freezing; for the majority of European Russia the average is only somewhat higher.

In summer, warm, moist air from the Atlantic is able to push as far east as central Siberia, under the influence of a prevailing low pressure system. The result is moisture-bearing air that delivers fairly high amounts of precipitation. Russia's short growing season relies heavily upon this rainfall for it crops; unfortunately, distribution of the moisture in many areas is often irregular and unpredictable. Droughts are not uncommon, especially in early summer; on the other hand, heavy rains in middle and late summer may compromise harvesting. In the east, late summer Pacific air can bring monsoon-like rainfall, with disastrous effects.

Overall, lack of sunshine characterizes the Russian climate. Overcast skies are the rule, especially in winter. Moscow typically experiences 23 days of cloud cover in December; indeed, sunless winter days are the rule in virtually all of the Russian nation.

Russia's climate zones lie in easily distinguishable belts that run east-to-west across the whole country. In the far north, Novaya Zemlya and Severnaya Zemlya and numerous smaller Arctic islands experience a polar desert climate. Below this, a tundra climate predominates for at least 60 mi (100 km) south, extending up into the steep mountain slopes of the far east. Next a broad subarctic zone passes southward as far as St. Petersburg in the west, crosses the Urals, and takes in nearly all the rest of Siberia. Last is a wide belt of cold, dry steppe climate starting at the Black Sea, crossing the North Caucasian Plain, moving through the lower Volga Valley and the southern Urals into Siberia.

Temperature

Lying in the continental climate zone, Moscow experiences temperatures in a range from 3° to 16° F (-16° to -9° C) in January and from 55° to 73° F (13° to 23° C) in July. In the far southeastern city of Vladivostok, temperatures range from 0° to 13° F (-18° to -11° C) in January and from 60° to 71° F (16° to 22° C) in July. None of these ranges describe especially severe conditions. However, the northeast Siberian city of Verkhoyansk has recorded an absolute temperature range (the difference between the hottest and the coldest temperatures) of 188° F (105° C). This is easily the greatest temperature range of any location on earth. July temperatures in Verkhoyansk average 56° F (13° C), but have reached 98° F (37° C). Winter temperatures have dipped to nearly -90° F (-32° C).

Rainfall

Most of Russia experiences only modest precipitation, but averages vary by region. On the Great European Plain, averages decrease from more than 30 in (80 cm) in the west to less than 16 in (40 cm) on the Caspian Sea coast. Siberia uniformly sees annual precipitation ranging from 20 to 32 in (500 to 800 mm), although amounts are generally less than 12 in (300 mm) in extreme northeastern Siberia. In high elevations precipitation totals may reach 40 in (1000 mm) or more, but in the valleys they may average less than 12 in (300 mm).

Plains

The Ural Mountains separate two vast plains, the Great European Plain and the even larger West Siberian Plain. Both of these so-called plains contain a wide variety of terrain, including vast forests, swamps, and stretches of tundra. However there are also many areas of grassland and farmland, especially in the Great European Plain.

Steppe

The steppe is a broad band of nearly treeless, grassy plains that extend across Hungary, Ukraine, southern Russia, and Kazakhstan before ending in Manchuria. Although historically presented as the typical Russian landscape, the steppe in Russia proper is in fact small, located mainly northwest of the Greater Caucasus Mountains and across the southern Volga Valley, the southern Urals, and parts of western Siberia. Isolated pockets of steppe can also be found in the mountain valleys of southeastern Siberia. Moderate temperatures and normally adequate levels of sunshine and moisture give the steppe zone relatively favorable conditions for agriculture, although precipitation can be unpredictable, sometimes even catastrophically dry.

Tundra

About 10 percent of Russia is tundra, a treeless, marshy plain that lies along Russia's northernmost zone. The tundra stretches from the Finnish border to the Bering Strait, then extends south along the Pacific coast to the Kamchatka Peninsula. The North Siberian and Kolyma Lowlands are entirely made up of tundra. Only mosses, lichens, dwarf willows and shrubs can grow on the permafrost and survive the long, harsh, sunless winters. In summer dusk comes at midnight and dawn follows within minutes. The powerful Siberian rivers that cut across the tundra toward the Arctic Ocean do a poor job of draining the region, due to partial and intermittent thawing. The most important physical process at work in the tundra is frost weathering, a vestige of the glaciation that shaped it during the last ice age.

Forests

Russia's vast forests comprise nearly one-fourth of the world's total forested area and contribute almost a third of its softwood timber. Located chiefly in Siberia, they cover over two-fifths of Russia's total territory, with two distinct areas: in the north is the taiga—a large, mainly

Population Centers – Russia

(1998 POPULATION)

Name	Population	Name	Population
Moscow	8,298,000	Omsk	1,158,000
St. Petersburg	4,695,000	Kazan'	1,092,000
Novosibirsk	1,402,000	Ufa	1,087,000
Nizhniy Novgorod	1,362,000	Chelyabinsk	1,086,000
Yekaterinburg	1,271,000	Perm	1,018,000
Samara	1,171,000	Rostov-na-Donu	1,006,000

SOURCE: "Major Cities of Russia (at the end of 1998)." *Handbook Russia 2000.* State Committee of the Russian Federation on Statistics.

Administrative Districts – Russia

2001 POPULATION ESTIMATES

Name	Population	Area (sq mi)	Area (sq km)	Capital
Central	36,700,000	250,965	650,000	Moscow
North West	14,400,000	647,669	1,677,900	St. Petersburg
South	21,500,000	227,490	589,200	Rostov-na-Donu
Privolzhsky	31,800,000	400,772	1,038,000	Nizhniy Novgorod
Urals	12,600,000	690,694	1,788,900	Yekaterinburg
Siberian	20,700,000	1,974,824	5,114,800	Novosibirsk
Far East	7,100,000	2,399,959	6,215,900	Khabarovsk

SOURCE: *Geodata,* 1989 edition and State Committee of the Russian Federation on Statistics. Cited on Geohive. http://www.geohive.com (accessed 22 May 2002).

coniferous forest—and in the south there is a much smaller area of mixed forest.

The taiga extends across the Ural Mountains and covers most of Siberia. Much of the land beneath it is permafrost. Although primarily coniferous, the taiga also offers birch, poplar, aspen, willow, and other deciduous trees in some places. Significant numbers of fir, birch, and other trees are also present in the extreme northwestern part of the European region. Firs predominate eastward to the western Urals, although some regions are almost exclusively birch. On the West Siberian Plain, the taiga trees are mostly pine, but along the southern fringes birch dominate. Much of the Central Siberian Upland and the mountains of eastern Siberia are covered by larch trees, a type of deciduous conifer.

The central portion of the Great European Plain between St. Petersburg and the Ukrainian border features a mixed forest of both conifers and deciduous trees. Oak, beech, maple, and hornbeam are the primary broadleaves. Moving south, the mixed forest passes through a narrow zone of forest-steppe 95 mi (150 km) wide, on average, before giving way to a zone of true steppe.

HUMAN POPULATION

Most Russians live in what's called the "fertile triangle" between the Baltic and Black seas and the southern Urals. The country's overall population density is 22 persons per sq mi (9 persons per sq km), but the distribution is very uneven. In European Russia density averages about 65 persons per sq mi (25 persons per sq km). Of course the heaviest concentrations are in sprawling metropolitan areas, but by contrast more than a third of the nation—Siberia and large tracts of northern European Russia—has fewer than 3 people per sq mi (1.2 people per sq km). About 78 percent of Russia's inhabitants live in the cities; 22 percent live in rural areas. Most of the large cities are in Europe, but there are also some in southern Siberia, most notably Novosibirsk, Omsk, and Vladivostok.

NATURAL RESOURCES

Russia is believed to have nearly half the world's coal reserves and probably has the largest petroleum reserves of any nation. The largest coal deposits, although undeveloped, are in central and eastern Siberia; developed fields lie in western Siberia, the region near Europe, the area around Moscow, and in the Urals. The oil fields in western Siberia and the Volga-Urals region are the nation's largest. Russia possesses some 40 percent of the world's natural gas reserves. Deposits are found along the Arctic coast in Siberia and in the northern Caucasus region.

In an area known as the Kursk Magnetic Anomaly, near Ukraine, are iron-ore deposits so vast they affect the earth's magnetic field. Russia also has abundant quantities of iron alloys (nickel, tungsten, cobalt, and molybdenum), and minor deposits of manganese can be found in the Urals. Nonferrous metals such as copper, lead, zinc, mercury, and aluminum are abundant as well. Some of the largest gold reserves in the world are located in Siberia and the Urals.

Potassium and magnesium salt deposits for the manufacture of chemicals are plentiful, as are phosphate ores, rock salt, and sulfur. High-grade limestone used for making cement is also found in many parts of the country.

Altogether, Russia may well be the most mineral-rich nation in the world. However, many of its deposits are located in remote areas with prohibitive climates, making them exceedingly expensive to extract. In addition to its mineral resources, Russia has large industrial and agricultural sectors.

FURTHER READINGS

Blaney, John W. "Environmental and Health Crises in the Former Soviet Union." In *The Successor States to the USSR,* edited by John W. Blaney, 134-142. Washington, D.C.: Congressional Quarterly, 1995.

Feshbach, Murray. *Ecological Disaster: Cleaning Up the Hidden Legacy of the Soviet Regime.* New York: Twentieth Century Fund, 1995.

Heleniak, Tim. "The Projected Population of Russia in 2005," *Post-Soviet Geography* 35, 10 (October 1995): 608-614.

Lydolph, Paul E. *Geography of the U.S.S.R.* New York: John Wiley and Sons, 1964.

McCauley, Martin. *The Soviet Union 1917-1991.* 2d ed. London: Longman, 1993.

Rwanda

- **Area:** 10,169 sq mi (26,338 sq km) / World Rank: 147
- **Location:** Southern and Eastern Hemispheres, south-central Africa, bordering Uganda to the north, Tanzania to the east, Burundi to the south, and Democratic Republic of the Congo to the west and northwest
- **Coordinates:** 2°00′S, 30°00′E
- **Borders:** 555 mi (893 km) / Burundi, 180 mi (290 km); Democratic Republic of the Congo, 135 mi (217 km); Tanzania, 135 mi (217 km); Uganda, 105 mi (169 km)
- **Coastline:** None
- **Territorial Seas:** None
- **Highest Point:** Mt. Karisimbi, 14,826 ft (4,519 m)
- **Lowest Point:** Rusizi River, 3,117 ft (950 m)
- **Longest Distances:** 154 mi (248 km) NE-SW; 103 mi (166 km) SE-NW
- **Longest River:** Kagera, 430 mi (692 km)
- **Largest Lake:** Kivu, 1,025 sq mi (2,665 sq km)
- **Natural Hazards:** Drought; volcanic activity
- **Population:** 7,312,756 (July 2001 est.) / World Rank: 90
- **Capital City:** Kigali, central Rwanda
- **Largest City:** Kigali, population 286,000 (2000 est.)

OVERVIEW

Rwanda is a small, landlocked country located south of the equator in east-central Africa. Much of the countryside is covered by grasslands and small farms extending over rolling hills, but there are also areas of swamps and rugged mountains, including volcanic peaks in the northwest border area. The divide between two of Africa's great watersheds, the Congo and Nile Basins, extends from north to south through western Rwanda at an average elevation of almost 9,000 ft (2,743 m). On the western slopes of this Congo-Nile ridgeline, the land slopes abruptly toward Lake Kivu in the Great Rift Valley on the western border of the country. The eastern slopes are more moderate, with rolling hills extending across the central uplands, at gradually reduced altitudes, to the plains, swamps, and lakes of the eastern border region.

Rwanda can be divided into five regions from west to east: (1) the narrow Great Rift Valley region along or near Lake Kivu, (2) the volcanic Virunga Mountains and high lava plains of northwestern Rwanda, (3) the Congo-Nile Ridge, (4) the rolling hills and valleys of the central plateaus, which slope eastward from the Congo-Nile Ridge, and (5) the savannas and marshlands of the eastern and southeastern border areas, which are lower in altitude, warmer, and drier than the central upland plateaus.

Rwanda lies on the African Tectonic Plate, along the Great Rift Valley that has been caused by the movement of the Arabian Plate along the African Plate. The Great Rift Valley is lined with lakes, volcanoes, and gorges, and is bordered by mountains on both sides.

MOUNTAINS AND HILLS

Mountains

Rising from high lava plains in the northwest corner of Rwanda are the Virunga Mountains, Rwanda's only mountain range. They consist of five volcanic peaks, of which two still emit smoke and steam. The highest of these is Mr. Karisimbi, which rises to over 14,000 ft (4,267 m). Three similar peaks lie across the border in the Democratic Republic of the Congo.

Plateaus

East of the Virunga Mountains lies the Central Plateau, with an average altitude of 4,700 ft (1,432 m). Covered by rolling hills, it becomes progressively lower in elevation as it extends toward the eastern border. The land in this region has been intensively farmed and grazed, resulting in considerable erosion and soil depletion.

Hills and Badlands

The rolling hills covering much of the Central Plateau have given Rwanda the nickname "Land of a Thousand Hills."

INLAND WATERWAYS

Lakes

Rwanda has many lakes. The largest, Lake Kivu, is located in the midst of the volcanic peaks in the Virunga Mountains and forms part of the border with the Democratic Republic of the Congo. The lake has a surface area

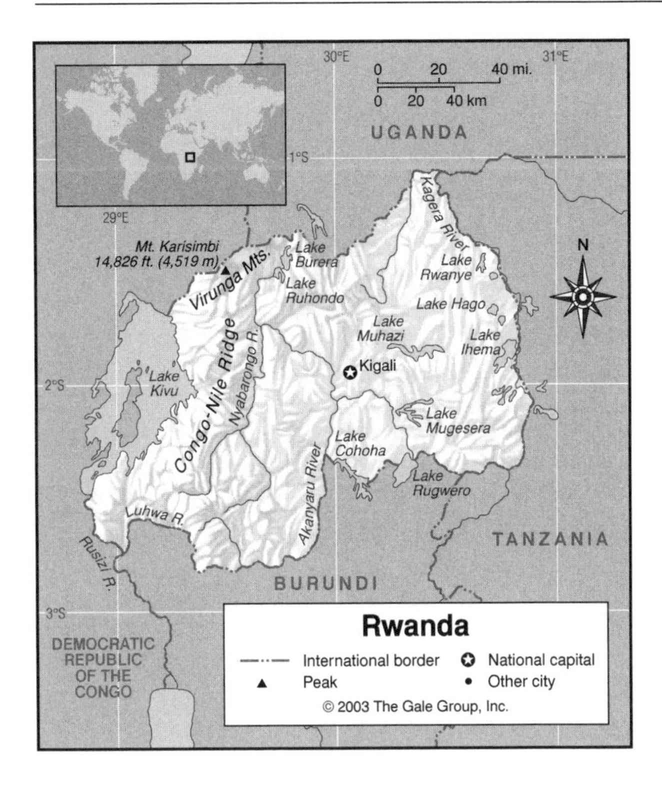

of 1,025 sq mi (2,665 sq km) and reaches a maximum depth of 1,558 ft (475 m), though its average depth is half that. Lake Cohoha and Lake Rugwero lie in Rwanda's southeast, partly in Burundi. There are also eight sizable lakes that lie entirely within Rwanda: Lakes Rwehikama, Ihema, Muhazi, Mugesera, Hago, and Rwanye in the east, and Lakes Ruhondo and Burera in the north.

Rivers

Most of Rwanda's rivers are in the eastern part of the country. The Kagera River in the east forms the boundary with Tanzania and part of the boundary with Burundi. With a total length of 430 mi (692 km), the Kagera is the longest river in Rwanda. The Nyabarongo River and its tributaries drain much of the Central Plateau. In the west, the Ruzizi flows southward from Lake Kivu along the border with the Democratic Republic of Congo, into Burundi, and on to Lake Tanganyika. In the south, the Luhwa and Akanyaru Rivers form parts of the boundary with Burundi.

Wetlands

Swampland is found among the savannas of the east and southeast.

THE COAST, ISLANDS, AND THE OCEAN

Rwanda is landlocked and has no oceanic coast. Its only significant island is the Île Gombo, located in Lake Kivu. The island has a surface area of only 3.5 sq mi (9 sq km).

CLIMATE AND VEGETATION

Temperature

High altitudes keep the climate moderate in much of Rwanda despite its proximity to the equator. In addition, trade winds from the Indian Ocean moderate the temperatures on the Central Plateau, where the annual average is 70°F (21°C). Temperatures in the mountains of the northwest are lower, especially at night, yet they average near 90°F (32°C) in parts of the eastern lowlands.

Rainfall

Average annual rainfall can range from as little as 30 in (76 cm) in the eastern lowlands to 70 in (179 cm) in the mountains. The average on the Central Plateau is about 45 in (114 cm).

Grasslands

The northeast is primarily savanna, while the central plateau consists of hilly grasslands.

Forests and Jungles

Very little of Rwanda's original forestland remains. Some reforestation has been carried out, much of it with eucalyptus trees. Dense vegetation is found near the shores of Lake Kivu in the west. Trees, shrubs, and lichens grow in the Virunga Mountains, and papyrus is found in the swamps of the eastern lowlands.

HUMAN POPULATION

Rwanda is the most densely populated country in Africa, yet more than 90 percent of the population in Rwanda is rural. The heaviest concentrations of people are found on the Central Plateau, in areas with an altitude between 5,000 and 7,500 ft (1,524 and 2,286 m). This central region has a population density of 1,104 people per sq mi (426 people per sq km). The western third has a population density of 796 per sq mi (308 per sq km), while the eastern third is the least populous, with a density of 383 per sq mi (148 per sq km).

NATURAL RESOURCES

Rwanda has few mineral resources or industry. Those that it does have include tin and tungsten ores, gold, and methane gas. The economy of Rwanda depends upon its agriculture, with coffee and tea being its leading exports. Most of the country's power is supplied through hydroelectricity provided by its many rivers.

FURTHER READINGS

Carr, Rosamond Halsey, and Ann Howard Halsey. *Land of a Thousand Hills: My Life in Rwanda.* Rockland, Mass.: Compass Press, 2000.

Destexhe, Alain. *Rwanda and Genocide in the Twentieth Century.* Trans. Alison Marschner. New York: New York University Press, 1995.

Dinar, Ali. *Rwanda Page.* http://www.sas.upenn.edu/African_Studies/Country_Specific/Rwanda.html (Accessed June 4, 2002).

Harelimana, Frodualđ. *Rwanda: Society and Culture of a Nation in Transition.* Corvallis, Ore.: Harelimana, 1997.

Murphy, Dervla. *Visiting Rwanda.* Dublin: Lilliput Press, 1998.

Nyankanzi, Edward L. *Genocide: Rwanda and Burundi.* Rochester, Vt.: Schenkman Books, 1998.

Saint Kitts and Nevis

- **Area:** 101 sq mi (261 sq km)/ World Rank: 195
- **Location:** Northern and Western hemispheres, in the Eastern Caribbean Sea between Puerto Rico and Trinidad
- **Coordinates:** 17°20′N, 62°45′W
- **Borders:** None
- **Coastline:** 84 mi (135 km)
- **Territorial Seas:** 12 NM
- **Highest Point:** Mount Misery, 3,793 ft (1,156 m)
- **Lowest Point:** Sea level
- **Longest Distances:** 23 mi (37 km) N-S; 5 mi (8 km) E-W
- **Longest River:** No perennial rivers of significant length
- **Largest Lake:** Great Salt Pond
- **Natural Hazards:** Hurricanes; earthquakes
- **Population:** 38,756 (July 2001 est.) / World Rank: 195
- **Capital City:** Basseterre, located on the western coast of Saint Kitts
- **Largest City:** Basseterre, population 11,600 (2002 est.)

OVERVIEW

Shaped like an exclamation mark, the popular tourist destinations of Saint Kitts and Nevis lie in the northern part of Leeward Islands in the Eastern Caribbean, with Barbuda in the northeast and Antigua to the southwest. They are volcanic islands separated by a two-mile-wide channel known as the Narrows. The larger island, Saint Kitts, has a dormant volcano, a slat lake, and tropical forests and is dotted by several bays along its southern region. Nevis is also home to a dormant volcano, as well as rich forests and boasts numerous black and white sand beaches. The islands are known for their lush vegetarian and gently warm tropical climate.

Saint Kitts and Nevis are located on the Caribbean Tectonic Plate where it meets the North American Plate. The subduction of the Atlantic Plate below the Caribbean Plate caused the islands to rise out of the ocean.

GEO-FACT

Historians believe that the Portuguese explorer, Christopher Columbus, may have thought the clouds surrounding the summit of Mount Nevis were snow cover, since he named the island Nevis and the Spanish word for snow is *nieve.*

MOUNTAINS AND HILLS

Mountains

The major mountain range on the island of Saint Kitts runs through the middle of the island in the southwesterly direction. Rainforests surround the higher slopes while plots of sugarcane cover the foothills. Mount Misery (Mt. Liamuiga) has the highest peak on the island at 3,793 ft (1,156 m). The circular island of Nevis slopes to its highest peak, Nevis Peak, which has an elevation of 3,232 ft (985 m) and is often capped in white clouds. Both Mt. Misery and Mt. Nevis are dormant volcanoes. Nevis itself is considered to be one giant volcano. Earthquakes in the surrounding areas confirm that both of these volcanoes will likely erupt again.

Plateaus

The southern peninsula on Saint Kitts consists of many low hills and expansive reaches of flat terrain. There is little vegetation in this area, since it does not receive as much rainfall as the forests in the higher elevations. Cacti are prevalent, and the area is subject to erosion.

INLAND WATERWAYS

Lakes

The baseball-shaped Great Salt Pond located near the southeastern tip of Saint Kitts is the only lake on the islands of significant size. Its waters have a higher salt concentration than those of the surrounding Caribbean Sea.

Rivers

Most of the rivers on Saint Kitts and Nevis have dried significantly over the course of history. Those that do remain are small and drain from the mountain ranges in the wet season, drying up in most parts if not completely in the dry season. Of significance are the Wingfield and Cayon rivers, which during the wet seasons will flow almost to the Caribbean.

Wetlands

Large swamps and marshes of all kinds can be found on the southern peninsula of Saint Kitts, with red and white mangrove as the dominant species in this area. Black vervet monkeys are common here, especially around the aptly-named Monkey Hill.

THE COAST, ISLANDS, AND THE OCEAN

Oceans and Seas

Saint Kitts and Nevis are located in the Caribbean Sea, separated by a two-mile-wide channel known as the Narrows. While there are coral reefs in the Caribbean around the islands, none near the islands are of significant size. The highest concentration of these reefs is near Nag's Head and the southwestern coast of Saint Kitts. The coral reefs on Saint Kitts, notably those near Sandy Point Bay, are rich in marine life. Barracuda, eels, rays, and sea turtles, as well as various corals and sponges, may be found in these waters.

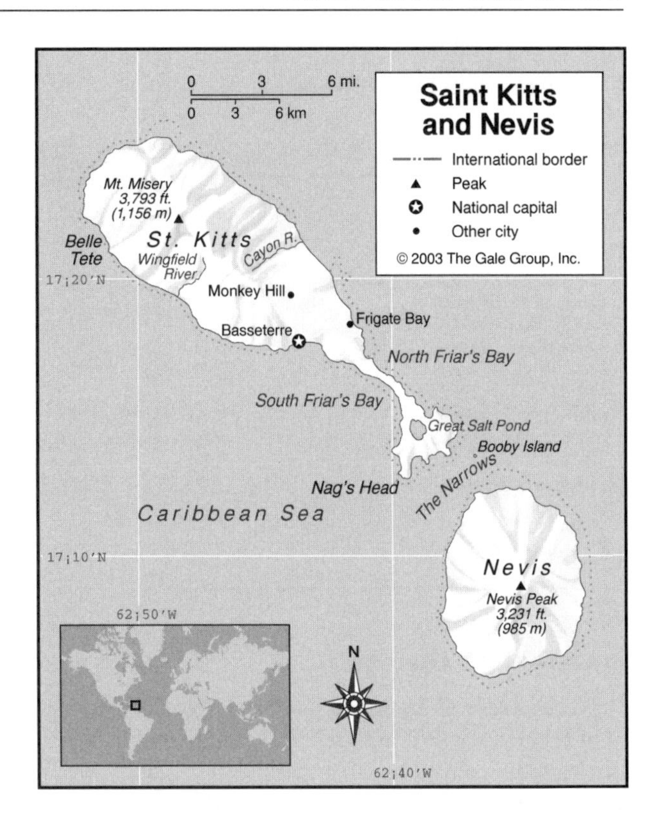

Major Islands

Located in the channel separating the islands of Saint Kitts and Nevis is the small Booby Island, an area rich in fish, especially snappers.

The Coast and Beaches

The coastlines of the islands are rather jagged and indented, providing for many bay and beach areas. Located on the southern tip of Saint Kitts are Majors Bay, Banana Bay, and Cockleshell Bay. Other bays line the coastline, including Half Moon Bay, Sandy Bay, Frigate and North Frigate Bays, and North and South Friar's Bays. There are two capes of interest: Belle Tete, on the northwestern shore of Saint Kitts, and Nag's Head at the end of the Frigate Bay Peninsula. The beaches on both islands range from smooth with white sand to coarse with black sand, with the best beaches found in southern bay area of Saint Kitts. The northern end of Saint Kitts has numerous black volcanic sand beaches. Nevis's most well known beach is Pinney's Beach, which boasts fine, white sand and is lined by coconut trees.

CLIMATE AND VEGETATION

Temperature

The temperatures recorded on Saint Kitts and Nevis change very little during the year, due to their close proximity to the equator. Year-round temperatures average 79°F (26°C) and rarely exceed 89°F (32°C).

Rainfall

Rainfall is greater and more frequent in higher elevations. Most rain falls between May and November, with an average annual rainfall of 43 in (109 cm). The summer months are especially humid; hurricanes are possible in the late summer and early fall months.

Grasslands

Fertile grasslands exist near the capital city of Basseterre, and are primarily used for crop cultivation. However, much of the natural vegetation has been removed.

Forests and Jungles

More than 240 species of trees are found in the forests that surround the islands' mountains. Common to both islands are lemon and palm trees, as well as hibiscus and tamarind. In recent years, hurricanes and deforestation have seriously affected the size of the rainforests.

HUMAN POPULATION

The island of Saint Kitts is more populated than Nevis, both in actual numbers and in population density. In 2002, the estimated population density of Saint Kitts was 445 inhabitants per sq mi (172 inhabitants per sq km) while the population density of Nevis was 246 per sq mi (95 per sq km). More than 25 percent of all the islands' inhabitants live in the capital city of Basseterre on Saint Kitts. Most of the country's population is of African descent.

Parishes – Saint Kitts and Nevis

Name	Area (sq mi)	Area (sq km)
Christ Church --Nichola Town	7.2	18.6
Saint Anne -- Sandy Point	4.9	12.8
Saint George -- Basseterre	11.1	28.7
Saint George -- Gingerland	7.1	18.5
Saint James -- Windward	12.0	31.1
Saint John -- Capisterre	9.6	24.8
Saint John -- Figtree	8.2	21.3
Saint Mary -- Cayon	5.8	15.1
Saint Paul -- Capisterre	5.3	13.8
Saint Paul -- Charlestown	1.4	3.5
Saint Peter -- Basseterre	8.0	20.7
Saint Thomas -- Middle Island	9.4	24.3
Saint Thomas -- Lowland	7.0	18.1
Trinity -- Palmetto Point	6.0	15.4

SOURCE: *Geo-Data: The World Geographical Encyclopedia*, 2nd ed. Detroit: Gale Research, 1989.

NATURAL RESOURCES

The islands of Saint Kitts and Nevis have limited natural resources apart from their fertile, arable lands, which support the cultivation of sugarcane, sea island cotton, peanuts, and coconut. Salt raking is done occasionally.

FURTHER READINGS

Gordon, Joyce. *Nevis: Queen of the Caribees.* London: Macmillan Caribbean, 1990.

Merrill, Gordon Clark. *The Historical Geography of St. Kitts and Nevis, the West Indies.* Mexico: Instituto Panamericano de Geografia e Historia, 1958.

Moll, Verna P. *Saint Kitts-Nevis.* Santa Barbara, Calif.: Clio Press, 1995.

Richardson, Bonham C. *Caribbean Migrants: Environment and Human Survival on Saint Kitts and Nevis.* Knoxville: University of Tennessee Press, 1983.

Saint Lucia

- **Area:** 239 sq mi (620 sq km) / World Rank: 183
- **Location:** Northern and Western Hemispheres, eastern Caribbean Sea, in the Windward Islands, south of Martinique, bordered by the Saint Lucia Channel, the North Atlantic Ocean, Saint Vincent Passage, and the Caribbean Sea.
- **Coordinates:** 13°53′N, 60°68′W
- **Borders:** None
- **Coastline:** 98 mi (158 km)
- **Territorial Seas:** 12 NM
- **Highest Point:** Mount Gimie, 3,117 ft (950 m)
- **Lowest Point:** Sea level
- **Longest Distances:** 27 mi (43 km) N-S; 14 mi (23 km) E-W
- **Longest River:** None of significant length
- **Natural Hazards:** Hurricanes; seismic activity
- **Population:** 158,178 (July 2001 est.) / World Rank: 180
- **Capital City:** Castries, northern Saint Lucia
- **Largest City:** Castries, population 53,000 (2000 est.)

OVERVIEW

Saint Lucia, located in the eastern Caribbean Sea between Martinique and Saint Vincent, is the second-largest of the Windward Islands. The interior of the volcanically formed island consists of mountains and hills, and is surrounded by a coastal strip. The cone-like twin peaks of the Gros Piton and Petit Piton are Saint Lucia's outstanding natural feature.

Saint Lucia is situated on the Caribbean Tectonic Plate. Evidence of past volcanic activity can be seen in the bubbling mud and escaping gases of the sulfur springs near the inactive crater at Soufrière.

MOUNTAINS AND HILLS

Mountains

Mountains occupy much of the interior, spanning the island from north to south. However, the southern half, which is geologically younger, is more mountainous, while the older northern half is hillier.

Although the highest elevation is in the south-central part of the island, where Mt. Gimie reaches a height of 3,117 ft (950 m), the country's best-known peaks are Gros Piton and Petit Piton, pyramids of volcanic rock that rise out of the ocean at Soufrière Bay on the southwest coast at respective elevations of 2,619 ft (798 m) and 2,461 ft (750 m).

INLAND WATERWAYS

Rivers

A number of small rivers flow outward from the central highlands to the coast. The principal ones are the Cul de Sac, Canelles, Dennery, Fond, Piaye, Doree, Canaries, Roseau, and Marquis.

THE COAST, ISLANDS, AND THE OCEAN

Oceans and Seas

Saint Lucia is located between the Atlantic Ocean and the Caribbean Sea. It is separated from Martinique to the

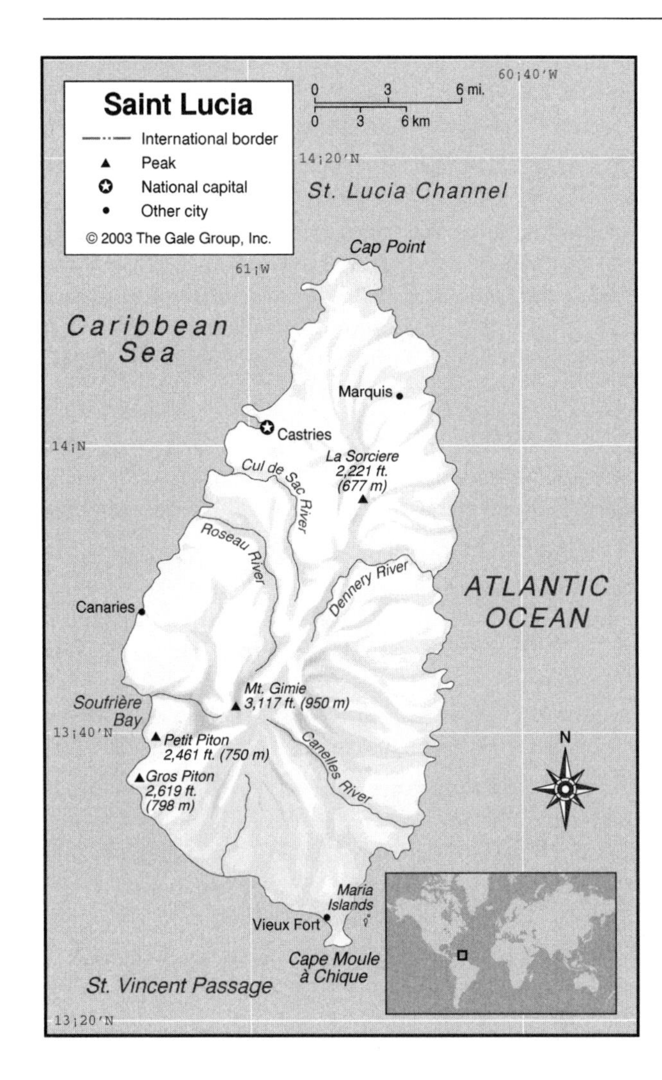

north by the Saint Lucia Channel, and from Saint Vincent to the south by the Saint Vincent Passage. Saint Lucia has two major ports: Castries and Vieux Fort. The harbor waters at the port of Castries are 27 ft (8 m) deep, but the underwater geography around the island varies drastically. There are extensive coral reefs, underwater cliffs, walls, and mountains surrounding Saint Lucia.

Major Islands

Other than the main island, Saint Lucia consists of the Maria Islands, located off the southeast coast. They are the site of a nature reserve.

The Coast and Beaches

The eastern coast has many small indentations, while the western coast is mostly smoother, with major indentations at the port of Castries in the northwest and Soufrière Bay in the southwest, where the peaks of Gros Piton and Petit Piton are located. The island has two major capes, Cap Point at the northern tip and Cape Moule à Chique at the southern. Saint Lucia is known for its many scenic beaches, some of which are covered with black volcanic sand.

Quarters – Saint Lucia

Name	Area (sq mi)	Area (sq km)	Capital
Anse-la-Raye/Canaries*	18.1	46.9	Anse-la-Rayel/Canaries
Castries	30.7	79.5	Castries
Choiseul	12.1	31.3	Choiseul
Dennery	26.9	69.7	Dennery
Gros Islet	39.2	101.5	Gros Islet
Laborie	14.6	37.8	Laborie
Micoud	30.9	80.0	Micoud
Soufrière	19.5	50.5	Soufrière
Vieux Fort	16.9	43.8	Vieux Fort

* Canaries is a city located within Anse-la-Raye with a special administrative status.

SOURCE: *Geo-Data: The World Geographical Encyclopedia,* 2nd ed. Detroit: Gale Research, 1989.

CLIMATE AND VEGETATION

Temperature

Saint Lucia's tropical climate is moderated by trade winds off the Atlantic. The mean temperature year round is about 80°F (27°C). Hurricanes are a hazard in the late summer months of June, July, and August.

Rainfall

Average annual rainfall ranges from about 50 in (127 cm) in the coastal areas to up to 150 in (381 cm) at higher elevations in the interior. The wet season is from June to September, and the dry season is from February to May.

Forests and Jungles

Dense rain forest is found on the higher mountain slopes. Lower down, the trees have largely been cleared for farming and timber production. Trees include many species of palm, bamboo, and giant ferns, as well as breadfruit, mangoes, coconut, and pawpaw. The forests are also rich in flowering plants, including orchids, anthurium, hibiscus, bird of paradise plants, and African tulip trees.

HUMAN POPULATION

More than one-third of the population lives in urban areas. The highest population density occurs in the region including Castries, northeast of the Cul de Sac River, where there are 1,619 people per sq mi (625 people per sq km). Other than this area, the northern half is less densely populated than the southern coastal region, where the density is 740 inhabitants per sq mi (286 inhabitants per sq km). The population density of the remainder of the island is 401 people per sq mi (155 people per sq km).

NATURAL RESOURCES

Saint Lucia's sandy beaches, scenic views, and pleasant climate form the basis for a lucrative tourist industry.

Other important natural resources are its forests, fertile soil, and geothermal potential from the sulfur springs and dormant volcanic craters. Other than tourism, the economy is based on agriculture: bananas are Saint Lucia's most important crop, followed by cocoa, spices, and coconuts.

FURTHER READINGS

Eggleston, George Teeple. *Orchids on the Calabash Tree.* New York: Putnam, 1962.

Ellis, G. *Saint Lucia: Helen of the West Indies.* London: Macmillan, 1988.

Interknowledge Corporation. *Saint Lucia: Simply Beautiful.* http://www.st-lucia.com (Accessed June 5, 2002).

Kingsolver, Barbara. *Homeland and Other Stories.* Rockland, Mass.: Wheeler Publishing, 1989.

Nieminen, Raija. *Voyage to the Island.* Washington, D.C.: Gallaudet University Press, 1990.

Philpott, Don. *Saint Lucia.* Lincolnwood, Ill.: Passport Books, 1996.

Saint Vincent and the Grenadines

- **Area:** 150 sq mi (389 sq km) / World Rank: 190
- **Location:** Islands in the Caribbean Sea, Northern and Western Hemispheres, part of the Windward Islands group of the Lesser Antilles, north of Grenada
- **Coordinates:** 13°15′N, 61°12′W
- **Borders:** None
- **Coastline:** 52 mi (84 km)
- **Territorial Seas:** 12 NM
- **Highest Point:** Soufrière, 4,049 ft (1,234 m)
- **Lowest Point:** Sea level
- **Longest Distances:** Saint Vincent Island, 18 mi (29 km) N-S; 11 mi (18 km) W-E
- **Longest River:** None of significant length
- **Natural Hazards:** Hurricanes, active volcanoes
- **Population:** 115,942 (July 2001 est.) / World Rank: 183
- **Capital City:** Kingstown, located on the southwest coast of St. Vincent
- **Largest City:** Kingstown, 27,000 (2000 est.)

OVERVIEW

Saint Vincent and the Grenadines (often simply called St. Vincent) is part of the Windward Islands group of the Lesser Antilles in the Caribbean Sea. Saint Vincent itself is a large volcanic island at the northern part of the country. The Grenadines are a chain of islets between Saint Vincent and Grenada, with half belonging to St. Vincent and the other to Grenada.

An active volcano, Soufrière, sits in the mountains in the north on Saint Vincent. The remainder of the island contains rugged land, except for lowlands in the interior and a valley that are home to tropical rainforests and Saint Vincent's best farmland, respectively. The Grenadines are generally rugged but low-lying.

MOUNTAINS AND HILLS

Mountains

Down its whole length, Saint Vincent is dominated by a volcanic range of mountains with four peaks at almost equal distance from each other: Soufrière (the highest point), Richmond, Grand Bonhomme, and St. Andrew. The Soufrière volcano also contains a crater lake that is 1 mi (1.6 km) wide. A rugged landscape with steep slopes comprises most of the remaining areas of Saint Vincent.

The Grenadines are formed by a volcanic ridge between Saint Vincent and Grenada that runs north to south. Mt. Tobaoi (1,010 ft / 308 m), the highest point in the Grenadines, is found on Union Island.

THE COAST, ISLANDS, AND THE OCEAN

Oceans and Seas

St. Vincent and the Grenadines is located between the Caribbean Sea to the west and the Atlantic Ocean to the east. Coral reefs surround the Grenadines. The Saint Vincent Passage is found north of that island; Martinique Channel separates the country from Grenada.

Major Islands

The island of Saint Vincent itself is by far the largest in the country, with an area of 133 sq mi (344 sq km). The Grenadines are a group of low-lying islands south of Saint Vincent, with wide beaches and coral reefs surrounding them. Union Island, Mayreau, Mustique, Canouan, Bequia, and many other unihabited rocks, reefs, and cays are part of the Grenadines that belong to St. Vincent.

The Coast and Beaches

Saint Vincent's east and west coasts are comprised of alternating rock cliffs and stretches of black sand beaches. The Grenadines have low-lying land, wide beaches, and shallow harbors and bays.

CLIMATE AND VEGETATION

Temperature

A tropical country, the average temperature for St. Vincent and the Grenadines is 79°F (26°C). September is

Population Centers – Saint Vincent and the Grenadines

(2001 POPULATION ESTIMATES)

Name	Population
Kingstown (capital)	28,000
Georgetown	22,000

SOURCE: "World Urbanization Prospects: The 2001 Revision." United Nations Population Division.

Census Divisions – Saint Vincent and the Grenadines

Name	Area (sq mi)	Area (sq km)
Barrouallie	14.2	36.8
Bridgetown	7.2	18.6
Calliaqua	11.8	30.6
Chateaubelair	30.0	77.7
Colonaire	13.4	34.7
Georgetown	22.2	57.5
Kingstown (city)	1.9	4.9
Kingstown (suburbs)	6.4	16.6
Layou	11.1	28.7
Marriaqua	9.4	24.3
Northern Grenadines	9.0	23.3
Sandy Bay	5.3	13.7
Southern Grenadines	7.5	19.4

SOURCE: *Geo-Data: The World Geographical Encyclopedia,* 2nd ed. Detroit: Gale Research, 1989.

the warmest month, with an average temperature of 81°F (27°C), and January is the coolest, with an average temperature of 77°F (25°C).

Rainfall

On Saint Vincent, yearly rainfall averages 91 in (231 cm), and in the mountainous regions, it averages more than 150 in (380 cm). The rainy season occurs between May or June through December. In most of the Grenadines, rainfall is the only source of fresh water.

Grasslands

The lowlands on Saint Vincent are covered with coconut and banana trees, and arrowroot. Some of the island's most fertile farmland is housed in the Mesopotamia Valley, which is northeast of Kingstown, the capital city.

Forests and Jungles

Forests and woodlands comprise 36 percent of St. Vincent and the Grenadines, with most of Saint Vincent's interior housing tropical rainforest.

HUMAN POPULATION

The majority of the population lives on Saint Vincent, the main island, and a little over half lives in urban areas. The Grenadines are sparsely populated, with many of the islands being completely uninhabited.

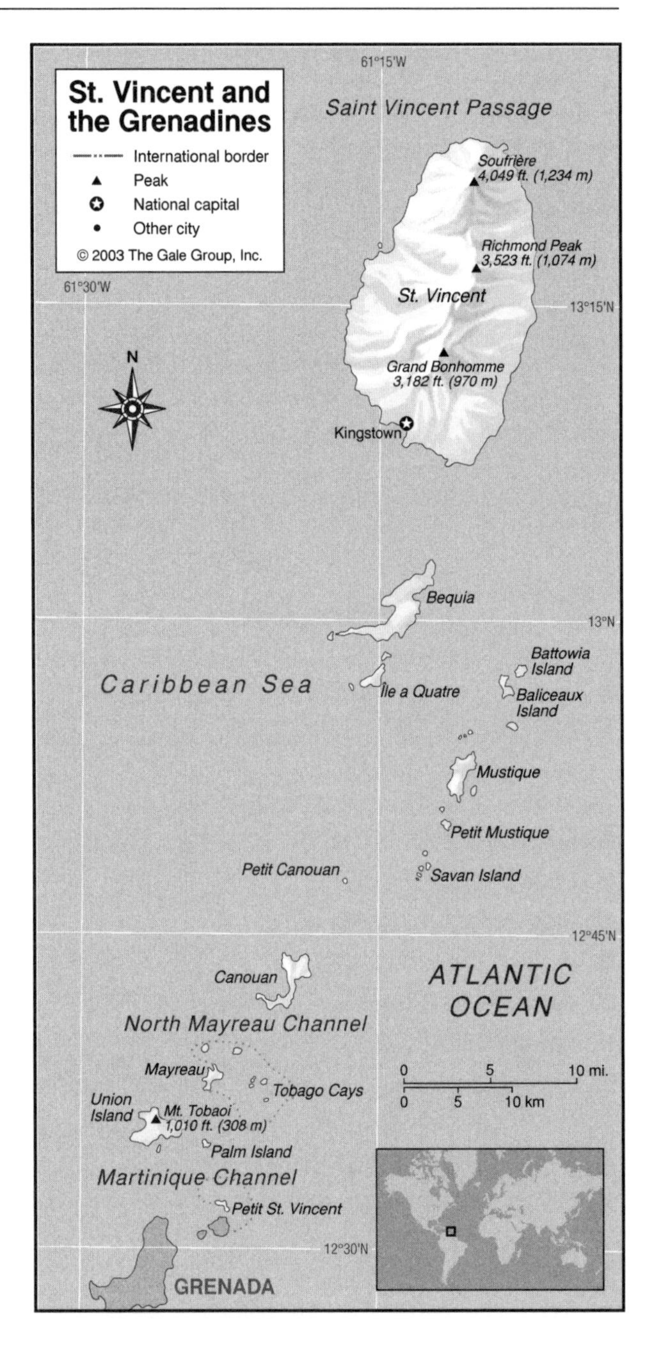

NATURAL RESOURCES

Cropland and hydropower are St. Vincent and the Grenadines' natural resources. Agriculture, especially bananas, is an important part of the economy. Arrowroot starch, eddoes and taro, vegetables, and tennis rackets are other exports. Tourism is also important to the country's economy.

FURTHER READINGS

Lonely Planet Guides. *Saint Vincent and the Grenadines.* http://www.lonelyplanet.com/destinations/caribbean/saint_vincent_and_the_grenadines/index.htm (accessed March 19, 2002).

Philpott, Don. *St. Vincent & Grenadines.* Lincolnwood, Ill.: Passport Books, 1996.

Scuba St. Vincent and the Grenadines. *Visions of Paradise.* http://www.scubasvg.com/islands/islandsintro.html (accessed March 19, 2002).

Walton, Chelle Koster. *Caribbean Ways: A Cultural Guide.* Westwood, Mass.: Riverdale, 1993.

WorldTravelGuide.Net. *Saint Vincent and the Grenadines.* http://www.worldtravelguide.net/data/vct/vct001.asp (accessed March 20, 2002).

Young, Virginia Heyer. *Becoming West Indian: Culture, Self, and the Nation in St.Vincent.* Washington, D.C.: Smithsonian Institution Press, 1993.

GEO-FACT

Robert Louis Stevenson, Scottish author of the classic adventure tale, *Treasure Island*, purchased property at the foot of Mount Vaea, near Apia on Upolu. He named the place Vailima, meaning "five waters," for the small streams that ran across the property. Here he built the home in which he spent the last five years of his life. He is buried on the island, and tourists often visit his gravesite.

Samoa

- **Area:** 1,104 sq mi (2,860 sq km) / World Rank:169
- **Location:** Southern and Western Hemispheres, Oceania, group of islands in the South Pacific Ocean, midway between Hawaii and New Zealand.
- **Coordinates:** 13°35′S, 172°20′W
- **Borders:** None
- **Coastline:** 250 mi (403 km)
- **Territorial Seas:** 12 NM
- **Highest Point:** Mauga Silisili, 6,093 ft (1,857 m)
- **Lowest Point:** Sea level
- **Longest Distances:** 93 mi (150 km) ESE-WNW; 24 mi (39 km) NNE-SSW
- **Longest River:** None of significant size
- **Natural Hazards:** Typhoons; earthquakes; tsunamis; volcanic activity
- **Population:** 179,058 (July 2001 est.) / World Rank: 177
- **Capital City:** Apia, on the northern shore of Upolu
- **Largest City:** Apia, population 33,000 (2000 est.)

OVERVIEW

Samoa (formerly Western Samoa) is located almost centrally in the Polynesian region of the South Pacific. It also comprises the last large area of land east of Australia in the central Pacific.

Samoa consists of the two large islands of Upolu and Savai'i and seven small islets, of which only Manono and Apolima are inhabited. The islands are volcanic, with coral reefs surrounding most of them. They have narrow coastal plains with volcanic, rocky, rugged mountains in the interior. The economy is primarily based on agriculture, but has also traditionally been dependent on development aid and family remittances from overseas. Tourism is an expanding sector.

Samoa is located on the Pacific Tectonic Plate just east of the Australian Tectonic Plate boundary, in the Pacific "Ring of Fire" directly northeast of the Tonga-Kermadec Trench. The plate boundaries give rise to many underwater volcanoes as the plates drift apart and magma flows through the gap. Some geologists believe that the islands themselves were formed by such rifting in the nearby Tonga-Kermadec Trench, and others believe that they were formed simply by magma breaking through a weak spot in the Pacific Plate. The activity of the plates near the Tonga-Kermadec Trench also causes earthquakes and tsunamis.

MOUNTAINS AND HILLS

Mountains

Rugged ranges are prevalent on both major islands, reaching 3,608 ft (1,100 m) on Upolu and 6,093 ft (1,857 m) on Savai'i. The significant peaks are Mauga Silisili—which is the highest point in Samoa at an elevation of 6,093 ft (1,857 m)—Mauga Loa (3,857 ft / 1,176 m), and Mauga Fito (Va'aifetu) (3,660 ft / 1,116 m). The islands are in an area of active volcanism, which has progressed westward. Savai'i, geologically the youngest island, last experienced eruptions from Matavanu 1905–10 and Mauga Mu in 1902. Other volcanoes on Savai'i are Mauga Afi, and Mauga Silisili. Volcanoes on Upolu consist of Mauga Ali'i and Mauga-o-Savai'i.

Plateaus

Savai'i's central volcanoes are surrounded by lava plateaus that give way farther down to hills and coastal plains. Upolu's central volcanic range, rising to Mauga Fito, slopes down on both sides to hills and coastal plains.

INLAND WATERWAYS

Lakes

Crater lakes are fed by rainfall that averages 118 in (300 cm) annually at Apia. On Upolu, there is a very deep lake, Lake Lanoto'o (Goldfish Lake) in the center of a volcanic crater. The lake is pea-green in color and full of wild goldfish that gather along the shores. The bottom of the lake has never been found.

Rivers

Both islands have numerous, swiftly flowing rivers with rapids and waterfalls. However, most of the rivers only flow during the wet season. Due to the porous nature of the volcanic soil, most of the water that does not evaporate quickly seeps into the ground.

Wetlands

In the quiet, swampy bay areas of Samoa red mangroves are common colonists. Vaiusu Bay in Apia is one of the largest areas of mangroves in Eastern Polynesia and the biggest fish breeding and feeding grounds in Samoa. Mangroves provide protected nursery areas for juvenile reef fishes, crustaceans, and mollusks, as well as nesting for coastal birds, and they contribute to higher water quality. Mangrove areas in Samoa have been used as garbage dumping grounds and extensive conservation programs are underway to reclaim them. A creature unique to this habitat is the flying fox (*Pteropus samoensis Peale*), a type of fruit bat, that is one of two mammals indigenous to Samoa. The fruit bat inhabits forests and swamps, by day roosting in trees in a noisy active colony, at dusk flying to fruit trees to feed.

THE COAST, ISLANDS, AND THE OCEAN

Major Islands

Savai'i, with an area of 665 sq mi (1,717 sq km), is the second highest island (6,094 ft / 1,858 m) in Polynesia outside of Hawaii and New Zealand. Upolu (434 sq mi / 1,125 sq km) is 3,749 ft (1,143 m) high. Apolima and Manono are located between the main islands of Savai'i and Upolu in the Apolima Strait. Apolima is a volcanic crater about 0.6 miles (1 km) wide and 1.1 mi (1.8 km) long. It has walls of basaltic rock that rise sharply out of the sea to heights of about 980 ft (300 m). A break in the wall of the volcanic crater allows a narrow channel with access to the sea. Only a handful of families live on Apolima. Manono, only a mile or two (3 km) in circumference, is a triangular-shaped island consisting mainly of coral sand. In the past, this island was home to the highest chiefs.

The Coast and Beaches

Coral reefs nearly surround the volcanic islands of Samoa, broken only in places where they are interrupted by constant wave action and lava flow. Of the 80 countries with coral reefs, Samoa ranks 58 in reef area with 490 sq mi (1,269 sq km), or 0.17 percent of the world's reefs. However, some reefs in Samoa are threatened by the agricultural industry, mining, construction, sewage, and over fishing and exploitation. Ports and harbors are located at Apia and Mulifanua on Upolu, and at Asau and Salelologa on Savai'i.

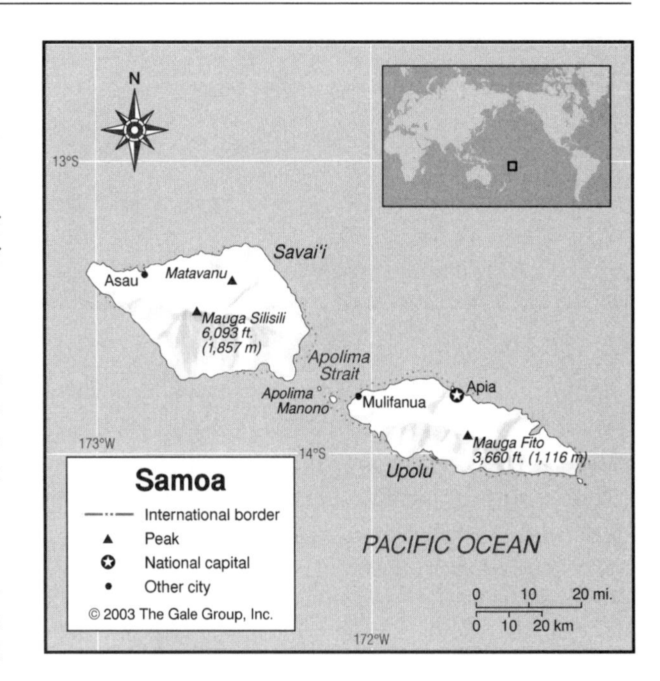

The southern shore of Upolu features a series of beaches. Toward the eastern end of the island are Aganoa Black Sand Beach and Salamuma Beach, both of which draw snorkellers interested in the coves and shallow waters. At the extreme eastern end of Upolu are spectacular turquoise reefs.

CLIMATE AND VEGETATION

Temperature

Samoa has a tropical marine climate. The hottest month is December, and the coldest is July. Due to the oceanic surroundings, the temperature ranges on the islands are not appreciable. The mean daily temperature is about 81°F (27°C) year round.

Rainfall

The dry season is May to October; the wet season is November to April. Rainfall averages 113 in (287 cm) annually, and the average yearly relative humidity is 83 percent. Because the interior of the islands is mountainous, there is also a considerable difference between the rainfall on the coast and of that in the jungle further inland. Average annual rainfall varies from 200 to 280 in (500 to 700 cm) on the southern windward side and 100 to 120 in (250 to 300 cm) on the leeward side. Trade

Islands – Samoa

Name	Area (sq mi)	Area (sq km)
Savai'i	659	1,707
Upolu	432	1,119

SOURCE: *Geo-Data: The World Geographical Encyclopedia*, 2nd ed. Detroit: Gale Research, 1989.

winds from the southeast are fairly constant throughout the dry season.

Forests and Jungles

Along the coasts there are mangrove forests. Inland, over lower slopes of the mountains, are the remaining lowland forests. Inland at higher elevations are rain forests, and higher elevations of Savai'i contain moss forest and mountain scrub. It is estimated that during the 3,000-year history of Samoa's human habitation about 80 percent of lowland forests have been lost.

HUMAN POPULATION

The main island of Upolu is home to nearly three-quarters of Samoa's population and has a density of about 275 people per sq mi (106 people per sq km). The population density on the larger Savai'i is only 73 inhabitants per sq mi (28 inhabitants per sq km). Most of the population is rural: in 2000, it was estimated that only 22 percent of the population lived in urban areas, with almost all of that number living in the capital of Apia. No other town has a population of more than 2,000, and only seven others have a population greater than 1,000.

NATURAL RESOURCES

Samoa's natural resources are its hardwood forests, fish, and hydropower. Besides fishing, the economy is dependent upon agriculture, with the main crops being taro, coconuts, and cocoa. Copra and pineapples are also grown.

FURTHER READINGS

Ahlburg, Dennis A., and Brown, Richard P. C. *Migrants' Intentions to Return Home and Capital Transfers: A Study of Tongans and Western Samoans in Australia.* Pacific Islands Development Program, Reports on Sustainable Development and Population Honolulu: East-West Center, 1997.

Antelope Internet Systems. *KavaRoot.com.* http://kavaroot.com/index.htm (Accessed May 19, 2002).

Chand, Kishore. *Samoa.* http://www.fao.org/waicent/faoinfo/agricult/agl/swlwpnr/y_pa/z_ws/ws.htm#overview (Accessed May 20, 2002).

Dahl, Arthur L. *Regional Ecosystems Survey of the South Pacific Area.* Noumea, New Caledonia: South Pacific Commission, 1980.

Eakin, Mark C. *State of the Reefs: Regional and Global Perspectives.* http://www.ogp.noaa.gov/misc/coral/sor/sor_contents.html#toc (Accessed May 20, 2002).

International Center for Island Studies. http://www.islandstudies.org/ (Accessed May 19, 2002).

National Museum of Natural History. *Global Volcanism Program.* http://www.volcano.si.edu/gvp/volcano/region04/index.htm (Accessed May 19, 2002).

National Oceanographic Data Center. *Coral Reefs and Associated Ecosystems.* http://www.nodc.noaa.gov/col/projects/coral/Coralhome.html (Accessed May 20, 2002).

Secretariat of the Pacific Community. *Pacific Island Region.* http://www.spc.org.nc/En/region.htm (Accessed May 19, 2002).

Tamua, Evotia. *Samoa.* Auckland, New Zealand: Pasifika Press, 2000.

Vaai, Saleimoa. *Samoa Faamatai and the Rule of Law.* Western Samoa: National University of Samoa, 1999.

San Marino

- **Area:** 24 sq mi (61 sq km) / World Rank: 202
- **Location:** Northern and Eastern Hemispheres, surrounded by Italy, in southern Europe.
- **Coordinates:** 43°46′N, 12°25′E
- **Borders:** 24 mi (39 km), all with Italy
- **Coastline:** None
- **Territorial Seas:** None
- **Highest Point:** Monte Titano, 2,477 ft (755 m)
- **Lowest Point:** Torrente Ausa, 180 ft (55 m)
- **Longest Distances:** 8 mi (13km) NE-SW; 6 mi (9 km) SE-NW
- **Longest River:** None of significant length
- **Natural Hazards:** None
- **Population:** 27,336 (July 2001 est.) / World Rank: 199
- **Capital City:** San Marino, west-central San Marino
- **Largest City:** San Marino, 5,000 (2000 estimate)

OVERVIEW

San Marino is a tiny landlocked country, located entirely within Italy, about 15 mi (24 km) southwest of the city of Rimini, in the Apennine Mountains between Italy's Marche and Romagna regions. It is Europe's third-smallest independent state (only Vatican City and

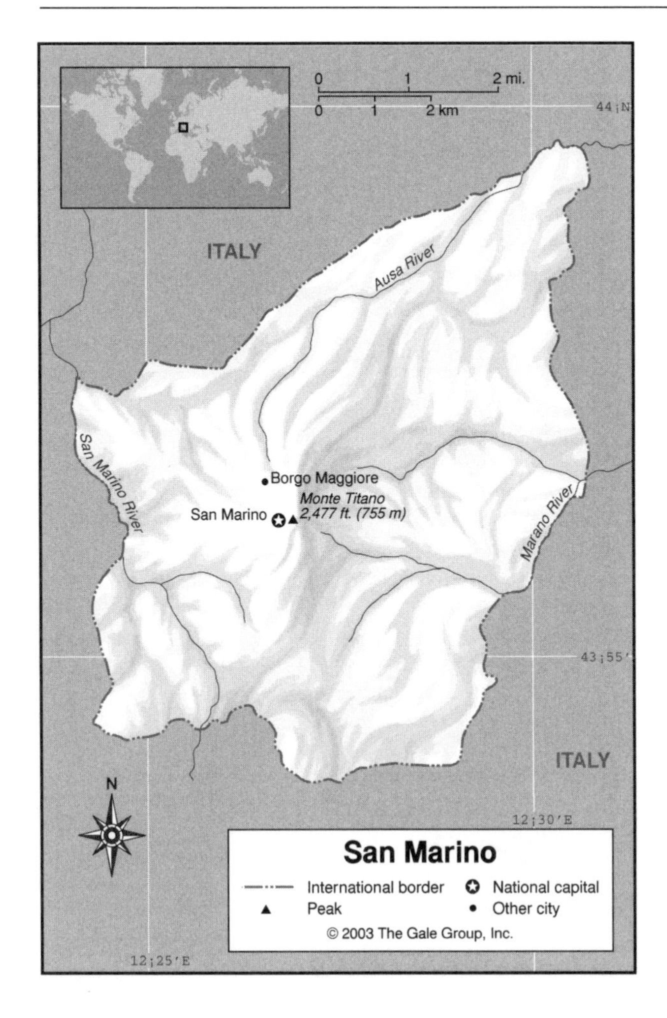

Monaco are smaller) and the world's second-smallest republic (after Nauru). San Marino is dominated by Mt. Titano, on whose slopes and crest most of the irregularly shaped country is situated. However, there is enough level land at the base of the mountain for agriculture. San Marino is situated on the Eurasian Tectonic Plate.

MOUNTAINS AND HILLS

The limestone peaks of Mt. Titano occupy the central part of the republic. There are three major peaks, with ancient fortifications on each one. The summit of Mt. Titano (2,477 ft / 755 m) commands a panoramic view of the Adriatic Sea, only 12 mi (19 km) away. Mt. Titano is bordered by hills to the southwest.

INLAND WATERWAYS

Rivers

San Marino lies largely within the basin of Italy's Marecchia River, into which the San Marino River drains, flowing northward and forming part of the republic's border with Italy. The Marano and Ausa rivers drain into the Adriatic Sea.

Castles – San Marino

Name	Area (sq mi)	Area (sq km)	Capital
Acquaviva	1.88	4.86	Acquaviva
Borgo Maggiore	3.48	9.01	Borgo
Citta	2.74	7.09	San Marino
Chiesanuova	2.11	5.46	Chiesanuova
Domagnano	'2.56	6.62	Domagnano
Faetano	2.99	7.75	Faetano
Fiorentino	2.53	6.56	Fiorentino
Montegiardino	1.28	3.31	Montegiardino
Serravalle / Dogano	4.07	10.53	Serravalle

SOURCE: *Geo-Data: The World Geographical Encyclopedia,* 2nd ed. Detroit: Gale Research, 1989.

THE COAST, ISLANDS, AND THE OCEAN

San Marino is landlocked.

CLIMATE AND VEGETATION

San Marino has the mild, temperate climate typical of northeastern Italy. Summer highs rarely rise above 79°F (26°C), and winter lows rarely fall below 19°F (7°C). Annual rainfall averages between 22 in (56 cm) and 32 in (80 cm).

There is some level land at the base of Mt. Titano, which is given over to agriculture. The mountain itself supports pasturelands as well as forests. Tree species include pine, oak, ash, elm, poplar, olive, and other varieties typical of northeastern Italy.

HUMAN POPULATION

San Marino is one of the world's most densely populated countries, with 1,139 people per sq mi (448 people per sq km) in 2001. More than 95 percent of the population is urban, living in the nation's nine small towns. The capital city of San Marino is located on the western slope of Mt. Titano, close to the summit. Commercial activity is centered on Borgo Maggiore, some 600 ft (183 m) below San Marino and connected to it by a winding road 1.5 mi (2.4 km) long.

NATURAL RESOURCES

San Marino's most important natural resources are its scenic landscape, which attracts tourists, and the limited area of arable land at the base of Mt. Titano. Building stone is its only mineral resource.

FURTHER READINGS

Carrick, Noel. *San Marino.* New York: Chelsea House, 1988.

Catling, Christopher. *Umbria, The Marches, and San Marino.* London: Black, 1994.

Cleary, Amelia. "My Country 'Tis So Wee: The Little Country of San Marino Puts Its Stamp on the World." *Washington Post* November 15, 2000: Style section, C15.

"Now, After 1,600 Years, Time to Join the World (San Marino to Become Member of the United Nations)." *New York Times* February 26, 1992.

São Tomé and Príncipe

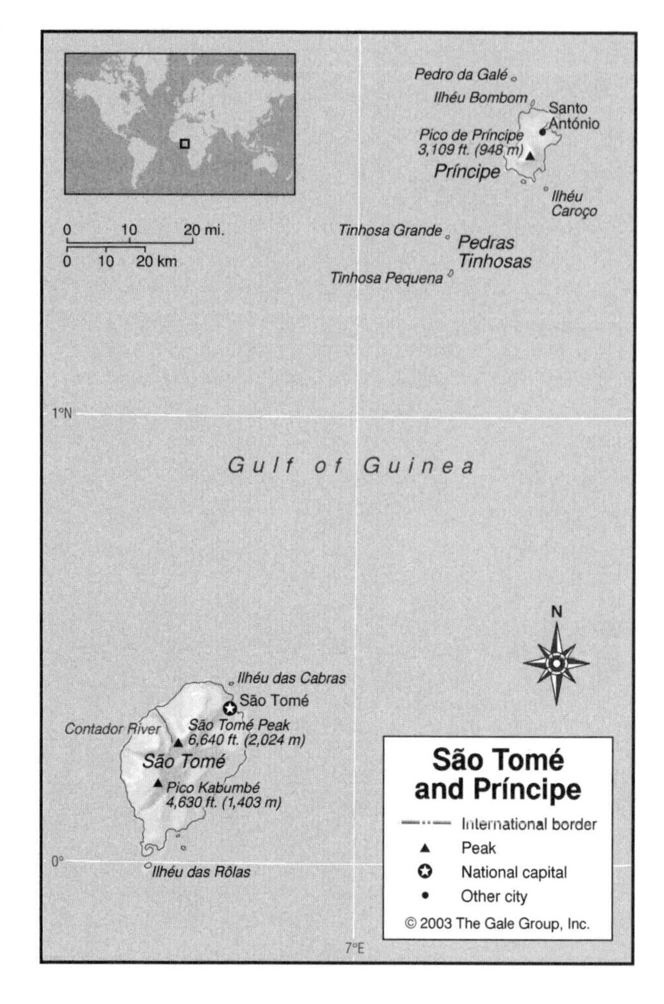

- **Area:** 386 sq mi (1,001 sq km) / World Rank: 175
- **Location:** Northern and Eastern Hemispheres, in the Gulf of Guinea, near the equator, west of the northern coast of Gabon on the African continent.
- **Coordinates:** 1°00′N, 7°00′E
- **Borders:** None
- **Coastline:** 130 mi (209 km)
- **Territorial Seas:** 12 NM
- **Highest Point:** São Tomé Peak, 6,640 ft (2,024 m)
- **Lowest Point:** Sea level
- **Longest Distances:** São Tomé: 30 mi (49 km) NNE-SSW / 18 mi (29 km) ESE-WNW; Príncipe: 13 mi (21 km) SSE-NNW / 9 mi (15 km) ENE-WSW
- **Longest River:** Not available
- **Natural Hazards:** Drought
- **Population:** 165,034 (July 2001 est.) / World Rank: 179
- **Capital City:** São Tomé, located on the northeast coast of the island of São Tomé
- **Largest City:** São Tomé, population 57,000 (2000 est.)

OVERVIEW

Africa's smallest country, São Tomé and Príncipe, is part of a chain of extinct volcanoes. The two main islands are São Tomé (330 sq mi/855 sq km) and Príncipe (42 sq mi/109 sq km). The country also includes the tiny Ilhéu Bombom, Ilhéu Caroço, and Ilhéu das Rôlas.

São Tomé and Príncipe's landscape is mostly mountainous. Rainforest covers other large areas of land, most of which, at higher elevations, give way to mountain-cloud forests. Most of the coastline is comprised of white sand beaches. Almost all of the population lives on the island of São Tomé.

The islands of São Tomé and Príncipe are located on the African Tectonic Plate.

MOUNTAINS AND HILLS

Mountains

The islands of São Tomé and Príncipe were once part of a chain of ocean volcanoes; these volcanoes are now extinct. Both São Tomé and Príncipe are mountainous. São Tomé's highest peaks are Pico de São Tomé (São Tomé Peak) at 6,640 ft (2,024 m) and Pico Kabumbé (Kabumbé Peak) at 4,630 ft (1,403 m). While there are 10 peaks that rise over 3,500 ft (1,067 m), many of the island's other peaks reach only a little more than half that height. Príncipe's highest elevation is Pico de Príncipe (Príncipe Peak) at 3,109 ft (948 m).

Plateaus

Príncipe features a large plateau that extends along the northwestern coast at elevations that reach 3,110 ft (948 m); the terrain of São Tomé also features a plateau, although it is smaller.

INLAND WATERWAYS

Rivers

The Contador River is located in the northwest, and its river valley is possibly the deepest in São Tomé. Several streams run down from the volcanic highlands into the Gulf of Guinea, the body of water that separates São Tomé and Príncipe from the mainland of Africa.

THE COAST, ISLANDS, AND THE OCEAN

Oceans and Seas

The Gulf of Guinea, the portion of the Atlantic Ocean that lies of the coast of West Africa, surrounds São Tomé and Príncipe.

Major Islands

São Tomé and Príncipe comprise an island nation. Small islets lie around the two main islands, and include Ilhéu das Rôlas, straddling the equator off the southern tip of São Tomé; Ilhéu Caroço, off the southern tip of Príncipe; and Bombom, off the northern coast of Príncipe. In the waters between the two main islands are Tinhosa Peqeuna, Pedras Tinhosas, and Tinhosa Grande. These islets are uninhabited.

The Coast and Beaches

Mostly untouched white sand beaches line most of the coasts and the country is attempting to develop a tourist industry around them. The main ports are São Tomé on the main island, and Santo António on Príncipe.

CLIMATE AND VEGETATION

Temperature

Lying near the equator, the islands' climate is tropical, and temperatures vary with the different altitudes. Temperatures in the coastal regions average 81°F (27°C), while the mountain areas average 68°F (20°C). The seasons are differentiated by precipitation changes rather than temperature changes.

Rainfall

The northern regions of São Tomé and Príncipe receive approximately 40 to 60 in (100 to 150 cm) of rain during the rainy season from October to May, while the most of the southern regions receive between 150 and 200 in (380 and 510 cm). The dry season occurs from early June to September.

Grasslands

In the northern region of São Tomé, there is a dry area where the climate resembles that of the savanna.

Forests and Jungles

Forestland and jungle covers most of the islands. Tropical rainforest changes to cloud-mountain forest above eleveations of 4,500 ft (1,370 m).

HUMAN POPULATION

Almost 95 percent of the population lives on São Tomé, and an estimated 47 percent lives in urban areas.

Districts – São Tomé and Príncipe

Name	Area (sq mi)	Area (sq km)	Capital
Aqua Grande	7	17	São Tomé
Cantagalo	46	119	Santana
Caué	103	267	São João Angolares
Lemba	88	229	Neves
Lobata	41	105	Guadalupe
Mé-zóchi	47	122	Trinidade
Paguê	55	142	São António

SOURCE: *Geo-Data: The World Geographical Encyclopedia,* 2nd ed. Detroit: Gale Research, 1989.

Santo António is the largest town on Príncipe. The population density is: São Tomé, 392 inhabitants per sq mi (151 per sq km) on São Tomé; 98 per sq mi (38 per sq km) on Príncipe.

NATURAL RESOURCES

São Tomé and Príncipe's natural resources are few; however, the island nation has considerable fish and hydropower resources. The economy, based on agriculture, was once based on the growth of sugarcane, but cocoa has since become the primary resource. Ninety percent of the country's food supply is imported.

FURTHER READINGS

Hodges, Tony. *São Tomé and Príncipe: from Plantation Colony to Microstate.* Boulder, Colo: Westview Press, 1988.

Iafrica.com. *São Tomé and Príncipe.* http://africa.iafrica.com/countryinfo/saotome/ (Accessed March 21, 2002).

LonelyPlanet. *São Tomé and Príncipe.* http://www.lonelyplanet.com/destinations/africa/sao_tome_and_principe/environment.htm (Accessed March 21, 2002).

Shaw, Caroline S. *São Tomé and Príncipe.* Santa Barbara, Calif.: Clio Press, 1994.

Saudi Arabia

- **Area:** 756,984 sq mi (1,960,582 sq km) / World Rank: 16
- **Location:** Northern and Eastern Hemispheres, Middle East, bordering the Persian Gulf in the east and the Red Sea in the west; north of Yemen, west of Oman, the United Arab Emirates, and Qatar; south of Kuwait, Iraq and Jordan.

- **Coordinates:** 25°00′ N, 45°00′ E
- **Borders:** 2,743 mi (4,415 km) / Iraq, 506 mi (814 km); Jordan, 452 mi (728 km); Kuwait, 138 mi (222 km); Oman, 420 mi (676 km); Qatar, 37.3 mi (60 km); United Arab Emirates, 284 mi (457 km); Yemen, 906 mi (1,458 km)
- **Coastline:** 1,640 mi (2,640 km)
- **Territorial Seas:** 12 NM
- **Highest Point:** Jabal Sawdā', 10,279 ft (3,133 m)
- **Lowest Point:** Sea level
- **Longest Distances:** 1,426 mi (2,295 km) ESE-WNW; 884 mi (1,423 km) NNE-SSW
- **Longest River:** There are no perennial rivers
- **Natural Hazards:** Frequent sand and dust storms
- **Population:** 22,757,092, including 5,360,526 non-nationals (July 2001 est.) / World Rank: 45
- **Capital City:** Riyadh, in the center of the country
- **Largest City:** Riyadh, population 3,600,000 (2002 est.)

OVERVIEW

The Kingdom of Saudi Arabia sits on the Arabian Tectonic Plate and constitutes about four-fifths of the Arabian Peninsula. It offers a land bridge connecting Africa with Eurasia. It is the third largest country in Asia after China and India. However, because several of its borders are incompletely demarcated, its precise area is difficult to specify. Saudi Arabia has the largest reserves of petroleum in the world, and ranks as the largest exporter of petroleum. Its extensive coastlines on the Persian Gulf and Red Sea provide great shipping opportunities (especially crude oil) through the Persian Gulf and Suez Canal. The country can be divided into six geographical regions: the Red Sea escarpment, from Hejaz in the north to 'Asir in the south; in the south, a coastal plain, the Tihamah, rises gradually from the sea to the mountains; the central plateau, Nejd, extends to the Tuwayq Mountains and further; and three sand deserts: the Ad Dahnā', the An-Nafūd, and, south of Nejd, the Rub′ al-Khali Desert, which is one of the largest sand deserts in world. Saudi Arabia's climate and geography has led to recent issues concerning the continuing encroachment of sand dunes on agricultural land, the preservation and development of water sources, and pollution and sanitation problems. At current rates of consumption, the nation's water supply may be exhausted in 10 to 20 years.

GEO-FACT

Saudi Arabia's borders are slowly being demarcated. A final border resolution with Qatar was achieved in March 2001. The location and status of the boundary with the United Arab Emirates is not final; the de facto boundary reflects a 1974 agreement. A June 2000 treaty delimited the boundary with Yemen, but final demarcation requires adjustments based on tribal considerations.

MOUNTAINS AND HILLS

Mountains

The Hejaz Mountains (3,000-9,000 ft / 910-2,740 m) rise sharply from the Red Sea and run in a north-south direction parallel to the Sea. Mount Lawz, at 8,464 ft (2,580 m), rises in the far north of the Hejaz in the vicinity of the Red Sea and Jordan. The northern range in the Hejaz seldom exceeds 6,888 ft (2,100 m) and gradually decreases to about 1,968 ft (600 m) around Mecca. Close to Mecca the Hejaz coastal escarpment is separated by a gap. In the plateau region of Nejd the Ajā' Mountains are just south of the An-Nafūd desert. The highest mountains (over 9,000 ft / 2,740 m) are in 'Asir in the south. This region extends along the Red Sea for 230 mi (370 km) and inland about 180 to 200 mi (290 to 320 km). Saudi Arabia's highest peak, Jabal Sawdā', is found here towering to a height of 10,276 ft (3,133 m).

Plateaus

East of Hejaz and 'Asir lies the central uplands, Nejd, a large mainly rocky plateau ranging from about 5,000 ft (1,520 m) in the west to about 2,000 ft (610m) in the east. The Nejd is scarred by extensive lava beds (*harrat*), evidence of fairly recent volcanic activity. A low plateau, Al-Hasa, to the east—average width, 100 mi (160 km); average altitude, 800 ft (240 m)—gives way to the low-lying Gulf region. The area north of the An-Nafūd, Badiyat ash Sham, is an upland plateau that is geographically part of the Syrian Desert. The Wādī as Sirhān, a large basin 984 ft (300 m) below the surrounding plateau, is a vestige of an ancient inland sea. For thousands of years the Wādī as Sirhān has served as a transit stop for caravan routes between the Mediterranean and the central and southern Saudi peninsula. East of the Ad Dahnā' lies the rocky, barren As-Summān Plateau, about 74 ft (120 km) wide and dropping in elevation from about 1,312 ft (400 m) in the west to about 787 ft (240 m) in the east. Separated from the As-Summān Plateau by the Ad Dahnā' is the Al-'Aramah Plateau, which runs right up to Riyadh.

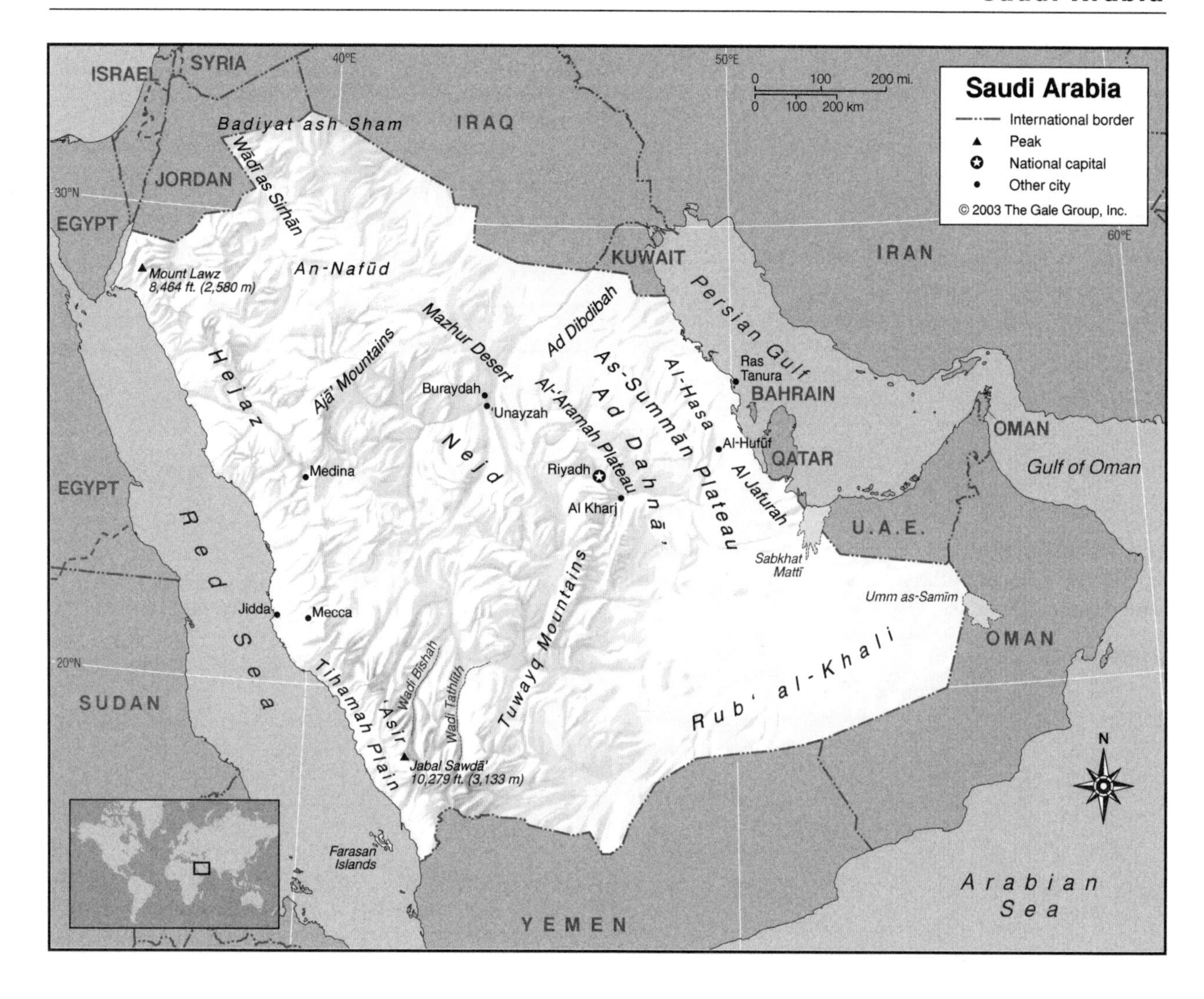

Hills and Badlands

The Tuwayq escarpment is a 496 mi (800 km) arc of spectacular limestone cliffs, plateaus, and canyons, eroded by wind and sand, that cuts across the Ad Dahnā' desert. Its steep west face rises anywhere from 328 to 820 ft (100 to 250 m) above the Nejd plateau. Many oases exist in this region. Buraydah, 'Unayzah, Riyadh, and Al Kharj are the most important. Large salt marshes (*sabkah*) are spread throughout the area.

INLAND WATERWAYS

Except for artesian wells in the eastern oases, there is no perennially existing water, either as lakes or flowing in rivers, in Saudi Arabia. In the northern Hejaz region, dry riverbeds (wadis) trace the courses of ancient rivers. The wadis will contain water for a brief period following significant rainfall, but the only real existence of water is in oases. Medina is the largest and most important oasis in the Hejaz region. In the southern 'Asir fertile wadis, Wadi Bīshah and Wadi Tathlīth allow for oasis agriculture. Eastern Arabia is also known as Al Ahsa, or Al-Hasa, after the largest oasis in the country, which actually encompasses two neighboring oases and the town of Al-Hufūf.

THE COAST, ISLANDS, AND THE OCEAN

Oceans and Seas

Saudi Arabia is bordered by two bodies of water: the Persian (Arabian) Gulf to the east, and the Red Sea to the west, the warmest and saltiest of the world's seas. The Persian Gulf is the marginal offshoot of the Indian Ocean that lies between the Arabian Peninsula and Iran, extending about 600 mi (970 km) from the Shatt al Arab delta to the Strait of Hormuz. The Gulf's width varies from a maximum of 210 mi (338 km) to a minimum of 34 mi (55 km) in the Strait of Hormuz, which opens into the Arabian Sea linking it with the Gulf of Oman. The shallow Gulf waters have very slow currents and limited tidal range. The Arabian Peninsula is a rock platform that was once continuous with northeast Africa. The Red Sea and the Gulf of Aden were formed when fissures opened, as the result of which a large trough, or rift valley, was formed and later occupied by the sea. The Saudi Arabian Red Sea coast is where the upward tilt of this platform is greatest. No perennial coastal rivers or streams flow into the Red Sea, and it is partially isolated from the open ocean. There are practically no natural harbors along the Red Sea. The Red Sea ecoregion is best known for the spectacular corals that live in the central and northern areas. Fewer coral species thrive in the Persian Gulf than in the Red Sea. Nevertheless, the entire Arabian Peninsula is fringed by some of the most beautiful coral reefs in the world, reefs which have been used by seagoing Arabs for thousands of years. The Arabian coral reefs have adapted to conditions not experienced by coral reefs elsewhere, such as high water temperature and salinity. Other than the reefs close to larger coastal cities and developments, most of these coral reefs in the Red Sea are healthy. However, reports from diving enthusiasts target coastal construction, dredging, and littering as expanding north and south from the Jidda area as having a significant impact on the reefs.

Major Islands

The principal islands of Saudi Arabia are found in an archipelago in the Red Sea, the Farasān Islands. The Farasan Islands number more than 120 and are fringed by beautiful pristine coral reefs, seagrass beds, and mangroves. Among the largest islands are Farasān al Kabir, 152 sq mi (394.7 sq km), altitude of 246 ft (75 m); Sajid, 60 sq mi (156 sq km); and Zufaf, 13 sq mi (33.2 sq km), altitude 377 ft (115 m); and two others, Sinafir and Tiran. All are uninhabited. The waters of the archipelago are occupied by important and endangered species such as turtles, whales, dolphins, and dugong (aquatic herbivorous mammals related to manatees). The beaches are settled as nesting grounds by turtles, and some are accredited as internationally and regionally important sites for waterbirds. Traditional Saudi maritime activities, such as artisanal fishing by Saudi nationals, pearl diving, and the annual gathering of harid parrotfish, still occur in the Farasān and nowhere else.

The Coast and Beaches

The Tihama Plain parallels the Red Sea coast. It is a salty tidal plain with the width averaging only about 40 mi (65 km). It is hot year round with a mean annual temperature 86°F (30.2°C) and mean annual rainfall of 1 in (2.5 cm). Jidda, on the Red Sea, is the chief port of entry for Muslim pilgrims going to Mecca. The flat, lowland coastal plain along the Persian Gulf is about 37 ft (60 km) wide. The north is the Ad Dibdibah gravel plain and the south is the Al Jafurah sandy desert. The Persian Gulf coast is extremely irregular and the land surface is unstable. Only the construction of long moles (breakwater) at Ras Tanura has opened the Gulf coast to seagoing tankers. In addition, on the Persian Gulf side, Saudi Arabia has infilled more than 40 percent of its coastline, wiping out half its mangroves; dredging and sedimentation are causing major ecological problems in these coastal habitats. Saudi Arabia's coast has no significant bays or capes.

Population Centers – Saudi Arabia

(2001 POPULATION ESTIMATES)

Name	Population
Riyadh (capital)	4,300,000
Jidda (Jeddah)	2,700,000
Mecca (Makkah)	1,500,000

SOURCE: Saudi Arabian Information Resource.

Administrative Areas – Saudi Arabia

Name	Area (sq mi)	Area (sq km)	Capital
Al Baha	3,830	9,921	Al Baha
Al Jouf	38,692	100,212	Sakākah
Asir	29,611	76,693	Abha
Hail	40,111	103,887	Hail
Jazan	4,506	11,671	Jīzān
Madinah	58,683	151,990	Medina (Al Madinah)
Makkah	59,123	153,128	Mecca (Makkah)
Najran	57,726	149,511	Najran
Qasim	22,412	58,046	Buraydah
Riyadh	156,077	404,240	Riyadh (Ar Riyād)
Tabouk	56,398	146,072	Tabouk

SOURCE: "Population (6–30 Years) by Enrollment Status in Administrative Areas 1413 A.H. (1992 A.D.)." Central Department of Statistics, Kingdom of Saudi Arabia.

CLIMATE AND VEGETATION

Temperature

Saudi Arabia's desert climate is generally very dry and very hot. However, in winter there can be frost and freezing temperatures. Day and night temperatures vary greatly. Two main climate extremes exist between coastal lands and the interior. The Red Sea coastal regions and Persian Gulf encounter high humidity and high temperatures, hot mists during the day and a warm fog at night. In the interior, daytime temperatures from May to September can reach 129°F (54°C) and are among the highest recorded anywhere in the world. The climate is more moderate from October through April with evening temperatures between 61° and 70°F (16° and 21°C). The prevailing winds are from the north. A southerly wind brings an increase in temperature and humidity, and a particular kind of storm known in the Gulf area as *kauf.* A strong northwesterly wind, the *shamal,* blows in late spring and early summer. In Eastern Arabia the *shamal* is especially severe, producing sand and dust storms that can decrease visibility to a few yards.

Rainfall

Average annual rainfall is 3.5 in (9 cm). A year's rainfall may consist of one or two torrential outbursts that flood the wadis and quickly disappear in the sand. Most rain falls from November to May. The eastern coast is noted for heavy fogs, and humidity can reach 90 percent. Between 10 and 20 in (3 and 5 cm) of rain falls in the mountainous 'Asir area, where there is a summer monsoon. Much of the Rub'al-Khali is considered "hyper-arid," without rainfall for more than 12 consecutive months.

Deserts

At least one-third of the total area of Saudi Arabia is sandy desert. The Rub'al-Khali (the Empty Quarter) in the south is the largest, with an area roughly 250,000 sq mi (647,500 sq km). It is considered the biggest continuous body of sand anywhere in the world. The Rub'al-Khali orients on a southwest to northeast axis. It is sand overlying gravel or gypsum plains and its depth can vary from zero to 820 ft (250 m). Its surface elevation varies from 2,624 ft (800 m) in the far southwest to near sea level in the northeast. Dune types vary, including moving crescent-shaped (barchan), longitudinal dunes more than a hundred of miles long, and enormous dune mountains. Salt flats like the Umm as-Samīm and the Sabkhat Mattī can harbor quicksands. The Rub'al-Khali extends into Qatar, the Abu Dhabi region of the United Arab Emirates, western Oman, and eastern Yemen. Its northern counterpart, the An-Nafūd, has an area of about 22,000 sq mi (57,000 sq km) with an elevation of about 3, 280 ft (1,000 m). It is covered by lengthy longitudinal dunes as much as 300 ft (90m) high. Dunes are separated by valleys of up to 10 ft (16 km) wide. The sand has a red tint from iron oxide. The Rub'al-Khali is connected to the An-Nafūd Desert by the Ad Dahnā' Desert, also called the river of sand. Its sands also have a reddish tint. The Ad Dahnā', with an average width of 35 mi (56 km) and an average altitude of 1,500 ft (460m), connects to the An-Nafūd Desert by way of the Mazhur Desert.

HUMAN POPULATION

Saudi Arabia's 2001 population was estimated to be about 22.7 million, including about 6.4 million resident foreigners. Until the 1960s, most of the population was nomadic or semi nomadic; due to rapid economic and urban growth, more than 95 percent of the population now is settled. Most of the population is concentrated in cities along the Persian Gulf and Red Sea coasts, as well as the densely populated internal oases, but much of the internal regions are uninhabited desert. Some cities and oases have densities of more than 2,600 people per sq mi (1,000 people per sq km).

NATURAL RESOURCES

Saudi Arabia's number one natural resources are petroleum and natural gas, with the northwestern region harboring the most. The Kingdom invested vast sums in the development of its oil industry, which generates revenues that finance the country's development and, at the same time, supports a non-oil industrial sector. Saudi Arabia's development plans include extensive exploration and survey for minerals of all kinds, hoping to exploit the deposits of iron ore, gold, uranium, bauxite, coal, iron, phosphate, tungsten, zinc, silver, and copper that exist. New mines at Al Amar, Al Ha 'ar, and Hamdhah and the magnesite deposit at Zharghat are coming into production.

FURTHER READINGS

Action Atlas Coral Reefs. *The Middle East.* http://www.motherjones.com/coral_reef/middle_east.html (Accessed May 20, 2002).

Lancaster, William. *The Rwala Bedouin Today.* Prospect Heights, Ill.: Waveland Press, 1997.

Long, David E. *The Kingdom of Saudi Arabia.* Gainesville: University Press of Florida, 1997.

Munton, P. "An overview of the ecology of the sands." *Journal of Oman Studies* Special Report 3, 1988: 231-240.

Nance, Paul J. *The Nance Museum: A Journey into Traditional Saudi Arabia.* St. Louis, MO: Nance Museum Pub., 1999.

NASA. *Gemini Earth Photographs.* "Ar-Rub' al-kahli." http://www.hq.nasa.gov/office/pao/History/SP-4203/phot01.htm (Accessed May 11, 2002).

United Nations. *Saudi Arabia.* http://www.un.int/saudiarabia/ (Accessed May 9, 2002).

US Department of Energy. *Saudi Arabia.* http://www.fe.doe.gov/international/saudiarabia.html (Accessed May 19, 2002).

US Department of Energy. Energy Information Administration. *Saudi Arabia: Environmental Issues.* http://www.eia.doe.gov/emeu/cabs/saudenv.html (Accessed May 20, 2002).

Senegal

- **Area:** 75,749 sq mi (196,190 sq km) / World Rank: 87
- **Location:** Northern and Western Hemispheres, on coastal West Africa, north of Guinea and Guinea-Bissau, west of Mali, southwest of Mauritania, and surrounding the Gambia.
- **Coordinates:** 14°00′N, 14°00′W
- **Borders:** 1,927 mi (3,101 km) / Gambia, 460 mi (740 km); Guinea, 205 mi (330 km); Guinea-Bissau, 210

mi (338 km); Mali, 260 mi (419 km); Mauritania, 505 mi (813 km)

- **Coastline:** 330 mi (531 km)
- **Territorial Seas:** 12 NM
- **Highest Point:** Unnamed feature near Nepen Diakha, 1,906 ft (581 m)
- **Lowest Point:** Sea level
- **Longest Distances:** 429 mi (690 km) SE-NW; 252 mi (406 km) NE-SW
- **Longest River:** Senegal, 1,015 mi (1,663 km)
- **Largest Lake:** Lac de Guiers, 60 sq mi (150 sq km)
- **Natural Hazards:** Lowlands seasonally flooded; periodic droughts; cold rains in north
- **Population:** 10,284,929 (July 2001 est.) / World Rank: 75
- **Capital City:** Dakar, located on Cap Vert peninsula facing the Atlantic Ocean
- **Largest City:** Dakar, population 2,077,000 (2000 est.)

OVERVIEW

Senegal is the westernmost part of a broad savanna extending across the Sahel. Most of the country lies upon a low sedimentary basin characterized by an expanse of flat and undulating plains with sparse grasses and woody shrubs, remarkable only for the near absence of natural landmarks or major changes in elevation. Broken terrain and steep slopes are found only in the extreme southeast.

Extensive riverine areas have been converted to farmland, especially in the Siné and Saloum River basins; the lowlands between Thiès and Kaolack yield significant peanut and other food crops. Beyond these areas, most of the land has little potential except as pasturage. Volcanic action created the Cap Vert peninsula and the nearby islets. Senegal lies on the African Tectonic Plate, and is slightly smaller than the state of South Dakota.

MOUNTAINS AND HILLS

Mountains

None of significant size.

Plateaus

In the extreme southeast, the Fouta Djallon plateau extends into Senegal from Guinea.

Hills

Except for the dunes in the coastal belt and several minor hills northwest of Thiès, the southeast is the only area with elevations of more than 300 ft (91 m) above sea level; even there, only a few ridges exceed 1,300 ft (396 m). The country reaches its highest point, 1,906 ft (581 m), near Nepen Diakha.

INLAND WATERWAYS

Lakes

The ebb and flow of the Senegal River is checked at some points by dikes that are opened to admit the fresh water and are later closed to impound it for use during the dry season and to exclude advancing salt water. A dam and a gate on the Taoué channel in northern Senegal 100 mi (160 km) inland control a shallow lake extending southward about 50 mi (80 km) and averaging about 8 mi (12 km) in width, known as the Lac de Guiers. At highest level the lake waters reach another 40 to 50 mi (64 to 80 km) southeastward into the Ferlo valley.

To the north of the Cap Vert peninsula lies Lac Rose, a shallow salt-water lake occupying a depression behind the coastal dunes. Organisms in the lake give it a pinkish color and villagers extract its salt for commercial purposes.

Rivers

Senegal's largest rivers—the Senegal, Siné, Saloum, Gambia, and Casamance—are sluggish, marsh-lined streams emptying into broad estuaries along the Atlantic Ocean. The Senegal rises in Guinea from the Bafing River, which is joined in eastern Mali by the Bakoye River. As it enters Senegal, it is joined from the south by the Falémé River. The average gradient is only a few inches per mile, and at high flood stage water spreads through a system of channels, sloughs, and adjacent lowlands until most of the valley is a sheet of water from which the tops of trees appear as green patches and villages stand out as isolated islands. As the long dry season sets in, ocean tides range nearly 300 mi (483 km) upstream, but following the rainy season, salty water is forced seaward and the system is filled with fresh water.

The Gambia, which rises in Guinea, receives the flow of a perennial river, the Koulountou, which also runs north from Guinea to join it near the Gambian border. Between the Gambia and Guinea-Bissau, the Casamance River drains a narrow basin less than 20 mi (32 km) wide, becoming a broad estuary 65 miles (104 km) from the sea, 6 mi (10 km) wide at the mouth.

Wetlands

The Saloum River and its major affluent, the Siné River, feed into an extensive tidal swamp just north of the Gambia. Only the lower reaches carry water all year, and these are brackish, as the tides penetrate far up the various channels through the swamp. Senegal's river valleys are subject to annual flooding.

THE COAST, ISLANDS, AND THE OCEAN

Oceans and Seas

Senegal faces the North Atlantic Ocean and has many major ports and harbors, the biggest being the capital city

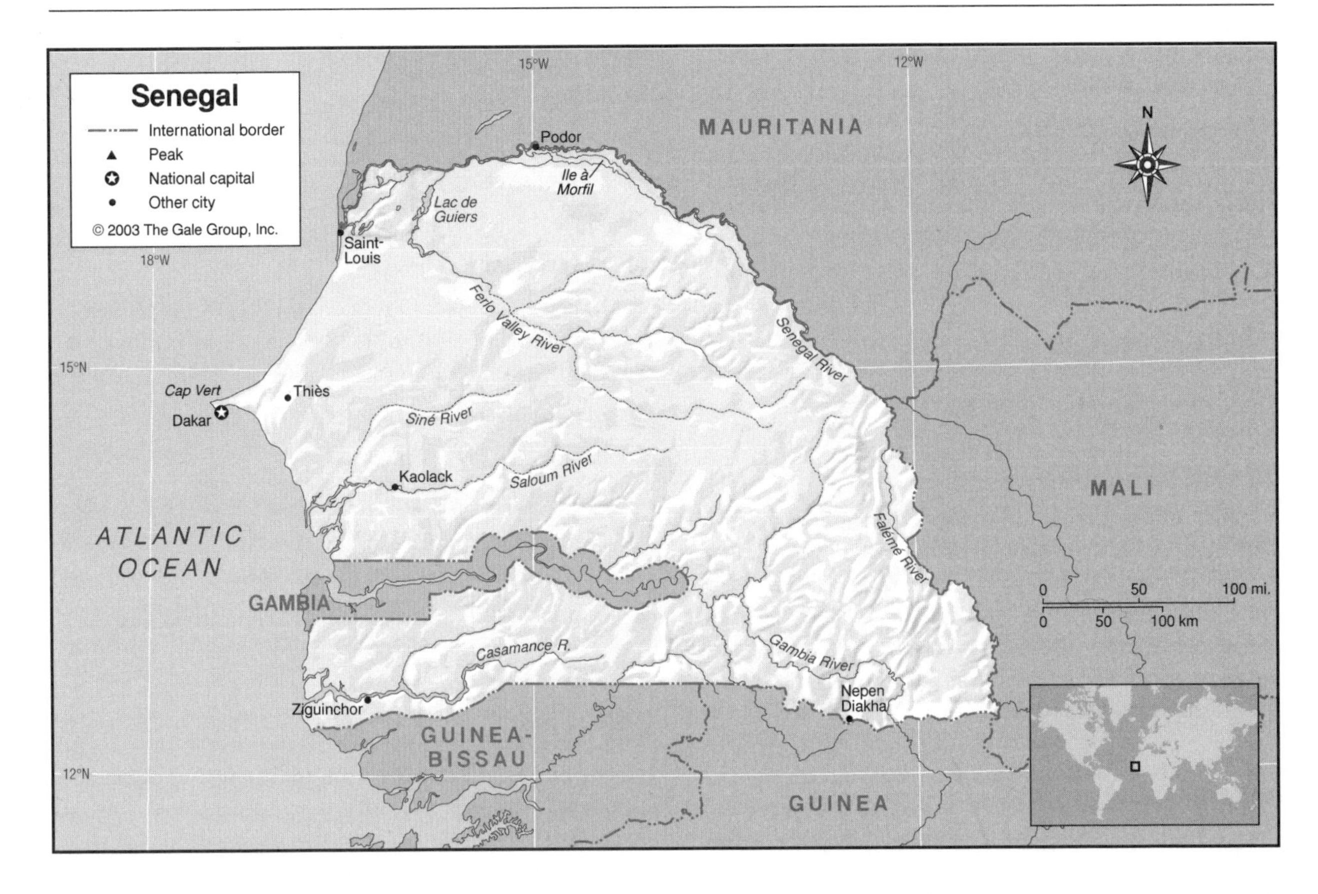

of Dakar. Other harbors up and down the coast are Kaolack, Matam, Podor, Richard Toll, Saint-Louis, and Ziguinchor. The North Atlantic provides Senegal with a great deal of rich fishing ground, and the fishing industry is a major component of Senegal's economy. Of some prevalence is the goblin shark, an animal with a peculiarly shaped body and of which little is known.

Major Islands

St. Louis, the former capital of colonial French West Africa, is located on an island near the mouth of the Senegal River. Ile de Gorée, once a slave transshipment point, is situated between the Cap Vert peninsula and the Petite Côte of the mainland. In the Senegal River valley above Dagana is the Ile à Morfil, a narrow island several hundred miles long between the river's main channel and the Doué channel on the opposite side. Senegal's estuaries contain many flat islands dividing numerous river channels.

The Coast and Beaches

The coastal belt north of the Cap Vert peninsula was formed by northeast trade winds, heavy surf, and the Canary Current, which moves southwestward along the coast. It is covered by small swamps or pools separated by very old dunes as high as 100 ft (30 m). The peninsula of Cap Vert itself is the westernmost point on the African coast. South of Dakar the coastal strip of sand beach narrows and is interrupted by a rocky promontory at Popenguine. In the Siné-Saloum, the coast is broken by the channels and islands of the Saloum River estuary. South of the Casamance River, creeks and estuaries are clogged by silt and sand in an area of salt flats.

CLIMATE AND VEGETATION

Temperature

Temperatures are lowest along the coast. At Dakar they vary from 79°F (26°C) to 63°F (17°C) from December to April, and from 86°F (30°C) to 68°F (20°C) from May to November.

Rainfall

The rainy season lasts from June to October, but rainfall is heavier and the rainy season is longer in the southern Casamance region than north of the Gambia. In the semi-arid extreme north, Podor has an average rainfall of 13 in (34 cm) while Ziguinchor, near the Guinea-Bissau border, has an average of 61 in (155 cm). Dakar averages 22 in (57 cm) of rain each year.

Forests and Savanna

Mangroves, thick forest, and oil palms characterize the coastal area of Casamance. This vegetation shades into wooded or open savanna in the central and eastern parts of the Casamance and throughout the Siné-Saloum. From Mauritania to the Gambia lies the Ferlo, a featureless expanse of savanna where dried tufts of grass, scrub, and thorn trees dominate over the long dry season.

Population Centers – Senegal

(1994 POPULATION ESTIMATES)

Name	Population
Dakar (capital)	1,641,358
Thiès	216,381
Kaolack	193,115
Ziguinchor	161,680
Saint-Louis	132,499

SOURCE: "Population of Capital Cities and Cities of 100,000 and More Inhabitants." United Nations Statistics Division.

HUMAN POPULATION

Most of the population is concentrated along coastal settlements. Nearly a quarter of the population lives in the capital city of Dakar. Overall, the country has a low population density of about 136 people per sq km (52 people per sq km). However, the eastern half of the country has few people, which means actual living conditions are more crowded than this figure would indicate.

NATURAL RESOURCES

Senegal's main resources are fish, phosphates, and iron ore. Off-shore hydrocarbon deposits are under exploration. More than half of the population makes a living working in either the fishing or agriculture industries. The main agricultural products that are produced in Senegal are peanuts, cotton, rice, and millet.

FURTHER READINGS

Africa South of the Sahara 2002. "Senegal." London: Europa Publishers, 2001.

Clark, Andrew F., and Lucie C. Phillips. *Historical Dictionary of Senegal.* 2nd ed. Metuchen, N.J.: Scarecrow Press, 1994.

Gellar, Sheldon. *Senegal: An African Nation between Islam and the West.* 2nd ed. Boulder, Colo: Westview Press, 1995.

Miller, Leanne. *The Goblin Shark.* http://www.umich.edu/~bio440/fishcapsules98/Mitsukurina.html (Accessed June 2002).

Seychelles

- **Area:** 176 sq mi (455 sq km) / World Rank: 186
- **Location:** An archipelago of more than 100 islands in the Indian Ocean off the east coast of Africa, Southern and Eastern Hemispheres, northeast of Madagascar
- **Coordinates:** 4°35′S, 55°40′E
- **Borders:** None
- **Coastline:** 305 mi (491 km)
- **Territorial Seas:** 12 NM
- **Highest Point:** Morne Seychellois, 2,992 ft (912 m)
- **Lowest Point:** Sea level
- **Longest Distances:** 17 mi (27 km) N-S, 7 mi (11 km) E-W; stretching 100 mi (1,200 km) from northeast to southwest
- **Longest River:** None of significant length
- **Natural Hazards:** None
- **Population:** 79,715 (July 2001 est.) / World Rank: 187
- **Capital City:** Victoria, on the largest island of Mahé at the northeastern end of the archipelago
- **Largest City:** Victoria, 40,000 (2000 metropolitan est.)

OVERVIEW

Seychelles is located in the Indian Ocean, 1,000 mi (1,600 km) east of the African continent and 700 mi (1,126 km) northeast of the island nation of Madagascar on the African Tectonic Plate. Its more than 100 islands fall into two categories: the core group of high-rising granitic islands, and a group of low coralline atolls making up the southwest part of the county.

MOUNTAINS AND HILLS

Forming the highest part of the Mascarene Ridge (a ridge that for much of its length is underwater in the Indian Ocean and runs in a north-south direction), the core group of islands making up the Seychelles are granitic peaks or ridges. On Mahé, Morne Seychellois (Mount Seychelles) reaches the highest point at 2,992 ft (912 m). The mountainous characteristics of the granitic islands are among the notable characteristics that appeal to tourists.

INLAND WATERWAYS

There are no rivers or lakes in Seychelles. Small streams drain the mountain slopes and there are small ponds on some of the islands.

THE COAST, ISLANDS, AND THE OCEAN

Oceans and Seas

The Seychelles archipelago is spread over approximately 150,000 sq mi (388,498 sq km) of the Indian Ocean east of Africa. Surrounding the islands are coral reefs.

GEO-FACT

The second largest of the Seychelles islands, Praslin, is the only place on earth where the coco-de-mer palm is found growing wild. The coco-de-mer has a huge double nut that resembles a human pelvis.

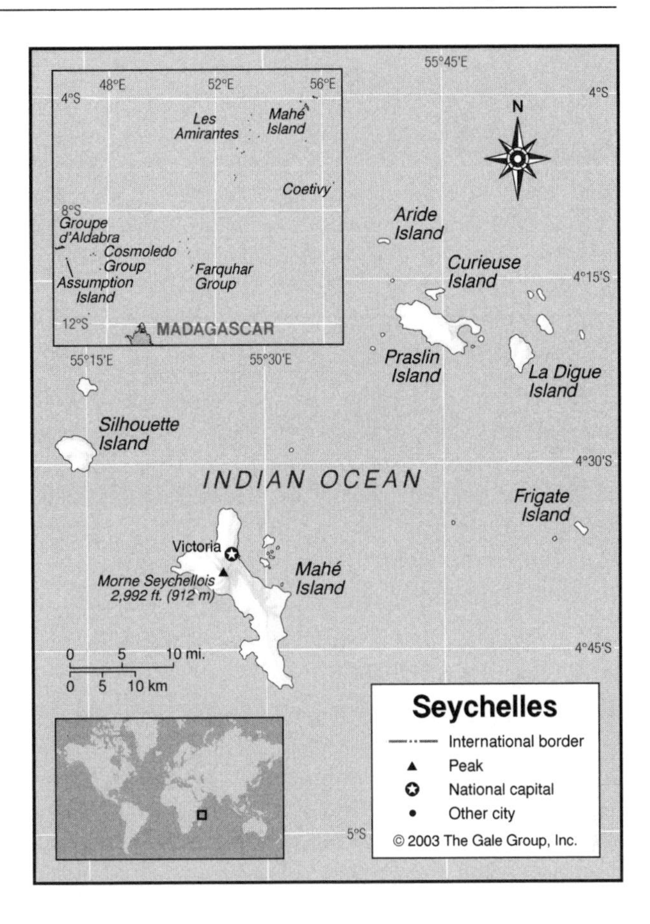

Major Islands

The count on the total number of islands varies depending upon what is considered an island. Some are merely sand cays and shoals barely above the high tide mark. There is general agreement that there are 32 granitic Seychelles islands. The remaining 70–90 islands are in the coralline group. The total land area of the granitic group is about 100 sq mi (259 sq km).

The largest granitic island is Mahé (56 sq mi/144 sq km). It is surrounded by coral reefs, and ringed by beaches featuring fine white sand. Other inhabited (or tourist destination) islands include Praslin (the second-largest island, lying northeast of Mahé); La Digue (east of Praslin); Frigate (directly east of Mahé and south of La Digue); and Silhouette (northwest of Mahé). The most northerly of the granitic islands is Aride, home to a bird sanctuary.

The Cosmoledo Group makes up the most south-westerly of the Seychelles. The coralline Aldabra, one of the Groupe d'Aldabra (Aldabra Group), is the world's largest atoll.

The Coast and Beaches

White, sandy beaches surround the granitic islands with flats of corals and shells behind them. Tar balls have washed up on the beaches for decades, indicating the possibility of undersea oil reserves.

CLIMATE AND VEGETATION

Temperature

Despite lying close to the equator, the trade winds keep the Seychelles' climate temperate. Coastal temperatures remain fairly constant at 81°F (27°C) throughout the year. Temperatures are generally lower at higher altitudes, especially at night. Humidity tends to be high, particularly in the coastal regions.

Rainfall

Average annual rainfall varies markedly across the islands of the Seychelles. The coastal regions on Mahé experience an annual rainfall of 93 in (236 cm), while the areas at higher altitude receive much more (140 in / 356 cm). The coral islands of the southwest, such as Aldabra and Assumption, experience much less rainfall, averaging about 20 in (50 cm) annually.

Generally, the period from May to October is slightly drier, although southeasterly winds bring brief rains every two to three days even during these months of the year. The northeasterly winds prevail from December through March, bringing heavier and more frequent rains.

Forests

Only 11 percent of Seychelles is considered forest land. Primary forests exist only on Praslin and Curieuse islands, which lie north of Mahé. These native forests of the coco-de-mer are now protected in small reserves. Coconut plantations have virtually replaced all broadleaf evergreen rain forests. There are other native tree species on the Seychelles that have adapted to the local conditions. Many forests are underplanted with fruit and spice plants, making good use of scarce land resources.

HUMAN POPULATION

The granitic Seychelles, including Mahé, Silhouette, Praslin, and La Digue, are situated in an area 35 mi (56 km) wide, and are home to the majority of Seychellois. Eighty percent of the population lives on the island of

Island Groups – Seychelles

Name	Area (sq mi)	Area (sq km)
Central Group	93	239
Outer Group	83	214

SOURCE: *Geo-Data: The World Geographical Encyclopedia*, 2nd ed. Detroit: Gale Research, 1989.

Mahé, almost two-thirds of that in the capital city of Victoria alone. The coralline Seychelles—including the Groupe d'Aldabra (Aldabra Islands), Coetivy, and the nearby Atoll de Cosmoledo (Cosmoledo Group), Atoll de Farquhar (Farquhar Group), and the Les Amirantes (Amirante Isles)—are inhabited mostly by a small number of temporary residents.

NATURAL RESOURCES

Fish, copra, coconuts, and cinnamon trees are the major natural resources, with copra and cinnamon being produced for export. Agriculture has long been the backbone of the country's economy, but it only contributes about 5 percent to the gross domestic product (GDP). Industrial fishing is in the early stages, accounting for 8 percent of exports and 1 percent of GDP.

Tar balls that have washed up on the country's beaches for decades have led to exploration for petroleum beginning in the 1970s, and the formation of a national oil company in the 1980s. As of 2002 however, no significant, exploitable, oil reserves had been identified.

FURTHER READING

Carpin, Sarah. *Seychelles.* Chicago: Passport Books, 1997.

Journey through Seychelles. Edison, N.J.: Hunter Publishing Co, 1994.

Ozanne, J.A.F. *Coconuts and Créoles.* London: P. Allan & Co., 1936.

Seychelles Magic/Seychelles Super Site. http://www.sey.net/ (Accessed April 10, 2002).

Seychelles Nation. http://www.seychelles-online.com.sc/ (Accessed April 10, 2002).

Travis, William. *Beyond the Reefs.* New York: Dutton, 1959.

Vine, Peter. *Seychelles.* London: Immel Publishing, 1992.

Sierra Leone

- **Area:** 27,699 sq mi (71,740 sq km) / World Rank: 119
- **Location:** Northern and Western Hemispheres, west coast of Africa, bounded on the north and east by Guinea, on the southeast by Liberia, and on the south and west by the Atlantic Ocean.
- **Coordinates:** 8°30′N, 11°30′W
- **Borders:** 595 mi (958 km) / Guinea, 405 mi (652 km); Liberia, 190 mi (306 km)
- **Coastline:** 250 mi (402 km)
- **Territorial Seas:** 200 NM
- **Highest Point:** Loma Mansa, 6,391 ft (1,948 m)
- **Lowest Point:** Sea level
- **Longest Distances:** 210 mi (338 km) N-S; 189 mi (304 km)
- **Longest River:** Rokel River, 270 mi (440 km)
- **Natural Hazards:** Harmattan winds; sandstorms; dust storms
- **Population:** 5,426,618 (July 2001 est.) / World Rank: 102
- **Capital City:** Freetown, located on the northern Atlantic Coast.
- **Largest City:** Freetown, population 699,000 (2000 est.)

OVERVIEW

Slightly smaller than the state of South Carolina, Sierra Leone, roughly circular in shape, is a compact country located in the southwestern part of the great bulge of West Africa. Lying between the seventh and tenth parallels north of the equator, it is bounded on the west by the Atlantic Ocean and inland by Guinea and Liberia. Its varied terrain includes the striking, mountainous Sierra Leone Peninsula, a zone of low-lying coastal marshland along the Atlantic Ocean, and a wide plains area extending inland to about the middle of the country. East of the plains the land rises to a broad, moderately elevated plateau from which emerge occasional hill masses and mountains.

Sierra Leone is located on the African Tectonic Plate.

MOUNTAINS AND HILLS

Mountains

The mountainous Sierra Leone Peninsula, on which Freetown is located, is 25 mi (40 km) long and about 10 mi (16 km) wide. Its unusual features bear no direct relationship to those of the adjacent coastal region. It consists mainly of igneous rocks that form with the nearby

Banana Islands the visible part of a much larger igneous mass submerged beneath the sea. Around the base of the mountains is a strip of land about 1 mi (1.6 km) wide consisting of hardpan. The Loma Mountains span the northeastern part of the country. The highest point in Sierra Leone, Mount Loma Mansa (Bintimani), rises to a height of 6,391 ft (1,948 m) in the Loma Mountains.

Plateaus

The plateau region, which encompasses roughly the eastern half of the country, consists mainly of a large uplifted area having elevations of above 1,000 ft (304 m) to about 2,000 ft (608 m). Several mountain masses rise above the relatively flat surface.

Hills and Badlands

In the region's southern section erosion has resulted in a large area of rolling terrain 40 mi (64 km) wide at certain points and having elevations between 500 and 1,000 ft (152 and 304 m). The western edge of the plateau exhibits different stages of erosion and in some places is characterized by steep-sided river valleys and highly dissected hills.

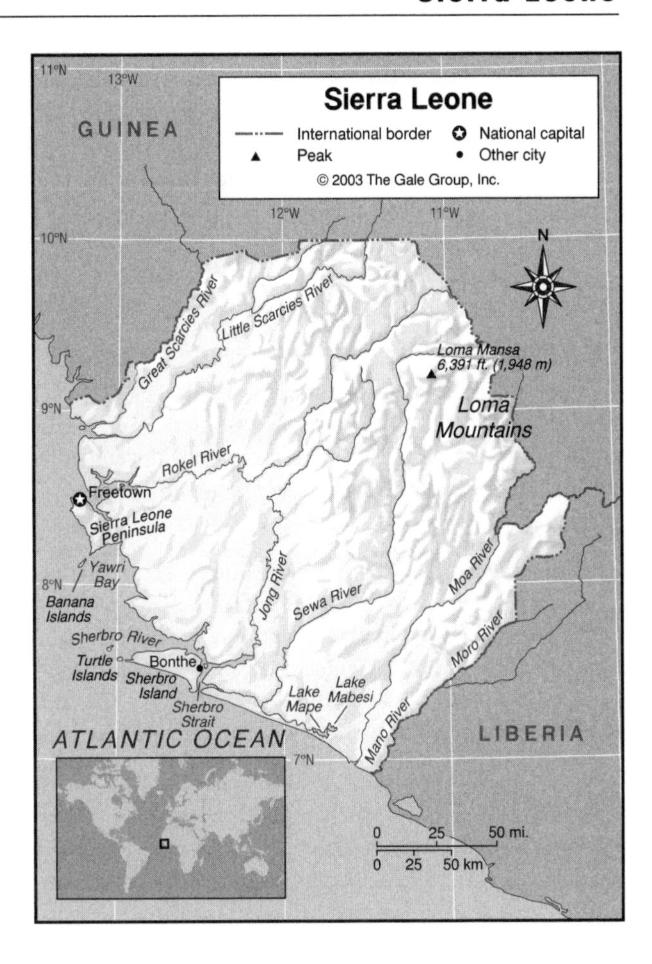

INLAND WATERWAYS

Lakes

There are several small lakes in Sierra Leone, most located in the south. The three largest and most important are Lake Sonfon, Lake Mabesi, and Lake Mape.

Rivers

Most of the rivers of Sierra Leone drain into the Atlantic Ocean; a few exceptions terminate at inland lakes. Of the numerous rivers, the most important ones are the Great and Little Scarcies in the north, and the Rokel in the central region. The Great Scarcies forms part of the northern border with Guinea. The Rokel River originates in the Loma Mountains and flows west to the Atlantic Ocean near Freetown. At 270 mi (440 km), the Rokel is the longest river in the country. Also important are the Mano and Moro Rivers, which form the southern border with Liberia. Other major rivers include the Jong, Sewa, Soa, and Moa. Most of the major rivers are navigable in the rainy season.

Wetlands

Mangrove swamps line much of the coast, behind which marine and freshwater swamps occupy large areas.

THE COAST, ISLANDS, AND THE OCEAN

Oceans and Seas

Sierra Leone is bounded on the southwest and west by the Atlantic Ocean and lies just north of the Grain Coast. There are oil and gas reserves under the ocean floor off the coast.

Major Islands

There are three major island groups off the coast of Sierra Leone: The Banana Islands, the Turtle Islands, and Sherbro Island. Sherbro Island is by far the largest. The city of Bonthe is located on this island.

The Coast and Beaches

The coast is very irregular, forming many bays, inlets, and peninsulas. The most significant features are Sierra Leone Peninsula, where Freetown is located, and Yawri Bay, which is located in the center of the coast just south of Sierra Leone Peninsula. Sherbro Island is separated from the mainland by Sherbro River on the north and Sherbro Strait on the east.

The coastal region covers a zone varying from about 5 to 25 mi (8 to 40 km) wide along the coast. Numerous estuaries whose river channels, as in the case of the Sierra Leone River, continue under the sea, characterize the region. Soils in this coastal stretch are relatively fertile and produce cash crops for the Freetown market.

CLIMATE AND VEGETATION

Temperature

Being so close to the equator, Sierra Leone has a tropical climate, and temperatures do not vary significantly

GEO-FACT

The emerald cuckoo, described as one of the most beautiful birds in Africa, is found in Sierra Leone. The emerald cuckoo is nearly extinct in the rest of West Africa.

year round. The mean temperature is about 81°F (27°C) on the coast and almost as high on the eastern plateau. There are two distinct seasons: the dry season, from November to April, and wet season over the rest of the year.

Rainfall

The prevailing winds from the southwest monsoon characterize the rainy season. Rainfall is greatest along the coast, especially in the mountains, where there is more than 230 in (580 cm) of rainfall annually. This compares with an average of approximately 125 in (315 cm) in the rest of the country. During the dry season, harmattan winds blow from the Sahara Desert bringing sandstorms and little rain.

Grasslands

About 25-35 percent of the land area, mostly in the north, consists of grasslands.

Forests and Jungles

Most of the Sierra Leone Peninsula's hills are included in a forest reserve established to halt erosion and preserve the watershed as a source of water supply for domestic purposes.

HUMAN POPULATION

The two largest ethnic groups are the Mende and the Temne, which together comprise about 60 percent of the population. The country is broken into four administrative divisions, roughly corresponding to the northern half of the country, the southeastern third, the southwestern third, and the Sierra Leone Peninsula. The Sierra Leone Peninsula—including the capital of Freetown—is by far the most densely populated, with a density of 4,799 people per sq mi (1,853 people per sq km). The population density in the north is 115 people per sq mi (44 people per sq km), while the southeast and southwest regions have densities of 202 per sq mi (78 per sq km) and 123 per sq mi (47 per sq km) respectively.

NATURAL RESOURCES

Sierra Leone has a rich store of mineral resources. The country was once the largest producer of diamonds in the world and remains in the top ten. Sierra Leone is also rich in such minerals as chrome, bauxite, and iron ore. However, the economy of the country is still heavily dependent on fishing and agriculture, where rice, coffee, cocoa, palm kernels and oil, and peanuts comprise the main crops.

Population Centers – Sierra Leone

(1990 POPULATION ESTIMATES)

Name	Population
Freetown (capital)	472,000
Koidu	82,000

SOURCE: Projected from United Nations Statistics Division data.

FURTHER READINGS

Ferme, Mariane. *The Underneath of Things: Violence, History, and the Everyday in Sierra Leone.* Berkeley: University of California Press, 2001.

Hirsh, John. *Sierra Leone: Diamonds and the Struggle for Democracy.* (International Peace Academy Occasional Paper Series). Boulder, Colo.: Lynne Rienner Publishers, 2000.

Richards, Paul. *Fighting for the Rain Forest: War, Youth, and Resources in Sierra Leone* (African Issues Series). Portsmouth, N.H.: Heinemann, 1996.

Sierra Leone Web. http://www.sierra-leone.org/index.html (Accessed June 10, 2002).

Singapore

- **Area:** 250 sq mi (648 sq km) / World Rank: 181
- **Location:** Northern and Eastern Hemispheres, Southeast Asia, an island off the tip of the Malay Peninsula, bordered on the north by the Johore Strait, and on the south by the Singapore Strait.
- **Coordinates:** 1°22′N, 103°48′E
- **Borders:** None
- **Coastline:** 120 mi (193 km)
- **Territorial Seas:** 3 NM
- **Highest Point:** Bukit Timah, 545 ft (166 m)
- **Lowest Point:** Sea level
- **Longest Distances:** 26 mi (42 km) ENE-WSW; 14 mi (23 km) SSE-NNW
- **Longest River:** Seletar, 9 mi (14 km)
- **Natural Hazards:** Monsoon storms

- **Population:** 4,300,419 (July 2001 est.) / World Rank: 117
- **Capital City:** Singapore, southern coast
- **Largest City:** Singapore, 4,300,419 (2001 metropolitan area est.)

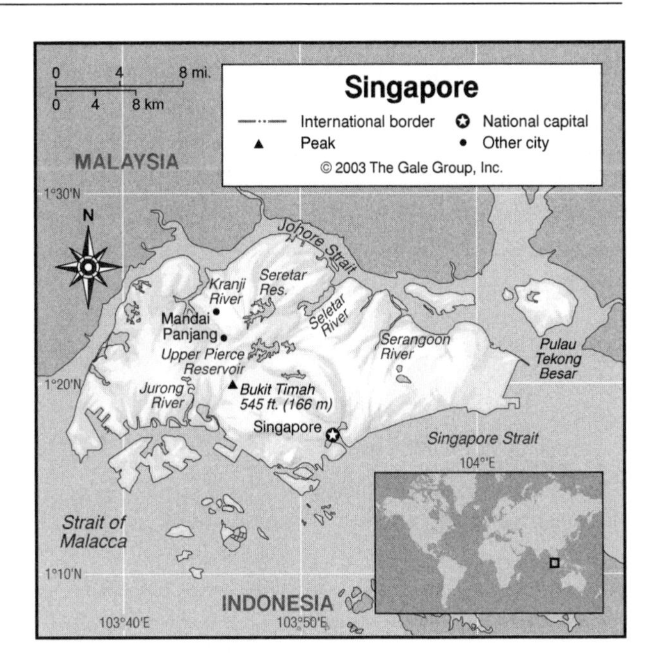

OVERVIEW

The Republic of Singapore consists of a main island and 63 islets just south of the tip of the Malay Peninsula in Southeast Asia. Singapore is often described as a city-state. The diamond-shaped main island, which accounts for all but about 15 sq mi (38 sq km) of the republic's area, is almost entirely urban. The island has three major geographic divisions: an elevated, hilly area in the center; a section of lower, rolling land to the west; and flatlands to the east. Land reclamation has added almost 6 sq mi (15 sq km) to the total territory since 1966, mostly along the southeast coast. Singapore is situated on the Eurasian Tectonic Plate.

MOUNTAINS AND HILLS

Plateaus

The highest land on Singapore is a ridge of rugged hills in the center of the island. The highest is Bukit Timah Hill, at 545 ft (165 m). The other hills include Mandai and Panjang. Lower ridges extend northwest-to-southeast in the western and southern parts of the island. The eastern part of the island is a low, eroded plateau.

INLAND WATERWAYS

Lakes

Fourteen reservoirs have been built on Singapore's rivers for flood control as well as private and industrial water use. Almost all the reservoirs are located in the center of the island or at the mouths of rivers on the northeastern or western coasts. Among the largest are Seretar and Upper Pierce in the center of the island.

Rivers

Singapore's rivers are all short, including its main river, which has the same name as the island itself. The Singapore River is important because it flows into the wide harbor on the southern island's southeastern coast. Other rivers include the Seletar (at 9 mi / 14 km, the longest on the island), Jurong, Kalang, Kranji, and Serangoon.

THE COAST, ISLANDS, AND THE OCEAN

Oceans and Seas

Singapore is located between the Indian Ocean and the South China Sea. It is bordered on the north by the Johore Strait, which separates it from the Malay Peninsula, on the southeast by the Singapore Strait, and on the southwest by the Strait of Malacca. A causeway less than 1 mi (3 km) long bridges the Johore Strait, connecting Singapore to the Malaysian state of Johore. The island historically owes much of its economic prominence to its strategic position at the eastern end of the Strait of Malacca, which is the shortest sea route between Southern and Eastern Asia. The coastal waters surrounding Singapore are generally less than 100 ft (30 m) deep.

Major Islands

After Singapore Island, the next-largest island is Pulau Tekong Besar to the northeast, whose area is only 7 sq mi (18 sq km).

The Coast and Beaches

The easternmost part of the coastline is smooth, but the rest has many indentations, of which the most important is the deep natural harbor at the mouth of the Singapore River on the southern coast.

CLIMATE AND VEGETATION

Temperature

Singapore has a humid, rainy, tropical climate, with temperatures moderated by the seas surrounding the islands. Temperatures are nearly uniform throughout the year, averaging 77°F (25°C) in January and 81°F (27°C) in June. Although the island lies between 1 and 2 degrees north of the equator, the maritime influences moderate the heat of the region. The all-time high temperature has been only 97°F (36°C).

Rainfall

Singapore is very humid, with heavy rainfall all year. Annual rainfall averages 93 in (237 cm). The northeast

Census Areas – Singapore

Name	Area (sq mi)	Area (sq km)
Central city area	3	8
City periphery	17	46
Suburbs	49	127
Outlying areas	169	437

SOURCE: *Geo-Data: The World Geographical Encyclopedia,* 2nd ed. Detroit: Gale Research, 1989.

monsoon that occurs between November and March brings the heaviest rainfall of the year.

Forests and Jungles

The lush vegetation that initially thrived in Singapore's warm, wet climate has largely been cleared for timber and urbanization. The greatest concentration of plant life is found in the Bukit Timah Nature Reserve, which has 200 acres (81 hectares) of primary rain forest and a wide variety of indigenous plants. There are mangrove swamps near the coast.

HUMAN POPULATION

The city of Singapore started out at the southern end of the main island, but increasing urbanization has expanded the metropolitan area to the point where the entire island is essentially a city-state, and is considered 100 percent urban. Singapore's population density is the second highest in the world at 17,202 people per sq mi (6,636 people per sq km; 2001 est.).

NATURAL RESOURCES

Due largely to its strategic geographic position, Singapore has a very strong and large economy. Its wealth is based on shipping, commerce, and manufacturing. Its greatest natural resource is the wide deepwater port on its southern shore. Its waters also support a fishing industry.

FURTHER READINGS

Fuller, Barbara. *Berlitz Discover Singapore.* Oxford, England: Berlitz Publishing, 1993.

Krannich, Ronald L., and Caryl Krannich. *The Treasures and Pleasures of Singapore and Malaysia: Best of the Best.* Manassas Park, Va.: Impact Publications, 1996.

Rowthorn, Chris, et al. *Malaysia, Singapore and Brunei.* Oakland, Calif.: Lonely Planet, 1999.

"Singapore and Malaysia." *Knopf Guides.* New York: Knopf, 1996.

Warren, William. *Singapore, City of Gardens.* Hong Kong: Periplus Editions, 2000.

Slovakia

- **Area:** 18,859 sq mi (48,845 sq km)/ World Rank: 129
- **Location:** Northern and Eastern Hemispheres, Central Europe, bordering Ukraine to the east, Hungary to the south, Austria to the southwest, the Czech Republic to the northwest, and Poland to the north
- **Coordinates:** 48°40′N, 19°30′E
- **Borders:** 842 mi (1,355 km) / Austria, 57 mi (91 km); Czech Republic, 134 mi (215 km); Hungary, 320 mi (515 km); Poland, 276 mi (444 km); Ukraine, 56 mi (90 km)
- **Coastline:** None
- **Territorial Seas:** None
- **Highest Point:** Gerlachovsky Peak, 8,711 ft (2,655 m)
- **Lowest Point:** Bodrok River, 308 ft (94 m)
- **Longest River:** Danube, 1,775 mi (2,857 km)
- **Natural Hazards:** None
- **Population:** 5,414,937 (July 2001 est.) / World Rank: 103
- **Capital City:** Bratislava, near the western border
- **Largest City:** Bratislava, 448,292 (2000 est.)

OVERVIEW

Slovakia (or Slovak Republic) occupies an area of Central Europe that constituted the eastern part of Czechoslovakia from 1918 to 1993. It is a landlocked country dominated by the western portion of the Carpathian Mountains, which extend over its northern and central regions. To the south are subsidiary mountain ranges, with distinct lowland areas in the southwest and east. The capital city of Bratislava is located on the Danube River, which flows through the country briefly in the west and along part of its southern border. Slovakia is situated on the Eurasian Tectonic Plate.

MOUNTAINS AND HILLS

Mountains

The portion of the Carpathian mountain system within Slovakia consists of a number of different ranges separated by valleys and river basins. The highest range is the High Tatras (or Vysoké Tatry), which extend in a narrow ridge along the border with Poland and have traditionally been a popular summer resort area. They include Slovakia's highest peak, Gerlachovsky. Snow persists at the higher elevations well into the summer months and all year long in some sheltered pockets. The tree line is at about 4,921 ft (1,500 m).

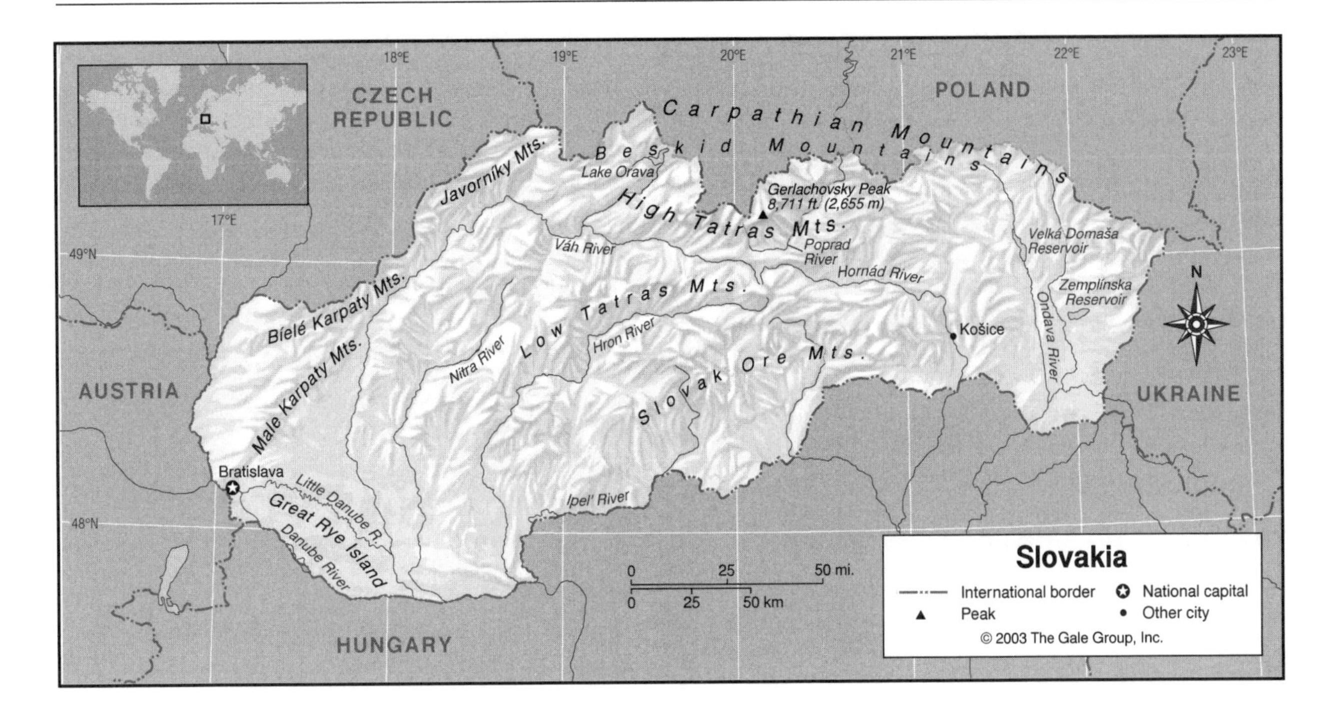

To the south, across the Váh River, the Low Tatras (Nízke Tatry) rise to elevations of 6,500 ft (1,981 m). Still farther south, across the Hron River, are the Slovak Ore (Slovenské Rudohrie) Mountains, whose mineral deposits were significant in supporting the industrialization of the region that occurred following World War II. These mountains of central Slovakia eventually give way to hills in the south-central part of the country.

In addition to the three major ranges in the center of the country, smaller ones rise in both the east and west. In the west, the Little Carpathian (Małe Karpaty) range rises in the southeast, near Bratislava, and several ranges, including the Bíelé Karpaty, Javorníky, and Beskid mountains, extend into the western part of the Czech Republic and southern Poland.

INLAND WATERWAYS

Lakes

Clear lakes dot the mountains of Slovakia. Many, such as Lake Orava and Lake Popradské, are associated with rivers. Lake Orava, at the Polish border, is one of the country's largest lakes; others include the manmade Zemplínska Reservoir and Velká Domaša Reservoirs.

Rivers

Most of Slovakia's rivers flow south into the Danube, which, together with the Morava, forms the country's southwestern border. From a point a few kilometers south of the Slovakian capital of Bratislava, the main channel of the Danube River demarcates the border between Slovakia and Hungary for about 108 mi (175 km). As it leaves Bratislava, the Danube divides into two channels. The main channel, the Danube proper, continues southward along the border with Hungary. The smaller channel, called the Little Danube, branches eastward and then southeast to meet the Váh River. The Váh continues south and converges with the Nitra and with the main branch of the Danube at Komárno. The Hron and Ipel' also flow south and enter the Danube before that river turns south into Hungary.

Slovakia's eastern rivers also tend to flow southward, and eventually enter the Danube. Among them are the Hornád and the Ondava. The Poprad, also in the east, is the only sizable river that flows northward, into Poland.

Wetlands

The corner of southeastern Slovakia between the Little Danube and the Danube, known as the Great Rye Island (Velky Litny Ostrov), is a marshland that was maintained for centuries as a hunting preserve for the nobility. Dikes and artificial drainage have made the land

GEO-FACT

There are a dozen caves open to the public in Slovakia. The Belian Cave in Tatra National Park is 5,778 ft (1,752 m) long; it is home to a natural auditorium where musical performances are staged, and provides a habitat for eight species of bat.

Population Centers – Slovakia

(1995 POPULATION ESTIMATES)

Name	Population
Bratislava (capital)	451,272
Košice	240,139

SOURCE: "Populations of Capital Cities and Cities of 100,000 and More Inhabitants.: United Nations Statistics Division.

cultivable for grain production, but it is still sparsely settled.

THE COAST, ISLANDS, AND THE OCEAN

Slovakia is landlocked.

CLIMATE AND VEGETATION

Temperature

Slovakia has a continental climate with sharp seasonal contrasts. Mean temperatures are 30°F (-1°C) in January and 70°F (21°C) in July. However, weather can vary considerably with elevation. The January average can be as low as 23°F (-5°C) in the mountains, where temperatures are colder than in the lowlands.

Rainfall

Like temperature, rainfall varies with elevation. The annual average in the lowlands is about 25 in (64 cm), but it can be more than twice as high in the High Tatras.

Grasslands

The lowlands in southwestern Slovakia, which belong to the Danube Basin, are part of the same structural depression as the Little Alföld (or Hungarian Plain) farther south in Hungary. The lowlands in the east are part of the Carpathian Depression.

Marsh grasses and reeds are abundant in Slovakia's southwestern lowland, which includes some original steppe grassland.

Forests and Jungles

About one-third of Slovakia is covered by forestland. Oak is found in the lowlands, and beech, spruce, pine, and mountain maple on the mountain slopes up to the timber line, with dwarf pines and low shrubs beyond. Still higher up, alpine meadows reach elevations of 7,500 ft (2,300 m), with grasses and wildflowers including glacial gentians, carnations, and androsace, and edelweiss. Many types of moss, lichens, and fungus are also found in the High Tatras. There are scenic national parks in both the High and Low Tatras.

HUMAN POPULATION

Population density is highest in the river valleys, and sparsest at the highest mountain elevations and in parts of the southern lowlands that are poorly drained. More than half the population is urbanized.

NATURAL RESOURCES

Natural resources include arable land, coal, iron, copper, lead, and zinc.

FURTHER READINGS

Brewer, Ted. *The Czech and Slovak Republics Guide.* New York: Open Road Publishing, 1997.

Humphreys, Rob, and Tim Nollen. *Czech and Slovak Republics: The Rough Guide.* New York: Penguin, 1998.

Husovská, 'Ludmilá. *Slovakia: Walking Through Centuries of Cities and Towns. Bratislava:* Priroda, 1997.

King, John, and Richard Nebesky. *Czech and Slovak Republics.* Oakland, Calif.: Lonely Planet, 1995.

Kirschbaum, Stanislav J. *A History of Slovakia: The Struggle for Survival.* New York: St. Martin's Press, 1995.

Slovakia, Heart of Europe. http://www.heartofeurope.co.uk/info.htm (Accessed June 24, 2002).

Slovenia

- **Area:** 7,820 sq mi (20,253 sq km) / World Rank: 153
- **Location:** Northern and Western Hemispheres, on the southeastern part of the European continent, south of Austria, southwest of Hungary, northwest of Croatia, and east of Italy.
- **Coordinates:** 46°00′N, 15°00′E
- **Borders:** 724 mi (1,165 km) / Austria, 205 mi (330 km); Hungary, 63 mi (102 km); Croatia, 311 mi (501 km); and Italy, 144 mi (232 km)
- **Coastline:** 29 mi (46.6 km)
- **Territorial Seas:** Not available.
- **Highest Point:** Mt. Triglav, 9,396 ft (2,864 m)
- **Lowest Point:** Sea level
- **Longest Distances:** 101 mi (163 km) N-S; 154 mi (248 km) E-W
- **Longest River:** Sava River, 452 mi (727 km)
- **Largest Lake:** Lake Cerknišco, 9.3 sq mi (24 sq km).
- **Natural Hazards:** Earthquakes occur in the east; occasional flooding of river valleys

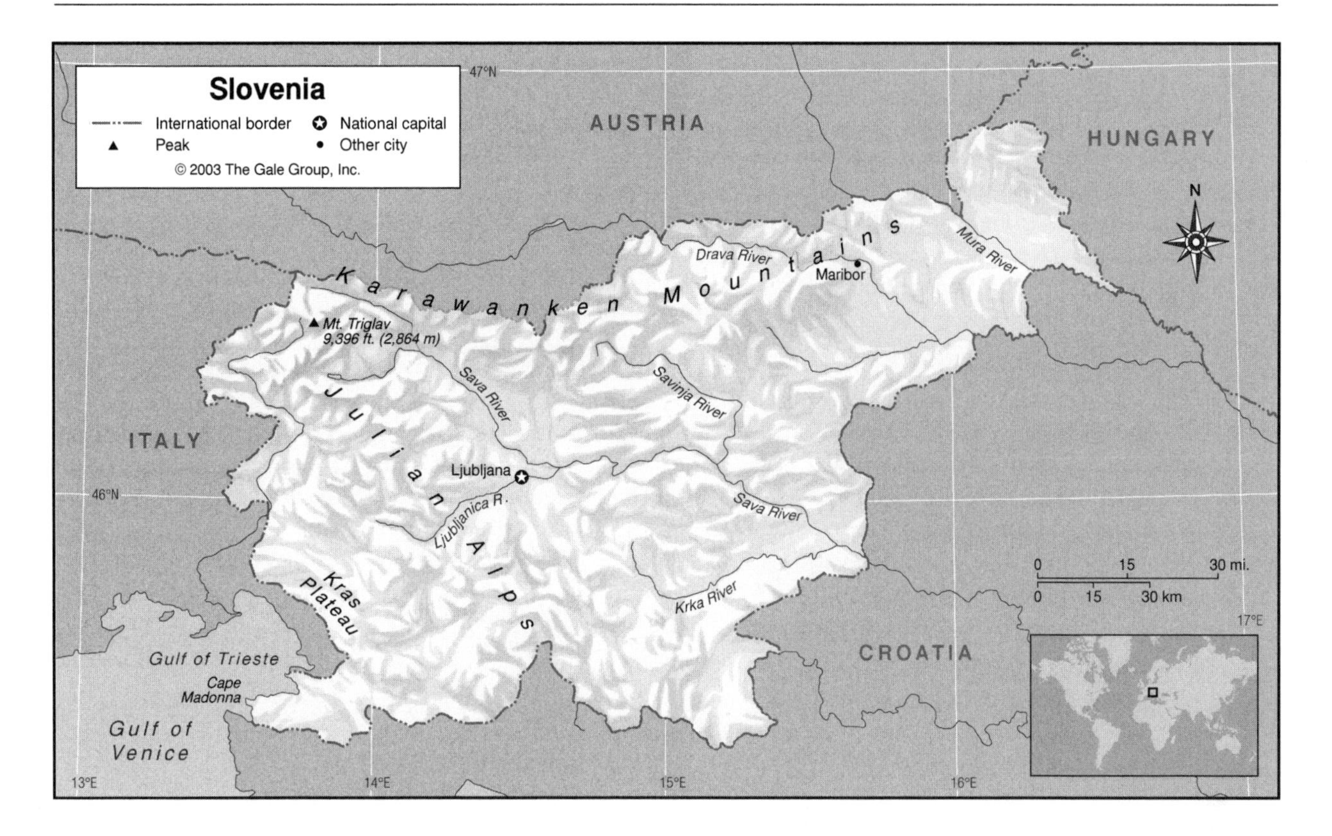

- **Population:** 1,930,132 (July 2001 est.) / World Rank: 143
- **Capital City:** Ljubljana, in the center of Slovenia
- **Largest City:** Ljubljana, 263,290 (June 2000)

OVERVIEW

Slovenia lies at the northwestern end of the Balkan Peninsula, at the intersection of Central Europe, the Mediterranean Sea, and the Balkans. It has a short coastline on the Adriatic Sea in the southwest, but a majority of Slovenia is covered by the Alps, principally in the north and south. In the east is the Pannonian Plain.

Geological fault lines are widespread in the mountains south of the Sava River. These structural seams in the earth's crust periodically shift, causing earth tremors and occasional earthquakes. Hundreds of tremors are recorded annually.

MOUNTAINS AND HILLS

Mountains

The northern and northwestern parts of Slovenia contain mountains resembling the higher Austrian Alps to the north, having sharp peaks and ridges. The Julian Alps are the highest of the three alpine ranges in Slovenia, occupying the northwestern third of the country. The Julian Alps are among the most rugged in Europe and contain many summits that exceed 5,900 ft (1,800 m). The six highest mountains in Slovenia are in the Julian Alps, with Mt. Triglav (9,394 ft / 2,864 m) the country's highest peak.

The Karawanken Mountains of the Alps run along the border with Austria; Mt. Stol (7,336 ft / 2,236 m) is its highest peak. The Kamnik-Savinja range lies south of the Karawankens; its highest peak is Mt. Grintovec (8,392 ft / 2,558 m). Eastward, these mountains possess less well-defined ridges, and their crests decrease in height to about 3,300 ft (1,000 m) in the vicinity of Maribor.

The Dinaric Alps parallel the coast in the southwest and range from 2,300 ft (700 m) to over 7,200 ft (2,200 m) in height. The flat depressions formed in the limestone hills vary considerably in size. Most are quite narrow, but some have been elongated to as much as 37 mi (60 km). The general surface configuration of the karst area is rocky, featuring many desolate cliffs that support little vegetation.

Hills and Plateaus

South of the northern Alps, the rough terrain in the west changes to hilly areas interspersed with flat valleys. One of the largest of these basins is located near Ljubljana. Farther to the south the terrain is elevated in the form of the Kras Plateau, which extends into the limestone ranges of the Dinaric Alps. This region, frequently referred to as karst or karstland, is distinctive because of the underground drainage channels that have been formed by the long-term seepage of water down through the soluble limestone. This action leaves the surface dry and over the years has formed many large depressions in the high coastal plateaus.

Canyons

Along the plateaus in the vicinity of Postonja and Cerknica in southwestern Slovenia, there are special karst-related depressions known as *poljes*. A polje is a large steep-walled enclosed depression with a flat floor across which may flow an intermittent or perennial stream. The stream originates at one end from a karst spring and disappears at the other end into a *ponor*, or cave entrance.

Hills and Badlands

Slovenia contains about 6,500 karst-formed caves, the largest of which is Postonja Cave, extending some 12.1 mi (19.5 km) underground. The Skocjan Caves of the southwest have 3.1 mi (5 km) of accessible caves and are listed as a UNESCO World Natural Heritage site.

INLAND WATERWAYS

Lakes

Slovenia's largest lake is Lake Cerknišco, which covers 9.3 sq mi (24 sq km) and, as a karst lake, fills and drains periodically. In 1971 a sluice was constructed at the entrance of the ponor in order to extend the lake's filled season from six to seven months per year. Slovenia also has 78 mineral and thermal springs, mostly in the Pannonian Plain.

Rivers

Formed at the confluence of the Sava Dolinka and Sava Bohinjka rivers, the Sava River is the central and longest river in Slovenia, flowing through the country for 137 mi (221 km). Its tributaries include the Trziska Bistrica, Savinja, Ljubljanica, and Krka rivers. After the Sava, the largest rivers in Slovenia are the Drava and Mura, both in the northeast. All of these rivers arise in the Alps of Slovenia, Austria, and Italy, then flow southeast into Croatia, eventually reaching the Danube.

THE COAST, ISLANDS, AND THE OCEAN

Slovenia has only 29 mi (46.6 km) of coastline, all on the Gulf of Venice, which is at the northern end of the Adriatic Sea. The only beaches are near Koper; the coast between Izola and Piran is lined with steep cliffs up to 260 ft (80 m) high. The sea around Cape Madonna near Piran reaches depths of 120 ft (37 m) and is a national marine reserve.

CLIMATE AND VEGETATION

Temperature

The average January and July temperatures in Ljubljana are 30°F (-1.1°C) and 68°F (19.9°C), respectively; in Maribor, 30°F (-1.3°C) and 67°F (19.6°C). Ljubljana annually has about 90 days with a minimum temperature below freezing and about 61 days with a maximum exceeding 77°F (25°C).

Population Centers – Slovenia

(2000 POPULATION ESTIMATES)

Name	Population
Ljubljana (capital)	270,506
Maribor	114,891
Kranj	51,805
Celje	49,313

SOURCE: Statistical Office of the Republic of Slovenia.

Rainfall

Ljubljana annually receives about 54.9 in (139 cm) of rain, with some 28 precent falling between April and June.

Pannonian Plain

Occupying the east and northeast is the Pannonian Plain, the most fertile farmland in the nation. The plain was once occupied by an ancient sea, which disappeared as the runoff from the surrounding mountains gradually filled it with rich alluvial deposits. It is a sedimentary region containing wide valley basins, alluvial plains, sandy dunes, and low, rolling hills covered with fertile loam. In general the area is low and flat. The major rivers are located on broad flood plains.

Forests

The government estimates that 56.5 percent of the land area is covered by forests. The largest protected area is Triglav National Park near Mt. Triglav. Another forested area can be found in the south, near Kočevje.

HUMAN POPULATION

Slovenia had 192 municipalities in 2000; an estimated 53% of the population inhabited urban areas that year, up from 48% in 1980.

NATURAL RESOURCES

Slovenia's major natural resources include coal (mined at Velenje), mercury (mined and smelted at Idrija), and timber.

FURTHER READINGS

Brân, Zoë. *After Yugoslavia*. Oakland, Calif.: Lonely Planet, 2001.

Fallon, Steve. *Slovenia*. Hawthorn, Australia: Lonely Planet, 1998.

Natek, Karel, *Discover Slovenia*, tr. Martin Cregeen Ljubljana: Cankarjeva zalozba, 1999.

Statistical Office of the Republic of Slovenia. *Statistical Yearbook of the Republic of Slovenia 2000*. http://www.gov.si/zrs/ (Accessed June 1, 2002).

Slovene Government website. http://www.gov.si/vrs/ang/ang-text/slovenia/uvod.html (Accessed June 1, 2002).

Solomon Islands

- **Area:** 11,000 sq mi (28,450 sq km) / World Rank: 143
- **Location:** Southern and Eastern Hemispheres, in the South Pacific region of Oceania, nearly 1,200 mi (1,900 km) northeast of Australia and about 300 mi (485 km) east of Papua New Guinea
- **Coordinates:** 8°00′S, 159°00′E
- **Borders:** None
- **Coastline:** 3,301 mi (5,313 km)
- **Territorial Seas:** 12 NM
- **Highest Point:** Mount Makarakomburu, 8,127 ft (2,447 m)
- **Lowest Point:** Sea level
- **Longest Distances:** 1,049 mi (1,688 km) ESE-WNW / 291 mi (468 km) NNE-SSW
- **Longest River:** None of significant length
- **Natural Hazards:** Earthquakes, volcanoes
- **Population:** 480,442 (July 2001 est.) / World Rank: 161
- **Capital City:** Honiara, which is located on the northern coast of Guadalcanal Island.
- **Largest City:** Honiara, 53,000 (2000 est.)

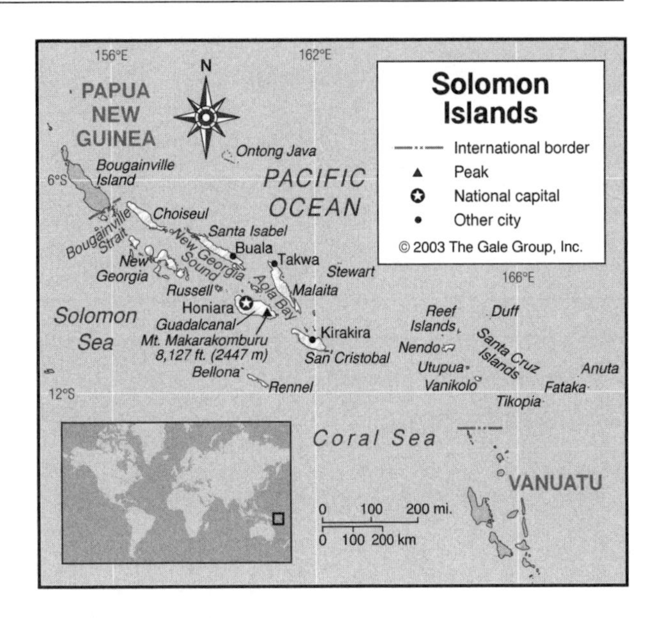

OVERVIEW

The Solomon Islands is an archipelago comprised a double chain of high continental islands formed from the exposed peaks of a submerged mountain chain. This chain extends from Bougainville Island in Papua New Guinea to the northern islands of Vanuatu. The Solomon Islands nation covers the central islands of this chain.

Almost all of the larger islands are volcanic in origin and are covered with steaming jungles and mountain ranges intersected by narrow valleys. Most of the smaller islands are low coral atolls. The Solomon Islands lie on the Transitional Zone along the edge of the Pacific and Australian Tectonic Plates, and earthquakes and volcanic activity are not uncommon.

MOUNTAINS AND HILLS

The five largest islands—Choiseul, New Georgia, Santa Isabel, Guadalcanal, and Malaita—are characterized by heavily forested mountain ranges. The terrain is very rugged, Mount Makarakomburu reaches 8,127 ft (2,447 m) on the southern end of Guadalcanal. Steep, narrow valleys intersect with the mountain ranges.

INLAND WATERWAYS

Short, narrow, and impassable, the rivers of the Solomon Islands are only navigable by canoe. Lagoons and mangrove swamps surround the islands at the coasts.

THE COAST, ISLANDS, AND THE OCEAN

Major Islands

The Solomon Islands nation is part of the archipelago of the same name. The largest island is Guadalcanal, which covers 2,047 sq mi (5,300 sq km). There are five other large islands, all in the western part of the chain: Choiseul, New Georgia, Santa Isabel, Malaita, and San Cristobal. Smaller islands include: Bellona, Duff, Gizo, Kolombangara, Ontong Java, Rennell, Savo, the Shortland Islands, Ranongga, Simbo, Rendova, Vangunu, Nggatoake, Russell, and Vella Lavella. To the east of these islands lie the Santa Cruz Islands part of the chain, which includes Santa Cruz, Nendo, Tikopia, Utupua, Vanikolo, Anuta, Fetaka, Duff, and the Reef Islands. In addition, there are approximately 992 islets, atolls, and reefs in the group.

Oceans and Seas

The Solomon Islands are mostly low-lying atolls that are surrounded by expanses of coral reefs. Unfortunately, much of the coral barrier is dead or dying. The Solomon Sea is southwest of the islands, the Coral Sea directly

Population Centers – Solomon Islands

(2002 POPULATION ESTIMATES)

Name	Population
Honiara (capital)	53,000
Gizo	8,000

SOURCE: Projected from United Nations Statistics Division data.

Provinces – Solomon Islands

1992 POPULATION ESTIMATES

Name	Population	Area (sq mi)	Area (sq km)	Capital
Central	19,898	237	615	Tulagi
Guadalcanal	103,266	2,069	5,358	Honiara
Isabel	17,061	1,597	4,136	Buala
Makira	26,070	1,231	3,188	Kirakira
Malaita	87,258	1,631	4,225	Auki
Temotu	16,867	346	895	Lata
Western	51,357	2,114	5,475	Gizo

SOURCE: Statistical Yearbook 1993, Solomon Islands.

south. To the north and east is the open Pacific Ocean. The Bougainville Strait lies between the northwestern islands of Choiseul, Vella Lavella, and the Shortland Islands, and Bougainville Island in Papua New Guinea. The New Georgia Sound and Aola Bay are sheltered areas of between the two lines of islands that make up the archipelago.

CLIMATE AND VEGETATION

Temperature

Because of cooling southeast trade winds off the surrounding seas, the temperatures of the islands are rarely extreme, despite it being tropical. November to March is the hottest period; from April to November it is cooler and drier. Normally, the daytime temperatures range from 77 to 90°F (25 to 32°C), with the nighttime ranging from 38 to 41°F (3 to 5°C).

Rainfall

The northwest monsoon, which brings warmer and wetter weather, lasts from November to March. Cyclones often start in the Coral Sea and the area of the Solomons, but often veer away from the Islands themselves. Annual average rainfall is 120 in (305 cm), and humidity is nearly 80 percent on average.

Grasslands

Guadalcanal Island contains the nation's only grassy plains of any extent, created by the alluvial deposits of the streams there.

Forests and Jungles

Rainforests and woodland cover about nine-tenths of the Solomon Island's area (approximately 6 million acres / 2.4 million ha). Important trees are teak, African and Honduras mahogany, balsa, Queensland maple, silky oak, black bean, and kuari. A significant environmental problem is deforestation, causing a related problem of soil erosion.

HUMAN POPULATION

Of the population (totaling 480,442), more than 90 percent are ethnic Melanesians. In 2001 the population was growing at a rate of 2.98 percent. It has been estimated that 20 percent of the population lives in urban areas, with the overall density (1996) being 39 per sq mi (15 per sq km); these figures vary from island to island. Malaita and Guadalcanal are the most populous islands. The Santa Cruz Islands are particularly sparsely populated.

NATURAL RESOURCES

Known mineral resources in the Solomon Islands include gold, bauxite, phosphates, lead, zinc, and nickel, but there has been little or no exploitation of them. Fishing is an important commercial activity both for export and for local consumption. Tuna and prawns are the primary fish products harvested. Timber and forest products are another important resource.

FURTHER READINGS

Bennett, Judith A. *Wealth of the Solomons: A History of a Pacific Archipelago, 1800–1978.* Honolulu: University of Hawaii Press, 1987.

Jack-Hinton, Colin. *The Search for the Islands of Solomon 1567–1838.* Oxford: Clarendon Press, 1969.

Newton Abbot, David and Charles Newton Abbot. *The Solomon Islands.* Harrisburg, PA: Stackpole Books, 1972.

Solomon Islands: A Travel Survival Kit. South Yarra, Victoria, Australia: Lonely Planet Publications, 1988.

Somalia

- **Area:** 246,201 sq mi (637,657 sq km) / World Rank: 43
- **Location:** Northern and Eastern Hemispheres, northeast coast of Africa, bordered by the Gulf of Aden to the north, the Indian Ocean to the east and south, Kenya to the southwest, Ethiopia to the west, and Djibouti to the northwest

- **Coordinates:** 10°00′N, 49°00′E
- **Borders:** 1,470 mi (2,366 km) / Djibouti, 36 mi (58 km); Ethiopia, 1,010 mi (1,626 km); Kenya, 424 mi (682 km)
- **Coastline:** 1,880 mi (3,025 km)
- **Territorial Seas:** 200 NM
- **Highest Point:** Mt. Shimbiris, 7,927 ft (2,416 m)
- **Lowest Point:** Sea level
- **Longest Distances:** 1,148 mi (1,847 km) NNE-SSW; 519 mi (835 km) ESE-WNW
- **Longest River:** Shabeelle, 1,250 mi (2,011 km)
- **Natural Hazards:** Drought, floods, dust storms
- **Population:** 7,488,773 (July 2001 est.) / World Rank: 89
- **Capital City:** Mogadishu, on the Indian Ocean coast
- **Largest City:** Mogadishu, 1,227,000 (2000 est.)

OVERVIEW

Somalia is located in easternmost Africa, sharing the Horn of Africa with Djibouti and Ethiopia. Coasts on the Gulf of Aden and Indian Ocean provide access to important sea routes. The land itself, however, is predominantly scrubland and desert. Only 13 percent of the land is arable, and there are few rivers or other dependable sources of fresh water. Somalia faces daunting food and water management issues that have often reached a state of crisis.

Sovereignty disputes are ongoing over the neighboring Ogaden region of Ethiopia. In the northwest, along the Gulf of Aden, the Republic of Somaliland, with some 3.5 million people, declared independence from Somalia in 1991. Somaliland has a functioning government but is not internationally recognized; within it is Puntland State, another autonomous region.

MOUNTAINS AND HILLS

Mountains

Somalia's only mountains, the Migiurtinia and Ogo ranges, are in the north, extending from Ethiopia and following the Gulf of Aden coast with a high escarpment until the cliffs form the tip of the Horn of Africa. Somalia's highest peak, Mt. Shimbiris, rises 7,926 ft (2,416 m) at the center of the northern range.

Plateaus

South of the mountains, the dry Somali Plateau continues from eastern Ethiopia's Ogaden region to become the Ogo Plateau, the Mudug Plain, and the Haud region of central/southwest Somalia. These plateau regions vary in height from 6,000 ft (1,829 m) in the Ogo to 1,640 ft (500 m) in the Haud.

Canyons

Throughout Somalia, soil erosion has caused gullies and canyons to appear. A lack of roads has led to trucks being driven across pasture land, eroding gullies in the dry soil. Deep ravines are also cut by seasonal watercourses.

Hills and Badlands

In the northern region called the Ogo, limestone hills 2,953 to 3,937 ft (900 to 1,200 m) high distinguish a rough terrain dissected with dried up streambeds. The hills are covered with scrub vegetation which is grazed by livestock and antelopes.

INLAND WATERWAYS

Lakes

Somalia lacks any permanent lakes. In the Haud, some basins are filled by rains and intermittent flood waters, creating temporary ponds. Somalia also has artificial ponds designed to capture precious seasonal waters for irrigation and drinking. Wells and springs are of great importance for Somalia's water supply.

Rivers

Somalia's two permanently flowing rivers, the Jubba (Gestro) and Shabeelle are used for irrigation but are not navigable by large boats. The Jubba and Shabeelle Rivers both have their sources in Ethiopia and run south through Somalia towards the Indian Ocean. The Jubba River is approximately 1,000 mi (1,610 km) long, and the Shabeelle River is 1,250 mi (2,011 km) long, of which 621 mi (1,000 km) is in Somalia. The Jubba River empties directly into the Indian Ocean in southern Somalia. To its north, the Shabeelle River flows towards the coast, then turns southeast following the coast, dwindling to its end in marshlands and sand flats. In times of heavy rain, the Shabeelle waters can meet those of the Jubba. The area between the two rivers is Somalia's most fertile region.

The Jubba/Shabeelle river system as well as the seasonal watercourses found in badly eroded, deforested, and desert terrain are highly vulnerable to sporadic flooding. In 1997 Somalia experienced its worst floods since 1961, with the Jubba and Shabeelle rivers overflowing their banks, and flooding recurred in 2002.

The two largest watercourses in northern Somalia are the seasonal Daror and Nugaaleed stream systems. Both are usually dry.

Wetlands

The wetlands of Somalia surround the outlet of the Jubba River and the lower reaches of the Shabeelle River, where swamp basins are the habitat of birds and reptiles. Some mangrove forests are still found in Somalia, especially along the Jubba outlet, but most have been destroyed by cutting for fuel and fodder.

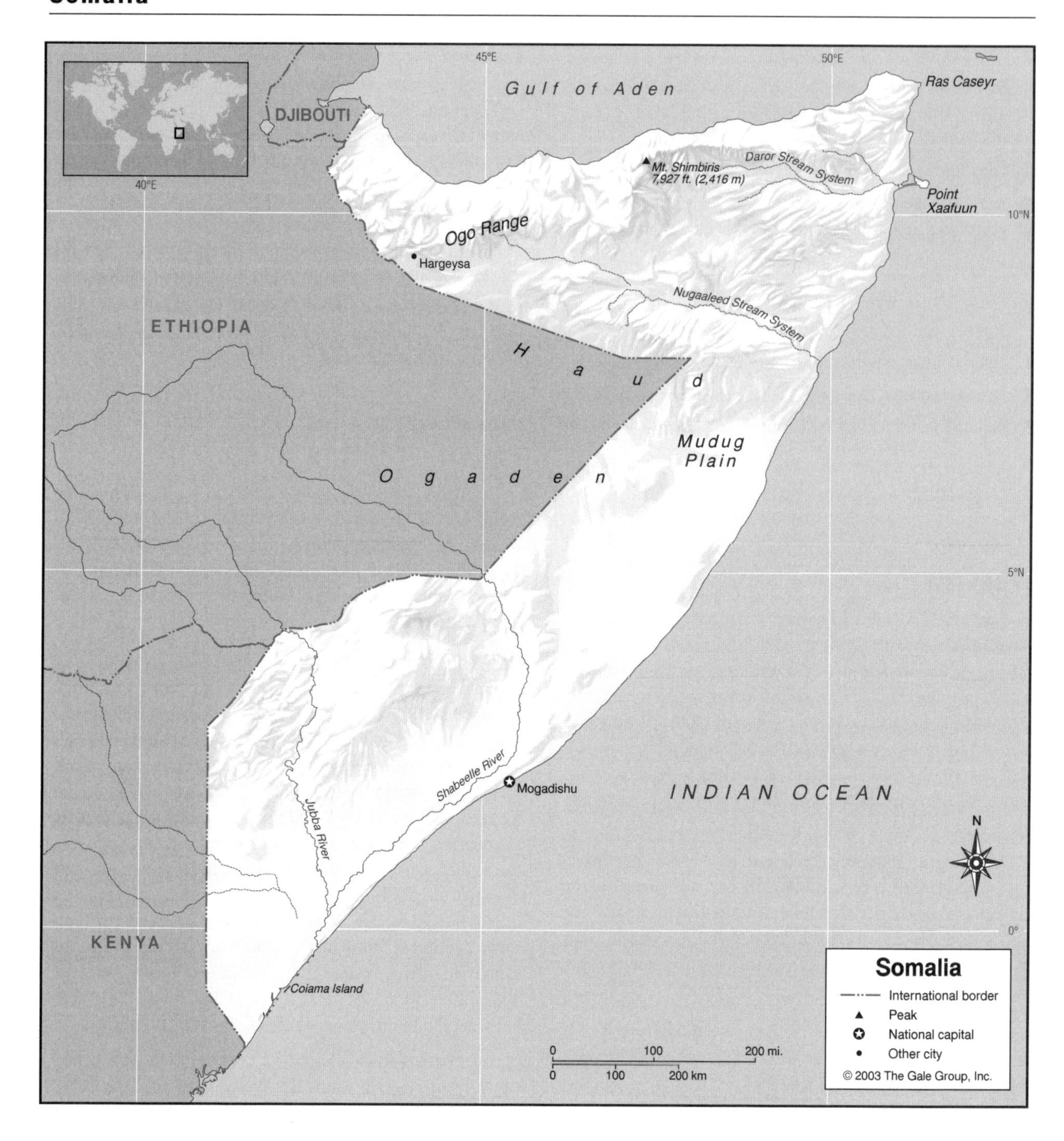

THE COAST, ISLANDS AND THE OCEAN

Oceans and Seas

Northern Somalia meets the Gulf of Aden, an inlet of the Indian Ocean, with Yemen across the Gulf. Because it leads to the Red Sea and Suez Canal, the Gulf of Aden is a crucial shipping lane, particularly for petroleum vessels, and oil tanker pollution affects Somalia's northern coast. The east coast of Somalia directly faces the Indian Ocean, where over-fishing by foreign trawler fleets is a problem.

Major Islands

The Bajuni are a 77 mi (125 km) coral reef chain of several small islands and many islets or rocks. They include Coiama (Somalia's largest island, 2.5 sq mi /6 sq km), Ngumi, the Ciovai pair, Ciula (inhabited), Daracas, and Ciandra. Most are barren and without permanent settlement.

The Coast and Beaches

Somalia has the second longest coastline in Africa (only South Africa's is longer.) The northern coast, along the Gulf of Aden, begins on the west at the border of Djibouti. Sandy beaches are interspersed with rocky cliffs, and the north coast lacks reefs. Ras Caseyr (Cape Guardafui) is a rugged headland where the north and east coasts meet. Due south of the Cape, the Point Xaafuun (Ras Hafun) promontory juts out. From there, the Indian Ocean coast runs south in a succession of sandy beaches with little indentation. In the southern stretch, from Mogadishu to the Kenya border, coral reefs are a barrier to the shore, which lacks natural harbors.

CLIMATE AND VEGETATION

Temperature

Somalia has an arid or semi-arid climate. In normal years there are four seasons, two with rain and two essentially without rain. December to March, the time of northeast monsoon winds, is a very dry season, with moderate temperatures in the north and hot temperatures in the south. April to June is a spring-like rainy season with hot temperatures. July to September, the time of southwest monsoon winds, is a dry and hot season. October to November is a humid, sporadically rainy season.

Somalia's average temperature is between 77° and 82° F (25° and 28° C.) Temperatures fall as low as 32° F (0° C) in the mountains of the north, and reach as high as 117° F (47° C) on the coasts.

Rainfall

In non-drought times, Somalia's average annual rainfall is only 11 in (28 cm) Droughts can strike Somalia when rainfall decreases even slightly. Their effects are worsened by factors such as over-grazing, erosion, disruptions of nomadic routes, and breakdowns in water access and food distribution. Major droughts ravaged Somalia in 1974-75, 1984-85, 1992, 1999, and 2001.

Flooding, worsened by the same factors, caused damage in 1997 and 2002. In such an arid land, a delicate ecological balance of rainfall, water absorption, and water storage is needed for the survival of humans, livestock and wildlife.

Grasslands

Up to 70 percent of Somalia is a scrubland ecosystem of coarse grass-patches and shrubs. This terrain is especially pervasive in the Haud Plateau region of the north, and throughout the south. The scrub vegetation receives minimal rain, but is resilient. Where there is water, as in the area between the Jubba and Shabeelle Rivers, good pastureland results.

The scrub grasslands are used by nomadic Somalis for their herds of camels, cattle, goats, and sheep. Much of Somalia's grassland is being lost to desertification as a result of over-grazing and the cutting of fodder grass for export to neighboring countries.

Deserts

Some 25 percent of Somalia is desert, usually consisting of sand or gravel mixed with some vegetation. The deserts run along most of Somalia's northern and central coasts and extend into the interior. Desertification is steadily claiming grassland and wooded areas across Somalia.

Regions – Somalia

Name	Area (sq mi)	Area (sq km)	Capital
Bakool	10,000	27,000	Xuddur
Banaadir	400	1,000	Mogadishu
Bari	27,000	70,000	Boosaaso
Bay	15,000	39,000	Baydhabo
Galguduud	17,000	43,000	Dhuusa Mareeb
Gedo	12,000	32,000	Garbahaarrey
Hiiraan	13,000	34,000	Beled Weyne
Jubbada Dhexe	9,000	23,000	Bu'aale
Jubbada Hoose	24,000	61,000	Kismaayo
Mudug	27,000	70,000	Gaalkacyo
Nugaal	19,000	50,000	Garoowe
Sanaag	21,000	54,000	Ceerigaabo
Shabeellaha Dhexe	8,000	22,000	Towhar
Shabeellaha Hoose	10,000	25,000	Marca
Togdheer	16,000	41,000	Burko
Woqooyi Galbeed	17,000	45,000	Hargeysa

SOURCE: *Geo-Data: The World Geographical Encyclopedia,* 2nd ed. Detroit: Gale Research, 1989.

On the Gulf of Aden coast, the Guban Desert is a hot, dry plain with a system of sandy seasonal watercourses. The arid Hobyo region extends north from Somalia's capital, Mogadishu, along the Indian Ocean coast. It is a desert with low vegetation that is a habitat for birds, reptiles, and antelopes. Areas of sand dunes along the Indian Ocean coast have been destabilized by over-grazing of the grasses that held the dunes in place.

Forests and Jungles

Somalia has only 1 percent forest cover left, which is mainly in the far south. Trees are cut for fuel, fodder, and livestock shelters, and there is very little reforestation. The southern forest includes eucalyptus, tall cactus, and mahogany. Trees that provide myrrh and frankincense are also native to Somalia. The north has some acacia scrub and savannah forest.

HUMAN POPULATION

Somalia has a high population growth rate of 3.4 percent (2001 estimate) and a low population density of 10.3 people per sq mi (4 people per sq km; 2001 estimate). Famine and warfare caused at least 300,000 deaths in Somalia in the catastrophic drought year of 1992.

Urban dwellers are an estimated 28 percent of Somalia's population (2000), despite instability and destroyed infrastructure in many of the cities and towns. Poverty is severe throughout urban and rural areas. Roughly 60 percent of the population is not only rural, but nomadic, moving from place to place in search of grazing land and water.

NATURAL RESOURCES

Livestock and agriculture are Somalia's major resources. The mineral resources of Somalia include uranium, iron, tin, gypsum, bauxite, copper, and salt, but few of them have been exploited. There is significant solar energy potential.

FURTHER READINGS

Adam, Hussein M., and Richard Ford, eds. *Mending Rips in the Sky: Options for Somali Communities in the 21st Century.* Lawrenceville N.J.: Red Sea Press, 1997.

Banadir.com. *Gateway to Somalia.* http://www.banadir.com (Accessed June 20, 2002).

D'Haem, Jeanne. *The Last Camel: True Stories About Somalia.* Lawrenceville N.J.: Red Sea Press, 1997.

Somali Environmental Protection and Anti-Desertification Organization (SEPADO). http://members.tripod.com/~sepado/ (Accessed June 20, 2002).

South Africa

- **Area:** 471,011 sq mi (1,219,912 sq km) / World Rank: 26
- **Location:** Southern and Eastern Hemispheres, at the southern tip of Africa; bordered by Namibia, Botswana, and Zimbabwe to the north; Mozambique, Swaziland, the Indian Ocean to the east; the Atlantic Ocean to the southwest; and the Indian Ocean to the south.
- **Coordinates:** 29°00′S, 24°00′E
- **Borders:** 2,952 mi (4,750 km) / Botswana, 1,143 mi (1,840 km); Lesotho, 565 mi (909 km); Mozambique, 305 mi (491 km); Namibia, 531 mi (855 km); Swaziland, 267 mi (430 km); Zimbabwe, 140 mi (225 km)
- **Coastline:** 1,739 mi (2,798 km)
- **Territorial Seas:** 12 NM
- **Highest Point:** Njesuthi Mountain, 11,181 ft (3,408 m)
- **Lowest Point:** Sea level
- **Longest Distances:** 1,132 mi (1,821 km) NE-SW; 662 mi (1,066 km) SE-NW
- **Longest River:** Orange River, 1,400 mi (2250 km)
- **Largest Lake:** Saint Lucia, 135 sq mi (350 sq km)
- **Natural Hazards:** Drought
- **Population:** 43,586,097 (July 2001 est.) / World Rank: 26
- **Capital City:** Pretoria, located in the northeastern part of the country
- **Largest City:** Cape Town, on southwestern coast, population 2,727,000 (2000 est.)

OVERVIEW

Comparatively, the area occupied by South Africa is slightly less than twice the size of the state of Texas. The country's general topography consists of a broad, centrally depressed plateau edged by a prominent escarpment overlooking marginal slopes that descend to the eastern, southern and western slopes. The mountainous edges of the plateau extend in a sweeping arc from the country's northeastern tip to its southwestern extremity. Collectively these edges are known as the Great Escarpment. The marginal zone tends to be more dissected than the plateau and in places exhibits acute relief features. Inland from the crest of the Great Escarpment the country consists generally of rolling plains dropping gradually to an altitude of about 2,952 ft (900 m) in the center. Within the plateau, there are a number of generally distinctive sub-regions. The largest of these is the Highveld (veld is Afrikaans for grassland), extending from Pretoria and encompassing all of the center of the country and reaching southwestward through to within a few hundred miles of the Atlantic Coast. The undulating land surface lies mostly between 3,937 and 5,906 ft (1,200 and 1,800 m) above sea level. Its northern limit is formed by the Witwatersrand Ridge, on which Johannesburg stands at 5,906 ft (1,800 m).

The Republic of South Africa lies on the African Tectonic Plate.

MOUNTAINS AND HILLS

Mountains and Plateaus

The Groote-Swartberge lies between the Great Karroo Range and the Little Karroo Range in the southern part of the country. Between the latter area and the coastal plain is another mountain range, the Langeberg. On the southern coast, just south of Cape Town, an isolated peak, Table Mountain, rises to about 3,563 ft (1,086 m). On the southwestern coast, the edge of the plateau is marked by the Roggeveld Mountains, a range of folded mountains, irregular in character and direction, which descends abruptly into a coastal plain.

The topography of South Africa consists primarily of a great plateau, which occupies about two-thirds of the country. The plateau reaches its greatest heights along the southeastern edge, which is marked by the Drakensberg Mountains, a range that is part of the Great Escarpment, which separates the plateau from coastal areas. The escarpment includes Njesuthi Mountain 11,178 ft

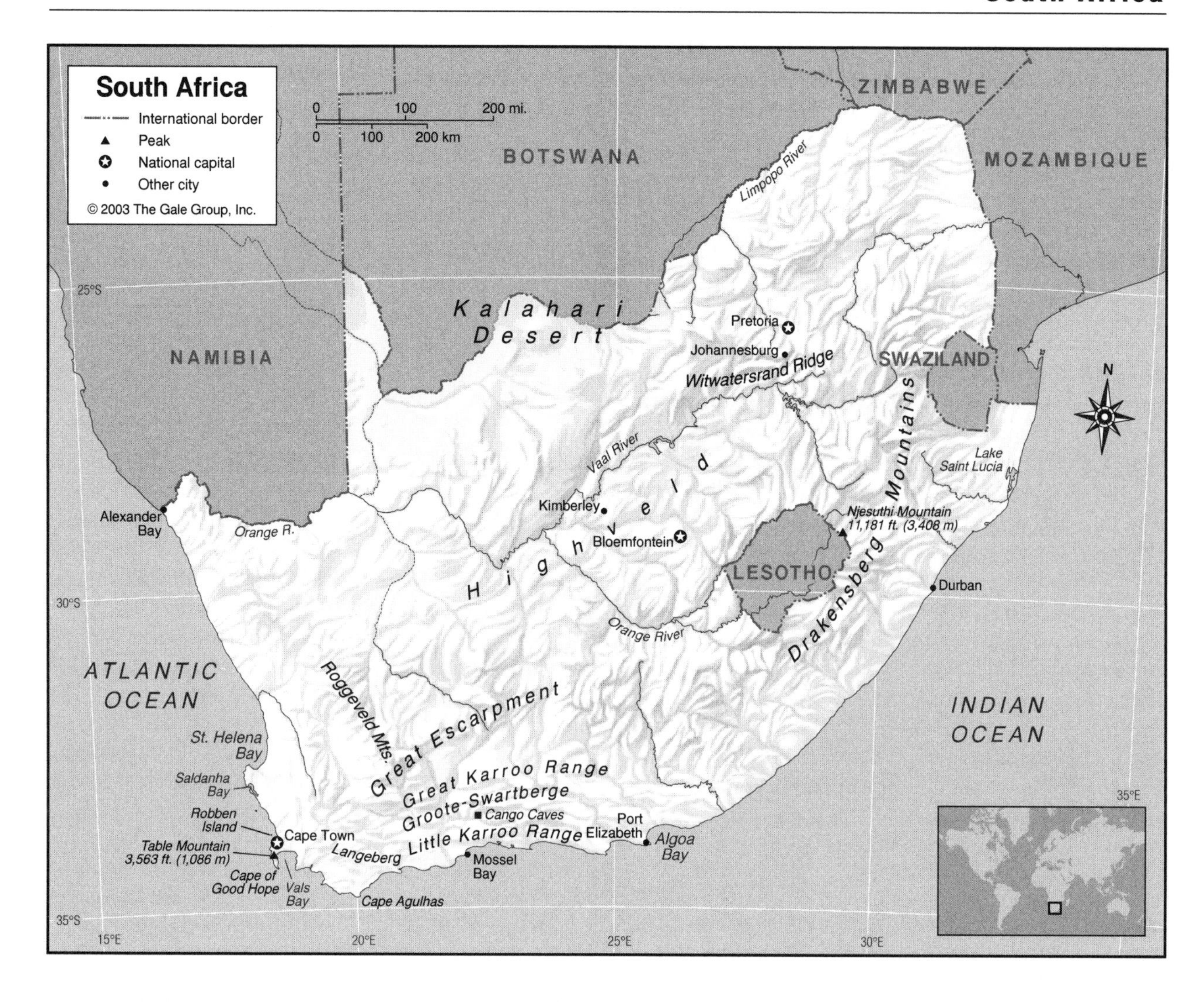

(3,408 m), the highest elevation in the country. Within the plateau three regions may be distinguished: the Highveld, the Bushveld, and the Middle Veld.

The Highveld, which covers most of the plateau, ranges in elevation between about 4,000 to 6,000 ft (about 1,200 and 1,800 m) and is characterized by level or gently undulating terrain. The northern limit of the Highveld is marked by a rock ridge, called the Witwatersrand, which includes the city of Johannesburg. North of the Witwatersrand is the Bushveld or Transvaal Basin. This section, much of which is broken into basins by rock ridges, slopes downward from east to west toward the Limpopo River. The Bushveld averages less than 4,000 ft (1,200 m) in height. The western section of the plateau, known as the Middle Veld (or Kaap Plateau), also slopes downward in a westerly direction. The elevation of the Middle Veld varies between about 2,000 to 4,000 ft (about 600 and 1,200 m).

Between the edge of the plateau and the eastern and southern coastline the land descends seaward in a series of abrupt grades, or steps. Along the eastern coast are two steps. The interior step is a belt of hilly country, called the Eastern Uplands. The exterior step is a low-lying plain, called the Eastern Lowveld. On the south, the steps, proceeding from the interior to the coast, are a plateau called the Great Karroo, or Central Karroo; a lower plateau called the Little Karroo, or Southern Karroo; and a low-lying plain.

Caves

The 20 million-year-old Cango Caves, located near Oudtshoorn in the Groote-Swartberge Mountains, are the longest underground cave sequence and have among the largest stalagmite formations in the world. Their underground area coves more than 3 mi (5 km) of widely branching caverns, interconnected tunnels, and deep pits, complete with magnificent limestone formations and colorfully illuminated sandstone formations.

INLAND WATERWAYS

Lakes

The largest lake in South Africa is Saint Lucia. Its surface area varies from about 115 sq mi (298 sq km) in the dry season to 135 sq mi (350 sq km) during the wet season, and its depth ranges from 3 to 8 ft (0.9 to 2.4 m). St. Lucia is a salt water lagoon located on the northeastern coast of the country near Sodwana Bay, separated from the Indian Ocean only by a narrow, seven-mile long (11.3 km) channel. The lake and surrounding area are home to a wide variety of animal species, and is the only place on earth where hippopotamuses, sharks, and crocodiles share the same waters.

Rivers

The chief rivers of South Africa are the Orange, Vaal, and Limpopo. The Orange River is the longest river in the country. It originates in Lesotho, flows in a northwestern direction, and empties into the Atlantic Ocean after a course of some 1,300 mi (2,100 km). The westernmost section of the Orange River forms the boundary between South Africa and Namibia. The Vaal River originates in the northeastern section of the country, near Swaziland. It flows in a southwestern direction to a point in the central portion of the country, where it joins the Orange River. The Limpopo River originates in the northeastern region and flows northwest to the Botswana border and then east along the borders of Botswana and Zimbabwe before entering Mozambique and continuing to the Indian Ocean. In general, the rivers of the country are irregular in flow. Many are dry during much of the year. Consequently, the rivers are of little use for navigation or hydroelectric power but are of some use for irrigation.

THE COAST, ISLANDS, AND THE OCEAN

Oceans and Seas

The country is bounded on the east by the Indian Ocean, on the south by the confluence of the Indian and Atlantic Oceans, on the west by the Atlantic Ocean. There are coral reefs off the eastern coast surrounding Sodwana Bay that attract divers from all over. Along this coast the continental shelf is generally narrow or entirely non-existant. Off the southern coast the continental shelf extends to form the large triangular Agulhas Bank, which has proven to be a source of oil and gas, while on the western coast it forms the Benguella Upwelling. The waters of the Indian Ocean off of South Africa's eastern coast have temperatures of 68°-77° F (20-25°C). The temperature of the Atlantic off the western coast is 48°-57° F (9°-14°C), while off the southern coast the waters average 61°-70°F (16°-21°C).

Major Islands

South Africa controls two small islands, Prince Edward and Marion, which lie some 1,200 mi (1,920 km) southeast of Cape Town. Other extraterritorial holdings include a number of small islands off the southwestern coast: Dassen, Robben and Bird Islands. All are inhabited except for Robben Island, which serves as the site of the country's maximum-security prison, once the site of Nelson Mandela's imprisonment.

The Coast and Beaches

South Africa has a rugged coastline with rocky shores and few sheltered bays or harbors. However, the high wave energy that the exposed coast is subjected to has formed sandy beaches in many places between the otherwise jagged shoreline. These beaches are usually backed by low sand dunes. The coastal belt of the west and south ranges between 500 and 600 ft (150 and 180 m) above sea level and is very fertile. Ports such as Cape Town play an important role in the economy of the country. Cape Town is also a popular vacation resort with natural scenic beauty and fine beaches.

The South African coastline begins at Alexander Bay on the Atlantic Ocean in the west and continues south rather uneventfully until it reaches St. Helena Bay. Here the coastline cuts back to the west. It then turns back east, carving out the tiny Saldanha Bay, then continues south, jutting out at the Cape of Good Hope peninsula, on which Cape Town is located. The coast cuts north around the peninsula then back to the south carving out Vals Bay before turning southeast. After rounding the southernmost point in Africa, Cape Agulhas, the coastlines heads to the northeast. It follows a slight indentation at Mossel Bay before turning around Port Elizabeth and forming a few indentations including the larger Algoa Bay. From here the smooth coastline heads due northeast with no features other than the St. Lucia estuary and Sodwana Bay in the northeast.

GEO-FACT

One of the biggest tourist's attractions in South Africa is Cape Town. It lies on the Atlantic west coast where the water may be icy but the sand is hot and the beaches are more sheltered from the renowned 'Cape Doctor' south-east wind than those on the east. The drive from Cape Town along Victoria Road to the western peninsula beaches of Clifton, Camps Bay, Llandudno, and Sandy Bay provides magnificent views of the jagged peaks of the twelve apostles. But one of the most spectacular marine drives in the world continues onto Chapman's Peak Drive, which skirts the Atlantic 1,960 ft (about 600 m) above sea level.

CLIMATE AND VEGETATION

Temperature

The climate of South Africa ranges from Mediterranean in the southwest to temperate in the interior plateau, to subtropical in the northeast. Snow is rare in South Africa, although winter frosts occur in the higher areas of the plateau. The average January temperature range in Durban, on a low-lying part of the northeastern coast, is 69° to 81°F (21° to 27°C). The corresponding temperature range in Johannesburg, in the north central area, is 58° to 78°F (14° to 26°C). Johannesburg, although closer to the equator than Durban, has a cooler summer largely because of its higher elevation some 5,470 ft (1,670 m) above sea level. The average January temperature range in Cape Town, on the southern coast, is 60° to 78°F (16° to 26°C); the Cape Town area is under the influence of cool winds from the South Atlantic. The winter temperature ranges follow the same regional pattern. The average July temperature range is 52° to 72°F (11° to 22°C) in Durban, 39° to 63°F (4° to 17°C) in Johannesburg, and 45° to 63°F (7° to 17°C) in Cape Town.

Rainfall

In general nearly all of South Africa enjoys a mild, temperate climate. Except for the extreme southwest, most of the country is under the influence of the easterly trade winds, which blow from over the Indian Ocean. Laden with moisture, these winds bring about 35 in (about 89 cm) of precipitation yearly to the Eastern Lowveld and the Eastern Uplands as far west as the Drakensberg. The Highveld receives about 15 to 30 in (about 38 to 76 cm) of precipitation annually, the amount diminishing rapidly toward the west. On the western coast rainfall is often as low as 2 in (5 cm) annually. The rainfall deposited by the trade winds occurs mainly between October and April. In the drier regions of the plateau, the amount of rainfall and the beginning of the rainy season vary greatly from year to year. The extreme southwest is under the influence of western winds originating over the Atlantic Ocean. This region annually receives about 22 in (about 56 cm) of rainfall, most of which occurs between June and September.

Grasslands

The grasslands of the central South African plateau region have dark to black soils, or *chernozems,* which are similar to those of the North American prairies. In the western areas, which receive less rain, the chernozem soils give way to poorer, chestnut-colored soils. In the south the soils are thin and often red. The soils in the northeast are reddish and yellowish.

Deserts

South Africa also includes a part of the Kalahari Desert in the northeast, generally covered with red soil and low-growing grasses and brush, except in the east, where large patches of sand are found. There is a section of the Namibian Desert in the west.

Forests

Forests, located mostly on the east, south, and southwest coasts, constitute a major feature of the vegetation of South Africa.

Population Centers – South Africa

(2000 POPULATION ESTIMATES)

Name	Population	Name	Population
Johannesburg	2,950,000	Pretoria (capital)	1,590,000
Cape Town	2,930,000	East Rand	1,552,000
Durban	2,391,000	Port Elizabeth	1,006,000

SOURCE: "Table A.12. Population of Urban Agglomerations with 750,000 Inhabitants or More in 2000, by Country 1950–2015. Estimates and Projections: 1950–2015." United Nations Population Division, World Urbanization Prospects: The 2001 Revision.

Provinces – South Africa

1996 CENSUS

Name	Population	Area (sq mi)	Area (sq km)	Capital
Eastern Cape	6,300,000	65,475	169,580	King Williams Town/Bisho
Free State	2,600,000	49,992	129,480	Bloemfontein
Gauteng	7,400,000	6,568	17,010	Johannesburg
KwaZulu-Natal	8,400,000	35,560	92,100	Pietermaritzburg/ Ulundi
Mpumalanga	2,800,000	30,691	79,490	Nelspruit
Northern Cape	840,000	139,703	361,830	Kimberely
Northern Province	4,900,000	47,842	123,910	Pietersburg/ Lebowa-Kgomo
North West	3,300,000	44,911	116,320	Mmabatho/ Mafeking
Western Cape	3,900,000	49,950	129,370	Cape Town

SOURCE: "Did You Know? Statistics South Africa." 1996 Census.

HUMAN POPULATION

South Africa has one of the world's most complex ethnic patterns. Furthermore, legal separation of the racial communities was a cornerstone of government policy throughout most of the twentieth century. This racial policy, often called apartheid but referred to in South African government circles as "separate development," created and maintained one of the most rigidly segregated societies in the world. During the 1970s and 1980s, enforcement of separatist policies eased, but the division of the population into four racial communities remained.

Until 1994 South Africa was divided into four provinces (Cape Province, Natal, Orange Free State, and Transvaal) and ten black homelands. Under the country's interim constitution, which took effect at the time of the country's first multiracial elections in April 1994, South Africa was divided into nine provinces. The black homelands, including those that were declared independent—Transkei, Bophuthatswana, Venda, and Ciskei—were dissolved and reincorporated into South Africa when the interim constitution took effect. According to a 1998 estimate, blacks formed the largest segment of the population, constituting 75.2 percent of the total. Whites accounted for 13.6 percent; Cape Coloreds 8.6 percent; and Asians 2.6 percent.

Another unusual feature of South Africa is that its governmental functions are split between three different cities, giving it in effect three different capitals. Pretoria is the administrative capital. The legislature sits in Cape Town. Bloemfontein is the site of South Africa's Supreme Court.

The population in South Africa is unevenly distributed. Because of the desert conditions in the west and the location of most resources in the east, this half of the country is much more densely populated. The western third of the country has a population density of only 28 inhabitants per sq mi (11 inhabitants per sq km). The northeastern region has a density of 128 people per sq mi (49 people per sq km), while the southeastern region is the most densely populated with 158 people per sq mi (61 people per sq km).

NATURAL RESOURCES

South Africa is very rich in mineral resources, concentrated mostly in the east. Gold, coal, and diamonds are the chief minerals. Gold is mined primarily in the Witwatersrand, the site of the richest gold field in the world, which was discovered in 1886. The gold in the Witwatersrand occurs in minute specks, invisible to the naked eye, in pebble beds called bankets, which are mined to depths below 10,000 ft (3,000 m). Uranium also is extracted commercially in the Witwatersrand. Vast and easily worked coal seams occur in the northeast between Lesotho and Swaziland.

Diamonds are another important source of mineral wealth. Most of the diamonds come from diamond fields near Kimberley, which were discovered in 1870. Surface workings were soon exhausted, but the diamonds were traced to their source rock and mined by large-scale methods. Other minerals found in South Africa include copper, nickel, platinum, asbestos, chromium, fluorite, phosphates, vanadium, tin, titanium, antimony, and manganese and iron ores.

FURTHER READINGS

Cohen, Robin et al. *African Islands and Enclaves.* Beverly Hills, Calif.: Sage, 1983.

Lamar, Howard, and Leonard Thompson, eds. *The Frontier in History: North America and Southern Africa Compared.* New Haven, Conn.: Yale University Press, 1981.

Mandela, Nelson. *The Struggle Is My Life.* New York: Pathfinder, 1986.

Nasson, Bill. *Abraham Esau's War: A Black South African War in the Cape, 1899-1902.* Cambridge: Cambridge University Press, 1991.

South Africa Tourism. *Discover South Africa.* http://satourweb.satour.com/index.cfm (Accessed June 12, 2002).

Spain

- **Area:** 194,897 sq mi (504,782 sq km) / World Rank: 52
- **Location:** Northern and Western Hemispheres; southwestern Europe, bordered by the Bay of Biscay, France, and Andorra to the north, the Mediterranean Sea to the east and south, the North Atlantic Ocean to the south, west, and northwest, and Portugal to the west.
- **Coordinates:** 40°00′N, 4°00′W
- **Borders:** 1,192 mi (1,918 km) / Andorra, 40 mi (63.7 km); France, 387 mi (623 km); Gibraltar, 0.7 mi (1.2 km); Morroco 10 mi (16.1 km), Portugal 754 mi (1,214 km)
- **Coastline:** 3,084 mi (4,964 km) / Mediterranean Sea, 1,038 mi (1,670 km); Atlantic and Bay of Biscay, 1,388 mi (2,234 km)
- **Territorial Seas:** 12 NM
- **Highest Point:** Pico de Teide, 12,198 ft (3,718 m)
- **Lowest Point:** Sea level
- **Longest Distances:** 764 mi (1,085 km) E-W; 590 mi (950 km) N-S
- **Longest River:** Tagus, 630 mi (1,010 km)
- **Natural Hazards:** Drought
- **Population:** 40,037,995 (July 2001 est.) / World Rank: 29
- **Capital City:** Madrid, centrally located
- **Largest City:** Madrid 4,072,000 (2000 est.)

OVERVIEW

Spain occupies the greater part of the Iberian Peninsula in southwestern Europe, which it shares with Portugal. Spain includes the Balearic Islands in the Mediterranean and Canary Islands in the Atlantic. Spain also exerts its control over five "places of sovereignty"

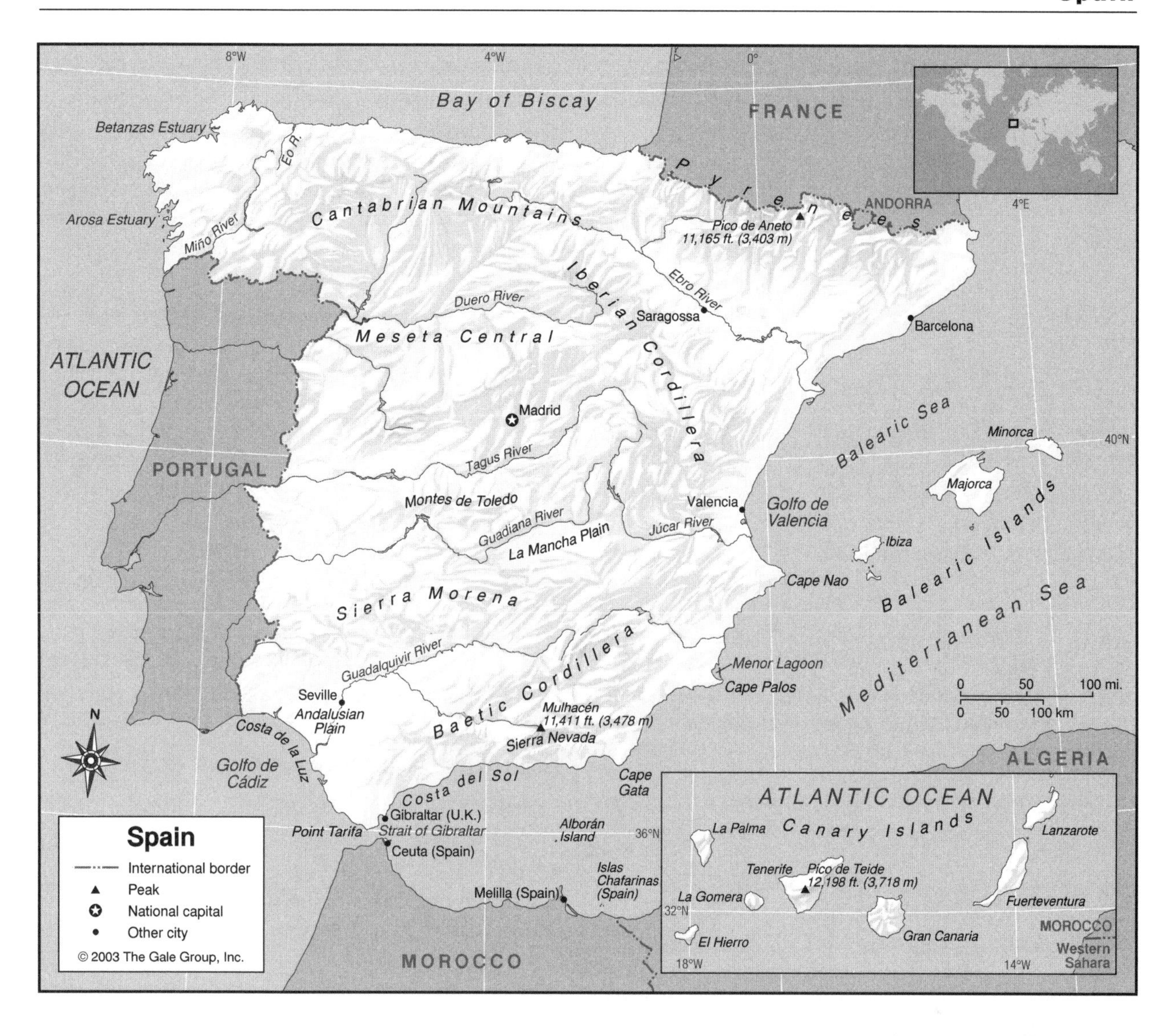

(plazas de soberania) on and off the coast of Morocco: the coastal enclaves of Ceuta and Melilla, which Morocco contests, and the islands of Penón de Alhucemas, Penón de Velez de la Gomera, and Islas Chafarinas. Spain has a centuries-old dispute with the United Kingdom over Gibraltar, a small enclave to the south of Spain.

Overall, Spain's terrain is mountainous, with major ranges running throughout the country. The Pyrenees are particularly noteworthy. One of Europe's most effective natural boundaries, the highest terrain of the main portion of this range marks the border with France. The tiny nation of Andorra is also located there. Most of the level land is in river valleys, along the coast, or on the Meseta Central (Central Mesa), the large plateau at the center of the country.

Topographically Spain is divided into four parts: the temperate region to the north and northwest, the marginal mountain ranges, the Meseta and the interior region, and the coastal regions. The categories are far from clear-cut. The temperate region, for example, includes significant portions of the mountains and coastal regions. Further, the Meseta contains two large low-lying river valleys and is traversed by several major mountain systems.

MOUNTAINS AND HILLS

Mountains

The Pyrenees extend across the country between the Bay of Biscay and the Mediterranean, a distance of about 260 mi (418 km). Their width averages 50 mi (80 km), with 80 mi (129 km) at the widest point. The French-Spanish border runs through these mountains, connecting six of the highest peaks. On the Spanish side three exceed 11,000 ft (3,353 m); the highest, Pico de Aneto, is 11,165 ft (3,403 m). The Pyrenees are very steep and rugged, with few passes.

In the north, the Cantabrian Mountains (Cordillera Cantábrica) extends across the country parallel to, and in some places adjacent to, the Bay of Biscay. There are drops exceeding 5,000 ft (1,524 m) within 20 mi (32 km) of the shore. Generally, peaks in the Cantabrian Mountains range from 7,000 to 8,000 ft (2,133 to 2,938 m).

The Iberian Cordillera (Sistema Ibérico) extend southeast from the center of the Cantabrian Mountains, reaching nearly to the Mediterranean coast. The nearly 8,000 sq mi (20,725 sq km) of this region are generally barren and rugged terrain. The Spanish call it the "area of difficulty," and it separates the Meseta Central from the northeastern river valleys.

The Baetic Cordillera (Sistema Penibético, Andalusian Mountains) of southwestern Spain extend from Cape Nao (Cabo de la Nao) to Gibraltar, a distance of some 360 mi (579 km). The most impressive part of this range is that closest to the coast, the Sierra Nevada, much of it desolate. Its 11,411 ft (3,478 m) peak, Mulhacén, is the highest point on the Iberian Peninsula. The coastal Sierra Nevada is separated from a sister range in the north by a geological fault line that runs roughly parallel to the coast. The northern range is equally forbidding, with the exception of a few exotic places, such as Granada. Further north is another, lower, chain, the Sierra Morena, with elevations between 500 and 2,000 ft (152 and 610 m).

Plateaus

The Meseta Central, the vast Spanish tableland, dominates central Spain from the Cantabrian Mountains in the north to the Sierra Morena in the south and from the Portuguese border in the west to the Iberian Cordillera in the east. Generally the Meseta has an elevation of between 2,000 and 2,500 ft (610 and 762 m), except in the river valleys. However, there are also many small mountains ranges within the Meseta, such as the Montes de Toledo and the Cordillera Carpetovetonica, the latter of which has elevations of up to 8,500 ft (2,591 m). In general, the Meseta gives way to higher land in the western part of the country between the basins of its three largest rivers.

Hills and Badlands

The land between the Bay of Biscay and the Cantabrian Mountains is hilly, having an average elevation of 2,000 ft (610 m). The verdant region is fairly broad in the west, narrower to the east where it is confined to the ocean-side slopes of the mountains.

INLAND WATERWAYS

Rivers

Spain has some 1,800 rivers and streams, of which only the Tagus (Río Tajo) is more than 600 mi (965 km) long; all but 90 are less than 60 mi (97 km) long. The Tagus, Duero, Guadiana, and the Guadalquivir all have their sources in the center of the country and drain to the west, into the Atlantic Ocean. The Ebro rises in the north and flows southeast between the Pyrenees and the Iberian Cordillera, into the Mediterranean. The Júcar, whose source lies in the southern Iberian Cordillera, also flows into the Mediterranean. The mountain rivers in the north all have short courses, owing to the nearness of their source to the sea. Those in the northwest are the longest, particularly the Miño (Minho). Many of them encounter the sea through deep estuaries extending from the mountains to the sea, similar to fjords.

Owing to scant and unpredictable rain, many of Spain's lesser riverbeds are dry most of the year. All the Meseta rivers are sluggish most of the year except for a few days each spring and fall, when the raging waters fill the riverbeds. Even the Tagus, the largest of the three, is variable in its volume of water. The Miño carries a volume of water equal to or greater than that of the Ebro, although the Miño's course is less than half as long and its basin covers only about a fifth as much area. South of the Meseta and the Sierra Morena and draining most of the Andalusian Plain, the Guadalquivir is the country's most consistent and valuable river, and Spain's only major river port, Seville (Sevilla), is located on it. The delta of the Guadalquivir is marshy and frequently saline.

THE COAST, ISLANDS, AND THE OCEAN

Oceans and Seas

The north of Spain faces the Bay of Biscay and the Atlantic Ocean. The southwest also has a coast on the Atlantic, in the form of the Golfo de Cádiz. The narrow Strait of Gibraltar between Spain and Morocco connects the Atlantic to the Mediterranean Sea to the east. The part of the Mediterranean east of Spain is often called the Balearic Sea, after the islands located there, and this part of Spain's coast is marked by the Golfo de Valencia.

Major Islands

The Baetic Cordillera extends underwater from southern Spain northeastward into the Mediterranean Sea and reappears as the Balearic Islands. The major islands of the archipelago are Majorca, Miñorca, and Ibiza, with Majorca by far the largest. Formentera and Cabrera are smaller islands in the Balearics. All of the islands are mountainous.

The Canary Islands are an archipelago of ten volcanic islands in the North Atlantic not far from Africa, about 823 mi (1,324 km) southwest of mainland Spain. They have been a possession of Spain for centuries and are considered a part of the country. Tenerife, Fuerteventura, Lanzarote, and Gran Canaria are the largest of the Canaries. La Palma, Gomera, Hierro, Graciosa, Lobos, and Alegranza make up the rest of the archipelago. The islands are predominantly mountainous and Pico de Teide (12,198 ft / 3,718 m) on Tenerife is Spain's tallest mountain.

The Coast and Beaches

The northern coast extends about 450 mi (724 km) from France to the northwestern corner of the country. The Cantabrian Mountains are never far from shore in this region, and the coast is generally even, marked only by occasional river estuaries. The largest of these, the Betanzas Estuary and Arosa Estuary, are in the most extreme northwest, fronting on the Atlantic Ocean proper.

Spain's "industry without smokestacks"—tourism—thrives on the beauty of its sun-drenched southern beaches. In fact, the southern Atlantic coast is called Costa de la Luz (Coast of Light) because of its bright sunshine. At the Strait of Gibraltar is Point Tarifa, the southernmost point in Europe. East of this is the narrow Costa del Sol (Sun Coast), which extends to Cape Gata (Cabo de Gata). The Costa Blanca, from Cape Gata to Cape Nao (Cabo de la Nao), has white and sunny beaches facing the warm Mediterranean Sea. Cape Palos forms the Menor Lagoon along this coastline.

CLIMATE AND VEGETATION

Temperature

Daytime summer temperatures reach the mid- to upper-90°s F (35° to 39° C) in the northern Meseta and are somewhat hotter in the south. Temperatures of 109° F (43° C) have been recorded in the Ebro basin. Nights are significantly cooler. In the northern Atlantic maritime region, temperatures are moderate. In the Mediterranean region, winter temperatures average between 50° and 55.4° F (10° and 13° C) and between 71.6° and 80.6° F (22° and 27° C) in summer.

Rainfall

Rainfall is highly irregular, but averages between 11.8 and 19.7 in (30 and 50 cm). The northern Meseta enjoys two rainy seasons—April to June and October to November. In the southern Meseta, the spring rainy season begins in March and is the wetter than the fall. The maritime northwest has abundant rainfall throughout the year with October through December being the wettest time of the year. The Mediterranean side receives the least rainfall with most precipitation coming in the fall and winter.

Grasslands

Grasslands are sparse in Spain, due to its semi-arid climate, and are mostly found in the valleys of the major rivers. The Andalusian Plain, in the valley of the Guadalquivir, is the largest and most important of these. It is the best farmland in the country, and the only low-lying area that permits easy entry from the sea. The La Mancha Plain is found near the Guadiana River. Grasses also cover parts of the Meseta Central and the high Sierra Nevada.

Population Centers – Spain

(2001 POPULATION ESTIMATES)

Name	Population	Name	Population
Madrid (capital)	2,882,860	Las Palmas de Gran Canaria	358,518
Barcelona	1,496,266	Murcia	357,166
Valencia	739,014	Bilbao	354,271
Seville (Sevilla)	700,716	Palma de Majorca	333,925
Saragossa (Zaragoza)	604,631		
Málaga	531,565		

SOURCE: Censo de Población y Viviendas 2001, INE (Instituto Nacional de Estadistica), Spain.

Autonomous Communities – Spain

2001 POPULATION ESTIMATE

Name	Population	Area (sq mi)	Area (sq km)	Capital
Andalucía	7,357,371	33,694	87,268	Seville (Sevilla)
Aragón	1,204,215	18,398	47,650	Saragossa (Zaragoza)
Asturias, Principado de	1,062,998	4,079	10,565	Oviedo
Baleares, Illes (Balearic Islands)	841,669	1,936	5,014	Palma de Majorca
Canarias	1,694,477	2,796	7,242	Santa Cruz de Tenerife
Cantabria	535,131	2,042	5,289	Santander
Castilla-La Mancha	1,760,516	30,591	79,230	Toledo
Castilla y León	2,456,474	36,368	94,193	Valladolid
Cataluña	6,343,110	12,328	31,930	Barcelona
Extremadura	1,058,503	16,063	41,602	Mérida
Galicia	2,695,880	11,365	29,434	Santiago de Compostela
La Rioja	270,400	1,944	5,034	Logroño
Madrid, Communidad de	5,423,384	3,087	7,995	Madrid
Murcia, Región de	1,197,646	4,370	11,317	Murcia
Navarra (Cdad. Foral de)	555,829	4,023	10,421	Pamplona
País Vasco	2,082,587	2,803	7,261	Vitoria-Gasteiz
Valenciana, Communidad	4,162,776	8,998	23,305	Valencia

SOURCE: Censo de Población y Viviendas 2001, INE (Instituto Nacional de Estadistica), Spain.

Deserts

Except in the north and northwest, the Meseta Central is substantially denuded and desert-like; scrub growth has replaced forests. Portions of the Mediterranean region are dry and desolate, especially the Baetic Cordillera, which receives dry, hot Leveche winds from the east or southeast originating over North Africa.

Forests

The Iberian peninsula is a botanical crossroads between Africa and Europe with a variety of natural vegetation and more than 8,000 species of plants, some of which originate in North Africa. Conifers predominate at higher elevations in the Meseta mountains and in northeastern Catalonia. Deciduous trees such as beech and oak predominate throughout northwestern 'wet Spain,' covering a rich and varied undergrowth dominated by ferns, gorse and heather. In the Pyrenees the deciduous zone extends from about 4,250 to 5,250 ft (1,295 to 1,600 m) while hardier pine species survive from 5,250 to 7,000 ft (1,600 to 2,133 m). Lowlands and coastal areas have mostly oak forests, to which pine, walnut, and chestnut have been added by humans.

HUMAN POPULATION

Since the death of President Franco in 1975, Spain has undergone a series of social transformations. The birth rate has fallen dramatically; the largely rural population has either moved to the cities or emigrated. In 1999 the population density was estimated at 209 people per sq mi (79 people per sq km), with 76 percent of the population living in urban areas. In 2001, the population growth rate was 0.1 percent, and life expectancy at birth was 78.9 years, equal to or better than advanced industrial societies. Forecasters predicted that by 2001 the number of people over 65 would exceed 15 percent of the population. Population growth is expected to stabilize at about 46 million in 2020. Ninety-nine percent of the population is nominally Roman Catholic. Seventy-four percent of the population speak Castilian Spanish, 17 percent speak Catalan, 7 percent Galician, and 2 percent Basque.

NATURAL RESOURCES

Virtually all types of minerals are found in Spain, but of more than 100 mined, only 17 are produced in significant quantities. These are iron, pyrites, copper, lead, tin, mercury, wolfram (metallic minerals), refractory argillite, bentonite, quartz, fluorous spar, glauberite, calcinated magnetite, sea salt, rock salt, and potassic and sepiolitic salts (nonmetallic minerals). The country has significant hydropower potential. Farming is also an important part of the economy.

FURTHER READINGS

Sí Spain. http://www.sispain.org/ (Accessed June 24, 2002).

Simonis, Damien. *Spain*. 3rd ed. Oakland, Calif.: Lonely Planet, 2001.

Smith, Angel. *Historical Dictionary of Spain*. Metuchen, N.J.: Scarecrow Press, 1996.

Solsten, Eric, and Sandra W. Meditz, eds. *Spain: A Country Study*. Area Handbook Series. Federal Research Division Library of Congress. Washington, D.C.: Department of the Army, 1988.

Sri Lanka

- **Area:** 25,332 sq mi (65,610 sq km) / World Rank: 122
- **Location:** An island in the Indian Ocean south of India, in the Northern and Eastern Hemispheres
- **Coordinates:** 7°00′N, 81°00′E
- **Borders:** None
- **Coastline:** 833 mi (1,340 km)
- **Territorial Seas:** 12 NM
- **Highest Point:** Pidurutalagala, 8,281 ft (2,524 m)
- **Lowest Point:** Sea level
- **Longest Distances:** 270 mi (435 km) N-S, 140 mi (225 km) E-W
- **Longest River:** Mahaweli, 206 mi (341 km)
- **Largest Lake:** Maduru Oya Reservoir, 24 sq mi (63 sq km)
- **Natural Hazards:** Floods, droughts, cyclones, tornadoes
- **Population:** 19,408,635 (July 2001 est.) / World Rank: 51
- **Capital City:** Colombo, on the southeast coast.
- **Largest City:** Colombo, 2,409,000 (2002 est.)

OVERVIEW

At one time the Indian Ocean island of Sri Lanka was part of the Indian subcontinent. Barely 18 mi (29 km) of shallow sea now separates the island from India. Situated on the Indian Tectonic Plate, the island is a teardrop-shaped mass of crystalline rock on which three levels can be distinguished: a coastal belt that rises from sea level to 100 ft (30 m); a belt of rolling plain corrugated with ridges rising to 500 ft (152 m) in the south; and, in the center, an irregularly shaped mass of hills and mountains having heights of over 6,000 ft (1,829 m).

MOUNTAINS AND HILLS

Mountains

The island's southwest is a series of ridges and valleys. Close to the sea the ridges are low and parallel to the coast, but inland they become mountain chains alternating with long, narrow depressions. The Sabaragamuwa Ridges cover nearly the entire southern region of the country.

The central highlands (also known as the hill country) are distinguished by high mountain walls. Elevations of more than 5,000 ft (1,524 m) above sea level are the rule; Adam's Peak, a pilgrimage destination, rises to 7,360 ft (2,243 m). The Piduru Ridges comprise the central mass of the hill country. This formidable, nearly inaccessible mountain fortress includes Sri Lanka's highest mountain, Pidurutalagala, with a summit of 8,281 ft (2,524 m).

The northernmost sections of the central highlands are the Dumbara, or Knuckles, group of mountains, including Knuckles Peak that rises to a height of 6,112 ft (1,863 m). The Dolosbage mountain group is separated by the valley of the Mahaweli River from the rest of the central highlands.

Plateaus

The Hatton Plateau is one of a series of high plains of the central highlands. Its height is between 3,000 and 4,000 ft (914 to 1,219 m) above sea level. The rivers that flow between its ridges ultimately form the Mahaweli. Nearly all of the Hatton Plateau is under tea cultivation. The ancient town of Kandy, a UNESCO World Heritage Site, is situated on the Kandy Plateau in the northwest central highlands. Horton Plains, a 12 sq mi (32 sq km) National Park in the southern central highlands is Sri Lanka's highest plateau at 6,988 ft (2,130 m).

Canyons

In the Dolosbage area of the central highlands, deep, narrow valleys lie between the ridges creating a rock maze. The Kandy Plateau is also cut by ridges and valleys and by the Mahaweli River gorge.

Hills and Badlands

The island's southeastern plain is interspersed with rounded hills that are the bare tops of eroded mountains. Gentler, grass-covered hills occur in the Uva Basin of the central highlands.

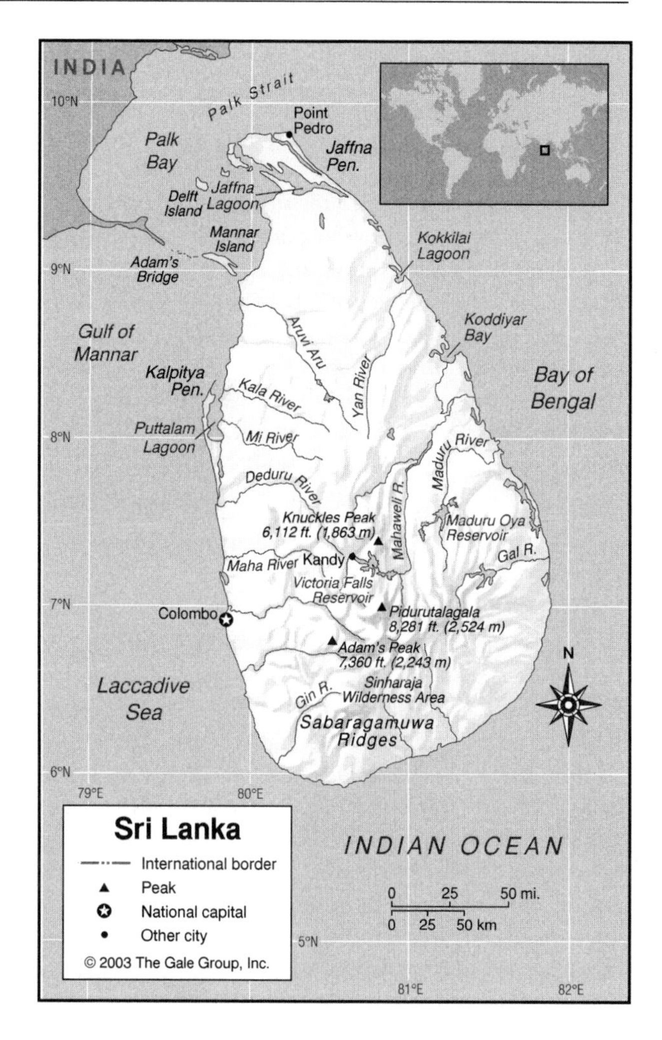

INLAND WATERWAYS

Lakes

Although Sri Lanka has few natural lakes, there are 12,000 bodies of water ranging from tiny ponds to human-made reservoirs several miles wide. The oldest of the traditional reservoirs, known as tanks, is believed to be Basawakkulam, more than 11 sq mi (30 sq km), built about 300 B.C. There are as many as 10,000 tanks of various sizes. There are also flood plain lakes, called villus, which are generally truncated river bends.

Sri Lanka's largest lake, Maduru Oya (24 sq mi/63 sq km), is a modern reservoir in the central highlands. Other large reservoirs include Randenigala (10 sq mi/27 sq km), Victoria Falls (9 sq mi/23 sq km), and Kotmale (4 sq mi/10 sq km). These huge highlands reservoirs were formed by the damming of the Mahaweli for irrigation, hydroelectricity, and water supply projects completed between 1977 and 1983. Sri Lanka has 46 large dams, and many smaller hydopower projects. Nature sanctuaries have been established around the reservoirs to protect the watersheds, but tens of thousands of people were displaced for the projects, and valuable agricultural land was submerged. The reservoirs are suffering siltation and drops in water level.

Rivers

The rivers of Sri Lanka rise in the high mountains and flow over the plateaus to the plains in a ring of waterfalls. A survey found 272 waterfalls on the island. There are sixteen principal rivers. The Mahaweli, which flows northeast from the central highlands for 206 mi (341 km), is the longest. With the exception of the 104 mi (167 km) long Aruvi Aru in the northwest, the other chief rivers range from 62 to 97 mi (100 to 156 km) in length. They are not useful for navigation, being too wild in the mountains and too shallow on the plains. Inland navigation is made possible by 153 mi (246 km) of canals. The Yan flows from the center of the island northeast to the Bay of Bengal. In the southeast, the relatively short Gal runs eastward from Gal Oya National Park to the ocean, and just north of it the Maduru runs to the coast near Batticaloa. The southern end of the island has the Gin River, and the northwestern region has rivers at nearly even intervals running westward from the center to the coast. From north to south they are the Aruvi Aru, Kala, Mi, Deduru, and Maha.

Wetlands

Sri Lanka's wetlands include freshwater marshes such as Muthurajawela, a peat bog between Colombo and Negombo; rare swamp forest, such as the Walauwa Watta Wathurana in the south near Ratnapura; and 42 salt water lagoons. Two sites in Sri Lanka are declared Wetlands of International Importance under the Ramsar Convention: Bundala National Park, a lagoon network abundant in waterbird species; and Annaiwilundawa Sanctuary, a 12th Century system of cascading water tanks.

THE COAST, ISLANDS AND THE OCEAN

Oceans and Seas

Sri Lanka lies in the northern Indian Ocean, with the Bay of Bengal to its east. The waters surrounding the island are so deep that Sri Lanka is almost unaffected by tidal variations. The Palk Strait and Palk Bay separate Sri Lanka's Jaffna Peninsula from India. To the south of Adam's Bridge, the Gulf of Mannar comes between Sri Lanka's northwest coast and India. Coral reefs extend around the Gulf of Mannar and sections of the southern and eastern coasts. Much of the coral is dying, however, from pollution, dynamite fishing, and changes in sea temperatures due to global warming. Sri Lanka is bordered on the southwest by the Laccadive Sea.

Major Islands

A few small islands extend from the north of Sri Lanka to the Indian mainland. Delft, 19 sq mi (50 sq km), and Velanai, 26 sq mi (68 sq km), are in Palk Bay. Mannar Island is part of Adam's Bridge, leading to India from the northwest.

The Coast and Beaches

The Jaffna Peninsula, a dry limestone extension, is Sri Lanka's northernmost region, with Point Pedro at its apex and Jaffna Lagoon to its south. Southwest of the Jaffna Peninsula, an elevated portion of the continental shelf forms the chain of rocky islands known as Adam's Bridge, nearly connecting Sri Lanka's northwest coast to India. Further south on the western coast, the Kalpitya Peninsula extends in a hook enclosing Puttalam Lagoon. The south/southwest coastline of Sri Lanka is famous for its many beautiful beaches, shared by tourist resorts and fishing communities. The southernmost point of the island is Dondra Head, marked with a lighthouse built in 1899. Koddiyar Bay indents the eastern coast, forming a superb natural harbor for the port of Trincomalee. A little further north Kokkilai Lagoon cuts into the coast near where the Yan River empties into the sea.

CLIMATE AND VEGETATION

Temperature

Sri Lanka receives the northeast monsoon in December-March and the southwest monsoon in June-October. In any one place, the temperature remains fairly constant year-round. The temperature in Colombo varies only from 77°F to 82.5°F (25°C to 28°C). The island's lowland areas have hot weather, with annual temperatures averaging 73°F to 88°F (23°C to 31°C), while the central mountains are cooler, averaging 57°F to 75°F (14°C to 24°C). Sri Lanka's humidity averages 70 percent to 90 percent.

Rainfall

A dry zone takes up three quarters of Sri Lanka, the northern and eastern regions, with an average annual rainfall of 50 to 75 in (127 to 190 cm), most of which comes from the northeast monsoon. The wet zone, the southwest region of the island, receives 100 to 200 in (254 to 508 cm) annual rainfall, mostly from the southwest monsoon.

Grasslands

Grasslands occur in the central highlands, the arid north, and along the eastern hills. The Uva Basin has distinctive wet grasslands called "patanas." Gal Oya, in the southeast, is a National Park, with tall grasses and monsoon forest. It has medicinal plants and is an elephant habitat. The Horton Plains are grasslands mixed with temperate forest, though they are dying off. Fires, livestock grazing, and invasive plant species degrade Sri Lanka's remaining grasslands.

Forests and Jungles

Sri Lanka is considered a biodiversity hot spot, as half of its species are endemic to the island. About 25 percent of Sri Lanka has forest cover of some type. About a fifth of that is tropical rainforest. Sinharaja, in the southern lowlands, is Sri Lanka's last significant primary rainforest, and has been declared a UNESCO World Heritage Site and Biosphere Reserve. Dry zone forests include thorn forests in the northwest and southeast, dry evergreen forests, and deciduous monsoon forests. The eastern slopes of the central highlands contain savannah forests, very susceptible to burning and droughts. Tropical evergreen rainforests are found at low and high elevations of the wet zone. Mangrove forests are declining along the coasts. Remaining forest cover exists mostly in disconnected patches of protected land.

HUMAN POPULATION

Sri Lanka has a very high population density, with 800 people per sq mi (309 per sq km). Consequently, family planning programs have been implemented to control population growth and this has stabilized the annual growth rate at 0.87 percent (2001), one of Asia's lowest rates. Thirty percent of Sri Lankans live in urban

Population Centers – Sri Lanka

(POPULATIONS FROM 2000 CENSUS)

Name	Population	Name	Population
Colombo (capital)	642,020	Negombo	121,933
Dehiwala-Mount Lavinia	209,787	Sri Jayawardanapura Kotte	115,826
Moratuwa	177,190	Kandy	110,049
Jaffna	135,000		

SOURCE: Department of Census and Statistics, Sri Lanka.

Districts – Sri Lanka

Name	Area (sq mi)	Area (sq km)	Capital
Amparai	1,778	4,604	Amparai
Anduradhapura	2,809	7,275	Anuradhapura
Badulla	1,090	2,822	Badulla
Batticaloa	1,017	2,633	Batticaloa
Colombo	268	695	Colombo
Galle	652	1,689	Galle
Hambantota	1,013	2,623	Hambantota
Jaffna	833	2,158	Jaffna
Kalutara	624	1,615	Kalutara
Kandy	833	2,158	Kandy
Kegalle	642	1,663	Kegalle
Kurunegala	1,844	4,776	Kurunegala
Mannar	778	2,014	Mannar
Matale	768	1,989	Mateale
Matara	481	1,247	Matara
Monaragala	2,188	5,666	Monaragala
Mullaitivu	798	2,066	Mullaitivu
Nuwara Eliya	555	1,437	Nuwara Eliya
Polonnaruwa	1,332	3,449	Polonnaruwa
Puttalam	1,172	3,036	Puttalam
Ratnapura	1,251	3,239	Ratnapura
Trincomalee	1,048	2,714	Trincomalee
Vavuniya	1,021	2,645	Vavuniya

SOURCE: *Geo-Data: The World Geographical Encyclopedia,* 2nd ed. Detroit: Gale Research, 1989.

areas (2001), and this number is expected to increase to 45 percent by 2015. One out of every eight Sri Lankans lives in the capital city of Colombo.

NATURAL RESOURCES

Being an island nation, Sri Lanka has developed a sizable fishing industry. It also produces and exports large amounts of clothing and textiles. The agricultural produce in the wet zones of the island include tea—which is grown heavily on the Hatton Plateau—rubber, coconuts, and potatoes. Sri Lanka also mines many mineral resources including graphite, heavy mineral sands, iron, salt, limestone, and clay. Precious gems are found in veins in the rock of the Sabaragamuwa Ridges south of Ratnapura. Sri Lanka has tapped into its many rivers through dams and waterfalls to produce hydropower, and also uses the monsoon winds to take advantage of wind energy.

FURTHER READINGS

Bradnock, Robert, and Rona Bradnock. *Sri Lanka Handbook.* Emeryville, Calif.: Avalon Travel Publishing, 2001.

Devendra, Tissa. *Sri Lanka, the Emerald Island.* Torrence, Calif.: Heian International, 2000.

Gunesekera, Romesh. *Reef.* New York: Riverhead Books, 1996.

Sri Lanka Wildlife Conservation Society. http://www.benthic.com/sri_lanka (accessed May 24, 2002).

Sri Lanka WWW Virtual Library. http://www.lankalibrary.com (accessed May 24, 2002).

Sudan

- **Area:** 967,499 sq mi (2,505,810 sq km) / World Rank: 11
- **Location:** Northern and Eastern Hemispheres, in northern Africa bordering the Red Sea to the northeast, south of Egypt and Libya, east of Chad and the Central African Republic, north of Uganda and Kenya, and west of Ethiopia and Eritrea
- **Coordinates:** 15°00′ N, 30°00′ E
- **Borders:** 4,776 mi (7,687 km) / Egypt, 791 mi (1,273 km); Ethiopia, 998 mi (1,606 km); Kenya, 144 mi (232 km); Uganda, 270 mi (435 km); Democratic Republic of the Congo, 390 mi (628 km); Central African Republic, 724 mi (1,165 km); Chad, 845 mi (1,360 km); Libya, 238 mi (383 km); Eritrea, 376 mi (605 km)
- **Coastline:** Red Sea, 530 mi (853 km)
- **Territorial Seas:** 12 NM
- **Highest Point:** Kinyeti, 10,456 ft (3,187 m)
- **Lowest Point:** Sea Level
- **Longest Distances:** 1,362 mi (2,192 km) SSE-NNW; 1,168 mi (1,880 km) ENE-WSW
- **Longest River:** Nile, total length 4,140 mi (6,670 km)
- **Largest Lake:** Lake Nubia, 373 sq mi (968 sq km)
- **Natural Hazards:** Dust storms, drought, floods
- **Population:** 36,080,373 (July 2001 est.) / World Rank: 33
- **Capital City:** Khartoum, central-northeast at the confluence of the Blue and White Nile
- **Largest City:** Khartoum, 2,700,000 (2000 est.)

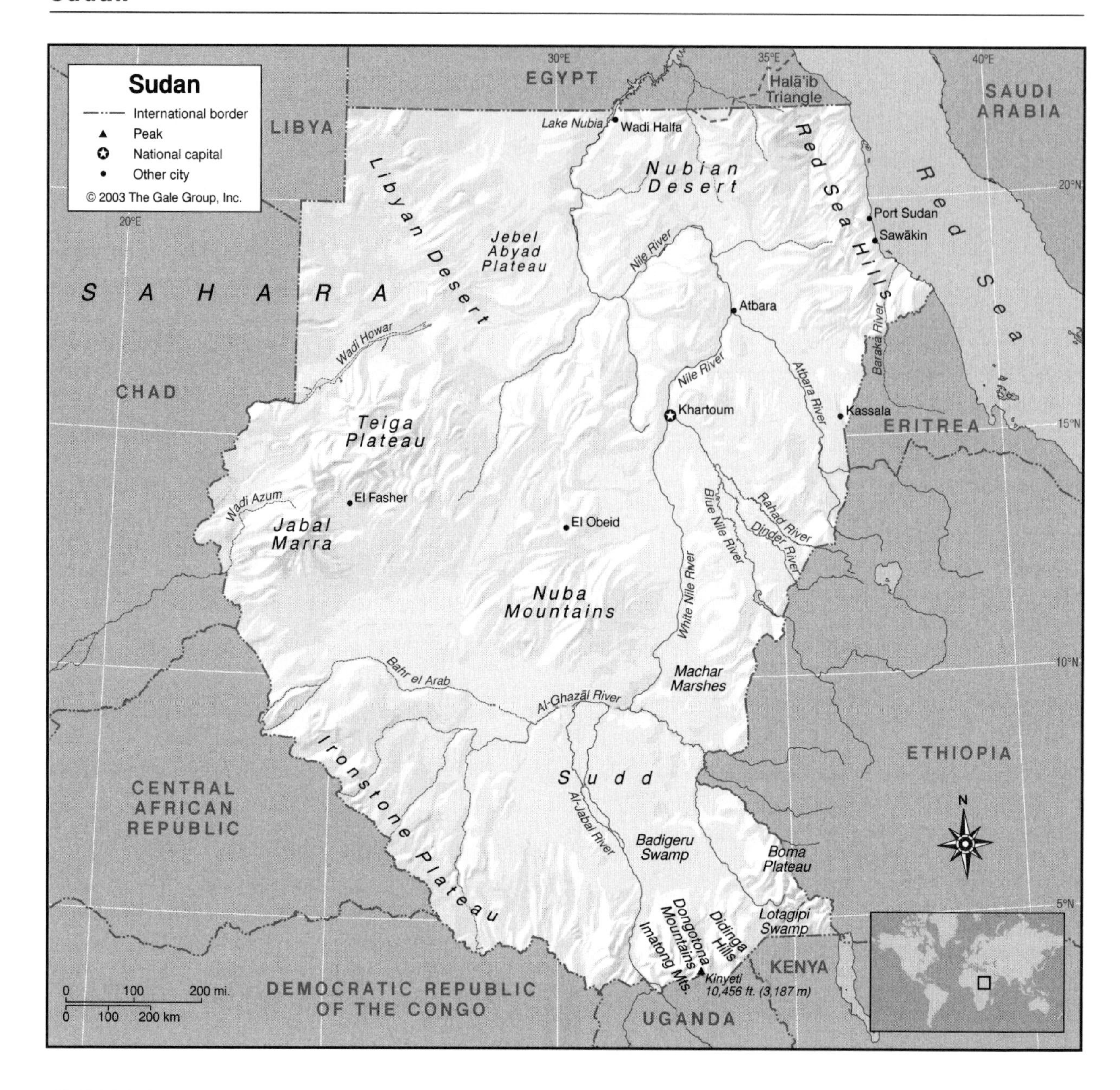

OVERVIEW

Sudan, the largest country in Africa, is an immense, sparsely populated plain, with plateaus or mountainous areas near the borders in the west, the southeast, and along the Red Sea coast in the northeast. It is slightly more than one-quarter the size of the United States.

The most prevalent landscape is semiarid savanna, a mixture of short grasses, scattered brush, and short trees. Daytime temperatures are high throughout the year, and the dry season ranges from three months in the relatively humid south to nine months in the capital city of Khartoum.

Narrow belts of irrigated cropland, no more than a few miles wide, bisect the northern savanna and desert along the main Nile River and along the White Nile, the Blue Nile, and the Atbara Rivers. They contrast sharply with the arid savanna or barren desert just beyond the limits of irrigation. Only 5 percent of the land is arable, 24 percent is meadows and pastures, 20 percent is forest and woodland, and 51 percent is semi-arid desert.

Sudan's administrative boundary with Kenya does not coincide with the international boundary, and Egypt asserts its claim to the "Halā'ib Triangle," a barren area of 8,026 sq mi (20,580 sq km) under partial Sudanese administration that is defined by an administrative boundary that supersedes the treaty boundary of 1899.

MOUNTAINS AND HILLS

Mountains

There are four mountain or upland zones. To the northeast near the coast lie the Red Sea Hills; in the west are the Marra, a mountain complex sloping to the border with Chad; and in central Sudan south of El Obeid are the Nuba Mountains, a relatively minor complex rising above the clay plains. The fourth zone includes the Imatong and Dongotona Mountains in the extreme south, along the Uganda border.

The Red Sea Hills are eroded outcroppings of base rock rising from a narrow coastal plain, the abruptness of their eastern slope giving rise to gushing torrents during winter rains blown in from the sea. The western slopes of the mountains incline more slowly toward the Nile and receive only light summer rains. North of the Atbara-Port Sudan railway, the hills extend into the desert, bare of vegetation except in the valleys. South of the railway, however, increased rainfall permits the growth of a few trees and thorny shrubs. The area is inhospitable and supports only semi-nomadic herders, who also cultivate hardy varieties of millet in the wetter valleys. They move their flocks laterally across the mountains or to higher or lower altitudes, depending upon the vagaries of the rainfall at various altitudes. The highest of the Red Sea Hills are above 7,000 ft (2,133 m).

The only major mountain range in western Sudan, the Marra, stands near the city of El Fasher. Rising above 10,000 ft (3,048 m) in elevation, this range forms part of the watershed between the Nile River and Lake Chad drainage basins. The Marra is of volcanic origin, and its valleys are relatively fertile. The upper elevations receive a slightly higher rainfall than the surrounding plains, and the relatively rich soil of the valley is more productive. Some of the rocks and peaks have a sculptured appearance resulting from the action of the rains upon the soft volcanic rock. Much of the eroded rock is deposited by streams on the desert floor below, but on the higher hillsides manmade terraces of ancient origin retain topsoil and water. Although cultivation is generally dependent upon the seasonal rains, some small valleys and terraces are irrigated with water from small perennial mountain streams.

The Nuba Mountains of central Sudan are scattered granitic masses, rising as much as 3,000 ft (914 m) above a flat clay plain. They are covered in many areas by variations of savanna vegetation. Some slopes were once terraced and then abandoned by subsistence farmers. Water is not as scarce in the mountains as in the surrounding plains. Wells are numerous in the open valleys, and a few short mountain streams continue to flow throughout the year.

The Imatong and Dongotona Mountains stand in the extreme south, the lower Didinga Hills flanking them to the east. The Imatongs are the highest mountains in Sudan, with peaks above 10,000 ft (3,048 m) including Mount Kinyeti, the highest point in the country, which rises to a height of 10,456 ft (3,187 m). The Dongotona Mountains, lying east of the Imatongs, reach a maximum height of about 8,300 ft (2,529 m). Both mountain chains have a considerable coverage of rain forest.

Plateaus

Plateau-like formations characterize the mountainous areas and their foothills, and therefore tend to rim the country serving as watersheds for the great Nile basin drainage. The best examples are found in the large Teiga Plateau north of the Marra in the west and the extensive Ironstone Plateau in the southwest. Near the Imatongs and Dongotona Mountains area in the southeast, on the border with Ethiopia, sits the Boma Plateau, the site of a national park. West of this region, north of the mountains and northeast of Ironstone Plateau, lower plateaus slope generally northward toward the Sudd. In the north, the Libyan Desert runs across the Jebel Abyad Plateau. Along the Red Sea coast in the northeast there are also some smaller plateaus.

INLAND WATERWAYS

Lakes

Sudan has few lakes; the largest are manmade, resulting from dams on the Blue Nile and Upper Nile Rivers. The backwaters of the Aswan Dam in Egypt form two distinct bodies of water: Lakes Nassar and Nubia. Nassar, the larger of the two, is completely within Egypt's borders. Lake Nubia, which begins in Egypt, extends into Sudan as far as the northern terminus of the Sudanese railway at Wadi Halfa. Its total surface area during the wet season is 373 sq mi (968 sq km).

Rivers

As the most distant and southernmost source of the Nile River, the White Nile, known in southern Sudan as the Bahr Al-Jabal, derives much of its water from the lake plateau of east-central Africa. These headwaters include the watersheds around Lake Victoria and Lake Albert, lying on or near the equator, where rainfall exceeds 50 in (127 cm) per year. Much of this water is lost to evaporation before it reaches the Nile tributaries, but a large volume is carried into the swamp areas, including the Sudd. Losses to evaporation are also heavy in this area. Partially for this reason, the annual input from the White Nile into the upper Nile at Khartoum is only one-fifth of that from the Blue Nile, but it is important because much of the White Nile water arrives during the months when the Blue Nile input is very low.

The Blue Nile rises at Lake Tana in the Ethiopian highlands and makes its way through the mountains for about 500 mi (804 km) before entering Sudan. Torrential summer rains draining into the fast-flowing Blue Nile from these highlands cause the seasonal flood on the lower reaches of the Blue Nile and on the upper Nile, floods upon which half of the people of Sudan are dependent. During flood-time the Blue Nile and its two major tributaries, the Dindar and the Ar Rahad, contribute 70 percent of the water of the upper Nile. During flood-times the flow of the Blue Nile may be 60 times that of its low water period, and 300 times its low stage during short periods of heavy flooding. During the low water stage on the upper Nile, however, the Blue Nile and the other eastern tributaries may contribute only 20 percent of the total flow.

An important tributary to the upper Nile is the Atbara River, similar in seasonal behavior to the Blue Nile and also originating in the mountains of Ethiopia. It traverses northwest across eastern Sudan and empties into the Nile at the town of Atbara. The gradient of the Nile from Khartoum to Wadi Halfa on the northern border of Sudan is considerably steeper than that of its 900 mi (1,448 km) course south of Khartoum. Along this lower reach are five of the Nile's six cataract areas of swift, rough water.

All perennial streams of significant size in Sudan are part of the Nile system. There are also numerous wadis, or intermittent streams, which flow only part of the year. Some drain into the Nile during the rainy season and stand empty at other times. Others drain into swamps that have no outlet to a river or disappear into the sands of an inland basin during the dry months. For example, the Wadi Howar and the Wadi Al-Ku, both originating in the Teiga Plateau region, disappear into the desert. Of similar origin, the Wadi Azum eventually reaches the Lake Chad drainage system to the west. Some of these intermittent streams carry large amounts of water during the rainy season and support local areas of agriculture. The Mareb—also known as the Gash, or Al-Qāsh in Sudan—and Baraka Rivers flow into northeast Sudan from the Eritrean highlands during the months of July, August, and September. The Mareb River provides water for important irrigation schemes north of Kassalā, and the Baraka feeds the Tawkar delta near the Red Sea Coast. The Bahr el Arab in southwestern Sudan is another important seasonal river.

Wetlands

Permanent swamps in the southern provinces and Upper Nile cover about 50,000 sq mi (129,500 sq km) where there is an excess of water for most of the year. This phenomenon is best characterized by the Sudd, a vast region of swamps and marshes covering an area of about 3,000 sq mi (7,770 sq km) and extending from Boma National Park several hundred miles northwestward to the Al-Ghazāl River, ending at the Machar Marshes near the Ethiopian border. The vast swamp and marsh area is as monotonous as the featureless plains farther north, but there is considerable variety of terrain and vegetation in the uplands south of the swamps, particularly near the Uganda and Kenya borders. The largest swamp in the Sudd, Badigeru Swamp, is located between the Al-Jabal and Boma National Park. Located in the southeast corner at the junction with Kenya and Ethiopia is Lotagipi Swamp.

GEO-FACT

Nubia, home to Africa's earliest black civilization (traced to 3100 BC), is an ancient designation for the region comprised by modern-day southern Egypt and northern Sudan; much of the territory was submerged under Lake Nasser when the Aswan Dam was built on the Nile River. The capital of Nubia in 600 BC was Meroe, near modern-day Khartoum. Pyramids, smaller and different in design from those in Egypt, still exist in the Nubian region of Sudan. The largest measures about 170 ft (51 m) at the base.

THE COAST, ISLANDS, AND THE OCEAN

Sudan faces the Red Sea, a major shipping route between the Mediterranean Sea and the Gulf of Aden and Indian Ocean. At its widest point it is only 205 mi (326 km). The Red Sea is rather deep, with an average depth of 1,640 ft (500 m). It reaches a maximum depth of 6,562 ft (2000 m) and features red coral reefs and extensive coral gardens. Natural harbors exist at Port Sudan (Bur Sudan) and Sawākin.

CLIMATE AND VEGETATION

Temperature

Sudan has an equatorial climate. The northern plains and desert region are hot and dry with maximum temperatures reaching 108°F (42°C) from March to June. November to February are the coolest months with average temperatures of 90°F (32°C), but sometimes going down to as low as 40°F (4.4°C) at night. Average temperatures in the central and southern regions are 80°F (26.7°C) and 85°F (29.4°C) respectively.

Rainfall

Rainfall increases from north to south. From less than 4 in (10 cm) in the north, the southern regions receive 30 to 50 in (76 to127 cm) of rain during the southern six-to-nine month wet season and produce a rich variety of tall grasses, shrubs, and trees. The lush vegetation in the south contrasts sharply with the deserts of Northern Province, where the occasional rains vanish in the parched sand and broad areas are devoid of either vegetation or people.

Savanna Grasslands

Much of Sudan's vegetation is a typical savanna mixture of grasses, thorny shrubs (sometimes called scrub), and scattered short trees. The vegetation varies from a lush mixture in the south, where rainfall is relatively heavy for as much as nine months of the year, to sparse grasses and shrubs on sandy soils in the region near Khartoum. Grass also covers the higher steppe region of the southeast.

Deserts

A line running east to Atbara and Port Sudan from the western frontier at 16° north latitude defines the approximate southern limit of desert, which covers the northern quarter of Sudan. The Libyan Desert extends into Sudan from the northwest; to the northeast the Nubian Desert covers the area between the Nile and the Red Sea Hills. These deserts are part of the larger Sahara Desert. From the confluence of the White and Blue Nile rivers near Khartoum, the Upper Nile winds northward through this desert area for a distance of 800 mi (1,287 km) inside Sudan and provides the only water for the narrow strips of cultivation along the riverbanks. In the area from Atbara to Wadi Halfa on the Egyptian frontier almost no rain falls; Wadi Halfa is often completely rainless for years at a time. The settlements along the Nile depend for their livelihood on various types of irrigation or inundating.

The hinterland west of the Nile supports only a few Arab nomads who, with their camels, sheep, or goats, cover great expanses of the parched country in search of grazing, usually south of 18° to 19° north latitude, where a little rain occurs during most years and grass or browse springs to life. Water is available only in scattered oases, such as Al Atrun in the western desert and Well No. 6 on the railway between Wadi Halfa and Abu Hamand. Terrain in this northern desert consists of broad areas of sand and flintrock with occasional hills and outcroppings of basalt, granite, and limestone, often surrounded by banks of sand deposited by the wind.

Plains

The topography of the country outside the mountains and the Nile valley is generally devoid of contrast, and the flat plain making up most of its huge area is distinguishable more by range of vegetation than by peculiarities of terrain. The plain, extending some 500 to 600 mi (804 to 965 km) from east to west and more than 1,000 mi (1,609 km) in its north-south axis, is a part of the broad savanna belt that begins at the southern edge of the Sahara Desert and extends across the African continent. For thousands of square miles the only features relieving the monotony of the Sudanese plain are low rolling hills, sometimes referred to locally as mountains, or extensive sand dunes created thousands of years ago and partially or entirely fixed by vegetation. Soils are composed mainly of clay, much of which is impermeable and difficult to cultivate, or sand that contains little clay or humus.

Population Centers – Sudan

(1993 POPULATION ESTIMATES)

Name	Population	Name	Population
Khartoum (capital)	924,505	Omdurman	228,778
Port Sudan	305,385	El Obeid	228,096
Kassalā	234,270	Medani	218,714

SOURCE: Population of Capital Cities and Cities of 100,000 and More Inhabitants." United Nations, Statistics Division.

States – Sudan

2000 POPULATION ESTIMATES

Name	Population	Area (sq mi)	Area (sq km)	Capital
Bahr al Jabal	2,256,942	8,863	22,956	Juba
Blue Nile State	633,129	17,700	45,844	Al-Damazin
East Equatorial	1,234,486	31,869	82,542	Kapoita
Gedarif	1,414,531	29,059	75,263	Gedarif
Gezira State	3,310,928	9,024	23,373	Wad Medani
Jungoli	n.a.	46,781	121,164	Bor
Kassala	1,433,730	14,174	36,710	Kassalā
Khartoum	4,740,290	8,549	22,142	Khartoum
Lakes State	n.a.	15,535	40,235	Rumbek
Noirth Kordufan	1,439,930	71,545	185,302	El Obeid
North Bahr al Gazal	n.a.	12,957	33,558	Awil
North Darfur	1,409,894	114,448	296,420	Al-Fashir
Northern	578,376	134,633	348,697	Dongula
Red Sea State	709,637	84,912	219,920	Port Sudan
River Nile State	895,893	47,152	122,123	Al-Damar
Sennar	1,132,758	14,612	37,844	Sennar
South Darfur	2,708,007	49,151	127,300	Nyala
South Kordufan	1,066,117	30,683	79,470	Kadugli
Unity	n.a.	139	360	Bantio
Upper Nile State	1,342,943	30,028	77,773	Malakal
Warap	n.a.	11,980	31,027	Warap
West Bahr al Gazal	n.a.	36,255	93,900	Wau
West Darfur	1,531,682	30,680	79,460	Geneina
West Equatorial	n.a.	30,398	78,732	Yambio
West Kordufan	1,078,330	43,001	111,373	Al-Fula
White Nile State	1,431,701	11,742	30,411	Rabak

SOURCE: "States of the Republic." *Facts About Sudan.* Sudan Ministry of Foreign Affairs.

Forests and Jungles

In the central and northern parts of the country, acacia and desert scrub dot the semi-arid and desert landscape. Areas of closed, broad-leaf tropical forests and rain forests are found in the well-watered southwest, and the Imatong and Dongotona Mountain Ranges have considerable rainforest coverage.

HUMAN POPULATION

Sudan has a very low population density—only 37.6 people per square mi (14.5 people per sq km). However, most of the population lives in or around the capital city of Khartoum and the surrounding Nile River valley. A large portion also lives in the southern mountain regions. Much of the rest of the country is sparsely populated or uninhabited, especially the desert. In 2000, about 36 percent of the population lived in urban areas.

NATURAL RESOURCES

Eighty percent of the population works in the agriculture business exploiting Sudan's ability to grow sugar, cotton, and other textiles. Sudan's climate is also suitable for the growth of wheat, sesame, mangos, and peanuts. Besides relying heavily upon agriculture, Sudan also has small reserves of other natural resources. It currently produces 185,000 barrels per day of petroleum, most of which it exports as crude. The country also has small reserves of mineral deposits, such as iron ore, copper, chromium ore, zinc, tungsten, mica, silver, and gold. Due to dams and the cataracts of the Nile, Sudan is also able to exploit hydropower.

FURTHER READINGS

Africa South of the Sahara 2002. Sudan. London: Europa Publishers, 2001.

ArabNet. *Sudan.* http://www.arab.net/sudan/sudan_contents.html (Accessed May 2002).

LexicOrient. *Sudan.* http://lexicorient.com/m.s/sudan/index.htm (Accessed May 2002).

Lobban, Jr. Richard A., Robert S. Kramer, and Carolyn Fluehr-Lobban. *Historical Dictionary of the Sudan.* 3rd Ed. Metuchen, N.J.: Scarecrow, 2002.

Moorehead, Alan. *The Blue Nile.* New York: HarperTrade, 2000.

Williams, Martin A.J. and D.A. Adamson. *Land Between Two Niles*: Quaternary Geology and Biology of the Central Sudan. Salem, N.H.: MBS, 1982.

Suriname

- **Area:** 63,038 sq mi (163,270 sq km) / World Rank: 92
- **Location:** Located in the Northern and Western Hemispheres, on the northern part of the South American continent, bordered by the Atlantic Ocean on the north, French Guiana on the east, Brazil on the south, Guyana on the west
- **Coordinates:** 4°00′N, 56°00′W
- **Borders:** 1,058 mi (1,707 km) total / Brazil, 371 mi (597 km); French Guiana, 317 mi (510 km); Guyana, 372 mi (600 km)
- **Coastline:** 239 mi (386 km)
- **Territorial Seas:** 12 NM
- **Highest Point:** Juliana Top, 4,034 ft (1,230 m)
- **Lowest Point:** Sea level
- **Longest Distances:** 411 mi (662 km) NE-SW / 303 mi (487 km) SE-NW
- **Longest River:** Courantyne River, 475 mi (764 km)
- **Natural Hazards:** None
- **Population:** 433,998 (July 2001 est.) / World Rank: 164
- **Capital City:** Paramaribo, located on the Atlantic coast
- **Largest City:** Paramaribo, 112,000 (2000 est.)

OVERVIEW

The smallest independent country on the South American continent, Suriname is divided into several distinct natural regions: a coastal plain, a region of forested mountains, and high savanna in the southwest. Of the three, the mountains are by far the largest, covering roughly three-quarters of the country. Seven significant rivers run through the country, all flowing into the Atlantic Ocean in the north.

Located near the equator, Suriname's climate is tropical, and rainfall varies throughout the different regions. The daily trade winds that blow in from the Atlantic Ocean influence the country's temperatures. Suriname is located on the South American Tectonic Plate.

MOUNTAINS AND HILLS

The mountainous rain forest region that covers 75 percent of Suriname has only been partially explored. It consists of a number of chains, with the terrain gradually rising to the country's highest elevation, Juliana Top (4,034 ft / 1,230 m), in the Wilhelmina Gebergte at the center of the country. The Van Asch Van Wijck Mountains make up the rest of the central mountain chain, which is connected to the Tumuc-Humac Mountains along the Brazilian border by the southern Eilerts de Haan Mountains. Other ranges include the Kayser and Bakhuis Mountains in the west and the Oranje and Lely Mountains in the east.

INLAND WATERWAYS

Lakes

W. J. van Blommestein Lake, the largest in Suriname, is man-made. It is a result of the Afobaka Dam, which was built in the 1960s across the Surinam River in the east-central region. The dam generates electricity for the processing of bauxite, one of the country's natural resources.

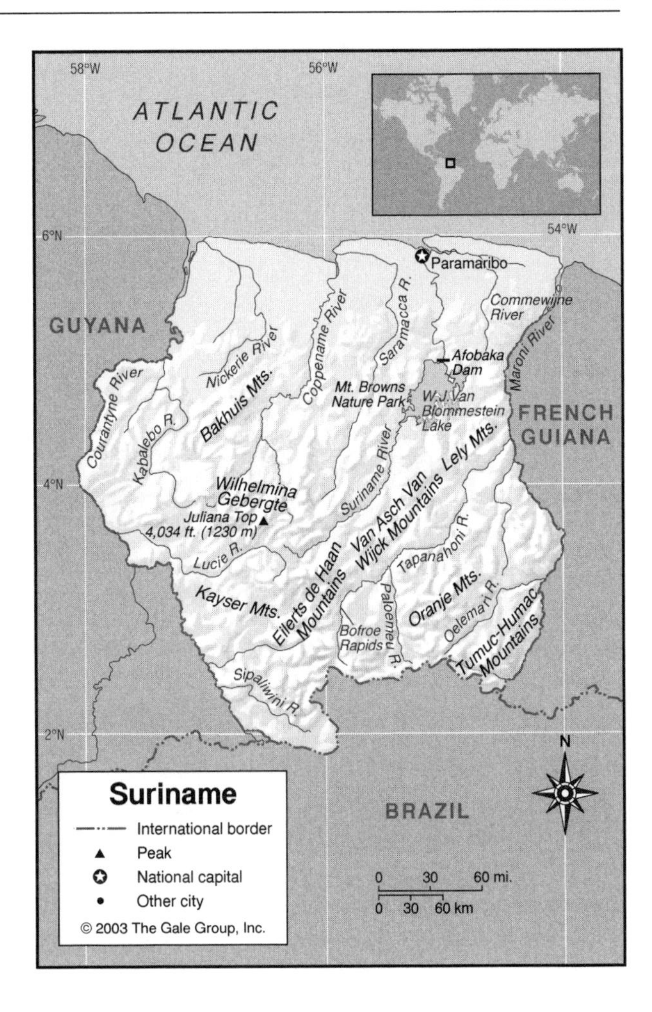

Rivers

Numerous rivers dissect the land, all interconnected by a remarkable system of channels. In the central part of the country the principal rivers are the Nickerie, the Coppename, the Saramacca, the Suriname, and the Commewijne. The largest river in the country is the Courantyne (Corantjin), which marks the border with Guyana. Major tributaries of the Courantyne in Suriname are the Sipaliwini, Lucie, and Kabalebo. Along the eastern border with French Guiana is another large river, the Maroni, with its tributaries the Tapanahoni, Paloemeu, and Oelemari. All of the rivers flow northward into the Atlantic Ocean, with many rapids and waterfalls.

Wetlands

Large portions of the coastal plain are swampland, as most of this area lies at sea level. Mud banks and other deposits from slow-moving rivers in their delta stage also contribute to the swamps. Some of these swamps have been drained to make land available for farming.

THE COAST, ISLANDS, AND THE OCEAN

The Atlantic Ocean is located along Suriname's northern region. The shape and make-up of the coastline constantly changes slowly because of the deposits from Suriname's numerous rivers. Ocean currents and wind push the river deposits to form unevenly shaped mud banks and ridges along the coast.

CLIMATE AND VEGETATION

Temperature

Suriname's climate is tropical and moist, as the country is located near the equator. Temperatures range from 82-90°F (28-32°C) during the day, and nighttime temperatures are as low as 70°F (21°C), which is the result of the northeast trade winds that blow in from the Atlantic all year.

Rainfall

Annual rainfall in Paramaribo, the capital city, is approximately 90 in (230 cm). Most rainfall occurs in the mountains in the southern region, and it varies along the coast. Annually, the western region receives 76 in (193 cm), while the eastern area receives 95 in (241 cm). Suriname experiences two wet seasons, and two dry. A long rainy season occurs from April to August, and is followed by a long dry season from August to November. Another rainy season occurs from December to February, but it is shorter and less rainy, and is followed by a shorter, drier season from February to March.

Grasslands

The coastal plains in the north cover about 16 percent of the country. Much of this is swampland, but some areas have been drained for fertile farmland. In the far south, past the mountain ranges, grassy savannas are scattered throughout the forests.

Forests and Jungles

Approximately 80 percent of the country is covered by tropical rain forest. This is essentially all of the country south of the coastal plains, with the exception of some small savannas in the south. The rain forest is considered is one of the best-preserved on earth. In the late 1990s the Central Suriname Wilderness Nature Reservation was created, setting aside about 10 percent of the country as a protected area.

HUMAN POPULATION

A little over half of the population lives in cities. The coastal plain, where the capital is located, supports a dense population because of its moderately fertile soils. Nearly two-thirds of the population lives within 50 mi (80 km) of the capital city of Paramaribo. Other cities tend to be located along the banks of Suriname's rivers.

NATURAL RESOURCES

Suriname is one of the world's leading producers of bauxite, a claylike ore that is the source of aluminum,

Administrative Districts – Suriname

Name	Area (sq mi)	Area (sq km)	Capital
Brokopondo	2,843	7,364	Brokopondo
Commewijne	908	2,353	Nieuw Amsterdam
Coronie	1,507	3,902	Totness
Marowijne	1,786	4,627	Albina
Nickerie	2,067	5,353	Nieuw Nickerie
Para	2,082	5,393	Onverwacht
Paramaribo	71	183	Paramaribo
Saramacca	1,404	3,636	Groningen
Sipaliwini	50,412	130,567	Paramaribo
Wanica	171	442	Lelydorp

SOURCE: *Geo-Data: The World Geographical Encyclopedia*, 2nd ed. Detroit: Gale Research, 1989.

with large deposits in the east-central region. The country's other natural resources are timber, hydropower, fish, kaolin, shrimp, gold, as well as small amounts of nickel, copper, platinum, and iron ore.

FURTHER READINGS

Goslinga, Cornelis C. *A Short History of the Netherlands Antilles and Suriname.* Norwell, MA: Kluwer Academic Press, 1978.

Hoefte, Rosemarijn. *Suriname.* Santa Barbara: Clio Press, 1990.

LonelyPlanet.com. *Suriname.* http://www.lonelyplanet.com/destinations/south_america/suriname/ (accessed March 6, 2002).

Wooding, Charles J. *Evolving Culture: A Cross-cultural Study of Suriname, West Africa, and the Caribbean.* Washington, D.C.: University Press of America, 1981.

"World Watch: Paramaribo." *Time International,* June 29, 1998, Vol. 150, No. 44, p. 14.

Swaziland

- **Area:** 6,704 sq mi (17,363 sq km) / World Rank: 156
- **Location:** Southern and Eastern Hemispheres, southern Africa, between Mozambique and South Africa
- **Coordinates:** 26°30′ S, 31°30′ E
- **Borders:** 332 mi (535 km) / Mozambique 65 mi (105 km); South Africa 267 mi (430 km)
- **Coastline:** None
- **Territorial Seas:** None
- **Highest Point:** Emlembe, 6,109 ft (1,862 m)
- **Lowest Point:** Great Usutu River, 69 ft (21 m)
- **Longest Distances:** 109 mi (176 km) N-S; 84 mi (135 km) E-W
- **Longest River:** Great Usutu River, 135 mi (217 km)
- **Natural Hazards:** None
- **Population:** 1,104,343 (July 2001 est.) / World Rank: 152
- **Capital City:** Mbabane, located in west-central region
- **Largest City:** Mbabane, population 67,200 (2002 est.)

OVERVIEW

Landlocked Swaziland is located in Southern Africa, nearly surrounded by South Africa. It is topographically part of the South African Plateau and is divided west to east into four well-defined regions of nearly equal breadth. From the high veld in the west, averaging 3,500 to 3,900 ft (1,050 m to 1,200 m) in elevation, there is a step-like descent eastward through the middle veld 1,475 ft to 1,970 ft (450 m to 600 m) to the low veld 490 to 980 ft (150 m to 300 m). To the east of the low veld is the Lebombo Range 1,475 to 2,700 ft (450 to 825m), which separates the country from the Mozambique coastal plain. Swaziland is slightly smaller than the state of New Jersey.

MOUNTAINS AND HILLS

Mountains

On the west side of the country is the high veld, which rises to 6,070 ft (1,850 m). Mt. Emlembe is located on the northwestern border with South Africa. Lion's cavern, a cavern containing an ancient mine, is found in the northwest. In the east, the Lebombo Mountains offer an undulating plateau rising high above the Lebombo plain from a striking escarpment.

Plateaus

Swaziland occupies the eastern edge of the South African plateau where it breaks apart and drops to the Mozambican plain on thc Indian Ocean.

INLAND WATERWAYS

Rivers

Swaziland is well-watered with four large rivers flowing eastward across it into the Indian Ocean. These are the Komati (source in South Africa) and the Mbuluzi (or Umbeluzi) Rivers in the north, the Great Usutu (or Lusutfu) River (source in South Africa) in the center and, the Ngwavuma River in the south.

Swaziland's highest waterfall, Malolotja Falls (about 3,000 ft/100 m high), is found in the Malolotja Nature Reserve about 12 mi (19 km) northwest of Mbabane. The reserve, at almost 5,000 ft (1,500 m) elevation, has more than twenty waterfalls as the Malolotja River flows down

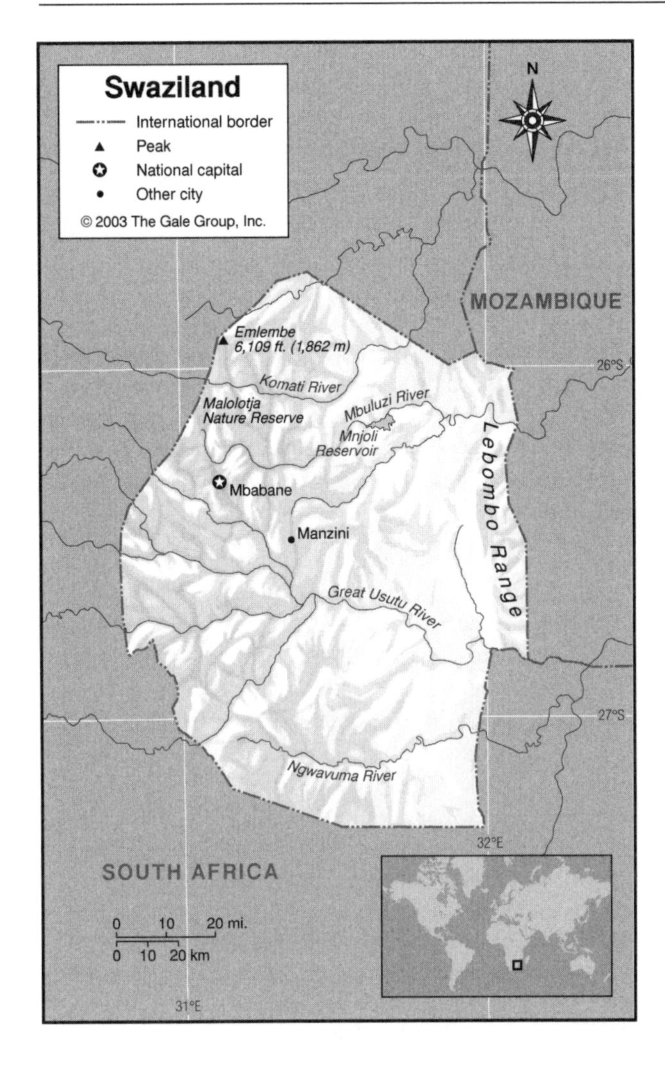

from the highest elevations to join the Komati River at about 3,000 (900m).

THE COAST, ISLANDS, AND THE OCEAN

Swaziland is landlocked.

CLIMATE AND VEGETATION

Temperature

Temperatures vary from as low as 27°F (-3°C) on the highlands in winter, to 108°F (42°C) in summer in the low veld. Temperatures rise and the climate warms as altitude drops. In Mbabane the average temperature ranges from 43 to 66°F (6 to 17°C) in June to 59 to 77°F (15 to 25°C) in January or February

Rainfall

The high veld region has a humid temperate climate and receives 55 in (140 cm) of rain annually. The Lebombo plain and middle veld are warmer and drier and receive only about 33 in (85 cm) per year. The nearly

Population Centers – Swaziland

(1997 POPULATION ESTIMATES)

Name	Population
Manzini	50,000
Mbabane (capital)	40,000

SOURCE: Projected from United Nations Statistics Division data.

Districts – Swaziland

POPULATIONS FROM 1997 CENSUS

Name	Population	Area (sq mi)	Area (sq km)	Capital
Hhohho	247,539	1,378	3,569	Mbabane
Lebombo	190,617	2,296	5,947	Siteki
Manzini	276,636	1,571	4,068	Manzini
Shiselweni	198,084	1,459	3,780	Nhlangano

SOURCE: Common Country Assessment, 1997, Swaziland .

tropical low veld receives an average of 24 in (60 cm) of rain annually. The wettest period of the year is from October to March when violent rainstorms may occur.

Grasslands

Swaziland is covered almost entirely by grasslands, savanna, and mixed scrub.

Forests

Swaziland's high veld has the largest man-made forests of conifers and eucalyptus in Africa.

HUMAN POPULATION

The Nkosi Dlamini, the royal clan, dominates the more than 70 clans in the country. The population is growing slightly and is densest on the middle veld, with more than 150 people per sq mi (58 people per sq km). About 36 percent of the population is urban, but only three cities have populations that are exceed 20,000. Protestants comprise 55 percent of the population, Muslims 10 percent, Roman Catholics 5 percent, and indigenous beliefs 30 percent. English is the official language for government business, but siSwati is widely spoken.

NATURAL RESOURCES

With its weather and land, most of the country is suited for agriculture, and agriculture makes up the lives of nearly two-thirds of the population. The forests of the high veld offer good sources of lumber. Corn, cotton, and citrus fruits are grown, but the primary exports are sugar, wood pulp, and soft-drink concentrate. Swaziland has great potential for hydropower with its many rivers and the Mnjoli Reservoir. The country is also home to large coal reserves, deposits of asbestos, low grade iron ore, clay, cassiterite, and gold in the northwest. Diamond deposits used to provide income but the easily accessible reserves have been depleted.

FURTHER READINGS

Africa South of the Sahara 2002. "Swaziland." London: Europa Publishers, 2001.

Booth, A.R. *Historical Dictionary of Swaziland.* Metuchen, N.J.: Scarecrow Press, 1975.

Gills, D.H. *The Kingdom of Swaziland.* Westport, CT: Greenwood Publishing Group, 1999.

World Gazetteer. *Swaziland.* http://www.world-gazetteer.com/t/t_sz.htm (Accessed June 6, 2002).

Sweden

- **Area:** 173,732 sq mi (449,964 sq km) World Rank: 56
- **Location:** Northern and Eastern Hemispheres, in Scandinavia, Northern Europe, bordering Norway to the west and north, Finland to the northeast, Baltic Sea in the east and south, North Sea in the southwest.
- **Coordinates:** 62°00′N, 15°00′E
- **Borders:** 1,370 mi (2,205 km) / Finland, 364 mi (586 km); Norway, 1,006 mi (1,619 km)
- **Coastline:** 2000 mi (3,218 km)
- **Territorial Seas:** 12 NM
- **Highest Point:** Kebnekaise, 6,926 ft (2,111 m)
- **Lowest Point:** Sea level
- **Longest Distances:** 978 mi (1,574 km) N-S 310 mi (499 km) E-W
- **Longest River:** Göta, 447 mi (720 km)
- **Largest Lake:** Lake Vänern, 1,387 sq mi (3,593 sq km)
- **Natural Hazards:** Ice floes in the Baltic Sea
- **Population:** 8,875,053 (July 2001 est.) / World Rank: 82
- **Capital City:** Stockholm, located near the southeast coast with the Baltic Sea
- **Largest City:** Stockholm, 755,305 (2001 est.)

OVERVIEW

The largest of the Scandinavian countries and the fourth-largest country in Europe, Sweden is one of the countries located farthest from the equator. It extends from north to south at roughly the same latitude as Alaska, with about 15 percent of its total area situated north of the Arctic Circle. The entire western and northern shores of the Baltic Sea and its large inlet the Gulf of Bothnia are along the Swedish coast. Two large islands in the Baltic, Gotland and Öland, are Swedish territory. The Baltic is connected to the North Sea by the straits that separate Sweden from Denmark to the south and west. Sweden's other borders are with Norway in the west and north, and Finland in the northeast.

The most notable of Sweden's geographical features is its length—indeed, the Swedes speak of *vart avlanga land* (our long, drawn-out land). It shares this and many other features with its western twin in Scandinavia, Norway, but Sweden is a land of lower altitudes and less dissected relief than Norway.

Four topographical divisions can be discerned in the country, although they are of unequal size. The largest is Norrland, the northern three-fifths of Sweden. Characterized by a landscape of hills and mountains, forests, and large river valleys, it stretches roughly from the lower reaches of the Dal River northward. Svealand, or central Sweden, constitutes the second region; it is made up of lowlands dotted with thousands of lakes, including the country's largest, Lake Vänern. Småland in the south is the third region, it is an area of forested hills. The fourth region is in southernmost part of the country and is known as Skåne (Scania), a continuation of the fertile plains of Denmark and northern Germany.

MOUNTAINS AND HILLS

Mountains

Norrland, the northern region of Sweden, covers about 60% of Sweden's territory, and includes the areas of highest elevation. The western highlands of Norrland follow the Norwegian frontier and rise to elevations of over 6,000 ft (1,818 m), of which the highest is Kebnekaise 6,926 ft (2,111 m). The terrain slopes to the southeast, away from the Kölen (Kjølen) Mountains along the border with Norway, to the Gulf of Bothnia. The flow of rivers in this region have incised the surface and leveled much of the terrain to a plateau. There are a number of small icefields in the far northern reaches above 66° N latitude.

Plateaus

Fulufjäll, a 22-mi (35-km) long, 9-mi (15-km) wide sandstone plateau in the center of the country near the Norwegian border, rises to 3,300 ft (1,000 m), with steep slopes and forested ravines all around it.

Småland

Småland in southeastern Sweden is an area of low highlands, with elevations generally less than 500 ft (152 m). It separates the plains of Skåne in the southernmost part of the country from the more extensive lowlands of Svealand to the north.

INLAND WATERWAYS

Lakes

Sweden has nearly 100,000 lakes. They are found throughout the country, but central Sweden in particular is a shatter zone of lakes and plains. The four largest lakes in the country are found here: Vänern, Vättern, Hjälmaren, and Mälaren. Vänern (1,387 sq mi / 3,593 sq km) and Vättern (738 sq mi / 1911 sq km) are among the four largest lakes in Europe. Vänern has an outlet to the west by way of the Göta River. It claims Sweden's largest catchment area. The Trollhättan Falls (108 ft / 33 m) on the Göta River are indicative of the change in level between Vänern and the lowlands along the Skagerrak in the west. Lake Mälaren (440 sq mi / 1140 sq km) lies only about 2 ft (0.6 m) above the average level of the Baltic Sea. The capital city of Stockholm is located along the strait that connects the lake to the sea. Archaeological evidence suggests that this lakes and plains region was the core of early Swedish settlements.

The depressions of the Norrland region are filled by lakes, most of which lie somewhat more than 1,000 ft (305 m) above the level of the Baltic. The largest of these, located in the Western Highlands, are the Torn Träsk (122 sq mi / 317 sq km) in the north, the Storsjön (176 sq mi / 456 sq km) in the south, and between them, the interconnected trio of Hornavan, Uddjaur, and Storavan (255 sq mi / 660 sq km).

The largest lake in southern Sweden, lying at 469 ft (142 m) above sea level with a depth of 111 ft (37 m), is Lake Bolmen (71 sq mi / 184 sq km).

Rivers

The rivers flowing in Norrland (northern Sweden) include the Torne, the Lule, the Skellefte, the Göta, the Ume (and its tributary, the Vindel), the Ångerman, the Ljungan, and the Dal. All flow generally southeast from the high elevations along the border with Norway, frequently punctuated by waterfalls and rapids, until they flow outward to the Gulf of Bothnia. The Torne and its tributaries form the border with Finland. The Göta River cuts through rocky back country into the lowlands of Svealand. The Klar, flowing south from Norway to Lake Vänern, has been used by foresters for floating logs for decades and is a favorite spot for recreational rafting.

The rivers flowing in the southern and western part of the country are shorther than those of the north. They include the Viskan, Ätran, Nissan, and Lagan, all well known for their abundant salmon. A second Göta River drains Lake Vänern into the Kattegat. It is part of the so-called Göta Canal. Built in the early 1800s, this is a 383 mi (613 km) long waterway from Göteborg to Stockholm, formed by linking lakes and other natural waterways with a series of canals. The system never had any real economic purpose, and is now used primarily by tourists. Several dozen locks compensate for the 330-ft (90-m) difference in elevation between the two cities.

Wetlands

North of the Arctic Circle lies a region of wetland and tundra landscape, with large peat marshes covering 40 percent of the land.

Population Centers – Sweden

(2001 POPULATION ESTIMATES)

Name	Population	Name	Population
Stockholm (capital)	754,948	Västerås	127,799
Göteborg (Gothenburg)	471,267	Örebro	124,873
Malmö	262,397	Norrköping	122,896
Uppsala	191,110	Helsingborg	118,512
Linköping	134,039	Jönköping	117,896

SOURCE: Statistics Sweden.

Countiess – Sweden

2001 POPULATION ESTIMATES

Name	Population	Area (sq mi)	Area (sq km)	Capital
Blekinge	150,017	1,178	3,051	Karlskrona
Dalarna	277,010	11,736	30,396	Falun
Gävleborg	278,171	7,615	19,722	Gävle
Gotland	57,412	1,225	3,173	Visby
Halland	276,653	2,205	5,711	Halmstad
Jämtland	128,586	20,925	54,197	Östersund
Jönköping	327,824	4,344	11,252	Jönköping
Kalmar	234,697	4,504	11,666	Kalmar
Kronoberg	176,582	3,639	9,424	Växjö
Norrbotten	254,733	40,883	105,886	Luleå
Örebro	273,137	3,591	9,301	Örebro
Östergötland	412,363	4,490	11,629	Linköping
Skåne	1,136,571	4,381	11,346	Malmö
Södermanland	257,220	2,541	6,581	Nyköping
Stockholm	1,838,882	2,620	6,785	Stockholm
Uppsala	296,627	2,773	7,183	Uppsala
Värmland	273,933	7,479	19,371	Karlstad
Västerbotten	254,818	22,853	59,194	Umeå
Västernorrland	245,078	8,932	23,134	Härnösand
Västmanland	257,957	2,549	6,603	Västerås
Västra Götaland	1,500,857	9,785	25,343	Göteborg

SOURCE: Statistics Sweden.

THE COAST, ISLANDS, AND THE OCEAN

Oceans and Seas

Lying to the east and south is the Baltic Sea, which is linked to the North Sea by narrow and shallow straits. The channel of water separating Denmark and Sweden and linking the Kattegat with the Baltic Sea is Öresund. The Kattegat lies along the southwest shore of Sweden; as it reaches the northernmost extent of Denmark, the Kattegat Strait flows into the Skagerrak, a triangular body of water that lies between Norway, Sweden, and Denmark. The Kattegat and Skagerrak are considered part of the North Sea. The Gulf of Bothnia, between Sweden and Finland, is the northernmost extension of the Baltic Sea.

Major Islands

Like other Scandanavian countries, Sweden has many islands. The archipelago of Stockholm shows the most intense concentration of islands, the outermost of which are separated from their Finnish counterparts by the Åland Sea. In contrast, the western coast archipelago of Bohusian is a skerry (rocky reef) zone where the ice, waves, and winds have left the skerries bald in appearance.

Of all the Swedish islands, Gotland (1,225 sq mi / 3,173 sq km) is the largest, and occupies a special and central place. Although it has a plateau appearance and is skirted with limestone cliffs, it has some of the finest beaches in the Baltic. Its principal town is Visby. Öland Island, not far off of Sweden's southeastern coast, is the second largest island at 519 sq mi (1,344 sq km).

The Coast and Beaches

The Bothnian coastal plain merges almost imperceptibly into the sea. Both the littoral and estuaries are crowded with islands. The Bothnian coast may be divided into three sections—lower, middle, and upper—of which the middle extends from Örnsköldsvik to Skellefteå. The area around Örnsköldsvik is designated as the High Coast, a UNESCO World Heritage site because of the geological process of uplift that is ongoing there. After the ice retreated from Sweden 9,600 years ago, geologists believe the land was about 940 ft (285 m) lower than it is today. In some areas, the land rises as much as 3 ft (nearly 1 m) per century.

Skäne differs from much of Sweden not merely in its structure and geological history but also in the fact that its coast is free of islands. The shore in the south is characterized by steep and rocky cliffs.

The island of Gotland has some of the finest beaches in the Baltic region; it is also rimmed in places by dramatic limestone cliffs.

CLIMATE AND VEGETATION

Temperature

Because of the influence of the ocean current known as the North Atlantic Drift and the prevailing air currents, Sweden's average temperatures are warmer than similar northern countries that lie further inland. In winter, the average temperature in southern Sweden in winter is 26°F (-3°C); in summer, 64°F (18°C). Norrland (northern Sweden) is much colder, with a winter season that extends for up to eight months, with snow remaining on the ground for about six months.

Rainfall

Annual rainfall averages 24 in (61 cm). The western part of the country along the border with Norway experiences the country's heaviest rainfall.

Grasslands

Throughout Svealand, the region dotted by numerous lakes, are extensive plains such as Uppland (centered on Uppsala), Västmanland, and Narke. Väster- and Öster-Götland (East and West Götland, not to be confused with the island of Gotland) are also grassland regions. South of Lake Vättern lie the faulted landscapes of Skäne, which, although fertile, and resembling the Danish plains across the Öresund, have areas of much more pronounced relief.

Tundra

The extreme north of Norrland is comprised of arctic tundra.

Forests and Jungles

Except for the northernmost arctic region, which is mostly treeless, much of Sweden's terrain is forested. Mountain birch forests cover a large part of Norrland; south of this region are coniferous forests, with Scots pine and Norway spruce. Further south stands of birch, aspen, beech, oak, ash, and elm are distributed across the territory. In the southwest, Norway spruce and pine predominate. The major forests surrounding Lake Vänern provide raw materials for the pulp and paper mills in the region.

HUMAN POPULATION

Most Swedes (about 85 percent) live in urban areas. The overall population density is 51 people per sq mi (20 people per sq km), but the vast majority of the population is concentrated in the southern two-fifths of the country, particularly in Svealand and Småland. The few population centers of the north tend to be found near the coast or on major rivers.

NATURAL RESOURCES

Significant iron and other metallic ore deposits are found in Norrland. In central Sweden, the region around Bergslagen developed as a center of Swedish industry because of other ore deposits. Sweden is a leading producer of arsenic, which is mined near the Skellefte River in Norrland. Sweden's vast forests support the large forest products industry.

FURTHER READINGS

Alderton, Mary. *Sweden.* London: A. & C. Black, 1995.

The Arctic Connection. *Arctic Sweden.* http://www.arcticconnection.com/Countries/sweden.shtml (Accessed July 18, 2002).

The High Coast. http://www.highcoast.net/ (Accessed July 18, 2002).

Frommer's . . .Sweden. New York: Macmillan, 1999.

Williams, Brian. *Guide to Sweden.* Jackson, Tenn.: Davidson, 2000.

Switzerland

- **Area:** 15,942 sq mi (41,290 sq km) / World Rank: 135
- **Location:** Northern and Eastern Hemispheres, in central Europe bordering Germany to the north, Liechtenstein and Austria to the east, Italy to the south and southeast, and France to the west and northwest.
- **Coordinates:** 47°00′N, 8°00′E
- **Borders:** 1,151 mi (1,852 km) / Austria, 102 mi (164 km); France, 356 mi (573 km); Italy, 460 mi (740 km); Liechtenstein, 25 mi (41 km); Germany, 208 mi (334 km)
- **Coastline:** None
- **Territorial Seas:** None
- **Highest Point:** Dufourspitze, 15,203 ft (4,634 m)
- **Lowest Point:** Lake Maggiore, 640 ft (195 m)
- **Longest Distances:** 216 mi (348 km) E-W; 137 mi (220 km) N-S
- **Longest River:** Rhine, 820 mi (1,320 km)
- **Largest Lake:** Lake Geneva, 224 sq mi (581 sq km)
- **Natural Hazards:** Floods; avalanches; landslides
- **Population:** 7,283,274 (July 2001 est.) / World Rank: 91
- **Capital City:** Bern, west-central Switzerland
- **Largest City:** Zürich, north-central Switzerland, population 984,000 (2000 est.)

OVERVIEW

Switzerland is a small, mountainous, landlocked country in Central Europe, famous for its picturesque Alpine vistas. It has three distinct geographical regions: the various branches of the Alps, comprising 60 percent of the country's total territory and extending over the southern part of the country; the Jura Mountains in the northwest (comprising 10 percent of the total territory); and the Mittelland (comprising 30 percent of the total territory) between. Switzerland is a federation of highly autonomous and ethnically, religiously, and linguistically distinct cantons. This structure has been in many ways influenced by the geography of the country, with villages and cantons cut off from each other by high mountains or deep valleys.

Switzerland is also an important European watershed—several of Europe's major rivers, including the Rhone and the Rhine, rise in the Swiss Alps. The majority of the population lives in the Mittelland, a plateau region of slightly rolling hills, meadowlands, winding valleys, and hundreds of lakes. Few villages are further than 15 mi (24 km) from a lake.

Switzerland lies on the Eurasian Tectonic Plate.

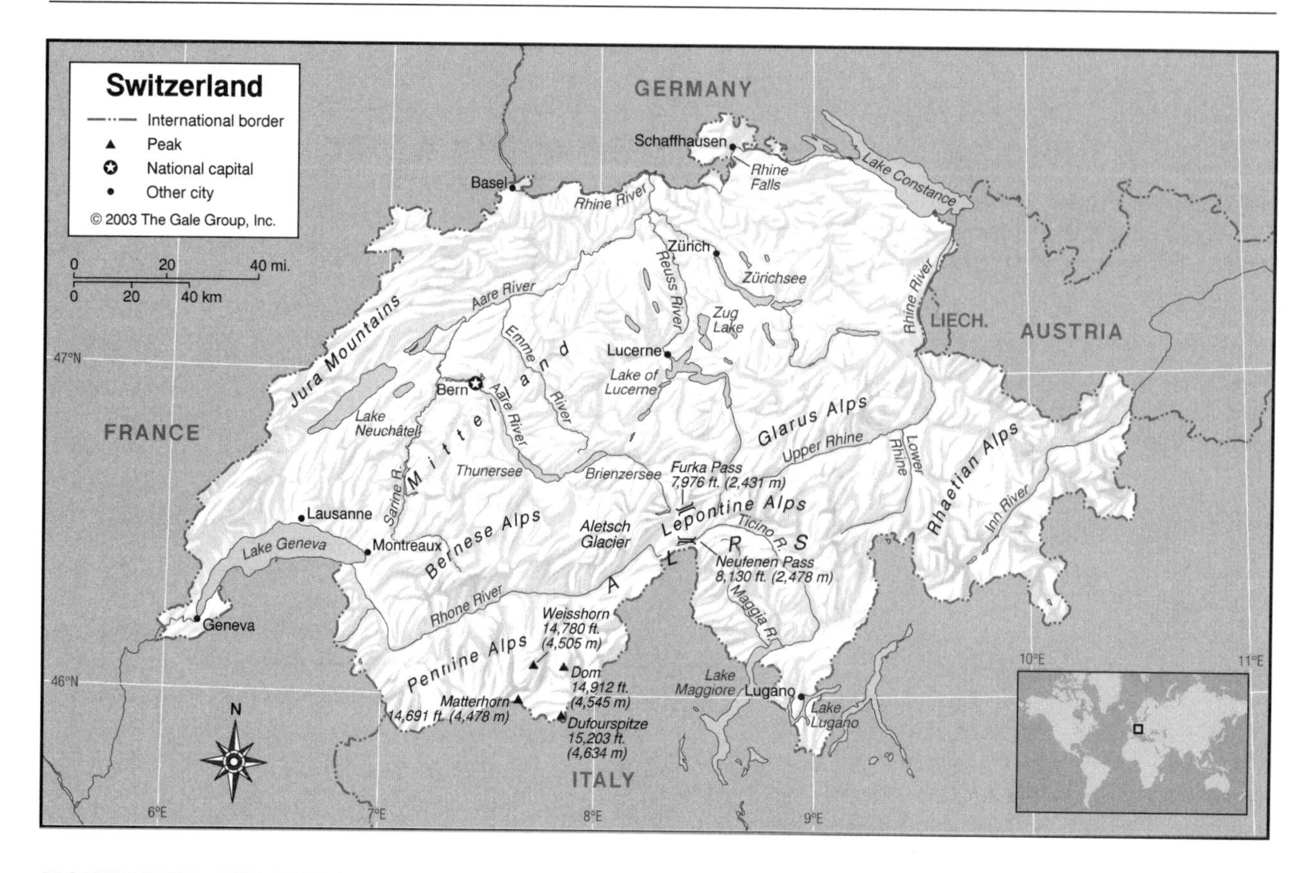

MOUNTAINS AND HILLS

Mountains

The Alps, the largest chain of mountains in Europe, cover three-fifths of Switzerland. The country contains the central part of the Alps, roughly one-fifth of the total range. The Swiss Alps are divided into different groups lengthwise by the Rhone and Rhine Valleys and crosswise by the Reuss and Ticino River Valleys. The main subdivisions are the Bernese Alps and Pennine Alps in the west, the Lepontine Alps in the center, and the Glarus Alps and Rhaetian Alps in the east. Their mean altitude is around 5,577 ft (1,700 m), but some 100 summits average 13,123 ft (4,000 m) or higher. Among the most dramatic peaks are the Dufourspitze on the Monte Rosa Massif, which is Switzerland's highest peak at 15,203 ft (4,634 m); the Weisshorn (14,780 ft / 4,505 m); the Dom (14,912 ft / 4,545 m); and the Matterhorn, the most famous Swiss peak, which has an elevation of 14,691 ft (4,478 m). All of these are located in the Pennine Alps.

The rugged Alpine landscape is rich and varied. Its slopes are cultivated up to about 5,000 ft (1,524 m). Forest predominates from that level to the tree line, which is above 7,000 ft (2,134 m), and pastureland from there until nearly 10,000 ft (3,048 m). Erosion has carved out valleys, terraces, and peaks whose continuing evolution makes the Alps a constantly changing region. In addition, there are more than 1,000 glaciers in the Alps, covering some 1,158 sq mi (3,000 sq km) in all. The largest is the Aletsch Glacier, which is 15 mi (24 km) in length. The Swiss Alps have more glaciers than the rest of the range. Although some of the glaciers are still advancing, the majority have been retreating since the twentieth century. The Alps are also the scene of some 10,000 avalanches per year, most of them occurring between February and April. Protection from this natural hazard is afforded by both forests and manufactured barriers. Measurements taken along the central region show that the Alps continue to grow by an average of one millimeter per year, this growth being balanced, however, by erosion. The subalpine (or "Pre-Alps") region on the northwest fringe of the Alps has a less complex structure than the main range. Many of its peaks reach heights of about 6,562 ft (2,000 m).

The Jura Mountains stretch across the northwestern part of the country, from Geneva in the west to Schaffhausen, and into western France. They form 160 mi (257 km) of the Swiss/French border. These mountains, covered by forest or pastureland, have a much less complex structure than the Alps and are also much lower. The mean altitude of these generally rounded mountains is 2,296 ft (700 m), but they include some peaks that rise to around 5,249 ft (1,600 m); the highest peak in the Swiss Jura is Mt. Tendre, which reaches 5,508 ft (1,679 m). The Jurassic Period was named for the Jura Mountains, whose many fossils date to that geological era.

Switzerland's central location in Europe has made it a transit point for European trade for centuries and over that time many mountain passes have been developed. Among the dozens of such passes, three of the longest are the Umbrail Pass from St. Maria to Bormio (Italy), which is 21 miles (33 km) long at an elevation of 8,202 ft (2,500 m); the Nufenen Pass from Ulrichen to Airolo is 22 mi (36 km) long at an elevation of 8,130 ft (2,478 m); and the Furka Pass from Gletsch to Hospental, at 7,976 ft (2,431 m) and 17 mi (28 km) in length.

Plateaus

At a mean altitude of 1,903 ft (580 m) the Mittelland, or Central Plateau, stretches from Lake Geneva to Lake Constance and forms a natural setting for settlement and commercial activity. The plateau was formed by the erosion from seas and lakes during the geological development of the Alps and the Jura Mountains. Vast amounts of sand, gravel, and pebbles brought down into hollows by the mountainside torrents were compressed to form new layers of rock. This fertile region is the country's agricultural heartland and home to most of its population.

Erosion has also created plateaus within the Jura Mountains, of which the most extensive is the Franches-Montagnes Plateau, which lies to the east of the border with France.

INLAND WATERWAYS

Lakes

Lakes are a striking feature of the Swiss landscape: no part of the country is farther than 9 mi (15 km) from a lake. A series of picturesque lakes stretches across the northern half of the country at the edges of the Mittelland and the subalpine region, from Lake Geneva (Lake Leman) in the west to Lake Constance (the Bodensee) in the northeast. They were formed during the Ice Age, when glacial ice or moraines created depressions and basins.

Lake Geneva is shaped like an arc, with the city of Montreaux at its easternmost point, and the city of Geneva at its westernmost. As the lake dips slightly southward toward Geneva, it is nearly bisected at a point the Swiss call the Nernier, dividing the lake into the Grand Lac (Large Lake) to the east, and the Petit Lac (Small Lake) to the south and west, the area immediately surrounding Geneva. With an area of 224 sq mi (581 sq km), Lake Geneva is Switzerland's largest lake, while Lake Neuchâtel, with an area of 83 sq mi (215 sq km), is the largest entirely within Swiss borders. At the far end of the Mittelland on the German border is Lake Constance, which is Switzerland's second largest lake at 208 sq mi (540 sq km). The other lakes in the Mittelland include Thunersee, Brienzersee, Lake of Lucerne, Zug, and Zürichsee. Lakes Lugano and Maggiore lie south of the Alps, at the Italian border. Switzerland also boasts of hundreds of smaller lakes, mainly in the Alps. There are few lakes in the Jura.

GEO-FACT

Though it has no ocean coast, two-thirds of Switzerland's boundary consists of natural barriers—either mountains, rivers, or lakes.

Rivers

Two of Europe's major rivers, the Rhone and the Rhine, have their sources in the Swiss Alps, within less than 20 mi (32 km) of each other. The Rhone originates from the Rhone Glacier in the Alps near Lake Geneva, from which it flows 505 mi (813 km) through France and into the Mediterranean Sea at Arles. The Rhone is a mostly mountainous river, cutting through numerous valleys in a wild course. The Rhine is one of the most important waterways in continental Europe. Its headwaters are in the Swiss Alps (at the confluence of the Upper Rhine and Lower Rhine Rivers) from which it flows 865 mi (1,391 km) to the North Sea. The river forms the border of Switzerland and Liechtenstein and Austria and then flows into Lake Constance on the border with Germany. The Rhine forms a nearly complete demarcation between Switzerland and Germany until it reaches Basel at which point it heads north out of Switzerland. The river flows south of the Swiss-German border briefly in the north-easternmost section of Zürich Canton where it forms Rhine Falls, a striking waterfall and one of the more dramatic in Europe. Of Switzerland's rivers, the Rhine has both the greatest total length and the greatest length within Swiss borders (233 mi / 375 km).

Other important rivers rise in the central Alps, including the Inn, which flows to Austria and the Danube; the Maggia, which originates from Lake Maggiore; the Ticino, a tributary of the Po that flows southward to Italy and empties into the Adriatic; and the Aare, which flows northward through Bern to join the Rhine at the German border. The Aare is the largest river entirely within Switzerland. Other rivers that help drain the Mittelland are the Sarine, the Emme, and the Reuss.

THE COAST, ISLANDS, AND THE OCEAN

Switzerland is landlocked.

CLIMATE AND VEGETATION

Temperature

Switzerland is in a climatic transitional zone, subject to Atlantic (western), Arctic (northern), continental (eastern), and Mediterranean (southern) influences. In addition, there is considerable variation due to differences in altitude. The Mittelland has warm, pleasant summer temperatures between 65°F and 70°F (18°C and 21°C), while temperatures in the mountains are cooler at high elevations but hotter in the valleys. In autumn and winter, fog is common at lower elevations, while higher altitudes enjoy the dry, sunny weather that has traditionally made the Swiss Alps a popular site for spas and sanatoriums. The average annual temperature for the country is 48°F (9°C). The canton of Ticino located south of the Alps has a Mediterranean climate. However, winter lows can fall below 32°F (0°C) in any part of the country.

One feature peculiar to Switzerland is the Foehn, a warm wind that blows through the Alpine valleys to the central lowlands, most often in the spring. The air current is generated when the low temperatures in the north coincide with high temperatures in the south. When the wind blows down the northern slopes, its temperature rises substantially, sucking up all moisture, for which reason the Foehn is described as the "Sahara Air." In general, Switzerland's prevailing winds are westerly.

Rainfall

Rainfall increases with altitude, ranging from 21 in (53 cm) in the Rhone Valley to 67 in (170 cm) in the city of Lugano, located in the mountains on the Lake of Lugano in the southern tip of Switzerland. Areas that are located near each other but have sharply contrasting elevations can also have sharp differences in rainfall. More than 75 percent of the country averages more than 40 in (102 cm) annually. The higher the elevation, the greater percentage of the total falls in the form of snow. At Alpine elevations of greater than 11,500 ft (3,505 m), all precipitation falls as snow.

Forests and Jungles

Switzerland's vegetation varies along with its climate. Deciduous forest predominates at lower elevations, and evergreens at higher ones. Oak and beech, as well as pines, grow both in the Mittelland and in the Jura. Spruce, larch, and pine are found at the higher Alpine elevations. Chestnut trees, as well as cypress, fig, and other typically Mediterranean species, grow in the region south of the Alps. On the highest slopes are Alpine meadows filled with wildflowers, including saxifrage, gentians, blue thistle, aster, Alpine roses and pansies, and others species, in particular the flower most closely associated with Switzerland: the edelweiss.

The Swiss have a history of maintaining their forests through conservation and replanting. However, this natural resource is now threatened by acid rain, which has damaged as many as 40 percent of the trees on the slopes of the country's mountains.

Population Centers – Switzerland

(1994 POPULATION ESTIMATES)

Name	Population
Zürich	342,804
Geneva (Genève)	172,138
Basel	175,809
Bern (capital)	128,872
Lausanne	116,917

SOURCE: "Population of Capital Cities and Cities of 100,000 and More Inhabitants." United Nations Statistics Division.

Cantons – Switzerland

2001 POPULATION ESTIMATES

Name	Population	Area (sq mi)	Area (sq km)	Capital
Aargau	550,900	542	1,405	Aarau
Appenzell Ausser-Rhoden	53,200	94	243	Herisau
Appenzell Inner-Rhoden	15,000	66	172	Appenzell
Basel	261,400	14	37	Basel
Berne	947,100	2,335	6,049	Berne
Fribourg	239,100	645	1,670	Fribourg
Genève	414,300	109	282	Geneva
Glarus	38,300	264	684	Glarus
Graubünden	185,700	2,744	7,106	Chur
Jura	69,100	323	837	Delémont
Lucerne	350,600	576	1,492	Lucerne
Neuchâtel	166,500	308	797	Neuchâtel
Nidwalden	38,600	107	276	Stans
Obwalden	32,700	189	491	Sarnen
Saint Gallen	452,600	778	2,014	Saint Gallen
Schaffhausen	73,400	115	298	Schaffhausen
Schwyz	131,400	351	908	Schwyz
Solothurn	245,500	305	791	Solothurn
Thurgau	228,200	391	1,013	Frauenfeld
Ticino	311,900	1,085	2,811	Bellinzona
Uri	35,000	416	1,076	Altdorf
Valais	278,200	2,018	5,226	Sion
Vaud	626,200	1,243	3,219	Lausanne
Zug	100,900	92	239	Zug
Zürich	1,228,600	668	1,729	Zürich

SOURCE: Swiss Federal Statistical Office, Neuchâtel.

HUMAN POPULATION

The overall population density of Switzerland is 474 people per sq mi (183 people per sq km). Switzerland's population centers are concentrated in the Mittelland, home to three-quarters of the population and the major cities of Zürich, Geneva, Lucerne, and the capital Bern. In the Jura Mountains the major city is Basel. Few significant cities can be found in the Alps, though small tourist centers like Davos and St. Moritz are world-famous.

NATURAL RESOURCES

The major rivers that have their source in the Swiss Alps provide an abundant source of hydropower. Switzerland is also rich in timber resources, and the country's scenic mountains, lakes, and valleys are the basis for a thriving tourism industry. Salt is among the most important minerals, although mining as a whole does not play a prominent role in the Swiss economy.

FURTHER READINGS

Insight guides. *Switzerland.* Singapore: APA Publications, 2001.

Lambert, Anthony J. *Switzerland: Rail, Road, Lake: The Bradt Travel Guide.* Guilford, Conn.: Globe Pequot Press, 2000.

New, Mitya. *Switzerland Unwrapped: Exposing the Myths.* London: I.B. Tauris, 1997.

Renouf, Norman. *Daytrips Switzerland: 45 One Day Adventures by Rail, Car, Bus, Ferry or Cable Car.* Norwalk, Conn.: Hastings House, 1999.

Steinberg, Jonathan. *Why Switzerland?* Cambridge: Cambridge University Press, 1996.

TRAMsoft Ambühler & Müller. *Information about Switzerland.* http://www.about.ch/ (Accessed June 7, 2002).

Syria

- **Area:** 71,498 mi (185,180 sq km) / World Rank: 88
- **Location:** Northern and Eastern Hemispheres, Middle East, bordering the Mediterranean Sea, south of Turkey, west of Iraq, north of Jordan, east of Israel and Lebanon.
- **Coordinates:** 35°00′N, 38°00′E
- **Borders:** 1,400 mi (2,253 km) / Iraq, 376 mi (605 km); Israel, 47 mi (76 km); Jordan, 233 mi (375 km); Lebanon, 233 mi (375 km); Turkey, 511 mi (822 km)
- **Coastline:** 120 mi (193 km)
- **Territorial Seas:** 35 NM
- **Highest Point:** Mount Hermon, 9,232 ft (2,814 m)
- **Lowest Point:** Unnamed location, 656 ft (200 m) below sea level
- **Longest Distances:** 493 mi (793 km) ENE-WSW; 268 mi (431 km) SSE-NNW
- **Longest River:** Euphrates, 2,235mi (3,596 km)
- **Natural Hazards:** Sandstorms
- **Population:** 16,728,808 (July 2001 est.) / World Rank: 56
- **Capital City:** Damascus, located in the southwestern part of the country
- **Largest City:** Damascus, 2,335,000 (2000)

OVERVIEW

Syria is located in Western Asia, north of the Arabian Peninsula in the Middle East. It consists of a fairly narrow set of mountain ranges in the west, which gives way to a broad plateau that slopes gently toward the east, bisected by the Euphrates River valley. Syria's western mountain slopes catch moisture-laden sea winds from the west and are thus more fertile and more heavily populated than the eastern slopes, which receive only hot, dry winds blowing across the desert.

Northeast of the Euphrates River, which originates in the mountains of Turkey and flows diagonally across Syria into Iraq, is the fertile Al Jazīrah region watered by the tributaries of the Euphrates. Oil and natural gas discoveries in the extreme northeastern portion of the Al Jazīrah have significantly enhanced the region's economic potential. Syria is extensively irrigated, with 28 percent of the land arable, 3 percent dedicated to permanent crops, 46 percent utilized as meadows and pastures, and only 3 percent forest and woodland. Other uses constitute the remaining 20 percent.

Control over the Golan Heights in the extreme southwest is disputed with Israel. Israel occupied the Golan Heights during a 1967 war with Syria and has remained there since. Israel annexed the Golan Heights in 1981, but Syria and other countries do not recognize this action and Syria continues to demand their return.

MOUNTAINS AND HILLS

Mountains

The An Nuşayrīyah Mountains (Jabal an Nuşayrīyah), a range paralleling the coast in the northwest, averages just over 3,976 ft (1,212 m) in height; the highest peak, Nabi Yunis, is about 5,167 ft (1,575 m). The An Nuşayrīyah terminates before reaching the Lebanese border and the Anti-Lebanon Mountains, leaving a corridor called the Homs Gap. For centuries the Homs Gap has been a favorite trade and invasion route from the coast to the country's interior and to other parts of Asia. Eastward, the line of the An Nuşayrīyah is separated from the Jabal az Zawiyah range and the plateau region by the Al Ghab depression, a fertile, irrigated trench crossed by the meandering Orontes River.

Inland and farther south, the Anti-Lebanon Mountains rise to peaks of over 8,858 ft (2,700 m) on the Syrian-Lebanese frontier and spread in spurs eastward toward the plateau region. The eastern slopes have little rainfall and vegetation and merge eventually with the desert.

In the southwest the lofty Mount Hermon (Jabal ash Shaykh), also on the border between Syria and Lebanon, descends to the Hawran Plateau. All but the lowest slopes of Mount Hermon are uninhabited, however. Volcanic

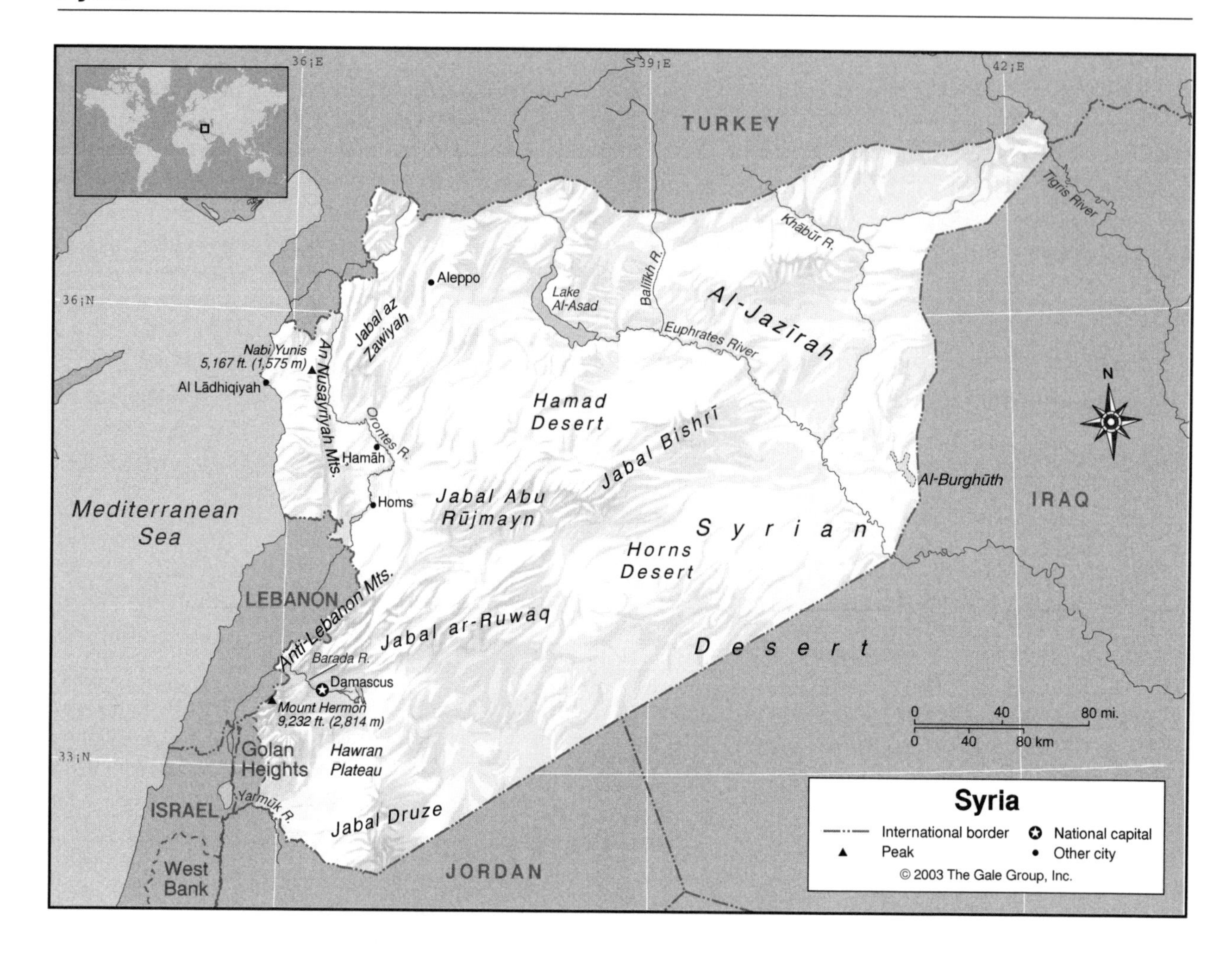

cones, some of which reach over 2,952 ft (900 m), intersperse the open, rolling, once-fertile Hawran Plateau south of Damascus and east of the Anti-Lebanon Mountains. Southeast of the Hawran lies the high volcanic region of the Jabal Druze range, home of the country's Druze population. The entire eastern plateau region is intersected by a low chain of mountains, the Jabal ar-Ruwāq, the Jabal Abū Rujmayn, and the Jabal Bishrī, extending northeastward from the Jabal Druze to the Euphrates River.

Plateaus

Hawran Plateau, frequently referred to as the Hawran, is a broad, expansive steppe lying east of the Anti-Lebanon mountains. The Hawran receives rain-bearing winds from the Mediterranean.

Hills

Along the coast, parallel to the Mediterranean, is situated a range of high hills that moderates the humidity and cooler temperatures coming off the water, restricting this effect to the narrow coastal belt. East of the Orontes River lie several ranges of hills fanning out gradually to the southwest.

INLAND WATERWAYS

Lakes

In the west the eastern slopes of various ridges descend to small streams that run dry or form salt lakes in shallow inland basins. The backwaters of the Euphrates dam built in 1973 upstream from Ar Raqqah constitute Lake Al-Asad (Buhayrat al Assad), a body of water about 50 mi (80 km) long and averaging five miles (8 km) in width.

Rivers

The country's waterways are of vital importance to its agricultural development. The longest and most important river is the Euphrates, which represents more than 80 percent of the country's water resources. Its main left-bank tributaries, the Balīkh and the Khābūr, are both major rivers and also rise in Turkey. The right-bank tributaries of the Euphrates, however, are small seasonal streams called wadis. The Tigris River flows along the northeastern border for a short distance.

Throughout the plateau region east of Damascus, oases, streams, and a few interior rivers that empty into swamps and small lakes provide water for local irrigation.

Population Centers – Syria

(1994 POPULATION ESTIMATES)

Name	Population	Name	Population
Damascus (capital)	1,549,000	Hamah	273,000
Aleppo	1,542,000	Al-Raqqa	138,000
Homs	558,000	Deir El-Zor	133,000
Al Lādhiqiyah (Latakia)	303,000	Al-Kamishli	113,000

SOURCE: United Nations Statistics Division.

Governorates – Syria

Name	Area (sq mi)	Area (sq km)	Capital
al-Hasakah	9,009	23,334	al-Hasakah
al-Lādhiqiyah	887	2,297	Al Lādhiqiyah
al-Qunaytirah	719	1,861	al-Qunaytirah
ar-Raqqah	7,574	19,616	ar-Raqqah
as-Suwayda'	2,143	5,550	as-Suwayda'
Dar'ā	1,440	3,730	Dar'ā
Dayr az-Zawr	12,765	33,060	Dayr az-Zawr
Rif Dimashq	6,962	18,032	Damascus
Halab	7,143	18,500	Aleppo
Hamāh	3,430	8,883	Hamah
Hims	16,302	42,223	Homs
Idlib	2,354	6,097	Idlib
Dimashq	41	105	-
Tartūs	730	1,892	Tartūs

SOURCE: *Geo-Data: The World Geographical Encyclopedia*, 2nd ed. Detroit: Gale Research, 1989.

The most important of these is the Barada, a river that rises in the Anti-Lebanon Mountains and disappears into the desert. The Barada creates the Al Ghutah Oasis, the site of Damascus. This verdant area, some 11.5 sq mi (30 sq km) has enabled Damascus to prosper since earliest times.

Areas in the Al Jazīrah have been brought under cultivation with the waters of the Khābūr River (Nahr al Khābūr). The Sinn, a minor river in the northwest, is used to irrigate the area west of the An Nuşayrīyah, while the Orontes waters the area east of these mountains. In the south the springs that feed the upper Yarmūk are diverted for irrigation of the Hawran.

Underground water reservoirs that are mainly natural springs are tapped for both irrigation and drinking. The Al Ghab region is richest in underground water resources and contains some nineteen major springs and underground rivers that have a combined yield of thousands of liters per minute.

Wetlands

East of the An Nuşayrīyah, the coastal range, lies the flat-bottomed Al Ghab depression. The snake-like Orontes river meanders there, creating flooding during winter and marshes in summer. There are salt flats in the northeast, including Rawdah and Al-Burghūth.

THE COAST, ISLANDS, AND THE OCEAN

Oceans and Seas

Syria has a short, narrow, coast on the Eastern Mediterranean Sea, stretching south from the Turkish border to Lebanon. Sand dunes cover this littoral, and its flatness is broken only by lateral promontories running down from the mountains to the sea.

CLIMATE AND VEGETATION

Temperature

East of the Anti-Lebanon ridges Syria demonstrates a typically continental style climate with hot days reaching temperatures of 100° F (38° C) or 109° F (43° C). By contrast, nights are cool and winters are fairly cold with temperatures falling to frost levels. The coastal hills enjoy a moderate climate along the Mediterranean and on the highest peaks snow may be found from late December to April.

Rainfall

Syria's average annual rainfall is less than 10 in (250 mm), but as much as 39 in (1,000 mm) of rain falls on the coastal plains and mountains, and on parts of the steppe east of the Homs Gap. Between eight and 15 in (200 mm to 380 mm) is not uncommon on the southern steppe of the fertile crescent. Rainfall diminishes greatly in the eastern desert, but increases in the extreme east on the Zagros mountains

Grasslands

The steppes of the western side of the Jabal Druze are part of the great fertile crescent arc, and unless cultivated are covered with seasonal grasses. The coastal strip also is home to wild grasses and shrubs such as tamarisk and buckthorn.

Deserts

Most of eastern Syria is part of the Syrian Desert, which is barren except for where rivers allow irrigated cultivation. All of the country west of the Euphrates south of the central mountain ranges is part of the barren desert region called Hamad. North of the mountains and east of the city of Homs is another barren area known as the Horns Desert, which has a hard-packed dirt surface. Even the Al Jazīrah 'island' land between the Euphrates and Tigris rivers is predominately desert.

Forests

The Anti-Lebanon mountains contain forests of Aleppo pine and Syrian oak.

HUMAN POPULATION

Syria is one of the most densely populated countries in the Middle East, but with one of the most skewed population distributions. Roughly 80 percent of the population lives in the west. Along the Mediterranean coastline, in the Aleppo area, and in the greater Damascus area, the population density exceeds 648 people per sq mi (250 people per sq km). By contrast, vast desert areas in the east between the Jordanian and Turkish borders are virtually uninhabited, as most of the eastern population is found along the Euphrates and its tributaries.

NATURAL RESOURCES

Petroleum and natural gas, in Al Jazīrah, are Syria's main resource. Syria also has phosphates, chrome and manganese ores, asphalt, iron ore, rock salt, marble, gypsum, and hydropower.

FURTHER READINGS

ArabNet. *Syria.* http://www.arab.net/syria/syria_contents.html (Accessed June 24, 2002).

Collelo, Thomas, ed. *Syria: A Country Study.* Area Handbook Series. Federal Research Division Library of Congress. Washington, D.C.: Department of the Army, 1987.

Copeland, Paul W. *The Land and People of Syria.* Rev. ed. Philadelphia: Lippincott, 1972.

Devlin, John F. *Syria: A Profile.* London: Croom Helm, 1982.

"Syria." *The Middle East and North Africa 2002.* 48th ed. London: Europa Publications, 2001.

Taiwan

- **Area:** 13,892 sq mi (35,980 sq km) [with offshore islands] / World Rank: 137
- **Location:** Northern and Eastern Hemispheres, East Asia off China's southeastern coast in the Western Pacific Ocean, bordered by the Philippine Sea and Taiwan Strait.
- **Coordinates:** 23°30′N, 121°00′E
- **Borders:** None
- **Coastline:** 973 mi (1,566 km)
- **Territorial Seas:** 12 NM
- **Highest Point:** Yü Shan, 13,114 ft (3,997 m)
- **Lowest Point:** Sea level
- **Longest Distances:** 245 mi (394 km) NNE-SSW; 89 mi (144 km) ESE-WNW
- **Longest River:** Choshui, 120 mi (190 km)
- **Largest Lake:** Sun Moon Lake, 4.5 sq mi (9 sq km)
- **Natural Hazards:** Earthquakes; typhoons
- **Population:** 22,370,461 (July 2001 est.) / World Rank: 46
- **Capital City:** Taipei, northern Taiwan
- **Largest City:** Taipei, 2,880,000 (2000 est. metropolitan population)

OVERVIEW

Taiwan is an island in the Pacific Ocean, approximately 100 mi (161 km) off mainland China's southeastern coast. It lies to the north of the Philippines and southeast of the Ryukyu islands of Japan. Taiwan also has jurisdiction over a number of islands in the Taiwan Strait, including the Pescadores, Quemoy, and Mat-Su island groups, and a few in the Pacific.

Taiwan's political status is unique and complicated. While in practice it is self-governing, it does not claim to be an independent country, nor is it officially recognized as one. Historically, the island is a part of China. However, when Communist revolutionaries took over China in 1947, Taiwan remained outside of their control. The pre-revolutionary national government of China, the Nationalists, established itself on the island, and for many years maintained that they, not the Communist government on the mainland, were the legitimate government of all of China. In practice, however, the Communists have been widely recognized as the real government of China since the 1970s. The mainland government insists that Taiwan still belongs to China and is in revolt against them.

High, rugged mountains and foothills occupy about two-thirds of the island, extending in a north-south direction from its northern tip to its southern extremity. On the east coast the mountains mostly drop precipitously to the Pacific Ocean. Near the center of the coast, however, a narrow rift valley intervenes between the central range and a lower, but also steep, coastal range. In the west the high mountains are succeeded by foothills that gradually give way to flat alluvial plains where the many short, rapid, mountain streams become meandering rivers. Extensive cultivation is undertaken on the slopes of the foothills, but most of the island's cultivable land lies in the west on the coastal plains.

Taiwan is situated on the Eurasian Tectonic Plate. The island lies near the border of the Philippine Plate, putting it in the Pacific Rim "Ring of Fire." More than 200 minor earthquakes are recorded each year.

GEO-FACT

Taipei's stormy, humid climate has given rise to the saying "The weather in Taipei is like a stepmother's temper."

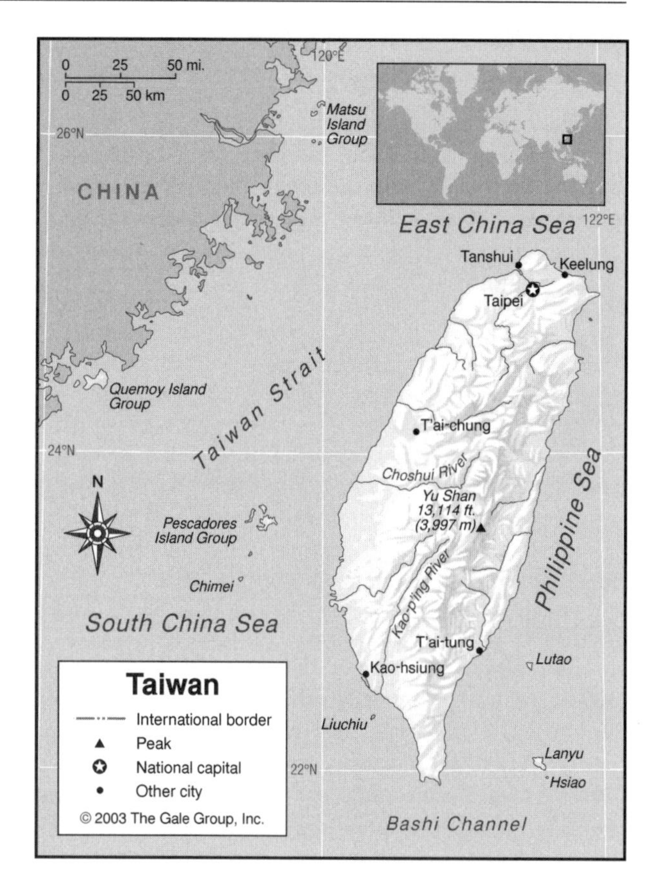

MOUNTAINS AND HILLS

Mountains

The Central Range, Taiwan's dominant geographical feature, spans the length of the island along a north-south axis. It has more than 60 peaks with elevations of over 10,000 ft (3,048 m). The highest is Yü Shan, located near the center of the island. In the far north, detached from the main mountain system, is a short range of volcanic origin called Tatun Shan, which rises to over 4,000 ft (1,219 m).

Plateaus

The hills that border the Central Range on the west descend to a rolling, terraced plateau with average elevations of 330 to 1,640 ft (101 to 500 km).

To the west of this plateau are the coastal plains.

Hills and Badlands

The foothills of the Central Range, which lie mostly to the west, have average elevations of 4,000 to 5,000 ft (1,219 to 1,524 m). In addition, there are a number of separate hills averaging about 5,000 ft (1,524 m).

INLAND WATERWAYS

Lakes

Two of Taiwan's major lakes are Coral Lake in the southwest and, near the center of the island, Sun Moon Lake, which is said to have once been two separate lakes, called Sun Lake and Moon Lake.

Rivers

Taiwan's rivers flow across the long, narrow island, rising in the Central Range and descending to the coasts, so they are all short. Two of the major rivers depart from this pattern: the Tanshui drains northward toward Taipei, and the Kao-p'ing drains southward toward the southeastern coast. The third major river is the Choshui, which drains westward across the mountains and through the coastal plain at one of its widest points.

THE COAST, ISLANDS, AND THE OCEAN

Oceans and Seas

Taiwan borders the Pacific Ocean to the east, the Taiwan Strait to the west, the East China Sea to the north, the South China Sea to the southeast, and the Bashi Channel of the Philippine Sea to the south.

Major Islands

The major island group associated with Taiwan are the Pescadores (Penghu Archipelago), 64 islands located roughly 25 miles (40 km) west of the main island. They have a total land area of 49 sq mi (127 sq km), spread out over more than 600 sq mi (1,554 sq km). The Quemoy (or Kinmen) and Mat-Su island groups are both located less than 2 mi (3 km) from the Chinese mainland and are populated mostly by military personnel. The Pescadores are relatively flat coral reefs that support some agriculture. The main island of the Quemoy group is rocky and boulder strewn, but it is also partially arable. Mat-Su, made up of masses of igneous rocks, supports no agriculture.

Taiwan's other islands include Lan-yü, or Orchid Island, and Lü tao (Green Island) southeast of the main island; Ch'i-Mei Yü to the west, and Hsiao Liu-Chiu Yü to the southwest.

The Coast and Beaches

The Central Range plunges abruptly to the sea along the eastern coast, except for an area about 100 mi (161 km) long north of T'ai-tung, where the T'ai-tung Rift Valley and, farther east, a short coastal ridge are located.

On the western side of the island, a coastal plain meets the sea in a band of swamps and tidal flats.

The coast is fairly smooth, except for deep indentations at the mouths of the Kao-p'ing River in the south and the Tanshui River in the north, and the deltas of several rivers in the southwest. The major deepwater ports are located at Keelung in the north and Kao-hsiung, in the Haochiung Bay in the south.

CLIMATE AND VEGETATION

Temperature

Taiwan's subtropical climate, warmer in the south and cooler in the north, is modified by Pacific breezes. Average temperature readings for January are 61°F (16°C) in the north and 68°F (20°C) in the south, while the average July temperature in both regions is 82°F (28°C).

Rainfall

Rainfall in Taiwan is generally heavy, averaging about 100 in (250 cm) annually and much higher in some regions. The northeast, or "winter," monsoon brings heavy rains to the northern part of the island between October and March, while the southwest, or "summer," monsoon brings rain to the south between May and September. The summer months also bring dangerous typhoons and cyclones.

Grasslands

A coastal plain of varying widths extends along the entire western coastline, from north to south. This plain, drained by numerous rivers, is Taiwan's agricultural heartland.

Forests and Jungles

Plant life on Taiwan is similar to that in neighboring China and the Philippines and varies with altitude. About half of Taiwan is forested, with most forests located in the central range. Mangroves and bamboo grow in the tidal flats and other parts of the coast. Tropical palm and broadleaf evergreens and acacias are found at elevations of up to 2,000 ft (610m), as well as teak, pandanus, and camphor. Mixed evergreen and deciduous forest including maple, cypress, cedar, and pine grows at altitudes between 2,000 and 7,000 ft (610 m and 2,134 m), and forests of fir, juniper, and other conifers beyond that.

HUMAN POPULATION

Taiwan has one of the highest population densities in the world, 1,742 people per sq mi (673 people per sq km) in 2000. Roughly 90 percent of the population inhabits the plateaus and plains to the west of the Central Range. The twentieth century saw significant rural-to-urban migration, and urban dwellers accounted for more than three-fourths of the population by the end of the century.

Population Centers – Taiwan

(1998 POPULATION ESTIMATES)

Name	Population
Taipei (capital)	2,639,939
Kao-hsiung	1,462,302
T'ai-chung	917,788
Tainan	721,832

SOURCE: Bureau of Statistics, Taiwan.

NATURAL RESOURCES

Taiwan's economy is based on commerce and manufacturing. It has modest reserves of coal, oil, natural gas, marble, limestone, and dolomite.

FURTHER READINGS

Copper, John Franklin. *Taiwan: Nation-state or Province?* Boulder, Colo.: Westview Press, 1996.

Fetherling, Doug. *The Other China: Journeys around Taiwan.* Vancouver: Arsenal Pulp Press, 1995.

Kemenade, Willem van. *China, Hong Kong, Taiwan, Inc.: The Dynamics to a* New *Empire.* New York: Vintage Books, 1998.

Rubinstein, Murray A., ed. *Taiwan: A New History.* Armonk, N.Y.: M.E. Sharpe, 1999.

Storey, Robert. *Taiwan.* Hawthorn, Vic.: Lonely Planet, 1998.

Tajikistan

- **Area:** 55,251 sq mi (143,100 sq. km) / World Rank: 94
- **Location:** Northern and Eastern Hemispheres; Central Asia, east of Uzbekistan, south of Kyrgyzstan, west of China, and north of Afghanistan.
- **Coordinates:** 39°00′N, 71°00′E
- **Borders:** 2,269 mi (3,651 km) / Afghanistan, 749 mi (1,206 km); China, 257 mi (414 km); Kyrgyzstan, 541 mi (870 km); Uzbekistan, 721 mi (1,161 km)
- **Coastline:** None
- **Territorial Seas:** None
- **Highest Point:** Qullai Ismoili Somoni, 24,590 ft (7,495 m)
- **Lowest Point:** Banks of the Syr Dar'ya, 984 ft (300 m)
- **Longest Distances:** 434 mi (700 km) E-W; 217 mi (350 km) N-S
- **Longest River:** Amu Dar'ya, 1,578 mi (2,539 km)

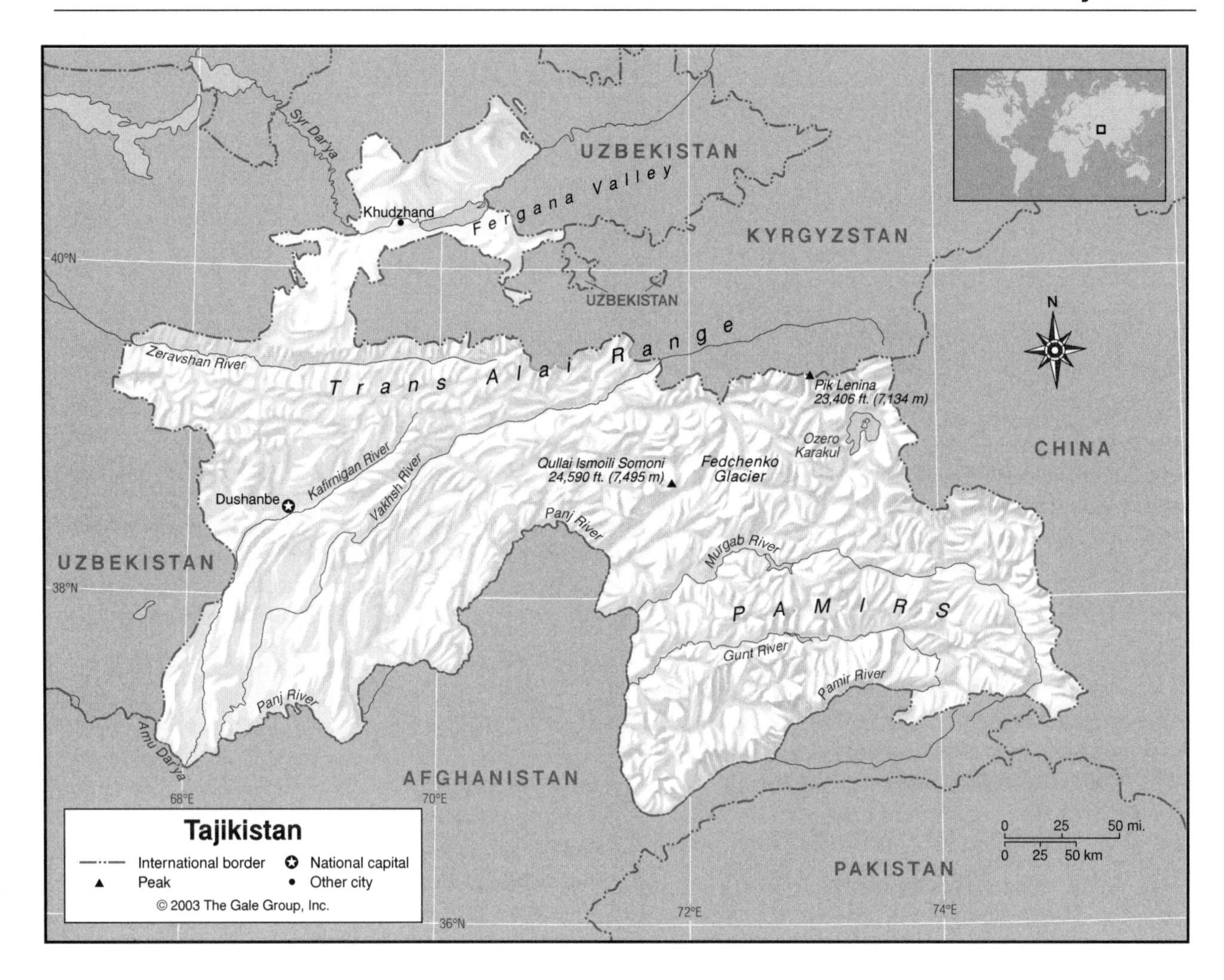

- **Largest Lake:** Ozero Karakul, 147 sq mi (380 sq km)
- **Population:** 6,578,681 (July 2001 estimate) / World Rank: 95
- **Capital City:** Dushanbe, in the west
- **Largest City:** Dushanbe, 664,000 (2000 est.)

OVERVIEW

Tajikistan is the smallest nation in Central Asia. It is dominated by mountains, including some of the world's highest peaks, with the Pamirs in the south and the Trans Alai range in the north. Tajikistan's mountainous terrain is also notable for its many glacier-fed rivers. The massive Fedchenko glacier, at more than 270 sq mi (700 sq km), is the largest glacier in the world outside of the polar regions.

Elevations in northwest and southwest Tajikistan are generally lower than in the rest of the country. The most notable lowland feature is the fertile Fergana Valley in the far north, whose soils of rich river deposits make the valley ideal for agriculture.

Tajikistan's climate is primarily continental, with hot summers and mild winters; but elevation plays a big role. The highest temperatures and near subtropical conditions prevail in the lowlands of the southwest, while semiarid to polar conditions characterize the high mountains all year.

Tajikistan is on the Eurasian Tectonic Plate, not far north of its border with the Indian Tectonic Plate. It lies on a seismic belt that is active throughout southeastern Central Asia. Earthquakes are common and can be devastating.

MOUNTAINS AND HILLS

Mountains

Nearly half of Tajikistan has elevations exceeding 9,800 ft (3,000 m). The Pamirs are the dominant mountain range and are among the highest mountains in the world, with an extraordinary mean elevation of 13,000 ft (3,965 m). Centered in southern Tajikistan they branch off in every direction, connecting with other great mountain ranges like the Tian Shan.

Qullai Ismoili Somoni (Communism Peak) in the Pamirs is the highest mountain in the country at 24,590 ft (7,495 m). Pik Lenina (Lenin Peak), in the Trans Alai range of the Pamirs in northeastern Tajikistan is also one of the world's highest at 23,406 ft (7,134 m). Many other peaks exceed 20,000 ft (6,096 m).

Plateaus

A portion of Tajikistan lies on the high Godesberg plateau, which also stretches into China and Afghanistan. This great plateau is considered part of the Pamirs system.

Hills

In the west, about a third of Tajikistan is comprised of foothills and steppes. The nation's lowest elevations are found in the southwestern river valleys and in the northern spur region that crosses the Fergana Valley.

INLAND WATERWAYS

Lakes

The majority of Tajikistan's lakes were formed by glaciers and are found in the eastern Pamirs. Ozero Karakul (Lake Karakul) is the largest, located in the northeast at an elevation of 13,000 ft (4,000 m). A salt lake, Karakul is essentially lifeless.

Rivers

Tajikistan's mountainous terrain has created an extensive network of rivers, but several large Central Asian rivers originating elsewhere cross the country as well, most notably the Syr' Darya (Sirdaryo—Syr River) and the Amu' Darya (Amu River).

The Amu' Darya is Central Asia's largest river, at 1,580 mi (2,540 km). It originates in the form of it's upper tributary, the Panj, along the Tajikistan-Afghan border. In Tajikistan, the Amu' Darya runs for 571 mi (921 km). The Vakhsh (Surkohb) River and the Kafirnigan River are two large tributaries of the Amu' Darya that run northeast to southwest in great valleys across western Tajikistan. Of these, the Vakhsh is second in length in Tajikistan only to the Amu' Darya. East of the capital Dushanbe, at the city of Norak (Nurek), the Vakhsh has been dammed to produce hydroelectric power and to facilitate irrigation. One of the world's highest, this dam forms the Norak Reservoir and serves Tajikistan's principal manufacturing center with water and power.

After the Amu' Darya, the Syr' Darya in northern Tajikistan is the second longest river in Central Asia, with a total length of 1,488 mi (2,400 km). The Syr' Darya flows through the country only briefly, traversing the Fergana Valley in northern Tajikistan for 121 mi (195 km). Another major northern river, the Zeravshan, crosses Tajikistan east-to-west for 196 mi (316 km); its total length is 484 mi (781 km).

Melting snow and melting glaciers cause Tajikistan's rivers to run high in the spring and summer, respectively. In summer, the glacial runoff is a critical aid to irrigation in Tajikistan's northern and western valleys.

THE COAST, ISLANDS, AND THE OCEAN

Tajikistan is landlocked.

CLIMATE AND VEGETATION

Temperature

Average temperatures vary significantly by region in Tajikistan. A continental climate predominates in the lowlands, with hot summers and cold winters. Khudzhand in the Fergana Valley has an average July temperature of 81° F (27° C) and a January average of 34° F (-1° C). Extreme temperatures in summer can reach 118° F (48° C), with strong dust storms in the semiarid areas. By contrast, the eastern mountains have average July temperatures of 50° F (10° C) or less, and January temperatures of -4° F (-20° C). Winter temperatures in the eastern Pamirs have experienced extremes of -76° F (-60° C).

Rainfall

For most of Tajikistan, the average annual precipitation ranges between 28 in (700 mm) and 63 in (1,600 mm). Although generally meager in the lowlands and mountains, sudden, substantial amounts of precipitation are known to cause devastating landslides. Winter and spring are the chief snowfall/rainfall seasons; summer and fall can be drought-stricken. Heaviest precipitation rates occur at the Fedchenko Glacier, where levels of 88 in (223 cm) per year have been recorded. The lowest annual averages are in the eastern Pamirs, with annual rainfall of less than 4 in (10 cm).

Grasslands

Western Tajikistan has some scattered areas of steppe that rise into the foothills of its mountains. The steppe vegetation features low shrubs and drought-resistant grasses, but also broad fields of wild poppies and even tulips.

Forests

Only about 4 percent of Tajikistan is forested, but the country's woodlands are noted for their beauty and abundant wildlife. Trees are primarily coniferous, although the lower slopes feature ancient stands of walnut trees. Wildlife is diverse and includes many varieties of mountain sheep and goat, notably the rare markhor as well as the Siberian horned goat. The lower mountain regions are home to brown bears, wolves, wild boar, lynx, and the endangered snow leopard; deer, wolves, foxes, and badgers can be found on the steppes.

Population Centers – Tajikistan

(1993 POPULATION ESTIMATES)

Name	Population
Dushanbe (capital)	528,600
Khudzhand (Kodzhent)	163,00

SOURCE: "Population of Capital Cities and Cities of 100,000 and More Inhabitants." United Nations Statistics Division.

HUMAN POPULATION

With a total population of 6,578,681 (July 2001 estimate), Tajikistan has a population density of 120 persons per sq mi (46 persons per sq km). The greatest density occurs in the northern and western lowlands. Only about 33 percent of the people live in urban areas. The largest city is the capital, Dushanbe, located in the Hisor Valley in western Tajikistan, with a population of 664,000. The next largest city is Khudzhand (called Leninabad from 1936 to 1991), population 163,000, situated in the Fergana Valley of northern Tajikistan.

NATURAL RESOURCES

Tajikistan's mineral resources include uranium, mercury, brown coal, lead, zinc, antimony, tungsten, silver, and gold. Petroleum reserves are limited. A small number of factories produce chemicals, fertilizers, cement, and vegetable oil. The nation's many rivers offer excellent sources for hydroelectric power, but Tajikistan's mountainous terrain severely limits the amount of land available for cultivation to only six percent.

FURTHER READINGS

Gleason, Gregory. "The Struggle for Control over Water in Central Asia: Republican Sovereignty and Collective Action." *RFE/RL Report on the USSR* [Munich], 3, 25 (June 21, 1991): 11-19.

Gretsky, Sergei. "Civil War in Tajikistan and Its International Repercussions." *Critique*, Spring 1995: 3-24.

Rubin, Barry M. "The Fragmentation of Tajikistan." *Survival* 35, 4 (Winter 1993-1994): 71-91.

Tajikistan Travel. *Adventure on the Roof of the World.* http://www.traveltajikistan.com/ (Accessed June 27, 2002).

Tanzania

- **Area:** 364,900 sq mi (945,087 sq km) / World Rank: 32
- **Location:** Southern and Eastern Hemispheres, in eastern Africa, bordering the Indian Ocean, south of Uganda and Kenya, east of Democratic Republic of the Congo, southeast of Burundi and Rwanda, and north of Mozambique, Malawi, and Zambia.
- **Coordinates:** 6°00′S, 35°00′E
- **Borders:** 2,114 mi (3,402 km) / Uganda, 246 mi (396 km); Kenya, 478 mi (769 km); Mozambique, 470 mi (756 km); Malawi, 295 mi (475 km); Zambia, 210 mi (338 km); Burundi, 280 mi (451 km); Rwanda, 135 mi (217 km)
- **Coastline:** 885 mi (1,424 km)
- **Territorial Seas:** 12 NM
- **Highest Point:** Mount Kilimanjaro, 19,341 ft (5,895 m)
- **Lowest Point:** Sea level
- **Longest Distances:** 760 mi (1,223 km) N-S; 740 mi (1,191 km) E-W
- **Longest River:** Ruvuma, 437 mi (704 km)
- **Largest Lake:** Lake Victoria, 26,828 sq mi (69,484 sq km)
- **Natural Hazards:** Drought; flooding on the central plateau
- **Population:** 36,232,074 (July 2001 est.) / World Rank: 32
- **Capital City:** Dar es Salaam, central Indian Ocean coast (note: presently moving to Dodoma)
- **Largest City:** Dar es Salaam, population 1,747,000 (2000 est.)

OVERVIEW

Tanzania, lying between one and twelve degrees south of the equator, is about the size of Texas, including 8,000 sq mi (20,650 sq km) of inland water. Most of the country, rising steadily toward the west, consists of extensive rolling plains demarcated by the Rift Valley, a series of immense faults creating both depressions and mountains. Much of it is above 3,000 ft (900 m) and some above 5,000 ft (1,500 m). A small portion, including the islands and the coastal plains, lies below 600 to 700 ft (about 200 m). The landscape is extremely varied, changing from coastal mangrove swamps to tropical rain forests and from rolling savannas and high arid plateaus to mountain ranges.

Four major ecological regions can be distinguished: high plateaus, mountain lands, lakeshore region, and coastal belt and islands. The mountain ranges and the

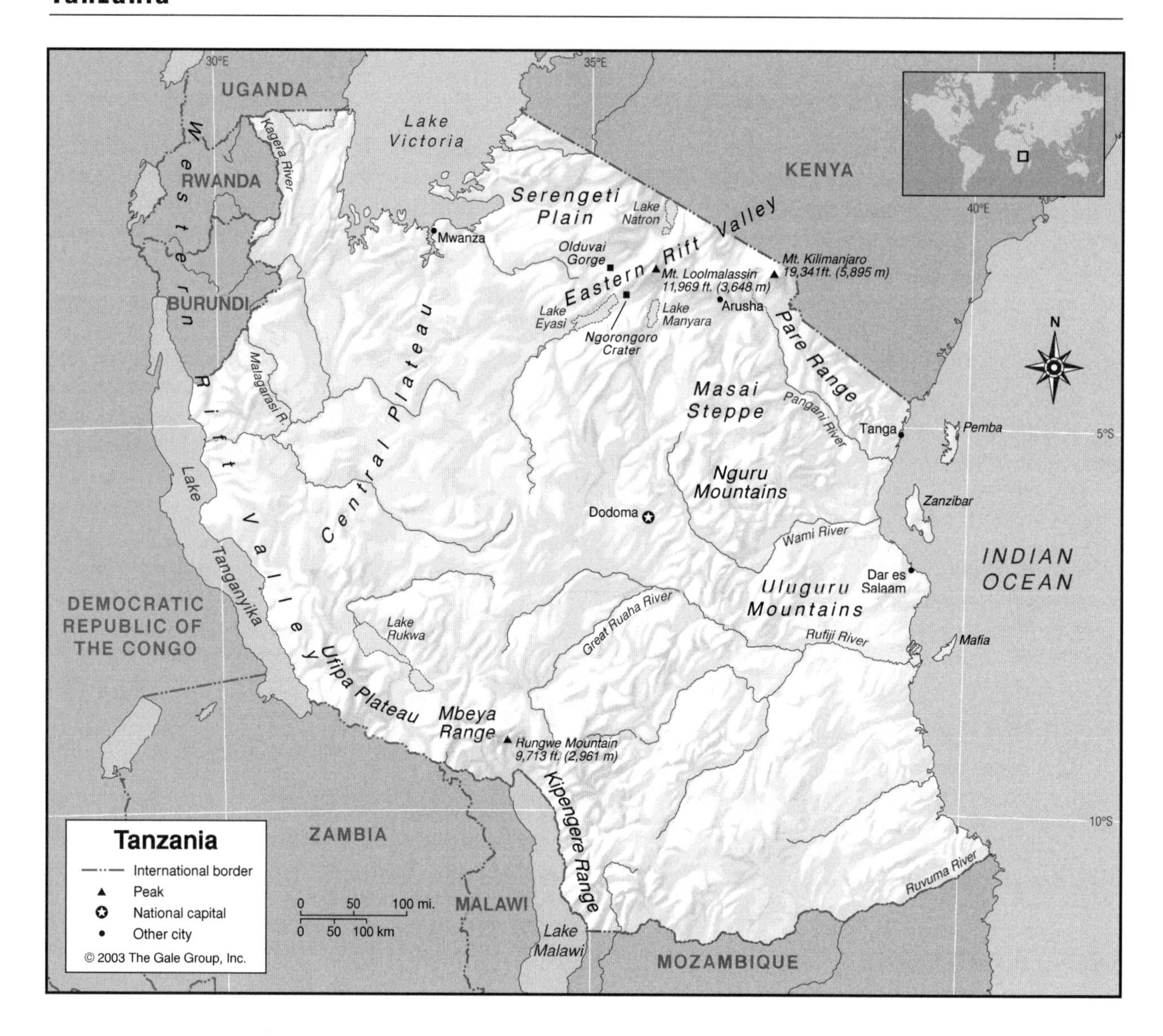

area around Lake Victoria (Victoria Nyanza) receive generous amounts of rain, but vast plateau areas in the center of the country are so dry that they cannot support significant cultivation activity, and tsetse fly infestation precludes animal husbandry. About 5 percent of the land is arable, 1 percent dedicated to permanent crops, 40 percent utilized as meadows and pastures, and 47 percent for forest and woodland.

MOUNTAINS AND HILLS

Mountains

One of three major mountainous zones extends inland from Tanga to near Lake Manyara. It includes the Usambara and Pare ranges, which together form a wedge-shaped mass reaching a height of almost 7,550 ft (2,300 m) and the Northern Highlands, which contain Mount Kilimanjaro and Mount Meru. Mount Kilimanjaro, the highest point in Africa, rises in two peaks united by a saddle to an ultimate height of 19,341 ft (5,895 m). The so-called glaciers on top of Kibo, the higher peak, are rapidly decaying remains of a former, more extensive icecap. Meru, the lower peak, rises to 14,960 ft (4,560 m).

The second zone stretches from the western shore of Lake Natron southward in a series of isolated mountains and mountain chains. They are interspersed with lakes and craters and connected with the northern part of the Eastern Rift. Between Lake Natron and Lake Manyara are the Winter Highlands, a volcanic region containing Mount Loolmalassin and the Ngorongoro Crater, roughly 60 to 70 mi (100 to 110 km) wide, in which is found one of the heaviest concentrations of wildlife in Africa. The shores of Lake Manyara and the nearby Serengeti Plain also teem with wildlife. West of the crater lies Olduvai Gorge, where the paleontological explorations of the late Louis S. B. Leakey and his associates led to the hypothesis that the earliest forms of man may have originated in East Africa.

The third major mountainous region stretches from the Nguru Mountains, about halfway between Dodoma and Dar es Salaam, and the Uluguru Mountains, farther south, to the Kipengere Range, which descends sharply toward the eastern shore of Lake Malawi (Nyasa). Around the northern shore of Lake Malawi the Mbeya Range, including Rungwe Mountain at 9,713 ft (2,961 m), completes the mountains of the south.

Plateaus

The high plateaus are characterized by monotonous undulating terrain cut slightly by mostly intermittent rivers. There are two major plateaus, the Central Plateau and the Eastern Plateau. The Central Plateau lies between the two branches of the Rift Valley. Its vast expanse forms a huge uplifted basin. Elevation varies from roughly 3,000 to 5,900 ft (900 to 1,800 m) above sea level, the greater part lying at about 4,000 ft (1,200 m). It is a hard dry plain dotted with granitic outcrops.

The Eastern Plateau is in effect a series of lower plateaus descending gradually to the coastal lowlands. In the north it consists primarily of the Masai Steppe, an extensive semiarid plain of more than 26,000 sq mi (almost 70,000 sq km). Varying from just under 800 to 3,500 ft (250 to over 1,000 m) above sea level, the steppe is near desert with vast areas of dry bush and scanty grass. The Makonde Plateau in the extreme southeast is a poorly watered tableland of about 1,200 sq mi (3,100 sq km).

A smaller plateau, the Ufipa Plateau, occupies the southwestern corner of Tanzania wedged between the Mbeya Mountains, Lake Rukwa, and Lake Tanganyika. The Ufipa Plateau consists mainly of highland swamp with some grassland and forest cover.

Hills

In the southeast, the terrain is broken and toward the coast is characterized by outcrops of isolated hill masses rising sharply from the surrounding land. On the western side of Zanzibar are several ridges that reach more than 200 ft (60 m) high; Masingini Ridge at 390 ft (119 m) is the highest point on the island. Pemba Island is hilly, and its highest point is 311 ft (95 m).

INLAND WATERWAYS

Lakes

Tanzania's lakes provide transportation, are a source of food and livelihood, and offer abundant water supplies for irrigation. With a surface area of 24,300 sq mi (62,940 sq km), Lake Victoria is the largest lake in Africa and the second largest freshwater lake on the globe. It is located in the north, shared also by Uganda and Kenya. About half of the lake is in Tanzania. Along the western coast, Lake Tanganyika, the world's second deepest lake, has a precipitous coastline and a few poor harbors. Found in the south, Lake Malawi also has poor harbors. To the east of Lake Tanganyika, Lake Rukwa is small and shallow and tends to be brackish. A series of small lakes in the northern part of the country all have salty water—Lake Natron is commercially exploited for salt and soda. Other lakes in the Eastern Rift Valley include Lake Eyasi and Lake Manyara.

Rivers

The country's rivers drain into four major basins. Five important rivers and a number of minor ones in the eastern third of Tanzania enter the Indian Ocean directly, including the Pangani, Wami, and Rufiji. Forming most of the southern border with Mozambique is the Ruvuma River, the longest river in Tanzania. The Ruvuma originates just east of Lake Malawi in the hills near Songea, and runs west before shortly coursing counterclockwise to head almost due east to the Indian Ocean, where it ends 437 miles (704 km) later. Other streams around Lake Nyasa empty into the lake and reach the Indian Ocean via the Zambezi River in Mozambique. A number of short rivers (except for the longer Kagera River in northwestern Tanzania) drain into Lake Victoria and ultimately via the Nile River into the Mediterranean Sea. Several rivers in western Tanzania, the longest of which is the Malagarasi, drain into Lake Tanganyika and ultimately via the Congo River into the Atlantic Ocean. Streams in the north-central and southwestern sections empty into smaller lakes and interior basins, with the notable exception of the Great Ruaha, which originates in the Mbeya Mountains and flows northeast to the center of the country before turning southwest and eventually feeding into the Rufiji.

Wetlands

Tanzania's lakes and swamps represent 5.8 percent of the total land surface, but this number excludes seasonally inundated flood plains and riverine marshes. The Sagara Swamp, which forms most of western Tanzania, is a huge floodplain with an area of 6,415 sq mi (16,614 sq km). It includes the Moyowosi Game Reserve and is home to many species of wildlife.

THE COAST, ISLANDS, AND THE OCEAN

Oceans and Seas

Tanzania faces the Indian Ocean on its eastern border. The continental shelf off the coast is relatively narrow, usually only 5 to 6 miles (8 to 10 km) wide, but extends about 25 mi (40 km) off the shore of the islands of Zanzibar and Mafia. There are many fringing reef systems offshore, but the further out the better developed and more diversified they get, with the best being off the Tanga coast and that of the offshore islands.

Major Islands

The islands of Tanzania are basically coral. Zanzibar, separated from the mainland by a channel 22 mi (35 km) wide at its narrowest point, is the largest coralline island on the African coast. It is about 50 mi (80 km) long and 25 mi (40 km) wide with a total area of 640 sq mi (1,657 sq km).

Pemba, north of Zanzibar, is smaller; it is 42 mi (67 km) long and 14 mi (22 km) wide with a total area of 380 sq mi (984 sq km). Its topography varies with small steep hills and valleys. Mafia, 27 mi (43 km) long and more than 9 mi (14 km) wide, is a low island situated about halfway down the coast south of Tanzania opposite the mouth of the Rufiji River.

The Coast and Beaches

The coastal belt is narrow in the north and south, averaging between 10 and 40 mi (16 and 60 km) in breadth. It is broader in the center near the lowlands of the Rufiji River valley where it almost reaches the Uluguru mountains.

The 500 mi (800 km) coast is difficult to approach because of numerous coral reefs and shifting sandbars at the mouths of rivers. The land slopes sufficiently toward the coast to cause most rivers to be unnavigable because of rapids.

Much of Tanzania's coastline consists of palm-fringed sandy beaches. The best beaches are located on the islands of Zanzibar and Mafia, but a particularly good stretch on the mainland is a 20-mile (32-km) strip beginning at Dar es Salaam and continuing south.

Population Centers – Tanzania

(1995 POPULATION ESTIMATES)

Name	Population	Name	Population
Dar es Salaam	1,747,000	Tanga	188,000
Mwanza	233,000	Zanzibar	158,000
Dodoma	204,000	Arusha	140,000

SOURCE: Projected from 1988 census, Tanzania Bureau of Statistics.

CLIMATE AND VEGETATION

Temperature

Being just south of the equator, the climate of Tanzania is mostly tropical, but gives way to temperate in the highlands. The coastal area is tropical and humid with average temperatures of about 81°F (27°C), while further inland the central plateau is hot and drier; temperatures vary by season and time of day. In the more temperate highlands, days are warm, but nights bring brisk temperatures.

Rainfall

From November to December and from March through May, the north enjoys rainy seasons, while the south has only one season of rain, from November to March. On the coast, annual rainfall averages 40 to 76 in (100 to 193 cm), but only 20 to 30 in (50 to 76 cm) on the central plateau. The eastern section of Lake Victoria receives 30 to 40 in (75 to 100 cm), and the western side 80 to 90 in (200 to 230 cm).

The islands receive heavy rains in April and May, and lighter rains in November and December. Drier weather occurs during the alternating monsoons, which blow from the northeast from December to March, and from the southwest from June to October.

GEO-FACT

Tanzania contains both the highest point in Africa, Mount Kilimanjaro, and the lowest, which is the floor of Lake Tanganyika.

Grasslands

The highlands are unique for their grasses and heath; about a third of the country is covered with wooded grassland savanna. Two thirds of Zanzibar Island is covered with bush and grass.

The northern portion of the Central Plateau slopes gently downward to form the large shallow depression containing Lake Victoria, which lies at an elevation of about 3,700 ft (1,180 m). On the lakeshore are large flooded inlets. The gradual slope of the land permits agricultural development not possible along the steep embankments of Lakes Tanganyika and Nyasa. The area is densely populated, and the people have a close cultural affinity with those living in the Uganda and Kenya portions of the Lake Victoria basin.

Forests and Jungles

Closed-forest hardwoods are found in the Lake Victoria basin and softwoods in mountain areas. Bush, thickets, and stunted forests dominate the drier central areas, while mangroves colonize the coastal strip. Rainforest conditions prevail on the southern slopes of Mount Kilimanjaro between 5,600 and 9,500 ft (1,700 and 2,900 m).

HUMAN POPULATION

The most densely populated areas are found in elevated and well-watered regions including the Usambara Mountains, Kilimanjaro and Meru, the Lake Victoria basin, the Southern Highlands, and the coast around Tanga and Dar es Salaam. The south is very sparsely populated. About 28 percent of the population lives in urban areas (2000 est.). In 2001, the population was growing at a rate of 2.61 percent. Christians comprise 45 percent of the population; Zanzibar is more than 99 percent Muslim.

NATURAL RESOURCES

Tanzania is heavily dependant on agriculture, and most of its industries are based on refining and exporting agricultural products including corn, sisal, cotton, tobacco, coffee, sugar, rice, and coconuts. Natural gas has been discovered in the Rufiji Delta and its exploitation looks promising. Other mineral deposits in Tanzania that may prove profitable in the future include tin, phosphates, iron ore, coal, diamonds, gemstones, gold, and nickel. With its many rivers, hydropower also presents major potential. The Great Ruaha River is the site of a hydroelectric station, and the Pangani River, which rises in the northeastern highlands, has three hydroelectric stations.

FURTHER READINGS

ABC/Kane Productions International, Inc. "Ngorongoro: Africa's Cradle of Life." *The Living Edens.* http://www.pbs.org/edens/ngorongoro/ (Accessed May 2002).

Africa South of the Sahara 2002. "Tanzania." London: Europa Publishers, 2001.

Asch, Lisa, and Peter Blackwell. *Tanzania.* Lincolnwood, Ill.: Passport Books, 1997.

Koornhof, Anton. *The Dive Sites of Kenya and Tanzania: Including Pemba, Zanzibar, and Mafia.* Lincolnwood, Ill.: Passport Books, 1997.

Spectrum Guide to Tanzania. New York: Interlink Books, 1998.

UNEP World Conservation Monitoring Centre. *World Atlas of Coral Reefs.* http://www.unep-wcmc.org (Accessed June 2002).

United Republic of Tanzania. *The Tanzania National Website.* http://www.tanzania.go.tz/ (Accessed May 2002).

Thailand

- **Area:** 198,457 sq mi (514,000 sq km) / World Rank: 51
- **Location:** Northern and Eastern Hemispheres, in Southeast Asia, bordering the Andaman Sea and the Gulf of Thailand in the south; southwest of Laos, northwest of Cambodia, north of Malaysia, southeast of Myanmar
- **Coordinates:** 15°00′N, 100°00′E
- **Borders:** 3,022 mi (4,863 km) / Laos, 1,090 mi (1,754 km); Cambodia, 499 mi (803 km); Malaysia, 314 mi (506 km); Myanmar (Burma), 1,118 mi (1,800 km)
- **Coastline:** 2,000 mi (3,219 km) / Gulf of Thailand, 1,684 mi (2,710 km); Andaman Sea, 316 mi (509km)
- **Territorial Seas:** 12 NM
- **Highest Point:** Doi Inthanon, 8,451 ft (2,576 m)
- **Lowest Point:** Sea level
- **Longest Distances:** 1,024 mi (1,648 km) N-S; 485 mi (780 km) E-W
- **Longest River:** Mekong River, 2,700 mi (4,350 km)
- **Largest Lake:** Thale Sap Songkla, 401 sq mi (1,040 sq km)
- **Natural Hazards:** Flooding, droughts, typhoons, land subsidence, landslides
- **Population:** 61,797,700 (July 2001 est.) / World Rank: 19
- **Capital City:** Bangkok, on the Chao Phraya River near the Gulf of Thailand
- **Largest City:** Bangkok, population 7,610,000 (2001est.)

OVERVIEW

Thailand lies on the Eurasian Tectonic Plate at the center of continental Southeast Asia. Features of the terrain include mountain ranges, an alluvial central plain, and an upland plateau. The mountains of southern China and northern Thailand extends down to a fertile central plain formed by the mighty Chao Phraya river. Settlement tended to concentrate in the Chao Phraya valley, with its fertile floodplains and tropical monsoon climate so ideally suited to wet-rice cultivation, rather than in the marginal uplands and mountains of the northern region, or in the arid Khorat Plateau. From the north-central mass of the country, the very narrow Malay Peninsula extends to the south, shared in part with Myanmar (Burma) and Malaysia. Numerous islands are scattered off both of the peninsula's coasts, and Thailand's part of the continental shelf extends to a depth of 656 ft (200 m).

MOUNTAINS AND HILLS

Mountains

Mountain chains cover most of northern Thailand and also rise along the western border with Myanmar (Burma) down to form the spine of the Malay Peninsula. The north consists of an area of high mountains incised by steep river valleys and upland areas that border the central plain. Doi Inthanon, a 8,451 ft (2,576 m) limestone peak, is Thailand's highest mountain.

Thailand's frontier mountain chains include the Tanen and Doi Luang Ranges, extensions of the Himalayan Foothills, in the north; the limestone peaks of the Dawna and Bilauktaung Ranges in the west; and the Dan-

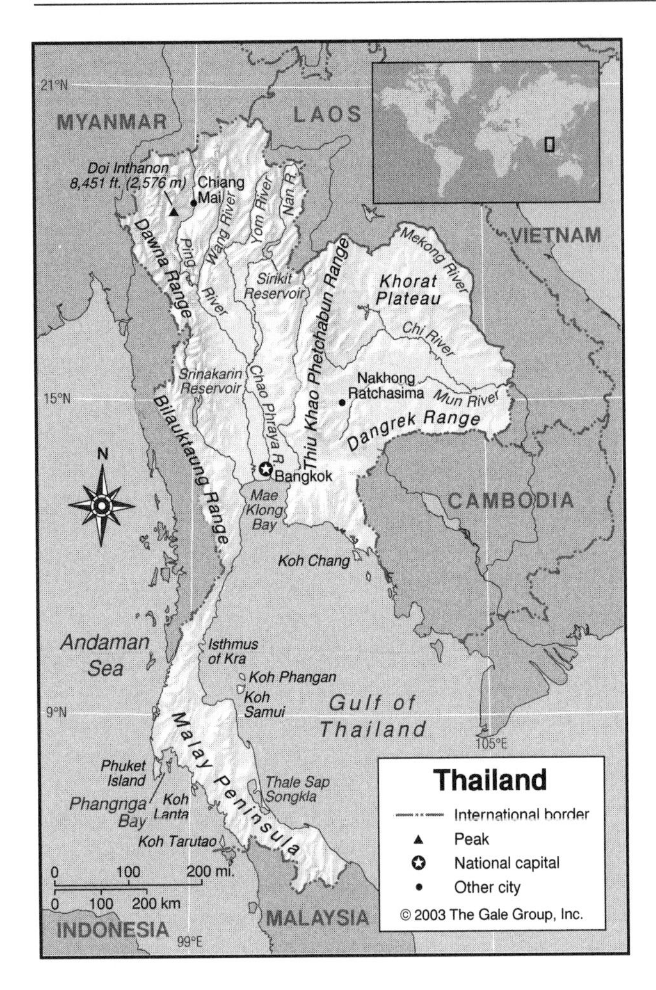

grek and Chanthaburi Ranges in the east along the Cambodian border. The Thiu Khao Phetchabun Range runs north-south down the middle of the country, setting off the Khorat plateau. The southern peninsular region has rolling to mountainous terrain unbroken by large rivers.

Plateaus

Northeast Thailand consists mainly of the dry Khorat Plateau, which is beset with many ecological problems, especially poor soils. This upland plateau, 200 to 700 ft (60 to 210 m) above sea level, is a gently rolling region of low hills and shallow lakes, drained almost entirely by the Mekong River via the Mun River. Mountains ring the plateau on the west and south, and the Mekong River delineates much of the eastern rim. Phu Kadueng, a national park in the north, is a 4,462 ft (1,360 m) high mesa, 23 sq mi (60 sq km) in surface area, that has wooded slopes and savannah (mixed grassland and forest) at the top. Phu Wiang and Phu Keaw are other mesas in the north.

Canyons

The Chaem River forms the narrow, rocky Ob Luang Gorge in the northwest. Phae Muang Phi ("City of Ghosts") is a canyon of labyrinthine rock formations sculpted by erosion in the north, near the town of Phrae. The small erosion-formed Sao Din Canyon is also located in the north, in the Nan Valley.

Hills and Badlands

Hill regions in Thailand include the countryside surrounding the northern city of Chiang Mai; the gem mining region of the southeast near Cambodia; and the picturesque limestone outcroppings along the southern peninsula and on the islands.

INLAND WATERWAYS

Lakes

Thale Sap Songkla (401 sq mi / 1,040 sq km) is Thailand's largest inland body of water. A lagoon lake on the southern peninsula, with a small inlet from the Gulf of Thialand, Thale Sap Songkla has a mixture of fresh and some brackish water. Two sanctuaries for waterbirds surround the lagoon lake's perimeters. Bung Nong Han is a 12 sq mi (32 sq km) freshwater lake in northeast Thailand.

Thailand has several huge reservoirs, including Srinakarin (300 sq mi / 419 sq km) near the Bilauktaung mountains, Khao Laem (150 sq mi / 388 sq km), Bhumiphol (116 sq mi / 300 sq km), Sirikit (100 sq mi / 260 sq km) in the north on the Nan River, and Rajjaprabha (64 sq mi / 165 sq km.) The reservoirs were formed by enormous environmentally and socially controversial dam projects. The country has 28 medium sized and large dams, constructed for irrigation diversion, domestic and industrial water supply, and generation of electric power.

Rivers

The Mekong River flows along much of Thailand's border with Laos. Approximately 2,700 mi (4,350 km) in length, it is the longest river in Southeast Asia. The eastern and some of the northern part of Thailand are drained by it. The Mun River, 400 mi (644 km) is the largest river within the northeast. The Mun and its Chi tributary empty into the Mekong. Rapids and falls in Laos and Cambodia prevent navigation down the Mekong from Thailand to the South China Sea.

The Chao Phraya 143 mi (230 km) and its tributaries drain an estimated one-third of the nation's territory. Together with part of the watershed of the Mekong, the two account for the importance of hydrology in the country's economic development. The network of rivers and human-made canals not only supports wet rice cultivation but provides vitally important transport waterways.

Northern Thailand's many rivers, including the Ping, Wang, Yom, and Nan, which unite in the lowlands to form the Chao Phraya watershed, provide water resources for numerous small irrigation schemes that support intensive rice cultivation in the alluvial valleys. The central plain is the lowland area drained by the Chao

Phraya, the country's principal river, and the smaller streams that feed into the delta near Bangkok. The highly developed irrigation systems of the central region support a concentrated population. The region is dominated by the Chao Phraya and its tributaries, with their waterborne transport and annual floods, and by the vast patchwork of cultivated paddy fields that stretches across it. Sprawling metropolitan Bangkok, the country's focal point of trade, transportation, and industrial activity, is situated on the southern edge of the plains region at the head of the Gulf of Thailand and encloses part of the delta of the Chao Phraya system.

Wetlands

Thailand's wetlands are being diminished by pollution, shrimp farms, and the cutting of mangroves for charcoal manufacture and firewood. Thailand has several sites designated as Wetlands of International Importance under the Ramsar Convention on Wetlands. They are: Bung Khong Long, in the north near Nong Khai, with several endemic fish species; Don Hai Lot, a rare ecosystem of inter-tidal mudflats in the south on Mae Klong Bay; the Krabi Estuary; Princess Sirindhorn Wildlife Sanctuary (Pru To Daeng), a large and very biodiverse peat swamp forest near Narathiwat and the Malaysian border; Kuan Ki Sian, a varied freshwater ecosystem near Thale Sap Songkla; and Nong Bong Kai (Chiang Saen Lake), an important bird habitat in the north.

THE COAST, ISLANDS AND THE OCEAN

Oceans and Seas

The shoreline of southern Thailand meets the Andaman Sea of the Indian Ocean to the west, and the Gulf of Thailand (formerly the Gulf of Siam) of the Pacific Ocean to the east. The offshore depths in the Gulf range from 98 to 262 ft (30 to 80 m). Thailand has 822 sq mi (2,130 sq km) of coral reefs. An estimated 96 percent of Thailand's coral reefs are considered "threatened," as they are endangered by dynamite fishing, pollution, oil spills, shrimp farming, and tourist activities.

Major Islands

Thailand's three largest islands are Phuket, 210 sq mi (543 sq km) in the Andaman Sea; Koh Samui, 93 sq mi (240 sq km) in the Gulf of Thailand off the Malay Peninsula; and Koh Chang, 85 sq mi (219 sq km) in the Gulf of Thailand off the southeast coast. Other islands in the Andaman Sea include the nine-island Similian group; the twin islands of Koh Phi-Phi; Koh Lanta; and the Turatao group, a Marine Park composed of 50 small islands. Other islands in the Gulf of Thailand are Koh Samet, a National Park off the southeast coast; and Koh Tao and Koh Phangan, both off the peninsula. Many of the islands have been developed for tourism purposes, and some are protected parks.

The Coast and Beaches

The Isthmus of Kra, which is just 15 mi (24 km) wide, connects the north-central mass of Thailand to its southern peninsula. There have been proposals for digging a canal through it, or building a superhighway across the isthmus, in order to use it as a transport channel between the Andaman Sea and the Gulf of Thailand, and thereby the Indian and Pacific Oceans. Thailand's Andaman Sea coastline, on the west side of the peninsula, extends south from the Myanmar (Burma) border to the Malaysian border, with many small islands. The large island of Phuket lies below a promontory that shelters the Andaman Sea's Phangnga Bay. Many small islands with dramatic limestone formations and caves attract visitors to Phangnga Bay.

The Gulf of Thailand coastline contains Mae Klong Bay, which indents into the country reaching its apex at the Chao Phraya outlet near Bangkok. The Gulf coast extends eastward to the Cambodian border, and extends southwest from Mae Klong Bay to the Malaysian border. The shoreline and islands on both the east and west coasts are graced with excellent beaches and harbors for fishing boats.

CLIMATE AND VEGETATION

Temperature

Most of Thailand has a tropical monsoon climate, with an equatorial climate affecting the southern peninsula. Three seasons are experienced: the rainy season from May to October, when the southwest monsoon arrives; the cool season from October to March, during the northeast monsoon; and the hot season from March to May. The country's average annual temperature is 83°F (28°C), with the average temperature in Bangkok varying from 77°F to 86°F (25°C to 30°C). Thailand's humidity averages 82 percent, dropping to 75 percent during the hot season.

Rainfall

The average annual rainfall in Thailand is 55 in (140 cm). Areas close to the sea receive more rain than inland areas. Northeast Thailand lies in the rain shadow of Indochina's mountains, and is very prone to droughts and chronic water shortages. Typhoons sometimes strike Thailand's south. In 2001, Typhoon Usagi caused flooding and massive mudslides on deforested hillsides, which left 176 people dead in southern Thailand. Global warming also threatens Thailand with changes in rainfall patterns and the possibility of major coastal flooding.

Grasslands

In the dry northeast scrub grassland is prevalent. Invasive imperata grasslands are common in the north where repeated burning of forests for agricultural clearing has taken place, though local and foreign aid groups

Population Centers – Thailand

(1990 POPULATION ESTIMATES)

Name	Population	Name	Population
Bangkok (metropolitan area, capital)	5,876,000	Songkhla	243,000
Nakhong Ratchasima	278,000	Nonthaburi	233,000
		Khon Kaen	206,000
		Chon Buri	187,000

SOURCE: "Population of Capital Cities and Cities of 100,000 and More Inhabitants." United Nations Statistics Division.

are attempting to reforest some of these areas. Wild and domesticated bamboo grows everywhere in Thailand, while cultivated ground cover include vetiver and kenaf. Khao Yai National Park, about 124 mi (200 km) north of Bangkok, has natural grasslands that are an important tiger, elephant, and deer habitat.

Forests and Jungles

The vegetation of Thailand is especially diverse, with more than 15,000 plant species, most of which are found in forest settings. Many types of orchids grow in Thailand's forests. During the twentieth century, Thailand's forests were extensively logged for teak and other hardwoods. The Thai government banned all logging in 1989 after disastrous floods were caused by deforestation. Thailand's deforestation comes from "slash and burn" agricultural clearing, frequent forest fires, and illegal timber trade. Approximately 25 percent forest cover remains; very little of it is primary rainforest. Types of forest in Thailand include mangrove (less than 1 percent), monsoon (5 to 9 percent), evergreen rainforest (10 percent), montane, and conifer. Tree plantations for commercial species such as eucalyptus and rubber exist but are environmentally controversial.

HUMAN POPULATION

Thailand's population growth rate is a very low 0.91 percent (2001) due to effective family planning programs. Population density in 2000 was 313 people per sq mi (121 people per sq km). Some 40 percent of Thais live in urban areas (2000), a segment that is projected to rise to 53 percent by 2010. More than 12 percent of the entire population lives in Bangkok There are substantial refugee populations from Myanmar (Burma) and other neighboring countries.

NATURAL RESOURCES

Thailand's mineral resources include tin—mined extensively in the south—iron ore, tungsten, tantalum, lead, gypsum, lignite, and fluorite. Precious gemstones are also found in the southeast near Cambodia. Being tropical, Thailand produces much rice, sugarcane, pineapples, and coconuts. It also grows corn, rubber, and soybeans, and has a sizable fishing economy. There are significant natural gas reserves in the Gulf of Thailand, and its many rivers make Thailand a large producer of hydropower through its dams and reservoirs. Thailand also brings in money through its tourism industry.

FURTHER READINGS

Boyd, Ashley J. *Thailand's Coral Reefs*. Bangkok: White Lotus, 1995.

Cubitt, Gerald, and Belinda Stewart-Cox. *Wild Thailand*. Cambridge MA: MIT Press, 1995.

Komar, Vitaly, and Alexander Melamid. *When Elephants Paint*. New York: HarperCollins, 2000.

Kuneepong, Parida. *Thailand, Gateway to Land and Water Information*. http://www.ldd.go.th/FAO/z_th/th.htm (accessed May 24, 2002).

Thai Society for the Conservation of Wild Animals. http://www.tscwa.org (accessed May 24, 2002).

Winichakul, Thongchai. *Siam Mapped: A History of the Geo-Body of a Nation*. Honolulu: University of Hawaii, 1994.

Togo

- **Area:** 21,925 sq mi (56,785 sq km) / World Rank: 125
- **Location:** Northern and Eastern Hemispheres, West Africa, bordering Benin to the east, the Atlantic Ocean to the south, Ghana to the west, and Burkina Faso to the north.
- **Coordinates:** 8°00′N, 1°10′E
- **Borders:** 1,023 mi (1,647 km) / Benin, 400 mi (644 km); Burkina Faso, 78 mi (126 km); Ghana, 545 mi (877 km)
- **Coastline:** 35 mi (56 km)
- **Territorial Seas:** 30 NM
- **Highest Point:** Mt. Agou, 3,235 ft (986 m)
- **Lowest Point:** Sea level
- **Longest Distances:** 317 mi (510 km) N-S; 87 mi (140 km) E-W
- **Longest River:** Oti, 340 mi (550 km)
- **Natural Hazards:** Drought; harmattan winds
- **Population:** 5,153,088 (July 2001 est.) / World Rank: 108
- **Capital City:** Lomé, on the southern coast
- **Largest City:** Lomé, population 662,000 (2000 est.)

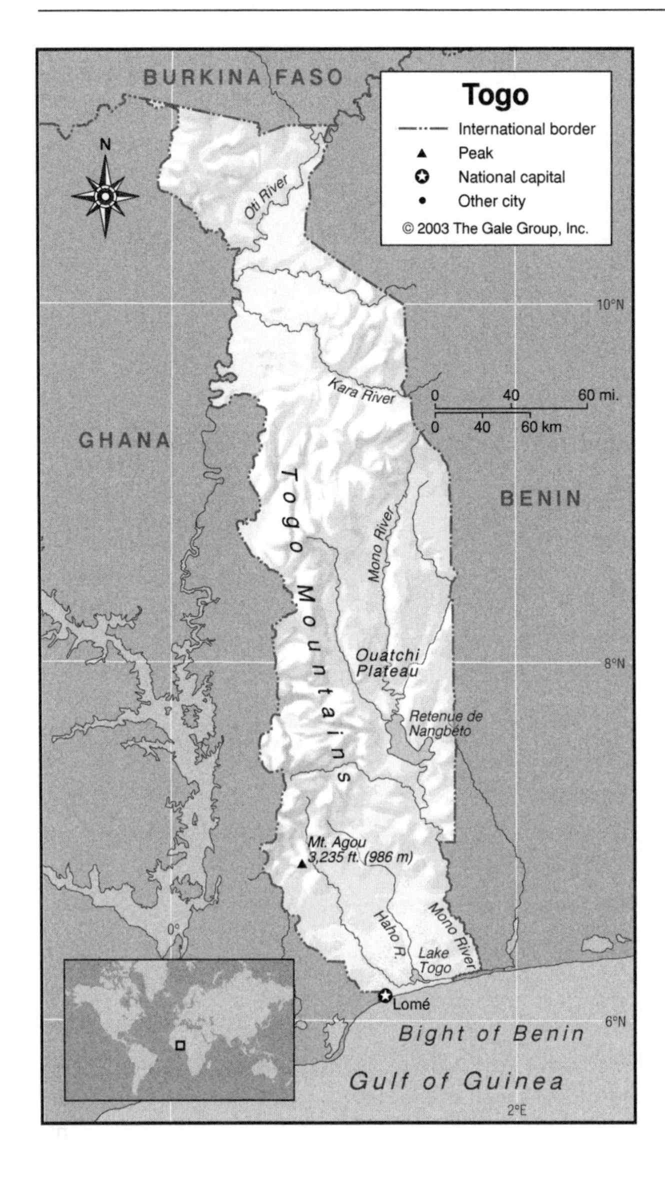

OVERVIEW

Togo is a long, narrow country in West Africa, sandwiched between Ghana and Benin. Its dominant physical feature is a chain of low mountains that stretches across the country trending southwest to northeast. Several different types of terrain lie north and south of these mountains. At the southernmost end is a narrow coastal strip, bordered by the low Ouatchi Plateau, which, in turn, gives way to the higher plateau that rises to the mountains. North of the Togo Mountains is yet another plateau, drained by the Oti River and crossed from southwest to northeast by granite escarpments. Togo is situated on the African Tectonic Plate.

MOUNTAINS AND HILLS

Mountains

The Togo Mountains, which cross Togo from southwest to northeast, belong to a mountain system that extends from the Atakora Mountains in Benin to Ghana's Akwapim Hills. Togo's highest peak, Mt. Agou, is located at the southern edge of these mountains rising to a height of 3,235 ft (986 m).

Plateaus

Togo has three different plateaus. The Ouatchi Plateau, which borders the coastal strip, is a transitional belt of reddish, lateritic clay soil. At elevations of between 200 and 300 ft (61 and 91 m), it extends some 20 mi (32 km) from the edge of the coastal region to a higher plateau drained by the Mono River. This second plateau stretches northward to the edge of the Togo Mountains. North of the mountains, a sandstone plateau traversed by granite ridges in the northwest is drained by the Oti River.

INLAND WATERWAYS

Lakes

Lake Togo is the largest of the inland lagoons lining Togo's coast. The largest lake in the country is a reservoir on the Mono River, the Retenue de Nangbéto.

Rivers

The Mono River flows north to south, traversing more than half the length of Togo before flowing into the Gulf of Guinea. Together with its tributaries, it drains most of Togo south of the central mountain chain. The primary river north of the mountains is the Oti River, a major tributary of the Volta River, and Togo's longest river with a total length of 340 mi (550 km). Besides the Mono and Oti, Togo's two other major rivers are the Kara, which crosses the Togo Mountains in the north, and the Haho River in the south, which drains into Lake Togo.

THE COAST, ISLANDS, AND THE OCEAN

Togo is bounded on the south by the Bight of Benin, which is within the Gulf of Guinea. Togo's coast is flat, low-lying, and narrow. It is fringed with sandy beaches separated from the rest of the land by lagoons and tidal flats, which give this area a swampy character. The beaches are not generally good for swimming because of the strong undertow—there is only one spot that is safe due to protection from a natural coral reef. However, fishing is possible from the shoreline or by boat. Marlin and sailfish are not unusual catches. Whales can often be seen from offshore boats.

CLIMATE AND VEGETATION

Temperature

Located only eight degrees north of the equator, Togo has a tropical climate. The northernmost part of the country, which is farther from the coast, has the greatest variations in temperature. The average high and low temperatures in the northern town of Mango are 95°F (35°C) and 59°F (15°C), compared with 86°F (30°C) and 73°F (23°C) in Lomé, which is on the southern coast.

Rainfall

Togo's climate, while moist, is drier than those of its neighbors on the Gulf of Guinea. The coast receives an annual average of about 31 in (78 cm), although it has two rainy seasons, one between April and early August, and a second, lesser one in October and November. The plateau region to the north has only the April to August rainy season but still averages 40 in (100 cm) of rainfall annually. The heaviest rainfall is in the Togo Mountains, which receive an average of around 60 in (150 cm) per year.

Grasslands

Togo's plateaus are covered by savanna grassland, with baobob and other large tree species found in the southern plateau only. Greenery flourishes in these areas during the rainy season, but the vegetation is reduced to dried-out grasses and shrubs when the weather turns dry.

Forests and Jungles

There are areas of tropical rain forest in the southwestern section of the Togo Mountains. Mangroves and dense patches of reeds grow in the coastal swamps and lagoons.

HUMAN POPULATION

The coastal plain region is by far the most densely populated, with a population density of 725 people per sq mi (280 people per sq km). The central mountainous and plateau region, by far the largest region, has a population density of about 160 people per sq mi (62 people per sq km) while the northern savanna region has a density of 172 people per sq mi (66 people per sq km). About one-third of the population is urban. The rest lives in small, widely scattered rural villages.

NATURAL RESOURCES

Togo's climate and the quality of its arable land allow for the cultivation of a variety of crops, on which its economy heavily depends. The most significant cash crops are cotton, coffee, and cocoa. Togo is also a major producer of phosphates; other mineral resources include marble and limestone.

FURTHER READINGS

Curkeet, A. A. Togo: *Portrait of a West African Francophone Republic in the 1980s*. Jefferson, N.C.: McFarland, 1993.

Knoll, Arthur J. *Togo Under Imperial Germany, 1884-1914: A Case Study in Colonial Rule*. Stanford, Calif.: Hoover Institution Press, 1978.

Mbendi Information for Africa. *Togo*. http://www.mbendi.co.za/land/af/to/p0005.htm (Accessed June 13, 2002).

Packer, George. *The Village of Waiting*. New York: Vintage Books, 1988.

Tonga

- **Area:** 289 sq mi (748 sq km) / World Rank: 178
- **Location:** An archipelago in the South Pacific Ocean, in the Southern and Western Hemispheres, Oceania, between Hawaii and New Zealand.
- **Coordinates:** 20°00′S, 175°00′W
- **Borders:** None
- **Coastline:** 260 mi (419 km)
- **Territorial Seas:** 12 NM
- **Highest Point:** Kao Island, 3,389 ft (1,033 m)
- **Lowest Point:** Sea level
- **Longest Distances:** 392 mi (631 km) NNE-SSW; 130 mi (209 km) ESE-WNW
- **Longest River:** None of significant length
- **Natural Hazards:** Earthquakes; volcanic activity; cyclones
- **Population:** 104,227 (July 2001 est.) / World Rank: 184
- **Capital City:** Nuku'alofa, northern Tongatapu
- **Largest City:** Nuku'alofa, population 40,000 (2000 est.)

OVERVIEW

Tonga, also known as the Friendly Islands, is an archipelago consisting of 171 islands in the South Pacific Ocean. Tonga is about one-third of the way from New Zealand to Hawaii. The nearest island groups are the Nieu Islands to the east, the Kermadec Islands to the south, Fiji to the west, and Wallis and Futuna to the north.

There are two types of natural division among Tonga's islands: between east and west, and between north and south. From north to south, the islands are clustered in three major groups: Vava'u to the north, Ha'apai in the middle, and Tongatapu to the south. There is also a smaller, more remote group, called the Niuas, farther north, as well as various individual islands both to the north and south. The islands of all the groups, from north to south, can be viewed as roughly falling into two parallel rows. The islands in the western row are purely volcanic in origin; those in the eastern row consist of submerged volcanoes capped by coral and limestone formations.

Tonga is situated where the Indo-Australian and Pacific Tectonic Plates converge. A long underwater channel called the Tonga Trench, which reaches from Tonga to New Zealand, is located at the meeting point of the two plates. The region's continuing seismic activity created a new island, called Metis Shoal, in 1995.

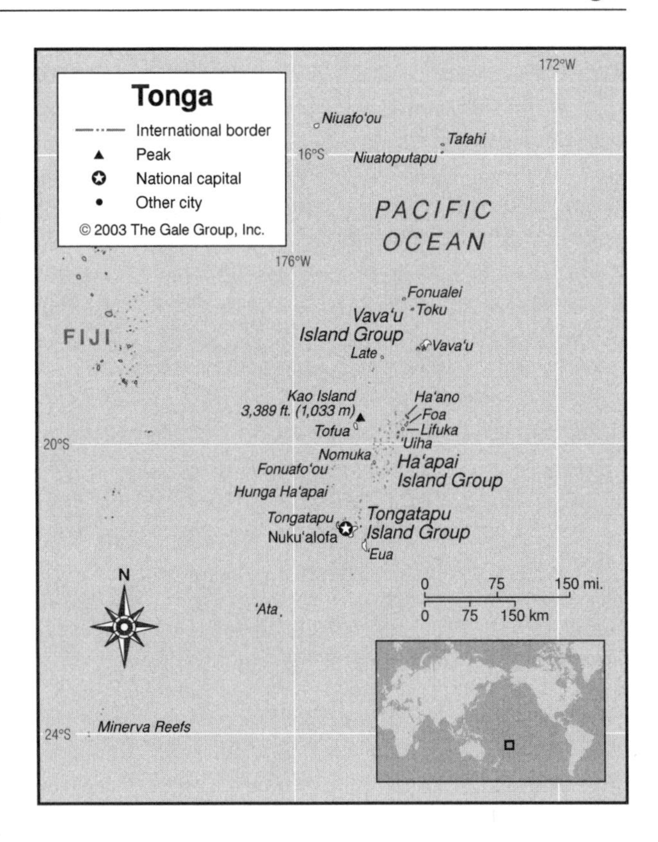

MOUNTAINS AND HILLS

Mountains

Tonga's islands are the tops of submerged volcanoes, four of which are still active on the islands of Tofua and Niuafo'ou. The islands to the west, which are purely volcanic, have risen above sea level thanks to volcanic activity alone, while the peaks to the east, which have been covered by coral, were originally lower.

Tonga's highest point is on Kao Island, in the central Ha'apai group, an altitude of 3,389 ft (1,033 m). A volcanic ridge on the island of 'Eua, the second-largest island in the Tongatapu group, rises to 1,078 ft (329 m).

Hills and Badlands

Hills rising to elevations between 500 and 1,000 feet (152 and 305 km) are found on islands in the Vava'u group.

INLAND WATERWAYS

Lakes

There are lakes on the islands of Vava'u, Nomuka, Tofua, and Niuafo'ou, some of which have waters that are very good for swimming, but none of which are of significant size.

Rivers

Tonga has no rivers. The island of 'Eua has creeks, and there is a single stream on Niuatoputapu.

THE COAST, ISLANDS, AND THE OCEAN

Oceans and Seas

The South Pacific Ocean surrounding Tonga is very seismically active. Tonga Trench, which is 35,400 ft (10,800 m) deep, has one of the greatest ocean depths in the world. Many of the islands are formed of coral reefs, and there are many other submerged reefs in the surrounding waters, including Minerva Reefs at the islands' southern end.

Major Islands

Tongatapu, with an area of 99 sq mi (256 sq km), is the largest single island and the site of Tonga's capital. The northernmost island group, Vava'u, has thirty-four islands; the Ha'apai group in the middle has thirty-six. The Tongatapu group to the south is composed of the island of Tongatapu, one other major island ('Eua), two much smaller ones, and a number of reefs.

The Coast and Beaches

Tonga's islands are fringed by coral reefs, which are gradually being eroded by rainwater. There are many white sandy beaches and magnificent swimming, diving, and snorkeling locations.

CLIMATE AND VEGETATION

Temperature

Most of Tonga is far enough from the equator to have a pleasant subtropical climate moderated by trade winds.

GEO-FACT

Being immediately west of the International Dateline, Tonga is the first nation to greet each new day, leading to the saying "Tonga is where time begins." Because of this, tourists flocked to the islands on December 31, 1999, to be the first to greet the new millennium.

Departments – Tonga

1996 POPULATION ESTIMATES

Name	Population	Area (sq mi)	Area (sq km)	Capital
'Eua	4,900	33.7	87.4	Ohonua
Ha'apai	8,100	42.5	110.0	Pangai
Niuas	2,000	27.7	71.7	Hihifo
Tongatapu	66,600	100.6	260.5	Nuku'alofa
Vava'u	15,800	46.0	119.2	Neiafu

SOURCE: *Geo-Data,* 1989 ed., and Johan van der Heyden, Geohive, http://www.geohive.com (accessed July 2002).

There are only two real seasons: the warmer season, from December to May, and the cooler season from May to December. Temperatures range from 60°F to 70°F (16°C to 21°C) in the coolest months of June and July, and average 80°F (27°C) in December, the hottest month.

Rainfall

Rainfall and humidity increase from south to north. Average annual rainfall ranges from 63 in (160 cm) in Tongatapu, to 87 in (221 cm) in Vava'u, to 101 in (257 cm) in Niuatoputapu.

Forests and Jungles

Original forest is found on the islands of the Vava'u group, as well as several islands in the Ha'apai group. The volcanic ridge on the island of 'Eua is the site of a forest reserve, and this island has the greatest number and variety of trees in Tonga. Tree species include coconut palm, paper mulberry, tavahi, toi, and other tropical varieties, as well as bushes and a wide array of tropical flowers, including frangipani, hibiscus, and datura.

HUMAN POPULATION

Tongatapu is home to two-thirds of the population, and this island is the most densely populated at 672 people per sq mi (259 people per sq km). The rest of the inhabited islands have a total density of 208 people per sq mi (80 per sq km). Of Tonga's 171 islands, only 45 are inhabited. Nearly half the population lives in urban areas.

NATURAL RESOURCES

Tonga's primary natural resources are its fertile soil and the plentiful fish in its waters. Its economy is heavily dependent on agriculture, with its main exports other than fish being coconuts, squash, bananas, and vanilla beans. It lacks in any mineral resources, but with its beautiful beaches and coral reefs it has developed a significant tourist industry.

FURTHER READINGS

Ellem, Elizabeth Wood. *Queen Salote of Tonga: The Story of an Era 1900-1965.* Auckland, New Zealand: Auckland University Press, 1999.

Fletcher, Matt, and Nancy Keller. *Tonga.* London: Lonely Planet, 2001.

Grijp, Paul van der. *Islanders of the South: Production, Kinship and Ideology in the Polynesian Kingdom of Tonga.* Tr. Peter Mason. Leiden: KITLV Press, 1993.

Rutherford, Noel, ed. *Friendly Islands: A History of Tonga.* New York: Oxford University Press, 1977.

Stanley, David. *Tonga-Samoa Handbook.* Emeryville, Calif.: Moon Publications, 1999.

Tonic. *Tonga: The Kingdom of Ancient Polynesia.* http://www.vacations.tvb.gov.to/index.htm (Accessed June 11, 2002).

Trinidad and Tobago

- **Area:** 1,980 sq mi (5,128 sq km) / World Rank: 167
- **Location:** Northern and Western Hemispheres, off the northeast coast of South America, two islands in the Atlantic Ocean northeast of Venezuela.
- **Coordinates:** 11°00′N, 61°00′W
- **Borders:** None
- **Coastline:** 225 mi (362 km)
- **Territorial Seas:** 12 NM
- **Highest Point:** Mount Aripo, 3,085 ft (940 m)
- **Lowest Point:** Sea level
- **Longest Distances:** Trinidad: 89 mi (143 km) N-S; 38 mi (61 km) E-W / Tobago: 26 mi (42 km) NE-SW; 7.5 mi (12 km) NW-SE
- **Longest River:** Ortoire River, 31 mi (50 km)
- **Largest Lake:** None of significant size
- **Natural Hazards:** None
- **Population:** 1,169,682 (July 2001 est.) / World Rank: 151
- **Capital City:** Port-of-Spain, located on the Gulf of Paria coast on Trinidad
- **Largest City:** Port-of-Spain, 52,000 (2000 metropolitan est.)

OVERVIEW

Trinidad and Tobago are two islands situated on the continental shelf of South America and are geographically but not geologically part of the West Indies. Trinidad, the larger of the two, is at some points within sight of the Venezuelan coast and was once a part of the mainland. Tobago, a few miles northeast of Trinidad, is part of a sunken mountain chain related to the continent. Trinidad, second largest of the Commonwealth Caribbean islands, is roughly rectangular in shape with lateral peninsular extensions at the northeast, northwest, and southwest corners. Smaller Tobago lies to the northeast of Trinidad and is separated from its sister island by a channel about 20 mi (32 km) in width. Both islands sit on the South American Tectonic Plate.

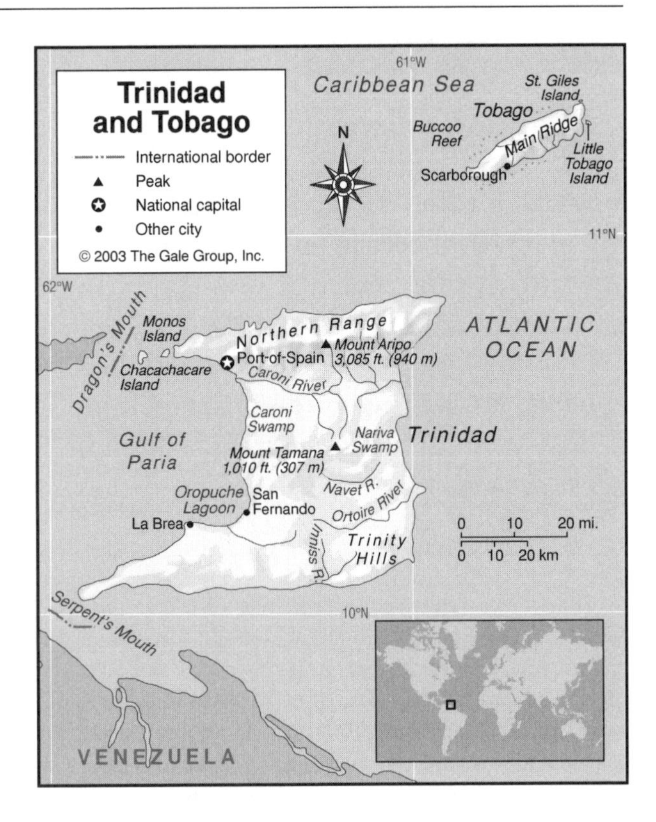

MOUNTAINS AND HILLS

Trinidad, the larger island, has one moderately high and two low mountain chains. The principal mountain system is the Northern Range, a rugged chain that covers the entire northern portion of the island. It includes the highest point in the country, Mount Aripo (Cerro del Aripo), with an elevation of 3,085 ft (940 m). The Northern Range is geologically an outlier of the Venezuelan portions of the Andes Mountains. Extending on a slant from northeast to southwest across the middle of the island, the Central Range has average elevations of 200 to 500 ft (61 to 152 m) and a maximum elevation at Mount Tamana, 1,010 ft (307 m). Along the southern coast the low and discontinuous Southern Range reaches a maximum elevation of a little less than 1,000 ft (304 m) in the Trinity Hills of the southeast.

Tobago, the smaller island, is generally mountainous. It has an uneven terrain dominated by the Main Ridge, a series of mountains near the northeast coast about 18 mi (29 km) long with elevations reaching a maximum of about 1,800 ft (548 m). South of the Main Ridge on Tobago are lower hills in which rivers have cut numerous deep and fertile valleys, and the southwestern part of the island consists of an extensive and fairly level coral platform.

INLAND WATERWAYS

Rivers

Rivers on Trinidad are numerous but short. Small rivers and streams are numerous on Tobago, but flooding and erosion are less serious on this island than on Trinidad because the upper slopes of the mountains retain much of their original forest cover. The longest river of the islands, the Ortoire, extends 31 mi (50 km) eastward to the Atlantic Ocean in the south. The second longest river, the Caroni at 25 mi (40 km), runs westward to the Gulf of Paria in the north. The Navet river begins in the dead center of the island and flows east to the ocean. Flowing to the southern coast is the Inniss. The only notable river on Tobago is the Courland River, which runs westward into the Caribbean Sea between the coral platform and the Main Ridge (a series of mountains near the northeastern coast).

Wetlands

There are no natural lakes, but extensive swamps occur along the eastern, southern, and western coasts on Trinidad. Some are mangrove swamps separated from the sea by wide sandbars. The most extensive of the swamplands are the Caroni Swamp and the Oropuche Lagoon on the Gulf of Paria, and the Nariva Swamp on the Atlantic coast to the east. The waters of most rivers and streams ultimately drain through these swamplands.

THE COAST, ISLANDS, AND THE OCEAN

Oceans and Seas

The Trinidad and Tobago islands are surrounded by the Caribbean Sea to the north and west, and the Atlantic Ocean to the east. In the Caribbean, southwest of Tobago, the Buccoo Reef houses coral gardens.

Major Islands

The Chacachacare and Monos islands and most of the remainder of the numerous small islands close to the Trinidad shoreline are located in or near the Dragon's Mouth, a narrow strait located in the Gulf of Paria. Tobago has several small satellite islands; the largest are Little Tobago Island and St. Giles, or Melville, Island.

The Coast and Beaches

On the north coast of Trinidad the shoreline is heavily indented, and the bays are rockbound. There is

GEO-FACT

Trinidad is home to Pitch Lake, the largest natural asphalt bog in the world. It covers 114 acres (46 hectares), is 250 ft (76 m) deep in the center, and is the world's leading source of natural asphalt. Commercially mined since the 19th century, the lake has a noxious odor—burbling, hissing, and occasionally spitting fire—yet it still attracts nearly 20,000 tourists per year.

no coastal plain between tidewater and the steep mountain cliffs. On the south the water is shallow, and the bays are narrow. The east coast is bordered by the Atlantic Ocean, providing several beaches. On the west, the land slopes gently from the Gulf of Paria to an interior of fertile hills and plains. An oval-shaped body of water, the Gulf of Paria separates Trinidad from Venezuela on the South American continent. The Gulf of Paria has narrow straits on the north and south named the Daron's Mouths and the Serpent's Mouth, respectively.

The town of Scarborough on Tobago is the only important port, but there are several small harbors, and the coastline is indented by numerous inlets and sheltered beaches.

CLIMATE AND VEGETATION

Temperature

The temperature varies little throughout the year. For the entire nation, the annual average temperature is 70°F (21°C). In Port-of-Spain, the capital, the minimum average temperature in January is 68°F (20°C) and the maximum is 86°F (30°C). In July, the temperature ranges from 73 to 88°F (23 to 31°C). In Trinidad's Northern Range, a decrease in temperature is caused by a corresponding increase in elevation, and nighttime temperatures are usually cool. For the most part, Tobago is cooler than Trinidad, owing to the more constant northeast trade winds.

Rainfall

Annual rainfall exceeds 100 in (250 cm) in Trinidad's northern and central hill areas, and on Tobago. In certain areas, the rainfall exceeds 150 in (380 cm). A large amount of hilly sections receive 80 in (200 cm) or more of rain, while in the lowlands the average is below 65 in (165 cm), and in other sections it is below 50 in (125 cm). The wet season occurs between June and December, and a relatively dry season occurs from January to May, but it is not a season of drought because rain still falls every few days in most areas.

Population Centers – Trinidad and Tobago

(1994 POPULATION ESTIMATES)

Name	Population
San Fernando	60,000
San Juan	55,000
Port-of-Spain (capital)	44,000
Arima	29,000
Marabella	27,000

SOURCE: Projected from United Nations Statistics Division data.

Grasslands

On the southern flank of the Northern Range on Trinidad, the mountains are deeply indented by river valleys as they slope gently to the broad Caroni Plain, the most extensive of the country's lowlands. The Central Range marks the southern limit of the Caroni Plain, and to the south of the Central Range the Naparima Plain on the west and the Nariva Plain on the east are the island's other major lowland areas. Throughout the lowlands the terrain ranges from flat to gently undulating.

Most of the limited amount of level land on Tobago occurs in the southwest, although narrow patches of coastal plain are found elsewhere, most notably around the mouth of the Courland River.

Forests and Jungles

About 31 percent of the land, or 398,000 acres (161,000 hectares), is covered by forests, with four-fifths of the forestland owned or administered by the government. Much of this land, however, is located in the hills areas and is inaccessible for exploitation for lumber.

HUMAN POPULATION

Most of the population lives on Trinidad, and about 74 percent lives in urban areas. The capital city, Port-of-Spain, its suburbs, and the area within 10 mi (16 km) of the suburbs houses one-third of the population. San Fernando, also on Trinidad, is the second most important town. The eastern half of Trinidad and the island of Tobago are sparsely populated. The overall population density is roughly 556 people per sq mi (215 people per sq km).

NATURAL RESOURCES

Petroleum, natural gas, and asphalt are Trinidad and Tobago's chief natural resources. On the southwestern coast of Trinidad, at La Brea, is Pitch Lake, the world's largest natural supply of asphalt. The islands also export large amounts of bananas, coffee, cocoa, sugar, and citrus fruits. Tourism is another contributor to the economy.

FURTHER READINGS

Bereton, Bridget. *A History of Modern Trinidad.* Portsmouth, N.H.: Heinemann Educational Books, 1981.

Columbus Publishing. *WorldTravelGuide.Net.* http://www.travel-guides.com/data/tto/tto.asp (Accessed March 12, 2002).

Lonely Planet Publications. *Lonely Planet Online.* http://www.lonelyplanet.com/destinations/caribbean/trinidad_and_tobago/index.htm (Accessed March 11, 2002).

O'Donnell, Kathleen, and Harry S. Pefkaros. *Adventure Guide to Trinidad & Tobago.* Edison, N.J.: Hunter Publishing, 1996.

Tourism and Industrial Development Company of Trinidad and Tobago Limited (TIDCO). http://www.visittnt.com/General/about/geography.html (Accessed March 12, 2002).

Winer, Lise. *Trinidad and Tobago.* Philadelphia: J. Benjamins, 1993.

Tunisia

- **Area:** 63,170 sq mi (163, 610 sq km) / World Rank: 91
- **Location:** Northern and Western Hemispheres, in North Africa to the south and west of the Mediterranean Sea, bordered by Libya on the southeast and Algeria on the west
- **Coordinates:** 34°00′N, 9°00′E
- **Borders:** 884 mi (1,424 km) total / Algeria, 600 mi (965 km); Libya, 285 mi (459 km)
- **Coastline:** 713 mi (1,148 km)
- **Territorial Seas:** 12 NM
- **Highest Point:** Mt. Ash-Sha′nabī, 5,065 ft (1,544 m)
- **Lowest Point:** Chott el Gharsa, 56 ft (17 m) below sea level
- **Longest Distances:** 492 mi (792 km) N-S / 217 mi (350 km) E-W
- **Longest River:** Medjerda, 230 mi (360 km)
- **Largest Lake:** Chott el Djerid, 1,900 sq mi (5,000 sq km; approximate maximum area, dry during summer)
- **Natural Hazards:** None
- **Population:** 9,705,102 (July 2001 est) / World Rank: 81
- **Capital City:** Tunis, located on the Mediterranean Sea coast
- **Largest City:** Tunis, 1.9 million (2000 metropolitan est.)

GEO-FACT

Tunisia has been a civilized land since ancient times. The ancient city of Carthage, located near Tunis, fought many wars with Rome before being conquered and becoming part of the Roman Empire. Roman ruins can be found throughout the country. El-Jem, an ancient, well-preserved colosseum (almost as big as the one in Rome) is located on a plateau about 210 km (130 mi) south of the capital city Tunis. It is believed to have been built between 230 A.D. and 238 A.D., and it is estimated that the seating capacity was 30,000. Two long underground passageways that held the animals and gladiators that were going to compete in the arena are still accessible.

OVERVIEW

Tunisia juts into the Mediterranean Sea on the northern coast of the African continent. Together with Algeria, Morocco, and the northwestern portion of Libya known historically as Tripolitania, Tunisia makes up the Maghreb, a region in which fertile coastal lands give way to the great Atlas mountain chain of North Africa and, finally, to the interior expanses of the Sahara Desert. It is located on the extreme northern edge of the African Tectonic Plate; parts of the northern coast are on the neighboring Eurasian Plate.

Tunisia can be divided into northern, southern, and central regions, determined in part by topography and quality of the soils, and, in particular, by the incidence of rainfall, which decreases progressively from north to south. The Mediterranean Sea influences the climate in the north, and the Sahara Desert influences the south.

MOUNTAINS AND HILLS

Mountains

The Atlas mountain system, which begins in southwestern Morocco, terminates in northeastern Tunisia. The principal mountain chain, the Dorsale, slants northeastward across the country from the Algerian border to Cape Bon. Mount Ash-Sha′nabī near the Algerian border in this range is the country's highest point at 5,064 ft (1,544 m), but elevations average less than 984 ft (300 m) and rarely exceed 3,280 ft (1,000 m). The Dorsale is cut

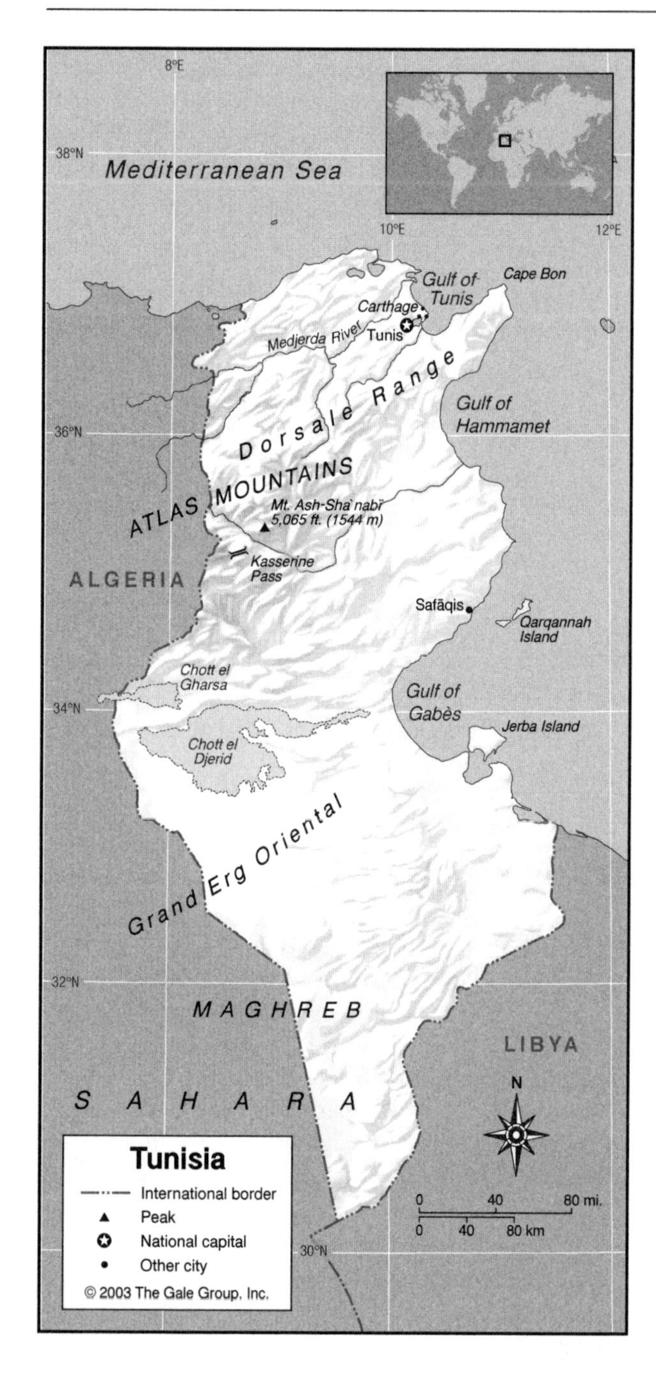

by several transverse depressions, among them the Kasserine (Al Qasrayn) Pass.

Plateaus and Badlands

Northern Tunisia, a generally mountainous region that comprises one-fourth of the country, is sometimes referred to as the Tell. It is a heavily populated area of high ground located close to the Mediterranean Sea. The region is bisected from east to west by the Medjerda River and is divided into subregions made up of the Medjerda Valley and the several portions of the Tell.

The western part of central Tunisia along the border with Algeria is moderately elevated and known as the High Steppes. In the desert region of the south there are many hills, small plateaus, and other forms of elevated terrain.

INLAND WATERWAYS

Lakes

Two large chotts or shatts (salt lakes) are located in Tunisia's southern region: the Chott el Djerid (the largest lake in the country) and the Chott el Gharsa (the lowest point). Chott el Djerid is dry during half the year but is flooded to form a shallow salt lake during the winter months.

Rivers

The most important river system in Tunisia, the Medjerda, rises in Algeria and drains into the Gulf of Tunis. It is also the only river that flows perenially, while other watercourses are seasonal. Even the volume of flow of the Medjerda in June and July is less than one-twelfth of that in February. In the central Tunisian steppes, occasional watercourses flow southward out of the Dorsale after heavy rains but evaporate in salt flats without reaching the sea.

THE COAST, ISLANDS, AND THE OCEAN

Tunisia is bordered by the Mediterranean Sea on the north and east. In the north the shoreline is indented by the Gulf of Tunis. Cape Bon forms the southeastern shore of this gulf with the coast curving sharply to the south at its point. Immediately to the south of it is the Gulf of Hammamet. Further to the south is the largest of Tunisia's gulfs, the Gulf of Gabès. Jerba and Qarqannah islands are located there. The eastern shoreline is smooth and sandy, and the northern shoreline is rocky.

CLIMATE AND VEGETATION

Temperature

Along the Mediterranean coast, temperatures are moderate—the average temperature is 64°F (18°C). In the interior south, containing the Sahara Desert, it is very hot. The summer season in the north (May-September) is hot and dry. In the winter months (October-April) the climate is mild with frequent rains. Temperatures at the capital city of Tunis range from 43°F (6°C) to 57°F (14°C) in January, and 70°F (21°C) to 91°F (33°C) in August.

Rainfall

Rainfall reaches a high of 59 in (150 cm) in the northern part of the country, while the extreme south rainfall averages less than 8 in (20 cm) per year.

Grasslands

Tunisia's most fertile land is in the north, on the plateaus of the Tell. The Medjerda River valley is particularly

Governorates – Tunisia

2000 POPULATION ESTIMATES

Name	Population	Area (sq mi)	Area (sq km)	Capital
Aryānah	678,900*	602	1,558	Aryānah
Bājah	317,300	1,374	3,558	Bājah
Banzart	518,500	1,423	3,685	Banzart
Bin `Arūs	446,000	294	761	Bin `Arūs
Jundūbah	424,900	1,198	3,102	Jundūbah
al-Kāf	280,400	1,917	4,965	al-Kāf
Madanīyīn	421,800	3,316	8,588	Madanīyīn
al-Mahdīyah	366,400	1,145	2,966	al-Mahdīyah
al-Munastīr	413,400	393	1,019	al-Munastīr
Nābul	632,700	1,076	2,788	Nābul
Qābis	332,300	2,770	7,175	Qābis
Qafsah	327,600	3,471	8,990	Qafsah
al-Qasrayn	416,300	3,114	8,066	al-Qasrayn
al-Qayrawān	563,200	2,591	6,712	al-Qayrawān
Qibilī	141,400	8,527	22,084	Qibilī
Safāgis	808,700	2,913	7,545	Safāqis
Sīdī Bū Zayd	398,600	2,700	6,994	Sīdī Bū Zayd
Silyānah	255,300	1,788	4,631	Silyānah
Sūsah	492,500	1,012	2,621	Sūsah
Tatāwin	147,300	15,015	38,889	Tatāwin
Tawzar	96,300	1,822	4,719	Tawzar
Tūnis	929,500	134	346	Tunis
Zaghwān	154,100	1,069	2,768	Zaghwān

*Includes Manouba, which was formed after the Aryanah governorate was split into two parts in September 2000.

SOURCE: National Statistics Institute, Tunisia.

heavily cultivated. This is the only region that receives adequate rainfall. South of the Tell and the Dorsale Mountains is central Tunisia, a region of generally poor soils and sparse rainfall. Its interior consists of a semiarid grasslands known as the High Steppes and the Low Steppes, the former occurring at greater elevations near the Algerian border. (The term steppes was used by the French to define the semiarid interior highlands of North Africa, and the area has little other than its name in common with the better known steppes of Central Asia.) Rainfall is scant in the steppes, but the dew is heavy and provides sufficient water for modest plant life. Parts of the region have been cultivated for cereals, olives and other tree crops. Esparto grass, the characteristic vegetation of the steppe region, covers more than one-fourth of the country. The coast of the central region, called the Sahel, has a more humid climate and more closely resembles northern Tunisia than the rest of this region.

Deserts

In the southern region, the land rises to form the plateaus, tablelands, and occasional eroded hills that make up the Tunisian portion of the Sahara Desert. Southern Tunisia commences with an area where elevations are lower and where the landscape is marked by numerous shatts (salt marshes or lakes) that lie below sea level. The Grand Erg Oriental, the edge of the Saharan dunes, is interrupted by the flat-topped Monts des Ksour. Fringed by lagoons and salt flats, the narrow gravelly coast of this southern region is sparsely settled, and the interior is almost totally barren except for a few nomads and inhabitants of oases, such as Nefta and Tozeur that occur along a line of springs at the foot of an interior escarpment. Surface water is otherwise extremely scarce.

HUMAN POPULATION

The Sahel (coast or shore) and portions of northern Tunisia combine to make up the economic, political, and cultural heartland of the country. About two-thirds of the population lives in urban areas; hardly anyone lives in the interior of the southern region, with the exception of a few oases cities. Almost all Tunisians are of Arabic ethnicity and are Muslim.

NATURAL RESOURCES

Tunisia's natural resources include iron ore, phosphates, petroleum, lead, zinc, and salt. Iron, petroleum, and phosphates are the most significant, but Tunisia's diverse economy does not depend on them. Tunisia's farmland produces a variety of Mediterranean foodstuffs, most notably olives and olive oil.

FURTHER READINGS

ArabNet. *Tunisia Geography.* http://www.arab.net/tunisia/geography/tunisia_geography.html (accessed March 7, 2002).

Brown, Roslind Varghese. *Tunisia.* New York: Marshall Cavendish, 1988.

Darke, Diana. *Passport's Illustrated Travel Guide to Tunisia.* Chicago: Passport Books, 1996.

LonelyPlanet.com. *Tunisia.* http://www.lonelyplanet.com/destinations/africa/tunisia/index.htm (accessed March 7, 2002).

Nelson, Harold D., ed. *Tunisia: A Country Study.* 3rd edition. Washington D.C.: Department of the Army, 1988.

Wilkinson, Stephan. "North Africa: Looking for an Oasis." *Conde Nast Traveler.* March 1995, Vol. 32, No. 3, p. 72+.

Turkey

- **Area:** 301,382 sq mi (780,580 sq km) / World Rank: 38
- **Location:** Northern and Eastern Hemispheres, with territory in both Europe and Asia, bordering Bulgaria to the northwest; Greece to the west; Iraq and Syria to the south; Armenia, Azerbaijan, and Iran to the east; and Georgia to the northeast.

- **Coordinates:** 39°00′N, 35°00′E
- **Borders:** 1,632 mi (2,627 km) / Armenia, 167 mi (268 km); Azerbaijan, 6 mi (9 km); Bulgaria, 149 mi (240 km); Georgia, 157 mi (252 km); Greece, 128 mi (206 km); Iran, 310 mi (499 km); Iraq, 206 mi (331 km); Syria, 511 mi (822 km)
- **Coastline:** 4,474 mi (7,200 km)
- **Territorial Seas:** 6 NM in the Aegean Sea; 12 NM in the Black and Mediterranean Seas
- **Highest Point:** Mount Ararat, 16,949 ft (5,166 m)
- **Lowest Point:** Sea level
- **Longest Distances:** 994 mi (1,600 km) SE-NW; 404 mi (650 km) NE-SW
- **Longest River:** Euphrates, 2,235 mi (3,596 km)
- **Largest Lake:** Lake Van, 2,545 sq mi (3,713 sq km)
- **Natural Hazards:** Severe earthquakes in the north
- **Population:** 66,493,970 (July 2001 est.) / World Rank: 16
- **Capital City:** Ankara, slightly northwest of the center of the country
- **Largest City:** Istanbul, in the northwest on the Sea of Marmara, 8,141,163 (1997 est.)

OVERVIEW

Turkey straddles the continents of Asia and Europe and consists of regions in each. About 3 percent is in Europe; it is part of the European region known as Thrace. It is separated from the Asian portion of Turkey by the Bosporus (İstanbul Boğazi, Karadeniz Bogazi), the Sea of Marmara (Marmara Denizi), and the Dardanelles (Çanakkale Boğazi), which are a series of waterways that connect the Black Sea with the Aegean Sea. The rest of the country is located in Asia, mostly on the peninsula of Asia Minor, the westernmost extension of the continent. This region is also called Anatolia, or simply Asiatic Turkey.

The terrain is structurally complex, and may be categorized into five regions: the Black Sea region in the north; Sea of Marmara region in the northwest; the Aegean Sea region in the far west; the Mediterranean Sea region in the south; and the Anatolian Plateau region in the country's center. All of the regions share a generally mountainous terrain, and there are many large lakes and rivers found throughout the country.

Turkey is located on the Eurasian Tectonic Plate. However, the boundaries with the Arabian Tectonic Plate and the African Tectonic Plate are located on or near the southern borders of the country. There is also a major fault line running along the northern part of Asia Minor. As a result of all this, the country is subject to a very high level of seismic activity. The earthquakes cause massive damage to buildings and, especially if they occur at night during the winter months, numerous deaths and injuries. For example, a violent earthquake in Erzincan the night of December 28–29, 1939, devastated most of the city and caused an estimated 160,000 deaths; more recently, an August 1999 quake centered on İzmit (Kocaeli) in the far west near the Sea of Marmara resulted in more than 17,000 deaths. Earthquakes of moderate intensity frequently continue with sporadic shocks over periods of several days or even weeks. The most earthquake-prone region centers on an arc that stretches from the general vicinity of the Sea of Marmara (Marmara Denizi) to the area north of Lake Van (Van Gölü) on the border with Georgia and Armenia.

MOUNTAINS AND HILLS

Mountains

Except for a relatively small segment along the Syrian border that is a continuation of the Arabian Platform, Turkey is part of the great Alpine-Himalayan mountain belt. The intensive folding and uplifting of this mountain belt during the Tertiary Period was accompanied by strong volcanic activity and intrusions of igneous rock material, followed by extensive faulting in the Quaternary Period. As a result, mountain ranges can be found throughout most of the country.

The most important mountain range in the south is that of the Taurus Mountains (Toros Dağlari). They run along the entire Mediterranean coast and extend far inland to the border with Iran, and include many peaks of over 10,000 ft (3,048 m). Smaller mountain ranges surround the Taurus on all sides, including the Aydin, Nur, Tahtali, Karagöl, and Mardin Mountains.

Another series of mountain ranges runs along the northern coast on the Black Sea. Principal among them are the Köroğlu, Küre, and Pontic Mountains. In the Marmara region of the northwest, the highest peak is Ulu Dağ (Mount Olympus), which rises to 8,392 ft (2,543 m) and provides a center for winter sports. Further east, the mountains rise as high as 12,897 ft (3,931 m) at Mt. Kaçkar (Kaçkar Dagi).

The nation's highest peak is the extinct volcano Mount Ararat (Buyuk Agri Dagi), which rises to 16,949 ft (5,166 m) in the far east near the border with Iran. To its southwest is a 12,857-ft (3,896-m) peak known as Little Mount Ararat; a plateau of lava covers the territory between the two peaks.

Plateaus

A large central massif, the Anatolian Plateau, composed of uplifted blocks and downfolded troughs, covered by recent deposits and giving the appearance of a plateau with rough terrain, is wedged between two the

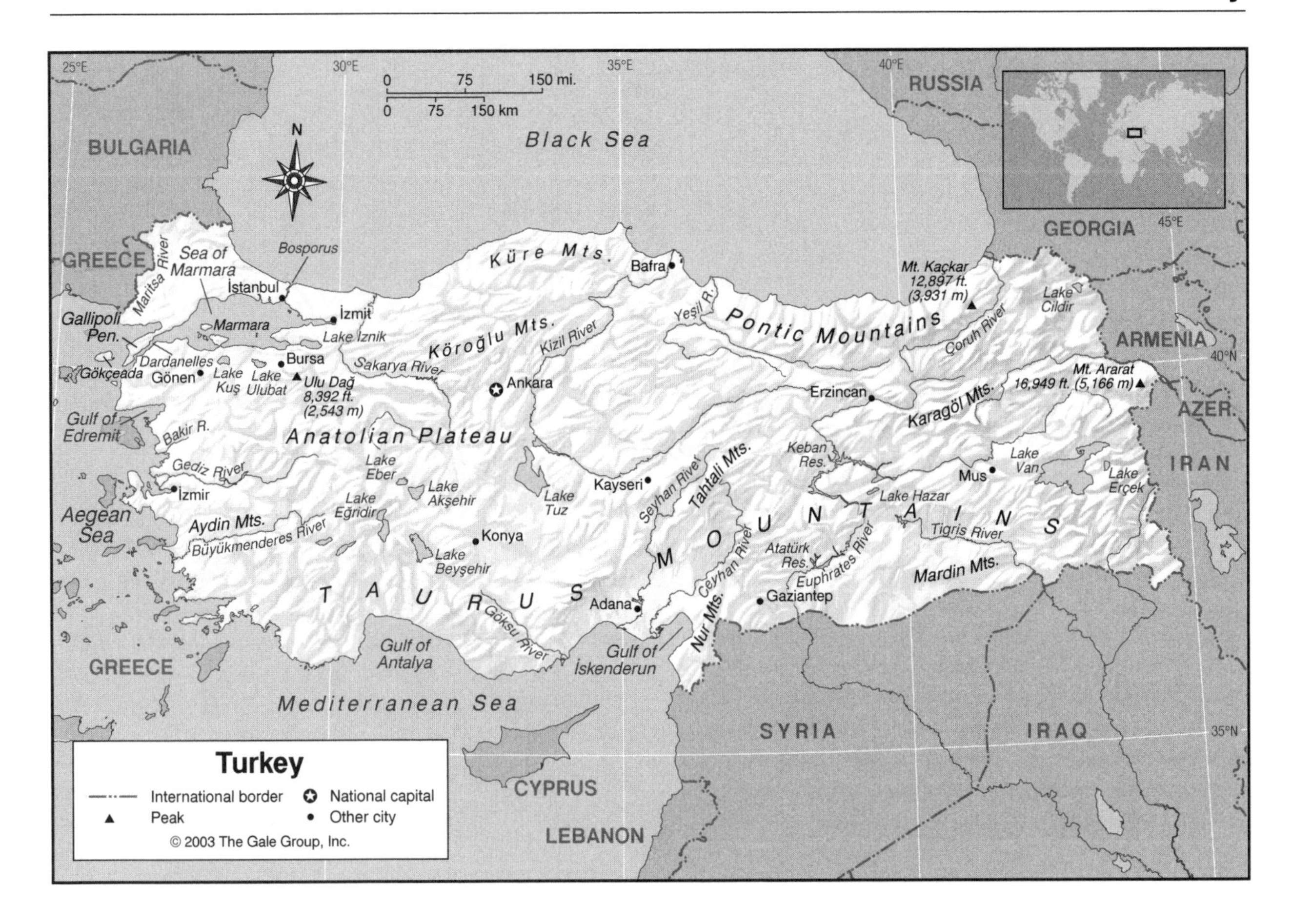

northern and southern mountain ranges. This plateau, with altitudes rising from west to east (from 1,980 to 3,960 ft / 600 to 1,200 m), is the heartland of the country. Except in the northwest, the mountains act as formidable barriers between the coastal regions and the plateau. It is crossed by many rivers, and is also the site of several large lakes, and has extensive plains.

Between Lake Tuz (Tuz Gölü) and Kayseri lies the tourist center of Ügrüp, where ancient dwellings were carved into the rock faces. In this region, the exposed rock has been eroded into strange formations called fairy chimneys. They resemble mushrooms, inverted cones, and obelisks, and have been carved by various civilizations through history to provide living space.

Canyons

In the central Anatolian region, the Melendiz River has eroded the Ihlara Valley to produce a deep canyon. The walls of the canyon have been carved to form Byzantine chapels, featuring frescoes. Also in this same area, underground villages were carved into the rock where early Christians hid to avoid persecution.

Hills and Badlands

There are regions of moderate hills in Thrace and in the far east along the border with Syria.

INLAND WATERWAYS

Lakes

The largest lake in the country, Lake Van (Van Gölü; 2,545 sq mi / 3,713 sq km) lies near the border with Iran. Other lakes in this eastern region include Ercek, Cildir, and Hazar. Turkey's second-largest lake, the shallow and salty Lake Tuz (Tuz Gölü) lies in central Anatolia directly south of Ankara. Lakes Akşehir and Eber lie west of Lake Tuz. Further to the southwest, in the Taurus Mountains west of Konya, are Lakes Beyşehir and Eğridir. Lying around the Sea of Marmara are numerous small lakes, the largest of which are Kuş, Ulubat, and Iznik.

Thermal Springs

Southwest of the Sea of Marmara region lies Gönen, where springs bubble from deep under ground, reaching the Earth's surface at about 180°F (82°C). Gönen has been the site of therapeutic mineral baths since the fifth century. In and around Bursa, thermal springs and therapeutic baths may also be found.

Rivers

The Euphrates (Firat) River has its source in central eastern Turkey. The longest river in Turkey, and in all of the Middle East, the Euphrates flows west initially, then curves south, crosses the Taurus Mountains, and enters Syria. It eventually flows southeast through Iraq and into

GEO-FACT

According to the Old Testament of the Bible, Noah's Ark landed on Mount Ararat (Buyuk Agri Dagi) in what is now eastern Turkey when the waters of the great flood subsided.

the Persian Gulf. There are two large reservoirs on the Euphrates in Turkey, the Keban and the Atatürk. The Tigris (Dicle) river also has its source in Turkey, somewhat south and west of that of the Euphrates in the Taurus Mountains. It follows a southeasterly path and soon exits Turkey for Iraq, where, hundreds of miles later, it joins the Euphrates shortly before reaching the Persian Gulf.

The longest river that flows completely within Turkey is the Kizil (Kizilirmak, Halys; 847 mi / 1,355 km), which follows a twisting path through central Anatolia. It forms a broad half-circle just east of Ankara, flowing first southwest but eventually curving all the way to the northeast to empty into the Black Sea at the headland of Bafra. Other rivers that empty into the Black Sea are the Yeşil in the east and the Sakarya in the west. The Çoruh River, renowned for its white-water rafting, rises in the mountains of eastern Turkey and reaches the Black Sea through neighboring Georgia.

The Gediz, Kücükmenderes, and Büyükmenderes Rivers flow westward to the Aegean Sea in Anatolia. The Maritsa River also flows into the Aegean in Europe, and marks most of Turkey's border with Greece. The Seyhan, Ceyhan, and Göksu rivers flow southward into the Mediterranean Sea. Lying 47 mi (76 km) south of Kayseri is the Kapuzbasi waterfall, which features a 230-ft (70-m) cascade fed by seven underground springs.

Wetlands

Turkey has extensive wetlands, most of which provide protected habitat for birds. The most important wetland area forms part of Kuscenneti National Park near Lake Kuz, where the habitat supports more than 225 bird species, with visits by an estimated 3 million migratory birds; Kuscenneti was established as a national park in 1959. Eleven other parks protect wetland bird habitats.

THE COAST, ISLANDS, AND THE OCEAN

Oceans and Seas

Turkey has coastlines on four different seas: the Black Sea, the Sea of Marmara, the Aegean Sea, and the Mediterranean Sea. The Black Sea is to the north of Anatolia, which makes up its entire southern coast. The Aegean Sea is west of Anatolia. The two are connected through the Sea of Marmara. This is a small inland sea lying between Asian and European Turkey in the far northwest of Turkey. It has a surface area of about 4,382 sq mi (11,350 sq km). The Dardanelles connects the Sea of Marmara to the Aegean Sea in the west, while the Bosporus connect it to the Black Sea to the northeast. The great city of İstanbul (Constantinople, Byzantium) is located on the Haliç (Golden Horn) estuary of the Bosporus. These two straits and the Sea of Marmara itself are what separate Europe and Asia. To the south of Anatolia is the Mediterranean Sea, which all of Turkey's other neighboring seas ultimately feed into.

Major Islands

There are numerous islands off the west coast in the Aegean Sea, but almost all of them belong to Greece. One of the few exceptions is the island of Gökçeada (İmroz), Turkey's largest, not far from the Dardanelles. It is covered with pine and olive trees, and surrounded by sparkling clear water. There is also an archipelago of nine small islands in the Sea of Marmara, where wealthy Turks have summer homes.

The Coast and Beaches

There are narrow coastal lowlands along the Black Sea and Mediterranean coasts. The western, Aegean, coastline is extremely irregular, with dramatic mountain faces rising perpendicularly from the sea and many islands just off shore (most of them Greek territory). The Gallipoli Peninsula extends southwest from Thrace to form the northern side of the Dardanelles. Along the west coast bordering the Aegean Sea, there are a number of inlets, including the Gulf of Edremit. This protected gulf encloses clear Aegean waters, and the shores lining the gulf feature sandy beaches surrounded by olive groves. The Gulf of Antalya indents the middle of the southern coast, and the Gulf of İskenderun marks the southeastern edge of Asia Minor.

CLIMATE AND VEGETATION

Temperature

The southern part of Turkey enjoys a Mediterranean climate, with a mean annual temperature of 63° to 68°F (17° to 20°C). In İstanbul temperatures average 40°F (4°C) in winter, and 81°F (27°C) in summer. The northern area along the Black Sea is slightly cooler, with a mean annual temperature range from 57° to 60°F (14° to 16°C); in winter, the average is 45°F (7°C) and in summer, 69°F (23°C). The central plateau region experiences wider daily and seasonal temperature variation, with cold winters and hot summers, yielding an annual mean temperature in the range of 46° to 54°F (8° to 12°C). The eastern region has higher elevations and temperatures are cooler overall, with the yearly mean in the range of 39° to

Population Centers – Turkey

(1995 POPULATION ESTIMATES)

Name	Population	Name	Population
İstanbul	7,774,169	Bursa	1,016,760
Ankara (capital)	2,837,937	Gaziantep	730,435
İzmir (Smyrna)	2,017,699	Konya	584,785
Adana	1,066,544	Antalya	502,269

SOURCE: "Population of Capital Cities and Cities of 100,000 and More Inhabitants." United Nations, Statistics Division.

48°F (4° to 9°C); winters can be severe in this region, with 120 days of snow cover and minimum temperatures of -4° to 3°F (-30° to -38°C); the average temperature in winter is 21°F (-13°C), and in summer, 63° (17°C).

Rainfall

Adequate rainfall occurs along the Mediterranean coast and the western coast of the Aegean Sea (23 to 51 in / 58 to 130 cm). The region bordering the Black Sea is also well-watered, with annual rainfall in the range of 28 to 87 in (71 to 220 cm). The Taurus Mountains along the Mediterranean prevents rain from reaching the heart of the country, which is therefore much drier, with annual rainfall between 22 to 28 in (56 to 71 cm).

Grasslands

The Ergene Plain is a lowland region in Thrace, extending along rivers that discharge into the Aegean Sea or the Sea of Marmara. There are many grassland areas in Anatolia. To the east and south of the Sea of Marmara, fertile plains stretch from west to east, following the flow of the Gediz, Bakir, and Kücük Menderes rivers. At an elevation of about 2,967 ft (899 m) around Tuz Gölü (Salt Lake) there are grassland plains. Relatively flat land is also found to the east of Konya, and south of Ankara. A fertile broad valley lies west of Lake Van, centered on Mus.

Forests

Slightly more than 10 percent of Turkey is forest, most of which lies in protected national reserves or parks. Forests are found in the mountainous areas near the Black Sea, Sea of Marmara, Aegean Sea, and the Mediterranean Sea. Small pine forests are found in central Anatolia, but the most common forest species is oak.

HUMAN POPULATION

Turkey is a populous country, with an overall population density of 221 people per sq mi (85 people per sq km) despite its large size. The highest population density is found around the Sea of Marmara in the northwest, where the former capital of İstanbul is located. There are also many people living along the coasts and in the central plains regions.

NATURAL RESOURCES

Although Turkey's mineral resources are extensive, only a few have been fully exploited as of the early twenty-first century. In the late 1990s, Turkey was the world's primary producer of boron; and was among the primary producers of limestone, marble, perlite, and pumice. It also mined significant deposits of coal, chromite, and copper.

FURTHER READINGS

"Big Western Turkey Cities Hit by Quake." *The New York Times* August 17, 1999, A9.

Embassy of Turkey. http://www.turkey.org/countryprofile/ (Accessed July 18, 2002).

Facaros, Dana. *Turkey.* London: Cadogan, 2000.

Focus Online Magazine. "Turkey the Key." http://www.focusmm.com/tanamenu.htm (Accessed July 18, 2002).

Karpat, Kermit H., ed. *Ottoman Past and Today's Turkey.* Boston: Brill, 2000.

Kinzer, Stephen. *Crescent and Star: Turkey Between Two Worlds.* New York: Farrar, Straus and Giroux, 2001.

Turkmenistan

- **Area:** 188,456 sq mi (488,100 sq km) / World Rank: 53
- **Location:** Northern and Eastern Hemispheres, in Central Asia, bordering the Caspian Sea to the west, Kazakhstan to the northwest, Uzbekistan to the north and northeast, Afghanistan to the southeast, and Iran to the southwest.
- **Coordinates:** 40°00′N, 60°00′E
- **Borders:** 2,321 mi (3,736 km) / Afghanistan, 462 mi (744 km); Iran, 616 mi (992 km); Kazakhstan, 235 mi (379 km); Uzbekistan, 1,007 mi (1,621 km)
- **Coastline:** None
- **Territorial Seas:** None
- **Highest Point:** Gora Ayribaba, 10,299 ft (3,139 m)
- **Lowest Point:** Vpadina Akchanaya, 266 ft (81 m) below sea level
- **Longest River:** Amu Dar'ya, 1,580 mi (2,540 km)
- **Largest Lake:** Lake Sarygamysh, 309 sq mi (800 sq km)
- **Natural Hazards:** Earthquakes; sandstorms

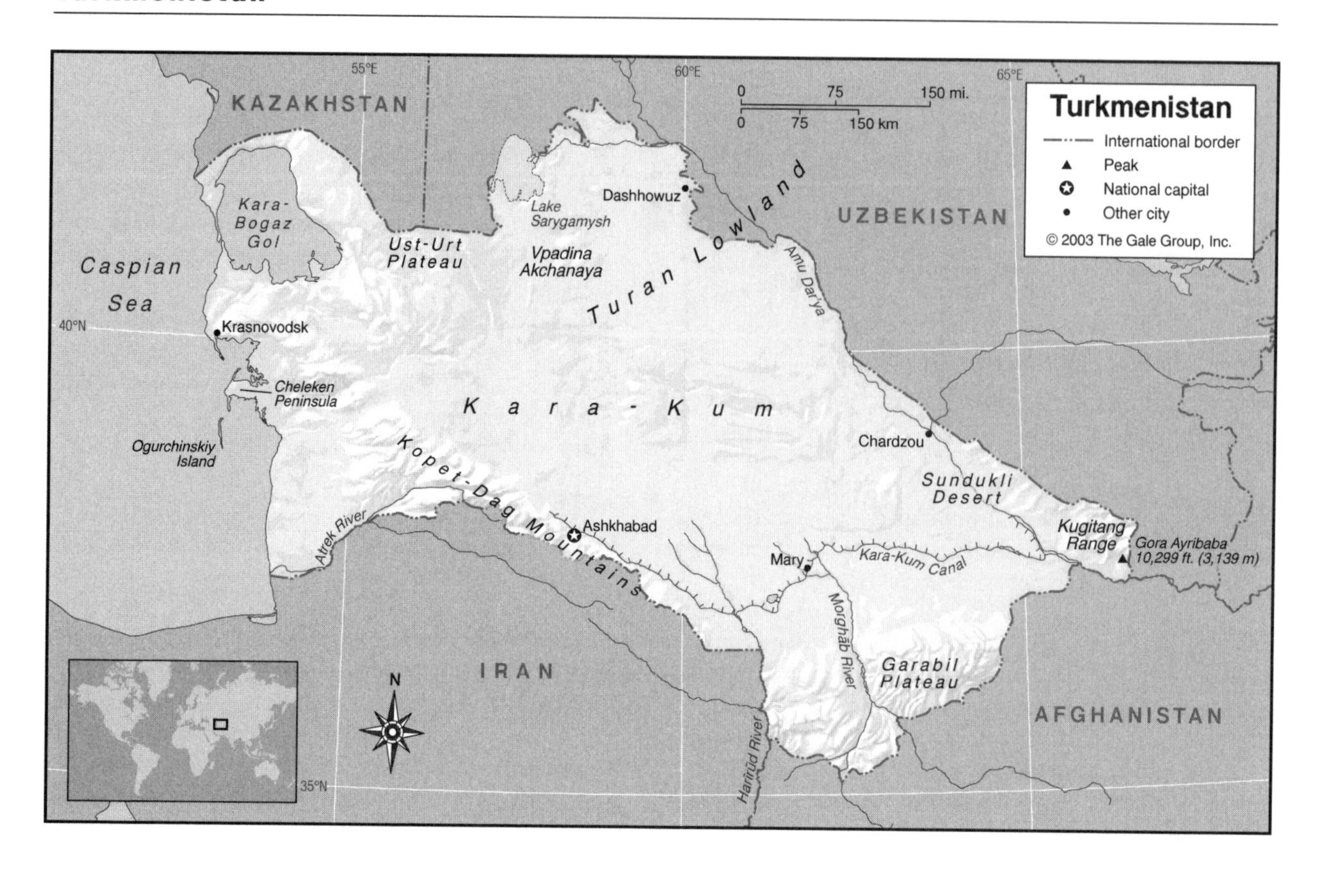

- **Population:** 4,603,244 (July 2001 est.) / World Rank: 113
- **Capital City:** Ashkhabad, south-central part of the country, near Iran border
- **Largest City:** Ashkabad, 462,000 (2000 est.)

OVERVIEW

Turkmenistan is a desert country characterized by a subtropical, desert climate with little fresh water. Its terrain is mostly low and flat, with nearly all of the western and central portions of the country—some 80 percent—covered by the great Kara-Kum (Garagum) Desert. The Kara-Kum is itself just a part of the Turan, a vast area of desert and steppe extending throughout Central Asia.

The desert gives way to mountains in the south; the eastern region is a plateau, the Garabil. Although Turkmenistan is considered landlocked, it borders the saltwater Caspian Sea on the west. Turkmenistan is on the Eurasian Tectonic Plate. Violent earthquakes are frequent in the mountains of the south.

MOUNTAINS AND HILLS

Mountains

The highest point in the country is Gora Ayribaba (Mount Ayribaba; 10,295 ft / 3,139 m), which is found in the small part of the Kugitang Range of mountains that extends across the border from Afghanistan in the east. The Kopet-Dag Mountains rise south of the Kara-Kum Desert and lie across the Turkmenistan-Iran border. Their highest point in Turkmenistan is Mt. Shahshah, 9,554 ft (2,912 m). Frequent, violent earthquakes occur in these mountains. To the north the Kopet-Dag has a chain of foothills, which feature a belt of oases fed by its mountain streams. Otherwise Turkmenistan is notably low in elevation, averaging only 1,640 ft (500 m) above sea level or less. The lowest point occurs in the Vpadina Akchanaya (Akdzhakaya Depression) of north-central Turkmenistan, 266 ft (81 m) below sea level.

Plateaus

Two plateaus occupy portions of Turkmenistan. The larger is the Garabil Plateau, which lies in the eastern portion of the country near the Afghanistan border. In the north the fringes of the Ust-Urt (Ustyurt) plateau extend across the border.

INLAND WATERWAYS

No significant rivers originate in Turkmenistan. The Atrek River flows along part of the border with Iran, emptying into the Caspian Sea. Tajikistan's Amu Dar'ya (Amu River) flows westward into Turkmenistan from the nation's Uzbekistan-Afghanistan border country, and forms sections of the border with each of those countries. It flows northwest near the country's northeastern border, and feeds Lake Sarykamysh in the north.

Two other significant rivers are the Morghāb and the Harīrūd (Tejen), both of which originate in Afghanistan. These waters flow northwest into Turkmenistan before drying up in the desert.

Water is diverted from all three rivers to fill the Soviet-built Kara-Kum Canal, which runs east to west across southern Turkmenistan. More than 870 mi (1,400 km) long, the Kara-Kum Canal is one of the longest canals in the world. It starts at the Amu Dar'ya near the Afghanistan border and extends across Turkmenistan to Krasnovodsk on the Caspian Sea. About 188 mi (300 km) of the canal is an enclosed aqueduct, but for the majority of its length it remains uncovered. Small river craft can navigate the canal for nearly half its length. The Soviets began construction of the canal in 1954, completing more than half of it by 1962 from the Amu Dar'ya to the nation's capital, Ashkhbad; but the remainder was not completed until 1986.

The Kara-Kum Canal provides irrigation water to most of southern Turkmenistan. In the northeast, other lesser canals channel the Amu Dar'ya's waters to irrigate portions of the country along the Uzbekistan border. Without these river-fed canal systems, Turkmenistan would have limited freshwater resources for cultivating crops or even providing drinking water. However, the diversion of so much water has been a major contributor to the drying up of the Aral Sea (the body of water that the Amu Dar'ya drains into).

Population Centers – Turkmenistan

(1990 CENSUS OF POPULATION)

Name	Population
Ashkhabad (capital)	407,000
Chardzou	164,000
Nebitdag	150,000
Tashauz	114,000

SOURCE: United Nations Statistics Division.

Divisions – Turkmenistan

1995 POPULATION ESTIMATES

Name	Population	Area (sq mi)	Area (sq km)	Capital
Ahal-Ashkhabad City	1,327,500	36,680	95,000	Ashkhabad, Annau*
Balkan	424,700	53,282	138,000	Nebitdag
Dashhowuz	1,059,800	28,417	73,600	Dashhowuz
Lebap	1,034,700	36,216	93,800	Chardzhou
Mary	1,146,800	33,513	86,800	Mary

* Ashkhabad is located within the Ahal Division and is the capital of its own subdivision, Annau is the capital of the rest of the Ahal Division.

SOURCE: U.S. and Foreign Commercial Service and the U.S. Dept. of State, 1999. Cited on Johan van der Heyden, Geohive, http://www.geohive.com (accessed May 2002).

THE COAST, ISLANDS, AND THE OCEAN

Turkmenistan is landlocked with no ocean coasts or islands. It does lie along the Caspian Sea, which forms its entire western border for 1,096 mi (1,768 km). There is a deep, almost circular indentation in the coast called the Kara-Bogaz Gol (Garabogazkol Gulf), with the Cheleken Peninsula further south. Strictly speaking the Caspian Sea is a very large, landlocked, saltwater lake, not a sea of ocean water.

CLIMATE AND VEGETATION

Temperature

Turkmenistan's desert continental climate features exceedingly hot summers followed by cold winters, and temperature ranges are fairly uniform for the country as a whole. In January, they range from 21° F to 41° F (-6° to 5° C); in July the range is generally between 81° and 90° F (27° and 32° C).

Rainfall

Annual rainfall amounts can vary locally, from 3 to 16 in (8 to 40 cm), but nearly two-thirds of Turkmenistan receives 6 in (15 cm) of precipitation or less.

Grasslands

In the northeast, where the great Turan lowland dips into Turkmenistan, there is steppe, a semiarid, grassy plain.

Deserts

Covering an area of about 110,000 sq mi (284,900 sq km), the Kara-Kum desert is one of the world's largest sand deserts. It extends westward from the Amu Dar'ya almost to the Caspian Sea and stretches from the Ust-Urt Plateau in the north to the Kopet-Dag Mountains in the south. The Kara-Kum occupies almost all of the country.

The name "Kara-Kum" means "black sand" in the Turk language, which aptly characterizes the coloration of much of this vast wasteland. The desert's chief features are rolling sand dunes as well as extensive regions of hard-packed clay and rock. Little in the way of vegetation can be found there, although in the southeast steppe areas some bushes and flowering plants do survive.

East of the Amu Dar'ya is the Sundukli Desert. This desert is an extension of the Kyzyl Kum Desert of Uzbekistan.

HUMAN POPULATION

With a population of 4,603,244 (July 2001 estimate), Turkmenistan is the least populated of the five Central

Asian countries that were formerly part of the Soviet Union. Most of its citizens live along waterways (rivers and canals) and the oasis country of the southern foothills of the Kopet-Dag Mountains. The Kara-Kum Desert is too inhospitable to support much human population. Overall the population density averages 24 persons per sq mi (9 persons per sq km). Roughly 45 percent of Turkmenistan's people live in cities, with the largest being located along the Kara-Kum Canal and the Amu Dar'ya.

NATURAL RESOURCES

Turkmenistan possesses significant quantities of mineral resources, including coal, salt, sulfur, and magnesium. In addition, oil and natural gas reserves within the country's territorial waters along the Caspian Sea are considered to be substantial. The Kara-Kum desert may also yield significant quantities of petroleum resources.

FURTHER READINGS

Mandelbaum, Michael, ed. *Central Asia and the World: Kazakhstan, Uzbekistan, Tajikistan, Kyrgyzstan, and Turkmenistan.* New York: Council on Foreign Relations Press, 1994.

Micklin, Phillip P. "The Water Management Crisis in Soviet Central Asia". *The Carl Beck Papers in Russian and East European Studies, No. 905.* Pittsburgh: University of Pittsburgh Press, 1992.

Nichol, James, and Leah Titerence. "Turkmenistan: Basic Facts." *CRS Report for Congress.* Washington: Library of Congress, Congressional Research Services, March 16, 1993.

Turks and Caicos Islands

Overseas Territory of the United Kingdom

- **Area:** 166 sq mi (430 sq km) / World Rank: 189
- **Location:** Northern and Western Hemispheres, in the Caribbean Sea; two island groups in the North Atlantic Ocean, southeast of The Bahamas.
- **Coordinates:** 21°45′N, 71°35′W
- **Borders:** None
- **Coastline:** 242 mi (389 km)
- **Territorial Seas:** 12 NM
- **Highest Point:** Blue Hills, 161 ft (49 m)
- **Lowest Point:** Sea level
- **Longest River:** None of significant length
- **Natural Hazards:** Subject to hurricanes
- **Population:** 18,122 (July 2001 est.) / World Rank: 202
- **Capital City:** Cockburn Town (on Grand Turk)
- **Largest City:** Cockburn Town, 4,900 (2002 est.)

OVERVIEW

Turks and Caicos Islands are a British dependency consisting of more than 30 islands forming the southeastern end of the Bahamas chain of islands about 90 mi (145 km) north of Haiti. Eight of the islands are inhabited: Grand Turk, Salt Cay, South Caicos, Middle Caicos, North Caicos, Providenciales, Pine Cay, and Parrot Cay. The islands consist of a low, flat limestone terrain with extensive marshes and mangrove swamps, and the north shore of Middle Caicos features a network of limestone cave formations. The islands are located on the North American Tectonic Plate, near where this plate borders the Caribbean Tectonic Plate, but not close enough to draw any threat of major or frequent seismic activity.

MOUNTAINS AND HILLS

The islands do not feature any mountains or significant hills, being mostly flat limestone. There are some areas of small hills, however. The highest point on the islands is only 161 ft (49 m), in the Blue Hills on the island of Providenciales, one of the largest and oldest settlements on the islands.

INLAND WATERWAYS

There are relatively few inland waterways, and little fresh water. On North and Middle Caicos there is a limited underground water supply but most islanders rely on cisterns to gather fresh water. However there are numerous salt ponds known as salinas, which are believed to have been connected to the ocean at some point in history. Wind and tidal patterns caused shifts in the shoreline and land mass to the extent that salinas, once bays, are now inland. Refreshed by seasonal tide activity, the salinases continue to collect salt. In areas of low elevation they form salt-saturated swamps. The salt concentration—up to four times that of the open ocean—varies according to rainfall, tidal patterns, and general climatic conditions.

THE COAST, ISLANDS, AND THE OCEAN

Oceans and Seas

The islands are surrounded by the Caribbean Sea, and to the north by the Atlantic Ocean. The ocean waters around the islands feature one of the world's most extensive coral barrier reefs, 65 mi (105 km) across and 200 mi (322 km) long. Running roughly southwest to northeast, a 7,000-ft (2,100-m) deep trench known as the Turks Island or Columbus Passage separates the Turks Islands group from the Caicos Islands group; the trench, ranked as one of the best diving spots in the world, is known to scuba divers as "The Wall." The islands are separated from the most southeastern Bahamian islands by a 30-mi (48-km) section of ocean known as the Caicos Passage.

Major Islands

The islands are divided into two groups. The Turks Islands (including Grand Turk, Salt Cay, Pine Cay, Parrot Cay, Ambergris Cay, Bush Cay, Big Sand Cay) lie to the east, and are more widely scattered and are separated from the island of Hispaniola by the Mouchoir Passage and from the Caicos Islands by Turks Island Passage. The islands of the Caicos Islands group (including West Caicos, Providenciales, North Caicos, Middle Caicos, East Caicos, and South Caicos) lie fairly close together, separated by narrow channels.

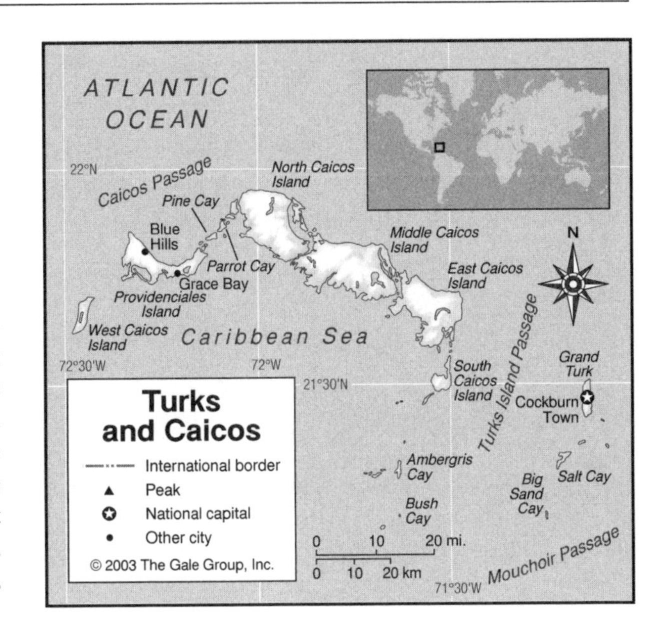

The Coast and Beaches

The islands have many beautiful white sandy beaches that attract numerous tourists. Most of these are found on the north and west shores, facing the open ocean. Some of the islands, such as West Caicos, have coastlines that drop abruptly to deep water. Middle Caicos, North Caicos, and Providenciales all feature a shoreline that zigzags in and out with many bluffs and small coves. Grace Bay on Providenciales is a 5-mile (8-km) long strip where most of the tourist resorts are located. Grand Turk Island features numerous bays on its eastern coast. South Caicos features Belle Sound and also has a formidable harbor in Cockburn Harbor, the only town on the island.

CLIMATE AND VEGETATION

Temperature

Situated just south of the tropic of Cancer, the Turks and Caicos Islands have a climate that is marine tropical, moderated by trade winds. It is usually sunny and relatively dry. The average temperature is 83°F (28°C) year round, with winter temperatures averaging 77°F (25°C) and summer readings averaging 90°F (32°C). The humidity is usually low, making the island heat more tolerable.

Rainfall

Rainfall averages 21 in (53 cm) per year, with slightly more precipitation falling during the summer. Hurricanes strike the islands frequently, with major hurricanes occurring on average at a rate of one every 10 to 15 years.

HUMAN POPULATION

The majority of the population is descended from African slaves who were brought to Providenciales Island to grow cotton. The rest of the population (less that 5 percent) consists of British, American, French, Canadian, and Scandinavian expatriates. Most of the islands have only one town. More than one-quarter of the population lives in the capital of Cockburn Town. Providenciales, the only town on the island of the same name, is the most tourist-oriented island with the most resorts and residential developments for retirees.

NATURAL RESOURCES

Having little or no arable land or mineral deposits, the most important resources of the Turks and Caicos Islands are tourism and fishing. The two chief industries are tourism and collecting sea sponges. The leading exports are sponges and shellfish, including the spiny lobster and conchs. Another leading source of income for the Turks and Caicos economy is revenue from customs receipts and offshore financial services.

FURTHER READINGS

Baker, Chrstopher. *Lonely Planet Bahamas Turks & Caicos.* Lonely Planet: 1998.

Cable & Wireless. *Turks and Caicos Islands Gateway-TCI Mall.* http://www.tcimall.tc/index.htm (Accessed June 2002).

Davies, Julia, and Phil Davies. *The Turks & Caicos Islands: Beautiful by Nature.* New York: Macmillan Educational Corp, 2000.

Palmer, Charles. *Living in the Turks and Caicos.* Atlanta, Ga.: Protea Publishing, 2001.

Pavlidis, Stephen. *The Turks and Caicos Guide: A Crusing Guide to the Turks and Caicos Islands.* Port Washington, Wis.: Seaworthy Publications, 1999.

Sadler, Herbert. *Turks Islands Landfall.* Turks and Caicos Islands: Marjorie Sadler, 1997.

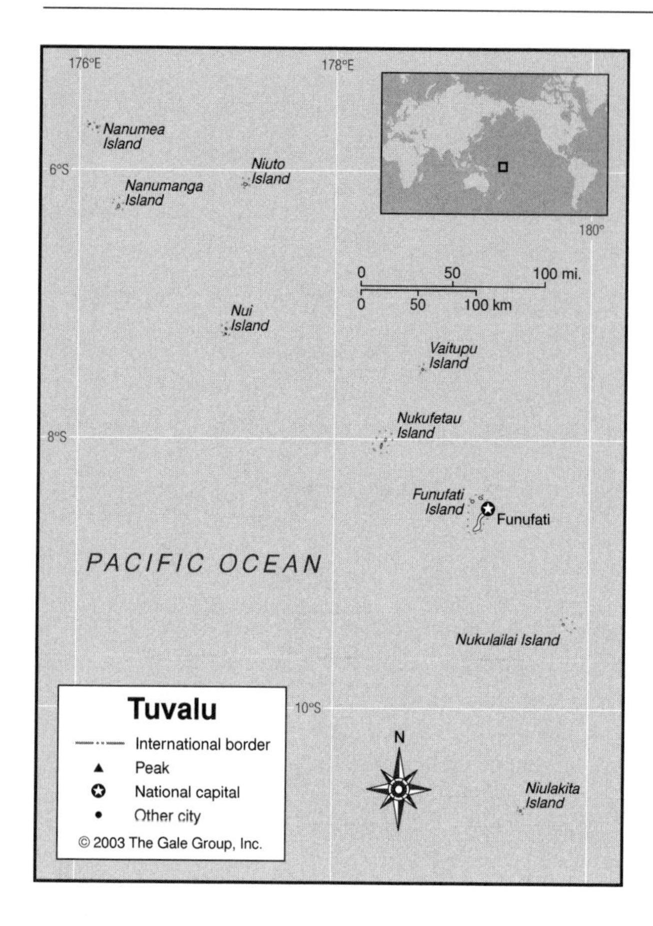

Tuvalu

- **Area:** 10 sq mi (26 sq km) / World Rank: 204
- **Location:** Southern and Eastern Hemispheres, in Oceania; island group consisting of nine coral atolls in the Southwestern Pacific Ocean, about equidistant from Hawaii and Australia.
- **Coordinates:** 8°00′S, 178°00′E
- **Borders:** None
- **Coastline:** 15 mi (24 km)
- **Territorial Seas:** 12 NM
- **Highest Point:** Unnamed location, 16 ft (5 m)
- **Lowest Point:** Sea level
- **Longest River:** The country has no rivers
- **Natural Hazards:** Cyclones; flooding due to changes in sea level
- **Population:** 10,991 (July 2001 est.) / World Rank: 205
- **Capital City:** Funafuti, on the east coast of the island of Funafuti
- **Largest City:** Funafuti, population 5,100 (2002 est.)

OVERVIEW

Tuvalu (formerly the Ellice Islands) is one of the smallest and most remote countries on Earth. Located just south of the equator on the Pacific Tectonic Plate, Tuvalu consists of a cluster of nine low-lying coral islands, plus islets. These remote atolls lie in a 370-mi-long (595-km) chain extending over some 500,000 sq mi (1,300,000 sq km) of ocean. Too remote and too small to develop a tourist industry, Tuvalu is ranked by the United Nations as among the least-developed countries.

MOUNTAINS AND HILLS

Tuvalu consists of very low-lying coral atolls, none of which is more than 16 ft (5 m) above sea level. As a result of these low levels the islands are very sensitive to changes in sea level.

INLAND WATERWAYS

There are no rivers, lakes, or streams on the islands. Five of the atolls do enclose sizable lagoons, but there is still no fresh water available other than the amount of rainfall that can be caught and stored.

THE COAST, ISLANDS, AND THE OCEAN

Major Islands

Tuvalu's islands are coral reefs on the outer arc of ridges formed by pressure from the Central Pacific against the ancient Australian landmass. All the islands are low lying with elevations of no greater than 16 feet (5 m) above sea level. The islands in the chain are Funafuti, Nanumea, Nanumanga, Niulakita (formerly uninhabited), Niuto, Nui, Nukufetau, Nukulailai and Vaitupu.

The Coast and Beaches

Coral reefs on five islands enclose sizeable lagoons. Funafuti and Nukufetau are the only islands with natural harbors for oceangoing ships. If the sea level rises significantly in the twenty-first century, most of these islands will be completely submerged.

CLIMATE AND VEGETATION

Temperature

Tuvalu has a tropical climate with little seasonal variation. The annual mean temperature is 86°F (30°C), moderated by easterly trade winds that blow from March to November.

Rainfall

Tuvalu is very wet—annual rainfall averages more than 140 in (355 cm). Westerly gales bring heavy rain

Islands – Tuvalu

Name	Area (sq mi)	Area (sq km)
Funafuti	0.91	2.36
Nanumanga	1.00	2.59
Nanumea	1.38	3.57
Niulakita	0.16	0.41
Niuto	0.82	2.12
Nui	1.27	3.29
Nukufetau	1.18	3.06
Nukulailai	0.64	1.66
Vaitupu	1.89	4.90

SOURCE: *Geo-Data: The World Geographical Encyclopedia,* 2nd ed. Detroit: Gale Research, 1989.

from November to March. Although the islands lie north of the main cyclone belt, Funafuti was devastated in 1894, 1972, and 1990.

Forests and Jungles

Coconut plantations have replaced most of Tuvalu's original vegetation of scrubby forest. However, its soils are poor and much of its vegetation has been cleared for fuel.

HUMAN POPULATION

Since the islands that make up the country are very small, Tuvalu's population density is high—about 1,055 residents per sq mi (407 residents per sq km) in 1999. In 2000 it was estimated that 52 percent of the population lived in urban areas, with almost half of the population living in the capital of Funafuti. All of the other towns on the islands have a population of less than a thousand. Only in recent times have the islands recovered from a population loss begun in the nineteenth century when many Tuvaluans were enslaved to work on plantations around the Pacific, Australia, and South America.

NATURAL RESOURCES

There are no known mineral resources in Tuvalu and very little exports. Tuvaluans make intensive use of its limited resources: fish and coconuts. Copra is the only cash crop. Currently, Tuvalu has garnered significant income from leasing its Internet addresses in its .tv domain, as well as selling the use of its telephone area code for "900" lines.

GEO-FACT

Of all nine islands in Tuvalu, Funafuti is the only one with an airport: a single grass strip that cannot even be used for jet aircraft.

FURTHER READINGS

Cannon, Brian. *Tuvalu Online.* http://www.tuvaluislands.com/ (Accessed May 6, 2002).

Lane, John. *Tuvalu: State of the Environment Report, 1993.* Western Samoa: SPREP, 1993.

Mueller-Dombois, Dieter, and F. Raymond Fosberg. *Vegetation of the Tropical Pacific Islands.* New York : Springer-Verlag, 1998.

Rodgers, K. A. 1991. "A Brief History of Tuvalu's Natural History." *South Pacific Journal of Natural Science* 11: 1-14.

Thaman, Randolph R. *Samoa, Tonga, Kiribati, and Tuvalu: A Review of Uses and Status of Trees and Forests in Land Use Systems with Recommendations for Future Actions.* Rome: Food and Agriculture Organization of the United Nations, 1995.

Uganda

- **Area:** 91,136sq mi (236,040 sq km) / World Rank: 82
- **Location:** Northern and Eastern Hemispheres, eastern Africa, west of Kenya, south of Sudan, east of the Democratic Republic of the Congo, and north of Rwanda and Tanzania
- **Coordinates:** 1°00′ N, 32°00′ E
- **Borders:** 1,676 mi (2,698 km) / Sudan, 270 mi (435 km); Kenya, 580 mi (933 km); Tanzania, 246 mi (396 km); Rwanda, 105 mi (169 km); Democratic Republic of the Congo, 475 mi (765 km)
- **Coastline:** None
- **Territorial Seas:** None
- **Highest Point:** Margherita Peak, 16,765 ft (5,110 m)
- **Lowest Point:** Lake Albert, 2,037 ft (621 m)
- **Longest Distances:** 489 mi (787 km) NNE-SSW; 302 mi (486 km) ESE-WNW
- **Longest River:** Nile, 4,160 mi (6,693 km)
- **Largest Lake:** Lake Victoria, 26,828 sq mi (69,484 sq km)
- **Natural Hazards:** None
- **Population:** 23,985,712 (July 2001 est.) / World Rank: 42
- **Capital City:** Kampala, central southern location on Lake Victoria
- **Largest City:** Kampala, 1,207,000 (2000 est.)

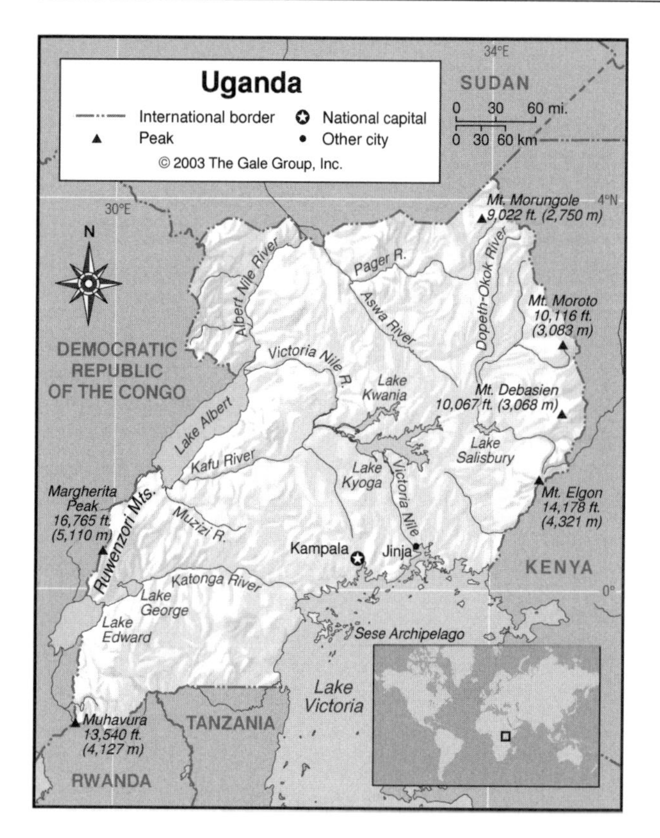

GEO-FACT

The Ruwenzori snow-capped 'Mountains of the Moon,' immortalized in Rider Haggard's novels, constitute Africa's highest mountain range.

OVERVIEW

Situated astride the equator, Uganda is about the size of the state of Oregon and lies on the great plateau of east-central Africa, at a point less than 500 mi (805 km) in a direct line from the Indian Ocean. Over most of the country the landscape is similar; a row of volcanoes along the east and the western rift valley system, flanked by highlands and mountains, offer sharp contrasts with the open vistas and level horizons of the central plateaus. Uganda lies at a relatively stable position in the middle of the African Tectonic Plate, however in recent geologic times the unwarping and faulting that created the western rift valley led to a piling up of waters in the relatively downwarped zone to the east that now forms the basin of Lake Victoria on the southern boundary. The country reaches its lowest points in the northwest in the valley of the Albert Nile on the Sudan border.

MOUNTAINS AND HILLS

Mountains

In the extreme southwest are the Mufumbiro volcanoes, of which only the northern slope is in Uganda. From these volcanic highlands an area of more than 5,000 ft (1,524 m) above sea level extends northeastward through Kigezi District into western Ankole District. The Mufumbiro Range includes the 11,960 ft (3,645 m) Mount Sabinio, the meeting point of Uganda, Rwanda, and Democratic Republic of the Congo. Its highest mountain is Muhavura at 13,540 ft (4,127 m).

These highlands are separated from the Ruwenzori Mountains, also known as the Mountains of the Moon, by a low valley containing Lake George and its outlet into Lake Edward, the Kazinga Channel. The Ruwenzori Range, skirting the western border with Zaire, is about 50 mi (80 km) long and rises into a number of peaks of more than 14,000 ft (4,267 m) of which the highest is Margherita Peak, at 16,763 ft (5,109 m). Above the 14,000 ft (4,267 m) level the mountains are capped with snow and large glaciers.

To the east, the approach to the Kenya borderlands is marked by volcanic centers and hills. Mount Elgon, between Sebei District and Kenya, is 14,178 ft (4,321 m) at its highest point. Mount Debasien, in Karamoja District, is 10,067 ft (3,068 m) while Mount Moroto, still further north, is 10,116 ft (3,083 m). Mount Morungole near the northeast border is 9,022 ft (2,750 m) and Mount Zulia in the extreme northeast is 7,048 ft (2,148 m) high. On the northern border are the southern outlines of the Imatong Mountains of the Sudan, all about 6,000 ft (1,828 m) high.

Plateaus

Between mountain masses east and west, Uganda's prominent relief feature is that of plateau, dissected by numerous rivers, swamps, and lakes. The plateau is fairly regular, maintaining an altitude in the range of 2,600–6,600 ft (800–2,000 m) above sea level. In the southwest this region is known as the Ankole, named after the native kingdom which used to occupy the land.

Hills and Uplands

West of the east border mountains are a number of other smaller mountain masses including the Labwor Hills, ranging from 5,900 to 8,300 ft (1,798 to 2,530 m). These hills are more or less isolated from each other, rising abruptly out of the plains.

INLAND WATERWAYS

Lakes

Lakes Albert, Edward, and George are troughs in the western rift valley system, while Lakes Victoria and Kyoga are shallow basins on the plateau. Lake Albert and

Population Centers – Uganda

(1991 CENSUS OF POPULATION)

Name	Population	Name	Population
Kampala (capital)	774,241	Entebbe	42,763
Jinja	65,169	Mbarara	41,031
Mbale	53,987	Soroti	40,970
Masaka	49,585	Gulu	38,297

SOURCE: Uganda Bureau of Statistics.

Districts – Uganda

POPULATIONS FROM 1991 CENSUS

Name	Population	Area (sq mi)	Area (sq km)	Capital
Adjumani	96,264			
Apac	454,504	2,510	6,490	Apac
Arua	637,941	3,020	7,830	Olaki
Bugiri	239,307			
Bundibugyo	116,566	900	2,340	Busaru
Bushenyi	579,137	2,080	5,400	Bumbaire
Busia	163,597			
Gulu	338,427	4,530	11,740	Bungatira
Hoima	197,851	3,820	9,900	Hoima
Iganga	706,476	4,939	12,792	Bulamogi
Jinja	289,476	297	768	Jinja
Kabale	417,218	668	1,730	Rubale
Kabarole	746,800	3,230	8,360	Karambe
Kalangala	16,371	3,501	9,067	
Kampala	774,241	76	197	Kampala
Kamuli	485,214	1,680	4,350	Namwendwa
Kapchorwa	116,702	670	1,740	Kaptanya
Kasese	343,601	1,240	3,200	Rukoki
Katakwi	144,597			
Kibale	220,261	1,639	4,246	
Kiboga	141,607	1,562	4,046	
Kisoro	186,681	281	728	
Kitgum	357,184	6,230	16,140	Labongo
Kotido	196,006	5,100	13,210	Kotido
Kumi	236,694	1,100	2,860	Kumi
Lira	500,965	2,810	7,250	Lira
Luwero	349,194	3,550	9,200	Luwero
Masaka	694,697	6,300	16,330	Kaswa Bukoto
Masindi	260,796	3,720	9,640	Nyangeya
Mbale	710,980	980	2,550	Bunkoko
Mbarara	798,774	4,180	10,840	Kakika
Moroto	174,417	5,450	14,110	Katikekile
Moyo	79,381	1,930	5,010	Moyo
Mpigi	913,867	2,400	6,220	Mpigi
Mubende	500,976	2,393	6,198	Bageza
Mukono	824,604	5,525	14,309	Kaguga Mukono
Nakasongola	100,497			
Nebbi	316,866	1,120	2,890	Nebbi
Ntungamo	289,222	794	2,056	
Pallisa	357,656	769	1,992	
Rakai	144,039	1,920	4,970	Byakabanda
Rukungiri	390,780	1,060	2,750	Kagunga
Sembabule	383,501			
Soroti	285,793	3,880	10,060	Soroti
Tororo	391,977	1,007	2,609	Sukulu

Districts with no area or capital listings are subdivisions of other districts, but their populations are tracked separately by UBOS.

SOURCE: "Population and Housing Census (number of persons), Region and District." UBOS (Uganda Bureau of Statistics).

Lake Edward are shared by Uganda and the Democratic Republic of the Congo, but Lake George, connected with Lake Edward by the Kazinga Channel, is wholly within Uganda.

All of the lakes are relatively shallow. The maximum depth recorded in Lake Victoria is 270 ft (82 m), that in Lake Albert, 168 ft (51 m), in Lake Edward, 384 ft (117 m), in Lakes Kyoga and Kwania, 24 ft (7.3 m), and in Lake George, 10 ft (3 m). The waters of Lake Albert are more extensively fished than those of Lake Victoria but less so than those of Lake Kyoga. There are many fish to be found in Lake George, and the catch from the more extensive waters of Lake Edward is similar in size to that from Lake George.

Lake Victoria is the second-largest freshwater lake in the world. Only Lake Superior in North America is larger. However, its 605 cubic miles (2,518 cubic kilometers) of water gives it only the seventh-largst volume. Lake Victoria contains numerous islands, has an indented coast with many deep gulfs, and has multiple tributaries that feed into it. Of Lake Victoria's 26,828 sq mi (69,484 sq km), 11,749 sq mi (20,430 sq km) are in Uganda, the remainder being divided between Kenya and Tanzania. Inside the lake there are many archipelagos, reefs, and more than 200 species of fish.

Rivers

The White Nile basin provides the country's main drainage. Lake Victoria overflows at a low point near Jinja to form the Victoria Nile, flowing into Lake Kyoga. The waters of the Lake Kyoga basin are carried by the Victoria Nile to Lake Albert and then, by the Albert Nile, to the north. From the Owen Falls, at Jinja, to the point at which the Albert Nile crosses the northern border with the Sudan, there is a descent of more than 1,700 ft (518 m), accomplished for the most part by a series of falls and rapids.

The innumerable other rivers are, for the most part, sluggish. Some are not much more than vegetation-covered swamps. The Katonga runs into a swamp at the northeast corner of Lake Victoria. The Kafu flows into the western end of Lake Kwania, but its headwaters connect with those of the Muzizi, flowing westward into the southern end of Lake Albert. Other major rivers are the Aswa, Pager, and Dopeth-Okok of the northeast and the Mpongo, a tributary of the Kafu. Only in the hill regions and on the slopes of the western rift valley are clear, running streams commonly found.

Wetlands

Most of Uganda's 2,000 sq mi (5,180 sq km) of swamp lie in the lowland area bordering the Nile. Lakes Kyoga and Kwania in the center of the country are surrounded by a large area of swamp. Lake Salisbury, to the northeast of Lake Kyoga, provides an outlet for the waters north of Mount Elgon to the Nile system. West of Lake Victoria, in the south, is a group of some six lakes connected by swamp. In the upland areas of the southwest a number of swampy areas have been reclaimed.

THE COAST, ISLANDS, AND THE OCEAN

Uganda has no borders with the ocean, however it does have many islands within Lake Victoria. The Sese archipelago, a chain of 62 islands in the lake off the coast southwest of Kampala, contains many inhabitants, most of them fishermen. The densely populated Ukerewe is the largest of the islands. It rises over 650 feet (200 m) above the lake's surface.

CLIMATE AND VEGETATION

Temperature

Being right on the equator, temperatures vary little on the plateau. At Lake Albert annual temperatures range only from 72° to 84°F (22° to 29°C). However, temperatures drop significantly in higher altitudes. At Kampala, for instance, the average extremes are 63° to 81°F (17° to 27°C).

Rainfall

While most of Uganda receives an annual rainfall of at least 40 in (140 cm), the northeast receives only 27 in (69 cm) per year. The areas around the lakes receive more rainfall on average. The city of Entebbe, on Lake Victoria, receives 64 in (162 cm).

Grasslands

Scattered patches of elephant grass dominate the southern reaches of the country, while long grasses colonize the western highlands. In the drier, northern savanna there is still mostly grasslands, but the grass is significantly shorter.

Forests and Jungles

Thick forest containing Mvuli trees in patches marks the south, while the western mountains are covered in parts with forests adapted to higher altitudes. Open woodlands, thorn trees, borassus palms, and scrub characterize the north.

HUMAN POPULATION

The population of Uganda is mostly rural with an overall density of 295 inhabitants per square mile (114 per sq km, 1999 est.). The capital city is the only one that has a population of more than 100,000 people. Much of Uganda's population is concentrated in the southern region surrounding Lake Victoria.

NATURAL RESOURCES

The western part of the country contains small quantities of minerals such as tin, tungsten, beryllium, and copper. Limestone and apatite are mined in the east. Soils and climatic conditions are excellent; about 78 percent of the land is potentially agriculturally productive, although only about 25 percent is used at any one time for cultivation and grazing. The rich soils of the south produce timber (Mvila tree), robusta coffee, tea, and bananas. About 30 percent of the land is forest and woodland. Hydroelectric power is found on the Nile at Jinja. Lakes Victoria and Katwe are rich in fish.

FURTHER READINGS

Africa South of the Sahara 2002. "Uganda." London: Europa Publishers, 2001.

Nzita, R., and Mbaga-Niwampa. *Peoples and Cultures of Uganda.* 2nd Ed. Kampala, Uganda: Fountain Publishers, 1995.

Pirouet, M.L. *Historical Dictionary of Uganda.* Metuchen, N.J.: Scarecrow Press, 1995.

ThinkQuest. *The Living Africa.* http://library.thinkquest.org/16645/contents.html (Accessed June 5, 2002).

Tourism in Uganda. http://www.visituganda.com/inside.htm (Accessed May 2002)

Ukraine

- **Area:** 233,090 sq mi (603,700 sq km) / World Rank: 45
- **Location:** Northern and Eastern Hemispheres, in Eastern Europe, bordering the Black Sea to the south, Belarus to the north, Moldova and Romania to the south and west, Russia to the east, and Hungary, Poland, and Slovakia to the west.
- **Coordinates:** 49°00′N, 32°00′E
- **Borders:** 2,832 mi (4,558 km) / Belarus, 554 mi (891 km); Hungary, 64 mi (103 km); Moldova, 583 mi (939 km); Poland, 266 mi (428 km); Romania, 330 mi (531 km); Russia, 979 mi (1,576 km); Slovakia, 56 mi (90 km)
- **Coastline:** 1,729 mi (2,782 km)
- **Territorial Seas:** 12 NM
- **Highest Point:** Hora Hoverlya, 6,762 ft (2,061 m)
- **Lowest Point:** Sea level
- **Longest Distances:** 818 mi (1316 km) E-W / 555 mi (893 km) N-S
- **Longest River:** Danube, 1,771 mi (2,850 km)
- **Natural Hazards:** None

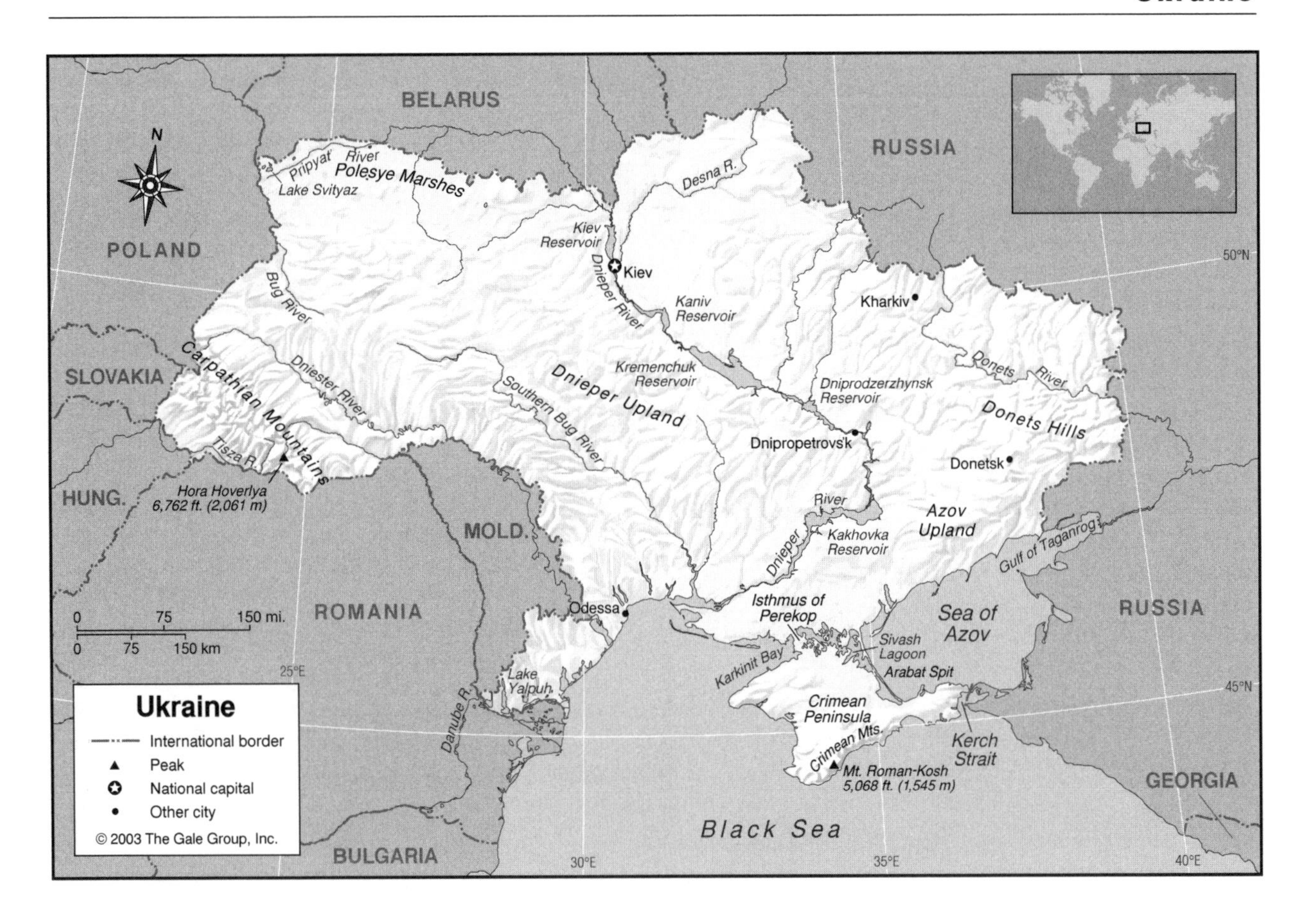

- **Population:** 48,760,474 (July 2001 est.) / World Rank: 24
- **Capital City:** Kiev, on the Dnieper in north-central Ukraine
- **Largest City:** Kiev, 2,932,000 (2002 est.)

OVERVIEW

Ukraine is the second largest country in Eastern Europe, after Russia. Due to its great size, Ukraine features a wide variety of terrain and climate conditions. The center of the country is predominantly a rolling upland plain, or steppe. This plain is crossed by many of Eastern Europe's major rivers, including the Dnieper (Dnipro), the Dniester, the Bug, the Donets, and the Tisza. Other, lower, plains are found along the Black Sea coast, and the southwestern corner of the country is part of the delta of the Danube. The Polesye (Polissya) Marshes are a lowland of swamps and wooded bogs in northern Ukraine, extending into Belarus. The Carpathian Mountains can be found in the west; other, lower, mountains are in the Crimean Peninsula (an autonomous republic considered part of Ukraine) and in the southeastern Donets region. Ukraine is located on the Eurasian Tectonic Plate.

MOUNTAINS AND HILLS

Ukraine has few mountains, they cover only about 5 percent of its total territory. The Carpathian Mountains in the extreme west are the highest in the country. It is here that Mount Hoverlya, the tallest in the country (6,762 ft / 2,061 m), can be found. The Crimean Mountains at the southern end of Crimea are also notable, reaching a height of 5,068 ft (1,545 m) at Mt. Roman-Kosh.

Outside of its mountains, Ukraine has several areas of hills and uplands. The most noteworthy are the Azov Upland north of the Sea of Azov, the Donets Hills, and the Dnieper Upland, which is the watershed between the Dnieper and the Southern Bug.

INLAND WATERWAYS

Lakes

More than 20,000 small lakes dot the Ukraine landscape, covering a total area of about 7,000 sq mi (18,139 sq km). The largest lakes in the country are all artificial, as the many dams on the Dnieper have created huge reservoirs. The Kremenchuk Reservoir and the Kakhovka Reservoir are the largest; the Kiev, Kaniv, and Dniprodzerzhynsk Reservoirs are also noteworthy. The largest natural lake is Lake Yalpuh (136 sq mi / 220 sq km) in the

Danube flood plain. Lake Svityaz (17 sq mi / 27 sq km) is a lake in Polesye Marshes of the northwest.

Rivers

Ukraine's most important river is the Dnieper (Dnipro). It flows south across the middle of the country for about 610 mi (980 km), curving first east, then west, then finally south again before entering the Black Sea. It flows for a total of 1,420 mi (2,290 km) from its source in Russia, making it the third longest river in Europe—only the Volga and the Danube rivers are longer.

More than half of the country's rivers are part of the Dnieper system, draining a vast area of nearly 200,000 sq mi (518,000 sq km). At its greatest width in its middle and lower reaches (approximately 1 mi / 2 km), it passes through Ukraine's most agriculturally developed and industrialized areas, where the river is used to ship grain, lumber, and metals. In Ukraine the river is entirely navigable, although it freezes during the winter. The capital city of Kiev is located on the upper Dnieper. There are numerous hydroelectric dams and large reservoirs all along the Dnieper in Ukraine. Important tributary rivers include Berezina, Desna, and Pripyat' (Pripet).

Ukraine's southwestern border with Romania is marked by the northernmost channel of the Danube. At 1,771 mi (2,850 km), it is the second longest river in Europe. Thus, while it flows through Ukraine only for a short distance before emptying into the Black Sea, the Danube is the longest river to pass through the country. The Danube has been a vital commercial and communications link since ancient times, connecting the interior of Eastern and Central Europe to the Black Sea.

The Dniester (Dnister) River originates in the Carpathian Mountains near Drohobych in western Ukraine. It then flows southeast for 870 mi (1,400 km) through western Ukraine and eastern Moldova (forming part of the border with that country), before emptying into the Black Sea southwest of Odessa. Its average width is 500 to 750 ft (152 to 229 m); near the mouth it reaches a maximum width of 1,400 ft (427 m) and also forms a broad, marshy lagoon, the Dnistrovskyy Lyman. For most of the year grain, vegetables, sunflower seeds, cattle, and lumber—all products of the Dniester River Basin—are shipped to the Black Sea and on to European and Asian markets. The Dniester Basin encompasses some 30,000 sq mi (77,700 sq km). In winter the river remains largely frozen.

The Donets River (631 mi / 1,015 km) has its source in Russia and flows south into Ukraine, then curves east across the easternmost part of the country and re-enters Russia. It is a tributary of Russia's Don River, which empties into the Sea of Azov. It has long been used as a transportation artery, and the Donets Basin is an important center of industry and population in Ukraine.

GEO-FACT

In April, 1986, a reactor at the Chernobyl nuclear power station in northern Ukraine experienced an explosion and core meltdown. Radioactive contamination spread through the air over northern Ukraine and southern Belarus and seeped into the ground, poisoning the water supply and making farmland toxic. The devastating effects of this accident on human health and the environment continue into the 21st century.

The Bug River (Western Bug) originates in western Ukraine and flows north, forming part of the border with Poland. Another river of the same name, the Southern Bug, rises in northwestern Ukraine and flows southeast, eventually emptying into the Black Sea near the mouth of the Dneiper. Navigation is limited to 100 mi (160 km) by shallow conditions and rough water. At 532 mi (856 km) in length, the Southern Bug is the longest river found entirely within Ukraine.

The Tisza river, noted for its abundance of fish, is formed by the confluence of the Black Tisza and the White Tisza rivers in the Ukraine's Carpathian Mountains. It then flows northeast into Romania before curving southwest, then south, running for a total of some 600 mi (970 km) before finally entering the Danube in northern Yugoslavia.

Wetlands

The Polesye (Polissya, Pripet) Marshes are a lowland in northern Ukraine and southern Belarus, located along the Pripyat' (Pripet) River and covering about 105 sq mi (270 sq km), making them the largest wetland in Europe. The land is mostly flat, sandy, bog soils, with a few low hills. Forests cover about a third of the marshes. The marshes range in elevation from 328 ft (100 m) above sea level in the northeast to 820 ft (250 m) above sea level in the south.

THE COAST, ISLANDS, AND THE OCEAN

The Black Sea and the Sea of Azov

All of Ukraine's coastline is on the Black Sea in the south. However, only the southwestern coast is on the Black Sea proper, The rest is on the Sea of Azov, an arm of the Black Sea that is formed by Ukraine's Crimean Peninsula. The coast on the Black Sea itself is a lowland area, with clayey soils. It is marked by the estuaries of the

Dnieper, Southern Bug, and Dniester Rivers, as well as the delta of the Danube in the southwest. Karkinit Bay indents the coast deeply, nearly separating the Crimean Peninsula from the mainland. On the far side of the Peninsula, the Kerch Strait connects the Black Sea to the Sea of Azov. The Sea of Azov is otherwise completely enclosed by Ukraine in the west and Russia in the east. It has an area of 14,517 sq mi (37,599 sq km). Its coastline in Ukraine consists of uplands and steppes. In the northeast it extends deeply into Russia in the form of the Gulf of Taganrog. In the west, the Sivash Lagoon nearly reaches Karkinit Bay in the Black Sea, separated only by the narrow Isthmus of Perekop.

The Crimean Peninsula

The Crimean Peninsula, also known as the Crimea, is an autonomous republic in southeastern Ukraine. The peninsula extends well into the Black Sea, measuring 110 mi (175 km) from north to south and 200 mi (320 km) east to west with a total area of 10,036 sq mi (25,993 sq km). The narrow Isthmus of Perekop joins it to the mainland in the north, and the Kerch Peninsula extends to the east, almost linking it with Russia. The Arabat Spit is a long spit of sand along the northeast coast of Crimea, helping to form the Sivash Lagoon.

The climate along the southern Crimean coast is mild and the land is scenic, with an abundance of vineyards, fruit orchards, and resorts. Although the southeastern section of the peninsula is mountainous, most of the interior is a flat plain or steppe. In contrast to the Mediteranian-like southern coast, the Crimean plains experience cold, windy winters and arid summers.

CLIMATE AND VEGETATION

Temperature

The climate of Ukraine is considered moderate and continental, with warm summers and cold winters. Along the southern Crimean coast the climate is Mediterranean— winters are more mild and wet while summers are hot and dry. In Kiev the July temperature averages 69°F (20°C); in January the average is 21°F (-6°C). Summers are warmer and winters are colder in eastern Ukraine, whose weather is influenced by large air masses from the steppes of Central Asia.

Rainfall

Ukraine's mild-to-moderate climate includes moderate levels of precipitation, with levels that average around 20 in (50 cm) per year, although the number varies by region. Rainfall is most frequent in summer; but it is highest in the Carpathian Mountains and lowest on the Black Sea coast, which proves favorable for the Crimean tourism industry.

Population Centers – Ukraine

(POPULATION ESTIMATE 1997)

Name	Population	Name	Population
Kiev (capital)	2,646,100	Odessa	1,086,700
Kharkiv (Kharkov)	1,615,000	Zaporozhye	899,500
Dnipropetrovs'k (Dnepropetrovsk)	1,185,500	Lvov	810,000
		Krivoi Rog	737,300
Donetsk	1,121,200	Lugansk	503,800

SOURCE: "Population of Capital Cities and Cities of 100,000 and More Inhabitants." United States Statistics Division.

Grasslands

Central Ukraine is characterized by mixed forest-steppe, with grasslands interspersed with deciduous trees, primarily oak. A true steppe zone (grassy plains) covers the lower third of the country, thinning out in the drier, more arid south. Along the southern Crimean coast lies a narrow Mediterranean zone of mixed shrubs, grasses, and evergreens.

Forests

Ukraine has well defined forest zones, with beech trees in the west; linden, oak, and pine in the north and northwestern swamps and meadows; and spruce in the northeast. About 18 percent of the country is forested, with tree cover most dense in the Carpathians and in the Polesye Marshes.

HUMAN POPULATION

With a population of 48,760,474 (July 2001 estimate), Ukraine has a population density of 209 persons per sq mi (81 persons per sq km). The population of Ukraine has been steadily declining over the last decade, with a loss nearing 1.7 million alone in the five-year period between 1997 and 2002. Approximately 68 percent of Ukrainians live in urban areas. The areas of densest settlement are in the Dnieper Lowlands of central Ukraine and the Donets Basin of the east, although large cities can be found throughout the country.

NATURAL RESOURCES

Ukraine is rich in both mineral resources and highly fertile agricultural land. Salt, sulfur, brimstone, ozocerite, graphite, titanium, magnesium, kaolin, nickel, and mercury are among its varied mineral deposits. In the southeast, the Donets Basin yields coal in abundance, while rich iron ore deposits are mined in the east central Kryvyy Rih area. In south central Ukraine, Nikopol on the Dnieper River has some of the world's largest deposits of manganese. Considerable oil and natural gas reserves have been found in the Carpathian foothills, the Donets Basin, and along the Crimean coast. Some 58 percent of Ukraine is arable land that supports a viable agricultural economy, with exceptionally fertile, black *chernozem* soils predominating in the central and southern regions.

FURTHER READINGS

Embassy of Ukraine. http://www.ukremb.com/ (accessed July 9, 2002).

Frydman, Roman, Andrzej Rapaczynski, John S. Earle, et al., eds. *The Privatization Process in Russia, Ukraine, and the Baltic States.* New York: Central European University Press, 1993.

Magocsi, Paul Robert. *Ukraine: A Historical Atlas.* Toronto: University of Toronto Press, 1985.

Subtelny, Orest. *Ukraine: A History.* Toronto: University of Toronto Press, 1988.

United States. U.S. Department of State. *Background Notes: Ukraine.* Washington, D.C.: Government Printing Office, 2001.

Welcome to Ukraine. http://www.ukraine.org/ (accessed July 9, 2002).

United Arab Emirates

- **Area:** 32,000 sq mi (82,880 sq km) / World Rank: 116
- **Location:** Northern and Eastern Hemispheres, in the Middle East, bordered on the east by Oman and the Gulf of Oman, on the north by the Persian Gulf, and on the west and south by Saudi Arabia.
- **Coordinates:** 24°00′N, 54°00′E
- **Borders:** 539 mi (867 km) / Oman, 255 mi (410 km); Saudi Arabia, 284 mi (457 km)
- **Coastline:** 819 mi (1,318 km)
- **Territorial Seas:** 12 NM
- **Highest Point:** Mount Yibir, 5,010 ft (1,527 m)
- **Lowest Point:** Sea level
- **Longest Distances:** 338 mi (544 km) NE-SW / 224 mi (361km) SE-NW
- **Longest River:** There are no perennial rivers
- **Natural Hazards:** Sand and dust storms
- **Population:** 2,407,460 (July 2001 est.) / World Rank: 137
- **Capital City:** Abu Dhabi, on the Persian Gulf coast
- **Largest City:** Abu Dhabi, population 928,000 (2000 est.)

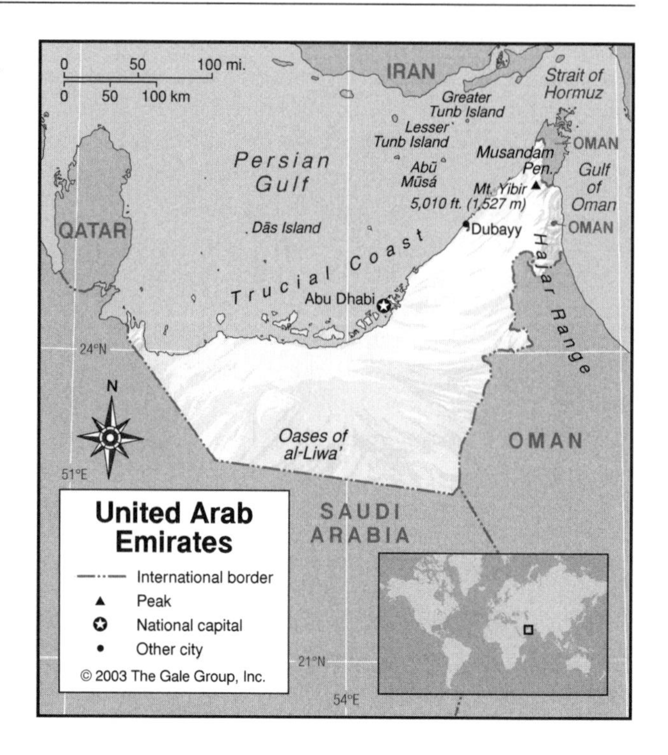

OVERVIEW

Seven emirates (states) make up the United Arab Emirates (UAE): Abu Dhabi, Dubayy, Ash Shāriqah, Ra's al Khaymah, Al Fujayrah, Umm al Qaywayn, and 'Ajmān. The location and status of the UAE's boundary with Saudi Arabia is not final. The UAE's boundary with Oman has not been bilaterally defined; the northern section in the Musandam Peninsula is an administrative boundary.

The UAE consists mainly of sandy inhospitable desert. On the west an enormous sebkha, or salt flat, extending southward for almost 70 mi (112 km) delimits it. The Trucial Coast in the east is made up of mud flats, sand spits, shallow seas, and lagoons. The eastern boundary runs northward over gravel plains and high dunes until it almost reaches the Hajar Mountains. The UAE is located on the Arabian Tectonic Plate.

MOUNTAINS AND HILLS

Mountains

The UAE's eastern region comprises barren, rugged mountains, part of neighboring Oman's Hajar Range. The highest peak in the country, Mount Yibir with a height of 5,010 ft (1,527 m), is located in this region. Behind Ra's al Khaymah and separating Al Fujayrah from the Persian Gulf the northern ridges of Hajar rise rapidly to 3,000 ft (900m), then go down to a narrow coastal plain along the Gulf of Oman.

Plateaus

The main gravel plain extends inland and southward from the coast of Ra's al Khaymah to Al-'Ayn and beyond.

INLAND WATERWAYS

There are no perennial lakes or rivers in the UAE, as not enough water falls to support them. However, there are small areas of wetlands. The Oases of al-Liwa' run in an arc along the dunes in the south at the edge of the vast Rub'al-Kali, the Empty Quarter. The flora of coastal marshes, swamps, and mangroves stands are salt-loving vegetation.

THE COAST, ISLANDS, AND THE OCEAN

Oceans and Seas

The UAE is at a strategic location along southern approaches to the Strait of Hormuz, which connects the Persian Gulf and the Gulf of Oman and is a vital transit point for world crude oil. The UAE is well provided with port facilities. Dubayy's Port Rashid is one of the largest artificial harbors in the Middle East.

Major Islands

There are many islands, most of which are owned by Abu Dhabi. These include Dās, the site of oil operations, and Abū Mūsá in the Persian Gulf exploited for oil and red oxide. Abū Mūsá is jointly claimed and administered by Iran and the UAE. Iran occupies two other islands in the Persian Gulf that are also claimed by the UAE, Lesser Tunb (Ţunb aş Şughrá) and Greater Tunb (Ţunb al Kubrá). The UAE has garnered significant diplomatic support in the region protesting these Iranian actions.

The Coast and Beaches

The flat coastal strip that makes up most of the UAE has an extensive area of *sebkha* subject to flooding. The alluvial flats on the Gulf of Oman south of Sharjah are a continuous, well-watered fertile littoral strip, known as the Batinah Coast. The Batinah runs between the mountains and the sea and continues into Oman.

CLIMATE AND VEGETATION

Temperature

The climate is arid and subtropical. The months between May and October are extremely hot, with shade temperatures of between 100° and 120°F (39° and 49°C). Humidity on the coast can exceed 85 percent. Winter temperatures can fall as low as 36°F (2°C) but average between 63° and 68°F (17° and 20°C). It is cooler in the eastern mountains.

Rainfall

Normal annual rainfall is from 2 to 4 in (5 to 10 cm), with considerably more in the mountains. Annual average rainfall in the mountain region is 5 to 8 in (14 to 20 cm) and along the east cost 4 to 5 in (10 to 14 cm). The wettest months are February and March. Prevailing winds, including the cool Shamal from the northeast and the Khamsin from the south, produce sandstorms. Influenced by monsoons, they vary by season and location.

Population Centers – United Arab Emirates

(1995 CENSUS OF POPULATION)

Name	Population
Dubayy (Dubai)	669,181
Abu Dhabi (capital)	398,695
Sharjah	320,095
al-Ain	225,970
Ajman	114,395

SOURCE: Ministry of Planning, UAE.

Emirates – United Arab Emirates

2000 POPULATION ESTIMATES

Name	Area (sq mi)	Area (sq km)	Population	Capital
Abu Dhabi	26,000	67,350	1,186,000	Abu Dhabi
Ajman	100	250	174,000	Ajman
Dubai	1,500	3,900	913,000	Dubayy (Dubai)
Fujairah	450	1,150	98,000	Fujairah
Ras al-Khaimah	650	1,700	171,000	Ras al-Khaimah
Sharjah	1,000	2,600	520,000	Sharjah
Umm al-Quwain	300	750	46,000	Umm al-Quwain

SOURCE: Ministry of Planning, United Arab Emirates.

Deserts

More than two thirds of the total area of the UAE is sandy desert land, running from the westernmost tip of Abu Dhabi east to the land border with Oman and the Gulf of Oman. The desert foreland extends from Ra's al Khaymah and Al-Ayn towards the west to the coast. The Westerly Desert is located in the northern edge of the Empty Quarter of Saudi Arabia and extends towards the borders of Abu Dhabi.

HUMAN POPULATION

In 2000 it was estimated that 86 percent of the population lived in urban areas, concentrating in each emirate's capital. About 80 percent of the UAE's population originate from outside its borders. This population is very unevenly distributed. Of the emirates, Abu Dhabi, Dubayy, and Ash Shāriqah are by far the most populous, with about 84 percent of the entire population—the city of Abu Dhabi alone has about 40 percent of the population. Settlements are concentrated in the coastal region, with a few also scattered on the Saudi Arabian border. The interior is largely uninhabited.

NATURAL RESOURCES

Petroleum and natural gas are the United Arab Emirates natural resources. Since 1973, the UAE has undergone a profound transformation from an impoverished region of small desert principalities to a modern state with a high standard of living. At present levels of production, oil and gas reserves should last more than 100 years. These fuel resources have brought the country wealth; the UAE has an open economy with a high per capita income: $22,800 (2000 est.). However, the fortunes of the economy fluctuate with the prices of oil and gas.

FURTHER READINGS

Crocetti, Gina L. *Culture Shock! United Arab Emirates.* Portland, Oreg.: Graphic Arts Center Pub. Co., 1996.

Etisalat. *UAE Pages.* http://www.uae.org.ae/general/contents.htm (accessed May 11, 2002).

Johnson, Julia. *United Arab Emirates.* Philadelphia: Chelsea House Publishers, 2000.

Kay, Shirley. *Seafarers of the Gulf.* Dubai: Motivate Pub., 1992.

Ministry of Information and Culture. *UAE Interact.* http://www.uaeinteract.com/default.asp (accessed May 9, 2002).

Trident Press Ltd. *Arabian Wildlife.* http://www.arabianwildlife.com/main.htm (accessed May 9, 2002).

United Kingdom

- **Area:** 94,526 sq mi (244,820 sq km) / World Rank: 78
- **Location:** Northern Hemisphere divided by the Prime Meridian between the Eastern and Western Hemispheres, in Western Europe, between the North Atlantic Ocean and the North Sea, northwest of France; Northern Ireland borders Ireland to the south.
- **Coordinates:** 54°00′N, 2°00′W
- **Borders:** 224 mi (360 km), all with Ireland
- **Coastline:** 7,723 mi (12,429 km)
- **Territorial Seas:** 12 NM
- **Highest Point:** Ben Nevis, 4,406 ft (1,343 m) / Mt. Paget, 9,626 ft (2,934 m), on South Georgia Island
- **Lowest Point:** Fenland, 13 ft (4 m) below sea level
- **Longest Distances:** 600 mi (965 km) N-S; 300 mi (485 km) E-W (Great Britain only)
- **Longest River:** Severn, 220 mi (354 km)
- **Largest Lake:** Lough Neagh, 154 sq mi (400 sq km)
- **Natural Hazards:** None
- **Population:** 59,647,790 (July 2001 est.) / World Rank: 20
- **Capital City:** London, in southern England on the Thames River
- **Largest City:** London, 6,962,319 (1994 census)

OVERVIEW

The United Kingdom (U.K.) is located on an archipelago off the northwestern coast of Europe, the British Isles. The major islands in the British Isles are Great Britain (often simply called Britain) and Ireland; numerous smaller islands are found nearby. Only the northern part of Ireland belongs to the United Kingdom, with the rest of the island being the Republic of Ireland. The British Isles lie between the Atlantic Ocean on the north and northwest and the North Sea on the east. Great Britain, the largest of the islands and the heartland of the country, is separated from the continent by the Strait of Dover and the English Channel, 21 mi (34 km) wide at its narrowest point, and from the Ireland by the Irish Sea, North Channel, and St. George's Channel. No place in the United Kingdom is more than 75 mi (120 km) from the sea.

The United Kingdom has four primary regions: England (50,337 sq mi / 130,373 sq km), Wales (8,018 sq mi / 20,767 sq km), and Scotland (30,415 sq mi / 78,775 sq km), all on the island of Great Britain; and Northern Ireland (5,452 sq mi / 14,120 sq km), on the island of Ireland. Each of these regions has its own geography.

England is found on the southern half of Great Britain, with Wales extending off in the west. It is composed mostly of rolling hills. The highest elevations are found in the north. In the northwest, a region known as the Lake District includes a number of small lakes and the terrain reaches higher elevations in a range known as the Cumbrian Mountains. In the north-central region, there are limestone hills known as the Pennine Chain. In the southwest, a peninsula with low plateaus and granite outcroppings makes up the region known as the West Country.

Scotland is the northern half of Great Britain, and is primarily mountainous. The Highlands occupy almost the entire northern half of the country and contain the highest peaks in the United Kingdom. South of the Highlands are the Central Lowlands, with an average elevation of 500 ft (152 m) and containing the valleys of the Tay, Forth, and Clyde Rivers. Beyond this are the Southern Uplands, with moorland cut by many valleys and rivers.

Wales is a region of rugged hills and mountains, with extensive tracts of high plateau and shorter stretches of mountain ranges deeply dissected by river valleys. It extends west from England. The Cambrian Mountains occupy almost the entire area and include Wales' highest point, Mt. Snowdon (3,560 ft / 1,085 m). There are nar-

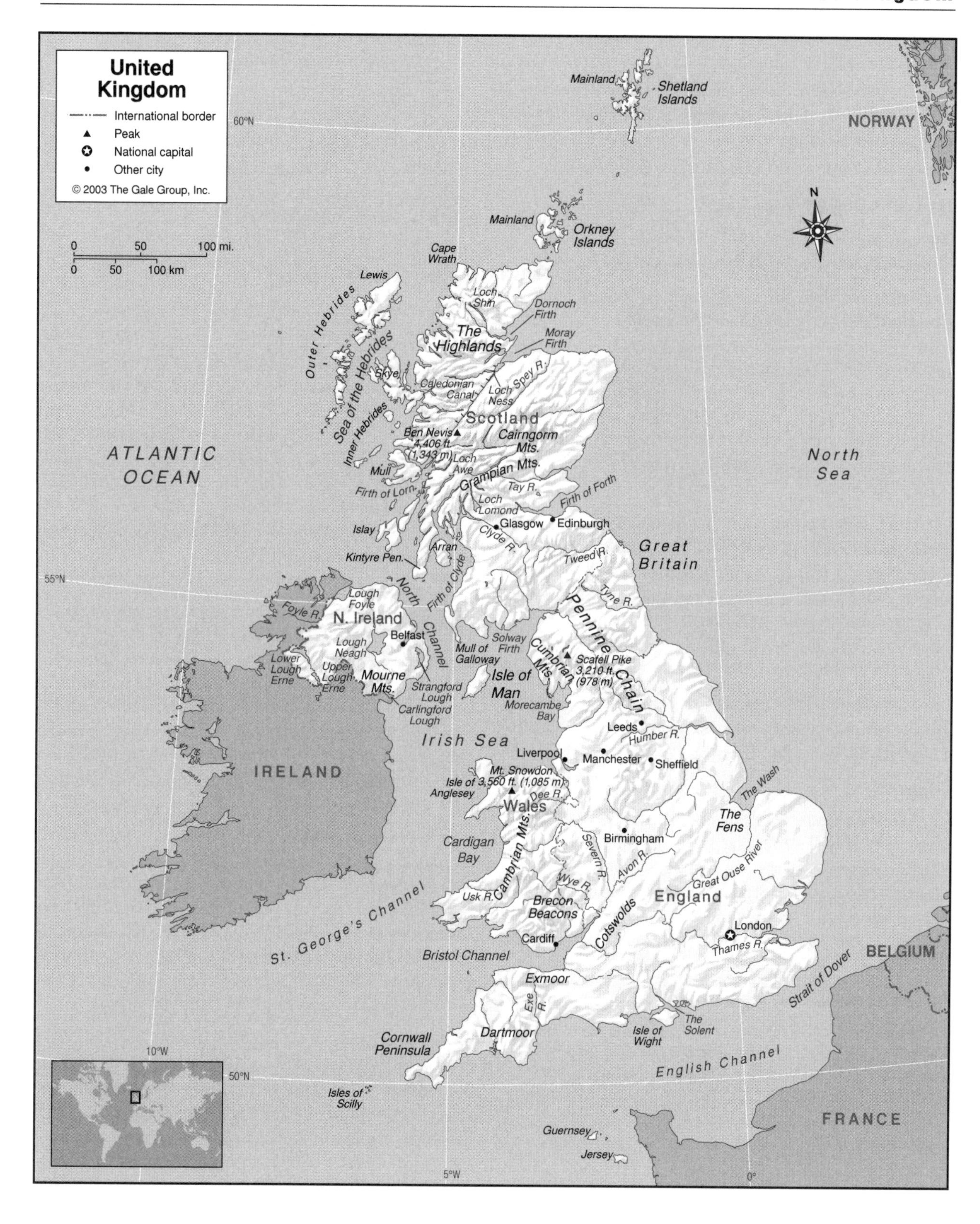

row coastal plains in the south and west and small lowland areas in the north, including the valley of the Dee.

Northern Ireland consists mostly of low-lying plateaus and hills, generally from 500 to 600 ft (152 to 183 m) above sea level. The Mourne Mountains mark the southeast. At the center of the region is Lough Neagh, a large lake.

There are several island groups and hundreds of individual islands in the vicinity of Great Britain and Ireland that are part of the United Kingdom. The best known among them are the Orkney Islands, the Shetland Islands, the Outer Hebrides, Skye, Mull, Islay, Arran, and the Isles of Scilly. In addition, the United Kingdom has dependencies scattered all around the world, including islands in every ocean except the Arctic.

MOUNTAINS AND HILLS

Mountains

The United Kingdom has no tall mountains by world standards, but there are many lower yet rugged ranges. The Highlands of Scotland are dominated by the Grampian Mountains and their subsidiary mountain ranges, including the Cairngorm Mountains. Ben Nevis (4,406 ft / 1,343 m), the highest peak in the United Kingdom, is found in this region, and there are more than 40 peaks that rise higher than 3,000 ft (900 m). At the southern end of Scotland are the Southern Uplands, with peaks of 2,750 ft (838 m).

The Cumbrian Mountains are the highest mountains in England. They are located in the northwestern Lake District. Scafell Pike (3,210 ft / 978 m) is the highest peak in the range. Farther south, the Cambrian Mountains occupy most of Wales and house its highest peak, Mt. Snowdon. The Black Mountains and Brecon Beacons located in the southern Wales are glacial mountains.

The Mournes Mountains in southeastern Northern Ireland include twelve rounded peaks. Slieve Donard (2,796 ft / 847 m) is the tallest.

Plateaus

The West Country of England, located on the southwestern Cornwall Peninsula, is the site of Exmoor and Dartmoor, low plateaus with granite projections. The Cairngorm Plateau in Scotland, located adjacent to the mountains of the same name, is a broad desert-like barren region with an elevation of more than 4,000 ft (1,220 m).

Hills and Badlands

Most of England consists of low plains and rolling downs, particularly in the south and the southeast, where the land does not rise higher than 1,000 ft (305 m) at any point. Running from east to west on the Scottish border are a series of sandstone ridges known as the Cheviot Hills, and from north to south from the Scottish border to central England is the Pennine Chain. South of the Pennines lie the Central Midlands, a plains region with low, rolling hills and fertile valleys. Southern England is the site of three ranges of low hills, the Cotswolds in the west and the North and South Downs in the east.

The Rannock moor lies in the center of Scotland, at an elevation of 1,000 ft (303 m). It is the source for most rivers in the country, and consists of peat bogs lying over granite bedrock, with rocky outcroppings. Foothills surround the mountains of Scotland and Wales.

The majority of Northern Ireland consists of low plateaus and hills, generally between 500 and 600 ft (150 to 180 m) high. In the east small hills called "drumlins" surround the area of Strangford Lough; some drumlins are partially submerged in the lough, and rise above the water's surface as small islands.

INLAND WATERWAYS

Lakes

The largest lake in the U.K. is Lough Neagh (Lake Neagh, 153 sq mi/396 sq km). It lies in the center of Northern Ireland. Because it is so shallow (30 ft / 8.9 m), the water remains muddy at all times, stirred by the strong prevailing winds in the area. Fishing for eel and salmon is an important activity on Lough Neagh; thousands of tons of eel are caught there and exported each year. Southwest of Lough Neagh are the Upper and Lower Lough Erne, two large lakes on the Erne River that extend across the country and into Ireland.

Scotland is a region of many lakes, called lochs. Loch Lomond (27 sq mi / 70 sq km) lies in this region, at the foot of Highlands. It is the largest lake on Great Britain. Loch Ness (22 sq mi / 57 sq km), famous for its legendary Loch Ness monster, lies further north. Like many of the lochs, Loch Ness was created by glacial activity and is very deep. Its mean depth is 433 ft (132 m), with a maximum depth of 754 ft (230 m). The long, narrow Loch Awe (15 sq mi/39 sq km) lies northwest of Loch Lomond. Loch Shin and Loch Loyal are located in the far north.

There are no large lakes in England or Wales. However, on the northwest coast of England, near the border with Scotland, there is a region of small, picturesque lakes called the Lake District. Within the confines of the Lake District National Park, there are 15 lakes: Bassenthwaite, Ullswater, Derwent, Loweswater, Crummock, Thirlmere, Buttermere, Ennerdale, Haweswater, Grasmere, Rydal, Wast Water, Devoke Water, Coniston, and Windermere.

Rivers

Rivers are plentiful throughout the United Kingdom, but most are short as the sea is always nearby. The longest rivers are found in England and Wales. The Severn is the longest in the nation (220 mi/352 km). It flows east from its source in Wales before curing south and finally west, emptying into the Bristol Channel. The Thames (200 mi / 322 km), England's best-known river and the second longest in the U.K., is navigable for about 135 mi (216 km), with more than 40 locks. Other English and Welsh rivers include the Humber, Tees, Tyne, and Great Ouse in the east, and the Avon, Wye, Dee, and the Exe in the west. The Usk, the longest river entirely within Wales, flows for 85 mi (136 km) in southern Wales. An extensive series of

GEO-FACT

The Channel Tunnel is a set of tunnels underneath the Strait of Dover that connects southeastern England to northeastern France. Dug between 1988 and 1991, it opened for use in 1994. When construction began it was the most expensive engineering project ever attempted. The final cost was $21 billion. At 31 mi (50 km) long it is among the longest tunnels on Earth; 24 mi (38 km) are under the English Channel itself. The Channel Tunnel actually consists of three tunnels: two main tunnels for rail traffic with a smaller maintenance tunnel between them. They are buried an average of 148 ft (45 m) beneath the ocean floor. Traveling from one end of the tunnel to the other takes roughly 35 minutes.

canals in England connects many of its southern rivers and cities together.

Scotland's river system is largely separate from that of England, although the Tweed (96 mi / 154 km) flows west along part of the border with England. The two major rivers of Scotland's central lowland are the River Clyde and the River Forth. The River Clyde (106 mi/170 km) has its source in the Southern Uplands of Scotland; it flows northwest, past Glasgow, into the Firth of Clyde. The River Forth (116 mi / 187 km) flows in an easterly direction to the North Sea at the Firth of Forth; strong tides ebb and flow, affecting the waters of the River Forth for half its length (53 mi / 85km). A canal runs across Scotland to connect the Clyde and the Forth. Scotland's longest river, the River Tay (117 mi/188 km), is further north, also flowing east to the Firth of Tay and the North Sea, near Dundee. Another major river is the Spey (110 mi/177 km), which flows north from the Northern Highlands to the North Sea. The Caledonian Canal cuts all the way across northwestern Scotland.

Northern Ireland's major rivers are the Erne, with Upper and Lower Lough Erne along its course, and the Foyle, which marks part of the border with Ireland. A canal connects Lough Neagh with the Irish Sea.

Wetlands

Wetlands are common throughout the United Kingdom, but are particularly common in East Anglia. This area northeast of London along the east coast is predominantly marshy, with expansive habitat for wading and sea birds. The Fens, in the northeast around the North Sea inlet, The Wash, feature vast expanses of flat land reclaimed by artificial drainage from its original marsh state. The lowest point in the U.K. is found in the fenland region. Estuaries on the south coast, such as the one at Poole Bay, feature wetlands habitats. Northern Ireland's landscape includes many peat bogs, swamps, and fens. Some coastal areas include marshy and rocky areas that support a wide range of wildlife. The region around Inverness, in Scotland, and on the Hebrides, Shetland, and Orkney Islands, supports rich wetland habitats. The Dyfi National Nature Reserve is a wetlands area in western Wales, with sand dunes, an estuary, and a peat bog, Cors Fochno. The Isles of Scilly off the southwest coast of Great Britain have been designated as protected wetlands.

THE COAST, ISLANDS, AND THE OCEAN

Oceans and Seas

The United Kingdom is surrounded by water. The North Sea lies between Scotland and England; about 125 mi (200 km) off the coast of Dundee, Scotland, lies the Devil's Hole, a series of deep trenches in the North Sea. The Devil's Hole reaches depths of 760 ft (230 m), compared with the average of 260 to 300 ft (80 to 90 m) in the surrounding water.

The Irish Sea lies between the island of Great Britain and Ireland. The northern part of the Irish Sea is known as the North Channel, while the southern part is St. George's Channel. South of Ireland and west of the southernmost tip of Great Britain is the Celtic Sea. The reefs that lie in the waters off the Northern Ireland coast make sailing in this area of the Irish Sea dangerous. Northwest of Great Britain is the Sea of the Hebrides. Beyond that sea and its islands is the open waters of the North Atlantic Ocean.

The English Channel lies along the southern coast of Great Britain, separating it from the mainland of Europe and the northwest coast of France. The English Channel connects the Atlantic Ocean to the west and the North Sea to the northeast, and is one of the world's busiest waterways, with more than 500 vessels plying its waters daily. The waters of the English Channel are often rough because of the combination of strong currents and heavy winds. The coast and English Channel are periodically shrouded in dense fog. The narrowest point of the English Channel lies between Dover in the southeast Great Britain and Calais, France; the waters there are known as the Strait of Dover.

Major Islands

The United Kingdom is found in the British Isles, an enormous collection of islands northwest of the European landmass. By far the largest is the island of Great Britain (88,150 sq mi / 228,300 sq km). Great Britain is the largest island in Europe. Ireland is the second largest of the British Isles, Northern Ireland is located on the northern part of this island.

Several smaller archipelagos near Great Britain are a part of the United Kingdom. The most extensive are the Hebrides, off the northwest coast of Scotland. The Outer Hebrides include the large islands of Lewis, Harris, and North and South Uist. The principal islands of the Inner Hebrides, which are located closer to shore and extending further south, are Skye, Mull, and Islay. The Orkney Islands are a smaller archipelago, located just north of Scotland. Hoy and Mainland Islands are the largest of the group. Much further north in the North Sea are the Shetland Islands. The largest island here is also called Mainland; it has a highly irregular shape with many peninsulas, some of which enclose St. Magnus Bay. Other Shetland Islands include Yell and Unst. The Isles of Scilly are located at the other end of the country, off the southwest tip of England in the Celtic Sea. They are a collection of 140 small, rocky, islands, most notably St. Martin's, St. Mary's, and St. Agnes.

Besides these archipelagos, there are many relatively isolated islands, large and small, near Great Britain. The Isle of Wight lies in the English Channel just off the southcentral coast of England. The Isle of Anglesey is off Wales in the Irish Sea. Arran is off the coast of Scotland, southeast of the Kintyre Peninsula.

Several islands near Great Britain are crown dependencies, meaning that they are technically not a part of the United Kingdom but instead belong to that country's royal family. In practice, they are specially administered by the U.K. government. The Isle of Man is the largest (221 sq mi / 572 sq km). It lies in the Irish Sea between the Cumbrian coast of England and Northern Ireland. The Channel Islands are located south of Great Britain in the English Channel, near the coast of France. They consist of Jersey (64 sq mi / 116 sq km), Guernsey (30 sq mi / 78 sq km), and the smaller islands of Alderney, Sark, Herm, Jethou, Brechou, and Lihou. Guernsey is mostly level with low hills in southwest. Jersey, southeast of Guernsey, is a gently rolling plain with low rugged hills along the north coast.

Overseas Dependencies

The United Kingdom has numerous dependencies scattered around the entire world. These are overseas territory and dependencies of the United Kingdom itself, not the royal family. Many of the largest are in or near the Caribbean Sea. The British Virgin Islands are the eastern half of the Virgin Islands, a group of small coral and volcanic islands (see entry on British Virgin Islands) in the eastern Caribbean. Anguilla is a flat, dry, scrub-covered coral island in the Leeward Islands (see entry on Anguilla), not far east of the British Virgin Islands. Southeast of Anguilla, beyond Antigua and Barbuda, is the single island of Montserrat (39 sq mi / 100 sq km). Montserrat is rugged and volcanically active. Eruptions in 1995 and 1996 forced the evacuation of much of the population. The Turks and Caicos Islands are located further north and west, in the Atlantic Ocean. They are a series of small limestone islands at the southeastern end of the Bahamas (see entry on Turks and Caicos). South of Cuba, in the Caribbean Sea, another small archipelago dependency of the U.K. can be found, the Cayman Islands (see entry on Cayman Islands). All of these dependencies are sparsely populated; the Cayman Islands' population of roughly 35,000 is the largest among them.

Other dependencies are scattered around the world. Bermuda is an archipelago of coral islands about 580 mi (935 km) east of the United States in the North Atlantic Ocean. It has long been a favorite destination of tourists (see entry on Bermuda). A group of widely scattered volcanic islands in the South Atlantic Ocean take their name from the island of Saint Helena, where their capital is located. Other islands in this dependency include Ascencion, Tristan da Cunha, Inaccessible Island, Gough, and the Nightingale Islands. Collectively they have an area of 273 sq mi (440 sq km). Queen Mary's Peak reaches 6,758 ft (2,060 m) on Tristan da Cunha.

Further south in the Atlantic Ocean, about 300 mi (483 km) east of the southern tip of South America, are the Falkland Islands (Islas Malvinas). They consist of two large main islands, East and West Falkland, and a number of smaller islands that surround them. Together they have an area of 4,700 sq mi (12,173 sq km), making them the largest of the United Kingdom's dependencies, but they have a population of only a few thousand people. The islands have irregular coastlines and low, rugged, mountains; Mt. Usborne is the highest peak at 2,312 ft (705 m). The Falkland Islands are also claimed by Argentina, and were the scene of fighting between the two countries in 1982. South Georgia and the South Sandwich Islands, located in two chains about 620 mi (1000 km) east of the Falkland Islands, are another dependency of the United Kingdom that is claimed by Argentina. These islands are mostly volcanic in origin and include some active volcanoes. Mt. Paget on South Georgia Island is the tallest mountain in all of the United Kingdom's possessions (9,626 ft / 2,934 m).

The United Kingdom has three other dependencies of note. The Chagos Archipelago, coral islands in the northern Indian Ocean, is the British Indian Ocean Territory. Its main island is Diego Garcia, the site of a large military base. The Pitcairn Islands are a group of small, low, volcanic islands in the south central Pacific Ocean. Gibraltar is not really an island, but rather a small peninsula (2.5 sq mi / 6.5 sq km) projecting south off of Spain's Mediterranean coastline. A British possession since 1713, Gibraltar occupies a very strategic location at the Straits of Gibraltar that connect the Mediterranean Sea to the Atlantic Ocean. It is also the site of the famous Rock of Gibraltar (1,398 ft / 426 m).

The Coast and Beaches

The coasts of both Great Britain and Northern Ireland are very irregular, with many long peninsulas and deep bays, firths, and loughs. The most even part of the nation's coastline is the eastern coast of England. Much of this part of the coast is less than 5 m (15 ft) above sea level, and protected by embankments against inundations from gales and unusually high tides. Even here the coast is marked by the estuaries of the Humber and the Thames, with The Wash jutting into the coast between them.

Along the southeast coast, the Strait of Dover is bordered by white chalk cliffs that rise to 825 ft (250 m). The faces of these famous white cliffs of Dover are scarred from various battle and shipwreck events through several centuries of British history. Several short promontories, including Dungeness and Beachy Head, mark England's southern coast. Between the main island and the Isle of Wight is The Solent. The whole of southwestern England is a peninsula, Cornwall, which extends 75 mi (120 km) west into the Atlantic, separating the English Channel in the south from the Bristol Channel to the north. Its coast is rugged and rocky. White clay, used in producing fine china, is quarried in this region.

Between Wales and England, the Celtic Sea flows into the Bristol Channel. The west coast of Wales curves around Cardigan Bay, a wide bay at the east edge of St. George's Channel, with the Lleyn Peninsula at its northern end. The coastline features rugged cliffs, coves, wooded estuaries, and sandy beaches. The Pembrokeshire Coast Path, the country's only coastal national park, extends more than 167 miles (269 km). North of Wales is the open Irish Sea. Further east are Liverpool Bay and Morecambe Bay on England's northwestern coast.

As the coast approaches Scotland, both in the west and the east, it becomes even more irregular than in the rest of the country. Scotland's entire coast was deeply incised by glacial activity, which has left behind towering cliffs and headlands as well as numerous bays and inlets, the deep and narrow lochs and wide firths of Scotland. Along the sparsely populated west coast, sea inlets form lochs edged by mountains. In some areas, the terrain opens to broad open beaches. Many river estuaries serve as fine harbors. Tides are very strong in the narrower inlets. The broad Solway Firth marks the end of England's northwestern coast and the beginning of Scotland. It is separated from the North Channel by a long, narrow peninsula, ending in the Mull of Galloway. Further north are two more great firths, the Firth of Clyde and the Firth of Lorn, with another long peninsula, Kintyre, between them. Both firths extend deep into the country in multiple arms, and both are connected to the North Sea on the far side of Scotland by canals. Further north on the western coast there are numerous narrower but still lengthy inlets. Cape Wrath marks the northwestern end of Great Britain.

The eastern coast of Scotland features fewer of the small, narrow inlets so predominate in the west, but instead has two deep, broad, indentations, with a headland between them. The northernmost ends in Moray and Dornoch Firths. Further south is the Firth of Forth, which plunges nearly straight into the island, reaching deep into the central lowlands region.

Along the east coast of Northern Ireland is a large sea inlet known as the Strangford Lough. Seawater rushes through the narrow gap, causing powerful, fast-moving currents twice every day with the changing tides. Strangford comes from a Scandinavian word for "violent fjord." Other notable loughs on Northern Ireland's coast are Lough Foyle at the northwestern border, and Carlingford Lough along the southeastern border. The Giant's Causeway is an unusual natural feature resembling a huge ramp, about 900 ft (274 m) across and extending 500 ft (150 m) into the ocean.

CLIMATE AND VEGETATION

Temperature

Warmed by the North Atlantic Drift, the United Kingdom enjoys a temperate climate, with the temperature rarely exceeding 90°F (32°C) in the summer months or dropping below 14°F (-10°C) in the winter. During the winter, monthly temperatures range from 37°F (3°C) to 41°F (5°C). Summertime temperatures range from 54°F to 61°F (12 to 16°C).

Rainfall

Rainfall is lightest along the eastern and southeastern coasts, and heaviest on the western and northern heights, where precipitation can exceed 150 in (380 cm). Average annual rainfall across the country is just above 40 in (100 cm), with rain distributed evenly throughout the year.

Plains

Except for the hilly areas in the north and west, and some forested areas, most of England is naturally rolling grassland. This is particularly true of the Midlands part of England. However, most of England's open land is now highly developed and urbanized.

Forests

Most of England's forestland has been destroyed, with new natural forest areas remaining. The government has

Population Centers – United Kingdom

(2000 ESTIMATES)

Name	Population	Name	Population
London, England	7,640,000	Tyneside (Newcastle upon Tyne), England	981,000
Birmingham, England	2,272,000	Liverpool, England	915,000
Manchester, England	2,252,000		
Leeds, England	1,433,000		

SOURCE: "Table A.12. Population of Urban Agglomerations with 750,000 Inhabitants or More in 2000." United Nations Population Division, *World Urbanization Prospects: The 2001 Revision.*

Regions – United Kingdom

2000 POPULATION ESTIMATES

Name	Population	Area (sq mi)	Area (sq km)	Capital
England	49,997,100	50,363	130,439	London
Northern Ireland	1,697,800	5,452	14,120	Belfast
Scotland	5,114,600	30,418	78,783	Edinburgh
Wales	2,946,200	8,019	20,768	Cardiff

SOURCE: "Mid-2000 UK Populations Estimates." National Statistics: the Official UK Statistics Site." (9 May 2002). [Online] Available http://www.statistics.gov.uk/ (accessed 3 June 2002).

made efforts to plant coniferous trees, but this has changed the soil acidity and led to further destruction. Most of the country's forestland is found in England. The area known as New Forest, located between Southampton and Bournemouth, is the largest preserved forestland in England. Its beech and oak trees cover 145 sq mi (376 sq km). It was set aside as a hunting area by William the Conqueror in 1079. The area retains many of the old medieval practices. For example, local inhabitants, referred to as "commoners" are granted right to graze cattle, horses, and pigs on forest land.

HUMAN POPULATION

Although there are many ethnic and religious minorities, the majority of the population is Caucasian and Christian, mainly Anglican. Approximately nine-tenths of the population live in urban areas. Greater London is the most heavily and most densely populated part of the country. Another area of dense population with many major cities is found along England's Irish Sea coast, extending into the center of the country. Scotland's Central Lowlands is another center of population, though not nearly so densely settled as those in England. The primary concentration of people in Wales is along the southern coast; in Northern Ireland it is in the east and southeast. The English comprise about 81.5 percent of the population, followed by Scottish (9.6 percent), Irish (2.4 percent) Welsh (1.9 percent), and Ulster (1.8 percent). The remaining 2.8 percent of the population is composed mainly of West Indian, Indian, and Pakistani people.

NATURAL RESOURCES

There are few mineral resources in the United Kingdom. Coal was historically the country's most important resource, with extensive deposits throughout northern England, but much of this ore has been exhausted. However, the United Kingdom has control over major oil deposits in the North Sea. Other minerals include limestone, dolomite, kaolin, ball clay, gypsum, sand, and gravel. The economy of the United Kingdom is among the world's largest and most highly developed, with important manufacturing, commercial, and service sectors.

FURTHER READINGS

Botting, Douglas. *Wild Britain: A Traveller's Guide.* New York: Interlink Books, 2000.

Gresko, Marcia S. *Scotland.* Woodbridge, Conn.: Blackbirch Press, 2000.

Lake District National Park Authority Online. http://www.lake-district.gov.uk/ (Accessed July 23, 2002).

Northern Ireland Tourist Board. *Discover Northern Ireland.* http://www.discovernorthernireland.com/ (Accessed July 23, 2002).

Norwich, John Julius. *England & Wales.* New York: Knopf, 2000.

Scotland. New York: Knopf, 2001.

UK 2002: The Official Yearbook of the United Kingdom of Great Britain and Northern Ireland. Norwich: Stationery Office, 2001.

VisitScotland. http://www.visitscotland.com/ (Accessed June 23, 2002).

Welcome to Scotland. http://www.geo.ed.ac.uk/home/scotland/scotland.html (Accessed July 23, 2002).

United States of America

- **Area:** 3,717,813 sq mi (9,629,091 sq km) / World Rank: 4
- **Location:** Northern and Western Hemispheres, bordering Canada to the north, Mexico to the south, the Atlantic Ocean to the east, the Gulf of Mexico to the

southeast, and the Pacific Ocean to the west. The exclave of Alaska borders Canada in the east.

- **Coordinates:** 38°00′N, 97°00′W
- **Borders:** 7,593 mi (12,219 km) / Canada, 5,526 mi (8,893 km); Mexico, 2,067 mi (3,326 km)
- **Coastline:** 12,380 mi (19,924 km)
- **Territorial Seas:** 12 NM
- **Highest Point:** Mt. McKinley, 20,322 ft (6,194 m)
- **Lowest Point:** Death Valley, 282 ft (86 m) below sea level
- **Longest Distances:** 2,897 mi (4,662 km) ENE-WSW / 2,848 mi (4,583 km) SSE-NNW
- **Longest River:** Missouri, 2,466 mi (3,968 km)
- **Largest Lake:** Lake Superior, 31,820 sq mi (82,732 sq km)
- **Natural Hazards:** Floods; hurricanes; tornadoes; forest fires; earthquakes
- **Population:** 278,058,881 (July 2001 est.) / World Rank: 3
- **Capital City:** Washington, D.C., mid-Atlantic coast
- **Largest City:** New York City, northern Atlantic coast, 8,008,278 (2000 census)

OVERVIEW

The United States of America (United States, U.S.A., U.S.), the world's third-largest country, occupies the central part of the North American continent, between Canada and Mexico. It extends about 1,200 mi (1,900 km) north to south between these two countries and spans the width of the continent from the Atlantic Ocean to the Pacific. In addition to the forty-eight conterminous states (and the District of Columbia) contained within this area, Alaska, at the northwestern edge of the continent, and Hawaii, an island state in the Pacific Ocean about 2,500 mi (4,000 km) west of North America, are integral parts of the U.S.A. It is the second largest country on the continent, Canada being larger. The United States also has many overseas dependencies and possessions. Most notable among these are Puerto Rico and the U.S. Virgin Islands in the Caribbean Sea; and Guam, American Samoa, and the North Mariana Islands, in the Pacific Ocean.

In its broadest topographic outline, the conterminous U.S. comprises a large central lowland—accounting for close to half its total area—bordered on the east and west by highlands. The western highland area, which begins with the Rocky Mountains, is by far the more extensive of the two, accounting for about one-third the total area of the country. The band of highlands on the east, which is lower and less extensive, consists of the Appalachian Mountains. The lowland between them is dominated by the Mississippi River and its tributaries, with the Great Lakes to the north. The western part of this lowland is known as the Great Plains. East and south of the Appalachian Mountains are coastal plains.

The conterminous United States can be broken down into the following major geographic regions:

A low-lying coastal plain extends for more than 2,000 mi (3,200 km) along the eastern and southeastern fringes of the country, encompassing the coasts of both the Atlantic Ocean and the Gulf of Mexico. The plain is narrow in New England but reaches a maximum width of about 200 mi (320 km) farther south. Its terrain is mostly flat but includes ridges and low hills, as well as large tracts of marshland.

The Appalachian Mountains extends northeast to southwest from Maine to Alabama, a length of some 1,200 mi (1,000 km). This varied region includes mountains, plateaus, and valleys. Among its subregions are the Piedmont Plateau, the Blue Ridge Mountains, the Appalachian Plateaus, the Adirondacks, and the St. Lawrence Valley. The area of this region is roughly equal to that of the Atlantic coastal plain.

The central lowland, with an area of about 585,000 sq mi (1,515,000 sq km), is the country's agricultural heartland. Stretching from the Appalachians to the Great Plains, and from the northern border to the southern section of the coastal plain, it is the country's largest region, accounting for about 15 percent of its total area and extending over 16 states. This region, which includes the Great Lakes, ranges in elevation from under 1,000 ft (305 m) to around 2,000 ft (600 m), gradually getting higher from east to west and encompassing flat plains, rolling land, woodlands, and valleys.

Extending northward from central Texas to the Canadian border, in an unbroken band with an average width of about 500 mi (805 km), the Great Plains represent a continuation of the central lowland. Continuing that region's gradual westward rise in elevation, from 2,000 ft (600 m) to over 5,000 ft (1,500 m), they cover an area of approximately 450,000 sq mi (1,165,000 sq km). The Great Plains include such varied subregions as the Missouri Plateau, the Black Hills, the High Plains, the Colorado Piedmont, and the Llano Estacado.

West of the Great Plains are the Rocky Mountains. They are characterized by many high peaks, rugged relief, extensive forests, and spectacular scenery. This is the principal mountain range of North America, extending across the continent from north to south in a vast series of parallel ridges. They are part of a continuous chain of mountains stretching from Alaska to Patagonia, the backbone of the Western Hemisphere. Within the United States they constitute the Continental Divide, with rivers

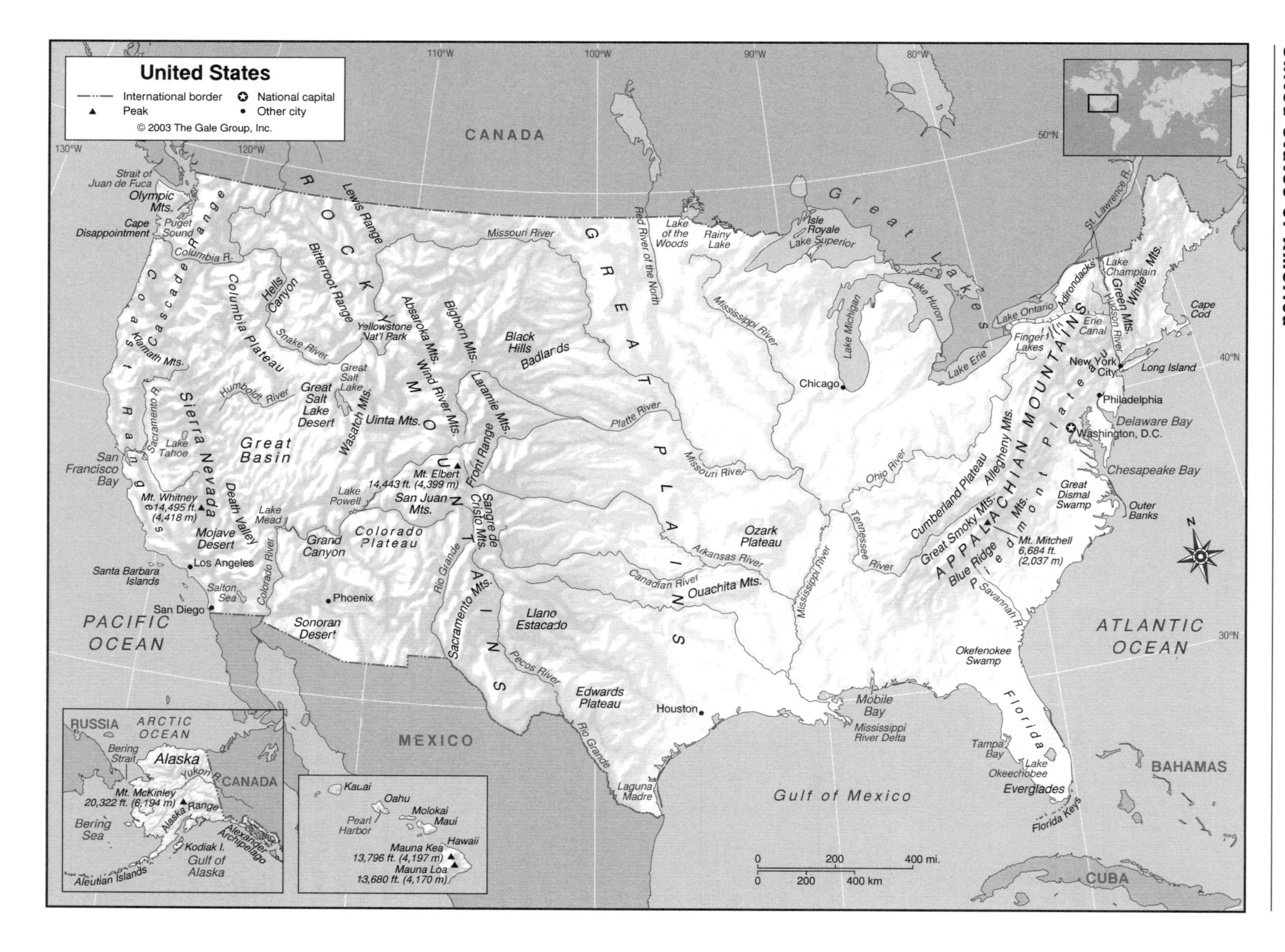
United States
International border
Peak
National capital
Other city
© 2003 The Gale Group, Inc.
CANADA
MEXICO
PACIFIC OCEAN
ATLANTIC OCEAN
Gulf of Mexico
BAHAMAS
CUBA
N
50°N
40°N
30°N
80°W
90°W
100°W
110°W
120°W
130°W
0 200 400 mi.
0 200 400 km
Great Lakes
Lake Superior
Isle Royale
Lake Michigan
Lake Huron
Lake Erie
Lake Ontario
St. Lawrence R.
Adirondacks
Lake Champlain
Green Mts.
White Mts.
Hudson River
Erie Canal
Finger Lakes
New York City
Long Island
Cape Cod
Philadelphia
Delaware Bay
Washington, D.C.
Chesapeake Bay
Outer Banks
Great Dismal Swamp
Piedmont Plateau
APPALACHIAN MOUNTAINS
Allegheny Mts.
Mt. Mitchell 6,684 ft. (2,037 m)
Blue Ridge Mts.
Great Smoky Mts.
Cumberland Plateau
Savannah R.
Okefenokee Swamp
Florida
Tampa Bay
Lake Okeechobee
Everglades
Florida Keys
Ohio River
Tennessee River
Chicago
Mississippi River
Mobile Bay
Mississippi River Delta
Ozark Plateau
Arkansas River
Ouachita Mts.
Missouri River
Canadian River
Houston
Laguna Madre
Rio Grande
Edwards Plateau
Llano Estacado
Pecos River
Lake of the Woods
Rainy Lake
Red River of the North
GREAT PLAINS
Platte River
Black Hills
Badlands
Missouri River
Bighorn Mts.
Laramie Mts.
Front Range
Sangre de Cristo Mts.
Sacramento Mts.
Mt. Elbert 14,443 ft. (4,399 m)
San Juan Mts.
ROCKY MOUNTAINS
Wind River Mts.
Absaroka Mts.
Yellowstone Nat'l Park
Uinta Mts.
Wasatch Mts.
Great Salt Lake
Great Salt Lake Desert
Lewis Range
Bitterroot Range
Snake River
Hells Canyon
Colorado Plateau
Lake Powell
Grand Canyon
Phoenix
Sonoran Desert
Lake Mead
Colorado River
Great Basin
Humboldt River
Columbia Plateau
Death Valley
Mojave Desert
Los Angeles
Salton Sea
San Diego
Sierra Nevada
Mt. Whitney 14,495 ft. (4,418 m)
Lake Tahoe
Sacramento R.
Klamath Mts.
Coast Ranges
Cascade Range
Columbia R.
Puget Sound
Olympic Mts.
Strait of Juan de Fuca
Cape Disappointment
San Francisco Bay
Santa Barbara Islands
Kauai
Oahu
Molokai
Maui
Hawaii
Pearl Harbor
Mauna Kea 13,796 ft. (4,197 m)
Mauna Loa 13,680 ft. (4,170 m)
RUSSIA
ARCTIC OCEAN
Bering Strait
Bering Sea
Alaska
Mt. McKinley 20,322 ft. (6,194 m)
Alaska Range
Yukon R.
CANADA
Kodiak I.
Gulf of Alaska
Alexander Archipelago
Aleutian Islands

to their east feeding into the Atlantic and the Gulf of Mexico, rivers to their west emptying into the Pacific, or never reaching the sea at all. Together with the Wyoming Basin they cover about 10 percent of the country.

The Intermontane Plateau region, which lies between the Rocky Mountains and the Pacific coastal mountains, covers the largest area within the western highland region: about 550,000 sq mi (1,424,000 sq km). It encompasses vast expanses of desert, as well as river basins, valleys, mountains, canyons, and hills. Much of this region has no drainage to the sea.

The Pacific mountain system parallels, and in some places extends to, the western coast, covering an area of about 200,000 sq mi (518,000 sq km) and including several different mountain ranges as well as a variety of other physical features, such as valleys and deserts.

Clearly demarcated among the foregoing large regions are two smaller but topographically distinct ones. The Superior Upland, an area bordering Lake Superior and extending 75,000 sq mi (194,000 sq km) over parts of Minnesota, Wisconsin, and Michigan, is a continuation of the Precambrian Shield of Canada that rises to elevations of between 1,000 and 2,000 ft (305 and 610km). The Interior Highlands, which include the Ozark Plateaus and the Ouachita Mountain region, have average elevations as high as 2,600 ft (792 m). With an area of less than 100,000 sq mi (250,000 sq km), they include parts of Missouri, Arkansas, and Oklahoma.

Separated from the rest of the country by Canada, Alaska is the largest state, with a land area about one-fifth that of the other forty-nine states combined. It is characterized by extremes of elevation, from Mount McKinley, at 20,320 ft (6,193 m) the highest point in North America, to the Aleutian Trench just off shore, at 25,000 ft (7,620 m) below sea level the lowest point bordering North America. Alaska has seven topographical regions: the southeastern coastal mountains; the glaciered coast; south-central Alaska; the Alaska Peninsula and the Aleutian Islands; interior Alaska; the Seward Peninsula and the Bering Coast Uplands; and the Arctic Slope.

The Hawaiian Islands are basaltic volcanoes near the middle of the Pacific Ocean along a northwest-trending ridge that divides two oceanic deeps, each of which descends more than 18,000 ft (5,486 m) below sea level. The United States also has a number of overseas island territories and dependencies in the Caribbean Sea and the Pacific Ocean, covering a total area of approximately 4,000 sq mi (10,360 sq km). Puerto Rico is the largest of these.

All of the United States lies on the North American Tectonic Plate, except for a narrow strip in southern California and the Hawaiian Islands, which are on the Pacific Plate. The Juan de Fuca Plate is located not far off the Pacific coast. Certain regions of the United States are subject to either severe or moderate seismic activity. One seismic belt, which extends along parts of the Rocky Mountains, has been the site of two major earthquakes, one at Helena, Montana, in 1925 and the other at Hebgen Lake in 1959. Earthquakes are more frequent on the Pacific Coast, along fault lines in the area where the North American and Pacific Tectonic Plates meet. The Pacific coastal area experienced more than 200 earthquakes between the mid-nineteenth and mid-twentieth century, including highly destructive ones in 1857, 1872 and 1906. Earthquakes are also frequent in two belts of seismic activity in Alaska. Volcanoes can be found all along the Pacific coast and in Alaska and Hawaii.

MOUNTAINS AND HILLS

Mountains make up one-quarter of the country. The central plains are flanked on the east by the Appalachian Mountains and on the west by the Rocky Mountain system. More high mountains can be found along the Pacific coast, and in Alaska and Hawaii.

The Rocky Mountains

The Southern Rockies are the highest in the chain, with many peaks of over 14,000 ft (4,267 m). They can be crossed only through high passes, all above 9,000 ft (2,743 m). These mountains consist of a series of ranges, among them the Laramie, Front, Sangre de Cristo, San Juan, and Sacramento Mountains. The highest peak in the Rocky Mountains is Mt. Elbert (14,433 ft / 4,399 m) in the Southern Rockies. The Middle Rockies include some of the most impressive mountains in the western United States, but only some parts are higher than 11,000 ft (3,353 m). Major ranges include the Bighorn, Absaroka, Wind River, Uinta and Wasatch. The Northern Rockies are the lowest part of the chain, rarely exceeding 11,000 ft (3,353 m). The Bitterroot and Lewis Ranges are the largest in this part of the Rockies.

The Appalachian Mountains

The Appalachians are the major mountain range of the eastern U.S.A. Although they are neither as high nor as rugged as the Rocky Mountains, with more rounded landforms, they are very extensive and were a serious impediment to travel in the early history of the country. They enter the United States from Canada in the northeast and extend southwest most of the way to the Gulf of Mexico. Only 15 peaks exceed 2,400 ft (731 m), with the highest being Mt. Mitchell (6,684 ft / 2,037 m).

The Appalachian Highlands consist of several distinct ranges. The Great Smoky Mountains in the southern part of the Appalachians are a national park. The Blue Ridge Mountains extend from Georgia to Pennsylvania. In places they consist of a single ridge; elsewhere, a complex of closely spaced ridges. The Allegheny Mountains are parallel to the Blue Ridge. The Adirondack Mountains in northern New York State have a domelike structure more than 100 mi (161 km) in diameter. The Appalachians are at their highest average elevation in New England, particularly the Green Mountains in Vermont, the Hoosac Mountains in Massachusetts, and the White Mountains of New Hampshire and Maine. Isolated mountains that rise above the general level of the surrounding terrain are called monadnocks.

Pacific Mountains

The Pacific mountain system comprises some 200,000 sq mi (518,135 sq km) along the western coast of the U.S. It includes granitic mountains, such as the Sierra Nevada and Klamath mountains; volcanic mountains such as the Cascade Range; folded and faulted mountains such as the Transverse Ranges, the Olympic Mountains, and the Oregon Coast Range; and dome mountains, such as the Marysville Buttes of the California Central Valley. These mountains are among the highest, roughest, and most scenic in the United States. The major divisions of the Pacific mountain system are the Cascade and Sierra Mountains, the Coast Ranges, and the Lower California Peninsular Range.

The Sierra Nevada is a huge block mountain about 350 mi (563 km) long and roughly 60 mi (97 km) wide. Not a single river crosses it in its entire length. The Cascades extend north from the Sierra Nevada all the way to Canada. These high mountains form the western borders of the great plateaus of the western U.S., with the Rocky Mountains in the east. Mt. Whitney, in the Sierra Nevada, is the highest point in the United States outside of Alaska at 14,495 ft (4,418 m).

The Coast Ranges are a series of mountains along the Pacific coastline, including the California coast ranges, the Transverse ranges, the Klamath Mountains, the Oregon coast range, and the Olympic Mountains in Washington. The highest ranges in the Transverse Ranges are the San Gabriel and the San Bernardino Mountains, both located along the San Andreas fault. The Oregon Coast Range consists of irregular hills and low mountains along the coast of Oregon and southwestern Washington. The highest summits are lower than 4,000 ft (1,219 m). The Lower California Peninsular Range is located mostly in Mexico, but extends across the border into the extreme southwest of the United States.

Alaska's major mountain ranges are found in the south-central part of the state: Chugach, Wrangell, Talkeetna, the Alaska Range, and the Aleutian Range. The highest mountains in the U.S. are all in Alaska. The north and south peaks of Mt. McKinley (Denali), at 20,320 ft (6,193 m) and 19,470 ft (5,934 m), respectively, are the highest peaks on the North American continent. Lower mountains, the Brooks Range, are found in the northern part of Alaska.

The Hawaiian Islands in the Pacific Ocean are all volcanic in origin and have mountainous interiors. The island of Hawaii itself is the site of two great volcanoes, Mauna Kea (13,796 ft / 4,197 m) and Mauna Loa (13,680 ft / 4,170 m).

Plateaus

About a quarter of the country rests on plateaus. The eight major ones are the Piedmont, Appalachian, and interior low plateaus in the east; the Ozark Plateau, Edwards Plateau, and the Llano Estacado in the central U.S.A.; and the Colorado and Columbia Plateaus, both of which belong to the Intermontane Plateau region of the west.

The Piedmont stretches along the eastern edge of the Appalachian Mountains south from the Hudson River to the end of the range. It has a maximum width of about 125 mi (201 km). Altitudes range between 100 and 1,000 ft (30 and 305 m). The Appalachian Plateaus are a series of plateaus west of the Appalachian Mountains, the most famous of which is the Cumberland Plateau. They too begin near the Hudson River and run south along the mountains, with their highest elevations, roughly 3,000 ft (914 m), near the Hudson. The interior low plateaus are another series of plateaus, found to the west of the Appalachian Plateaus. They are less well defined and have lower elevations than the Appalachian Plateaus, and run parallel to the Appalachians for 600 mi (965 km) from Alabama to Ohio

The Ozark Plateaus, often called the Ozarks, are found southwest of the junction of the Mississippi and Missouri Rivers in the south central United States. They are sometimes referred to as mountains, but for most of their area have average elevations between only 1,000 and 1,500 ft (300 and 460 m). The Edwards Plateau and Llano Estacado are both elevated areas of the Great Plains found along the southeastern edge of the Rocky Mountains.

The Colorado Plateau, between the Southern Rockies, Sierra Nevada, and the Great Basin, is the most colorful part of the United States, with spectacular geological features. It is primarily semi-arid, with the Sonoran Desert bordering it to the south. The general plateau surface is higher than 5,000 ft (1,524 m), and some peaks reach 11,000 ft (3,353 m). Striking topographic features include volcanoes, cinder cones, volcanic necks, mesas, and dome mountains. The Grand Canyon is located in the southwestern part of this region.

The Columbia Plateau is located on the far side of the Great Basin from the Colorado Plateau, in the north. It too is characterized by a semiarid climate and dry canyons, and is dominated by the Columbia and Snake Rivers and their plains. Yellowstone National Park is located on the Yellowstone Plateau east of the Snake River Plain, at the southeastern edge of the Columbia Plateau, where volcanic activity is still strong.

Canyons and Depressions

The country's most dramatic canyons are in the Intermontane Region between the Rocky Mountains and the Pacific coastal mountains. The major rivers of these regions have carved deep canyons in the plateaus.

In the Colorado Plateau, the Colorado River passes through the picturesque Grand Canyon, one of the country's most famous natural features. It is more than 1 mi (1,600 m) deep and 217 mi (349 km) long. Farther north in the same plateau region is the Canyonlands area of southeastern Utah, which is dominated by multiple canyons, some of which are more than 2,000 ft (600 m) deep. The pinnacles and spires of red rock in southwestern Utah's Bryce Canyon are among the most remarkable sights in the country.

In the Columbia Plateau is the single deepest canyon in the United States: Hells Canyon, carved by the Snake River. Its average depth is 6,600 ft (2,000 m), and it extends for 125 mi (200 km). In the same region, the Columbia River has carved canyons that reach depths of 2,000 ft (600 m).

The Great Basin is a vast area in the western U.S. that has no drainage to the ocean. This is because, although it is well above sea level in most places, the surrounding mountains and plateaus are higher still. The Great Salt Lake is found in the northeastern part of the Great Basin. Death Valley, the lowest point in North America, is along the southwestern edge of the Great Basin. Much of the Basin is desert.

Hills and Badlands

There are foothills associated with all of the major mountain ranges of the United States. The best-known region outside of these mountains designated as hilly are the Black Hills of the northern Great Plains. These are actually dome mountains; some rise to elevations of over 2,000 ft (610 m). The craggy, heavily eroded peaks are forested at lower elevations but bare at the top, with steeply rising pinnacles and other rock formations. East of the Black Hills are the White River Badlands, where erosion has sculpted buttes, spires, and other fantastic shapes into the plateau. The badlands continue to erode, on average one inch per year.

INLAND WATERWAYS

Lakes

The five Great Lakes make up the world's largest group of freshwater lakes, and Lake Superior has the greatest surface area of any freshwater lake on Earth, (31,800 sq mi / 82,362 sq km). Lake Huron is the next largest. Lake Michigan, the third largest, is the only one of the Great Lakes found entirely within the United States, the others are shared with Canada. It has an area of 22,178 sq mi (57,441 sq km) and a maximum depth of 923 ft (281 m). All three of these lakes drain into Lake Erie, which in turn empties into Lake Ontario. The St. Lawrence River flows northwest out of Lake Ontario into the Atlantic Ocean.

Outside of the Great Lakes, the next largest in the country is the much smaller Great Salt Lake in Utah, with an area of about 2,300 sq mi (5,957 sq km) and a maximum depth of 15 m (48 ft). Surprisingly, the arid western United States is a land of lakes, both dry and perennial, including not only the Great Salt Lake, but also Sevier Lake, Utah Lake, and Carson Sink in the Great Basin; and Lake Chelan, Crater Lake, Lake Tahoe, and Yosemite Lake in the mountains.

Florida has many lakes; Lake Okeechobee, to the north of the Everglades, covers 1,943 sq km (750 sq mi). Minnesota is also known for its lakes, including the Red Lakes, and also Lake of the Woods and Rainy Lake on the

GEO-FACT

Founded in 1872, Yellowstone National Park was the first national park ever established anywhere in the world, and is one of the oldest nature reserves. It is 3,472 sq mi (8,987 sq km) in area, covering parts of the Yellowstone Plateau and the Northern Rockies. Yellowstone is situated above a geothermal hotspot, a place where the magma normally found deep within the Earth has made its way close to the surface. As a result the entire park and surrounding region is very seismically active. In prehistoric times the park was the site of incredibly powerful volcanic eruptions. An eruption 600,000 years ago left a crater 50 mi (32 km) long and 30 mi (19 km) wide. This volcano, since buried, is still technically active. The most famous attractions of the park, its hot springs and geysers, are a result of all of this volcanic activity. Yellowstone National Park is thought to contain roughly 10,000 hot springs and geysers, more than half of all such features on Earth.

Canadian border. New York is home to the Finger Lakes, as well as Lake Champlain. The so-called Salton Sea is a salt lake in southern California that was formed unintentionally by run-off water from the Colorado River and from farming.

Rivers

With few exceptions, the rivers to the east of the Rocky Mountains drain into the Atlantic Ocean and the Gulf of Mexico; those to the west drain into the Pacific. There are a few rivers in the Great Basin that never reach any ocean, the largest of which is the Humboldt. The Red River of the North in the northern Great Plains drains north into Canada.

There are many short rivers east of the Appalachian Mountains that flow into the Atlantic. Even the longest of these flows for only several hundred miles. Chief among them is the Hudson River, which is linked via the Erie Canal to the Great Lakes. The Connecticut, Roanoke, and Savannah are other notable rivers of this region.

Most of the central United States drained by the Mississippi River and its tributaries. The Mississippi is one of the world's great rivers in terms of both volume and length (2,348 mi/4,127 km). It flows south across the country, somewhat east of its center, and empties into the Gulf of Mexico in a great delta, The most important of its tributaries are the Arkansas, Ohio, and Missouri Rivers. The Arkansas (1,450 mi / 2,333 mi) flows east from the Rocky Mountains across the center of the country, cutting through the Ozarks. It too has many tributaries, including the Cimarron and the Canadian Rivers. The Ohio River (975 mi / 1,569 km) forms at the confluence of three smaller rivers at Pittsburgh, in the Appalachians. It then flows west, and gradually south before merging with the Mississippi. Its largest tributary, the Tennessee River (652 mi / 1,049 km), is one of the most important in the eastern United States in its own right, with many hydroelectric dams and reservoirs along its course.

The largest of the Mississippi's tributaries by far is the Missouri, which at 2,466 mi (3,968 km) is longer than the Mississippi itself. The Missouri is the longest river in the country and the second-longest in North America. It has its source in the Northern Rockies, and flows east across the northern Great Plains before making a great curve south. It then flows southeast and meets the Mississippi near the middle of that river's course. The length of the Mississipppi-Missouri system from the source of the Missouri to the mouth of the Mississippi is 3,860 mi (6,211 km) making it the world's third-longest river system. The Missouri's many tributaries include the Bighorn, James, and Platte.

The rivers of the southern Great Plains are among the few in the central United States that reach the Gulf of Mexico directly, without flowing into the Mississippi first. The Rio Grande, along the border with Mexico, is one of the longest in the country at 1,885 ft (3,033 m). Its major tributary is the Pecos. At the far side of the central plains, the St. Lawrence River flows northeast out of Lake Ontario and into Canada.

The principal river of the Colorado Plateau is the Colorado River (1,450 mi / 2,350 km) itself. The Colorado flows southwest, and receives all of the other large rivers in the region, including the Green, San Juan, and Gila, before exiting into Mexico and eventually the Pacific Ocean. The Colorado has been dammed in several locations, forming the great Lake Mead and Lake Powell reservoirs. It also flows through the famous Grand Canyon. The Columbia Plateau is dominated by the Columbia River and its large tributary, the Snake. As with the Colorado, both rivers have dug deep canyons, and are the site of major reservoirs, including Franklin D. Roosevelt Lake behind Grand Coulee Dam. The Sacramento, Klamath, and Willamette are wide rivers of the Pacific mountain region, although they are short and drain small areas.

In Alaska the Yukon is the longest river (1,979 mi / 3,185 km). It is navigable almost all of the way to its headwaters in Canada. Its principal tributaries are the Koyukuk, Tanana, and Porcupine rivers. The Kuskokwim, Alaska's other major river, rises in the Alaska Range and flows southwest.

Wetlands

The Atlantic coastal plain, more than half of which is below 500 ft (152 m), is noted for its low topographic relief and extensive marshy tracts. Marshes begin to appear in some areas as the plain widens between New York City and North Carolina. Farther south, there is a network of lagoons, sea channels, and salt marshes between the shoreline of the Carolinas and Georgia, as well as the Great Dismal Swamp in North Carolina and Virginia.

The Okefenokee Swamp in northeastern Florida, with an area of around 700 sq mi (1,813 sq km), is the largest single swamp in North America. Occupying the tip of the Florida peninsula, south of Lake Okeechobee is the vast network of swamps and marshes known as the Everglades. Near the middle of the Gulf Coast shoreline is are the swamplands, mud flats, and bayous of the Mississippi River Delta. There are also swamplands in Upper Michigan and numerous large swamps in the deltas of Alaska's lower Yukon and lower Kuskokwim rivers.

THE COAST, ISLANDS, AND THE OCEAN

Oceans and Seas

The conterminous United States is bordered on the east by the Atlantic Ocean, on the southeast by the Gulf of Mexico, and on the west by the Pacific Ocean.

Off the Atlantic coast the continental shelf is more than 100 mi (161 km) wide; beyond this, the ocean floor plunges to depths of more than 2 mi (3.2 km).

The continental shelf along most of the Pacific coast is quite narrow. North of Point Conception it is barely 50 mi (80 km) wide, but south of this point the width increases to 150 mi (241 km). Two major mountain ridges extend about 1,500 mi (2,414 km) westward from the coast into the Pacific Ocean.

Alaska is bordered on the north by the Beaufort Sea and the Arctic Ocean, on the west by the Chukchi Sea, the Bering Strait, and the Bering Sea, and on the south by the Gulf of Alaska. The southern shores of the Alaska Peninsula and the Aleutian Islands are bordered by oceanic trenches. The Hawaiian Islands lie in the North Pacific Ocean.

Major Islands

There are few large islands off the coast of the conterminous United States. Long Island near the mouth of the Hudson River is the largest (1,723 sq mi / 4,462 sq km). The Florida Keys are a series of small islands arcing southwest from the south coast of Florida into the Gulf of Mexico. There are numerous smaller islands in Chesapeake Bay, the Outer Banks, and off the northeastern coast. The largest islands off the Pacific coast are the Santa Barbara Islands, with the only other islands of any size found in Puget Sound. Isle Royale is a large island and nature preserve in Lake Superior.

The Hawaiian Islands are basaltic volcanoes near the middle of the Pacific Ocean along a northwest-trending ridge that divides two oceanic deeps, each of which descends more than 18,000 ft (5,486 m) below sea level. There are five large islands—Hawaii, Maui, Molokai, Oahu, Kauai—with four smaller islands close by—Kahoolawe, Lanai, Kaula, and Niihau. These make up the southeastern part of the archipelago; a number of smaller islands extend in a line to the northwest.

Hawaii itself, the most easterly and largest of the islands (4,021 sq mi / 10,414 sq km), is the top of a truly enormous undersea mountain. It has four volcanic peaks, two more than 13,000 ft (3,962 m) in altitude. Cooled lava covers a significant part of the island, due mostly to the frequent eruptions of Mauna Loa. Red Hill on Maui reaches 10,023 ft (3,055 m), but otherwise elevations in the archipelago are not nearly as high as on Hawaii, although all of the islands are hilly and home to volcanoes, especially toward the east. The coastlines are mostly rocky and rough. At the exposed northeastern shores erosion has produced sea cliffs 3,000 ft (914 m) high. Only Oahu and Niihau have large coastal plains. There is only one harbor, Pearl Harbor, west of Honolulu (the state capital and principal city) on Oahu.

There are many islands located off the Alaskan coast. The southern part of Alaska is the site of the coastal Alexander Archipelago. Further east is the Alaska Peninsula, with the Aleutian Islands extending from its tip. Volcanic in origin, they form an arc consisting of more than 75 volcanoes extending 1,500 mi (2,414 km) from Mount Spurr opposite Cook Inlet, to Buldir Volcano between the Rat Islands and the Near Islands. Kodiak Island and St. Lawrence Island are another two large islands near Alaska.

Island Dependencies

Except for Puerto Rico (see entry on Puerto Rico), U.S. territories and dependencies consist of very small islands. These include American Samoa, Guam, the U.S. Virgin Islands, Baker Island, Howland Island, Jarvis Island, Johnston Atoll, Kingman Reef, Midway Islands, Navassa Island, Palmyra Atoll, and Wake Island.

The U.S. Virgin Islands comprises over 50 islands about 40 mi (64 km) east of Puerto Rico along the Anegada Passage in the Caribbean Sea, with a total land area of roughly 136 sq mi (352 sq km). Only three of the islands are important in size: St. Croix, St. Thomas, and St. John. The terrain is mostly hilly to rugged and mountainous, with little level land. About three-fourths of St. John is a national park. Navassa is a small island, about 2 sq mi (5.2 sq km) in area, between Haiti and Jamaica in the Caribbean. It has a central limestone plateau ringed by vertical white cliffs, with a terrain that is mostly exposed rock and some grassland.

All other U.S. dependencies are located in the Pacific Ocean. Baker Island, Howland Island, and Jarvis Island are all small islands of the Line Islands group, located south of Honolulu in the North Pacific Ocean. All three are coral islands. Johnston Atoll comprises two small islands, Johnston Island and Sand Island, west-southwest of Honolulu in the North Pacific Ocean. Their total land area is about 1 sq mi (2.8 sq km). The islands are flat, with a maximum elevation of 4 meters. Kingman Reef is a small, flat coral island with an area of less than 1 sq mi (2.6 sq km), about 993 mi (1,600 km) south-southwest of Honolulu. Palmyra Atoll is a small island grouping the Central Pacific with a land area of about 5 sq mi (13 sq km) about 993 mi (1,600 km) south-southwest of Honolulu. It comprises about 50 small islets covered with dense vegetation.

The Midway Islands are a group of low, flat coral islands, the largest of which are Eastern Island and Sand Island, with a total land and water area of 2 sq mi (5.2 sq km), about 1,459 mi (2,350 km) west-northwest of Honolulu. They are the westernmost islands of the Hawaiian Islands, but are not a part of that state. Wake Island, 2,298 mi (3,700 km) west of Honolulu, is an atoll about 3 sq mi (7 sq km) in area, consisting of three small coral islands formed on top of an underwater volcano. The central lagoon is a former crater and the islands are part of the rim.

American Samoa is part of the Samoan Archipelago, found about 2,300 mi (3,700 km) southwest of Hawaii. It is made up of seven islands—Tutuila, Tau, Olosega, Ofu, Aunuu, Rose, and Swain's—with a total land area of 77 sq mi (199 sq km). Most of the islands are volcanic with rugged peaks and limited coastal plains; two are coral atolls.

Guam is the southernmost and largest of the Mariana Islands, about 1,347 mi (2,170 km) west of Honolulu. It has a total land area of 209 sq mi (541 sq km). The island is of volcanic origin and is surrounded by coral reefs.

The Coastline

The Atlantic coastline can be divided into three sections. Large peninsulas characterize the northern (or embayed) section, which stretches from the northeastern end of the country halfway down the coast to Chesapeake Bay. This is the largest bay of the embayed section of coast, but other notable bays are found further north, including the New York and Delaware Bays. The complex New England coastline includes the long, bent "arm" of Cape Cod and a number of islands.

South of the embayed section is the Sea Islands section, a region of coastal lagoons and islands. The Outer Banks are the most famous and extensive of these. Coastal land in this region is generally low and swampy. The final segment of the eastern coast is the smooth, sandy, east coast of the Florida Peninsula.

The southern, or Gulf, coast has multiple indentations in its eastern section, including Tampa and Mobile Bays. The irregularly shaped Mississippi delta juts out in the middle, and the shoreline to the west is smoother. The westernmost part of the Gulf coast, in the Corpus Christi area, is the site of the Laguna Madre, a distinctive inland waterway that has a nearly identical counterpart—called by the same name—just south of the U.S. border along the coast of Mexico. These narrow strips of water are two of only three coastal lagoons in the world that are hypersaline (i.e. saltier than the ocean).

The Pacific shoreline is straight and fully exposed to the surf, without barrier beaches or lagoons. There are two major indentations in the Pacific Coast. Puget Sound is a deep inlet with numerous islands at the northwestern corner of the conterminous United States; it is connected to the open ocean by the Strait of Juan de Fuca. San Francisco Bay is an inlet near the middle of the coast where the Sacramento River exits into the sea. The Columbia River also has a broad estuary further north, near Cape Disappointment.

The coast of Alaska is deeply embayed to the west, southwest, and south. Indentations in the south include Prince William Sound. The Alaska Peninsula extends southwest into the ocean, with Bristol Bay to the north. Further north is the Seward Peninsula. Point Barrow on the Arctic Ocean coast is the northernmost point in the country.

The Shore

The coast of northern New England coast is rocky, while the Atlantic coast south of New England is a plain with extensive sandy beaches. In some stretches of coastline south of New York City, the beaches border marshland areas. In the Sea Islands section of the Atlantic coastal plain, the islands off the coast have attractive sandy beaches facing the ocean. Sandy beaches also rim much of the Gulf Coast, except for the Mississippi Delta area, where marshes, swamps, and bayous extend to the sea.

Unlike the Atlantic coast, the Pacific Coast has plains in only about half a dozen places. Much of the coast is mountainous, with steep bluffs, elevated marine terraces, and fewer beaches than the Atlantic and Gulf coasts. A narrow coastal plain rings Puget Sound.

Alaska's coast is mostly low-lying in the north and west and mostly mountainous in the south and in both panhandles. The Hawaiian Islands are ringed with narrow coastal plains.

CLIMATE AND VEGETATION

Temperature

Although the conterminous United States lies within the temperate zone of the Northern Hemisphere, between the Tropic of Cancer and 50° north latitude, there are wide variations in climate, including extremes in temperature and violent weather disturbances. The influence of the Atlantic and Pacific Oceans is limited by the mountain systems that intervene between the coasts and the interior both on the east and west, and also by the fact that air masses move primarily from west to east across the country. Thus, the states along the eastern seaboard have a continental climate despite their proximity to the Atlantic Ocean, as do the states of the Midwest, across which the same air masses move. These parts of the country have hot, humid summers and cold, snowy winters, except for the southeastern region, where winter weather is tempered by latitude. The mean annual temperature in Miami, Florida, is 76°F (24°C), while that in Boston is 51°F (11°C).

In the country's vast central lowlands, the tendency toward sharp contrasts and sudden changes is augmented by the interplay between cold arctic air masses that blow down from Canada and warm air masses moving north from the Gulf of Mexico. This combination can bring sudden shifts in temperature, summer thunderstorms, and more violent disturbances, including tornadoes and blizzards. The areas closest to the center of the country have some of the most extreme weather, including frigid winter temperatures. The northern Great Plains has seen summer highs of 121°F (49°C) and winter lows of -60°F (-51°C).

In contrast to the continental climate experienced in much of the country, the west coast, with its proximity of the Pacific Ocean, has a maritime climate with warm summers and mild winters that are rainy and overcast in the northwest but can include clear weather in the southern parts of California. Seattle, in the northern part of the Pacific coast has average temperatures of 39°F (4°C) in January and 65°F (18°C) in July. Farther south along the coast, Los Angeles averages 56°F (13°C) in January and 69°F (21°C) in July. Some of the maritime influence from the Pacific reaches the plateau region between the coastal ranges and the Rockies, but there are greater contrasts in temperature and much less precipitation.

The panhandle region of southern Alaska has a mild maritime climate, while the interior of the state has extremes of both heat and cold. The far north, within the Arctic Circle, has a uniformly frigid arctic climate. By contrast, Hawaii has a stable, even climate with temperatures averaging 73°F (23°C) in January and 80°F (27°C) in July.

Rainfall

Average annual rainfall is more than 40 in (100 cm) in an area covering roughly the eastern 2/5ths of the country, extending furthest west in the south. The prairie and Great Plains states to the north and west are considerably drier, with average rainfall as low as 18 in (46 cm) per year, dropping to 10 in (25 cm) in the northern plains.

In the Rocky Mountains, precipitation varies according to altitude, with higher elevations receiving more rain. The deserts to the west of Rockies are the driest parts of the country, although other parts of the intermontane region receive considerably heavier rainfall. The unevenly distributed precipitation of the region ranges from annual averages of 3 in (8 cm) in Yuma, Arizona to 30 in (76 cm) in higher parts of Arizona and New Mexico, to as much as 60 in (152 cm) even farther north, in central Idaho and Washington state. Annual rainfall in the Pacific coastal area varies widely with latitude, from 1.78 in (4.52 cm) in Death Valley in the south, to more than 140 in (356 cm) in the Olympic Mountains of Washington state.

In Alaska, the southern arc of the Aleutian islands and the panhandle have a wet maritime climate, while the interior is, on the whole, quite dry, in spite of its snow. Hawaii is generally moderately rainy (28 in / 71cm annually), but with very heavy rainfall at higher elevations. Mt. Waialeale, on the island of Kauai, can receive as much as 460 in (1,168 cm) of precipitation per year.

Grasslands

Almost half the country consists of plains. The Atlantic coastal plain, with a width of 100 to 200 mi (161 to 322 km), extends along the Atlantic seaboard south of the Hudson River and wraps around the Appalachian Mountains to follow the Gulf coast. It makes up about 10 percent of the country's land area, and, particularly in the south, contains the most fertile land in the country.

Between the Rocky Mountains and the Appalachian Mountains is a vast area of grasslands and plains, which extends from the coastal plains of the south well into northern Canada. West of the Mississippi, where rainfall is lightest, the rolling prairies are known as the Great Plains.

The vast central plain extends northward toward Canada and southward to the coastal plain bordering the Gulf of Mexico. Other extensive plains occur in the structural basins of the western mountains, such as the broad Central Valley of the Sacramento and San Joaquin rivers in California, and the Wyoming Basin.

Both tall and short grasses are found in the Great Plains and the Intermontane Plateau area between the Rockies and the Pacific coast. Species include buffalo, bunch, needle, wheat, and mesquite.

Deserts

The Colorado Plateau and the Great Basin are both characterized by arid areas of bare rock, and sparse vegetation elsewhere. Included in this region is the Great Salt Lake Desert, Death Valley, and the Sonoran Desert system, which extends southward into Mexico and also includes the Mojave Desert in southern California. The true deserts of the southwest support only scrub and annuals that appear only intermittently, after rainfall. Brush and bunchgrass are found in semidesert areas.

Tundra

Vast areas of Alaska are permanently frozen, and the state has thousands of glaciers. The area bordering the Arctic Ocean is a nearly featureless and permanently frozen coastal plain. Further south the lowlands thaw at the surface during the summer, but permafrost remains further down.

Forests and Jungles

A band of evergreen forest extends along the northern border from the east coast all the way to the Great Plains, with species such as pine, spruce, hemlock, and balsam. Farther south, this gives way to a mixed evergreen and deciduous forest that includes oak, maple, beech, walnut, ash, linden, and sycamore. Cypress and white cedar grow in the swamp areas of the Atlantic coastal plain. The most spectacular forests are found on the Pacific coast, which is famous for its giant redwoods, Douglas firs, and giant sequoias. The Olympia Mountains are the site of a temperate rainforest.

Population Centers – United States of America

(2000 POPULATION)

Name	Population	Name	Population
New York City	8,008,000	Phoenix, Arizona	1,321,000
Los Angeles, California	3,695,000	San Diego, California	1,223,000
Chicago, Illinois	2,896,000	Dallas, Texas	1,189,000
Houston, Texas	1,954,000	San Antonio, Texas	1,145,000
Philadelphia, Pennsylvania	1,518,000	Detroit, Michigan	951,000

SOURCE: "No. 34. Incorporated Places with 100,000 or More Inhabitants in 2000—Population, 1970 to 2000, and Land Area, 2000." U.S. Census Bureau, Statistical Abstract of the United States: 2001.

States – United States of America

1999 POPULATION ESTIMATES

Name	Population	Area (sq mi)	Area (sq km)	Capital
Alabama	4,370,000	51,705	133,915	Montgomery
Alaska	620,000	591,004	1,530,693	Juneau
Arizona	4,778,000	114,000	295,259	Phoenix
Arkansas	2,551,000	53,187	137,754	Little Rock
California	33,145,000	158,706	411,047	Sacramento
Colorado	4,056,000	104,091	269,594	Denver
Connecticut	3,282,000	5,018	12,997	Hartford
Delaware	745,000	2,044	5,294	Dover
Florida	15,111,000	58,664	151,939	Tallahassee
Georgia	7,788,000	58,910	152,576	Atlanta
Hawaii	1,185,000	6,471	16,760	Honolulu
Idaho	1,252,000	83,564	216,430	Boise
Illinois	12,128,000	57,871	149,885	Springfield
Indiana	5,943,000	36,413	94,309	Indianapolis
Iowa	2,869,000	56,275	145,752	Des Moines
Kansas	2,654,000	82,277	213,096	Topeka
Kentucky	3,961,000	40,409	104,659	Frankfort
Louisiana	4,372,000	47,752	123,677	Baton Rouge
Maine	1,253,000	33,265	86,156	Augusta
Maryland	5,172,000	10,460	27,091	Annapolis
Massachusetts	6,175,000	8,284	21,455	Boston
Michigan	9,864,000	97,102	251,493	Lansing
Minnesota	4,776,000	86,614	224,329	St. Paul
Mississippi	2,769,000	47,689	123,514	Jackson
Missouri	5,468,000	69,697	180,514	Jefferson City
Montana	883,000	147,046	380,847	Helena
Nebraska	1,666,000	77,355	200,349	Lincoln
Nevada	1,809,000	110,561	286,352	Carson City
New Hampshire	1,201,000	9,279	24,032	Concord
New Jersey	8,143,000	7,787	20,168	Trenton
New Mexico	1,740,000	121,593	314,924	Santa Fe
New York	18,197,000	52,735	136,583	Albany
North Carolina	7,651,000	52,669	136,412	Raleigh
North Dakota	634,000	70,702	183,117	Bismarck
Ohio	11,257,000	44,787	115,998	Columbus
Oklahoma	3,358,000	69,956	181,185	Oklahoma City
Oregon	3,316,000	97,073	251,418	Salem
Pennsylvania	11,994,000	46,043	119,251	Harrisburg
Rhode Island	991,000	1,212	3,139	Providence
South Carolina	3,886,000	31,113	80,582	Columbia
South Dakota	733,000	77,116	199,730	Pierre
Tennessee	5,484,000	42,144	109,152	Nashville
Texas	20,044,000	266,807	691,027	Austin
Utah	2,130,000	84,899	219,887	Salt Lake City
Vermont	594,000	9,614	24,900	Montpelier
Virginia	6,873,000	40,767	105,586	Richmond
Washington	5,756,000	68,139	176,479	Olympia
West Virginia	1,807,000	24,231	62,758	Charleston
Wisconsin	5,250,000	66,215	171,496	Madison
Wyoming	480,000	97,809	253,324	Cheyenne
District of Columbia	519,000	69	179	-

SOURCE: "No. 19. Resident Population—States: 1991 to 1999." U.S. Census Bureau, Statistical Abstract of the United States: 2001.

Small trees including dogwood, hawthorn, and wild cherry grow in many parts of the country. Some common species of the estimated 20,000 species of native flowering plants are dandelions, wild rose, black-eyed Susan, columbine, aster, phlox, and forget-me-not. Bamboos and ferns dominate the tropical woodlands of Hawaii.

The Everglades in southern Florida, which cover an area 50 mi (80 km) wide and 100 mi (160 km) long, are the country's last sizable subtropical wilderness. Sawgrass covers much of the region, which is inundated by a sheet of water 6 in (15 cm) deep. Fern-draped hardwood trees, including mahogany and other species, are found on its hammocks (small islands). Willows, cypress, mangroves, and slash pines are among the other trees that grow in the region, which is known for its diverse wildlife, including 347 bird species.

HUMAN POPULATION

In 2001 the overall population density of the United States was estimated at 75 people per sq mi (29 per sq km). However, the population of the United States is unevenly distributed across the country. The major population concentrations are along the northeast Atlantic and the southwest Pacific coastal regions. The Mid-Atlantic coastal region between New York City and Washington, D.C., is the most densely populated of all. Lesser concentrations of people can be found in the Midwest between the Great Lakes, Mississippi, and Ohio Rivers; Florida; and the rapidly growing southwest. The population is lowest in the arid interior of the west, the northern Great Plains, and in Alaska. More than three-quarters of the U.S. population is urban. This includes those people living in the suburbs around major cities, the growth of which was one of the major U.S. population trends of the last half of the twentieth century. Major cities can be found scattered throughout most of the country.

NATURAL RESOURCES

The United States has an abundance of almost every natural resource. Oil and natural gas are found in the south and off shore in the Gulf of Mexico. There are also major deposits in Alaska. There are huge coal reserves in both the Appalachian and Rocky Mountains. Iron and copper can be found in both of these ranges, as well as in the northern Great Lakes region. Gold and silver are mined in the west and Alaska. Large amounts of timber are harvested from the nation's forests, particularly those in the west. The many large rivers are a source of abundant water and hydropower. Huge tracts of farmland make the United States one of the world's leading producers of food.

FURTHER READINGS

Brinkley, Douglas. *The Majic Bus: An American Odyssey.* New York: Anchor Books, 1994.

Horwitz, Tony. *Confederates in the Attic: Dispatches from the Unfinished Civil War.* New York: Pantheon Books, 1998.

McPhee, John A. *Coming into the Country.* New York: Farrar, Straus & Giroux, 1977.

———. *Basin and Range.* New York: Farrar, Straus & Giroux, 1981.

Sierra Club. http://www.sierraclub.org (accessed July 12, 2002).

U.S. Geological Survey. *USGS.* http://www.usgs.com (accessed July 12, 2002).

U.S. National Park Service. *ParkNet.* http://www.nps.gov (accessed July 12, 2002).

Uruguay

- **Area:** 68,039 sq mi (176,220 sq km) / World Rank: 90
- **Location:** Southern and Western Hemispheres, in southern South America, on the Atlantic Ocean, bordering Brazil to the northeast and Argentina to the west.
- **Coordinates:** 30°00′S, 56°00′W
- **Borders:** 972 mi (1,564 km); Brazil, 612 mi (985 km); Argentina, 612 mi (579 km)
- **Coastline:** 410 mi (660 km)
- **Territorial Seas:** 12 NM
- **Highest Point:** Cerro Catedral, 1,686 ft (514 m)
- **Lowest Point:** Sea level
- **Longest Distances:** 345 mi (555 km) NNW-SSE; 313 mi (504 km) ENE-WSW
- **Longest River:** Uruguay, 1,000 mi (1,610 km)
- **Largest Lake:** Embalse del Río Negro, 4,000 sq mi (10,360 sq km, approximate)
- **Natural Hazards:** Drought; flooding
- **Population:** 3,360,105 (2001) / World Rank: 128
- **Capital City:** Montevideo, on the southern Atlantic coast
- **Largest City:** Montevideo, population 1,344,839 (2001)

OVERVIEW

Small Uruguay forms a flat wedge between its giant neighbors, Brazil and Argentina. Well-watered grasslands predominate, with elevations rising into hills in the north. The southern coast is bordered by the great Río de la Plata estuary and the Atlantic Ocean, and on the west the country is separated from Argentina by the Río Uruguay. Eastern Uruguay is marked by swamps and lagoons.

MOUNTAINS AND HILLS

Plateaus

The interior of Uruguay is a low, broken plateau, which is a transition from the pampas of Argentina to the hilly uplands of southern Brazil. The geological foundation of most of this region is made up of gneiss, sandstone, and granite, and an extension of the basaltic plateau of Brazil reaches southward west of the Río Negro.

Hills and Badlands

Cerro Catedral (1,686 ft / 514 m), near the southern coast, is Uruguay's highest point. Uruguay's interior plateau features ranges of low hills that become more prominent in the north as they merge into the highlands of southern Brazil. The most important of Uruguay's *cuchillas* (hill ranges) are the Grande Range and the Haedo Hills. Only in these and in the Santa Ana Hills along the Brazilian frontier do altitudes exceed 600 ft (183 m) with any frequency.

INLAND WATERWAYS

Lakes

Embalse del Río Negro, a reservoir formed by the Río Negro dam in the central part of the country, is the largest artificial lake in South America, with a surface area of more than 4,000 sq mi (10,359 sq km). Other reservoirs are Lake Palmar, also on the Río Negro, and Lake Salto Grande on the Uruguay.

Lagoons appear along the eastern coast. The largest is Lagoa Mirím (Laguna Merín), which extends across the border into Brazil. Some of the lagoons are fresh water, but most are brackish due to direct tidal connection with the Atlantic.

Rivers

The largest of Uruguay's rivers is the Uruguay itself, at 1,000 mi (1,620 km). The river flows for 270 mi (435 km) through the country. It marks the entire western boundary with Argentina and extends farther to the north as a portion of the Argentina-Brazil frontier. It has low banks, and floods sometimes inundate large areas. Industrial

and mining pollution affect the Uruguay and other rivers in the country.

The Uruguay merges with the Parana River to form the Río de la Plata. The Río de la Plata is a vast estuary of the Atlantic Ocean, and is saline except at its western extremity, where the Parana and Uruguay gush enormous quantities of fresh water into it.

The Río Negro rises in southern Brazil, then bisects Uruguay as it flows southwestward to join the Uruguay. It is the most important river to flow through Uruguay, rather than along its borders, and is the site of several major reservoirs. Its principal tributaries are the Río Yi and Tacuarembó. Smaller rivers are found throughout the country, with the Cuareim and Yaguarón flowing along parts of the border with Brazil.

Wetlands

The Atlantic coastal plain contains marshes as well as lagoons. These coastal wetlands, the habitat of birds, fish and seals, are threatened by rice cultivation, pine plantations, and livestock grazing. The Banados del Este y Franja Costera, more than 1,679 sq mi (4,350 sq km) of wetlands on the Atlantic coast, is a UNESCO-designated World Biosphere Reserve. It includes Uruguay's part of Lagoa Mirím, as well as Laguna Negra and Laguna Castillos.

THE COAST, ISLANDS AND THE OCEAN

The Coast and Beaches

Uruguay's coast is entirely on the Atlantic Ocean. The curve of the coastline is characterized by beaches and rocky headlands. The eastern coast is marked by swamps and lagoons. It then curves west and leaves the open Atlantic, running for more than 200 mi (322 km) along the Río de la Plata estuary to reach the mouth of the Uruguay.

At the center of the southern coastline, the city of Montevideo has nine beaches on the Atlantic. East of Montevideo is Punta del Este, a peninsular beach resort. National parks protect stretches of beach along the eastern coast.

Major Islands

A few small islands lie off the coast of Uruguay. Isla de Lobos, 0.16 sq mi (0.4 sq km) with one of the largest sea lion populations in the world, is offshore from the town of Punta del Este.

CLIMATE AND VEGETATION

Temperature

Uruguay has a temperate climate with four seasons: summer, December-March; autumn, April-June; winter, July-August; spring, September-November. Average temperatures are 63°F (17°C) in spring; 77°F (25°C) in summer; 64°F (18°C) in autumn; and 54°F (12°C) in winter. Winds often sweep across Uruguay from the Atlantic Ocean, and the "pampero" is a cold winter wind from Argentina.

Rainfall

Most of Uruguay's rain falls in the winter months of July and August. The yearly average precipitation is 41 in (105 cm). Humidity averages 65 percent. Although freezing temperatures occur, snow is rare.

Grasslands

A prairie ecosystem dominates Uruguay, where vast expanses of undulating grasslands cover more than 90 percent of the country. The grasslands are extensions of Argentina's pampas. High potassium content in the soil produces nutritious grasses for livestock, mainly cattle and sheep. Grazing accounts for more than three quarters of land use in Uruguay.

Forests and Jungles

Forests cover just 3 percent of the land, and they are found mostly along river banks. Plantations of eucalyptus and pine for pulpwood production have been replacing grasslands in some areas. The tree plantations are controversial because of their water consumption and soil depletion.

Departments – Uruguay

2000 POPULATION ESTIMATES

Name	Population	Area (sq mi)	Area (sq km)	Capital
Artigas	77,909	4,605	11,928	Artigas
Canelones	493,485	1,751	4,536	Canelones
Cerro Largo	85,666	5,270	13,648	Melo
Colonia	125,428	2,358	6,106	Colonia del Sacramento
Durazno	57,507	4,495	11,643	Durazno
Flores	25,520	1,986	5,144	Trinidad
Florida	68,491	4,022	10,417	Florida
Lavalleja	62,596	3,867	10,016	Minas
Maldonado	144,688	1,851	4,793	Maldonado
Montevideo	1,380,962	205	530	Montevideo
Paysandú	116,944	5,375	3,922	Paysandú
Río Negro	54,253	3,584	9,282	Fray Bentos
Rivera	104,127	3,618	9,370	Rivera
Rocha	72,421	4,074	10,551	Rocha
Salto	125,396	5,468	14,163	Salto
San José	103,499	1,927	4,992	San José de Mayo
Soriano	84,024	3,478	9,008	Mercedes
Tacuarembó	87,602	5,961	15,438	Tacuarembó
Treinta y Tres	51,573	3,679	9,529	Trienta y Tres

SOURCE: INE (Instituto Nacional de Estadística), Uruguay.

HUMAN POPULATION

Uruguay has a higher population density than most of South America: 47 people per sq mi (18 people per sq km). This figure is deceptive, however, as almost half of the entire population is concentrated in Montevideo, the capital and only large city. As a consequence, much of the rest of the country is sparsely populated. This is reflected by the fact that 91 percent of the population lives in cities.

NATURAL RESOURCES

Uruguay's primary natural resource is its grazing land, which is very well suited to the raising of livestock and also supports significant agriculture. Fishing is another important resource. Hydroelectricity provides more than 95 percent of Uruguay's power needs. Mineral resources are limited, but there is some mining of granite, gems, and marble.

FURTHER READINGS

Arrarte, Carlos Perez, and Guillermo Scarlato. "The Laguna Merín Basin of Uruguay: From Protecting Natural Heritage to Managing Sustainable Development." *Cultivating Peace.* Ottawa, Ontario: International Development Research Center, 1999.

Box, Ben. *South American Handbook.* Lincolnwood Ill.: Passport Books, 1999.

Bridal, Tessa. *The Tree of Red Stars.* Minneapolis: Milkweed Editions, 1997.

Carrere, Ricardo, and Larry Lohmann. *Pulping the South: Industrial Tree Plantations in the World Paper Economy.* London: Zed Books, 1996.

Verdesio, Gustavo. *Forgotten Conquests: Rereading New World History from the Margins.* Philadelphia: Temple University Press, 2001.

Uzbekistan

- **Area:** 172,741 sq mi (447,400 sq km) / World Rank: 57
- **Location:** Northern and Eastern Hemispheres, in Central Asia, north of Turkmenistan and Afghanistan, west of Tajikistan and Kyrgyzstan, and south and east of Kazakhstan.
- **Coordinates:** 41°00′N, 64°00′E
- **Borders:** 3,866 mi (6,221 km) / Afghanistan, 85 mi (137 km); Kazakhstan, 1,369 mi (2,203 km); Kyrgyzstan, 683 mi (1,099 km); Tajikistan, 721 mi (1,161 km); Turkmenistan, 1,007 mi (1,621 km)
- **Coastline:** None
- **Territorial Seas:** None
- **Highest Point:** Adelunga Toghi, 14,111 ft (4,301 m)
- **Lowest Point:** Sariqarnish Kuli, 39 ft (12 m) below sea level
- **Longest River:** Amu Dar'ya, 1,580 mi (2,540 km)
- **Largest Lake:** Aral Sea, 12,000 sq mi (31,080 sq km, estimate)
- **Natural Hazards:** Earthquakes; drought
- **Population:** 25,155,064 (July 2001 est.) / World Rank: 41
- **Capital City:** Tashkent, in the northeast
- **Largest City:** Tashkent, 2,495,000 (2000 est.)

OVERVIEW

Uzbekistan's varied terrain includes high mountains and semiarid grasslands in the east, and lowlands and a predominantly flat plateau region in the west. In the center lies the vast Kyzyl Kum, one of the world's largest desert. It is a hot, dry country with long summers and mild winters.

Despite its desert character, Uzbekistan is crossed by many large rivers, including the Syr Dar'ya, the Amu Dar'ya, and the Zeravshan. Their highly irrigated river valleys support Uzbekistan's agricultural economy and most of its human inhabitants. A portion of the fertile Fergana Valley crosses eastern Uzbekistan.

Uzbekistan is located on the Eurasian Tectonic Plate. The entire country is tectonically active. Earthquakes are frequent and can be severe, particularly in the east. A 1966 earthquake destroyed much of the capital city of Tashkent.

Nearly 40 percent of western Uzbekistan is occupied by the Qoraqalpogh Autonomous Republic (known also as Qoraqalpoghistan or Karakalpakstan). Uzbekistan's constitution accords Qoraqalpoghistan a self-governing status with its own legislature, supreme court, and local governments. Nevertheless, the Uzbekistan central government still exercises considerable control in the republic's affairs.

MOUNTAINS AND HILLS

Mountains

In the east and northeast, Uzbekistan is predominantly mountainous. In the northeast, the Tian Shan extends into the country from the east. Further south, on the far side of the Fergana Valley, are the Alai Mountains, a part of the Pamirs. Both ranges are tall, reaching up to 14,111 ft (4,301 m) at Adelunga Toghi, and rise even higher further to the east in Kyrgyzstan and Tajikistan.

Plateaus

West and south of the Aral Sea is the Ustyurt (Ust' Urt) Plateau. This is a well defined upland, broken by occasional small mountain ridges. It extends west from the shores of the Aral Sea to the Caspian Sea shoreline in Kazakhstan. Its area is roughly 77,220 sq mi (200,000 sq km).

INLAND WATERWAYS

Lakes

The southern half of the Aral Sea is located in northwestern Uzbekistan, with the rest in Kazakhstan. The water is salty, and the lake's size led to its being called a sea, but with no outlet to the ocean it is technically a lake. As recently as the 1960s the Aral Sea was the world's fourth largest lake. Since then its area has been shrinking at an alarming rate. The use of water from the primary rivers that feed the lake for irrigation have left the lake only half its former size. Large islands have surfaced in the middle of the Aral Sea where none were formerly found, and the northernmost part of the lake (in Kazakhstan) has been separated from the south by dry land.

Lake Aydarkul in eastern Uzbekistan is the largest fresh water lake in the country. Lake Sarygamysh extends into the country from Turkmenistan in the southwest.

Rivers

There are three significant rivers in Uzbekistan, the Amu Dar'ya, Syr Dar'ya, and Zeravshan. All of these rivers originate in the high mountains east of Uzbekistan. The Amu Dar'ya is the largest of the three. It flows west along the southern border with Afghanistan, then curves northwest into Turkmenistan. Further north it becomes the border between Turkmenistan and Uzbekistan, then flows just within Uzbekistan's borders for a time. Near the city of Nukus it turns north and spreads out into a delta. The delta feeds into the southern Aral Sea, although diversion of water from the Amu Dar'ya has been so extensive that the river often dries out before reaching the Aral.

The Syr Dar'ya enters the country from Kyrgyzstan in the northeast and flows west through the fertile Fergana Valley. It cuts across the spur of northern Tajikistan, then turns north back through Uzbekistan and into Kazakhstan. It eventually empties into the northern Aral Sea.

The Zeravshan enters the country from the mountains of Tajikistan to the east, then arcs across the south-

GEO-FACT

Recognized as one of the worst ecological disasters in the world, the evaporation of the Aral Sea has come about as a result of massive irrigation withdrawals from the Amu Dar'ya and the Syr Dar'ya that feed the sea. Starting in the early 1960s, the goal was to increase cotton yields dramatically in Central Asia, an effort requiring enormous amounts of irrigation. By 2002 the Aral Sea had shrunk dramatically; more than half of its basin is dry and salt-encrusted. The impact on the region's ecosystem has been devastating. In addition to the lake itself, the Amu Dar'ya and Syr Dar'ya deltas have also dried up, destroying much of the plant and wildlife habitat, even driving some species into extinction. Salt from the exposed lake bed is picked up by winds and carried as far as 250 mi (400 km) away, accelerating the process of transforming arable land into desert (desertification). Furthermore, chemical pesticides, carried along with the salt, are believed to have contributed to human respiratory illnesses and even some cancers.

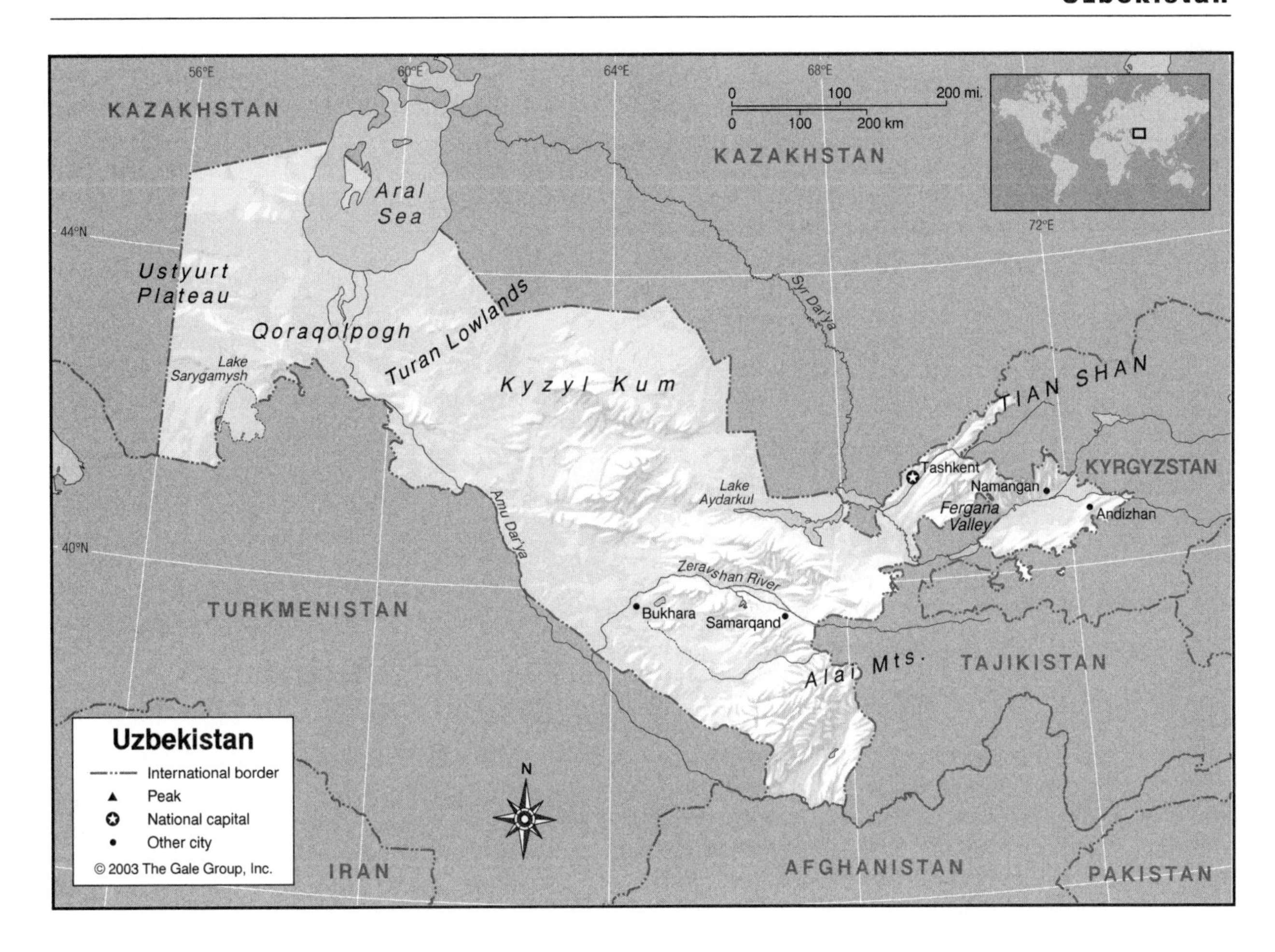

east portion of the country. The Zeravshan was once the Amu Dar'ya's largest tributary, joining it in Turkmenistan. Now it does not even reach the Amu Dar'ya; instead, it expires in the desert near the city of Bukhara, drained to nothing for purposes of irrigation.

Besides rivers, Uzbekistan has extensive canal systems, built mostly during the Soviet period. The Amu-Bukhara canal is the most notable, but there are many others.

THE COAST, ISLANDS, AND THE OCEAN

Oceans and Seas

Uzbekistan is landlocked with no ocean coasts or islands. It does encompass the southern half of the Aral Sea, with 260 mi (420 km) of shoreline. However, the Aral Sea is technically a landlocked, saltwater, lake, not an ocean or sea.

CLIMATE AND VEGETATION

Temperature

Uzbekistan has a continental climate, meaning that it has definite seasonal variations as well as significant day and nighttime differences. July (summer) high temperatures are generally between from 79° to 90°F (26° to 32°C) but can get much higher. January highs are usually between 21° to 36°F (-6° to 2°C).

Rainfall

Most precipitation falls in the months of March and April; droughts are common in Uzbekistan's long, hot summers. Although snow falls regularly in the winter months, it seldom amounts to any significant measure and soon melts. Overall, precipitation is light, with only the best watered areas receiving more than 12 in (30 cm) annually.

Grasslands

The Fergana Valley in the northeast is a fertile area, largely devoted to agriculture. Outside of this area, most of the country is desert. Semi-arid grasslands, or steppe, can be found in the west as part of the Turan or on the Ustyurt Plateau.

Deserts

The Kyzyl Kum desert occupies an immense area some 115,000 square mi (298,000 sq km) in extent, making it the largest desert in Central Asia. It extends southeast of the Aral Sea, between the valleys of the Amu Dar'ya and Syr Dar'ya, meaning that the bulk of it is located in Uzbekistan. It is mostly covered with red sand,

Population Centers – Uzbekistan

(1990 CENSUS OF POPULATION)

Name	Population	Name	Population
Tashkent (capital)	2,094,000	Andizhan	297,000
Samarqand (Samarkand)	370,000	Bukhara	228,000
		Fergana	198,000
Namangan	312,000		

SOURCE: "Population of Capital Cities and Cities of 100,000 and More Inhabitants." United Nations Statistics Division.

which is the meaning of its name. It is an extremely arid and inhospitable area. Another desert, the Mirzachol, lies southwest of the capital, Tashkent

HUMAN POPULATION

Of the former Soviet republics, only Russia and Ukraine are more populous than Uzbekistan. With a population of 25,155,000 (July 2001 estimate), its average population density is 136 persons per sq mi (52 persons per sq km). The population is densest in the fertile Fergana Valley. Most other large cities are found near the Zeravshan or Amu Dar'ya. The Kyzyl Kum Desert and Ustyurt Plateau have few inhabitants. Overall, about 42 percent Uzbekistan's inhabitants live in cities. The capital city of Tashkent, with a metropolitan population estimated at 2,495,000 in 2000, is the largest city in Central Asia.

NATURAL RESOURCES

Minerals are found in abundance in Uzbekistan, including rich deposits of gold, silver, uranium, copper, zinc, lead, tungsten, lithium, and molybdenum. Anthracite coal mining is also a thriving industry. Large reserves of oil and natural gas have been discovered in the Fergana Valley, and production and export are increasing every year.

Although only 10 percent of the land in Uzbekistan is arable, the country's economy depends heavily upon agriculture. Much rich farmland is found at the bases of the mountains and along the river valleys. Uzbekistan is one of the world's largest exporters of cotton and cotton seed, although in the interests of land management greater focus is being placed on grain production.

FURTHER READINGS

Ferdinand, Peter, ed. *The New States of Central Asia and Their Neighbors.* New York: Council on Foreign Relations Press, 1994.

Feshbach, Murray, and Alfred Friendly, Jr. *Ecocide in the USSR: Health and Nature under Seige.* New York: Basic Books, 1992.

Horton, Scott, and Tatiana Geller. "Investing in Uzbekistan's Natural Resources Sector." *Central Asia Monitor 1* (1996): 25-35.

Lubin, Nancy. "Uzbekistan." In Philip R. Pryde, ed., *Environmental Resources and Constraints in the Former Soviet Republics,* 289-306. Boulder, Colo.: Westview Press, 1995.

Micklin, Philip. "The Aral Sea Crisis: Introduction to the Special Issue." *Post-Soviet Geography,* 33, 5 (May 1992): 269-82.

Nichol, James. "Uzbekistan: Basic Facts." *CRS Report for Congress,* May 28, 1996.

Vanuatu

- **Area:** 4,710 sq mi (12,200 sq km) / World Rank: 159
- **Location:** Southern and Eastern Hemispheres, in Oceania, group of islands in the South Pacific Ocean, northeast of Australia.
- **Coordinates:** 16°00′S, 167°00′E
- **Borders:** None
- **Coastline:** 1,570 mi (2,528 km)
- **Territorial Seas:** 12 NM
- **Highest Point:** Mount Tabwemasana, 6,158 ft (1,877 m)
- **Lowest Point:** Sea level
- **Longest River:** None of significant length
- **Natural Hazards:** Typhoons; volcanic eruptions; minor earthquakes; mudslides; heavy rainfall
- **Population:** 192,910 (July 2001 est.) / World Rank: 176
- **Capital City:** Port-Vila, on the island of Éfaté
- **Largest City:** Port-Vila, on Éfaté, population 30,000

OVERVIEW

Vanuatu, formerly the Anglo-French Condominium of the New Hebrides, is a Y-shaped Melanesian archipelago of more than 80 islands, stretching between west of Fiji and northeast of New Caledonia. The entire chain is the result of active volcanism as the Australian and Pacific Plates converge at a rate of 3.5 in (9 cm) per year, uplifting Vanuatu around 1.5 in (4 cm) per year. Lying along the Pacific Ring of Fire, there are active volcanoes

GEO-FACT

Beach rock on Espíritu Santo has naturally welded itself to human garbage along the shore, including the remains of WWII machinery and tens of thousands of Coca-Cola and 7-Up bottles.

on Tanna, Ambrim, and Lopevi. Seventy of the islands are inhabited.

MOUNTAINS AND HILLS

Most of the islands are rugged and mountainous with cultivated narrow coastal plains. The principal peak, Mount Tabwemasana, rises to a height of 6,158 ft (1,877 m) on Espíritu Santo. Other significant peaks include the 4,166 ft (1,270 m) high Mount Maroum on Ambrim, and Mount Tukosmera, which reaches 3,556 ft (1,084 m) on Tanna.

INLAND WATERWAYS

Because the islands are generally very small, there are no rivers or lakes of significant size. However, many small streams do drain the mountains, including the Jourdain, Sarakana, and Wamb Rivers. Some small lakes do exist in extinct volcanic craters and other low-lying areas, including Lake Manaro Ngoro, Lake Manaro Lakua, Lake Voui, and Lake Siwi.

THE COAST, ISLANDS, AND THE OCEAN

Oceans and Seas

The Pacific Ocean surrounding the islands contains many coral reefs just off the rocky shores plunging to hundreds of meters below the surface, as well as underwater volcanoes.

Major Islands

The larger islands are of volcanic origin overlaid with limestone formations; the smaller ones are coral and limestone. The thirteen major islands are Torres Islands (Îles Torres), Bank Islands (Îles Banks—Mota Lava, Sola, Gaua), Espíritu Santo, Ambae, Maéwo, Pentecost, Malakula, Ambrim, Epi, Tongoa, Éfaté, Erromango, Aniwa, Tanna, Fortuna, and Aneityum. The largest islands are Espíritu Santo, Malakula, and Éfaté. Vanuatu makes a disputed claim to Matthew and Hunter Islands east of New Caledonia. "Ownership" of these would considerably extend Vanuatu's Maritime Economic Zone.

The Coast and Beaches

The beach rock is an unusual aspect of local geology. Calcium carbonate leaches from decayed shells and zooplankton skeletons to the littoral zone of the beaches. When water evaporates the calcium carbonate cements everything within the littoral zone together, forming beach rock.

CLIMATE AND VEGETATION

Temperature

Vanuatu's climate is tropical, moderated by southeast trade winds from about May to September each year. It is hot with humidity averaging 83 percent all year. Average midday temperatures in Port-Vila range from 77°F (25°C) in winter to 84°F (29°C) in summer.

Rainfall

Rainfall averages about 94 in (239 cm) per year, and as high as 160 in (406 cm) per year in the northern islands. During November–April the islands are threatened by tropical cyclones.

Forests and Jungles

Lowland forests cover the southeastern, or windward, sides of Vanuatu's islands. At approximately 1,640 ft (500 m) montane forests begin. Threatened by the logging industry, hardwood forests cover 75 percent of the land area.

HUMAN POPULATION

The majority of the population lives in some 2,000 small villages. The only major city is Port-Vila, which has a population of about 30,000. No other village has more than 2,000 inhabitants. Only 20 percent of the Vanuatuans live in urban areas.

NATURAL RESOURCES

Forests, manganese, and fish are Vanuatu's major natural resources. Both tuna and bonito are frozen and exported to Japan and the United States. Tourism is a developing industry. The absence of personal and corporate income taxes and a shipping registry under a "flag of convenience" have made Vanuatu an offshore financial center.

FURTHER READINGS

Bonnemaison, Joël. *The Tree and the Canoe: History and Ethnogeography of Tanna.* Honolulu: University of Hawaii Press, 1994.

Food and Agriculture Organization of the UN. Fisheries. http://www.fao.org/fi/default.asp (accessed May 6, 2002).

Jolly, Margaret. *Women of the Place: Kastom, Colonialism, and Gender in Vanuatu.* Philadelphia: Harwood Academic Publishers, 1994.

Kilham, Christopher. *Kava: Medicine Hunting in Paradise.* Rochester, Vt.: Park Street Press, 1996.

Rodman, Margaret. *1947-Houses Far from Home: British Colonial Space in the New Hebrides.* Honolulu: University of Hawai'i Press, 2001.

Seach, John. *Volcano Live.* http://www.volcanolive.com/contents.html (accessed May 6, 2002).

Vatican City

The Holy See

- **Area:** 0.17 sq mi (0.44 sq km) / World Rank: 207
- **Location:** Northern and Eastern Hemispheres, in southern Europe, surrounded by Italy.
- **Coordinates:** 41°54′N, 12°27′E
- **Borders:** 2 sq mi (3.2 sq km), all with Italy
- **Coastline:** None
- **Territorial Seas:** None
- **Highest Point:** unnamed location, 246 ft (75 m)
- **Lowest Point:** unnamed location, 62 ft (19 m)
- **Longest River:** None
- **Largest Lake:** None
- **Natural Hazards:** None
- **Population:** 890 (July 2001 est.) / World Rank: 206
- **Capital City:** Vatican City
- **Largest City:** Vatican City, 890 (2001 est.)

OVERVIEW

Vatican City (the Holy See; the Vatican) is an urban, landlocked, enclave of Rome, Italy. It is the world's smallest state, located on the west bank of the Tiber River and to the west of the Castel Sant'Angelo. On the west and south it is bounded by the Leonine Wall. Vatican City is the administrative center of the Roman Catholic Church and the home of the Pope. It contains the following buildings and landmarks: St. Peter's Square; St. Peter's Basilica, the largest Christian church in the world; a quadrangular area north of the square containing administrative buildings and the Belvedere Park; the pontifical palaces to the west of Belvedere Park; and the Vatican Gardens, which occupy about half the area.

Certain areas in and near Rome are designated as extraterritorial to the Holy See, and receive tax exemption from the Italian government because of their association with the Vatican. Outside Rome, the sites include Castel Gandolfo, the pope's summer villa and its environs (almost 100 acres / 40 hectares) and Santa Maria de Galeria (about 1,037 acres / 420 hectares), some 12 mi (19.3 km) from Rome.

MOUNTAINS AND HILLS

The principality is built on a slight hill, but the variation in elevation is less than 200 ft (60 m).

INLAND WATERWAYS

Italy's Tiber River flows nearby, but there are no waterways within the Vatican City itself.

THE COAST, ISLANDS, AND THE OCEAN

Vatican City is landlocked.

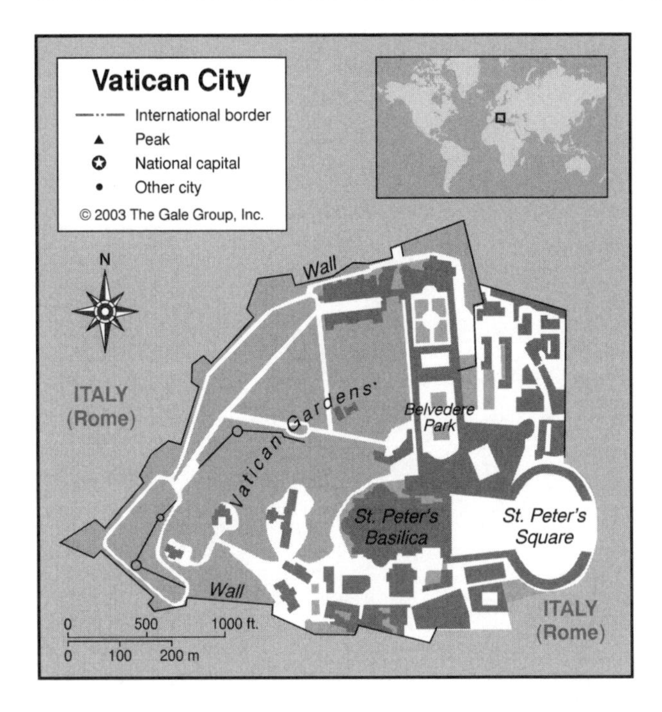

CLIMATE AND VEGETATION

Temperature

Vatican City's climate may be characterized as temperate. The temperature in January averages 45°F (7°C) and in July, 75°F (24°C). The gardens of Vatican City, featuring cultivated plants such as orchids, are world-renowned.

Rainfall

The rainy season is winter (September to mid-May), while there is very little rainfall from May to September. The annual rainfall averages 20 in (50 cm).

HUMAN POPULATION

Due to its small size and unusual nature, Vatican City does not have natives in the usual sense of the word. Its population is made up entirely of clergymen and other functionaries of the Roman Catholic Church who were born elsewhere.

NATURAL RESOURCES

Vatican City has no exploitable natural resources.

FURTHER READINGS

Hirst, Michael, et al. *The Sistine Chapel: A Glorious Restoration.* New York: H.N. Abrams, 1994.

Hutchinson, Robert J., *When in Rome: A Journal of Life in Vatican City.* New York: Doubleday, 1998.

McDowell, Bart. *Inside the Vatican.* Washington, D.C.: National Geographic Society, 1993.

Pietrangeli, Carlo, ed. *Paintings in the Vatican.* Boston: Little Brown, 1996.

Reese, Thomas J. *Inside the Vatican: The Politics and Organization of the Catholic Church.* Cambridge, Mass.: Harvard University Press, 1996.

Vatican: The Holy See. http://www.vatican.va/index.htm (accessed June 19, 2002).

Venezuela

- **Area:** 352,144 sq mi (912,050 sq km) / World Rank: 34
- **Location:** Northern and Western Hemispheres, on the northern coast of South America, south of the Caribbean Sea, west of Guyana, north and west of Brazil, and north and east of Colombia.
- **Coordinates:** 8°00′N, 66°00′W
- **Borders:** 3103 mi (4993 km) / Brazil, 1,367 mi (2,200 km); Colombia, 1,274 mi (2,050 km); Guyana, 462 mi (743 km)
- **Coastline:** 1,740 mi (2,800 km)
- **Territorial Seas:** 12 NM
- **Highest Point:** Pico Bolívar, 16,427 ft (5,007 m)
- **Lowest Point:** Sea level
- **Longest Distances:** 924 mi (1,487 km) WNW-ESE / 730 mi (1,175 km) NNE-SSW
- **Longest River:** Orinoco, 1,600 mi (2,574 km)
- **Largest Lake:** Lake Maracaibo, 6,300 sq mi (16,316 sq km)
- **Natural Hazards:** Floods; mudslides; drought
- **Population:** 23,916,810 (July 2001 est.) / World Rank: 43
- **Capital City:** Caracas, close to the center of the Caribbean Coast
- **Largest City:** Caracas, population 1,976,000 (2000 est.)

OVERVIEW

Venezuela occupies a large and varied region of northern South America, with a Caribbean coast, extensions of the Andes Mountains, rainforests, and grassy plains. Geographers divide Venezuela into four regions: the Maracaibo Lowlands, the Northern Mountains, the Orinoco Lowlands, and the Guiana Highlands. Venezuela has conflicting land claims with Guyana and conflicting maritime claims with Colombia.

Venezuela is situated on the South American Tectonic Plate. The northern shoreline, however, is the border between this plate and the Caribbean Plate. The South

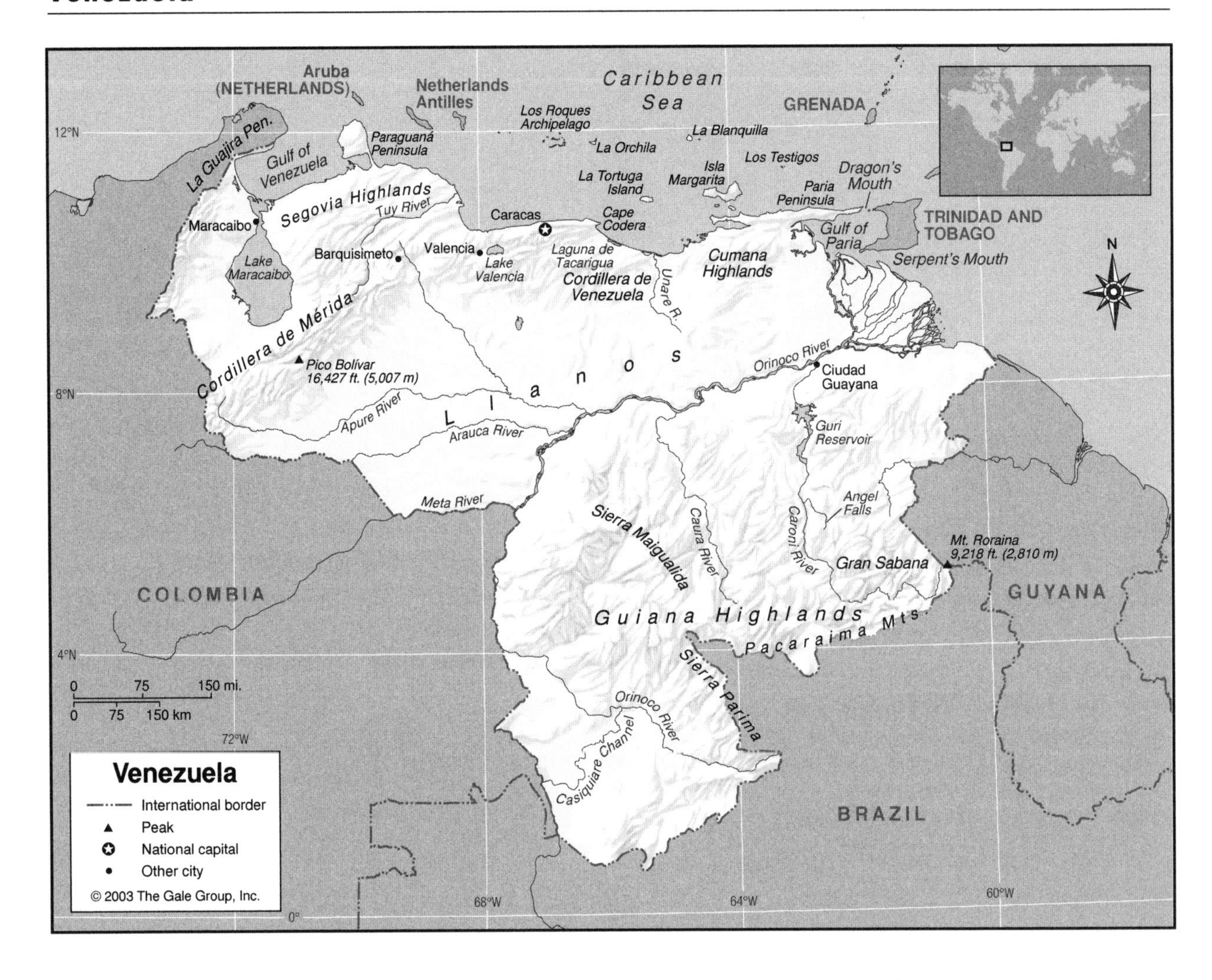

American Tectonic Plate is slowly sliding westward while the Caribbean Plate is sliding eastward. Over millions of years, the action of these plates has caused the formation of rocky cliffs on the Caribbean Coast as well as myriad fault lines running through North Central Venezuela. The major fault, the San Sebastian Fault, is the border between the two plates, on which earthquakes and landslides have occurred, the last major one hitting Caracas in 1967.

MOUNTAINS AND HILLS

Mountains

The Northern Mountains and their spur ranges extend from the Colombian border on the west to the coastal Paria Peninsula on the east. The Andes rise in Venezuela as the Cordillera de Mérida, containing permanently snowcapped peaks. The highest mountain in Venezuela, Bolívar Peak, at over 16,400 ft (24,998 m), is located in this chain.

The Cordillera de Mérida extends nearly to the Caribbean Coast. Then a coastal range, the Cordillera de Venezuela runs eastward. This range, where altitudes average over 5,000 ft (1,524 m) and peaks reach from 7,000 to 9,000 ft (2,133 to 2,743 m), is flanked on the north by narrow coastal plains except at points where the mountain slopes descend directly to the Caribbean. Part of the Cordillera de Venezuela terminates at Cape Codera on the Caribbean, but remnants of a parallel range continue eastward, ending near the Unare River.

Farther eastward the Cumana Highlands (also called the Eastern Highlands) rise in a broad block and extend eastward to terminate near the Gulf of Paria. At the core of the Cumana Highlands peaks reach 8,000 ft (2,438 m), but most of the system is made up of relatively low, dissected uplands.

In the south are the Guiana Highlands, an area with many mountain ranges. The Sierra Parima and Pacaraima Mountains form the southeastern borders with Brazil, extending south and east respectively from a common point of origin. The Sierra Parima reach heights of 5,000 ft (1,524 m) while Mount Roraina in the Pacaraima Mountains towers to 9,218 ft (2,810 m). The Sierra

Maigualida form an arc in the center of southern Venezuela.

Plateaus

In the northwest, the Cordillera de Mérida chain broadens northward to form the Segovia Highlands, consisting of heavily dissected plateaus decreasing in altitude from 6,000 ft (1,828 m) at their southern extremity to 600 ft (183 m) in the north before descending to the coastline.

The Guiana Highlands, rising almost immediately south of the Orinoco River, are considered to be the oldest land areas of the country, and erosion over the centuries has caused unusual formations. Comprising about 57 percent of the national territory, the 200,000 sq mi (517,988 sq km) highlands consist principally of plateau areas scored by swiftly running tributaries of the Orinoco. The most conspicuous topographical feature of the region is the Gran Sabana, a deeply eroded high plateau some 14,000 sq mi (36,260 sq km) that rises deep in the interior in abrupt cliffs of up to 2,500 ft (762 m) in height. From its rolling surface emerge massive perpendicular, flat-topped bluffs, called "tepuis." The loftiest tepui, Mount Roraima at the Venezuela-Brazil-Guyana tri-point, is over 9,000 ft (2,743 m) above sea level.

Canyons

Dramatic river canyons cut through the Canaima region of the Guiana Highlands. Among them are Devil's Canyon, at the foot of Angel Falls in southern eastern Venezuela; Kavac Canyon, one of the world's narrowest with a depth of 400 ft (122 m) but a width of only 4 ft (1.2 m wide); and Hacha Canyon. Cordillera de Mérida also contain river canyons, such as Santa Catalina Canyon near Merída.

Hills and Badlands

Hill regions of Venezuela include Tachira (a coffee-growing area in the west), the Sierra de San Luis in the northwest, Isla Margarita, and the Paria Peninsula, as well as parts of the south. The capital, Caracas, is surrounded by urbanized, deforested hillsides that are vulnerable to landslides.

INLAND WATERWAYS

Lakes

Lake Maracaibo, covering about 6,300 sq mi (16,316 sq km), is the largest body of water in Latin America. In the north it is directly connected with the Gulf of Venezuela by an island-dotted narrows some 25 mi (40 km) in length. The lake has an average depth of 30 ft (9 m) and is navigable to its southern end. The connection with the sea makes the lake's waters brackish in parts.

Second in importance among Venezuela's hundreds of lakes is Lake Valencia (142 sq mi / 369 sq km) located southwest of Caracas in the heart of the country's best agricultural lands. Originally the lake drained southward toward the Orinoco, but forest clearing on surrounding mountain slopes and overplanting of adjacent level ground caused its waters to subside to a point where it was left without a surface outlet. Lake Valencia and Lake Maracaibo are both badly polluted by sewage and industrial waste.

Other lakes include the large, mercury-contaminated Guri Reservoir on the Canaima River, and other reservoirs formed by hydroelectric dams, as well as numerous small mountain lakes in the Cordillera de Mérida. The coastal lowlands are also scattered with lagoons.

Rivers

Although there are more than 1,000 rivers in Venezuela, the river systems are dominated by the Orinoco. Flowing first west, then north, then east for 1,600 mi (2,574 km) to the Atlantic Ocean from its source in the Guiana Highlands at the Brazilian border, the Orinoco carries an enormous amount of water and is among the greatest rivers in the world in terms of volume. It is 5 mi (8 km) wide in some areas. Its flow varies substantially by season. When the river is low, Atlantic tidal effects can reach Ciudad Bolívar, 260 mi (418 km) upstream.

The Orinoco River system—together with its 436 tributaries including the Arauca, Apure, Meta, Guaviare, and Ventuari—provides drainage for about four-fifths of the country. It gathers the interior runoff from the Northern Mountains, most of the water from the Guiana Highlands, and the seasonal waters of the extensive great plains (*llanos*). As the Orinoco passes through the central part of southern Venezuela, it divides its waters. Through the Casiquiare Channel it sends one-third of its volume through the Negro River to the Amazon River along navigable waterways.

Most of the rivers rising in Venezuela's Cordillera de Mérida flow southeastward to the Apure River, a tributary of the Orinoco. From its headwaters in the Cordil-

GEO-FACT

Angel Falls, the highest waterfall in the world at 3,212 ft (979 m), including a straight drop of 2,647 ft (807 m), is a spectacular sight in Venezuela's Guiana Highlands. Its waters plunge from the 232 sq mi (600 sq km) Auyán Tepuy, considered the abode of spirits by local Pemon Indians. The waterfall is named after United States bush pilot Jimmie Angel, who revealed its existence to the world in 1935.

lera de Mérida, the Apure crosses the *llanos* in a generally eastward direction. There are also rivers that flow north from the Cordillera de Mérida into Lake Maracaibo and the Caribbean, including the Tuy River, which drains the country's most prosperous agricultural lands.

The other major Venezuelan river is the fast-flowing Caroni, which originates in the Gran Sabana and flows northward to join the Orinoco at Cuidad Guayana. Major hydroelectric projects have been established on its course. The enormous Guri hydroelectric project on the Caroni flooded large forest areas, and there is environmental controversy over plans to run power cables from the Guri project through the Cainama National Park to Brazil. An environmentally controversial mega-dam project for electricity export has been proposed for the Caura River, in the central Guiana Highlands.

Wetlands

Venezuela has five sites designated as Wetlands of International Importance under the Ramsar Convention on Wetlands. The Los Roques Archipelago is a group of 45 small islands surrounding a lagoon, with coral reefs and mangroves. Ciénaga de Los Olivitos is a coastal saltmarsh area, a significant bird habitat, which is threatened by salt mining. Cuare, Laguna de la Restinga, and Laguna de Tacarigua are also coastal wetlands, with mangroves, bird and turtle populations.

Other wetlands include the mudflats of the Orinoco Delta, with more than 70 outlets spread out over 9,000 sq mi (23,300 sq km); and the *llanos* grasslands north of the Orinoco, which are seasonally inundated with river waters.

THE COAST, ISLANDS AND THE OCEAN

Oceans and Seas

Venezuela's north shore meets the Caribbean Sea of the Atlantic Ocean, with a long coastline, coral reefs, and many islands. The coral reefs have been damaged by siltation and tourist development, and a surface "slick" of unidentified origin killed marine life on Morrocoy National Park in 1996. The Orinoco River empties directly into the Atlantic Ocean.

Major Islands

Seventy-two islands belong to Venezuela. The most important by far is Isla Margarita, 412 sq mi (1,067 sq km). Though rocky and receiving little rainfall, it is nevertheless heavily populated and intensively farmed. The other islands vary from coral atolls, to sandbars, to rocks. The 85 sq mi (220 sq km) La Tortuga is located 55 mi (88 km) west of Margarita. The most distant, the tiny island of Aves, is situated 300 mi (483 km) north of Isla Margarita. Morrocoy National Park, a wildlife preserve, is a small archipelago off the eastern coast.

The Coast and Beaches

Venezuela boasts the Caribbean's longest coastline. Nature refuges and tourist resort areas are interspersed along the rocky coast. The western section begins at the Gulf of Venezuela, which extends south to Lake Maracaibo. The Gulf is defined by the La Guajira (Colombia) and Paraguaná Peninsulas. The central coast has sand beaches and rocky cliffs as it undulates gently around to the Paria Peninsula. The Gulf of Paria on the eastern coast is formed by the Paria Peninsula and enclosed by the neighboring island nation of Trinidad and Tobago. Connecting the Gulf of Paria to the ocean are two straits that separate Trinidad from Venezuela. These are the Dragon's Mouth on the north and the Serpent's Mouth of the south. Further east, near the Guyana border, is the delta of the Orinoco River.

CLIMATE AND VEGETATION

Temperature

With a tropical climate, Venezuela has little seasonal variation in temperature, but there is considerable variation based on altitude, with much cooler weather in the Andean heights of the northwest than on the plains. The average temperatures in the lowland and plains areas below 2,625 ft (800 m) is 79° to 83°F (26° to 28°C). In the 2,625–6,560 ft (800–2,000 m) range the average temperatures are between 54° and 77°F (12° and 25°C), while at altitudes of 6,560–9,840 ft (2,000–3,000 m) the average is 48° to 52°F (9° to 11°C). In the high mountains (above 9,840 ft / 3,000 m) there are permanent snowfields and annual temperatures average below 46°F (8°C).

Rainfall

Two basic seasons occur in Venezuela: a wet season, commonly referred to as winter, from May through November, and a dry season, or summer, from December through April. The average annual rainfall in Venezuela is 32 in (81 cm), with more rain in the mountains and less on the Caribbean coast and islands. Humidity averages 50–60 percent. Heavy rains cause periodic flooding; a recent major episode occurred in December 1999 when mudslides destroyed settlements on deforested river banks and hillsides in northern Venezuela, killing as many as 30,000 people.

Grasslands

North of the Orinoco, the *llanos* grasslands cover about 115,800 sq mi (300,000 sq km). These plains, broken by low mesas, are used for cattle grazing. The rivers and streams winding through the *llanos* seasonally overflow their banks, turning the grasslands into wetlands, which then gradually dry out. These alternately wet and dry grasslands form an extraordinary wildlife habitat with many species of birds (ibis, herons, storks), mammals (capybaras, pumas), and reptiles (anacondas, caimans).

In the Gran Sabana, south of the Orinoco in the Guiana Highlands, along the Brazilian frontier, grasslands surround the forested "tepui" tabletop mountains.

Forests and Jungles

Venezuela suffered the loss of more than 8 percent of its forests during the 1980s, some 23,166 sq mi (60,000 sq km), an area larger than Costa Rica. The deforestation resulted mainly from agricultural and ranching expansion, and also from urbanization, pollution, and logging. Sixty percent of the natural forest north of the Orinoco River was destroyed.

At present, 48 percent of Venezuela still has forest cover, which survives mostly in the northeast around the Orinoco Delta, the southeast, and the south. Mining and logging operations, legal and illegal, continue to deforest the Guiana Highlands, where much of the remaining natural forest is found. Venezuela's watersheds have been affected by deforestation, which has caused siltation of rivers, lakes, and reservoirs. The control of forest areas by indigenous peoples has been decreased by political decisions in the south, which still contains biodiverse primary tropical rainforest. Northern cloud forests (found at higher elevations) and coastal mangrove forests are also disappearing under urban and tourist industry pressures.

Efforts are being made to protect the remaining forests, with 35 percent of Venezuela's land officially regulated, and 29 percent of Venezuela designated as national park land. Huge forest parks include the Biosphere Reserve of the Upper Orinoco-Casiquiare (32,046 sq mi / 83,000 sq km) in the south, which is the world's largest protected tropical rainforest, and Canaima National Park (11,583 sq mi / 30,000 sq km) in the Guiana Highlands.

HUMAN POPULATION

In 2000, an estimated 87 percent of the population was urban and as much as 90 percent is concentrated north of the Orinoco River in the western Andean Region and along the coast, living in the coastal plains and mountain valleys. In the region south of the Orinoco and Apure Rivers the population density is less than 10 people per sq mi (4 people per sq km). Divided into thirds, the area north of these rivers has a population density of a little more than 60 per sq mi (23 per sq km) in the east (including Margarita Island), 231 per sq mi (89 per sq km) in the central, and 126 per sq mi (49 per sq km) in the west.

NATURAL RESOURCES

Venezuela has some of the world's largest petroleum reserves, as well as important deposits of bauxite, coal, gold, diamonds, iron, manganese, and ferro-nickel. Resources also include hydroelectric power, livestock, fishing, agriculture, and tourist attractions.

Population Centers – Venezuela

(2000 POPULATION ESTIMATES)

Name	Population	Name	Population
Caracas (capital)	1,976,000	Maracay	459,000
Maracaibo	1,373,000	Barcelona/Puerto La Cruz	411,000
Barquisimeto	876,000	San Cristóbal	329,000
Valencia	832,000	Ciudad Bolivar	308,000
Ciudad Guayana	692,000		
Petare	521,000		

SOURCE: INE (Instituto Nacional de Estadistica), Venezuela.

States – Venezuela

2000 POPULATION ESTIMATES

Name	Population	Area (sq mi)	Area (sq km)	Capital
Amazonas	100,325	67,900	175,750	Puerto Ayacucho
Anzoátegui	1,140,400	16,700	43,300	Barcelona
Apure	467,000	29,537	76,500	San Fernando
Aragua	1,481,000	2,700	7,014	Maracay
Barinas	583,500	13,600	35,200	Barinas
Bolívar	1,307,000	91,900	238,000	Cuidad Bolívar
Carabobo	2,106,000	1,795	4,650	Valencia
Cojedes	262,000	5,700	14,800	San Carlos
Delta Amacuro	138,000	15,500	40,200	Tucupita
Distrito Federal	2,285,000	745	1,930	Caracas
Falcón	747,700	9,600	24,800	Coro
Guárico	638,600	25,091	64,986	San Juan de Los Morros
Lara	1,581,000	7,600	19,800	Barquisimeto
Mérida	745,000	4,400	11,300	Mérida
Miranda	2,607,000	3,070	7,950	Los Teques
Monagas	599,800	11,200	28,900	Maturín
Nueva Esparta	377,700	440	1,150	La Asunción
Portuguesa	830,000	5,900	15,200	Guanare
Sucre	824,800	4,600	11,800	Cumaná
Táchira	1,031,200	4,300	11,100	San Cristóbal
Trujillo	587,300	2,900	7,400	Trujillo
Vargas	309,000	n.a.	n.a.	La Guaira
Yaracuy	519,000	2,700	7,100	San Felipe
Zulia	3,210,000	24,400	63,100	Maracaibo

SOURCE: INE (Instituto Nacional de Estadistica), Venezuela.

FURTHER READINGS

Coronil, Fernando. *The Magical State: Nature, Money and Modernity in Venezuela.* Chicago: University of Chicago Press, 1997.

Dominic Hamilton. *Venezuela Voyages.* http://www.venezuelavoyage.com (accessed April 29, 2002).

Global Forest Watch. "The State of Venezuela's Forests: A Case Study of the Guayana Region." Washington, D.C.: World Resources Institute and Global Forest Watch, 2002.

Jordan, Tanis and Jordan, Martin. Angel Falls: A South American Journey. New York: Kingfisher Books, 1995.

Miranda, Marta. *All That Glitters is Not Gold: Balancing Conservation and Development in Venezuela's Frontier Forests.* Washington, D.C.: World Resources Institute, 1998.

Murphy, Alan, and Day, Mick. *Venezuela Handbook.* Lincolnwood, Ill.: Passport Books, 2001.

World Resources Institute. http://www.wristore.com/allthatgliti.html (accessed 29 April 2002).

GEO-FACT

Out of seven new species of mammals identified worldwide in the twentieth century, two were found in a nature reserve in the northwest corner of Vietnam in the 1990s: the giant muntjac (a type of barking deer) and the Vu Quang ox.

Vietnam

- **Area:** 127,244 sq mi (329,560 sq km) / World Rank: 67
- **Location:** Northern and Eastern Hemispheres, in southeast Asia, bordering China to the north, the South China Sea to the east and south, the Gulf of Thailand to the southwest, and Cambodia and Laos to the west.
- **Coordinates:** 16°00′N, 106°00′E
- **Borders:** 2,883 mi (4,639 km) / Cambodia, 763 mi (1,228 km); China, 796 mi (1,281 km); Laos, 1,324 mi (2,130 km)
- **Coastline:** 2,140 mi (3,444 km)
- **Territorial Seas:** 12 NM
- **Highest Point:** Fan-si-pan, 10,312 ft (3,143 m)
- **Lowest Point:** Sea level
- **Longest Distances:** 1,025 mi (1,650 km) N-S / 373 mi (600 km) E-W
- **Longest River:** Mekong, 2,800 mi (4,506 km)
- **Natural Hazards:** Typhoons; flooding
- **Population:** 79,939,014 (July 2001 est.) / World Rank: 14
- **Capital City:** Hanoi, northern Vietnam
- **Largest City:** Ho Chi Minh City, southern Vietnam, population 3,678,000 (2000 metropolitan est.)

OVERVIEW

Vietnam is a long, narrow country at the eastern edge of the Indochinese Peninsula in Southeast Asia. It has four major topographical divisions: the Red River Delta in the north; the Mekong Delta in the south; the Annamese Cordillera, a mountain system that spans nearly the entire length of the country; and the central lowlands, a narrow coastal plain between the mountains and the sea in the central part of the country.

Vietnam's shape has led people to compare it with the poles that the Vietnamese sling over their shoulders to carry rice—the country's long, narrow middle is the pole, and the river deltas on the north and south are the baskets of rice at either end. At its narrowest point, the country is only 31 mi (50 km) wide.

Vietnam is situated on the Eurasian Tectonic Plate.

MOUNTAINS AND HILLS

Mountains

Mountains account for three-fourths of Vietnam's terrain. The country's Annamese Cordillera mountain system has two major branches. One branch, projecting southward from Yunnan Province in China, extends along the country's entire border with Laos and, except at the northeastern tip of Laos, separates the Red River Basin from that of the Mekong River. Elevations along this branch range from 3,000 to 10,000 ft (914 to 3,048 m). The northern portion of this branch, called the Hoang Lien Mountains, includes Vietnam's highest peak, Fan-si-pan, which rises in the extreme northwest to a height of 10,312 ft (3,143 m).

The southern part, called the Truong Son, continues for more than 750 mi (1,207 km) along Vietnam's boundary with Laos and part of its boundary with Cambodia until it reaches the Mekong Delta, where it terminates about 50 mi (80 km) north of Ho Chi Minh City. The Truong Son is irregular in height and form and has numerous eastward spurs, which divide the coastal strip into separate sections. Its peaks range in height from about 5,000 ft (1,524 m) to 8,521 ft (2,597 m).

The second branch of Vietnam's mountains, sometimes referred to as the Northern Highlands, extends along the border with China, terminating in a series of islands northeast of Haiphong in the Gulf of Tonkin.

Plateaus

Within the wider, southern portion of the Truong Son is a plateau area known as the Central Highlands that covers approximately 20,000 sq mi (51,800 sq km) and consists of two distinct parts. The northern part, called the Dac Lac Plateau, varies in elevation from about 600 to 1,600 ft (182 to 487 m), with a few peaks rising much higher. The southern portion of the Central Highlands rises to elevations of over 3,000 ft (914 m) in many places. The hill city of Da Lat is in the center of this area.

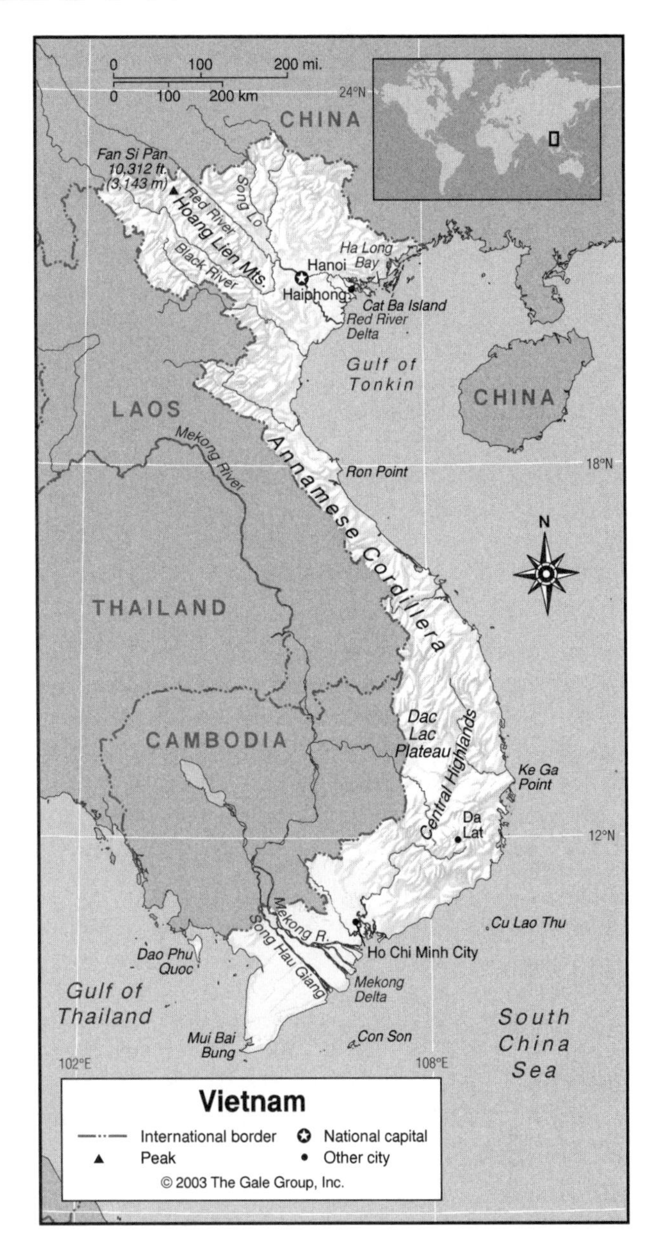

INLAND WATERWAYS

Lakes

There are many small lakes scattered across Vietnam; however, lakes and waterfalls are predominantly found in the Central Highlands. Lake Ba Be is one of the few lakes in northern Vietnam. It has a surface area of 1.7 sq mi (4.5 sq km). The lake is surrounded by limestone hills in which the Nang River has dug caves, including the famous Puong Grotto. Ho Tay (West Lake), located in western Hanoi city, is also one of Vietnam's largest lakes. It has a surface area of 1.6 sq mi (4.1 sq km).

Rivers

The Red River, located in the far north, has a total length of about 725 mi (1,167 km). Its two major tributaries, the Song Lo and the Black River, give it a large flow volume during the rainy season—as much as 800,000 cu ft (22,653 cu m) per second, or twice as much as the maximum flow of the Nile River in Egypt.

The 2,800 mi (4,506 km) long Mekong is one of the great rivers of the world. From its source in the high plateau of Tibet, not far from the headwaters of the Yangtze of China, it flows through Tibet and China to the northern border of Laos. There it separates Myanmar from Laos and, farther downstream, Laos from Thailand. Flowing through Cambodia, it bifurcates at the capital, Phnom Penh. The southern branch, the Song Hau Giang, flows directly to the sea; the larger northern branch splits into four parts about 50 mi (80 km) before reaching the sea. The Mekong Delta, with an area of approximately 26,000 sq mi (67,340 sq km), was built up by the five branches of the Mekong, as well as three smaller rivers. These rivers carry so much sediment to the sea that the coastline in the south is advancing as much as 250 ft (76 m) per year.

In addition to Vietnam's two major rivers, a number of shorter rivers and streams rise in the Annamese Cordillera and flow to the sea. Several, including the Saigon River, flow across the lowlands between the Mekong Delta and the southern edge of the cordillera.

THE COAST, ISLANDS, AND THE OCEAN

Oceans and Seas

Vietnam is bordered on the northeast by the Gulf of Tonkin, on the southeast by the South China Sea, and on the southwest by the Gulf of Thailand. Coral reefs surround Vietnam's coastline on all sides including its offshore islands—more than 90 percent of the entire coastal length. However, many of these reefs are in poor condition due to the use of fishing practices involving explosives and electric shock.

Major Islands

Vietnam has a number of offshore islands in Ha Long Bay to the north, in the South China Sea to the east, and near the Mekong River Delta in the south. The largest of the islands in the north is Cat Ba, with an area of 137 sq mi (355 sq km). Ha Long Bay is also the site of some 3,000 uninhabited islets formed of limestone or dolomite. Dao Phu Quoc is a large Vietnamese island in the Gulf of Thailand.

Jurisdiction over several of the islands off the coast of Vietnam, including the Spratly Islands to the southeast, has been disputed by neighboring countries, including Malaysia, the Philippines, Indonesia, China, and Taiwan.

The Coast and Beaches

Vietnam's S-shaped coast snakes from the Gulf of Thailand in the southwest, to the South China Sea, then up to the Gulf of Tonkin in the northeast. It is heavily indented at the mouth of the Mekong River in the south, with another major indentation at Haiphong Harbor on the Red River Delta in the north. There are moderate indentations at the mouths of other rivers along the coast. The southern tip of Vietnam, the cape of Mui Bai Bung, marks the most southeastern point of the Gulf of Thailand. Moving north along the coast there are several other capes including Ke Ga Point and Ron Point. Much of the shore along the northern half of the coastal plain is fringed by a narrow line of sand dunes backed by an intensively cultivated flat fertile area.

CLIMATE AND VEGETATION

Temperature

The southern part of the country is warmer than the north. The average annual temperature in Ho Chi Minh City is 81°F (27°C), compared with 74°F (23°C) in Hanoi. However, the Central Highlands are cooler although they are in the south, thanks to their higher elevation. The average annual temperature at Da Lat, in the highlands, is 70°F (21°C).

Rainfall

Average annual rainfall ranges from 68 in (172 cm) in Hanoi to as much as 160 in (406 cm) or more in the mountains. Violent typhoons strike the central coastal region between July and November, causing serious damage to crops and property, as well as loss of life.

Grasslands

The low, level plain of the Mekong Delta, nowhere more than 10 ft (3 m) above sea level, is very fertile, and cultivated land extends to the immediate shoreline in the vicinity of the river mouths. The Red River Delta is a flat, triangular region, smaller but more intensively developed than the Mekong River Delta. Almost entirely built up of alluvium, the delta was formerly an extension of the Gulf of Tonkin, which has since been filled by the deposits of the rivers that run into the basin. The central lowlands consist of a fertile, narrow, coastal strip along the eastern slopes of the Truong Son Mountains. Tropical grasses grow throughout the lowlands, and there is a large savanna area in the southwest.

Forests and Jungles

Vietnam's diverse ecosystems are estimated to support as many as 12,000 plant species. Tree species range from conifers to hardwoods like teak and ebony to bamboo, palm, and mangroves. Tropical broadleaf rain forest

Provinces – Vietnam

1999 POPULATION ESTIMATES

Name	Population	Area (sq mi)	Area (sq km)	Capital
An Giang	2,049,039	1,349	3,493	Long Xuyen
Ba Ria – Vung Tau	800,568	759	1,965	Vung Tau
Bac Can (Kan)	275,520	1,851	4,795	Bac Kan
Bac Giang	1,492,191	1,474	3,817	Bac Giang
Bac Lieu	736,325	959	2,485	Bac Lieu
Bac Ninh	941,389	308	799	Bac Ning
Ben Tre	1,296,914	859	2,225	Ben Tre
Binh Dinh	1,461,046	2,346	6,076	Qui Nhon
Binh Duong	716,427	105	272	Thu Dau Mot
Binh Phuoc	653,644	2,624	6,796	Dong Zoai Town
Binh Thuan	1,047,040	3,066	5,864	Phan Thiet
Ca Mau	1,117,829	2,009	5,204	Ca Mau
Can Tho	1,811,100	1,145	2,965	Can Tho
Cao Bang	490,700	3,261	8,445	Cao Bang
Da Nang	684,131	364	942	Da Nang
Dak Lak	1,776,331	7,645	19,800	Buon Me Thuot
Dong Nai	1,989,541	2,264	5,864	Bien Hoa
Dong Thap	1,564,977	1,309	3,391	Csa Dec
Gia Lai	971,920	6,259	16,212	Plei Ku
Ha Giang	602,700	3,024	7,831	Ha Giang
Ha Nam	791,618	318	823	Phu Ly
Ha Noi	2,672,122	356	921	Hanoi
Ha Tay	2,386,770	829	2,148	Ha Dong
Ha Tinh	1,269,013	2,337	6,054	Ha Tinh
Hai Duong	1,649,779	641	1,661	Hai Duong
Hai Phong	1,672,992	580	1,503	Haiphong
Hoa Binh	757,637	1,781	4,612	Hoa Binh
Hung Yen	1,068,705	346	895	Hung Yen
Khanh Hoa	1,031,262	2,030	5,257	Nha Trang
Kien Giang	1,494,433	2,455	6,358	Rach Gia
Kon Tum	314,042	3,836	9,934	Kon Tum
Lai Chau	588,666	6,586	17,068	Lai Chau
Lam Dong	966,219	3,835	9,933	Da Lat
Lang Son	704,643	3,161	8,187	Lang Son
Lao Cai Son	594,637	3,108	8,050	Lao Cai
Long An	1,306,202	1,681	4,355	Tan An
Nam Dinh	1,888,400	644	1,669	Nam Dinh
Nghe An	2,858,265	6,321	16,371	Vinh
Ninh Binh	884,080	536	1,388	Ninh Binh
Ninh Thuan	503,048	1,323	3,427	Phan Rang-Thap cham
Phu Tho	1,261,500	1,338	3,465	Viet Tri
Phu Yen	786,972	1,038	5,278	Tuy Hoa
Quang Binh	793,863	3,083	7,984	Dong Hoi
Quang Nam	1,372,424	4,264	11,043	Hoi An
Quang Ngai	1,190,006	1,999	5,177	Quang Ngai
Quang Ninh	1,004,461	2,293	5,938	Hong Gai
Quang Tri	573,331	1,771	4,588	Dong Ha
Soc Trang	1,183,820	1,232	3,191	Soc Trang
Son La	881,383	5,586	14,468	Son La
Tay Ninh	965,240	1,556	4,029	Tay Ninh
Thai Binh	1,785,600	577	1,495	Thai Binh
Thai Nguyen	1,046,163	1,455	3,769	Thai Nguyen
Thanh Hoa	3,467,609	4,300	11,138	Thanh Hoa
Thanh Pho Ho Chi Minh	5,037,155	807	2,090	Ho Chi Minh City
Thua Thien – Hué	1,045,134	1,934	5,009	Hué
Tien Giang	1,605,147	918	2,377	My Tho
Tra Vinh	965,712	915	2,369	Tra Vinh
Tuyen Quang	675,110	2,240	5,801	Tuyen Quang
Vihn Long	1,010,486	571	1,478	Vihn Long
Vinh Phuc	1,091,973	526	1,362	Vinh Yen
Yen Bai	679,700	2,629	6,808	Yen Bai

SOURCE: General Statistical Office, Government of Vietnam.

Population Centers – Vietnam

(1992 POPULATION ESTIMATES)

Name	Population	Name	Population
Ho Chi Minh City	3,015,743	Qui Nhon	163,385
Hanoi (capital)	1,073,760	Vung Tau	145,145
Haiphong	783,133	Rach Gia	141,132
Da Nang	382,674	Long Xuyen	132,681
Buonmathuot	282,095	Thai Nguyen	127,643
Nha Trang	221,331	Hon Gai	127,484
Hué	219,149	Vinh	112,455
Cantho	215,587	Mytho	108,404
Campha	209,086	Da Lat	106,409
Nam Dinh	171,699		

SOURCE: "Population of Capital Cities and Cities of 100,000 and More Inhabitants." United Nations Statistics Division.

and deciduous forest is found in the mountains of the both the northern and central parts of the country. Bamboo grows on the lower slopes. The southernmost tip of the Mekong Delta, known as the Mui Bai Bung (Ca Mau Peninsula), is covered with dense jungle, and its shoreline is lined with mangroves.

In the last half of the twentieth century, Vietnam's forested area was reduced by 50 percent as a result of logging, herbicides used during the war of the 1960s and 1970s, and slash-and-burn agriculture. By the late 1990s, forests covered only one-fifth of the country, but a system of national parks and nature reserves had been established to protect the rich and diverse ecosystems that remained. Nearly half of Cat Ba Island has been turned into a national park, with evergreen rain forests, waterfalls, mangrove swamps, and abundant wildlife, including wild boar, dolphins, waterfowl, and 200 species of fish.

HUMAN POPULATION

An estimated 80 percent of Vietnamese are rural dwellers. More than two-thirds of the population lives in the northern and southern river delta regions, which have population densities of 648 people per sq mi (250 people per sq km) and 1,137 people per sq mi (439 people per sq km) respectively. The Central Highlands has a population density of 328 people per sq mi (127 people per sq km) while the narrow lowlands region has a density of 532 people per sq mi (206 people per sq km).

NATURAL RESOURCES

Vietnam's fertile land and forests are important natural resources. It is also rich in mineral resources. Coal reserves are abundant in the north, and mining has been expanding since the war. Vietnam also has reserves of manganese, bauxite, and phosphates. There are offshore oil and gas reserves, though they have yet to be fully exploited.

FURTHER READINGS

Brownmiller, Susan. *Vietnam: Encounters of the Road and Heart.* New York: HarperCollins, 1994.

Clifford, Geoffrey. *Vietnam: The Land We Never Knew.* San Francisco: Chronicle Books, 1989.

Hunt, Christopher. *Sparring with Charlie: Motorbiking Down the Ho Chi Minh Trail.* New York: Anchor Books, 1996.

Maitland, Derek. *Insider's Vietnam, Laos, and Cambodia Guide.* Edison, N.J.: Hunter Publishing Inc., 1995.

Vietnam National Administration of Tourism. *Vietnam-tourism.com.* http://www.vietnamtourism.com/e_pages/e_index.htm (Accessed June 13, 2002).

Warmbrunn, Erika. *Where the Pavement Ends: One Woman's Bicycle Trip Through Mongolia, China, and Vietnam.* Seattle: Mountaineers Books, 2001.

Virgin Islands

Dependency of the United States of America

- **Area:** 136 sq mi (352 sq km) / World Rank: 191
- **Location:** Northern and Western Hemispheres, between the Caribbean Sea and the North Atlantic Ocean, east of Puerto Rico.
- **Coordinates:** 18°20′N, 64°50′W
- **Borders:** None
- **Coastline:** 117 mi (188 km)
- **Territorial Seas:** 12 NM
- **Highest Point:** Crown Mountain, 1,555 ft (474 m)
- **Lowest Point:** Sea level
- **Longest River:** None of significant length
- **Natural Hazards:** Hurricanes; droughts; floods; earthquakes
- **Population:** 122,211 (July 2001 est.) / World Rank: 182
- **Capital City:** Charlotte Amalie, south-central St. Thomas island
- **Largest City:** Charlotte Amalie, 11,000 (2000 est.)

OVERVIEW

The Virgin Islands (Virgin Islands of the United States) were purchased from Denmark by the U.S.A. in 1917. The islands are organized as an unincorporated territory of the United States under the jurisdiction of the Office of Insular Affairs of the Department of the Inte-

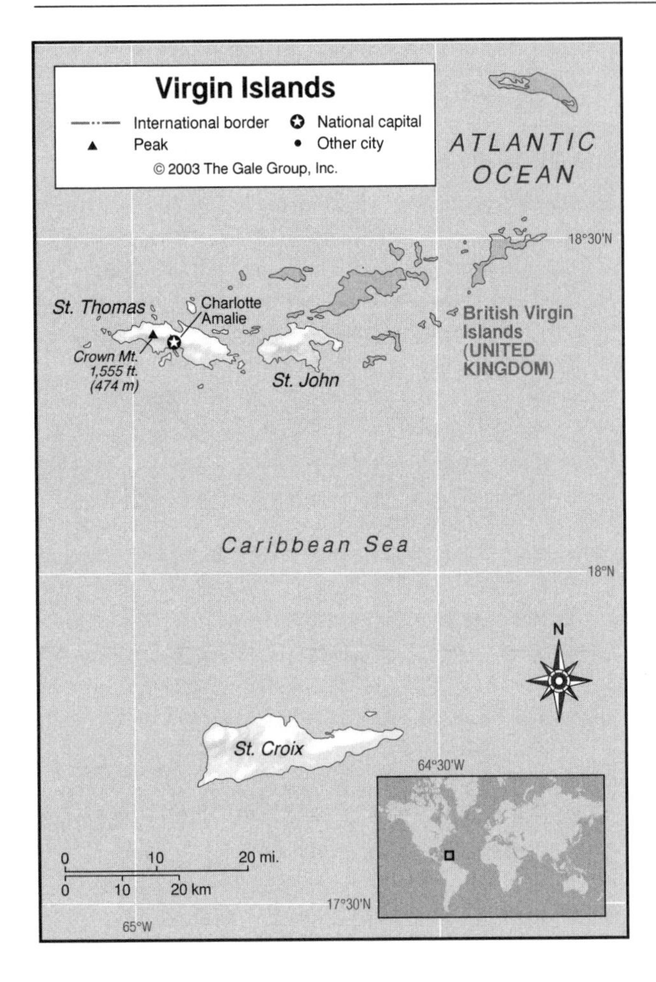

rior. They were formerly known as the Danish West Indies. The British Virgin Islands, located further east, are part of the same island chain (see entry on British Virgin Islands). The U.S. portion of the archipelago consist of 68 islands, about 40 mi (64 km) east of Puerto Rico. They are the westernmost islands of the Lesser Antilles. Most of the Virgin Islands are very small. St. Croix, St. John, and St. Thomas are the major islands.

Altogether, the land area of the islands is twice that of Washington, DC. Although rarely affected by hurricanes, the islands are subject to frequent severe droughts, floods, earthquakes, and lack of natural fresh water resources. Arable land accounts for 15 percent of the territory, 6 percent is under permanent cultivation, 26 percent is permanent pastureland, and 6 percent is forest and woodland; 47 percent is given to other uses. About three-fourths of St. John is a national park.

MOUNTAINS AND HILLS

While the small coral islands are relatively flat, the larger volcanic islands are steep and hilly. Elevations can exceed 1,000 ft (305 m). The highest point is Crown Mountain, (1,555 ft / 474 m) on St. Thomas.

INLAND WATERWAYS

Due to their small sizes and often steep and hilly terrain, the Virgin Islands lack rivers, streams, or freshwater lakes of any consequence. However, salt ponds and marshes dot the flat coraline islands.

THE COAST, ISLANDS, AND THE OCEAN

The Virgin Islands are situated in the Caribbean Sea near the Anegada Passage, which handles heavy shipping traffic for the Panama Canal. Only three of the islands are important in size: St. Croix, St. Thomas and St. John. St. Croix is the largest at 84 sq mi (218 sq km). St. Thomas is the next largest (28 sq mi / 73 sq km) and is the site of one of the best natural deepwater harbors in the Caribbean. St. John is 20 sq mi (52 sq km) in size. The rest of the Virgin Islands are islets. All of the islands are fringed by coral reefs.

CLIMATE AND VEGETATION

Temperature

The climate is subtropical, tempered by easterly trade winds, and has little seasonal variation. Temperatures range from 70° to 90°F (21° to 32°C) year-round, with relatively low humidity.

Rainfall

The rainy season is from May to November. Prolonged periods of rain are rare, but the islands are subject to occasional hurricanes.

Grasslands

Some of the slopes and tablelands are covered with grasses and scrub.

Forests and Jungles

Rainforest is found on St. Croix, and covers the upper third of St. John, three-fifths of which is national park. Some 800 species of plants are found in the islands.

HUMAN POPULATION

The islands have no first-order administrative divisions as defined by the US Government, but Saint Croix, Saint John, and Saint Thomas constitute second-order units. Almost all of the population lives on St. Croix and St. Thomas, with most of the rest on St. John. In 2001, the population growth rate was 1.06 percent.

NATURAL RESOURCES

The Virgin Island's economy relies primarily on tourism. Its greatest natural resource is the subtropical beauty of its islands, which includes sun, sand, sea, surf, and coral reefs. A substantial portion of the natural surroundings is preserved in the Virgin Islands National Park covering 14,689 acres (5,945 hectares). The Virgin Islands also produce rum and bay rum.

FURTHER READINGS

Sullivan, Lynne. *Adventure Guide to the Virgin Islands*. 5th ed. Edison, N.J.: Hunter Publishing Inc., 2001.

Government of the US Virgin Islands. http//www.gov.vi/ (accessed July 17, 2002).

Here.VI. http://www.here.vi/ (accessed July 17, 2002)

Yemen

- **Area:** 203,850 sq mi (527,970 sq km) / World Rank: 50
- **Location:** Northern and Eastern Hemispheres, on the southwestern Arabian Peninsula, bordering Oman to the east and Saudi Arabia to the north.
- **Coordinates:** 15°00′N, 48°00′E
- **Borders:** 1085 mi (1746 km) / Oman, 179 mi (288 km); Saudi Arabia, 906 mi (1,458 km)
- **Coastline:** 1,184 mi (1,906 km)
- **Territorial Seas:** 12 NM
- **Highest Point:** An-Nabī Shu'ayb, 12,336 ft (3,760 m)
- **Lowest Point:** Sea level
- **Longest Distances:** 336 mi (540 km) N-S / approximately 777 mi (1,250 km) SW-NE
- **Longest River:** Wadi Hadhramawt, 149 mi (240 km)
- **Largest Lake:** Marib Reservoir, approximately 12 sq mi (30 sq km)
- **Natural Hazards:** Droughts; earthquakes; sandstorms; flash floods
- **Population:** 18,078,035 (July 2001 est.) / World Rank: 54
- **Capital City:** Sanaa, in the western highlands
- **Largest City:** Sanaa, population 629,000 (2000)

OVERVIEW

Yemen wraps around the southwest corner of the Arabian Peninsula. Southern Yemen, where the city of Aden is located, and North Yemen, where the city of Sanaa is located, were independent countries until they unified to become Yemen in 1990. Yemen has five principal geographic regions: the Tihama coastal plain, the mountainous interior, the high plateau and the Hadhramawt and Al Mahra uplands, the "Empty Quarter" interior desert, and offshore islands.

MOUNTAINS AND HILLS

Mountains

Yemen's interior is quite mountainous, with ranges along a north-south axis parallel to the Red Sea and along an east-west axis parallel to the Gulf of Aden. The mountains, which include extinct volcanoes, reach 8,000 ft (2,438 m) in the extreme west and gradually taper off to the east. Average elevations in the interior mountains are from 7,000 to 10,000 ft (2,133 to 3,048 m). The ranges are characterized by rocky spars and sharp, steep ridges, which have rendered access to the interior difficult.

There are western, central and eastern ranges. The western mountains, although steep, are terraced for intensive agriculture. The central mountain range begins in the vicinity of the old city of Ta'izz and includes Arabia's highest peak, An-Nabī Shu'ayb, which rises to 12,336 ft (3,760 m). Yemen's capital, Sanaa, is located in one of the largest basins of the central range, at an elevation of 7,874 ft (2,400 m). The eastern highlands have heights of 2,500 to 3,500 feet (762 m to 1,067 m).

Plateaus

Yemen's eastern region occupies the irregular southern end of the Arabian plateau, which is formed by ancient granites and partly covered by sedimentary limestones and sand. The central highlands of Yemen are broken into plateaus ranging in height from 4,000 to 10,000 ft (1,200 to 3,000 m). The Harra plateau, north of Sanaa, is a spectacular landscape of lava rock, sandstone striations, and extinct volcanic cones.

Canyons

Yemen's canyons include the Al Guedam canyon in the mountains north of Sanaa; Wadi Dahero canyon on Socotra island; and the Bir Maqsur limestone crevasse, also on Socotra. Deep eroded ravines known as "wadis," cut by extinct or seasonally flowing rivers, fissure much of Yemen's interior.

Hills and Badlands

Yemen's eastern mountains slope down into hills that merge with the sands of the Rub' al-Khali desert. Other hill areas include the Hadhramawt and Al Mahra uplands in the east. Throughout Yemen, the foothills of mountain ranges are terraced for farming.

INLAND WATERWAYS

Lakes

Yemen lacks natural freshwater lakes. A dam built in 1986 near the site of the ancient Marib dam in western Yemen produced a 12 sq mi (30 sq km) reservoir, but the

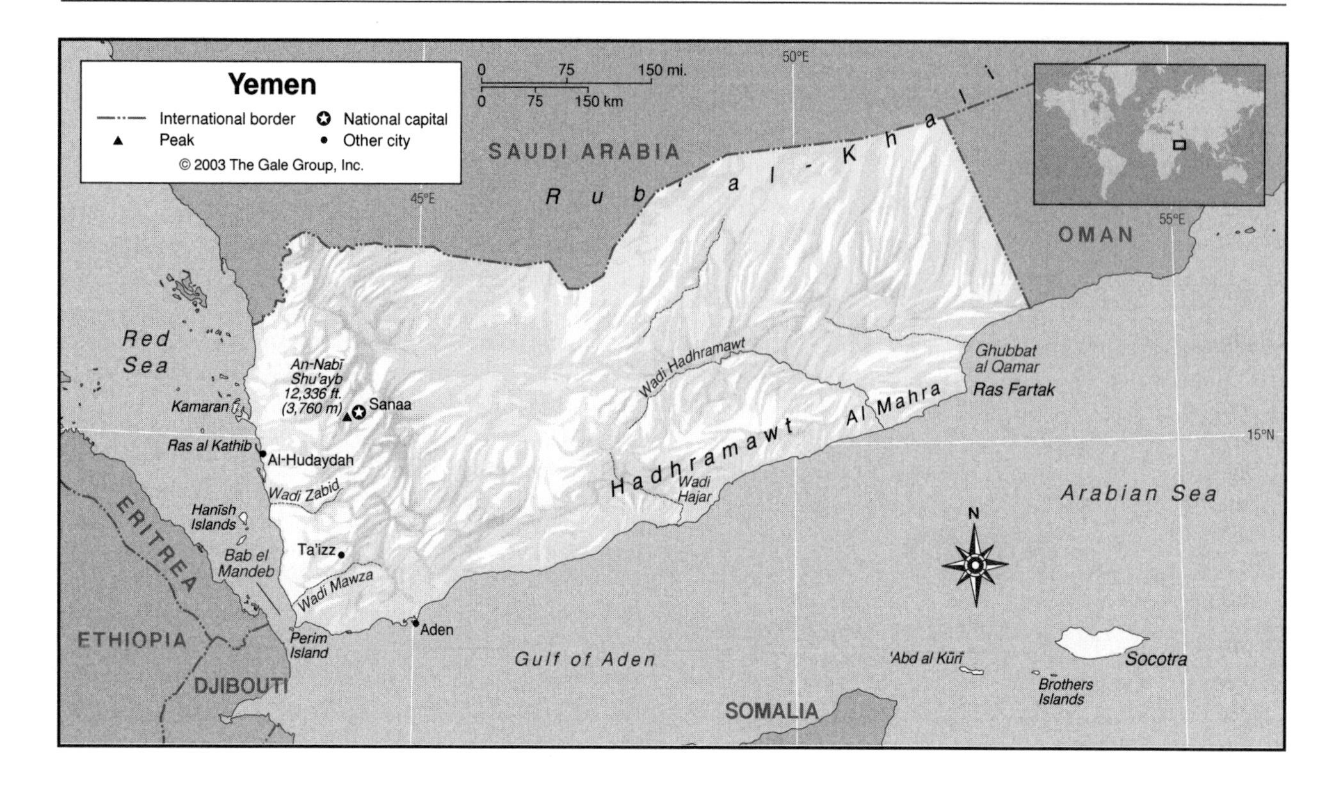

water levels have been depleted and it has been afflicted with algae blooms. There are some small brackish lagoons along the coast and several hot springs in the highlands.

Rivers

Yemen's highlands are interspersed with deep wadis, or riverbeds. The wadis are usually quite dry with little vegetation but many will fill dramatically during times of heavy rains. Rainfall drains through seven major wadis that can flow west as far as the Red Sea: Rima, Rasyan, Mawr, Surdud, Siham, Zabid and Mawza. Wadis that drain south into the Gulf of Aden from the eastern regions include Hajar, Jahr, Warazan, and Yemen's longest, Wadi Hadhramawt (149 mi / 240 km). The valley of Wadi Hadhramawt extends from the central part of the country southeastward to the Gulf of Aden. Surrounded by desolate hills and desert, the upper and middle parts of the Hadhramawt, with their alluvial soil and seasonal floodwaters, are relatively fertile and are inhabited by a farming population; the lower eastern part of the valley, which turns southward to the sea, is barren and largely uninhabited.

Wetlands

The valleys of Wadi al-Malih and Wadi Warazan near the city of Ta'izz, and Wadi Zabid near the port of Al Hudaydah, contain marshes that are decreasing in area due to demands on the groundwater, agricultural conversion, and grazing. Wastewater lagoons north of Ta'izz and northeast of Al Hudaydah, and a treated-sewage outflow area west of the city of Aden, have become important bird habitats. Mudflats, sandbars, and mangroves form wetlands ecosystems along the Red Sea coast.

THE COAST, ISLANDS AND THE OCEAN

Oceans and Seas

The Red Sea meets Yemen to the west, and the Gulf of Aden of the Arabian Sea meets Yemen to the south; all of these are extensions of the Indian Ocean. One of the world's most important shipping lanes, the Bab el-Mandeb, connects the Red Sea and the Gulf of Aden southwest of Yemen, with Djibouti and Eritrea on the far side. Some 5 percent of Yemen's coast has coral reefs, with particularly diverse marine habitats in the Red Sea.

Major Islands

Yemen has more than 115 islands, including the significant islands of Perim and Kamaran in the Bab el-Mandeb and the Hanīsh Islands further north. Yemen also possesses the 1,400 sq mi (3,626 sq km) island of Socotra in the Gulf of Aden. Socotra has numerous endemic species, with intact land and marine ecosystems. The Brothers are a chain of small islands near Socotra.

The Coast and Beaches

The north-south stretch of Yemen's Red Sea coast is interrupted by the promontories of Ras al Shi'b (15 mi / 25 km) and, near the port of Al Hudaydah, Ras al Kathib (19 mi / 30 km). The town of Turba marks the corner on the Bab el-Mandeb strait where the Red Sea and Gulf of Aden coasts converge. Yemen's Gulf of Aden coast runs from southwest to northeast. To the east of Turba is the excellent natural harbor of Aden. Coastal plains follow on the Gulf of Aden, with sandy beaches including Ras Sharma and Dhobbah, which are nesting sites for endangered green turtles. The coast curves inward at Ras Fartak point, forming the Ghubbat al Qamar before Yemen's border with Oman.

Population Centers – Yemen

(1993 POPULATION ESTIMATES)

Name	Population
Sanaa (Sana'a, capital)	926,595
Aden	400,783
Ta'izz (Taiz)	290,105
Al-Hudaydah (Hodeidah)	246,068

SOURCE: "Population of Capital Cities and Cities of 100,000 and More Inhabitants." United Nations Statistics Division.

CLIMATE AND VEGETATION

Temperature

Yemen has a very hot semitropical climate, with temperatures that can reach as high as 129°F (54°C). The average temperature varies over the two basic seasons, ranging from 72°F (22°C) in summer to 57°F (14°C) in winter. The Red Sea coast is particularly hot and humid. The interior mountain regions experience frosts in winter. Sandstorms often appear in summer and winter as winds sweep across Yemen.

Rainfall

Monsoon rains drench much of otherwise dry Yemen twice each year, in March-May and July-September. In the southwest corner of the country there is more consistent rain, with constant fog along the coast. Yemen's average annual rainfall is 20 in to 36 in (51 to 91 cm), with great regional variation: less than 5 in (12 cm) on the coastal lowlands, contrasting with 39 in (100 cm) in highlands above 9,842 ft (3,000 m).

Grasslands

Scrub grasslands, with sparse ground cover and shrubbery, are common throughout Yemen. This type of terrain, about 30 percent of Yemen's area, is used for raising livestock, and over-grazing is an environmental threat.

Deserts

Inland from the mountains and north of the Wadi Hadhramawt valley, gravel deserts transition into the sand dune deserts of the Rub al-Khali, or Empty Quarter, which extends across the border from Saudi Arabia. Even in this inhospitable region, oases are inhabited during the rainy season. Productive salt pans are found in the Rub al-Khali.

On the Gulf of Aden coast there is a coastal fog desert ecosystem, with vegetation that then gives way to the Tihama desert. The Tihama is a narrow, hot, humid, almost waterless strip that extends along the seacoast and occupies approximately 10 percent of the country.

Forests and Jungles

Most of Yemen's forests were destroyed during the twentieth century, by agricultural clearing and cutting for fuel wood, fodder and timber. Yemen had less than 4 percent forest cover as of 1998. Remaining forests are found in the mountains, along some wadis, and on Socotra island. They include old-growth stands of juniper such as Jebal Bura, an ancient forest sanctuary in the eastern highlands, as well as some recently planted juniper groves.

Jebel al Houf, on the far eastern coast, has Yemen's largest forest (77 sq mi /200 sq km), in an area where mountains trap monsoon moisture to create a foggy, misty zone; it is protected by the local community. Efforts are being made to preserve the forests of the mountainous Utma region of the central highlands, which include medicinal and fragrant trees.

HUMAN POPULATION

Yemen's population growth rate is a high 3 percent per year (1999). An estimated 38 percent of Yemen's people are city or town dwellers (2000), while the rest of the population lives in small farming, herding, and fishing villages, taking maximum advantage of the 6 percent of the land that supports agriculture. Yemen's human settlements occupy a great array of terrains, including looming mountains and baking hot deserts. The national population density is 85 people per sq mi (33 people per sq km).

NATURAL RESOURCES

Yemen has lucrative petroleum reserves. Other resources include rock salt, marble, coal, nickel, copper, semiprecious stones, and ocean fishing. Agriculture is important, although water supply problems are constant. As a transition zone between the Asian and African bioregions, Yemen has an especially high biodiversity of flora and fauna.

FURTHER READINGS

Hansen, Eric. *Motoring with Mohammed: Journeys to Yemen and the Red Sea.* New York: Vintage, 1992.

MacKintosh-Smith, Tim. *Yemen: The Unknown Arabia.* New York: Overlook Press, 2001.

Stark, Freya. *The Southern Gates of Arabia: A Journey in the Hadhramawt.* London: John Murray, 1940.

Wald, Peter. *Yemen.* London: Pallas Athene, 1996.

Wetlands International: *Republic of Yemen.* http://www.wetlands.org/inventory&/MiddleEastDir/YEMEN.htm (accessed July 6, 2002).

Yemen Gateway. http://www.al-bab.com/yemen/about.htm (accessed July 6, 2002).

Yugoslavia

- **Area:** 39,518 sq mi (102,350 sq km) / World Rank: 108
- **Location:** Northern and Western Hemispheres, in southeastern Europe, south of Hungary, west of Romania and Bulgaria, north of the Former Yugoslav Republic of Macedonia and Albania, and east of Croatia and Bosnia and Herzegovina.
- **Coordinates:** 44°00′N, 21°00′E
- **Borders:** 1,396 mi (2,246 km) / Albania, 178 mi (287 km); Bosnia and Herzegovina, 327 mi (527 km); Bulgaria, 198 mi (318 km); Croatia, 166 mi (266 km); Hungary, 94 mi (151 km); Former Yugoslav Republic of Macedonia, 137 mi (221 km); Romania, 296 mi (476 km).
- **Coastline:** 124 mi (199 km)
- **Territorial Seas:** Not available
- **Highest Point:** Mt. Daravica, 8,714 ft (2,656 m)
- **Lowest Point:** Sea level
- **Longest Distances:** 306 mi (492 km) N-W / 235 mi (378 km) E-W
- **Longest River:** Danube River, 1,729 mi (2,783 km)
- **Largest Lake:** Lake Scutari, 150 sq mi (400 sq km).
- **Natural Hazards:** Earthquakes
- **Population:** 10,677,290 (July 2001 est.) / World Rank: 69
- **Capital City:** Belgrade, in north central Yugoslavia.
- **Largest City:** Belgrade, 1,168,454 (2001 census)

OVERVIEW

Yugoslavia, in southern Europe, covers the middle of the Balkan Peninsula and extends westward to meet the Adriatic Sea. The southern half of Yugoslavia is rugged and mountainous, but in the north is the Danube River basin and the southern extent of the Pannonian Plain.

When Yugoslavia was originally formed in 1918 (known at that time as the Kingdom of Serbs, Croats, and Slovenes), it was much larger than it is currently. Starting in 1991, many of the republics that had made up Yugoslavia broke away and became independent countries. The Former Yugoslav Republic of Macedonia, Bosnia and Herzegovina, Slovenia, and Croatia were all part of Yugoslavia. Modern Yugoslavia consists of the two remaining republics: Serbia, making up the eastern 86 percent of the country; and coastal Montenegro, which occupies the southwestern 14 percent. Within Serbia there are two nominally autonomous provinces: Kosovo (4,203 sq mi / 10,887 sq km) in the south; and Vojvodina (8,303 sq mi / 21,506 sq km) in the north. Kosovo was the scene of much bloodshed and ethnic conflict in the late 1990s, prompting intervention by the NATO military alliance and the deployment of NATO and Russian peacekeepers.

Located on the Eurasian Tectonic Plate, Yugoslavia is seismically active. Two parallel thrust fault lines extend from northwest to southeast Montenegro. Serbia has thrust fault lines on either side of the Velika Morava and Južna Morava river basins. There is also a tectonic contact line along the eastern border with Romania. These structural seams in the earth's crust periodically shift, causing earth tremors and occasional destructive earthquakes. In 1979, a major earthquake severely damaged towns and villages in the southern highlands and also caused great devastation along the coast.

MOUNTAINS AND HILLS

Mountains

About half of Serbia is covered by mountains. Serbia is ringed by the Dinaric Alps on the west, the Sar Mountains and the North Albanian Alps (or Prokletije) on the south, and the Balkan Mountains and (just across the border in Romania) the Transylvanian Alps on the east. There are many peaks exceeding 6,000 feet (1,800 m) above sea level, including 13 mountains that are over 7,870 ft (2,400 m).

Nearly all of Montenegro is mountainous. The name *Montenegro* (Black Mountain) is believed to come from the thick "black" forests that once covered the area. The high Dinaric Alps of Montenegro rise steeply from the Adriatic coastline, separating a narrow ribbon of coastal plain only 1 to 6 mi (2 to 10 km) wide from the interior.

The four highest peaks in Yugoslavia are all in Serbia: Daravica, 8,714 ft (2,656 m); Crni Vrh, 8,481 ft (2,585 m); Gusan, 8,330 ft (2,539 m); and Bogdaš, 8,311 ft (2,533 m). Bobotov Kuk, 8,275 ft (2,522 m), the next highest mountain, lies in Montenegro.

Canyons

Tara Canyon follows the Tara River along Montenegro's northwestern border with Bosnia and Herzegovina. At 4,265 ft (1,300 m), Tara Canyon is Europe's deepest canyon. There are also canyons in the Piva and Morača

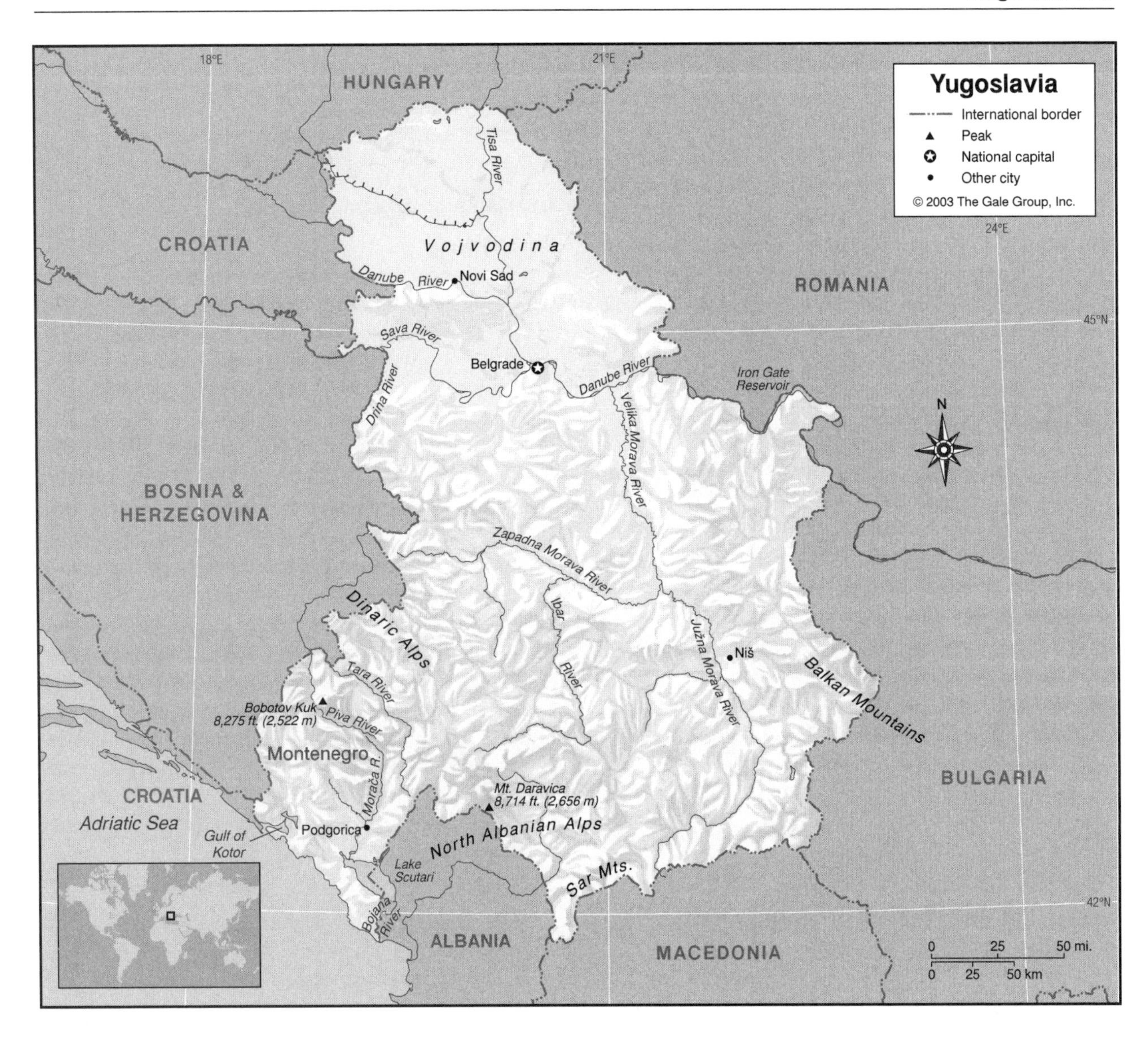

river basins of Montenegro that are about 3,940 ft (1,200 m) deep.

INLAND WATERWAYS

Lakes

Lake Scutari (Skadar), only 4 mi (7 km) from the Adriatic coast in Montenegro, is the one of 40 lakes in Montenegro and is by far the largest lake in Yugoslavia (as well as in the entire Balkan region). Covering approximately 150 sq mi (400 sq km), about two-thirds of the lake lies within Yugoslavia's borders, with the rest in Albania. Although its surface area is large, its average depth is only 16 ft (5 m).

High mountains rise to the southwest of the lake, while to the northeast is a wide swamp. Although Scutari is adjacent to the Adriatic Sea, there are about 30 spots, known as *oke* (singular: *oko*) where its bed is under sea level and groundwater springs forth from the bottom of the lake. The Morača River is the largest stream flowing into Lake Scutari.

Rivers

Most of Yugoslavia's rivers flow eastward towards the Black Sea basin. Yugoslavia's most important river is the Danube, which forms part of the border with Croatia, then flows across northern Serbia and then along the border with Romania. The Danube River is the second longest river in Europe, rising in Germany and flowing southeast into the Black Sea. The total length of the Danube River is 1,729 mi (2,783 km), of which 365 mi (588 km) flows through Yugoslavia. As the only major European river to flow west to east, the Danube has long been an important commercial and military transportation route, and is a vital link between Yugoslavia and the rest of Europe.

Along Yugoslavia's northeastern border with Romania the Danube flows through the Iron Gate. This is a

GEO-FACT

At Krivosije, above the Bay of Kotor, the average annual rainfall is 207 in (532 cm), making it the rainiest place in Europe. However, since the porous limestone of the region offers negligible water retention, the amount of groundwater available for potable usage in the vicinity is limited.

gorge with rapids where the Danube cuts through the Transylvanian Alps. In 1972, the joint Yugoslav-Romanian Iron Gate Dam, with its two hydroelectric plants, was completed at the gorge. With the help of this dam and other engineering feats the Danube is navigable throughout Yugoslavia.

The Danube's main tributaries in Yugoslavia are the Tisa, Sava, and Morava rivers. The Tisa River is 600 mi (966 km) long, of which 103 mi (168 km) flows through Yugoslavia. It enters the country from Hungary and flows south across the Pannonian Plain to the Danube.

The Sava River is 587 mi (945 km) long, entering the country from Bosnia and Herzegovina and flowing east for 128 mi (206 km) before meeting the Danube at Belgrade (Beograd). The Drina is a major tributary of the Sava, and makes up part of Yugoslavia's border with Bosnia and Herzegovina.

The Zapadna Morava (191 mi / 308 km), flowing eastward, and the Južna Morava (183 mi / 295 km), flowing towards the north, merge to form the Velika Morava (115 mi/185 km) near the center of the country. The Morava Rivers and their tributary the Ibar drain the mountainous areas of central and southern Serbia and flow northward to join the Danube east of Belgrade.

THE COAST, ISLANDS, AND THE OCEAN

With the independence of Croatia, Yugoslavia lost most of its coastline. Only 124 mi (199 km) of Adriatic coastline remains. The coast is indented with numerous bays and coves. The largest, and most impressive is the Gulf of Kotor, which is the world's southernmost fjord. The Dinaric Alps rise very close to the water in most places. Only 32 mi (52 km) of coast can be considered beach. Velika plaza (Long Beach) at Ulcinj has the longest continuous stretch of sandy beach, at 8 mi (13 km).

Since the coastline is so rugged, access to the sea from inland is difficult. The port of Bar and the Gulf of Kotor are the main access points. At about 125 mi (200 km), the Adriatic Sea is at its widest between Yugoslavia and southern Italy. This portion of the Adriatic is also the deepest, reaching some 4,360 ft (1,330 m) at a point about 75 mi (120 km) southwest of the Gulf of Kotor.

CLIMATE AND VEGETATION

Temperature

Yugoslavia's climate varies greatly from one part of the country to another, due to the many mountain ranges. The climate of Serbia is on the whole continental, with cold, dry winters and warm, humid summers. The Pannonian Plains have cold winters with hot and dry summers. In Vojvodina, July temperatures average 70°F (21°C), and temperatures in January average about 32°F (0°C).

The Adriatic coast has a more Mediterranean climate, but the Dinaric Mountains prevent the Mediterranean weather from penetrating inland Montenegro. The average seaside July temperatures are between 74.1°F (23.4°C) and 78.1°F (25.6°C). Summers are usually long and dry, winters short and mild. Intense summer heat penetrates the Bojana River valley, over the Lake Scutari basin, and upstream along the Morača River. Podgorica, on the Morača, is the warmest city in Yugoslavia, with July temperatures averaging 79.5°F (26.4°C). The absolute maximum can sometimes reach 104°F (40°C). The average January temperatures are around 41°F (5°C), with the absolute minimum of 14°F (-10°C).

Rainfall

Annual precipitation in Serbia ranges from 22 to 75 inches (56 to 190 cm), depending on elevation and exposure. Heavy rains in spring and autumn frequently cause floods. Snow is rare along the Montenegrin coast and in the Lake Scutari basin, but just inland, in the region of high limestone mountains, the climate is typically subalpine, with cold, snowy winters and mild summers. Snow can be found well into summer, and sometimes even year round, on the highest mountain peaks.

Grasslands

Occupying northern Serbia is the Pannonian Plain and the low-lying plains of Vojvodina, where the Danube River is joined by two of its major tributaries, the Sava and Tisa rivers. The region is mostly flat, with some low hills, and contains fertile soils used for farmland and grazing. The Pannonian Plain is in an ancient dry seabed. It is filled with rich alluvial deposits, forming fertile farmland and rolling hills. Kosovo, at the southern end of the country, covers a montane basin with high plains.

Forests and Jungles

The government estimated the total forested area at 6,249,000 acres (2,529,000 hectares), of which 4,902,000 acres (1,984,000 hectares) were in Serbia and 1,347,000 acres (545,000 hectares) were in Montenegro. Yugoslavia's

Autonomous Republics – Yugoslavia

1998 POPULATION ESTIMATES

Name	Population	Area (sq mi)	Area (sq km)	Capital
Kosovo I Metohija (Kosovo-Metohija)	2,222,000	4,203	10,887	Pristina
Montenegro (Crna Gora)	647,000	5,333	13,812	Podgorica
Serbia	5,780,000	21,609	55,968	Belgrade
Vojvodina	1,968,000	8,303	21,506	Novi Sad

SOURCE: Federal Statistical Office, Yugoslavia.

nine national parks cover 618,700 acres (250,400 hectares), of which 64 percent is in Serbia. The forests in Serbia are composed of about 170 broadleaf and 35 coniferous species of trees and shrubs.

HUMAN POPULATION

Yugoslavia's official population was 10,633,000 in 2000, of whom 9,979,000 lived in Serbia and 654,000 resided in Montenegro. Of the total inhabiting Serbia, 2,291,000 lived in Kosovo and 1,946,000 resided in Vojvodina. Serbia's largest cities are Belgrade, Niš, Kragujevac, and Èaèak. Montenegro's largest city is Podgorica, and Vojvodina's largest city is Novi Sad. Priština (33,305) is the largest city in Kosovo.

NATURAL RESOURCES

The fertile plains of Vojvodina supply grain and sugar beets, while the hilly central areas of Serbia specialize in dairy, fruit, and livestock. Yugoslavia has deposits of oil, gas, coal, antimony, copper, lead, zinc, nickel, gold, pyrite, and chrome. Iron and copper deposits are located in central Serbia; coal, lead, and zinc reserves are found in Kosovo; and oil reserves are located in Vojvodina.

FURTHER READINGS

Anzulovic, Branimir. *Heavenly Serbia: from Myth to Genocide.* New York: New York University Press, 1999.

Brân, Zoë. *After Yugoslavia.* Oakland, Calif.: Lonely Planet, 2001.

Federal Republic of Yugoslavia. http:// www.gov.yu (accessed June 14, 2002).

Malcolm, Noel. *Kosovo: A Short History.* New York: HarperPerennial, 1999.

Radovanovic, Ivana. *The Iron Gates Mesolithic.* Ann Arbor, Mich.: International Monographs in Prehistory, 1996.

Zambia

- **Area:** 290,586 sq mi (752,614 sq km) / World Rank: 40
- **Location:** Southern and Eastern Hemispheres, in southern Africa, bordering the Democratic Republic of the Congo to the north and west, Tanzania to the northeast, Angola to the west, Namibia and Zimbabwe to the south, Mozambique to the south and east, and Malawi to the east.
- **Coordinates:** 15°00′S, 30°00′E
- **Borders:** 3,519 mi (5,664 km) / Angola, 690 mi (1,110 km); Democratic Republic of the Congo, 1,199 mi (1,930 km); Malawi, 520 mi (837 km); Mozambique, 260 mi (419 km); Namibia, 145 mi (233 km); Tanzania, 210 mi (338 km); Zimbabwe, 495 mi (797 km)
- **Coastline:** None
- **Territorial Seas:** None
- **Highest Point:** Location in Mafinga Hills, 7,549 ft (2,301 m)
- **Lowest Point:** Zambezi River, 1,079 ft (329 m)
- **Longest Distances:** 749 mi (1,206 km) E-W / 506 mi (815 km) N-S
- **Longest River:** Zambezi, 1,700 mi (2,735 km)
- **Largest Lake:** Lake Tanganyika, 12,700 sq mi (32,893 sq km)
- **Natural Hazards:** Tropical storms
- **Population:** 9,770,199 (July 2001 est.) / World Rank: 80
- **Capital City:** Lusaka, central southern Zambia
- **Largest City:** Lusaka, 1,200,000 (2002 est.)

OVERVIEW

Zambia, a vast country slightly larger than Texas, is situated in the tropical south-central portion of Africa, atop a plateau ranging from 3,000 to 4,500 feet (910 to 1,370 m) above sea level. Highest elevations are found in the northeast of the country, which is home to the Muchinga Mountains. The mountains and plateaus recede as the land is cut by the Luangwa River in the east and the Kafue River in the west, both tributaries of the Zambezi, which flows to the south of the country through the wondrous Victoria Falls and the artificial Lake Kariba. Several natural lakes, Mweru, Bangweulu, and Tanganyika, are located in the north of the country near the border with Tanzania. Nineteen national parks dot the entire landscape, in efforts to preserve the country's diverse wildlife.

Zambia is located on the African Tectonic Plate, right along the Great Rift Valley.

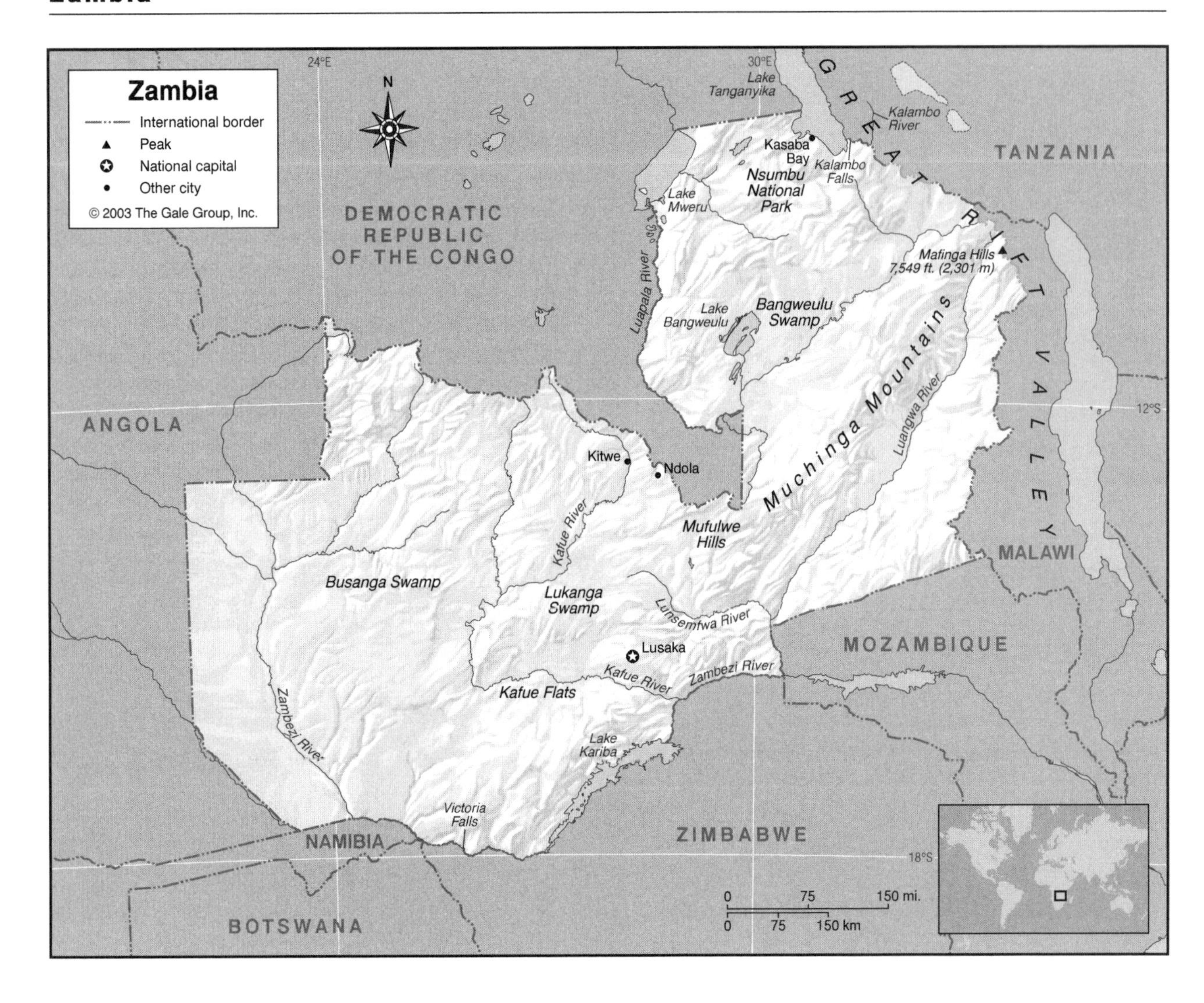

MOUNTAINS AND HILLS

Mountains

The highest points in Zambia are found in the north-east corner on the country, along the borders with Tanzania and Malawi. Most significant are the Mbala Highlands near Tanzania, the Mafingi Mountains and the Copperbelt Highlands near Malawi, and the Muchinga Mountain Range. The highest point in Zambia, an unnamed location at 7,549 ft (2,301 m), is located in the Mafinga Hills.

Plateaus

Most of Zambia lies on a portion of the great plateau that dominates central and southern Africa's landmass. Some of the plateau is undulating, some relatively flat, but most of it ranges between 2,952 and 4,921 ft (900 and 1,500 m), the higher sections above 3,937 ft (1,200 m), occurring, for the most part, in the north. Vegetation on the plateau is primarily open woodland or savanna, dominated by acacia and baobob trees. Also prevalent are thorn trees and bushes. Vegetation becomes more dry and sparse as one travels south, but in some northeastern areas, tropical flora such as orchids can be found.

Hills and Badlands

The significant areas of lower land are rift valleys in the east (the Luangwa River Valley) and in the south (the middle Zambezi River Valley), both bounded by escarpments. The Luangwa Valley hosts a variety of birds, including the kingfisher, lark, stork, red-billed quela, and even the occasional ostrich. Safari wildlife is also prevalent in this area, including the giraffe, zebra, rhinoceros, elephant, baboon, hyena, and lion.

INLAND WATERWAYS

Lakes

Zambia has three significant natural lakes, the Bangweulu, Mweru, and Tanganyika, which are all located in the northern reaches of the country near the borders with Tanzania and the Democratic Republic of the Congo. Lake Tanganyika is the largest overall. Only the southern end of this long, narrow, lake is within Zambia. Lake

Mweru is a much smaller and shallower fresh water basin on the border with the Democratic Republic of the Congo. Both Lakes Mweru and Tanganyika are known for their abundance of fish; more than 300 species of fish live in their waters. Lake Bangweulu is the smallest of the three northern lakes, but is the largest found entirely within Zambia, with a surface area of 3,000 sq mi (9,840 sq km). It is drained by the Luapala River.

Zambia also boasts an artificial lake on its southern border, Lake Kariba, which was formed with the creation of a dam on the Zambezi River in 1960. With a surface area of 2,124 sq mi (5,500 sq km), Kariba is one of the largest artificial lakes in the world.

Rivers

Most of Zambia's streams, except for those in part of the eastern lobe, ultimately drain into the Indian Ocean via the Zambezi River and its main tributaries. In addition to those streams that enter the Zambezi directly, there are three main tributary systems, those of the Kafue, Luangwa, and Lunsemfwa Rivers.

With a total length of 1,700 mi (2,735 km), the Zambezi River is the longest river in Zambia. The upper Zambezi, running roughly from north to south, has a low gradient, and the area through which it passes is marked by floodplains and swamps. After turning eastward, the Zambezi flows over Victoria Falls and through the middle Zambezi Valley, much of it occupied by the great artificial Lake Kariba. Much of the upper Kafue River is also characterized by a low gradient, and extensive swamps are common. The Lunsemfwa drains a small portion of Zambia between near the Luangwa Valley and joins the Luangwa River at a steeper gradient than most of Zambia's streams.

The flow of all watercourses in Zambia is affected by the clear demarcation between rainy and dry seasons. Most small streams dry up sometime between May and October, and even the larger rivers show a substantial difference between maximum (occurring variously between February and May) and minimum discharges.

The famous Victoria Falls, straddling the border between Zimbabwe and Zambia, are majestic waterfalls that reach 350 ft (106 m) high at their maximum and nearly a mile (nearly 1.5 km) wide. The average flow over Victoria Falls is 38,000 cu ft per sec (1,090 cu m per sec), though it varies widely from year to year.

In addition to the world-renowned Victoria Falls, there are two other significant waterfalls located in Zambia. Ngonye or Sioma Falls, located just 185 mi (300 km) from Victoria Falls, features horseshoe-shaped drops that carry over 10,000 cubic feet (283 cu m) of water per second. East of the Nsumbu National Park are the Kalambo Falls, which flow from the Kalambo River into Lake Tanganyika. These falls are the second highest continuous waterfalls on the continent, plunging 725 ft (221 m).

GEO-FACT

The world famous Victoria Falls are referred to by local Zambians as *Mosi-oa-Tunya*, which means "the smoke that thunders." Thrill seekers often flock to the falls for white water rafting, jet boating, canoeing, and bungee jumping. The rapids beneath the falls are among the safest in the world, due in part to their expansive width.

Wetlands

Lake Bangweulu and several smaller bodies of water are part of the Bangweulu Swamp complex, the largest swamp area in the country. The main swamp is permanently flooded; the periphery, a belt about 24 mi (40 km) wide, is flooded during and immediately after the rainy season. These are the most important wetland areas in the country, housing a wealth of bird life, including the rare shoebill, which looks much like the extinct dodo bird. Also found here are more than 150 species of reptiles and thousands of different insect types, many of which are peculiar to the region. This region also features small islands composed of floating vegetation that appear during the wet seasons. In the western part of the country, northwest of Lusaka, is a large marsh area that includes Busanga Swamp and Lukanga Swamp.

THE COAST, ISLANDS, AND THE OCEAN

The Coast and Beaches

Zambia is landlocked, with no direct access to the ocean. The coast of Lake Tanganyika is marked by Kasaba, Nkamba, and Ndole Bays.

CLIMATE AND VEGETATION

Temperature

Because of its high altitude, most of the country enjoys a pleasantly temperate climate. Low-lying areas such as the valleys of the Zambezi, Luangwa, and Kwafe Rivers and the shores of the country's lakes, have the highest temperatures in the country. The hottest months of the year are August through October, when daily temperatures often reach a high of 86° to 89°F (30° to 32°C). The months of May through July are only slightly cooler, with temperatures ranging from 63° to 79°F (17° to

Population Centers – Zambia

(1990 CENSUS OF POPULATION)

Name	Population	Name	Population
Lusaka (capital)	982,362	Chingola	162,954
Ndola	376,311	Mufulira	152,944
Kitwe	338,207	Luanshya	146,275
Kabwe	166,519		

SOURCE: United Nations Statistics Division.

Provinces – Zambia

POPULATIONS FROM 1990 CENSUS

Name	Population	Area (sq mi)	Area (sq km)	Capital
Central	721,000	36,446	94,395	Kabwe
Copperbelt	1,427,000	12,096	31,328	Ndola
Eastern	966,000	26,682	69,106	Chipata
Luapula	525,000	19,524	50,567	Mansa
Lusaka	987,000	8,454	21,896	Lusaka
Northern	855,000	57,076	147,826	Kasama
North-Western	388,000	48,582	125,827	Solwezi
Southern	907,000	32,928	85,283	Livingstone
Western	607,000	48,798	126,386	Mongu

SOURCE: Central Statistics Office, Zambia. Cited by Johan van der Heyden, Geohive, http://www.geohive.com (accessed June 2002).

26°C). At night, however, temperatures may drop as low as 41°F (23°C).

Rainfall

The rainy season is long, beginning in the middle of November and lasting until April, throughout which heavy tropical storms are prevalent. Rainfall is generally highest in the northern provinces of Zambia, decreasing as one travels south. Average annual rainfall is about 50 in (125 cm) in the north and only 30 in (75 cm) in the south. The capital city of Lusaka receives approximately 32 in (81cm) of rainfall each year.

Grasslands

The Kafue Flats, through which the Kafue River passes after turning east, are too poorly drained to provide land for crop cultivation but furnish good pasture for livestock. The wide, grassy areas are known as *dambos* and feed area wildebeests, zebras, and buffalo. The Flats are also home to lions and cheetahs, who feed on these animals.

Forests and Jungles

The most common woodlands are of a tropical savanna type, and are found primarily on the plateau areas that dominate the northeast of the country. Thin forests grow in the north and the east of the country, with more lush savanna regions found along the plateaus. The southwest of the country near Angola has forests of Zambian teak. The western province has many deciduous trees tall and dense grass undergrowths. Higher rainfall regions support eucalyptus trees. The Chipya region in the west, which extends into Angola, houses evergreen trees, which grow continuously since the area remains well-watered year round.

HUMAN POPULATION

Almost half of the population lives in urban areas; the majority resides in or near the capital of Lusaka. Nearly 99 percent of the population of Zambia is of Bantu descent, although a few Europeans and people of other nationalities reside in the country as well. Among the groups of Bantu, the Bemba group is the most prevalent, accounting for approximately 37 percent of Zambia's population. Members of the Bemba group mainly reside in the northeast and in the Copper Belt region of the north-central area. Members of the Tongo group inhabit the south of Zambia, while the Nyanja live in the east and the Lozi reside in the west.

NATURAL RESOURCES

Zambia is rich in copper, which provides the country with much of its export economy. The copper is found in the central highlands region north of the Lukanga Swamp and northwest of the Mufulwe Hills. The country is also rich in cobalt, of which it is the second-largest producer in the world. Lead and zinc mines are located near the Kabwe River. Coal, emeralds, amethysts, silver, gold, selenium, and feldspar are also found in lesser amounts throughout the country. Hydropower is also a rich resource, providing for 70 percent of the country's electricity needs. Hydroelectric power is tapped from the Kafue River and Victoria Falls.

FURTHER READINGS

Burdette, Marcia. *Zambia: Between Two Worlds*. Boulder, Colo.: Westview Press, 1988.

Cox, Thorton. *Traveler's Guide to East Africa*. New York: Hastings House, 1980.

Holmes, Timothy. *Zambia*. New York: Benchmark Books, 1998.

McIntyre, Chris. *Guide to Zambia*. Chalfont St. Peter: Bradt, 1999.

Naipal, Shiva. *North of South: An African Journey*. New York, Penguin Books, 1996.

Zimbabwe

- **Area:** 150,804 sq mi (390,580 sq km) / World Rank: 61
- **Location:** Southern and Eastern Hemispheres, in southern Africa, bordered in the east by Mozambique, in the south by South Africa, in the west by Botswana, and in the north by Zambia.
- **Coordinates:** 20°00′S, 30°00′E
- **Borders:** 1,905 mi (3,066 km) / Botswana, 505 mi (813 km); Mozambique, 765 mi (1,231 km); South Africa, 140 mi (225 km); Zambia, 495 mi (797 km)
- **Coastline:** None
- **Territorial Seas:** None
- **Highest Point:** Inyangani, 8,504 ft (2,592 m)
- **Lowest Point:** Junction of the Runde and Save Rivers, 531 ft (162 m)
- **Longest Distances:** 529 mi (852 Km) WNW-ESE; 710 mi (1,223 Km) NNE-SSW
- **Longest River:** Zambezi River, total length 1,650 mi (2,650 km)
- **Largest Lake:** Lake Kariba, 3,000 sq mi (7,770 sq km)
- **Natural Hazards:** Drought
- **Population:** 11,365,366 (July 2001 est.) / World Rank: 66
- **Capital City:** Harare, located in the central northeast of the country
- **Largest City:** Harare, population 1,800,000 (2000 est.)

OVERVIEW

Most of Zimbabwe is comprised of a granite plateau, with more than 75 percent of it lying between 2,000 and 5,000 ft (600 and 1,500 m). Much of the plateau surface is an extensive veld, a rolling plain covered with a mixture of grasses and open woodlands. The high plateau, known as the Highveld, is approximately 400 mi (650 km) long by 50 mi (80 km) wide. The high altitude portion of the country is a central ridge forming the country's watershed, with streams flowing southeast to the Limpopo and the Sabi Rivers and northwest into the Zambezi River. Only the largest rivers in Zimbabwe maintain a year-round flow of water. The Highveld slopes gently downward from a central upland region through a Middleveld region to considerably lower plains areas—the Lowveld—near the country's borders.

Zimbabwe is located on the African Tectonic Plate.

MOUNTAINS AND HILLS

Mountains

In north-central Zimbabwe, the broad expanse of Highveld breaks up into several arms, one of which joins a lesser watershed ridgeline, extending southeastward from the Harare region to the mountains around Umtalis along the eastern border. The eastern mountain complex is the highest in the country. Most peaks are between 6,000 and 8,000 ft (1,828 and 2,368 m) in elevation; the loftiest, Mount Inyangani at 8,504 ft (2,592 m), is the tallest mountain in Zimbabwe. The other arm extends north from Harare as the Umvukwe Range, which meets the Zambezi Escarpment in the far north.

Plateaus

The plateaus of Zimbabwe are divided into three sections, the Highveld (high altitude), Middleveld (medium altitude), and the Lowveld (low altitude). The Highveld stretches northeast to southeast at elevations of 4,000 to 5,500 ft (1,219 to 1,675 m), reaching Mount Inyangani. Its surface features vary from relatively smooth or rolling to rough, almost mountainous terrain. The Middleveld (middle altitude) areas are located on either side of the Highveld, and range from 2,000 to 4,000 ft (600 to 1,200 m) in height.

Hills and Badlands

Both the Highveld and Lowveld regions contain rocky hills and buttes known locally as Kopjes (Hills). The central high altitude areas are marked by a massive extrusion of ancient lava that extends from the northeast to the southwest for 300 mi (482 km) called the Great Dike Hills. Only a few miles wide in most segments, it rises as much as 1,500 ft (457 m) above the surrounding Highveld in a series of eroded ridges. Various segments of this long series of ridges are known as the Doro Range, the Selondi Range, the Mashona Hills, and the Umvukwe Range. The Umvukwe is the northernmost and highest segment of the Great Dike, reaching to more than 5,600 ft (1,706 m) above sea level.

INLAND WATERWAYS

Lakes

On the Zambezi River about 300 mi (483 km) downstream from Victoria Falls, a dam, completed in 1959, created Lake Kariba. Approximately 300 mi (483 km) in length, it is one of the largest artificial lakes in the world. Lake Kariba, with an area of 3,000 sq mi (7,770 sq km) lies on the border between Zimbabwe and Zambia and is the largest lake in Zimbabwe.

Rivers

Except for a small area in the southwest, Zimbabwe's territory is drained by three rivers that flow east to the Indian Ocean via Mozambique. Two of them originate outside Zimbabwe—the Zambezi along the Zambian border and the Limpopo in South Africa. The headwaters of the Sabi, the third major river, are south of Harare on the eastern slopes of the Highveld.

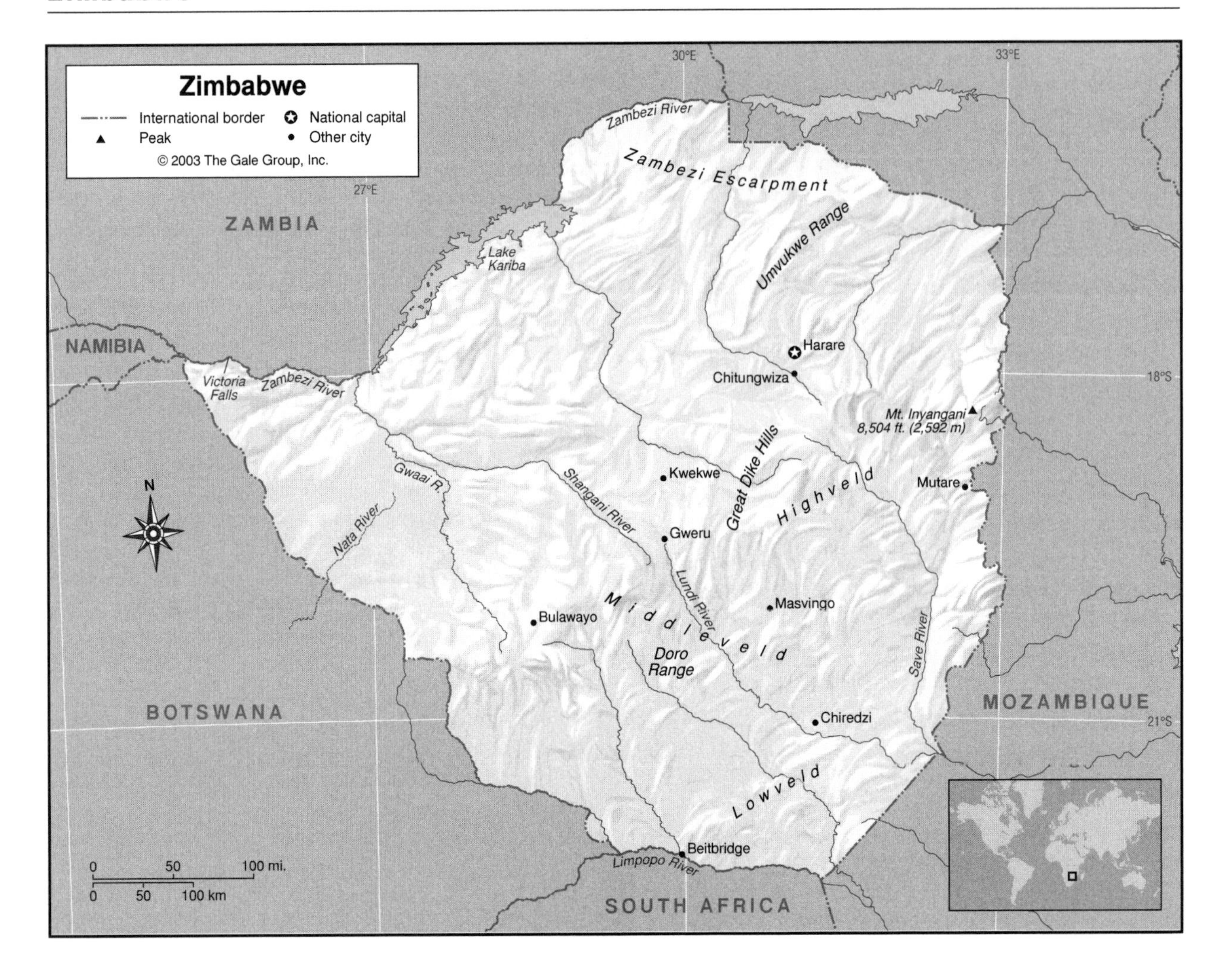

The Zambezi water system, made up of the Zambezi and its tributaries, collect runoff waters from nearly all of the land in the western, northern, and northeastern areas of the country northwest of the central ridgeline of the Highveld. The Zambezi, which marks much of the northern border with Zambia, is the longest of all African rivers flowing to the Indian Ocean. The Zambezi is already carrying a heavy volume of water as it approaches the westernmost part of Zimbabwe. In this area the river drops over Victoria Falls, a cataract 350 ft (106 m) high at its maximum and nearly a mile (nearly 1.5 km) wide. The average flow over Victoria Falls is 38,000 cu ft per sec (1,090 cu m per sec), though this amount varies widely from year to year.

Runoff from most of the eastern and southeastern slopes of the central ridgeline flows generally southeastward through numerous tributaries of the Limpopo and the Sabi rivers. Although it is shorter and carries a lesser volume of water than the Zambezi, the Limpopo, forming the border between Zimbabwe and South Africa, is also one of Africa's major rivers. Its gradient is relatively even except in the Beitbridge area, where it flows through an extensive gorge marked by rapids and several waterfalls. Numerous tributaries carry a heavy volume of water from upper levels of the veld, where annual rainfall is heavier than in the Lowveld areas near the Limpopo River itself.

Southeastern Zimbabwe is drained by the Sabi River and its tributaries. Near the town of Chiredzi, it is joined by the Lundi River, which collects water from the Gweru area. It turns eastward at the Mozambique border, eventually reaching the Indian Ocean.

In an area of low rainfall near the southwestern border with Botswana, the Nata River and a few other small intermittent streams carry the annual runoff to an internal drainage basin in Botswana.

THE COAST, ISLANDS, AND THE OCEAN

Zimbabwe is landlocked.

CLIMATE AND VEGETATION

Temperature

The temperature of Zimbabwe is greatly affected by the altitude. The eastern high altitudes are cooler than the lower areas. Average temperatures in the high altitudes range from 54° to 55°F (12° to 13°C) in the winter and are around 75°F (24°C) in the summer. In the lower altitudes, temperatures are usually 11°F (6°C) higher than the higher altitude areas.

Rainfall

The summer rainy season lasts from November to March. It is followed by a transitional season, during which temperature and rainfall decreases. The cool dry season follows, lasting from mid-May to mid-August. Finally, there is a warm dry season, which lasts until the onset of the rainy season.

Altitude also affects the rainfall in Zimbabwe. The eastern mountainous regions receive more than 40 in (100 cm) annually. The capital city of Harare receives approximately 32 in (81 cm) of rainfall per year. The southern and southwestern regions of the country receive very little rain. Seasonal water shortages are commonplace in the southwestern region.

Grasslands

The highest area of grass and woodlands, the Highveld, is between 50 and 100 mi (80 to 160 km) wide and extends from northeast to southwest for 400 mi (643 km) across the center of the country. On its northwestern and southeastern slopes, the Highveld merges imperceptibly into the medium altitude wooded grasslands (Middleveld), generally defined as the area between 3,000 and 4,000 ft (914 and 1,219 m) in elevation. In the southeast, the Middleveld is no more than 75 mi (120 km) wide in most areas, narrowing to less than 50 mi (80 km) northeast of Masvingo. On the opposite slope of the Highveld in western and northwestern Zimbabwe, the Middleveld covers a much larger area, having a breadth of 150 mi (210 km) or more in the Gwaai and Shangani River Valleys northwest of Bulawayo.

Below 2,000 ft (600 m) are areas called the Lowveld, made up of wide grassy plains. In the southeast the Lowveld, which is generally considered to include the land below 3,000 ft (914 m), extends from the nominal edge of the Middleveld to the southern and southeastern borders, covering nearly one-fifth of the territory. In the northwest and the north the Lowveld is divided into three major sections, partially separated by escarpments and local ranges of hills. These sections slope directly to the Zambezi River or to the shoreline of Lake Kariba.

Forests

Most of the country is savanna, though the mountainous east supports tropical evergreen and hardwood forest. Tree species include mahogany, teak, knobtorn, and baobab. Exotic trees such as eucalyptus and saligna gum are used for power transmission poles. The people of Zimbabwe depend on trees as their main source of fuel, light, and heat. The rapid deforestation in Zimbabwe has caused severe soil erosion and river siltation.

Population Centers – Zimbabwe

(1992 CENSUS OF POPULATION)

Name	Population
Harare	1,189,103
Bulawayo	621,742
Chitungwiza	274,912
Mutare	131,367
Gweru	128,037

SOURCE: United Nations Statistics Division.

HUMAN POPULATION

The overall population density is 78 people per sq mi (30 people per sq km). Approximately 35 percent of the people lived in urban areas in 2000 as compared to 22 percent recorded in 1980. Between 60 percent and 70 percent of the population live in the rural areas of the country despite the poor land conditions and frequent droughts.

Africans make up 98 percent of the population of Zimbabwe and are mainly related to the two major Bantu-speaking groups, the Shona and the Ndebele. The Shona group makes up about 71 percent of the population and the Ndebele comprises 16 percent of the population. Other groups account for approximately 11 percent of the African populace, which includes groups like the Tonga in the north and the Sotho, Venda, and Hlengwe along the southern border. Whites make up 1 percent of the non-African population. Asian and peoples of mixed ancestry make up the remaining 1 percent.

NATURAL RESOURCES

Zimbabwe is rich in many natural resources, the major ones being gold, chrome, coal, copper, and nickel. Reserves of chromium ore, asbestos, iron ore, vanadium, lithium, tin, and platinum are also found in Zimbabwe. Agriculture is also important to Zimbabwe's economy, with major exports being cotton, tobacco, sugar, and tea. However, Zimbabwe's commercial farmers continue to be mostly white, which leaves most black Africans as subsistence farmers. Some black Africans own their farms, but most of them hold land in poor and densely populated areas.

FURTHER READINGS

Fromentin, Eughne. *Between Sea and Sahara: An Algerian Journal.* Tr. Blake Robinson. Athens: Ohio University Press, 1999.

Matowanyika, Joseph Zano Zvapera. *The History of Land Use in Zimbabwe from 1900.* http://www.lead.org/lead/training/international/zimbabwe/1997/papers/d1matowan.html (accessed March.15, 2002).

McCrea, Barbara, and Tony Pinchuck. *Zimbabwe: The Rough Guide.* 3rd ed. London and New York: Rough Guides, 1997.

Ranger, Terence. *Voices from the Rocks: Nature, Culture & History in the Matopos Hills of Zimbabwe.* Bloomington: Indiana University Press, 1999.

WORLD RANKINGS APPENDIX

Continents by area

Rank	Continent	Area (sq mi)	Area (sq km)
1	Asia	17,139,445	44,391,162
2	Africa	11,677,239	30,244,049
3	North America	9,361,791	24,247,039
4	South America	6,880,706	17,821,029
5	Antarctica	5,500,000	14,245,000
6	Europe	3,997,929	10,345,636
7	Australia	2,967,909	7,686,884

Islands by area

All measurements are approximate

Rank	Island	Continent	Body of water	Area (sq mi)	Area (sq km)
1	Greenland	N. America	Atlantic Ocean	840,000	2,175,600
2	New Guinea	Oceania	Pacific Ocean	305,000	790,000
3	Borneo	Asia	South China Sea	285,000	737,000
4	Madagascar	Africa	Indian Ocean	226,657	587,040
5	Baffin	N. America	Baffin Bay	196,000	507,000
6	Sumatra	Asia	Andaman Sea	164,000	425,000
7	Honshu	Asia	Pacific Ocean	88,000	228,000
8	Great Britain	Europe	North Sea	84,400	219,000
9	Victoria	N. America	Viscount Melville Sound	83,900	217,000
10	Ellesmere	N. America	Arctic Ocean	75,800	196,000
11	Sulawesi (Celebes)	Asia	Celebes Sea	67,400	174,000
12	South Island (New Zealand)	Oceania	Pacific Ocean	58,200	151,000
13	Java	Asia	Indian Ocean	50,000	129,000
14	North Island (New Zealand)	Oceania	Pacific Ocean	44,200	114,000
15	Newfoundland	N. America	Atlantic Ocean	42,000	109,000
16	Cuba	N. America	Caribbean Sea	40,500	105,000
17	Luzon	Asia	Pacific Ocean	40,400	105,000
18	Iceland	Europe	Atlantic Ocean	39,769	103,000
19	Mindanao	Asia	Pacific Ocean	36,500	94,600
20	Ireland	Europe	Atlantic Ocean	32,500	84,100
21	Hokkaido	Asia	Pacific Ocean	30,100	78,000
22	Sakhalin	Asia	Sea of Okhotsk	29,500	76,400
23	Hispaniola	N. America	Atlantic Ocean	29,200	75,600
24	Banks	N. America	Arctic Ocean	27,000	70,000
25	Sri Lanka	Asia	Indian Ocean	25,332	65,610
26	Tasmania	Australia	Indian Ocean	24,900	64,400
27	Devon	N. America	Baffin Bay	21,300	55,200
28	Novaya Zemlya	Europe	North Kara Sea	18,900	48,900
29	Grande de Tierra del Fuego	S. America	Atlantic Ocean	18,700	48,400

[continued]

Islands by area

All measurements are approximate

Rank	Island	Continent	Body of water	Area (sq mi)	Area (sq km)
30	Marajo	S. America	Atlantic Ocean	18,500	48,000
31	Alexander	Antarctica	Bellingshausen Sea	16,700	43,200
32	Axel Heiberg	N. America	Arctic Ocean	16,700	43,200
33	Melville	N. America	Viscount Melville Sound	16,300	42,100
34	Southampton	N. America	Husdon Bay	15,900	41,200
35	West Spitsbergen	Europe	Arctic Ocean	15,300	39,500
36	New Britain	Oceania	Bismarck Sea	14,600	37,800
37	Taiwan	Asia	Pacific Ocean	13,892	35,980
38	Kyushu	Asia	Pacific Ocean	13,800	35,700
39	Hainan	Asia	South China Sea	13,100	34,000
40	Prince of Wales	N. America	Viscount Melville Sound	12,900	33,300
41	Novaya Zemlya	Europe	Barents Sea	12,800	33,300
42	Vancouver	N. America	Pacific Ocean	12,100	31,300
43	Timor	Asia	Timor Sea	10,200	26,300
44	Sicily	Europe	Mediterranean	9,810	25,400
45	Somerset	N. America	Lancaster Sound	9,570	24,800
46	Sardinia	Europe	Mediterranean	9,190	23,800
47	Bananal	S. America	Araguaia River	7,720	20,000
48	Halmahera	Asia	Molucca Sea	6,950	18,000
49	Shikoku	Asia	Pacific Ocean	6,860	17,800
50	Ceram	Asia	Banda Sea	6,620	17,200
51	New Caledonia	Oceania	Coral Sea	6,470	16,700
52	Bathurst	N. America	Viscount Melville Sound	6,190	16,000
53	Prince Patrick	N. America	Arctic Ocean	6,120	15,800
54	North East Land	Europe	Barents Sea	5,790	15,000
55	Flores	Asia	Flores Sea	5,520	14,300
56	Oktyabrskoy Revolyutsii	Asia	Arctic Ocean	5,470	14,170
57	Sumbawa	Asia	Indian Ocean	5,160	13,400
58	King William	N. America	Queen Maud Gulf	5,060	13,100
59	Samar	Asia	Pacific Ocean	5,050	13,100
60	Negros	Asia	Sulu Sea	4,900	12,700
61	Palawan	Asia	South China Sea	4,550	11,800
62	Kotelnyy	Asia	Arctic Ocean	4,500	11,700
63	Panay	Asia	Sulu Sea	4,450	11,500
64	Bangka	Asia	Java Sea	4,370	11,320
65	Ellef Ringnes	N. America	Arctic Ocean	4,360	11,300
66	Bolshevik	Asia	Arctic Ocean	4,350	11,270
67	Sumba	Asia	Indian Ocean	4,310	11,200
68	Bylot	N. America	Baffin Bay	4,270	11,100
69	Jamaica	N. America	Caribbean Sea	4,243	10,990
70	Dolak	Asia	Arafura Sea	4,160	10,800
71	Hawaii	Oceania	Pacific Ocean	4,040	10,500
72	Viti Levu	Oceania	Pacific Ocean	4,010	10,400
73	Cape Breton	N. America	Atlantic Ocean	3,980	10,300

[continued]

Islands by area

All measurements are approximate

Rank	Island	Continent	Body of water	Area (sq mi)	Area (sq km)
74	Bougainville	Oceania	Pacific Ocean	3,880	10,000
75	Mindoro	Asia	South China Sea	3,760	9,730
76	Prince Charles	N. America	Foxe Basin	3,680	9,520
77	Kodiak	N. America	Pacific Ocean	3,670	9,510
78	Cyprus	Asia	Mediterranean	3,571	9,250
79	Komsomolets	Asia	Arctic Ocean	3,480	9,010
80	Buru	Asia	Banda Sea	3,470	9,000
81	Corsica	Europe	Mediterranean	3,370	8,720
82	Puerto Rico	N. America	Atlantic Ocean	3,350	8,680
83	New Ireland	Oceania	Pacific Ocean	3,340	8,650
84	Disco	N. America	Davis Strait	3,310	8,580
85	Chiloe	S. America	Pacific Ocean	3,240	8,390
86	Crete	Europe	Mediterranean	3,190	8,260
87	Anticosti	N. America	Gulf of St. Lawrence	3,070	7,940
88	Wrangel	Asia	Chukchi Sea	2,820	7,300
89	Leyte	Asia	Visayan Sea	2,780	7,210
90	Zealand	Europe	Baltic Sea	2,710	7,020
91	Cornwallis	N. America	Barrow Strait	2,700	7,000
92	Wellington	S. America	Trinidad Gulf	2,610	6,750
93	Iturup (Etorofu)	Asia	Pacific Ocean	2,600	6,720
94	Prince of Wales	N. America	Pacific Ocean	2,590	6,700
95	Graham	N. America	Pacific Ocean	2,460	6,360
96	East Falkland	S. America	Atlantic Ocean	2,440	6,310
97	Melville	Asia	Timor Sea	2,400	6,220
98	Novaya Sibir	Asia	East Siberian Sea	2,390	6,200
99	Kerguelen	Antarctica	Indian Ocean	2,320	6,000
100	Andros	N. America	Grand Bahama Bank	2,300	5,960
101	Prince Edward Island	N. America	Gulf of St. Lawrence	2,180	5,660
102	Bali	Asia	Indian Ocean	2,170	5,620
103	Guadalcanal	Oceania	Solomon Sea	2,170	5,630
104	Vanua Levu	Oceania	Pacific Ocean	2,140	5,530
105	Coats	N. America	Hudson Bay	2,120	5,500
106	Santa Ines	S. America	Pacific Ocean	2,120	5,500
107	Lombok	Asia	Indian Ocean	2,100	5,430
108	Chicagof	N. America	Gulf of Alaska	2,080	5,400
109	Madura	Asia	Java Sea	2,040	5,290
110	Amund Ringnes	N. America	Peary Channel	2,030	5,260
111	MacKenzie King Island	N. America	Hazen Strait	1,950	5,050
112	Edge	Europe	Barents Sea	1,940	5,030
113	Faddeyevskiy	Asia	Arctic Ocean	1,930	5,000
114	Espiritu Santo	Asia	Coral Sea	1,930	5,000
115	Caviana	S. America	Atlantic Ocean	1,920	4,970
116	Grande de Gurupa	S. America	Amazon River	1,880	4,860
117	Vendsyssel -Thy	Europe	Skagerrak	1,810	4,680
118	Malaita	Oceania	Pacific Ocean	1,750	4,530
119	Long	N. America	Atlantic Ocean	1,723	4,462
120	Kangaroo	Australia	Indian Ocean	1,680	4,350
121	West Falkland	S. America	Atlantic Ocean	1,680	4,350
122	Hoste	S. America	Pacific Ocean	1,590	4,110
123	Santa Isabel	Oceania	Pacific Ocean	1,550	4,010
124	Berkner	Antarctica	Weddell Sea	1,500	3,880
125	Isabela	S. America	Pacific Ocean	1,450	3,760
126	South Georgia	S. America	Atlantic Ocean	1,450	3,760
127	Euboea	Europe	Aegean Sea	1,410	3,650
128	Majorca	Europe	Mediterranean	1,400	3,630
129	Makira (San Cristóbal)	Oceania	Pacific Ocean	1,350	3,500
130	Vaygach	Europe	Kara Sea	1,310	3,380
131	Adelaide	Antarctica	Bellingshausen Sea	1,270	3,300
132	Waigeo	Asia	Pacific Ocean	1,250	3,250
133	Kolguyev	Europe	Barents Sea	1,240	3,200
134	Riesco	S. America	Strait of Magellan	1,200	3,110
135	Gotland	Europe	Baltic Sea	1,160	3,000
136	Fyn	Europe	Baltic Sea	1,150	2,980
137	Zemlya Georga	Europe	Barents Sea	1,120	2,900
138	Zemlya Aleksandry	Europe	Barents Sea	1,080	2,800
139	Manitoulin	N. America	Lake Huron	1,068	2,766
140	Sarema	Europe	Baltic Sea	1,030	2,680
141	Magdalena	S. America	Moraleda Channel	998	2,580
142	Reunion	Africa	Indian Ocean	970	2,512
143	Navarino	S. America	Nassau Bay	955	2,470
144	Ross	Antarctica	Ross Sea	888	2,300
145	Clarence	S. America	Strait of Magellan	863	2,240
146	Hinn	Europe	Norwegian Sea	849	2,200
147	Bioko	Africa	Bight of Bonny	779	2,020
148	Charcot	Antarctica	Bellingshausen Sea	772	2,000

[continued]

Islands by area

All measurements are approximate

Rank	Island	Continent	Body of water	Area (sq mi)	Area (sq km)
149	Tenerife	Africa	Atlantic Ocean	745	1,930
150	Mauritius	Africa	Indian Ocean	720	1,860
151	Skye	Europe	Sea of the Hebrides	643	1,670
152	Fuerte Ventura	Africa	Atlantic Ocean	642	1,663
153	Zanzibar	Africa	Indian Ocean	640	1,656
154	Lesbos	Europe	Aegean Sea	629	1,630
155	Gran Canaria	Africa	Atlantic Ocean	592	1,530
156	Mexiana	S. America	Atlantic Ocean	592	1,530
157	Sao Luis	S. America	Atlantic Ocean	465	1,200
158	Grande Comore	Africa	Indian Ocean	443	1,147
159	Sao Tiago	Africa	Atlantic Ocean	384	995
160	Pemba	Africa	Indian Ocean	380	984
161	Santa Cruz	S. America	Pacific Ocean	358	927
162	Margarita	S. America	Pacific Ocean	355	920
163	Dahlac	Africa	Red Sea	347	900
164	São Tomé	Africa	Gulf of Guinea	330	854
165	Lanzarote	Africa	Atlantic Ocean	302	782
166	Santo Antao	Africa	Atlantic Ocean	302	782
167	Madeira	Africa	Atlantic Ocean	286	741
168	La Palma	Africa	Atlantic Ocean	256	662
169	Fernandina	S. America	Pacific Ocean	253	655
170	Boa Vista	Africa	Atlantic Ocean	239	620
171	San Salvador	S. America	Pacific Ocean	199	515
172	Djerba	Africa	Gulf of Gabes	197	510
173	San Cristobal	S. America	Pacific Ocean	185	480
174	Fogo	Africa	Atlantic Ocean	184	476
175	Nzwani	Africa	Mozambique Channel	164	424
176	Mayotte	Africa	Mozambique Channel	144	374

Oceans and seas by area

All measurements are rounded to the nearest thousand

Rank	Name	Area (sq mi)	Area (sq km)
1	Pacific Ocean	60,060,000	155,557,000
2	Atlantic Ocean	29,638,000	76,762,000
3	Indian Ocean	26,469,000	68,556,000
4	Southern Ocean	7,848,000	20,327,000
5	Arctic Ocean	5,427,000	14,056,000
6	Coral Sea	1,850,000	4,791,000
7	Arabian Sea	1,492,000	3,864,000
8	South China Sea (Nan Hai)	1,423,000	3,685,000
9	Weddell Sea	1,080,000	2,796,000
10	Caribbean Sea	1,063,000	2,753,000
11	Mediterranean Sea	971,000	2,515,000
12	Tasman Sea	900,000	2,331,000
13	Bering Sea	890,000	2,305,000
14	Bay of Bengal	839,000	2,173,000
15	Sea of Okhotsk	614,000	1,590,000
16	Gulf of Mexico	596,000	1,544,000
17	Gulf of Guinea	592,000	1,533,000
18	Barents Sea	542,000	1,405,000
19	Norwegian Sea	534,000	1,383,000
20	Gulf of Alaska	512,000	1,327,000
21	Hudson Bay	476,000	1,233,000
22	Greenland Sea	465,000	1,205,000
23	Bellinghausen Sea	430,000	1,110,000
24	Amundsen Sea	400,000	1,036,000
25	Arafura Sea	400,000	1,036,000
26	Philippine Sea	400,000	1,036,000
27	Sea of Japan	378,000	979,000
28	Mozambique Channel	376,000	975,000
29	Ross Sea	370,000	958,000
30	East Siberian Sea	361,000	936,000
31	Scotia Sea	347,000	900,000
32	Kara Sea	341,000	883,000
33	Labrador Sea	309,000	800,000
34	East China Sea (Dong Hai / Tung Hai)	290,000	752,000
35	Solomon Sea	278,000	720,000
36	Laptev Sea	270,000	700,000

[continued]

Oceans and seas by area

All measurements are rounded to the nearest thousand

Rank	Name	Area (sq mi)	Area (sq km)
37	Baffin Bay	268,000	695,000
38	Banda Sea	268,000	695,000
39	Drake Passage	240,000	620,000
40	Timor Sea	237,000	615,000
41	Andaman Sea	232,000	601,000
42	North Sea	232,000	601,000
43	Davis Strait	230,000	596,000
44	Chukchi Sea	225,000	582,000
45	Great Australian Bight	187,000	484,000
46	Beaufort Sea	184,000	476,000
47	Celebes Sea	182,000	472,000
48	Black Sea	178,000	461,000
49	Red Sea	175,000	453,000
50	Java Sea	167,000	433,000
51	Sulu Sea	162,000	420,000
52	Yellow Sea (Huang Hai)	161,000	417,000
53	Baltic Sea	147,000	382,000
54	Gulf of Carpentaria	120,000	310,000
55	Molucca Sea	119,000	307,000
56	Persian Gulf	93,000	241,000
57	Gulf of Thailand	92,000	239,000
58	Gulf of St. Lawrence	92,000	239,000
59	Bismarck Sea	87,000	225,000
60	Gulf of Aden	85,000	220,000
61	Makassar Strait	75,000	194,000
62	Ceram Sea	72,000	187,000

Bays, gulfs, inlets and passages by area

All measurements are rounded to the nearest thousand

Rank	Name	Continent	Country	Area (sq mi)	Area (sq km)
1	Bay of Bengal	Asia	India/Myanmar/Sri Lanka	839,000	2,173,000
2	Gulf of Mexico	N. America	U.S.A./Cuba/Mexico	596,000	1,544,000
3	Gulf of Guinea	Africa	Cameroon/Nigeria/ Equatorial Guinea	592,000	1,533,000
4	Gulf of Alaska	N. America	U.S.A./Canada	512,000	1,327,000
5	Hudson Bay	N. America	Canada	476,000	1,233,000
6	Baffin Bay	N. America	Greenland/Canada	268,000	695,000
7	Great Australia Bight	Australia	Australia	187,000	484,000
8	Gulf of Carpentaria	Australia	Australia	120,000	310,000
9	Persian Gulf	Asia	Bahrain/Iran/Qatar/ Saudi Arabia/U.A.E.	93,000	240,000
10	Gulf of St. Lawrence	N. America	Canada	93,000	240,000
11	Gulf of Thailand	Asia	Cambodia/Thailand	92,000	239,000
12	Gulf of Aden	Asia and Africa	Somalia/Yemen	85,000	220,000
13	Gulf of Oman	Asia	Iran/U.A.E/Oman	70,000	181,000
14	Gulf of California	N. America	Mexico	62,000	161,000
15	Gulf of Tonkin	Asia	Vietnam/China	45,000	117,000
16	Gulf of Bothnia	Europe	Finland/Sweden	45,000	117,000
17	Bight of Benin	Africa	Benin	26,000	67,000
18	Strait of Malacca	Asia	Malaysia/Indonesia	25,000	65,000
19	Gulf of Boothia	N. America	Canada	23,000	60,000
20	Amundsen Gulf	N. America	Canada	23,000	60,000
21	Gulf of Finland	Europe	Finland/Estonia	12,000	30,000

Ocean trenches by depth

Rank	Name	Ocean	Depth (ft)	Depth (m)
1	Mariana Trench	Pacific	38,635	11,784
2	Philippine Trench	Pacific	37,720	11,505
3	Tonga Trench	Pacific	37,166	11,336
4	Izu Trench	Pacific	36,850	11,239
5	Kermadec Trench	Pacific	34,728	10,592

[continued]

Ocean trenches by depth

Rank	Name	Ocean	Depth (ft)	Depth (m)
6	Kuril Trench	Pacific	34,678	10,577
7	New Britain Trench	Pacific	31,657	9,655
8	Puerto Rico Trench	Atlantic	31,037	9,466
9	Bonin Trench	Pacific	29,816	9,094
10	Japan Trench	Pacific	29,157	8,893
11	South Sandwich Trench	Atlantic	28,406	8,664
12	Palau Trench	Pacific	27,972	8,531
13	Peru-Chile Trench	Pacific	27,687	8,445
14	Yap Trench	Pacific	27,552	8,403
15	Aleutian Trench	Pacific	26,775	8,166
16	Roanche Gap	Atlantic	26,542	8,095
17	Cayman Trench	Atlantic	26,519	8,088
18	New Hebrides Trench	Pacific	25,971	7,921
19	Ryukyu Trench	Pacific	25,597	7,807
20	Java Trench	Indian	24,744	7,547
21	Diamantina Trench	Indian	24,249	7,396
22	Mid America Trench	Pacific	22,297	6,801
23	Brazil Basin	Atlantic	22,274	6,794
24	Ob Trench	Indian	21,785	6,644
25	Vema Trench	Indian	19,482	5,942
26	Agulhas Basin	Indian	19,380	5,911
27	Ionian Basin	Mediterranean	17,306	5,278
28	Eurasia Basin	Arctic	16,122	4,917

Rivers by length

All measurements are approximate

Rank	Name	Continent	Country	Length (mi)	Length (km)
1	Nile	Africa	Egypt, Sudan, Uganda	4,160	6,693
2	Amazon	S. America	Brazil, Colombia, Peru, Venezuela	3,900	6,280
3	Mississippi-Missouri	N. America	U.S.A.	3,860	6,211
4	Chang Jiang (Yangtze or Yangtse)	Asia	China	3,434	5,525
5	Ob'-Irtysh	Asia	Kazakhstan, Russia	3,335	5,380
6	Paraná	S. America	Argentina, Brazil, Paraguay	3,030	4,870
7	Huang He (Huang-ho or Yellow)	Asia	China	2,903	4,671
8	Irtysh	Asia	Kazakhstan, Russia	2,760	4,441
9	Lena	Asia	Russia	2,734	4,400
10	Amur	Asia	China, Russia	2,719	4,350
11	Congo (Zaire)	Africa	Angola, Dem. Rep. of the Congo, Rep. of the Congo	2,700	4,344
12	Mackenzie	N. America	Canada	2,635	4,290
13	Mekong	Asia	Cambodia, China, Laos, Myanmar, Thailand, Vietnam	2,600	4,200
14	Niger	Africa	Benin, Guinea, Mali, Niger, Nigeria	2,594	4,184
15	Yenisey	Asia	Russia	2,566	4,129
16	Missouri	N. America	U.S.A.	2,466	3,968
17	Mississippi	N. America	U.S.A.	2,348	3,787
18	Volga	Europe	Russia	2,293	3,689
19	Ob'	Asia	Russia	2,270	3,650
20	Euphrates	Asia	Iraq, Syria, Turkey	2,235	3,596
21	Purus	S. America	Brazil, Peru	2,100	3,380
22	Madeira	S. America	Brazil	2,013	3,241
23	Lower Tunguska	Asia	Russia	2,000	3,220
24	Indus	Asia	Pakistan	1,988	3,200
25	São Francisco	S. America	Brazil	1,988	3,199
26	Yukon	N. America	Canada, U.S.A.	1,980	3,180
27	Rio Grande	N. America	Mexico, U.S.A.	1,885	3,034
28	Brahmaputra (Jamuna)	Asia	Bangladesh, China, India	1,800	2,900
29	Danube	Europe	Austria, Bulgaria, Croatia, Germany, Hungary, Romania, Ukraine, Slovakia, Yugoslavia	1,775	2,857
30	Salween	Asia	China, Myanmar	1,770	2,849

[continued]

Rivers by length

All measurements are approximate

Rank	Name	Continent	Country	Length (mi)	Length (km)
31	Darling	Australia	Australia	1,702	2,739
32	Tocantins	S. America	Brazil	1,677	2,698
33	Nelson	N. America	Canada	1,660	2,671
34	Vilyuy	Asia	Russia	1,650	2,650
35	Zambezi	Africa	Angola, Mozambique, Namibia, Zambia, Zimbabwe	1,650	2,650
36	Murray	Australia	Australia	1,609	2,589
37	Paraguay	S. America	Argentina, Brazil, Paraguay	1,584	2,549
38	Amu Dar'ya	Asia	Afghanistan, Tajikistan, Turkmenistan, Uzbekistan	1,580	2,540
39	Kolyma	Asia	Russia	1,562	2,513
40	Ganges	Asia	Bangladesh, India	1,560	2,510
41	Ishim	Asia	Kazakhstan, Russia	1,520	2,450
42	Ural	Asia	Kazakhstan, Russia	1,510	2,430
43	Japurá	S. America	Brazil, Colombia	1,500	2,414
44	Arkansas	N. America	U.S.A.	1,460	2,350
45	Colorado	N. America	U.S.A.	1,450	2,330
46	Dnieper	Europe	Belarus, Russia, Ukraine	1,420	2,290
47	Negro	S. America	Brazil, Colombia, Venezuela	1,400	2,250
48	Ubangi	Africa	Central African Rep., Dem. Rep. of the Congo, Rep. of the Congo	1,400	2,253
49	Aldan	Asia	Russia	1,390	2,240
50	Columbia-Snake	N. America	Canada, U.S.A.	1,390	2,240
51	Syr Dar'ya	Asia	Kazakhstan, Kyrgyzstan, Uzbekistan	1,370	2,200
52	Araguaia	S. America	Brazil	1,366	2,198
53	Olenek	Asia	Russia	1,350	2,170
54	Irrawaddy	Asia	Myanmar	1,350	2,170
55	Kasai	Africa	Angola, Dem. Rep of the Congo	1,338	2,153
56	Ohio-Allegheny	N. America	U.S.A.	1,310	2,109
57	Tarim	Asia	China	1,300	2,090
58	Orange	Africa	Lesotho, Namibia, South Africa	1,300	2,090
59	Orinoco	S. America	Venezuela	1,281	2,061
60	Shabeelle	Africa	Ethiopia, Somalia	1,250	2,011
61	Xingu	S. America	Brazil	1,230	1,979
62	Columbia	N. America	Canada, U.S.A.	1,214	1,953
63	Mamoré	S. America	Bolivia	1,200	1,931
64	Tigris	Asia	Iraq, Turkey	1,180	1,900
65	Northern Dvina	Europe	Russia	1,160	1,870
66	Don	Europe	Russia	1,153	1,860
67	Angara	Asia	Russia	1,151	1,852
68	Kama	Europe	Russia	1,120	1,800
69	Indigirka	Asia	Russia	1,112	1,789
70	Pechora	Europe	Russia	1,112	1,789
71	Limpopo	Africa	Botswana, South Africa, Mozambique	1,100	1,770
72	Salado	S. America	Argentina	1,110	1,770
73	Guaporé	S. America	Bolivia, Brazil	1,087	1,749
74	Tobol	Asia	Kazakhstan, Russia	1,042	1,677
75	Snake	N. America	U.S.A.	1,038	1,670
76	Red	N. America	U.S.A.	1,018	1,638
77	Sénégal	Africa	Guinea, Mali, Mauritania, Senegal	1,015	1,663
78	Churchill	N. America	Canada	1,000	1,613
79	Jubba	Africa	Ethiopia, Somalia	1,000	1,613
80	Okavango	Africa	Angola, Botswana	1,000	1,613
81	Pilcomayo	S. America	Argentina, Bolivia, Paraguay	1,000	1,613
82	Uruguay	S. America	Uruguay	1,000	1,613
83	Volta	Africa	Burkina Faso, Ghana	992	1,600

Rivers by drainage area

In thousands of sq mi and sq km. All measurements are approximate

Rank	Name	Continent	Country	Drainage basin (sq mi)	Drainage basin (sq km)
1	Amazon	S. America	Brazil, Colombia, Peru, Venezuela	2,375	6,150
2	Congo (Zaire)	Africa	Angola, Dem. Rep. of the Congo, Rep. of the Congo	1,476	3,822
3	Nile	Africa	Egypt, Sudan	1,293	3,349
4	Mississippi	N. America	U.S.A.	1,247	3,230
5	Paraná	S. America	Argentina, Brazil Paraguay	1,197	3,100
6	Ob'-Irtysh	Asia	Kazakhstan, Russia	1,154	2,990
7	Yenisey	Asia	Russia	996	2,580
8	Lena	Asia	Russia	961	2,490
9	Niger	Africa	Benin, Guinea, Mali, Niger, Nigeria	808	2,092
10	Amur (Hei-lung chiang or Heilong Jiang)	Asia	China, Russia	716	1,855
11	Chang Jiang (Yangtze or Yangtse)	Asia	China	705	1,827
12	Mackenzie	N. America	Canada	690	1,787
13	Saint Lawrence	N. America	Canada, U.S.A.	550	1,424
14	Volga	Europe	Russia	525	1,360
15	Zambezi	Africa	Angola, Mozambique, Namibia, Zambia, Zimbabwe	514	1,331
16	Madeira	S. America	Brazil	463	1,200
17	Indus	Asia	Pakistan	450	1,165
18	Nelson	N. America	Canada	437	1,132
19	Shatt al-Arab*	Asia	Iran, Iraq, Syria, Turkey	427	1,105
20	Paraguay	S. America	Argentina, Paraguay	425	1,100
21	Murray	Australia	Australia	408	1,057
22	Negro (Guainía)	S. America	Brazil	386	1,000
23	Ganges	Asia	Bangladesh, India	368	952
24	Kasai	Africa	Angola, Dem. Rep. of the Congo	349	904
25	Orinoco	S. America	Venezuela	340	880
26	Orange	Africa	Lesotho, Namibia, South Africa	330	855
27	Yukon	N. America	Canada, U.S.A.	328	850
28	Para	S. America	Brazil	323	836
29	Danube	Europe	Austria, Bulgaria, Croatia, Germany, Hungary, Romania, Ukraine, Slovakia, Yugoslavia	315	816
30	Mekong	Asia	Cambodia, China, Laos, Myanmar, Thailand, Vietnam	313	811
32	Okavango	Africa	Angola, Botswana	303	78531
	Salado	N. America	Argentina	309	800
33	Ubangi	Africa	Central African Rep., Dem. Rep. of the Congo, Rep. of the Congo	298	773
34	Huang He (Huang-ho or Yellow)	Asia	China	297	771
35	Ob'	Asia	Russia	295	765
36	Aldan	Asia	Russia	281	729
37	Shari	Africa	Cameroon, Chad	270	700
38	Columbia	N. America	Canada, U.S.A.	258	668
39	Kolyma	Asia	Russia	248	648
40	Colorado	N. America	Mexico, U.S.A.	247	640
41	São Francisco	S. America	Brazil	241	623
42	Guapore (Iténez)	S. America	Bolivia, Brazil	232	600
43	Brahmaputra	Asia	Bangladesh	224	580
44	Panjnad	Asia	India, Pakistan	206	533
45	Ohio-Allegheny	N. America	U.S.A.	204	528
46	Sungari	Asia	China	202	524
47	Kama	Europe	Russia	196	507
48	Dnieper	Europe	Belarus, Russia, Ukraine	195	504
49	Lower Tunguska	Asia	Russia	183	473
50	Tapajós	S. America	Brazil	179	463
51	Volta	Africa	Burkina Faso, Ghana	154	398
52	Northern Dvina	Europe	Russia	138	357

*The Shatt al-Arab is the confluence of the Euphrates and Tigris Rivers.

Rivers by discharge rate

All measurements are approximate

Rate	River	Continent	Country	Discharge rate (cu ft per sec)	Discharge rate (cu m per sec)
1	Amazon	S. America	Brazil, Colombia, Peru, Venezuela	6,180,000	175,000
2	Congo (Zaire)	Africa	Angola, Dem. Rep. of the Congo	1,377,000	39,000
3	Negro (Guainía)	S. America	Brazil	1,236,000	35,000
4	Chang Jiang (Yangtze or Yangtse)	Asia	China	1,137,000	32,190
5	Orinoco	S. America	Venezuela	890,000	25,200
6	Paraná	S. America	Argentina, Brazil, Paraguay	809,000	22,900
7	Madeira	S. America	Brazil	770,000	21,800
8	Brahmaputra	Asia	Bangladesh, China, India	678,000	19,200
9	Yenisey	Asia	Russia	622,000	17,600
10	Lena	Asia	Russia	586,000	16,600
11	Mississippi	N. America	U.S.A.	531,000	15,040
12	Mekong	Asia	Cambodia, China, Laos, Myanmar, Thailand, Vietnam	501,000	14,200
13	Irrawaddy	Asia	Myanmar	447,000	12,660
14	Zhu (Pearl)	Asia	China	441,000	12,500
15	Ganges	Asia	Bangladesh, India	440,000	12,460
16	Ob'-Irtysh	Asia	Russia	438,000	12,400
17	St. Lawrence	N. America	Canada, U.S.A.	355,000	10,050
18	Salween	Asia	Myanmar	353,000	10,000
19	Kasai	Africa	Rep. of the Congo, Dem. Rep. of the Congo	351,000	9,951
20	Amur (Hei-lung chiang or Heilong Jiang)	Asia	China, Russia	349,000	9,860
21	Mackenzie	N. America	Canada	316,000	8,940
22	Para	S. America	Brazil	305,000	8,630
23	Volga	Europe	Russia	285,000	8,060
24	Magdalena	S. America	Colombia	283,000	8,000
25	Ohio	N. America	U.S.A.	272,000	7,710
26	Solo	Asia	Indonesia	268,000	7,600
27	Ubangi	Africa	Central African Rep., Rep. of the Congo, Dem. Rep. of the Congo	265,000	7,500
28	Zambezi	Africa	Angola, Mozambique, Namibia, Zambia, Zimbabwe	250,000	7,070
29	Caqueta	S. America	Brazil	247,000	7,000
31	Indus	Asia	Pakistan	235,000	6,640
32	Yukon	N. America	Canada, U.S.A.	223,000	6,310
33	Danube	Europe	Austria, Bulgaria, Croatia, Germany, Hungary, Romania, Ukraine, Slovakia, Yugoslavia	221,000	6,250
34	Araguaia	S. America	Brazil	217,000	6,140
35	Tapajos-Juruena	S. America	Brazil	212,000	6,000
36	Niger	Africa	Benin, Guinea, Mali, Niger, Nigeria	201,000	5,700
37	Uruguay	S. America	Argentina, Brazil, Uruguay	194,000	5,490
38	Columbia	N. America	Canada, U.S.A.	194,000	5,490
39	Branco	S. America	Brazil	191,000	5,400
40	Aldan	Asia	Russia	177,000	5,010
41	Putumayo	S. America	Brazil	177,000	5,010
42	Ob	Asia	Russia	174,000	4,920
43	Atrato	S. America	Colombia	173,000	4,900
44	Caroni	S. America	Venezuela	168,000	4,750
45	Ogooue	Africa	Gabon	165,000	4,670
46	Fly	Asia	Papua New Guinea	157,000	4,450
47	Paraguay	S. America	Argentina, Brazil, Paraguay	155,000	4,400
48	Chindwin	Asia	Myanmar	141,000	4,000
48	Jurua	S. America	Brazil	141,000	4,000
48	Pechora	Europe	Russia	141,000	4,000
49	Northern Dvina	Europe	Russia	120,000	3,400
49	Fraser	N. America	Canada	120,000	3,400
50	Lower Tunguska	Asia	Russia	117,000	3,300
51	Kama	Europe	Russia	99,000	2,800

Lakes by area

All measurements are approximate

Rank	Name	Continent	Country	Area (sq mi)	Area (sq km)
1	Caspian Sea	Asia	Azerbaijan, Iran, Kazakhstan, Russia, Turkmenistan	143,000	371,000
2	Superior	N. America	Canada, U.S.A.	31,820	82,732
3	Victoria	Africa	Uganda, Tanzania, Kenya	26,828	69,484
4	Aral Sea	Asia	Kazakhstan, Uzbekistan	24,900	64,500
5	Huron	N. America	Canada, U.S.A.	23,000	59,570
6	Michigan	N. America	U.S.A.	22,400	58,020
7	Tanganyika	Africa	Burundi, Dem. Republic of the Congo, Tanzania, Zambia	12,700	32,020
8	Baikal	Asia	Russia	12,160	31,500
9	Great Bear	N. America	Canada	12,095	31,328
10	Great Slave	N. America	Canada	11,030	28,570
11	Erie	N. America	Canada, U.S.A.	9,920	25,690
12	Winnipeg	N. America	Canada	9,420	24,390
13	Malawi	Africa	Malawi, Mozambique, Tanzania,	8,680	22,490
14	Ontario	N. America	Canada, U.S.A.	7,440	19,240
15	Balkhash	Asia	Kazakhstan	7,030	18,200
16	Ladoga	Russia	Russia	7,000	18,130
17	Maracaibo	S. America	Venezuela	5,020	13,010
18	Chad	Africa	Cameroon, Chad, Niger, Nigeria	4,000–10,000	10,360–25,900
19	Embalse del Río Negro	S. America	Uruguay	4,000	10,360
20	Patos	S. America	Brazil	3,920	10,153
21	Onega	Europe	Russia	3,750	9,720
22	Eyre	Australia	Australia	3,668	9,500
23	Volta	Africa	Ghana	3,276	8,485
24	Titicaca	S. America	Bolivia, Peru	3,200	8,288
25	Nicaragua	S. America	Nicaragua	3,150	8,160
26	Athabasca	N. America	Canada	3,060	7,940
27	Reindeer	N. America	Canada	2,570	6,650
28	Smallwood Reservoir	N. America	Canada	2,500	6,460
29	Turkana (Rudolf)	Africa	Ethiopia, Kenya	2,473	6,405
30	Issyk Kul	Asia	Kyrgyzstan	2,360	6,100
31	Torrens	Australia	Australia	2,230	5,780
32	Albert	Africa	Dem. Republic of the Congo, Uganda	2,160	5,590
33	Vanern	Europe	Sweden	2,160	5,580
34	Netilling	N. America	Canada	2,140	5,540
35	Winnipegosis	N. America	Canada	2,070	5,370
36	Nasser	Africa	Egypt, Sudan	2,026	5,248
37	Bangweulu	Africa	Zambia	1,930	5,000
38	Chott el Djerid	Africa	Tunisia	1,930	5,000
39	Urmia	Asia	Iran	1,879	4,868
40	Nipigon	N. America	Canada	1,870	4,850
41	Gairdner	Australia	Australia	1,840	4,770
42	Manitoba	N. America	Canada	1,800	4,660
43	Kyoga	Africa	Uganda	1,710	4,430
44	Khanka	Asia	China, Russia	1,700	4,400
45	Saimaa	Europe	Finland	1,700	4,403
46	Mweru	Africa	Dem. Republic of the Congo	1,680	4,350
47	Great Salt	N. America	U.S.A.	1,680	4,350
48	Qinghai (Koko)	Asia	China	1,625	4,209
49	Woods	N. America	Canada	1,580	4,100
50	Taymyr	Asia	Russia	1,540	3,990
51	Nasser	Africa	Egypt	1,522	3,942
52	Orumiyeh	Asia	Iran	1,500	3,880
53	Dubawnt	N. America	Canada	1,480	3,830
54	Van	Asia	Turkey	1,430	3,710
55	Tana	Africa	Ethiopia	1,390	3,600
56	Peipus	Europe	Estonia, Russia	1,386	3,555
57	Uvs	Asia	Mongolia	1,300	3,366

Lakes by depth

Rank	Lake	Continent	Country	Depth (ft)	Depth (m)
1	Baikal	Asia	Russia	5,315	1,621
2	Tanganyika	Africa	Burundi, Tanzania, Dem. Republic of the Congo, Zambia	4,825	1,471
3	Caspian Sea	Asia	Azerbaijan, Iran, Kazakhstan, Russia, Turkmenistan	3,363	1,025
4	Malawi	Africa	Malawi, Tanzania, Mozambique	2,316	706
5	Issyk Kul	Asia	Kyrgyzstan	2,303	702
6	Great Slave	N. America	Canada	2,015	614
7	Matana	Asia	Indonesia	1,936	590
8	Crater	N. America	U.S.A.	1,932	589
9	Toba	Asia	Indonesia	1,736	529
10	Hornindals	Europe	Norway	1,686	514
11	Sarez	Asia	Tajikistan	1,657	505
12	Tahoe	N. America	U.S.A.	1,645	501
13	Chelan	N. America	U.S.A.	1,605	489
14	Kivu	Africa	Rwanda, Dem. Republic of the Congo	1,575	480
15	Quesnel	N. America	Canada	1,560	475
16	Sals	Europe	Norway	1,522	464
17	Adams	N. America	Canada	1,500	457
18	Mjøsa	Europe	Norway	1,473	449
19	Manapuri	Oceania	New Zealand	1,453	443
20	Poso	Asia	Indonesia	1,444	440
21	Nahuel Huapi	S. America	Argentina	1,437	438
22	Dead Sea	Asia	Israel, Jordan	1,421	433
23	Tazawa	Asia	Japan	1,394	425
24	Great Bear	N. America	Canada	1,356	413
25	Como	Europe	Italy	1,352	412
26	Superior	N. America	Canada, U.S.A.	1,333	406
27	Hawea	Asia	New Zealand	1,286	392
28	Wakatipu	Asia	New Zealand	1,240	378
29	Suldals	Europe	Norway	1,234	376
30	Maggiore	Europe	Italy, Switzerland	1,221	372
31	Fyres	Europe	Norway	1,211	369
32	Chilko	N. America	Canada	1,200	366
33	Pend Oreille	N. America	U.S.A.	1,200	366
34	Shikotsu	Asia	Japan	1,191	363
35	Powell	N. America	Canada	1,174	358
36	Llanquihue	S. America	Chile	1,148	350
37	Garda	Europe	Italy	1,135	346
38	Towada	Asia	Japan	1,096	334
39	Wanaka	Asia	New Zealand	1,086	325
40	Bandak	Europe	Norway	1,066	325
41	Telestskoya	Asia	Russia	1,066	325
42	Eutsuk	N. America	Canada	1,060	323
43	Atitlan	N. America	Guatemala	1,050	320
44	Lunde	Europe	Norway	1,030	314
45	Geneva	Europe	France, Switzerland	1,017	310
46	Morar	Europe	Scotland	1,017	310
47	Kurile	Asia	Russia	1,004	306
48	Walker	N. America	U.S.A.	1,000	305
49	Titicaca	S. America	Bolivia, Peru	997	304
50	Argentino	S. America	Argentina	984	300
51	Iliamna	N. America	U.S.A.	980	299
52	Tyrifjorden	Europe	Norway	968	295
53	Lugano	Europe	Italy, Switzerland	945	288
54	Takla	N. America	Canada	941	287
55	Ohrid	Europe	Albania, Yugoslavia	938	286
56	Atlin	N. America	Canada	930	283
57	Nuyakuk	N. America	U.S.A.	930	283
58	Michigan	N. America	U.S.A.	923	285
59	Harrison	N. America	Canada	916	279
60	Te Anau	Oceania	New Zealand	906	276

Lakes by volume

Rank	Lake	Continent	Country	Volume (cu mi)	Volume (cu km)
1	Caspian Sea	Asia	Azerbaijan, Iran, Kazakhstan, Russia, Turkmenistan	18,882	78,707
2	Baikal	Asia	Russia	5,517	22,995
3	Tanganyika	Africa	Burundi, Tanzania, Democratic Republic of the Congo, Zambia	4,391	18,304
4	Superior	N. America	Canada, USA	2,921	12,174
5	Malawi	Africa	Malawi, Tanzania, Mozambique	1,473	6,140
6	Michigan	N. America	USA	1,169	4,874
7	Huron	N. America	Canada, USA	858	3,575
8	Victoria	Africa	Kenya, Tanzania, Uganda	604	2,518
9	Great Bear	N. America	Canada	550	2,292
10	Great Slave	N. America	Canada	542	2,258
11	Issyk Kul	Asia	Kyrgyzstan	420	1,725
12	Ontario	N. America	Canada, USA	369	1,539
13	Aral Sea	Asia	Kazakhstan, Uzbekistan	348	1,450
14	Ladoga	Russia	Russia	218	905
15	Titicaca	S. America	Bolivia, Peru	198	827
16	Erie	N. America	Canada, USA	111	462
17	Winnipeg	N. America	Canada	89	370
18	Hovsgol	Asia	Mongolia	88	367
19	Kivu	Africa	Rwanda Congo, Dem. Republic of	80	333
20	Onega	Europe	Russia	70	285
21	Maracaibo	S. America	Venezuela	67	280

Glaciers, ice shelves, and related features, by area

Only named features are included in this list. All measurements are approximate. Also note that glaciers and similar features can vary in size over time.

Rank	Name	Continent	Country	Area (sq mi)	Area (sq km)
1	Antarctic Ice Shelf*	Antarctica	Antarctica	3,805,235–4,619,957	9,855,570–11,965,700
2	Greenland Ice Sheet	N. America	Greenland	659,087	1,707,038
3	Novaya Zemlya Ice Sheet	Asia	Russia	8,301–9,112	21,500–23,600
4	Prince of Wales Ice Field	N. America	Canada	7,247	18,770
5	Mer de Glace Agassiz	Europe	Switzerland	6,216	16,100
6	Devon Island Ice Cap	N. America	Canada	5,019	13,000
7	Southern Patagonian Ice Field	S. America	Argentina-Chile	5,019	13,000
8	Bering Glacier	N. America	U.S.A.	3,800	9,800
9	Austfonna Ice Cap	Europe	Svalbard (dependency of Norway)	3,248	8,413
10	Vatnajökull Ice Cap	Europe	Iceland	3,166–3,297	8,200–8,538
11	Akaioa Ice Cap	N. America	Canada	2,799	7,250
12	Malaspina Glacier**	N. America	U.S.A.	847–3,089	2,195–8,000
13	Barnes Ice Cap	N. America	Canada	2,292	5,935
14	Bering Glacier	N. America	Canada-U.S.A.	2,008	5,200
15	Axel Heiberg Island Ice Cap	N. America	Canada	1,969	5,100
16	Northern Patagonian Ice Field	S. America	Argentina-Chile	1,622	4,200
17	Hubbard Glacier	N. America	U.S.A.	1,313	3,400
18	Olav V Land Ice Field	Europe	Svalbard (Norway)	1,158	3,000
19	Kallstenius Ice Field	N. America	Greenland	1,050	2,720
20	Vestfonna Ice Cap	Europe	Svalbard (Norway)	967	2,505
21	Juneau Icefield	N. America	U.S.A.	703	1,820
22	Skeidararjökull	Europe	Iceland	665	1,722
23	Asgardfonna Ice Cap	Europe	Svalbard (Norway)	635	1,645
24	Edgeoyjokulen Ice Cap	Europe	Svalbard (Norway)	512	1,300

[continued]

Glaciers, ice shelves, and related features, by area

Only named features are included in this list. All measurements are approximate. Also note that glaciers and similar features can vary in size over time.

Rank	Name	Continent	Country	Area (sq mi)	Area (sq km)
25	Breidamerkurujökull	Europe	Iceland	489	1,266.5
26	Glacier Bruggen (Pio XI)	S. America	Argentina-Chile	488	1,265
27	Hinlopenbreen	Europe	Svalbard (Norway)	482–483	1,248–1,250
28	Negribreen	Europe	Svalbard (Norway)	456	1,180
29	Brasvellbreen	Europe	Svalbard (Norway)	448	1,160
30	Etonbreen	Europe	Svalbard (Norway)	413–440	1,070–1,140
31	Columbia Glacier	N. America	U.S.A.	422–425	1,093–1,100
32	Siachen Glacier	Asia	India-Pakistan	386	1,000
33	Langjökull	Europe	Iceland	368	953
34	Leighbreen	Europe	Svalbard (Norway)	357	925
35	Holtedahlfonna and Isachsenfonna	Europe	Svalbard (Norway)	347	900
36	Fedchenko Glacier	Asia	Tajikistan	347	900
37	Nabesna Glacier	N. America	U.S.A.	316	819
38	Jostedalsbreen	Europe	Norway	315	815
39	Kvitoyjokulen	Europe	Svalbard (Norway)	272	705
40	Stonebreen	Europe	Svalbard (Norway)	270	700
41	Myrdals Glacier	Europe	Iceland	268.3	695
42	Kronebreen	Europe	Svalbard (Norway)	267.5	693
43	Taku Glacier	Europe	Svalbard (Norway)	259	671
44	Hochstetterbreen Glacier	Europe	Svalbard (Norway)	224.3	581
45	Kahiltna Glacier	N. America	U.S.A.	223.9	580
46	Barentsjokulen	Europe	Svalbard (Norway)	220	571
47	Kennicott Glacier	N. America	U.S.A.	213	551
48	Shokalsky Glacier	Asia	Kazakhstan-Kyrgyzstan	210.4	545
49	Balderfonna	Europe	Svalbard (Norway)	209.7	543
50	Baird Glacier	N. America	U.S.A.	205	532
51	Muldrow Glacier	N. America	U.S.A.	199	516
52	Baltoro Glacier	Asia	Pakistan	70	181
53	Aletsch Glacier	Europe	Switzerland	66	171

* Antarctic glaciers are difficult, if not impossible, to list by their surface area, because Antarctica is an effectively continuous ice mass and there is no easy way to determine the boundaries of a given glacier.

** Scientists do not agree on the size of this glacier; area measurements reported reflect the range of research findings as of 2002.

Deserts by area

All measurements are approximate

Rank	Name	Continent	Country	Area (sq mi)	Area (sq km)
1	Sahara	Africa	Algeria, Chad, Egypt, Libya, Mali, Mauritania, Morocco, Niger, Sudan, and Tunisia	3,475,000	9,000,000
2	Arabian*	Asia	Saudi Arabia, Kuwait, Qatar, the United Arab Emirates, Oman, Yemen, Jordan, Syria, Iraq	900,000	2,330,000
3	Gobi	Asia	China, Mongolia	500,000	1,300,000
4	Kalahari	Africa	Botswana, Namibia, South Africa	360,000	930,000
5	Great Victoria	Australia	Australia	134,652	348,750
6	Taklimakan (Takla Makan)	Asia	China	125,000	320,000
7	Sonoran	N. America	United States of America, Mexico	120,000	310,000
8	Kara-Kum	Asia	Kazakhstan, Turkmenistan	115,830	300,000
9	Kyzyl Kum	Asia	Kazakhstan, Uzbekistan	115,000	297,850
10	Namib	Africa	Namibia, South Africa	110,000	285,000
11	Great Sandy	Australia	Australia	103,185	267,250
12	Somali	Africa	Somalia	100,000	260,000
13	Thar	Asia	India, Pakistan	90,000	233,000
14	Tanami	Australia	Australia	71,235	184,500
15	Atacama	S. America	Chile, Peru	70,000	180,000
16	Simpson	Australia	Australia	68,150	176,500
17	Gibson	Australia	Australia	60,230	156,000
18	Little Sandy	Australia	Australia	43,050	111,500

* Two deserts are commonly referred to by this name. This entry refers to the deserts of the Arabian Peninsula and not the Arabian Desert of Egypt, which is part of the Sahara.

Mountain peaks by height and continent

All measurements are approximate. Note that many mountains have multiple peaks, which will appear separately in the table.

Rank	Name	Country	Elevation (ft)	Elevation (m)
Africa				
1	Kibo (Mt. Kilimanjaro)	Tanzania	19,340	5,895
2	Mawensi (Mt. Kilimanjaro)	Tanzania	17,100	5,210
3	Batian (Mt. Kenya)	Kenya	17,058	5,203
4	Nelion (Mt. Kenya)	Kenya	17,020	5,190
5	Margherita Peak (Mt. Stanley)	Dem. Rep. of the Congo, Uganda	16,756	5,110
6	Alexandra Peak (Mt. Stanley)	Dem. Rep. of the Congo, Uganda	16,700	5,094
7	Albert Peak (Mt. Stanley)	Dem. Rep. of the Congo	16,690	5,090
8	Savoia Peak (Mt. Stanley)	Uganda	16,330	4,981
9	Elena Peak (Mt. Stanley)	Uganda	16,300	4,972
10	Elizabeth Peak (Mt. Stanley)	Uganda	16,170	4,932
11	Phillip Peak (Mt. Stanley)	Uganda	16,140	4,923
12	Moebius Peak (Mt. Stanley)	Uganda	16,130	4,920
13	Vittorio Emanuele (Mt. Speke)	Uganda	16,040	4,892
14	Ensonga (Mt. Speke)	Uganda	15,960	4,868
15	Johnston (Mt. Speke)	Uganda	15,860	4,834
16	Edward (Mt. Baker)	Uganda	15,890	4,846
17	Umberto (Mt. Emin)	Dem. Rep. of the Congo	15,740	4,798
18	Semper (Mt. Baker)	Uganda	15,730	4,795
19	Kraepelin (Mt. Emin)	Dem. Rep. of the Congo	15,720	4,791
20	Iolanda (Mt. Gessi)	Dem. Rep. of the Congo	15,470	4,751
21	Bottego (Mt. Gesi)	Dem. Rep. of the Congo	15,418	4,699
22	Sella (Mt. Luigi)	Dem. Rep. of the Congo	15,178	4,626
23	Ras Deshen	Ethiopia	15,157	4,620
24	Weismann (Mt. Luigi)	Dem. Rep. of the Congo	15,157	4,620
25	Okusoma (Mt. Luigi)	Dem. Rep. of the Congo	15,020	4,578

[continued]

Mountain peaks by height and continent

All measurements are approximate. Note that many mountains have multiple peaks, which will appear separately in the table.

Rank	Name	Country	Elevation (ft)	Elevation (m)
Antarctica				
1	Vinson	Antarctica	16,860	5,142
2	Tyree	Antarctica	16,290	4,968
3	Shinn	Antarctica	15,750	4,800
4	Gardner	Antarctica	15,370	4,690
5	Epperly	Antarctica	15,100	4,600
Asia				
1	Everest (Zhumulangma Feng)	Nepal, China	29,030	8,850
2	K2	China, Pakistan	28,251	8,611
3	Kanchenjunga	India, Nepal	28,169	8,586
4	Lhotse	China, Nepal	27,890	8,500
5	Makalu	China, Nepal	27,824	8,481
6	Kanchenjunga, south peak	India, Nepal	27,800	8,479
7	Kanchenjunga, west peak	India, Nepal	27,620	8,424
8	Lhotse Shar	China, Nepal	27,500	8,388
9	Dhaulagiri	Nepal	26,813	8,172
10	Manslu	Nepal	26,775	8,155
11	Cho Oyu	China, Nepal	26,750	8,150
12	Nanga Parbat I	Pakistan	26,660	8,130
13	Masherbrum I	Pakistan	26,610	7,810
14	Annapurna I	Nepal	26,500	8,080
15	Gasherbrum I	Pakistan	26,470	8,070
16	Broad, highest peak	Pakistan	26,400	8,050
17	Gasherbrum II	Pakistan	26,360	8,030
18	Gosainthan	China	26,290	8,010
19	Broad, middle peak	Pakistan	26,250	8,000
20	Gasherbrum III	Pakistan	26,090	7,950
21	Annapurna II	Nepal	26,040	7,940
22	Gasherbrum IV	Pakistan	26,000	7,930
23	Gyachung Kang	China, Nepal	25,990	7,927
24	Nanga Parbat II	Pakistan	25,950	7,910
25	Kangbachen	India, Nepal	25,930	7,909
26	Manslu, east pinnacle	Nepal	25,900	7,900
27	Distaghil Sar	Pakistan	25,870	7,890
28	Nuptse	Nepal	25,850	7,880
29	Himachuh	Nepal	25,800	7,860
30	Khiangyang Kish	Pakistan	25,760	7,850
31	Ngojumba Ri	China, Nepal	25,720	7,847
32	Dakura	Nepal	25,710	7,842
33	Masherbrum II	Pakistan	25,660	7,826
34	Nanda Devi, west peak	India	25,650	7,823
35	Nanga Parbat III	Pakistan	25,650	7,823
36	Rakaposhi	Pakistan	25,550	7,793
37	Batura Mustagh I	Pakistan	25,540	7,790
38	GasherbrumV	Pakistan	25,500	7,770
39	Kamet	China, India	25,440	7,760
Europe				
1	El'brus (Elborus), west peak	Russia	18,481	5,633
2	El'brus (Elborus), east peak	Russia	18,360	5,590
3	Shkhara	Georgia, Russia	17,064	5,205
4	Dykh, west peak	Russia	17,050	5,200
5	Dykh, east peak	Russia	16,900	5,150
6	Koshtan	Russia	16,880	5,148
7	Pushkina	Russia	16,730	5,100
8	Kazbek, east peak	Georgia	16,526	5,040
9	Dzhangi	Georgia	16,520	5,039
10	Katyn	Georgia, Russia	16,310	4,975
11	Shota Rustaveli	Georgia, Russia	16,270	4,962
12	Mizhirgi, west peak	Russia	16,170	4,932
13	Mizhirgi, east peak	Russia	16,140	4,923
14	Kundyum-Mizhirgi	Russia	16,010	4,880
15	Gestola	Georgia, Russia	15,930	4,860
16	Tetnuld	Georgia, Russia	15,920	4,850
17	Mont Blanc, main peak	France, Italy	15,772	4,810
18	Dzhimariy	Georgia	15,680	4,780
19	Adish	Georgia, Russia	15,570	4,749
20	Courmayer (Mont Blanc)	France, Italy	15,577	4,748
21	Ushba	Georgia	15,450	4,710
North America				
1	McKinley (Denali), south peak	U.S.A.	20,323	6,194
2	Logan, central peak	Canada	19,550	5,959
3	Logan, west peak	Canada	19,470	5,930
4	McKinley (Denali), north peak	U.S.A.	19,470	5,930
5	Logan, east peak	Canada	19,420	5,920
6	Pico de Orizaba	Mexico	18,701	5,700
7	Logan, north peak	Canada	18,270	5,570
8	Saint Elias	U.S.A., Canada	18,010	5,490
9	Popocatepetl	Mexico	17,887	5,452
10	Foraker	U.S.A.	17,400	5,300
11	Ixtacihuatl	Mexico	17,342	5,286
12	Queen	Canada	17,300	5,270
13	Lucania	Canada	17,150	5,230
14	King	Canada	16,970	5,170
15	Steele	Canada	16,640	5,070
16	Bona	U.S.A.	16,500	5,033
17	Blackburn, highest peak	U.S.A.	16,390	5,000
18	Blackburn, southeast peak	U.S.A.	16,290	4,968
19	Sanford	U.S.A.	16,240	4,950
20	Wood	Canada	15,880	4,840
Oceania				
1	Puncak Jaya	Indonesia	16,503	5,033
2	Daam	Indonesia	16,150	4,926
3	Pilimsit	Indonesia	15,750	4,800
4	Trikora	Indonesia	15,580	4,752
5	Mandala	Indonesia	15,420	4,700
6	Wisnumurti	Indonesia	15,080	4,590
7	Yamin	Indonesia	14,860	4,530
8	Wilhelm	Papua New Guinea	14,793	4,509
9	Kubor	Papua New Guinea	14,300	4,360
10	Herbert	Papua New Guinea	14,000	4,270
South America				
1	Aconcagua	Argentina	22,835	6,960
2	Ojos del Salado, southeast peak	Argentina, Chile	22,573	6,880
3	Bonete	Argentina	22,550	6,870
4	Pissis	Argentina	22,240	6,780
5	Mercedario	Argentina	22,210	6,770
6	Huascarán, south peak	Peru	22,204	6,768
7	Llullaillaco	Argentina, Chile	22,100	6,730
8	Libertador	Argentina	22,050	6,720
9	Ojos del Salado, northwest peak	Argentina, Chile	22,050	6,720
10	Tupungato	Argentina, Chile	21,900	6,670
11	Gonzalez, highest peak	Argentina, Chile	21,850	6,664
12	Huascarán, north peak	Peru	21,840	6,661
13	Muerto	Argentina, Chile	21,820	6,655
14	Yerupaja, north peak	Peru	21,760	6,630
15	Incahuasi	Argentina, Chile	21,700	6,610
16	Galan	Argentina	21,650	6,600
17	Tres Cruces	Argentina, Chile	21,540	6,560
18	Gonzalez, north peak	Argentina, Chile	21,490	6,550
19	Sajama	Bolivia	21,463	6,542
20	Yerupaja, south peak	Peru	21,380	6,510
21	Chimborazo	Ecuador	20,681	6,267

[continued]

Volcanoes by height

All measurements are approximate.

Rank	Name	Continent	Country	Elevation (ft)	Elevation (m)
1	Tupungato	S. America	Chile	22,310	6,800
2	Tipas	S. America	Argentina	21,845	6,660
3	Cerro el Condor	S. America	Argentina	21,425	6,532
4	Antofallo	S. America	Argentina	20,008	6,100
5	Guallatiri	S. America	Chile	19,882	6,060
6	Lascar	S. America	Chile	19,652	5,990
7	Cotopaxi	S. America	Ecuador	19,344	5,896
8	Kilimanjaro	Africa	Tanzania	19,341	5,895
9	El Misti	S. America	Peru	19,031	5,801
10	Pico de Orizaba	N. America	Mexico	18,702	5,700
11	Tolima	S. America	Colombia	18,425	5,616
12	Popocatépetl	N. America	Mexico	17,887	5,450
13	Yucamani	S. America	Peru	17,860	5,444
14	Sangay	S. America	Ecuador	17,159	5,230
15	Tungurahua	S. America	Ecuador	16,684	5,085
16	Cotacachi	S. America	Ecuador	16,250	4,939
17	Purace	S. America	Colombia	15,604	4,756

[continued]

Volcanoes by height

All measurements are approximate.

Rank	Name	Continent	Country	Elevation (ft)	Elevation (m)
18	Klyuchevskaya	Asia	Russia	15,584	4,750
19	Kronotskaya	Asia	Russia	15,580	4,749
20	Shiveluch	Asia	Russia	15,580	4,749
21	Pichincha	S. America	Ecuador	15,173	4,625
22	Karasimbi	Africa	Dem. Rep. of the Congo	14,873	4,507
23	Rainier	N. America	USA	14,410	4,395
24	Wrangell	N. America	USA (Alaska)	14,163	4,317
25	Colima	N. America	Mexico	13,993	4,265
26	Tajumulco	N. America	Guatemala	13,845	4,220
27	Mauna Kea	Oceania	USA (Hawaii)	13,796	4,205
28	Mauna Loa	Oceania	USA (Hawaii)	13,680	4,170
29	Cameroon	Africa	Cameroon	13,353	4,070
30	Tacana	N. America	Guatemala	13,300	4,053
31	Kerintji	Asia	Indonesia	12,483	3,805
32	Erebus	Antarctica	Antarctica	12,448	3,794
33	Fuji	Asia	Japan	12,388	3,776
34	Fuego	N. America	Guatemala	12,346	3,763
35	Agua	N. America	Guatemala	12,307	3,751
36	Rindjani	Asia	Indonesia	12,224	3,726
37	Pico de Teide	Africa	Spain (Canary Is.)	12,198	3,718
38	Tolbachik	Asia	Russia	12,077	3,682
39	Semeru	Asia	Indonesia	12,060	3,676
40	Ichinskaya	Asia	Russia	11,800	3,621
41	Atitlan	N. America	Guatemala	11,650	3,551
42	Torbert	N. America	USA (Alaska)	11,450	3,480
43	Nyirangongo	Africa	Dem. Rep. of the Congo	11,365	3,465
44	Kroyakskaya	Asia	Russia	11,336	3,456
45	Irazu	S. America	Costa Rica	11,260	3,432
46	Slamet	Asia	Indonesia	11,247	3,428
47	Spurr	N. America	USA (Alaska)	11,137	3,385
48	Lautaro	S. America	Chile	11,120	3,380
49	Sumbing	Asia	Indonesia	11,060	3,371
50	Raung	Asia	Indonesia	10,932	3,332
51	Etna	Europe	Italy	10,902	3,323
52	Baker	N. America	USA	10,778	3,285
53	Lassen	N. America	USA	10,492	3,187
54	Dempo	Asia	Indonesia	10,390	3,158
55	Sundoro	Asia	Indonesia	10,367	3,151
56	Agung	Asia	Indonesia	10,337	3,142
57	Prahu	Asia	Indonesia	10,285	3,137
58	Llaima	S. America	Chile	10,245	3,125
59	Redoubt	N. America	USA (Alaska)	10,197	3,108
60	Tjiremai	Asia	Indonesia	10,098	3,078
61	One-Take	Asia	Japan	10,056	3,067
62	Nyamulagira	Africa	Dem. Rep. of the Congo	10,026	3,056
63	Iliamna	N. America	USA (Alaska)	10,016	3,053
64	Ardjuno-Welirang	Asia	Indonesia	9,968	3,038
65	San Pedro	N. America	Guatemala	9,902	3,020
66	Gede	Asia	Indonesia	9,705	2,958
67	Zhupanovsky	Asia	Russia	9,705	2,958
68	Apo	Asia	Philippines	9,692	2,954
69	Merapi	Asia	Indonesia	9,551	2,911
70	Marapi	Asia	Indonesia	9,479	2,891
71	Geureudong	Asia	Indonesia	9,459	2,885
72	Bezymianny	Asia	Russia	9,449	2,882
73	Shishaldin	N. America	USA (Alaska)	9,372	2,856
74	Tambora	Asia	Indonesia	9,350	2,850
75	Villarrica	S. America	Chile	9,318	2,840
76	Fogo	Africa	Cape Verde	9,281	2,829
77	Ruapehu	Oceania	New Zealand	9,175	2,796
78	Peuetsagoe	Asia	Indonesia	9,115	2,780
79	Paricutin	N. America	Mexico	9,100	2,775
80	Big Ben	Antarctica	Heard Island (dependency of Australia)	9,006	2,745
81	Balbi	Oceania	Papua New Guinea	8,999	2,743
82	Avachinskaya	Asia	Russia	8,987	2,741
83	Melbourne	Antarctica	Antarctica	8,957	2,732
84	Poas	N. America	Costa Rica	8,872	2,704
85	Papandajan	Asia	Indonesia	8,744	2,665
86	Piton de la Faournaise	Africa	Reunion (dependency of France)	8,626	2,631

[continued]

Volcanoes by height

All measurements are approximate.

Rank	Name	Continent	Country	Elevation (ft)	Elevation (m)
87	Pacaya	N. America	Guatemala	8,367	2,552
88	Mt. St. Helens	N. America	USA	8,366	2,550
89	Asama	Asia	Japan	8,300	2,530
90	Pavlof	N. America	USA (Alaska)	8,261	2,518
91	Veniaminof	N. America	USA (Alaska)	8,220	2,507
92	Mayon	Asia	Philippines	8,077	2,462
93	Sinabung	Asia	Indonesia	8,066	2,460
94	Yake Dake	Asia	Japan	8,049	2,455
95	Tandikat	Asia	Indonesia	7,993	2,438
96	Canalaon	Asia	Philippines	7,984	2,435
97	Shoshuenco	S. America	Chile	7,941	2,422
98	Idjen	Asia	Indonesia	7,823	2,386
99	Izalco	N. America	El Salvador	7,828	2,386
100	Karthala	Africa	Comoros	7,746	2,361
101	Alaid	Asia	Russia	7,669	2,339
102	Bromo	Asia	Indonesia	7,636	2,329
103	Griggs	N. America	USA (Alaska)	7,597	2,317
104	Ulawun	Oceania	Papua New Guinea	7,546	2,300
105	Sibajak	Asia	Indonesia	7,541	2,300
106	Ngauruhoe	Oceania	New Zealand	7,515	2,291
107	Beerenberg	Europe	Jan Mayen Is. (dependency of Norway)	7,470	2,277
108	Guntur	Asia	Indonesia	7,377	2,249
109	Bamus	Oceania	Papua New Guinea	7,338	2,248
110	Puyehue	S. America	Chile	7,331	2,236
111	Chokai	Asia	Japan	7,314	2,229
112	Butak Petarangan	Asia	Indonesia	7,285	2,222
113	Galunggung	Asia	Indonesia	7,113	2,168
114	Mageik	N. America	USA (Alaska)	7,098	2,165
115	Douglas	N. America	USA (Alaska)	7,016	2,140
116	Amburombu	Asia	Indonesia	6,964	2,124
117	San Miguel	N. America	El Salvador	6,957	2,120
118	Tokachi	Asia	Japan	6,814	2,077
119	Chiginagak	N. America	USA (Alaska)	6,777	2,067
120	Azuma	Asia	Japan	6,775	2,065
121	Queen Mary's Peak	Africa	Tristan da Cunha (dependency of the United Kingdom)	6,760	2,060
122	Katmai	N. America	USA (Alaska)	6,715	2,047
123	Kukak	N. America	USA (Alaska)	6,688	2,040
124	Makushin	N. America	USA (Alaska)	6,678	2,035
125	Calbuco	S. America	Chile	6,567	2,003
126	Pogromni	N. America	USA (Alaska)	6,564	2,002
127	Niigata Yakeyama	Asia	Japan	6,557	2,000
128	Tongariro	Oceania	New Zealand	6,516	1,986
129	Zheltovskaya	Asia	Russia	6,403	1,953
130	Kaba	Asia	Indonesia	6,400	1,952
131	Sangeang Api	Asia	Indonesia	6,390	1,949
132	Nasu	Asia	Japan	6,289	1,917
133	Rincon de la Vieja	S. America	Costa Rica	6,282	1,916
134	Hudson	S. America	Chile	6,246	1,905
135	Trident	N. America	USA (Alaska)	6,111	1,864
136	Awu	Asia	Indonesia	6,102	1,860
137	Martin	N. America	USA (Alaska)	6,098	1,860
138	Soputan	Asia	Indonesia	5,994	1,827
139	Tiatia	Asia	Russia	5,964	1,819
140	Manam	Oceania	Papua New Guinea	5,925	1,807
141	Tanaga	N. America	USA (Alaska)	5,921	1,806
142	Siau	Asia	Indonesia	5,830	1,778
143	El Viejo (San Cristobal)	N. America	Nicaragua	5,721	1,745
144	Great Sitkin	N. America	USA (Alaska)	5,705	1,740
145	Kelud	Asia	Indonesia	5,679	1,731
146	Cleveland	N. America	USA (Alaska)	5,672	1,730
147	Batur	Asia	Indonesia	5,630	1,717
148	Askja	Europe	Iceland	4,954	1,510
149	La Soufriere	N. America	Guadeloupe (dependency of France)	4,813	1,467
150	Lopevi	Oceania	Vanuatu	4,744	1,447
151	Pelée	N. America	Martinique (dependency of France)	4,583	1,397

[continued]

Volcanoes by height

All measurements are approximate.

Rank	Name	Continent	Country	Elevation (ft)	Elevation (m)
152	Catarman	Asia	Philippines	4,367	1,332
153	Vesuvius	Europe	Italy	4,190	1,277
154	Soufrière	S. America	St. Vincent and the Grenadines	4,048	1,234

Waterfalls by height

All measurements are approximate. If a waterfall has multiple cascades they are listed separately.

Rank	Name	Continent	Country	Height (ft)	Height (m)
1	Angel (upper falls)	S. America	Venezuela	2,648	807
2	Utigord	Europe	Norway	2,625	800
3	Monge	Europe	Norway	2,539	774
4	Mtarazi (Mutarazi)	Africa	Mozambique, Zimbabwe	2,500	760
5	Itatinga	S. America	Brazil	2,060	628
6	Cuquenán (Kukenaam)	S. America	Guyana, Venezuela	2,000	610
7	Kahiwa	N. America	U.S.A. (Hawaii)	1,750	533
8	Tysse (Tusse)	Europe	Norway	1,749	533
9	Maradalsfos	Europe	Norway	1,696	517
10	Ribbon	N. America	U.S.A.	1,612	491
11	Roraima	S. America	Guyana	1,500	457
12	Della	N. America	Canada	1,445	440
13	Yosemite, Upper	N. America	U.S.A.	1,430	436
14	Gavarnie	Europe	France	1,385	422
15	Tugela (highest falls in chain)	Africa	South Africa	1,350	411
16	Krimml	Europe	Austria	1,250	380
17	Silver Strand	N. America	U.S.A.	1,170	357
18	Basaseachic	N. America	Mexico	1,020	311
19	Staubbach	Europe	Switzerland	980	299
20	Vettis	Europe	Norway	902	275
21	King George VI	S. America	Guyana	850	260
22	Wallaman	Oceania	Australia	850	260
23	Takakkaw	N. America	Canada	838	254
24	Hunlen	N. America	Canada	830	253
25	Jog (Gersoppa)	Asia	India	830	253
26	Skykje	Europe	Norway	820	250
27	Sutherland, Upper	Oceania	New Zealand	815	248
28	Sutherland, Middle	Oceania	New Zealand	751	229
29	Kaieteur	S. America	Guyana	741	226
30	Wollomombi	Oceania	Australia	726	220
31	Kalambo	Africa	Tanzania, Zambia	704	215
32	Fairy	N. America	U.S.A.	700	213
33	Feather	N. America	U.S.A.	640	195
34	Maletsunyane	Africa	Lesotho	630	192
35	Bridalveil	N. America	U.S.A.	620	189
36	Multnomah	N. America	U.S.A.	620	189
37	Panther	N. America	Canada	600	183
38	Voringfoss	Europe	Norway	597	182
39	Nevada	N. America	U.S.A.	594	181
40	Angel, Lower	S. America	Venezuela	564	172
41	Augrabies (Aughrabies)	Africa	South Africa	480	146
42	Tully	Oceania	Australia	450	137
43	Helmcken	N. America	Canada	450	137
44	Nachi	Asia	Japan	430	131
45	Tequendama	S. America	Colombia	427	130
46	Bridal Veil	N. America	Canada	400	122
47	Illilouette	N. America	U.S.A.	370	113
48	Yosemite, Lower	N. America	U.S.A.	320	98
49	Twin	N. America	Canada	260	80

Countries and selected dependencies, by area

Rank	Country	Continent	Area (sq mi)	Area (sq km)	Percent of world land area
1	Russia	Asia/Europe	6,592,735	17,075,200	11
2	Antarctica	Antarctica	5,405,000	14,000,000	9.4
3	Canada	N. America	3,851,788	9,976,140	6.7
4	U.S.A.	N. America	3,717,792	9,629,091	6.4
5	China	Asia	3,705,386	9,596,960	6.4
6	Brazil	S. America	3,286,470	8,511,965	5.7
7	Australia	Australia	2,967,893	7,686,850	5.1
8	India	Asia	1,269,338	3,287,590	2.3
9	Argentina	S. America	1,072,157	2,776,890	1.9
10	Kazakhstan	Asia	1,049,150	2,717,300	1.8
11	Sudan	Africa	967,493	2,505,810	1.7
12	Algeria	Africa	919,590	2,381,740	1.6
13	Congo, Dem. Rep. of the	Africa	905,563	2,345,410	1.6
14	Greenland	N. America	840,000	2,175,600	1.5
15	Mexico	N. America	761,606	1,972,550	1.3
16	Saudi Arabia	Asia	756,984	1,960,582	1.3
17	Indonesia	Asia	741,096	1,919,440	1.3
18	Libya	Africa	679,358	1,759,540	1.2
19	Iran	Asia	636,293	1,648,000	1.1
20	Mongolia	Asia	604,247	1,565,000	1.0
21	Peru	S. America	496,223	1,285,220	0.9
22	Chad	Africa	495,755	1,284,000	0.9
23	Niger	Africa	489,189	1,267,000	0.8
24	Angola	Africa	481,350	1,246,700	0.8
25	Mali	Africa	478,764	1,240,000	0.8
26	South Africa	Africa	471,008	1,219,912	0.8
27	Colombia	S. America	439,733	1,138,910	0.8
28	Ethiopia	Africa	435,184	1,127,127	0.8
29	Bolivia	S. America	424,162	1,098,580	0.7
30	Mauritania	Africa	397,953	1,030,700	0.7
31	Egypt	Africa	386,660	1,001,450	0.7
32	Tanzania	Africa	364,879	945,037	0.6
33	Nigeria	Africa	356,667	923,768	0.6
34	Venezuela	S. America	352,143	912,050	0.6
35	Namibia	Africa	318,694	825,418	0.6
36	Pakistan	Asia	310,401	803,940	0.5
37	Mozambique	Africa	309,494	801,590	0.5
38	Turkey	Asia	301,382	780,580	0.5
39	Chile	S. America	292,258	756,950	0.5
40	Zambia	Africa	290,584	752,614	0.5
41	Myanmar	Asia	261,969	678,500	0.5
42	Afghanistan	Asia	250,000	647,500	0.4
43	Somalia	Africa	246,199	637,657	0.4
44	Central African Republic	Africa	240,534	622,984	0.4
45	Ukraine	Europe	233,089	603,700	0.4
46	Botswana	Africa	231,803	600,370	0.4
47	Madagascar	Africa	226,656	587,040	0.4
48	Kenya	Africa	224,961	582,650	0.4
49	France	Europe	211,208	547,030	0.4
50	Yemen	Asia	203,849	527,970	0.4
51	Thailand	Asia	198,455	514,000	0.3
52	Spain	Europe	194,896	504,782	0.3
53	Turkmenistan	Asia	188,455	488,100	0.3
54	Cameroon	Africa	183,567	475,440	0.3
55	Papua New Guinea	Oceania	178,703	462,840	0.3
56	Sweden	Europe	173,731	449,964	0.3
57	Uzbekistan	Asia	172,741	447,400	0.3
58	Morocco	Africa	172,413	446,550	0.3
59	Iraq	Asia	168,753	437,072	0.3
60	Paraguay	S. America	157,046	406,750	0.3
61	Zimbabwe	Africa	150,803	390,580	0.3
62	Japan	Asia	145,882	377,835	0.3
63	Germany	Europe	137,846	357,021	0.2
64	Congo, Rep. of	Africa	132,047	342,000	0.2
65	Finland	Europe	130,127	337,030	0.2
66	Malaysia	Asia	127,316	329,750	0.2
67	Vietnam	Asia	127,243	329,560	0.2
68	Norway	Europe	125,181	324,220	0.2
69	Côte d'Ivoire	Africa	124,502	322,460	0.2
70	Poland	Europe	120,728	312,685	0.2
71	Italy	Europe	116,305	301,230	0.2
72	Philippines	Asia	115,830	300,000	0.2
73	Ecuador	S. America	109,483	283,560	0.2
74	Burkina Faso	Africa	105,869	274,200	0.2
75	New Zealand	Oceania	103,737	268,680	0.2
76	Gabon	Africa	103,347	267,667	0.2

[continued]

Countries and selected dependencies, by area

Rank	Country	Continent	Area (sq mi)	Area (sq km)	Percent of world land area
77	Guinea	Africa	94,926	245,857	0.2
78	United Kingdom	Europe	94,525	244,820	0.2
79	Ghana	Africa	92,100	238,540	0.2
80	Romania	Europe	91,699	237,500	0.2
81	Laos	Asia	91,428	236,800	0.2
82	Uganda	Africa	91,135	236,040	0.2
83	Guyana	S. America	83,000	214,970	0.1
84	Oman	Asia	82,031	212,460	0.1
85	Belarus	Europe	80,154	207,600	0.1
86	Kyrgyzstan	Asia	76,640	198,500	0.1
87	Senegal	Africa	75,749	196,190	0.1
88	Syria	Asia	71,498	185,180	0.1
89	Cambodia	Asia	69,900	181,040	0.1
90	Uruguay	S. America	68,039	176,220	0.1
91	Tunisia	Africa	63,170	163,610	0.1
92	Suriname	S. America	63,039	163,270	0.1
93	Bangladesh	Asia	55,598	144,000	0.1
94	Tajikistan	Asia	55,251	143,100	0.1
95	Nepal	Asia	54,363	140,800	0.1
96	Greece	Europe	50,942	131,940	0.1
97	Nicaragua	N. America	49,998	129,494	0.1
98	Eritrea	Africa	46,842	121,320	0.1
99	Korea, North (Democratic People's Republic of)	Asia	46,540	120,540	0.1
100	Malawi	Africa	45,745	118,480	0.1
101	Benin	Africa	43,483	112,620	0.1
102	Honduras	N. America	43,278	112,090	0.1
103	Liberia	Africa	43,000	111,370	0.1
104	Bulgaria	Europe	42,822	110,910	0.1
105	Cuba	N. America	42,803	110,860	0.1
106	Guatemala	N. America	42,042	108,890	0.1
107	Iceland	Europe	39,769	103,000	0.1
108	Yugoslavia	Europe	39,517	102,350	0.1
109	Korea, South (Republic of)	Asia	38,023	98,480	0.1
110	Hungary	Europe	35,919	93,030	0.1
111	Portugal	Europe	35,672	92,391	0.1
112	Jordan	Asia	35,637	92,300	0.1
113	French Guiana	S. America	35,135	91,000	0.1
114	Azerbaijan	Asia	33,436	86,600	0.1
115	Austria	Europe	32,378	83,858	0.1
116	United Arab Emirates	Asia	32,000	82,880	0.1
117	Czech Republic	Europe	30,450	78,866	0.1
118	Panama	N. America	30,193	78,200	0.1
119	Sierra Leone	Africa	27,699	71,740	0.05
120	Ireland	Europe	27,135	70,280	0.05
121	Georgia	Asia	26,911	69,700	0.05
122	Sri Lanka	Asia	25,332	65,610	0.04
123	Lithuania	Europe	25,174	65,200	0.04
124	Latvia	Europe	24,938	64,589	0.04
125	Togo	Africa	21,925	56,785	0.04
126	Croatia	Europe	21,831	56,542	0.04
127	Bosnia and Herzegovina	Europe	19,741	51,129	0.03
128	Costa Rica	N. America	19,730	51,100	0.03
129	Slovakia	Europe	18,859	48,845	0.03
130	Dominican Republic	N. America	18,815	48,730	0.03
131	Bhutan	Asia	18,147	47,000	0.03
132	Estonia	Europe	17,462	45,226	0.03
133	Denmark	Europe	16,638	43,094	0.03
134	Netherlands	Europe	16,033	41,526	0.03
135	Switzerland	Europe	15,942	41,290	0.03
136	Guinea-Bissau	Africa	13,946	36,120	0.02
137	Taiwan	Asia	13,892	35,980	0.02
138	Moldova	Europe	13,067	33,843	0.02
139	Belgium	Europe	11,780	30,510	0.02
140	Lesotho	Africa	11,720	30,355	0.02
141	Armenia	Asia	11,506	29,800	0.02
142	Albania	Europe	11,100	28,748	0.02
143	Solomon Islands	Asia	10,985	28,450	0.02
144	Equatorial Guinea	Africa	10,831	28,051	0.02
145	Burundi	Africa	10,745	27,830	0.02
146	Haiti	N. America	10,714	27,750	0.02
147	Rwanda	Africa	10,169	26,338	0.02

[continued]

Countries and selected dependencies, by area

Rank	Country	Continent	Area (sq mi)	Area (sq km)	Percent of world land area
148	Macedonia	Europe	9,781	25,333	0.02
149	Belize	N. America	8,867	22,966	0.02
150	Djibouti	Africa	8,494	22,000	0.01
151	El Salvador	N. America	8,124	21,040	0.01
152	Israel	Asia	8,019	20,770	0.01
153	Slovenia	Europe	7,820	20,253	0.01
154	Fiji	Oceania	7,054	18,270	0.01
155	Kuwait	Asia	6,880	17,820	0.01
156	Swaziland	Africa	6,704	17,363	0.01
157	East Timor	Asia	5,640	14,609	0.01
158	Bahamas	N. America	5,382	13,940	0.01
159	Vanuatu	Oceania	4,710	12,200	0.01
160	Qatar	Asia	4,416	11,437	0.01
161	Gambia, The	Africa	4,363	11,300	0.01
162	Jamaica	N. America	4,243	10,990	0.01
163	Lebanon	Asia	4,015	10,400	0.01
164	Cyprus	Asia	3,571	9,250	0.01
165	Puerto Rico	N. America	3,515	9,104	0.01
166	Brunei	Asia	2,228	5,770	0.004
167	Trinidad and Tobago	S. America	1,980	5,128	0.003
168	Cape Verde	Africa	1,557	4,033	0.003
169	Samoa	Oceania	1,104	2,860	0.002
170	Luxembourg	Europe	998	2,586	0.002
171	Comoros	Africa	838	2,170	0.001
172	Mauritius	Africa	718	1,860	0.001
173	Guadeloupe	N. America	687	1,780	0.001
174	Martinique	N. America	425	1,100	0.0007
175	São Tomé and Príncipe	Africa	386	1,001	0.0007
176	Netherlands Antilles	N. America	371	960	0.0006
177	Dominica	N. America	291	754	0.0005
178	Tonga	Oceania	289	748	0.0005
179	Kiribati	Oceania	277	717	0.0005
180	Micronesia	Oceania	271	702	0.0005
181	Singapore	Asia	250	647.5	0.0004
182	Bahrain	Asia	239	620	0.0004
183	St. Lucia	N. America	239	620	0.0004
184	Andorra	Europe	181	468	0.0003
185	Palau	Oceania	177	458	0.0003
186	Seychelles	Africa	176	455	0.0003
187	Antigua and Barbuda	N. America	171	442	0.0003
188	Barbados	N. America	166	430	0.0003
189	Turks & Caicos Islands	N. America	166	430	0.0003
190	Saint Vincent and the Grenadines	N. America	150	389	0.0003
191	Virgin Islands	N. America	136	352	0.0002
192	Grenada	N. America	131	340	0.0002
193	Malta	Europe	122	316	0.0002
194	Maldives	Asia	115	300	0.0002
195	St. Kitts and Nevis	N. America	101	261	0.0002
196	Cayman Islands	N. America	100	259	0.0002
197	Aruba	S. America	75	193	0.0001
198	Marshall Islands	Oceania	70	181.3	0.0001
199	Liechtenstein	Europe	62	160	0.0001
200	British Virgin Islands	N. America	58	150	0.0001
201	Anguilla	N. America	35	91	0.0001
202	San Marino	Europe	24	61.2	--
203	Bermuda	N. America	23	58.8	--
204	Tuvalu	Oceania	10	26	--
205	Nauru	Oceania	8.1	21	--
206	Monaco	Europe	0.7	1.95	--
207	Vatican City	Europe	0.17	0.44	--

Countries and selected dependencies, by population

Rank	Country	Continent	Population (July 2001)
1	China	Asia	1,273,111,290
2	India	Asia	1,029,991,145
3	U.S.A.	N. America	278,058,881
4	Indonesia	Asia	228,437,870
5	Brazil	S. America	174,468,575
6	Russia	Asia/Europe	145,470,197
7	Pakistan	Asia	144,616,639
8	Bangladesh	Asia	131,269,860
9	Japan	Asia	126,771,662
10	Nigeria	Africa	126,635,626
11	Mexico	N. America	101,879,171
12	Germany	Europe	83,029,536
13	Philippines	Asia	82,841,518
14	Vietnam	Asia	79,939,014
15	Egypt	Africa	69,536,644
16	Turkey	Asia	66,493,970
17	Iran	Asia	66,128,965
18	Ethiopia	Africa	65,891,874
19	Thailand	Asia	61,797,700
20	United Kingdom	Europe	59,647,790
21	France	Europe	59,551,227
22	Italy	Europe	57,679,825
23	Congo, Democratic Republic of the (Zaire)	Africa	53,624,718
24	Ukraine	Europe	48,760,474
25	Korea, Republic of (South Korea)	Asia	47,904,370
26	South Africa	Africa	43,586,097
27	Myanmar	Asia	41,994,678
28	Colombia	S. America	40,349,388
29	Spain	Europe	40,037,995
30	Poland	Europe	38,633,912
31	Argentina	S. America	37,384,816
32	Tanzania	Africa	36,232,074
33	Sudan	Africa	36,080,373
34	Algeria	Africa	31,736,053
35	Canada	N. America	31,592,805
36	Kenya	Africa	30,765,916
37	Morocco	Africa	30,645,305
38	Peru	S. America	27,483,864
39	Afghanistan	Asia	26,813,057
40	Nepal	Asia	25,284,463
41	Uzbekistan	Asia	25,155,064
42	Uganda	Africa	23,985,712
43	Venezuela	S. America	23,916,810
44	Iraq	Asia	23,331,985
45	Saudi Arabia	Asia	22,757,092
46	Taiwan	Asia	22,370,461
47	Romania	Europe	22,364,022
48	Malaysia	Asia	22,229,040
49	Korea, Democratic People's Republic of (North Korea)	Asia	21,968,228
50	Ghana	Africa	19,894,014
51	Sri Lanka	Asia	19,408,635
52	Mozambique	Africa	19,371,057
53	Australia	Australia	19,357,594
54	Yemen	Asia	18,078,035
55	Kazakstan	Asia	16,731,303
56	Syria	Asia	16,728,808
57	Côte d'Ivoire	Africa	16,393,221
58	Madagascar	Africa	15,982,563
59	Netherlands	Europe	15,981,472
60	Cameroon	Africa	15,906,500
61	Chile	S. America	15,328,467
62	Ecuador	S. America	13,183,978
63	Guatemala	N. America	12,974,361
64	Cambodia	Asia	12,491,501
65	Burkina Faso	Africa	12,272,289
66	Zimbabwe	Africa	11,365,366
67	Cuba	N. America	11,184,023
68	Mali	Africa	11,008,518
69	Yugoslavia	Europe	10,677,290
70	Greece	Europe	10,623,835
71	Malawi	Africa	10,548,250
72	Angola	Africa	10,366,031
73	Niger	Africa	10,355,156
74	Belarus	Europe	10,350,194
75	Senegal	Africa	10,284,929
76	Czech Republic	Europe	10,264,212
77	Belgium	Europe	10,258,762

[continued]

Countries and selected dependencies, by population

Rank	Country	Continent	Population (July 2001)
78	Hungary	Europe	10,106,017
79	Portugal	Europe	10,066,253
80	Zambia	Africa	9,770,199
81	Tunisia	Africa	9,705,102
82	Sweden	Europe	8,875,053
83	Dominican Republic	N. America	8,581,477
84	Bolivia	S. America	8,300,463
85	Austria	Europe	8,150,835
86	Azerbaijan	Asia	7,771,092
87	Bulgaria	Europe	7,707,495
88	Guinea	Africa	7,613,870
89	Somalia	Africa	7,488,773
90	Rwanda	Africa	7,312,756
91	Switzerland	Europe	7,283,274
92	Chad	Africa	7,114,400
93	Haiti	N. America	6,964,549
94	Benin	Africa	6,590,782
95	Tajikistan	Asia	6,578,681
96	Honduras	N. America	6,406,052
97	El Salvador	N. America	6,237,662
98	Burundi	Africa	6,223,897
99	Israel	Asia	5,938,093
100	Paraguay	S. America	5,734,139
101	Laos	Asia	5,635,967
102	Sierra Leone	Africa	5,426,618
103	Slovakia	Europe	5,414,937
104	Denmark	Europe	5,252,815
105	Libya	Africa	5,240,599
106	Finland	Europe	5,175,783
107	Jordan	Asia	5,153,378
108	Togo	Africa	5,153,088
109	Papua New Guinea	Oceania	5,049,055
110	Georgia	Asia	4,989,285
111	Nicaragua	N. America	4,918,393
112	Kyrgyzstan	Asia	4,753,003
113	Turkmenistan	Asia	4,603,244
114	Norway	Europe	4,503,440
115	Moldova	Europe	4,431,570
116	Croatia	Europe	4,334,142
117	Singapore	Asia	4,300,419
118	Eritrea	Africa	4,298,269
119	Puerto Rico	N. America	3,937,316
120	Bosnia and Herzegovina	Europe	3,922,205
121	New Zealand	Oceania	3,864,129
122	Ireland	Europe	3,840,838
123	Costa Rica	N. America	3,773,057
124	Lebanon	Asia	3,627,774
125	Lithuania	Europe	3,610,535
126	Central African Republic	Africa	3,576,884
127	Albania	Europe	3,510,484
128	Uruguay	S. America	3,360,105
129	Armenia	Europe	3,336,100
130	Congo, Republic of the	Africa	3,258,400
131	Liberia	Africa	3,225,837
132	Panama	N. America	2,845,647
133	Mauritania	Africa	2,747,312
134	Jamaica	N. America	2,665,636
135	Mongolia	Asia	2,654,999
136	Oman	Asia	2,622,198
137	United Arab Emirates	Asia	2,407,460
138	Latvia	Europe	2,385,231
139	Lesotho	Africa	2,177,062
140	Bhutan	Asia	2,049,412
141	Macedonia, The Former Yugoslav Republic of	Europe	2,046,209
142	Kuwait	Asia	2,041,961
143	Slovenia	Europe	1,930,132
144	Namibia	Africa	1,797,677
145	Botswana	Africa	1,586,119
146	Estonia	Europe	1,423,316
147	Gambia, The	Africa	1,411,205
148	Guinea-Bissau	Africa	1,315,822
149	Gabon	Africa	1,221,175
150	Mauritius	Africa	1,189,825
151	Trinidad and Tobago	S. America	1,169,682
152	Swaziland	Africa	1,104,343
153	Fiji	Oceania	844,330
154	East Timor	Asia	779,567
155	Qatar	Asia	769,152

[continued]

Countries and selected dependencies, by population

Rank	Country	Continent	Population (July 2001)
156	Cyprus	Asia	762,887
157	Guyana	S. America	697,181
158	Bahrain	Asia	645,361
159	Comoros	Africa	596,202
160	Equatorial Guinea	Africa	486,060
161	Solomon Islands	Oceania	480,442
162	Djibouti	Africa	460,700
163	Luxembourg	Europe	442,972
164	Suriname	S. America	433,998
165	Guadeloupe	N. America	431,170
166	Martinique	N. America	418,454
167	Cape Verde	Africa	405,163
168	Malta	Europe	394,583
169	Brunei Darussalam	Asia	343,653
170	Maldives	Asia	310,764
171	Bahamas, The	N. America	297,852
172	Iceland	Europe	277,906
173	Barbados	N. America	275,330
174	Belize	N. America	256,062
175	Netherlands Antilles	N. America	212,226
176	Vanuatu	Oceania	192,910
177	Samoa	Oceania	179,058
178	French Guiana	S. America	177,562
179	São Tomé and Príncipe	Africa	165,034
180	Saint Lucia	N. America	158,178
181	Micronesia, Federated States of	Oceania	134,597
182	Virgin Islands, U.S.	N. America	122,211
183	Saint Vincent and the Grenadines	N. America	115,942
184	Tonga	Oceania	104,227
185	Kiribati	Oceania	94,149
186	Grenada	N. America	89,227
187	Seychelles	Africa	79,715
188	Marshall Islands	Oceania	70,822
189	Dominica	N. America	70,786
190	Aruba	N. America	70,007
191	Andorra	Europe	67,627
192	Antigua and Barbuda	N. America	66,970
193	Bermuda	N. America	63,503
194	Greenland	Europe	56,352
195	Saint Kitts and Nevis	N. America	38,756
196	Cayman Islands	N. America	35,527
197	Liechtenstein	Europe	32,528
198	Monaco	Europe	31,842
199	San Marino	Europe	27,336
200	British Virgin Islands	N. America	20,812
201	Palau	Oceania	19,092
202	Turks and Caicos Islands	N. America	18,122
203	Anguilla	N. America	12,132
204	Nauru	Oceania	12,088
205	Tuvalu	Oceania	10,991
206	Holy See	Europe	890
207	Antarctica	Antarctica	0

Countries and selected dependencies, by boundary length

Borders plus coastline

Rank	Name	Total boundaries (mi)	Total boundaries (km)
1	Canada	156,949	252,684
2	Russia	35,743	57,614
3	Indonesia	35,718	57,490
4	Greenland	27,333	44,087
5	Australia	22,831	36,735
6	China	22,905	36,671
7	Philippines	22,499	36,289
8	United States	19,991	32,172
9	Japan	18,479	29,751
10	Norway	15,156	24,440
11	Brazil	13,752	22,182
12	India	13,084	21,103
13	Antarctica	11,230	17,968
14	New Zealand	9,404	15,134
15	Kazakstan	9,284	14,976
16	Greece	9,250	14,886

[continued]

Countries and selected dependencies, by boundary length

Borders plus coastline

Rank	Name	Total boundaries (mi)	Total boundaries (km)
17	Argentina	9,106	14,654
18	Mexico	8,617	13,868
19	United Kingdom	7,720	12,789
20	Chile	7,834	12,606
21	Congo, Democratic Republic of the (Zaire)	6,684	10,781
22	Turkey	6,142	9,827
23	Italy	5,911	9,532
24	Colombia	5,758	9,212
25	Iran	5,355	8,620
26	Sudan	5,306	8,540
27	Venezuela	5,277	8,493
28	Mongolia	5,070	8,161.9
29	Vietnam	5,023	8,083
30	Thailand	5,022	8,082
31	Peru	4,491	7,950
32	Croatia	4,886	7,863
33	Pakistan	4,859	7,820
34	Myanmar	4,840	7,806
35	South Africa	4,686	7,548
36	Denmark	4,587	7,382
37	Malaysia	4,563	7,344
38	Algeria	4,553	7,341
39	Ukraine	4,576	7,340
40	Mali	4,661	7,243
41	Saudi Arabia	4,383	7,055
42	Mozambique	4,373	7,041
43	Spain	4,272	6,881.8
44	Angola	4,215	6,798
45	Bolivia	4,190	6,743
46	Uzbekistan	4,117	6,641
47	France	3,925	6,316
48	Libya	3,823	6,153
49	Micronesia, Federated States of	3,789	6,112
50	Germany	3,732	6,007
51	Papua New Guinea	3,711	5,972
52	Chad	3,700	5,968
53	Mauritania	3,622	5,828
54	Niger	3,540	5,697
55	Congo, Republic of the	3,518	5,673
56	Zambia	3,496	5,627
57	Afghanistan	3,428	5,529
58	Sweden	3,370	5,423
59	Namibia	3,373	5,396
60	Somalia	3,350	5,391
61	Solomon Islands	3,301	5,313
62	Ethiopia	3,300	5,311
63	Central African Republic	3,233	5,203
64	Laos	3,151	5,083
65	Eritrea	3,009	5,015
66	Egypt	3,109	5,014
67	Cameroon	3,095	4,993
68	Iceland	3,098	4,988
69	Nigeria	3,044	4,900
70	Madagascar	3,000	4,828
71	Tanzania	2,997	4,826
72	Bangladesh	2,995	4,820
73	Estonia	2,744	4,427
74	Ecuador	2,556	4,247
75	Korea, Democratic People's Republic of (North Korea)	2,590	4,168
76	Botswana	2,488	4,013
77	Kenya	2,473	3,982
78	Paraguay	2,436	3,920
79	Kyrgyzstan	2,410	3,878
80	Morocco	2,391	3,852.9
81	Finland	2,327	3,754
82	Turkmenistan	2,322	3,736
83	Cuba	2,316	3,735
84	Guinea	2,311	3,719
85	Iraq	2,291	3,689
86	Yemen	2,270	3,652
87	Tajikistan	2,264	3,651
88	Senegal	2,257	3,632
89	Bahamas, The	2,201	3,542
90	Oman	2,154	3,466
91	Gabon	2,135	3,436
92	Poland	2,099	3,379

[continued]

Countries and selected dependencies, by boundary length

Borders plus coastline

Rank	Name	Total boundaries (mi)	Total boundaries (km)
93	Burkina Faso	1,983	3,192
94	Belarus	1,925	3,098
95	Zimbabwe	1,905	3,066
96	Honduras	1,894	3,050
97	Panama	1,892	3,045
98	Cambodia	1,873	3,015
99	Portugal	1,868	3,007
100	Nepal	1,818	2,926
101	Guyana	1,811	2,921
102	Malawi	1,789	2,881
103	Azerbaijan	1,751	2,813
104	Romania	1,698	2,733
105	Uganda	1,676	2,698
106	Korea, Republic of (South Korea)	1,656	2,651
107	Ghana	1,635	2,632
108	Tunisia	1,597	2,572
109	Austria	1,588	2,562
110	Vanuatu	1,567	2,528
111	Syria	1,520	2,446
112	Yugoslavia	1,520	2,445
113	Uruguay	1,381	2,224
114	United Arab Emirates	1,355	2,185
115	Liberia	1,345	2,164
116	Bulgaria	1,343	2,162
117	Nicaragua	1,330	2,141
118	Benin	1,308	2,110
119	Suriname	1,297	2,093
120	Haiti	1,268.7	2,046
121	Guatemala	1,251	2,017
122	Hungary	1,248	2,009
123	Costa Rica	1,204	1,929
124	Czech Republic	1,169	1,881
125	Switzerland	1,151	1,852
126	Ireland	1,124	1,808
127	Georgia	1,098	1,771
128	Togo	1,058	1,703
129	Latvia	1,043	1,681
130	Jordan	1,022	1,645
131	Taiwan	973	1,566
132	Dominican Republic	977	1,563
133	French Guiana	970	1,561
134	Palau	944	1,519
135	Bosnia and Herzegovina	919	1,479
136	Netherlands	918	1,478
137	Belgium	900	1,451
138	Moldova	864	1,389
139	Lithuania	850.4	1,372
140	Sierra Leone	250	1,360
141	Slovakia	842	1,355
142	Sri Lanka	833	1,340
143	Israel	795	1,279
144	Armenia	778	1,254
145	Slovenia	753	1,211.6
146	Kiribati	709	1,143
147	Fiji	702	1,129
148	Albania	672	1,082
149	Bhutan	668	1,075
150	Guinea-Bissau	667	1,074
151	Jamaica	634	1,022
152	Burundi	605	974
153	Cape Verde	598	965
154	Kuwait	598	963
155	Lesotho	565	909
156	Belize	559	902
157	Rwanda	555	893
158	El Salvador	530	852
159	Equatorial Guinea	517	835
160	Djibouti	511	822
161	Gambia, The	510	820
162	East Timor	492	792
163	Macedonia, The Former Yugoslav Republic of	465	748
164	Lebanon	422	679
165	Cyprus	403	648
166	Maldives	400	644
167	Qatar	387	623
168	Brunei Darussalam	337	541
169	Swaziland	332	535
170	Côte d'Ivoire	2,254	515
171	Puerto Rico	311	501
172	Seychelles	305	491
173	Tonga	260	419
174	Samoa	250	403
175	Turks and Caicos Islands	241	389
176	Netherlands Antilles	232.33	374.2
177	Marshall Islands	230	370
178	Trinidad and Tobago	225	362
179	Luxembourg	221	356
180	Martinique	217	350
181	Comoros	211	340
182	Guadeloupe	197.4	316.2
183	Malta	157	253
184	São Tomé and Príncipe	130	209
185	Singapore	120	193
186	Virgin Islands	116.7	188
187	Mauritius	110	177
188	Cayman Islands	100	160
189	Saint Lucia	98	158
190	Antigua and Barbuda	95	153
191	Dominica	92	148
192	Saint Kitts and Nevis	84	135
193	Bahrain	78	126
194	Grenada	75	121
195	Andorra	74.6	120.3
196	Bermuda	64	103
197	Barbados	60	97
198	Saint Vincent and the Grenadines	52	84
199	British Virgin Islands	50	80
200	Liechtenstein	47	76
201	Aruba	42.6	68.5
202	Anguilla	38	61
203	San Marino	24	39
204	Nauru	18.6	30
205	Tuvalu	15	24
206	Monaco	5.2	8.5
207	Vatican City	1.2	3.2

[continued]

SELECTED SOURCES

The World Rankings tables in *Geo-Data: The World Geographical Encyclopedia*, 3rd ed. were based on a large number of sources. A key source of information for all of these tables are the country entries appearing elsewhere in the book. Each of these entries has its own set of sources, and it would be impossible to list all of them here. However, some sources which cover a large number of tables or were particularly useful are listed below.

Altai Republic Territory and Geography. http://www.altai-republic.com/territory/resp_altay_eng.htm (accessed September 2002).

Atlas of Canada. http://atlas.gc.ca/site/english/ (accessed July–August 2002).

AUGLIG: Geoscience Australia: Spatial Information for the Nation. http://www.auslig.gov.au (accessed July–August 2002).

CIA World Factbook 2002. http://www.odci.gov/cia/publications/factbook (accessed July–August 2002).

Federal Research Division, Library of Congress. Country Studies. http://lcweb2.loc.gov/frd/cs/cshome.html#toc (accessed July–August 2002).

Geo-Data: The World Geographical Encyclopedia, 2nd ed. Detroit: Gale Research, 1989.

Hoelzle, M. and W. Haeberli. World Glacier Inventory. World Glacier Monitoring Service and National Snow and Ice Data Center/World Data Center for Glaciology, Boulder, CO, 1999. http://www.nsidc.org/data/glacier_inventory/ (accessed September 2002).

Internal Lake Environment Committee. LakeNet. http://www.worldlakes.org (accessed July–August 2002).

The Living Desert: World Desert. http://www.livingdesert.com/worlddes.htm (accessed July–August 2002).

Merriam-Webster's Geographical Dictionary, Third Edition. Springfield, Mass.: Merriam-Webster, Inc., 1997.

Munro, David, ed. *Oxford Dictionary of the World.* New York: Oxford University Press, 1995.

Peakware World Mountain Encyclopedia. http://www.peakware.com/encyclopedia/index.htm (accessed September 2002).

River Systems of the World. http://www.rev.net/~aloe/river (accessed July–August 2002).

United States Geological Survey. http://www.usgs.gov/ (accessed September 2002);

World Atlas. http://www.graphicmaps.com (accessed July–August 2002).

GLOSSARY

Aboriginal □ Someone or something that is the first or earliest known of its type in a country or region.

Acid rain □ Rain (or snow) that has become slightly acid by mixing with industrial air pollution.

Alluvium □ Clay, silt, sand or gravel deposted by running water such as a stream or river.

Alluvial plain □ Flatlands containing deposits of alluvium.

Antarctic □ Relating to the southernmost part of the Earth.

Aquatic □ Of or relating to the water, particularly the animals and plants that live there.

Aqueduct □ A bridge-like structure that carries water over obstacles, usually man-made.

Aquifer □ An underground layer of porous rock, sand, or gravel that holds water.

Arable land □ Land that is naturally suitable for cultivation by plowing and is used for growing crops.

Archipelago □ A group of islands, or a body of water containing many islands.

Arctic □ Relating to the northernmost part of the Earth, or anything that is frigidly and invariably cold.

Arid □ Extremely dry, particularly applied to regions of low rainfall where non-irrigated agriculture is impossible and there is little natural vegetation.

Artesian well □ A type of well where underground pressure forces water to the surface.

Asphalt □ A substance containing high concentrations of hydrocarbons that is found in naturally occurring beds, and is also produced as a by-product of petroleum refinining. Used in road building.

Atmosphere □ The air surrounding the Earth's surface.

Atoll □ An island consisting of a strip or ring of coral surrounding a central lagoon.

Avalanche □ A swift sliding of an accumulation of snow or ice down a mountain.

Badlands □ Eroded and barren land.

Barren land □ Unproductive land, partly or entirely treeless.

Barrier island □ An island formed by wave and tidal action, oriented parallel to the shore and serving to protect the shore from ocean wave action.

Barrier reef □ A coral reef that lies parallel to the coast, often forming a lagoon along the shore.

Basalt □ Black or nearly black dense rock, usually formed by the solidification or magma or from some other high-temperature geological event.

Basin □ A depression on land or on the ocean floor. Usually relatively broad and gently sloped, as compared to a trench, canyon, or crater.

Bay □ An indented body of water surrounded by land on three sides.

Bayou □ A stagnant or slow-moving body of water.

Beach □ Sediment deposited through the action of waves and the process of erosion along the shoreline of a large body of water.

Bedrock □ Solid rock lying under loose earth.

Biosphere □ All living organisms on Earth.

Bluff □ Elevated area with a broad, steep cliff face.

Bog □ Wet, soft, and spongy ground where the soil is composed mainly of decayed or decaying vegetable matter.

Bora ☐ A very cold wind blowing from the north in the Adriatic Sea region.

Biome ☐ Ecological community of an area, such as tropical rainforest or desert.

Broadleaf forest ☐ A forest composed mainly of broadleaf (deciduous) trees, as opposed to a coniferous forest.

Butte ☐ An elevated, flat-topped area, similar to but smaller than a plateau or mesa.

Caldera ☐ A crater formed by the eruption of a volcano.

Canal ☐ An artificial waterway, either to connect two bodies of water or for irrigation.

Canyon ☐ A deep gorge cut by a river, usually found in arid regions characterized by plateaus.

Cape ☐ A projection of land that extends away from a landmass out into a body of water.

Catchment ☐ Area that collects water.

Cave ☐ Hollow places in the Earth produced by the combined action of fracture of rock layers and the eroding action of ground water.

Cay ☐ Low lying island or reef, also called a Key.

Caucasus ☐ Region of southeast Europe between the Black and Caspian seas.

Channel ☐ A narrow body of water that connects two larger areas. An area where water flows through a narrow restricted path.

Chapparal ☐ Dense, shrubby vegetation that grows in regions with hot, dry summers and cooler winters, such as in the southwest United States and northwest Mexico.

Cliff ☐ A high, vertical face of rock.

Climate ☐ Weather conditions pertaining to a specific area.

Coastal belt ☐ A coastal plain area of lowlands and somewhat higher ridges that run parallel to the coast.

Coastal plain ☐ A fairly level area of land along the coast of a land mass.

Coniferous forest ☐ A forest consisting mainly evergreen trees such as pine, fir, and cypress trees.

Conifers ☐ Plants, mostly evergreen trees and shrubs, that produce cones.

Continent ☐ One of the seven major land masses of the Earth.

Continental climate ☐ A climate typical of the interior of a continent. Particulars can vary widely depending on the region, but in general areas with a continental climate have greater variations in temperature, both on a daily and seasonal basis, than in areas with a maritime climate.

Continental divide ☐ The geographic line separating the drainage basins of a continent.

Continental shelf ☐ A shallow submarine plain extending from the coast of a continent coast into the sea, and varying in width; typically ends in a steep slope to the ocean floor.

Coral reef ☐ A ridge in the warm-water areas of the ocean made up of the limestone and calcium deposits of coral animals.

Cordillera ☐ A continuous ridge, range, or chain of mountains.

Crater ☐ A bowl-shaped depression on the surface of the Earth, generally with relatively deep, steep, sides. The most common type of crater is a caldera, formed by volcanic eruption.

Cultivable land ☐ Land that can be prepared by plowing for the production of crops.

Cyclone ☐ A weather occurrence that involves the wind blowing spirally around and in towards a center, creating a powerful storm. In the northern hemisphere, the cyclonic movement is usually counter-clockwise, and in the southern hemisphere, it is clockwise.

Dam ☐ A structure built across a river that restricts its flow, causing a reservoir to form behind it. Dams are often used to generate hydropower.

Deciduous forest ☐ A forest consisting of trees that shed their leaves on a yearly basis (broadleaf) as opposed to those (coniferous) that retain them.

Deciduous species ☐ Any species of tree or shrub that sheds or casts off a part of itself after a definite period of time, commonly used to describe trees that lose their leaves on a yearly basis.

Deforestation ☐ The removal or clearing of a forest, usually to enable the land to be used for another purpose, such as agriculture or settlements.

Delta ☐ Triangular-shaped deposits of soil formed at the mouths of large rivers. They are formed out of the silt carried by the river, and have the effect of forcing the river to split into distributary channels, sometimes over a very wide area.

Density ☐ Number of units in a specific area.

Depression ☐ A point on the Earth's surface that forms a hollow, or that has sunken or fallen in. Any place where the Earth's surface is lower than the surrounding terrain.

Desert ☐ A dry land area with little precipitation and sparse vegetation.

Desertification ☐ The process wherby land that supports vegetation gradually becomes desert as a result of climatic changes, land mismanagement, or both.

Dike ☐ An artificial riverbank built up to control the flow of water.

Discontiguous ☐ Not connected to or sharing a boundary with.

Distributary ☐ A stream which branches off from a river and never rejoins it, flowing indepdently into another body of water.

Doldrums ☐ An area near the equator characterized by variable winds and periods of calm.

Dormant volcano ☐ A volcano that has not exhibited any signs of activity for an extended period of time.

Drumlin ☐ A rounded, oval hill rarely more than 250 ft (75 m) high or more than a half mile (1 km) long.

Dry climate ☐ Weather patterns characterized by little rainfall and dramatic temperature differences between day and night.

Dune ☐ A mound or ridge of loose, wind-blown sand.

Earth ☐ Fifth-largest planet in the solar system; its orbit is third from the sun, its circumference measured at the equator is 24,900 mi (40,064 km), and measured around the poles is 24,860 mi (40,000 km). The diameter at the equator is 7,926 mi (12,753 km), and from pole to pole, 7,900 mi (12,711 km).

Earthquake ☐ Shaking or other movement of the earth that is caused by tectonic shifts or volcanic activity.

Easterlies ☐ Winds or air currents blowing more or less consistently form east to west.

Eastern Hemisphere ☐ The eastern half of the earth's surface, as divided by the Prime Meridian and 180th meridian.

Ecology ☐ The branch of science that studies the relationship between organisms and their environments.

Eddy ☐ An air or water current that follows a course different from that of the main flow and usually has a swirling circular motion.

Enclave ☐ A political or cultural unit within a region that is distinct and different from the territories that surround it is an enclave within those territories. An enclave within one country or culture is often an exclave in respect to another country or culture.

Endangered species ☐ A plant or animal species whose existence as a whole is threatened with extinction.

Endemic ☐ Anything that is native to or restricted to a specific place, especially if it is unique to or characteristic of that place.

Equator ☐ An imaginary line running around the middle of the Earth halfway between the North and South Poles, identified as 0° latitude, which divides the Northern and Southern Hemispheres.

Erosion ☐ Changes in the shape of the earth's surface as a result of wind, water, or ice.

Escarpment ☐ A steep slope that separates areas of different elevations.

Estuary ☐ The region where a river and a large lake or sea meet, with their waters gradually blending into each other.

Exclave ☐ Part of a country that is separated from the larger, main portion of the country by foreign territory. An exclave is considered an enclave in respect to the foreign country or countries that surround it.

Fault ☐ An area of weakness in the Earth's crust where the rock formation splits, allowing the opposing sides shift and adjust to relieve stresses, sometimes causing earthquakes. Most commonly found along the boundaries between tectonic plates. Also called a fault line.

Fauna ☐ Animal life.

Fen ☐ Wet, soft, and spongy ground where the soil is composed mainly of decayed or decaying vegetable matter and is fed by surrounding soils and groundwater; fens are similar to bogs but have higher nutrient levels.

Fjord ☐ A relatively narrow arm of the sea that indents deeply into the land, with generally steep slopes or cliffs on each side.

Flood ☐ The flow of excessive quantities of water over land that is generally above water.

Flood plain ☐ An area of low-lying land bordering a stream of water where floods, and the resulting deposits of alluvium, occur frequently.

Flora ☐ Vegetation.

Front ☐ Weather term describing the point at which two air masses interact.

Game reserve ☐ An area of land reserved for wild animals that are hunted for sport or for food.

Geothermal energy ☐ Energy derived from the heat that constantly and naturally radiates out from the center of the Earth. Also used to desrcibe the radiation itself.

Geyser ☐ A hot spring that periodically erupts through an opening in the surface of the Earth, spewing boiling water and steam.

Glacier ☐ A large body of permanent ice on the Earth's surface.

Gorge ☐ A deep ravine with steep, rocky walls, through which a stream flows.

Grassland ☐ An area where the vegetation is mostly grasses and other grass-like plants, often providing a transition between forests and deserts.

Groundwater ☐ Water located below the earth's surface, the source from which wells and springs draw their water.

Growler ☐ A semi-submerged iceberg that poses a sea hazard.

Gulf ☐ Inlet of the sea formed by a concave indentation in the coastline.

Gulf Stream ☐ Warm ocean current flowing from roughly the Gulf of Mexico northeast along the coast of North America, then east toward Europe.

Hardpan ☐ A layer of hardened clay soil, usually underlying a thin layer of topsoil.

Hardwoods ☐ Deciduous trees, such as cherry, oak, maple, and mahogany, that produce very hard, durable, and therefore valuable, lumber.

Harmattan ☐ An intensely dry, dusty wind felt along the coast of Africa between Cape Verde and Cape Lopez. It prevails at intervals during the months of December, January, and February.

Headland ☐ Slightly elevated land lying along or jutting into a body of water.

Headstream ☐ Stream that forms the source of a river.

Headwater ☐ Source of a stream or river.

Heath ☐ Uncultivated land with low shrubs.

Hemisphere ☐ Any half of the globe—north, south, east or west—as divided by the Prime Meridian and 180° longitude, or by the equator.

Hill ☐ A rounded area of elevation rising more or less prominently above the surrounding, flatter landscape. Hill is generally used to describe an elevation of no more than 1,000 ft (300 m) above the surrounding land, but this can vary greatly depending on the region.

Humboldt Current ☐ A cold ocean current that runs north from Antarctica along the west coast of South America, primarily from June to November.

Hurricane ☐ A tropical storm with winds over 74 mph.

Hydropower ☐ Electricty generated by daming rivers and using the water to turn turbines. Also called hydroelectric power.

Hydrosphere ☐ Pertaining to the area of water along the Earth's surface.

Iceberg ☐ A massive block of ice floating independently in water that has broken from a glacier or an ice shelf through a process known as calving.

Ice caps ☐ Ice sheets covering less than 19,000 sq mi (50,000 sq km).

Ice sheets ☐ Very large glaciers found in polar and sub-polar regions, generally occupying high and relatively flat regions.

Ice shelves ☐ Sheets of ice that extend over the sea and float on the water, typically ranging from approximately 500–3,500 ft (200–1000 m) thick. The Arctic Ocean is partly covered by ice shelves, and the continent of Antarctica is almost completely surrounded by them.

Inlet ☐ A narrow passage through which water from an ocean or other large body passes, usually into a bay or lagoon.

International Date Line ☐ An arbitrary line at about the 180th meridian that designates where one day begins and another ends.

Island ☐ Land entirely surrounded by water, of relatively small area compared to the land mass of a continent.

Isthmus ☐ A narrow strip of land bordered by water on two side and connecting two larger bodies of land on the others, such as two continents, a continent and a peninsula, or two parts of an island.

Japan Current ☐ A warm current in the Pacific Ocean.

Karst ☐ Region of limestone characterized by underground streams and caverns.

Key ☐ A low flat island consisting of coral or sand, coral fragments, shell fragments, and other debris deposited on a coral flat, just above the high water level. Also called a Cay.

Labrador Current ☐ A North Atlantic current that flows southward from polar waters along the east coast of Canada.

Lagoon ☐ A shallow body of water, often connected with, or barely separated from, a nearby larger body of water.

Lake ☐ An inland body of standing water of considerable size.

Landlocked country ☐ A country that does not have direct access to the sea; it is completely surrounded by other countries.

Landslide ☐ A flow of muddy soil or loose rock that is usually triggered by heavy rainfall in areas where the terrain is steep.

Latitude ☐ An imaginary line running around the Earth, parallel to the equator. In geography, another name for a parallel. The Earth is divided into two sets of lines of latitude, each of 90 degrees, starting at the equator, 0° latitude, and extending north or south to the poles.

Lava ☐ Molten rock (magma) that has been poured out on the Earth's surface, usually through a volcano.

Leeward ☐ The direction identical to that of the prevailing wind.

Littoral ☐ The area between the high water and low water marks of a shore or coastal region.

Loam ☐ Light soil consisting of clay, silt, and sand.

Loess ☐ A windblown accumulation of fine yellow clay or silt.

Longitude ☐ An imaginary line that extends along the surface of the Earth directly from one pole to another. Another name for a meridian. The Earth is divided into 360 degrees of longitude, with 0° being designated the Prime Meridian.

Maghreb ☐ Region in northwest Africa made up of Algeria, Morocco, and Tunisia.

Magma ☐ Rock beneath the Earth's surface that has been melted by the heat of the Earth's interior. When magma breaches the Earth's surface it is known as lava.

Mangrove ☐ A tree which abounds on tropical shores in both hemispheres. Characterized by its numerous roots which arch out from its trunk and descend from its branches. Mangroves form thick, dense growths along the tidal muds, reaching lengths hundreds of miles long.

Marsh ☐ An area of soggy land, usually covered wholly or in part by shallow water, and containing aquatic vegetation.

Maquis ☐ Scrubby, thick underbrush, typical of that found along the coast of the Mediterranean Sea.

Marginal land ☐ Land that is difficult or unattractive for cultivation.

Marine life ☐ The life that exists in, or is formed by the sea.

Maritime climate ☐ The climate and weather conditions typical of areas bordering large bodies of water. Details vary from one region to another, but generally areas close to water have more even temperatures (lower high temperatures and higher low temperatures) than areas with a continental climate.

Massif ☐ Central part of a mountain, or the dominant part of a range of mountains.

Mean temperature ☐ The air temperature unit measured by the National Weather Service by adding the maximum and minimum daily temperatures together and diving the sum by 2.

Mediterranean climate ☐ A wet-winter, dry-summer climate with a moderate annual temperature range, as is typically experienced by countries along the Mediterranean Sea.

Meridian ☐ See longitude.

Mesa ☐ An isolated, elevated, flat-topped area of land, typically larger than a butte but smaller than a plateau.

Mistral ☐ In southern France, a cold, dry, northerly wind.

Monsoon ☐ Seasonal change in the wind direction of Southeastern Asia, leading to wet and dry seasons. The monsoon develops when there is a significant difference in temperature between the ocean and the land. Monsoon is sometimes used to describe any heavy rainfall.

Moor ☐ A poorly drained open area containing peat and heath.

Moist tropical climate ☐ A weather pattern typical to the tropics, known for year-round high temperatures and large amounts of rainfall.

Moraine ☐ A deposit of rocky earth deposited by a glacier.

Mountain ☐ A lofty elevation of land, generally at least higher than 1,000 ft (300 m), but varying greatly depending on the surrounding terrain, with little surface area at its peak; commonly formed in a series of ridges or in a single ridge known as a mountain range.

Natural harbor ☐ A protected portion of a sea or lake along the shore resulting from the natural formations of the land.

Nature preserve ☐ An area where one or more species of plant and/or animal are protected from harm, injury, or destruction.

Nilas ☐ A thin sheet of ice on the surface of the sea, such as the sea around Antarctica, that moves with the waves but does not break apart.

Northern Hemisphere ☐ The northern half of the Earth's surface, as divided by the equator.

Nunatak ☐ A mountain that rises above the ice sheet in Antarctica.

Oasis ☐ Originally, a fertile spot in the Libyan Desert where there is a natural spring or well and vegetation; now refers to any fertile tract in the midst of a wasteland.

Ocean ☐ The interconnected bodies of saltwater that cover almost three-fourths of the Earth's surface.

Oceania ☐ A name used to refer collectively to the islands that lie in the central and south Pacific Ocean—including the islands of Micronesia (Kiribati and Marshall Islands, and neighboring islands), Melanesia (Vanuatu and Fiji, and neighboring islands), and Polynesia (Hawaii, Tonga, Tuvalu, and neighboring islands)—that are located far from any continent.

Pampa ☐ Grass-covered plain of South America.

Panhandle ☐ A long narrow strip of land projecting like the handle of a frying pan.

Parallel ☐ See latitude.

Peneplain ☐ A flat land surface that has been subjected to severe erosion.

Peninsula ☐ A body of land surrounded by water on three sides.

Permafrost ☐ A frozen layer of soil that never thaws.

Petroglyph ☐ Ancient human carvings on rock, often found in caves.

Piedmont ☐ The terrain at the base of a mountain or range of mountains.

Plain ☐ An expansive area free of major elevations and depressions.

Plateau ☐ A relatively flat area of elevated land.

Plate tectonics ☐ A set of theories about the Earth's structure used by many geologists to explain why landmasses and oceans are arranged as they are, and why seismic activity occurs. According to plate tectonics the Earth's surface, including the bottoms of the oceans, rests on a number of large tectonic plates. These plates are slowly moving across the interior layers of the Earth. Where they grind against each other, earthquakes and other seismic activity occurs, and the shape of the land gradually changes.

Polar circles ☐ A circle around the Earth distinguishing the frigid polar zones from the temperate zones. The Earth has two polar circles, the Arctic Circle in the north and the Antarctic Circle in the south.

Polar climate ☐ A humid, severely cold climate controlled by arctic air masses, with no warm or summer season.

Polar regions ☐ The areas surrounding the northernmost (north pole) or southernmost (south pole) points on the Earth, marked by extreme, year-round, cold temperatures.

Polder ☐ Low land recovered from a body of water and protected by dikes or embankments.

Pole, geographic ☐ The extreme northern and southern points of the Earth's axis, where the axis intersects the spherical surface.

Pole, magnetic ☐ Either of two points on the Earth's surface, close to the geographic north pole and south poles, where the magnetic field is most intense.

Pond ☐ A small body of still, shallow water.

Prairie ☐ An area of level grassland that occurs in temperate climate zones.

Prime Meridian ☐ The meridian designated as zero degrees in longitude that runs through Greenwich, England, site of the Royal Observatory. All other longitudes are measured from this point .

Proved reserves ☐ The quantity of a recoverable mineral resource (such as oil or natural gas) that is still in the ground.

Rainforest ☐ Forest of tall trees with a high, leafy canopy where the annual rainfall is at least 100 in (254 cm).

Ravine ☐ A narrow, deep valley usually forming the channel for a stream.

Reef ☐ String of rocks or coral formations, usually on a sandy bottom, that are barely submerged.

Reforestation ☐ Systematically replacing forest trees lost due to fire or logging.

Reservoir ☐ A lake that was formed artificially by a dam.

Ring of Fire ☐ Region of seismic activity roughly outlined by a string of volcanoes that encircles the Pacific Ocean.

River ☐ A substantial stream of water following a clear channel as it flows over the land.

Riverine ☐ Related to a river or the banks of a river.

Roundwood ☐ Timber used as poles or in similar ways without being sawn or shaped.

Sahel ☐ Semi-arid plains region in Africa, bordering the Sahara Desert on the north, stretching approximately from the Atlantic Ocean to the Red Sea.

Salinization ☐ An accumulation of soluble salts in soil. This condition is common in irrigated areas with desert climates, where water evaporates quickly in poorly drained soil due to high temperatures. Sever salinization renders soil poisonous to most plants.

Salt pan ☐ An area of land in a sunny region that is periodically submerged in shallow water, usually due to tides or seasonal floods. The sun causes the shallow water to evaporate and leave the salt it contained behind on the ground. Also called a salt flat.

Sand bar ☐ A deposit of sedimentary material in a river, lake, or other body of water that lies in shallow water.

Savanna ☐ A treeless or near treeless plain of a tropical or subtropical region dominated by drought-resistant grasses.

Sea ☐ A body of salt water that is connected to (and therefore a part of) the ocean.

Season ☐ Regular variations in weather patterns that occur at the same times every year.

Seasonal ☐ Dependant on the season. The flow of rivers and volume of lakes often varies greatly between seasons, as can vegetation.

Sedimentary rock ☐ Rock, such as sandstone, shale, and limestone, formed from the hardening of material deposits.

Seismic activity ☐ Relating to or connected with an earthquake or earthquakes in general.

Semiarid ☐ A climate where water and rainfall is relatively scarce but not so rare as to prohibit the growth of modest vegetation. Semiarid areas are often found around arid deserts, and semiarid land is sometimes called a desert itself.

Shoal ☐ A shallow area in a stream, lake, or sea, especially a sand bank that lies above water at low tide or during dry periods.

Sierra ☐ A chain of hills or mountain.

Silt ☐ Fine gravely inorganic material, between fine sand and coarse clay particles in character, that is carried the flow of a river and deposited along its banks. Silt is generally very fertile soil.

Sirocco ☐ Hot, sand-laden wind in Algeria.

Skerry ☐ A rocky island.

Slough ☐ A marshy pond that occurs in a river inlet.

Softwoods ☐ Coniferous trees, whose wood density as a whole is relatively softer than the wood of those trees referred to as hardwoods.

Sound ☐ A wide expanse of water, usually separating mainland from islands or connecting two large bodies of water; often lies parallel to the ocean coastline.

Southern Hemisphere ☐ The southern half of the Earth's surface, as divided by the equator.

Spring ☐ Water flowing from the ground through a natural opening.

Stalactites and stalagmites ☐ Deposits of calcium carbonite formed in a cavern or cave; stalactites hang down from the ceiling like icicles, and stalagmites rise from the floor.

Steppe ☐ A flat, mostly treeless, semiarid grassland, marked by extreme seasonal and daily temperature variations. Although sometimes used to describe other areas, the term applies primarily to the plains of southeastern Europe and Central Asia.

Strait ☐ Narrow body of water connecting two larger bodies of water.

Stream ☐ Any flowing water that moves generally downhill from elevated areas towards sea level.

Subarctic climate ☐ A high latitude climate of two types: continental subarctic, which has very cold winters, short, cool summers, light precipitation and moist air; and marine subarctic, a coastal and island climate with polar air masses causing large precipitation and extreme cold.

Subcontinent ☐ A land mass of great size, but smaller than any of the continents; a large subdivision of a continent.

Subtropical climate ☐ A middle latitude climate dominated by humid, warm temperatures and heavy rainfall in summer, with cool winters and frequent cyclonic storms.

Taiga ☐ An area of open forest made up of coniferous trees.

Tectonic ☐ Relating to the structure of the Earth's crust.

Tectonic plate ☐ According to the theory of plate tectonics, the outer layer of the Earth consists of a series of large plates of rock called tectonic plates. The largest plates have entire oceans or continents on their surface.

Temperate zone ☐ The parts of the Earth lying between the tropics and the polar circles. The northern temperate zone is the area between the tropic of Cancer and the Arctic Circle. The southern temperate zone is the area between the tropic of Capricorn and the Antarctic Circle. Temperate zones are marked by the greatest seasonal variations in temperature.

Terraces ☐ Successive areas of flat lands.

Terrain ☐ General characteristics of the Earth's surface in a region, including its characteristic vegetation.

Tidal bore ☐ A distinctive type of wave that travels up a shallow river or estuary on the incoming tide. It is a dramatic phenomenon that occurs in few places in the world; the incoming tidal waters are flowing against the river's current.

Tidal wave ☐ See Tsunami.

Tide ☐ The rise and fall of the surface of a body of water caused by the gravitational attraction of the sun and moon.

Timber line ☐ The point of high elevation on a mountain above which the climate is too sever to support trees.

Topography □ The elevations and depressions of a region are its topography; also, the study of such features.

Tornado □ A violent, whirling wind that form a funnel-shaped cloud that moves in a path over the surface of the Earth.

Torrid zone □ The part of the earth's surface that lies between the tropics, so named for the warm, humid, character of its climate.

Trade winds □ Winds that consistently blow from the northeast and southeast toward the equator.

Transcaucasia □ Region in southeast Eruope between the Black and Caspian seas and south of the Caucasus Mountains.

Trench □ Steep-sided section of the ocean floor where the water is very deep.

Tributary □ Any stream that flows into another larger stream.

Tropic of Cancer □ A latitudinal line located 23°27' north of the equator, the northernmost point at which the sun can shine directly overhead.

Tropic of Capricorn □ A latitudinal line located 23°27' south of the equator, the southernmost point at which the sun can shine directly overhead.

Tropical monsoon climate □ One of the tropical rainy climates; it is sufficiently warm and rainy to produce tropical rainforest vegetation, but also has a winter dry season.

Tsunami □ A powerful, massive, and destructive ocean wave caused by undersea earthquake or volcanic eruption.

Tundra □ A nearly level treeless area whose climate and vegetation are characteristically arctic due to its position near one of the poles; the subsoil is permanently frozen.

Typhoon □ Violent hurricane occurring in the region of the South China Sea, usually in the period from July through October.

Valley □ An elongated depression through which a stream of water usually flows.

Vegetation □ Plants.

Veldt □ In South Africa, an unforested or thinly forested tract of land or region, a grassland.

Volcano □ A hole or opening, usually at the peak of a cone-shaped mountain, through which molten rock and superheated steam erupt from the interior of the Earth.

Wadi □ Dry stream bed, usually in a desert region in southwest Asia or north Africa.

Waterfall □ A steep natural descent of water flowing over a cliff or precipice to a lower level.

Watershed □ An area of shared water drainage, that is, all the rainfall drains into a common river or lake system.

Wave □ The alternate rising and falling of ridges of water, generally produced by the action between the wind and the surface of a body of water.

Weather □ Atmospheric conditions at a given place and time.

Western Hemisphere □ The western half of the earth's surface divided by the Prime Meridian at 180°.

West Indies □ Islands lying between North America and South America made up of the Greater Antilles (Cuba, Haiti, Dominican Republic, Jamaica, and Puerto Rico) and the Lesser Antilles (Virgin Islands, Trinidad and Tobago, Barbados), and Bahamas.

Wildlife sanctuary □ An area of land set aside for the protection and preservation of animals and plants.

Windward □ Facing into the prevailing wind, or lying closest to the direction from which the wind is blowing.

INDEX

Page numbers in bold indicate primary treatment of a country; those in italics indicate illustrations; and the letter *t* following a page number denotes a table. The index is sorted word by word. Entries are in natural language order with the exception of country names, which are sorted by their most recognizable name as in the rest of the book (for example: Korea, Democratic People's Republic of). Variant and abbreviated forms of names are listed as cross references.

G

H

I

Index

L

M

N

O

Q

S

T

U

V

W

X

Y

Z